A TEXTBOOK
OF PHYSICAL
CHEMISTRY

third edition

A TEXTBOOK OF PHYSICAL CHEMISTRY

third edition

Arthur W. Adamson

University of Southern California

ACADEMIC PRESS COLLEGE DIVISION

Harcourt Brace Jovanovich, Publishers

Orlando   San Diego   San Francisco   New York   London
Toronto   Montreal   Sydney   Tokyo   São Paulo

Academic Press, Inc.
Orlando, Florida 32887

United Kingdom Edition published by
Academic Press, Inc. (London) Ltd.
24/28 Oval Road, London NW1 7DX

ISBN: 0-12-044255-8
Library of Congress Catalog Card Number: 85-70837

Printed in the United States of America

He thought he saw a Royal Beast
    Cavorting with a Mate:
He looked again, and found it was
    A P. Chem. Text sedate.
"The choice is clear," he said, "to whom
    The book to dedicate!"

                      with apologies

3

Contents

chapter 2   Kinetic molecular theory of gases

chapter 3   Some additive physical properties of matter

chapter 4 Chemical thermodynamics. The first law of thermodynamics

chapter 5 Thermochemistry

chapter 6   The second and third laws of thermodynamics

chapter 7 Chemical equilibrium

chapter 8 Liquids and solids

chapter 9   Solutions of nonelectrolytes

chapter 10 Dilute solutions of nonelectrolytes. Colligative properties

chapter 11 Heterogeneous equilibrium

chapter 12 Solutions of electrolytes

chapter 13 Electrochemical cells

chapter 14 Kinetics of gas-phase reactions

chapter 15 Kinetics of reactions in solution

chapter 16 Wave mechanics

chapter 17 Molecular symmetry and bonding

chapter 18 Wave mechanics and bonding

chapter 19 Molecular spectroscopy and photochemistry

chapter 20 The solid state

chapter 21 Colloids and macromolecules

chapter 22 Nuclear chemistry and radiochemistry

Preface

A Textbook of Physical Chemistry serves as an intermediate-level textbook for the standard year-long course in physical chemistry, usually taken by chemistry and engineering majors in their junior year. Shorter texts are available, in which the presentation is kept to a fairly elementary level. Longer books try for considerable mathematical and conceptual elegance. I have tried here to steer a middle course. The book can be made more or less demanding, however, by suitable inclusion or omission of sections (including those in small print) in the various chapters. Care has been taken that otherwise easier sections do not rely on some previous, more advanced material.

I have held to the precept that physical chemistry is not a descriptive subject but a quantitative one epitomizing the scientific method. With a few inevitable exceptions, all equations and relationships have been derived from first principles and in sufficient detail that the student should not be left mystified. No mathematical preparation beyond calculus is required. Partial differentiation is explained in detail so that the student need not have prior exposure to it, and the same is true of such topics as matrix multiplication (which appears in the chapter on group theory and symmetry).

I have written in as plain and straightforward a manner as I know how. The student audience that I am addressing has always been in my mind.

It is appropriate to make some comparisons with the previous, second edition. The sequence of topics is about the same. The same general structure is retained, namely that of a main section to each chapter, followed by a section called "Commentary and Notes" and a "Special Topics" section. The Commentary and Notes sections are intended to add qualitative insight to topics covered in the main portion of the chapter and to provide interesting descriptive material. Biographies of outstanding scientists and such topics as the thermochemistry of nutrition are included. In general, no problems are written on the material of these sections. The Special Topics sections treat certain subjects that can be covered at the instructor's option but that are not essential to the main body of material. The presentation in the Special Topics sections is rigorous, and problems on them are included. Several sections that were Special Topics in the second edition have been moved into the main body of the text, however, in order to improve the continuity of presentation.

As in the second edition, the statistical mechanical treatment of the first and second laws of thermodynamics follows the phenomenological presentation of these laws. Instructors preferring to treat statistical thermodynamics as a single unit may do so by setting aside Sections 4-11 and 6-11 for later presentation.

There is some 20 percent turnover of material to allow the inclusion of such new subjects as Fourier transform spectroscopy, more detail on the RRKM treatment of reaction rates, and some discussion of oscillatory reactions. The Carathéodory theorem is included as an alternative to the standard Carnot cycle development of the entropy concept. The postulates of wave mechanics are now treated explicitly and in some detail. Contemporary data have replaced older material in many instances, and there are many new figures, including some computer graphics where they have appeared useful.

There is some 80 percent turnover in the Exercises and Problems sections (exercises and problems are now numbered in a single sequence). As before, answers are given to the exercises. A manual of worked-out answers to both exercises and problems, prepared by Edwin Joseph Bohać, Jr., of the Georgia Institute of Technology, is available from the publisher on request.

Both the traditional cgs/esu and the newer SI systems of units are used, often in parallel. (Table 3-3 gives a useful set of factors that allows conversion from cgs/esu to SI equations and vice versa.) After all, the chemist or engineer who is unacquainted with *both* systems is seriously handicapped. Some current literature, especially from overseas, is in SI units, but much appears mostly in cgs or a mixture of cgs and SI quantities; and, of course, the vast store of past literature and data compilations is in the older system. I add that I am not enamored of the SI system, considering it to be neither particularly relevant nor convenient to physical chemistry [see *J. Chem. Ed.*, *55*, 634 (1978)].

The preparation of this edition was carried out in part while I was a guest at the University of Hawaii and would have been completed while I was a guest at the University of Canterbury in Christchurch, New Zealand, except for the intervention of a change in personal plans. I am very grateful to both chemistry departments for their hospitality, both institutionally and as individual faculty members. My secretary, Kay Siu, has been indispensable in many aspects; and, as before, I thank my wife for her help in proofreading—help that makes the book (if not the subject) partly hers. My special thanks go to the several reviewers who provided valuable suggestions for improvements in the second edition, and to those who helped greatly with suggestions and corrections at the manuscript stage.

A TEXTBOOK
OF PHYSICAL
CHEMISTRY
third edition

chapter 1
Ideal and
nonideal gases

1-1   Introduction

Physical chemistry comprises the quantitative and theoretical study of the properties of the elements in their various states of combination. The definition is a sweeping one—it includes the behavior and the structure of individual molecules as well as all the various kinds of intermolecular interactions. At one time physical chemistry was considered to be a part of physics; and physics, yet earlier, lay within the formal discipline of natural philosophy. This historical relationship is reflected in the name "doctor of philosophy" for the highest degree in science. The name should not be considered as a purely archaic one, however; the chemist and other scientists *are* philosophers in that they inquire into the underlying causes of natural behavior.

Science, or the second philosophy as it is sometimes called, has progressed far indeed; moreover, its development shows no signs of slackening. For example, we need not go back very far in time—say thirty years—to observe that many topics in this book were once either unknown or at the research frontier. During this thirty year period an avalanche of facts has been compressed by the physical chemist into tables of standard data and into far-reaching empirical relationships, and the great theories of physics and chemistry have been made more precise and more capable of treating complex situations. New phenomena—the natural world is still full of surprises—have been discovered, measured, and then fitted within a theoretical framework. The same processes are going on today—the student thirty years from now will no doubt be confronted with much material not to be found in present texts.

The textbook of physical chemistry has never been easy to assimilate (or to write!)—there is so much to cover and there are so many important things to emphasize. The major empirical laws must be described and the great theories of molecular dynamics, statistical thermodynamics, and wave mechanics must be treated in sufficient detail to provide both a real application and a foundation for more advanced work. Furthermore, throughout the book the tone of the writing should be quantitative, not descriptive; the student should experience the scientific method at work.

The material that follows has been written in as plain and direct a way as this writer knows. Much attention has been given to its organization. The student should read the Preface carefully; it describes the philosophy of the book, its structure, and various practical aspects of its use. One point should be mentioned here. It is assumed that the student has taken a modern course in introductory college chemistry and that he is reasonably familiar with the gas laws, simple thermodynamics and the concept of chemical equilibrium, and the elements of chemical kinetics. Such background material is generally reviewed briefly early in each chapter. The student is also assumed to be comfortable with the qualitative language of wave mechanics and chemical bonding, although no detailed background in these subjects is required.

Both cgs (centimeter-gram-second) and SI (Système Internationale or meter-kilogram-second-ampere) units will be used. The former system is the traditional one in chemistry. However, the better features of the newer SI system are gaining use in the United States. The two systems of units are discussed in an introductory way in the Commentary and Notes section of this chapter, and in more detail in the corresponding section of Chapter 3 (Section 3-CN-1). Also, some conversion tables are given on the inside of the front cover.

We proceed now to the topic of this chapter. In keeping with the above assumptions, we will not belabor the ideal gas law or its simple applications. We do show, in Section 1-3, how the law is obtained, but with the purpose of demonstrating how it is used to define a temperature scale. The procedure provides a beautiful illustration of the scientific method; a quantity such as temperature is a very subtle one in its ultimate "meaning," yet we are able to define it exactly and unambiguously. The rest of the chapter deals with the behavior of nonideal gases and with critical phenomena. Some previous experience with this subject is assumed and the material is therefore covered rather briefly. The main emphasis is on the van der Waals equation because it is so widely used for the qualitative treatment of real gases. A glimpse of the more rigorous, modern approach is given in the Special Topics section.

1-2   Equations of state

A system at equilibrium may be described by the macroscopic properties of volume v, pressure P, and temperature t. (Temperature is defined for the moment by means of some arbitrarily chosen thermometer, and we neglect the need to specify what magnetic, electric, or gravitational fields are present.) That is, all other properties of the system are determined if these variables are specified. The *equation of state* of a system is just the functional relationship

$$v = mf(P, t) \tag{1-1}$$

where m is the mass present and $f(P, t)$ is some function of pressure and temperature. If V denotes the molar volume, an alternative form of Eq. (1-1) is

$$V = f(P, t) \tag{1-2}$$

As a matter of convenience, an equation of state usually is written for a pure chemical substance; if a mixture is involved, then composition is added as a variable.

Note that P, t, V, and density $\rho = m/v$ are *intensive* quantities. That is, their value does not depend on the amount of material present. Total volume v and mass m are *extensive* quantities. The latter gives the amount of the system and the former is proportional to m as indicated by Eq. (1-1). It is customarily assumed that an equation of state can always be written in a form involving only intensive quantities as in Eq. (1-2). This expectation is more a result of experience than a fundamental requirement of nature. We know, for example, that if a sufficiently small portion of matter is sampled, then its intensive properties *will* depend on its mass. In fact, one way of taking this aspect into account is by adding a term for the surface energy of the system. Such a term is ordinarily negligible and will not be considered specifically until Chapter 8.

To resume the original line of discussion, an equation of state describes a range of equilibrium conditions for a substance. That is, we require Eq. (1-2) to be obeyed over an appreciable range of the variables and that its validity be independent of past history. Suppose, for example, that some initial set of values (P, t) determines a molar volume V_1, and that P and t are then varied arbitrarily. It should be true that if they are returned to the original values, then V returns to V_1.

An equation such as Eq. (1-2) is a *phenomenological* one; it summarizes empirical observation and involves only variables that are themselves experimentally defined. Such relationships are often called *laws* or *rules*. In contrast, *theories* or *hypotheses* draw on some postulated model or set of assumptions and may not be and in fact usually are not entirely correct. A phenomenological relationship, however, merely reflects some aspect of the behavior of nature, and *must* therefore be correct (within the limits of the experimental error of the measurement).

1-3 Development of the concept of an ideal gas; the absolute temperature scale

The first reported reasonably quantitative data on the behavior of gases are those of Robert Boyle (1662). Some of his results on "the spring of air" are given in Table 1-1; they show that for a given temperature, the Pv product is essentially constant. Much later, in 1787, Charles added the observation that this constant was a function of temperature. At this point the equation of state for all gases was observed to be

$$PV = f(t) \tag{1-3}$$

Very accurate contemporary measurements add some important refinements. A selection of such results is given in Table 1-2, and we now see that not only does the PV product depend on pressure at constant temperature, but it does so in different ways for different gases. The data can be fitted to the equation

$$PV = A(t) + b(t)P + c(t)P^2 + \cdots \tag{1-4}$$

where t in parentheses is a reminder that the coefficients A, b, c, etc., are temperature dependent. The important observation is that while b, c, and so forth depend also on the nature of the gas, the constant A does not. As $P \rightarrow 0$, Eq. (1-4) becomes

Robert Boyle: 1627–1691

As the son of the Earl of Cork, he was born to wealth and nobility. While residing at Oxford, he discovered "Boyle's law," first identified methyl alcohol and phosphoric acid, and noted the darkening of silver salts by light. In "The Sceptical Chymist," he attacked the alchemical notion of the elements, giving an essentially modern definition. A founder of the Royal Society. (From H. M. Smith, "Torchbearers of Chemistry," Academic Press, New York, 1949.)

Two of Boyle's Experiments

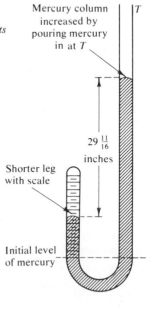

Mercury column increased by pouring mercury in at T

T

$29\frac{11}{16}$ inches

Shorter leg with scale

Initial level of mercury

FIGURE 1-1
On the left: A demonstration that a paddle wheel of feathers fell rapidly in a vacuum, and without turning. Boyle was seeing if air had some "subtle" component that could not be removed. On the right: How Boyle obtained the data of Table 1-1. (From "Robert Boyle's Experiments in Pneumatics" J. B. Conant, ed., Harvard Univ. Press, 1950.)

v	P (in. Hg)	Pv
48	$29\frac{2}{16}$	1400
44	$31\frac{15}{16}$	1405
40	$35\frac{5}{16}$	1412
28	$50\frac{5}{16}$	1409
16	$87\frac{14}{16}$	1406
12	$117\frac{9}{16}$	1411

TABLE 1-1
"The Spring of Air"
(Boyle, 1662)[a]

[a]Note Fig. 1-1.

TABLE 1-2
Isothermal P–V data for various gases at 0°C

V (liter mole^{-1})	P (atm)	PV (liter atm mole^{-1})

O$_2$

22.3939	1.00000	22.3939
29.8649	0.75000	22.3987
44.8090	0.50000	22.4045
89.6384	0.25000	22.4096

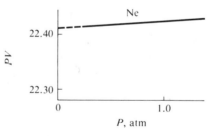

Ne

22.4280	1.00000	22.4280
33.6360	0.66667	22.4241
67.6567	0.33333	22.4189

CO$_2$

22.2643	1.00000	22.2642
33.4722	0.66667	22.3148
44.6794	0.50000	22.3397
67.0962	0.33333	22.3654
89.5100	0.25000	22.3897

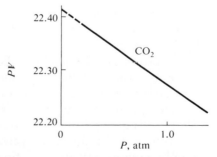

$$PV = A(t) \tag{1-5}$$

where, at 0°C, $A(t) = 22.4140$ for V in liter mole^{-1} and P in atmospheres. Note that Eq. (1-5) is a *limiting law,* that is, it gives the behavior of real gases in the limit of zero pressure.

If now the value of $A(t)$ is studied as a function of temperature, one finds the approximate behavior to be (as did Gay-Lussac around 1805)

$$A = j + kt \tag{1-6}$$

The values of j and k depend on the thermometer used; moreover, the temperature dependence of A is not exactly linear. Specifically, the results obtained using a mercury thermometer are not quite the same as those obtained using an alcohol thermometer. It is both arbitrary and inconvenient to have a temperature scale tied to the way some specific substance, such as mercury or alcohol, expands with temperature. The constant A, however, is a universal one, valid for all gases and we therefore use Eq. (1-6) as the defining equation for temperature.

The limiting gas law thermometer, or, as it is usually called, the *ideal gas thermometer,* is commonly based on a centigrade scale; the specific defining statements are as follows:

1. $t = 0$ at the melting poing of ice, at which temperature $A = 22.4140$;
2. $t = 100$ at the normal (1 atm) boiling point of water, at which temperature $A = 30.6197$;
3. intervening t values are defined by $A = j + kt$.

On combining these conditions, we have

$$j = 22.4140 \qquad k = 0.082057$$

(using liter mole^{-1} and atmosphere units). Equation (1-6) then becomes

$$A = 22.4140 + 0.082057t = 0.082057(273.15 + t) \tag{1-7}$$

where t is now the familiar temperature in degrees Centigrade (or Celsius).

The next step is obvious. Clearly Eq. (1-7) takes on a yet simpler and more rational form if a new temperature scale is adopted such that $T = 273.15 + t$°C. We then have

$$A = 0.082057T \tag{1-8}$$

or, inserting the definition of A into Eq. (1-5), we obtain

$$\lim_{P \to 0} PV = RT \tag{1-9}$$

T is called the absolute temperature and R is the gas constant, whose numerical value depends on the choice of units. Some useful sets of units and consequent R values are given in Table 1-3.

The procedure for obtaining Eq. (1-9) has been described in some detail not only because of the importance of the equation, but also because the procedure itself provides a good example of the scientific method. We have taken the phenomenological observation of Eq. (1-5), noticed the approximate validity of Eq. (1-6), and then defined our temperature scale so as to make Eq. (1-9) exact. In effect, the procedure provides an operational, that is, an unambiguous experimental, definition of temperature. At no point has it been

Units[a]	R	**TABLE 1-3**
liter atmosphere mole^{-1} K^{-1}	0.082057	
cubic centimeter atmosphere mole^{-1} K^{-1}	82.057	
joule mole^{-1} K^{-1}	8.3141	
erg mole^{-1} K^{-1}	8.3141 $\times$ 10^7	
calorie mole^{-1} K^{-1}	1.9872	

[a]Degree absolute is denoted by K.

in the least necessary to understand or to explain why gases should behave this way or what the fundamental meaning of temperature is.

To summarize, Eq. (1-9) is an equation obeyed (we assume) by all gases in the limit of zero pressure. As Boyle and Charles observed, it is also an equation of state that is approximately obeyed by many gases over a considerable range of temperature and pressure.

At this point it is convenient to introduce the concept of a hypothetical gas that obeys the equation

$$PV = RT \tag{1-10}$$

under all conditions. Such a gas we call an *ideal gas*. It is important to keep in mind the distinction between Eq. (1-9) as an exact limiting law for all gases and Eq. (1-10) as the equation for an ideal gas or as an approximate equation for gases generally. This type of distinction occurs fairly often in physical chemistry, such as, for example, in the treatment of solutions.

Equation (1-10) relates three variables: P, V, and T. The equation therefore corresponds to a surface in three-dimensional space, and a portion of this surface is shown in Fig. 1-2. As shown in the figure, one may take sections

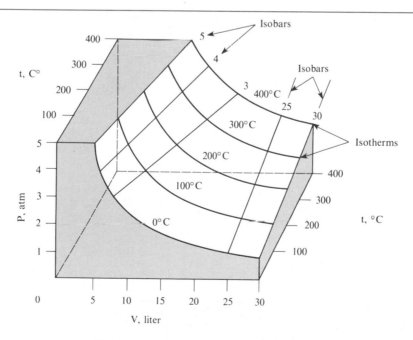

FIGURE 1-2
Plot of Eq. (1-10) showing isothermal, isobaric, and isochoric (or isosteric) sections.

at constant temperature, called *isotherms;* these are hyperbolas in the case of an ideal gas. Alternatively, one may take sections at constant pressure, called *isobars;* as a plot of V vs. T at constant P, an isobar is a straight line in the case of an ideal gas, of slope R/P. Finally, sections at constant volume are called *isochors* (or *isosteres*); the plot of P vs. T at constant V is again a straight line, but now of slope R/V.

1-4   The ideal gas law and related equations

Equation (1-10) can be put in various alternative forms, such as

$$Pv = nRT \qquad (n = \text{number of moles}); \tag{1-11}$$

$$Pv = \frac{m}{M}RT \qquad (M = \text{molecular weight}); \tag{1-12}$$

$$PM = \rho RT \qquad (\rho = \text{density}). \tag{1-13}$$

Equation (1-13) tells us, for example, that the molecular weight of any gas can be obtained approximately if its pressure and density are known at a given temperature. Furthermore, since the ideal gas law is a limiting law, the limiting value of P/ρ as pressure approaches zero must give the exact molecular weight of the gas. In effect, by writing Eq. (1-4) in the form

$$\frac{P}{\rho} = \frac{Pv}{m} = \frac{RT}{M} + \frac{\beta P}{M} + \frac{\gamma P^2}{M} + \cdots \tag{1-14}$$

one notes that the intercept of Pv/m (or P/ρ) plotted against P must give RT/M for any gas. Such a plot is illustrated schematically in Fig. 1-3.

EXAMPLE   The density of a certain hydrocarbon gas at 25°C is 12.20 g liter^{-1} at $P = 10$ atm and 5.90 g liter^{-1} for $P = 5$ atm. Find the molecular weight of the gas and its probable formula.

At 10 atm, P/ρ is $10/12.20 = 0.8197$, and at 5 atm, it is $5/5.90 = 0.8475$. Linear extrapolation to zero pressure gives $P/\rho = 0.8753$. Hence $M = RT/(P/\rho) = (0.082057)(298.15)/(0.8753) = 27.95$ g mole^{-1}. The probable formula is C_2H_4.

FIGURE 1-3
Variation of P/ρ with P for a hypothetical nonideal gas.

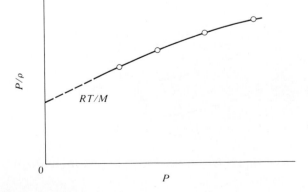

Convert the data above to SI units and rework the problem. See Commentary and Notes   **EXAMPLE**
section.

   The SI unit of force is the newton, N; this force gives an acceleration of 1 m sec^{-2} to
1 kg. The SI unit of pressure is the pascal, Pa; 1 Pa is 1 N per m^2. Thus

$$1 \text{ atm} = (0.760 \text{ m Hg})(13.5951 \text{ g cm}^{-3})(10^{-3} \text{ kg g}^{-1})(10^6 \text{ cm}^3 \text{ m}^{-3})$$

$$(9.80662 \text{ m sec}^{-2}) = 1.01325 \times 10^5 \text{ Pa or N m}^{-2}$$

Also,

$$1 \text{ g liter}^{-1} = 1 \text{ kg m}^{-3}$$

The problem now reads that the density is 12.20 kg m^{-3} at $P = 1.01325 \times 10^6$ Pa and
is 5.90 kg m^{-3} at $P = 5.06625 \times 10^5$ Pa. The respective P/ρ values are 83,053 and 85,869
Pa m^3 kg^{-1}, and the value extrapolated to zero pressure is 88,685 Pa m^3 kg^{-1}. The molecular
weight is thus

$$M = (8.31441)(298.15)/(88685) = 0.02795 \text{ kg mole}^{-1}$$

   Note that in the SI system, molecular weights are a thousandfold smaller in numerical
value than in the cgs system. This is because the unit of mass is the kilogram, while
Avogadro's number remains the same.

   It is generally useful to have an accepted standard condition of state for a
substance. Often this is 25°C and 1 atm pressure. In the case of gases, an
additional, frequently used condition is that of 0°C and 1 atm pressure. This
state will be referred to as the *STP* state (standard temperature and pressure).

1-5   Mixtures of ideal gases; partial pressures

So long as the discussion about gases deals with a single chemical species it
is immaterial whether volume is put on a per mole or a per unit mass basis.
If the amount of gas, expressed in either way, is doubled at constant temperature and pressure, the volume must also double. Suppose, however, that
a container holds 1 g of hydrogen at STP, and 1 g of oxygen is added. The
STP volume of the mixture will not be doubled; it would be, however, if 16
g of oxygen were added. (This last statement is strictly true only if the condition is the limiting one of zero pressure rather than 1 atm.)

   We are involved at this point with another observation, embodied in the
statement known as *Avogadro's hypothesis*, which says that equal volumes of
gases at the same pressure and temperature contain equal numbers of moles.
Again this is really a limiting law statement, exact only in the limit of zero
pressure, but approximately correct for real gases. In effect, the constant A
of Eq. (1-4) is a universal constant only if V is volume per mole, and not, for
example, per gram of gas. A more general form is thus

$$Pv = nA + nbP + ncP^2 + \cdots \tag{1-15}$$

Since A is independent of the nature of the gas, n is simply the total number
of moles of gas, irrespective of whether there is a mixture of species present.
The corresponding ideal gas law is given by Eq. (1-11).

   We now define the partial pressure of the ith species in a mixture of ideal
gases by the equation

$$P_i v = n_i RT \tag{1-16}$$

Dividing Eq. (1-16) by Eq. (1-11) gives

$$\frac{P_i}{P} = \frac{n_i}{n}$$

or

$$P_i = x_i P \tag{1-17}$$

where x_i denotes the mole fraction of the ith species. Since the sum of all mole fractions must by definition equal unity, it further follows that

$$\sum P_i = P \tag{1-18}$$

that is, the total pressure of a mixture of ideal gases is given by the sum of the partial pressures of the various species present. This is a statement of *Dalton's law*.

A useful quantity which can now be defined is the average molecular weight of a gas mixture, given by

$$M_{av} = \frac{m}{n} \tag{1-19}$$

where m and n are respectively the total mass and number of moles present. Further, we have

$$M_{av} = \frac{\sum m_i}{n} = \frac{\sum n_i M_i}{n} = \sum x_i M_i \tag{1-20}$$

It also follows, on combining Eqs. (1-13) and (1-18), that

$$PM_{av} = \rho RT \tag{1-21}$$

Thus the procedure illustrated by Fig. 1-3 will, for a mixture of gases, give the average molecular weight.

A complication may now arise. In order to determine the exact average molecular weight of a gas mixture, it is necessary to extrapolate P/ρ to the limit of zero pressure, yet it can happen that the composition of the mixture is itself dependent on the pressure. Thus, gaseous N_2O_4 will actually consist of a mixture of NO_2 and N_2O_4 in amounts given by the equilibrium constant for the process

$$N_2O_4 = 2NO_2$$

$$K_P = \frac{P_{NO_2}^2}{P_{N_2O_4}} \tag{1-22}$$

The proportion of NO_2 present will increase as the total pressure decreases, so that M_{av} now varies with pressure. In effect, one now knows the species and hence their molecular weight and K_P is the unknown. Equation (1-22) for K_P is exact only for ideal gases, however, and the following procedure is necessary. One first determines M_{av} for a series of total pressures using Eq. (1-21). Each determination provides a value for K_P, assuming ideal gas behavior, and these values are then plotted against pressure. The true K_P is given by the intercept at zero pressure. See Problem 1-17.

1-6   Partial volumes; Amagat's law

The partial volume v_i, of a component of a gaseous mixture is defined as the volume that component would occupy were it by itself at the pressure and temperature of the mixture:

$$v_i = \frac{n_i RT}{P} \tag{1-23}$$

Since $\Sigma n_i = n$, it follows that

$$\sum v_i = v \tag{1-24}$$

and, further, that

$$v_i = x_i v \tag{1-25}$$

Equation (1-24) is a statement of *Amagat's law of partial volumes*, and although its derivation assumes ideal gas behavior, the equation is often more closely obeyed by real gases than is its counterpart involving partial pressures, Eq. (1-17).

1-7   Deviations from ideality—critical behavior

The equation of state of an actual gas is given in one form by Eq. (1-4),

$$PV = A(T) + b(T)P + c(T)P^2 + \cdots \tag{1-26}$$

where $b(T)$, $c(T)$, and so on are not only functions of temperature, but also are characteristic of each particular gas. A form that is more useful for theoretical purposes is the following:

$$PV = A(T)\left[1 + \frac{B(T)}{V} + \frac{C(T)}{V^2} + \cdots\right] \tag{1-27}$$

or $\tag{1-28}$

$$\frac{PV}{RT} = 1 + \frac{B(T)}{V} + \frac{C(T)}{V^2} + \cdots$$

This type of equation is known as a *virial equation*, and $B(T)$ and $C(T)$ are called the second and third virial coefficients, respectively. This form is more useful than Eq. (1-4) because molar volume is a measure of the average distance between molecules and an expansion in terms of V is thus an expansion in terms of intermolecular distance. The virial coefficients can then in turn be estimated by means of various theories for intermolecular forces of attraction and repulsion.

The left-hand term of Eq. (1-28), PV/RT, is called the compressibility factor Z and its deviation from unity is a measure of the deviation of the gas from ideal behavior. Such deviations are small at room temperature for cryoscopic gases, that is, low-boiling gases such as argon and nitrogen, until quite high pressures are reached, as illustrated in Fig. 1-4, but can become quite large for relatively higher-boiling ones, such as carbon dioxide. Figure 1-5 shows that for nitrogen at $t_2 = 50°C$ (curve 2), the plot of the compressibility factor Z against P increases steadily with increasing pressure, but at a lower tem-

FIGURE 1-4
Variation of
compressibility factor with
pressure.

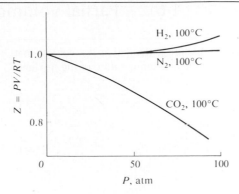

FIGURE 1-4
Variation of
compressibility factor with
pressure.

perature, it first decreases. At one particular temperature, the plot of Z versus P approaches the $Z = 1$ line asymptotically as P approaches zero. This is known as the *Boyle temperature*. The analytical condition is

$$\left(\frac{\partial Z}{\partial P}\right)_T \rightarrow 0 \qquad \text{as} \qquad P \rightarrow 0 \tag{1-29}$$

The partial differential sign, ∂, and the subscript, T, mean that the derivative of Z is taken with respect to P with the temperature kept constant. A gas at its Boyle temperature behaves ideally over an exceptionally large range of pressure essentially because of a compensation of intermolecular forces of attraction and repulsion.

The gas of a substance which can exist in both the gas and liquid states at a given temperature is often distinguished from gases generally by being called a vapor. Clearly, as a vapor is compressed at constant temperature, condensation will begin to occur when the pressure of the vapor has reached the vapor pressure of the liquid. The experiment might be visualized as involving a piston and cylinder immersed in a thermostat bath; the enclosed space contains a certain amount of the substance, initially as vapor, and the piston is steadily pushed into the cylinder. The arrangement is illustrated in Fig. 1-6. At the point of condensation, reduction in volume ceases to be accompanied by a rise in pressure; more and more vapor simply condenses to liquid at constant pressure P^0. Eventually all the vapor is condensed, and

FIGURE 1-5
Variation of
compressibility factor with
temperature and pressure
for nitrogen.

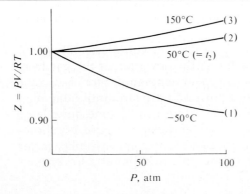

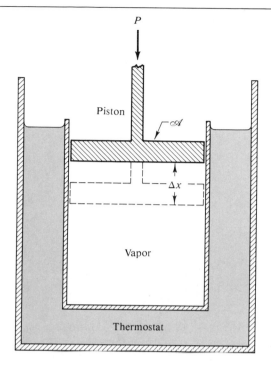

FIGURE 1-6
Compression and eventual condensation of a vapor. $\mathscr{A}$ is the area of the piston.

the piston now rests against liquid phase; liquids are generally not very compressible, and now great pressure is needed to reduce the volume further. The plot of P versus V corresponding to this experiment is shown in Fig. 1-7, where P^0 denotes the vapor pressure of the liquid and V_ℓ its molar volume. The plots are for constant temperature and are therefore isotherms.

As further illustrated in Fig. 1-7, at some higher temperature the isotherm will lie above the previous one, and the horizontal portion representing condensation will be shorter. This is because, on the one hand, $P^{0\prime}$ is larger than P^0, so the molar volume of the vapor at the condensation point is smaller, and on the other hand, the liquid expands somewhat with temperature, so $V_\ell{}'$ is greater than V_ℓ. One can thus expect, and in fact does observe, that at some sufficiently high temperature the horizontal portion just vanishes. This temperature is called the *critical temperature* T_c, and the isotherm for T_c is also shown schematically in Fig. 1-7. The broken line in the figure gives the locus of the end points of the condensation lines, and hence encloses the region in which liquid and vapor phases coexist.

There is not only a critical temperature, but also a *critical point,* which is the vestigial point left by the condensation line as it just vanishes; alternatively, the critical point is the maximum of the broken line of the figure. This point then defines a *critical pressure* P_c and a *critical volume* V_c as well as T_c. The critical temperature can also be considered as the temperature above which we speak of a gas rather than of a vapor. Compression of a gas (that is, in this context, a gaseous substance above its critical temperature) results not in condensation, but only in a steady increase in pressure, as illustrated by the curve labeled T'' in Fig. 1-7.

Figure 1-8 displays a perspective view of the $P–V–T$ surface corresponding

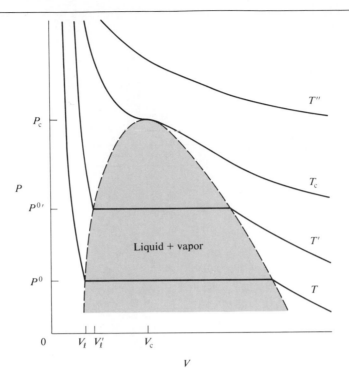

FIGURE 1-7
P–V isotherms for a real vapor.

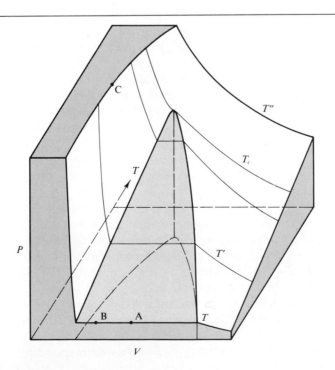

FIGURE 1-8
Perspective view of the *P–V–T* surface for a vapor showing condensation and critical behavior. See Problem 1-17.

to Fig. 1-7, and shows the isotherms for T, T', T_c, and T''. Isobars and isochors are not shown; these will no longer be straight lines as in Fig. 1-2, since we are not dealing with an ideal gas. However, at sufficiently large volumes, the P–V–T surface for any substance will approach that for an ideal gas. That is, Fig. 1-8 if extended to yet larger molar volumes will take on the appearance of Fig. 1-2.

At the other extreme, that of low temperatures and especially of small volumes, one has the liquid phase. The isotherms are then given by the coefficient of compressibility of the liquid, β, defined as

$$\beta = -\frac{1}{V}\left(\frac{\partial V}{\partial P}\right)_T \qquad (1\text{-}30)$$

Values of β for liquids are small, about 10^{-5} atm^{-1}. Thus for small changes in volume the slope of the P–V isotherm for a liquid will be approximately $-(1/V\beta)$; for water at 20°C it is about 1.2×10^6 atm liter^{-1}. Thus the curves in this region of Fig. 1-7 are nearly vertical lines. The isobars are given by the coefficient of thermal expansion α defined as

$$\alpha = \frac{1}{V}\left(\frac{\partial V}{\partial T}\right)_P \qquad (1\text{-}31)$$

Values of α are likewise small, about 10^{-4} K^{-1}, and the slope of the V–T isobar for a liquid will thus be $V\alpha$ for small changes in V. For water $V\alpha$ is about 8.0×10^{-6} liter K^{-1}. Consequently isobars corresponding to the liquid phase appear as nearly horizontal lines.

The preceding digression was intended to help fix characteristic general features of the typical P–V–T relationship for a real substance, insofar as vapor and liquid phases are involved. At the moment, however, we are primarily interested in the vapor and gas regions, and for these there is an important observation known as the *principle of corresponding states*. The intermolecular forces of attraction and repulsion which determine deviations from ideality also determine the conditions for condensation and, in particular, the values of T_c, P_c, and V_c. It is therefore perhaps not surprising that if the equation of state for a gas or vapor is written in the form

$$\frac{V}{V_c} = f\left(\frac{P}{P_c}, \frac{T}{T_c}\right) \qquad (1\text{-}32)$$

the function f turns out to be nearly independent of the substance. The quantities P/P_c, V/V_c, and T/T_c are known as the reduced variables and are denoted by P_r, V_r, and T_r, the reduced pressure, reduced volume, and reduced temperature, respectively.

This statement about Eq. (1-32) is essentially a statement of the *principle of corresponding states*. Alternatively, the principle affirms that all gases at a given P_r and T_r have the same V_r. A corollary is that gases or vapors in corresponding states have the same value for Z, the compressibility factor. Figure 1-9 may be used to obtain a fairly good value for the compressibility factor and hence for V if P and T are known, for any substance whose critical constants are also known.

The critical constants for a selection of substances are given in Table 1-4.

FIGURE 1-9
Hougen–Watson chart for the calculation of pressure, volume, and temperature relations at high pressure. (From O. A. Hougen and K. M. Watson, "Chemical Process Principles," Part II. Copyright 1959, Wiley, New York. Used with permission of John Wiley & Sons, Inc.)

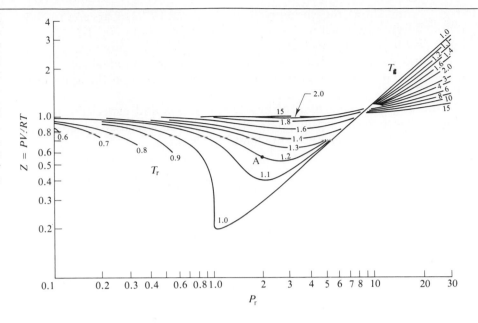

TABLE 1-4	Substance	Melting point (K)	Boiling point (K)	T_c (K)	P_c (atm)	V_c (cm^3 mole^{-1})
Critical constants and related physical properties[a]	He	<1	4.6	5.2	2.25	61.55
	Ne	24.5	27.3	44.75	26.86	44.30
	H$_2$	14.1	20.7	33.2	12.8	69.68
	O$_2$	54.8	90.2	154.28	49.713	74.42
	N$_2$	63.3	77.4	125.97	33.49	90.03
	Cl$_2$	172.2	238.6	417.1	76.1	123.4
	CO	74	81.7	134.4	34.6	90.03
	NO	109.6	121.4	177.1	64	57.25
	CO$_2$	216.6[b]	194.7	304.16	72.83	94.23
	H$_2$O	273.2	373.2	647.3	218.5	55.44
	NH$_3$	195.5	239.8	405.5	112.2	72.02
	CCl$_4$	250.2	349.7	556.25	44.98	275.8
	CH$_4$	90.7	109.2	190.25	45.6	98.77
	C$_2$H$_2$	191.4	189.2	308.6	61.65	112.9
	C$_2$H$_4$	104.1	169.5	282.8	50.55	126.1
	CH$_3$OH	175.3	338.2	513.1	78.50	117.7
	C$_2$H$_5$OH	155.9	351.7	516.2	62.96	167.2
	CH$_3$COOH	289.8	391.1	594.7	57.11	171.2
	C$_6$H$_6$	278.7	353.3	561.6	47.89	256.4

[a]Critical constants from E. A. Moelwyn-Hughes, "Physical Chemistry." Pergamon, Oxford, 1961; melting and boiling points from "Handbook of Chemistry and Physics," 51st ed. Chemical Rubber Publ., Cleveland, Ohio, 1970.
[b]At 5.2 atm.

EXAMPLE

Suppose that we wish to find the molar volume of ammonia gas at 212°C and 224 atm pressure. Then P_r and T_r are 224/112 = 2.0 and 485/4.05 = 1.2. From Fig. 1-9, point A, the value of Z for $P_r = 2$ and $T_r = 1.2$ is 0.57. The molar volume of the ammonia is then $V = 0.57RT/P = (0.57)(0.0821)(485)/(224)$ or $V = 0.101$ liter.

1-8 Semiempirical equations of state. The van der Waals equation

The relative success of the principle of corresponding states, as illustrated in the use of the chart of Fig. 1-9, suggests that it should be possible to find a not too complicated analytical expression for the function $V = f(P, T)$. In fact quite a number of such functions have been proposed, some of which are given in the Commentary and Notes section at the end of the chapter. Such functions, being analytical, are in many ways more convenient than a graph such as Fig. 1-9; they permit more precise (although not necessarily more accurate) calculations. If the function is so constructed that its form and the constants it contains have at least an approximate physical meaning, then it also provides a basis for seeing physically why different gases differ in their critical and nonideal behavior.

A semiempirical equation that meets the preceding criteria fairly well is the *van der Waals equation*, which may be assembled as follows.

First, one recognizes that molecules take up space, so that the volume occupied by a gas is only partly free space. It thus seems reasonable to replace V in the ideal gas law by the free-space volume $V - b$, where b is the effective volume occupied by a mole of molecules. This volume b is not the actual molar volume, but is the so-called *excluded volume*. The point involved is illustrated in Fig. 1-10. In the case of two identical spherical molecules the center of an approaching molecule A cannot come closer than a distance $2r$ (r being the radius) to the center of another like molecule B. Thus the excluded volume is $(4\pi/3)(2r)^3$. The effect is a mutual one, however, and further thought indicates that, per molecule, the excluded volume should be four rather than eight times the volume per molecule. One thus expects b to be something like four times the molar volume, but clearly this expectation is approximate since molecules are not impenetrable and in general are not spherical. Further, with increasing gas density there will be an increasing number of molecules in mutual proximity, with further sharing of the excluded volume. The effective b value should thus diminish.

As an approximation, however, we neglect the foregoing complication and take b to be a constant, and thus obtain the corrected equation

$$P(V - b) = RT \tag{1-33}$$

Next the pressure exerted by a gas must originate, on the molecular scale, as a result of a bombardment of the walls of the container by the molecules of the gas. There must be some mutual attractive force between the molecules, however; the fact that a vapor will condense to a liquid is clear enough evidence of this. As a consequence one expects that the actual pressure observed should be less than that for an ideal gas, where such attractive forces are not present. In the van der Waals equation this correction takes the form

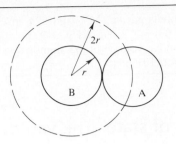

FIGURE 1-10
Illustration of excluded
volume.

of a correction a/V^2 applied to the observed pressure. The complete equation is then

$$\left(P + \frac{a}{V^2}\right)(V - b) = RT \qquad (1\text{-}34)$$

The exact form of this last correction can only be defended as an approximation. V is a measure of the average volume per molecule and hence of the cube of the average distance apart of molecules. V^2 is then proportional to r^6, where r is this average distance. There are a number of indications that the potential energy of attraction between molecules does vary as the inverse sixth power of their distance of separation (see Chapter 8, Section 8-ST-1). We can thus see that the correction term to the pressure should somehow depend inversely on V and that the actual $1/V^2$ dependence used is not unreasonable.

The van der Waals equation may be put in the form of a virial equation. On solving Eq. (1-34) for P and then multiplying both sides by V/RT one obtains

$$Z = \frac{PV}{RT} = \frac{1}{1 - (b/V)} - \frac{a}{RTV} \qquad (1\text{-}35)$$

The first term on the right can be expanded in a power series in b/V, and on collecting terms we have

$$Z = 1 + \left(b - \frac{a}{RT}\right)\frac{1}{V} + \left(\frac{b}{V}\right)^2 + \cdots \qquad (1\text{-}36)$$

The second and third virial coefficients are thus

$$B(T) = b - \frac{a}{RT} \qquad C(T) = b^2$$

Since b/V is usually a small number in the case of a gas, the cubic and higher terms of Eq. (1-36) can be neglected. An approximate form of Eq. (1-36) valid for small pressures and hence large V is obtained when only the first two terms on the right are kept and V is replaced by RT/P:

$$Z = 1 + \frac{P}{RT}\left(b - \frac{a}{RT}\right) + \left(\frac{b}{RT}\right)^2 P^2 \qquad (1\text{-}37)$$

Differentiation of Eq. (1-37) gives

$$\lim_{P \to 0} \left(\frac{\partial Z}{\partial P} \right)_T = \frac{1}{RT} \left(b - \frac{a}{RT} \right) \tag{1-38}$$

Recalling the discussion in Section 1-7 on the Boyle temperature [see Eq. (1-29)], we conclude that

$$T_B = \frac{a}{bR} \tag{1-39}$$

The physical meaning assigned to the a and b coefficients confirms the earlier analysis that at the Boyle temperature there is a balance between intermolecular attraction, measured by a, and intermolecular repulsion, measured by the excluded volume b.

The van der Waals equation allows calculation of isotherms such as those shown schematically in Fig. 1-7. This is best done by solving Eq. (1-34) for P,

$$P = \frac{RT}{V - b} - \frac{a}{V^2} \tag{1-40}$$

Then, for a given choice of a and b a value of P can easily be found for each of a series of values of V. The isotherms of Fig. 1-11 were computed by this procedure for water with $a = 5.72$ liter^2 atm mole^{-2} and $b = 0.0319$ liter mole^{-1}. (See the next section for a discussion of the problem of choosing van der Waals constants.)

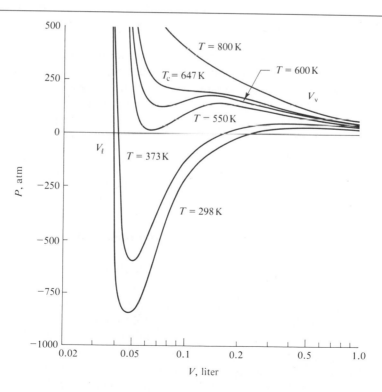

FIGURE 1-11
Isotherms calculated from the van der Waals equation ($a = 5.72$ liter^2 atm mole^{-2}, $b = 0.0319$ liter mole^{-1}).

The most obvious aspect of Fig. 1-11 is that many of the isotherms show a maximum and a minimum; this is to be expected since Eq. (1-34) [or (1-40)] is a cubic equation in volume:

$$PV^3 - (Pb + RT)V^2 + aV - ab = 0 \qquad (1-41)$$

For a given P there should in general be three roots or values of V. However, for any given choice of a and b there will be a particular value of T for which these three roots become equal. Above this value of T two roots become imaginary, leaving one real root. At high temperatures, then, the isotherms of Fig. 1-11 look much like Fig. 1-7 for a real substance. The problem is to rationalize the region showing a maximum and minimum in Fig. 1-11 with the region showing a horizontal line in Fig. 1-7. This is done as follows.

For a real substance isothermal compression across the flat portion of an isotherm corresponds to conversion of vapor to liquid at constant pressure. The amount of mechanical work done, as in the piston and cylinder arrangement of Fig. 1-6, is given by

$$w = \text{work} = \int_{V_\ell}^{V_v} P \, dV \qquad (1-42)$$

Notice that if a piston under pressure P sweeps a volume dV, then, as shown in the figure, the total force acting on the piston is $f = P\mathcal{A}$ and this force acts through a distance dx, where $dV = \mathcal{A} \, dx$. Thus the integral of Eq. (1-42) corresponds to $\int_{x_1}^{x_2} f \, dx$ and indeed gives the work done. The limits of integration for Eq. (1-42) are from the molar volume of the vapor when condensation just starts, V_v, to the molar volume of the liquid when condensation is just completed, V_ℓ. Since P^0 is constant, the work is just

$$w = P^0(V_v - V_\ell) \qquad (1-43)$$

We turn now to the van der Waals equation; referring to Fig. 1-11, it seems clear that the section labeled V_v must represent the molar volume of the gaseous state of the substance, while that labeled V_ℓ should correspond to the liquid state. The van der Waals equation connects these two branches with the section showing a maximum and a minimum, but a real substance takes the short cut of direct condensation when P reaches P^0 as illustrated in Fig. 1-12. We would like to know how to locate the horizontal line of this short cut, and hence the liquid vapor pressure P^0.

We can regard the route taken by the van der Waals equation and that given by the short cut as alternate paths requiring the *same amount of work*. That is, we require the integral $\int_{V_\ell}^{V_v} P \, dV$ to be the same along the curved path *abcd* in Fig. 1-12 and along the straight-line path *ad*,

$$w = P^0(V_v - V_\ell) = \int_{V_\ell}^{V_v} P \, dV \text{ (curve)} \qquad (1-44)$$

Graphically, this amounts to equating the two differently shaded areas in the figure; it also amounts to requiring that the net area between the line *ad* and the curve *abcd* be zero. Figure 1-13 repeats Fig. 1-11, but with horizontal lines added, as located by the preceding criterion. It is thus possible to interpret the van der Waals equation so as to obtain liquid vapor pressures or P^0 values.

There are some further interesting aspects to the above considerations. The

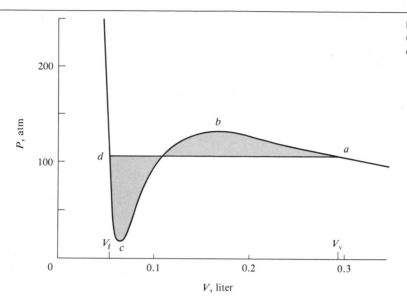

FIGURE 1-12
Condensation and the van der Waals equation.

section *abcd* of the van der Waals isotherm of Fig. 1-12 represents an unstable situation. Thus along the portion *ab* the pressure of the vapor is greater than the condensation pressure for the liquid state. It is actually possible to compress vapors beyond the condensation pressure; the system is unstable toward condensation, but if the vapor is free of dust, the first appearance of liquid droplets may be delayed. The effect is known as *supersaturation* and occurs because a small liquid drop has a higher vapor pressure than does the bulk liquid, by virtue of having a large surface-to-volume ratio and consequently

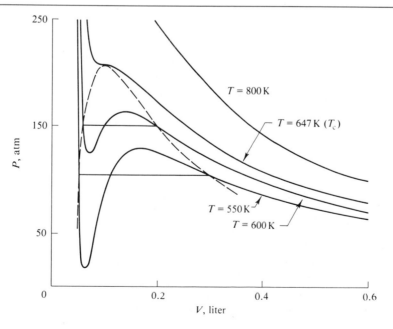

FIGURE 1-13
Figure 1-11 with condensation lines added.

an appreciable added energy due to the surface energy. The section dc is also metastable; here liquid is under less pressure than its vapor pressure and should spontaneously form vapor bubbles. Liquid surrounding a small cavity exerts a lower vapor pressure than normal, however, again because of a surface tension effect. This relationship connecting size of droplet or bubble, surface tension, and vapor pressure is given by the Kelvin equation, discussed in Section 8-9.

As a further point note that in Fig. 1-11 the lowest van der Waals isotherms reaches negative pressures. The implication is that a liquid can exist under tension, as a metastable condition. This, too, has been verified, and for water a tensile strength of as much as 100 atm has been found (see Commentary and Notes section). Finally, the section bc must represent a totally unstable region as opposed to a metastable one, since it calls for volume to increase with increasing pressure.

The van der Waals equation, although fairly simple algebraically, thus describes not only nonideal gas behavior, but also condensation and regions of vapor and liquid metastability; and, of course, it can be used for calculation of the coefficients of compressibility and thermal expansion for both liquids and gases. As discussed in the following section it also predicts critical phenomena and is consistent with the principle of corresponding states. The simplicity of the equation, the wide range of properties which can be treated, and the rather straightforward physical meaning assigned to the a and b constants have combined to make the van der Waals equation by far the one most commonly used in approximate applications.

1-9   The van der Waals equation, critical phenomena, and the principle of corresponding states

As noted in the preceding section, for a given set of a and b values there will be one single temperature at which the van der Waals equation will have three equal roots. At this temperature the equivalent straight-line section will have diminished to a point; this is the critical temperature for a van der Waals substance. The van der Waals critical point may be related to the a and b constants. Perhaps the most convenient way is as follows. At the point of three equal roots the maxima and minima must have just merged. This means that at this point the isotherm for T_c must be horizontal and, moreover, have an inflection point (as illustrated in Fig. 1-11).

The mathematical statements of these conditions are that $(\partial P/\partial V)_T = 0$ and $(\partial^2 P/\partial V^2)_T = 0$. On applying the indicated differentiations to Eq. (1-40), we have

$$\left(\frac{\partial P}{\partial V}\right)_T = 0 = \frac{-RT_c}{(V_c - b)^2} + \frac{2a}{V_c^3} \tag{1-45}$$

$$\left(\frac{\partial^2 P}{\partial V^2}\right)_T = 0 = \frac{2RT_c}{(V_c - b)^3} - \frac{6a}{V_c^4}$$

On solving Eqs. (1-40), (1-45), and (1-46) simultaneously we find

$$T_c = \frac{8a}{27bR} \qquad V_c = 3b \qquad P_c = \frac{a}{27b^2} \tag{1-47}$$

Alternatively, we have

$$b = \frac{RT_c}{8P_c} \qquad a = \frac{27(RT_c)^2}{64P_c} \tag{1-48}$$

Finally, the expressions for V_c, P_c, and T_c may be combined to give

$$P_cV_c = \tfrac{3}{8}RT_c \tag{1-49}$$

The van der Waals equation also conforms to the principle of corresponding states. From Eqs. (1-47), $P = (a/27b^2)\, P_r$, $V = 3bV_r$, and $T = (8a/27bR)\, T_r$, and substitution into the van der Waals equation (1-34) then yields

$$\left(P_r + \frac{3}{V_r^2}\right)\left(V_r - \frac{1}{3}\right) = \frac{8}{3}T_r \tag{1-50}$$

As required by eq. (1-32), we now have a relationship connecting V_r, P_r, and T_r which contains no constants specific to the particular substance.

Table 1-5 gives pairs of van der Waals constants for a number of substances. Different sources give somewhat different values for these constants, however. They may be obtained in various ways. One is from the critical constants, with Eqs. (1-47). Another is by a best fitting of the van der Waals equation to the gas (as opposed to the vapor) portion of the compressibility chart of Fig. 1-9. The constants can also be obtained from the coefficients of compressibility and thermal expansion for a liquid, and so on. Since the van der Waals equation is still only an approximate equation, each method will yield somewhat different a and b values. Any one set will then be best suited for calculations around that region of conditions for which the set was obtained

TABLE 1-5

Van der Waals constants for gases[a]

Substance	a (liter² atm mole⁻²)	b (liter mole⁻¹)
He	0.03412	0.02370
Ne	0.2107	0.01709
H_2	0.2444	0.02661
O_2	1.360	0.03183
N_2	1.390	0.03913
Cl_2	6.493	0.05622
CO	1.485	0.03985
NO	1.340	0.02789
CO_2	3.592	0.04267
H_2O	5.464	0.03049
NH_3	4.170	0.03707
CH_4	2.253	0.04278
C_2H_2	4.390	0.05136
C_2H_4	4.471	0.05714
C_2H_6	5.489	0.06380
CH_3OH	9.523	0.06702
C_2H_5OH	12.02	0.08407
CH_3COOH	17.59	0.1068
C_6H_6	18.00	0.1154

[a]Landolt-Bornstein, "Physical Chemistry Tables." Springer, Berlin, 1923.

and will be apt to give poor results when used in calculations for some quite different pressure and temperature region.

As an example, the a and b values for water used in calculating Fig. 1-11 give a good fit to P–V–T data for water well above its critical temperature and pressure. They are appreciably different from the ones calculated from the critical point for water (since V_c is 55 cm^3 mole^{-1}, this would give $b = 18.5$ cm^3 as compared to 31.9 cm^3 used for Fig. 1-11.) One result is that while one would expect the 25°C curve of Fig. 1-11 to cross the $P = 0$ line at about 18 cm^3 mole^{-1}, the molar volume of liquid water, the calculated curve does so at 31.9 cm^3 mole^{-1}. Clearly, a different set of a and b values would better represent this region of the P–V–T plot for water. It is as a consequence of such quantitative deficiencies of the van der Waals equation that various more elaborate analytical equations of state have been proposed. Some of these are mentioned in the Commentary and Notes section.

COMMENTARY AND NOTES

A section of this type occurs at the end of the main portion of most chapters. The purposes are, first, to provide some qualitative commentary on interesting but less central aspects of the chapter material and, second, to supply, for reference purposes, additional quantitative results. The latter will ordinarily be presented without derivation or much discussion.

1-CN-1   Systems of units

It is appropriate to comment at this point on the rather mixed state of affairs in the matter of systems of units (see Adamson, 1978). The standard metric or cgs system leads to the erg and the dyne as the units of energy and of force, respectively, and through the mechanical equivalent of heat, to the calorie as an alternative unit of energy. Force is defined by means of Newton's first law,

$$f = ma \tag{1-51}$$

where a denotes acceleration, and the dyne is that force required to accelerate 1 g by 1 cm sec^{-2}. The dimensions of dyne and erg are g cm sec^{-2} and g cm^2 sec^{-2}, respectively. The unit of volume is the cubic centimeter, with the liter or 1000 cm^3 as a secondary unit.

The mks (meter–kilogram–second) system, now incorporated by a set of international commissions into what is called the SI system (Système Internationale d'Unités) makes several changes. The unit of force is still given by Eq. (1-51), but in terms of kg m sec^{-2}, and is now called the newton (N). One N is 10^5 dyn. The unit of energy is in kg m^2 sec^{-2}, or the joule (J). One J is 10^7 erg. (See McGlashan, 1968).

In addition to critically reevaluating the numerical values of fundamental constants, the SI commissions recommended that use of the calorie be dropped, as well as special names for subunits such as the liter, micron (10^{-6} m), and angstrom (10^{-8} cm). A special set of prefixes was adopted for the designation of multiples or fractions of the primary units. These are tabulated on the inside cover. For example, nano- means 10^{-9}; the closest SI unit to the angstrom is the nanometer (nm) or 10^{-7} cm.

Avogadro's number was not changed, however, and molecular weights therefore *have* changed. Thus the molecular weight of O$_2$ is 0.03200 kg mole^{-1} in the SI system.

SI units of particular relevance to this chapter are those for force, pressure, and concentration. The former is discussed above. The SI unit of pressure is the pascal, a force of one newton per m^2. Since the newton is 10^5 dyne, one pascal or Pa corresponds to 10 dyne cm^{-2}, or 1 × 10^{-5} bar (the bar is 1 × 10^6 dyne cm^{-2}). Alternatively, 1 bar corresponds to 1 × 10^5 Pa. Also, 1 atm is 1.01325 bar or 1.01325 × 10^5 Pa. (Note the Example in Section 1-4.) The SI units of volume and of concentration are m^3 and mol m^{-3}. A cubic meter is a rather large volume, and these units are not very convenient in size. Often, however, volume and concentration will be expressed in dm^3 and mol dm^{-3}; since the dm^3 is *exactly* one liter, this usage is equivalent to giving volume in liters and concentration in mole liter^{-1}. Remember, however, that in working problems in SI units one must convert to m^3 or to mol m^{-3}.

At the time of this writing the cgs and related conventional units remain in principal usage in U.S. technical journals in chemistry, although our National Bureau of Standards has supported use of the SI system. Great Britain officially requires the use of SI units in British journals; usage is mixed in other European countries. Textbooks of physical chemistry in the United States are now recognizing both systems, as is done here. There appears to be a slow movement toward complete adoption of the SI system.

Since this is a commentary section, an opinion is permissible. The cgs system and associated secondary units have three important characteristics. (1) They are decimalized (unlike the English units such as feet and inches). (2) Units are defined operationally, that is, in terms of fundamental laws such as Newton's law. (3) Commonly measured quantities come out to be of the order of unity. Thus the density of water is about 1 g cm^{-3} and its heat capacity about 1 cal g^{-1} K^{-1}; the size of an atom is about 1Å; pressure at sea level is about 1 atm; the molecular weight of hydrogen is about 1 g mole^{-1}, and so on.

The SI system retains (1) but departs from (2) (see Section 3-CN-1) and (3). With respect to (3), for example, the density of water becomes about 10^3 kg m^{-3} (or 10^{-3} kg cm^{-3}), and its heat capacity about 4×10^6 J K^{-1} m^{-3}; atoms are around 0.1 nm in size; sea level pressure becomes about 1×10^5 Pa; the atomic weight of hydrogen is 10^{-3} kg mole^{-1}.

Criterion (3) alone suggests that the cgs and related units will continue to be used. Convenience is imporant in science as it is elsewhere. (Consumers in the United States are being urged to think in terms of liters rather than quarts; they will be yet more reluctant to buy milk in terms of cubic meters or decimeters or to joule-count in watching their diet!)

The situation with respect to electrical and magnetic units is even more complicated. It is here that the SI system departs from criterion (2) above. This aspect is discussed in detail in Section 3-CN-1 since topics involving electrical units are not taken up until Chapter 3.

1-CN-2   Other equations of state

A number of semiempirical equations of state have found use. These tend to be of the form of the van der Waals equation, but with improvements designed to allow for a temperature dependence of a and b. Some of these are

Clausius equation:

$$\left[P + \frac{a}{T(V + c)^2} \right](V - b) = RT \qquad (1\text{-}52)$$

Berthelot equation:

$$\left(P + \frac{a}{TV^2} \right)(V - b) = RT \qquad (1\text{-}53)$$

Dieterici equation:

$$P(V - b) = RTe^{-a/RTV} \qquad (1\text{-}54)$$

Beattie–Bridgman equation:

$$P = \frac{RT(1 - \epsilon)}{V^2}(V + B) - \frac{A}{V^2} \qquad (1\text{-}55)$$

where $A = A_0[1 + (a/V)]$, $B = B_0[1 - (b/V)]$, and $\epsilon = c/VT^3$, and A_0 and B_0 are constants. This particular equation is designed for gases at high pressures.

Benson–Golding equation:

$$\left(P + \frac{a}{V^{5/3}T^{2/3}} \right)(V - bV^{-1/2}) = RT. \qquad (1\text{-}56)$$

(See Glasstone, 1946; Weston, 1950; Benson and Golding, 1951.)

1-CN-3   Some general comments. The law of the rectilinear diameter

Some of the various properties that an equation of state should in principle be able to predict were mentioned in the discussion of the van der Waals equation. A brief elaboration is worthwhile. For example, the idea of a tensile strength for a liquid may seem unexpected. The experimental problem, of course, is that one cannot simply pull on a column of liquid as one might on a rod of solid material. What one actually does is to fill a capillary tube with liquid at some elevated temperature and then seal the tube. On cooling, the liquid should contract, but to do so a bubble of vapor would have to form, and if the liquid is free of dust or if the cooling is rapid enough, the column of liquid remains intact and therefore under tension. One calculates this tension from the coefficient of compressibility, knowing how much the liquid has been forced to expand in order to keep filling the capillary at the lower temperature. Negative pressure may be applied mechanically, but less easily. This situation does occur, however, with a boat propeller; liquid behind the rotating blades is momentarily under tension. An important practical problem is to avoid *cavitation*, or the formation of vapor bubbles; the sudden collapse of such bubbles not only hampers the propeller but can pluck out metal grains to roughen and eventually destroy the surface.

Another property which can be calculated from an equation of state is the surface tension of a liquid. We ordinarily think of pressure as a scalar of non-directional quantity, but in the case of a crystalline, nonisotropic solid, application of a uniform pressure will distort the crystal. To avoid this, we would have to exert different pressures on each crystal face. In general, then, pressure can be treated as a set of stresses or directional vectors.

Since a liquid does not support stress, the pressure around any portion of a liquid must be isotropic. This is not true, however, in the surface region. The pressure normal to the surface must indeed be the same as the general pressure throughout the system. However, the pressure parallel to the surface varies through the interface. Analysis shows that the surface tension γ can be calculated if the difference between the pressure normal to the surface and that parallel to the surface is known as a function of distance through the interface. The equation is

$$\gamma = \int_{\text{vapor phase}}^{\text{liquid phase}} (P - p)\,dx \qquad (1\text{-}57)$$

where P is the general, isotropic pressure, p is the local pressure component parallel to the

surface, and x is distance normal to the surface. If we can, by some analysis, calculate how the density or molar volume varies across a liquid–vapor interface, then use of an equation of state such as the van der Waals equation allows a calculation of p as a function of x, and hence of the surface tension (see Tolman, 1949).

A final brief consideration concerns the determination of the critical point of a substance. The reality of a critical point can be seen by means of the following type of experiment. A capillary tube is evacuated and then partly filled with liquid, the remaining space containing no foreign gases but only vapor of the substance in question; the tube is then sealed. On heating, opposing changes take place. The liquid phase increases its vapor pressure, and the vapor density increases as vaporization occurs; this acts to diminish the volume of liquid phase. On the other hand, the liquid itself expands on heating. If just the proper degree of filling of the capillary was achieved, these two effects will approximately balance, and the liquid–vapor meniscus will remain virtually fixed in position as the capillary is heated. A temperature will then be reached at which the meniscus begins to become diffuse and then no longer visible as a dividing surface. At this temperature the system often shows opalescence; the vapor and liquid densities are so nearly the same and their energy difference is so small that fluctuations can produce transient large liquid-like aggregates in the vapor and *vice versa* in the liquid. There is still an average density gradient. However, at a slightly higher temperature, perhaps 5–10 K more, the system becomes essentially uniform. This last is the critical temperature; knowing the amount of substance and the volume of the capillary, one also knows the critical molar volume.

This type of visual experiment, although quite interesting, does not allow a very precise determination of the critical point. An alternative procedure makes use of a series of isotherms such as are shown in Fig. 1-7. However, while the broken line joining the end points of the condensation lines can be fairly well established, its exact maximum point is hard to fix exactly. This locus may alternatively be plotted as temperature versus the equilibrium vapor and liquid densities ρ_v and ρ_ℓ as illustrated Fig. 1-14. A useful observation, known as the *law of the rectilinear diameter*, states that the average density $\rho_{av} = (\rho_\ell + \rho_v)/2$ is a linear function of

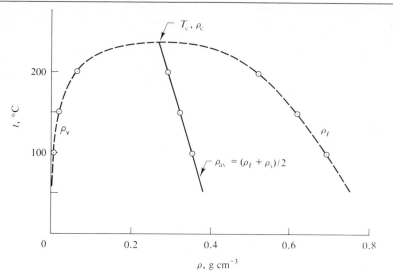

FIGURE 1-14
Illustration of the law of the rectilinear diameter.

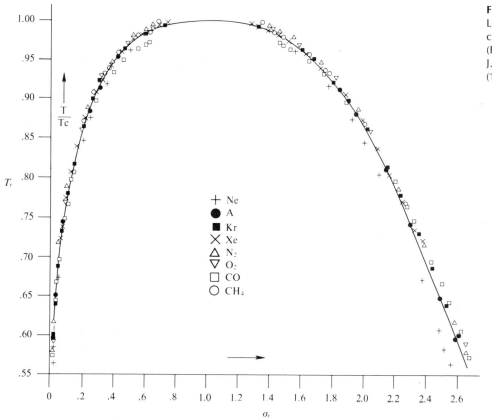

FIGURE 1-15
Liquid-vapor co-existence curve in reduced form. (From E.A. Guggenheim, J. Chem. Phys., **13**, 253 (1945).)

temperature, as also illustrated in the figure. The essentially straight and nearly vertical line of ρ_{av} versus T makes an easily defined intersection with the curved line of densities. This intersection then gives the critical temperature and density.

As might be expected, the principle of corresponding states is obeyed fairly well. Figure 1-15 is, in a sense, a sequel to Fig. 1-9; data for a variety of substances fall nearly on the same curve if plotted as T_r vs. ρ_r, where $\rho_r = \rho/\rho_c$ and ρ_c is the density at the critical temperature.

SPECIAL TOPICS

The Special Topics section at the end of each chapter takes up either more specialized or more advanced aspects of the chapter material. The intention is to provide a section, separated from the main body of material, which can be considered if time in the course permits, or as self-study on the part of the interested student. Although the material in the main body of succeeding chapters will not draw on previous Special Topics sections, subsequent Special Topics sections may make use of the results of the preceding ones.

1-ST-1 Use of a potential function for nonideal gases

The van der Waals and other equations of state cited in this chapter are illustrations of a semi-empirical approach in which the goal is to obtain a definite functional form for the equation

of state. Contemporary theoretical chemists now pursue the line of using a detailed expression for the intermolecular potential between two molecules to obtain values for the virial coefficients of Eq. (1-28). The potential function will be somewhat approximate or semiempirical, but the ensuing and generally quite elaborate theoretical development may be rigorous.

If the mutual potential energy between two molecules as a function of the separation r is denoted by $\phi(r)$, then a statistical mechanical derivation, which is beyond the scope of this text, gives the following equation for the second virial coefficient (see Hirschfelder et al., 1954):

$$B(T) = 2\pi N_A \int_0^\infty (1 - e^{-\phi(r)/kT})\, r^2\, dr \qquad (1\text{-}58)$$

where N_A is Avogadro's number. According to the Boltzmann Principle, discussed in Chapter 2, the exponential term, $e^{-\phi(r)/kT}$, is the relative probability of a molecule having a given po-

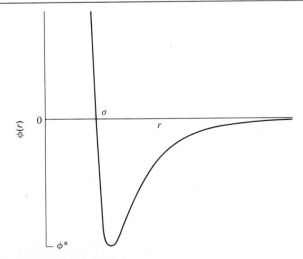

FIGURE 1-16
Variation of potential energy as two molecules approach each other.

tential energy. The constant k is known as the Boltzmann constant, and is just the gas constant per molecule.

If now molecules experience mutual attractive and repulsive forces, $\phi(r)$ can be expected to have a form of the type illustrated in Fig. 1-16. That is, as two molecules approach each other they will at first be attracted and then, at small separations, repelled. Correspondingly, their mutual potential energy is zero at infinite separation and diminishes to a maximum negative value at some separation r_0. At $r = \sigma$, the potential energy is just zero again; and for $r < \sigma$, it is positive and very rapidly increasing.

The average total energy of the pair of molecules can be written

$$\epsilon_{\text{tot}} = \epsilon_{r \to \infty} + \phi(r) \qquad (1\text{-}59)$$

Thus ϵ_{tot} diminishes to a minimum at r_0 and then rises rapidly as r is decreased. In terms of the Boltzmann principle, the effect is to make separation distances around $r = r_0$ relatively more probable and separation distances of $r < \sigma$ relatively less probable than for an ideal gas.

A second effect of $\phi(r)$ is on the ability of molecules to exert a pressure. Perhaps a helpful although very crude physical explanation is as follows. Two molecules at a distance r_0 apart have less energy than the average, and were they to separate without any energy being supplied, they would end up as separate molecules making less than the average contribution to the pressure. Conversely, a pair of molecules at $r < \sigma$ should make a greater than average contribution to the pressure. Thus the presence of intermolecular forces affects both the distri-

bution of intermolecular distances and the expected pressure of the now nonideal gas.

The way in which Eq. (1-58) works can be illustrated by the following example. Consider the gas molecules to be hard spheres so that the potential energy plot is as shown in Fig. 1-17. That is, there are no attractive forces, and $\phi(r)$ jumps to infinity at $r = \sigma$. The integral of Eq. (1-58) can then be written in two parts:

(a) at $r > \sigma$,

$$\phi(r) = 0 \qquad e^{-\phi(r)/kT} = 1$$

$$\int_{\sigma}^{\infty} (1 - e^{-\phi(r)/kT})\, r^2\, dr = 0$$

(b) at $r < \sigma$,

$$\phi(r) = \infty \qquad e^{-\phi(r)/kT} = 0$$

$$\int_{0}^{\sigma} (1 - e^{-\phi(r)/kT})\, r^2\, dr = \tfrac{1}{3}\sigma^3$$

and

$$B(T) = \tfrac{2}{3}\pi N_A \sigma^3 \qquad (1\text{-}60)$$

Since σ corresponds to a molecular diameter, $B(T)$ is just four times the volume of a mole of molecules, or has essentially the same meaning as the van der Waals constant b. Considering only the second virial coefficient, we see that the equation of state of the hard-sphere gas is then

$$\frac{PV}{RT} = 1 + \frac{B(T)}{V} = 1 + \frac{b}{V} \qquad (1\text{-}61)$$

If the approximation is made that $1/V = P/RT$, then Eq. (1-61) reduces to Eq. (1-33), $P(V - b) = RT$.

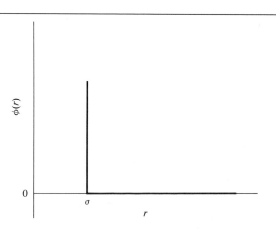

FIGURE 1-17
Hard-sphere potential plot.

The hard-sphere model provides only a poor approximation to real gases, just as Fig. 1-17 is a most crude approximation to Fig. 1-16. Theoreticians make use of more realistic potential functions than the hard-sphere one. However, the determination of really accurate functions is a wave mechanical problem that has not been fully solved as yet. What one usually does is to take a semiempirical form chosen both for its probable approximate correctness and for mathematical convenience. A commonly used such form is the *Lennard–Jones potential*

$$\phi(r) = \frac{\alpha}{r^6} + \frac{\beta}{r^{12}} \tag{1-62}$$

As mentioned in Section 1-8, the attractive potential between molecules is expected to vary as $1/r^6$ at least for large separations; the first term on the right of Eq. (1-62) assumes this attractive potential to apply at all distances. The second term on the right is undoubtedly incorrect theoretically but constitutes a mathematically convenient way of providing a steeply rising repulsive potential.

The effect of introducing both attractive and repulsive potentials is to make $B(T)$ a complicated quantity as far as physical significance is concerned. It is now temperature-dependent and, moreover, may be positive or negative. A calculated curve for $B(T)$ as a function of temperature using the Lennard–Jones potential is given in Fig. 1-18. Here B is given in terms of b_0, where b_0 is the hard-sphere value of $2\pi N_A \sigma^3 / 3$; T^* is a reduced temperature, defined as kT/ϕ^*. As shown in Fig. 1-16, ϕ^* is the minimum

in the potential curve. The quantities σ and ϕ^* are natural ones to use in connection with the Lennard–Jones potential, since this function takes on a rather simple form in terms of them:

$$\phi(r) = -4\phi^* \left[\left(\frac{\sigma}{r}\right)^{12} - \left(\frac{\sigma}{r}\right)^{6} \right] \tag{1-63}$$

One may use Fig. 1-18 to calculate $B(T)$ for various gases if their parameters σ and ϕ^* are known, and some of these values are given in Table 1-6. Notice that the existence of a single curve for $B(T)$ is an illustration of the principle of corresponding states. Here B/b_0 and kT/ϕ^* are the reduced variables. The Lennard–Jones potential becomes a rather poor approximation, however, as one goes to diatomic and polyatomic molecules, especially those which are polar and nonspherical, and the treatment as outlined is much too simple to give a good representation of the $P–V–T$ behavior of such gases over more than a narrow range of conditions. A more accurate potential function would, for example, include the dependence on the relative angular orientation of two approaching molecules.

The second virial coefficient, as indicated by the form of Eq. (1-58), is determined by the form of the potential energy of interaction between two molecules. The third virial coefficient involves the mutual interaction potential for molecules taken three at a time. Although many such calculations have been made, they are obviously quite complicated and the reader is referred to advanced texts at this point.

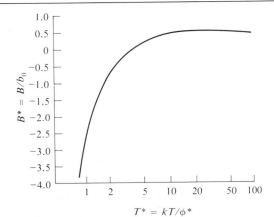

FIGURE 1-18

Reduced plot for the calculation of $B(T)$ of Eq. (1-69) assuming a Lennard–Jones potential. (Adapted from J.O. Hirschfelder, C.F. Curtiss, and R.B. Bird, "Molecular Theory of Gases and Liquids," corrected ed. Copyright 1964, Wiley, New York. Used with permission of John Wiley & Sons. Inc.)

$$B^* = B/b_0$$

$$T^* = kT/\phi^*$$

Gas	ϕ^*/k (K)	σ (Å)	b_0 (cm^3 mole^{-1})
Ne	34.9	2.78	27.10
Ar	119.8	3.40	49.80
Kr	171	3.60	58.86
Xe	221	4.10	86.94
N_2	95.1	3.70	63.78
O_2	118	3.46	52.26
CH_4	148.2	3.82	70.16
CO_2	189	4.49	113.9

TABLE 1-6
Lennard–Jones parameters from second virial coefficients[a]

[a]From J. O. Hirschfelder, C. F. Curtiss, and R. B. Bird, "Molecular Theory of Gases and Liquids," corrected ed., p. 165. Wiley, New York, 1964.

The references at the end of each chapter are generally to specialized monographs from which more detailed information can be obtained on the subject or subjects of the chapter. See the Preface for some additional comments.

GENERAL REFERENCES

BERRY, R.S., RICE, S.A., AND ROSS, J. (1980). "Physical Chemistry." Wiley, New York.

GLASSTONE, S. (1946). "A Textbook of Physical Chemistry," 2nd ed. Van Nostrand–Reinhold, Princeton, New Jersey. A good although partially outdated general reference.

HIRSCHFELDER, J.O., CURTISS, C.F., AND BIRD, R.B. (1964). "Molecular Theory of Gases and Liquids," corrected ed. Wiley, New York. An excellent advanced treatise on the statistical mechanical approach to the properties of gases.

MOELWYN-HUGHES, E.A. (1961). "Physical Chemistry," 2nd ed. Pergamon, Oxford. A useful intermediate-level general text.

PARTINGTON, J.R. (1949). "An Advanced Treatise on Physical Chemistry," Vol. 1. Longmans, Green, Boston, Massachusetts. A very detailed reference.

In addition, a collection of worked-out examination questions is available: ADAMSON, A.W. (1969). "Understanding Physical Chemistry," 2nd ed. Benjamin, New York.

CITED REFERENCES

ADAMSON, A.W. (1978). *J. Chem. Ed.*, **55**, 634.

BENSON, S.W., AND GOLDING, R.A. (1951). *J. Chem. Phys.* **19**, 1413.

GLASSTONE, S. (1946). "Textbook of Physical Chemistry," Van Nostrand–Reinhold, Princeton, New Jersey.

HIRSCHFELDER, J.O., CURTISS, C.F., AND BIRD, R.B. (1964). "Molecular Theory of Gases and Liquids," corrected ed. Wiley, New York.

MACDOUGALL, F.H. (1936). *J. Amer. Chem. Soc.* **58**, 2585.

MCGLASHAN, M.L. (1968). "Physico-Chemical Quantities and Units." Royal Inst. of Chem. Publ. No. 15, London.

WESTON, F. (1950). "An Introduction to Thermodynamics. The Kinetic Theory of Gases, and Statistical Mechanics," Addison-Wesley, Reading, Massachusetts.

EXERCISES AND PROBLEMS

As noted in the Preface, each section of this type is divided into three parts. The first consists of *Exercises*, or very straightforward illustrations of the textual material. The *Problems* section contains more demanding and often longer applications of the same material; and, as the name indicates, *Special Topics Problems* draw on the Special Topics section of this chapter. Numbers given to one significant figure are to be taken as exact. Problems marked with an asterisk require fairly lengthy computations.

EXERCISES

1-1   Calculate the molar volume V and the density of ethane gas at STP, assuming ideal behavior.

Ans. $V = 22.414$ liter, $\rho = 1.342 \times 10^{-3}$ g cm^{-3}.

1-2   Repeat Exercise 1-1, but for the conditions of 25°C and 1.5×10^5 Pa.

Ans. $V = 1.6526 \times 10^{-2}$ m^3 (or 16.526 dm^3);
$\rho = 1.8196$ kg m^{-3} (or 1.8196×10^{-3} kg dm^{-3}).

1-3   Calculate V and ρ for dry air (av. molecular weight = 29 g mole^{-1}) at STP. Repeat the calculation for air saturated with water vapor at 25°C and 1 atm total pressure. Assume ideal behavior.

Ans. (a) $V = 22.414$ liter, $\rho = 1.294 \times 10^{-3}$ g cm^{-3};
(b) $V = 24.47$ liter, $\rho = 1.171 \times 10^{-3}$ g cm^{-3}.

1-4   In the Dumas method one determines the molecular weight of a gas by a direct measurement of its density. A glass bulb weighs 20.2000 g when evacuated, 125.4000 g when filled with water at 25°C, and 20.2171 g when filled with a hydrocarbon gas at 25°C and 100 Torr pressure. Calculate the molecular weight of the gas, assuming ideal gas behavior.

Ans. 30.1 g mole^{-1}.

1-5   The amount 2.731 g of N_2O_4 is introduced into a 2-liter flask at 25°C. Partial dissociation into NO_2 occurs, and the equilibrium pressure is 0.8623 atm. Calculate the degree of dissociation, α, and the value of K_P [Eq. (1-22)].

Ans. $\alpha = 0.1875$, $K_P = 0.1257$ atm.

1-6   Calculate the partial volumes of H_2O, O_2, and N_2 in air saturated with water vapor at 50°C and at 1 atm total pressure. Assume ideal behavior and one mole of total gas.

Ans. $V_{H_2O} = 3.228$ liter, $V_{O_2} = 4.658$ liter, $V_{N_2} = 18.63$ liter.

1-7   Derive the van der Waals equation for n moles of gas.

Ans. $[P + (an^2/v^2)](v - nb) = nRT$.

1-8   Calculate the second and third virial coefficients for NH_3 assuming it to be a van der Waals gas.

Ans. $B(T) = 0.03707 - (50.82/T)$ liter; $C(T) = 1.37 \times 10^{-3}$ liter^2.

1-9   What is the Boyle temperature of NH_3 assuming it to be a van der Waals gas?

Ans. 1370 K.

1-10   Tables 1-4 and 1-5 come from different sources and are not necessarily consistent. Calculate the van der Waals constants for H_2O from its critical point.

Ans. $a = 5.447$ liter^2 atm mole^{-2},
$b = 0.0304$ liter mole^{-1} [Eq. (1-48)], 0.0185 liter mole^{-1} [Eq. (1-47)].

1-11   A hot air balloon has a volume of 100 m^3 and is filled with air at 100°C. Ambient temperature and pressure are 25°C and 1 atm, respectively. (a) Calculate the lift of the balloon in kg. (b) Suppose that the balloon were filled with He at ambient temperature. Calculate the lift in kg. (Assume ideal gas behavior.)

Ans. (a) 23.8 kg, (b) 102.2 kg.

1-12   Suppose that in case (b) of Exercise 1-11 the mass attached to the balloon were 50 kg (that is, the balloon fabric, etc., and its payload). Would the balloon be able to rise to an altitude where ambient temperature and pressure were −40°C and 0.25 atm, respectively?

Ans. No (the lift is only 32.7 kg).

1-13   Fifty moles of NH_3 is introduced into a two-liter cylinder at 25°C. Calculate the pressure if (a) the gas is ideal and (b) it obeys the van der Waals equation.

Ans. (a) 612 atm, (b) 5740 atm.

Using Fig. 1-9, calculate the molar volume of CO_2 at 100°C and 100 atm pressure. Compare this with the ideal gas volume.   **1-14**

Ans. Fig. 1-9: 0.21 liter; ideal gas: 0.306 liter.

What is the critical temperature of a van der Waals gas for which P_c is 100 atm and b is 50 cm³   **1-15**
mole⁻¹?

Ans. 487.5 K.

Jovians, if such intelligent beings exist, might know about the behavior of gases and their extrapolated   **1-16**
behavior in the limit of zero pressure, Eq. 1-5. Their units of mass and length would, of course, be different from ours, and hence their units of pressure and volume; let us, to be specific, call them "spring," S, and "box," B, respectively. Their A values A_J, would thus be given in S B units. Since hydrogen and methane are prevalent substances, they define their absolute or Jovian temperature, T_J, in terms of the A_J values at the two triple points. A_J at the triple point of hydrogen is found to be 65.32 S B, and A_J at the triple point of methane is 422.7 S B. They also use a centigrade scale, assigning 100°J to the above temperature interval. Calculate T_J for the triple point of water. (On *our* centigrade scale, the triple points of hydrogen, methane, and water are -259.14°C, -182.48°C, and 0.01°C, respectively.) Also, calculate their value for the gas constant, R, in S B T_J^{-1} units.

PROBLEMS

Referring to Fig. 1-8, make a semiquantitative plot of (a) an isochor going through point A, (b) an   **1-17**
isochor going through point B, and (c) an isobar going through point C.

As illustrated in Fig. 1-19, a 150 cm long pipe (of cross-section 20 cm²) is lowered into water until   **1-18**
the closed end of the pipe is flush with the general water level. Calculate the distance x to which the water will rise in the pipe. Assume ambient conditions at 25°C and 1 bar pressure, and ideal gas behavior.

A mixture of 1 mole of H_2 and 2 of O_2 is at 25°C and 20 liter volume. Calculate the partial pressure   **1-19**
and partial volume of the H_2 and the O_2. Give pressures in both atmospheres and newtons per square meter. Assume ideal gas behavior.

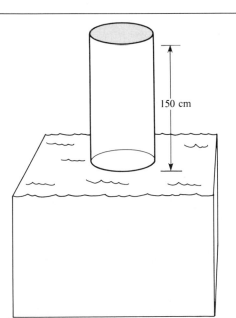
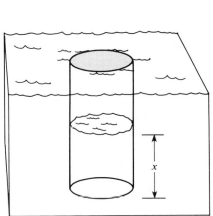

FIGURE 1-19
Illustration for Problem
1-18.

150 cm

x

1-20   The mixture of Problem 1-19 is exploded by means of a spark (the container is a strong one) and reaction to form water goes to completion. The mixture is returned to 25°C. Calculate, assuming ideal gas behavior, the partial pressures and volumes for all species present (remember that some of the water may condense).

1-21   Bulb A, of 400 cm³ volume, contains 0.3 mole of ideal gas A and is thermostatted at 0°C. Bulb B contains 0.4 mole of ideal gas B at a pressure of 2×10^6 N m⁻²; it is thermostatted at 50°C. A connection between the two bulbs is opened so that the gases equilibrate to uniform pressure. Calculate the final pressures of gases A and B.

1-22   A tank of compressed nitrogen gas has a volume of 50 liters; the pressure is 2000 atm initially (at 25°C). Owing to a faulty valve, gas is leaking out at a rate proportional to the difference between the pressure inside the tank and the pressure outside (1 atm). The initial rate of leakage is 0.5 g of gas per second. If we assume that the process continues isothermally at 25°C, how long will it take for half the gas initially present in the tank to leak out?

1-23   The McLeod gauge (see Fig. 1-20) is a device enabling one to make a manometric measurement of very low pressures (down to 10^{-7} Torr). The device is operated as follows. Initially the mercury level is below point a so that the entire apparatus is at the uniform low pressure P_1 which is to be measured. By raising the reservoir B, the mercury level is raised past point b and then further until the meniscus in tube A is at the level c. Once the mercury passes b, the gas in the bulb C is trapped and as the mercury level is raised further, this gas is compressed into the capillary tube D and the meniscus in the capillary reaches level d when the level in tube A reaches c. The distance between c and d is now related to the value of P_1. If V denotes the volume (in cubic centimeters) of bulb C and if the capillary tube is of total length d and of uniform radius r (in millimeters), then if x denotes the distance between c and point d, derive the relationship between x and P_1. In the case of a particular McLeod gauge, V is 250 cm³, d is 10 cm, and r is 0.5 mm; calculate x for P_1 equal to 10^{-5}, 10^{-4}, 10^{-3} Torr, respectively.

1-24   Bulb A, of 500 ml volume, initially contains N_2 at 1.4 atm pressure and 25°C; bulb B, of 600 ml volume, initially contains O_2 at 0.5 atm pressure and at 0°C. The two bulbs are then connected so that there is free passage of gas back and forth between them and the assembly is then brought to a uniform temperature of 20°C. Calculate the final pressure.

1-25   A simple thermometer consists of a bulb filled with an essentially ideally behaving gas and connected to a manometer or pressure gauge. The volume of the gas is kept constant and its temperature is

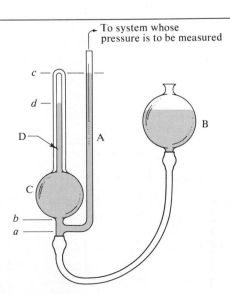

FIGURE 1-20
Schematic for a McLeod gauge.

To system whose pressure is to be measured

measured in terms of its pressure. Suppose that the pressure is 75 mm Hg when the bulb is at 100°C. Calculate the pressure reading at 101°C, at 25°C, and at 26°C.

The vapor of acetic acid contains single and double molecules in equilibrium as shown by the reaction $(CH_3COOH)_2 = 2CH_3COOH$. At 25°C and 0.020 atm pressure, the Pv product for 60 g of acetic acid vapor is 0.541 RT, and at 40°C and 0.020 atm, it is 0.593RT. Calculate the fraction of the vapor forming single molecules at each temperature and the value for the equilibrium constant at each temperature, $K_P = P^2_{CH_3COOH}/P_{(CH_3COOH)_2}$ (see MacDougall, 1936).   **1-26**

A certain gas has a Z value of 0.7 at 200 atm and 0°C. Calculate V_c, P_c, and T_c for this gas, assuming that it obeys the van der Waals equation, and that the value for the constant a is 2.253 liter^2 atm mole^{-2}.   **1-27**

Derive the value of T_r such that   **1-28**

$$\frac{d(P_rV_r)}{dP_r} = 0 \text{ as } P_r \to 0,$$

that is, the value of T_r at the Boyle temperature of a van der Waals gas.

Derive an expression for the coefficient of thermal expansion α,   **1-29**

$$\alpha = \frac{1}{V}\left(\frac{\partial V}{\partial T}\right)_P,$$

for a gas that follows (a) the ideal gas law and (b) the van der Waals equation.

Calculate the pressure versus volume isotherm for CH_3OH at 325 K using the van der Waals equation. Plot the resulting curve (up to 30 liter mole^{-1}). (a) Indicate on the graph how you would estimate the vapor pressure of methanol at this temperature. (b) Obtain the slope dV/dP (at constant T) for $P - 1$ atm, and calculate the coefficient of compressibility of the liquid at this pressure: **1-30***

$$\beta = -\frac{1}{V}\left(\frac{\partial V}{\partial P}\right)_T$$

Compare the result with an experimental value. (c) Estimate the tensile strength of methanol (liquid) at this temperature.

Calculate the ratio P(actual) to P(ideal) for CH_4 at -20°C and a volume of 1.50 liter mole^{-1}. Assume van der Waals behavior.   **1-31**

A nonideal gas is at 0°C and 300 atm pressure; its T_r and P_r values are those of point A in Fig. 1-9. Calculate the van der Waals constants for this gas and its critical volume; assume the gas obeys the van der Waals equation.   **1-32**

The ratio P/ρ obeys the equation $P/\rho = 5.161 \times 10^4 - 2.672 \times 10^{-3} P + 1.822 \times 10^{-11} P^2$ for a certain gas at 0°C; P is in N m^{-2} and ρ is in kg m^{-3}. Calculate (a) the coefficients if P is in atm and ρ is in g liter^{-1}, and (b) the molecular weight of the gas.   **1-33**

Calculate the van der Waals constants a and b for the gas of Problem 1-33.   **1-34**

Let n^0 be the number of moles of N_2O_4 introduced into a liter volume at 25°C. Partial dissociation into NO_2 occurs, and the equilibrium pressure is recorded. The data are   **1-35**

$n^0(\times 10^3)$	6.28	12.59	18.99	29.68
P(atm)	0.2118	0.3942	0.5719	0.8550

[Adapted from F. H. Verhoek and F. Daniels, *J. Amer. Chem. Soc.* **53,** 1250 (1931).] Calculate each K_P [Eq. (1-22)] and the true K_P by extrapolation to zero pressure.

The curve for $T/T_c = 0.8$ in the Hougen–Watson chart (Fig. 1-9) ends abruptly. Reproduce this curve in a sketch, and show by means of a dotted line what a continuation of it should look like. Also sketch in for reference the complete curve shown for $T/T_c = 1$.   **1-36**

1-37 The following data are obtained for a certain gas at 298.2 K:

P (atm)	0.6097	0.8121	0.9737	1.2155
ρ/P (g liter^{-1} atm^{-1})	5.3834	5.3892	5.3940	5.4018

Calculate the molecular weight by the extrapolation method and find the value of $B(T)$.

1-38* Make a plot of V versus P at 25°C for a substance which obeys the van der Waals equation and whose critical temperature and pressure values are those for water. The plot should extend over the range from liquid to gaseous state so as to show the minimum and maximum in pressure that the equation predicts. Estimate from the plot (making your procedures clear) (a) the tensile strength of liquid water, (b) the compressibility of liquid water (compare with the experimental value), and (c) the vapor pressure of water (compare with the experimental value).

1-39* Calculate P versus V for CH_3OH using the van der Waals equation. Do this for 40°C intervals between 0°C and 200°C and plot the results. Recast the data in terms of compressibility factor Z in terms of P_r for various T_r, and plot these curves.

1-40 Make a semiquantitative plot of (a) isobars for P around $P^{0\prime}$, and (b) isosteres for V around $V_\ell{}'$. Use the curves of Figs. 1-7 and 1-8 as a guide.

1-41 Using the Hougen–Watson chart (Fig. 1-9) and the critical constants (Table 1-4) for CO, obtain the value for the second virial coefficient for CO at 25°C. It is suggested that a graphical method be used.

1-42 A certain gas obeys the van der Waals equation with $a = 10^7$ atm cm^6 mole^{-2} and $b = 100$ cm^3 mole^{-1}. Calculate the volume of four moles of the gas when the pressure is 5 atm and the temperature is 300°C.

SPECIAL TOPICS PROBLEMS

1-43 Calculate $B(T)$ for CH_3OH at 25°C (a) using its van der Waals constants and (b) assuming the hard-sphere model and using only the density of liquid methanol.

1-44 Verify Eq. (1-63). That is, show that ϕ^* is the minimum potential energy and that σ is the value of r when ϕ is zero.

1-45 Using Table 1-4 and Fig. 1-18, calculate the compressibility factor for CO_2 at 5 atm and 25°C and at 5 atm and 200°C.

1-46 The value of $B(T)$ for Xe gas is -130.2 cm^3 mole^{-1} at 298.2 K and -81.2 cm^3 mole^{-1} at 373.2 K. Find the value of ϕ^*/k which, using Fig. 1-18, will reproduce this ratio of $B(T)$ values, and from this ϕ^* and b_0.

1-47 Derive Eq. (1-63) from Eq. (1-62).

1-48 Make an approximate calculation of $B(T)$ for methane at 25°C as follows. Use. Eq. (1-58) but assume that $\phi = \infty$ for $r < \sigma$ and that $\phi(r) = -\alpha/r^6$ for $r > \sigma$; and also that $e^{-\phi(r)/kT}$ can be approximated by $(1 - \phi(r/kT))$. Use the appropriate ϕ^*/k and σ values from Table 1-6. Compare your result with the value estimated from Fig. 1-18.

chapter 2
Kinetic molecular theory
of gases

2-1   Introduction

The treatment of ideal and nonideal gases in Chapter 1 is carried out largely from a phenomenological point of view. Behavior is described in terms of the macroscopic variables P, V, and T, although some molecular interpretation was included in the discussion of the a and b parameters of the van der Waals equation (Section 1-8) and in the Special Topics section. We take up here the detailed model of a gas, that is, the kinetic molecular theory of gases. In this model, a gas is considered to be made up of individual molecules, each having kinetic energy in the form of a random motion. The pressure and the temperature of a gas are treated as manifestations of this kinetic energy. In its simplest form, kinetic theory assumes that the molecules experience no mutual attractions.

The elementary picture is the familiar one of a molecule having an average velocity u and bouncing back and forth between opposite walls of a cubical container. With each wall collision a change in momentum $2mu$ occurs, where m is the mass of a molecule. If the side of the container is l, the frequency of such collisions is $u/2l$, and the momentum change per second imparted to the wall, that is, the force on it, is mu^2/l. The pressure, or force per unit area, becomes mu^2/l^3 or $mu^2/\mathbf{v}$. The quantity u refers to the velocity component in some one direction, and the total velocity squared, c^2, is $c^2 = u^2 + v^2 + w^2$, where v and w are the components in the other two directions; on the average these components should be equal, and so we conclude that $u^2 = c^2/3$, and obtain the final equation

$$P\mathbf{v} = \tfrac{1}{3}mc^2 \tag{2-1}$$

where $\mathbf{v}$ denotes the volume per molecule. Per mole, this becomes

$$PV = \tfrac{1}{3}Mc^2 \tag{2-2}$$

where M is the molecular weight of the gas.

One now takes note that the simple picture corresponds, for real gases, to the limiting condition of zero pressure, for which the phenomenological law is the ideal gas equation

$$PV = RT \tag{2-3}$$

and the relationship which is intuitively expected to exist between kinetic energy and temperature is simply

$$RT = \tfrac{1}{3}Mc^2 = \tfrac{2}{3}(\text{kinetic energy}) \tag{2-4}$$

the molar kinetic energy being $\tfrac{1}{2}Mc^2$. Alternatively, one writes

$$c^2 = \frac{3\,RT}{M} \tag{2-5}$$

or

$$c^2 = \frac{3kT}{m} \tag{2-6}$$

where k is the gas constant per molecule, called the *Boltzmann constant*.

This treatment is unsatisfactory at some points. Clearly, it is unrealistic to take all molecules as having the same velocity. Even occasional intermolecular collisions must eventually bring about a distribution of velocities, and since we are describing a theory for time-invariant or equilibrium properties of a gas, we should be dealing with steady state velocity distributions. The quantity c^2 must then really be some kind of average quantity. The argument that one should use $\tfrac{1}{3}c^2$ for the velocity component squared in some one direction is plausible but is not a proof. The sections that follow take up a more elaborate but more satisfying way of obtaining not only the above results, but much additional information as well.

There remain a number of important properties of a gas which cannot be explained unless a finite molecular size is specifically assumed. The model at this point becomes one of molecules that act as hard spheres. By "hard" we mean that they behave like spheres of definite radius r and that their collisions are elastic so that kinetic energy as well as momentum is conserved. Beyond this lie more advanced treatments which allow for the presence of attractive as well as of repulsive forces between molecules. Some aspects of these will be taken up in the Special Topics section.

The immediate task, however, is the treatment of velocity distributions. Here, the central assumption will be that of the Boltzmann principle, namely that the probability of a molecule having an energy ϵ is given by

$$p(\epsilon) = Ae^{-\epsilon/kT} \tag{2-7}$$

where A is a constant. Equation (2-7) can be obtained in various ways, one of which is given in the next section (and another in Section 4-10).

2-2   The Boltzmann distribution law

The Boltzmann principle is central to all of the statistical aspects of physical chemistry, and we proceed to obtain the corresponding relationship, the Boltzmann distribution law, in as plain and simple a manner as seems possible. Later, in Section 4-10, a somewhat more elegant derivation will be given as an introduction to the treatment of statistical thermodynamics.

We consider a system that is isolated (no exchange of matter, heat, or work with its surroundings), so that its total energy **E** is constant and the total number of molecules N is also constant. The molecules making up this system can have various energy states; let them be called ϵ_1, ϵ_2, ϵ_3, and so on.

There will be many ways in which the N molecules could be assigned specific energies so as to give the same total energy **E**. For example, let **E** be 20 units of energy and N be ten molecules, and suppose that there are three possible states of one, three, and five units of energy. Possible distributions are given in Table 2-1. In this case there are three different ways of satisfying the two requirements of fixed N and fixed **E**.

We will be looking for the most probable distribution among energy states, and therefore need to assign a relative probability or statistical weight to each distribution. This statistical weight is obtained as follows. If we were just putting N molecules in as many boxes, their permutations would be $N!$. However, we consider only molecules in different energy states as distinguishable. That is, the N_1 molecules in the ϵ_1 box are taken to be indistinguishable, as are the N_2 molecules in the ϵ_2 box, and so on. We must then divide out the permutations that should not be present. Thus in the case of system 2, there are 6! ways of permuting the molecules in the ϵ_1 box, 3! ways for the ϵ_2 box, and 1! or one way for the ϵ_3 box. The distinguishable permutations, which give the desired statistical weight, are thus $10!/(6!)(3!)(1!) = 840$.

The general statements of the preceding conditions are then

$$N = \sum N_i = \text{constant} \tag{2-8}$$

$$\mathbf{E} = \sum N_i \epsilon_i = \text{constant} \tag{2-9}$$

$$W = \frac{N!}{N_1! N_2! N_3! \cdots} = \frac{N!}{\Pi_i N_i!} \tag{2-10}$$

where N_i denotes the number of molecules in the ith state of energy ϵ_i, and W is the statistical weight or probability of the particular distribution. To repeat, the denominator of Eq. (2-10) serves to take out those permutations that do not count because of the indistinguishability of molecules in the same energy state.

One further point completes the basic picture. If N is a very large number, for example, Avogadro's number, it turns out that W will peak very sharply at some one distribution. That is, there will be some set of N_1, N_2, and so on values giving the largest W, and relatively small departures from this proportion will cause W to drop sharply. This largest W is called W_{max}. Thus the set of requirements of Eqs. (2-8)–(2-10) acts to define a most probable distri-

	Number in given state[b]			
SYSTEM	$\epsilon_1 = 1$	$\epsilon_2 = 3$	$\epsilon_3 = 5$	
1	5	5	0	
2	6	3	1	
3	7	1	2	

TABLE 2-1
Ways of distributing molecules between states[a]

[a]Assuming ten molecules and total system energy of 20 units.
[b]Energy in arbitrary units.

bution and one which is assumed, in the case of a large number of molecules, to be *the* distribution.

The preceding conditions and assumptions are in fact sufficient to give the immediate precursor to the Boltzmann distribution law, namely the conclusion that

$$p(\epsilon) = \text{(constant)}\ e^{-\beta\epsilon} \tag{2-11}$$

where $p(\epsilon)$ is the probability that the molecule has energy ϵ. The constant β will be identified with $1/kT$ in Section 2-7, when a consequence of Eq. (2-11) is compared with the ideal gas law.

We can obtain Eq. (2-11) as follows. Since we are considering a system at equilibrium, the distribution will be one for which W is at a maximum—that is, we expect the equilibrium distribution to be the one that is the most probable. We now imagine that a small redistribution δ of molecules takes place, subject to the restriction that neither N nor the total energy of the system changes,

$$\sum_i \delta N_i = 0 \tag{2-12}$$

$$\sum_i \epsilon_i\ \delta N_i = 0 \tag{2-13}$$

Since W is at a maximum, it also follows that δW must be zero. It is convenient at this point to take the logarithm of Eq. (2-10):

$$\ln W = \ln N! - \sum_i \ln N_i!$$

and write as the condition

$$\delta(\ln W) = 0 = \sum_i \delta(\ln N_i!) \tag{2-14}$$

since $\delta(\ln N!) = 0$. We are dealing with very large numbers and it is permissible to replace factorials by Stirling's approximation,

$$\ln x! = x \ln x - x \tag{2-15}$$

Equation (2-14) becomes

$$0 = \sum_i \delta(N_i \ln N_i - N_i)$$

$$= \sum_i \left[N_i \frac{\delta N_i}{N_i} + (\ln N_i)\ \delta N_i - \delta N_i \right]$$

or

$$\sum_i (\ln N_i)\ \delta N_i = 0 \tag{2-16}$$

Equations (2-12)–(2-14) impose three different conditions on the δN's and are to be obeyed even though the system makes small, arbitrary fluctuations. A way of handling such a situation is Lagrange's method of undetermined multipliers (see Appendix B). We add the three conditions, but in some ratio which is to be determined; that is, we write

$$\sum_i (\alpha + \beta\epsilon_i + \ln N_i)\ \delta N_i = 0 \tag{2-17}$$

where Eqs. (2-12) and (2-13) have been multiplied by the undetermined coefficients α and β, respectively.

Equation (2-17) can be written out in detail as a sum of terms in δN_1, δN_2, δN_3, and so on:

$$(\alpha + \beta\epsilon_1 + \ln N_1)\,\delta N_1 + (\alpha + \beta\epsilon_2 + \ln N_2)\,\delta N_2$$
$$+ (\alpha + \beta\epsilon_3 + \ln N_3)\,\delta N_3 + \cdots = 0 \quad (2\text{-}18)$$

Now, if there were only two states, say ϵ_1 and ϵ_2, Eqs. (2-8) and (2-9) would absolutely fix the distribution (for example, system 1 of Table 2-1). With three states the most probable population of ϵ_3 could be varied, but given δN_3, this would then determine δN_1 and δN_2. With a larger number of states, δN_3, δN_4, and so on could be chosen arbitrarily, and this would then fix δN_1 and δN_2. If we elect to choose values for α and β such that the terms of Eq. (2-18) in δN_1 and δN_2 are zero, which can be done since α and β are adjustable constants, then the equation reduces to the requirement that the sum of all the terms in δN_3, δN_4, and so on must be zero. Since the variations δN_3, δN_4, and so on are arbitrary, the only way for this requirement to be generally true is for each term separately to be zero. The general condition is therefore

$$\alpha + \beta\epsilon_i + \ln N_i = 0 \qquad\qquad\qquad (2\text{-}19)$$

or

$$N_i = e^{-\alpha}e^{-\beta\epsilon_i} = (\text{constant})\, e^{-\beta\epsilon_i} \qquad\qquad (2\text{-}20)$$

which is Eq. (2-11).

This result, namely that the most probable number of molecules in a state of energy ϵ_i is proportional to $e^{-\beta\epsilon_i}$ may seem startling in that it is obtained on so general a basis. To repeat, it is a consequence of the restrictions of Eqs. (2-12)–(2-14) plus the assumption that molecules will find that energy distribution having the greatest statistical weight as given by the permutation formula of Eq. (2-10).

Later in this chapter (Section 2-7) we note that $\beta = 1/kT$, so that the Boltzmann distribution law becomes

$$N_i = A\, e^{-\epsilon_i/kT} \qquad\qquad\qquad (2\text{-}7)$$

where A is a constant. If we sum over all states,

$$N = \sum_i N_i = A \sum_i e^{-\epsilon_i/kT} \qquad\qquad (2\text{-}21)$$

The fraction x_i of molecules in a given state is then

$$x_i = N_i/N = \frac{e^{-\epsilon_i/kT}}{\displaystyle\sum_i e^{-\epsilon_i/kT}} \qquad\qquad (2\text{-}22)$$

The denominator of Eq. (2-22) is called the *partition function*, **Q**,

$$\mathbf{Q} = \sum_i e^{-\epsilon_i/kT} \qquad\qquad\qquad (2\text{-}23)$$

Q plays a central role in statistical thermodynamics, as we shall see in Chapter 4. Since there may be two or more states of identical energy, a more general statement is

$$Q = \sum_i g_i e^{-\epsilon_i/kT} \qquad (2\text{-}24)$$

where g_i is the number of times the ith energy level occurs, and is known as the *degeneracy*.

2-3   The barometric equation

A very simple application of the Boltzmann distribution law is to the case of a column of ideal gas in a uniform gravitational field. The potential energy of a molecule of mass m at an elevation h is just mgh, where g is the acceleration due to gravity. From Eq. (2-7), the ratio of the numbers of molecules at elevations h_2 and h_1 is just

$$N_2/N_1 = e^{-mg(h_2-h_1)/kT} \qquad (2\text{-}25)$$

if the column is at a uniform temperature. If we take $h_1 = 0$ (i.e., sea level), then

$$N_h = N_0 e^{-mgh/kT} \qquad (2\text{-}26)$$

For an ideal gas at constant temperature, N is proportional to concentration, C, and molecular weight M is Avogadro's number times m, or $M = N_A m$. Also noting that $R = N_A k$, an alternative and more useful form of Eq. (2-26) is

$$C_h = C_0 e^{-Mgh/RT} \qquad (2\text{-}27)$$

Since C is proportional to pressure at constant temperature, it follows that

$$P_h = P_0 e^{-Mgh/RT} \qquad (2\text{-}28)$$

Equation (2-28) is known as the *barometric equation*. Note that if pressure is in atm, then $P_0 = 1$ atm.

EXAMPLE   As an application of Eq. (2-28), consider a column of air of $M_{av} = 29$ g mole^{-1}, $T = 298$ K, $g = 981$ cm sec^{-2}, and $P_0 = 1$ atm at $h = 0$. The exponential term must be dimensionless, so R must now be in ergs K^{-1} mole^{-1} and h in centimeters. One then finds

$$P_h = \exp\left[\frac{-(29)(981)h}{(8.31 \times 10^7)(298)}\right] = \exp(-1.149 \times 10^{-6}h)$$

Thus if $h = 1$ km, or 10^5 cm

$$P_h = e^{-0.1149} = 0.891$$

Note that in the SI system, $M_{av} = 0.029$ kg mole^{-1}, $g = 9.81$ m s^{-2}, h is in meters, and R should be in J K^{-1} mol^{-1}.

It is worth taking a moment to discuss some of the mathematical aspects of an exponential equation such as Eq. (2-28). Figure 2-1 gives the plot of P vs. h for the conditions of the above Example. Note that at $h = 6.03$ km, $P_h/P_0 = 1/2$ or $P = 0.5$ atm. The "half-height" or $h_{1/2}$ (the height for the pressure to decrease by a factor of two) is independent of the value of P_0. Thus, starting at 6.03 km, the pressure will decrease by half again in another 6.03 km, and so will be 0.25 at $h = 12.06$ km, and so on, as illustrated in the figure.

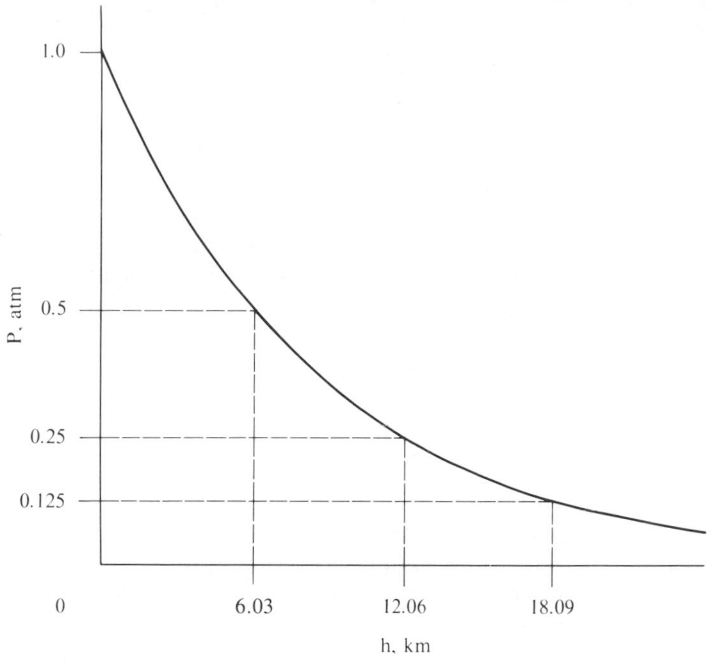

FIGURE 2-1
Decrease of barometric pressure with altitude for air at 298 K.

Equation (2-28) is of the general form **EXAMPLE**

$$y = y_0 e^{-kx} \tag{2-29}$$

or

$$\ln (y/y_0) = -kx. \tag{2-30}$$

Now, $x_{1/2}$ is the value of x for $y/y_0 = 1/2$, so

$$\ln(1/2) = -kx_{1/2}$$

or

$$kx_{1/2} = 0.6931 \tag{2-31}$$

Thus, in the example just discussed, $h_{1/2} = 0.6931/0.1149 = 6.03$ km.

A further point is as follows. In view of the laws for mixtures of ideal gases discussed in Section 1-5, Eq. (2-28) actually applies separately to each component of a mixture. Thus each component of the earth's atmosphere has its own separate barometric distribution (assuming ideal gas behavior), with the consequence that the pressures and hence concentrations of the lighter components decrease less rapidly with altitude than do those of the heavier ones.

If the ratio of nitrogen to oxygen pressures is 4 at sea level (i.e., 80 mole percent N_2), **EXAMPLE** what should the ratio be at 10 km assuming constant g and a constant temperature of 25°C? If we write Eq. (2-28) for each gas, and take the ratio, then

$$(P_{N_2}/P_{O_2})_{h=10 \text{ km}} = 4 \exp \left\{ - \left[\frac{M_{N_2}gh}{RT} - \frac{M_{O_2}gh}{RT} \right] \right\}$$

$$= 4 \exp \left[-(28.01 - 31.99) \frac{981 \times 1 \times 10^6}{8.31 \times 10^7 \times 298} \right]$$

$$= 4.68$$

Thus the components of air should be appreciably fractionated at 10 km, and the use of an average molecular weight of 29 for air in the previous example was an approximation.

In obtaining Eq. (2-28), it was assumed that g and T were constant. The latter restriction is an especially severe one. An alternative approach to the barometric equation, given in Section 4-ST-1, is easily adopted to handle more complicated conditions.

2-4   The distribution of molecular velocities*

We start by applying Eq. (2-7) to the case of a one-dimensional gas, that is, to a system of molecules having only kinetic energy due to motion along one direction in space, say the x-direction. Other forms of energy than kinetic are not of concern here, and it is assumed that the probability of a given molecule having a certain kinetic energy is independent of the kinetic energies of other molecules. In effect, we assume ideal gas-like behavior. The Boltzmann equation then gives

$$p(u) = A \exp \left(-\frac{mu^2}{2kT} \right) \tag{2-32}$$

where u is the velocity in the x direction (and could be either positive or negative), $\epsilon = \frac{1}{2} mu^2$, and A is constant (not the same as in Eq. 2-21).

Kinetic energy states are so closely spaced that we have an essentially continuous distribution of velocities, and we are interested not so much in the probability of some exact velocity u but rather in the fraction $\Delta x(u)$ of molecules having a small range of u values, where x denotes mole fraction. If we make this range small enough, that is, between u and $u + du$, then $\Delta x(u)$ becomes $dx(u)$ and

$$p(u) = dx(u)/du \tag{2-33}$$

If the total number of molecules is Avogadro's number, then an alternative form of Eq. (2-32) is

$$\frac{dN(u)}{N_A} = A \left[\exp\left(-\frac{mu^2}{2kT} \right) \right] du \tag{2-34}$$

The proportionality constant A can be evaluated by the requirement that the sum of probabilities for all possible velocities must be unity:

$$\int \frac{dN(u)}{N_A} = 1 = A \int_{-\infty}^{\infty} \left[\exp\left(-\frac{mu^2}{2kT} \right) \right] du \tag{2-35}$$

The integral of Eq. (2-35) can be put in a standard form, that is,

---

*We will use "velocity" and "speed" as synonymous; in a stricter usage, "speed" is a scalar quantity and "velocity" is a vector one.

$$\int_0^\infty \left[\exp(-a^2x^2) \right] dx = \frac{\sqrt{\pi}}{2a} \tag{2-36}$$

where in the present case $a = (m/2kT)^{1/2}$. It is entirely arbitrary whether we call a direction plus or minus, and, of course, the sign of u vanishes when we write u^2; the distribution must therefore be symmetric around $u = 0$. We therefore have

$$\int_{-\infty}^\infty \left[\exp\left(-\frac{mu^2}{2kT} \right) \right] du = 2 \int_0^\infty \left[\exp\left(-\frac{mu^2}{2kT} \right) \right] du = \left(\frac{2\pi kT}{m} \right)^{1/2}$$

The constant A, called the normalization constant, is thus $(m/2\pi kt)^{1/2}$ and Eq. (2-34) becomes

$$\frac{dN(u}{N_A} = \left(\frac{m}{2\pi kT} \right)^{1/2} \left[\exp\left(-\frac{mu^2}{2kT} \right) \right] du \tag{2-37}$$

Equation (2-37) is known as the *Maxwell–Boltzmann* equation and was obtained in about 1860 by J.C. Maxwell and L. Boltzmann, assuming a classical continuum of velocities. Quantum mechanics, which came later, yields virtually the same result because translational energy states are so very closely spaced (see Section 4-12).

The graph of Eq. (2-37), that is, of $(1/N_A)dN(u)/du$ versus u, is shown in Fig. 2-2 for the case of a gas of molecular weight 28, such as nitrogen, at 25°C and 1025°C. The coefficient of u^2 can be written $M/2RT$ and is thus equal to $28/(2)(8.314 \times 10^7)(298.1) = 5.649 \times 10^{-10}$ at 25°C and 1.297×10^{-10} at 1025°C, with u in centimeters per second. The coefficient A can likewise be written $(M/2\pi RT)^{1/2}$ and has the values 1.34×10^{-5} and 0.644×10^{-5} sec cm^{-1} at the

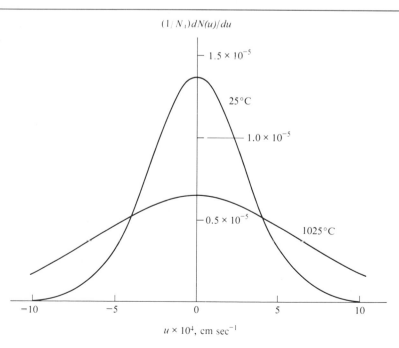

$(1/N_1)dN(u)/du$

1.5 × 10⁻⁵

25°C

1.0 × 10⁻⁵

0.5 × 10⁻⁵

1025°C

−10   −5   0   5   10

$u \times 10^4$, cm sec^{-1}

FIGURE 2-2
Molecular velocity distribution in one dimension for N_2 at 25°C and at 1025°C.

two temperatures, respectively. The distribution is symmetric, of course, and the most probable velocity is zero. The physical explanation of this last observation is that a molecule has an equal chance of gaining a velocity increment in either the positive or the negative direction. Finally, the spread of the distribution, as measured, for example, by the width at half-maximum, increases with increasing temperature.

It is significant that Eq. (2-37) has the same functional form as the *normal error function*. This last may be written in the form

$$p(x) = \frac{1}{\sigma(2\pi)^{\frac{1}{2}}} \exp\left(-\frac{x^2}{2\sigma^2}\right) \tag{2-38}$$

Here, $p(x)$ is the probability of observing an error or departure x from the mean, and the quantity σ is called the *standard deviation* (from the mean). Equation (2-38) follows from Eq. (2-37) if x is set equal to u, and σ^2 to kT/m.

The normal error function is usually derived from consideration of the case of a measurement that is subject to cumulative error arising from the random effect of a large number of small errors, each of which may be positive or negative. For a large number of such measurements σ^2 has the alternative meaning of being the average of the squares of the deviations from the mean.

Equation (2-37) is thus an "error" curve for velocities in that it gives the chance of finding some given velocity or deviation from the most probable velocity. That is, if repeated samplings of a gas are made, the error curve gives the frequency with which different specific velocities are to be observed. If a large number of such samplings are made, then σ^2 gives the average of the squares of the observed velocities, that is, the mean square velocity. Thus

$$\sigma^2 = \overline{u^2} = \frac{kT}{m} \tag{2-39}$$

Equation (2-37) gives the velocity distribution for a one-dimensional ideal gas, a rather fictitious substance. However, it also gives the distribution of the x-velocity components of a three-dimensional ideal gas. As we will see in Section 2-6, this can be a very useful quantity.

The case of a two-dimensional gas follows in a straightforward manner. We now allow velocities along the x and y directions, given by u and v. The probability function $p(u,v)$ is now the product of the two separate functions, so that the Maxwell-Boltzmann equation in two dimensions is

$$\frac{dN(u,v)}{N_A} = A\left\{\exp\left[-\frac{m(u^2 + v^2)}{2kT}\right]\right\} du\ dv \tag{2-40}$$

where $dN(u, v)/N_A$ is now the fraction of molecules having velocity components between u and $u + du$ and v and $v + dv$. Equation (2-40) factors into two integrals, each analogous to that of Eq. (2-35), with the result that on carrying out the normalization procedure, one finds that $A = m/2\pi kT$, and the complete distribution equation becomes

$$\frac{dN(u,v)}{N_A} = \frac{m}{2\pi kT}\left\{\exp\left[-\frac{m(u^2 + v^2)}{2kT}\right]\right\} du\ dv \tag{2-41}$$

The distribution function of Eq. (2-41) now requires a three-dimensional plot for its display, as illustrated in Fig. 2-3. Again the maximum is at u and

$(1/N_0)\, dN(u, v)/du\, dv$

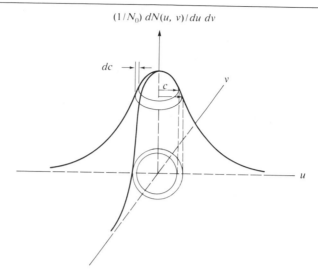

FIGURE 2-3
Molecular velocity
distribution for a gas in
two dimensions.

v equal to zero, and, of course, it is symmetric with respect to positive and negative velocities.

It is ordinarily of more interest to deal with the net velocity c than with the separate components. That is, although the u and v velocity components represent independent ways in which the molecule can have kinetic energy, it is the net velocity of the molecule and its total kinetic energy $mc^2/2$ that are needed for most applications. The speed or net velocity c is

$$c^2 = u^2 + v^2$$

$$(2\text{-}42)$$

The sum of all the possible ways in which the net velocity can increase from c to $c + dc$, that is, by adding increments du and dv to various combinations of u and v, is given by the area of the annulus shown in Fig. 2-3. This argument is illustrated in more detail in Fig. 2-4. The effect is that $du\, dv$ can be replaced by $2\pi c\, dc$, so that the distribution law becomes

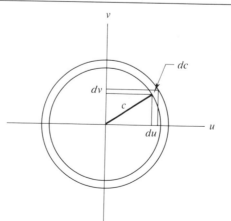

FIGURE 2-4
Projection of the annulus
of Fig. 2-3 onto the u,v
plane.

$$\frac{dN(c)}{N_A} = \frac{m}{kT}\left[\exp\left(-\frac{mc^2}{2kT}\right)\right]c\,dc \tag{2-43}$$

The two-dimensional distribution law for c, Eq. (2-43), can now be plotted on an ordinary graph, since only the net velocity is involved; also we have lost the information as to whether c is positive or negative and plot only its magnitude. Some sample graphs, again for a gas of molecular weight 28, are shown in Fig. 2-5. Notice that there is now a nonzero most probable velocity, in contrast to the situation in Fig. 2-3. The reason is that, while the probability of individual velocity components u and v decreases with their increasing values, the number of ways in which a given velocity magnitude c can be made up of the u and v components increases in proportion to c. For a while the latter effect overrides the former. Notice also that the most probable c value increases with increasing temperature and that the half-width increases with increasing temperature.

The extension to three dimensions follows the same series of steps. The basic distribution law is

$$\frac{dN(u,\,v,\,w)}{N_A} = A\left\{\exp\left[-\frac{m(u^2 + v^2 + w^2)}{2kT}\right]\right\}du\,dv\,dw \tag{2-44}$$

where w is the added velocity component in the z direction. This again can be factored into three integrals, so that the normalization constant becomes the cube of that for the one-dimensional gas, to give

$$\frac{dN(u,v,w)}{N_A} = \left(\frac{m}{2\pi kT}\right)^{3/2}\left\{\exp\left[-\frac{m(u^2 + v^2 + w^2)}{2kT}\right]\right\}du\,dv\,dw \tag{2-45}$$

Again the chief interest is in the net velocity, now given by $c^2 = u^2 + v^2 + w^2$, and the number of ways in which a velocity increment dc can be obtained is related to the separate increments by

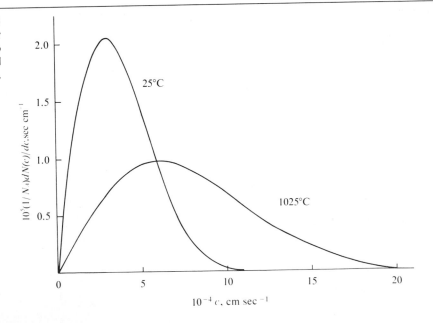

25°C

1025°C

$10^5(1/N_0)dN(c)/dc$, sec cm^{-1}

$10^{-4}c$, cm sec^{-1}

$$4\pi c^2\, dc = du\, dv\, dw$$

Note that $4\pi c^2 dc$ is the volume of a spherical shell of radius c and thickness dc.

The distribution law in net velocity c is then

$$p(c) = \frac{dN(c)}{N_A dc} = 4\pi\left(\frac{m}{2\pi kT}\right)^{3/2}\left[\exp\left(-\frac{mc^2}{2kT}\right)\right]c^2. \tag{2-46}$$

This is the Maxwell–Boltzmann distribution equation in three dimensions.

Calculate $p(c)$ for O_2 at 25°C and $c = 1 \times 10^5$ cm sec^{-1}. The ratio $m/2kT = M/2RT = (31.999)/(2)(8.31441 \times 10^7)(298.15) = 6.4542 \times 10^{-10}$ and $mc^2/2kT = 6.4542$, so that $\exp(-mc^2/2kT) = 1.5739 \times 10^{-3}$. Substitution into Eq. (2-46) gives   **EXAMPLE**

$$p(c) = 4\pi(6.4542 \times 10^{-10}/\pi)^{3/2}(1.5739 \times 10^{-3})(1 \times 10^5)^2 = 5.824 \times 10^{-7}$$

In SI units, $M/2RT = (0.031999)/(2)(8.31441)(298.15) = 6.4542 \times 10^{-6}$, $mc^2/2kT = 6.4542$ (or the same, as it must be since the exponential is dimensionless and therefore cannot depend on the choice of system of units). Then

$$p(c) = 4\pi(6.4542 \times 10^{-6}/\pi)^{3/2}(1.5739 \times 10^{-3})(1 \times 10^3)^2 = 5.824 \times 10^{-5}$$

Note that $p(c)$ is a hundredfold larger in the SI calculation. This is because the velocity interval dc is 100 times larger when c is in m sec^{-1} than when it is in cm sec^{-1}.

Equation (2-46) is graphed in Fig. 2-6 and looks much like the one for a two-dimensional gas. The most probable velocity, that is, the velocity at the maximum of the distribution, is larger, however. This most probable velocity c_p may be evaluated analytically by setting $dp(c)/dc$ equal to zero. On doing this, one obtains from Eq. (2-46) the condition

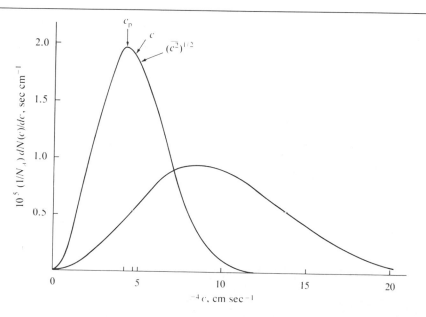

FIGURE 2-6
Three-dimensional velocity distribution for N_2 at 25°C and at 1025°C.

$$-\frac{m}{2kT}\left[\exp\left(-\frac{mc_p^2}{2kT}\right)\right](2c_p)(c_p^2) + \left[\exp\left(-\frac{mc_p^2}{2kT}\right)\right](2c_p) = 0$$

or

$$c_p^2 = \frac{2kT}{m} \tag{2-47}$$

As a final comment on the general character of these distribution laws, we note that the probability of a given velocity rapidly decreases with increasing c owing to the exponential term. There is, however, always a finite probability for any given velocity, no matter how large. For example, the fraction of molecules per unit velocity increment having a velocity of $5c_p$ is only about 10^{-10} of that having velocity c_p itself. Such a number, although small, can be quite significant when chemical reaction rates are being treated.

2-5   Average quantities from the distribution laws

It is a common experience to take the average of some property for a collection of objects. For example, the average value of a coin in a bag of mixed coins would be obtained by dividing the total value by the number of coins. In more detail, the operation is one of dividing two sums:

$$\bar{s} = \frac{\Sigma\, s_i N_i}{\Sigma\, N_i} \tag{2-48}$$

That is, the product of the value s_i of the ith type of coin times the number N_i of such coins is summed over all types and is then divided by $\Sigma\, N_i$, the total number of coins. The average of any property could be obtained in this way; if s denotes the mass of a given type of coin, then $\bar{s}$ would now be the average mass of a coin, and so on.

If there is a very large number of objects with many gradations of the property s, then these summations can be approximated by an integration procedure. The principal change is that N_i is now replaced by a probability or distribution function $N(s)$, where $dN(s)$ gives the chance of finding objects with s lying between s and $s + ds$:

$$\bar{s} = \frac{\int_0^\infty s\, dN(s)}{\int_0^\infty dN(s)} = \frac{1}{N_0}\int_0^\infty s\, dN(s) \tag{2-49}$$

The integral $\int_0^\infty dN(s)$ simply represents the sum of all objects and must therefore equal N_0, their total number.

We can apply Eq. (2-49) to obtain two important types of velocity averages for the three-dimensional case. The first is simply the average velocity $\bar{c}$ given by

$$\bar{c} = \frac{1}{N_A}\int_0^\infty c\, dN(c)$$

or, from Eq. (2-46),

$$\bar{c} = 4\pi\left(\frac{m}{2\pi kT}\right)^{3/2}\int_0^\infty\left[\exp\left(-\frac{mc^2}{2kT}\right)\right]c^3\, dc \tag{2-50}$$

The integral may be evaluated as follows. We let $x = mc^2/2kT$, and so obtain

$$\bar{c} = 4\pi\left(\frac{m}{2\pi kT}\right)^{3/2} 2\left(\frac{kT}{m}\right)^2 \int_0^\infty x e^{-x}\, dx \tag{2-51}$$

The integral is now a standard one known as a gamma function* and its value is just unity. The resulting expression for $\bar{c}$ is thus

$$\bar{c} = \left(\frac{8kT}{\pi m}\right)^{1/2} \tag{2-52}$$

The second average is that of c^2; this is given by

$$\bar{c^2} = 4\pi\left(\frac{m}{2\pi kT}\right)^{3/2} \int_0^\infty \left[\exp\left(-\frac{mc^2}{2kT}\right)\right] c^4\, dc \tag{2-53}$$

When we let $x = mc^2/2kT$ the integral reduces to one of the form $\int_0^\infty x^{3/2} e^{-x}\, dx$, again a gamma function, and one whose value is $\frac{3}{4}\sqrt{\pi}$ [see the footnote following Eq. (2-51)]. On working through the algebra, we find

$$\bar{c^2} = \frac{3kT}{m} \tag{2-54}$$

or, alternatively, the average molar kinetic energy E is then

$$E = \frac{3}{2}RT \tag{2-55}$$

Note that the right-hand sides of Eqs. (2-6) and (2-54) are identical. Thus the average velocity referred to in the introductory section is really the square root of the average of velocities squared, $(\bar{c^2})^{1/2}$, or the root mean square velocity. See also Section 2-7.

To summarize this section, the three characteristic velocities for a gas in three dimensions are

$$\text{most probable} \quad c_p \quad = \left(\frac{2kT}{m}\right)^{1/2} \qquad \text{[Eq. (2-47)]},$$

$$\text{average} \quad \bar{c} \quad = \left(\frac{8kT}{\pi m}\right)^{1/2} = 1.128 c_p \qquad \text{[Eq. (2-52)]},$$

$$\text{root mean square} \ (\bar{c^2})^{1/2} \ = \left(\frac{3kT}{m}\right)^{1/2} = 1.225 c_p \qquad \text{[Eq. (2-54)]}.$$

These three velocities are located on one of the distributions of Fig. 2-6 to show their relative positions.

---

*Integrals of the type $\int_0^\infty x^{n-1} e^{-x}\, dx$ occur frequently in physical chemistry; this integral is called the gamma function $\Gamma(n)$. Standard tables of $\Gamma(n)$ are available for $0 < n < 1$ and $1 < n < 2$; some important values are $\Gamma(1/2) = \sqrt{\pi}$ and $\Gamma(1) = 1$. Also, a very useful relationship is $\Gamma(n + 1) = n\Gamma(n)$ for n a positive integer.

2-6   Some applications of simple kinetic molecular theory. Collision frequency on a plane surface and Graham's law

An important quantity given by simple kinetic molecular theory is the frequency with which molecules hit a plane surface. This frequency is, for example, central to the kinetic treatment of adsorption-desorption processes or, more generally, to those of evaporation and condensation.

The relationship can be derived fairly simply as follows. Since all directions in space should be equivalent, the result should be independent of direction; we can therefore assume that the molecules impinging on a plane surface are doing so from the x direction, so that only their velocity components in that direction need be considered. The distribution law for such x components is given by Eq. (2-37):

$$\frac{d\mathbf{n}(u)}{\mathbf{n}_0} = \left(\frac{m}{2\pi kT}\right)^{1/2}\left[\exp\left(-\frac{mu^2}{2kT}\right)\right]du$$

where $\mathbf{n}$ and $\mathbf{n}_0$ denote numbers of molecules per unit volume, $\mathbf{n}_0$ being the total concentration; $d\mathbf{n}(u)/\mathbf{n}_0$ is again the fraction of molecules having velocity between u and $u + du$. (We use $\mathbf{n}$ rather than n since the latter symbol has previously been used to denote number of moles.)

As illustrated in Fig. 2-7, if we consider a particular group of molecules of velocity component u_1, the ones that will reach the surface in the next second will be those within the distance x_1, where $x_1 = u_1$ (unit time). Per unit area of surface, the total number of molecules of this velocity group reaching the surface per second will then be $\mathbf{n}(u_1)u_1$ (since u_1 times the unit area gives the volume containing the successful molecules of that velocity group). For some other group of, say, lower velocity u_2 the number of that group reaching the surface per second will be $\mathbf{n}(u_2)u_2$. In effect, then, we need to weight the distribution equation by u to get $dZ(u)$, the number of molecules per unit volume reaching the wall whose velocity component lies between u and $u + du$:

$$dZ(u) = \mathbf{n}_0\left(\frac{m}{2\pi kT}\right)^{1/2}\left[\exp\left(-\frac{mu^2}{2kT}\right)\right]u\,du \qquad (2\text{-}56)$$

We obtain Z, the total surface collision frequency per unit area, by integrating over all acceptable values of u, namely from zero to plus infinity, where a

FIGURE 2-7
Volumes swept out in unit time by molecular velocity groups u_1 and u_2, $u_1 > u_2$.

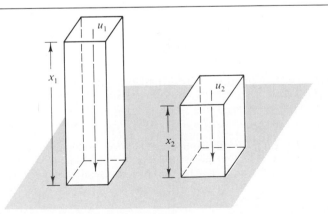

plus velocity is one toward the surface. On carrying out this integration, we obtain

$$Z = \mathbf{n}\left(\frac{kT}{2\pi m}\right)^{1/2} \tag{2-57}$$

(the subscript to the concentration $\mathbf{n}$ is no longer needed).

Equation (2-57) can be phrased in some alternative ways. On referring to Eq. (2-52) for the average velocity $\bar{c}$, we see that

$$Z = \tfrac{1}{4}\mathbf{n}\bar{c} \tag{2-58}$$

Alternatively, the kinetic molecular theory at this level of sophistication implies ideal gas behavior, and since the concentration $\mathbf{n}$ is then equal to P/kT, we have

$$Z = P\left(\frac{1}{2\pi mkT}\right)^{1/2} = PN_A\left(\frac{1}{2\pi MRT}\right)^{1/2} \tag{2-59}$$

Finally, the mass collision frequency Z_m follows if Z is multiplied by the mass per molecule:

$$Z_m = \rho\left(\frac{kT}{2\pi m}\right)^{1/2} \tag{2-60}$$

where ρ is the density.

The preceding discussion provides a theoretical basis for understanding a conclusion reached in 1829, known as *Graham's law*. Graham studied the rate of *effusion* of gases, that is, the rate of escape of gas through a small hole or orifice. He found that for a given temperature and pressure difference, the rate of effusion of a gas is inversely proportional to the square root of its density. For two different gases, then, we have

$$\frac{n_1}{n_2} = \left(\frac{\rho_2}{\rho_1}\right)^{1/2} \tag{2-61}$$

where n denotes the number of moles escaping per unit time. Since the comparison is at a given pressure and temperature, an alternative form of Graham's law is

$$\frac{n_1}{n_2} = \left(\frac{M_2}{M_1}\right)^{1/2} \tag{2-62}$$

where M is molecular weight.

Graham's law follows directly from Eq. (2-59) if it is assumed that in an effusion experiment the rate of passage of molecules through the hole is proportional to the rate at which they would be hitting the area of surface corresponding to the area of the hole. Effusion rates are thus often assumed to be given directly by Z. There is a problem in that, unless the hole is very small, there may be both a pressure and a temperature drop in its vicinity, which introduces a correction term dependent on the effusion rate and hence on the molecular weight of the gas.

It is also important in effusion that the flow be molecular, that is, the molecules should escape directly through the hole without collisions with the sides of the hole or with each other. Should such collisions occur, then some molecules will be reflected back into the vessel whence they came. The limiting case in which many molecular collisions occur as the gas flows through a channel is one of *diffusion*, a much slower process than effusion; and the limiting case in which molecules make many collisions with the sides of the hole, but not with each other, is called *Knudsen flow*, again a slower process than effusion.

EXAMPLE   A sample calculation is appropriate to illustrate the use of units and to give an order-of-magnitude appreciation of Z. Consider the case of a water surface at 25°C and in equilibrium with its vapor. At equilibrium the rates of evaporation and of condensation must be equal, and we can calculate the latter with the assumption that every vapor molecule hitting the surface sticks to become part of the liquid phase. The vapor pressure of water at 25°C is 23.76 Torr or 0.0313 atm or 3.17×10^4 dyn cm^{-2} or 3.17×10^3 N m^{-2}. The moles of collisions per square centimeter per second Z_n is given by Eq. (2-59) as

$$Z_n = 3.17 \times 10^4 \left[\frac{1}{(2)(3.141)(18)(8.314 \times 10^7)(298.15)} \right]^{1/2}$$

$$= 0.0189 \text{ mole cm}^{-2} \text{ sec}^{-1}$$

The frequency Z is then 1.14×10^{22} molecules cm^{-2} sec^{-1}. In the SI system (see Section 1-CN-1), P becomes 3.17×10^3 N m^{-2}, R is 8.314 J mole^{-1}, and N_A is 6.023×10^{23} molecules mole^{-1}, so

$$Z_n = 3.17 \times 10^3 \left[\frac{1}{(2)(3.141)(0.018)(8.314)(298.15)} \right]^{1/2}$$

$$= 189 \text{ mole m}^{-2} \text{ sec}^{-1} = 1.14 \times 10^{26} \text{ molecule m}^{-2} \text{ sec}^{-1}.$$

This result carries the implication that the evaporation rate is also 1.14×10^{22} molecule cm^{-2} sec^{-1}. Since a liter of water is about 55 moles, some *three liters* of water evaporate from and condense on a square meter of water surface every second! Also, an individual water molecule occupies about 10 Å^2 area or 10^{-15} cm^2 and the evaporation rate from this area is then 1.14×10^7 molecules sec^{-1}. Thus the lifetime of a surface water molecule must be about 10^{-7} sec, so that the water surface is far from quiescent on a molecular scale.

2-7   A rederivation of the ideal gas law

We now can repeat the derivation of the ideal gas law in a more rigorous manner. The velocity component u is taken to be perpendicular to the wall, as before, and the element of pressure contributed by molecules of velocity between u and $u + du$ is given by the momentum change times $dZ(u)$, or $dP = 2mu \, dZ(u)$. Therefore,

$$P = 2 \, mn_0 \int_0^\infty \left(\frac{m}{2\pi kT} \right)^{1/2} \exp\left[-\frac{mu^2}{kT} \right] u^2 \, du \tag{2-63}$$

Now, $\overline{u^2}$ is given by the integral

$$\overline{u^2} = \int_{-\infty}^\infty \left(\frac{m}{2\pi kT} \right)^{1/2} \exp\left[-\frac{mu^2}{kT} \right] u^2 \, du \tag{2-64}$$

which, on evaluation, gives

$$\overline{u^2} = kT/m \tag{2-65}$$

Notice that $\overline{u^2}$ is just one third of $\overline{c^2}$. Returning to Eq. (2-63), we can replace $\mathbf{n}_0$ by $1/\mathbf{v}$, where $\mathbf{v}$ is the overall volume per molecule, and replace the integral by $\overline{u^2}/2$, so that on rearrangement we get

$$P\mathbf{v} = (2m)\overline{u^2}/2) = m\,\overline{u^2} = \tfrac{1}{3}m\overline{c^2} \tag{2-66}$$

or

$$PV = \tfrac{1}{3}M\overline{c^2} \tag{2-67}$$

We have thus obtained Eq. (2-2), but with c^2 identified as $\overline{c^2}$, the average of velocities squared.

  An important point is the following one. The derivation of the Boltzmann equation in Section 2-2 led to Eq. (2-11), in which the exponent could only be said to involve a constant, β, times ϵ. The entire development of the preceding and present sections could have been carried out with this indeterminate form of the Boltzmann distribution equation. Equation (2-54) would then have led to the result that $\overline{c^2} = 3/\beta m$, for example. We would then combine Eq. (2-66) with the ideal gas law, to obtain

$$P\mathbf{v} = kT = \tfrac{1}{3}m\overline{c^2} \tag{2-68}$$

On replacing $\overline{c^2}$ by $3/\beta m$, β would be found to be $1/kT$. A formal treatment would have proceeded in such a manner, but this would have involved deriving a number of equations with β and then restating them later with β replaced by $1/kT$.

  It is appropriate to make an additional observation at this point. The kinetic energy in one dimension is $\tfrac{1}{2}m\overline{u^2}$ and, from Eq. (2-65), this is just $\tfrac{1}{2}kT$. Since the three velocity components, u, v, and w, are independent, the above conclusion applies to each. Thus:

x component: $\epsilon^{kin}_{1-dim} = \tfrac{1}{2}m\overline{u^2}$

x and y components: $\epsilon^{kin}_{2-dim} = \tfrac{1}{2}m\overline{u^2} + \tfrac{1}{2}m\overline{v^2}$

x, y, and z components: $\epsilon^{kin}_{3-dim} = \tfrac{1}{2}m\overline{u^2} + \tfrac{1}{2}m\overline{v^2} + \tfrac{1}{2}m\overline{w^2} = \tfrac{1}{2}m\overline{c^2}$

Alternatively, the contribution from each component to the kinetic energy is $\tfrac{1}{2}kT$, which adds up to the $\tfrac{3}{2}kT$ (or $\tfrac{3}{2}RT$ per mole) given by Eq. (2-55).

2-8   Bimolecular collision frequency and mean free path

There are a number of important properties of a gas that we cannot explain without specifically invoking a molecular size. Clearly, the frequency of intermolecular collisions is one of these properties, as is the related quantity, the mean free path between collisions. It is less obvious perhaps, but equally true, that the explanations of properties such as viscosity, thermal conductivity, and diffusion rates also require the assumption of some finite molecular diameter.

  The model that will be assumed at this point is that of a spherical molecule that can be treated as having a definite radius r and which undergoes elastic collisions. Using this "hard-sphere" model one neglects the fact that molecules are actually "soft," in that colliding molecules approach closer in a violent collision than in a mild one, and one also neglects intermolecular forces of attraction. Treatments not involving these approximations properly belong in advanced treatises. The derivations, even with these assumptions, will not be given here in full detail [see, for example, Moelwyn-Hughes (1961)]. Rigorous treatments will generally modify the derivations given here by only a small numerical factor, however.

A. Bimolecular collision frequency

The frequency of collisions between like molecules, Z_{11}, may be derived on a very simple basis. As illustrated in Fig. 2-8 we imagine a molecule of radius r_1 moving with an average velocity c_1. As it moves it will contact, that is, collide with, any second molecule lying within a cylinder of radius $2r_1$ (see also Fig. 1-10).

The volume swept out in each second is thus $\pi(2r_1)^2\bar{c}_1$, or $\pi\sigma_1^2\bar{c}_1$, where $\sigma_1 = 2r_1$ and is the collision diameter of the molecule, which is assumed to be spherical. If $\mathbf{n}_1$ is the concentration of molecules, then the average number of collisions per second is $\pi\sigma_1^2\bar{c}_1\mathbf{n}_1$. This gives Z_1, the frequency with which a given molecule collides with others. A more rigorous treatment takes into account the relative motion of molecules, to give

$$Z_1 = 2^{1/2}\pi\sigma_1^2\bar{c}_1\mathbf{n}_1 = 4\sigma_1^2\left(\frac{\pi kT}{m}\right)^{1/2}\mathbf{n}_1 \tag{2-69}$$

The reciprocal of Z_1 is sometimes called the *mean free time*, τ. If we could form a gas with a non-equilibrium velocity distribution, say all molecules having $c = (\bar{c^2})^{1/2}$, the equilibrium distribution should eventually be established as molecules underwent random collisions with each other. The experiment can be "carried out" by means of a computer, and the finding is that only a few mean free times are needed for the equilibrium velocity distribution to be approached fairly closely.

The total collision frequency for all molecules in a unit volume is obtained by multiplying Z_1 by $\mathbf{n}_1$, and dividing by two (so as not to count collisions twice):

$$Z_{11} = \pi\left(\frac{1}{2}\right)^{1/2}\sigma_1^2\bar{c}_1\mathbf{n}_1^2 = 2\sigma_1^2\left(\frac{\pi kT}{m}\right)^{1/2}\mathbf{n}_1^2 \tag{2-70}$$

Z_{11} gives the collisions per unit volume per second.

FIGURE 2-8
Molecular collision diameter.

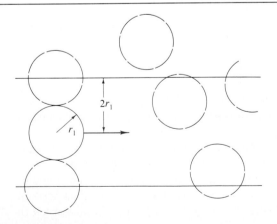

EXAMPLE

Calculate Z_1 and Z_{11} for oxygen at 25°C and 1 atm pressure; σ is 3.61 Å. The quantity $(\pi kT/m)$ or $(\pi RT/M)$ is (π) (8.3144×10^7) $(298.15)/(32.00) = 2.4337 \times 10^9$, and the square root is 4.933×10^4. Let C denote concentration in mol cm^{-3}, $C = P/RT = (1)/(82.057)$ $(298.15) = 4.087 \times 10^{-5}$ and $\mathbf{n} = N_A C = 2.461 \times 10^{19}$ molecule cm^{-3}. Substitution into Eqs. (2-69) and (2-70) gives

$$Z_1 = (4)(3.61 \times 10^{-8})^2(4.933 \times 10^4)(2.461 \times 10^{19})$$

$$= 6.328 \times 10^9 \text{ collisions s}^{-1}$$

$$Z_{11} = Z_1 C/2 = (6.328 \times 10^9)(4.087 \times 10^{-5})/(2)$$

$$= 1.293 \times 10^5 \text{ moles of collisions cm}^{-3} \text{ s}^{-1}$$

In SI units, $(\pi kT/m) = (\pi)(8.3144)(298.15)/(0.03200) = 2.4337 \times 10^5$, and the square root is 493.3. C, now in mol m^{-3}, is $(1.0133 \times 10^5)/(8.3144)(298.15) = 40.87$, so

$$Z_1 - (4)(3.61 \times 10^{-10})^2(493.3)(40.87) - 1.051 \times 10^{-14} \text{ moles of collisions s}^{-1}$$

$$\text{or } 6.328 \times 10^{-1} \text{ (as before)}$$

The calculation of Z_{11} is left as an exercise (will it be the same as the cgs value?).

Certain shortcuts and alternative routes are illustrated in the calculations. In evaluating kT/m, the equivalent ratio RT/M was used. In obtaining C, it was convenient to use P in atm and the gas constant in cm^3 atm K^{-1} mol^{-1}.

Although the derivation for collisions between like molecules gave a result which is correct except for the small correction factor, a somewhat more elaborate approach is needed to obtain the correct form for Z_{12}, the collision frequency between unlike molecules.

We now have two kinds of molecule, 1 and 2, and the frequency with which a single molecule of type 1 will collide with molecules of type 2 is

$$Z_{1(2)} = 2 \sqrt{2} \, \sigma_{12}^2 \left(\frac{\pi kT}{\mu_{12}}\right)^{1/2} \mathbf{n}_2 \tag{2-71}$$

Here $\mathbf{n}_2$ is the concentration of species 2 in molecules per unit volume, σ_{12} is the average collision diameter, $(\sigma_1 + \sigma_2)/2$, and μ_{12} is the reduced mass,[*]

$$\mu_{12} = \frac{m_1 m_2}{m_1 + m_2} \tag{2-72}$$

The frequency of all collisions between the two types of molecule is then

$$Z_{12} = 2 \sqrt{2} \, \sigma_{12}^2 \left(\frac{\pi kT}{\mu_{12}}\right)^{1/2} \mathbf{n}_1 \mathbf{n}_2 \tag{2-73}$$

There are actually several Z quantities for the case of a mixture of gases 1 and 2. Let $Z_i' = 4\sigma_i^2(\pi kT/m_i)^{1/2}$, $i = 1$ or 2, and $Z_{12}' = 2\sqrt{2}\sigma_{12}^2 \, (\pi kT/\mu_{12})^{1/2}$. There are four kinds of molecular collision frequencies:

$$
\begin{array}{ll}
Z_{1(1)} = Z_1'\mathbf{n}_1 & Z_{2(2)} = Z_2'\mathbf{n}_2 \\
Z_{1(2)} = Z_{12}'\mathbf{n}_2 & Z_{2(1)} = Z_{12}'\mathbf{n}_1
\end{array}
\tag{2-74}
$$

[*]In some respects, it is more logical to use $m_{12} = 2m_1 m_2/(m_1 + m_2)$ rather than μ_{12}, since m_{12} would reduce to just m if $m_1 = m_2$. See Problem 2-43.

There are then three kinds of bimolecular collision frequencies (collisions per unit volume per second):

$$Z_{11} = \tfrac{1}{2}Z_1'\mathbf{n}_1^2 \quad Z_{22} = \tfrac{1}{2}Z_2'\mathbf{n}_2^2 \quad \text{and} \quad Z_{12} \text{ (or } Z_{21}) = Z_{12}'\mathbf{n}_1\mathbf{n}_2 \tag{2-75}$$

The total collision frequency is the sum of these three. The extension to mixtures of more than two components is straightforward.

B. Mean free path

As the name implies, the mean free path λ of a molecule is the average distance traveled between collisions. As a slightly intuitive definition, λ is just the mean velocity divided by the collision frequency, and so it follows from Eq. (2-69) that

$$\lambda = \frac{\bar{c}}{Z_1}$$

or $\hspace{8cm}$ (2-76)

$$\lambda = \frac{1}{\sqrt{2}\,\pi\sigma^2\mathbf{n}}$$

Equation (2-76) applies to a gas consisting of a single molecular species, and since they are not needed, the subscripts of Eq. (2-69) have been dropped. For an ideal gas the concentration in molecules per unit volume, $\mathbf{n}$, is equal to P/kT, and an alternative form of Eq. (2-76) is

$$\lambda = \frac{kT}{\sqrt{2}\,\pi\sigma^2 P} \tag{2-77}$$

EXAMPLE

Calculate λ for oxygen at 25°C and 1 atm. Using Eq. (2-76) and $\mathbf{n}$ from the example of the preceding section, $\lambda = 1/(2)^{1/2}(\pi)\,(3.61 \times 10^{-8})^2(2.461 \times 10^{19}) = 7.018 \times 10^{-6}$ cm.

In SI units, and using Eq. (2-77), $\lambda = (1.3807 \times 10^{-23})(298.15)/(2)^{1/2}(\pi)\,(3.61 \times 10^{-10})^2(1.0133 \times 10^5) = 7.016 \times 10^{-8}$m.

Mean free paths in a mixture may also be formulated. Thus $\lambda_1 = \bar{c}_1/[Z_{1(1)} + Z_{1(2)}]$ and $\lambda_2 = \bar{c}_2/[Z_{2(2)} + Z_{2(1)}]$. Again, the extension to mixtures of more than two components is straightforward.

2-9 Transport phenomena; viscosity, diffusion, and thermal conductivity

Transport phenomena, as the name implies, refer to the transport or spatial motion of some quantity. In the case of *viscosity* the quantity transported is momentum: in *diffusion* one deals with molecular transport or the drift of molecules from one place to another; in *thermal conductivity* one deals with the flow or transport of heat down a temperature gradient. The three quantities have much in common and the kinetic molecular theory treatment of them leads to somewhat similar equations.

A. Viscosity

The *coefficient of viscosity* η of a fluid is defined as a measure of the friction that is present when adjacent layers of the fluid are moving at different speeds. For example, if a fluid is flowing between parallel plates, as illustrated in Fig.

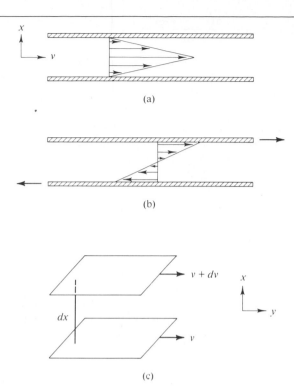

FIGURE 2-9
Velocity profiles and defining model for viscosity.

2-9(a), we ordinarily assume that the material immediately adjacent to the walls is stationary; the material farther away then moves increasingly rapidly. The arrows in the figure give this velocity profile. Alternatively, if the fluid is stationary but the walls are moving with equal and opposite velocities, the velocities of various layers of the fluid are as shown in Fig. 2-9(b). If we now consider two layers of fluid parallel to the walls and separated by distance dx, then, as indicated in Fig. 2-9(c), their velocities will differ by dv or alternatively by $(dv/dx)dx$. If the area of each layer is $\mathscr{A}$, then the frictional drag or force f between the layers is given by η and

$$f = \eta \mathscr{A} \frac{dv}{dx} \tag{2-78}$$

Equation (2-78) is in fact the defining equation for the coefficient of viscosity and is known as *Newton's law of viscosity*. In the cgs system the unit of viscosity is the *poise* (abbreviated P, and equivalent to grams per centimeter per second). The unit is a relatively large one; ordinary gases have viscosities of 10^{-3}–10^{-4} P and liquids have viscosities of around 10^{-2} P. The SI unit of viscosity (not yet given a name) is ten times larger than the cgs unit, the dimensions being kg m^{-1} s^{-1}. The reciprocal of viscosity is called the *fluidity*.

The experiment corresponding directly to the defining equation (2-78), namely that of measuring the drag between oppositely moving plates, is not a convenient one. More often one determines instead the pressure drop required to produce a given fluid flow rate down a cylindrical tube. The appropriate equation is

$$\eta = \frac{\pi(P_1 - P_2)r^4 t}{8Vl} \tag{2-79}$$

where $P_1 - P_2$ is the pressure drop down the tube, usually a capillary, of radius r and length l; and V is the volume of fluid flowing in time t. In the case of gases V is a strong function of pressure, and in application of the equation V should be evaluated at the average pressure $(P_1 + P_2)/2$. Equation (2-79) is known as the *Poiseuille* equation; a detailed derivation is given in Section 8-ST-2. Alternatively, one may observe the rate of damping of a torsion pendulum suspended in the gas.

The coefficient of viscosity of a gas may be derived theoretically, and the derivation that follows represents one of the triumphs of early kinetic molecular theory. The experimental observation was that the viscosity of a gas does not vary appreciably with pressure; this was very hard to understand on an intuitive basis, since it seemed that the denser a gas, the greater should be its viscosity. Kinetic molecular theory provided the explanation.

In simple theory molecules of a gas do not interact except by means of collisions, and these occur on the average only after a free flight distance given by the mean free path λ. The theoretical picture is then one of molecules moving in free flight back and forth between adjacent layers of gas, the layers being separated on the average by the distance λ. As illustrated in Fig. 2-10, a molecule of layer 1 will have a velocity component v corresponding to the average flow velocity of the gas in that layer. On arriving at layer 2 the molecule then experiences a collision which delivers the directional momentum mv to the molecules of that layer. Since these have on the average a net velocity component of $v + \Delta v$, corresponding to the layer velocity of layer 2, the net momentum loss realized in layer 2 is $m\,\Delta v$. The same momentum transfer occurs with opposite sign when a counterpart molecule of layer 2 arrives at layer 1.

If we consider each layer to be of area $\mathscr{A}$, then by the surface collision formula (2-58), the number of molecules arriving each second at layer 2 from layer 1 will be $\frac{1}{4}n\bar{c}\mathscr{A}$, and likewise the number of counterpart molecules ar-

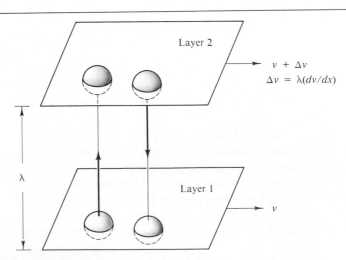

FIGURE 2-10
Model for calculation of gas viscosity.

Layer 2

$v + \Delta v$
$\Delta v = \lambda(dv/dx)$

λ

Layer 1

v

riving at layer 1 from layer 2. The total momentum transport per second is then

$$\frac{d(mv)}{dt} = \frac{2\mathscr{A}}{4}\,\mathbf{n}\bar{c}m\Delta v \tag{2-80}$$

Where the factor of two takes into account the two-way traffic of molecules. By Newton's law a portion of matter, in this case a layer, acted on by a force shows a momentum change with time, and $d(mv)/dt$ due to the interchange of momentum by the gas molecules will appear as an equivalent force f acting between the layers. Also, since Δv is given by $\lambda(dv/dx)$, we can write

$$f = \mathscr{A}\left(\frac{\mathbf{n}\bar{c}m\lambda}{2}\right)\frac{dv}{dx} \tag{2-81}$$

Equation (2-81) has the same form as the defining equation for viscosity, Eq. (2-78), and on comparing the two equations, we conclude that

$$\eta = \tfrac{1}{2}\mathbf{n}\bar{c}\lambda m \tag{2-82}$$

Further, since $\mathbf{n}m$ gives the density of the gas, we have

$$\eta = \tfrac{1}{2}\rho\bar{c}\lambda \tag{2-83}$$

A yet more instructive form is obtained on elimination of λ between Eqs. (2-83) and (2-76):

$$\eta = \frac{m\bar{c}}{2\sqrt{2}\pi\sigma^2} \tag{2-84}$$

or, using Eq. (2-52) for $\bar{c}$

$$\eta = \frac{1}{\pi\sigma^2}\left(\frac{kTm}{\pi}\right)^{1/2} \tag{2-85}$$

The simplified derivation given here takes the molecules as moving perpendicularly to the planes of shear (Fig. 2-10). Actually, of course, molecules have a distribution of directions of motion; those headed from one plane towards the other but on a slant path will, on the average, make a collision before reaching the second plane. The rigorous treatment, which takes this detail into account, yields Eq. (2-85) but with a correction factor of $5\pi/16$. A useful reduction is given by Herschfelder et al.:

Example

$$\eta = 2.669 \times 10^{-5}\,(MT)^{1/2}/\sigma^2 \tag{2-86}$$

where η is in poise and σ is in Å.

Equation (2-85) (or Eq. (2-86)) shows explicitly that although gas viscosity depends on $T^{1/2}$, it is independent of any change in pressure or density at constant temperature. The physical rationale of this result is that, whereas the frequency of molecules moving back and forth between layers increases in proportion to the pressure, the mean free path, or effective distance between layers, decreases in proportion to the pressure and the two effects just cancel with respect to the momentum transport per unit distance.

A treatment of the viscosity of a nonideal gas is discussed in Section 2-ST-1.

EXAMPLE   Calculate η for oxygen at 25°C and 1 atm. We will use Eq. (2-83) and draw on the examples of the preceding section. Thus $\rho = MC = (32.00)(4.087 \times 10^{-5}) = 1.308 \times 10^{-3}\,\text{g cm}^{-3}$, and $\lambda = 7.018 \times 10^{-6}$ cm. Further, $\bar{c} = (8kT/\pi M)^{1/2} = [(8)(8.314 \times 10^{7})(298.15)/(\pi)(32.00)]^{1/2} = 4.442 \times 10^{4}$ cm sec^{-1}. Then $\eta = (1/2)(1.308 \times 10^{-3})(4.442 \times 10^{4})(7.018 \times 10^{-6}) = 2.038 \times 10^{-4}$ P (or g cm^{-1} sec^{-1}).

In SI units, $\rho = (0.03200)(40.87) = 1.308$ kg m^{-3}, $\bar{c}$ is 4.442×10^{2} m sec^{-1}, and λ is 7.016×10^{-8} m. The viscosity is now 2.038×10^{-5} kg m^{-1} sec^{-1}; the SI unit of viscosity (not yet named) is thus ten times larger than the cgs unit.

B. Diffusion

Diffusion is a process of spatial drift of molecules due to their kinetic motion, as impeded by random collisions. The physical picture is that a molecule makes a free flight until a collision with another molecule causes a change in direction and velocity. The process is often referred to as a random walk, the analogy being to a person taking successive steps but with each step unrelated in direction to the preceding one (the alternative scientific colloquialism is the "drunkard's walk"). As illustrated in Fig. 2-11, a given molecule will thus drift away from its original position in the course of time. If there is a concentration gradient, then the statistical effect will be one of a net average drift from a more concentrated to a more dilute region. The matter is discussed in more detail in Section 10-7, but the defining phenomenological equation, known as *Fick's law*, is

$$J = -\mathscr{D}\,\frac{d\mathbf{n}}{dx} \tag{2-87}$$

where J is this net drift expressed in molecules crossing unit area per second, and $d\mathbf{n}/dx$ is the concentration gradient in the drift direction. The coefficient $\mathscr{D}$ is known as the *diffusion coefficient*; in the cgs system its units are square centimeters per second; in SI, $\mathscr{D}$ is given in m²s^{-1}, so the unit is 10^{4} larger than the cgs one. Diffusion coefficients for gases are around unity, and for liquids are about 10^{-5} or less, in cm² s^{-1}.

An important equation due to Einstein allows us to evaluate $\mathscr{D}$ in terms of gas kinetic theory. We consider a situation in which there is a concentration gradient in one direction only, as in the case of diffusion along the length of a long tube or cell. As indicated in Fig. 2-11, there will be some average displacement x of a molecule in time t. The direction of this displacement will be random, but as a simplification we take the direction to be along the concentration gradient for half of the molecules, and opposite to the concentration gradient for the other half. We now imagine a reference cross-section of unit area and at right angles to the concentration gradient, as shown in Fig. 2-12, and seek to count the number of molecules passing through this

FIGURE 2-11
Random walk.

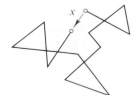

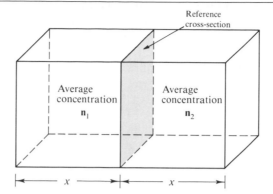

FIGURE 2-12
Here n_1 and n_2 are average concentrations.

cross-section in time t. (Imagine a fish ladder with an observation window so that one can count the fish passing a given point.) In our simplified picture, half of all molecules within a distance x of the reference cross-section will pass through it by diffusion during a time interval t. The total number of molecules within distance x to the left of the cross-section is just $n_1 x$, where n_1 is the average concentration in that volume. The flow from left to right is therefore $\frac{1}{2}n_1 x$. Similarly, the flow from right to left is $\frac{1}{2}n_2 x$, where n_2 is the average concentration in the volume $n_2 x$ to the right of the reference cross-section. In the cgs system, these flows are in molecules per square centimeter per second.

Since $n_2 = n_1 + x(dn/dx)$, where dn/dx is the concentration gradient, the *net* flux across the reference plane is

$$J = \frac{1}{t}\left[\frac{1}{2}\,n_1 x - \frac{x}{2}\left(n_1 + x\,\frac{dn}{dx}\right)\right] = -\frac{x^2}{2t}\frac{dn}{dx} \tag{2-88}$$

Comparison with Eq. (2-87) gives

$$\mathscr{D} = \frac{x^2}{2t} \tag{2-89}$$

Equation (2-89) allows a very simple derivation of $\mathscr{D}$ from kinetic molecular theory. Although the displacement x can be the net result of many random steps, it can also be put equal to the smallest step which is random with respect to succeeding ones. This smallest step is the mean free path distance λ. During such a step the molecule is moving with velocity $\bar{c}$, so the average time between steps is just $\lambda/\bar{c}$. On substituting $x = \lambda$ and $t = \lambda/\bar{c}$ into Eq. (2-89) we get

$$\mathscr{D} = \frac{1}{2}\lambda\bar{c} \tag{2-90}$$

Drawing on the preceding example, for oxygen at 25°C and 1 atm, $\mathscr{D} = (1/2)(7.018 \times 10^{-6})$ $(4.442 \times 10^4) = 0.156$ cm^2 s^{-1}. Alternatively, in SI units, $\mathscr{D} = 1.56 \times 10^{-5}$ m^2 s^{-1}. **EXAMPLE**

Equation (2-90) applies to the interdiffusion of molecules of a single kind. Such a diffusion is called *self-diffusion*, and can be measured experimentally by the use of isotopic labeling. That is, in the experiment corresponding to Fig. 2-12, there can be no overall pressure gradient since otherwise there would be a wind or viscous flow of gas. However, labeling the molecules on, say, the left of the reference plane makes possible the measurement of their net diffusion migration across it even though the total concentration of all molecules is uniform.

As with viscosity, the rigorous treatment of the hard sphere model makes a detailed analysis of the averaging over all velocity directions. Again, the effect is only to change the numerical coefficient. Equation (2-90) becomes (see Hirschfelder et al.):

$$\mathcal{D} = \frac{3\pi}{16}\lambda\bar{c} = \left(\frac{3}{8}\right)\left(\frac{kT}{m}\right)^{1/2}\left(\frac{1}{\pi\sigma^2}\right)\left(\frac{1}{\mathbf{n}}\right) \tag{2-91}$$

The case of interdiffusion of molecules in a mixture of gases will not be treated here, because of its complexity. The general situation is that now each type of molecule has its own mean free path and intrinsic diffusional drift. Because of the differing drift rates, a pressure gradient does develop until balanced by an opposite bulk flow or wind. The observed interdiffusion or mutual diffusion coefficient is determined by the intrinsic diffusion rate as modified by the bulk flow.

C. Thermal conductivity

The phenomenological equation for thermal conductivity, or the equation analogous to Newton's law for viscous flow, Eq. (2-78), and to Fick's law for diffusional flow, Eq. (2-87), is attributed to Fourier and is

$$J = -\kappa\frac{dT}{dx} \tag{2-92}$$

where J is now the flow of heat per second across unit area perpendicular to a temperature gradient dT/dx. The quantity κ is the *coefficient of thermal conductivity*.

The kinetic molecular theory analysis is quite similar to that for viscosity. Again the picture is one of molecules moving back and forth between planes a distance λ apart, but now planes 1 and 2 of Fig. 2-10 are at different temperatures. Molecules from plane 1 carry energy ϵ_1 and those from plane 2 carry energy ϵ_2 and if the collision at the end of a mean free path results in complete energy exchange, the energy transported is then $\lambda\,d\epsilon/dx$. Following the same arguments as for Eqs. (2-80) and (2-81), we write

$$-J = \frac{d\epsilon}{dt} = \frac{2\mathbf{n}c\lambda}{4}\frac{d\bar{\epsilon}}{dx}$$

$$J = -\frac{\mathbf{n}\bar{c}\lambda}{2}\frac{d\bar{\epsilon}}{dT}\frac{dT}{dx} \tag{2-93}$$

The thermal conductivity coefficient is, on comparison with Eq. (2-92),

$$\kappa = \frac{\mathbf{n}\bar{c}\lambda}{2}\frac{d\epsilon}{dT} \tag{2-94}$$

The quantity $d\epsilon/dT$ is the molecular heat capacity of the gas (at constant volume) $\mathbf{c}_v$, or mc_v, where c_v is the heat capacity per unit mass. Equation (2-94) then becomes

$$\kappa = \tfrac{1}{2}\mathbf{n}\bar{c}\lambda m c_v \tag{2-95}$$

or, invoking the equation for viscosity, Eq. (2-82),

$$\kappa = \tfrac{1}{2}\rho\bar{c}\lambda c_v = \eta c_v \tag{2-96}$$

The molar heat capacity, C_v, for oxygen at 25°C is 5.03 cal K^{-1} mole^{-1} so that c_v is 0.157 cal K^{-1} g^{-1}. Drawing on the example following Eq. (2-86), we have $\kappa = (2.038 \times 10^{-4})(0.157) = 3.20 \times 10^{-5}$ cal K^{-1} cm^{-1} sec^{-1}. In SI units, $c_v = (0.157)(4.184)(10^3) = 657$ J kg^{-1} and $\kappa = (2.038 \times 10^{-5})(657) = 0.0134$ J K^{-1} m^{-1} sec^{-1}.   **EXAMPLE**

As with viscosity and diffusion, the exact hard sphere treatment gives the same form of equation as does our simple treatment, with a slightly different numerical coefficient (see Hirschfelder et al.). The correct equation for κ is:

$$\kappa = \left(\frac{25}{32}\right)\left(\frac{\pi k T}{m}\right)^{1/2}\left(\frac{1}{\pi\sigma^2}\right) c_V \tag{2-97}$$

2-10   Summary of kinetic molecular theory quantities

A variety, perhaps a bewildering variety, of quantities have been derived from the kinetic molecular model for gases. The following discussion is intended to summarize these in an organized way so that a general picture can emerge. First are the fundamental Boltzmann distribution functions and the various average velocities derived from them. Immediate applications are the identification of the average kinetic energy with $\tfrac{3}{2}kT$ and the derivation of the surface or wall collision frequency and its application to effusion.

Up to this point the finite size of molecules has been accepted, but has played no direct role in the treatments. That is, bimolecular collisions act to redistribute momenta and energies but do not change the overall statistical situation. Next, however, molecules are specifically considered to be hard spheres of definite radius r or collision diameter σ, which now permits the calculation of the various types of collision frequencies and mean free paths.

The three transport phenomena discussed involve the presence of a gradient in pressure, in concentration, or in temperature. The systems are no longer in equilibrium but it is assumed that the gradients are not such as to perturb the general distribution equations significantly. It is then possible to calculate the rate of mass flow, of concentration change, and of heat flow due to these gradients. The effect of intermolecular attractive forces is disregarded in all of these derivations.

The principal relationships are assembled in Table 2-2, together with the approximate values for the various quantities for a typical gas. The table can then be used, for example, to estimate the numerical value of a quantity for some other type of gas or condition.

Finally, in Table 2-3 are collected the experimentally measured values for the viscosity, self-diffusion, and thermal conductivity coefficients of some actual gases. These provide a means for testing the accuracy of the hard-sphere model.

TABLE 2-2
Summary of kinetic molecular theory quantities[a]

Symbol	Quantity	Formula	Equation number	Approximate value[b]
$p(c)$	Three-dimensional distribution law	$4\pi(m/2\pi kT)^{3/2}\,[\exp(-mc^2/2kT)]c^2$	(2-46)	—
c_p	Most probable velocity	$(2kT/m)^{1/2}$	(2-47)	3.94×10^4 cm s^{-1}
$\bar{c}$	Average velocity	$(8kT/\pi m)^{1/2}$	(2-52)	4.44×10^4 cm s^{-1}
$(\overline{c^2})^{1/2}$	Root mean square velocity	$(3kT/m)^{1/2}$	(2-54)	4.82×10^4 cm s^{-1}
Z	Surface collision frequency	$\frac{1}{4}\mathbf{n}\bar{c}$	(2-58)	0.453 moles cm^{-2} s^{-1}
Z_1	Collision frequency for a molecule	$4\sigma_1{}^2(\pi kT/m)^{1/2}\mathbf{n}_1$	(2-69)	6.33×10^9 collisions s^{-1}
λ	Mean free path	$1/\sqrt{2}\pi\sigma^2\mathbf{n}$	(2-76)	7.02×10^{-6} cm
Z_{11}	Bimolecular collision frequency, like molecules	$2\sigma_1{}^2(\pi kT/m)^{1/2}\mathbf{n}_1{}^2$	(2-70)	1.29×10^5 moles cm^{-3} s^{-1}
Z_{12}	Bimolecular collision frequency, unlike molecules	$2\sqrt{2}\sigma_{12}^2(\pi kT/\mu_{12})^{1/2}\,\mathbf{n}_1\mathbf{n}_2$	(2-73)	—
η	Viscosity coefficient	$\frac{1}{2}\rho\bar{c}\lambda$	(2-83)	2.04×10^{-4} P
$\mathscr{D}$	Self-diffusion coefficient	$\frac{1}{2}\bar{c}\lambda$	(2-90)	0.156 cm^2 s^{-1}
κ	Thermal conductivity coefficient	$\frac{1}{2}\rho\bar{c}\lambda c_V$	(2-96)	3.20×10^{-5} cal cm^{-1} s^{-1} K^{-1}

[a]Definition of symbols; m, mass per molecule; $\mathbf{n}$, molecules per unit volume; μ_{12}, reduced mass; ρ, density; c_V, heat capacity per unit mass; σ, collision diameter or $2r$.

[b]Calculated for a gas at 25°C and 1 atm, of molecular weight 32, and collision radius 3.61 Å.

TABLE 2-3
Transport coefficients for various gases at 1 atm[a]

Gas	Collision diameter σ (Å)	Temperature (K)	Self-diffusion coefficient $\mathscr{D}$ (cm^1 s^{-1})	Viscosity $10^4\eta$ (P)	Coefficient of thermal conductivity $10^4\,\kappa$ (cal cm^{-1} s^{-1} K^{-1})
He	2.18	100	0.270[b]	0.951	1.744
		300	1.669[b]	1.981	3.583
Ar	3.64	100	0.023	0.815	0.154
		300	0.186	2.271	0.422
N$_2$	3.75	100	0.027	0.698	0.232[b]
		300	0.205	1.663	0.602[b]
		500	0.495[b]	2.657	0.876[b]
O$_2$	—	273.2	0.18	—	0.58
		280	—	1.959	—
CH$_4$	—	273.2	0.206	—	0.73
		280	—	1.052	—
		353.2	0.318	—	—
CO$_2$	—	273.2	0.0970	—	0.35
		280	—	1.402	—
HCl	—	295	0.1246	—	—

[a]From J.O. Hirschfelder, C.F. Curtiss, and R.B. Bird, "Molecular Theory of Gases and Liquids," corrected ed. Wiley, New York, 1964.

[b]Estimated values.

COMMENTARY AND NOTES

2-CN-1 Verification of the kinetic molecular theory

The Maxwell–Boltzmann distribution laws are, of course, verified indirectly by the agreement of experiment with various derived quantities, such as effusion rates or transport coefficients. A direct verification requires sampling of a gas to obtain the actual frequency of occurrence of various molecular velocities, that is, to obtain an experimental measurement of a velocity distribution.

A schematic arrangement for doing this is illustrated in Fig. 2-13. Molecules effusing from a small hole in a reservoir are first required to travel in a certain direction by means of a succession of slits. The collimated beam now encounters a pair of disks mounted coaxially and rotating. Each disk has a slit in it and one sets the relative position of the slits and the speed of rotation. If the disks rotate at ω revolutions per second and the slits are θ degrees apart, then the second slit comes into position θ/ω 360 sec after the first one. Only those molecules in the beam that take θ/ω 360 sec to travel the distance between the disks will get through the slits. One then varies the time-of-flight requirement and, by means of a detector, determines the relative number of molecules of various velocities.

Early attempts (see Herzfeld and Smallwood, 1951) gave qualitative agreement, and more recent experiments using a considerably more sophisticated approach than that of Fig. 2-13 have given rather good agreement, as shown in Fig. 2-14 for the case of a vapor of potassium atoms.

Clearly, the kinetic molecular theory of gases is well established. It may seem at this point rather unnecessary even to discuss its validity. However, the question of what is meant by the "validity" of a theory has subtleties that are worth illustrating in the present context.

A theory will always start with some sort of model. The model may be a physical picture; it may be a set of mathematical premises (as in wave mechanics). The first question to ask about a theory is whether there is evidence that the model is qualitatively correct. In the case of the kinetic molecular theory, we do have much evidence that matter comes in small units or molecules and, from the time-of-flight experiments mentioned earlier, that the molecules of a gas do have a velocity distribution.

The second question is whether the theory leads to the correct form or parametric dependence of the quantity treated. That is, does it give the phenomenological limiting law and does it give the observed approximate dependence of the coefficient in question on the state variables of the system? Again, kinetic molecular theory meets the test. It leads to relationships matching the defining equations for the various transport processes and correctly predicts the approximate dependence of η, $\mathscr{D}$, and κ on the state of the gas. It has been mentioned, for example, that one of the early triumphs of kinetic molecular theory was its prediction that viscosity should be independent of pressure.

The third, and in this discussion, final question to be asked of a theory is whether it predicts precisely how the theoretically derived quantity or coefficient varies with conditions. It is here that theories generally fail—even the best ones—presumably because the model is not adequate in detail.

Kinetic molecular theory, as developed in terms of the hard-sphere model, is no excep-

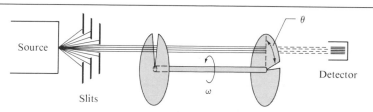

FIGURE 2-13
Schematic velocity selector apparatus.

Source

Slits

θ

ω

Detector

FIGURE 2-14
Experimental verification
of the Maxwell-Boltzmann
distribution of molecular
velocities for potassium
atoms around 900 K. The
plot is of relative intensity
observed by the detector
vs. reduced velocity, that
is, velocity relative to the
most probable velocity.
(From R.C. Miller and P.
Kusch, Phys. Rev., **99,** 1314
(1955).)

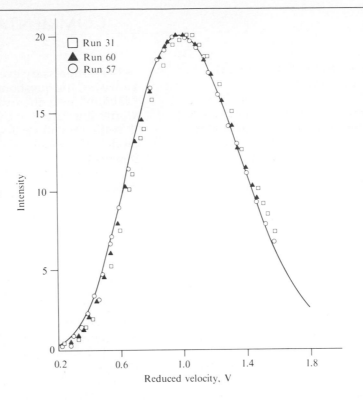

FIGURE 2-14
Experimental verification of the Maxwell-Boltzmann distribution of molecular velocities for potassium atoms around 900 K. The plot is of relative intensity observed by the detector vs. reduced velocity, that is, velocity relative to the most probable velocity. (From R.C. Miller and P. Kusch, Phys. Rev., **99,** 1314 (1955).)

tion. Thus, as illustrated by the data of Table 2-4, the values of σ from viscosity and from diffusion measurements are not exactly the same. The theoretical parametric dependence is not exactly obeyed either. Equation (2-85) requires that viscosity be independent of pressure and although this is true of dilute gases, it ceases to be so for dense ones. The equation further predicts that viscosity should be proportional to $T^{1/2}$ and this is not quite correct, even for dilute gases. Thus the viscosity of oxygen at 200 K is 1.48×10^{-3} P and at 800 K is 4.12×10^{-3} P, considerably more than twice as much.

Such defects may be handled by recognizing that molecules do experience mutual attraction. By adding a suitable attractive potential, but still assuming hard spheres, Sutherland obtained the equation

$$\eta = \frac{\eta^0}{1 + (S/T)} \qquad (2\text{-}98)$$

TABLE 2-4
Molecular diameters (Å)[a]

Gas	From gas viscosity	From self-diffusion	From van der Waals b constant[b]
Ar	3.64	3.47	5.7
N_2	3.75	3.48	5.0
O_2	3.61	3.35	2.6
CH_4	4.14	3.79	5.1
CO_2	4.63	4.28	5.1

[a]The experimental transport coefficients are from measurements at 0°C and 1 atm. The values from viscosity and self-diffusion are as given by J.O. Hirschfelder, C.F. Curtiss, and R.B. Bird, "Molecular Theory of Gases and Liquids," corrected ed., p. 545. Wiley, New York, 1964.
[b]The van der Waals b constant is taken to be $V_c/3$.

where S is an empirical constant. Equation (2-98) is quite successful in fitting viscosity data.

The yet more sophisticated modern approach is to use a detailed potential function for both attractive and repulsive forces, including, for diatomic and polyatomic molecules, a dependence on the relative orientation of approaching molecules. By now, the mathematical problems have become formidable and approximate computational methods have to be used.

In effect, one says at this point that the theory is exactly correct and that any quantitative defects can in principle be corrected by a suitably elaborate potential function. This is the terminal stage of development of a major theory. Its validity is taken for granted, and effort is concentrated on refinements of the model and on ways of avoiding mathematical approximations.

2-CN-2   Some more about the kinetic molecular theory

The simple kinetic molecular theory treatment of transport phenomena leads to expressions for the viscosity coefficient η, self-diffusion coefficient $\mathcal{D}$, and thermal conductivity coefficient κ in which the only unknown is the collision diameter σ. The equations are summarized in Table 2-2.

The experimentally measured coefficients may then be used in calculations of molecular diameters. Some such results are given in Table 2-4 along with values calculated from the van der Waals constant b. The agreement is not bad, and this speaks well for the merits of the hard-sphere model as a first approximation. We find that the greatest discrepancy occurs with σ values from thermal conductivity; part of the difficulty is that with polyatomic molecules the heat capacity reflects both internal energy and kinetic energy, although it is largely the latter that is transferred in a collision. As a result c_v in Eq. (2-95) becomes a semiempirical quantity itself, along with σ.

In order to calculate σ from the kinetic molecular theory one must use the theoretical expression for λ, which requires knowing the value of **n,** the concentration expressed as molecules per unit volume. This in turn implies a knowledge of Avogadro's number. Alterna-

tively, one may use the van der Waals constant b, or some other independent measurement of molar volume, and, with some algebraic manipulation, combine this information with a measured viscosity or self-diffusion coefficient to obtain a value for Avogadro's number. Approximately correct answers result.

There are some aspects to the transport phenomena which provide an important illustration of the workings of the scientific method.

We start with an empirical observation, or law, which in this case is of the form

$$\text{rate of transport} = (\text{constant})(\text{driving force}),$$

that is, rate of momentum transport =

$$(\eta)(\text{velocity gradient}),$$
$$\text{rate of mass transport} =$$
$$(\mathcal{D})(\text{concentration gradient}),$$
$$\text{rate of heat transport} =$$
$$(\kappa)(\text{temperature gradient}). \tag{2-99}$$

The driving force in each instance is given by a gradient of an intensive property: velocity, concentration, or temperature. The quantity transported is an extensive property closely related to the intensive one: momentum, mass, or heat. Rate processes generally obey this form, particularly in systems not too far from equilibrium; the detailed elaboration of this observation by Lars Onsager won him a Nobel prize.

Second, having observed empirically that a relationship of the type of Eq. (2-99) is approximately obeyed by systems generally, and appears to be exactly obeyed under some limiting condition, one then uses the equation as a defining equation for the proportionality constant η, $\mathcal{D}$, or κ. The equations involve only macroscopic properties of the system and are thus phenomenological in nature. The definitions are unambiguously phrased in terms of experiment and are thus operational definitions (see Bridgman, 1946). The scientific method provides this procedure for defining quantities exactly without having to first supply any kind of theoretical explanation.

We have a similar situation with the ideal gas law. As a limiting law, we have the phenomenological statement that $PV = $ constant for any gas at a given temperature and in fact used this to define an absolute temperature scale.

Now the statement $PV/T = R$ is not exactly correct for a real gas. One could proceed by tabulating "R" values for real gases under various conditions. It is, of course, more convenient to use instead the deviation from unity of the compressibility factor PV/RT, or to use a virial expansion. A similar situation applies to transport phenomena. The defining equations (2-99) are not ordinarily obeyed exactly. Thus diffusion rates generally depend not just on the concentration gradient, but also on the actual level of concentration. As with the compressibility factor, we must report values of η, $\mathcal{D}$, or κ for a specified state of the gas.

The task of theory is then not only to explain the limiting laws but also to predict how the coefficients which they define should vary with the state of the system.

SPECIAL TOPICS

2-ST-1 Use of a Lennard–Jones potential function

It was mentioned in the preceding Commentary and Notes section that more elaborate treatments of transport phenomena make use of a potential function that allows for attractive forces, such as the widely used Lennard–Jones function described in the Special Topics section of Chapter 1. The function is [Eq. (1-62)]

$$\phi(r) = -\frac{\alpha}{r^6} + \frac{\beta}{r^{12}}$$

where the first term represents the attractive part of the potential and the second that of the repulsive part. Thus molecules are considered also to be somewhat "soft" in that the repulsion rises rapidly but not infinitely rapidly as two molecules come to within their collision diameter σ. Equation (1-62) is plotted in Fig. 1-

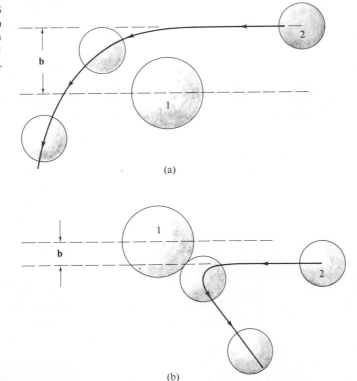

FIGURE 2-15
Molecular trajectories in the presence of an attractive–repulsive potential.

b

1

2

(a)

1

b

2

(b)

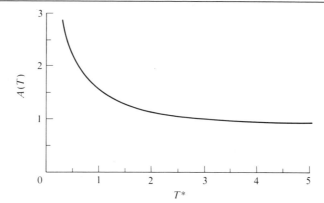

FIGURE 2-16
Viscosity parameter A(T) [Eq. (2-100)] as a function of T*. (Data from J.O. Hirschfelder, C.F. Curtiss, and R.B. Bird, "Molecular Theory of Gases and Liquids," corrected ed. Wiley, New York, 1964.)

16, in which are given the two characteristic quantities σ and ϕ^*. As a reminder, σ is now the internuclear separation at which ϕ is just zero and ϕ^* is the value of ϕ at the potential energy minimum.

In the more advanced treatment (see, for example, Hirschfelder et al., 1964), it is necessary to consider in detail the trajectories of approaching molecules for various impact parameters **b**. As illustrated in Fig. 2-15, the *impact parameter* is the minimum internuclear distance that a pair of approaching molecules would reach were there no interaction between them. The figure also illustrates the point that as a result of the attractive potential, molecule 2 may swing behind molecule 1 (taken as stationary) and thus be deflected. In the treatment of viscosity, for example, we need detailed calculations to obtain the average momentum exchange for all possible impact parameters, using the Lennard–Jones potential function. This averaging integration has been carried out, and the result can be expressed as a correction to Eq. (2-85):

$$\eta = \frac{1}{\pi\sigma^2}\left(\frac{kTm}{\pi}\right)^{1/2}\frac{1}{A(T)} \qquad (2\text{-}100)$$

where $A(T)$ is this integral.

The function $A(T)$ obeys the principle of corresponding states in that it may be tabulated in terms of the reduced temperature $T^* = kT/\phi^*$. A plot of $A(T)$ versus T^* is given in Fig. 2-16. Actual viscosity data are now used in the calculation of the values of ϕ^*/k and σ which give the best fit for a given gas, and some of these values are collected in Table 2-5.

TABLE 2-5

Parameters for the Lennard–Jones potential from viscosity coefficients[a]

Gas	ϕ^*/k (K)	σ (Å)
Ar	124	3.42
Kr	190	3.61
N_2	91.5	3.68
O_2	113	3.43
CH_4	137	3.82
CO_2	190	4.00

[a]From J.O. Hirschfelder, C.F. Curtiss, and R.B. Bird, "Molecular Theory of Gases and Liquids," corrected ed. Wiley, New York, 1964.

EXAMPLE　Calculate the corrected viscosity for CO_2 at 0°C. From Table 2-5 we find $\phi^*/k = 190$ and hence $T^* = 273/190 \simeq 1.44$, and read from Fig. 2-16 that $A(T) \simeq 1.35$. Also, from Table 2-5, $\sigma = 4.00$ Å and substitution into Eq. (2-85) gives

$$\eta_{ideal} = \frac{1}{(\pi)(4.00 \times 10^{-8}(6.023 \times 10^{23})} \left[\frac{(8.314 \times 10^7)(273.2)(44.01)}{\pi} \right]^{1/2}$$

$$= 1.86 \times 10^{-4} \text{ P.}$$

The corrected viscosity is $1.86 \times 10^{-4}/1.35$ or 1.38×10^{-4} P, as compared to the experimental value of 1.42×10^{-4} P.

Note that $A(T)$ is not a function of pressure or of gas density. This is because the gas has been assumed to be sufficiently dilute that it is only during actual bimolecular collisions that the potential function exerts an influence. Alternatively, one is neglecting trimolecular collisions. Note also that the σ parameters of Table 2-5 are not quite the same as the corresponding ones given in Table 1-6. The latter were obtained when a Lennard–Jones potential treatment was fitted to the observed second virial coefficients of the gases. This discrepancy illustrates the point that even the more advanced treatments are approximate and retain a semiempirical aspect.

GENERAL REFERENCES　BERRY , R.S., RICE, S.A. AND ROSS, J. (1980). "Physical Chemistry." Whiley, New York.

HIRSCHFELDER, J.O., CURTISS, C.F., AND BIRD, R.B. (1964). "Molecular Theory of Gases and Liquids," corrected ed. Wiley, New York.

KAUZMANN, W. (1966). "Kinetic Theory of Gases." Benjamin, New York.

CITED REFERENCES　BRIDGMAN, P. (1946). "The Nature of Modern Physics." Macmillan, New York.

HERZFELD, K.F., AND SMALLWOOD, H. (1951). "A Treatise on Physical Chemistry" (H.S. Taylor and S. Glasstone, eds.), 3rd ed., Vol. 2, Van Nostrand–Reinhold, Princeton, New Jersey.

HIRSCHFELDER, J.O., CURTISS, C.F., AND BIRD, R.B. (1964). "Molecular Theory of Gases and Liquids." Wiley, New York, p. 527, 545.

MOELWYN-HUGHES, E.A. (1961). "Physical Chemistry." Pergamon, Oxford.

EXERCISES

The student is reminded to make a consistent choice of units in carrying out the following calculations. Take as exact numbers given to one significant figure.

2-1　Calculate W for systems 1 and 3 of Table 2-1.

Ans. $W_1 = 252$; $W_3 = 360$.

2-2　What is $h_{1/2}$ for argon—that is, the elevation at which the pressure of argon in the atmosphere is half of its sea level value? Assume 20°C .

Ans. 4.31 km.

2-3　A good vacuum for many purposes might still have a pressure of 10^{-5} Pa. Treating air as a single gas of molecular weight 29, at what elevation will this pressure be found? Assume g to be constant at the standard value and temperature to be uniform at -70°C.

Ans. 136.9 km.

2-4　Calculate the fraction of H_2 molecules per unit velocity interval, $(1/N_A)dN(u)/du$, having velocity $u = 0$ if $PV = 10$ liter atm.

Ans. 5.627×10^{-6}.

2-5

Calculate the fraction of O_2 molecules per unit velocity interval, $(1/N)_A \, dN(u)/du$, having velocity u such that their kinetic energy in one dimension is equal to kT at 25°C. Repeat the calculation in two and in three dimensions where, in each case, the velocity c is such that the corresponding kinetic energy is kT at 25°C.

Ans. 0.527×10^{-5}, 1.87×10^{-5}, 2.11×10^{-5}.

2-6

Repeat the calculations in Exercise 2-5, but taking the one-, two-, and three-dimensional velocities to be 2×10^5 cm sec^{-1}.

Ans. 8.790×10^{-17}, 1.584×10^{-15}, 9.084×10^{-15}.

2-7

Calculate the values of c_p, $\bar{c}$, and the root mean square velocity for argon gas at (a) 25°C and (b) 200°C.

Ans. (a) 3.523×10^4, 3.975×10^4, 4.315×10^4
and (b) 4.438×10^4, 5.008×10^4, and 5.435×10^4 (all in cm sec^{-1}).

2-8

Consider N_2 gas at 77 K and 1×10^4 Pa. Calculate (a) $\bar{c}$, (b) **n,** (c) Z in moles per square meter per second hitting a surface, and (d) Z_m in kg per square meter per second hitting a surface.

Ans. (a) 241.2 m s^{-1}, (b) 9.406×10^{24} molecules m^{-3},
(c) 9.420×10^2 mol m^{-2} s^{-1}, and (d) 26.40 kg m^{-2} s^{-1}.

2-9

Referring to Exercise 2-8, if every N_2 molecule that strikes the surface sticks and each N_2 molecule occupies 0.16 nm^2 area, how long should it take for an initially clean surface to be 1% covered with N_2?

Ans. 1.102×10^{-10} s.

2-10

An evacuated bulb has a pinhole of 1×10^{-0} m^2 area. How many moles of air ($M_{av} = 0.029$ kg mole^{-1}) should leak into the bulb per minute at 25°C?

Ans. 2.861×10^{-3} mole min^{-1}.

2-11

An astronaut's spacesuit has a pinhole leak. If the suit contains air of normal composition, what should be the composition of the gas escaping into space from the pinhole, according to Graham's law?

Ans. 18.96% oxygen.

2-12

Calculate λ in meters for N_2 at 250 K and 2×10^6 N m^{-2} pressure. Take $\sigma = 0.37$ nm.

Ans. 2.837×10^{-9} m.

2-13

For good insulation the pressure in the evacuated space of a Dewar or thermos flask should be reduced to the point that the mean free path is greater than the distance between the walls. What should the pressure of air be at 25°C if the mean free path is to equal 0.5 cm? Take the average molecular diameter to be 3.70 Å.

Ans. 1.336×10^{-5} atm.

2-14

Calculate the mean free path of argon gas at 10^{-11} atm (an obtainable pressure under modern ultrahigh-vacuum conditions) and 0°C, taking σ to be 3.64 Å.

Ans. 6.32×10^5 cm.

2-15

Calculate the frequency of $O_2 - N_2$ collisions in air at 25°C. Take the collision diameters to be 3.61 Å for O_2 and 3.75 Å for N_2.

Ans. 4.452×10^4 mole cm^{-3} sec^{-1}.

2-16

The viscosity of air at 0°C is 1.71×10^{-4} P. What should the volume flow rate of air be, in liter sec^{-1} through a capillary tube 10 m long and 0.1 mm in radius if the pressure drop is 15 atm?

Ans. 3.490×10^{-4} liter sec^{-1}.

2-17

Calculate σ_{av} for air using information supplied in Exercise 2-16.

Ans. 3.76 Å. (We use 1.71×10^{-4} P as the viscosity of air at 0°C, and Eq. (2.86).)

2-18   Calculate the self-diffusion coefficient for helium at STP taking σ to be 2.18 Å. Also calculate the viscosity and the thermal conductivity coefficient taking C_v to be 3.00 cal K^{-1} $mole^{-1}$.

Ans. $\mathscr{D} = 1.059$ cm^2 sec^{-1}, $\eta = 1.891 \times 10^{-4}$ P,
$\kappa = 1.417 \times 10^{-4}$ cal K^{-1} cm^{-1} sec^{-1}.

2-19   Referring to Exercise 2-18 for data, how far should a helium atom at STP diffuse in 1 sec? How long should it take to diffuse 1 cm?

Ans. 1.46 cm, 0.472 sec.

2-20   Repeat the calculations of Exercise 2-18 in SI units.

Ans. $\mathscr{D} = 1.059 \times 10^{-4}$ m^2 s^{-1}, $\eta = 1.891 \times 10^{-5}$ kg m^{-1} s^{-1},
$\kappa = 5.930 \times 10^{-2}$ J K^{-1} m^{-1} s^{-1}.

PROBLEMS

Problems marked with an asterisk require fairly lengthy computations.

2-21   Extend the example of Table 2-1 to find the possible distributions and their probability weights if we have 30 molecules with an average energy of 2 units per molecule.

2-22*   Consider a system of 20 molecules that may be distributed among states of energy one, two, three, and four units. List the possible distributions having an average energy of two units and calculate W for each.

2-23*   How accurate is Sterling's approximation? Compare the actual values of $x!$ with those from Eq. 2-15 for $n = 10, 50, 100, 300$. (If you have a programmable calculator, set it to calculate $\ln(n!)$ rather than $n!$; otherwise you will run out of exponent capability). A more exact expression is

$$\ln(x!) = \tfrac{1}{2} \ln (2\pi) + (x + \tfrac{1}{2}) \ln x - x + \frac{1}{12x} - \frac{1}{360x^3} \cdots \tag{2-101}$$

(see G. Arfken, "Mathematical Methods for Physicists," Academic, New York, 1968, p. 363.) How good is Eq. (2-101)?

2-24   At sea level, the composition of air is 80 mole % nitrogen and 20 mole % oxygen. Estimate the altitude (in miles) at which, according to the barometric formula, the air should contain only 15 mole % oxygen. Assume 0°C.

2-25   Assume that a body of air ($M_{av} = 29.0$ g $mole^{-1}$) at 25°C is in barometric equilibrium with g constant at 980 cm sec^{-2}. Calculate the mass of air contained in a column 1 square mile in area and 1 mile high. The pressure is 0.9 atm at the base of the column. What would your answer be if air had a molecular weight of 58.0 g $mole^{-1}$?

2-26   Calculate the exact value of c at the crossing point of the two distributions of Fig. 2-6.

2-27*   Calculate enough values of $1/N_A)dN(c)/dc$ for methane at 100°C to make a reasonably accurate plot of probability versus velocity. Verify by a graphical or an analytical method that the plot is normalized. Locate c_p, $\bar{c}$, and $(\overline{c^2})^{1/2}$ on the plot.

2-28   Make the change of variable $E = \tfrac{1}{2}mc^2$ to derive from Eq. (2-43) the expression for the fraction dN/N_A of molecules having energy in two dimensions between E and $E + dE$, as a function of E. Calculate the fraction of molecules of a gas which have an energy equal to or greater than ten times their average kinetic energy (which is $3kT/2$).

2-29   Calculate c_p, $\bar{c}$, and $(\overline{c^2})^{1/2}$ for Ar gas at 100°C and at 1000°C.

2-30   Calculate $p(c)$ for $c = \bar{c}$ in the case of N_2 gas at 25°C.

2-31   Derive the expressions for $\bar{c}$ and $\overline{c^2}$ for a two-dimensional gas.

Derive the expression for the average kinetic energy of molecules escaping through a pinhole by effusion. Note that Eq. (2-56) gives the probability of a molecule hitting the pinhole (if multiplied by the area of the pinhole). Your answer will be larger than the average kinetic energy in one dimension. Why?
$\quad$ **2-32**

Methane gas is present in a flask at 25°C and pressure P. The flask has a pinhole which may be regarded as a cylindrical hole of 2×10^{-4} cm² area and 0.20 cm length. Calculate the rate of escape of the methane assuming that the process is one of (a) effusion and (b) diffusion, and for $P = 1$ atm and $P = 10^{-7}$ atm. Express your answer in moles of gas per second.
$\quad$ **2-33**

A flask contains a 3:1 mole ratio of H_2 to He at 100°C and 2 atm total pressure, and has a pinhole of 1×10^{-4} cm² area. Calculate (a) the total moles of gas escaping per second and (b) the total mass of gas escaping per second.
$\quad$ **2-34**

Calculate the mass per second of CO_2 striking each square mm of a leaf in air containing CO_2 at a partial pressure of 100 Pa and at 25°C.
$\quad$ **2-35**

What quantity of heat in calories per centimeter length per second will be lost through conduction by gas molecules by a filament 0.2 mm in diameter heated electrically to 200°C in a light bulb containing argon at 0.01 mm pressure, the wall temperature of the bulb being 25°C? [*Note:* This pressure may be assumed to be low enough so that gas molecules travel from filament to wall and back again without hitting each other on the way. Therefore argon atoms hit the filament, then leave it with a kinetic energy corresponding to 200°C, and fly to the wall to be "cooled down" to 25°C. For steady-state conditions, the rate of leaving the filament should be the same as the rate of hitting it.]
$\quad$ **2-36**

In studying the state of adsorbed molecules, high vacuum surface chemists typically expose a clean surface to a specified low pressure of the adsorbate gas, for a specified time. A unit that has come into use is the *Langmuir*. One Langmuir, L, is an exposure of 1×10^{-6} torr-second. A smooth surface of tungsten is exposed to 3 L of CO gas. What fraction of the surface should be covered if every molecule of CO hitting the surface sticks? Take the molecular area of CO to be 16 Å² and the temperature to be 100°C. Comment on your answer. The actual degree of surface coverage was about 0.6.
$\quad$ **2-37**

Calculate the collision frequency for H_2 at STP. Take the collision diameter to be 2.5 Å.
$\quad$ **2-38**

Calculate the collision frequency for Cl_2 at STP. Take the collision diameter to be 3.5 Å.
$\quad$ **2-39**

Calculate the initial rate for the reaction $H_2 + Cl_2 = 2\,HCl$ in an equimolar mixture at STP assuming that reaction occurs with each collision (note Problems 2-38 and 2-39). The reaction is actually quite slow; why might this be?
$\quad$ **2.40**

The collision diameter for argon is 3.64 Å. Calculate Z_{Ar-Ar} at STP. The average concentration of molecules which are in the act of colliding may be estimated as given by their collision frequency times the collision lifetime of about 10^{-13} sec. Calculate the frequency, in moles of events per liter per second, with which such a collision pair will undergo collision with an argon atom—that is, the frequency of triple collisions.
$\quad$ **2-41**

The molar volume V of a substance may be calculated from the liquid density at the normal boiling point. If the molecules are assumed to be spherical and to be close-packed in the liquid phase, then V can be related to σ and N_A. Equation (2-85) may now be restated in a form that involves only experimentally measurable quantities, with N_A an *unknown*. Derive this form and calculate N_A from the liquid density of CH_4 (see a handbook) and its gas viscosity (Table 2-3). Comment on why your calculated N_A differs from the correct value. Note: close-packed spheres fill 74% of space.
$\quad$ **2-42**

Consider a mixture of gases 1 and 2. Show that the expressions for $Z_{1(1)}$, $Z_{1(2)}$, $Z_{2(2)}$, and $Z_{2(1)}$ are given by a single general expression for $Z_{i(j)}$ with $i = 1$ or 2 and $j = 1$ or 2 if m_{ij} is used instead of μ_{ij} (see footnote following Eq. (2-73)).
$\quad$ **2-43**

Verify Eq. (2-86).
$\quad$ **2-44**

2-45 The coefficient for the interdiffusion of two chemical species is given by

$$\mathcal{D}_{12} = 2.6280 \times 10^{-3} \frac{[T^3(M_1 + M_2)/2M_1M_2]^{1/2}}{P\sigma_{12}^2} \ (cm^2 \ sec^{-1}),$$

where M denotes molecular weight. This is an equation of convenience in that the units of P and $\mathcal{D}_{12}$ are not in fundamental cgs or SI units, but in some commonly used units. Deduce what these units must be, and whether M is in g mole^{-1} or in kg mol^{-1}. Calculate $\mathcal{D}_{12}$ for the interduffusion of oxygen and nitrogen in air at STP.

2-46 Three of the *experimentally measurable* quantities listed in Table 2-2 may be combined as a dimensionless product (or quotient). Find one such combination, calculate its theoretical values, and compare with experiment.

2-47 Calculate C_V

for methane at STP from the data of Table 2-3.

2-48 The viscosity of H_2 at STP is 8.40×10^{-5} P. Calculate the mean free path.

2-49 Suppose we have an equimolar mixture of $^{13}CH_4$ and $^{13}CD_4$ (species A and B, respectively) at STP. Calculate Z_A and $Z_{A(B)}$. Explain if the answers are different. Assume that σ is 4.14 Å for both species.

2-50 We have a mixture of H_2 and O_2 at 25°C, the respective partial pressures being $\frac{2}{3}$ atm and $\frac{1}{3}$ atm. Calculate the moles per liter per second of (a) $H_2 - H_2$, (b) $H_2 - O_2$, and (c) $O_2 - O_2$ collisions. Calculate also the average distance traveled by a hydrogen molecule before making a collision with (d) another hydrogen molecule, (e) an oxygen molecule, and (f) any molecule. The cross sections for H_2 and O_2 are 2.74 Å and 3.61 Å, respectively.

SPECIAL TOPICS PROBLEMS

2-51 Calculate the viscosity of CH_4 at 280 K using Eq. (2-100) and compare your result with the experimental value.

2-52 The viscosity of Xe is 2.235×10^{-4} P at 290 K and 3.954×10^{-4} P at 550 K. Devise a trial and error method so as to calculate reasonably matching values of ϕ^*/k and σ.

chapter 3
Some additive physical properties of matter

3-1   Introduction. The principle of additivity

We take up in this chapter several physical properties of matter, especially of liquids and gases, properties which would constitute a diversion if thrust into one or another of the main chapters of this text. The properties of light absorption, index of refraction and dielectric constant behavior, and magnetochemistry are too useful to be relegated to some terminal chapter, however, and so are presented here.

These topics sound rather diverse, but they have two aspects in common. First, the effects all arise from the interaction of an electric or a magnetic field with molecules, interactions which, to a first approximation, are determined by the separate behavior of atoms or small groups of atoms within a molecule. The second aspect in common is that in each case the measured property can be put on a molar basis and, when this is done, the property is approximately *additive*. That is, the molar property can be formulated as a sum of contributions from various parts of the molecules present and, if a mixture of species is present, as a mole fraction weighted sum of contributions from the components of the mixture.

The additivity principle is central to large areas of analytical chemistry, and for this reason alone it deserves some special emphasis. To be more specific as to what additivity means, let $\mathscr{P}$ denote the molar property in question. Then, if the molecular formula is $A_a B_b C_c \cdots$, additivity means that

$$\mathscr{P} = a\mathscr{P}_A + b\mathscr{P}_B + c\mathscr{P}_C + \cdots \tag{3-1}$$

where $\mathscr{P}_A$ and so on denote the atomic properties. For a mixture of species the average molar property is

$$\mathscr{P}_m = x_1\mathscr{P}_1 + x_2\mathscr{P}_2 + \cdots = \sum_i x_i\mathscr{P}_i \tag{3-2}$$

where $\mathscr{P}_i$ is the molar property of the ith species and x denotes mole fraction. This last attribute is widely used in chemical analysis. The result is that by means of a physical measurement one can determine the composition of a mixture or, as in chemical kinetics, follow changes in composition with time.

We have already encountered some additive properties. Mass is strictly additive (neglecting relativity), so molar mass or molecular weight obeys Eqs. (3-1) and (3-2) exactly. Volume at constant pressure and temperature is additive for ideal gases, but usually only approximately so in the case of liquids.

The discussion of cgs vs. SI units, begun in Section 1-CN-1, is continued in Section 3-CN-1. In the area of electrical units there are some major differences between the two systems, including differences in philosophy!

3-2   Absorption of light

The term "light" will be used in the present context to mean electromagnetic radiation generally. That is, the radiation need not be of wavelength corresponding to the visible region: It could be ultraviolet, infrared, or microwave radiation. Such radiation can interact with matter to show various effects such as refraction, scattering, and absorption. It is only the last that we discuss at the moment. The phenomenon of optical activity is taken up in Section 19-ST-2.

When light is absorbed its energy is converted into some other form, and we accept that this conversion can occur only in the discrete amount given by the quantum of energy of the light. It is part of the quantum theory of light that this quantum of energy is given by $h\nu$, where h is Planck's constant and ν is the frequency of the light. The usual consequence of absorption of a light quantum is that some particular molecule gains energy $h\nu$ and is thereby put in some higher energy state, that is, in some excited state.

A point of importance here is that although light can transfer energy only in units of quanta, it otherwise behaves as a wave. One consequence is that the train of waves passing through a region of matter may or may not interact with any of the molecules present. There is only a certain probability that interaction will occur. The derivation of the absorption law is then based on a probability analysis.

Let k_ℓ be the probability that light of a given wavelength will be absorbed per unit length of matter; $k_\ell \delta$ is then the probability of absorption in some small distance δ. The probability that the light is not absorbed is $1 - k_\ell \delta$. Further, the probability that light will not be absorbed in a second increment of distance δ is $(1 - k_\ell \delta)^2$ and the probability of it not being absorbed in n such distances becomes $(1 - k_\ell \delta)^n$.

Now let ℓ be the total distance, $\ell = n\delta$. We want to keep the same path length ℓ, but to take the limit of $n \to \infty$, that is, the limit of an infinite number of infinitesimal lengths δ:

$$\lim_{n \to \infty} \left(1 - \frac{k_\ell \ell}{n} \right)^n = e^{-k_\ell \ell} \tag{3-3}$$

(It should be recognized that $\lim_{n \to \infty}[1 + (1/n)^n] = e$.) The probability that light will not be absorbed in distance ℓ is then $e^{-k_\ell \ell}$. If light of intensity I_0 is incident on a portion of matter, then the probable intensity after a distance ℓ is

$$I = I_0 e^{-k_\ell \ell} \tag{3-4}$$

I/I_0 is the fraction of light transmitted through, that is, not absorbed by, a sample of thickness ℓ.

Equation (3-4) is a statement of Lambert's law, where k_ℓ is the *linear absorption coefficient* (also called the Napierian extinction coefficient *b*). However, since the absorption process is a molecular one, it is somewhat more rational to recognize that, per unit area of target, a depth ℓ will contain $\ell\rho/m$ molecules, and to write Eq. (3-4) as

$$\frac{I}{I_0} = e^{-k_N N} \qquad k_N = \frac{k_\ell m}{\rho} \tag{3-5}$$

where ρ and m are respectively the density and the mass per molecule of the material, N is the number of molecules traversed in depth ℓ per unit cross section, and k_N is now the molecular absorption coefficient.

A useful physical picture is the following. Rather than talk about the probability of absorption by a molecule, let us assign to it an *effective* cross-sectional area σ. That is, we regard the molecule as behaving like a totally opaque object of area σ. As illustrated in Fig. 3-1, the collection of N molecules presents a target area of $N\sigma$ per unit total cross-section of the incident beam of radiation and thus occludes fraction $N\sigma$ of the beam. It follows that $k_N = \sigma$. Keep in

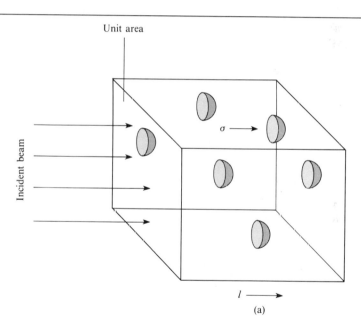

(a)

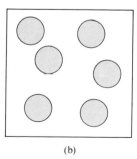

(b)

FIGURE 3-1
(a) Perspective view of a beam of radiation incident on a depth ℓ of material containing N molecules, each of cross section σ. (b) View along the path of the beam, illustrating that the effective occluded area is $N\sigma$.

mind that σ bears no particular relationship to the actual molecular size; actual σ values may range from being somewhat larger than, say, the van der Waals area of the molecule to very much smaller values. In the case of nuclear reactions (see Chapter 22), σ values are often around 10^{-24} cm^2, or of the order of a nuclear diameter squared.

An important case for the physical chemist is that in which the absorbing species is present in solution. If we can assume that k_N is not affected by concentration, then, since $N = N_A C \ell / 1000$, where C is in moles per liter, that is, in units of molarity M,

$$\frac{I}{I_0} = e^{-\alpha C \ell} \qquad \alpha = \frac{k_N N_A}{1000} \tag{3-6}$$

The constant α is called the *molar absorption coefficient*, and Eq. (3-6) is a statement of the combined Beer–Lambert law for light absorption.

Finally, a common modification of Eq. (3-6) is

$$\frac{I}{I_0} = 10^{-\epsilon C \ell} \qquad \epsilon = \frac{\alpha}{2.303} \tag{3-7}$$

If we now define the *optical density* D as equal to $\log(I_0/I)$, we have

$$D = \epsilon C \ell \tag{3-8}$$

The quantity D is alternatively called the *absorbance* and is denoted by the symbol A. Also, the *extinction coefficient* ϵ is alternatively called the *molar absorptivity*.

Returning to the derivation that led to Eq. (3-3), if more than one type of interacting species is present, then the probability k_ℓ of absorption per unit distance will now be given by a sum over all species, $k_{\ell,\text{tot}} = \Sigma k_{\ell,i}$. The same applies to the other absorption coefficients. In particular, for a solution of more than one species, we get

$$D = \ell \sum \epsilon_i C_i \tag{3-9}$$

Often the product $\epsilon \ell C$ for the solvent is abstracted from the sum to give

$$D = D_i + \ell \sum \epsilon_i C_i \tag{3-10}$$

Optical density or absorbance is thus an additive property of mixtures, to the extent that changing concentration does not change the character of the species present. It may also be additive with respect to chromophoric groups within a molecule; this is especially common in the case of infrared spectra where vibrational excitations of the various types of bonds contribute somewhat independently to the overall absorption. It should be emphasized that ϵ is very much a function of wavelength so that any statement of an ϵ value must be accompanied by the wavelength to which it pertains.

EXAMPLE   A 0.01 M solution of $Cr(NH_3)_5(NCS)^{2+}$ (as the perchlorate salt) absorbs 88.79% of incident light of 500 nm wavelength when placed in a 1-cm spectrophotometer cell. The solvent medium absorbs somewhat at this wavelength; a 10-cm cell filled with solvent transmits 31.6% of incident light at 500 nm.

On standing, the reaction

$$Cr(NH_3)_5(NCS)^{2+} + H_2O = Cr(NH_3)_5(H_2O)^{3+} + NCS^-$$
$$\quad\quad A \quad\quad\quad\quad\quad\quad\quad\quad\quad\quad\quad\quad B$$

occurs and the reaction goes to completion in several weeks at 45°C. The per cent transmission of this final solution through a 1-cm cell is 44.64. How far had the reaction proceeded after 75 hr at 45°C if, after this time, the optical density of the solution in the 1-cm cell was 0.650? Also calculate the extinction coefficients of the complex ions A and B.

First, the various statements should be reduced to optical densities for a 1-cm path length. For the original solution, $D_0 = \log(I_0/I) = -\log(0.1121) = \log(8.92) = 0.950$. For the solvent, $D_8 = -\log(0.316) = \log(3.16) = 0.500$, or, for a 1-cm path length, $D_8 = 0.050$. The net optical density due to the starting complex is then $0.950 - 0.050 = 0.900$, and the extinction coefficient is therefore $\epsilon_A = 0.900/0.01 = 90.0$ liter mole^{-1} cm^{-1}. After complete reaction, $D_\infty = -\log(0.4464) = \log(2.24) = 0.350$. The net optical density due to the product is then $0.350 - 0.050 = 0.300$ and the extinction coefficient of the product is $\epsilon_B = 0.300/(1 \times 0.01) = 30.0$ liter mole^{-1} cm^{-1}.

The net optical density after 75 hr is $0.650 - 0.050 = 0.600$. Then, by Eq. (3-10)

$$0.600 = (1)(90.0C_A + 30.0C_B) = 60.0C_A + 0.300$$

(since $C_B = 0.01 - C_A$), or $C_A = 0.300/60.0 = 0.0050$ M. The reaction had thus proceeded to 50%, or by one half-life.

The solution of the previous example would transmit 12.58% of incident light (assuming no solvent absorption) with a 1-cm path length, corresponding to an ϵ_A value of 90 liter mole^{-1} cm^{-1}. Calculate the corresponding values of k_ℓ, k_N, σ, and α.

Since $\ln(I/I_0) = -2.0727$, it follows from Eqs. (3-4), (3-5), (3-6), and (3-7) that

$$2.0727 = k_\ell\ell = k_N N = \sigma N = \alpha C\ell = 2.303 \,\epsilon C\ell$$

In this example, $\ell = 1$ cm and $N = (6.022 \times 10^{23})(0.01)(1)/(1000) = 6.022 \times 10^{18}$ molecule cm^{-2}. Then $k_\ell = 2.0727$ cm^{-1}, $k_N = \sigma = (2.0727)/(6.022 \times 10^{-18}) = 3.4419 \times 10^{-19}$ cm^2, and $\alpha = (2.0727)/(0.01)(1) = 207.27$ liter mole^{-1} cm^{-1}. Notice that the cross-section σ corresponds to an effective molecular diameter of about 0.06 Å, or far less than the van der Waals diameter, which would be something like 6 Å for the species involved, $Cr(NH_3)_5(NCS)^{2+}$. This is just an alternative way of expressing a rather small absorption probability per molecule.

In SI units, distances should be in meters and concentrations in mol m^{-3}. More often than not, however, the cgs and mole liter^{-1} units are used, but with liter replaced by its equivalent, dm^3.

EXAMPLE

3-3 Molar refraction

A second type of interaction of light with a medium is refraction, measured by the index of refraction n, which gives the ratio of the velocity of light in vacuum to that in the substance. The actual experimental measurement is based on the bending of a light ray as it passes from air into the medium.

The theory involves the interaction of the oscillating electric field of electromagnetic radiation with the various characteristic frequencies for electrons in an atom or collection of atoms, and leads, in first order, to the equation

$$n = 1 + \sum \frac{a_i}{\nu_0^2 - \nu^2} \tag{3-11}$$

where the sum is over the various characteristic frequencies v_0 for a given frequency of radiation v, and the a_i's are constants. Ordinarily, v is small compared to any v_0, so the index of refraction does not vary much with the wavelength of light used, although for n to be purely characteristic of the atom, v should be zero. Often, however, a standard wavelength, such as that of the yellow emission from a sodium lamp, is used. There is an important exception to this situation. If the compound absorbs light this means that there is some transition between states of differing energy. This energy corresponds to a frequency hv_0 and, as a consequence, the index of refraction will vary rapidly as the frequency of the light used, v, approaches v_0, that is, through a region of strong light absorption. For most small molecules, however, v_0 is rather large; in fact, hv_0 often corresponds approximately to the ionization energy of the molecule. Equation (3-11) is approximate and n does not actually go to infinity at $v = v_0$.

An early treatment by H. A. Lorentz and L. V. Lorenz in 1880, as well as more modern ones, considers that the index of refraction should vary with the density of a medium of given atomic composition according to the relationship

$$\mathbf{R} = \frac{n^2 - 1}{n^2 + 2} \frac{M}{\rho} \tag{3-12}$$

Here M/ρ is the molar volume and $\mathbf{R}$ is a scalar quantity called the *molar refraction* or *molar refractivity*. If molecules are treated as perfectly conducting spheres and light of infinite wavelength is used in measuring n, then an advanced treatment shows that $\mathbf{R}$ is the actual molar volume of the molecule.

There is an alternate way of treating this limiting situation. An imposed electric field $\mathbf{F}$ will in general displace the electrons in an atom somewhat to produce a dipole moment as illustrated in Fig. 3-2. A charge separation occurs in the initially electrically symmetric atom, equivalent to having small equal and opposite charges $\delta+e$ and $\delta-e$ separated by distance d. The dipole moment μ is defined as

$$\mu = \delta ed \tag{3-13}$$

where e is the charge on the electron. The traditional unit for dipole moment is the debye (D), defined as 10^{-18} esu cm. Since d is of the order of 1 Å and e has the value 4.803×10^{-10} esu, dipole moment quantities generally come

---

FIGURE 3-2
Molecular dipole moment induced by an electric field.

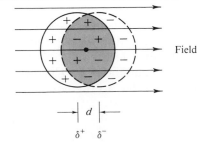

out as some small number of debye units. (In the SI system, dipole moment is given in coulomb-meter units.)

The induced dipole moment is proportional to the electric field,

$$\mu_{ind} = \mathbf{m}_\alpha = \alpha \mathbf{F} \tag{3-14}$$

where the proportionality constant α is called the polarizability and the symbol **m** is used for moments induced by various individual effects. Only one such effect is present in Eq. (3-14), so that here μ_{ind} and $\mathbf{m}_\alpha$ are the same. The symbol μ is reserved for the experimentally observed dipole moment; if it is induced, the subscript "ind" may be used. In general, μ_{ind} may be the sum of more than one term [as in Eq. (3-20)]. In the cgs system, the field is in esu volt per centimeter, the dipole moment in esu-centimeter, and α is then in cubic centimeters. Usual values of α are about 10^{-24} cm^3 or about actual atomic volumes. In the SI system, polarizability has the dimensions of A^2 s^4 kg^{-1} (or C^2 m^2 J^{-1}, which is equivalent), and **F** is in V m^{-1}. It turns out that to write an equation valid in SI units, α must be replaced by $\alpha_{SI}/4\pi\epsilon_0$, where ϵ_0 is the permittivity of vacuum (see Section 3-CN-1), $\epsilon_0 = 8.842 \times 10^{-12}$ C^2 J^{-1} m^{-1}. A cgs polarizability of 1×10^{-24} cm^3 becomes $(4\pi)(8.842 \times 10^{-12})$ $(10^{-2})^3$ $(1 \times 10^{-24}) = 1.111 \times 10^{-40}$ C^2 m^2 J^{-1} in SI units. Often, α_{SI} is divided by $4\pi\epsilon_0$ and reported in m^3. Some representative polarizability values are given in Table 3-1.

The conducting sphere model provides a connection between index of refraction at infinite wavelength, n_∞ [$\nu = 0$ in Eq. (3-11)] and the polarizability of the sphere, again given in Lorentz and Lorenz. This relationship is

$$\mathbf{R} = \frac{n^2 - 1}{n^2 + 2} \frac{M}{\rho} = \frac{4}{3} \pi N_A \alpha \tag{3-15}$$

Substance	Polarizability (Å^3)	Dipole moment (D)
He	0.204	—
Ar	1.62	—
H$_2$	0.802	—
N$_2$	1.73	—
Cl$_2$	4.50	—
HCl	2.56	1.05
HBr	3.49	0.807
HI	5.12	0.39
CO$_2$	2.59	0.00
H$_2$O	1.44	1.82
NH$_3$	2.15	1.47
CH$_3$Cl	—	1.87
C$_6$H$_6$	25.1	0.00
C$_6$H$_5$Cl	—	1.70
p-C$_6$H$_4$Cl$_2$	—	0.00
o-C$_6$H$_4$Cl$_2$	—	2.2
m-C$_6$H$_4$Cl$_2$	—	1.4

TABLE 3-1
Dipole moments and polarizabilities[a]

[a]From **R** values.

where N_A is Avogadro's number. Thus α corresponds to the cube of the radius of the sphere. While Eq. (3-14) is the same in SI, Eq. (3-15) becomes

$$\mathbf{R} = \frac{n^2 - 1}{n^2 + 2}\frac{M}{\rho} = N_A\alpha_{SI}/3\epsilon_0 \tag{3-16}$$

As suggested by the summation of Eq. (3-11), molar refraction is, ideally, an additive property; empirically, it is very nearly one. Thus values of **R** can be assigned to individual atoms in a molecule. Somewhat better results are obtained if molar polarizations are assigned to types of bonds rather than to individual atoms. See Glasstone (1946) and Moelwyn–Hughes (1961).

EXAMPLE   Calculate the molar refractions for two isomeric liquids at 20°C, acetic acid and methyl formate. The respective molar volumes M/ρ are $60.05/1.0491 = 57.24$ cm³ mole⁻¹ and $60.05/0.9742 = 61.64$ cm³ mole⁻¹; and the refractive indices with the sodium D line are 1.3721 and 1.3433, respectively. Substitution into Eq. (3-12) gives molar refractions of 13.013 cm³ mole⁻¹ for acetic acid and 13.033 cm³ mole⁻¹ for methyl formate. Thus **R** value can distinguish between isomers. The respective values from the atomic refractions are 12.972 and 13.090 cm³ mole⁻¹ (and those from the bond refractions, 12.770 and 12.782 cm³ mole⁻¹, but now for infinite wavelength).

Use of Eq. (3-15) gives 22.7×10^{-24} cm³ and 24.4×10^{-24} cm³ for the polarizabilities of acetic acid and methyl formate, respectively.

3-4   Molar polarization; dipole moments

It was stressed in the preceding section that, on theoretical grounds, the index of refraction should be for zero-frequency light in the calculation of molar refraction. As an alternative, one may directly impose an external low- or zero-frequency electric field on the medium by having it between the plates of a capacitor. The capacitance of a parallel plate capacitor is

$$C_0 = \frac{q}{V} = \frac{q}{\mathbf{E}_0 d} \tag{3-17}$$

where V is the applied potential difference and $\mathbf{E}_0$ is the resulting field if the separation of the plates is d. In the presence of a medium other than vacuum, this field is reduced from $\mathbf{E}_0$ to $\mathbf{E} = \mathbf{E}_0/D$, where D is the dielectric constant of the medium, and we now have

$$C = \frac{q}{\mathbf{E}d} = C_0 D \tag{3-18}$$

The effect of the medium is thus to increase the capacitance from C_0 to C. The value of C_0 is related to the dimensions of the capacitor,

$$C_0 = \epsilon_0\mathscr{A}/d \tag{3-19}$$

where $\mathscr{A}$ is the area of the plates, and C_0 is in farads.

From a molecular point of view, the presence of the field induces a proportional dipole moment in the molecules as in Eq. (3-14). At the low frequencies which characterize dielectric constant measurements there are now two (main) contributions to μ_{ind}. As before there is the dipole moment arising from the polarizability of the molecule, $\mathbf{m}_\alpha$, and there is also the effective moment $\mathbf{m}_\mu$ resulting from the partial net alignment by the field of any permanent molecular dipoles that are present. This second type of induced dipole moment is again proportional to the field acting on the molecules (as discussed later) and the proportionality constant is called the orientation polarizability α_μ. Equation (3-14) becomes

$$\mu_{\text{ind(tot)}} = \mathbf{m}_\alpha + \mathbf{m}_\mu = (\alpha + \alpha_\mu)\mathbf{F} \tag{3-20}$$

or

$$\mathbf{m}_{\text{tot}} = \alpha_{\text{tot}}\mathbf{F} \tag{3-21}$$

If the molecules are far apart, as in a dilute gas, the total induced polarization simply acts to reduce the apparent field $\mathbf{E}_0$; from an analysis of the situation using Eq. (3-19) one finds

$$\mathbf{E} = \mathbf{E}_0 - \mathbf{I}/\epsilon_0 \tag{3-22}$$

where $\mathbf{I}$ is the total polarization per unit volume, $\mathbf{I} = \mathbf{m}_{\text{tot}}\mathbf{n}$ and $\mathbf{n}$ denotes concentration in molecules per unit volume.

More generally, however, the degree of polarization of each molecule is affected by the field of the induced polarization of the other molecules, and it is necessary to take this into account. This corrected, or net effective local field $\mathbf{F}$ is now given by

$$\mathbf{F} = \mathbf{E} + \mathbf{I}/3\epsilon_0 \tag{3-23}$$

or, from Eq. (3-22), since $\mathbf{E}_0 = D\mathbf{E}$

$$\mathbf{F} = \tfrac{1}{3}(D + 2)\mathbf{E} \tag{3-24}$$

On eliminating $\mathbf{E}$, we obtain

$$\mathbf{F}\left(1 - \frac{3}{D + 2}\right) = \mathbf{I}/3\epsilon_0 = \mathbf{m}_{\text{tot}}\mathbf{n}/3\epsilon_0$$

and, on eliminating $\mathbf{F}$ by means of Eq. (3-21), we obtain

$$\frac{D - 1}{D + 2} = \mathbf{n}\alpha_{\text{tot}}/3\epsilon_0 \tag{3-25}$$

Since $\mathbf{n} = N_A\rho/M$, the final form is

$$\mathbf{P} = \left(\frac{D - 1}{D + 2}\right)\left(\frac{M}{\rho}\right) = N_A\alpha_{\text{tot}}/3\epsilon_0 \tag{3-26}$$

Equation (3-26) is in SI units. $\mathbf{P}$ is called *molar polarization* and the contribution to $\mathbf{P}$ from the polarizability α of the molecules is denoted $\mathbf{P}_\alpha$. Theoretical analysis shows the $\mathbf{P}_\alpha$ should be equal to $\mathbf{R}$, the molar refraction, provided the latter is the limiting value from indices of refraction at infinite wavelength, $\mathbf{R}_0$.

The contribution of α_μ [Eq. (3-20)] to the total polarizability is quite important if the molecules have a nonzero dipole moment μ. This will be the case if bonds between unlike atoms are present, as in NO or HCl, provided there is no internal cancellation of such bond dipoles, as happens in CH_4 or CO_2 owing to molecular symmetry. The relationship between α_μ and the permanent molecular dipole moment μ is as follows.

If a molecule has a permanent dipole moment, it will tend to be oriented along the direction of the local electric field **F**. That is, the energy ϵ of the dipole in the field would be zero if it were oriented perpendicular to the field and it would be μF if it were oriented in line with the field. For some intermediate orientation

$$\epsilon = -\mathbf{F}\mu\cos\theta \tag{3-27}$$

where, as shown in Fig. 3-3(a), θ is the orientation angle defined such that $\mu\cos\theta$ gives the component of μ in the field direction. At 0 K all dipoles should orient completely, but at any higher temperature thermal agitation will prevent this and the probability of a given orientation is proportional to the Boltzmann factor $e^{-\epsilon/kT}$. At infinite temperature the effect of the field becomes negligible; all orientations are then equally probable and the average net dipole moment of a collection of molecules is zero.

What is now needed is the average orientation or, more precisely, the average of $\mu\cos\theta$, which gives the average net dipole moment $\mathbf{m}_\mu$. The procedure followed is that discussed in Section 2-4; that is, one writes the averaging equation

$$\mathbf{m}_\mu = \frac{N\int_0^\pi (\mu\cos\theta)\, e^{(F\mu\cos\theta)/kT}(2\pi\sin\theta\, d\theta)}{N\int_0^\pi e^{(F\mu\cos\theta)/kT}(2\pi\sin\theta\, d\theta)} \tag{3-28}$$

where N is the number of molecules and $(2\pi\sin\theta\, d\theta)$ gives $d\omega$, the element of solid angle.

In the polar coordinate system used (shown in Fig. 3-3b) an element of solid angle is given by

$$d\omega = \sin\theta\, d\theta\, d\phi$$

Integration over all directions gives

$$\int d\omega = \int_0^{2\pi}\int_0^\pi \sin\theta\, d\theta\, d\phi = \left| -\cos\theta\,\right|_0^\pi \left|\,\phi\,\right|_0^{2\pi} = 4\pi \tag{3-29}$$

In the present case the energy is not affected by the ϕ angle, so this has been integrated out in Eq. (3-28) to give the factor 2π.

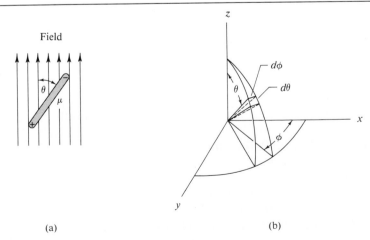

FIGURE 3-3
(a) Component of dipole moment in the direction of a field. (b) Polar coordinate system.

(a)

(b)

The integrals of Eq. (3-28) may be evaluated as follows. Let $a = \mu F/kT$ and $x = \cos \theta$, so $dx = \sin \theta \, d\theta$; then

$$\frac{\mathbf{m}_\mu}{\mu} = \frac{\int_1^{-1} e^{ax} x \, dx}{\int_1^{-1} e^{ax} \, dx}$$

The integrals are now in a standard form, and algebraic manipulation gives*

$$\frac{\mathbf{m}_\mu}{\mu} = \frac{e^a + e^{-a}}{e^a - e^{-a}} - \frac{1}{a} = \coth a - \frac{1}{a} \tag{3-30}$$

Under ordinary experimental conditions $\mu F/kT$ is small compared to unity, and on expansion of the expression $(\coth a - 1/a)$, the first term is $a/3$. To a first approximation we have

$$\mathbf{m}_\mu = \left(\frac{\mu^2}{3kT} \right) \mathbf{F} \tag{3-31}$$

Comparison with Eq. (3-20) gives

$$\alpha_\mu = \frac{\mu^2}{3kT} \tag{3-32}$$

The expression for $\mathbf{P}$ [Eq. (3-26)] may now be written explicitly as

$$\mathbf{P} = N_A \alpha_{SI}/3\epsilon_0 + (N_A/3\epsilon_0) \frac{\mu^2}{3kT} \tag{3-33}$$

The corresponding equation in cgs/esu units is:

$$\mathbf{P} = \frac{4}{3} \pi N_A \alpha + \frac{4}{3} \pi N_A \frac{\mu^2}{3kT} \tag{3-34}$$

Thus the molar polarization $\mathbf{P}$ as obtained from dielectric constant measurements should vary with temperature according to the equation

$$\mathbf{P} = a + b/T \tag{3-35}$$

where

$$a = \mathbf{P}_\alpha = \mathbf{R} = N_A \alpha/3\epsilon_0$$

and

$$b = (N_A/3\epsilon_0) \frac{\mu^2}{3k}$$

It is an approximation to equate $\mathbf{P}_\alpha$ to $\mathbf{R}$, but if $\mathbf{R}$ is determined from a refractive index measured well away from an absorption band, the approximation usually is an acceptable one.

Equation (3-33) is known as the Debye equation, and typical plots of $\mathbf{P}$ versus $1/T$ are shown in Fig. 3-4. Equation (3-33) is valid for gases. In liquids, especially associated ones, the molecular dipoles may not be able to respond freely to the applied field.

---

*$\cosh x = \frac{1}{2}(e^x + e^{-x})$ and $\sinh x = \frac{1}{2}(e^x - e^{-x})$; $\coth x = (\cosh x)/\sinh x$.

FIGURE 3-4
Variation of molar
polarization with
temperature.

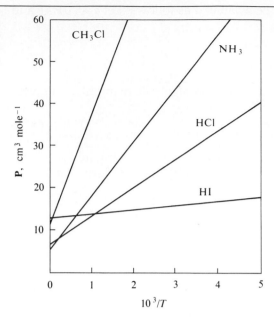

EXAMPLE Some sample calculations are now in order! First, the statement that $\mathbf{F}\mu/kT$ is ordinarily a small quantity can be verified. A capacitor might be charged to as much as 6000 V cm^{-1} or 6×10^5 V m^{-1}. We can take this as the approximate local field $\mathbf{F}$, and if μ is in the usual range of about 1 debye, then in SI units $\mu = (1 \times 10^{-18})(10/c)(10^{-10}/10^{-8})$ where c is the cgs velocity of light, or $\mu = (1 \times 10^{-18})(10/2.998 \times 10^{10})(10^{-2}) = 3.34 \times 10^{-30}$ C m. The maximum energy in the field is then

$$\epsilon = \mu\mathbf{F} = (3.34 \times 10^{-30})(6 \times 10^5) = 2.00 \times 10^{-24} \text{ J}$$

If T is about 300 K, then kT is $(1.38066 \times 10^{-23})(300)$ or about 4×10^{-21} J, and ϵ/kT is about 5×10^{-4}, and is indeed a small number.

Turning next to an application of the Debye equation, Eq. (3-33), we find that the dielectric constant of gaseous HCl at 1 atm pressure is reported (International Critical Tables) to be 1.0026 at 100°C and 1.0046 at 0°C. The molar volume is RT/P assuming ideal gas behavior, or $(8.3144)(T)/(1.01325 \times 10^5)$, giving 3.062×10^{-2} m^3 mol^{-1} and 2.241×10^{-2} m^3 mol^{-1} at the two temperatures respectively. From Eq. (3-26), the corresponding $\mathbf{P}$ values are 2.651×10^{-5} m^3 mol^{-1} and 3.431×10^{-5} m^3 mol^{-1}. We now solve Eq. (3-35) as a pair of simultaneous equations for the two temperatures, to obtain $a = 5.205 \times 10^{-6}$ m^3 mol^{-1} and $b = 7.950 \times 10^{-3}$ m^3 K mol^{-1}. Then

$$\alpha = 3\epsilon_0 a/N_A = (3)(8.854 \times 10^{-12})(5.20 \times 10^{-6})/(6.022 \times 10^{23})$$
$$= (2.29 \times 10^{-40} \text{ C}^2 \text{ m}^2 \text{ J}^{-1}$$
$$\mu^2 = 9\epsilon_0 kb/N_A = (9)(8.854 \times 10^{-12})(1.38066 \times 10^{-23})(7.95 \times 10^{-3})/$$
$$(6.022 \times 10^{23}) = 1.452 \times 10^{-59} \text{ c}^2\text{m}^2$$

whence $\mu = 3.81 \times 10^{-30}$ C m.

Better values for α and μ are 2.84×10^{-40} C^2 m^2 J^{-1} and 3.50×10^{-30} C m, respectively. The use of just two points in the calculation leads to some error.

The above calculations may, of course, be carried out using cgs-esu units. Thus

$$\mu\mathbf{F}/kT = (1 \times 10^{-18})(6000/300)/(1.38066 \times 10^{-16})(300) = 5 \times 10^{-4} \text{ (as before)}$$

Also, the molar volumes are now 3.062×10^4 cm^3 mole^{-1} and 2.241×10^4 cm^3 mole^{-1} at 100°C and 0°C, respectively, and the corresponding **P** values are 26.51 cm^3 mol^{-1} and 34.31 cm^3 mol^{-1}. Simultaneous solution of equations (3-35) written for the two temperatures now gives 5.20 cm^3 mol^{-1} for the intercept and 7.95×10^3 cm^3 K mol^{-1} for the slope, whence $\alpha = 2.06 \times 10^{-24}$ cm^3 and $\mu = 1.14$ debye.

Dipole moments may also be determined from measurements at a single temperature if a dielectric constant measurement is combined with one of index of refraction. The latter gives the molar refraction and hence $\mathbf{P}_\alpha$ directly, for use in Eq. (3-35). Indices of refraction, however, are usually measured with visible light, as with the sodium D line, and some correction may be necessary.

This general approach may be extended to solutions. The molar polarization of a solution is an additive property, so that for a solution

$$\mathbf{P} = \frac{D-1}{D+1} \frac{x_1 M_1 + x_2 M_2}{\rho} = x_1 \mathbf{P}_1 + x_2 \mathbf{P}_2 \tag{3-36}$$

where x denotes mole fraction. $\mathbf{P}_1$ is obtained from measurements on the pure solvent and $\mathbf{P}_2$ can then be calculated for the solute. To discount solute–solute interactions, the resulting $\mathbf{P}_2$ values are usually extrapolated to infinite dilution. To obtain $\mathbf{R}_2$, one applies the same procedure to molar refractions from index of refraction measurements and then proceeds as before to calculate a dipole moment for the solute. Not only is this procedure somewhat approximate, but the solvent will often have a polarizing effect on the solute as a result of solvation interactions. Consequently $\mathbf{P}_2$ and corresponding μ values will not in general be exactly the same as those obtained from the temperature dependence of the dielectric constant of the pure solute vapor. In the case of nonvolatile, solid substances, however, the solution procedure may be the best one available for estimating molecular dipole moments.

3-5   Dipole moments and molecular properties

It is conventional and very useful to regard a molecular dipole moment as an additive property. One assigns individual dipole moments to each bond and the addition, of course, is now a vectorial one.

In the case of diatomic molecules the bond moment is just the measured dipole moment of the molecule. One can proceed a step further. If the internuclear distance is known from crystallographic or spectroscopic measurements, then from the definition of dipole moment $\mu = \delta ed$ [Eq. (3-13)], the fractional charge δe on each atom can be calculated. This is, in effect, the degree of ionic character of the bond.

Some dipole moment values are given in Table 3-1 and the calculation may be illustrated as follows. The bond length HCl is 1.275 Å and the dipole moment is 1.05 D (debye). The fraction of ionic character δ is then $1.05 \times 10^{-18}/(1.275 \times 10^{-8})(4.80 \times 10^{-10}) = 0.17$. The H—Cl bond is thus said to have 17% ionic character. The similarly obtained values for HF, HBr, and HI are 45%, 11%, and 4%, respectively. This set of values has been used to relate the electronegativity difference between atoms and the degree of ionic character of the bond (see, for example, Douglas and McDaniel, 1965).

Turning to triatomic molecules, Table 3-1 shows a zero dipole moment for CO_2. This is not because the individual C=O bonds are nonpolar, but because the two bond dipoles cancel, as illustrated in Fig. 3-5(a). This arrangment of two dipoles end-to-end corresponds to a quadrupole (see Special Topics) and the experimental quadrupole moment of -4.3×10^{-26} esu cm^2 for CO_2, when combined with the known C=O distance, indicates that each oxygen atom carries a charge of about $0.3e$.

Water, on the other hand, shows a net dipole moment because the bond angle is not 180° and the two H—O dipoles fail to cancel. As illustrated in Fig. 3-5(b), the actual angle is 105° and the observed moment is the vector sum of the two bond moments.

The situation becomes increasingly complicated with polyatomic molecules. The net dipole moment of ammonia is now the vector sum of three N—H bond moments, while for the planar BF_3 molecule the three bond moments cancel to give zero molecular dipole moment. The same cancellation occurs for CH_4, which is unfortunate, since one thus gets no information about the C—H bond moment. An indirect estimate (Partington, 1954) gives -0.4 for μ_{C-H} (carbon is negative), and this now permits the calculation of other bond moments involving carbon. Thus μ_{C-Cl} can now be obtained from the dipole moment of CH_3Cl and a knowledge of the bond angles.

A few such bond moments are collected in Table 3-2. It must be remembered, however, that the resolution of a molecular moment into bond moments is somewhat arbitrary. Thus one could assign a moment to the lone electron pairs in water or ammonia as well as to the bonds. Also the dipole moment of a molecule is merely the first term in an expansion of the function which describes the complete radial and angular distribution of electron density in a molecule (see Special Topics). This function, however, is in general not known or can only be estimated theoretically, and the bond moment procedure just given is adequate for many purposes.

Finally, molecular dipole moments can provide useful qualitative information. Thus the zero moment for p-$C_6H_4Cl_2$ must mean that the C—Cl bonds are in the plane of the ring, while the nonzero moment for p-$C_6H_4(OH)_2$ (of 1.64 D) indicates that the O—H bonds lie out of the plane of the ring. The series of p-, m-, and o-isomers of $C_6H_4Cl_2$ can be identified since the order given should be that of increasing dipole moment. As an example from inorganic chemistry, the isomers of square planar PtA_2B_2 complexes can be distinguished on the basis of their dipole moments. That of the *trans* isomer should be small or zero, whereas that of the *cis* isomer should be fairly large.

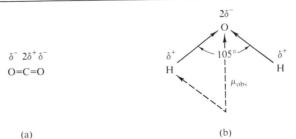

FIGURE 3-5
(a) A molecule with zero dipole moment (but nonzero quadrupole moment). (b) Vector addition of bond dipole moments for water.

(a)

(b)

Bond	Moment (D)	Bond	Moment (D)	
O—H	−1.53	C—Cl	1.56	**TABLE 3-2**
N—H	−1.31	C—O	0.86	Some bond dipole
C—H	(−0.40)[b]	C=O	2.4	moments[a]
C—F	1.51	C≡N	3.6	

[a]The sign of the dipole moment indicates that of the first-named atom.
[b]Assumed value.

As a specific example, the isomers of $Pt[P(C_2H_5)_3]_2Br_2$ have dipole moments of zero and 11.2 D.

COMMENTARY AND NOTES

3-CN-1 Systems of units

We continue here the discussion of Section 1-CN-1 on systems of units. Three criteria were stated as desirable for a system of units: (1) It should be decimalized, (2) units should be defined operationally, that is, in terms of fundamental laws, and (3) commonly measured quantities should come out to be of the order of unity (see also Adamson, 1978. It was noted that both the cgs and the SI systems meet criterion (1) but that SI units are often inconvenient, as in the case of the Pa as the unit of pressure and the m^3 as the unit of volume. We now take a look at electrical units, and encounter criterion (2).

Much of physical chemistry concerns itself with electrical forces between atoms and molecules, such as van der Waals forces, and forces between charged particles such as ions, electrons, and nuclei. The latter two are central to the statement of the potential energy in dealing with the wave mechanics of atoms and molecules. This is essentially a matter of electrostatics, for which the fundamental law is Coulomb's law. In the cgs-esu system, this law is used to *define* charge, just as Newton's law is used to define force. We thus write

$$f = q_1q_2/r^2 \tag{3-37}$$

The esu unit of charge, the statcoulomb, is defined operationally by the statement that unit charges repel with a force of 1 dyne if at a distance of 1 cm. If the medium is other than vacuum, the force is reduced by the factor D, to $f = q_1q_2/Dr^2$, where D is the dielectric constant.

Electrostatic units are not convenient in those areas of physical chemistry involving electrochemistry and electrode reactions. The "practical" units are now the coulomb (C) for quantity of electricity, the ampere, or $C\ s^{-1}$, as the unit of current and ordinary volt, V, as the unit of potential. C and V are related in that it requires 1 J of work to transport 1 C across a potential difference of 1 V. These units are not in rational relationship to Coulomb's law, and to use them in connection with this law, one now writes

$$f = \frac{q_1q_2}{4\pi\epsilon_0 r^2} \tag{3-38}$$

where ϵ_0 is a conversion factor known as the "permittivity of vacuum," $\epsilon_0 = 10^7/(4\pi c^2) = 8.85419 \times 10^{-12}\ C^2\ J^{-1}\ m^{-1}$. The factor of 4π is put into the definition of ϵ_0, and taken out again in Eq. (3-38); other equations such as (3-19) and (3-22), which would have a factor of 4π or of $(1/4\pi)$ in the esu system, now do not have such factors when written in SI units. To return to Eq. (3-38), force f is now in newtons, charge is in coulombs, and r is in meters.

Notice that we are dealing with more than just alternative choices of units; *the actual equations are different*, such as Eq. (3-37) vs. Eq. (3-38). The same is true for relationships involving dipole moment and polarizability. The situation is summarized in Table 3-3.

	Parameter in Cgs/Esu	Parameter in SI	Parameter in Cgs/Esu	Parameter in SI
TABLE 3-3 Correspondences for converting from Cgs/Esu to SI equations[a]	V	$(4\pi\epsilon_0)^{1/2}V$	μ	$\mu/(4\pi\epsilon_0)^{1/2}$
	q	$q/(4\pi\epsilon_0)^{1/2}$	α	$\alpha/(4\epsilon\epsilon_0)$
	$\mathbf{F}$	$(4\pi\epsilon_0)^{1/2}\mathbf{F}$	$\mathbf{H}$	$(4\pi/\mu_0)^{1/2}\mathbf{H}$
			$\mathbf{B}$	$(4\pi/\mu_0)^{1/2}\mathbf{B}$

[a]See Appendix in J.D. Jackson, *Classical Electrodynamics*, Wiley, New York, 1962. Also, Bryce Crawford, private communication.

EXAMPLE   We convert Eq. (3-37) to Eq. (3-38) by replacing each q by $q/(4\pi\epsilon_0)^{1/2}$. Equation (3-14) is the same in both systems. Thus

$$\mu_{esu} = \alpha_{esu}\mathbf{F}_{esu}$$

and, using Table 3-3,

$$\frac{\mu_{SI}}{(4\pi\epsilon_0)^{1/2}} = \frac{\alpha_{SI}}{4\pi\epsilon_0}\,\mathbf{F}_{SI}(4\pi\epsilon_0)^{1/2}$$

and the $(4\pi\epsilon_0)$ quantities cancel out. On the other hand, Eq. (3-15), written for cgs/esu units, becomes Eq. (3-16) on following the instruction in Table 3-3 to replace α_{esu} by $\alpha_{SI}/(4\pi\epsilon_0)$.

As an interesting and instructive analogy, consider the following hypothetical situation. Astronomers make much use of the equation for gravitation attraction,

$$f = G\frac{m_1 m_2}{r^2} \tag{3-39}$$

where G is the gravitational constant, $G = 6.6720 \times 10^{-8}$ cm^3 s^{-2} g^{-1}, and force is in dynes if m is in grams and r is in centimeters. The conversion factor G is needed because Eq. (3-39) is not in rational relationship to Newton's law,

$$f = ma \tag{3-40}$$

where a is acceleration. Equation (3-40) is our *defining* equation for force, that is, it requires unit force to accelerate unit mass by unit acceleration.

Suppose, now, that astronomers control the definitions of units, and decide that the really fundamental law is that of Eq. (3-39), so they write it as

$$f = \frac{m_1 m_2}{r^2} \tag{3-41}$$

That is, the new unit of force, let's call it the kepler, $\mathfrak{K}$, is now defined by Eq. (3-41); unit masses separated by unit distance experience unit gravitational attractive force. The new unit, $\mathfrak{K}$, now displaces the newton, with the consequence that Eq. (3-40) must be written

$$f = \epsilon_{\mathfrak{K}} ma \tag{3-42}$$

where $\epsilon_{\mathfrak{K}}$ might be called the inertial "permittivity." Astronomers wishing to calculate the force on a mass under acceleration would now use Eq. (3-42). Other equations, such as Coulomb's law, would now contain $\epsilon_{\mathfrak{K}}$ if conventional units of charge are retained. The physical chemist, however, might well balk at this hypothetical new system, and prefer to stay with Newton's law as the defining equation for force. It would be awkward to deal with a variety of equations containing $\epsilon_{\mathfrak{K}}$, just as it is awkward to introduce the factor $(1/4\pi\epsilon_0)$ into Coulomb's law if SI units are to be used. The calculation of the numerical value of $\epsilon_{\mathfrak{K}}$ is left as an exercise (Exercise 3-16).

In magnetochemistry, the traditional units are the electromagnetic or Gaussian ones. A magnetic field $\mathbf{H}$ in vacuum gives a magnetic induction or flux density $\mathbf{B}$, with $\mathbf{B} = \mathbf{H}$. In the presence of matter, one has $\mathbf{B} = \mu\mathbf{H}$, where μ is the permeability. Alternatively,

$$\mathbf{B} = (1 + \chi)\,\mathbf{H} \tag{3-43}$$

where χ is called the magnetic susceptibility. The unit of **H** is the oersted and the unit of **B** is the gauss. In SI units, Eq. (3-43) becomes

$$B = \mu_0 (1 + \chi) H \qquad (3\text{-}44)$$

where μ_0 is the permeability of vacuum, $\mu_0 = 4\pi \times 10^{-7}$. **H** is in amperes per meter and B is in tesla, T, where $1\,T = 10^4$ gauss.

As has been remarked (Jackson, 1962), the mks (i.e., SI) system has the virtue of overall convenience in practical, large-scale phenomena, especially in engineering applications. The Gaussian (cgs) system is more suitable for microscopic problems involving electrodynamics of individual charged particles. The present writer concurs, and the latter system is therefore emphasized in this textbook.

SPECIAL TOPICS

3-ST-1 The charge distribution of a molecule

The wave mechanical picture of a molecule is that of an electron density distribution; the charge density at each volume element (or the probability of an electron in that element) is given by the square of the psi function ψ for that molecule. Recalling the coordinate system of Fig. 3-3, we see that this charge density is in general some complicated function of r, θ, and ϕ, $\rho(r, \theta, \phi)$. Now, just as a wavy line can be approximated by a series of terms in cosines and sines, that is, by a Fourier expansion, so can a function in three dimensions be approximated by an expansion in spherical harmonics. In fact, it turns out that the wave mechanical solutions for the hydrogen atom consist of one or another spherical harmonic (Section 16-7). These are just the functions s, p_x, p_y, p_z, d_{z^2}, $d_{x^2-y^2}$, d_{xy}, d_{xz}, d_{yz}, and so on.

An s function is spherically symmetric and is everywhere positive (or, alternatively, everywhere negative). The sum or integral over a charge distribution given by the s function,

$$q_e = \sum_i q_i = \int \rho(r, \theta, \phi)\,d\tau \qquad (3\text{-}45)$$

where $d\tau$ is the element of volume, simply gives a charge q_e due to the electron cloud behaving as though it were centered at the nucleus (Fig. 3-6a). The algebraic sum of q_e and the nuclear charge gives the net charge q. An atom or molecule whose charge distribution is given by an s function behaves like a point charge q. The energy of interaction with a unit charge q_0 is

$$\epsilon = \frac{q_0 q}{r} = qV \qquad (3\text{-}46)$$

Here V is the potential of the test charge,

$$V = \frac{q_0}{r} \qquad (3\text{-}47)$$

A p function has positive and negative lobes, as illustrated in Fig. 3-6(b). The sum over such a charge distribution is zero according to Eq. (3-45). However, the sum or integral

$$\mu = \sum_i q_i r_i = \int r\rho(r,\theta,\phi)\,d\tau \qquad (3\text{-}48)$$

is not zero. This now corresponds to a dipole moment. Next, an electron distribution as given by the d_{z^2} function looks as in Fig. 3-6(c). There is now no dipole moment, but Q is nonzero as given by

$$Q = \sum_i q_i r_i^2 = \int r^2 \rho(r,\theta,\phi)\,d\tau \qquad (3\text{-}49)$$

where Q is called the quadrupole moment.

Thus for a molecule such as HCl, the first approximation, or s contribution, says the molecule is neutral. The second term in the expansion, or the p_z contribution, gives the dipole moment. In this approximation the atoms can be regarded as having net charges δ^+ and δ^- separated by the bond distance d. The third term in the series expansion would be the contribution of the d_{z^2} term. This recognizes that the electrons are not spherically disposed around each atom, but concentrate to some extent between the atoms, that is, form a bond. In terms

FIGURE 3-6
(a) Charge distribution
according to an *s* function.
A monopole. (b) Charge
distribution according to a
p function. A dipole. (c)
Charge distribution
according to a d$_{z^2}$
function. A quadrupole.

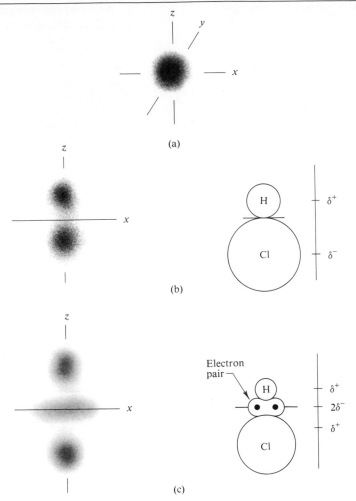

FIGURE 3-6
(a) Charge distribution according to an *s* function. A monopole. (b) Charge distribution according to a *p* function. A dipole. (c) Charge distribution according to a d$_{z^2}$ function. A quadrupole.

of point charges, a quadrupole consists of two equal, opposing dipoles.

In the case of a neutral diatomic molecule like HCl, the complete charge distribution is given by the sum of the dipole and quadrupole contributions. With more complicated molecules yet higher moments would be needed. A number of quadrupole moments are known; the ones that are have been obtained from analysis of spectroscopic data. A few values are given in Table 3-4.

Equation (3-47) gives the potential V due to a point charge. The potential energy of a molecule having a charge, a dipole moment, and a quadrupole moment can be expressed in terms of the electric potential at that point and its derivatives:

$$\epsilon = Vq + \mu\frac{dV}{dx} + \frac{Q}{2}\frac{d^2V}{dx^2} \qquad (3\text{-}50)$$

or

$$\epsilon = Vq + \mu\mathbf{F} + \frac{Q}{2}\frac{d\mathbf{F}}{dx} \qquad (3\text{-}51)$$

where $\mathbf{F}$ is the field. It is assumed that the dipole and quadrupole are aligned in the x direction, and dV/dx gives the field $\mathbf{F}$ in this direction. The verification of Eq. (3-51) is left as an exercise.

Substance	Quadrupole moment $\times\ 10^{26}$ (esu cm^2)
H_2	0.65
O_2	-0.4
N_2	-1.4
CO_2	-4.3
C_2H_6	-0.8
C_2H_4	2.0

TABLE 3-4
Some quadrupole moments[a]

[a]From A.D. Buckingham, R.L. Disch, and D.A. Dunmur, *J. Amer. Chem. Soc.* **90**, 3104 (1968).

3-ST-2 Magnetic properties of matter

The magnetic properties of matter are often of interest to the chemist. In some cases, important information about the electronic structure of a molecule can be obtained.

We measure a quantity called the molar magnetic susceptibility χ_M, which is rather analogous to molar polarization in the electrical situation. A convenient way of making this measurement is by means of a Gouy magnetic balance, illustrated in Fig. 3-7. An electromagnet establishes a field H_0 and one measures the resulting change in weight of a sample. The sample is contained in a tube which is suspended between the pole pieces of the magnet, with the bottom of the sample at the centerline. The hollow lower portion of the tube compensates for the action of the field on the tube itself. The suspension is attached to a sensitive balance, and one measures the change in

pull that occurs when the magnetic field is turned on.

As illustrated in Fig. 3-8(b) the sample may be diamagnetic, in which case the magnetic field within it is less than H_0 and magnetic lines of force deflect away from the sample. Alternatively, as in Fig. 3-8(c), the sample may be paramagnetic and show the reverse effects. In the first case the pull decreases when the field is applied; and in the second case it increases. A straightforward derivation shows that the change in force f on the sample is given by

$$f = \tfrac{1}{2}\mathcal{A}\chi_M \frac{\rho}{M} H_0{}^2 \qquad (3\text{-}52)$$

where $\mathcal{A}$ is the cross-sectional area of the tube, ρ is the sample density, and M is the molecular weight.

From the molecular point of view, χ_M is made up of two contributions: that due to the inherent magnetic polarizability α and that due to the alignment in the field of any permanent

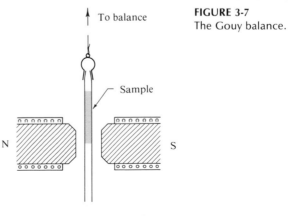

FIGURE 3-7
The Gouy balance.

To balance

Sample

N   S

FIGURE 3-8
Lines of magnetic force in
(a) vacuum, (b) a
diamagnetic substance,
and (c) a paramagnetic
substance.

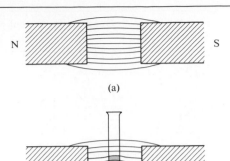

N   S

(a)

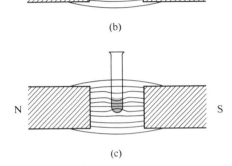

N   S

(b)

N   S

(c)

magnetic moments present. Just as with dipolar molecules in an electric field, the alignment of permanent molecular moments is opposed by the randomizing effect of thermal agitation. The actual equation is

$$\chi_M = N_A \left(\alpha + \frac{\mu_{M^2}}{3kT} \right) \qquad (3\text{-}53)$$

where μ_M is the permanent molecular magnetic moment.

As in the electrical case α describes the tendency of the applied field to induce an opposing field in an otherwise homogeneous medium. A physical description of the effect is that the applied field causes the electrons of the atoms to undergo a precession; this precession gives rise to an electric current which in turn generates an opposing magnetic field. We call the effect diamagnetism, and report a molar diamagnetic susceptibility $N_A\alpha$ or χ_α. Pascal observed around 1910 that χ_α's are approximately additive, although as with molar polarization or refraction, corrections are needed for special types of bonds. Some atomic and bond susceptibilities are given in Table 3-5.

	Molar		Molar
TABLE 3-5 Some molar diamagnetic susceptibilities[a] — Atom or bond	susceptibility $\times 10^6$	Ion	susceptibility $\times 10^6$
H	-2.93	Na^+	-5
C	-6.00	NH_4^+	-11.5
N (open chain)	-5.55	Fe^{2+}	-13
N (ring)	-4.61	Cr^{3+}	-11
O (alcohol,		Co^{3+}	-10
ether)	-4.61	Cl^-	-26
O (carbonyl)	-1.72	NO_3^-	-20
Cl	-20.1	SO_4^{2-}	-40
C=C	5.5		

[a]From P.F. Selwood, "Magnetochemistry," 2nd ed. Wiley (Interscience), New York, 1956.

As a numerical illustration, suppose the magnet to provide a field $\mathbf{H}_0$ of 10,000 Oe, $\mathscr{A}$ to be 0.1 cm^2, and the sample to consist of water at 25°C. The value of χ_α for water is -13×10^{-6} and $\chi_\alpha(\rho/M)$ is then -0.722×10^{-6} cm^{-3}. Substitution into Eq. (3-52) gives   **EXAMPLE**

$$f = -\tfrac{1}{2}(0.1)(0.722 \times 10^{-6})(10^4)^2 = -3.61 \text{ dyn or } -3.7 \text{ mg.}$$

The force is negative, meaning that the sample will be 3.7 mg lighter when the field is on.

Permanent magnetic moments arise first of all from the orbital motion of electrons. For a single electron this moment is related to its angular momentum and therefore to the orbital quantum number l, where $l = 0$ for an s electron, $l = 1$ for a p electron, and so forth. Analysis gives

$$\mu = \frac{eh}{4\pi mc}[l(l + 1)]^{1/2} \qquad (3\text{-}54)$$

where the quantity $(eh/4\pi mc)$ is known as the Bohr magneton μ_B. Substitution of the numerical values for Planck's constant h, the mass of the electron m, and the velocity of light c gives

$\mu_B = 9.274 \times 10^{-21}$ erg Oe^{-1} (and 9.274×10^{-24} A m^{-2} in the SI system).

The second contribution to the permanent magnetic moment is that from the spin of the electron itself. Orbital motion is quantized in units of angular momentum $h/2\pi$ and spin angular momentum is quantized in units of $\tfrac{1}{2}h/2\pi$. Since the quantity $h/2\pi$ appears in both cases, it is not surprising that the spin magnetic moment is related to μ_B:

$$\mu_S = 2\mu_B [s(s + 1)]^{1/2} \qquad (3\text{-}55)$$

where s is the spin quantum number.

The orbital contribution, Eq. (3-54), is often

FIGURE 3-9
A modern superconducting magnet installed at the Japanese National Research Institute for Metals. Shown is the Nb_3Sn outer magnet, which gives a central field of 135,000 gauss. With an inner V_3Ga set of coils, the field becomes 175,000 gauss. In use, the whole unit is immersed in a He cryostat at 4.2K. (Courtesy Intermagnetics General Corp.)

small; this is because orbital motions of electrons may be so tied into the nuclear configuration of the atom that they are unable to line up with an applied field. As a fortunate consequence only the spin contribution is then important. If more than one unpaired electron is present, the individual spins combine vectorially,

$$\mu_S = 2\mu_B[S(S + 1)]^{1/2} \qquad (3\text{-}56)$$

where S is just one-half times the number of unpaired electrons. The molar susceptibility is then

$$\chi_M = \frac{4\mu_B{}^2 N_A}{3kT} [S(S + 1)]^{1/2} \qquad (3\text{-}57)$$

The total observed χ_M is the sum of the negative diamagnetic contribution $N_A\alpha$ and the positive contribution from Eq. (3-57).

Equation (3-57) often applies adequately to elements through the first long row of the periodic table, and therefore allows a determination of the number of unpaired electrons present per molecule. Thus a ferrous compound should have four unpaired electrons on the iron and therefore a moment of $[2(2 + 1)]^{1/2}\mu_B$ or $4.90\mu_B$. The observed value is $5.25\mu_B$.

Modern magnetochemistry deals with a great variety of effects in addition to ordinary susceptibility measurements. These range from studies of transport phenomena in superconductors to magnetic field effects in spectroscopy. Figure 3-9 shows a contemporary electromagnet in which superconducting material is used for the coils. This allows very high fields to be obtained with a minumum of energy expenditure (and of heating).

GENERAL
REFERENCES   HALLIDAY, D., AND RESNICK, R. (1962). "Physics." Wiley, New York.
SELWOOD, P. (1956). "Magnetochemistry," 2nd ed. Wiley (Interscience), New York.

CITED
REFERENCES   ADAMSON, A.W. (1978). *J. Chem. Ed.*, **55,** 634.
DOUGLAS, B.E., AND McDANIEL, D.H. (1965). "Concepts and Models of Inorganic Chemistry." p. 84, Blaisdell, Waltham, Massachusetts.
GLASSTONE, S. (1946). "Textbook of Physical Chemistry." Van Nostrand–Reinhold, Princeton, New Jersey.
HANTZSCH, A., AND DÜRIGEN, F. (1928). *Z. Phys. Chem.* **136,**1. International Critical Tables, National Research Council, McGraw-Hill, N.Y., 1926–1933.
JACKSON, J.D. (1962). "Classical Electrodynamics." Wiley, New York.
McGLASHAN, M.L. (1968). Physico-Chemical Quantities and Units. Royal Inst. of Chem. Publ. No. 15, London.
MOELWYN–HUGHES, E.A. (1961). "Physical Chemistry." Pergamon, Oxford.
PARTINGTON, J.R. (1954). "An Advanced Treatise on Physical Chemistry," Vol. 5, p. 476. Longmans, Green, New York.
RINGBOM, Z. (1938). *Anal. Chem.* **115,** 332.
SUGDEN, S. (1924). See GARNER, F.B., AND SUGDEN, S.S. (1929), *J. Chem. Soc.* **1929,** 1298.

EXERCISES

3-1   Benzene has a strong absorption at 180 nm. Estimate the molar extinction coefficient, assuming that the cross-section for light absorption is given by the physical size of the molecule, using the van der Waals b constant and taking the molecule to be approximately spherical. (The actual value is 10^5 M^{-1} cm^{-1}.)

Ans. 1.1×10^6 M^{-1} cm^{-1}.

3-2   How closely does $(1 - x/n)^n$ approach e^{-x}, that is, what is the value of $(1 - x/n)^n e^x$ for $n = 3, 10, 100$? Take x to be (a) 0.1 and (b) 2.

Ans. We evaluate the ratio $(1 - x/n)^n/e^{-x}$.
(a) 0.998, 0.9995, 0.9999. (b) 0.274, 0.793, 0.980.

Gaseous bromine has an extinction coefficient of 120 liter mole^{-1} cm^{-1} at 400 nm. Calculate (a) α, (b) k_N (c) k_ℓ, and (d) the percentage of incident light of this wavelength which would pass through a 1.5-cm cell filled with Br$_2$ at 100 Torr pressure and 25°C.   3-3

Ans. (a) 276.4, (b) 4.589 $\times$ 10^{-19}, (c) 1.486 cm^{-1}, (d) 10.76%.

The extinction coefficient for azoethane is 9.5 M^{-1} cm^{-1} at 28,000 cm^{-1} (this is the wave number $\bar{\nu}$, where $\bar{\nu} = 1/\lambda$). Calculate the optical density of a 1.5 $\times$ 10^{-3} M solution for a 2 cm cell, assuming that the solvent, ethanol, has an optical density of 0.05 cm^{-1} at this wavelength.   3-4

Ans. 0.1285.

The optical density of a 0.02 M solution of Cr(NH$_3$)$_6^{3+}$ is 0.589 at 462.5 nm with a 0.5-cm cell; that of the aqueous solvent alone is 0.200. Calculate the extinction coefficient for this complex ion and the percentage of incident light absorbed by the solution.   3-5

Ans. 38.9 liter mole^{-1} cm^{-1}, 74.2%.

The extinction coefficients of benzophenone and of naphthalene are the same at 300 nm, namely 500 liter mole^{-1} cm^{-1}. Naphthalene does not absorb at 350 nm, but benzophenone does, with an extinction coefficient of 110 liter mole^{-1} cm^{-1}. Neglecting solvent absorption, calculate the composition of a mixed solution of these two species if the optical density in a 1-cm cell is 0.500 at 300 nm and 0.030 at 350 nm.   3-6

Ans. 2.73 $\times$ 10^{-4} M benzophenone and 7.27 $\times$ 10^{-4} M naphthalene.

Referring to Exercise 3-6, a solution 2.73 $\times$ 10^{-4} M in benzophenone *alone* is placed in a 1-cm spectrophotometer cell, and a solution 7.27 $\times$ 10^{-4} M in naphthalene *alone* is placed in a second 1-cm cell. If now the two cells are lined up, as illustrated in Fig. 3-10, and a beam of 300 nm light is used, what will be the percentage transmission through the two cells together? To what optical density does this correspond (notice how this value compares with that for the combined solution of Exercise 3-6)? Does your answer depend on the order in which the two cells are lined up? (Neglect solvent absorption.)   3-7

Ans. 31.62%, 0.500.

Calculate the molar refraction of ammonia if the index of refraction of the gas at STP is 1.000373.   3-8

Ans. 5.57 cm^3 mol^{-1}.

The molar refraction of HI is 13.74 cm^3 mole^{-1}. Calculate the index of refraction of HI at 25°C and 0.1 atm pressure and the polarizability of the HI molecule.   3-9

Ans. n = 1.0000842, α = 5.45 $\times$ 10^{-24} cm^3.

The index of refraction of liquid water is 1.333. Calculate the molar refraction and the polarizability of water.   3-10

Ans. **R** = 3.70 cm^3 mol^{-1}, α = 1.47 $\times$ 10^{-24} cm^3.

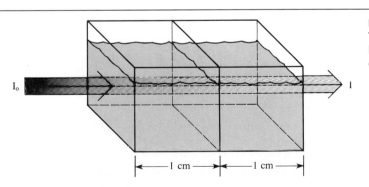

I_o — I

|← 1 cm →|← 1 cm →|

FIGURE 3-10
Two 1-cm square cells back to back in illustration of Exercise 3-7.

3-11 The dielectric constant of liquid water is 78.54 at 25°C. Calculate the molar polarization of water, and in conjunction with Exercise 3-10, the dipole moment of water. Explain why your value for μ is incorrect.

Ans. **P** = 17.3 cm^3 mol^{-1}, μ = 0.817 D.

3-12 As an alternative calculation to Exercise 3-11, the dielectric constant of water vapor at 1 atm varies from 1.005471 at 384.3 K to 1.004124 at 444.7 K. Calculate the dipole moment of water from this information.

Ans. 1.84 D.

3-13 The dielectric constant of chloroform gas at 120°C and 1 × 10^5 Pa pressure is 1.0042. Given that the dipole moment is 3.41 × 10^{-30} C m, calculate the polarizability of chloroform and the dielectric constant of liquid chloroform.

Ans. α = 1.302 × 10^{-39} C^2 m^2 J^{-1}, D(liq) = 4.94.

3-14 Calculate the bond length in HI.

Ans. 2.03 Å.

3-15 Calculate the dipole moment for the H—O bond in H$_2$O from information in Section 3-5.

Ans. 1.49 D.

3-16 Calculate the value of $\epsilon_{\mathcal{H}}$ in Eq. (3-42) for (a) cgs units of mass and acceleration and (b) SI units.

Ans. (a) 1.4988 × 10^7 g s^2 cm^{-3}, (b) 1.4988 × 10^{10} kg s^2 m^3.

PROBLEMS

3-17 The accompanying table gives the extinction coefficients for ferrocyanide ion and for ferricyanide ion at various wavelengths.

Wavelength (nm)	Ferrocyanide ϵ (M^{-1} cm^{-1})	Ferricyanide ϵ (M^{-1} cm^{-1})
280	970	1100
290	550	1350
300	300	1550
310	285	1350
320	315	1100
330	320	800
340	260	520
350	165	300
360	130	320
370	70	430
380	40	590
390	15	750
400	7	890
410	0	1000
420	0	1010
430	0	870
440	0	650
460	0	250

(a) Make a semilogarithmic plot (on semilog graph paper) of extinction coefficient versus wavelength for these two species. (b) Calculate the percentage of light of 350 nm absorbed by a 0.0005 M solution of each species in a 2-cm cell. (c) The reaction

$$2Fe(CN)_6^{3-} + H_2O + CN^- = 2Fe(CN)_6^{4-} + CNO^- + 2H^+$$

if being followed spectrophotometrically; CN^- and CNO^- do not absorb in the region of wavelengths covered in the accompanying table. The solution initially is 0.001 M in ferricyanide; after 10 min the optical density of the solution (in a 1-cm cell) is 0.906 at 320 nm. Calculate the percentage of the ferricyanide that has reacted.

The transmittance of a 0.02 M solution of bromine in chloroform is 56% for light of a certain wavelength and a path length of 1 mm. Calculate the extinction coefficient and the percentage of incident light that would be absorbed if the path length were 0.5 cm. **3-18**

Cis- and trans-azobenzene have the same extinction coefficient of 4.5 $\times$ 10^3 M^{-1} cm^{-1} at 270 nm (this is called an isosbestic point); at 330 nm the extinction coefficient of the cis compound is 700 M^{-1} cm^{-1} while that of the trans compound is 2.4 $\times$ 10^5 M^{-1} cm^{-1}. These values are for 95% ethanol as solvent (which does not absorb appreciably at these wavelengths). In an analytical procedure, a solution of azobenzene of unknown isomer composition is found to have D_{270} = 0.0945 and D_{330} = 1.4505. The path length is not known. Calculate the percentage of the azobenzene that is in the cis form. **3-19**

Calculate the absorption cross-section for trans-azobenzene at 330 nm (see Problem 3-19). **3-20**

At a wavelength of 500 nm the optical density of a 0.0200 M solution of the complex ion $[Co(NH_3)_5H_2O]^{3+}$ is 0.940, while that of a 0.00200 M solution of the ion $[Co(NH_3)_5(SCN)]^{2+}$ is 0.356. At 345 nm, the two optical densities are 1.060 and 1.33, respectively. The optical densities at 500 and 345 nm are then determined for a solution containing a mixture of the two complex ions and found to be 0.319 and 0.611, respectively. Calculate the concentrations of the complex ions in this solution. The same cell is used throughout, but the path length is not known. **3-21**

The amount of a certain substance A is to be estimated by measuring the optical density for a solution of A, using a spectrophotometer. The intensity of the transmitted light can be measured within an error of 1.0% of the initial intensity I_0, that is, $\delta I/I_0$ = 0.01. At what value for D will the relative error in the concentration of A be minimized? What is that error? (Ringbom, 1938) Note: It is necessary to obtain an expression for $\delta D/D$ and then to set $d(\delta D/D)/dD$ equal to zero, thus getting the minimum value of $\delta D/D$. The solution of A can then be made up so that the error in D and hence in concentration is a minimum. **3-22**

The molar refraction **R** is 1.643 for oxygen in an ether group (R-O-R). It is 6.818 for methane and 13.279 for dimethyl ether. Calculate the value of **R** for diethyl ether. (All values are in cm^3 mol^{-1}.) **3-23**

Given that the molar refractions for methane, ethane, n-propionic acid, and methyl propionate are 6.818, 11.436, 17.590, and 22.326, respectively, calculate the atomic refractions for H, C, carbonyl oxygen (C = O), hydroxyl oxygen (−OH), and ether oxygen (−O−R). Use only the above information plus the fact that the molar refraction for β-hydroxy propionic acid is 19.115. **3-24**

Show that at low pressures, the index of refraction of a gas should obey an equation of the form $n = 1 + bP$, where b is a constant at a given temperature. For a particular gas, b = 4.674 $\times$ 10^{-4} atm^{-1} at 25°C. Calculate the polarizability of the gas. **3-25**

The molar polarization of a certain vapor is found to obey the equation $P = 400 \times 10^{-6} + 18 \times 10^{-3}/T$ (in m^3 mol^{-1}). Assuming ideal gas behavior, calculate the index of refraction and the dielectric constant of the vapor at STP. What molecular properties can be obtained? Calculate their values. **3-26**

The accompanying data are reported for solutions of ethyl ether in cyclohexane at 20°C [from Glasstone, 1946). **3-27**

Mole fraction ether	Dielectric constant	Density (g cm^{-3})
0.04720	2.109	0.7751
0.08854	2.178	0.7720
0.12325	2.246	0.7691
0.17310	2.317	0.7664

The molar refraction for the ether is 22.48 cm³. Using the extrapolation method, calculate the molar polarization of the solvent and of the ether and the dipole moment of the latter.

3-28 The accompanying data have been obtained for aqueous LiCl solutions at 19°C (Hantzsch and Dürigen, 1928).

Grams of salt per 100 g solution	Density (g cm⁻³)	Index of refraction
0	0.9984	1.3334
3.44	1.0180	1.3397
6.77	1.0357	1.3462
9.99	1.0524	1.3523
16.12	1.0862	1.3644

Calculate the molar refraction of water at 19°C and, by the extrapolation method, that of LiCl(aq).

3-29 The polarizability of CCl_4 is 10.5 Å³. Calculate the index of refraction and the dielectric constant for the gas at STP and for liquid cabon tetrachloride.

3-30 The angle of inclination of the N—H bonds to the molecular axis is 111° in the case of NH_3. The molecular dipole moment is 1.48 D; calculate the dipole moment for a N—H bond.

3-31 The dipole moment of SO_2 is 1.61 D, the length of each S—O bond is 1.45 Å, and the bonds have approximately 25% ionic character. Calculate the bond angle, that is, the angle between the two S—O bonds.

3-32 The dipole moment of water is 6.17×10^{-30} C m and the two H—O bonds are at 105° to each other and each have 25% ionic character. Calculate the H—O bond length in angstroms.

3-33 The dipole moment of HCl is 3.60×10^{-30} C m and the bond length is 0.130 nm. Calculate the fractional ionic character to the HCl bond, that is, the charge on each atom expressed as a fraction of the electronic charge.

3-34 Write Eqs. (3-20) to (3-24) for the cgs-esu system.

3-35 An additive property that has had some historical importance is that known as the *parachor* [see O. R. Quale, *Chem. Rev.* **53**, 439 (1953)]. The parachor, [P], is a molar volume in which surface tension is used to correct for intermolecular forces, $[P] = \gamma^{1/4}M/\rho$. Calculate [P] for octane, heptane, and ethanol and, from these, the group parachors for CH₃—, CH₂—, and —OH.

3-36 Calculate the Coulomb force between two electrons 3 Å apart using (a) the cgs-esu system and (b) the SI system. (c) Derive, that is, verify, the value given for ϵ_0 in the Commentary and Notes section.

SPECIAL TOPICS PROBLEMS

3-37 Carbon monoxide has a dipole moment of 0.10 D and a quadrupole moment of -1.3×10^{-26} esu cm². Calculate the potential energy of interaction of CO with a Na⁺ ion 15 Å away. Assume that C is negative in CO and that the molecular axis points toward the ion, that is, Na⁺ · · · C—O. Express your result as ϵ/kT for 25°C.

3-38 Derive Eq. (3-48).

3-39 The energy of adsorption of polar molecules on ionic solids is attributed in part to ion–dipole interaction. Suppose a silica–alumina catalyst to have sites of charge −1 widely enough separated that each adsorbed H_2O interacts with only one site. Taking the dipole moment of H_2O to be 1.83 D, calculate the ion–dipole interaction energy as a function of distance from the surface from 3 to 15 Å assuming that the water molecule is oriented so that its dipole is (a) parallel to the surface and (b) perpendicular to the surface. Express your results as ϵ/kT at 25°C.

The magnetic field in a Gouy apparatus is calibrated by measurement of the pull for a sample of Mohr's salt. The tube is 3 mm in diameter and the packing of the solid is such that its density in the tube is 3.5 g cm^{-3}. Calculate $\mathbf{H}_0$ if the pull is 40 mg. Assume 25°C.   **3-40**

The molar susceptibility of benzene is -54.8×10^{-6} in cgs units. What force should be observed in a Gouy balance if a 0.5-cm-diameter tube is used with a field of 8000 Oe? Assume 25°C.   **3-41**

Calculate χ_M for the Cr^{3+} ion and for the complex ion Cr(NH$_3$)$_6^{3+}$, allowing for the diamagnetic corrections. What increase in weight should be observed if a 0.01 M solution of Cr(NH$_3$)$_6$Cl$_3$ in water is placed in a 1-cm-diameter tube and weighed in a Gouy balance for which $\mathbf{H}_0$ is 9500 Oe (allow for solvent). Assume 25°C.   **3-42**

The following results are obtained for a 0.5 M solution of a nickel coordination compound. With a field of 10,000 Oe and a tube of 0.1 cm^2 area, the net pull (after allowing for solvent) is 12.3 mg. The diamagnetic correction for the nickel compound itself is -2.0×10^{-4} in cgs units. Calculate the number of unpaired electrons per nickel atom. Assume 25°C.   **3-43**

chapter 4
Chemical thermodynamics.
The first law of
thermodynamics

Modern chemical thermodynamics has two interpenetrating structures. The first structure is that of classical thermodynamics, which is based on a set of far-reaching phenomenological laws. The detailed logical elaboration of these laws permits their precise application to very complicated chemical and physical situations and leads to relationships and conclusions that are otherwise far from obvious. However, all these relationships and conclusions are themselves phenomenological; that is, they deal with macroscopic, operationally defined quantities. They contain nothing that is not already implicit in the laws of thermodynamics themselves.

This aspect of thermodynamics has a strength and a weakness. The strength lies in the fact that since the laws rest entirely on experimental observation, they are correct to the extent of the accuracy and generality of such observations. The weight of evidence is by now so massive that we have no doubt that the laws are valid within our present scientific ken. Should some totally new reach be achieved, the laws might be expanded, but their present application would not be altered.

There is a not too old example of just such a situation. The first law of thermodynamics affirms the conservation of energy. In the absence of external fields (electrical, magnetic, or gravitational) this amounts to saying that chemical energy, heat energy, and work or mechanical energy may be interconverted but that their sum is constant for an isolated system. The discovery of radioactivity and, later, of nuclear fission showed that vast amounts of energy could be released by matter–energy conversion. The first law now has to include nuclear energy, or, alternatively, to affirm that the mass of an isolated system remains constant, recognizing that total mass includes energy mass. The point, however, is that the discovery of the whole new domain of nuclear energy in no way altered the laws of thermodynamics with respect to phenomena not involving nuclear changes. No new discovery, however enormous, is expected to do otherwise.

The weakness of classical thermodynamics stems from the fact that, being phenomenological, it deals only with the average behavior of a large number of molecules. It provides no detail about individual molecular behavior. Also, while it can relate macroscopic properties, it cannot predict them separately. Thus thermodynamics provides a relationship between the heat capacity of a gas and its coefficients of compressibility and thermal expansion, but it cannot predict the actual heat capacity of any particular gas. It relates vapor pressure and heat of vaporization, but is unable to say what the actual vapor pressure of a given liquid will be.

The second structure is that of statistical thermodynamics. Here one applies the Boltzmann principle to formulate the distribution of molecules among all of their possible energy states, much as is done in gas kinetic theory, and the various average quantities so computed can be identified with classical thermodynamic ones. It is now necessary to provide a great deal of detailed information about energy states, and in simple cases it is possible to do this through the use of wave mechanics. For example, statistical mechanics can give the actual heat capacity of a particular dilute gas with some precision.

In principle, then, statistical thermodynamics not only leads to the relationships of classical thermodynamics but also provides values for all the thermodynamic properties of individual substances. The practice is not so easy. Wave mechanics does not provide accurate energy states for any but the simplest of molecules, let alone for the multitudinous interactions among molecules in a condensed state. Even for gases the intermolecular potential functions that determine nonideality are as yet semiempirical. The full statistical mechanical treatment of liquids is barely within reach for a monatomic one such as argon, but otherwise seems hopelessly complex.

A consequence of the complexity of rigorous treatments is that various simplifying assumptions are introduced. These may take the form of neglecting certain categories of energy states, of making some structural assumptions that simplify the listing of such states, or of various mathematical approximations to make the computations feasible. In other words, statistical mechanics as practiced is very often based on simplified models and is essentially semiempirical. Herein lies a danger. Because the results appear in classical thermodynamic language, there is a tendency to imbue them with the full status of classical, phenomenological thermodynamics. This is at times quite easy to do if the simplifying assumptions are not stated explicitly but are implicit in some detail of mathematics or in the manner of counting of energy states.

However, because the two types of thermodynamics are so interpenetrating, the practice in this text will be to present the statistical treatments along with or closely following the classical ones. Use will be made of wave mechanical results, some of which are derived only later. For the present such results are introduced simply as accepted statements about the spacings of different kinds of energy levels.

4-2   Temperature and the zeroth law of thermodynamics

The first, second, and third laws of thermodynamics will receive a good deal of attention in this and succeeding chapters. There is also a zeroth law. The name, although peculiar sounding, is logical enough since the zeroth law properly precedes the others, yet was not stated formally until after the first and second laws. Had history occurred in a different order, we would instead be talking about the first, second, third, and fourth laws of thermodynamics!

The *Zeroth Law of Thermodynamics* is stated as follows:

Two systems that are separately in thermal equilibrium with a third system are in thermal equilibrium with each other.

In statements such as the above, words are used rather precisely. A *system* is that part of the universe which is under consideration, usually a very small part, such as some gas, liquid, or solid in a container in the laboratory. The astronomer, however, might want to consider the earth and its atmosphere, or the sun and its planets, or the galaxy, as a system (e.g., the *solar system*). A system is bounded, whatever its size; one may put it in a real or an imaginary box and regard everything within the box as the system.

If we bring two systems up to each other, that is, bring the boxes that contain them into close proximity, we can distinguish among several possibilities. The walls of the boxes may be immovable and electrically nonconducting, so that no mechanical or electrical work can be exchanged. They may be impermeable to matter, so that no material can be exchanged. Finally, the walls may be perfect insulators so that no heat can be exchanged. Systems whose walls have all these properties are said to be *isolated*. That is, no matter and no form of energy can be exchanged between them. Also, of course, any such system is isolated from the rest of the universe. (Less grandly but more usually, this means it is isolated from the rest of the laboratory.)

We can imagine, however, walls that do not allow exchange of work or of matter, but which are thermally conducting so that heat can pass between the two systems. Such a wall is called *diathermal* (*dia* meaning through). It is a matter of experience that two systems brought into proximity but separated by a diathermal wall may change with time, but will eventually come to equilibrium, that is, reach a condition of no further change. Such a pair of systems is said to be in *thermal equilibrium*.

The zeroth law states that if, for example, system A is in thermal equilibrium with system C and if system B is also in thermal equilibrium with system C, then systems A and B are in thermal equilibrium with each other. Clearly, there is *some* property of the systems that is associated with heat flow. We know this property as temperature. Bear in mind that the systems can be entirely different from each other in size and in chemical composition and remain that way; it is their *temperature* that becomes uniform at thermal equilibrium.

The law makes a very important statement about temperature. It is a statement that does not depend on the nature of a system or on how temperature is to be measured, provided we use the same "thermometer" on all systems. As discussed in Section 1-3, the natural thermometer to use is the ideal gas law thermometer, since this is based on the common limiting behavior of all gases. It is the limiting value of A (Eq. 1-5) that defines our temperature. The statement is that all systems in thermal equilibrium with each other will show the same A value on an ideal gas thermometer. (The actual Kelvin *scale* is merely a convenient arbitrary one, based on dividing the difference in A values at the ice point and the normal boiling point of water into one hundred equal parts; note Problem 1-16.)

4-3   The mechanical equivalent of heat

The origins of the first law of thermodynamics, or the conservation of energy, lie in the 18th and early 19th centuries, with the developing realization that heat and work are interconvertible in fixed ratio. The first clear, although qualitative, realization of this interconvertibility is attributed to Count Rumford in about 1780. At one stage in his career he was able to observe the great amount of heat liberated in the operation of boring cannon. His observations convinced him that the heat resulted from the dissipation of mechanical work and was not some substance released by the iron chips. For example, he noted that a dull tool produced no chips, yet about the same amount of heat. He thus attacked the then prevalent idea that heat was a substance (called caloric). The story of Count Rumford the man is a fascinating one, incidentally; read Section 4-CN-1!

It remained for James Joule to make the conclusion quantitative in the 1840's. One of Joule's experiments consisted in churning up water in a container by means of a paddle driven by a falling weight. He observed a temperature rise and from this calculated the mechanical equivalent of heat.

At first, the calorie was defined as that amount of heat required to raise the temperature of 1 g of water by 1°C. Since the heat capacity of water varies with temperature, a more precise definition made the temperature interval 14.5°C to 15.5°C (the 15-degree calorie). Now, however, the calorie is defined in terms of the joule,

$$1 \text{ cal} = 4.184 \text{ J (exactly)} \tag{4-1}$$

As discussed in Section 1-CN-1, the joule is the amount of work corresponding to the action of unit force, the newton, over unit distance, the meter; the newton, in turn, is defined as that force that gives 1 kg an acceleration of 1 m sec^{-2}. In the cgs system gram and centimeter are substituted for kilogram and meter, respectively; the unit of work is the erg; $1 \text{ J} = 10^7$ erg; and the unit of force is the dyne, $1 \text{ N} = 10^5$ dyn. In Joule's experiment the work done by the falling weight would be given by the product of the force acting on it (mass times acceleration due to gravity) and the distance it fell.

4-4   Energy and the first law of thermodynamics

The interconvertibility of heat and work makes it possible to define energy as the ability to produce heat or to do work. To be more precise, we state that

$$\Delta E = q - w \tag{4-2}$$

where ΔE is the internal energy change and

$$q = \text{heat absorbed by the system,} \tag{4-3}$$

$$w = \text{work done by the system.} \tag{4-4}$$

The quantities q and w will be treated as algebraic ones. That is, if heat is produced by the system, q will be negative, and likewise w, if work is done on the system. These sign conventions are such that ΔE is negative for the system if it furnishes overall heat plus work to its surroundings.*

We know, of course, that energy appears to reside in a system in different ways. We think of *thermal energy*—a system that releases heat or does work usually becomes cooler; the energy content of a system must then be a function of its temperature. However, if ice absorbs heat at 0°C or if two blocks of ice are rubbed together (as was done in 1799 by Humphry Davy), melting, but not necessarily any change in temperature, occurs. There is thus a difference in the energy content of ice and of water. We call this a latent energy or *latent heat.* Further, an electrochemical cell can produce work, as in the starting of a car by means of a lead storage battery. There is no particular temperature or phase change, but now a chemical reaction has occurred. Evidently the products contain a different amount of energy than do the reactants. We thus also have *chemical energy.*

In view of these various forms of energy and our lack of complete understanding of them, might it be possible that ΔE for a system could depend on the path taken in accomplishing the change? We can examine the implication of this possibility as follows. We begin with a system in some state A and take it to state B, with internal energy change ΔE_{AB}. There should in general exist some state C such that we can take state A to state C with ΔE_{AC} and then take state C to state B with ΔE_{CB}. We have thus taken state A to state B by the alternative path *via* state C, with overall energy change ΔE_{ACB}. Recalling the discussion of Section 1-2, the word *state* has a quite specific meaning. The state of a system is its condition as described by its measurable macroscopic properties (pressure, density, temperature, etc.); two systems are in the same state if all their properties are the same. This is our *thermodynamic* state.†

Resuming the hypothetical experiment, suppose that indeed $\Delta E_{AB} \neq \Delta E_{ACB}$. If we reverse the direction of the second sequence, then $\Delta E_{BCA} = -\Delta E_{ACB}$ (since both the qs and ws will necessarily change signs), and, as illustrated in Fig. 4-1, we can set up the cyclic sequence: A → B, B → C → A. By our supposition, $\Delta E_{cycle} = \Delta E_{AB} - \Delta E_{ACB}$ and $\Delta E_{cycle} \neq 0$. That is, we have returned the system to its exact original state A, but find an energy change, so that $w_{cycle} \neq q_{cycle}$. If, for example $w_{cycle} > q_{cycle}$, work has been done by the system without a compensatory absorption of heat. The cycle could be repeated indefinitely so that we would, in effect, have a machine that produced more work than it absorbed heat energy. Further, if we degraded some of the work into heat just sufficient to balance the heat absorption, the net effect would be a machine that produced work without drawing on any energy source. This would be a lovely machine indeed! (If it happened that $w_{cycle} < q_{cycle}$, we could simply reverse the direction of the cycle.)

---

*Some choices have been made here. Internal energy is sometimes designated by the symbol U, the better to distinguish internal from other types of energy. There is an alternative convention regarding w, which defines w as the work done *on* the system, so that Eq. (4-2) would read ΔE (or ΔU) = $q + w$.

†It is a regrettable source of confusion that the term "state" is also used to designate a microscopic condition, usually of a particular molecule, as in a *spectroscopic* state. Unfortunately, our vocabulary in this area has become sufficiently codified that we are stuck with this duality of meaning.

FIGURE 4-1
A cyclic process.

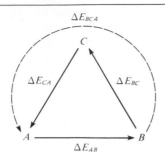

FIGURE 4-1
A cyclic process.

The possibility of net work being supplied by a system was for a while intensively tested. Machines that might accomplish this feat are known, of course, as perpetual motion machines. Some of them in the literature, including the early patent literature, are quite persuasive. Also, the idea was an acceptable one to the literary mind—read the voyage to Laputa in Gulliver's Travels and the description of how the floating island worked. Figure 4-2 shows one of these persuasive perpetual motion machines.

Our experience by now is that perpetual motion machines are *not* possible, and we make the conclusion specific by stating that $w_{tot} = q_{tot}$ for *any* cyclic process. This means that $\Delta E = 0$ for any cyclic process and, therefore, that

FIGURE 4-2
A machine for making energy.

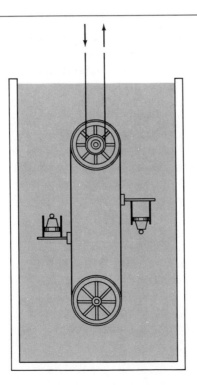

An endless belt has a number of small piston and cylinder units attached, of which two are shown. In the case of the units on the left side, the weight on the piston acts to expand the gas in the cylinder. With the units on the right side, the weight acts to compress the gas.

The whole assembly is immersed in a tank of water. Since the left-hand units have a larger gas volume than the right-hand ones, they are more buoyant. The endless belt thus rotates clockwise.

A secondary belt runs off a pulley on the upper wheel, to deliver useful power. This marvelous machine is yours for the making! What do you think of it?

Appearances are deceptive! Although there is a buoyancy difference which generates a net work per cycle, this work is just balanced by the work of displacing water as the unit swings around the bottom wheel.

ΔE for *any process* must be independent of path (i.e., in the above example, $\Delta E_{AB} = \Delta E_{ACB}$). Alternatively put, our experience requires that internal energy change as defined by Eq. (4-2) is entirely specified when the initial and final states of the system are specified. It is this attribute of E that gives the First Law of Thermodynamics its enormous importance. Conversely, without this attribute, Eq. (4-2) would be merely a trite and inconsequential statement.

It should be noted that Eq. (4-2) gives only an energy difference; it provides no absolute scale of energy. Ordinarily, we pick some standard or reference state as having an assumed energy, usually zero, and then obtain the energy of all other states through Eq. (4-2). In this sense, then, we speak of energy as a *state function;* we mean that changes in it are defined if the change of state is defined.

The various preceding statements are ways of affirming the First Law of Thermodynamics. To summarize, the alternative statements are:

$\Delta E = q - w$.
The internal energy of an isolated system is constant. (4-5)
The internal energy change in a cyclic process is zero.
Internal energy is a state function.

The expression "internal energy" is the correct one for E; it emphasizes that we are referring to the energy content of a definite portion of matter. When indicated by a capital E internal energy will be on a molar basis.*

Returning to the example of Fig. 4-1, we recall that the requirement to satisfy our experience with respect to perpetual motion machines is that $q_{tot} = w_{tot}$. Thus, per cycle, it would be allowable for work to be done on the system and an equivalent amount of heat evolved. It would also be allowable for heat to be absorbed and an equivalent amount of work to be done. The first situation encounters no objections, but the second turns out to have a restriction, again as a result of experience. The restriction is stated by the Second Law of Thermodynamics, taken up in Chapter 6.

As a final comment, the first law has so far been justified in terms of negative arguments, that is, in terms of the lack of observations of certain behavior. There is a similar aspect to the second law. However, both lead to relationships that have been abundantly verified. There is thus a great amount of direct evidence for the validity of the laws of thermodynamics.

4-5 Mathematical properties of state functions. Exact and path-dependent differentials

The conclusion that the internal energy change is zero for a cyclic process or that E is a state function carries a further implication. Referring once again to Fig. 4-1, we see that the change in energy on going from A directly to B, ΔE_{AB}, must be the same as that on going from A to C and then from C to B, ΔE_{ACB}. Otherwise, the first law of thermodynamics would be violated. We conclude therefore that ΔE cannot depend on the path taken or that, for some small change, dE is independent of path.

---

*We will use ordinary capital letters to denote quantities per mole, sans serif capitals for quantities for an arbitrary amount of system, and small capitals for quantities per molecule, thus: E (energy per mole), **E** (energy), and ᴇ (energy per molecule).

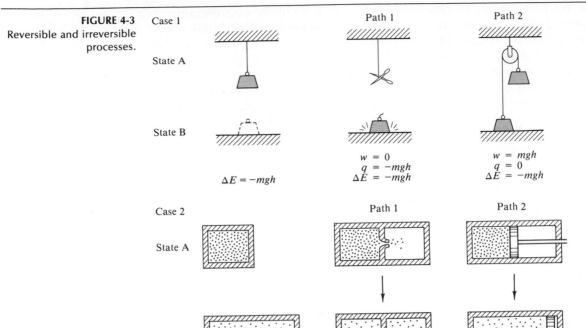

FIGURE 4-3
Reversible and irreversible processes.

We know, however, the q and w do depend on path. This is a matter of common experience, examples of which are illustrated in Fig. 4-3 and as follows.

Case 1 *State A:* mass m at height h.

State B: mass m at $h = 0$.

Path 1: The object is dropped in free fall and gains kinetic energy mgh, which is dissipated as heat when it strikes the stop at $h = 0$. No work is done, and so $w = 0$, but $q = -mgh$. Thus $\Delta E = q - w = -mgh$.

Path 2: The object is counterbalanced by a very slightly lesser mass, which is then raised from $h = 0$ to h. The work done is now $w = mgh$, and $q = 0$, as no significant kinetic energy is developed. Again, ΔE (for the object) is $q - w$ or $-mgh$.

Case 2 *State A:* Gas at T_1, P_1, and V_1.

State B: Gas at T_1, P_2, and V_2.

Path 1: The expansion is allowed to occur by rupture of a diaphragm, temperature being kept constant by means of a thermostat bath. No work is done, $w = 0$, $q = \Delta E$.

Path 2: The expansion occurs against a piston. The work done is (force on piston) × (displacement) or $w = \int f\, dx = \int P\mathscr{A}dx$, where $\mathscr{A}$ is the area of the piston. Hence $w > 0$, $q = \Delta E + w$.

Thus one of the very important corollaries of the first law is that while ΔE depends only on the initial and final states, it is given experimentally by the difference between two quantities q and w which separately do depend on the path taken in going from the initial to the final state. As will be seen in the ensuing material, we take advantage of this situation by choosing paths of convenience between two states to facilitate calculation of q and w, knowing that the resulting ΔE will be independent of our choice of path. In differential form Eq. (4-2) becomes

$$dE = \delta q - \delta w \tag{4-6}$$

where the lower case deltas are used to emphasize that the path must be specified for q and w.

There is a mathematical parallel, which can be developed as follows. First, the state of a system is defined by any two of the state variables V, P, and T. That is, if we know two of these, the equation of state (for example, the ideal gas law) gives the third. Since E depends only on the state, it follows that E is some function of any two of the state variables, and as a matter both of convention and of convenience we pick V and T and write $E = f(V,T)$. (For the present it is assumed that no chemical changes can occur; otherwise chemical composition would also have to be specified.) The general mathematical situation is that of a dependent variable y which is some function of variables u and u, $y = f(u, v)$.

A schematic representation of a surface given by $y = f(u, v)$ is shown in Fig. 4-4. We can define two slopes: $(\partial y/\partial u)_v$ and $(\partial y/\partial v)_u$, where the partial differential sign ∂ signifies that the subscript variable is held constant. As shown in the figure, these slopes are those of lines of constant v and constant u, respectively. Now suppose we wish to go from a point y_1 to a point y_2. Two possible paths would be that marked ab and that marked cd. Accordingly, Δy can be expressed in two ways:

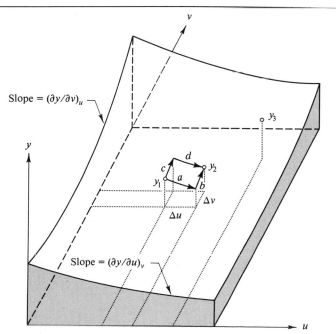

FIGURE 4-4
Surface illustrating total and partial differentials.

$$\Delta y = \left(\frac{\Delta y}{\Delta u}\right)_{v_1} \Delta u + \left(\frac{\Delta y}{\Delta v}\right)_{u_2 = u_1 + \Delta u} \Delta v$$

(a) $\qquad$ (b)

$$\Delta y = \left(\frac{\Delta y}{\Delta v}\right)_{u_1} \Delta v + \left(\frac{\Delta y}{\Delta u}\right)_{v_2 = v_1 + \Delta v} \Delta u$$

(c) $\qquad$ (d)

In the limit of very small differences, $\Delta y = dy$, and $(\Delta y/\Delta u)_{v_1} = (\Delta y/\Delta u)_{v_1 + \Delta v} = (\partial y/\partial u)_v$. A similar relation exists for the other slope. The resulting mathematical statement is

$$dy = \left(\frac{\partial y}{\partial u}\right)_v du + \left(\frac{\partial y}{\partial v}\right)_u dv \tag{4-7}$$

and the corresponding one for E is

$$dE = \left(\frac{\partial E}{\partial V}\right)_T dV + \left(\frac{\partial E}{\partial T}\right)_v dT \tag{4-8}$$

If dy is integrated between y_1 and some value y_3, the integral is simply $y_3 - y_1$, and, of course, is independent of the path. However, the integrals $\int_{u_1}^{u_3} (\partial y/\partial u)_v \, du$ and $\int_{v_1}^{v_3} (\partial y/\partial v)_u \, dv$ are path dependent. By way of illustration, let $y = uv^2$, so that

$$dy = v^2 \, du + 2uv \, dv \tag{4-9}$$

Let us integrate between the points $(u = 2, v = 1)$ and $(u = 4, v = 3)$, as indicated in Fig. 4-5. First,

$$\Delta y = y_2 - y_1 = 36 - 2 = 34$$

and this result must be independent of the integration path. Two possible paths are as follows.

Path 1:

(a) $(2, 1)$ to $(4, 1)$ $\qquad v^2 \int du \quad = (1)^2 (4 - 2) = 2$

(b) $(4, 1)$ to $(4, 3)$ $\qquad 2u \int v \, dv = 8[(3^2/2) - (1^2/2)] = 32$

Sum: $2 + 32 = 34$

Path 2:

(c) $(2, 1)$ to $(2, 3)$ $\qquad 2u \int v \, dv = 4[(3^2/2) - (1^2/2)] = 16$

(d) $(2, 3)$ to $(4, 3)$ $\qquad v^2 \int du \quad = 9(4 - 2) = 18$

Sum: $16 + 18 = 34$

Thus although the two integrals $v^2 \int du$ and $2u \int v \, dv$ depend on path, their sum does not.

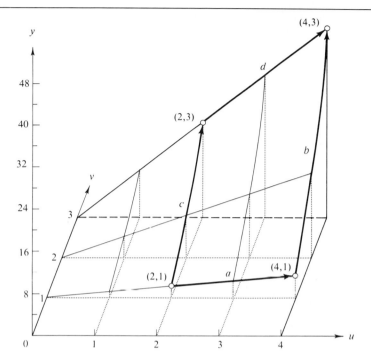

FIGURE 4-5
The function $y = uv^2$.

The differential dy in Eq. (4-7) is known as an *exact differential;* its integral is independent of path. However, this is not necessarily true of every expression of the form

$$dy = M\,du + N\,dv \qquad (4\text{-}10)$$

For example, the expression ($v\,du + 2uv\,dv$) is not an exact differential. It is possible to test an expression such as Eq. (4-10) on this point. Referring to Eq. (4-7), for an exact differential, we have

$$\left(\frac{\partial M}{\partial v}\right)_u = \left(\frac{\partial^2 y}{\partial u\,\partial v}\right) \qquad \left(\frac{\partial N}{\partial u}\right)_v = \left(\frac{\partial^2 y}{\partial v\,\partial u}\right)$$

Most thermodynamic functions are continuous and single-valued, so the order of partial differentiation makes no difference. It follows that

$$\left(\frac{\partial M}{\partial v}\right)_u = \left(\frac{\partial N}{\partial u}\right)_v \qquad (4\text{-}11)$$

The expression ($v^2\,du + 2uv\,dv$) meets the test of Eq. (4-11) whereas ($v\,du + 2uv\,dv$) does not, and is therefore not an exact differential.

Where the differential is known to be exact, Eq. (4-11) is very useful in providing an additional relationship. Such a relationship is known as an *Euler* or a *cross-differentiation equation* and will occur fairly often in Chapter 6.

Mathematics provides two additional very useful equations. Since u and v are independent variables du and dv may be chosen independently. In particular they may be chosen so that $dy = 0$. Equation (4-7) then becomes

$$0 = \left(\frac{\partial y}{\partial u}\right)_v du + \left(\frac{\partial y}{\partial v}\right)_v dv$$

or expressing the constancy of y in equation,

$$\left(\frac{\partial u}{\partial v}\right)_y = -\frac{(\partial y/\partial v)_u}{(\partial y/\partial u)_v} \tag{4-12}$$

Thus

$$\left(\frac{\partial V}{\partial T}\right)_E = -\frac{(\partial E/\partial T)_V}{(\partial E/\partial V)_T} \tag{4-13}$$

EXAMPLE   Let $V = f(P,T)$. Then

$$dV = \left(\frac{\partial V}{\partial P}\right)_T dP + \left(\frac{\partial V}{\partial T}\right)_P dT \tag{4-14}$$

If the gas is ideal, $(\partial V/\partial P)_T = -RT/P^2$ and $(\partial V/\partial T)_P = R/P$; by Eq. (4-12), $(\partial P/\partial T)_V$ is given by

$$\left(\frac{\partial P}{\partial T}\right)_V = -\frac{(\partial V/\partial T)_P}{(\partial V/\partial P)_T} = -\frac{R/P}{-RT/P^2} = \frac{P}{T}$$

The derivative $(\partial P/\partial T)_V$ may be found directly. Thus $P = RT/V$, so $(\partial P/\partial T)_V = R/V = P/T$. The result is the same as by the indirect method.

Second, if there is some quantity z which is a function of u and v, then one can impose the condition on du and dv that z be constant:

$$dy_z = \left(\frac{\partial y}{\partial u}\right)_v du_z + \left(\frac{\partial y}{\partial v}\right)_u dv_z$$

This constancy is usually indicated in partial differential form by dividing through by du_z or by dv_z; thus

$$\left(\frac{\partial y}{\partial v}\right)_z = \left(\frac{\partial y}{\partial u}\right)_v \left(\frac{\partial u}{\partial v}\right)_z + \left(\frac{\partial y}{\partial v}\right)_u \tag{4-15}$$

For example, since $P = f(V,T)$, on applying the condition of P constant to Eq. (4-8), we obtain

$$\left(\frac{\partial E}{\partial T}\right)_P = \left(\frac{\partial E}{\partial V}\right)_T \left(\frac{\partial V}{\partial T}\right)_P + \left(\frac{\partial E}{\partial T}\right)_V \tag{4-16}$$

4-6   Heat and work for various processes

A. Evaluation of q; heat capacity   As emphasized earlier, the value of q for a change of state depends on the path, so the path must always be specified. There is, of course, an infinite number of possible types of paths that a system might follow in going from one state to another, but we find it very convenient, both experimentally and theoretically, to emphasize a few special ones. We thus make much use of the following types of process.

1. *Adiabatic.* $q = 0$. The system is insulated so that no heat can enter or leave.

2. *Isochoric.* The system is constrained to a definite volume. We introduce the coefficient $(\partial q/\partial T)_V$ or, as a convenient nomenclature, specify the path by writing q_v, and hence dq_v/dT. This type of coefficient is called a heat capacity (the term *specific heat* refers to heat capacity per unit mass), C. For a constant-volume path C becomes C_V:

$$C_V = \frac{dq_V}{dT}$$

Further, under constant-volume conditions no mechanical work w can be done. As a consequence $\Delta E = q_V$, or $dE_V/dT = C_V$. Alternatively, by Eq. (4-8) we can write

$$\left(\frac{\partial E}{\partial T}\right)_V = C_V = \frac{dq_V}{dT} \tag{4-17}$$

3. *Isobaric.* The system is maintained at a particular pressure during the process. Following the preceding formalism we have

$$C_P = \frac{dq_P}{dT} \tag{4-18}$$

B. Work

Mechanical work is given by force f acting through a distance x, or $w = \int f\, dx$. The work to lift mass m by height h is $(mg)h$. The joule corresponds to a force of one newton acting over a distance of one meter, etc. We call such work *mechanical* work. There is an alternative way of expressing mechanical work that is very useful in physical chemistry. Imagine, as illustrated in Fig. 4-6, a gas contained in a piston and cylinder arrangement, and at pressure P. The force on the piston is $P\mathcal{A}$, where $\mathcal{A}$ is the area of the piston. If the gas is allowed to expand, it will do mechanical work

$$w = \int f\, dx = \int (P\mathcal{A})dx \tag{4-19}$$

However, $\mathcal{A}\, dx$ is just dV, the volume change, and Eq. (4-19) may alternatively be written

$$w = \int P\, dV \tag{4-20}$$

There are other kinds of work. Electrical potential is defined in terms of the work to transport unit charge q across the potential difference V (see Section 3-CN-1). Thus $w = \int V\, dq$. This would apply both to the work done by an electrochemical cell and to the work of charging a condenser. As another

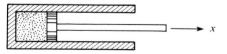

FIGURE 4-6
Illustration of work through the expansion of a gas.

example, surface tension γ is defined in terms of the work to increase the surface area of a system by unit amount. Therefore, in this case $w = \int \gamma \, d\mathcal{A}$.

In each of the above cases, work is given by an integral of the type $w = \int y \, dx$ where y is an *intensive* quantity and x is an *extensive* one. An *intensive* property is not specifically related to the size of a system; it can be measured in some small region. Thus pressure, electrical potential, surface tension, and temperature are intensive properties or variables. An *extensive* property depends on the size or amount of the system; volume, area, and charge are examples. In this chapter we will mostly limit ourselves to mechanical work expressed in terms of Eq. (4-19) or Eq. (4-20). It should be kept in mind, however, that w in Eq. (4-2) can occur in a variety of forms and contexts.

C. Reversible and irreversible work

We return to Eq. (4-20) and its illustration in Fig. 4-6. If w is to be applicable to the first law equation, Eq. (4-2), the pressure in Eq. (4-20) must be the pressure felt by the piston, that is, the external pressure P_{ext} indicated in Fig. 4-7(a), and

$$w = \int P_{ext} dV \tag{4-21}$$

However, while the value of P_{ext} would be important in obtaining the work output of a machine, it is not necessarily the same as the pressure in the gas. For example, the cylinder might contain a gas at 1 atm pressure confined by a piston held in place by means of a stop as shown in Fig. 4-7(b). For simplicity,

FIGURE 4-7
Reversible (a) and irreversible (b) expansions. (c) Illustration of Eq. (4-20).

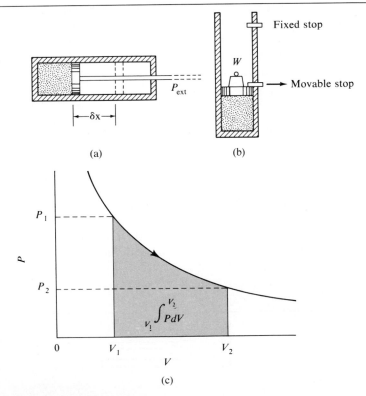

we can assume the piston and cylinder are kept at a constant temperature. The pressure P_{ext} is now defined by some weight W placed on the piston. This weight might be such as to make $P_{ext} = 0.1$ atm. The stop is pulled out and the piston moves back to the second stop, thereby increasing the volume by ΔV. The work done is $0.1\Delta V$ liter atm. Were w such as to make $P_{ext} = 0.4$ atm, then w would be $0.4\Delta V$ liter atm, and so on.* If 0.4 atm is the final equilibrium pressure of the gas, this last is the greatest weight that could be used if the piston is to reach the second stop.

Clearly, the maximum possible value of w would be obtained if P_{ext} (or W in the figure) were steadily adjusted so as always to be only slightly, or, in the limit, infinitesimally less than the pressure of the gas, which is the pressure P. Not only do we now have the maximum work, but since $P_{ext} = P$, we can find how P_{ext} must vary with the volume. That is, the equation of state of the substance gives us P as a function of T and V, $P = f(T,V)$, so Eq. (4-21) becomes

$$w_{max} = \int P \, dV = \int f(T, V)dV \qquad (4-22)$$

[note Fig. 4-6(c)].

We speak of the condition of $P_{ext} = P$ as giving the *reversible work*. Thus from Fig. 4-6(a) if P_{ext} is slightly less than P, expansion occurs, while if P_{ext} is made slightly more than P, compression occurs. Thus the direction of the process can be reversed by an infinitesimal change in P_{ext}. Reversible processes will be of central importance in the development of chemical thermodynamics, since by means of them we can calculate w if we know the equation of state of the substance and since they give the limiting or maximum value of w.

We may next consider w in terms of the various standard processes.

1. *Adiabatic.* $q = 0$. We do not know w directly, but, by the first law, it follows that

$$dE = -\delta w = -P \, dV \qquad (4-23)$$

2. *Isochoric.* $w = 0$. This follows since $dV = 0$, as noted previously. Then

$$dE = C_v \, dT \qquad (4-24)$$

3. *Isobaric.* $w = P\Delta V$. Since pressure is constant, it may be put in front of the integral of Eq. (4-21). Also,

$$dE = C_P \, dT - P \, dV \qquad (4-25)$$

Where only reversible mechanical work is involved, the first law evidently can be written in the form

$$dE = \delta q - P \, dV \qquad (4-26)$$

The various special cases then follow from Eq. (4-26). It should be emphasized that whenever P rather than P_{ext} is used, the process must be a reversible

*We can suppose that in each case friction prevents the piston and weight from gaining any appreciable kinetic energy and that the frictional heat is in this case dissipated into the heat reservoir that maintains constant temperature. In the alternative, adiabatic case, the heat would be dissipated into the gas. [See D. Kivelson and I. Oppenheim, *J. Chem. Ed.*, **43**, 273 (1966) for a discussion of this type of experiment.]

one. Also, of course, it is assumed that only pressure-volume mechanical work is involved.

EXAMPLE   Suppose that one mole of liquid water at 25°C and 1000 atm pressure is to be heated by 1°C (a) at constant volume and (b) at constant pressure. To keep the calculation simple, we take C_V to be constant at 17.3 cal K^{-1} $mole^{-1}$ and C_p to be constant at 18.0 cal K^{-1} $mole^{-1}$. The water will expand by 8.6×10^{-3} cm^3 $mole^{-1}$ per degree.

(a) $\Delta E = C_V \Delta T = 17.3$ cal. (b) $\Delta E = C_p \Delta T - P \Delta V = 18.0 - (1000)(8.6 \times 10^{-3})$ $(1.987/82.057) = 17.8$ cal. Note that had we taken P to be 1 atm, the $P \Delta V$ term would have been negligible. The term is in general important for a gas, but we will see in Section 4-8 that an interesting simplification is possible in the case of an ideal gas.

In conclusion to this section, we return briefly to the matter of reversibility. The requirement can be generalized. In Fig. 4-3, the weight is lowered reversibly by just counterbalancing it. An electrochemical cell operates reversibly if its emf is just balanced by an opposing, externally applied one. Referring to the integral $w = \int y \, dx$, the generalization is that the intensive variable must be just counterbalanced. This is another way of saying that the system experiencing work must be at equilibrium at all stages of the process.

4-7   Enthalpy. An alternative form of the first law

The first law takes on a particularly simple form for an isochoric process if only mechanical work is considered, that is, dE is simply $C_V dT$. Actual experiments are more often done under constant pressure conditions, however, and the following alternative statement of the first law is very useful. We define a new molar quantity H called the *enthalpy*:

$$H = E + PV \tag{4-27}$$

Since the volume and pressure of a system are determined solely by its state, as is E, H must also be a state function. The differential form of Eq. (4-27) is

$$dH = dE + P \, dV + V \, dP \tag{4-28}$$

and, using Eq. (4-26), we have

$$dH = \delta q + V \, dP \tag{4-29}$$

We then have the following special cases, corresponding to the standard processes.

1. *Adiabatic.* Since $q = 0$ (and δq is zero for all increments along the path), it follows that

$$dH = V \, dp \tag{4-30}$$

2. *Isochoric.*

$$dH = dE + V \, dP = C_V \, dT + V \, dP \tag{4-31}$$

3. *Isobaric.* Since dP is zero, Eq. (4-29) becomes $dH = \delta q_P$ or

$$dH = C_P \, dT \quad \text{or} \quad \left(\frac{\partial H}{\partial T}\right)_P = C_P \tag{4-32}$$

Note the parallel between Eq. (4-24) and Eq. (4-32).

Since H is a state function, its value depends on those of the state variables, P, V, and T, any one of which can be eliminated by means of the equation of state of the system. In the case of E, it is natural to pick V and T as the independent variables; in the case of H, the natural choice is P and T. Since dH is an exact differential, we have

$$dH = \left(\frac{\partial H}{\partial P}\right)_T dP + \left(\frac{\partial H}{\partial T}\right)_P dT \tag{4-33}$$

The introduction of the enthalpy function allows the following derivation of an important relationship for the difference between C_P and C_V:

$$C_P - C_V = \left(\frac{\partial H}{\partial T}\right)_P - \left(\frac{\partial E}{\partial T}\right)_V$$

or, from the definition of H, Eq. (4-25),

$$C_P - C_V = \left(\frac{\partial E}{\partial T}\right)_P + P\left(\frac{\partial V}{\partial T}\right)_P - \left(\frac{\partial E}{\partial T}\right)_V \tag{4-34}$$

On eliminating $(\partial E/\partial T)_P$ by means of Eq. (4-16), we obtain as the result

$$C_P - C_V = \left[P + \left(\frac{\partial E}{\partial V}\right)_T\right]\left(\frac{\partial V}{\partial T}\right)_P \tag{4-35}$$

There are some interesting aspects to this last equation. The heat capacity difference $(C_P - C_V)$ can be thought of as made up of two types of contributions. The first is the work against the external pressure P due to the change in volume on heating given by $P(\partial V/\partial T)_P$. The second is the work against the attractive forces between the molecules of the substance, given by $(\partial E/\partial V)_T$ $(\partial V/\partial T)_P$. The quantity $(\partial E/\partial V)_T$ has the dimensions of pressure, and since it gives the measure of these attractive forces, it is called the *internal pressure* P_{int}.

As we learn in the next section, the internal pressure is zero for an ideal gas, and for most real gases it is small compared to P. We can evaluate P_{int} by means of a thermodynamic relationship derived later, in Section 6-ST-1:

$$\left(\frac{\partial E}{\partial V}\right)_T = T\left(\frac{\partial P}{\partial T}\right)_V - P \tag{4-36}$$

By means of this expression, we find that for a van der Waals gas,

$$P_{int} = \left(\frac{\partial E}{\partial V}\right)_T = \frac{a}{V^2} \tag{4-37}$$

Also, E for a van der Waals gas can be obtained by integrating Eq. (4-37), to obtain

$$E(V,T) = -\frac{a}{V} + f(T) \tag{4-38}$$

The first term represents the contribution of attractive forces, and the second turns out to be the same as for an ideal gas, $\frac{3}{2}RT$.

EXAMPLE Calculate P_{int} and E for CO_2 at 100°C and 100 atm pressure. From Exercise 1-14, $V = 0.21$ liter, so $P_{int} = 3.592/(0.21)^2 = 81.5$ atm, a not insignificant value. E becomes $(-3.592)/(0.21) + \frac{3}{2} RT = 28.83$ liter atm or 698 cal mole^{-1}.

In the case of liquids (or solids) the internal pressure is usually much greater than the external pressure (usually around atmospheric). For common liquids P_{int} is around 3000 atm (see Table 8-2). Liquids, in effect, are held together by their internal pressure, that is, by their intermolecular attraction.

Finally, a very useful relationship is obtained by using Eq. (4-36) to eliminate the coefficient $(\partial E/\partial V)_T$ from Eq. (4-35):

$$C_P - C_V = T\left(\frac{\partial P}{\partial T}\right)_V \left(\frac{\partial V}{\partial T}\right)_P \tag{4-39}$$

or, since

$$\left(\frac{\partial P}{\partial T}\right)_V = -\frac{(\partial V/\partial T)_P}{(\partial V/\partial P)_T} = \frac{\alpha}{\beta} \quad \text{[Eq. (4-14)]}$$

where α is the coefficient of thermal expansion and β the coefficient of compressibility,

$$\alpha = \frac{1}{V}\left(\frac{\partial V}{\partial T}\right)_P \qquad \beta = -\frac{1}{V}\left(\frac{\partial V}{\partial P}\right)_T \tag{4-40}$$

the final form of Eq. (4-39) becomes

$$C_P - C_V = \frac{\alpha^2 VT}{\beta} \tag{4-41}$$

EXAMPLE Calculate $(C_P - C_V)$ for water liquid at 25°C. The coefficient of thermal expansion is 4.8×10^{-4} K^{-1} and that of compressibility is 4.6×10^{-5} atm^{-1}. We find $(C_P - C_V) = (4.8 \times 10^{-4})^2 (18)(298.15)/(4.6 \times 10^{-5}) = 26.9$ cm^3 atm K^{-1} mol^{-1} or 0.65 cal K^{-1} mol^{-1}.

4-8 Applications of the first law to ideal gases

A. Internal energy change for an ideal gas

The application of the first law to the case of an ideal gas not only provides simple illustrations of its use but turns out to be of considerable importance generally. To anticipate a little, the reason for this last statement is that ideal gas behavior is the limiting behavior of all real gases, so that the results provide the reference condition for treating nonideal behavior. Further, any solid or liquid is in principle in equilibrium with its vapor; knowledge of the thermodynamic properties of the vapor may therefore be used as an indirect path to treatment of the solid or liquid state. Finally, most gases are approximately ideal, so the thermodynamics of the ideal gas is often used directly to give approximate results, usually of acceptable accuracy, for real gases and vapors at ordinary pressures.

We owe again to Joule a fundamental piece of information—this time about an ideal gas. His experiment, described in 1843 and illustrated in Fig. 4-8,

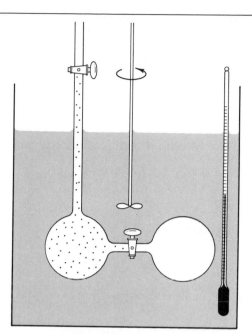

FIGURE 4-8
Joule's experiment.

consisted of allowing air under pressure (up to 10 or 20 atm) to expand from one vessel into another. There was no change in the temperature of the gas. Alternatively, the two bulbs were immersed in a water bath; no change in the temperature of the bath could be observed when the expansion occurred. The significance of this result is that since no work was done, $w = 0$, and since no temperature change occurred, $q = 0$; therefore ΔE was zero.

More accurate later experiments have shown that Joule's experiment does in fact lead to small temperature changes (see Special Topics) but that in the limit of ideal gas behavior, his conclusion was correct. The conclusion, in partial differential form, is

$$\left(\frac{\partial T}{\partial V}\right)_E = 0 \quad \text{(ideal gas)} \tag{4-42}$$

(Equations valid specifically for an ideal gas will be so labeled.) It follows from Eq. (4-13) that

$$\left(\frac{\partial E}{\partial V}\right)_T = 0 \quad \text{(ideal gas)} \tag{4-43}$$

and, from Eq. (4-8), that

$$dE = \left(\frac{\partial E}{\partial T}\right)_V dT \quad \text{(ideal gas)}$$

or

$$\frac{dE}{dT} = \left(\frac{\partial E}{\partial T}\right)_V = C_V \quad \text{(ideal gas)} \tag{4-44}$$

This last, very important result says that for an ideal gas the change in energy with temperature is always given by $C_V dT$, *regardless of the actual path*; that is, regardless of whether V is constant or not.

A similar situation holds for enthalpy. It follows from the defining equation for H that

$$\left(\frac{\partial H}{\partial P}\right)_T = \left(\frac{\partial E}{\partial P}\right)_T + \left(\frac{\partial (PV)}{\partial P}\right)_T$$

The last term is zero for an ideal gas; also

$$\left(\frac{\partial E}{\partial P}\right)_T = \left(\frac{\partial E}{\partial V}\right)_T \left(\frac{\partial V}{\partial P}\right)_T$$

(a partial derivative may be expanded in this manner if the same quantity is held constant in all terms), and since $(\partial E/\partial V)_T$ is zero, it follows that

$$\left(\frac{\partial E}{\partial P}\right)_T = 0 \quad \text{(ideal gas)} \tag{4-45}$$

The result is therefore that

$$\left(\frac{\partial H}{\partial P}\right)_T = 0 \quad \text{(ideal gas)} \tag{4-46}$$

Then, from Eqs. (4-33) and (4-32),

$$\frac{dH}{dT} = \left(\frac{\partial H}{\partial T}\right)_P = C_P \quad \text{(ideal gas)} \tag{4-47}$$

Thus for an ideal gas the change in enthalpy with temperature is always given by $C_P dT$, *regardless of the actual path*.

We can also relate C_P and C_V. By Eq. (4-35),

$$C_P - C_V = (P + 0)\frac{R}{P}$$

or

$$C_P - C_V = R \quad \text{(ideal gas)} \tag{4-48}$$

B. Various processes for an ideal gas

The equations for various processes that were developed in Sections 4-6 and 4-7 are general in that they apply to reversible processes for any substance, including, of course, an ideal gas. Several of these relationships can be developed further, however, with the use of Eqs. (4-44) and (4-47) and the ideal gas law. We will summarize here only those forms special for an ideal gas.

1. *Adiabatic.* Since dE is always $C_V dT$ for an ideal gas, Eq. (4-23) becomes

$$C_V dT = -P\,dV = -RT\,d(\ln V) \quad \text{(ideal gas)}$$

or

$$C_V d(\ln T) = -R\,d(\ln V) \quad \text{(ideal gas)} \tag{4-49}$$

Alternatively, $d(\ln P) + d(\ln V) = d(\ln T)$ for an ideal gas, so, on eliminating $d(\ln V)$ and remembering that $C_P = C_V + R$, we get

$$C_P d(\ln T) = R \, d(\ln P) \quad \text{(ideal gas)} \tag{4-50}$$

If the heat capacities are independent of temperature (see further below), the equations can be integrated directly to give

$$C_V \ln \frac{T_2}{T_1} = R \ln \frac{V_1}{V_2} \quad \text{(ideal gas)} \tag{4-51}$$

$$C_P \ln \frac{T_2}{T_1} = R \ln \frac{P_2}{P_1} \quad \text{(ideal gas)} \tag{4-52}$$

$$PV^\gamma = \text{constant} \quad \gamma = \frac{C_P}{C_V} \quad \text{(ideal gas)} \tag{4-53}$$

The last form is obtained as follows. Equation (4-51) may be written as $(T_2/T_1) = (V_1/V_2)^{R/C_V} = (V_1/V_2)^{\gamma-1}$. Since $(T_2/T_1) = (P_2 V_2/P_1 V_1)$, elimination gives $(P_2/P_1) = (V_1/V_2)^\gamma$ or $P_1 V_1^\gamma = P_2 V_2^\gamma$. Equation (4-53) is a generalization of this last conclusion.

2. *Isochoric.* From Eq. (4-47), $dH = C_P dT$ (as always for an ideal gas).
3. *Isobaric.* When only pressure-volume work is involved, the general expression for δw is $\delta w = P \, dV$, or, for an isobaric process, $w = P\Delta V$. In the case of an ideal gas, we can also write

$$dw = P \, dV = R \, dT \quad \text{(ideal gas)} \tag{4-54}$$

and, since R is a constant,

$$w = P\Delta V = R\Delta T \quad \text{(ideal gas)} \tag{4-55}$$

4. *Isothermal.* The reversible isothermal or constant-temperature process was not considered explicitly in Sections 4-6 and 4-7, since it leads to no specialized forms in the general case. However, for an ideal gas if dT is zero, then $dE = 0$ and $dH = 0$. It follows that $\delta q = \delta w$, and, further, that

$$\delta q = \delta w = P \, dV = (RT/V)dV = RT \, d(\ln V) \quad \text{(ideal gas)} \tag{4-56}$$

Since temperature is constant, integration gives

$$q = w = RT \ln(V_2/V_1) \quad \text{(ideal gas)} \tag{4-57}$$

Again since temperature is constant, we have from the ideal gas law that $d(\ln V) = -d(\ln P)$ and so

$$\delta q = \delta w = -RT \, d(\ln P) \quad \text{(ideal gas)} \tag{4-58}$$

and

$$q = w = RT \ln (P_1/P_2) \quad \text{(ideal gas)} \tag{4-59}$$

5. *A general path.* A reversible change of state need not be restricted to occur along some one of the above types of paths. Combinations of paths may be used, of course, as in the example of Section 4-8D. In addition, however, a path may be defined by means of some imposed constraint such as $P = g(V)$ along the path. A case that has some application to heat engines is the constraint

$$PV^a = k \tag{4-60}$$

where a and k are constants. In the case of an ideal gas, it turns out that the heat capacity along such a path is constant and is given by

$$C_{\text{path}} = C_V - \frac{R}{a-1} \qquad (4\text{-}61)$$

Notice that Eq. (4-53) is a special case of Eq. (4-60), with $a = \gamma$.

C. Heat capacity of an ideal gas

This subject is treated in more detail in Section 4-9. For the present, it is useful to consider qualitatively a special case of the general heat capacity behavior for an ideal gas. The special case is that of an ideal, monatomic gas. Such a gas can have only kinetic energy; its molar energy is then, from Eq. (2-4),

$$E = \tfrac{3}{2}RT$$

The heat capacity of the gas is therefore

$$C_v = \tfrac{3}{2}R \quad \text{(ideal monatomic gas)} \qquad (4\text{-}62)$$

and C_P, of course, is $(C_V + R)$, or

$$C_P = \tfrac{5}{2}R \quad \text{(ideal monatomic gas)} \qquad (4\text{-}63)$$

Gases, even though ideal, will in general have a higher heat capacity than that given here. If the molecule has more than one atom, then as the temperature is raised energy goes into vibration and rotation as well as into increased kinetic energy. The degree to which this happens increases with increasing temperature; some typical heat capacity curves are given in Fig. 4-9. It is convenient for many calculational purposes to represent this temperature variation by means of a polynomial in temperature:

$$C_P = a + bT + cT^{-2} + \cdots \qquad (4\text{-}64)$$

and some values for such coefficients are given in Section 5-6.

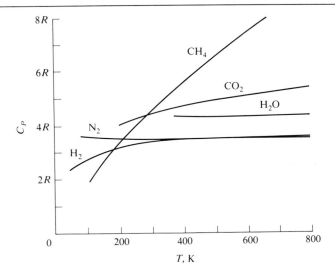

FIGURE 4-9
Heat capacity C_P as a function of temperature for various gases.

A number of equations have been introduced in this section for processes involving an ideal gas. The following example is intended to illustrate their application in a way that will also bring out important features of the first law of thermodynamics. We will use one mole of an ideal, monatomic gas as the working fluid—we can then use the constant values of C_V and C_P of $\frac{3}{2}R$ and $\frac{5}{2}R$, respectively, and thus simplify the calculations.

The gas initially is in state 1 at 0°C and 10 atm pressure, and hence at 2.241 liter volume. It is put through the following reversible cycle, cycle A in Fig. 4-10.

State 1 to state 2: Isothermal expansion to 1 atm. Then

$$V_2 = 22.41 \text{ liter} \qquad T_2 = 273.15 \text{ K}$$

$$w = RT \ln \frac{P_1}{P_2} = R(273.15)(2.303)(1) = 629.0R$$

$$q = w = 629.0R \qquad \Delta E = \Delta H = 0$$

The results can all be expressed in units of $(R)(K)$; we can obtain the actual values in calories by putting $R = 1.987$ cal K^{-1} mole^{-1}, or in joules by using $R = 8.314$ J K^{-1} mole^{-1}.

State 2 to state 3: Isobaric cooling to 2.24 liter. Then

$$T_3 = \frac{V_3}{V_2}T_2 = 27.315 \text{ K} \qquad P_3 = 1 \text{ atm}$$

$$w = R\,\Delta T = R(27.32 - 273.15) = -245.8R$$

D. Some calculations for an ideal, monatomic gas

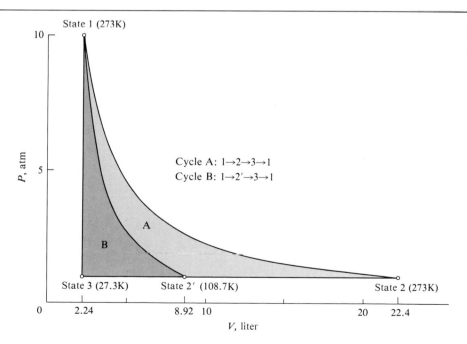

FIGURE 4-10
Two reversible ideal gas cycles (see text).

$$q = C_P \, \Delta T = \tfrac{5}{2}R(-245.8) = -614.5R$$

$$\Delta E = C_V \, \Delta T = \tfrac{3}{2}R(-245.8) = -368.8R$$

$$\Delta H = C_P \, \Delta T = q = -614.6R$$

State 3 *back to state* 1: Isochoric heating to 273.15 K, 10 atm, and 2.241 liter. Then

$$w = 0 \qquad q = C_V \Delta T = \tfrac{3}{2}R(273.15 - 27.32) = 368.8R$$

$$\Delta E = q = 368.8R \qquad \Delta H = C_P \, \Delta T = 614.6R$$

The results for cycle A are summarized in the top part of Table 4-1. Notice that for the cycle $q = w$ and ΔE and ΔH are zero. A further point is that since $w = \int P \, dV$, if a path is plotted as P versus V, then w is given by the area under that path. The area is positive for a left-to-right direction of progress along the path and negative for a right-to-left one. As a result the total work associated with a closed path or cycle is just the area enclosed by the cycle.

Next consider a second reversible cycle B using the same starting point; this is also shown in Fig. 4-10. The steps are now as follows.

State 1 *to state* 2′: Adiabatic expansion to 1 atm. Then $\log(T_2'/T_1) = (2/5) \log(P_2'/P_1) = -0.400$ or $T_2' = 0.3981 T_1 = 108.74$ K. Then

$$V_2' = (108.7/273.15)(10/1)(2.241) = 8.92 \text{ liter.}$$

Further,

$$q = 0$$

$$\Delta E = C_V \, \Delta T = \tfrac{3}{2}R(108.74 - 273.15) = -246.6R$$

$$w = -\Delta E = 246.6R$$

$$\Delta H = C_P \, \Delta T = \tfrac{5}{2}R(108.7 - 273.15) = -411.0R$$

State 2′ *to state* 3: Isobaric cooling to 2.24 liter; $T_3 = 27.31$ K and $P_3 = 1$ atm, as before. Then

$$w = R \, \Delta T = R(27.32 - 108.74) = -81.4R$$

$$q = C_P \, \Delta T = \tfrac{5}{2}R(-81.4) = -203.5R$$

	Change of state	Process	q	w	ΔE	ΔH
TABLE 4-1 Calculations for the two cycles of fig. 4-10[a]	Cycle A					
	1 to 2	Isothermal	629.0	629.0	0	0
	2 to 3	Isobaric	−614.5	−245.8	−368.8	−614.6
	3 to 1	Isochoric	368.8	0	368.8	614.6
	Cycle		383.3	383.3	0	0
	Cycle B					
	1 to 2′	Adiabatic	0	246.6	−246.6	−411.0
	2′ to 3	Isobaric	−203.5	−81.4	−122.1	−203.5
	3 to 1	Isochoric	368.7	0	368.7	614.5
	Cycle		165.2	165.2	0	0

[a]All values in terms of R(K).

$$\Delta E = C_V \, \Delta T = \tfrac{3}{2}R(-81.4) = -122.1R$$

$$\Delta H = q = -203.5R$$

State 3 *back to state* 1: Same as for cycle A.

This second set of results is assembled in the bottom part of Table 4-1. Again ΔE and ΔH are zero for the cycle, but notice in Fig. 4-10 that the P–V plot of the adiabatic expansion $1 \rightarrow 2'$ is steeper than that of the isothermal one $1 \rightarrow 2$. The physical reason for this is that in the adiabatic expansion the energy for the work done is supplied by a cooling of the gas. The temperature thus drops steadily during the expansion and the volume at each stage is less than at the corresponding pressure during the isothermal expansion. As a result the area under the P–V plot for the adiabatic process is smaller than that under the one for the isothermal process, and this is in conformity with the lesser amount of work done. The further consequence is that less work is done by cycle B than by cycle A, and a correspondingly smaller amount of heat is absorbed.

Notice also that while q and w are different for the sequences $1 \rightarrow 2 \rightarrow 3$ and $1 \rightarrow 2' \rightarrow 3$, the values of ΔE and ΔH are the same. We thus have a further illustration of the point that q and w are dependent on the path taken between two states, but ΔE and ΔH are not.

4-9 Molecular basis for heat capacities. The equipartition principle

It is pointed out in the preceding section that since the internal energy for an ideal monatomic gas consists only of its kinetic energy, C_V is therefore $\tfrac{3}{2}R$. Also, in Section 2-CN-1, it was noted that the heat capacity for a one- and a two-dimensional gas is $C_V = \tfrac{1}{2}R$ and R, respectively. Thus each independent velocity component has associated with it a molar kinetic energy of $\tfrac{1}{2}RT$ and corresponding heat capacity of $C_V = \tfrac{1}{2}R$. These three independent modes of translation in three dimensions reflect the fact that three coordinates are needed to specify the position of a particle in space and, by means of their time derivatives, the velocity of a particle in space.

We speak of the number of degrees of freedom f of a system as the number of variables that must be specified to fix the position of each particle or the velocity of each particle. For one mole of particles there are $3N_A$ degrees of freedom. Consider next the case of an ideal gas composed of diatomic molecules. There are now two moles of atoms and so, per mole of molecules, $f = 6N_A$. To generalize, a gas consisting of n atoms per molecule will have $3nN_A$ degrees of freedom per mole, or $f = 3n$ per molecule.

Returning to the case of the gas consisting of diatomic molecules, although one could specify the positions of all atoms by means of $6N_A$ coordinates per mole, there is a more rational and very useful alternative scheme. One recognizes that the gas in fact consists of molecular units of two atoms each. Of the six degrees of freedom required to locate two atoms, one first uses three to locate the center of mass of the molecule. One now needs to specify the positions of the two atoms relative to their center of mass. As illustrated in Fig. 4-11 two angles will be needed or, alternatively, the degrees of rotation around the x axis and around the y axis. Rotation about the molecular or z axis does not change the nuclear positions and hence contributes no infor-

FIGURE 4-11
Coordinates needed to
specify the rotation of a
diatomic molecule.

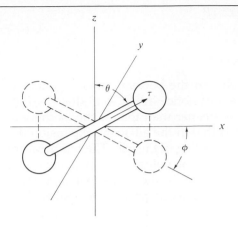

mation. (In the case of a nonlinear molecule, however, all three rotations would be needed.) If now we also specify the distance between the atoms, the description of their position in space is completely fixed. Thus six coordinates are also needed in the alternative scheme. In general, the number of coordinates sufficient to describe the positions of a set of n atoms cannot depend on the scheme used but is always equal to the $3n$ degrees of freedom.

As just noted, each degree of translational freedom implies an energy of $\frac{1}{2}RT$ per mole and a heat capacity contribution of $\frac{1}{2}R$ per mole. In the alternative scheme of describing the positions and velocities of atoms, the rotational degrees of freedom must likewise contribute $\frac{1}{2}R$ each to C_V. It might be supposed that the same should be true for those degrees of freedom describing interatomic distances. Here, however, a new factor enters. The atoms of a molecule are held together by the attractive forces of chemical binding. They execute vibrations, and their interatomic distances are therefore some periodic function of time. Thus in the case of the diatomic molecule of Fig. 4-11, dr/dt is also a function of time. It is zero at the end of a vibrational stretch or compression and a maximum in between; consequently dr/dt is insufficient to specify the total energy of the vibration, and one needs to state the potential energy as well as the kinetic energy for each value of r.

The conclusion from this analysis is that each vibrational degree of freedom should contribute twice the usual amount to the heat capacity, or that the total heat capacity associated with vibrations should be Rf_{vib}, where f_{vib} denotes the number of vibrational degrees of freedom. This same conclusion is obtained in Section 4-14 as a result of a quantum statistical mechanical treatment.

In summary, we have the following results.

Atoms per molecule	n
Total number of degrees of freedom per molecule	$f = 3n$
Translational degrees of freedom per molecule	$f_{\text{trans}} = 3$
Rotational degrees of freedom per molecule	f_{rot}
Linear molecule	$f_{\text{rot}} = 2$
Nonlinear molecule	$f_{\text{rot}} = 3$

Vibrational degrees of freedom per molecule

Linear molecule

$$f_{\text{vib}}$$
$$f_{\text{vib}} = 3n - (3 + 2)$$
$$= 3n - 5$$

Nonlinear molecule

$$f_{\text{vib}} = 3n - (3 + 3)$$
$$= 3n - 6$$

Heat capacity per mole

$$(f_{\text{trans}} + f_{\text{rot}}) (\tfrac{1}{2}R) + f_{\text{vib}}R$$

Some representative applications are given in Table 4-2.

This treatment of heat capacity is based on two principal assumptions. The first is that potential energy contributions are present only for vibrations within the molecules and not for those between molecules. The gas is therefore assumed to be ideal. The second assumption is that each degree of freedom has associated with it a heat capacity contribution of $\tfrac{1}{2}R$ (or R, if a vibration). The principle which states this assumption is known as the *principle of the equipartition of energy*. It is a principle based on classical mechanics and, as will be seen in Sections 4-13 and 4-14, it can be in serious error. A glance at Fig. 4-9 shows that, at best, the equipartition values of heat capacities are only approached at high temperatures. A closer look suggests that for molecules around room temperature the equipartition contributions from translation and rotation have been reached and that the main discrepancy lies with the vibrational contribution. An important function of the statistical mechanical treatments that follow is a detailed explanation of how and why heat capacities vary with temperature.

4-10   Statistical mechanics

It was stated in the Introduction to this chapter that our practice will be to follow a given area of classical or phenomenological thermodynamics with the corresponding statistical mechanical treatment. The two approaches not only are separately very useful but also complement each other in a way that

Molecule	f	f_{trans}	f_{rot}	f_{vib}	Equipartition heat capacity per mole in units of R	
					C_V	C_P
Ar	3	3	0	0	3/2	5/2
N_2	6	3	2	1	7/2	9/2
CO_2 (linear)	9	3	2	4	13/2	15/2
H_2O (nonlinear)	9	3	3	3	12/2	14/2
CH_4	15	3	3	9	24/2	26/2
Monatomic crystalline solid	3	0	0	3	6/2	~6/2
MX type of crystalline solid	6	0	0	6	12/2	~12/2

TABLE 4-2
Applications of the equiparitition principle

provides increased understanding of both. The full, rigorous discipline of statistical thermodynamics will not be needed for the applications we will make, fortunately, but some description of the general foundation of statistical mechanics is desirable.

We consider a system containing, say, N molecules. N may be a large number, but not necessarily. The volume is fixed and so is N, but the system is in contact with a large heat reservoir at some temperature T. Due to fluctuations the energy of the system may change from time to time, but we identify the time average energy as a thermodynamic property. The problem of tracing such variations over time is too formidable to consider, and we take an alternative approach. We imagine a very large number $\mathcal{N}$ of identical systems, all in thermal equilibrium. This collection is called a *canonical ensemble*, and it is a fundamental postulate of statistical mechanics that the average properties of the set of systems constituting the ensemble are the same as the time average properties of any one of them, in the limit of $\mathcal{N} \rightarrow \infty$.*

We can obtain the form of the Boltzmann equation from a consideration of the properties of a canonical ensemble. There will be a probability p_i of finding a system in the ensemble that has energy E_i; p_i will in general be some function of E_i, $p_i = g(E_i)$. Further, the probability p_j of finding a system with energy E_j will be $p_j = g(E_j)$. Taking two systems together, the probability of one having E_i and the other having E_j must be proportional to $p_i p_j$ or equal to some constant times the product $g(E_i)g(E_j)$. However, the two systems taken together have energy $(E_i + E_j)$, and we must also be able to write $p_{ij} = g(E_i + E_j)$. We now have the situation that

$$g(E_i)g(E_j) = Ag(E_i + E_j) \tag{4-65}$$

where A is a proportionality constant. The function that has the required behavior is of the form

$$g(E) = Ae^{-\beta E} \tag{4-66}$$

where β is a constant and the negative sign is required to avoid the catastrophe of infinite energy having infinite probability. Equation (4-66) is the same as Eq. (2-20), and it will be recalled that comparison of a consequence of Eq. (2-20) with the ideal gas law identified the quantity β as $1/kT$.

To return to the canonical ensemble, the situation becomes very simple for the case to be considered, namely that of the ideal gas. In this case, each molecule is essentially independent of the others and is thus its own system. The ensemble thus consists of N, usually N_A, molecules in volume V and in thermal equilibrium with surroundings at temperature T.

---

*There are two other types of ensembles. In the *microcanonical ensemble* each system has the same energy rather than the same temperature; and in the *grand canonical ensemble*, the systems, while at the same temperature, may exchange matter with each other so that composition may fluctuate. We will not deal specifically with either in this text.

4-11   Statistical mechanical treatment of first law quantities

As noted in the preceding section, it is sufficient for the present to regard our ensemble as being made up of non-interacting molecules, each of which has various translational, rotational, and vibrational energy states. According to the Boltzmann principle, the probability of a molecule being in some energy state ϵ_i is proportional to $e^{-\epsilon_i/kT}$. As was noted in Section 2-2, the sum over all energy states is given the special name of *partition function*, $\mathbf{Q}$:

A. The partition function. The average energy per molecule

$$\mathbf{Q} = \sum_i g_i e^{-\epsilon_i/kT} \tag{4-67}$$

The factor g_i allows for the possibility that the ith energy level may occur more than once, called *degeneracy*; that is, if a molecule can have an energy ϵ_i in more than one way, that energy level is said to be *degenerate*. (The use of g could be avoided by listing an energy level as many times as it occurs.)

We can now write an expression for the average energy of a molecule, following the usual averaging procedure, given by Eq. (2-48)

$$\bar{\epsilon} = \frac{\sum \epsilon_i g_i e^{-\epsilon_i/kT}}{\sum g_i e^{-\epsilon_i/kT}} \tag{4-68}$$

As noted in the preceding section, it is a fundamental postulate of statistical mechanics that the average energy $\bar{\epsilon}$ which a particular molecule should have is also the average energy per molecule, E, in a large collection of molecules. Now, from Eq. (4-67), $d\mathbf{Q}/dT = \sum_i(\epsilon_i/kT^2)g_i e^{-\epsilon_i/kT}$; consequently Eq. (4-68) can be written in the form

$$E = kT^2\frac{d\mathbf{Q}/dT}{\mathbf{Q}} = kT^2\frac{d(\ln \mathbf{Q})}{dT} \tag{4-69}$$

or, per mole,

$$E = RT^2\frac{d(\ln \mathbf{Q})}{dT} \tag{4-70}$$

Equation (4-70) may be written in the form

$$E = -R\left[\frac{\partial(\ln \mathbf{Q})}{\partial(1/T)}\right]_V \tag{4-71}$$

since $(-1/T^2)dT = d(1/T)$. Constancy of volume, implicit in the derivation, is now stated explicitly. Since $C_V = (\partial E/\partial T)_V = -T^{-2}[\partial E/\partial(1/T)]_V$, we obtain from Eq. (4-71)

$$C_V = \frac{R}{T^2}\left[\frac{\partial^2(\ln \mathbf{Q})}{\partial(1/T)^2}\right]_V \tag{4-72}$$

Enthalpy may also be expressed in terms of the partition function,

$$H = E + Pv = kT^2\left[\frac{\partial(\ln \mathbf{Q})}{\partial T}\right]_V + kTv\left[\frac{\partial(\ln \mathbf{Q})}{\partial v}\right]_T \qquad H = N_A\mathrm{H} \tag{4-73}$$

The derivation is best made using the second law of thermodynamics, however, and is therefore postponed until Section 6-9.

B. Separation of the partition function into translational, rotational, and vibrational parts

The energy ε_i of the ith state of a molecule will in general consist of the sum of the translational, rotational, and vibrational energies. In addition, if the state possesses electronic excitation, then this is added, too. As is discussed in the material immediately following, each type of energy is quantized. A molecule thus might be in the nth translational energy level, in the tenth rotational level, and in the second vibrational level, as a specific example. To a good first approximation, however, the energy of, say, the tenth rotational state is independent of what translational or vibrational state the molecule is in. If this approximation is made, a very convenient factoring of the partition function becomes possible.

Suppose that just two kinds of energy states are possible, denoted by ϵ_i' and ϵ_i''; these might, for example, be translational and rotational states. The molecule may then be in state ϵ_1', ϵ_2', ϵ_3', ... with respect to the first kind of energy and at the same time in state ϵ_1'', ϵ_2'', ϵ_3'', ... with respect to the second kind of energy. Possible total energy states would then be $(\epsilon_1' + \epsilon_1'')$, $(\epsilon_1' + \epsilon_2'')$, $(\epsilon_1' + \epsilon_3'')$,... , $(\epsilon_2' + \epsilon_1'')$, $(\epsilon_2' + \epsilon_2'')$, $(\epsilon_2' + \epsilon_3'')$,... , $(\epsilon_3' + \epsilon_1'')$,... . The corresponding partition function would then be

$$Q = e^{-(\epsilon_1' + \epsilon_1'')/kT} + e^{-(\epsilon_1' + \epsilon_2'')/kT} + \cdots + e^{-(\epsilon_2' + \epsilon_1'')/kT} + \cdots$$

(We may avoid using the weighting factors g_i simply by listing each level as many times as it is degenerate.) This equation can be written as

$$Q = (e^{-\epsilon_1'/kT} e^{-\epsilon_1''/kT}) + (e^{-\epsilon_1'/kT} e^{-\epsilon_3''/kT}) + \cdots + (e^{-\epsilon_2'/kT} e^{-\epsilon_1''/kT}) + \cdots$$

It is now possible to group terms:

$$Q = e^{-\epsilon_1'/kT}(e^{-\epsilon_1''/kT} + e^{-\epsilon_2''/kT} + \cdots) + e^{-\epsilon_2'/kT}(e^{-\epsilon_1''/kT} + e^{-\epsilon_2''/kT} + \cdots)$$
$$+ e^{-\epsilon_3'/kT}(e^{-\epsilon_1''/kT} + e^{-\epsilon_2''/kT} + \cdots) + \cdots$$

If we make the stated assumption, namely that a given ϵ'' state has an energy value that is independent of which ϵ' state is involved and vice versa, then the sums in parentheses are identical, and the whole expression reduces to the produce of two sums. Thus

$$Q = \left(\sum_i e^{-\epsilon_i'/kT}\right)\left(\sum_i e^{-\epsilon_i''/kT}\right)$$

or, returning to the use of g_i weighting factors,

$$Q = \left(\sum_i g_i' e^{-\epsilon_i'/kT}\right)\left(\sum_i g_i'' e^{-\epsilon_i''/kT}\right)$$

or

$$Q = Q'Q''.$$

The assumption of the mutual independence of energy states therefore allows the partition function to be expressed as a product of the separate partition functions for each kind of state. We can write

$$Q = Q_{\text{trans}}Q_{\text{rot}}Q_{\text{vib}}Q_{\text{elec}} \tag{4-74}$$

We can now write

$$\left[\frac{\partial(\ln \mathbf{Q})}{\partial\,(1/T)} \right]_V = \left[\frac{\partial(\ln \mathbf{Q}_{\text{trans}})}{\partial(1/T)} \right]_V + \frac{d(\ln \mathbf{Q}_{\text{rot}})}{d(1/T)} + \frac{d(\ln \mathbf{Q}_{\text{vib}})}{d(1/T)} + \frac{d(\ln \mathbf{Q}_{\text{elec}})}{d(1/T)} \quad (4\text{-}75)$$

In our case of non-interacting molecules, only translational energy levels depend on V, and the partial differentiation signs can be dropped for the other terms. It follows from Eq. (4-75) and Eq. (4-71) that

$$E = E_{\text{trans}} + E_{\text{rot}} + E_{\text{vib}} + E_{\text{elec}} \quad (4\text{-}76)$$

The approximation thus leads to the reasonable conclusion that the total average energy can be regarded as the sum of the separate average energies of translation, rotation, etc. It further follows from Eq. (4-72) that C_V may also be written as a sum of corresponding contributions:

$$C_V = C_{V(\text{trans})} + C_{V(\text{rot})} + C_{V(\text{vib})} + C_{V(\text{elec})} \quad (4\text{-}77)$$

The same is true for H and for C_P.

We will neglect partition functions due to a sum over electronic excited states because the energy of such states is usually so large that they are not populated at ordinary temperatures. We do this by setting $\mathbf{Q}_{\text{elec}}$ equal to unity, on the following basis. If the first and higher electronic excited states are large in energy compared to kT, then all but the first term in $\mathbf{Q}_{\text{elec}}$ will be negligible, and

$$\mathbf{Q}_{\text{elec}} = e^{-\epsilon_0/kT}$$

where ϵ_0 is the energy of the first electronic state. In general, we do not know absolute energy values, only relative ones, and we can set $\epsilon_0 = 0$ as our reference point. However, even if ϵ_0 were assigned some nonzero value, the consequence would simply be to add this value to E. Thus, referring to Eq. (4-69), we obtain

$$E_{\text{elec}} = (kT^2)\frac{\epsilon_0}{kT^2} = \epsilon_0$$

Our neglect of $\mathbf{Q}_{\text{elec}}$ then amounts, at the most, to regarding the absolute energy of the first electronic state as making a constant contribution to E. We ordinarily do set $\epsilon_0 = 0$ for electronic energy, but in the case of vibrational energy, we will use a nonzero value for energy of the first state (see Section 4-14).

4-12   Translational partition function for an ideal gas

A result from the wave mechanical treatment of a particle in a box, which will not be derived until Section 16-5B, gives the sequence of energy states that corresponds to a particle which is free of any interaction with other particles. In this respect, the picture is analogous to that used in Section 2-3 for the simple kinetic theory derivation for an ideal gas. The wave mechanical result is

$$\epsilon_n = \frac{n^2 h^2}{8a^2 m} \quad (4\text{-}78)$$

where n is a quantum number taking on all integral values, h is Planck's constant, m is the mass of the particle, and a is the length of the box. Equation (4-78) is actually for a one-dimensional box, that is, translation along only one direction in space is considered. It allows us to obtain the partition function for translation in one dimension, but, in view of the preceding analysis, it is evident that

$$Q_{\text{trans(3-dim)}} = Q_{\text{trans}(x)}Q_{\text{trans}(y)}Q_{\text{trans}(z)} = Q^3_{\text{trans(1-dim)}} \tag{4-79}$$

The desired partition function for three-dimensional translation will therefore be just the cube of that for the one-dimensional case.

We obtain the one-dimensional Q from the summation

$$Q_{\text{trans(1-dim)}} = \sum_{n-1}^{n=\infty} \exp\left(-\frac{n^2h^2}{8a^2mkT}\right) \tag{4-80}$$

In this case, an excellent approximation results if integration is substituted for the summation. The reason is that the quantity $h^2/8a^2m$ is a very small one compared to any ordinary value of kT.

EXAMPLE   Suppose that the dimension a is 1 cm and that we are dealing with a molecule of molecular weight 30 g mole^{-1}. The quantity $h^2/8a^2m$ is then $(6.63 \times 10^{-27})^2(6.02 \times 10^{23})/(8)(1^2)(30) = 1.1 \times 10^{-31}$ erg, as compared to the value of kT at 25°C of $(1.38 \times 10^{-16})(298) = 4.1 \times 10^{-14}$ erg molecule^{-1}. The ratio is about 3×10^{-18}!

As a consequence, the exponential terms in Eq. (4-80) will not begin to diminish from unity until extremely large n values, and we are dealing almost with a continuum of energy states. In integral form, Eq. (4-80) becomes

$$Q_{\text{trans(1-dim)}} = \int_0^\infty \exp\left(-\frac{n^2h^2}{8a^2mkT}\right)\, dn \tag{4-81}$$

which is of the standard form $\int_0^\infty \exp(-b^2x^2)dx$. The result is

$$Q_{\text{trans(1-dim)}} = \frac{2\pi mkT)^{1/2}}{h}a \tag{4-82}$$

Finally, we have

$$Q_{\text{trans(3-dim)}} = \frac{(2\pi mkT)^{3/2}}{h^3}v \tag{4-83}$$

where the volume v replaces a^3.

Equation (4-83) gives the translational partition function per molecule, and for the present purposes it may be abbreviated as

$$Q_{\text{trans}} = \alpha T^{3/2} \tag{4-84}$$

Insertion of this result into Eq. (4-69) gives

$$E_{\text{trans}} = kT^2\left\{\frac{\partial[\ln(\alpha T^{3/2})]}{\partial T}\right\}_v$$

or

$$E_{trans} = \tfrac{3}{2}kT$$

and

$$E_{trans} = \tfrac{3}{2}RT \qquad (4\text{-}85)$$

This is the same result as obtained from the kinetic molecular theory of an ideal gas. Also, $C_{V(trans)}$ is $\tfrac{3}{2}R$, which is just the equipartition value.

Further, $[\partial(\ln Q_{trans})/\partial v]_T$ is just $1/v$, so by Eq. (4-73)

$$H_{trans} = E_{trans} + (kTv)\frac{1}{v} = E + kT$$

or

$$H_{trans} = E_{trans} + RT$$

This is the result expected for an ideal gas, from Eq. (4-27) and the ideal gas law.

It may seem that all of the labor in dealing with the translational partition function has been rather useless in that it has produced a simple, already known result. We have, however, brought together classical and statistical thermodynamics and, moreover, Eq. (4-83) will be used in Section 6-9 to obtain a very important new equation.

An important point is that Q_{trans} depends on the volume v per molecule and thus on the macroscopic condition of the gas. As will be seen in the following sections, the rotational and vibrational partition functions do not have this dependence. They are computed purely on the basis of the spacing of energy states of the individual molecule and do not involve v. Often, then, Q_{rot} and Q_{vib} are grouped separately as giving the *internal* or intramolecular partition function, $Q_{int} = Q_{rot}Q_{vib}$, so that $Q = Q_{trans}Q_{int}$.

4-13   The rotational partition function

The rotational energy states for a diatomic molecule are given by wave mechanics as

$$\epsilon_{rot} = J(J + 1)\frac{h^2}{8\pi^2 I} \qquad (4\text{-}86)$$

where J is the rotational quantum number, which may have the integral values $1, 2, 2, \dots$, and I is the moment of inertia, given by

$$I = \mu r^2 \qquad \mu = \frac{m_1 m_2}{m_1 + m_2} \qquad (4\text{-}87)$$

where μ is the reduced mass [Eq. (2-72)] and r is the interatomic distance (Section 16-ST-2). The partition function for rotation is

$$Q_{rot} = \sum_{J=O}^{\infty} (2J + 1)\exp\left[-J(J + 1)\frac{h^2}{8\pi^2 I kT} \right] \qquad (4\text{-}88)$$

where the factor $(2J + 1)$ is the degeneracy g_J of each J state. The degeneracy for $J > 1$ arises from the number of ways the total angular momentum of a molecule can be quantized with respect to spatial orientation.

Equation (4-88) is for the special case of an A–B diatomic molecule, such as CO. The more general expression for **Q** is given in Section 4-CN-2.

The summation called for by Eq. (4-88) cannot be made in closed form but, analogous to the case with translational energy states, the rotational energy interval is often small enough compared to kT that no great error is made if the summation is replaced by an integration.

EXAMPLE   Consider the case of $H^{35}Cl$ at 25°C. The reduced mass is $35 \times 1/36 = 0.972$ g mol^{-1} and the bond length is 1.28 Å, giving a moment of inertia of $(0.972)(1.28 \times 10^{-8})^2 / (6.022 \times 10^{23}) = 2.64 \times 10^{-40}$ g cm^2.

The quantity $h^2/8\pi^2 I$ is then $(6.63 \times 10^{-27})^2/(8)(3.14)^2(2.64 \times 10^{-40}) = 2.11 \times 10^{-15}$ erg. Recalling from the previous example that kT at 25°C is 4.1×10^{-14} erg molecule^{-1}, we have a ratio of 0.051 so that rotational energy level spacings are about 1/20th of kT. In this particular case, the summation of Eq. (4-88) can safely be replaced by an integral.

The above approximation will not always be an acceptable one, however. Note Problem 4-40.

Equation (4-88) becomes

$$\mathbf{Q}_{rot} = \int_0^\infty (2J + 1) \exp\left[-J(J + 1)\frac{h^2}{8\pi^2 IkT}\right]dJ \tag{4-89}$$

The quantity $h^2/8\pi^2 Ik$ has the dimension of temperature and is called the *characteristic rotation temperature* θ_{rot}. If we let $y = J(J + 1)$, the integral reduces to a standard one, and the result is simply

$$\mathbf{Q}_{rot} = \frac{T}{\theta_{rot}} \tag{4-90}$$

An expression for the general case of a polyatomic molecule is given in the Commentary and Notes section, but it should be mentioned here that for A–A-type molecules $\mathbf{Q}_{rot}$ is reduced by a symmetry factor of one-half from the value for A–B-type molecules given by Eq. (4-90).

We can now calculate some thermodynamic quantities. Since $d(\ln \mathbf{Q})/dT$ is just $1/T$, we obtain from Eq. (4-69) that

$$\mathrm{E}_{rot} = kT \qquad \text{or} \qquad E_{rot} = RT \tag{4-91}$$

The heat capacity $C_{V(rot)}$ is then R. Thus as in the case of translational motion, the result is the same as that obtained from the equipartition principle. There is no dependence of $\mathbf{Q}_{rot}$ on volume, and therefore $H_{rot} = E_{rot}$.

That the equipartition values are obtained for E and for C_V is evidently a consequence of using the integration approximation, that is, the limiting case in which θ_{rot} is much smaller than T. At the other extreme, where $\theta_{rot} \gg T$, only the first rotation state will be populated and the situation will be similar to that considered for electronic excited states. Then E_{rot} becomes just the absolute energy of this first rotational state, taken to be zero; in any event, being constant, $E_{0,rot}$ now makes no contribution to the heat capacity. Thus the quantum mechanical treatment predicts that at a sufficiently low temperature the rotational contribution to the heat capacity will disappear. This effect is indeed observed and is illustrated in Fig. 4-9 in the case of H_2. Below about 200 K the heat capacity begins to drop increasingly rapidly from $\frac{7}{2}R$ for C_P to $\frac{5}{2}R$, the value for translation only. By about 50 K this limit is essentially reached. The quantum treatment thus shows that the equipartition value for rotational heat capacity is a maximum, limiting one and that the actual value may be significantly lower. In the case of translation, however, the energy level spacings are so very close together that deviation from the equipartition value for translational heat capacity is never observed for dilute or ideally behaving gases.

4-14 The vibrational partition function

As in the preceding two sections, the wave mechanical formulation of the spacing between states is merely presented for use at this point (see Section 16-5). In the case of vibration the first-order treatment for a diatomic molecule is based on the assumption of a harmonic oscillator. The result is

$$\epsilon_{\text{vib}} = (v + \tfrac{1}{2})h\nu_0 \tag{4-92}$$

where v is the vibrational quantum number, having values 0, 1, 2,... , and ν_0 is the characteristic frequency, related to the reduced mass μ and the force constant k by

$$\nu_0 = \frac{1}{2\pi}\left(\frac{k}{\mu}\right)^{1/2} \tag{4-93}$$

In the harmonic oscillator, the restoring force on a particle displaced by distance x is simply $k\,x$, where x is the displacement from the minimum potential energy position. Although more elaborate treatments are available (see Commentary and Notes), the one given here is adequate for the first few vibrational energy states.

The vibrational partition function is then

$$\mathbf{Q}_{\text{vib}} = \sum_{\nu} e^{-\epsilon_{\text{vib}}/kT} = e^{-h\nu_0/2kT} + e^{-3h\nu_0/2kT} + \cdots \tag{4-94}$$

This particular series can be summed as follows. Let $x = \exp(-h\nu_0/2kT)$; Eq. (4-94) reduces to a series whose sum is given by the binomial theorem:

$$\mathbf{Q}_{\text{vib}} = x(1 + x^2 + x^4 + \cdots) = \frac{x}{(1 - x^2)}$$

or

$$\mathbf{Q}_{\text{vib}} = \frac{\exp(-h\nu_0/2kT)}{1 - \exp(-h\nu_0/kT)} = \frac{\exp(-\theta_{\text{vib}}/2T)}{1 - \exp(-\theta_{\text{vib}}/T)} \tag{4-95}$$

where, as with rotational states, we define a characteristic vibration temperature as $\theta_{\text{vib}} = h\nu_0/k$. Differentiation gives

$$\frac{d(\ln \mathbf{Q}_{\text{vib}})}{dT} = \frac{\theta_{\text{vib}}}{2T^2} + \frac{\theta_{\text{vib}}}{T^2}\frac{\exp(-\theta_{\text{vib}}/T)}{1 - \exp(-\theta_{\text{vib}}/T)}$$

and, by Eq. (4-69),

$$\text{E} = \tfrac{1}{2}h\nu_0 + kT\frac{\theta_{\text{vib}}/T}{\exp(\theta_{\text{vib}}/T) - 1} \tag{4-96}$$

and

$$E = \tfrac{1}{2}N_A h\nu_0 + RT\, \frac{\theta_{vib}/T}{\exp(\theta_{vib}/T) - 1} \tag{4-97}$$

In the limit of very low T the second term of Eq. (4-97) drops out and $E = \tfrac{1}{2}N_0 h\nu_0$. This energy of the lowest vibrational state is called the zero-point energy. At the other extreme, where $\theta_{vib} \ll T$, Eq. (4-97) reduces to

$$E = \tfrac{1}{2}N_A h\nu_0 + RT \tag{4-98}$$

In this limit the heat capacity contribution is thus

$$C_{V(vib)} = R \tag{4-99}$$

which is the equipartition value for one degree of vibrational freedom. Thus the equipartition contribution of R per degree of freedom to vibrational heat capacity arises naturally from the mathematical treatment. As with $\mathbf{Q}_{rot}$ there is no dependence of $\mathbf{Q}_{vib}$ on volume, so $H_{vib} = E_{vib}$.

The general expression for C_V is obtained by differentiation of Eq. (4-97). This gives

$$C_{V(vib)} = R \left(\frac{\theta_{vib}}{T}\right)^2 \frac{\exp(\theta_{vib}/T)}{[\exp(\theta_{vib}/T) - 1]^2} \tag{4-100}$$

The expression reduces to zero for $\theta_{vib} \gg T$ and to R for $\theta_{vib} \ll T$. It is fortunate that it is possible to sum Eq. (4-93) explicitly since the usual situation is one lying between these extremes.

Force constants [f in Eq. (4-93)] are about 10^5 dyn cm^{-1}, and for a molecule like HCl, μ is about unity per mole, so ν_0 is about 10^{14} sec^{-1} and $h\nu_0/k$ is about 5×10^3 K. Thus θ_{vib} may correspond to several thousand degrees Kelvin. For HCl, $\nu_0 = 8.98 \times 10^{13}$ sec^{-1} so that $\theta_{vib} = 4.308 \times 10^3$ K. The energy of the first vibrational transition, from $v = 0$ to $v = 1$, which is just $h\nu_0$, is thus $4308/298$ or $14.5kT$ at 25°C. Thus the relative probability of even the first excited vibrational state being populated is only $e^{-14.5}$, or about 10^{-6}, at 25°C. Very little energy is present in the form of vibrational excitation, and the contribution to heat capacity is therefore quite small. Equation (4-100) is plotted in Fig. 4-12 as $C_{V(vib)}$ versus T/θ_{vib}, and the result confirms that vibrations contribute very little to the heat capacity of HCl around room temperature. On the other hand, for a more weakly bonded, heavy-atom molecule such as I_2, ν_0 is only $6.16 \times 10^{12} 5^{-1}$, or less than one-tenth that of HCl. Now $\theta_{vib} = 307$ K and from Fig. 4-12 the vibrational heat capacity contribution for I_2 approaches at the equipartition value at room temperature.

The varying extent to which vibrations contribute to heat capacity is also illustrated in Fig. 4-8. Although there is only one vibrational degree of freedom for a diatomic molecule, the number increases rapidly with molecular complexity. For a polyatomic molecule, the total vibrational partition function will be factorable, in the manner of Eq. (4-74), into a product of the separate ones for all of the independent vibrational modes:

$$\mathbf{Q}_{tot(vib)} = \prod_i \mathbf{Q}_{vib,i} \tag{4-101}$$

Such independent modes are called *normal modes* (see Section 19-5). The quantitative result is that a simple function such as that plotted in Fig. 4-12 is no

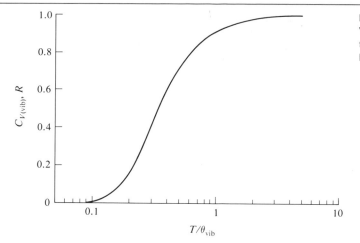

FIGURE 4-12
Variation of $C_{V(vib)}$ with temperature according to Eq. (4-100).

longer applicable, but, qualitatively, the result is much the same as for diatomic molecules. A reasonable rule of thumb for polyatomic molecules is that about 20% of the equipartition vibrational heat capacity will be developed at room temperature (and all of the equipartition translational and rotational contributions).

EXAMPLE

Estimate the molar heat capacity C_P of propane, C_3H_8, at 25°C. On the basis of the rule just given, the translational contribution is $\frac{3}{2}R$, and the rotational contribution is likewise $\frac{3}{2}R$ (the molecule is nonlinear). This leaves $(3 \times 11 - 6)$ or 27 vibrational degrees of freedom, for which the equipartition contribution to the heat capacity would be $27R$, of which only 20% should be present. We thus assemble C_P as $R + \frac{3}{2}R + \frac{3}{2}R + (0.2)(27)R = 9.4R$ or 18.7 cal K^{-1} mol^{-1}. The experimental value is 19.7 cal K^{-1} mol^{-1}.

COMMENTARY AND NOTES

4-CN-1 Count Rumford—The story of a strange man

There is not room in a text such as this one to delve very extensively into scientific history, but we will present some from time to time. The following article by E.C. Krauss is relevant to the history of the mechanical equivalent of heat; it also makes fascinating reading!

Scientist and Scoundrel, and Good in Both Lines*
E. C. Krauss

A bit of little-known history about the U.S. Military Academy has recently come to light

*Copyright, 1955, Los Angeles Times. Reprinted by permission.

after being buried in old records for generations. The story is that West Point had a narrow escape: it was almost established under the care of a man who had been both one of the world's most brilliant scientists and a scoundrel without a shred of character.

This curious individual, who might be called attractive in a revolting sort of way, was Benjamin Thompson; born at Woburn, Mass.; Sir Benjamin by action of His Majesty, King George III, and Count Rumford, a noble of the Holy Roman Empire.

When he was at the British court in the late 1790's, he proposed to Rufus King, our Minister, that he start an American military academy similar to one he had established for the Grand Duke of Bavaria. King knew him only as a man of apparently brilliant attainments and recommended him to President John Adams, who was on the point of naming him—would have named him but for a dispute over rank, Thompson asking to be made a full general, equal to Washington.

At this point, British officials who knew about the seamy side of Thompson's career gave King an earful about him that made the appointment impossible. Minister King gave Thompson a choice: he would tell the world why Thompson was not acceptable, and thus bring all the scandal to light; or Thompson would be publicly offered the appointment but must refuse it. The latter course, which saved face both for the United States and for Thompson, was agreed upon. And West Point started (in 1801) under somebody else.

What was the scandal? This being a family newspaper, not all of it can be told. However, Thompson started early: in Portsmouth, N. H., where he was teaching school, he won the heart of a rich widow of 32 and set up, at 19, as the local squire and a major in the militia. This was 1774 and Thompson was run out of town after being accused of being a spy for Gen. Gage, British commander in Boston. He was, but the patriots couldn't prove it at the time.

In Boston, he tried for a year to get a commission in the Colonial Army, but the New Hampshire patriots stopped that; so he went over openly to the British side. Finally Gen. Gage sent him to London as a member of a committee to explain to Lord George Germain, Colonial Secretary, the evacuation of Boston. He became a favorite of Lord George at once and was made undersecretary for Carolina and Georgia. In 1779 he took a cruise with the British Navy and in 1781 made a deal with a man named Lutterloh and a Frenchman named LaMotte to sell British naval secrets to the French.

The plot was detected; Lutterloh confessed and turned King's evidence; but Lord George would not permit the prosecution, or even the exposure of Thompson. LaMotte was convicted, hanged, drawn and quartered, and Thompson was sent to America as a lieutenant colonel of Dragoons. But the war was about over.

Thompson—he was Sir Benjamin by this time—then went to Bavaria where his great talents won him quick favor. He became Minister of War, Minister of the Interior and Royal Scientist; and he was also a spy for the British. However, at the Bavarian court he carried on further scientific experiments. At London, he had devised apparatus for measuring the force of gunpowder, and practically invented the science of ballistics. In Bavaria he disproved the "caloric theory" of heat, devised a new method of boring cannon, invented the cookstove (all cooking had been done on open fires) and reinvented the central heating system which the ancient Cretans had invented millenniums before. He invented the drip coffee pot and was the inventor of baking powder.

One of his achievements in the realm of statesmanship was to arrest the 2600 beggars of Munich and put them to work making clothing for the army.

Meanwhile he was grafting and stealing with both hands, as he had been in London; it was estimated that he stole the equivalent of much more than $1,000,000 during his public career—an incredible fortune for that day.

Thompson broke with the British spy system in 1785 and in 1791 was made a count. He went back to London in 1795 in the guise of a distinguished scientist, to read papers before the Royal Society and remained a year. On his return to Munich he was made Prime Minister; but he made so many enemies that he had to flee to London two years later, pausing only to appoint himself Bavarian Minister to the Court of St. James's.

The British court refused to accept him as such, giving the excuse that the King could not well receive as a foreign envoy one of his own subjects.

When the West Point deal fell through, Thompson, or Count Rumford, drifted to Paris where he became very chummy with Napoleon, who admired his artillery studies. And if the Americans did not want him as a military instructor, Napoleon did and the count established the famous school of St. Cyr, the French West Point.

At 60, he married the widow of the great French chemist Antoine Lavoisier—he who was guillotined when the court ruled that "the republic has no need of scientists." Mme. Lavoisier had a handsome fortune. But she cared for society and Count Rumford didn't; at their divorce he took half her inheritance.

His science kept equal pace with his scoundrelism; he invented the photometer for measuring light, first used the term "candle power" as a standard, and also invented the steam radiator.

The count lived in easy circumstances till 1814 and remembered his birthplace by endowing a professorship at Harvard—which is still in force.

Here was surely one of the strangest mixtures of good and evil the world ever saw.

4-CN-2 Additional aspects of statistical mechanical treatments

It is correctly suggested by the statistical mechanical treatment that it is the partition function for translation that determines ideal gas behavior. It gives the same value of E as from simple kinetic molecular theory, the same relationship between H and E, and leads to the ideal gas law. A gas, however, may be ideal and yet exhibit complex heat capacity behavior owing to the quantization of rotational and vibrational energy levels. Thus, by neglecting intermolecular forces, we make the gas ideal but are still free to consider in detail the structure of individual molecules and the intramolecular forces present. Alternatively put, in advanced kinetic molecular theory the emphasis is on treating intermolecular forces, including suitable modifications to Q_{trans}, while in this chapter we have been more interested in Q_{int}.

The treatment presented of Q_{int} was introductory, and a few of the complications might be mentioned here. First, Q_{rot} was derived only for the case of an A–B-type molecule. The general formula is

$$Q_{rot} = \frac{1}{\pi\sigma} \left(\frac{8\pi^3 IkT}{h^2} \right)^{n/2} \qquad (4\text{-}102)$$

where n denotes the number of independent rotation axes (two for a linear molecule* and three otherwise) and σ is a symmetry number. This last is the number of indistinguishable po-

sitions into which the molecule can be turned by rotations.

For an A–A molecule, $\sigma = 2$, for example, while for NH_3, $\sigma = 3$, and for CH_4, $\sigma = 12$. A nonlinear molecule will in general have three independent moments of inertia, and in such a case, I is their geometric mean, $I = (I_A I_B I_C)^{1/3}$. Equation (4-87) gives the definition of I for the diatomic molecule; in the general case $I_A = \Sigma m_i d_i^2$, where d_i is the perpendicular distance of mass m_i from rotation axis A. Corresponding equations give I_B and I_C for axes B and C. The three mutually perpendicular axes pass through the center of mass and usually are oriented to conform with the molecular symmetry.

With vibrations two types of complications enter immediately. An experimental one is that in the case of polyatomic molecules there are a number of vibrational modes, and the observed absorption spectrum will show transitions between various combinations of the different vibrational excited states. Therefore it can be difficult to analyze such a spectrum in order to obtain the separate fundamental frequencies.

The second complication is that the assumption of harmonic oscillator behavior is not strictly correct. As illustrated in Fig. 4-13(a) the assumption leads to a parabolic potential energy function with equally spaced vibrational energy states. The actual situation is that shown in Fig. 4-13(b), where for a diatomic molecule the interatomic potential resembles somewhat the Lennard–Jones form used for intermolecular potentials (Fig. 1-16), although the theoretical basis for the attractive part is now one of chemical bonding rather than van der Waals forces. At any rate, the effect is to make the potential well asymmetric, or to introduce anharmonicity in the vibrational behavior. The successive states become closer and closer in energy until a definite final state is reached beyond which dissociation occurs. In the case of H_2, there are 15 such bound states before dissociation. Fortunately, at the temperatures of most heat capacity calculations only the first few vibrational levels are important, so that the anharmonicity effect is usually of little consequence.

However, with the development of the space program it has become necessary to deal in practice with energies and heat capacities of small molecules at very high temperatures. These quantities are important in determining

---

*One might ask why, in the case of a linear molecule, there is not a rotation axis through the nuclei. In principle there is, but since most of the mass is that of the nuclei, which are very small in size, the moment of inertia for this axis is also very small. We see from Eq. (4-86) that the energy spacing of the corresponding rotational states will be very large—so large that their contribution to the partition function can be neglected. Note Exercise 4-18.

FIGURE 4-13
Potential wells for (a) a
harmonic and (b) an
anharmonic oscillator.

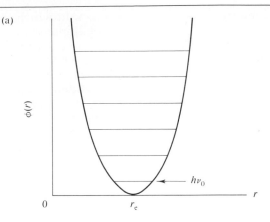

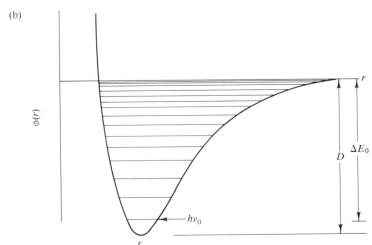

the practical thrust developed by a given fuel. Here, anharmonicity may seriously complicate a theoretical calculation of the vibrational contributions. (In addition, Q_{elec} may have to be considered.) We can also understand why gases are not likely ever to reach vibrational equipartition. For them to do so the temperature must be large compared to θ_{vib}, which means that the average molecule is in a high vibrational state. Since there are only a limited number of states before the dissociation limit, an appreciable fraction of the molecules will be dissociated and the experimental heat capacity will no longer be that of the original molecule only, but rather that of a mixture of it and its dissociation products.

An important exception to this situation oc-

curs with some liquids and especially with atomic crystals such as metals. This is discussed in more detail in the Special Topics section, but, briefly, one may find heat capacities closely approaching the vibrational equipartition limit. The qualitative explanation is that the vibrations are now between atoms that are held in a lattice and thus cannot dissociate.

A final point to mention is that the assumption of the independence of energy states which led to Eq. (4-74) is not fully valid. The approximation is very good with respect to Q_{trans}, but there is some interdependence of vibrational and rotational energy levels. For example, a molecule in an excited vibrational state will have a somewhat different moment of inertia from one in the ground or $v = 0$ state.

SPECIAL TOPICS

4-ST-1 The barometric equation revisited

The simple barometric equation for an ideal gas in a gravitational field is obtained in Section 2-3 as an application of the Boltzmann equation:

$$P_h = P_0 e^{-Mgh/RT} \qquad (2\text{-}28)$$

There is an alternative derivation that makes it easy to vary the assumptions and conditions. We consider a column of fluid in a gravitational field as illustrated in Fig. 4-14. If the column is of unit cross section, the pressure (force per unit area) at level h must be just the total weight of the column above that level. The change in pressure dP between h and $h + dh$ is then just the weight of fluid contained between the two levels:

$$dP = -\rho g\, dh \qquad (4\text{-}103)$$

Equation (4-103) is general. If the fluid is a liquid that is assumed to be incompressible, then the density ρ is independent of h and integration gives

$$P_h = P_0 - \rho gh \qquad (4\text{-}104)$$

where P_0 is the pressure at $h = 0$. This is just a statement of the variation of hydrostatic pressure with height.

If, however, the fluid is taken to be an ideal gas, then use of Eq. (1-13) gives

$$dp = -\frac{PM}{RT} g\, dh \qquad (4\text{-}105)$$

Rearrangement gives

$$\frac{dP}{P} = d(\ln P) = -\left(\frac{Mg}{RT}\right) dh \qquad (4\text{-}106)$$

and, if we assume g and T to be constant, we obtain on integration

$$\ln (P/P_0 = -Mgh/RT \qquad (4\text{-}107)$$

which is the same as Eq. (2-28) above.

Other, more realistic assumptions are possible. If we are considering extreme altitudes, then g will decrease noticeably with h, and its functional dependence on h can be introduced into Eq. (4-105) *before* integration. Temperature generally decreases with increasing h, and if its functional dependence on h is known, this, too, can be inserted in Eq. (4-105). Finally, the gas need not be taken to be ideal, and some more complicated function than (PM/RT) may be used for ρ in the precursor equation, Eq. (4-104). Several problems illustrate these possibilities.

It is especially appropriate to the material of

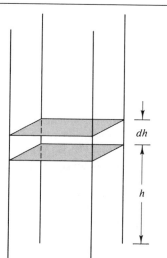

FIGURE 4-14
The barometric effect.

EXAMPLE   Consider air of average M about 29 g mol^{-1} and of C_P about $\frac{7}{2}R$. We find $dT/dh = -(29)(980)/(3.5)(8.31 \times 10^7) = -9.8 \times 10^{-5}$ K cm^{-1}. Temperature should thus decrease by about 10°C per km elevation.

---

this chapter to consider another type of constraint. A large column of air will have such a small area to volume ratio that heat transport across its boundaries will be negligible. Such a large column is, in effect, an adiabatic system. We now have, for the case of an ideal gas, the additional condition

$$d(\ln P) = (C_P/R)\, d(\ln T) \tag{4-50}$$

and combination of Eqs. (4-50) and (4-106) gives the very simple result

$$dT = -(Mg/C_P)\, dh \tag{4-108}$$

If we again assume g to be constant,

$$T = T_0 - (Mg/C_P)h \tag{4-109}$$

Equation (4-109) is known as the barometric adiabat; it predicts that temperature should decrease linearly with altitude.

The behavior predicted by the above example is fairly realistic for normal meteorological conditions. Under *inversion* conditions, however, air temperature may *increase* with altitude for a while, before resuming its normal behavior. When this happens in the Los Angeles area, for example, the effect is to put a "lid" over the city, above which smog and other pollutants cannot rise.

4-ST-2   The Joule–Thomson effect

It was mentioned in Section 4-8A that Joule's experiment of allowing free expansion of a gas does in fact lead to small temperature changes. While no mechanical work is done by the gas on the surroundings, for any real gas there is work done against the intermolecular forces of attraction. This is another manifestation of the internal pressure of a gas discussed in Section 4-7. The expansion, moreover, is essentially

adiabatic, so q is zero, and the process is therefore one in which E does not change. The work against the internal pressure thus is balanced by a temperature change in the gas, and we define the *Joule coefficient* as

$$J = \left(\frac{\partial T}{\partial V}\right)_E \tag{4-110}$$

Since by Eq. (4-12)

$$\left(\frac{\partial T}{\partial V}\right)_E = -\frac{(\partial E/\partial V)_T}{(\partial E/\partial T)_V} = -\frac{1}{C_V}\left(\frac{\partial E}{\partial V}\right)_T \tag{4-111}$$

we can write, using Eq. (4-36),

$$J = -\frac{1}{C_V}\left[T\left(\frac{\partial P}{\partial T}\right)_V - P\right] = -\frac{P_{\text{int}}}{C_V} \tag{4-112}$$

The Joule coefficient thus provides a direct measure of the internal pressure P_{int}.

The measurement of the Joule coefficient is a difficult one, and alternative experimental arrangements have been devised to obtain closely related quantities. An important such arrangement is the Joule–Thomson (Kelvin) experiment illustrated in Fig. 4-15. One allows a gas to expand through a porous plug so that entering gas may be kept at pressure P_1 and exiting gas at pressure P_2. This is shown schematically in the figure as a double piston and cylinder arrangement. The porous plug is made of thermally insulating material so that q is zero. The gas is initially at P_1 and V_1 and the final state is at P_2 and V_2. Since $q = 0$, $E = -w = P_1V_1 - P_2V_2$, so

$$E_2 - E_1 = P_1V_1 - P_2V_2 \quad \text{or} \quad H_2 = H_1$$

The experiment is therefore at constant H. The change in temperature of the gas is measured and the *Joule–Thomson coefficient* is defined as

---

EXAMPLE   Referring to the example following Eq. (4-38), P_{int} was found to be 81.5 atm for CO_2 at 100°C and 100 atm pressure, using the van der Waals equation. For CO_2, C_V may be guessed to be about 7 cal K^{-1} mol^{-1} or about 290 cm^3 atm K^{-1} mol^{-1}, whence $J = -0.28$ K mol cm^{-3}. At STP, the corresponding calculation gives a J value of about -2.6×10^{-5} K mol cm^{-3}. Thus J is rather small under ordinary conditions.

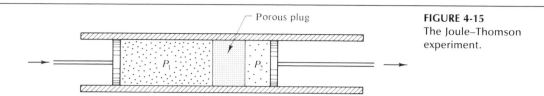

FIGURE 4-15
The Joule–Thomson
experiment.

$$\mu = \left(\frac{\partial T}{\partial P}\right)_H \qquad (4\text{-}113)$$

By the nature of the experiment ΔP is always negative, so a positive μ corresponds to cooling on expansion. Around room temperature, all gases except hydrogen, helium, and neon exhibit such cooling. At a sufficiently low temperature all gases do. The Joule–Thomson coefficient also depends on pressure, and the point at which $\mu = 0$, known as the *Joule–Thomson inversion point*, is then a function of pressure and temperature.

The Joule–Thomson coefficient may also be calculated from the equation of state of the gas. Since by Eq. (4-12)

$$\left(\frac{\partial T}{\partial P}\right)_H = -\frac{(\partial H/\partial P)_T}{(\partial H/\partial T)_P} = -\frac{1}{C_P}\left(\frac{\partial H}{\partial P}\right)_T \qquad (4\text{-}114)$$

an equation analogous to Eq. (4-36), but again derived later (Section 6-ST-1), gives

$$\left(\frac{\partial H}{\partial P}\right)_T = -T\left(\frac{\partial V}{\partial T}\right)_P + V \qquad (4\text{-}115)$$

and combination of this with Eq. (4-114) yields

$$\mu = \frac{1}{C_P}\left[T\left(\frac{\partial V}{\partial T}\right)_P - V\right] \qquad (4\text{-}116)$$

The van der Waals equation may be used for the calculation of μ and it gives moderately good agreement with experiment.

The Joule–Thomson effect is generally larger than the Joule effect. For CO_2 at STP, $\mu = 1.30$ K atm^{-1} (as compared to 0.73 K atm^{-1} calculated from the van der Waals equation), and for O_2, $\mu = 0.31$ K atm^{-1}. The Joule–Thomson effect may be produced in a cyclic operation, and machines that do this are widely used in the liquefaction of gases. In practice, the expansion may be through a throttle rather than a porous plug.

4-ST-3   The heat capacity of a solid

This subject might have been assigned to the chapter on the crystal structure of solids, but is treated here instead since it ties in very closely with the discussion of heat capacities. Before taking up the quantum mechanical aspect we first consider an extension of the equipartition principle.

In the case of a crystalline solid in which each lattice position is occupied by a single atom, there can be no translation (other than of the crystal as a whole!) and no rotation. The $3N_A$ degrees of freedom per mole are thus entirely vibrational, and one predicts an equipartition molar heat capacity of $3R$ as noted in Table 4-2. This prediction explains an old observation, known as the *law of Dulong and Petit*, to the effect that the molar heat capacity of a monatomic crystalline solid is about 6.2 cal K^{-1} mol^{-1}. This law is obeyed by a number of metals; for example, the heat capacities for Cu, Fe, and Ag are 5.9, 6.2, and 6.0 cal K^{-1} mol^{-1}, respectively, around room temperature.

The Dulong–Petit law finds useful application in metallurgy. The composition of a binary alloy can be estimated just from its heat capacity per gram, since division into 6.2 gives the average molecular weight. Table 4-2 includes an entry for an MX-type crystal, such as NaCl. The point here is that the equipartition value per mole is now doubled since there are two moles of atoms per formula weight.

The heat capacity of a copper–tin alloy is 0.0576 cal K^{-1} g^{-1}. Calculate its composition.   **EXAMPLE**
Ans. The average molecular weight is $6.2/0.0576 = 107.6$ g mol^{-1}; hence $107.6 = 63.5x_{Cu} + 118.7(1 - x_{Cu})$, or $x_{Cu} = 0.200$, where x denotes mole fraction.

While it is not appropriate here to cover the statistical mechanical treatment of solids in much detail, there is a simple and useful extension of the material of Section 4-14 to the case of a crystalline solid in which the lattice sites are occupied by atoms (or monatomic ions). If the $3N_A$ vibrations can be viewed as independent oscillators, then the total partition function is just the product of $3N_A$ individual ones, as in Eq. (4-101), or $Q_{vib}^{3N_A}$, where Q_{vib} is given by Eq. (4-95). The effect is simply to multiply the

expression for E [Eq. (4-96)] by $3N_A$, with the result that the molar heat capacity is just three times that given by Eq. (4-100):

$$C_{V(crystal)} = 3R \left(\frac{\theta_{vib}}{T}\right)^2 \frac{\exp(\theta_{vib}/T)}{[\exp(\theta_{vib}/T) - 1]^2}$$
(4-117)

This treatment, due to Einstein, leads to a predicted variation of $C_{V(crystal)}$ with T/θ_{vib}, which is the same as that shown in Fig. 4-12 except

FIGURE 4-16

Heat capacities of various crystalline solids as a function of temperature. (a) The best-fitting curves according to the Einstein equation (4-117), dashed lines. (b) Scaled by the best choice of θ_D to give a fit to the theoretical Debye curve (solid line). [Adapted from S. M. Blinder, "Advanced Physical Chemistry." Copyright 1969, Macmillan, New York.]

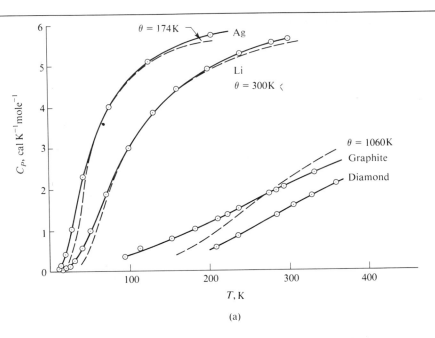

(a)

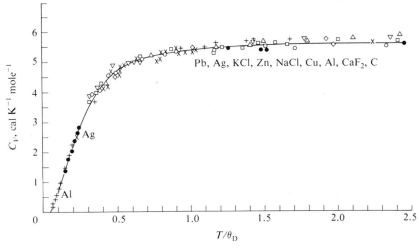

(b)

that the heat capacity scale is now from zero to 3R. The high-temperature limit is 3R, or the equipartition value (and that given by the Du-long–Petit law). Experimental results agree fairly well with the *Einstein equation,* as shown by the broken lines in Fig. 4-16(a). In each case θ_{vib} is empirically determined so as to give the best fit. Significant deviation sets in at low temperatures, however, and the theory was considerably improved by Debye. He assumed a range of possible ν_0 values up to a maximum ν_{max}, using the same distribution function as for blackbody radiation (Section 16-ST-1). This ν_{max} is simply a cutoff at the point where a total of $3N_A$ frequencies is reached and is an empirical parameter of the theory. The characteristic temperature is now defined as $h\nu_{max}/k$ and is thus also an empirical quantity. For the low-temperature region the Debye treatment predicts that the heat capacity should be proportional to T^3, and this provides a valuable limiting law. It is used, for example, in extrapolating experimental heat capacity values toward 0 K when calculating thermodynamic quantities.

BERRY, R.S., RICE, S.A., AND ROSS, J. (1980). "Physical Chemistry." Wiley, New York.

BLINDER, S.M. (1969). "Advanced Physical Chemistry." Macmillan, New York.

HIRSCHFELDER, J.O., CURTISS, C.F., AND BIRD, R.B. (1964). "Molecular Theory of Gases and Liquids." Wiley, New York.

GENERAL REFERENCES

EXERCISES

Show whether or not the following expressions are exact differentials. (a) $3v^3du + 3uv^2dv$, (b) $(6uv + 4v^2)du + (3u^2 + 8uv)dv$, (c) $4xy\,dx + 4x^2dy$. **4-1**

Ans. (a) No, (b) yes, (c) no.

Given the function $y = u(u + v)$, evaluate (a) $(\partial y/\partial u)_v$ and (b) $(\partial y/\partial v)_u$. Calculate Δy, $\int (\partial y/\partial u)_v du$, and $\int (\partial y/\partial v)_u dv$ for (c) path $(u_1,v_1) \rightarrow (u_2, v_1)$ followed by $(u_2, v_1) \rightarrow (u_2,v_2)$ and (d) path $(u_1,v_1) \rightarrow (u_1,v_2)$ followed by $(u_1,v_2) \rightarrow (u_2,v_2)$, where $u_1 = 2, u_2 = 5, v_1 = 3$, and $v_2 = 6$. **4-2**

Ans. (a) $2u + v$, (b) u,
(c) $\Delta y = 45$, $\int (\partial y/\partial u)_v\,du = 30$, $\int (\partial y/\partial v)_u\,dv = 15$,
(d) $\Delta y = 45$, $\int (\partial y/\partial v)_u\,dv = 6$, $\int (\partial y/\partial u)_v\,du = 39$.

For a right circular cylinder of radius r and length l, the area is $\mathcal{A} = 2\pi rl$ and the volume is $V = \pi r^2 l$. (a) Evaluate the coefficients $(\partial\mathcal{A}/\partial r)_l$, $(\partial\mathcal{A}/\partial l)_r$, $(\partial r/\partial l)_{\mathcal{A}}$, $(\partial V/\partial r)_l$, $(\partial V/\partial l)_r$. (b) Evaluate $(\partial\mathcal{A}/\partial V)_l$ by expressing $\mathcal{A}$ as a function of l and V only and then carry out the indicated partial differentiation. (c) Using Eqs. (4-13) and (4-15), express $(\partial\mathcal{A}/\partial V)_l$ in terms involving only partial derivatives listed in (a) and thus evaluate $(\partial\mathcal{A}/\partial V)_l$ indirectly. **4-3**

Ans. (a) $2\pi l$, $2\pi r$, $-\mathcal{A}/2\pi l^2$, $2\pi rl$, πr^2. (b) and (c) $1/r$.

Express $(\partial H/\partial T)_V$ in terms of derivatives that you can evaluate from the ideal gas law. **4-4**

Ans. $(\partial H/\partial T)_V = (\partial H/\partial T)_P + (\partial H/\partial P)_T(\partial P/\partial T)_V$
$= C_P + 0 = C_P.$

A certain gas obeys the equation $P(V - b) = RT$. (a) Evaluate $(\partial V/\partial T)_P$ and $(\partial V/\partial P)_r$. Obtain $(\partial P/\partial T)_v$ by (b) use of Eq. (4-12) and (c) direct differentiation. **4-5**

Ans. (a) R/P, $-RT/P^2$, (b) and (c) $R/(V - b)$.

For conditions around 20°C and atmospheric pressure, the coefficient of thermal expansion of liquid benzene is 1.24×10^{-3} K^{-1} and the coefficient of compressibility is 9.67×10^{-10} m^2 N^{-1}. The molar volume of benzene may be found in handbooks. Calculate $(C_P - C_V)$ and $(\partial E/\partial V)_T$. **4-6**

Ans. $(C_P - C_V) = 41.4$ J K^{-1}, $(\partial E/\partial V)_T = 3.76 \times 10^8$ Pa.

4-7   The coefficient of thermal expansion for Al is $(1/V)(\partial V/\partial T)_P = 2.4 \times 10^{-5}$ °C^{-1}, and $(\partial E/\partial V)_T$ is estimated at 1.0×10^5 atm. Calculate $C_P - C_V$ for Al at 25°C and atmospheric pressure.

Ans. 0.58 cal K^{-1} mol^{-1}.

4-8   One mole of an ideal, monatomic gas at STP undergoes an isochoric heating to 25°C. Calculate (a) P, (b) q, (c) w, (d) ΔE, and (e) ΔH.

Ans. (a) 1.09 atm, (b) 74.5 cal, (c) 0, (d) 74.5 cal, (e) 124 cal.

4-9   The same gas as in Exercise 4-8 undergoes an isobaric heating to 25°C. Calculate the same quantities as before, plus (f) the final volume.

Ans. (a) 1 atm, (b) 124 cal, (c) 49.7 cal, (d) 74.5 cal, (e) 124 cal, (f) 24.5 liter.

4-10   The same gas as in Exercises 4-8 and 4-9 undergoes a reversible adiabatic compression such that the final temperature is 25°C. Calculate quantities (a)–(f).

Ans. (a) 1.24 atm, (b) 0, (c) −74.5 cal, (d) 74.5 cal, (e) 124 cal, (f) 19.7 liter.

4-11   An empirical equation useful for gases at high pressures is $PV = bRT$, where b is a constant. Evaluate $(C_P - C_V)$ for such a gas.

Ans. bR.

4-12   Referring to Exercise 4-11, C_V and C_P are 14 and 17 J K^{-1}, respectively. Calculate the work to compress one mole of the gas isothermally at 120°C from 10 bar to 20 bar pressure.

Ans. −818 J.

4-13   Five moles of an ideal gas of $C_P = 6.5$ cal K^{-1} mol^{-1} are compressed isothermally and reversibly at 100°C from 0.1 to 5 atm. Calculate (a) the initial and final volumes, (b) q, (c) w, (d) ΔE, and (e) ΔH.

Ans. (a) 1531 liter, 30.62 liter, (b) −14,500 cal, (c) −14,500 cal, (d) 0, (e) 0.

4-14   Show what the value of *a* must be in Eq. (4-61) for a reversible isobaric process.

Ans. a = 0.

4-15   Calculate the equipartition heat capacity C_P for (a) O_3, (b) Xe, (c) HCl, and (d) C_2H_4, assuming ideal gas behavior, and C_V for (e) diamond and (f) NaCl (solid).

Ans. (a) $7R$, (b) $2\frac{1}{2}R$, (c) $4\frac{1}{4}R$, (d) $16R$, (e) $3R$, (f) $6R$.

4-16   A foam plastic has cells of 0.1 mm dimension. Calculate (a) the quantum number n for the one-dimensional translational state whose energy would be equal to kT at 25°C in the case of argon gas and (b) $\mathbf{Q}_{\text{trans(3-dim)}}$ at 25°C assuming the cells to be spheres of radius 0.1 mm.

Ans. (a) 7.1×10^6, (b) 1.02×10^{21}.

4-17   The rotational constant B_e of a linear molecule is defined as $B_e = h/8\pi^2 cI$, where c is the velocity of light; $B_e = 1.93$ cm^{-1} for CO. Calculate (a) the moment of inertia I for CO, (b) the quantum number J for the rotational state whose energy is equal to kT at 25°C, (c) the characteristic rotational temperature, and (d) $\mathbf{Q}_{\text{rot}}$ at 25°C.

Ans. (a) 1.45×10^{-39} g cm^2, (b) about 10, (c) 2.78 K, (d) 107.

4-18   We don't worry about rotation about the internuclear axis of a linear molecule. To illustrate the situation, calculate the energy of the $J = 1$ state relative to kT at 25°C for ^{35}Cl—^{35}Cl rotating on the bond axis. Nuclei are about 10^{-12} cm in diameter, and the moment of inertia of a sphere is about mr^2 where m is the mass of the sphere and r is its radius.

Ans. $\epsilon/kT \simeq 5 \times 10^5$.

4-19   The vibrational characteristic temperature is 3084 K for CO. Calculate (a) $h\nu_0$, (b) the force constant k , (c) the energy of the $v = 2$ state, (d) $C_{V(\text{vib})}$ at 25°C, (e) $\mathbf{Q}_{\text{vib}}$ at 25°C, and (f) ϵ_{vib} at 25°C.

Ans. (a) 4.257×10^{-13} erg, (b) 1.86×10^6 dyn cm^{-1},
(c) 1.06×10^{-12} erg (or $25.8kT$ at $25°C$), (d) $3.44 \times 10^{-3}\ R$,
(e) 5.675×10^{-3}, (f) 2.13×10^{-13} erg (or $5.2kT$).

PROBLEMS

Show that

$$(\partial P/\partial T)_H = -\frac{(\partial H/\partial T)_P}{(\partial H/\partial P)_T}$$

4-20

Derive the relationship $(\partial u/\partial y)_v\ (\partial v/\partial u)_y\ (\partial y/\partial v)_u = -1$. This is a useful alternative form to one of the equations in the text. It need not be memorized; just notice that u, v, and y occur once in each numerator, denominator, and subscript. Verify the relationship for $y = f(V,T)$, where the function f is for an ideal gas.

4-21

Show that if work w is done on an otherwise isolated system, the internal energy change of the system is exactly w.

4-22

Derive Eq. (4-37).

4-23

The area $\mathcal{A}$ of a right cone is given by $\pi r(r^2 + h^2)^{1/2}$, where r is the radius of the base and h is the altitude, and the volume is $(\pi/3)r^2h$. Evaluate $(\partial\mathcal{A}/\partial r)_h$, $(\partial\mathcal{A}/\partial h)_r$, $(\partial r/\partial h)_\mathcal{A}$, $(\partial V/\partial r)_h$, $(\partial V/\partial h)_r$, and $(\partial r/\partial h)_V$. Evaluate $(\partial\mathcal{A}/\partial V)_h$ by expressing $\mathcal{A}$ as a function of h and V only and then carrying out the indicated partial differentiation. Finally, evaluate $(\partial\mathcal{A}/\partial V)_h$ in terms of the differentials given here only, using the various partial differential relationships given in the chapter.

4-24

Derive the equation $C_P - C_V = -(\partial P/\partial T)_V[(\partial H/\partial P)_T - V]$.

4-25

Suppose that for a certain gas $(\partial E/\partial V)_T = 0$, but $P(V - b) = RT$. Calculate $(\partial H/\partial V)_T$ and $C_P - C_v$.

4-26

One mole of liquid water is vaporized reversibly at $75°C$. Calculate ΔE of vaporization and the reversible work. ΔH of vaporization is 10.5 kcal mol^{-1}.

4-27

A laboratory experiment for determining the heat capacity ratio γ is as follows. Air pressure is built up in a large bottle to some value P_2, atmospheric pressure being P_1. The stopper is removed so that air rushes out of the bottle in an essentially adiabatic expansion which cools the gas in the bottle. The bottle is stoppered and the air inside is allowed to warm up to the original ambient temperature T, at which point the pressure in the bottle rises to P_3. Derive an expression relating γ to P_1, P_2, and P_3, assuming ideal gas behavior. Calculate C_P for the gas if P_1, P_2, and P_3 are 1, 1.5, and 1.1 atm, respectively.

4-28

The heat capacity ratio, γ, is 1.20 for a certain ideal gas. By what factor does the pressure change if the volume is doubled in a reversible adiabatic expansion?

4-29

One mole of an ideal monatomic gas undergoes the following processes: 1. Adiabatic expansion from P_1, V_1, T_1 to P_2, V_2, T_2. 2. Return to initial state by the straight line path shown in the accompanying diagram. Calculate ΔE, ΔH, q, and w for each step and for the cycle if $P_1 = 4$ atm, $T_1 = 0°C$, and $V_2 = 3V_1$.

4-30

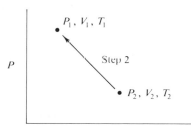

4-31   One mole of an ideal monatomic gas may be taken from the initial condition $P_1 = 2 \times 10^5$ Pa, $V_1 = 10$ dm^3 to the final condition $P_2 = 0.4 \times 10^5$ Pa, $V_2 = 1.5$ dm^3 by either one of the following paths: Path 1. (a) decrease in volume at constant pressure followed by (b) decrease in pressure at constant volume. Path 2. (a) decrease in pressure at constant volume followed by (b) decrease in volume at constant pressure.
For each path calculate ΔE, q, and w. If the gas is taken from P_1, V_1 to P_2, V_2 by path 1 and returned to the initial state by path 2, what are the values of ΔE, q, and w for the cycle?

4-32   Calculate ΔE, ΔH, q, and w when 1 mole of an ideal monatomic gas initially at 0°C and 2 atm is taken to a final pressure of 15 atm by the reversible path defined by the equation PV^3 = constant. Calculate (by means of a derivation) the heat capacity along this path, that is dq/dT for the path.

4-33   One hundred grams of oxygen at 25°C are held by a piston under 20 atm pressure. The pressure is suddenly released to 10 atm and the gas expands adiabatically. If C_v for oxygen is 4.95 cal K^{-1} mol^{-1}, calculate the final temperature of the gas. What are ΔE and ΔH for the process? Assume the gas is ideal.

4-34   Consider the hypothetical experiment of Fig. 4-7 in which a gas expands against a fixed weight of 0.1 atm. Suppose the gas to be one mole of an ideal monatomic one initially at STP, and that there is enough friction that the piston moves slowly, that is, negligible kinetic energy of motion develops. Calculate the final temperature of the gas and the amount of frictional heat developed if the expansion is (a) isothermal and (b) adiabatic with the frictional heat returned continuously to the gas.

4-35   One mole of CH$_4$ at 100°C at 80 atm is compressed isothermally to 100 atm. The pressure is then reduced to 80 atm by cooling at constant volume. Finally, the original state is regained by warming at constant pressure. Assuming that CH$_4$ obeys the van der Waals equation with constants as given in Table 1-5, calculate q and w for each step and for the cycle.

4-36   Referring to Exercise 4-19, calculate the heat capacity C_v for CO for various temperatures between 25°C and 1000°C and plot the results as C_v versus t.

4-37   The probability of a molecule being in a given rotational state at a given temperature increases at first with J because of the degeneracy factor, then decreases because the exponential factor takes over. Derive the expression for the most probable J value for an A—B diatomic molecule as a function of temperature. The rotational temperature for CO is 2.8 K. What is the most probable rotational level for CO at 0°C?

4-38   Calculate the heat capacity C_v for CO at 1500°C given that θ_{vib} is 3084 K. At what temperature would I_2 have the same heat capacity if its θ_{vib} is 307 K?

4-39   Calculate the vibrational heat capacity $C_{V(vib)}$ for CH$_3$I at 25°C given that the fundamental vibration frequencies ν_0 are $\bar{\nu}_1 = 2970$ cm^{-1}, $\bar{\nu}_2 = 1252$ cm^{-1}, $\bar{\nu}_3 = 533$ cm^{-1}, $\bar{\nu}_4 = 3060$ (2) cm^{-1}, $\bar{\nu}_5 = 1440$ (2)cm^{-1}, $\bar{\nu}_6 = 880$ (2) cm^{-1}. Frequency is given in terms of wave number $\bar{\nu}$, or the reciprocal of the wavelength corresponding to the ν_0; here (2) indicates that two vibrations have the same frequency. Each frequency may be assumed to make its independent contribution to the heat capacity.

4-40   The rotational temperature is 66 K for HD. Calculate the average rotational energy at 30 K and at 40 K and estimate the rotational contribution to the heat capacity of HD at 35 K. Note that the integrated form of the partition function cannot be used.

SPECIAL TOPICS PROBLEMS

4-41   A more general form for the earth's gravitational attraction is that the acceleration experienced by something of mass M is (GE)/r^2 where GE is the geocentric gravitational constant, GE = 3.99×10^{14} m^3 s^{-2}, and r is the distance from the center of the earth. We still take $h = 0$ at the surface of the earth, for which r_0 is 6.38×10^6 m. If we still assume constant temperature and ideal gas behavior, show that to a good approximation Eq. (4-107) becomes

$$1n\ (P/P_0)\ =\ -(Mg/RT)h(1\ -\ h/r_0)$$

(4-118)

Calculate P/P_0 for an ideal gas of molecular weight 29 g mol^{-1} at $h = 2$ km, but assuming that T is 25°C at $h = 0$ and diminishes by 10°C per km.

4-42

Derive the integrated form of the barometric equation for a gas obeying the equation $P(V - b) = RT$. Assume g and T constant.

4-43

Derive the equation

4-44

$$C_V\ =\ C_P\left[1\ -\ \mu\left(\frac{\partial P}{\partial T}\right)_V\right]\ -\ V\left(\frac{\partial P}{\partial T}\right)_V$$

An alternative Joule–Thomson type of experiment is to measure the quantity of heat that must be supplied when the expansion occurs at constant temperature. This gives the *isothermal* Joule–Thomson coefficient $\phi = (\partial H/\partial P)_T$. Relate ϕ to μ and calculate ϕ for CH_4 at STP assuming it to be a van der Waals gas and looking up any additional data needed.

4-45

Using Eq. (4-109), derive an expression of μC_P for a van der Waals gas. Estimate the value of μ for CH_4 at 1 atm and 25°C.

4-46

The Debye characteristic temperature $\theta_D = 86$ K for Pb. Estimate the heat capacity of Pb at -100°C and at 25°C. Calculate θ_{vib} for Pb for the best-fitting Einstein equation.

4-47

The heat capacity of Pt is (in calories per gram) 0.00123 at -255°C, 0.0261 at -152°C, 0.0324 at 20°C, and 0.0365 at 750°C. Estimate the Debye characteristic temperature θ_D and θ_{vib} for the best-fitting Einstein equation. Plot the data as heat capacity versus temperature along with the two theoretical curves.

4-48

chapter 5
Thermochemistry

5-1   Introduction

The preceding chapters have dealt with the physical properties and changes of state of pure substances or fixed mixtures of substances. We now consider the application of the first law of thermodynamics to chemical changes—a subject called *thermochemistry*. The subject is of great practical importance since it deals with the organization of the large amount of data concerning energies of chemical reactions. It also provides the foundation for obtaining chemical bond energies and hence for some aspects of theoretical chemistry.

This chapter also serves to introduce the general notation used in the application of thermodynamics to chemical change. The balanced equation for a chemical reaction is in the strict sense merely a particular affirmation that atoms are conserved. Thus the equation

$$aA + bB + \cdots = mM + nN + \cdots \qquad (5\text{-}1)$$

is no more than a statement that substances A, B,..., if taken in the molar amounts $a, b,...,$ will collectively contain the same number of each kind of atom as will substances M, N,..., taken in the amounts of $m, n,....$ The equation does not predict that the chemical change will in fact occur, or that if it does occur, other types of reactions may not also be present.

From the point of view of the first law, each substance has an internal energy and an enthalpy which are functions of the state of that substance. Thus $E_A = f_A(V_A, T_A)$, $E_B = f_B(V_B, T_B)$, and so on, and similarly $H_A = f_A'(P_A, T_A)$, $H_B = f_B'(P_B, T_B)$, and so on. To be complete, therefore, Eq. (5-1) should specify the state of each substance:

$$aA(V_A \text{ or } P_A, T_A) + bB(V_B \text{ or } P_B, T_B) + \cdots$$
$$= nN(V_N \text{ or } P_N, T_N) + mM(V_M \text{ or } P_M, T_M) + \cdots \quad (5\text{-}2)$$

(It is sufficient to give T and either P or V since the three are related by the equation of state for the substance.) The substances are not necessarily at the same temperature and pressure or at the same temperature and molar volume. We then use the symbol Δ as an operator which may be applied to any characteristic extensive property $\mathscr{P}$ such as $E, H,$ or V to signify the following operation:

$$\Delta \mathscr{P} \text{ means } (m\mathscr{P}_\mathrm{M} + n\mathscr{P}_\mathrm{N} + \cdots) - (a\mathscr{P}_\mathrm{A} + b\mathscr{P}_\mathrm{B} + \cdots) \tag{5-3}$$

or

$$\Delta \mathscr{P} = \sum_{\text{products}} n_\mathrm{p}\mathscr{P}_i - \sum_{\text{reactants}} n_\mathrm{r}\mathscr{P}_i$$

where n_p and n_r are the stoichiometric coefficients in the balanced equation. The property $\mathscr{P}$ is for the indicated substance in that state specified by the chemical equation. Thus the value of $\Delta \mathscr{P}$ must always refer to a definite, completely described equation. In the case of reactions involving solutions, chemical composition must be specified; the detailed thermochemistry of solutions is deferred until Chapter 9.

As an illustration, the reaction by which water is formed from hydrogen and oxygen might be written

$$\mathrm{H}_2 \text{ (gas, 25°C, 1 atm)} + \tfrac{1}{2}\mathrm{O}_2 \text{ (gas, 25°C, 1 atm)} = \mathrm{H}_2\mathrm{O} \text{ (liq, 25°C, 1 atm)} \tag{5-4}$$

For this reaction ΔH is

$$\Delta H = H(\mathrm{H}_2\mathrm{O}) - H(\mathrm{H}_2) - \tfrac{1}{2}H(\mathrm{O}_2) \tag{5-5}$$

where the individual enthalpies are per mole of the substance in the indicated state. Note that the phase (gas, liquid, solid) of the substance must be specified if there is any possible ambiguity.

Unlike the situation with some extensive quantities, such as volume or mass, we have no absolute values for the internal energy or enthalpy of a substance. Although the first law permits us to calculate changes in E or H, its application never produces absolute values. The same is true in thermochemistry; ΔE and ΔH give the changes in internal energy and enthalpy that accompany a chemical reaction, and while either may be expressed in the form of Eq. (5-3), we do not know the separate E or H values. It is partly a consequence of this situation that much use is made of standard or reference states. Two systems of standard states are in use. The first is one of convenience, 1 atm pressure and, if so specified, 25°C. In the case of a reaction for which the reactants and products are all in a standard state, such as Eq. (5-4), one writes ΔE^0_{298} or ΔH^0_{298}, where the superscript zero means that the pressure is one atm and the subscript gives the temperature chosen. As will be seen, a large body of thermochemical data is reported on this basis.

The second system takes as the standard state the substance devoid of any thermal energy, that is, at 0 K. This is a more rational as well as a very useful approach and is developed in the Special Topics section. Its implementation does require either extensive knowledge of heat capacity data or sufficient spectroscopic information to allow evaluation of the various partition functions.

5-2 Measurement of heats of reaction: Relationship between ΔE and ΔH

The practice of thermochemistry involves the measurement of the heat absorbed or evolved when a chemical reaction occurs. That is, the determination is one of q in the first law statements (assuming only pressure-volume work):

$$dE = \delta q - P\,dV \tag{5-6}$$

$$dH = \delta q + V\,dP \tag{5-7}$$

For simplicity, one attempts to choose conditions such that either volume is constant, so that no work is done, in which case $\Delta E = q_V$, or pressure is constant, in which case $\Delta H = q_P$. In either event, if heat is evolved, the reaction is said to be *exothermic*; q is negative and likewise the corresponding ΔE or ΔH. If heat is absorbed, the reaction is said to be *endothermic*, and q and ΔE or ΔH are positive.

   The equipment used to make such measurements is the *calorimeter*; it may be of either the *constant-pressure* or the *constant-volume* type. A simple illustration of the former is shown in Fig. 5-1(a). Insulation is provided by a Dewar flask, and one observes the temperature change when the reactant dissolves in or reacts with the liquid or solution. Research calorimeters of this type are more elaborate, of course, and with the use of a thermistor or a thermocouple, temperature changes of as little as $10^{-5}\,°C$ can be recorded. An electrical

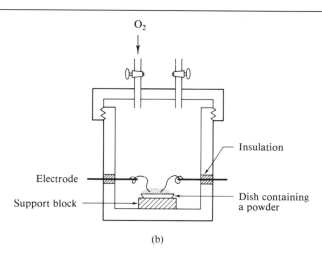

(b)

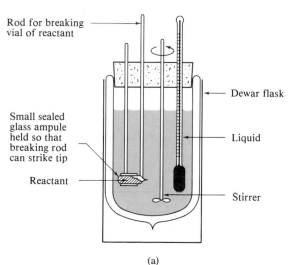

(a)

FIGURE 5-1
(a) Simple calorimeter such as might be used for measuring a heat of solution or of mixing. ΔH is measured since the contents are always at atmospheric pressure.
(b) Constant-volume combustion calorimeter.

heating element allows the introduction of an accurately known amount of heat so that the heat capacity of the calorimeter can be determined. Since the calibration is electric, it is conventional to record calorimetric heats in joules rather than in calories.

The *bomb* calorimeter, illustrated in Fig. 5-1(b), is of the constant-volume type. Typically, a substance is ignited in oxygen by means of a spark and the temperature rise of the calorimeter is measured. The heat of reaction, corresponding in this case to a ΔE, is again obtained through the known heat capacity of the calorimeter.

These calorimeters are *adiabatic;* insulation confines the heat of reaction, which is given by the product of the temperature change and the heat capacity of the assembly. The temperature change should be small for several reasons. Heat losses through conduction and radiation are reduced and thus the error in correcting for them. One ultimately wishes to report the results as a standard enthalpy or energy change, such as a ΔH^0_{298} or a ΔE^0_{298}, for a reaction in which reactants and products are at the *same* temperature, and the necessary corrections may be troublesome and introduce error if much heating has occurred in the actual reaction.

Heat loss may be virtually eliminated by making the temperature difference between the calorimeter and its immediate surroundings essentially zero. The *ice* calorimeter used by Lavoisier and Laplace in 1780 [see Hoch and Johnston (1961)] is a classic example. Here, the reaction vessel, uninsulated, is packed in ice in a container having a drain spigot. The whole unit is packed in an outer layer of ice. The heat of reaction melts some of the inside layer of ice and is given by the amount of water collected from the drain. Since the inside and outside layer are both at 0°C, there is no net heat flow between them.

In a modern version, the reaction vessel is in thermal contact with a large heat reservoir, such as a block of metal, through a layer of thermoelectric semiconductor material. The temperature difference across the layer is given accurately by the potential difference and integration over the time of the experiment gives the total heat flow and hence the heat generated by the reaction. At no time, however, is there more than a very small *temperature difference* between the reaction vessel and the heat reservoir.

Alternatively, the temperature of the heat reservoir may be kept the same as that of the calorimeter vessel, by means of a heater in the reservoir that is electronically controlled. The calorimeter is now adiabatic even though the temperature may rise during the reaction; there is *no* heat flow between the calorimeter vessel and the heat reservoir since the two are always at the same temperature.

> Calorimetry may also be used to measure *light* energy, the device being called a *bolometer.* The measurement is again one of temperature change, this time due to the energy delivered by the absorbed light. For example, one micromole of light quanta (one microeinstein of light) at 500 nm wavelength corresponds to 0.2392 J. Pulsed lasers may deliver this much light in a single nanosecond pulse, and here a *ballistic* calorimeter may be used. The light pulse is absorbed by a thermoelectric cavity and one sees a rapid increase in output voltage, followed by a decline as heat loss to the surroundings occurs. The instrument can be calibrated so that the maximum voltage gives the pulse energy.

To return to conventional calorimetry, the constant-volume type is generally used when a reactant or product is gaseous, as in combustion reactions.

Since the measured heat is a q_V, one obtains a ΔE of reaction, and it is usually desirable to convert the result to a ΔH of reaction. It follows from the definition of H that for a reaction

$$\Delta H = \Delta E + \Delta(PV) \tag{5-8}$$

and further

$$\Delta H^0_{298} = \Delta E^0_{298} + P\,\Delta V \qquad P = 1 \text{ atm} \tag{5-9}$$

where ΔV is given by Eq. (5-3). Molar volumes of liquids and solids are not large and their contribution to the ΔV of a reaction is usually of the order of a few cubic centimeters; the $P\,\Delta V$ term of Eq. (5-9) would then amount to only about 1 cal. If gases are involved, however, the correction can be quite appreciable. If we let Δn_g denote the number of moles of gaseous products minus the moles of gaseous reactants, then the $P\,\Delta n$ term becomes $(\Delta n_g)PV$. If the gases are nearly ideal, then an approximate form of Eq. (5-8) is

$$\Delta H \simeq \Delta E + (\Delta n_g)RT \tag{5-10}$$

Thus in the case of Eq. (5-4)

$$\Delta H^0_{298} = \Delta E^0_{298} - \tfrac{3}{2}(8.3144)(298.1) = \Delta E^0_{298} - 3.72 \text{ kJ cal}$$

The $P\,\Delta V$ term for $H_2O(l)$ amounts to only 18 cm^3 atm or about 1.7 J, so neglecting it introduces only a small error.

Constant-pressure calorimeters are especially convenient for reactions involving solutions. The process may be one of the dissolving of a substance in a solvent, in which case a *heat of solution* is obtained. Or a solution may be diluted by the addition of more solvent, so that a *heat of dilution* is obtained. Heats of chemical reactions may be studied by mixing two reacting solutions in the calorimeter. Since the calorimeter is open to the atmosphere or is at some constant pressure, the result gives ΔH directly. Where high precision and accuracy are not required, the simple arrangement shown in Fig. 5-1(a) may be entirely adequate. If the heat of reaction is small, however, as for example is often the case with heats of dilution, the apparatus becomes much more sophisticated [see Barthel (1975)].

The principal chemical requirement in calorimetry is that the measured q must be assignable to a definite process. This means that the products must be well defined and preferably should result from a single chemical reaction. Correction for incomplete reaction can be made, but it is very desirable that the reaction go to completion. In addition, the reaction should be fairly rapid; otherwise maintenance of the adiabatic condition of the calorimeter becomes very difficult. Most of the reactions of organic chemistry fail to meet one or another of these criteria, and for this reason organic thermochemistry deals mainly with combustion and hydrogenation; these processes generally can be made to go rapidly and cleanly. Heats of solution and dilution offer no great difficulty, and many inorganic reactions are easy to study. A good example of the latter would be a heat of neutralization of an acid by a base. On the other hand, coordination compounds often react too slowly for good calorimetry and are not easy to burn to well-defined products.

5-3   Some enthalpies of combustion, hydrogenation, and solution

The general type of calorimeter used for measuring heats of combustion was described in the preceding section. With excess pure oxygen present, most organic compounds will burn cleanly to carbon dioxide and water. Halogens or nitrogen present in a compound may appear as a mixture of the element and its oxides, and the exact proportions of such products may have to be determined for each experiment. In working problems in this text, however, it will be sufficient for the reader to assume that only the free element is formed, that is, Cl_2, Br_2, I_2, or N_2.

Some typical enthalpies of combustion are collected in Table 5-1. They are generally quite large, and since it is the difference among various enthalpies of combustion that usually is wanted (see next section), a very high degree of precision is desirable. This has indeed been achieved; most of the results in the table are accurate to a few tens of calories or better. Notice that compounds containing oxygen, such as ethanol and acetic acid, have lower enthalpies of combustion than the corresponding hydrocarbons. In a sense such compounds are already partly "burned."

The hydrogenation of unsaturated hydrocarbons is another reaction that has been used. One procedure is to pass a mixture of hydrogen and the

TABLE 5-1
Some enthalpies of reaction[a]

Substance	Reaction	ΔH^0_{298} kcal	ΔH^0_{298} kJ
Combustion[b]			
$H_2(g)$	$H_2 + \frac{1}{2}O_2 = H_2O$	-68.317	-285.84
C(graphite)	$C + O_2 = CO_2$	-94.052	-393.51
C(diamond)	$C + O_2 = CO_2$	-94.502	-395.40
$CO(g)$	$CO + \frac{1}{2}O_2 = CO_2$	-67.636	-282.99
$CH_4(g)$	$CH_4 + 2O_2 = CO_2 + 2H_2O$	-212.86	-890.36
$C_2H_2(g)$	$C_2H_2 + 2\frac{1}{2}O_2 = 2CO_2 + H_2O$	-310.62	-1299.63
$C_2H_4(g)$	$C_2H_4 + 3O_2 = 2CO_2 + 2H_2O$	-337.23	-1410.97
$C_2H_6(g)$	$C_2H_6 + 3\frac{1}{2}O_2 = 2CO_2 + 3H_2O$	-372.82	-1559.88
$C_3H_8(g)$	$C_3H_8 + 5O_2 = 3CO_2 + 4H_2O$	-530.61	-2220.07
$C_6H_6(g)$	$C_6H_6 + 7\frac{1}{2}O_2 = 6CO_2 + 3H_2O$	-789.08	-3301.51
$C_2H_5OH(l)$	$C_2H_5OH + 3O_2 = 2CO_2 + 3H_2O$	-326.71	-1366.95
$CH_3COOH(l)$	$CH_3COOH + 2O_2 = 2CO_2 + 2H_2O$	-208.5	-872.36
Hydrogenation			
$C_2H_4(g)$	$C_2H_4 + H_2 = C_2H_6(g)$	-32.747	-137.01
$cis\text{-}CH_3CH = CHCH_3(g)$	$C_4H_8 + H_2 = C_4H_{10}(g)$	-28.570^c	-119.54
$trans\text{-}CH_3CH = CHCH_3(g)$	$C_4H_8 + H_2 = C_4H_{10}(g)$	-27.621^c	-115.57
Solution			
$H_2SO_4(l)$	$H_2SO_4 + \infty H_2O = \text{solution}$	-22.99	-96.19
$H_2SO_4(l)$	$H_2SO_4 + 50H_2O = \text{solution}$	-17.53	-73.34
$HCl(g)$	$HCl + \infty H_2O = \text{solution}$	-17.96	-75.14
$NaOH(s)$	$NaOH + \infty H_2O = \text{solution}$	-10.246	-42.87
$NaCl(s)$	$NaCl + \infty H_2O = \text{solution}$	0.930	3.89
$NaC_2H_3O_2(s)$	$NaC_2H_3O_2 + \infty H_2O = \text{solution}$	-4.3	-18.0

[a]Values from F.A. Rossini et al., eds. (1952). Tables of Selected Values of Chemical Thermodynamic Properties. Nat. Bur. Std. Circ. No. 500, and the more recent NBS Technical Note 270-3, D.D. Wagman, W.H. Evans, V.B. Parker, I. Halow, M. Bailey, and R.H. Schumm, eds., 1968.

[b]Combustion products are $CO_2(g)$ and $H_2O(l)$.

[c]For 355 K, from A.B. Kistiakowsky and co-workers, *J. Amer. Chem. Soc.* **57,** 65, 876 (1935).

hydrocarbon through an adiabatic calorimeter which contains a platinum catalyst. The rate of heating of the calorimeter is then determined for a known flow rate of the gases and, if necessary, the effluent gas is analyzed so that the degree of reaction may be found. As illustrated by the values in Table 5-1 enthalpies of hydrogenation are generally much smaller than those of combustion. It is now possible to obtain fairly accurate differences in enthalpies of hydrogenation of isomeric compounds.

The third type of experimental heat of reaction which is listed in the table is that of solution. These are known as *integral* enthalpies of solution (see Section 9-ST-1) and the reaction consists in mixing one mole of the pure substance with the indicated amount of water so that a solution is formed. Notice that a ΔH of solution may be large and negative, as in the case of sulfuric acid, or small, as with sodium acetate. It may be positive, as with sodium chloride. Generally when heat is absorbed on dissolving, the value is small in magnitude. The heat of solution also depends on the final concentration obtained. As might be expected, the value (disregarding sign) tends to be larger for the limiting case in which an infinitely dilute solution is formed than for that involving some higher concentration. This last point is illustrated by the data for sulfuric acid and further in Table 5-3.

5-4   Combining ΔH or ΔE quantities

Table 5-1 provides a very small sample of the many individual reactions and types of reactions that have been studied experimentally. A complete tabulation of such individual results would be both clumsy and difficult to use. The first step toward systematization, taken in Table 5-1, is to reduce all values to a standard pressure and temperature, the latter usually being 25°C (see the next section for the manner of doing this). The second and very important step is to report results as standard heats of formation (also discussed in the next section). The basis for doing this is that, according to the first law, E and H are state functions, so that ΔE and ΔH for a given overall process are independent of the path taken. The principle was formulated independently by Hess in about 1840 and is sometimes known as *Hess's law of constant heat summation.*

The principle may be illustrated as follows. Consider the following two reactions:

$$C_2H_6(g) + 1\tfrac{1}{2}O_2(g) = CH_3COOH(l) + H_2O(l) \qquad \Delta H^0_{298} = -688.3 \text{ kJ} \qquad (5\text{-}11)$$

$$CH_3COOH(l) + 2O_2(g) = 2CO_2(g) + 2H_2O(l) \qquad \Delta H^0_{298} = -871.5 \text{ kJ} \qquad (5\text{-}12)$$

These correspond to first oxidation of ethane to acetic acid and then burning of the acetic acid to carbon dioxide and water. The net result of the two steps is simply the combustion of ethane:

$$C_2H_6(g) + 3\tfrac{1}{2}O_2(g) = 2CO_2(g) + 3H_2O(l) \qquad \Delta H^0_{298} = -1560 \text{ kJ} \qquad (5\text{-}13)$$

It is a common procedure in chemistry to obtain the net result of a series of chemical steps by the algebraic summation of the corresponding equations. Thus Eq. (5-13) results from the addition of Eqs. (5-11) and (5-12) and the cancellation of $CH_3COOH(l)$ since it appears once on each side of the equality sign. This procedure is always possible with balanced chemical equations, and the result must also be a balanced equation. The further conclusion of importance here is that the ΔH^0_{298} quantities combine similarly. If we write each according to its detailed meaning as given by Eq. (5-3) we have

$$\Delta H^0_{298}[\text{Eq. (5-11)}] + \Delta H^0_{298}[\text{Eq. (5-12)}]$$

$$= \left[\begin{matrix} H_{\text{CH}_3\text{COOH}} \\ (l,\ 25°\text{C},\ 1\ \text{atm}) \end{matrix} + \begin{matrix} H_{\text{H}_2\text{O}} \\ (l,\ 25°\text{C},\ 1\ \text{atm}) \end{matrix} - \begin{matrix} H_{\text{C}_2\text{H}_6} \\ (g,25°\text{C},\ 1\ \text{atm}) \end{matrix} - \begin{matrix} 1\frac{1}{2}H_{\text{O}_2} \\ (g,\ 25°\text{C},\ 1\ \text{atm}) \end{matrix} \right]$$

$$+ \left[\begin{matrix} 2H_{\text{CO}_2} \\ (g,\ 25°\text{C},\ 1\ \text{atm}) \end{matrix} + \begin{matrix} 2H_{\text{H}_2\text{O}} \\ (g,\ 25°\text{C},\ 1\ \text{atm}) \end{matrix} - \begin{matrix} H_{\text{CH}_3\text{COOH}} \\ (l,\ 25°\text{C},\ 1\ \text{atm}) \end{matrix} - \begin{matrix} 2H_{\text{O}_2} \\ (g,\ 25°\text{C},\ 1\ \text{atm}) \end{matrix} \right]$$

$$= \left[\begin{matrix} 2H_{\text{CO}_2} \\ (g,\ 25°\text{C},\ 1\ \text{atm}) \end{matrix} + \begin{matrix} 3H_{\text{H}_2\text{O}} \\ (g,\ 25°\text{C},\ 1\ \text{atm}) \end{matrix} - \begin{matrix} H_{\text{C}_2\text{H}_6} \\ (g,\ 25°\text{C},\ 1\ \text{atm}) \end{matrix} - \begin{matrix} 3\frac{1}{2}H_{\text{O}_2} \\ (g,\ 25°\text{C},\ 1\ \text{atm}) \end{matrix} \right]$$

The quantities $H_{\text{CH}_3\text{COOH}}$ cancel since each is for the same substance in the same state, and the result is just ΔH^0_{298} for the net equation (5-13). The sum of -688.3 kJ and -871.5 kJ is indeed -1560 kJ.

The general conclusion is that ΔH (or ΔE) quantities add (or subtract) as do the corresponding equations provided that the substances that are cancelled are in the same state. This last requirement will necessarily be met if standard ΔH's are used.

The lack of dependence of ΔH and ΔE on path is useful in the following ways. It allows the calculation of a ΔH or ΔE for a process that is difficult or impossible to carry out directly or which simply has not yet been measured. As an example, one can determine experimentally the heats of combustion of graphite and of diamond:

$$\text{C(graphite)} + \text{O}_2(g) = \text{CO}_2(g) \qquad \Delta H^0_{298} = -393.51\ \text{kJ} \qquad (5\text{-}14)$$

$$\text{C(diamond)} + \text{O}_2(g) = \text{CO}_2(g) \qquad \Delta H^0_{298} = -395.40\ \text{kJ} \qquad (5\text{-}15)$$

We may subtract Eq. (5-15) from Eq. (5-14) to get

$$\text{C(graphite)} = \text{C(diamond)} \qquad \Delta H^0_{298} = (-393.51) - (-395.40) = 1.89\ \text{kJ} \qquad (5\text{-}16)$$

The conversion of graphite to diamond (or *vice versa*) is not a feasible laboratory reaction, but by combining the two enthalpies of combustion one obtains the desired ΔH^0_{298}.

Enthalpies of combustion may, in fact, be used quite generally to obtain enthalpies of other reactions. Thus the reaction

$$\text{C}_2\text{H}_4(g) + \text{H}_2(g) = \text{C}_2\text{H}_6(g) \qquad (5\text{-}17)$$

may be written as

$$\text{C}_2\text{H}_4(g) + 3\text{O}_2(g) = 2\text{CO}_2(g) + 2\text{H}_2\text{O}(l) \qquad \Delta H^0_{298} = -1410.97\ \text{kJ} \qquad (5\text{-}18)$$

plus

$$\text{H}_2(g) + \tfrac{1}{2}\text{O}_2(g) = \text{H}_2\text{O}(l) \qquad \Delta H^0_{298} = -285.84\ \text{kJ} \qquad (5\text{-}19)$$

minus

$$\text{C}_2\text{H}_6(g) + 3\tfrac{1}{2}\text{O}_2(g) = 2\text{CO}_2(g) + 3\text{H}_2\text{O}(l) \qquad \Delta H^0_{298} = -1559.88\ \text{kJ} \qquad (5\text{-}20)$$

Then ΔH^0_{298} for Eq. (5-17) must be $(-1410.97) + (-285.84) - (-1559.88) = -136.93$ kJ. Notice that the result is not exactly the same as the directly determined value of -137.01 kJ. As mentioned earlier, the problem is that in using enthalpies of combustion we are combining very large numbers to give a small net result. Even small percentage errors in the former propagate to give a large percentage error in the result.

The example may be generalized. We may obtain ΔH^0_{298} for any reaction by adding the standard enthalpies of combustion of the reactants and subtracting the sum of those of the products:

$$\Delta H^0_{298} = \sum_{\text{reactants}} n_r \Delta H^0_{c,298} - \sum_{\text{products}} n_p \Delta H^0_{c,298} \qquad (5\text{-}21)$$

where $\Delta H^0_{c,298}$ denotes a standard enthalpy of combustion. As an illustration, ΔH^0_{298} for the (unlikely) reaction

$$2CH_4(g) + CH_3COOH(l) = C_3H_8(g) + CO_2(g) + 2H_2(g)$$

is given by

$$2\Delta H^0_{c,298}\,(CH_4) + \Delta H^0_{c,298}\,(CH_3COOH) - \Delta H^0_{c,298}\,(C_3H_8) - 2\Delta H^0_{c,298}\,(H_2)$$

There is no term for CO_2 since by definition its enthalpy of combustion is zero. A table of standard enthalpies of combustion thus implies a knowledge of ΔH^0_{298} values for all reactions that can be formulated as a combination of combustion reactions.

5-5   Enthalpies of formation

A. Standard enthalpy of formation for compounds

As may be gathered from the foregoing, experimental enthalpies of reaction have been obtained for a wide variety of processes—combustion, hydrogenation, solution, etc. From the experimental point of view, the process is usually chosen on the basis of convenience and feasibility. It would be very complicated and a nuisance to *use* such a welter of types of results, and what is done to take advantage of Hess's law to reduce everything to a standard type of reaction, namely the standard enthalpy of formation from the elements. Each element involved is taken to be in its stable chemical form at 25°C and 1 atm pressure. For example, the reaction giving the standard enthalpy of formation of $H_2O(l)$ is

$$H_2(g, 298\ K, 1\ atm) + \tfrac{1}{2}O_2(g, 298\ K, 1\ atm) = H_2O(l, 298\ K, 1\ atm) \qquad (5\text{-}22)$$

and $\Delta H^0_{f,298}(H_2O, l)$ is -285.84 kJ mol^{-1}. The standard enthalpy of formation of $CO_2(g, 298\ K, 1\ atm)$ is

$$C(\text{graphite}, 298\ K, 1\ atm) + O_2(g, 298\ K, 1\ atm) = CO_2(g, 298\ K, 1\ atm)$$
$$\Delta H^0_{f,298}(CO_2), g) = -393.51 \text{ kJ mol}^{-1} \qquad (5\text{-}14)$$

(Notice that the above two reactions could be studied experimentally; they correspond to the enthalpies of combustion of hydrogen and of graphite, respectively.)

We can now combine reactions (5-14), (5-18) and (5-22):

$$2CO_2(g) + 2H_2O(l) = C_2H_4(g) + 3O_2(g) \qquad \Delta H^0_{298} = 1410.97 \text{ kJ}$$

$$2C(\text{graphite}) + 2O_2(g) = 2CO_2(g) \qquad \Delta H^0_{298} = 2(-393.51) = -787.02 \text{ kJ}$$

$$2H_2(g) + O_2(g) = 2H_2O(l) \quad \Delta H^0_{298} = 2(-285.84) = -591.68 \text{ kJ} \qquad (5\text{-}23)$$

$$2H_2(g) + 2C(\text{graphite}) = C_2H_4(g) \quad \Delta H^0_{298} = 1410.97 - 787.02 - 571.68 = 52.27 \text{ kJ}$$

The result is just the enthalpy of formation of $C_2H_4(g)$, $\Delta H^0_{f,298} = 52.27$ kJ. Standard enthalpies of formation at 298 K have been compiled in this manner for a large number of substances, both organic and inorganic. Extensive tables have been compiled, and a sampling is given in Table 5-2. The table includes values for unstable forms of some elements, such as O_3 and C(diamond), and in some cases also gives values for both gaseous and liquid states of a compound, as for water. The difference between two such values is simply the enthalpy of vaporization.

Having tables of enthalpies of formation, we can use Hess's law to obtain ΔH^0_{298} for *any* reaction involving listed compounds. Suppose, for example, that our reaction is compound A going to product B + 2C. We then write

$$(\text{elements in compound A}) = A \qquad \Delta H^0_{f,298}(A)$$

$$(\text{elements in compound B}) = B \qquad \Delta H^0_{f,298}(B)$$

$$2(\text{elements in compound C}) = 2C \qquad 2\Delta H^0_{f,298}(C)$$

If we now subtract the first equation from the second two, we have

$$A + (\text{elements in B}) + 2(\text{elements in C}) = B + 2C + (\text{elements in A})$$

TABLE 5-2 Standard enthalpies of formation[a]		$\Delta H^0_{f,298}$			$\Delta H^0_{f,298}$	
Substance		kcal mol^{-1}	kJ mol^{-1}	Substance	kcal mol^{-1}	kJ mol^{-1}
AgCl(s)		−30.36	−127.03	$H_2O(l)$	−68.317	−285.84
$Br_2(g)$		7.34	30.71	$H_2O(g)$	−57.789	−241.79
C(graphite)		(0.000)	(0.000)	HF(g)	−64.79	−271.1
C(diamond)		0.4532	1.896	HCl(g)	−22.063	−92.31
$CaCO_3(s)$		−228.45	−1206.87	HBr(g)	−8.66	−36.23
CaO(s)		−151.9	−635.5	HI(g)	6.20	25.94
CO(g)		−26.416	−110.52	KCl(s)	−104.175	−435.87
$CO_2(g)$		−94.052	−393.51	NaCl(s)	−98.232	−411.03
$CH_4(g)$		−17.89	−74.85	$NH_3(g)$	−11.04	−46.19
$C_2H_2(g)$		54.19	226.73	NO(g)	21.600	90.37
$C_2H_4(g)$		12.50	52.30	$NO_2(g)$	8.091	33.85
$C_2H_6(g)$		−20.24	−84.68	$N_2O_4(g)$	2.58	10.8
$C_3H_8(g)$		−24.82	−103.85	$O_3(g)$	34.0	142.3
$C_6H_6(l)$		11.718	49.028	P(g)	75.18	314.6
$C_6H_6(g)$		19.820	82.93	$PCl_3(g)$	−73.22	−306.4
$C_2H_5OH(l)$		−66.356	−277.63	$PCl_5(g)$	−95.35	−398.94
$CH_3COOH(l)$		−116.4	−487.0	S(rhombic)	(0.000)	(0.000)
$CCl_4(l)$		−33.3	−139.3	S(monoclinic)	0.071	0.30
$Fe_2O_3(s)$		−196.5	−822.2	$SO_2(g)$	−70.96	−296.9
Glycine				$SO_3(g)$	−94.45	−395.2
$\quad H_2NCH_2COOH(s)$		−126.33	−528.56			

[a]Data from F.A. Rossini et al., eds. (1952). Tables of Selected Values of Chemical Thermodynamic Quantities. Nat. Bur. Std. Circ. No. 500. U.S. Gov't. Printing Office, Washington, D.C.

Since, by definition of a balanced equation, the sum of the elements in compounds B and 2C must equal the sum of those in compound A, the sums of elements *must* cancel, and we are left with

$$A = B + 2C \qquad \Delta H_{298}^0 = \Delta H_{f,298}^0(B) + 2\Delta H_{f,298}^0(C) - \Delta H_{f,298}^0(A)$$

We generalize the situation by writing

$$\Delta H_{298}^0 = \sum_{\text{products}} n_p \, \Delta H_{f,298}^0 - \sum_{\text{reactants}} n_r \, \Delta H_{f,298}^0 \qquad (5\text{-}24)$$

[Notice that Eq. (5-24) is similar to Eq. (5-21) for combining enthalpies of combustion, but with the sums for reactants and products interchanged (why?).]

Calculate the enthalpy of hydrogenation of ethylene [Eq. (5-17)] using enthalpies of formation.

EXAMPLE

$$\Delta H_{298}^0 = \Delta H_{f,298}^0 (C_2H_6, \, g) - \Delta H_{f,298}^0 (C_2H_4, \, g)$$

$$= -84.68 - 52.30 = -136.98 \text{ kJ mol}^{-1}$$

There is no term for hydrogen because its heat of formation is zero by definition.

Some values of enthalpies of solution are included in Table 5-1. Such data may be incorporated into the standard enthalpy of formation scheme by assigning the entire heat of solution to the solute. For example, the heat of dissolving of $HCl(g)$ in water to give an infinitely dilute solution is, per mole of HCl, -75.14 kJ (Table 5-1). We use the notation *aq* to denote the condition of infinite dilution in water, and may now write

B. Standard enthalpies of aqueous solutes and of ions

$$\tfrac{1}{2}H_2(g) + \tfrac{1}{2}Cl_2(g) = HCl(g) \qquad \Delta H_{f,298}^0 = -92.31 \text{ kJ}$$

$$HCl(g) + aq = HCl(aq) \qquad \Delta H_{298}^0 = -75.14 \text{ kJ}$$

Then we have

$$\tfrac{1}{2}H_2(g) + \tfrac{1}{2}Cl_2(g) + aq = HCl(aq) \qquad \Delta H_{f,298}^0 (HCl, \, aq) = -167.44 \text{ kJ}$$

The same convention may be applied to solutions of any concentration. Thus the standard enthalpy for the following reaction is -73.85 kJ:

$$HCl(g) + 100H_2O = \left\{ \begin{array}{l} \text{solution of concentration} \\ m = 0.55 \text{ mole per 1000 g } H_2O \end{array} \right\}$$

$$\Delta H_{298}^0 = -73.85 \text{ kJ}$$

The standard enthalpy of formation of HCl in 0.55 molal solution is then $-73.85 + -92.31 = -166.16$ kJ mol^{-1}.

In the case of infinitely dilute solutions of electrolytes it is convenient to proceed a step further. For such solutions the ions have no mutual interaction and the state of any given ion is entirely independent of the nature of its counter ion. It is therefore possible to combine equations as in the following example:

(a) $\tfrac{1}{2}H_2(g) + \tfrac{1}{2}Cl_2(g) = H^+(aq) + Cl^-(aq) \qquad \Delta H_{298}^0 = -167.44$ kJ

(b) $Na(s) + \tfrac{1}{2}Cl_2(g) = Na^+(aq) + Cl^-(aq) \qquad \Delta H_{298}^0 = -407.10$ kJ

(c) $Na(s) + \frac{1}{2}Br_2(l) = Na^+(aq) + Br^-(aq)$ $\Delta H^0_{298} = -360.58$ kJ

(d) $= (a) + (c) - (b)$:

$\frac{1}{2}H_2(g) + \frac{1}{2}Br_2(l) = H^+(aq) + Br^-(aq)$ $\Delta H^0_{298} = -120.92$ kJ

$HCl(aq)$ has been written as $H^+(aq) + Cl^-(aq)$, and similarly for $NaCl(aq)$ and $NaBr(aq)$ not only because it is more nearly correct but also because it emphasizes that the $Cl^-(aq)$ of the HCl solution cancels the $Cl^-(aq)$ of the NaCl solution.

This procedure is unnecessarily clumsy. If the enthalpy of formation of some one ion were known, it would then be possible to calculate those of all other ions. Such a value is not known, since ionic chemical reactions always involve combinations of ions such that the equation is balanced electrically (see Section 13-CN-3 for a further comment). By the same token, however, one may arbitrarily assign a value to the standard enthalpy of formation of some one ion, and the values then calculated for the other ions must always occur in any thermochemical equation in such a way that the assumption cancels out. The convention is that $\Delta H^0_{f,298}(H^+, aq)$ is zero.

Some representative standard enthalpies of formation of solutes and of individual aqueous ions are given in Table 5-3. We may now obtain the values of ΔH^0_{298} for reactions (a), (b), (c), and (d) simply by adding the appropriate individual ion values.

5-6   Dependence of ΔH and ΔE on temperature

The temperature dependence of individual H and E values is given by the equations

$$dE = C_V\, dT \qquad [\text{Eq. (4-24)}]$$

$$dH = C_P\, dT \qquad [\text{Eq. (4-32)}]$$

TABLE 5-3
Standard enthalpies of formation of aqueous species[a]

Solute	$\Delta H^0_{f,298}$ kcal mol^{-1}	$\Delta H^0_{f,298}$ kJ mol^{-1}	Ion	$\Delta H^0_{f,298}$ kcal mol^{-1}	$\Delta H^0_{f,298}$ kJ mol^{-1}
HCl					
in ∞ H$_2$O	-40.023	-167.47	H$^+$	(0.000)	(0.000)
in 100H$_2$O	-39.713	-166.16	Li$^+$	-66.554	-278.46
H$_2$SO$_4$			Na$^+$	-57.279	-239.66
in ∞ H$_2$O	-216.90	-907.51	K$^+$	-60.04	-251.21
in 100H$_2$O	-211.59	-885.29	Ag$^+$	-25.31	-105.90
NaOH			OH$^-$	-54.957	-229.94
in ∞ H$_2$O	-112.236	-469.60	F$^-$	-78.66	-329.11
in 100H$_2$O	-112.108	-469.06	Cl$^-$	-40.023	-167.46
NaCl			Br$^-$	-28.90	-120.92
in ∞ H$_2$O	-97.302	-407.11	I$^-$	-13.37	-55.94
in 100H$_2$O	-97.250	-406.89	ClO$_4^-$	-31.41	-131.42
NH$_4$Cl			SO$_4^{2-}$	-216.90	-907.51
in ∞ H$_2$O	-71.76	-300.24	CO$_3^{2-}$	-161.63	-676.26
in 100H$_2$O	-71.63	-299.70	NH$_4^+$	-31.7	-132.5

[a]Data from F.A. Rossini et al., eds. (1952). Tables of Selected Values of Chemical Thermodynamic Quantities. Nat. Bur. Std. Circ. No. 500. U.S. Gov't. Printing Office, Washington, D.C.

for constant volume and constant pressure heating, respectively. In integral form these equations become:

$$E_2 = E_1 + \int_{T_1}^{T_2} C_V \, dT \tag{5-25}$$

$$H_2 = H_1 + \int_{T_1}^{T_2} C_P \, dT \tag{5-26}$$

As discussed in Section 4-8C, heat capacities are in general temperature-dependent and the experimental values are usually summarized in the form of a polynomial in temperature. Three terms suffice to represent data over a considerable range of temperature. Thus we write*

$$C_P = a + bT + cT^{-2} \tag{5-27}$$

The polynomial formulation is convenient in allowing easy integration of Eq. (5-26). Values of a, b, and c are given in Table 5-4 for a number of gases, liquids, and solids.

The procedure for calculating the change in a heat of reaction with temperature is quite analogous to the preceding. Let ΔH_2 be the enthalpy of reaction at T_2 and ΔH_1 that at T_1. The general reaction equation (5-2) may then be written

$$-\Delta H_2 = [aH_2(A) + bH_2(B) + \cdots] - [mH_2(M) + nH_2(N) + \cdots]$$

$$= \left\{ a\left[H_1(A) + \int_{T_1}^{T_2} C_P(A) \, dT \right] + b\left[H_1(B) + \int_{T_1}^{T_2} C_P(B) \, dT \right] + \cdots \right\}$$

$$- \left\{ m\left[H_1(M) + \int_{T_1}^{T_2} C_P(M) \, dT \right] + n\left[H_1(N) + \int_{T_1}^{T_2} C_P(N) \, dT \right] + \cdots \right\}$$

On collecting terms, we obtain

$$\Delta H_2 = \Delta H_1 + \int_{T_1}^{T_2} \Delta C_P \, dT \tag{5-28}$$

where the operator Δ in ΔC_P has its usual meaning. The coefficients a, b, and c in the series expression for each individual heat capacity may be combined according to Eq. (5-27) so that ΔC_P may be written

$$\Delta C_P = \Delta a + (\Delta b)T + (\Delta c)T^{-2} \tag{5-29}$$

A formal integration of Eq. (5-28) gives

$$\Delta H_2 = \Delta H_1 + (\Delta a)(T_2 - T_1) + \left(\frac{\Delta b}{2}\right)(T_2^2 - T_1^2) - (\Delta c)\left(\frac{1}{T_2} - \frac{1}{T_1}\right) \tag{5-30}$$

The procedure is most commonly applied to ΔH^0 values, that is, to ΔH quantities for 1 atm pressure, and the C_P values of Table 5-4 are for this standard pressure. Analogous equations involving C_V apply to the calculation of a ΔE_2 from a ΔE_1.

---

*A widely used alternative series is of the form $C_P = a + bT + cT^2$ (note Problem 5-28).

TABLE 5-4
Variation of C_P^0 with
temperature,[a]
$C_P^0 = a + bT + cT^{-2}$

Substance	a	$10^3 b$	$10^{-5} c$
Gases (from 298 K to 2000 K)			
Monatomic gases	4.97 (20.78)	0	0
H_2	6.52 (27.28)	0.78 (3.26)	0.12 (0.50)
O_2	7.16 (29.96)	1.00 (4.18)	-0.40 (-1.67)
N_2	6.83 (28.58)	0.90 (3.77)	-0.12 (-0.50)
CO	6.79 (28.41)	0.98 (4.10)	-0.11 (-0.46)
Cl_2	8.85 (37.03)	0.16 (0.67)	-0.68 (-2.85)
CO_2	10.57 (44.23)	2.10 (8.79)	-2.06 (-8.62)
H_2O	7.30 (30.54)	2.46 (10.29)	—
NH_3	7.11 (29.75)	6.00 (25.10)	-0.37 (-1.55)
CH_4	5.65 (23.64)	11.44 (47.86)	-0.46 (-1.92)
C_6H_6	2.46 (10.29)	60.2 (251.9)	—
Liquids (from melting point to boiling point)			
H_2O	18.04 (75.48)	—	—
I_2	19.20 (80.33)	—	—
NaCl	16.0 (66.9)	—	—
C_6H_6	8.00 (33.50)	80 (335)	
Solids (from 298 K to melting point or 2000 K)			
C (graphite)	4.03 (16.86)	1.14 (4.77)	-2.04 (-8.54)
Al	4.94 (20.67)	2.96 (12.38)	—
Cu	5.41 (22.64)	1.50 (6.28)	—
Pb	5.29 (22.13)	2.80 (11.72)	0.23 (0.96)
I_2	9.59 (40.12)	11.90 (49.79)	—
NaCl	10.98 (45.94)	3.90 (16.32)	—

[a]G.N. Lewis and M. Randall, "Thermodynamics," 2nd ed., (revised by K.S. Pitzer and L. Brewer). McGraw-Hill, New York, 1961. Values in cal K^{-1} mol^{-1}; values in parentheses in J K^{-1} mol^{-1}.

EXAMPLE Consider the reaction

$$CH_4(g) + 2O_2(g) = CO_2(g) + 2H_2O(g)$$

We find ΔH_{298}^0 from Table 5-2: $\Delta H_{298}^0 = 2(-241.79) - 393.51 - (-74.85) = -802.24$ kJ. We obtain ΔH_{373}^0 as follows, using Eq. (5-30): $\Delta a = 44.23 + 2(30.54) - 23.64 - 2(29.96) = 21.75$ J K^{-1}; $\Delta b = [8.79 + 2(10.29) - 47.86 - 2(4.18)]$

$\times\ 10^{-3} = -26.85 \times 10^{-3}$ J K^{-2}; $\Delta c = [-8.62 + 0 - (-1.92) - 2(-1.67)] \times 10^5 = -3.36 \times 10^5$ J K. Then

$$\Delta H^0_{373} = -802,240 + 21.75(373.15 - 298.15)$$

$$+ \frac{-26.85 \times 10^{-3}}{2}[(373.15)^2 - (298.15)^2]$$

$$- (-3.36 \times 10^5)\left(\frac{1}{373.15} - \frac{1}{298.15}\right)$$

$$= -802,240 + 729 = -801.51 \text{ kJ}.$$

Equations (5-25) and (5-26) are applicable to a change of temperature which is not accompanied by any phase change. As illustrated in Fig. 5-2, a substance may exist in some phase α below a certain temperature and in some phase β above that temperature. The most common situations are those in which α is a solid phase and β the liquid phase, or in which α is the liquid phase and β the gaseous one. If we take H_1 and T_1 as the reference points, then Eq. (5-26) is obeyed up to the temperature of the phase transition $T_{\alpha\beta}$. At this temperature additional heat is needed to bring about the transition and H increases by the enthalpy of the transition $\Delta H_{\alpha\beta}$. Above $T_{\alpha\beta}$, H again rises in accord with Eq. (5-26), although now the heat capacity is $C_P(\beta)$ rather than $C_P(\alpha)$. An entirely analogous analysis applies to the behavior of E.

The presence of phase transitions introduces a corresponding complication in the calculation of a ΔH_2 from a ΔH_1. In such a case Eq. (5-28) may still be used in the form

$$\Delta H_2 = \Delta H_1 + \sum_{\text{products}} n_p(H_2 - H_1) - \sum_{\text{reactants}} n_r(H_2 - H_1) \qquad (5\text{-}31)$$

If a reactant or product undergoes a phase transition, this must be allowed for in calculating the corresponding $(H_2 - H_1)$ term; otherwise Eq. (5-26) is used in the usual manner.

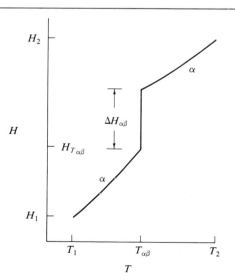

FIGURE 5-2
Variation of enthalpy with temperature for a substance undergoing a phase change.

COMMENTARY AND NOTES

5-CN-1 The thermochemistry of nutrition

The methods of thermochemistry have been applied extensively both to foods and to living systems. The overall processes that occur when food is metabolized may be summarized as follows. For sugars and starches,

$$(CH_2O)_n + nO_2 = nCO_2 + nH_2O(l)$$
$$\Delta H^0_{c,298} \simeq -4 \text{ kcal g}^{-1}$$

Animal fats are likewise converted to CO_2 and H_2O. Taking tripalmitin as a typical example,

$$2C_{51}H_{98}O_6 + 145O_2 = 102CO_2 + 98H_2O(l)$$
$$\Delta H^0_{c,298} \simeq -10 \text{ kcal g}^{-1}$$

In the case of proteins, the nitrogen appears in waste products, mainly urea. The combustion reaction is difficult to formulate precisely, but we can write in the case of glycine

$$2H_2NCH_2COOH + 3O_2$$
$$= CO(NH_2)_2(aq) + 3CO_2$$
$$+ 3H_2O(l) \quad \Delta H^0_{c,298} \simeq -2 \text{ kcal g}^{-1}$$

For an average protein $\Delta H^0_{c,298}$ is usually taken to be -4.4 kcal g^{-1}. In obtaining the "calorie" content of a food (in nutrition the word "calorie" actually means kilocalorie) it is first fractionated into the sugar + starch, fat, and protein components. One then applies the foregoing heats of combustion to obtain an overall calorie rating for the food.

The equations above do give the approximate net change that occurs in metabolism. One way of checking this conclusion is to measure the respiratory quotient, R.Q., which is the ratio of moles of CO_2 produced to moles of O_2 consumed by the subject. R.Q. should be about 1 for sugars and starches, about 0.7 for fats, and about 0.8 for proteins. In the case of a group of human subjects, the observed average R.Q. rose toward unity on a carbohydrate diet, and fell toward 0.7 on a fat diet.

In another approach, actual calorimetry is done on test subjects, to determine the rate of metabolic evolution of heat. The total oxidation occurring is calculated from the amount of CO_2 exhaled, supplemented by analyses of the excreta. The calculated heat production can then be compared with that from the whole body calorimetry. As an example, direct calorimetry on a group of men gave 2250 kcal as the heat produced in a 24-hour day under resting conditions, and 4690 kcal under working conditions. The calculated values were 2450 kcal and 4700 kcal, respectively. Thus most of the calorie content of a person's diet is accounted for chemically (see Gemmill and Brobeck, 1968). Note Fig. 5-3 for an interesting space-age application.

The finding that most of the energy content of food appears as work and heat is a reflection of the fact that the adult organism is not growing. Growing adds to the physical size, and the nutrients that go into new tissue cannot, of course, appear as combustion products. The adult uses food energy to do physical work and to keep the body in its nonequilibrium steady state. Examples of this last category include the energy needed for the regeneration of cells, to supply the heat dissipated by friction within the body, as in blood circulation, and to maintain body temperature.

SPECIAL TOPICS

5-ST-1 Explosions, flames, and rockets

The emphasis of this chapter has so far been on ΔE and ΔH quantities for reactions under standard conditions, and, in the preceding section, on the calculation of ΔE or ΔH at some temperature T_2 given the value at T_1. There is an interesting and important special case in which the chemical reaction occurs under essentially adiabatic conditions so that the heat of the reaction is confined to the system and

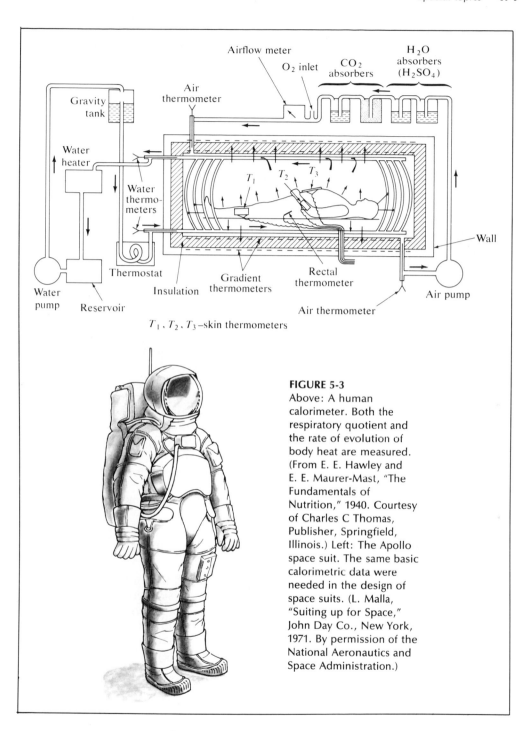

FIGURE 5-3
Above: A human calorimeter. Both the respiratory quotient and the rate of evolution of body heat are measured. (From E. E. Hawley and E. E. Maurer-Mast, "The Fundamentals of Nutrition," 1940. Courtesy of Charles C Thomas, Publisher, Springfield, Illinois.) Left: The Apollo space suit. The same basic calorimetric data were needed in the design of space suits. (L. Malla, "Suiting up for Space," John Day Co., New York, 1971. By permission of the National Aeronautics and Space Administration.)

therefore heats it as the reaction proceeds. This situation occurs in an explosion that takes place in an isolated system, such as is approximated by an open flame and in the combustion of rocket fuel.

The confined explosion is the simplest case for us to treat more quantitatively. We can do so in terms of a generalized combustion reaction of a hydrocarbon using a scheme similar to that of the example of Section 5-6:

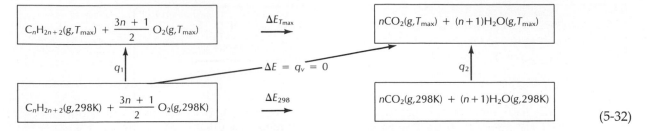

$$(5\text{-}32)$$

To simplify matters, we avoid phase changes by specifying all species to be gaseous. Also, since the system is to be at constant volume, we must deal with E's and C_V's. We are interested in the value of T_{max}, the maximum temperature attained under adiabatic conditions. This means that the overall process is the one indicated by the arrow labeled $\Delta E = q_V = 0$. The desired T_{max} may be calculated as follows. We allow the reaction to proceed at 298 K and apply the heat of reaction to warming the products, which means the q_2 must just equal ΔE_{298}:

$$0 = \Delta E_{298} + \int_{298\ K}^{T_{max}} [nC_V(CO_2) + (n+1)C_V(H_2O)]\ dT$$

or, in general,

$$0 = \Delta E_{T_1} + \int_{T_1}^{T_{max}} \sum_{products} n_p C_V\ dT \qquad (5\text{-}33)$$

The heat capacities should be known as a function of temperature, and on carrying out the integration, one obtains an equation which is then to be solved for T_{max}.

Fundamental hazards to calculations of this type are the following. Since ΔE is negative, we expect from Le Châtelier's principle that the equilibrium will shift progressively to the left with increasing temperature. This means that combustion reactions which go to completion at 25°C may fail to do so at some very high temperature. An even more serious difficulty is that other equilibria begin to be important at very high temperatures. Water may be partly dissociated into H, HO, and O radicals; carbon

dioxide begins to dissociate into CO and O_2 and further into C and O. Gases inert at room temperature, such as nitrogen, begin to enter into the reaction, and so on. All of these effects act either to reduce the amount of reaction or to expend some of the heat of the reaction in dissociating the products into fragments. The result is to reduce T_{max} below its simple theoretical value, and often considerably so.

The problem becomes a very acute one in the field of aerospace engineering. A rocket exhaust is essentially a flame which has reached some high temperature by virtue of the heat of reaction of the propellants, and it is essential to know both the actual T_{max} and the chemical composition of the exhaust if the thrust of the rocket is to be computed. The various dissociative equilibria mentioned here themselves depend on temperature, and the temperature reached depends on the extent of such dissociation. The consequence is that one must make a very involved series of successive approximations in order to arrive at a result.

5-ST-2 Chemical bond strengths

One of the important contributions of thermochemistry to physical chemistry and chemical physics is that it provides chemical bond energies. The definition of bond energy is simple in the case of a diatomic molecule; it is just the dissociation energy. Thus

$$H_2(g) = 2H(g) \qquad \Delta H^0_{298} = 435.9\ kJ \qquad (5\text{-}34)$$
$$HI(g) = H(g) + I(g) \qquad \Delta H^0_{298} = 298.7\ kJ \qquad (5\text{-}35)$$

EXAMPLE   Referring to the preceding example, ΔH^0_{298} for the combustion of methane is -804.24 kJ. The temperature reached in an explosion is given by applying ΔE_{298} (also -804.24 kJ in this case) to heating the products. Assuming for simplicity that C_V is $6R$ for both CO_2 and H_2O, the total C_V is $(3)(6)(8.314) = 150$ J K^{-1} and $\Delta T = 804,240/150 = 5360$ K, or $T_{max} = 5660$ K.

The H—H and H—I bond energies are then 435.9 and 298.7 kJ, respectively.

An alternative to the foregoing is to use ΔE_0 values for bond energies where, as indicated in Fig. 4-13, ΔE_0 is the energy to take a diatomic molecule from the lowest or zero-point vibrational state to the dissociation limit. Spectroscopy gives ΔE_0 values. There is undoubtedly some confusion in that spectroscopic ΔE_0 values and thermochemical ΔH^0_{298} values have been used somewhat interchangeably. The two values are not greatly different, however. For example, ΔE_0 for the atomization of hydrogen is 431.8 kJ, as compared to the ΔH^0_{298} of 436.0 kJ. The practice here will be to use the standard enthalpies.

Bond energies may also be obtained for polyatomic molecules. The enthalpy of atomization of methane is taken to be four times the average C—H bond energy:

$$CH_4(g) = C(g) + 4H(g)$$
$$\Delta H^0_{298} = 1662 \text{ kJ} \qquad (5\text{-}36)$$

The average C—H bond energy is then 415 kJ. The ΔH^0_{298} value is obtained by summing the following steps:

$$CH_4(g) = C(\text{graphite}) + 2H_2(g)$$
$$2H_2(g) = 4H(g)$$
$$\underline{C(\text{graphite}) = C(g)}$$
$$CH_4(g) = C(g) + 4H(g)$$

$$\begin{aligned}
\Delta H^0_{298} &= 74.89 \text{ kJ} \\
\Delta H^0_{298} &= 871.9 \text{ kJ} \\
\underline{\Delta H^0_{298} = 715.0 \text{ kJ}} \\
\Delta H^0_{298} &= 1662 \text{ kJ}
\end{aligned} \qquad (5\text{-}37)$$

Among these values the enthalpy of sublimation of graphite has been the most uncertain because of the great experimental difficulties involved; as a result, several values appear in the chemical literature. For this reason, different references may report somewhat different C—H bond energies. It should also be stressed that the value given here is the *average* C—H bond energy. The process of dissociating one hydrogen atom at a time to give CH_3, CH_2, CH, and then C will require somewhat different energies for each step; that for the first step is estimated to be 427 kJ, for example.

The H—O bond energy may be obtained from the sum of reactions

$$\begin{aligned}
H_2O(g) &= H_2(g) + \tfrac{1}{2}O_2(g) & \Delta H^0_{298} &= 241.8 \text{ kJ} \\
H_2(g) &= 2H(g) & \Delta H^0_{298} &= 436.0 \text{ kJ} \\
\underline{\tfrac{1}{2}O_2(g) = O(g)} & & \underline{\Delta H^0_{298} = 247.7 \text{ kJ}} \\
H_2O(g) &= 2H(g) + O(g) & \Delta H^0_{298} &= 925.5 \text{ kJ}
\end{aligned}$$
$$(5\text{-}38)$$

There are two H—O bonds in water, and so the average H—O bond energy is 463 kJ. The N—H bond energy follows from an analogous calculation involving ammonia.

We next assume that bond energies are additive, that is, that the strength of a given type of bond is independent of its chemical environment. For example, the enthalpy of dissociation of ethane into atoms can be calculated from its enthalpy of formation and the enthalpies of sublimation of graphite and of dissociation of hydrogen. We get

$$C_2H_6(g) = 2C(g) + 6H(g) \quad \Delta H^0_{298} = 2823 \text{ kJ}$$

Ethane consists of six C—H bonds and one C—C bond, and the assumption is that the sum of these bond dissociation energies should equal the overall enthalpy of atomization. The C—H bond is taken to be the same as the average value in methane, so that the C—C bond energy should be $2822 - 2493 = 329$ kJ. Repetition of the same type of calculation for propane leads to a C—C bond energy of 326 kJ, so the assumption is at least approximately correct. Application of the above procedure to methanol yields a C—O bond energy, and so on.

A number of such bond strengths are given in Table 5-5 along with some useful heats of sublimation of elements. Note that the value given there for the C—C and C—H bond energies are somewhat different than the ones just obtained. The figures in the table are offered as better average values for general use.

In some cases the discrepancy between an observed enthalpy and that calculated from bond energies appears to be real—that it, due to a specific, neglected factor. For example, one obtains from the enthalpies of formation of benzene and of atomization of graphite and hydrogen

$$C_6H_6(g) = 6C(g) + 6H(g)$$
$$\Delta H^0_{298} = 5515 \text{ kJ} \qquad (5\text{-}39)$$

TABLE 5-5
Enthalpies of formation of atoms and bond strengths[a]

Atom	$\Delta H^0_{f,298}$ [b]		Bond	ΔH^0_{298}	
	kcal mol⁻¹	kJ mol⁻¹		kcal mol⁻¹	kJ mol⁻¹
H	52.1	218.0	C—H	99.5	416.3
O	59.2	247.7	C—C	83	347
N	112.9	472.4	C=C	146	611
S	53.2	223.8	C≡C	198	828
P	75.2	314.6	C—N	72	301
C	170.9	715.0	C=N	147	615
F	18.3	76.6	C—O	85	356
Cl	29.0	121.3	C=O	175	732
Br	26.7	111.7	C—Cl	78	326
I	25.5	106.7	O—H	111	464
Na	26.0	108.8	O—O	33	138
K	21.5	90.0	N—H	93	389
Ca	46.0	192.4	N—N	39	163
Al	75.0	313.8	N=N	100	418
Ni	102	426.8	Cl—Cl	58	242
Fe	96.7	404.6	Br—Br	46	193
Ag	69.1	289.1	I—I	36	151

[a]The values are adapted from various sources. See NBS Circular 500 (Table 5-3); K.S. Pitzer, "Quantum Chemistry," Prentice-Hall, Englewood Cliffs, N.J., 1953; T. L. Cottrell, "The Strength of Chemical Bonds," Butterworth, London and Washington, D.C., 1958.
[b]From their standard states.

Benzene, if written in the Kekulé structure

would be assigned six C—H bonds, three C—C bonds, and three C=C bonds, which total 5372 kJ. The discrepancy of 143 kJ suggests that benzene is more stable than expected in terms of this structure. The modern explanation is that there are no fixed C=C bonds and that the electrons which might go into such bonds interact instead in a diffuse or delocalized way and are spread over the whole molecule. This difference between the estimated energy of a fixed bond structure and the actual energy of a delocalized bonding structure is called the *resonance energy*.

Although the assumption of additivity of bond strengths can lead to appreciable error,

it does provide a means of estimating enthalpies of formation of compounds not yet studied or difficult to study. In the case of flames, for example, one may need an estimate of the enthalpies of formation of various radicals. A similar situation occurs in chemical kinetics, where it is desirable to estimate the energy required to produce possible reaction intermediates (see Benson, 1968).

5-ST-3   Internal energy and enthalpy functions

The fact that we do not know absolute values for internal energies and enthalpies is reflected in our procedures in thermochemistry. Thus we tabulate standard enthalpies of formation of substances and not their absolute enthalpies. That is, we take the elements in their standard state as a point of reference. Since we are interested in ΔH or ΔE for a chemical reaction, only the differences between enthalpies or energies of formation are involved, and the choice of reference state cancels out. Were the absolute H and E values known for the elements, we could then convert all of the standard heats of formation to absolute values. However, such a set of values would lead to exactly the same ΔH or ΔE for a chemical process as before. (The proof of this statement is left as an exercise.)

The choice of reference state is therefore mainly a matter of convenience. Although the usual choice is well suited for the compilation of thermochemical data, there is an alternative, a more natural one from the point of view of statistical thermodynamics. This alternative may be introduced as follows. By Eq. (4-24),

$$\left(\frac{\partial E}{\partial T}\right)_V = C_V$$

where, it will be recalled, E is the average energy per molecule. Integration then gives

$$E = E_0 + \int_0^T C_V \, dT \tag{5-40}$$

The energy E_0 is the energy per molecule at 0 K and hence the energy in the lowest translational, rotational, vibrational, and electronic energy state. The first two we regard as zero, the third is the zero-point energy, $\frac{1}{2}h\nu_0$, and the fourth is unknown. In fact, $E_{0,\text{elec}}$ is essentially unknowable, since our experience, as embodied in the first law, is that we can determine only changes in energy and not absolute values. If we now define a quantity $E - E_0$, the *internal energy function*, we have

$$E - E_0 = \int_0^T C_V \, dT \tag{5-41}$$

and the quantity on the right is in principle experimentally determinable. It is, for example, actually $E - E_0$ that is given by Eq. (4-69),

$$E - E_0 = kT^2 \left[\frac{\partial(\ln \mathbf{Q})}{\partial T}\right]_V \tag{5-42}$$

or, per mole,

$$E - E_0 = RT^2 \left[\frac{\partial(\ln \mathbf{Q})}{\partial T}\right]_V \tag{5-43}$$

The statistical thermodynamic treatment in Chapter 4 showed that it is usually possible to factor a partition function into the separate translational, rotational, and vibrational contributions. Thus

$$E - E_0 = E_{\text{trans}} + E_{\text{rot}} + E_{\text{vib}} \tag{5-44}$$

Also, from the definition of H,

$$H - E_0 = (E - E_0) + PV \tag{5-45}$$

and

$$H^0 - E_0 = (E - E_0) + PV \tag{5-46}$$

where the zero superscript denotes that H is for the standard pressure of 1 atm; $H^0 - E_0$ is the standard enthalpy function.* For an ideal gas,

$$H^0 - E_0 = (E - E_0) + RT$$
$$\text{(ideal gas)} \tag{5-47}$$

The standard enthalpy function may be related to experiment as follows. For the general reaction

$$aA + bB + \cdots = mM + nN + \cdots$$

we can write ΔH^0 for some temperature T as

$$\Delta H_T^0 = \Delta(H^0 - E_0)_T + \Delta E_0 \tag{5-48}$$

If we know $\Delta(H^0 - E_0)_T$, ΔE_0 can be calculated from the experimental heat of reaction. Since ΔE_0 is the internal energy change when the reaction occurs with each species in its lowest energy state, it is a fundamental quantity for the reaction. Thus ΔH_T^0 has been written as the sum of ΔE_0, about which *nothing* can be calculated, and a term $\Delta(H^0 - E_0)_T$, which *can* be calculated. It is a way of concentrating the full capabilities of thermodynamics into a single term.

We can now proceed to examine the calculation of $(H^0 - E_0)_T$ for a substance. There are two independent ways of doing this. First, we can obtain $(E - E_0)_T$ and hence $(H^0 - E_0)_T$ by means of Eq. (5-41). This requires a complete knowledge of the heat capacity of the substance from 0 K up to the desired temperature. If phase changes occur in the process of heating the substance from 0 K, then the energies of these changes must be included, essentially in the manner indicated in Fig. 5-2. Alternatively, $(H^0 - E_0)_T$ could be obtained from a similar heating at constant pressure and from the variation of C_P with temperature and the enthalpies of any phase changes. This approach is discussed further in Section 6-ST-3.

The heat capacity approach does not really add anything over the discussion in Section 5-

*More often, this term is applied to $(H^0 - H_0^0)/T$ as discussed in Section 6-ST-3.

6 on the temperature dependence of ΔH and ΔE. It is of more importance that $(H^0 - E_0)_T$ can be calculated by the methods of statistical thermodynamics. Thus if the various energy states are known, $(H^0 - E_0)_T$ can be obtained directly through Eq. (5-44). By repeating the calculation for each substance involved in a chemical reaction, one then obtains $\Delta(H^0 - E_0)_T$ and, by means of Eq. (5-48), ΔE_0. Having ΔE_0, one can now calculate ΔH_T^0 for any temperature. There are some considerable further advantages to the use of enthalpy functions in dealing with applications of the second law of thermodynamics; these are discussed in the Special Topics section of the next chapter.

The preceding discussion illustrates several points. First we see again that the energy of an ideal gas is a function of temperature only; the calculations of $E - E_0$ did not require a value for the pressure. Second, the major contributions to $E - E_0$ were from translation and rotation. This will generally be true for small, strongly bonded molecules around 25°C (the discussion of Section 4-9) on the molecular interpretation of heat capacities is relevant at this point). Considerations such as this last one are often helpful in estimations of the vibrational contribution when quantitative information is not available. Thus the use of the enthalpy function allows an intermixing of phenomenological or classical thermodynamics and statistical thermodynamics.

EXAMPLE   To illustrate the application of statistical thermodynamics to thermochemistry, consider the reaction

$$3H_2 + N_2 = 2NH_3 \qquad \Delta H_{298}^0 = -92.38 \text{ kJ} \tag{5-49}$$

We assume that each gas is ideal and can apply the methods of Chapter 4 to the calculation of $E - E_0$ for each species. It is safe to assume that at 25°C the equipartition values will be reached with respect to the translational and rotational contributions. The calculations then center on obtaining E_{vib}; we use Eq. (4-96) and need the fundamental frequency or characteristic vibrational temperature θ_{vib} for each mode of vibration. The latter is given by $\theta_{vib} = h\nu_0/k$ or, if ν_0 is in cm^{-1}, $\theta_{vib} = hc\nu_0/k = 1.439\nu_0$. Then E_{vib} is [Eq. (4-97)]

$$E_{vib} = RT\frac{x}{e^x - 1} = RTf(x)$$

Hydrogen
    translation: $E_{trans} = \frac{3}{2}RT =$ 447 R
    rotation: $E_{rot} = RT =$ 298 R
    vibration: $\nu_0 = 4160.2$ cm^{-1}, $\theta_{vib} = 5987$ K,
                $x = 20.08$, $f(x) = 0.0000$:
                $E_{vib} = RTf(x) =$ 0

                                                          745   R

Nitrogen
    translation: same as for H_2 447 R
    rotation: same as for H_2 298 R
    vibration: $\nu_0 = 2230.7$ cm^{-1}, $\theta_{vib} = 3210$ K,
                $x = 10.77$, $f(x) = 0.0003$:
                $E_{vib} = 0.003RT =$ 0

                                                           745   R

Ammonia
    translation: same as for H_2 and N_2 447 R
    rotation: $E_{vib} = \frac{3}{2}RT =$ 447 R

    vibration: there are six vibrational degrees of freedom, but only four frequencies since two are doubly degenerate; thus

$\nu_1 = 3337$ cm^{-1}, from which
$f(x) = 0.0000$:

$\nu_2 = 950$ cm^{-1}, $f(x) = 0.047$:	$E_1 =$	0
$\nu_3 = 3414$ cm^{-1} (twice);	$E_2 = 0.047\ RT =$	14.0 R
$f(x) = 0.0000$:	$E_3 =$	0
$\nu_4 = 1627.5$ cm^{-1} (twice);		
$f(x) = 0.003$:	$E_4 = (0.003)2RT =$	1.8 R
		$\overline{910\ \ R}$

$$\Delta(E - E_0) = [2(910) - 745 - 3(745)]R = -1160\ R = -9645\ J$$

and

$$\Delta(H^0 - E_0) = -1160R - 2RT = -1756R = -14{,}600\ J$$

Finally, from Eq. (5-48),

$$\Delta E_0 = -92.38 - (-14.60) = -77.8\ kJ$$

Knowing ΔE_0, we could repeat the entire calculation for any other temperature to obtain a ΔH^0 for that temperature. The temperature could, for example, be one not easily accessible for direct experimental measurements.

GENERAL
REFERENCES

BARTHEL, J. (1975), "Thermometric Titrations." Wiley, New York.
COTTRELL, T.L. (1958). "The Strengths of Chemical Bonds." Butterworth, London and Washington, D. C. (Bond dissociation energies.)
ROSSINI, F.A., et al., eds. (1952). Tables of Selected Values of Chemical Thermodynamic Properties. Nat. Bur. Std. Circular No. 500. Also, EVANS, W.H., PARKER, V.B., HALOW, I., BAILEY, M., AND SCHUMM, R.H., eds., National Bureau of Standards Technical Note 270-3, 1968. (Sources for thermochemical data.)
STURTEVANT, J.M. (1959). "Physical Methods of Organic Chemistry" (A. Weissberger, ed.), 3rd ed., Vol. 1. Wiley (Interscience), New York. (Experimental calorimetry.)

CITED
REFERENCES

BARTHEL, J. (1975), "Thermometric Titrations." Wiley, New York.
BENSON, S.W. (1968), "Thermochemical Kinetics." Wiley, New York.
BROSEN, E.J., AND ROSSINI, F.D. (1946). J. Res. Nat. Bur. Std. A **36**, 274.
GEMMILL, C.L., AND BROBECK, J. R. (1968), "Medicinal Physiology," V.D. Mountcastle, ed. p. 473. C.V. Mosby Co., St. Louis.
HOCH, M., AND JOHNSTON, H.L. (1961). J. Phys. Chem. **65**, 856.
KNOWLTON, J.W., AND ROSSINI, F.D. (1949). J. Res. Nat. Bur. Std. A **43**, 113.
WAGMAN, D.D., KILPATRICK, W.J., TAYLOR, W.J., PITZER, K.S., AND ROSSINI, F.D. (1945). J. Res. Nat. Bur. Std. A **34**, 155.

EXERCISES

Remember to take as exact numbers given to one significant figure. Give your answers in kilocalories and in kilojoules. (It will be useful to get a feeling for typical numbers in both sets of units.)

Various types of carbon blacks (more or less amorphous carbon) are used as adsorbents. The heat of combustion of a particular carbon black is found to be 8030 cal g^{-1} in a constant volume combustion calorimeter, at 25°C. Calculate ΔH^0 for the process C(amorphous) = C(graphite).

5-1

Ans. -2.39 kcal mol^{-1}, -10.0 kJ mol^{-1}.

5-2   Calculate ΔH^0_{298} and ΔE_{298} for the reaction:

$$H_2O(l) + F_2(g) = 2HF(g) + \tfrac{1}{2}O_2(g)$$

Ans. $\Delta H^0_{298} = -61.26$ kcal or -256.3 kJ;
$\Delta E_{298} = -62.15$ kcal or -260.0 kJ.

5-3   A sample consisting of 0.200 g of $CH_3COOH(l)$ is ignited in a bomb calorimeter containing pure oxygen. The heat capacity of the calorimeter is 12,000 J °C^{-1}. What temperature increase should occur?

Ans. 0.242°C.

5-4   The heat capacity of a bomb calorimeter was first determined by electrical heating; the temperature rose by 0.890°C when a current of 2.500 A from a 12 V battery was passed through a resistance heater for 25 s. Combustion of 20 mg of styrene (C_8H_8) led to a temperature rise of 1.0097°C. Assume the temperature to be essentially constant at 25°C and neglect any additional heat capacity due to the reaction products. Assume also that the reaction is of $C_8H_8(g)$ and that the product water is present as liquid. Calculate ΔH^0_{298} of formation and of combustion of gaseous styrene, and the corresponding ΔE_{298} values.

Ans. $\Delta H^0_{c,298} = -4.4389 \times 10^3$ kJ (or -1.061×10^3 kcal) mol^{-1},
$\Delta E_{c,298} = -4.4315 \times 10^3$ kJ (or -1.059×10^3 kcal) mol^{-1};
$\Delta H^0_{f,298} = 147.5$ kJ (or 35.3 kcal) mol^{-1},
$\Delta E_{f,298} = 154.9$ kJ (or 37.0 kcal) mol^{-1}.

5-5   Calculate ΔH^0_{298} and ΔE_{298} for the incomplete combustion reaction

$$C_6H_6(g) + 1\tfrac{1}{2}O_2(g) = 6C(s) + 3H_2O(l)$$

Ans. $\Delta H^0_{298} = -224.77$ *kcal or* -940.5 *kJ;* $\Delta E = -223.30$ *kcal or*
-934.3 *kJ.*

5-6   Calculate ΔH^0_{298} for the combustion of glycine(s).

Ans. -162.4 kcal or -679.4 kJ.

5-7   Calculate ΔH^0_{298} per mole for the *cis* to *trans* isomerization of $CH_3CH = CHCH_3$.

Ans. -949 cal or -3.97 kJ.

5-8   A small catalytic hydrogenation unit has a heat capacity of 1200 J K^{-1}. What will its rate of heating be if an equimolar mixture of ethylene and hydrogen gases is passed through it at the rate of 2 millimoles of mixture per minute? Assume complete reaction.

Ans. 0.228 K min^{-1}.

5-9   Calculate ΔH^0_{298} for the process

$$Ag + HCl(aq) = AgCl(s) + \tfrac{1}{2}H_2(g)$$

The enthalpy of formation of AgCl(s) is -127.0 kJ.

Ans. 40.5 kJ or 9.67 kcal.

5-10   Obtain ΔH^0_{298} for the formation of $NH_4OH(aq)$.

Ans. -86.6 kcal or -362 kJ.

5-11   Calculate ΔH^0_{298} for the dilution process: NaCl (in 100 moles H_2O) = NaCl (aq).

Ans. -220 J or -53 cal.

5-12   Calculate ΔH^0_f of CH_4 at 400 K.

Ans. -77.9 kJ or -18.6 kcal.

Calculate ΔH_f^0 of $CH_3COOH(g)$ at 100°C given that ΔH_v is 405 J g^{-1} at the normal boiling point of 118°C and that the values of C_P are 2.30 J K^{-1} g^{-1} and 6.27 J K^{-1} g^{-1} for liquid and gaseous acetic acid, respectively.

5-13

Ans. -107.0 kcal mol^{-1} or -447.9 kJ mol^{-1}.

How many g of ice would be melted if 1 g of graphite were burned to CO_2 in an ice calorimeter? The heat of fusion of ice is 1.44 kcal mol^{-1}.

5-14

Ans. 97.9 g.

PROBLEMS

Problems marked with an asterisk require fairly lengthy computations.

The heat of combustion of ethylene glycol is $\Delta H_{c,298}^0 = -281.9$ kcal mol^{-1}. Calculate ΔH_{298}^0 for the reaction $(CH_2OH)_2(l) = C_2H_5OH(l) + \frac{1}{2}O_2(g)$ and $\Delta H_{f,298}^0$ for ethylene glycol. Assume the combustion is to $H_2O(l)$.

5-15

Calculate ΔH_{298}^0 and ΔE_{298} for the reaction

5-16

$$2H_2O(l) + CCl_4 + 3SO_2 = S(\text{monoclinic}) + CO_2 + 4HCl(g) + 2SO_3(g)$$

The heat of sublimation of graphite to carbon atoms has been estimated as 170 kcal mol^{-1}. The dissociation of molecular hydrogen into atoms, $H_2 = 2H$, has $\Delta H^0 = 103.2$ kcal mol^{-1}. From these data and the value for the heat of formation of methane, calculate ΔH^0 for

5-17

$$C(g) + 4H(g) = CH_4(g)$$

One-fourth of this value is a measure of the C—H bond energy in methane.

Calculate ΔH_{298} for the process C(graphite) = C(diamond) under a pressure of 10,000 atm. From Eq. (6-102), $(\partial H/\partial P)_T = V(1 - T\alpha)$, but we can make the approximation of neglecting the α term and also of assuming that V does not change significantly with pressure. The densities of graphite and of diamond are 2.25 and 3.51 g cm^{-3}, respectively.

5-18

At 25°C and constant pressure, the heat of formation of CO_2 from graphite is -94.272 kcal mol^{-1} and the heat of combustion of CO is -67.263 kcal mol^{-1}. Calculate ΔH for the formation of one mole of CO from C and CO_2 (at 1000 K) if the following equations are valid for the molar heat capacities:

5-19

for C (graphite) $C_P = 1.20 + 0.0050T - 1.2 \times 10^{-6}T^2$

for $CO_2(g)$ $C_P = 7.40 + 0.006T - 1.50 \times 10^{-6}T^2$

for CO(g) and $O_2(g)$ $C_P = 6.5 + 0.0010T$

ΔH_{298}^0 is 0.83 kJ for the reaction of $N_2O(g)$ with $NO_2(g)$ to give $N_2O_3(g)$ and is -206.47 kJ for the reaction of $N_2O(g)$ with $N_2O_3(g)$ to give NO(g). Using the data given and that in Table 5-2, calculate the standard enthalpies of formation of $N_2O(g)$ and $N_2O_3(g)$.

5-20

For the following reaction at 25°C, $\Delta H = -56$ kcal:

5-21

$$2HI(g, 1 \text{ atm}) + Cl_2(g, 1 \text{ atm}) = 2HCl(g, 1 \text{ atm}) + I_2(s)$$

Given that the heat of sublimation of $I_2(s)$ is 6.1 kcal mol^{-1} at 185°C; the heat capacities C_P for the gases HI, HCl, Cl_2, and I_2 may be taken to be 6.5 cal °C^{-1} mol^{-1} and independent of temperature; the heat capacity C_P for $I_2(s)$ is 12.6 cal °C^{-1} mol^{-1} (independent of T); calculate ΔH for the following reaction at 125°C:

$$2HI(g, 1 \text{ atm}) + Cl_2(g, 1 \text{ atm}) = 2HCl(g, 1 \text{ atm}) + I_2(g, 1 \text{ atm})$$

5-22   The $\Delta H_{c,298}^0$ is -1622 kJ for alanine, $CH_3CH(NH_2)COOH(s)$. Calculate ΔH_{298}^0 for the reaction: $C_2H_4(g) + NH_3(g) + CO_2(g) = CH_3CH(NH_2)COOH(s)$. Calculate also $\Delta H_{f,298}^0$ for alanine.

5-23   The heats of combustion at 25°C and 1 atm of cyclopropane, $[(CH_2)_3(g)]$, graphite, and hydrogen are -499.85 kcal mol^{-1}, -94.051 kcal mol^{-1}, and -68.317 kcal mol^{-1}, respectively [values for graphite and hydrogen are from Wagman *et al.* (1945)]. At 25°C, ΔH for the formation of propene, $CH_3CH = CH_2(g)$, from the elements is 4.879 kcal (Brosen and Rossini, 1946). Find ΔH for (a) the formation of one mole of cyclopropane from the elements, (b) the isomerization of one mole of cyclopropane at 25°C and 1 atm (Knowlton and Rossini, 1949).

5-24   Calculate ΔH_{298}^0 for the process: $H_2SO_4(aq) + 2NaCl(aq) = Na_2SO_4(aq) + 2HCl$.

5-25* Find a, b, and c values for the equation $C_P^0 = a + bT + cT^2$ that fit the data for NH_3 over the range 298 K to 2000 K. (Heat capacity data are often reported in this form). Use Eq. (5-27) and Table 5-4 to generate your data. Using calculated points at 200 K intervals, find the a, b, and c values giving the best least squares fit.

5-26* The heat capacity of methane is reported to fit the equation $C_P/R = \alpha + \beta T + \gamma T^2$ over the range 298 K to 1500 K, with $\alpha = 1.701$, $\beta = 9.080 \times 10^{-3}$, and $\gamma = -21.64 \times 10^{-7}$ with T in kelvins. Make a plot of C_P/R vs. T from 298 K to 1500 K using both the above equation and Eq. (5-27) with the constants given in Table 5-4. Also make a plot of $[(C_P/R)^* - (C_P/R)]/(C_P/R)^*$ over the same temperature range, where $(C_P/R)^*$ is the value given by Eq. (5-27) and (C_P/R) is the value given by the above equation.

5-27   Calculate ΔH^0 at 90°C for the combustion of $C_6H_6(g)$ to $CO_2(g)$ and $H_2O(g)$. Use only the following data. Substance (C_P in J K^{-1} mol^{-1}): $C_6H_6(l)$ (130); $C_6H_6(g)$ (100); $O_2(g)$ (30); $CO_2(g)$ (47); $H_2O(l)$ (75); $H_2O(g)$ (34). Also the enthalpies of vaporization are 34,000 and 44,000 J mol^{-1} for C_6H_6 and for H_2O, respectively, at their normal boiling points of 80°C and 100°C. The standard enthalpy of combustion of $C_6H_6(l)$ to $H_2O(l)$ is -3268 kJ mol^{-1} at 25°C.

5-28   A small photocalorimeter cell contains 25 cm^3 of a solution that is totally absorbing to light of 530 nm wavelength. Exposure of the cell to a beam of light of this wavelength for 35 s results in a change in temperature of 6.5 scale divisions on a chart recorder where scale divisions are proportional to temperature. A resistance heater of 10.0 Ω resistance produces a temperature change of 5.5 scale divisions when 0.25 A of current is passed through the resistance for a period of 15 s. Calculate the intensity of the light beam in einstein s^{-1} (an einstein is one mole of light quanta).

5-29   An aqueous solution of $Cr(NH_3)_6^{3+}$ is totally absorbing to 450 nm light. Ten cm^3 of solution at 25°C is irradiated for 10 s with an incident intensity of 8.00×10^{16} quanta s^{-1} and a temperature rise of 0.00950°C is observed. Calculate ΔH for the reaction $Cr(NH_3)_6^{3+} + H_2O = Cr(NH_3)_5(H_2O)^{3+} + NH_3(aq)$ if this is the photochemical reaction and the quantum efficiency is 0.36 (that is, 0.36 mole of reaction occurs for every mole of light quanta absorbed). Assume the heat capacity of the solution is that for water.

SPECIAL TOPICS PROBLEMS

5-30   Calculate the maximum temperature achieved by the mixture when 0.100 g of ethane gas is exploded in a constant-volume adiabatic (that is, insulated) calorimeter containing 10 g of air. The initial temperature is 25°C. Assume ideal gas behavior.

5-31* One-half mole of ethane at 25°C is exploded in a large excess, ten moles, of oxygen, burning completely to carbon dioxide and water. Calculate the final temperature under adiabatic conditions (a) assuming the explosion to be at constant pressure, (b) assuming it to be at constant volume.

5-32   Calculate ΔH_{298}^0 for formation of butadiene, using bond energies.

5-33   The standard heat of combustion of naphthalene (solid) is -1231 kcal mol^{-1} at 25°C (to CO_2 and liquid water) and the heat of sublimation is about 9 kcal mol^{-1}. Estimate the resonance energy for naphthalene.

ΔH^0_{298} is 71.4 kcal for the reaction $HI(g) = H(g) + I(g)$. C_P for HI is $5.00 + 1 \times 10^{-3}T$ in cal K^{-1} mol^{-1}. The bond energy of HCl at 25°C is 103 kcal mol^{-1}. Calculate (a) ΔH^0 and ΔE for the atomization of $HI(g)$ at 1000 K, and (b) ΔH^0_{298} for $HCl + I = HI + Cl$.

5-34

Calculate ΔE_0 for the reaction $H_2(g) + \frac{1}{2}O_2(g) = H_2O(g)$ using data from Tables 5-2 and 6-1; then calculate ΔH^0 at 3000°C. The fundamental frequencies for H_2O are: $\nu_1 = 3652$ cm^{-1}, $\nu_2 = 1595$ cm^{-1}, and $\nu_3 = 3756$ cm^{-1}.

5-35

Calculate ΔE_0 for the reaction: $CO(g) + \frac{1}{2}O_2(g) = CO_2(g)$ using data from Tables 5-2 and 6-1; then calculate ΔH^0 for the reaction at 1500°C. Compare your result with the value obtained using heat capacity data from Table 5-4 and the value of ΔH^0 at 25°C. For CO_2, a linear molecule, $\nu_1 = 1388.3$ cm^{-1}, $\nu_2 = 667.3$ cm^{-1} (twice), and $\nu_3 = 2349.3$ cm^{-1}.

5-36

chapter 6
The second and third laws of thermodynamics

6-1   Introduction

The second law of thermodynamics stems from general conclusions drawn from experience about the difference in nature between heat energy and work energy. The qualitative observation is that work energy can always be converted completely into heat energy but that the reverse is not possible in the absence of other change. Examples of the first part of the statement are common. The experiments of Count Rumford on the boring of cannon and Joule's experiments of heating water by means of a weight-driven paddle wheel were cited in connection with the first law. The braking of any motor reduces its work output to heat, the impact of a falling object converts its kinetic energy to heat, and so forth. In general any change or process capable of doing work may be arranged to produce instead the equivalent amount of heat, if only by providing sufficient friction.

As a further example, imagine an ideal gas contained in a cylinder with a piston, the whole in thermal contact with the surroundings, that is, at constant temperature. As illustrated in Fig. 6-1(a), the expansion might be reversible; the piston turns a wheel which drives a cam so designed that at each stage of the expansion the opposing force of the weight being lifted is only infinitesimally less than the force of the gas on the piston. The process is isothermal and, since the gas is ideal, $\Delta E = 0$ and $q = w$. Thus work w has been done at the expense of heat energy q from the surroundings, but with a change in state of the gas. The gas could be returned to its original state if the weight were made infinitesimally heavier. The work would now be $-w$ and the heat $-q$. The net for the cycle would be $w = 0$ and $q = 0$. Now consider the experiment depicted in Fig. 6-1(b). A valve in the piston is opened, allowing the gas to expand freely. ΔE is still zero and now $w = 0$, so that $q = 0$. However, to return the gas to its original state with the least possible work, the valve would have to be closed, and the minimum possible weight hung from the cam. Work would be $-w$ and heat $-q$; the net effect of this cycle would be that the surroundings have supplied work w, which has been entirely converted to heat. In the isothermal irreversible cycle, work has been converted to heat; in the isothermal reversible cycle there is no net conversion

FIGURE 6-1
(a) Reversible and (b)
irreversible expansions.

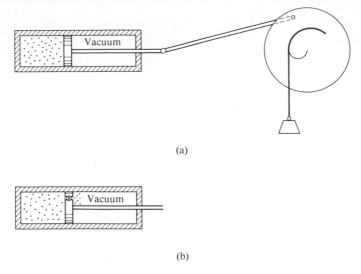

(a)

(b)

either way. This last represents the best we can do; we cannot find an isothermal cycle for our gas the net effect of which would be to convert heat into work.

One statement of the second law is simply a generalization of the above conclusion: *It is not possible to convert heat into work by means of an isothermal cyclic process.* This is a negative statement but one that summarizes much experience. Were the statement incorrect, it should be possible to devise a cycle, or essentially an engine, that operates isothermally and which supplies work energy at the expense of the heat energy of the surroundings. Such an engine could, for example, drive a ship at sea by drawing on the heat energy of the ocean. Engines of this type are known as *perpetual motion machines of the second kind,* and the above statement of the second law affirms the nonexistence of such an engine.

A second facet of observations such as those just referred to is that a process that will take place spontaneously can be made to do work. A gas expands spontaneously into a vacuum; and the arrangement of Fig. 6-1(a) illustrates a means of obtaining work out of the same change in state of the gas. If the process were carried out reversibly, the arrangement would give the maximum possible work. Another type of spontaneous process is the flow of heat from a higher to a lower temperature. As illustrated schematically in Fig. 6-2(a), a bar heated at one end has the other end in contact with a lower-temperature heat reservoir. The spontaneous change will be for heat to flow into the reservoir until the entire bar is at the reservoir temperature T_1. Alternatively, however, work could be obtained. As sketched in Fig. 6-2(b), the hot end of the bar might be used to convert some water into steam, which could then be introduced into a piston and cylinder setup. If the unit should be immersed in the cooler heat reservoir, the steam would condense, pulling down the piston and thus providing work. The end result would be that the same amount of energy would be lost by the bar as before, but now with some of it appearing as work rather than as heat at T_1.

A third type of spontaneous process is the chemical process. A battery

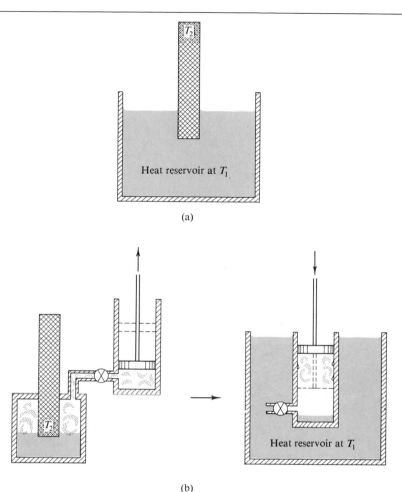

FIGURE 6-2
(a) Irreversible and (b) partially reversible use of heat at T_2.

(a)

(b)

contains chemicals that will react spontaneously if allowed to. The spontaneous reaction will occur if the chemicals are directly mixed or simply if the battery is short-circuited. The chemical energy that is released appears as heat. Alternatively, the battery may be made to run a motor and thus do work. The same chemical energy now appears partly as work.

The quantitative aspects of the second law have to do with specifying the maximum work that a spontaneous process can provide. This is not necessarily given by the internal energy change. For example, the isothermal expansion of an ideal gas provides work, but without an internal energy change of the gas. Next, in the case of heat flow from a hotter to a colder system we can see that the relative temperatures must play a role in determining how much work can be obtained. To pursue the point further, consider the coiled spring shown in Fig. 6-3(a). It can do a certain amount of work w' on uncoiling (as by lifting a weight). If, instead, the spring is placed in a heat reservoir at T_1 and then allowed to uncoil, it does no work, but an equivalent amount of frictional heat energy q_1 is added to the reservoir. Suppose, instead, that the spring had uncoiled in a reservoir at a higher temperature T_2 as illustrated in

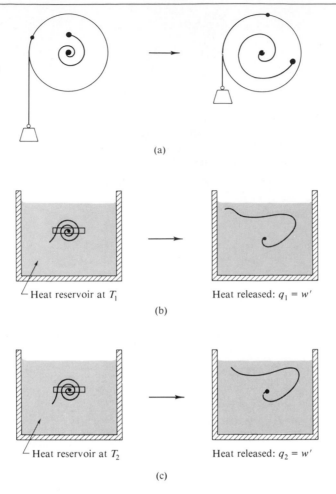

(a)

Heat reservoir at T_1 Heat released: $q_1 = w'$

(b)

Heat reservoir at T_2 Heat released: $q_2 = w'$

(c)

Fig. 6-3(c); it would then produce the same amount of heat as before, but at temperature T_2. Clearly the heat at T_2 could be used in the same way as the hot bar, namely, to do some work w by means of a heat engine operating between T_2 and T_1.

We can expect that if the operations are reversible the work done will be the maximum possible and thus be at a unique value. Also, in a qualitative way we would expect that the greater the difference between T_2 and T_1, the greater should be the fraction of q_2 that can appear as work. A more detailed analysis, as in the next two sections, leads to the conclusion that the measure of *unavailability* of heat energy for work is given by q_{rev}/T. In fact the increment $\delta q_{rev}/T$ turns out to have the properties of a state function; the integral from one state to another is independent of path. This is the central mathematical statement of the second law of thermodynamics and, accordingly, a new state function, *entropy*, is introduced, where

$$dS = \delta q_{rev}/T \qquad (6\text{-}1)$$

Again in a qualitative way, the example of the hot bar losing its heat to the lower-temperature heat reservoir involves an increase in entropy—successive "elements" of heat are being transferred from a higher to a lower temperature so that T in Eq. (6-1) decreases. The process is also a spontaneous one, and another way of generalizing the second law is to say that in a closed system (such as bar plus reservoir) the entropy increases in a spontaneous process.

The various introductory statements of the second law are then as follows:

It is not possible to convert heat into work by means of an isothermal cyclic process.

q_{rev}/T *is a state function.* (6-2)

The entropy of an isolated system will increase if spontaneous processes can occur.

The basis for the second law is a negative observation—our inability to construct a perpetual motion machine of the second kind. Similarly, the first law rests on our inability to construct a perpetual motion machine of the first kind—one that creates energy. As with the first law, there is also a positive side. The various implications of the second law lead to phenomenological relationships that have been abundantly verified. Both laws of thermodynamics are thus most solidly a part of our understanding of the behavior of nature.

6-2 A mathematical approach to the first and second laws of thermodynamics. The Carathéodory theorem.

A fairly standard derivation of the entropy function $dS = \delta q_{rev}/T$ and proof that S is a state function follows in the next section. There is a mathematical approach to the first and second laws of thermodynamics, however, which we outline in this section.

The approach rests heavily on the mathematical properties of exact and inexact differential expressions. We rewrite Eq. (4-10) as

$$dy = M(u,v)du + N(u,v)dv \qquad (6-3)$$

to emphasize that M and N are in general functions of the dependent variables u and v. Recall that Eq. (4-11) provides a test of whether Eq. (6-3) is an exact differential; if it is, then

$$\left(\frac{\partial M}{\partial v}\right)_u = \left(\frac{\partial N}{\partial u}\right)_v \qquad (4-11)$$

If the test holds, dy is a total differential, that is, the integral of dy is independent of path. In the thermodynamic context, this means that y is a state property and a change in Δy is fully determined by the initial and final states regardless of path. An important point, however, is that the functions $M(u,v)du$ and $N(u,v)dv$ need not each be an exact differential, and their integrals may *separately depend on the path of integration. This aspect was illustrated in the example of Section 4-5.*

The *mathematical* statement of the first law of thermodynamics is the affirmation that while δq and δw are inexact differentials individually, their algebraic sum $(\delta q - \delta w)$ *is* exact. That is, $dF = \delta q - \delta w$ and F is a state function.

There is also a mathematical way of formulating the second law of thermodynamics, and this has to do with the behavior of exact and inexact differentials. As was noted in Section 4-5, an expression $[M(u,v)du + N(u,v)dv]$ is not necessarily an exact differential. Thus $(v\,du + 2u\,dv)$ is not exact [and fails the test of Eq. (4-11)]. If, however, the expression is multiplied by v to obtain $(v^2du + 2uv\,dv)$, we now have an exact differential. A multiplying function $G(u,v)$ that converts an inexact differential to an exact one is known as an *integrating factor*. In the above case, v is an integrating factor for the inexact differential $(v\,du + 2u\,dv)$.

To be more specific, in the case of $dy = 0$, we have

$$M(u,v)du + N(u,v)dv = 0 \qquad (6\text{-}4)$$

If multiplication by some function $G(u,v)$ makes the equation integrable, then $G(u,v)$ is an integrating factor. On integration, one obtains some solution of the form $F(u,v) = C$, where C is a constant. In the above case, the solution to $v^2\,du + 2uv\,dv = 0$ is $uv^2 = C$. Thus, $F(u,v)$ is uv^2 and the solution $F(u,v) = C$ generates a family of curves, one for every value of C. It can be shown, in fact, that an integrating factor *must* exist for any differential expression in two variables, that is, of the type of Eq. (6-4) (see Blinder, 1969).

EXAMPLE   In general, if one integrating factor can be found, then a whole family will exist. In fact, $G(u,v)$ times *any function* of $F(u,v)$ will also be an integrating factor. In the above case, $G(u,v)$ is just v, and $F(u,v) = uv^2$. It will then be true that $v(uv^2)^n$ is also an integrating factor. Suppose that $n = -1$, and we have $v(uv^2)^{-1} = 1/uv$. Multiplication of the original expression $(v\,du + 2u\,dv)$ by $1/uv$ gives $(du/u + 2dv/v)$, which integrates to $\ln(uv^2)$. If $\ln(uv^2) = C$ then uv^2 also is a constant, and we have recovered the original $F(u,v)$ solution.

One can look at the reverse procedure. For an ideal gas, $V = RT/P$, hence $dV = -(RT/P^2)dP + (R/P)dT$. Clearly, the expression on the right is exact [and meets the test of Eq. (4-11)]. We can now factor out $(1/P)$ to write $(1/P)[-(RT/P)dP + R\,dT]$. The expression in brackets is *not* an exact differential, but one sees that $(1/P)$ is an integrating factor for it. Also, $(1/P)$ times any function of (RT/P) will be an integrating factor for $[-(RT/P)dP + R\,dT]$.

The case of three or more independent variables is an important one; for the present purpose nothing is gained by involving more than three such variables, and the equation analogous to Eq. (6-4) is:

$$M(u,v,w)du + N(u,v,w)dv + O(u,v,w)dw = 0 \qquad (6\text{-}5)$$

The expression on the left may or may not be exact, as determinable by cross-differentiation tests of the type of Eq. (4-11). If it is not exact, there may or *may not* exist an integrating factor. This is in contrast to the case of two variables, for which an integrating factor can always be found. If Eq. (6-5) is exact or if it has an integrating factor, a solution of the form $F(u,v,w) = C$ exists; graphically, this solution consists of a family of surfaces in the (u,v,w) coordinate system, as illustrated in Fig. 6-4. Consider now a line or path of change

---

FIGURE 6-4
Successive surfaces representing the solution $F(u,v,w) = $ constant for various values of the constant, assuming that Eq. (6-4) is integrable. Points P_1 and P_2 lie on one such surface (with the constant $= a$), and points P_3 and P_4 lie on different surfaces.

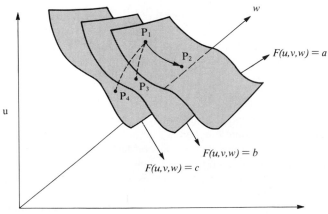

that lies on a given surface, $F(u,v,w) = a$, such as the line from P_1 to P_2, which lies on the surface $F(u,v,w) = a$. The surface is one for which $dy = 0$, which means this is true for each infinitesimal segment of the line from P_1 to P_2 and hence that y is constant along that line or path. It can be shown that surfaces of the type under discussion do not cross. As a consequence, no line satisfying the condition $F(u,v,w) = a$ can reach a point such as P_3 or P_4 in the figure, which lie on the surfaces $F(u,v,w) = b$ and $F(u,v,w) = c$, respectively.

A very important point is that if the expression of Eq. (6-5) is *not* integrable (is not an exact differential and has no integrating factor), then no solution function $F(u,v,w) = C$ exists and there are no solution surfaces. One can still construct paths from one point in (u,v,w) space to another that would satisfy Eq. (6-5), but the paths are no longer constrained to lie on some one surface. This case is illustrated by the dashed lines in Fig. 6-4, showing paths going from P_1 to P_3 and from P_1 to P_4. To repeat, such paths satisfy the constraint $dy = 0$, but are free of the constraint $F(u,v,w) = 0$.

How does all this bear on the second law of thermodynamics? We make the connection by considering y to be q_{rev} and write the identity

$$\delta q_{rev} \equiv M(u,v,w)du + N(u,v,w)du + O(u,v,w)dw \qquad (6\text{-}6)$$

where q_{rev} is heat absorbed in a reversible process, and u, v, and w (and in general other independent variables) define the state of the system. For any small *adiabatic* change, $\delta q_{rev} = 0$, and we recover Eq. (6-5). We know that q_{rev} does depend on path, as discussed in Chapter 4, so that q_{rev} is not a state function. The expression on the right of the identity sign of Eq. (6-6) is therefore *not* and exact differential. Might there be, however, an integrating factor? The first statement of the second law of thermodynamics given in statements (6-2), coupled with the above mathematical analysis, allows us to affirm that an integrating factor *must* exist. To reach this conclusion, we adopt the often used strategem of showing that the opposite conclusion cannot be true.

Suppose that the expression (6-5) is *not* integrable, i.e., does not have an integrating factor. This means, in terms of Fig. 6-4, that innumerable different adiabatic paths can exist, all passing through the point P_1. We can take the variables u and v to be T and V, respectively, and, to simplify matters, consider only paths in the T–V plane, as shown in Fig. 6-5. Out of the many adiabatic paths that would be permitted, we pick the two shown, where P_2 and P_3 are *isothermal* points. Since we are dealing with reversible q's, we can reverse the direction of one of the adiabatic paths so as to have a cycle: P_3 to P_2, P_2 to P_1, and P_1 back to P_3. Some work will be done in this cycle—as a minimum there will be pressure-volume work—and, by the first law of thermodynamics, an equivalent amount

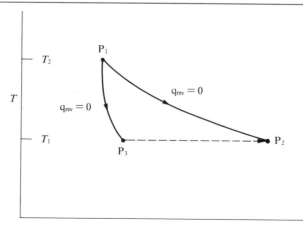

FIGURE 6-5
Possible paths in the T–V plane if Eq. (6-5) has no integrating factor.

of heat is absorbed. Because of the two adiabatic legs of the cycle, however, this heat is absorbed isothermally (at T_1 in the figure).

We thus have the situation in which heat is converted into work isothermally; the cycle would constitute a perpetual motion machine of the second kind, and would be in violation of the first of the forms (6-2) of the statement of the second law of thermodynamics. Such a cycle therefore *cannot* be possible. This means that possible adiabatic paths must lie on a single solution surface, e.g., points P_1 and P_2 in Fig. 6-4. It further means that points P_3, P_4, etc., are inaccessible by any reversible adiabatic path passing through P_1. For these conclusions to be valid, a function $F(u,v,w \ldots) = C$ *must* exist as a solution to Eq. (6-5), which means that the equation *is* integrable. We know that the expression itself is not directly integrable since it is not an exact differential; i.e., δq_{rev} in Eq. (6-6) depends on path. The summary conclusion is therefore that there *must* exist an integrating factor $G(T,V, \ldots)$ that makes Eq. (6-4) integrable.

The more concise and elegant statement of the above findings is known as Carathéodory's theorem:

> *If in every neighborhood of every point $P(u,v,w \ldots)$ there exist points which are inaccessible from P along a reversible adiabatic path, then δq_{rev} possesses an integrating factor.*

An essentially equivalent alternative statement is that one and only one reversible adiabat can pass through a given state of a system.

There will in general be any number of possible integrating factors, but a detailed consideration of how q and w are constrained produces as the simplest one: $G(T,V, \ldots) = 1/T$ (see Berry et al., 1980). The final conclusion is that $(1/T)\delta q_{rev}$ is exact and therefore that $\delta q_{rev}/T$ is a state function—a function that we call entropy, S. The Carathéodory theorem thus contains a mathematical statement of the second law of thermodynamics. Note that all of this has been quite general—we did not restrict ourselves to an ideal gas or to any particular type of substance or physical state.

The above has involved a somewhat specialized branch of mathematics, and has been a bit esoteric. We proceed in the next section to a fairly straightforward proof of the conclusion that $\delta q_{rev}/T$ is a state function, based on the analysis of heat engines.

6-3   The Carnot cycle—heat machines

One of the major advances in the early understanding of thermodynamics was published in 1824 by Sadi Carnot (1786–1832). Named after a Persian poet, Sadi could easily have been submerged by a brilliant and eminent father. Lazarre Carnot was a noted soldier, patriot, mathematician, and engineer. It was perhaps fortunate that during Sadi's twenties, his father was in exile for his republican beliefs. Sadi finished at the École polytechnique in Paris, had a period in the army, which was not to his liking, and left in 1820, on half-pension, to devote himself to studies. He moved in republican, radical circles and was perhaps doubly suspect by authorities because of his father.

The principal established science of the period was that of mechanics, pursued by mathematicians and engineers. Chemistry was an art, largely practiced by physicians. Sadi Carnot interested himself in a subject that had no established discipline, that of heat engines. The steam engine had been known for a century. The first devices were primitive ones in which a cylinder was alternately heated and cooled to cause a piston to move back and forth. About 1750, however, James Watt had added the major improvement of a condenser, thus allowing regular, reciprocating action and a great saving of heat energy (Fig. 6-6). The early engines were slow and clumsy and it was

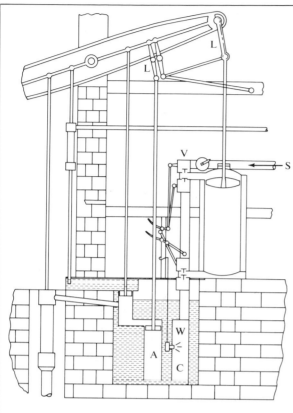

A burst of steam, S, is admitted to the cylinder, driving the piston by expansion. At the start of the return stroke, an air pump, A, has reduced the pressure in the condenser, C, and valves, V, change to allow spent steam to rush into C and be condensed by a spray of water, W. Note the linkage, L, which makes the piston rod travel on an accurate parallel to the cylinder axis.

FIGURE 6-6
A James Watt pumping engine, circa 1790. (From W. B. Snow, "The Steam Engine," American School of Correspondence, Chicago, 1909.)

In the earlier Newcomen engine, steam was admitted to the cylinder and a boy then turned on a water spray to cause condensation. The partial vacuum allowed atmospheric pressure to push the piston back. Watt's great invention was to have the condensation take place away from the cylinder so that the latter was not alternatively heated and cooled. Even the first Watt engines reduced threefold the fuel needed for a given job.

The term "horsepower" is due to Watt. Mine water pumps had been powered by horses hitched to a horizontal wheel whose rotation drove the pump. It was natural to speak of the number of horses that a steam pump could replace.

fortunate that they were in fact well suited for an important application—that of pumping water out of mines. With continuous operation and an easily measurable output (of water pumped), improvements could be reported quantitatively. The mechanical sophistication of steam-operated pumps was improving rapidly during Carnot's time, and he had access to a great many published accounts of performances.

The theory, that is, the fundamental science of steam engines was not at all understood, however. Thus while Carnot pére had been able in 1793 to formulate the theoretical conditions for the maximum efficiency of a source of mechanical energy (such as a paddle wheel), namely, no friction and a limit of zero velocity of moving elements, there was no comparable statement for heat engines. By around 1800 it had been established that the use of high pressure led to increased output, and for a while it was thought that there was no limit to the amount of work obtainable from a given amount of fuel. One only needed sufficiently high pressure. By 1820, however, this view was in doubt, as a result of practical experiences. High pressures were dangerous in use, moreover, and engineers had begun to lose enthusiasm for that approach.

In the meantime, the theory of heat had been progressing slowly. The invention of the thermometer in the 17th century had made the science of heat possible. Early chemists and physicians observed the phenomenon of latent heat, could measure specific heats, and could do calorimetry. An example of this last is the ice calorimeter of Lavoisier and Laplace in 1780 (Section 5-2). There evolved, quite naturally, the *caloric* theory, which treated heat as a substance that was always conserved. Caloric could be exchanged between substances but not destroyed. By 1798, however, Benjamin Thompson was proposing that heat could be generated mechanically through friction (see Section 4-2), but not until 1840 did Joule measure the mechanical equivalent of heat. Thus what we now understand, namely that heat and mechanical work are different forms of energy and are conserved only jointly, was obscure in Carnot's time.

Carnot, in fact, did not concern himself greatly about the *nature* of heat; the caloric theory was acceptable to him. What he did in his single and monumental work, "Réflexions sur la puissance motrice du feu," was to relate efficiency and *temperature.* He concluded that both a hot and a cold temperature were needed to obtain work from heat; he discussed the reversible heat engine; he calculated its maximum efficiency by means of what we now call the *Carnot cycle.* We will use this cycle shortly to obtain a definition of entropy and a statement of the second law of thermodynamics. Carnot appreciated, incidentally, that high-pressure steam engines were more efficient not because of pressure per se, but because they operated at a higher than usual temperature; he noted that water was not always the best fluid for a heat engine.

Carnot's treatise was largely ignored at the time. Engineers perceived nothing much new of practical importance. Physicians and chemists were not much interested in engines. Carnot himself seemed to have developed doubts about his theory, perhaps engendered by questions as to the nature of heat. Yet we cannot say what he might have produced further. He died in the great cholera epidemic of 1832, at the age of 46. His possessions were burned (a sanitary measure). He published nothing after his great treatise, and only a few private papers survive. In them, for example, is discussion of a mechanical equivalent of heat and of possible means of measuring it.

We proceed now to the Carnot cycle. This is a cycle containing the essential features of a heat engine in a very useful form. In a heat engine, a fluid is heated to a temperature T_2 by means of a heat source or reservoir at T_2, and then expands, doing work. Depending on the design, the expansion might be approximately adiabatic. We represent this expansion by an isothermal step followed by an adiabatic one, rather than as a single intermediate kind of path. As a result of the adiabatic step, the fluid cools to T_1. We now want to complete a cycle by returning to the initial state. It would not do to retrace exactly since we want net work to be done. The simplest thing is to reverse the order of the steps, that is, to have an isothermal compression at T_1, followed by an adiabatic compression to the original state at T_2. The advantage of this particular cycle is that heat exchange occurs only during the isothermal steps at T_2 and at T_1 rather than continuously over a range of temperature.

The preceeding description shows that the Carnot cycle corresponds to a real, although idealized, heat engine. We now further assume that the operating fluid is an ideal gas; later it will be shown that the conclusions must be valid for any substance.

The ideal gas cycle is shown in Fig. 6-7. We want to relate the heat absorbed during the isothermal expansion at T_2 and that evolved during the compression at T_1 to the net work done. To do this, we analyze each step, assuming one mole of gas.

Step ab Isothermal expansion at T_2 from V_a to V_b:

$$w_{ab} = RT_2 \ln \frac{V_b}{V_a} \qquad \Delta E_{ab} = 0 \qquad q_{ab} = q_2 = w_{ab}$$

Step bc Adiabatic expansion from T_2 to T_1 and from V_b to V_c:

$$q_{bc} = 0 \qquad -w_{bc} = \Delta E_{bc} = C_V \Delta T = C_V(T_1 - T_2)$$

(we assume a constant heat capacity),

$$C_V \ln \frac{T_1}{T_2} = -R \ln \frac{V_c}{V_b} \qquad \text{[ideal gas, Eq. (4-51)]}$$

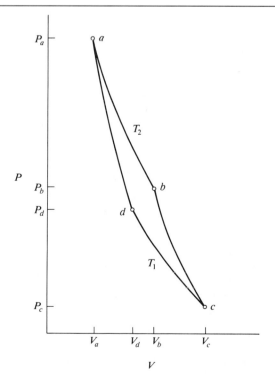

FIGURE 6-7
The Carnot cycle.

Step cd Isothermal compression at T_1 from V_c to V_d:

$$w_{cd} = RT_1 \ln \frac{V_d}{V_c} \qquad \Delta E_{cd} = 0 \qquad q_{cd} = q_1 = w_{cd}$$

Step da Adiabatic compression from T_1 to T_2 and from V_d to V_a:

$$q_{da} = 0 \qquad -w_{da} = \Delta E_{da} = C_V \Delta T = C_V(T_2 - T_1)$$

$$C_V \ln \frac{T_2}{T_1} = -R \ln \frac{V_a}{V_d} \qquad \text{[ideal gas, Eq. (4-51)]}$$

The two equations based on Eq. (4-51) provide a relationship between the four volumes:

$$C_V \ln \frac{T_1}{T_2} = -R \ln \frac{V_c}{V_b} = R \ln \frac{V_a}{V_d}$$

so that

$$\frac{V_b}{V_c} = \frac{V_a}{V_d} \qquad \text{or} \qquad \frac{V_b}{V_a} = \frac{V_c}{V_d} \tag{6-7}$$

We now assemble the terms for w_{tot}:

$$w_{tot} = RT_2\left(\ln \frac{V_b}{V_a}\right) + C_V(T_1 - T_2) + RT_1\left(\ln \frac{V_d}{V_c}\right) + C_V(T_2 - T_1)$$

The terms involving C_V cancel, and using Eq. (6-7), we get

$$w = w_{tot} = R(T_2 - T_1)\left(\ln \frac{V_b}{V_a}\right) \tag{6-8}$$

(the subscript is no longer needed for clarity). Since $q_2 = RT_2\ln(V_b/V_a)$, this further reduces to

$$w = \left(\frac{q_2}{T_2}\right)(T_2 - T_1) = q_2 \frac{T_2 - T_1}{T_2} \tag{6-9}$$

The efficiency of the heat engine is just w/q_2; this gives the work done per unit amount of heat energy supplied at T_2. The result is in accord with the qualitative statements of the preceding section. Thus the greater the difference between T_2 and T_1, the greater is the fraction of q_2 that can be converted to work. An additional point is that the ultimate in efficiency is reached if $T_1 = 0$ K (or $T_2 = \infty$ K); w is now equal to q_2. This result is restricted for the moment to the reversible Carnot cycle for an ideal gas but is otherwise quite general. Thus Eq. (6-9) is, of course, independent of the scale of the cycle, that is, of the number of moles of gas used, and of the number of cycles carried out. It is also independent of the value of C_V (or of the heat capacity ratio γ) and of the expansion ratio V_c/V_b, which determines the temperature ratio T_2/T_1.

There is an important corollary equation to Eq. (6-9). Since $\Delta E = 0$ for the cycle, it follows that

$$w = q_1 + q_2 \tag{6-10}$$

Combination of Eqs. (6-9) and (6-10) leads to a relation between q_1 and q_2:

$$\frac{q_1}{T_1} = -\frac{q_2}{T_2} \qquad\qquad (6\text{-}11)$$

We emphasize that Eq. (6-11) is derived for a reversible cycle, so that q_1 and q_2 are reversible heats.

Consider now the specific case of a Carnot heat engine operating between

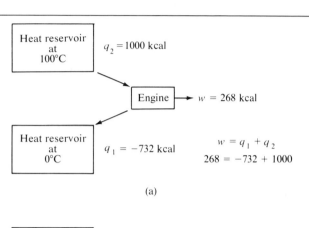

(a)

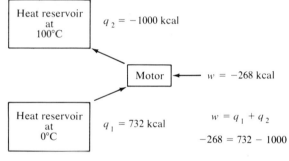

(b)

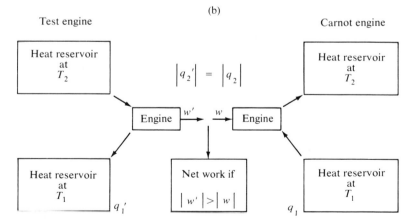

(c)

FIGURE 6-8
(a) The heat engine.
(b) The heat pump.
(c) Comparison of the ideal Carnot engine with a nonideal test engine.

100°C and 0°C. If 1000 kcal is absorbed at 100°C, the work done by the engine is (1000)(100)/(373) = 268 kcal; the efficiency is thus 26.8%. The remainder of the 1000 kcal, or 732 kcal, appears as heat delivered to the reservoir at 0°C. The analysis is shown schematically in Fig. 6-8(a).

The Carnot cycle is based on reversible steps and can therefore be operated in either direction with no change in the fundamental equations governing it. We could, for example, follow the cycle in a counterclockwise direction as illustrated in Fig. 6-8(b). It is now necessary to do work on the system. The net result is that by doing 268 kcal of work, 732 kcal have been taken from the reservoir at 0°C and this heat, plus the work energy, appears as 1000 kcal at 100°C. In this mode of operation, the Carnot cycle functions as a *refrigerator*. The 732 kcal abstracted at 0°C would, for example, correspond to the freezing of about 99 kg of ice. The relevant performance factor is now the ratio $|q_1|/|w|$, or 2.73.

The cycle, in this reverse mode of operation, is acting as a *heat pump*, transferring heat from a lower to a higher temperature. It could thus be used as a heating unit. If installed in a house with 0°C weather outside, the expenditure of 268 kcal of work would deliver 1000 kcal of heat at 100°C. The performance factor is, from this point of view, $|q_2|/|w|$, or 3.73. An advantage of using heat pumps now becomes apparent. The same 268 kcal of work could be converted directly into heat at 100°C (by means of an electric heater, for example), but now only 268 kcal of heat would result. Thus the performance factor of 3.73 is in fact the theoretical advantage factor of the heat pump over a direct heating system. In the case of the heating of a house, a more realistic set of temperatures might be 0°C and 25°C, in which case the performance factor rises to 11.9. Because of its greater initial capital cost the heat pump has not until recently been much used for heating. However, the heat pump has the ability to serve either for heating or for cooling, and this type of installation has now become relatively common. In the winter it provides space heating and in the summer, air conditioning.

6-4   Generalization of the Carnot cycle—the entropy function

The Carnot cycle would represent an important but rather specialized situation except that we can now bring in the second law statement about our inability to find a perpetual motion machine of the second kind. Recall that the Carnot cycle could represent a real engine, idealized only in the sense that we have reduced friction and related irreversibilities to the vanishing point, so that $w = w_{rev}$ and hence is the maximum possible. Ideal gas behavior is the limiting behavior of all real gases, so the use of an ideal gas is not completely unrealistic.

Let us suppose, then, that some other heat engine is to be matched against a Carnot engine. This second engine operates between the same two temperatures T_2 and T_1 and the scale of the Carnot engine is adjusted so that q_2 for it is equal to $-q_2'$ for the second or test engine. The situation is summarized as follows [see Fig. 6-8(c)].

Test engine   Absorbs heat q_2' at T_2, evolves heat q_1' at T_1, and produces w'.
Carnot engine   Evolves heat q_2 at T_2, $-q_2 = q_2'$; absorbs heat q_1 at T_1; and requires work w to be done on the cycle.

Since the heat absorption and evolution are balanced at T_2, q_2(net) = 0. Suppose that w' is greater than w, so that there is a net production of work by the combined machines. By the first law, $w_{net} = q_1 + q_1'$ (the q_2's cancel); since w_{net} is a positive number, more heat must have been absorbed at T_1 by the Carnot engine than evolved at T_1 by the test engine. That is, there must be a net absorption of heat at T_1. The overall result is therefore that heat energy at T_1 is converted into work, and we have a perpetual motion machine of the second kind. This is affirmed to be impossible by the second law, and we conclude that w' cannot be greater than w. The most efficient possible performance of the test engine, according to the second law, would be that $w' = w$, so that the combined engines would produce no net work and absorb no net heat at T_1.

The inescapable conclusion is that the Carnot efficiency is the maximum possible for any heat operating between the two temperatures. Equation (6-11) can thus be upgraded to the status of a general law of nature.

Now that we have established the generality of the conclusions from the Carnot cycle, it is very useful to put the cycle itself on a more abstract and general basis. Figure 6-9 depicts a reversible cycle between two states A and B. The paths A → B and B → A are arbitrary but can be approximated by a set of Carnot cycles. The interior portions cancel, and so the net result is to give the stepped path traced in heavy lines. Since Eq. (6-11) applies to each cycle, it must be true for the set of cycles that

$$\sum \frac{q_i}{T_i} = 0 \tag{6-12}$$

This sum can be divided into two parts, that for the terms that are made up by the contour from A to B and that for the terms made up by the contour of the path from B to A:

$$\sum_{A \to B} \frac{q_i}{T_i} + \sum_{B \to A} \frac{q_i}{T_i} = 0 \tag{6-13}$$

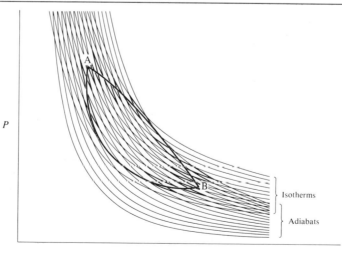

FIGURE 6-9
Approximating a general ideal gas cycle by means of a series of Carnot cycles.

Isotherms

Adiabats

P

V

Carnot cycles may be made as small as desired, so that in the limit we may reproduce the arbitrary path as closely as we please. For an infinite number of small steps, Eq. (6-13) takes an integral form:

$$\int_A^B \frac{\delta q_{rev}}{T} + \int_B^A \frac{\delta q_{rev}}{T} = 0 \tag{6-14}$$

or

$$\oint \frac{\delta q_{rev}}{T} = 0 \tag{6-15}$$

where $\oint$ denotes a cyclic integral, that is, one taken around a path, and the subscript has been added to q to emphasize again that the process must be reversible.

It is possible to go from A to B by some other reversible path, and to return from B to A always by the same path. Equation (6-14) does not depend on the specification of the path, and it must therefore be true that $\int_A^B \delta q_{rev}/T$ is independent of path. This conclusion, as well as the statement of Eq. (6-15), means that the quantity $\delta q_{rev}/T$ is a total or exact differential. Referring to Sections 4-5 and 6-2, we see that this is the defining property of a state function.

The recognition that $\delta q_{rev}/T$ is a state function is the principal quantitative contribution of the second law. We give this function its own name, *entropy* (Greek for change or a turning toward), denoted by the symbol S. The defining equation is

$$dS = \frac{\delta q_{rev}}{T} \tag{6-16}$$

To repeat, S, like E and H, is a thermodynamic quantity that depends only on the state of the system, so that ΔS, like ΔE and ΔH, is determined by the change of state, irrespective of the path. This last is an important point. The defining equation (6-16) allows us to calculate ΔS by setting up a reversible path and evaluating $\int \delta q_{rev}/T$. The resulting value of ΔS must be the same for that particular change in state regardless of the path used.

We can now write some combined first and second law equations. Thus Eq. (4-26) becomes

$$dE = T\,dS - P\,dV \tag{6-17}$$

and Eq. (4-29) becomes

$$dH = T\,dS + V\,dP \tag{6-18}$$

As with E and H, the practice will be to use S to denote the entropy per mole and s that per molecule. Entropy has the same units as the gas constant and heat capacity and ordinarily is expressed as calories or joules per kelvin per mole. The same alternative sets of units that are used for R may, of course, also be used for entropy. Some have considered that it is undignified for so important a thermodynamic quantity to have a composite dimension and have used the term EU, meaning entropy unit.

6-5   Calculations of ΔS for various reversible processes

In the absence of phase changes, q_{rev} for a change in temperature is given by

$$q_{rev} = \int C\, dT$$

where C is the heat capacity for the path, which may be some arbitrary one. The general expression for ΔS is therefore

$$\Delta S = \int \frac{C}{T}\, dT = \int C\, d(\ln T) \tag{6-19}$$

Usually, however, we consider only isochoric and isobaric paths, for which ΔS is given by

$$\Delta S = \int C_V\, d(\ln T) \tag{6-20}$$

$$\Delta S = \int C_P\, d(\ln T) \tag{6-21}$$

A. Changes of temperature

The heat capacity of liquid water between 20°C and 60°C can be represented by the equation: $C_P = 0.84092 + 3.467 \times 10^{-4}\, T + 4.77 \times 10^3/T^2$ (cal g^{-1} mol^{-1}). Find the entropy change when water is heated from 20°C to 60°C. Substitution into Eq. (6-21) gives $\Delta S_P = 0.84092$ ln $(333.15/293.15) + 3.467 \times 10^{-4}$ $(333.15 - 293.15) + (4.77 \times 10^3)(-1/2)(1/333.15^2 - 1/293.15^2) = 0.128$ cal K^{-1} mol^{-1}.

EXAMPLE

We now extend the considerations of Section 4-8 to include calculations of ΔS for the various standard processes. Previously, the process was required to be reversible in order that P_{ext} might be replaced by P, the pressure of the gas. We again require each process to be reversible, now both for that reason and because the defining equation for entropy, Eq. (6-16), requires the use of q_{rev}.

   Equation (6-17) takes a special form in the case of an ideal gas. We replace dE by $C_V dT$ [Eq. (4-44)] and P by RT/V:

$$C_V\, dT = T\, dS - \frac{RT}{V}\, dV$$

This rearranges to

$$dS = C_V\, d(\ln T) + R\, d(\ln V) \qquad \text{(ideal gas)} \tag{6-22}$$

or, since $d(\ln V) = d(\ln T) - d(\ln P)$,

$$dS = C_P\, d(\ln T) - R\, d(\ln P) \qquad \text{(ideal gas)} \tag{6-23}$$

   Equations (6-22) and (6-23) may be applied as follows (the results of Section 4-8B are included for completeness).

   Isochoric process:

$$w = 0 \qquad dE = C_V dT = dq_V \qquad dH = C_P\, dT$$

$$dS = \frac{dq_V}{T} = C_V\, d(\ln T) \qquad \Delta S = C_V \ln \frac{T_2}{T_1} \qquad \text{(ideal gas)} \tag{6-24}$$

B. Processes involving an ideal gas

Isobaric process:

$$dw = P\,dV = R\,dT \qquad w = R\,\Delta T$$

$$dE = C_V\,dT \qquad dH = C_P\,dT = dq_P \tag{6-25}$$

$$dS = \frac{dq_P}{T} = C_P\,d(\ln T), \qquad \Delta S = C_P \ln \frac{T_2}{T_1} \quad \text{(ideal gas)}$$

Isothermal process:

$$dq = dw = RT\,d(\ln V) = -RT\,d(\ln P)$$

$$q = w = RT \ln \frac{V_2}{V_1} = RT \ln \frac{P_1}{P_2}$$

$$dE = 0 \qquad dH = 0 \tag{6-26}$$

$$dS = R\,d(\ln V) \qquad \Delta S = R \ln \frac{V_2}{V_1} \qquad \Delta S = R \ln \frac{P_1}{P_2} \quad \text{(ideal gas)}$$

Adiabatic process:

$$dq = 0 \qquad dw = -dE \qquad dE = C_V dT \qquad dH = C_P dT$$

$$dS = \frac{dg}{T} = 0 \qquad \text{(ideal gas)}$$

$$C_V\,d(\ln T) = -R\,d(\ln V) \qquad \text{[ideal gas, Eq. (4-49)]} \tag{6-27}$$

$$C_P\,d(\ln T) = R\,d(\ln P) \qquad \text{[ideal gas, Eq. (4-50)]}$$

Note that Eqs. (4-49) and (4-50) follow from Eqs. (6-22) and (6-23) on setting $dS = 0$ for the adiabatic process.

An interesting and important result is obtained if we consider the overall entropy change for the gas plus its surroundings. By surroundings, we mean the rest of the system or apparatus that is involved. In the isochoric process, the surroundings gain (algebraically) $-q_v$, and therefore suffer an entropy change of $-\int dq_v/T$ or $-C_V \ln(T_2/T_1)$. The total entropy change for the gas plus surroundings is thus zero. The same result is obtained for the isobaric process, in which the surroundings gain $-dq_P/dT$ or $-C_P \ln(T_2/T_1)$ in entropy, and this just balances the entropy change of gas. The overall entropy change is again zero for the isothermal process, and, of course, for the adiabatic one since here both q and ΔS are zero.

The conclusion is that in a reversible process the *total* entropy change is zero for the *entire* system that is, for the system plus its surroundings. Since the entire system contains all the changes that occur, and therefore does not exchange heat with anything outside of itself, we speak of it as a *thermally isolated* system. Although the examples cited are for an ideal gas, the result can be shown to be quite general (Section 6-7):

$$\Delta S = 0 \qquad \text{for any reversible process in a thermally isolated system} \tag{6-28}$$

The Carnot cycle may be presented as an entropy versus temperature plot. Thus, referring to Section 6-3, $S_{ab} = R \ln(V_b/V_a)$; $S_{bc} = 0$; $S_{cd} = R \ln (V_d/V_c)$;

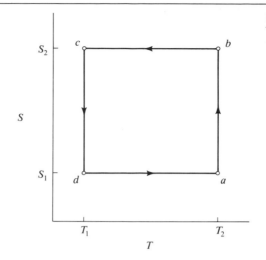

FIGURE 6-10
The Carnot cycle as an S versus T plot.

and hence, from Eq. (6-7), $S_{cd} = -S_{ab}$; $S_{da} = 0$. The corresponding plot for the cycle is shown in Fig. 6-10. Here, too, ΔS for the cycle plus its surroundings is zero. The entropy change for the ideal gas at T_2 is q_2/T_2, and is just balanced by the equal and opposite entropy change in the heat reservoir at T_2. Similarly, the entropy change at T_1, q_1/T_1, is again balanced by the equal and opposite entropy change in the reservoir at T_1.

In the case of a pure substance, that is, in the absence of chemical equilibria, a change of phase occurs sharply at a definite temperature for a given pressure. This means that at the particular T and P the two phases are in equilibrium; the process of converting one phase into the other is thus a reversible one. As noted in Section 5-6, there is a latent heat for a phase change, usually determined at constant pressure; this is a q_P and is equal to ΔH for the change. We have

C. Phase changes

$$\Delta S = \int \frac{dq_{\text{rev}}}{T} = \frac{1}{T} \int dq_{\text{rev}} = \frac{\Delta H}{T} \qquad (6\text{-}29)$$

Consider the process

EXAMPLE

$$\text{H}_2\text{O}(l, 100°\text{C}, 1 \text{ atm})$$
$$= \text{H}_2\text{O}(g, 100°\text{C}, 1 \text{ atm}) \qquad \Delta H^0_{373} = 9.720 \text{ kcal} \quad (6\text{-}30)$$

The process is reversible; since liquid and vapor are in equilibrium, an infinitesimal temperature change one way or the other will cause either evaporation or condensation. To carry out the process as written, we suppose that a heat reservoir infinitesimally above 100°C supplies the necessary heat. The entropy change is then $\Delta S = 9720/373 = 26.1$ cal K^{-1} mol^{-1}. The heat reservoir, naturally, loses this amount of heat at 100°C, and shows an entropy change of -26.1 cal K^{-1}; as with the ideal gas processes, the entropy change for the entire system is zero.

Calculation of the entropy change on freezing of a two-component system is more complicated. As an illustration, suppose that we have a 12.5% by volume aqueous ethylene glycol (antifreeze) solution. In such a solution there is about 0.048 mole of ethylene glycol per mole of water, and this produces a freezing point depression. Nothing happens at 0°C, but at -3.9°C ice begins to separate out, and the residual solution becomes more concentrated. By the time the mix has been cooled to -6.7°C, the solution is 17.0% ethylene glycol by volume, corresponding to 12.08 mol of water per mole of ethylene glycol. The original 0.048 mole remains in solution, since ice alone freezes out, and the corresponding amount of water in the 17.0% solution is $(0.048)(12.08) = 0.58$ mole. Thus 0.42 mole of ice has formed on cooling the solution from -3.9°C to -6.7°C.

The entropy change for the above process has three types of contributions. (a) The entropy change for cooling the solution, (b) that for freezing out the 0.42 mole of ice, and (c) that for cooling the ice as it is formed. Since the amounts of solution and of ice are changing with temperature, the actual calculation would involve setting up the appropriate differential equation and integrating over the temperature interval involved. See problem 6-42.

6-6   Calculation of ΔS for various irreversible processes

Irreversible processes are important because of their bearing on the question of equilibrium. In an irreversible process there is some imbalance—P_{ext} may be different from P or T_{ext} from T, or a chemical driving force may be present. In all cases the irreversible process is one that can occur spontaneously. By contrast, a system at equilibrium cannot be subject to spontaneous change. Thus the reversible process is associated with equilibrium and the irreversible one with a lack of equilibrium.

It is entirely possible to calculate ΔS for an irreversible process provided the initial and final states are given. It is only necessary to devise a reversible path for accomplishing the same change of state. As an example, suppose that one mole of an ideal gas initially at STP is allowed to expand into an evacuated space and thereby double its volume. Since the gas is ideal, there is no change in temperature, and the final state is 0.5 atm pressure and 44.8 liters. As illustrated in Fig. 6-11, the equivalent reversible process is an isothermal expansion against a piston. As given in the preceding section, $\Delta S = R \ln(V_2/V_1) = R \ln 2 = 0.693R$, and this same value applies to the irreversible process. Note that in this particular irreversible process the gas is in effect its own thermally isolated system, and that now the system entropy change is $0.693R$, or positive.

A second type of irreversible process would be one involving a phase change. Consider the process

$$H_2O(l, -10°C) = H_2O(s, -10°C) \tag{6-31}$$

The process is one that *could* be realized experimentally. In the absence of dust, a liquid may be supercooled below its freezing point. Water is, in fact, particularly prone to this, and in careful experiments it has been possible to supercool water to -40°C. Such supercooled water is unstable of course, and will eventually freeze. Since the process (6-31) is written as an isothermal one, we suppose that the water is in contact with a heat reservoir at -10°C; otherwise the latent heat released on freezing would warm the system.

In order to calculate ΔS for Eq. (6-31), we must produce the same change of state by means of reversible steps. This may be done as follows:

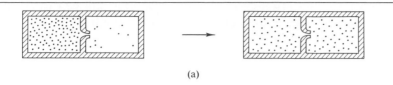

(a)

FIGURE 6-11
(a) Irreversible and
(b) reversible expansion
of a gas.

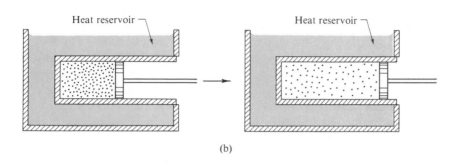

(b)

(a) $H_2O(l, -10°C) = H_2O(l, 0°C)$

(b) $H_2O(l, 0°C)$ $= H_2O(s, 0°C)$

(c) $H_2O(s, 0°C)$ $= H_2O(s, -10°C)$

Step (a) is reversible even though the liquid water is unstable toward freezing.

The entropy and heat changes for the three steps are as follows. (The heat capacities of solid and liquid water are 9.1 and 18 cal K^{-1} mol^{-1}, respectively, and the heat of fusion of water at 0°C is 1435 cal mol^{-1}.)

Step (a): $\Delta H = q_P = \int C_P \, dT = 18(273.2 - 263.2) = 180$ cal.

$$\Delta S = C_P \int d(\ln T) = 18 \ln(273.2/263.2) = 0.67 \text{ cal K}^{-1}$$

[by Eq. (6-21)].

Step (b): $q_P = \Delta H = -1435$ cal.

$$\Delta S = q_P/T = -1435/273.2 = -5.25 \text{ cal K}^{-1}$$

Step (c): $\Delta H = q_P = 9.1(263.2 - 273.2) = -91$ cal.

$$\Delta S = 9.1 \ln(263.2/273.2) = -0.34 \text{ cal K}^{-1}$$

The overall entropy change for the water is then $(0.67 - 5.25 - 0.34) = -4.92$ cal K^{-1}. The overall $\Delta H = (180 - 1435 - 90) = -1345$ cal; this is also q_P for the actual irreversible reaction, Eq. (6-31), and gives the heat absorbed by the heat reservoir at $-10°C$. The entropy change of the reservoir is therefore $1345/263.2 = 5.11$ cal K^{-1}. The overall entropy change is then $-4.92 + 5.11 = 0.19$ cal K^{-1}, or again positive.

These results suggest that we can expect ΔS to be positive for a system and its surroundings, that is, for a thermally isolated system, in which a spontaneous process occurs. The conclusion is correct and may be proved by reference to the Carnot cycle. An ideal gas may still be carried through the succession of states of the Carnot cycle even though one or more of the steps may now be irreversible. For example, the opposing pressure on the piston during the first isothermal expansion, P_{ext}, might be less than P. The expansion is then irreversible and spontaneous—it would not be reversed by an infinitesimal increase in P_{ext}. The process is still isothermal and still goes to the same final volume V_b in Fig. 6-7, so ΔS for the gas is the same as before. The heat reservoir at T_2 has supplied less heat than in the reversible process, however, and its entropy change, a negative number, will now be less than $|q_{2(rev)}/T|$. There will thus be an overall increase in entropy for the cycle plus its surroundings. Any irreversibility in the cycle must have the effect that either less heat is absorbed by the gas at T_2 or more heat is evolved at T_1. In either case, we find $\Delta S > 0$ for the gas plus the heat reservoirs.

It was shown in Section 6-4 that no actual cycle can do better than a Carnot cycle without violating the second law. Therefore, we can write the general statement that for any process in a thermally isolated system

$$dS \geqslant 0 \tag{6-32}$$

where the equal sign holds for a reversible process and the inequality sign for an irreversible or spontaneous process. The thermally isolated system is a special one, and Eq. (6-32) may be made yet more general. In each case examined the effect of irreversibility has been to decrease w (algebraically) and hence to decrease q (algebraically) relative to the reversible value. Since the entropy change is determined by q_{rev}, the general statement is

$$dS \geqslant \frac{\delta q}{T} \tag{6-33}$$

where the equal sign holds for q_{rev} and the inequality sign holds otherwise. In the thermally isolated system, δq is zero and we return to Eq. (6-32).

6-7   Free energy. Criteria for equilibrium

A. The Gibbs free energy G   Our preoccupation so far has been with entropy and heat quantities. In chemistry, we are also interested in the work that a process can perform, especially the chemical work. With the discovery that $\delta q_{rev}/T$ is a state function, we have a powerful tool for evaluating w_{rev}, or the maximum work permitted by certain important types of change in state. First, we have to recognize that other types of work than pressure–volume (i.e., mechanical) work are possible, such as chemical work, or work involving magnetic or electric fields. For the present purposes, however, the only form of nonmechanical work of interest to us will be chemical work. A good example of such work is that which a battery can provide. We therefore write that in general

$$\delta w_{rev} = P \, dV + \delta w'_{rev} \tag{6-34}$$

where $\delta w'_{rev}$ denotes all work other than pressure–volume work, or, for us, chemical work. The more general combined first and second law statement is then

$$dE = T \, dS - P \, dV - \delta w'_{rev} \tag{6-35}$$

It is now convenient to introduce a new thermodynamic quantity G defined as

$$G = H - TS \qquad (6\text{-}36)$$

Differentiation gives

$$dG = dH - T\,dS - S\,dT$$
$$= dE + P\,dV + V\,dP - T\,dS - S\,dT \quad (6\text{-}37)$$

Elimination of dE between Eqs. (6-35) and (6-37) gives

$$dG = -S\,dT + V\,dP - \delta w'_{\text{rev}} \qquad (6\text{-}38)$$

For a process at constant temperature and pressure Eq. (6-38) reduces to

$$dG = -\delta w'_{\text{rev}} \qquad (6\text{-}39)$$

and the usefulness of the G function is now apparent. The negative of dG gives the reversible and hence the maximum chemical work for a reaction at constant temperature and pressure. It is the free or available work energy, and G is called the *Gibbs free energy*, after an illustrious American scientist (see Commentary and Notes section).

An alternative type of free energy, the *Helmholtz free energy A*, is defined as

$$A = E - TS \qquad (6\text{-}40)$$

Differentiation gives

$$dA = dE - T\,dS - S\,dT \qquad (6\text{-}41)$$

and, in combination with Eq. (6-35), this gives

$$dA = -S\,dT - P\,dV - \delta w'_{\text{rev}} = S\,dT - dw_{\text{rev}} \qquad (6\text{-}42)$$

For a process at constant temperature, $-dA$ gives the maximum total work, mechanical plus chemical (and any other forms of work).

We can now derive two criteria for equilibrium, including chemical equilibrium. If Eq. (6-33), $dS \geqslant \delta q/T$, is considered in terms of the first law, we obtain

B. Criteria for equilibrium

$$dE \leqslant T\,dS - \delta w \qquad \text{or} \qquad dE \leqslant T\,dS - P\,dV - \delta w' \qquad (6\text{-}43)$$

where the equal sign applies for a reversible process and the inequality sign for an irreversible or spontaneous one. Equation (6-43) may be rearranged to yield

$$dS - \frac{dE + \delta w}{T} \geqslant 0 \qquad (6\text{-}44)$$

If the system is isolated, δq, and hence $dE + \delta w$, is zero. We thus regain Eq. (6-32):

$$dS \geqslant 0 \text{ (for an isolated system)}$$

The significance of this last statement with respect to equilibrium is that the entropy of an isolated system will increase until no further spontaneous processes are possible. As illustrated in Fig. 6-12(a), the entropy must then reach

FIGURE 6-12
(a) For an isolated system S is a maximum at equilibrium. (b) For a system at constant T and P, G is a minimum at equilibrium. Note Fig. 7-3 for a specific example.

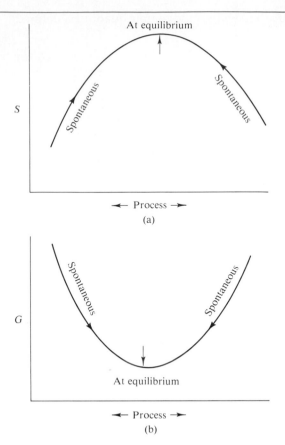

a maximum. At this point any change is a reversible one, and the system is in equilibrium.

Insertion of relationship (6-43) into Eq. (6-37) gives

$$- dG - S\, dT + V\, dP - \delta w' \geqslant 0 \tag{6-45}$$

If the processes are restricted to constant temperature and pressure, then

$$-dG \geqslant \delta w' \qquad \text{or} \qquad dG \leqslant -\delta w' \tag{6-46}$$

If there is no provision for the system to do chemical work (or any other kind of work than pressure–volume work), then $\delta w' = 0$, and

$$dG \leqslant 0 \qquad \text{(constant } T \text{ and } P, \text{ no chemical work)} \tag{6-47}$$

This would be an entirely different process than the constant volume one of Fig. 6-12(a), of course. The situation is now that the free energy of a system at constant temperature and pressure (and which may contain various reacting chemical species) will decrease until no further spontaneous processes are possible. As illustrated in Fig. 6-12(b), the free energy is then at a minimum. Any changes around this minimum are reversible, and the system is at equilibrium. The restriction of constant temperature and pressure may seem severe. It is not, however. Should some spontaneous fluctuation produce a local change in pressure or temperature, the reestablishment of uniform pressure and temperature equilibrium between the system and its surroundings would occur spontaneously. Thus no spontaneous irreversible changes at all can occur when G is at a minimum.

The condition of $\delta w' = 0$ is the usual one in chemical equilibrium. That is, we do not ordinarily arrange to extract chemical work. The usual form of Eq. (6-38) is therefore

$$dG = -S\,dT + V\,dP \tag{6-48}$$

This applies to reversible changes in temperature and pressure under conditions such that $\delta w' = 0$. (Just as $dE = T\,dS$ applies under constant–volume conditions where $\delta w = 0$.) Also, G is defined in terms of state functions [Eq. (6-36)] and hence is itself a state function; ΔG is independent of path and is determined only by the initial and final states. ΔG has a *direct physical* significance, however, only when the two states are at the same temperature and pressure. It then gives the maximum chemical work that is obtainable by the process at constant pressure and temperature that takes the system from the initial to the final state.

Like G, the Helmoltz free energy A is a state function by definition. The usual form of Eq. (6-42) is

$$dA = -S\,dT - P\,dV \tag{6-49}$$

As in the case of Eq. (6-48), Eq. (6-49) applies to reversible changes in temperature and volume under conditions such that $\delta w' = 0$.

The Gibbs free energy will be used throughout the rest of this text and its various applications will be illustrated as they come up. However, two very useful results may be written down immediately from Eq. (6-48). First, for a constant–pressure process, the change in G with temperature is $dG = -S\,dT$, that is,

C. Calculations of free energy changes

$$\left(\frac{\partial G}{\partial T}\right)_P = -S \tag{6-50}$$

or

$$\Delta G = -\int S\,dT \qquad \text{(pressure constant)} \tag{6-51}$$

For a constant–temperature process, the change of G with pressure is $dG = V\,dP$, or

$$\left(\frac{\partial G}{\partial P}\right)_T = V \tag{6-52}$$

and

$$\Delta G = \int V\,dP \qquad \text{(temperature constant)} \tag{6-53}$$

In the case of an ideal gas, $V - RT/P$ and Eq. (6-53) becomes

$$dG = \frac{RT}{P}\,dP = RT\,d(\ln P) \qquad \text{(ideal gas)} \tag{6-54}$$

or

$$\Delta G = RT \ln \frac{P_2}{P_1} \quad \text{(ideal gas)} \tag{6-55}$$

EXAMPLE (1) ΔG for the isothermal expansion of one mole of ideal gas from STP as the initial state to the final state of 0.5 atm is given by Eq. (6-55), $(1.987)(273.15) \ln(0.5/1)$ or -376.2 cal. Alternatively, by Eq. (6-36), $\Delta G = -T(\Delta S)$ since $\Delta H = 0$; $\Delta G = -(273.15)(R \ln 2) = -376.2$ cal.

(2) ΔG may be calculated for a process not at constant T. Consider the reversible isobaric expansion of one mole of N_2 from 1 atm at 298.15 K to twice the volume. T_2 is $T_1 V_2/V_1$, or 596.30 K. We can use Eq. (6-36) in the form $\Delta G = \Delta H - \Delta(TS) = \Delta H - (T_2 S_2 - T_1 S_1)$. S_1 is given later in Table 6-2 as 45.79 and using Eq. (6-21) and a C_P value of $(5/2)R$, $S_2 = 45.79 + (5/2)R \ln(596.30/298.15) = 49.23$ cal K^{-1} mole^{-1}. Thus $\Delta G = (5/2)R(596.30 - 298.15) - [(596.30)(49.23) - (298.15)(45.79)] = -14.22$ kcal. [As an exercise, show that the same result is obtained by using Eq. (6-51).]

(3) Finally, we may obtain ΔG for a process in which both T and P change. Consider a reversible adiabatic expansion of N_2 from the initial state of part (2) to 0.5 atm. From Eq. (4-52), we find $\ln(T_2/T_1) = (2/5)\ln(0.5/1)$, whence $T_2 = 225.96$ K. Since $q_{rev} = 0$, $\delta S = 0$, and $\Delta G = \delta H - S\Delta T = (5/2)R(225.96 - 298.15) + (45.79)(72.19) = 2947$ cal.

For none of the foregoing processes does ΔG have a simple physical meaning, since for none of them are both T and P constant. The value of the free energy quantity will become apparent when we look at phase changes and at chemical reactions.

It similarly follows from Eq. (6-49) that

$$\left(\frac{\partial A}{\partial V}\right)_T = -P \tag{6-56}$$

and

$$\left(\frac{\partial A}{\partial T}\right)_V = -S \tag{6-57}$$

These expressions are not suited to the usual applications of free energy in physical chemistry, but are very useful in theoretical treatments because of the relation of A to the partition function discussed in Chapter 4.

Free energy changes may be obtained for chemical reactions and are then used in much the same way as are enthalpies in thermochemistry. One may use either free energies of formation, as discussed in Chapter 7 (on chemical equilibrium), or the so-called free energy functions, described in the Special Topics section of this chapter.

6-8 Second law relationships

It is worthwhile to assemble the various basic equations before deriving a few relationships. These basic equations are as follows.

Statements of definition:

$H = E + PV$ [Eq. (4-27)]

$A = E - TS$ [Eq. (6-40)]

$G = H - TS$ [Eq. (6-36)]

Alternate forms of the combined first and second laws:

$$dE = T\,dS - P\,dV \qquad \text{[Eq. (6-17)]}$$

$$dH = T\,dS + V\,dP \qquad \text{[Eq. (6-18)]}$$

$$dA = -S\,dT - P\,dV \qquad \text{[Eq. (6-49)]}$$

$$dG = -S\,dT + V\,dP \qquad \text{[Eq. (6-48)]}$$

Immediate consequences of the definitions of entropy:

$$\Delta S = \frac{\Delta E}{T} \qquad (V \text{ and } T \text{ constant}) \tag{6-58}$$

$$\Delta S = \frac{\Delta H}{T} \qquad (P \text{ and } T \text{ constant}) \tag{6-59}$$

$$C_V = T\left(\frac{\partial S}{\partial T}\right)_V \qquad \text{[Eq. (6-20)]}$$

$$C_P = T\left(\frac{\partial S}{\partial T}\right)_P \qquad \text{[Eq. (6-21)]}$$

Criteria for equilibrium:

$$dS \geq 0 \text{ (isolated system)} \qquad \text{[Eq. (6-32)]}$$

$$dG \leq 0 \text{ (constant } T \text{ and } P) \qquad \text{[Eq. (6-46)]}$$

These relationships provide a number of secondary ones. For example, the alternative forms of the combined first and second laws are all exact differentials. This means that each generates an Euler or cross-differentiation relationship [Eq. (4-11)]. Two important cross-differentials come from the statements for dA and dG:

$$\left(\frac{\partial S}{\partial V}\right)_T = \left(\frac{\partial P}{\partial T}\right)_V \tag{6-60}$$

$$\left(\frac{\partial S}{\partial P}\right)_T = -\left(\frac{\partial V}{\partial T}\right)_P \tag{6-61}$$

A somewhat more complete picture is given in Section 6-ST-1.

Some applications of thermodynamic relationships are as follows. We can write Eq. (6-17) in the form:

$$\left(\frac{\partial E}{\partial V}\right)_T = T\left(\frac{\partial S}{\partial V}\right)_T - P \tag{6-62}$$

Use of Eq. (6-60) gives

$$\left(\frac{\partial E}{\partial V}\right)_T = T\left(\frac{\partial P}{\partial T}\right)_V - P \tag{6-63}$$

Equation (6-63) is called a *thermodynamic equation of state*. It provides a relationship among E and the state variables P, V, and T, which is valid for all substances. Its application to the calculation of internal pressures was discussed in Section 4-7. A completely parallel derivation beginning with the differentiation of Eq. (6-18) with respect to pressure at constant temperature and using Eq. (6-61) leads to a similar thermodynamic equation of state:

$$\left(\frac{\partial H}{\partial P}\right)_T = -T\left(\frac{\partial V}{\partial T}\right)_P + V \tag{6-64}$$

As discussed in Section 4-ST-2, Eq. (6-64) allows an evaluation of the Joule–Thomson coefficient from an equation of state such as the van der Waals equation.

We can now develop Eq. (4-35) further. The equation is

$$C_P - C_V = \left[P + \left(\frac{\partial E}{\partial V}\right)_T\right]\left(\frac{\partial V}{\partial T}\right)_P$$

On replacing $(\partial E/\partial V)_T$ by means of Eq. (6-63), we get

$$C_P - C_V = T\left(\frac{\partial P}{\partial T}\right)_V \left(\frac{\partial V}{\partial T}\right)_P \tag{6-65}$$

or, since

$$\left(\frac{\partial P}{\partial T}\right)_V = -\frac{(\partial V/\partial T)_P}{(\partial V/\partial P)_T} = \frac{\alpha}{\beta} \qquad [\text{Eq. (4-14)}]$$

where α is the coefficient of thermal expansion and β the coefficient of compressibility,

$$\alpha = \frac{1}{V}\left(\frac{\partial V}{\partial T}\right)_P \qquad \beta = -\frac{1}{V}\left(\frac{\partial V}{\partial P}\right)_T \tag{6-66}$$

the final form of Eq. (6-63) becomes

$$C_P - C_V = \frac{\alpha^2 VT}{\beta} \tag{6-67}$$

EXAMPLE   First of all, we can verify the value of $(C_P - C_V)$ for an ideal gas. Here, $\alpha = R/PV = 1/T$ and $\beta = -RT/VP^2 = 1/P$. Substitution into Eq. (6-67) yields $(C_P - C_V) = VP/T = R$, as expected.

It might be thought that $(C_P - C_V)$ would be small for a liquid or solid, but this is not necessarily the case. For benzene at 25°C, $\alpha = 1.24 \times 10^{-3}$ K^{-1} and $\beta = 9.58 \times 10^{-10}$ Pa^{-1}. Then $(C_P - C_V) = (1.24 \times 10^{-3})^2 (89 \times 10^{-6})(298)/(9.58 \times 10^{-10}) = 42.6$ J K^{-1} mol^{-1} or 5.12 R! The reason is that while a condensed phase may not expand much with temperature, as compared to a gas, the compressibility is much smaller than for a gas.

6-9   The third law of thermodynamics

A. Entropy and probability

We now take another look at entropy—in a way that will lead to the third law of thermodynamics as well as introduce the statistical mechanical approach.

Consider the situation shown in Fig. 6-13. We have a flask of volume V_1 connected to a second flask of equal volume, so that the combined volume V_2 is twice V_1. The flasks are kept at some uniform temperature. Suppose

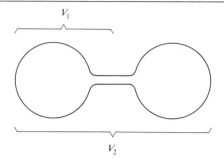

FIGURE 6-13

that a single molecule of an ideal gas is present in the system. Clearly, the chance of the molecule being in V_1 is $\frac{1}{2}$. If two molecules are present, their behavior will be independent of each other (ideal gas behavior) and the chance of both being in V_1 is $(\frac{1}{2})^2$. If N_A molecules are present, the chance that all of them will accidentally collect in V_1 must be $(\frac{1}{2})^{N_A}$.

The probability calculation may be generalized to the case of an arbitrary ratio of V_1 to V_2. The probability of all of the molecules being in V_1 is now $(V_1/V_2)^{N_A}$. As a further generalization, we can assign an individual probability weight p to a molecule being in a given volume. That is, p_1 for a molecule being in volume V_1 is proportional to V_1, and, similarly, p_2 is proportional to V_2. The ratio of p values will then give the relative probability of any two volume conditions for a molecule. Thus for the general process

State 1 (molecule in V_2) → State 2 (molecule in V_1)

the relative probability is V_1/V_2. For N_A molecules, the relative probability is then $(V_1/V_2)^{N_A}$. We can make the dependence on the number of molecules additive rather than exponential by introducing a new variable r, defined as $r = k \ln p$ (the reason for introducing the Boltzmann constant will be apparent shortly). Then for the process

state A (N_A molecules in V_2) → state B (N_A molecules in V_1),   (6-68)

we have

$$N_A \, \Delta r = k \ln \left(\frac{p_1}{p_2} \right)^{N_A} = k \ln \left(\frac{V_1}{V_2} \right)^{N_A}$$

or

$$N_A \, \Delta r = k N_A \ln \frac{V_1}{V_2} = R \ln \frac{V_1}{V_2} \qquad (6\text{-}69)$$

The right-hand side of Eq. (6-69) is the same as that of Eq. (6-26), so for isothermal volume changes of an ideal gas we can identify $N_A \, \Delta r$ with ΔS, or Δr with the entropy change per molecule. Alternatively, we see that the entropy of a state can be written

$$S = \alpha + k \ln p \qquad (6\text{-}70)$$

where p is the probability weight of the state, and α is a constant. The relation between entropy and probability was proposed by Boltzmann in 1896 [see also Eq. (6-77)].

Although Eq. (6-70) was obtained by consideration of volume changes for an ideal gas, the analysis can be made in quite general terms, where p would be the probability weight of a given state. In quantum theory, molecules are limited to definite energy states; p is now interpreted as the number of ways in which a molecule could, over a period of time, have a given average energy $\bar{\epsilon}$. Alternatively, for N equivalent molecules, p^N is proportional to the number of ways for their average total energy to be $N\bar{\epsilon}$.

There is one condition such that p should be unity. That is the situation in which only one arrangement of molecules in energy states is possible. For this to be true, there can be only one energy state and all molecules must be equivalent. The physical picture meeting this requirement is that of a perfect crystalline solid at 0 K. At 0 K all molecules must, by the Boltzmann principle, be in their lowest possible energy state, and if the solid is perfectly crystalline, all lattice positions are equivalent. The entropy per molecule should then be just the constant α of Eq. (6-70).

B. The third law of thermodynamics

The third law of thermodynamics has evolved out of considerations such as those just given. It adds one additional statement, namely, that the constant α is intrinsic to each element (α could, for example, include some nuclear probability weight). If we now consider a general chemical reaction

$$aA + bB + \cdots = mM + nN + \cdots$$

then ΔS for the reaction must be zero at 0 K. That is, no chemical reaction can change the number of each kind of atom, and therefore the constants α for each atom must cancel. Since we do not know the α's and since they must always cancel out, we simply proceed on the basis that they are zero. Acceptance of this assumption or reference point allows the third law to be stated in the following useful way:

The entropy of all perfect crystalline substances is zero at 0 K.

The term "perfect" in the above statement is an important one. As discussed in Section 6-CN-2, any disorder in a crystalline material will give it a residual entropy at 0 K, an entropy that shows up experimentally.

C. Application of the third law of thermodynamics

Since the third law affirms that $S_{0\,K} = 0$ for any perfect crystalline substance, we can calculate its entropy at some higher temperature, using the methods of thermochemistry. Suppose we want S_{298} for some gaseous substance. We start with the crystalline solid as close to 0 K as possible and determine the entropy change on warming it to its melting point:

$$\Delta S_1 = \int_{0\,K}^{T_f} \frac{C_P(s)}{T}\, dT$$

There is next an entropy contribution for the melting of the solid:

$$\Delta S_2 = \frac{\Delta H_f}{T_f}$$

where ΔH_f is the enthalpy of fusion. The liquid is then further warmed to its normal boiling point:

$$\Delta S_3 = \int_{T_f}^{T_b} C_P(l)d(\ln T)$$

We now add the entropy of vaporization:

$$\Delta S_4 = \frac{\Delta H_V}{T_b}$$

Finally, the gas is warmed to 298 K:

$$\Delta S_5 = \int_{T_b}^{298} C_P(g)d(\ln T)$$

The actual heat capacity data might appear as in Fig. 6-14. The data for the solid would look like one of the curves of Fig. 4-16. In practice, the actual data might extend down to some low temperature T_1, and would have to be extrapolated to 0 K, as indicated by the dashed portion of the curve. This extrapolation, in fact, would probably be made on the basis of the Debye treatment of the heat capacity of solids (Section 4-ST-3), which predicts that at low temperatures C_V should be proportional to T^3. Since the integral to obtain ΔS_1 is of C_P vs. ln T (and *not* T), the contribution to ΔS_1 from the extrapolated region is not necessarily negligible. The discontinuity in the figure at T_f corresponds to the freezing (or melting) point, and we pick up the heat capacity curve for the liquid. As suggested in the figure, heat capacities of liquids at their freezing points tend to be a bit lower than for the solid. This is possible if some degrees of vibrational freedom are converted to rotations in the liquid, which contributes $R/2$ rather than R to the heat capacity (see Section 4-8). The second jump in the curve is at the normal boiling point T_b. Notice that the heat capacity of the vapor is decidedly less than that of the liquid. Now some degrees of vibrational freedom have been converted both to translation and to rotation, each mode contributing only $R/2$ to C_P.

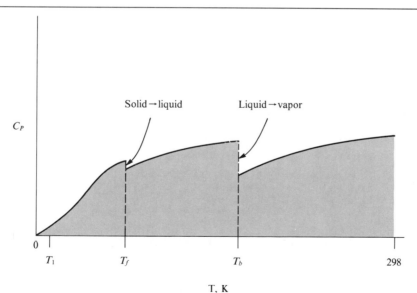

FIGURE 6-14
Hypothetical heat capacity curve for warming a substance from 0 K to 298 K.

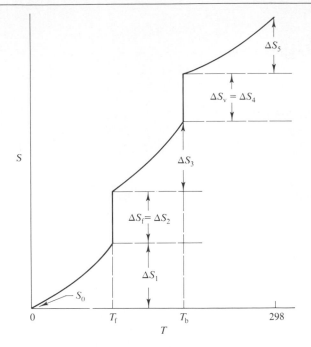

FIGURE 6-15
Sequence of steps
involved in obtaining a
third law entropy.

The corresponding development of the successive contributions to entropy is illustrated in Fig. 6-15, that is,

$$S_{298}^0 = S_0 + \Delta S_1 + \Delta S_2 + \Delta S_3 + \Delta S_4 + \Delta S_5$$

where S_0 is the entropy at 0 K and is affirmed to be independent of the chemical state of the substance or, in practice, zero. In the analogous calculation of the enthalpy changes on heating, shown in Fig. 5-2, the enthalpy at 0 K became equal to the internal energy E_0, but this latter depends on the chemical state. That is, ΔE_0, unlike ΔS_0, is definitely *not* zero for a chemical process at 0 K.

As may be imagined, the obtaining of data for experimental third law entropies is an arduous task. It has been done for a number of substances, however, and results are given later in Table 6-2. As is seen in the next section, statistical thermodynamics provides an alternative means for calculating S_{298}^0 values. The whole subject is discussed further in Section 6-CN-2.

6-10   Summary of the statements of the laws of thermodynamics

The remainder of this text will make much use of the results of thermodynamics, but in terms of specific applications. This seems an appropriate place to collect the various alternative general statements of the laws of thermodynamics.

First Law of Thermodynamics [Eq. (4-5)]:
$dE = \delta q - \delta w$.
The internal energy of an isolated system is constant.
The internal energy change of a cyclic process is zero.
Internal energy is a state function.

Second Law of Thermodynamics:
$dS = \delta q_{rev}/T; \; dS > \delta q_{irrev}/T.$
Entropy is a state function.

Corollaries are:
It is not possible to convert heat into work by means of an isothermal, cyclic process.
The entropy of an isolated system tends to a maximum.
The Carathéodory theorem (Section 6-2).

Third Law of Thermodynamics:
If the entropy of each element in some perfect crystalline state is taken as zero at 0 K, then the entropy of any perfect crystalline substance is also zero at 0 K.
ΔS for any isothermal process is zero at 0 K.

This last statement is sometimes known as the *Nernst heat theorem.* The various standard thermodynamic equations have been summarized in Section 6-8 and need not be repeated. A more general approach to obtaining such relationships is referred to in Section 6-ST-1.

6-11   Statistical mechanical treatment of second law quantities

Entropy may be expressed in terms of the partition function **Q,** using Eq. (4-72), which gives the statistical thermodynamic formulation for C_V, **A.** Entropy

$$C_V = \frac{R}{T^2}\left[\frac{\partial^2(\ln \mathbf{Q})}{\partial(1/T)^2}\right]_V$$

We substitute this expression for C_V into Eq. (6-20),

$$T\left(\frac{\partial S}{\partial T}\right)_V = C_V$$

The required integration is then made:

$$\int dS = R \int \frac{1}{T^3}\left[\frac{\partial^2(\ln \mathbf{Q})}{\partial(1/T)^2}\right]_V dT$$

or, since $d(1/T) = -dT/T^2$,

$$S = R \int \frac{1}{T}\left\{\frac{\partial[T^2\partial(\ln \mathbf{Q})/\partial T]}{\partial T}\right\}_V dT + \text{constant} \tag{6-71}$$

Equation (6-71) may be expanded into two terms,

$$S = 2R \int \left[\frac{\partial(\ln \mathbf{Q})}{\partial T}\right]_V dT + R \int T\left[\frac{\partial^2(\ln \mathbf{Q})}{\partial T^2}\right]_V dT + \text{constant}$$

and, on integrating the second term by parts, we obtain

$$S = R \ln \mathbf{Q} + RT\left[\frac{\partial(\ln \mathbf{Q})}{\partial T}\right]_V + \text{constant} \tag{6-72}$$

The last term in Eq. (6-72) is, by Eq. (4-70), the average energy divided by T:

$$S = \frac{E}{T} + R \ln \mathbf{Q} + \text{constant} \tag{6-73}$$

As 0 K is approached, only the first term of the partition function is important, and so $\mathbf{Q}$ reduces to $g_0 e^{-\epsilon_0/kT}$, where ϵ_0 is the energy of the lowest state and g_0 is its degeneracy. Then $\ln \mathbf{Q}$ becomes $-(\epsilon_0/kT) + \ln g_0$. The average energy E becomes $N_A\epsilon_0$ and we have

$$S_0 = \frac{N_A\epsilon_0}{T} - \frac{N_A k \epsilon_0}{kT} + N_A k \ln g_0 + \text{constant}$$

or

$$S_0 = N_A k \ln g_0 + \text{constant} \tag{6-74}$$

Equation (6-74) corresponds to Eq. (6-70) for 0 K and per mole. Thus the constant corresponds to α and $N_A k \ln g_0$ corresponds to the probability weight of the ground state. As already discussed, α will always cancel out in any process, and is conventionally set equal to zero. Although the ground state of a molecule may often be described as degenerate, at the limit of 0 K even the most minor perturbation would remove such degeneracy, so we can expect g_0 to be unity. It is thus essentially as an aspect of the third law of thermodynamics that we take S_0 to be zero. An interesting exception to this is mentioned in the Commentary and Notes section.

Ground state degeneracy does make a contribution to the entropy, however, above 0 K. This degeneracy is implicitly contained in the spectroscopic term symbol (Section 19-2A). It is 2 for atomic hydrogen, for example, because the odd electron has two possible orientations; we should add $R \ln 2$ to the translational entropy given by Eq. (6-104). Other common cases are (molecule, g_0): (Cl, 2), (O_2, 3), (NO, 4). For molecules with all electrons paired, $g_0 = 1$.

The resulting equation for S,

$$S = \frac{E}{T} + R \ln \mathbf{Q} \qquad \text{(distinguishable molecules)} \tag{6-75}$$

applies, as indicated, to the case where molecules in different energy states are distinguishable. This is true for the case of an ideal crystal and also for the *internal* energy states (rotation and vibration) of the molecules of an ideal gas.

As one way of seeing this reservation, we can now return to discuss the quantity p in Eq. (6-70) in more detail. It is usually called the *thermodynamic probability* and is the number of distinguishable ways in which molecules can be permitted among their most probable energy states. That is, p is the same as the quantity W_{max} that was described by Eq. (2-10). As a reminder, the number of ways of permuting N_A molecules is just $N_A!$, but these molecules populate energy states such that N_1 are in state ϵ_1, N_2 are in state ϵ_2, and so on, and we regard the molecules of a given energy state as indistinguishable. We must then divide out the numbers of ways of permuting N_1 molecules within state ϵ_1, of N_2 molecules within state ϵ_2, and so on. The net number of distinguishable permutations is then

$$W = \frac{N_A!}{N_1! \, N_2! \cdots N_i!} = \frac{N_A!}{\Pi_i \, N_i!} \tag{6-76}$$

where the populations of the various states are the *most probable ones;* for this case W is usually called W_{max}, or the *thermodynamic probability.* W_{max} is proportional to the probability of a state of the system but obviously is usually a very large number. It is not the same as the probability itself—this last is a number that does not exceed unity and can be thought of as W_{max} divided by a normalizing factor.

Equation (6-70) now becomes

$$S = \alpha + k \ln W_{max} \tag{6-77}$$

It is essentially this form of Eq. (6-70) that was proposed by Boltzmann. We proceed to evaluate W_{max} using Stirling's approximation; we also set α equal to zero. Thus

$$S = k[N_A \ln(N_A) - N_A] - k\sum_i [N_i \ln(N_i) - N_i]$$

or

$$S = kN_A \ln N_A - k\sum_i N_i \ln N_i \tag{6-78}$$

since $\sum N_i = N_A$. The second term of Eq. (6-78) can be handled as follows. We can write

$$N_i = N_A \frac{e^{-\epsilon_i/kT}}{\sum e^{-\epsilon_i/kT}} = \frac{N_A e^{-\epsilon_i/kT}}{\mathbf{Q}}$$

The summation term of Eq. (6-78) becomes

$$k \sum N_i \ln N_i = k \sum \left(N_i \ln N_A - \frac{N_i \epsilon_i}{kT} - N_i \ln \mathbf{Q} \right)$$

$$= kN_A \ln N_A - \frac{E}{T} - kN_A \ln \mathbf{Q}$$

and after the substitution, the terms $kN_A \ln N_A$ cancel to give

$$S = \frac{E}{T} + R \ln \mathbf{Q}$$

which is the same as Eq. (6-75).

It is the assumption of distinguishable molecules, on which Eq. (6-75) is based, that allows us to write the entropy for a mole of particles as just N_A times the entropy per particle. The situation is somewhat similar to that discussed in Section 4-11B except that we are now talking about the total partition function for a set of N_A molecules. If the energy states for each molecule are distinct, then the total partition function $\mathbf{Q}_{tot}$ can be written

$$\mathbf{Q}_{tot} = \mathbf{Q}_1\mathbf{Q}_2\mathbf{Q}_3 \cdots \mathbf{Q}_{N_A} = \mathbf{Q}^{N_A}$$

$\mathbf{Q}$ enters in the various expressions for thermodynamic quantities as $\ln \mathbf{Q}$, and since $\ln \mathbf{Q}_{tot} = N_A \ln \mathbf{Q}$, the effect is that the quantity per mole is always N_A times that per molecule. Strictly speaking, Eq. (6-75) should be written as

$$S = \frac{E}{T} + k \ln \mathbf{Q}_{\text{tot}} \tag{6-79}$$

but, for distinguishable molecules with $\mathbf{Q}_{\text{tot}} = \mathbf{Q}^{N_A}$, we regain Eq. (6-75).

B. Energy   The expression for energy E is given by Eq. (4-69), which we repeat here:

$$E = kT^2 \left[\frac{\partial(\ln \mathbf{Q}_{\text{tot}})}{\partial T} \right]_V = RT^2 \left[\frac{\partial(\ln \mathbf{Q})}{\partial T} \right]_V \tag{6-80}$$

An important point was noted in Section 5-ST-2 and should be restated here. The above energy is that developed on heating from 0 K and is the quantity of interest if we are dealing with a single substance. In the next chapter, however, we take up the changes in thermodynamic properties accompanying a chemical reaction, and must recognize that even at 0 K there will be an energy change, ΔE_0, for a chemical process. While we can determine energy *changes*, such as a ΔE_0, *absolute* energies are inaccessible. The first law of thermodynamics, for example, deals only with energy changes. It is convenient, although arbitrary, to attribute an E_0 to each chemical substance. Actual E_0 values will never be produced, but the formalism makes the handling of equations easier.

Equation (6-80) is therefore better written as

$$E = E_0 + RT^2 \left[\frac{\partial(\ln \mathbf{Q})}{\partial T} \right]_V \tag{6-81}$$

if we are going to apply statistical thermodynamics to chemical reactions [as is done in Section 5-ST-2; note Eq. (5-42)].

C. Free energy and related quantities   The expression for A follows from the defining equation, Eq. (6-40), $A = E - TS$. Substitution into Eq. (6-79) gives

$$A = E_0 - kT \ln \mathbf{Q}_{\text{tot}} \tag{6-82}$$

or, for distinguishable molecules,

$$A = E_0 - RT \ln \mathbf{Q} \tag{6-83}$$

In order to evaluate G and H, we need a statistical thermodynamic expression for pressure, and use Eq. (6-56) in combination with Eq. (6-83) to obtain

$$P = -\left(\frac{\partial A}{\partial V} \right)_T = kT \left[\frac{\partial(\ln \mathbf{Q}_{\text{tot}})}{\partial V} \right]_T = RT \left[\frac{\partial(\ln \mathbf{Q})}{\partial V} \right]_T \tag{6-84}$$

the last equality being for the case of distinguishable molecules. Eq. (6-84) is a general one, applicable to gases, liquids, and solids. Then, by Eq. (6-36) we have $G = H - TS = A + PV$, and

$$G = E_0 - kT \ln \mathbf{Q}_{\text{tot}} + kTV \left[\frac{\partial(\ln \mathbf{Q}_{\text{tot}})}{\partial V} \right]_T \tag{6-85}$$

or, for distinguishable molecules,

$$G = E_0 - RT \ln \mathbf{Q} + RTV \left[\frac{\partial(\ln \mathbf{Q})}{\partial V} \right]_T \tag{6-86}$$

Enthalpy is given by

$$H = E + PV = E_0 + RT^2 \left[\frac{\partial(\ln \mathbf{Q})}{\partial T} \right]_V + RTV \left[\frac{\partial(\ln \mathbf{Q})}{\partial V} \right]_T \qquad (6\text{-}87)$$

The derivatives of $\ln \mathbf{Q}$ that appear in the above equations are easy to evaluate for an ideal gas, and this is done immediately below. For liquids and solids, however, the energy states are largely but not entirely localized, and the statistical mechanical approach must be handled very carefully (see also Section 9-CN-3).

6-12   Statistical thermodynamic quantities for an ideal gas

Explicit statistical thermodynamic expressions can be written for the various thermodynamic functions if we are dealing with an ideal gas, but a new aspect must be discussed first. We follow Eq. (4-60) in writing $\mathbf{Q} = \mathbf{Q}_{trans} \mathbf{Q}_{rot} \mathbf{Q}_{vib} \mathbf{Q}_{elec}$. Ordinarily a substance is in its lowest electronic state, and $\mathbf{Q}_{elec}$ merely provides the term E_0 in the equations of the preceding section. We can make the further distinction

$$\mathbf{Q} = \mathbf{Q}_{trans} \mathbf{Q}_{int}, \qquad \mathbf{Q}_{int} = \mathbf{Q}_{rot} \mathbf{Q}_{vib} \qquad (6\text{-}88)$$

where $\mathbf{Q}_{int}$ is the partition function for energy states internal to the molecule.

The new aspect is as follows. The equations of the preceding section are written either in terms of $\mathbf{Q}_{tot}$ for N_A molecules, or, in the case of distinguishable molecules, $\mathbf{Q}_{tot} = \mathbf{Q}^{N_A}$ and the $(\ln \mathbf{Q}_{tot})$ terms could be replaced by $N_A \ln \mathbf{Q}$. The element of distinguishability is *not* present, however, in the case of translational energy states of an ideal gas. The translational energy of a molecule of an ideal gas is determined wave mechanically as that of a standing wave occupying the whole space. The translational energy states for a set of molecules are, in other words, not localized; the molecules are in this respect indistinguishable. The result is that when we take the product of the partition functions for each molecule, $\mathbf{Q}_{tot}$ has too high a multiplicity. We have over-counted the number of distinguishable energy states by the number of ways of permuting N_A molecules among themselves, or by the factor $N_A!$. For the ideal gas, then, we have

$$\mathbf{Q}_{tot} = \frac{1}{N_A!} \mathbf{Q}^{N_A} \qquad (6\text{-}89)$$

The factor $1/N_A!$ does not affect the average energy, since E involves only the derivative $\partial(\ln \mathbf{Q}_{tot})/\partial T = N_A \, \partial(\ln \mathbf{Q})/\partial T$. This complication could therefore be ignored until now. The entropy, however, depends on $\ln \mathbf{Q}_{tot}$ and not just on its derivative. The matter of indistinguishability has to do only with translational energy states, and we could write more specifically that

$$\mathbf{Q}_{tot} = \left[\frac{1}{N_A!} \mathbf{Q}_{trans}^{N_A} \right] [\mathbf{Q}_{int}]^{N_A} \qquad (6\text{-}90)$$

It is convenient, however, to retain the form of Eq. (6-89). Equation (6-79) should now be written

$$S = \frac{E}{T} + R \ln \mathbf{Q} - k \ln N_A! \qquad (6\text{-}91)$$

The last term is evaluated by means of Stirling's approximation, $\ln x! = x \ln x - x$, and Eq. (6-91) becomes

$$S = \frac{E}{T} + R \ln \mathbf{Q} - R \ln N_A + R \tag{6-92}$$

Alternatively, in terms of the separation indicated in Eq. (6-90),

$$S_{\text{trans}} = \frac{E_{\text{trans}}}{T} + R \ln \mathbf{Q}_{\text{trans}} - R \ln N_A + R \tag{6-93}$$

$$S_{\text{int}} = \frac{E_{\text{int}}}{T} + R \ln \mathbf{Q}_{\text{int}} \tag{6-94}$$

or, more specifically,

$$S_{\text{rot}} = \frac{E_{\text{rot}}}{T} + R \ln \mathbf{Q}_{\text{rot}} \tag{6-95}$$

$$S_{\text{vib}} = \frac{E_{\text{vib}}}{T} + R \ln \mathbf{Q}_{\text{vib}} \tag{6-96}$$

The definitions of A and G contain S, and are correspondingly altered from the expressions of the preceding section. Equation (6-82) becomes

$$A_{\text{trans}} = -RT \ln \mathbf{Q}_{\text{trans}} + RT \ln N_A - RT \tag{6-97}$$

$$A_{\text{rot}} = -RT \ln \mathbf{Q}_{\text{rot}} \qquad A_{\text{vib}} = -RT \ln \mathbf{Q}_{\text{vib}} \tag{6-98}$$

Equation (6-84) for P is unchanged, but Eq. (6-85) is now

$$G_{\text{trans}} = -RT \ln \mathbf{Q}_{\text{trans}} + RTV \left[\frac{\partial (\ln \mathbf{Q}_{\text{trans}})}{\partial V} \right]_T + RT \ln N_A - RT \tag{6-99}$$

$$G_{\text{rot}} = -RT \ln \mathbf{Q}_{\text{rot}} + RTV \left[\frac{\partial (\ln \mathbf{Q}_{\text{rot}})}{\partial V} \right]_T \tag{6-100}$$

$$G_{\text{vib}} = -RT \ln \mathbf{Q}_{\text{vib}} + RTV \left[\frac{\partial (\ln \mathbf{Q}_{\text{vib}})}{\partial V} \right]_T \tag{6-101}$$

Finally, Eq. (6-87) for H is unchanged. Notice that it is only entropy and quantities containing entropy that are affected by the matter of distinguishability vs. indistinguishability. As before, in assembling A_{tot} and G_{tot}, we will add E_0, the energy of the substance at 0 K.

We can now proceed to the obtaining of explicit expressions. That for S_{trans} is obtained by substitution for $\mathbf{Q}_{\text{trans}}$ [Eq. (4-83)],

$$\mathbf{Q}_{\text{trans}} = \frac{(2\pi mkT)^{3/2}}{h^3} V$$

in Eq. (6-93), remembering that for an ideal gas $E_{\text{trans}} = \frac{3}{2} R$, to obtain

$$S_{\text{trans}} = \frac{3}{2} R + R + R \ln \left[\left(\frac{2\pi mkT}{h^2} \right)^{3/2} \frac{V}{N^A} \right] \tag{6-102}$$

Note that we have used V, the volume per mole, in Eq. (4-83).

Equation (6-102) is known as the *Sackur–Tetrode equation*; it is one of the more valuable relations of statistical thermodynamics since it gives the absolute translational entropy of an ideal gas. Further manipulation puts the equation in the form

$$S_{trans} = \tfrac{5}{2} R + R \ln \left[\left(\frac{2\pi}{h^2} \right)^{3/2} \frac{1}{N_A{}^4} R^{5/2} \right] + R \ln \left(\frac{T^{5/2}M^{3/2}}{P} \right) \qquad (6\text{-}103)$$

(V has been replaced by RT/P). On evaluating the collections of constants, we see that Eq. (6-103) reduces to

$$S^0_{trans} \text{ (cal K}^{-1} \text{ mol}^{-1}) = R \ln(T^{5/2}M^{3/2}) - 2.31 \qquad (6\text{-}104)$$

for an ideal gas at 1 atm pressure. [The units of R as used in various parts of Eq. (6-103) must be chosen with care!] As an example, the entropy of argon at 298.1 K may be calculated as follows:

$$S_{298} = 1.987 \ln[(298.2)^{5/2}(39.95)^{3/2}] - 2.31 = 36.98 \qquad \text{cal K}^{-1} \text{ mole}^{-1}$$

EXAMPLE

A variation of the Sackur–Tetrode equation, important in surface chemistry, is the version for a two-dimensional gas. The physical situation is that of molecules adsorbed at an interface but able to move freely about on it. They thus behave as a two-dimensional gas, for which $\mathbf{Q}_{trans}$ can be found by interpolation between Eqs. (4-82) and (4-83):

$$\mathbf{Q}_{trans(2\text{-dim})} = \frac{2\pi mkT}{h^2} \sigma \qquad (6\text{-}105)$$

Insertion of this into Eq. (6-93) (remembering that E is now just RT) gives

$$S_{trans(2\text{-dim})} = 2R + R \ln \left(\frac{2\pi mkT}{h^2} \frac{\sigma}{N_A} \right) \qquad (6\text{-}106)$$

On substituting numerical values for the various constants, we obtain

$$S_{trans(2\text{-dim})} = R \ln (MT\sigma) + 65.8 \text{ (cal K}^{-1} \text{ mole}^{-1}) \qquad (6\text{-}107)$$

where σ denotes the area per molecule. As an example, argon at 25°C and 1 atm pressure has a concentration such that the molecules are, on the average, about 34 Å apart. A corresponding state for argon as a two-dimensional gas would then be one with $\sigma = (34)^2$ Å^2 per molecule. Substitution of this value into Eq. (6-107) gives

$$S_{Ar(2\text{-dim})} = 1.987 \ln[(39.95)(298.1)(1.16 \times 10^{-13})] + 65.8$$

$$= 25.3 \text{ cal K}^{-1} \text{ mole}^{-1}$$

Notice that this value is about two-thirds of that for the three-dimensional gas at 1 atm and 25°C. Why?

The thermodynamic functions A_{trans}, G_{trans}, and H_{trans} can be similarly obtained from Eqs. (6-97), (6-99), and (6-87) or, more simply, by using the equations of Section 6-8. The contributions from rotation and vibration are obtained by substitution of the appropriate partition function expressions into Eqs. (6-95) and (6-96). Thus

$$S_{rot} = R + R \ln \frac{T}{\theta_{rot}} \qquad (6\text{-}108)$$

for an A–B type molecule [from Eq. (4-90) and remembering that $E_{rot} = RT$]. A_{rot} and G_{rot} are the same since $\mathbf{Q}_{rot}$ does not depend on volume:

TABLE 6-1

Spectroscopic constants
for some diatomic
molecules[a]

Molecule	ν_0 (cm^{-1})	θ_{vib} (K)	θ_{rot} (K)
H_2	4160.2	5986	87.5
N_2	2330.7	3353	2.89
O_2	1556.4	2239	2.08
CO	2143.2	3084	2.78
$H^{35}Cl$	2885.9	4152	15.2
HBr	2559.3	3682	12.2
$^{35}Cl_2$	556.9	801.3	0.351
I_2	212.3	306.8	0.0538
HI	2230.1	3209	9.43

[a]Adapted from S.M. Blinder, "Advanced Physical Chemistry," p. 450, Macmillan, New York, 1969.

$$A_{rot} = G_{rot} = -RT \ln \frac{T}{\theta_{rot}} \qquad (6\text{-}109)$$

The vibrational partition function is given by Eq. (4-95): using it and E_{vib} from Eq. (4-97), we get

$$S_{vib} = R\left[\frac{\theta_{vib}/T}{e^{\theta_{vib}/T} - 1} - \ln(1 - e^{-\theta_{vib}/T}) \right] \qquad (6\text{-}110)$$

Notice that the $\frac{1}{2}h\nu_0$ term in the expression for E cancels out. Again, $A_{vib} = G_{vib}$, so

$$A_{vib} = G_{vib} = N_A\left(\frac{h\nu_0}{2}\right) + RT \ln(1 - e^{-\theta_{vib}/T}) \qquad (6\text{-}111)$$

As with E and H, the zero-point energy term $\frac{1}{2}h\nu_0$ is present.

The preceding equations, plus those of Sections 4-13 and 4-14, allow the calculation of the rotational and vibrational contributions to the various thermodynamic quantities provided θ_{rot} and θ_{vib} are known. These are generally obtained spectroscopically from an analysis of the spacings of rotational and vibrational energy states. These, in turn, are given by the absorption spectrum of the molecule, infrared for vibration and rotation and microwave for rotation only. The discussion of molecular spectroscopy as such is deferred until Chapter 19 but it is useful at this point to supply some of the results. These are given in Table 6-1 for some gaseous diatomic molecules. In the case of polyatomic molecules there is more than one degree of vibrational freedom and hence more than one θ_{vib}. Additional data for such molecules are given in the reference cited with the table.

We can illustrate the use of the preceding equations by calculating the thermodynamic quantities for N_2(g) at 25°C and 1 atm pressure, assumed to be an ideal gas. For translation, we have

$$E_{trans} = \tfrac{3}{2}RT = 889 \qquad \text{cal mol}^{-1}$$

$$S_{trans}^0 = 1.987 \ln(298.15)^{5/2}(28.01)^{3/2}] - 2.31 = 35.93 \quad \text{cal K}^{-1}\,\text{mol}^{-1}$$

Then

$$H_{trans}^0 = E_{trans} + RT = 1481 \quad \text{cal mol}^{-1}$$

$$G^0_{trans} = H^0_{trans} - TS^0_{trans} = -9232 \quad \text{cal mol}^{-1}$$

(The superscripts merely confirm that the pressure is 1 atm.)

For rotation, $\theta_{rot} = 2.89$ K; E_{rot} is then given by the equipartition value,

$$E_{rot} = RT = 592 \quad \text{cal mol}^{-1}$$

By Eq. (6-108),

$$S_{rot} = 1.987 + (1.987) \ln \frac{298.16}{2(2.89)}$$

$$= 9.82 \quad \text{cal K}^{-1} \text{ mol}^{-1}$$

(the factor of two enters because the symmetry number for N_2 is two). Then

$$H_{rot} = E_{rot} = 592 \quad \text{cal mol}^{-1}$$

$$G_{rot} = A_{rot} = E_{rot} - TS_{rot} = -2340 \quad \text{cal mol}^{-1}$$

Finally, for vibration, $\theta_{vib} = 3353$ K, so $\theta_{vib}/T = 11.24$. The consequence is that the vibrational contribution is negligible in all cases.

The total values are as follows:

$$E_{tot} = E_0 + 889 + 592 = E_0 + 1481 \quad \text{cal mol}^{-1}$$

(where E_0 represents the unknown absolute energy at 0 K),

$$H^0_{tot} = E_0 + 1481 + 592 = E_0 + 2073 \quad \text{cal mol}^{-1}$$

$$S^0_{tot} = 35.93 + 9.82 = 45.75 \quad \text{cal K}^{-1} \text{ mol}^{-1}$$

$$G^0_{tot} = E_0 - 9232 - 2336 = E_0 - 11,570 \quad \text{cal mol}^{-1}$$

Certain points of interest are brought out by the preceding calculation. First, the calculated S^0_{tot} is in quite good agreement with that obtained from heat capacity data using the third law. As cited in Table 6-2, the third law value is 45.90 cal K^{-1} mol^{-1}. It seems amazing that the calculated value, resting entirely on the partition functions for N_2 as an ideal gas, agrees with the thermal value obtained by considering the warming of crystalline nitrogen through its melting point and boiling point. The basic reason is that the entropy is zero for the perfect crystalline solid at 0 K and is also zero for the hypothetical ideal gas at 0 K. For that transition, ΔS is thus zero, and what we are observing is that entropy is a state function whose value is independent of path. The situation may be shown as follows:

N_2(ideal gas, 0 K, 1 atm)

$\uparrow \ \Delta S = 0 \qquad\qquad N_2(g,\ 298$ K, 1 atm)

N_2(crystal, 0 K) $\rightarrow$ liquid

A second point has to do with the large negative values for $(G^0 - E_0)$. These arise from the contribution of the $-TS^0$ term. Had the calculation been made at some lower temperature, the contribution of this term would have been smaller, not only absolutely, but relative to E and H^0. In effect, at low temperatures H^0 makes the principal contribution to G^0, while at high temperatures the $-TS^0$ term dominates.

Substance	Thermochemical (cal K⁻¹ mol⁻¹)	Calculated (cal K⁻¹ mol⁻¹)

TABLE 6-2
Thermochemical and statistical thermodynamic entropies at 298.1K[a]

Substance	Thermochemical $(\text{cal K}^{-1}\,\text{mol}^{-1})$	Calculated $(\text{cal K}^{-1}\,\text{mol}^{-1})$
Solid		
K	15.2	—
Ag	10.20	—
AgCl	23.00	—
Al	6.77	—
C(graphite)	1.37	—
Ca	9.95	—
CaO	9.5	—
Cd	12.37	—
Cu	7.97	—
Liquid		
Br_2	36.4	—
H_2O	16.73	—
Hg	18.17	—
C_6H_6	41.3	—
Gas		
CH_4	44.47	—
CO	46.20	47.3
CO_2	51.08	—
Cl_2	53.29	53.31
H_2	31.21	31.23
HCl	44.64	44.64
HBr	47.6	47.53
N_2	45.90	45.79
NO	50.34	50.43
O_2	49.01	49.03

[a]From G.N. Lewis and M. Randall, "Thermodynamics," 2nd ed. (revised by K.S. Pitzer and L. Brewer). McGraw-Hill, New York, 1961; and E.A. Moelwyn-Hughes, "Physical Chemistry," 2nd ed. Pergamon, Oxford, 1961.

COMMENTARY AND NOTES

6-CN-1   Statistical thermodynamics— ensembles—and J. Willard Gibbs

An account was given in Section 4-2 of a brilliant scientist, Benjamin Thompson or Count Rumford. Another brilliant scientist, also American by birth, was Josiah Willard Gibbs. The two men differed in all other respects, however, in about the most complete way imaginable.

J. Willard Gibbs was born in New Haven, Connecticut, on February 11, 1839, and died there in April, 1903. He was of English descent; we can trace his ancestry back to Sir Henry Gibbs of Warwickshire. It was a son of Sir Henry that emigrated to Boston in 1658, to establish the Gibbs family in the New England area. The family was academically oriented. After several generations of graduates from Harvard, we come to Josiah Gibbs, who graduated from Yale in 1809 and remained as Professor of Sacred Literature. This was the father of J. Willard Gibbs.

The young Gibbs entered Yale College in

1854 and graduated in 1858, receiving several prizes in Latin and mathematics; he received his doctorate in 1863 and was for a while appointed a tutor. He then spent several years abroad, in Paris, Berlin, and Heidelberg, with exposure to such eminent scientists as Magnus, Kirchhoff, and Helmholtz. In 1871 he was appointed Professor of Mathematical Physics at Yale College, which position he retained until his death. He never married, and lived with his sister and her family. As a person, he was modest and of a retiring disposition; his constitution was weak as a result of a childhood bout with scarlet fever. As a scientist he was thorough and meticulous, never publishing until he had worked and reworked even the most minor aspects of a paper. The one perhaps unfortunate consequence of this extreme care is that each word in his writings conveys a precise meaning and that no words are wasted; therefore, Gibbs' papers must be studied word by word—they cannot be speed-read!

Gibbs was one of the most brilliant scientists of the last century. He had the insight and power to perceive the most far-reaching logical consequences of fundamental postulates and had great inventive powers in mathematics. He made major contributions in vector analysis, as in the development of the theory of dyadics, and in the electromagnetic theory of light. He enjoyed geometry and graphical approaches. For physical chemists, Gibbs is almost the father of thermodynamics, both classical and statistical. He began publishing a series of lengthy articles in the *Transactions of the Connecticut Academy* in 1873. He developed in great depth the thermodynamic structure for dealing with heterogeneous equilibrium; he discovered the phase rule (Chapter 11) which bears his name and largely developed the thermodynamics of systems having an interface. We speak of the Gibbs adsorption equation in surface chemistry. He built much of the structure of statistical mechanics. Although his work was before the development of wave mechanics, statistical thermodynamic results do not require the use of quantum ideas—they are derivable and were derived by Gibbs on the basis of classical mechanics.

There were some delays in the general appreciation of Gibbs—perhaps because he chose to publish in a relatively obscure journal. Later, private copies of his papers were in circulation throughout the European scientific world and were avidly studied by his peers. Today the "Collected Works of J. Willard Gibbs" (Gibbs, 1931) remains virtually a bible for serious students of physical chemistry. By the late 1870's Gibbs had become an honored man, and he received a great variety of awards and honorary degrees during the following two decades. A major award of the American Chemical Society today is the J. Willard Gibbs medal.

Gibbs made much use of what are called "ensembles." As discussed in Section 4-10, an ensemble is simply the set of molecules, states, or other entities that is being examined. In statistical thermodynamics, it is the set that is being considered and over which summations are to be made. We have used, for example, a set called the *microcanonical* ensemble. This is the set of molecules, usually N_A in number, which occupies a specified volume and is at a specified total energy. As discussed in Section 2-2, if the energy is specified, there will be only certain ways in which the molecules can be distributed among the energy states. For most purposes the results obtained from the microcanonical set are the same as those from a *canonical* ensemble. This last corresponds to a collection of molecules occupying a specified volume and at a specified temperature. The system is in thermal contact with a heat reservoir and is not isolated as is the microcanonical set. The effect is to allow variations in average energy. However, if the number of molecules is large, this variation will be minute; the distribution will peak so sharply at the same place as that for the microcanonical ensemble of the same average energy that the difference is neglected unless the fluctuations themselves are being examined. The general derivations of thermodynamic quantities have so far been based on a canonical ensemble.

The third type of ensemble, known as the *grand canonical* ensemble, is a set of sets of molecules, each in thermal contact with the surroundings. The different sets of molecules may vary, however, in the number of molecules they contain. We speak of the grand canonical ensemble as an open system, meaning that molecules are allowed to enter or leave. This type of ensemble is used in statistical thermodynamics in calculations of a change in a thermodynamic property per mole of substance, primarily in dealing with mixtures or solutions. While the elaborate formality of the grand canonical ensemble will not be used here, it does constitute the statistical thermodynamic basis for treating partial molal quantities as discussed in Chapter 9.

6-CN-2   Additional comments on the third law of thermodynamics and on the attainment of 0 K

A corollary of the Nernst heat theorem (Section 6-10) is the statement: It is impossible to reduce the temperature of any system to absolute zero in a finite number of steps. To understand this statement, we need to consider how, in general, one can try to reach absolute zero. There is no lower-temperature heat reservoir now, and the only way to have a system cool itself is by a spontaneous adiabatic process. Suppose that the system consists of a hypothetical ideal gas; then since $dS \geq 0$, by Eq. (6-22),

$$C_V \ln \frac{T_2}{T_1} + R \ln \frac{V_2}{V_1} \geq 0$$

Let T_1 and V_1 be the starting condition, and now let the gas expand adiabatically to V_2; we want T_2 to be exactly 0 K. The term $C_V \ln(T_2/T_1)$ has now become minus infinity, and so the term $R \ln(V_2/V_1)$ must exceed plus infinity—clearly an impossibility!

The same type of analysis can be made in a more general way, but the added elaboration seems unnecessary. The conclusion remains the same: One can approach closer and closer to 0 K by some series of adiabatic processes but one cannot attain it in any finite number of steps.

The adiabatic expansion of an ideal gas is unrealistic as a practical procedure; all substances are solid near 0 K. A very imaginative method known as *adiabatic demagnetization*, developed by W. F. Giauque, is the one generally used. The procedure is to cool a paramagnetic material to as low a temperature as possible (liquid helium temperature) while in as strong a magnetic field as possible. The field is then turned off and the temperature drops. A temperature of 2×10^{-5} K has been reported (de Klerk et al., 1950). The physical reason that the temperature drops on demagnetization is roughly as follows. The magnetic field tends to align the molecular magnetic poles (discussed in Section 3-ST-2), reducing the entropy

of the system because of the greater order or lower thermodynamic probability of the system. It is, in effect, a restriction on the thermal motion of the molecules. On removal of the magnetic field, new energy states become available and more randomness or greater thermodynamic probability is sought. Since the system is adiabatic, the increase in entropy must be accompanied by a decrease in temperature. The formal analysis is not actually very different from that for the adiabatic expansion of a gas into a larger volume.

Although 0 K has not been reached exactly, nor is ever expected to be reached, it has been approached closely enough to suggest that heat capacity data can be extrapolated accurately from a few degrees Kelvin to 0 K by means of the Debye theory (Section 4-ST-3). The thermochemical procedure outlined in Section 6-9C has been carried out for a number of substances, with representative results as given in Table 6-2. As a specific example, an early value for the entropy of nitrogen at 25°C was obtained from the series of increments given in Table 6-3.

The third law of thermodynamics may be checked in two types of way. A most satisfying one is through statistical thermodynamics. If the final state is that of an ideal gas, then the equations of Section 6-12 may be used. The sample calculation given there led to S_{298}^0 for nitrogen of 45.75 cal K^{-1} mol^{-1} using the Sackur–Tetrode equation for an ideal gas and spectroscopic data giving the rotational and vibrational temperatures. The thermochemical value is 45.90 cal K^{-1} mol^{-1}. The agreement in this and several other cases is remarkably good, as shown in Table 6-2.

The second way of checking the third law is by means of a net chemical reaction. For example, combination of the separate thermochemical values for S_{298}^0 gives $\Delta S_{298}^0 = 4.78$ cal K^{-1} for the reaction

$$H_2 + Cl_2 = 2HCl$$

If now q_{rev} can be determined for the reaction, then ΔS is also given, by the second law, as q_{rev}/T. Several verifications of this type have been made, again with good agreement. Parenthetically, the usual procedure actually is to determine the equilibrium constant for the reaction and its temperature dependence (Section 7-4), but this amounts to the same thing.

Step	Entropy increment (cal K^{-1} mol^{-1})
0–10 K (by extrapolation)	0.458
10–35.61 K (integration of heat capacity data)	6.034
Crystal structure transition at 35.61 K	1.536
35.61–63.14 K (integration of heat capacity data)	5.589
Melting at 63.14 K	2.729
63.14–77.32 K (integration of heat capacity data)	2.728
Vaporization at 77.32 K	17.239
77.32–298.2 K (integration of heat capacity data)	9.37
Correction for gas nonideality	0.22
Total	45.9

TABLE 6-3
The entropy of N_2[a]

[a]W.F. Giauque and J.O. Clayton, *J. Amer. Chem. Soc.* **55,** 4875 (1933).

In the process of checking thermochemical and statistical thermodynamic entropies, occasional discrepancies have been found, of the order of a few calories per kelvin per mole, always with the thermochemical entropy too low. Instances where freezing produces a glass rather than a crystalline solid can be explained immediately on the basis of a residual entropy of disorder of the glassy structure. However, in cases such as CO, N_2O, and H_2O the crystals appear perfect, and the problem has been traced to a disorder on a molecular scale. Thus with CO the two atoms are so nearly the same that when the solid first crystallizes the end-to-end orientation of the CO is essentially random; that is, a crystal layer might appear thus:

CO OC OC CO CO OC CO

CO OC CO CO OC OC CO

The effect is that each CO has two possible positions, so for N_A molecules the thermodynamic probability is 2^{N_A} and a contribution of $R \ln 2$ is predicted by Eq. (6-70). As 0 K is approached, all the CO's should orient one way or the other but apparently they are prevented by an energy barrier from doing so in a finite time. Thus the thermochemical summations start with $R \ln 2$ or 1.4 cal K^{-1} mol^{-1} rather than with zero entropy and come out too low. The actual results are: spectroscopic entropy, 47.30; thermochemical entropy, 46.2. Adding 1.4 brings the thermochemical value to 47.6, or a little too high. Evidently the disorder is not complete.

Ice constitutes another type of example. Here one has a three-dimensional network of hydrogen bonds, and since hydrogen atoms could not be located by x-ray crystallography, for a long time it was uncertain whether they were positioned midway between oxygens or whether each hydrogen was closer to one oxygen than to another. A calculation on the latter supposition, assuming that each oxygen has two close and two distant hydrogens, but randomly chosen, leads to a thermodynamic probability of $(\frac{3}{2})^{N_A}$. This corresponds to an entropy contribution of $R \ln \frac{3}{2}$ or 0.806 cal K^{-1} mol^{-1}, which matches closely the discrepancy between the spectroscopic and thermochemical values. The two-position picture for hydrogen bonding in ice has since been confirmed by neutron diffraction measurements, which show (statistically) half a hydrogen atom at each of two positions between oxygens: $O - \frac{1}{2}H - \frac{1}{2}H - O$. The result is interpreted to mean that each oxygen has two close and two distant hydrogens but that the pattern throughout the crystal is random.

SPECIAL TOPICS

6-ST-1 Thermodynamic relationships

The various defining equations and statements of the combined first and second laws were summarized in Section 6-8. There are a number of thermodynamic relationships that stem from these and that can be produced in a fairly organized manner. We summarize such procedures here and use them to obtain a few of the more useful additional thermodynamic equations.

The various forms of the combined laws are repeated for convenience:

$$dE = T \, dS - P \, dV \qquad \text{[Eq. (6-17)]}$$

$$dH = T \, dS + V \, dP \qquad \text{[Eq. (6-18)]}$$

$$dA = -S \, dT - P \, dV \qquad \text{[Eq. (6-49)]}$$

$$dG = -S \, dT + V \, dP \qquad \text{[Eq. (6-48)]}$$

First, the equations lead directly to several simple coefficients. Thus

$$\left(\frac{\partial E}{\partial V}\right)_S = -P \qquad \left(\frac{\partial E}{\partial S}\right)_V = T$$

$$\left(\frac{\partial G}{\partial P}\right)_T = V \qquad \left(\frac{\partial G}{\partial T}\right)_P = -S \quad (6\text{-}112)$$

represent four out of the total of eight; the complete listing seems unnecessary.

Second, the equations are all exact differentials. This means that each generates an Euler or cross-differentiation [Eq. (4-11)]. We then have four new equations

$$\left(\frac{\partial T}{\partial V}\right)_S = -\left(\frac{\partial P}{\partial S}\right)_V \qquad (6\text{-}113)$$

$$\left(\frac{\partial T}{\partial P}\right)_S = \left(\frac{\partial V}{\partial S}\right)_P \qquad (6\text{-}114)$$

$$\left(\frac{\partial S}{\partial V}\right)_T = \left(\frac{\partial P}{\partial T}\right)_V \qquad \text{[Eq. (6-60)]}$$

$$\left(\frac{\partial S}{\partial P}\right)_T = -\left(\frac{\partial V}{\partial T}\right)_P \qquad \text{[Eq. (6-61)]}$$

These equations find a great deal of use in thermodynamic derivations.

Third, the four original statements of the combined laws generate a set of four equations based on the general procedure of Eq. (4-12). Thus

$$\left(\frac{\partial S}{\partial V}\right)_E = -\frac{(\partial E/\partial V)_S}{(\partial E/\partial S)_V} = \frac{P}{T} \qquad (6\text{-}115)$$

$$\left(\frac{\partial S}{\partial P}\right)_H = -\frac{(\partial H/\partial P)_S}{(\partial H/\partial S)_P} = -\frac{V}{T} \qquad (6\text{-}116)$$

$$\left(\frac{\partial T}{\partial V}\right)_A = -\frac{(\partial A/\partial V)_T}{(\partial A/\partial T)_V} = -\frac{P}{S} \qquad (6\text{-}117)$$

$$\left(\frac{\partial T}{\partial P}\right)_G = -\frac{(\partial G/\partial P)_T}{(\partial G/\partial T)_P} = \frac{V}{S} \qquad (6\text{-}118)$$

In addition, one may hold a dependent variable constant to give, for example,

$$\left(\frac{\partial G}{\partial P}\right)_V = -S\left(\frac{\partial T}{\partial P}\right)_V + V \qquad (6\text{-}119)$$

Finally, one can in a general way express $E = f(V, T)$, $H = f(P, T)$, $A = f(V, T)$, and $G = f(P, T)$ and obtain relationships such as Eq. (4-14). Rules for obtaining any desired partial differential coefficient have been worked out (Lewis and Randall, 1961).

6-ST-2 Thermodynamic treatment of a nonideal gas—fugacity

This topic is included for two reasons. First, gases are generally assumed to be ideal in this text, and it seems desirable to indicate in at least *one* place how to proceed if the assumption is to be avoided. Second, the procedure is a fairly typical one in chemical thermodynamics and will be used again in connection with nonideal solutions. In the case of an ideal gas, various thermodynamic relationships, such as Eq. (6-55), contain pressure as a variable by virtue of the ideal gas law. One now seeks to find an effective or thermodynamic pressure for a nonideal gas, defined so that it can be used in thermodynamic equations in the same way as actual pressure is in the case of an ideal gas.

In the case of an ideal gas, we can write Eq. (6-55) in the form

$$G = G^0 + RT \ln P \qquad (6\text{-}120)$$

where G^0 is the molar free energy in the reference state of unit pressure, usually 1 atm. As suggested by G. N. Lewis, let us define an effective pressure such that the same form of equation is exact for a nonideal gas. We call this effective pressure the *fugacity, f*:

$$G = G^0 + RT \ln f. \qquad (6\text{-}121)$$

Clearly, $f \to P$ as $P \to 0$ since all gases approach ideality in this limit. Application of Eq. (6-52) gives

$$\left(\frac{\partial G}{\partial P}\right)_T = V = RT \left[\frac{\partial(\ln f)}{\partial P}\right]_T \qquad (6\text{-}122)$$

or

$$RT\, d(\ln f) = V\, dP$$
$$\text{(at a given temperature)} \qquad (6\text{-}123)$$

Fugacity may be obtained experimentally by means of the following procedure. We define α as $V_{\text{ideal}} - V_{\text{actual}}$

$$\alpha = \frac{RT}{P} - V \qquad (6\text{-}124)$$

Substitution into Eq. (6-123) and integration at constant temperature gives

$$\int_0^P d(\ln f)$$
$$= \int_0^P d(\ln P) - \frac{1}{RT} \int_0^P \alpha\, dP \qquad (6\text{-}125)$$

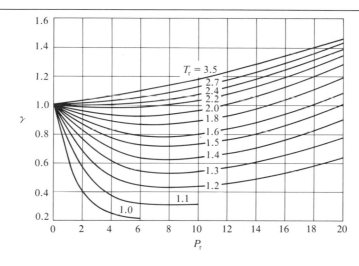

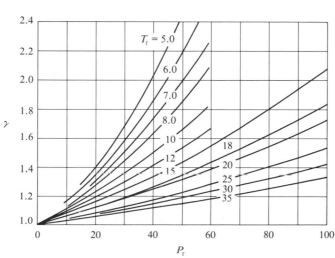

FIGURE 6-16
Activity coefficients for gases as a function of reduced pressure P_r for various reduced temperatures T_r. [From R. H. Newton, Ind. Eng. Chem., **27**, 302 (1935). Copyright 1935 by the American Chemical Society. Used by permission of the copyright owner.]

Since f and P become identical in the limit of zero pressure, Eq. (6-125) reduces to

$$\ln f = \ln P - \frac{1}{RT} \int_0^P \alpha \, dP \qquad (6\text{-}126)$$

The integral may, for example, be obtained graphically from a plot of α versus P. Where the nonideality is not large, a useful rule is that the observed pressure of a gas will be the geometric mean of the fugacity and the ideal gas law pressure given by RT/V:

$$P_{\text{actual}} = \left(\frac{RTf}{V} \right)^{1/2} \qquad (6\text{-}127)$$

The procedure for determining the fugacity of a gas in a mixture is similar, although more complicated. See Section 7-ST-1 for the treatment of equilibria using fugacities.

Note that neither the definition of fugacity nor the preceding derivations require that the substance be gaseous. We can obtain the fugacity of a solid or a liquid by integrating Eq. (6-123). The molecular weight used must be that of the gas at $P \to 0$, however, since this is the condition on which the definition of f is based.

Returning to the case of a nonideal gas, the law of corresponding states amounts to the observation that a plot of the compressibility factor Z against reduced pressure P/P_c for various reduced temperatures T/T_c gives a set of plots that are well obeyed by all gases. This is illustrated in Fig. 1-9. It should not be surprising that fugacities obey the law of corresponding states fairly well. Actually, it is more convenient to introduce a further quantity, the *activity coefficient* γ defined by $f_i = \gamma_i P_i$. The activity coefficient is the factor by which the fugacity differs from the actual pressure of the gas. Activity coefficients also obey the law of corresponding states and Fig. 6-16 shows a chart of γ versus P_r for various T_r. It is used in much the same way as in Fig. 1-9.

6-ST-3 The free energy function

We continue here the approach initiated in Section 5-ST-3. In that section the internal energy function $E - E_0$ was introduced, obtainable through either Eq. (5-40) or Eq. (5-43):

$$E - E_0 = \int_0^T C_V \, dt \qquad \text{or}$$

$$E - E_0 = RT^2 \left(\frac{\partial (\ln \mathbf{Q})}{\partial T} \right)_V$$

The nature of the E_0 quantity is also discussed in Section 6-11. As indicated above, the quantity $(E - E_0)$ can be determined either thermochemically or spectroscopically. Similarly, the enthalpy function $H^0 - E_0$ is defined by Eq. (5-46). It is now convenient to introduce a more symmetric function, $H^0 - H_0^0$, given by

$$H^0 - H_0^0 = \int_0^T C_P \, dT \qquad (6\text{-}128)$$

or, from Eq. (6-87)

$$H^0 - H_0^0 = RT^2 \left[\frac{\partial (\ln \mathbf{Q})}{\partial T} \right]_V$$

$$+ RTV \left[\frac{\partial (\ln \mathbf{Q})}{\partial V} \right]_T \qquad (6\text{-}129)$$

The *enthalpy function* may therefore also be determined either thermochemically if the substance is an ideal gas or from spectroscopically obtained values of θ_{vib} and θ_{rot} and the various partition functions given in Section 6-12. Recall that the superscript zero denotes 1 atm pressure and the subscript zero the value at 0 K.

Since by definition $G = H - TS$, we can write

$$\frac{G^0 - H_0^0}{T} = \frac{H^0 - H_0^0}{T} - S^0 \qquad (6\text{-}130)$$

(H_0^0 has merely been subtracted from both sides). The quantity $(G^0 - H_0^0)/T$ is called the *free energy function* and $(H^0 - H_0^0)/T$ the *enthalpy function*. As with the other functions,

---

EXAMPLE Calculate the fugacity of NH_3 at 350 K and 500 atm pressure. From Table 1-4, T_c and P_c are 304 K and 72.8 atm, hence $T_r = 350/304 = 1.15$, and $P_r = 500/72.8 = 6.87$. From Fig. 6-16, we read $\gamma = 0.35$ and thus the fugacity is $f = 500 \times 0.35 = 175$ atm.

$(G^0 - H_0^0)/T$ may be obtained purely from thermochemical data, including now a third law determination of S^0, or, in the case of an ideal gas, from the various partition function expressions.

We now consider data for some chemical reaction. For the general reaction

$$aA + bB + \cdots = mM + nN + \cdots$$

we can write ΔG^0 for some temperature T as

$$\frac{\Delta G^0}{T} = \frac{\Delta(G^0 - H_0^0)}{T} + \frac{\Delta H_0^0}{T} \qquad (6\text{-}131)$$

Similarly,

$$\Delta H^0 = \Delta(H^0 - H_0^0) + \Delta H_0^0 \qquad (6\text{-}132)$$

If $\Delta G^0/T$ can be evaluated for the reaction and $\Delta(G^0 - H_0^0)/T$ obtained either thermochemically or spectroscopically, then ΔH_0^0 can be determined. For this reaction ΔH_0^0 is a constant and Eq. (6-131) can be used for the calculation of ΔG^0 at any other temperature. As in thermochemistry the practice is to compile values of ΔH_0^0 for reactions in which a given substance is formed from the elements. One can then calculate ΔH_0^0 for any other reaction in the same way as with enthalpies of formation [Eq. (5-24)].

It is well at this point to summarize the rea-

It is well at this point to summarize the reasons for using the above functions. First, they provide a natural arrangement that facilitates the use of statistical thermodynamics. Once ΔH_0^0 has been determined for a reaction we can often calculate $\Delta(G^0 - H_0^0)/T$ for any desired temperature without recourse to further experimental data. In this way we can find ΔG_0 for a reaction at a temperature not experimentally accessible. Second, the free energy function $(G^0 - H_0^0)/T$ changes much more slowly with temperature than does G^0 itself. Tabulations of this function may then be for rather widely spaced temperature intervals. Tables 6-4 and 6-5 give some illustrative data of this type.

It will be seen later (Section 7-2) that ΔG^0 determines the equilibrium constant for a reaction, the equation being

$$\Delta G^0 = -RT \ln K \qquad (6\text{-}133)$$

where K is the equilibrium constant. Equation (6-131) may then be written in the form

$$\ln K = -\frac{1}{R}\left[\frac{\Delta(G^0 - H_0^0)}{T} + \frac{\Delta H_0^0}{T}\right] \qquad (6\text{-}134)$$

The ability to calculate a ΔG^0 for a reaction is equivalent to the ability to determine its equilibrium constant; hence the great importance of ΔG^0.

	$-(G^0 - H_0^0)/T$ (cal K^{-1} mol^{-1})					$H_{298}^0 = H_0^0$ (kcal mol^{-1})	ΔH_0^{0b} (kcal mol^{-1})
Element	298.15 K	500 K	1000 K	1500 K	2000 K		
Br$_2$(g)	50.85	54.99	60.80	64.31	66.83	2.325	8.37
Br$_2$(l)	25.5	—	—	—	—	3.240	0
C(graphite)	0.53	1.16	2.78	4.19	5.38	0.251	0
Cl$_2$(g)	45.93	49.85	55.43	58.85	61.34	2.194	0
F$_2$(g)	41.37	45.10	50.44	53.74	56.17	2.110	0
H$_2$(g)	24.42	27.95	32.74	35.59	37.67	2.024	0
I$_2$(g)	54.18	58.46	64.40	67.96	70.52	2.148	15.66
I$_2$(s)	17.18	—	—	—	—	3.154	0
N$_2$(g)	38.82	42.42	47.31	50.28	52.48	2.072	0
O$_2$(g)	42.06	45.68	50.70	53.81	56.10	2.07	0

TABLE 6-4

Free energies for the elements[a]

[a]From G.N. Lewis and M. Randall, "Thermodynamics," 2nd ed. (revised by K.S. Pitzer and L. Brewer), p. 680. McGraw-Hill, New York, 1961 (where additional values are given).

[b]Of formation from the element in its standard phase; H_0^0 includes the $\frac{1}{2}h\nu_0 N_A$ contribution from the zero-point vibrational energy.

TABLE 6-5

Free energies for various gaseous compounds[a]

Compound	$-(G^0 - H_0^0)/T$ (cal K^{-1} mol^{-1})					$H_{298}^0 - H_0^0$ (kcal mol^{-1})	ΔH_0^{0b} (kcal mol^{-1})
	298 K	500 K	1000 K	1500 K	2000 K		
CH$_4$	36.46	40.75	47.65	52.84	57.1	2.397	−15.99
C$_2$H$_2$	39.98	44.51	52.01	57.23	61.33	2.392	54.33
C$_2$H$_4$	43.98	48.74	57.29	63.94	69.46	2.525	14.52
C$_2$H$_6$	45.27	50.77	61.11	69.46	—	2.856	−16.52
C$_3$H$_8$	52.73	59.81	74.10	85.86	—	3.512	−19.48
C$_6$H$_6$	52.93	60.24	76.57	90.45	—	3.401	24.00
C$_2$H$_5$OH	56.20	62.82	75.28	85.15	—	3.39	−52.41
CO	40.25	43.86	48.77	51.78	53.99	2.073	−27.202
CO$_2$	43.56	47.67	54.11	58.48	61.85	2.238	−93.969
HBr	40.53	44.12	48.99	51.95	54.13	2.067	−8.1
HCl	37.72	41.31	46.16	49.08	51.23	2.065	−22.019
HI	42.40	45.99	50.90	53.90	56.11	2.069	6.7
H$_2$O	37.17	41.29	47.01	50.60	53.32	2.368	−57.107
NH$_3$	37.99	42.28	48.63	53.03	56.56	2.37	−9.37
NO	42.98	46.76	51.86	54.96	57.24	2.194	21.48
SO$_2$	50.82	55.38	62.28	66.82	70.2	2.519	−70.36
SO$_3$	51.89	57.14	66.08	72.40	77.1	2.77	93.06

[a]From G.N. Lewis and M. Randall, "Thermodynamics," 2nd ed. (revised by K.S. Pitzer and L. Brewer), p. 682. McGraw-Hill, New York, 1961.

[b]Of formation from the elements.

EXAMPLE The following calculation will demonstrate the use of the tables. Consider the reaction

$$3H_2 + N_2 = 2NH_3 \tag{6-135}$$

We used this in Section 5-ST-2 to illustrate the calculation of $H^0 - E_0$ from partition functions. We now want to obtain ΔG^0 (and hence the equilibrium constant) at 1000 K. First, ΔH_{298}^0 is just the standard enthalpy of formation of 2 moles of ammonia. From Table 5-2 we get $\Delta H_{298}^0 = -2(11.04) = -22.08$ kcal, and from Tables 6-4 and 6-5 we have

$$\Delta(H_{298}^0 - H_0^0) = 2(2.37) - 3(2.024) - 2.072 = -3.40 \quad \text{kcal}$$

Then from Eq. (6-132) we have

$$\Delta H_0^0 = -22.08 - (-3.40) = -18.68 \quad \text{kcal}$$

We next calculate $\Delta(G^0 - H_0^0)/T$ for 1000 K using the tables:

$$\frac{\Delta(G^0 - H_0^0)}{T} = -2(48.63) - [-3(32.74) - 47.31] = 48.27 \quad \text{cal K}^{-1}$$

or, multiplying by $T = 1000$ K, we get

$$\Delta(G^0 - H_0^0) = 48.27 \quad \text{kcal}$$

Finally, we have

$$\Delta G^0 = 48.27 - 18.68 = 29.59 \quad \text{kcal}$$

We can complete the calculation by obtaining the equilibrium constant. From Eq. (6-134) we have

$$\ln K = -\frac{29,590}{(1.987)(1000)} = -14.89$$

$$K = 3.41 \times 10^{-7}$$

This example could have been made yet more complete if we had carried out the evaluation of $G^0 - H_0^0$ for each substance at 1000 K using the appropriate partition functions. Essentially the same calculation was made in Section 6-9C for N_2 at 25°C.

GENERAL
REFERENCES

BLINDER, S.M. (1969), "Advanced Physical Chemistry." Macmillan, New York.
LEWIS, G.N., AND RANDALL, M. (1961). "Thermodynamics," 2nd ed. (revised by K.S. Pitzer and L. Brewer). McGraw-Hill, New York.

CITED
REFERENCES

BERRY, R.S., RICE, S.A., AND ROSS, J. (1980). "Physical Chemistry." Wiley, New York.
BLANDAMER, M.J. (1973). "Introduction to Chemical Ultrasonics." Academic, New York.
BLINDER, S.M. (1969). "Advanced Physical Chemistry." Macmillan, New York.
DE KLERK, D., STEENLAND, M.J., AND GORTER, C.J. (1950), *Physica* **16**, 571.
GIAUQUE, W.F., AND BLUE, R.W. (1936). *J. Amer. Chem. Soc.* **58**, 831.
GIBBS, J.W. (1931). "Collected Works of J. Willard Gibbs." Longmans, Green, New York.
LEWIS, G.N., AND RANDALL, M. (1961). "Thermodynamics," 2nd ed. (revised by K.S. Pitzer and L. Brewer), p. 667. McGraw-Hill, New York.

EXERCISES

Find two integrating factors for the expression $\left[-\dfrac{R}{P} dP + \dfrac{R}{T} dt \right]$. Show what $F(P,T)$ is.   **6-1**

Ans. One integrating factor is (T/P), and $F(P,T)$ is RT/P.
Other integrating factors would be $(T/P)(RT/P)^n$, $n = 1, -1, 2$, etc.

Find an integrating factor for $[v\, du + 3u\, dv]$ and thence $F(u,v)$.   **6-2**

Ans. $F(u,v) = uv^3$.

An ideal heat engine operates between $-50°C$ and $50°C$. How many joules of work does it deliver   **6-3**
per joule of heat energy absorbed at the upper temperature?

Ans. 0.310 J.

The same machine as in Exercise 6-3 is operated as a refrigerator. How much power (in watts) must   **6-4**
be supplied to the machine so that its low-temperature coils can maintain a cryostat bath at $-50°C$?
The bath gains 100 cal s^{-1} from its surroundings.

Ans. 188 W.

The same machine as in Exercise 6-3 is used as a heat pump. How much power must be supplied   **6-5**
to it if its hot coils are to replace a 40,000 Btu hr^{-1} house furnace? (This is a Canadian winter with
$-50°C$ outside, and the 50°C inside temperature will make the house rather warm!)

Ans. 3630 W.

The available cooling water for an ideal heat engine is 10°C. What must be the upper operating   **6-6**
temperature if the efficiency is to be 25% (and the lower temperature 10°C)?

Ans. 104°C.

One mole of an ideal monatomic gas is initially at STP and is put through the following alternative   **6-7**
reversible processes. For each process, calculate w, q, ΔE, ΔH, and ΔS. (a) Isochoric cooling to
$-100°C$. (b) Isothermal compression to 100 atm. (c) Isobaric heating to 100°C. (d) Adiabatic expansion
to 0.1 atm.

Ans. (a) $w = 0$, $q = -298$ cal, $\Delta E = q$,
$\Delta H = -497$ cal, $\Delta S = -1.36$ cal K^{-1} mol^{-1}.
(b) $w = q = -2500$ cal, $\Delta E = \Delta H = 0$, $\Delta S = -9.15$ cal K^{-1} mol^{-1}.
(c) $w = 199$ cal, $q = \Delta H = 497$ cal, $\Delta E = 298$ cal,
$\Delta S = 1.55$ cal K^{-1} mol^{-1}.
(d) $w = -\Delta E = 490$ cal, $q = \Delta S = 0$, $\Delta H = -817$ cal.

6-8 Assume that the fuel mix for a Diesel engine behaves like an ideal gas of heat capacity $C_V = 31.4$ J K^{-1} mol^{-1}. Each cylinder has a volume of 0.80 liter, and the diesel action amounts to an adiabatic heating of the fuel mixture to a temperature such that spontaneous combustion is rapid. If this temperature is 600°C and the initial fuel mixture is at 25°C and 1 atm pressure, calculate the compression ratio and the final pressure in Pa. Assume reversible processes.

Ans. 60.5 : 1, 18.2 MPa.

6-9 Two moles of an ideal monatomic gas initially at 1 atm and 25°C undergo an irreversible adiabatic expansion in which 200 cal of work is done. Calculate, if possible, the final state of the gas and q, ΔE, ΔH, and ΔS for the change.

Ans. $q = 0$, $T_f = 265$ K, $\Delta E = -200$ cal, $\Delta H = -333$ cal
(the final volume or pressure and ΔS cannot be obtained
without more knowledge of the details of the expansion).

6-10 Calculate the change in entropy if one troy ounce of silver is heated at constant volume from 300 K to 400 K. Note Fig. 4-16.

Ans. 0.495 cal K^{-1} mol^{-1}.

6-11 What is the change in entropy when one mole of gold freezes at its melting point of 1064°C? The heat of fusion is 15.8 cal g^{-1}.

Ans. 2.33 cal K^{-1}mol^{-1}.

6-12 What are the changes in energy, enthalpy, and entropy when one mole of water condenses at 100°C from vapor at 1 atm to liquid water?

Ans. $\Delta H = -9.717$ kcal mol^{-1}, $\Delta E = -8.976$ kcal mol^{-1},
$\Delta S = -26.04$ cal K^{-1} mol^{-1}.

6-13 Calculate from the data of Tables 5-4 and 6-2 the molar entropy of liquid water at 100°C and 1 atm pressure and of water vapor at 200°C and 1 atm pressure.

Ans. S^0(H$_2$O liq, 100°C) = 20.78 cal K^{-1} mol^{-1},
S^0(H$_2$O gas, 200°C) = 48.80 cal K^{-1} mol^{-1}.

6-14 Calculate ΔG for the isothermal expansion of one mole of water vapor at 25°C from 2600 Pa to 260 Pa. Assume ideal gas behavior.

Ans. -1360 cal.

6-15 Calculate the value of ΔG when one mole of benzene is taken from liquid at 1 atm and 80.1°C to vapor at 1 atm and 80.1°C and thence to vapor at 0.5 atm and 200°C. (Assume ideal gas behavior for the vapor.) C_P is given by $(8.00 + 0.080T)$ and $(2.46 + 0.0602T)$ for liquid and gaseous benzene, respectively, and S^0_{298} is 41.3 for benzene liquid, all values in cal K^{-1} mol^{-1}.

Ans. (a) $\Delta G = 0$, (b) $\Delta G = -9271$ cal mol^{-1} (the process is not
isothermal and so the negative sign does not imply spontaneity).

6-16 Show from Eq. (6-65) that $C_P - C_V = R$ for an ideal gas.

6-17 Calculate the coefficients of compressibility and of thermal expansion for an ideal gas.

Ans. $\beta = 1/P$, $\alpha = 1/T$.

6-18 Derive the value of $C_P - C_V$ for a gas obeying the equation $P(V - b) = RT$, where b is a constant. Comment on the result.

Ans. R.

6-19 Verify the constant -2.31 in Eq. (6-104); obtain two more significant figures.

6-20 Verify the constant in Eq. (6-107); obtain two more significant figures.

6-21 Calculate S_{trans}, A_{trans}, and G_{trans} of methane gas at 1 atm and 50°C.

Ans. $S^0_{trans} = 34.66$ cal K^{-1} mol^{-1}, $A_{trans} = -10,240$ cal mol^{-1},
and $G^0_{trans} = -9590$ cal mol^{-1}.

Calculate S_{vib}, A_{vib}, and G_{vib} at 50°C for (a) H_2 gas and (b) I_2 gas.   **6-22**

> Ans. (a) $S_{vib} = 1.58 \times 10^{-7}R$, $A_{vib} = G_{vib} = -2.91 \times 10^{-6} R$,
> (b)$S_{vib} = 1.089R$, $A_{vib} = G_{vib} - 158R$.

Calculate S_{rot}, A_{rot}, and G_{rot} for (a) H_2 at 50°C and (b) I_2 at 50°C. Explain why your answer to (a) may   **6-23**
be in error.

> Ans. (a) $S_{rot} = 2.31R$, $A_{rot} = G_{rot} = -422R$,
> (b) $S_{rot} = 9.70R$, $A_{rot} = G_{rot} = -2810R$.

PROBLEMS

Show whether the expression $[(2uw/v)du - (u^2w/v^2)dv + (u^2/v)dw]$ is a total differential. If it is,   **6-24**
find $F(u,v,w)$. Note: three cross-differential relationships patterned after that of Eq. (4-11) provide
a test.

Show whether the expression $[y\,dx - x\,dy]$ is an exact or total differential. If it is, find $F(x,y)$; if it is   **6-25**
not, find an integrating factor and the corresponding $F(x,y)$.

A reversible heat engine operates between 0°C and an upper temperature T_2 with 20% efficiency; it   **6-26**
produces 20,000 J per cycle. Calculate T_2, q_1, and q_2. Make a plot of S versus T for the Carnot cycle
corresponding to this engine. Also calculate the performance factors for the engine operating as a
refrigerator and as a heat pump.

A heat engine operates between 0°C and 1000°C and is ideal except that 20% of the work produced   **6-27**
is dissipated by friction at 0°C. How much heat must be supplied at 1000°C if 1000 J of useful work
is to be done by the machine?

A mercury vapor engine operates at a boiler pressure of 2 atm; the condenser is at 25°C. Calculate   **6-28**
the theoretical efficiency. Calculate also the theoretical efficiency of a steam engine operating at the
same boiler pressure and condenser temperature. The vapor pressure of mercury is given by log P
(mm Hg) $= 3.066 \times 10^3/T + 7.752$.

Gasoline has a heat of combustion of about 11.5 kcal/g and its density is about 6 lb per gallon; typical   **6-29**
cylinder and exhaust temperatures might be 2000°C and 600°C. On a level road, energy consumption
is due to friction of moving parts and to air resistance, and this has the effect of decelerating the car
at 2 mph per second at 40 mph. To maintain speed, energy from the engine counteracts this friction.
Assuming ideal heat engine behavior, how many miles per gallon should a 4000 lb automobile (say,
a Toronado) get?

A system experiences the following reversible constant-pressure process: **6-30**

$$(P_1 = 1 \text{ atm}, V_1 = 3 \text{ liter}, T_1 = 300 \text{ K}) \rightarrow (P_2 = 1 \text{ atm}, V_2 = 5 \text{ liter}, T_2 = 600 \text{ K})$$

The heat capacity of the system for this process is 25 cal K^{-1}; the entropy of the system is initially 30
cal K^{-1}. Calculate ΔE, ΔH, ΔS, ΔG, q, and w for the process.

Show that $(\partial E/\partial V)_T = 0$, for an ideal gas, using the first and second laws and related definitions.   **6-31**

Derive from the first and second laws of thermodynamics and related definitions: **6-32**

$$G - H = T\left(\frac{\partial G}{\partial T}\right)_P$$

Derive the equation   **6-33**

$$C_V = C_P\left[1 - \mu\left(\frac{\partial P}{\partial T}\right)_V\right] - V\left(\frac{\partial P}{\partial T}\right)_V$$

from the first and second laws and related definitions (μ is the Joule–Thomson coefficient).

6-34   Derive the equation

$$\left(\frac{\partial H}{\partial T}\right)_P = \left(\frac{\partial E}{\partial T}\right)_V + \left[P + \left(\frac{\partial E}{\partial V}\right)_T\right]\left(\frac{\partial V}{\partial T}\right)_P$$

from the first and second laws and related definitions.

6-35   The heat of fusion of sodium is 21.7 cal g^{-1} at its melting point of 97°C, and the heat capacities of solid and liquid sodium are both 0.28 cal K^{-1} g^{-1}. Liquids can generally be supercooled, and we consider the process:

Na(liq, 70°C, 1 atm) → Na(solid, 70°C, 1 atm)

Calculate ΔS and ΔG for this process. Note that ΔS will be negative, but this does not mean that supercooled liquid sodium will not spontaneously freeze, because the above process is not one that could occur in an isolated system. Write the process that *would* occur in an isolated system. Since conditions are now adiabatic, as the liquid sodium at 70°C starts freezing, the temperature of the system will rise. If it rises to 97°C, the final state will consist of both liquid and solid sodium. Determine what the final state will be and calculate ΔS for the process. Neglect mechanical work due to volume changes, and assume that $C_V = C_P$ and that $\Delta H = \Delta E$.

6-36   Given the process:

H_2O(solid, −15°C, 1 atm) → H_2O(liquid, −15°C, 1 atm)

and necessary thermal data from handbooks, calculate ΔS, ΔG, and ΔH. Note that ΔS is positive; explain whether this means that the process is spontaneous.

6-37   Calculate ΔS, ΔH, and ΔG for the process

18 g ice (0°C) + 200 g H_2O (50°C) = equilibrium system at t °C.

The process is carried out by placing the water in a Dewar flask and then dropping the ice in; it may be assumed that the flask is perfectly insulated.

6-38   Evaluate $\left(\frac{\partial E}{\partial V}\right)_T$ for a gas that obeys the equation $P(V - b) = RT$, where b is a constant.

6-39   One mole of an ideal monatomic gas initially at STP is expanded reversibly to a final pressure of 0.5 atm along the path PV^2 = constant (note Eq. (4-60)). Calculate ΔS for the process.

6-40   Calculate ΔG for the following changes of state for one mole of an ideal monatomic gas initially at 0°C and 1 atm pressure: (a) volume doubled at constant pressure; (b) pressure doubled at constant volume; (c) pressure doubled at constant temperature. Assume S_{273} at 1 atm to be 26 cal K^{-1} mol^{-1}.

6-41   Calculate ΔG and ΔA for three moles of an ideal gas that expands irreversibly from 2 atm at 25°C to 0.1 atm at 25°. Repeat the calculation for carbon dioxide, assuming it to obey the van der Waals equation. No work is done.

6-42   Following Eq. (6-30) there is a discussion of how entropy change might be calculated on cooling a solution. We can construct a numerical problem around the example given, namely an initial solution consisting of 1 mole of water and 0.048 mole of ethylene glycol, whose freezing point is −3.9°C. We ask for the entropy change on cooling to −6.7°C, at which point 0.42 mole of ice has formed, with the remaining water and ethylene glycol forming a solution of increased concentration. We make the calculation easier by assuming that the freezing point of this particular system obeys the equation $n_{H_2O} = (3.9)/(273 - T)$ where n_{H_2O} is the number of moles of water remaining in solution along with the 0.048 mole of ethylene glycol, and $n_{ice} = 1 - n_{H_2O}$. The heat capacity of liquid water and of ice are nearly the same, 18 cal K^{-1} mol^{-1}; the heat capacity of ethylene glycol is 33.7 cal K^{-1} mol^{-1}, and the heat of fusion of water from the solution is taken to be that for pure water, namely 1,436 cal mol^{-1}. To calculate ΔS one needs to set up the expression for dq/dT, where q is the heat evolved by the system on cooling. Three terms are involved, (a) that for cooling the solution, (b) that for cooling the ice present, and (c) that for the latent heat evolved as dn moles of water freeze over the interval dT. (A reasonable assumption to make is that the heat capacity of the solution is just the sum of the heat capacities due to the moles of water and of ethylene glycol present.)

Show that $C_P - C_V = 2aP/RT^2 + a^2P^2/R^3T^4$ for a van der Waals gas. [*Note:* this equation is obtained by making certain approximations. Start with the van der Waals equation in the form $PV = RT - (a/V) + bP + (ab/V^2)$, then neglect the term ab/V^2 and approximate a/V by aP/RT.] Using the van der Waals constants, calculate $C_P - C_V$ for CO_2 at 35°C and 10 atm.   **6-43**

A useful quantity in chemical ultrasonics is the isentropic compressibility, $\beta_S = -(1/V)(\partial V/\partial P)_S$. It can be shown, for example, that $\beta_S = 1/v^2\rho$, where v is the velocity of sound in the medium (M.J. Blandamer, 1973). Derive the relationship $\beta = \beta_s + \alpha^2TV/C_P$ where β is the ordinary compressibility, $\beta = -(1/V)(\partial V/\partial P)_T$, and α is the coefficient of thermal expansion. Hint: in a derivation of this type it usually is easier to start with the given equation and work backwards to a first or second law statement or to an identity.   **6-44**

The principal vibrational frequencies for CO_2 are $\nu_1 = 1388.3$ cm^{-1}, $\nu_2 = 667.3$ (twice), and $\nu_3 = 2349.3$ cm^{-1}. Calculate the vibrational partition functions for CO_2 at 25°C and at 50°C and make an approximate calculation of the vibrational contribution to the heat capacity at 37°C. Compare this with the experimental value. Also calculate the percentage of the equipartitional heat capacity that is present at 37°C. Finally, calculate the vibrational contribution to the entropy of CO_2 at 25°C.   **6-45**

The vibrational frequency for $^{79}Br^{81}Br$ is 321.1 cm^{-1}. Calculate (a) the separation of the first vibrational level from the ground state in cal mol^{-1}; (b) the vibrational contribution to the energy at 100°C; and (c) the vibrational contribution to the entropy at 100°C.   **6-46**

Calculate S_{trans}, S_{rot}, and S_{vib} for iodomethane, CH_3I, at 25°C. For this type of molecule,   **6-47**

$$\mathbf{Q}_{rot} = \frac{1}{\sigma}\left[\frac{\pi}{B^2A}\left(\frac{kT}{hc}\right)^3\right] \tag{6-136}$$

The two rotational constants, B and A, have the values 0.28 and 5.08 cm^{-1} respectively, and the vibrational frequencies are $\nu_1 = 2969.8$ cm^{-1}, $\nu_2 = 1251.5$ cm^{-1}, $\nu_3 = 532.8$ cm^{-1}, $\nu_4 = 3060.3$ cm^{-1} (twice), $\nu_5 = 1440.3$ cm^{-1} (twice), and $\nu_6 = 880.1$ cm^{-1} (twice).

Derive Eq. (6-136) from Eq. (4-102).   **6-48**

Giauque and Blue (1936) obtained the following data for hydrogen sulfide: **6-49**

T (K)	C_P (cal K^{-1} mol^{-1})	T (K)	C_P (cal K^{-1} mol^{-1})
20	1.25	110	11.79
30	2.48	120	13.27
40	3.56	126.2	(^{b})
50	4.56	130	13.25
60	5.51	140	13.33
70	6.43	150	13.45
80	7.31	170	13.92
90	8.26	187.6	(^{c})
100	9.36	190	16.21
103.5	(^{a})	210	16.31
105	11.25	212.8	(^{d})

^{a}Solid-state transition with $\Delta H = 368$ cal mol^{-1}.
^{b}Solid-state transition with $\Delta H = 121$ cal mol^{-1}.
^{c}Melting point, $\Delta H = 568$ cal mol^{-1}.
^{d}Normal boiling point, $\Delta H = 4463$ cal mol^{-1}.

Below 20 K the heat capacity obeys the law $C_P = aT^3$. Calculate the absolute entropy of hydrogen sulfide at 1 atm and 212.8 K.

An absorbed molecule is often thought to be confined or localized to a relatively small area around a particular active site of the adsorbent. Calculate the spacing between the $n = 1$ and $n = 2$ levels   **6-50** of translational energy for the case of argon molecules each confined to a two-dimensional box of 25 $\mathring{A}^2$ area, assuming 0°C. Explain whether the approximation made in obtaining Eq. (4-66) should still apply.

6-51 The rotational constant B for HCl is 10.59 cm^{-1}, where $B = h/8\pi^2cI$), and the fundamental vibrational frequency is 2886 cm^{-1}. Calculate the total entropy of HCl gas at STP, assuming it to be an ideal gas.

6-52 Calculate the molar entropy of Kr gas at STP assuming it to be ideal and using the Sackur–Tetrode equation. Also calculate the entropy per mole of Kr adsorbed on a surface assuming that it behaves as an ideal two-dimensional gas and that its state is 0°C and a surface concentration such that the average distance apart of molecules is the same as in the three-dimensional gas at STP.

6-53 One mole of an ideal monatomic gas expands adiabatically and reversibly from 400 K and 1×10^6 Pa to 273 K. Calculate ΔS, ΔE, ΔH, ΔG, q, and w.

6-54 One mole of an ideal monatomic gas at STP expands into an evacuated flask so that its volume triples. Calculate ΔS, ΔE, ΔH, ΔG, q, and w.

SPECIAL TOPICS PROBLEMS

Problems marked with an asterisk require fairly lengthy computations.

6-55 Assume that a gas obeys the equation $P(V - b) = RT$. Derive an expression for the fugacity of such a gas as a function of pressure and temperature.

6-56 Assume that a gas obeys the equation $(P + a/V^2)V = RT$. Derive an expression for the fugacity of such a gas as a function of pressure and temperature.

6-57* Assuming that NH$_3$ gas obeys the van der Waals equation, calculate the activity coefficient of the gas at 200°C from zero pressure up to 10 atm pressure.

6-58* The following data have been obtained for a certain gas at 25°C:

P, atm	1	5	10	20	30	40	50
Z	0.996	0.975	0.950	0.900	0.850	0.792	0.730

where Z is the compressibility factor. Calculate the fugacity of this gas at 25°C and 50 atm pressure. You may use a graphical integration procedure.

6-59 Estimate the activity coefficient of CO$_2$ at 300°C and 400 atm pressure.

6-60 Calculate ΔG^0 and the corresponding equilibrium constants at 298 K and 1000 K for the reaction

$$C_2H_4 + H_2O = C_2H_5OH$$

6-61 Verify the value of $(G^0 - H_0^0)/T$ at 298 K for N$_2$(g) given in Table 6-4 by evaluating the appropriate partition functions.

6-62 Calculate $\Delta S^0_{298\ K}$ for the reaction H$_2$ + Br$_2$ = 2 HBr.

6-63 Calculate K_p for reaction (6-135) at 1500 K.

chapter 7
Chemical equilibrium

7-1   Pre-thermodynamic ideas

Early chemists considered that different substances exhibited either empathies or antipathies towards each other. Those that reacted were thought to possess some kind of mutual affinity, and alchemical language was rich in phrases implying almost emotional attitudes on the part of various chemicals. Chemistry was then largely preparative in nature; reactions either occurred or did not occur. The concept of equilibrium gradually developed, however, with one of the first advances being made by C. L. Berthollet in 1801. Berthollet served as an advisor to Napoleon in an expedition to Egypt in 1799 and had observed that there were deposits of sodium carbonate in various salt lakes. He concluded that the usual reaction

$$Na_2CO_3 + CaCl_2 \rightarrow CaCO_3 + 2NaCl$$

could be reversed in the presence of a sufficiently high salt concentration, so that chemical affinity was more than an inherent attribute of a substance but also depended on its concentration.

Later, in the period 1850–1860, L. Wilhelmy and M. Berthelot added the important conclusion that chemical equilibrium resulted from a balance between forward and reverse reaction rates, rather than being a static condition. Their particular studies dealt with the hydrolysis of sugars and of esters, but the general conclusion was that for a reaction such as

$$A + B = M + N \tag{7-1}$$

the forward and reverse reaction rates must be equal at equilibrium. These rates were taken to be proportional to the concentrations, in what is known as the *law of mass action*. Thus for the preceding reaction

$$\text{forward rate} \quad = R_f = k_f(A)(B) \tag{7-2}$$
$$\text{backward rate} = R_b = k_b(M)(N) \tag{7-3}$$

where the parentheses denote concentrations. At equilibrium $R_f = R_b$, and on equating the respective expressions, we find

$$K = \frac{k_f}{k_b} = \frac{(M)(N)}{(A)(B)} \tag{7-4}$$

A more general formulation for a chemical reaction is

$$aA + bB + \cdots = mM + nN + \cdots \tag{7-5}$$

and the corresponding equilibrium constant becomes

$$K = \frac{(M)^m(N)^n \cdots}{(A)^a(B)^b \cdots} \tag{7-6}$$

The constant K is known, in this context, as the *mass action equilibrium constant.*

In more modern language, the contribution of the mass action approach was in establishing that, experimentally, an equilibrium system is one in dynamic balance. The important consequence is that a small change in condition, such as that caused by the addition or removal of one of the reactants or products, shifts the position of equilibrium accordingly. The same is true for a small change in temperature or pressure. Such changes are therefore reversible ones. Since they do not occur spontaneously, the free energy of the system is at a minimum, by the criterion for equilibrium of Section 6-7B.

One of the initial difficulties was in distinguishing between amount as such and concentration. Berthollet's reaction was between a concentrated NaCl solution and the $CaCO_3$ present in limestone or as a solid. The problem was in recognizing that the equilibrium depends on the amount of sodium chloride present, since this does affect its concentration, but does not depend on the amount of limestone, since this does *not* affect the concentration of $CaCO_3$. After a period of some confusion, it was realized that the mass action formulation had to be phrased in terms of "active masses" or concentrations. Another early problem lay in understanding the role of catalysts. If a catalyst can change the rate of a reaction, might it not also change the position of equilibrium? One of the important contributions of thermodynamics to chemical equilibrium studies is the conclusion that the free energy change for a chemical reaction is determined by the states of the reactants and products and is therefore independent of path. As a consequence, the position of equilibrium cannot be affected by a catalyst; some further analysis of this point is given in the Commentary and Notes section.

We proceed in the next section to develop the thermodynamic treatment of chemical equilibrium. For the present, the treatment is restricted to reactions involving gases only, or gases and pure solid phases. The thermodynamics is quite general, however, and is applied to solution equilibria in Chapter 12 with only minor modifications in terminology.

7-2   The thermodynamic equilibrium constant

The formal statement of the free energy change accompanying a chemical reaction such as that of Eq. (7-5) is

$$\Delta G = (mG_M + nG_N + \cdots) - (aG_A + bG_B + \cdots) \tag{7-7}$$

In the case of a gaseous species, assumed to be ideal, the molar free energy is given by Eq. (6-54). Integration at constant temperature T gives

$$G = G^0 + RT \ln (P/P^0) \text{ (ideal gas)} \tag{7-8}$$

G^0 is the molar free energy of the gas in its standard or reference state of unit pressure, ordinarily taken to be 1 atm. Since P^0 is unity, Eq. (7-8) is often simplified to

$$G = G^0 + RT \ln P \tag{7-9}$$

but it must be understood what the unit of pressure is.* Equations (7-8) and (7-9) apply to an ideal gas; for nonideal gases, fugacity f replaces P, as discussed in Section 7-ST-1.

Equation (7-7) becomes

$$\Delta G = (mG_M^0 + nG_N^0 + \cdots) - (aG_A^0 + bG_B^0 + \cdots) \\ + RT[(m\ln P_M + n\ln P_N + \cdots) - (a\ln P_A + b\ln P_B + \cdots)] \tag{7-10}$$

if we assume that only ideal gases are involved and that the process is written for some constant temperature T. Equation (7-10) reduces to

$$\Delta G = \Delta G^0 + RT\ln\frac{P_M^m P_N^n \cdots}{P_A^a P_B^b \cdots} = \Delta G^0 + RT \ln Q \tag{7-11}$$

where ΔG^0 is the free energy change when the reaction occurs with the reactants and products in their standard states of 1 atm and Q may be called the reaction quotient. The equation is a general one, in the sense that it gives the free energy change when the process occurs with pressures P_M, P_N, P_A, P_B, and so on, defining some arbitrary value of Q. See Section 7-CN-1 for a discussion of units.

The general case will not be one of chemical equilibrium, and reaction in one direction or the other will occur spontaneously so as to reduce the free energy of the system to a minimum. At this point ΔG will be zero for any small degree of reaction in either direction and the system will be in equilibrium. Equation (7-11) becomes

$$\Delta G^0 = -RT \ln K_P \tag{7-12}$$

where K_P is the equilibrium value of Q, and is the thermodynamic equilibrium constant for the reaction.

It should be appreciated that any set of individual pressures which combine to give the value of K_P will represent an equilibrium system. Alternatively, K_P is determined by ΔG^0, which depends only on the standard-state free energies of the reactants and products and is a constant for the system. However, ΔG^0 and hence K_P will vary with temperature, as discussed in Section 7-4. The value of ΔG^0 is a measure of the tendency of the reaction to occur. If it is large and negative, K_P is a large number, and at equilibrium the concentrations of products will be large compared to those of the reactants. Conversely, if ΔG^0 is a large positive number, K_P will be small, and the degree of reaction will be small at equilibrium. Finally, Eq. (7-12) relates equilibrium constants to the general body of thermodynamic data. Free energies for in-

*The U.S. National Bureau of Standards has been mandated to convert its tables to SI units. One of the less exciting jobs has been to convert G^0 values from a standard state of 1 atm to that of either 1 Pa or 1 bar.

dividual substances may be obtained from spectroscopic information through the methods of statistical thermodynamics so that equilibrium constants can be calculated indirectly (Section 7-7) or measured equilibrium constants can be given a thermodynamic interpretation.

The equilibrium constant for a reaction involving ideal gases may be expressed in several alternative forms. Since $P_i = (n_i/N)P$, where N and P are, respectively, the total moles present and the total pressure at equilibrium, it follows that

$$K_P = \frac{n_M{}^m n_N{}^n \cdots}{n_A{}^a n_B{}^b \cdots}\left(\frac{P}{N}\right)^{\Delta n} = K_n\left(\frac{P}{N}\right)^{\Delta n} \tag{7-13}$$

where Δn denotes the number of moles of products minus those of reactants. Since K_P is independent of pressure, it is evident that K_n must be a function of pressure. Therefore K_n is not a true equilibrium constant; it is useful, however, as an intermediate quantity in equilibrium constant calculations. Similarly, we have

$$K_P = \frac{x_M{}^m x_N{}^n \cdots}{x_A{}^a x_B{}^b \cdots}P^{\Delta n} = K_x P^{\Delta n} \tag{7-14}$$

where x denotes mole fraction; K_x, like K_n, is a function of pressure.

We may also express K_P in terms of concentration since $C = n/v = P/RT$. We obtain

$$K_P = \frac{C_M{}^m C_N{}^n \cdots}{C_A{}^a C_B{}^b \cdots}(RT)^{\Delta n} = K_C(RT)^{\Delta n} \tag{7-15}$$

where K_C, like K_P, is independent of pressure and is a true equilibrium constant. It is customary to specify units. Thus the units of K_P will usually be (atmosphere)$^{\Delta n}$ and those of K_C, (moles per liter)$^{\Delta n}$ (see Section 7-CN-1).

It is important to remember that an equilibrium constant is written for a specific reaction. Thus we have

$$2H_2 + O_2 = 2H_2O \qquad K_P = \frac{P_{H_2O}^2}{P_{H_2}^2 P_{O_2}}\text{ (atm)}^{-1}$$

$$H_2 + \tfrac{1}{2}O_2 = H_2O \qquad K_P' = \frac{P_{H_2O}}{P_{H_2}P_{O_2}^{1/2}}\text{(atm)}^{-1/2}$$

Halving the scale of a reaction halves the values of ΔG^0 and makes the new equilibrium constant the square root of the original one. A related aspect is that of adding or subtracting equations. One adds or subtracts ΔG^0 values, as with other Δ quantities, and therefore multiplies or divides the corresponding K_P's. For example,

$$\begin{aligned}\text{(a)}\quad & H_2O(g) + C(s) = CO(g) + H_2(g)\\ \text{(b)}\quad & \underline{H_2(g) + \tfrac{1}{2}O_2(g) = H_2O(g)}\\ \text{(c)} = \text{(a)} + \text{(b)}\quad & C(s) + \tfrac{1}{2}O_2(g) = CO(g)\end{aligned}$$

It follows that $\Delta G^0_{(a)} + \Delta G^0_{(b)} = \Delta G^0_{(c)}$ and $K_{P(a)}K_{P(b)} = K_{P(c)}$.

7-3   The determination of experimental equilibrium constants

A. Experimental procedures

Equations (7-13)–(7-15) provide alternative specifications for the quantities that must be determined for an equilibrium system. Since K_P varies with temperature, we assume that the temperature has a known, specified value. Several types of experimental approaches are used. A direct one is simply to analyze an equilibrium mixture by chemical means. It is necessary, of course, that the equilibrium not shift during the process of analysis, which means that the reaction must be slow. This will be the case if the equilibrium is being studied at some high temperature and samples of the equilibrium mixture are cooled suddenly, or "quenched." The effect is to freeze the system at its equilibrium composition, so that analyses may be made at leisure. Alternatively, if a catalyst is used to hasten equilibration, then removal of the catalyst makes chemical analysis possible. As an example, the reaction

$$H_2 + I_2 = 2HI \tag{7-16}$$

is rapid at high temperatures or in the presence of a catalyst such as platinum metal. A sample of the mixed gases, cooled rapidly to room temperature and away from the catalyst, is now just a mixture of nonreacting species. The I_2 and HI can be absorbed in aqueous alkali and analyzed and the hydrogen gas determined by gas analysis methods.

The overall composition of an equilibrium mixture is usually known from the amounts of material mixed. In the example here if the initial mixture consists of certain definite amounts of hydrogen and iodine, then analysis of the hydrogen in the equilibrium mixture suffices. The amount of hydrogen lost indicates the number of moles of HI produced and the amount of I_2 reacted and hence the number of moles of I_2 remaining. Equation (7-13) then gives K_P if the total pressure of the equilibrium mixture is known.

It is often possible to determine the equilibrium composition without disturbing the system. For example, the concentration of the various species may be found from some characteristic physical property. Thus I_2 vapor has a distinctive absorption spectrum and measurement of the optical density of an equilibrium mixture at a suitable wavelength gives the I_2 concentration directly. Such a measurement, in combination with the initial composition, allows calculation of the concentrations of H_2 and of HI, and hence of K_c. The equilibrium

$$N_2O_4 = 2NO_2 \tag{7-17}$$

can be followed by magnetic susceptibility measurements. Nitrogen dioxide is paramagnetic, while N_2O_4 is diamagnetic.

Methods such as the preceding are specifically tailored to the individual system; they have in common that the amounts or concentrations of specific species are determined. The degree of reaction may also be found from the P–v–T properties of the equilibrium mixture provided that Δn is not zero. Thus the density of an equilibrium mixture gives its average molecular weight:

$$M_{av} = \frac{\rho RT}{P} \tag{7-18}$$

In the case of Eq. (7-17), we have

$$M_{av} = 46x_{NO_2} + 92x_{N_2O_4} = 92 - 46x_{NO_2} \tag{7-19}$$

and so a determination of M_{av} gives x_{NO_2} and $x_{N_2O_4}$. Substitution into Eq. (7-14) gives K_P. This approach would not work for reaction (7-16); since $\Delta n = 0$ the average molecular weight does not change during the course of the reaction.

B. Effect of an inert gas on equilibrium

It might be supposed that, by definition, an inert gas should have no effect on an equilibrium. This is true in the sense that the presence of the inert gas does not affect K_P. Further, if one has an equilibrium mixture contained in a vessel at a given P and T, then introduction of an inert gas in no way affects the equilibrium composition (assuming ideal gas behavior). This is the intuitive expectation and it may also be shown to follow from Eq. (7-13). Addition of the inert gas increases both P and N but their ratio remains constant, and hence *so does* K_n. The numbers of moles of reactants and products remain the same.

If, however, inert gas is added, keeping the pressure constant, then N increases but not P. Therefore K_n must increase if Δn is positive and decrease if Δn is negative. In other words, if the number of moles increases with the degree of reaction, then the equilibrium must be shifted toward more reaction on dilution with the inert gas. Thus the degree of a reaction such as (7-17) should increase on dilution with an inert gas at constant pressure. Conversely, the degree of a reaction such as the following should decrease on such dilution:

$$3H_2 + N_2 = 2NH_3 \tag{7-20}$$

The effect of adding an inert gas at constant pressure is to increase the volume of the system and the results are just the same as if the equilibrium mixture were expanded into the larger volume. With no inert gas now present the total pressure decreases, and by Eq. (7-14), the mole fractions of products must increase if Δn is positive and decrease if Δn is negative.

C. Some sample calculations

The algebraic techniques for obtaining an equilibrium constant from experimental data generally require use of the stoichiometry of the reaction. In the experience of this writer, a good approach is to set up statements for the various mole numbers and first to evaluate K_n. Where intensive quantities such as density are given it may be useful to assume a certain amount of equilibrium mixture; the assumption should later cancel out. Also, a convention that is very helpful is to express by $n°$ the amount of a species that would be present were the mixture entirely unreacted.

Suppose that the equilibrium of Eq. (7-17) is studied. When 9.2g of NO_2 is introduced into a 36-liter flask at 25°C, the equilibrium pressure is found to be 0.1 atm. From the ideal gas law, $N = (0.100)(36)/(0.0821)(298.2) = 0.147$. The material and mole balance statements are

$$0.2 = n_{NO_2} + 2n_{N_2O_4} \qquad 0.147 = n_{NO_2} + n_{N_2O_4}$$

whence $n_{N_2O_4} = 0.053$, $n_{NO_2} = 0.094$, and

$$K_P = K_n \frac{P}{N} = \frac{(0.094)^2}{0.053} \frac{0.1}{0.147} = 0.113 \text{ atm} \tag{7-21}$$

Alternatively, only the equilibrium density and pressure might be known, 0.256 g liter^{-1} and 0.1 atm, respectively, in this case. From Eq. (7-18), $M_{av} = (0.256)(0.0821)(298.2)/(0.1) = 62.7$. We now write

$$62.7 = 46x_{NO_2} + 92x_{N_2O_4} \text{ or } 1.36 = x_{NO_2} + 2x_{N_2O_4}$$

whence $x_{NO_2} = 0.64$, $x_{N_2O_4} = 0.36$, and

$$K_P = K_x P = \frac{(0.64)^2}{0.36} 0.1 = 0.113 \text{ atm}$$

Reactions such as Eq. (7-17), in which a single species dissociates into products, are often characterized by a degree of dissociation α. We write

$$n_{N_2O_4} = n^\circ_{N_2O_4}(1 - \alpha) \qquad n_{NO_2} = n^\circ_{N_2O_4}(2\alpha) \qquad N = n^\circ_{N_2O_4}(1 + \alpha)$$

The equilibrium constant expression is

$$K_P = \frac{4n^\circ_{N_2O_4}\alpha^2}{n^\circ_{N_2O_4}(1 - \alpha) \, n^\circ_{N_2O_4}(1 + \alpha)} \frac{P}{} = \frac{4\alpha^2}{1 - \alpha^2} P = 0.113 \text{ atm} \tag{7-22}$$

For this example, since $P = 0.1$ atm, $\alpha = 0.470$. If the equilibrium pressure were 1 atm, then α would be 0.166. As expected from the preceding discussion, an increase in pressure shifts the equilibrium to the left.

Suppose, finally, that we take the original equilibrium mixture, with $P = 0.1$ atm, and add inert gas at constant pressure until the volume is 50 liters instead of the original 36 liters. The various statements of mole numbers become

$$n_{NO_2} = n_{NO_2} \qquad n_{N_2O_4} = n^\circ_{N_2O_4} - \tfrac{1}{2}n_{NO_2} = 0.1 - \tfrac{1}{2}n_{NO_2}$$

$$n_g = n_g(\text{inert gas}) \qquad N = n_g + 0.1 + \tfrac{1}{2}n_{NO_2}$$

The total number of moles N is also given by the ideal gas law, $N = (0.1)(50)/(0.0821)(298.2) = 0.204$. Substitution into Eq. (7-13) gives

$$K_P = 0.113 = \frac{n^2_{NO2}}{0.1 - \tfrac{1}{2}n_{NO_2}} \frac{0.1}{0.204}$$

On solving for n_{NO_2}, we obtain 0.105, or $\alpha = (0.105)/(2)(0.1) = 0.525$. Thus the degree of dissociation has increased from 0.470 to 0.525 on dilution with the inert gas.

A second illustration can be based on reaction (7-20). A mixture of hydrogen and nitrogen in a 3 : 1 mole ratio is passed over a catalyst at 500°C. $K_P = 2.49 \times 10^{-5}$ atm^{-2}. What should the total pressure be if the mole fraction of ammonia in the equilibrium mixture is to be 40%? Again it is convenient to set up a table of mole quantities, letting y denote the number of moles of ammonia present in a sample of equilibrium mixture which consists initially of three moles of hydrogen and one of nitrogen:

$$n_{NH_3} = y \qquad n_{H_2} = 3 - \tfrac{3}{2}y \qquad n_{N_2} = 1 - \tfrac{1}{2}y \qquad N = 4 - y$$

where n_{H_2} and n_{N_2} are given by the stoichiometry of the reaction. We want y/N to be 0.4, or $y/(4 - y) = 0.4$, whence $y = 8/7$ and $N = 20/7$ moles; n_{H_2} is then 9/7 and $n_{N_2} = 3/7$. Inserting these values into Eq. (7-13) gives

$$2.49 \times 10^{-5} = \frac{(8/7)^2}{(9/7)^3(3/7)}\frac{(20/7)^2}{P^2} = \frac{11.71}{P^2}$$

Recall that $\Delta n = -2$ in this case; note also that the denominators for the various mole numbers cancel out. The result obtained, $P = 686$ atm, is thus independent of the original choice of amount of reaction mixture. See, however, Section 7-ST-1.

As a final example, consider the reaction

$$PCl_5(g) = PCl_3(g) + Cl_2(g) \tag{7-23}$$

An equilibrium mixture consists of 1 mole of each species, in volume V and at temperature T. Three and one-third moles of Cl_2 are now added, keeping P and T constant. Calculate the number of moles of Cl_2 present when equilibrium is reestablished. Obtain also the ratio of the final to the original total volume. For the original mixture

$$K_P = \frac{(1)(1)}{1}\frac{P}{3} = \frac{P}{3}$$

We let y equal the number of moles of PCl_3 in the new equilibrium mixture, so that $1 - y$ gives the PCl_3 lost as a result of the shift in equilibrium to the left. The table of mole quantities is

$$n_{PCl_3} = y \qquad n_{PCl_5} = 1 + (1 - y) = 2 - y$$

$$n_{Cl_2} = 1 + 3\tfrac{1}{3} - (1 - y) = 3\tfrac{1}{3} + y \qquad N = 5\tfrac{1}{3} + y$$

We then have

$$K_P = \frac{P}{3} = \frac{y(3\tfrac{1}{3} + y)}{2 - y}\frac{P}{5\tfrac{1}{3} + y}$$

The pressure cancels out, and on solving for y we obtain $y = \tfrac{2}{3}$. The total number of moles present is then $5\tfrac{1}{3} + \tfrac{2}{3} = 6$, as compared to the original three moles. Since the pressure has remained constant, the volume must have doubled. Had no reaction occurred, the added Cl_2 would have increased the total number of moles to $3 + 3\tfrac{1}{3} = 6\tfrac{1}{3}$.

7-4   The variation of K_P with temperature

One of the valuable features of Eq. (7-12) is that it permits a thermodynamic treatment of the variation of K_P with temperature. According to Eq. (6-50),

$$\left(\frac{\partial G}{\partial T}\right)_P = -S$$

If this result is applied to each term in Eq. (7-7), we get

$$\left[\frac{\partial(\Delta G)}{\partial T}\right]_P = -\Delta S \tag{7-24}$$

Equation (7-23) holds for ΔG^0 as well, so we have

$$\frac{d(\Delta G^0)}{dT} = -\Delta S^0 \tag{7-25}$$

(the partial differentiation symbol is not needed since superscript zero indicates standard and hence constant pressure). We can thus obtain the standard entropy change for a reaction from the temperature coefficient of ΔG^0.

Alternatively, we have

$$\left[\frac{\partial(G/T)}{\partial T}\right]_P = -\frac{1}{T^2}G + \frac{1}{T}\left(\frac{\partial G}{\partial T}\right)_P = -\frac{1}{T^2}(G + TS) \tag{7-26}$$

By definition, $G = H - TS$, so Eq. (7-26) becomes

$$\left[\frac{\partial(G/T)}{\partial T}\right]_P = -\frac{H}{T^2} \tag{7-27}$$

Since a chemical reaction is written with each substance at the same temperature, Eq. (7-27) can be applied to Eq. (7-7) term by term to give

$$\left[\frac{\partial(\Delta G/T)}{\partial T}\right]_P = -\frac{\Delta H}{T^2} \tag{7-28}$$

and also

$$\frac{d(\Delta G^0/T)}{dT} = -\frac{\Delta H^0}{T^2} \tag{7-29}$$

Substituting the expression for ΔG^0 from Eq. (7-12) gives

$$\frac{d(\ln K_P)}{dT} = \frac{\Delta H^0}{RT^2} \tag{7-30}$$

The restriction of constant pressure is not required for ideal gases. Equation (7-29) is usually called the *van't Hoff equation*, after a famous Dutch physical chemist. Since $dT/T^2 = -d(1/T)$, an alternative form is

$$\frac{d(\ln K_P)}{d(1/T)} = -\frac{\Delta H^0}{R} \tag{7-31}$$

Equation (7-30) provides some immediate qualitative information. If ΔH^0 is negative, so that the reaction is exothermic, then K_P decreases with increasing temperature. Conversely, a positive ΔH^0 means that K_P increases with increasing temperature. More quantitatively, Eq. (7-31) indicates that a plot of $\ln K_P$ versus $1/T$ should be a straight line of slope equal to $-\Delta H^0/R$. Integration of Eq. (7-31) gives

$$\ln K_P = I - \frac{\Delta H^0}{RT} \tag{7-32}$$

where I is an integration constant, or if carried out between limits,

$$\ln \frac{K_{P,T_2}}{K_{P,T_1}} = \frac{\Delta H^0}{R}\left(\frac{1}{T_1} - \frac{1}{T_2}\right) \tag{7-33}$$

T(K)	α	K_P(atm)
439	0.124	0.0269
443	0.196	0.0329
462	0.244	0.0633
485	0.431	0.245
534	0.745	1.99
556	0.857	4.96
574	0.916	9.35
613	0.975	40.4

TABLE 7-1
The $PCl_5 = PCl_3 + Cl_2$
equilibrium[a]

[a]Adapted from E.A. Moelwyn-Hughes, "Physical Chemistry," 2nd revised ed., p. 998. Pergamon, Oxford, 1961.

By way of illustration, some data from Holland (1913) on the equilibrium for the dissociation of PCl_5, Eq. (7-23), are summarized in Table 7-1 and plotted according to Eq. (7-32) in Fig. 7-1. From the plot, $K_P = 10$ atm at $1/T = 1.73 \times 10^{-3}$ and 0.1 atm at $1/T = 2.15 \times 10^{-3}$ and the slope is therefore $(\ln 10 - \ln 0.1)/(1.73 - 2.15)(10^{-3}) = -1.096 \times 10^4$. Then

$$\Delta H^0 = (1.987)(1.096 \times 10^4) = 21{,}800 \text{ cal}$$

The foregoing integration assumed that ΔH^0 was itself independent of temperature. As discussed in Section 5-6, this is not strictly correct, since $\Delta C_P{}^0$ will not in general be zero. Equation (5-30) may be written in the form

$$\Delta H_T{}^0 = B + \Delta aT + \frac{\Delta bT^2}{2} - \frac{\Delta c}{T} \tag{7-34}$$

where B is a constant of integration. The formal integration of Eq. (7-30) gives

$$\ln K_P = -\frac{B}{RT} + \frac{\Delta a}{R} \ln T + \frac{\Delta b}{2R} T + \frac{\Delta c}{2R} \frac{1}{T^2} + I \tag{7-35}$$

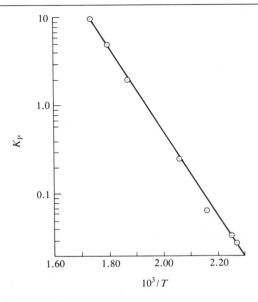

FIGURE 7-1
Variation with temperature of K_P for the equilibrium $PCl_5 = PCl_3 + Cl_2$ (data from Table 7-1).

In order to use Eq. (7-35), we must know ΔH^0 at some one temperature so as to determine B, and K_P at some one temperature so as to evaluate the integration constant I. In addition, of course, the coefficients Δa, Δb, and Δc for the temperature dependence of the heat capacity of reaction are needed. The application is tedious, although straightforward. A more elaborate but ultimately more convenient approach is described in Section 6-ST-3.

7-5   Entropy, enthalpy, and temperature

We can apply Eq. (6-36) to write

$$\Delta G^0 = \Delta H^0 - T\Delta S^0 \tag{7-36}$$

Equation (7-36) is instructive in that it tells us that ΔG^0 and hence K_P is determined by a balance between enthalpy and entropy. Consider, for example, the case of ΔG^0 negative or $K_P > 1$. This could arise primarily because of a negative standard enthalpy change, so that the reaction is downhill in energy. Alternatively, a negative ΔG^0 could stem primarily from a positive ΔS^0. Often the two terms oppose each other. Thus for the reaction

$$C_2H_4(g) + H_2(g) = C_2H_6(g) \tag{7-37}$$

ΔH^0_{298} is -32.74 kcal and so makes a negative contribution to ΔG^0, while ΔS^0_{298} is -28.84 cal K^{-1} and makes a positive contribution.

Reaction (7-20) provides an alternative example. Here, ΔH^0_{298} is 11.04 kcal and ΔS^0_{298} is 23.70 cal K^{-1}. The dissociation of ammonia is thus enthalpy hindered but entropy favored. The former term wins at 25°C— ΔG^0_{298} is 3.976 kcal. If, as a rough approximation, we take ΔH^0 and ΔS^0 not to vary much with temperature, we see that the factor T in Eq. (7-36) acts to make the dissociation increasingly favorable with increasing temperature. The same conclusion follows, of course, from Eq. (7-25). As a qualitative rule, ΔG^0 tends to be enthalpy dominated at low temperatures and entropy dominated at high temperatures.

The observation that ΔS^0_{298} is negative for reaction (7-37) and positive for reaction (7-20) is somewhat to be expected. For simple substances a substantial portion of ΔS^0_{298} is due to the translational contribution. If, as in reaction (7-37), two species combine to give a single product, one translational entropy contribution is lost and it is reasonable that ΔS^0_{298} be negative. On the other hand, in reaction (7-20) one reactant gives two moles total of product, so that one translational entropy contribution is gained, and it is now reasonable that ΔS^0_{298} be positive.

7-6   Gas-solid equilibria

A special case of heterogeneous equilibrium is that between a pure solid phase (or phases) and a gas (or mixture of gaseous products). Consider, for example, the reaction

$$CaCO_3(s) = CaO(s) + CO_2(g) \tag{7-38}$$

The expression for the ΔG of reaction is

$$\Delta G = G^0_{CO_2} + RT \ln P_{CO_2} + G^0_{CaO} - G^0_{CaCO_3} \tag{7-39}$$

(the free energy of a pure solid species at 25°C is just its G^0 value) or

$$\Delta G = \Delta G^0 + RT \ln P_{CO_2} \qquad (7\text{-}40)$$

At equilibrium $\Delta G = 0$ and

$$\Delta G^0 = -RT \ln P_{CO_2} \qquad (7\text{-}41)$$

The equilibrium constant for the reaction is simply $K_P = P_{CO_2}$. However $-RT$ $\ln K_P = \Delta G^0$, which involves the standard free energies of *all three* substances. The point is that unless CO_2 is in equilibrium with both $CaCO_3(s)$ and $CaO(s)$, its pressure will not correspond to K_P. With only one gaseous species present there is an all-or-nothing aspect to the equilibrium. If $CaO(s)$ is exposed to a pressure of CO_2 less than the K_P value, no $CaCO_3(s)$ forms at all. If the pressure is increased to the equilibrium value, then further addition of CO_2 results in conversion of $CaO(s)$ to $CaCO_3(s)$, the pressure remaining constant. Once the conversion is complete, the CO_2 pressure may increase again. The situation is illustrated graphically in Fig. 7-2(a), which shows the phases present for systems of various overall compositions and pressures. The system might, for example, be contained in a piston and cylinder arrangement at constant temperature. For system compositions lying between CaO and $CaCO_3$ if P is greater than K_P, then no gas phase is present at all but just a mixture of the two solids.

The corresponding temperature–composition diagram is shown schematically in Fig. 7-2(b), for a system under 1 atm pressure. The equilibrium pressure K_P is 1 atm at T_0. Below this temperature a system of overall composition x_1 consists of just CO_2 and $CaCO_3$. When heated to T_0, the $CaCO_3$ decomposes to give more CO_2, plus CaO, and above T_0 only the last two phases are present. A system of overall composition x_2 initially consists of $CaCO_3$ and CaO; again the former dissociates at T_0, and above T_0 only CO_2 and CaO are present. Diagrams of this type thus have the aspect of a phase map. They are discussed further in Chapter 8 (on liquids) and in a more formal way, in Chapter 11 (on the phase rule).

As another example of solid–gas equilibrium, K_P for the reaction

$$NH_4HS(s) = NH_3(g) + H_2S(g) \qquad (7\text{-}42)$$

is

$$K_P = P_{NH_3}P_{H_2S} \qquad (7\text{-}43)$$

FIGURE 7-2
The system CO_2–CaO. (a) Pressure versus composition. (b) Temperature versus composition.

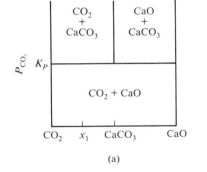

(a)

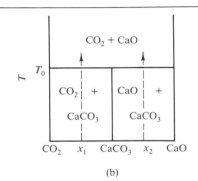

(b)

As in the preceding case, solid NH_4HS must be present in order for the ammonia and hydrogen sulfide pressures to be equilibrium pressures. One may also phrase K_P in terms of the total dissociation pressure above $NH_4HS(s)$. Since NH_3 and H_2S are formed in equal amounts, the pressure of each is half of the total pressure, so we have

$$K_P = \tfrac{1}{4}P^2 \tag{7-44}$$

A third example will serve to illustate an important point in stoichiometry. For the decompostion of methane $K_P = 20.5$ atm at 800°C:

$$CH_4(g) = C(s) + 2H_2(g) \qquad K_P = \frac{P_{H_2}^2}{P_{CH_4}} = 20.5 \tag{7-45}$$

Suppose that we initially have 3 mole of CH_4 in a 5-liter vessel at 800°C and wish to calculate the number of moles of each species at equilibrium. Let y be the number of moles of H_2 present. Then we have

$$n_{H_2} = y \qquad n_{CH_4} = 3 - \tfrac{1}{2}y \qquad N = 3 + \tfrac{1}{2}y$$

Also $P/N = RT/V = (0.0821)(1073)/5 = 17.6$ atm mol^{-1}. Substitution into Eq. (7-13) gives

$$20.5 = \frac{y^2}{3 - \tfrac{1}{2}y}\,17.6$$

from which $y = 1.60$. Then $N = 3.80$ and $P = 17.6N = 66.9$ atm. The number of moles of CH_4 are $3 - \tfrac{1}{2}y$ or 2.20, so 0.80 mole has decomposed, and n_C is therefore 0.80.

We next wish to compute how much hydrogen gas should be introduced into the same 5-liter vessel at 800°C in order to *just* convert 3 mole of carbon into methane. We first observe that there will then be 3 mole of methane, so that $P_{CH_4} = 3(17.6) = 52.8$ atm. Since the carbon is to have just disappeared, the hydrogen pressure will be the equilibrium pressure, or $P_{H_2}^2 = K_P\,P_{CH_4} = (20.5)(52.8)$, $P_{H_2} = 32.9$ atm, and $n_{H_2} = 32.9/17.6 = 1.87$ mole. The total number of moles of hydrogen required is then 1.87 plus the number of moles required to convert the carbon to methane, or plus an additional 6 mole, for a total of 7.87 mole. Thus the stoichiometry of the reaction must be considered as well as the equilibrium pressure.

7-7   Le Châtelier's principle

A principle put forth by H. Le Châtelier about 1890 reads as follows: *A change in a variable that determines the state of an equilibrium system will cause a shift in the position of the equilibrium in a direction tending to counteract the effect of the change in the variable.* We can see qualitatively that this principle must hold quite generally. Were the opposite to hold, namely, that a small change in a variable would cause a shift which magnified the effect of the change, then equilibrium systems would never be stable. A small fluctuation in condition would lead to a large change, which is contrary to observation.

Specific applications of Le Châtelier's principle to gas equilibria follow. First, it is observed in Section 7-3B that if an equilibrium system is diluted or expanded, then a shift occurs such as to increase the number of moles of gas present and therefore such as to oppose the effect of the expansion. That is, owing to the shift, the change in pressure on expansion is less than it would be otherwise. The examples of Sections 7-3C and 7-5 illustrated the point that addition of a product causes a shift in the equilibrium to the left, or such as to consume some of the added species. Thus addition of Cl_2 to the PCl_5–PCl_3–Cl_2 equilibrium causes more PCl_5 to form, and the addition of hydrogen to carbon causes the formation of methane.

The effect of temperature also obeys Le Châtelier's principle. As noted in Section 7-4, if heat is evolved by the reaction, then an increase in temperature shifts the equilibrium to the left, or in the direction such as to absorb heat and thus reduce the change in temperature that a given amount of heating would otherwise produce. Conversely, if the reaction absorbs heat, addition of heat to the system shifts the equilibrium to the right.

7-8 Free energy and entropy of formation

Free energies of formation are defined in the same way as are enthalpies of formation in Section 5-5, and a number of values are given in Table 7-2. The combined Tables 5-2 and 7-2 allow a calculation of entropies of formation from the relationship

$$\Delta G_f^0 = \Delta H_f^0 - T\Delta S_f^0 \tag{7-46}$$

The use of ΔG_f^0 values is similar to the use of ΔH_f^0 values.

TABLE 7-2
Standard free energies of formation[a]

Substance	$\Delta G_{f,298}^0$ (kcal mol^{-1})	Substance	$\Delta G_{f,298}^0$ (kcal mol^{-1})
AgCl(s)	−26.224	H_2O(l)	−56.6902
Br_2(g)	0.751	H_2O(g)	−57.6357
C(diamond)	0.6850	HCl(g)	−22.769
C(graphite)	(0.000)		
$CaCO_3$(s)	−269.78	HBr(g)	−12.72
CaO(s)	−144.36		
CO(g)	−32.8079	HI(g)	0.31
CO_2(g)	−94.2598	KCl(s)	−97.592
CH_4(g)	−12.140	NaCl(s)	−91.785
C_2H_2(g)	50.000	NH_3(g)	−3.976
C_2H_4(g)	16.282	NO(g)	20.719
C_2H_6(g)	−7.860	NO_2(g)	12.390
C_3H_8(g)	17.217	O_3(g)	39.06
C_6H_6(l)	29.756	P(g)	66.77
C_2H_6(g)	−7.860	PCl_3(g)	−68.42
C_2H_5OH(l)	−41.77	PCl_5(g)	−77.59
CH_3COOH(l)	−93.8	S(rhombic)	(0.000)
CCl_4(l)	−16.4	S(monoclinic)	0.023
Fe_2O_3(s)	−177.1	SO_2(g)	−71.79
Glycine, H_2NCH_2COOH(s)	−88.61	SO_3(g)	−88.52
Glycine(aq)	−89.1		
Glycylglycine(aq)	−117.3		

[a]Data from F.A. Rossini *et al.*, *Selected Values of Chemical Thermodynamic Quantities*, Nat. Bur. Std. Circ. No. 500. U. S. Govt. Printing Office, Washington, D.C., 1959; F.W. Carpenter, *J. Amer. Chem. Soc.* **82**, 1111 (1960).

The example used in Section 5-5A on the calculation of the enthalpy of hydrogenation of **EXAMPLE**
ethylene may be repeated using free energies of formation. The reaction is

$$C_2H_4(g) + H_2(g) = C_2H_6(g) \tag{7-47}$$

$\Delta G_{298}^0 = -7.860 - 16.282 = -24.142$ kcal. The corresponding ΔH_{298}^0 was previously
found to be -136.98 kJ or -32.74 kcal, and by Eq. (7-36), $\Delta S_{298}^0 = [-32,740$
$- (-24,142)]/298.15 = -28.84$ cal K^{-1}. Also, from Eq. (7-12), ln $K_P = 24,142/(1.987)$
(298.2) and $K_P = 4.96 \times 10^{17}$ atm^{-1}. (See Section 7-CN-1 regarding the use of units at this
point.)

If we ignore the complication of the ΔC_P^0 for the reaction and assume that ΔH^0 is
constant, then we may use Eq. (7-33) to obtain K_P at some other temperature, say 200°C.
We have

$$\ln \frac{K_{P,200°C}}{K_{P,25°C}} = -\frac{32,740}{(1.987)} \left(\frac{1}{298.2} - \frac{1}{473.2} \right)$$

$$= -20.43$$

whence $K_{P,200°C} = 6.62 \times 10^8$ atm. As expected from Le Châtelier's principle, K_P decreases with increasing temperature.

It is noted in Section 6-CN-2 that we may obtain absolute standard entropies
thermochemically, using heat capacity data close to 0 K and assuming the
third law of thermodynamics. A number of values so determined are given
in Table 6-2. The study of chemical equilibria interacts with such results in
two ways. First, the temperature dependence of measured equilibrium constants gives experimental values for ΔS^0 of reaction. These may then be compared with the values calculated from third law entropies, and one of the
major evidences of the validity of the third law is the excellent agreement
that has resulted. (The other is the agreement between thermochemical and
spectroscopic absolute entropies.)

The second interaction is that if ΔS^0 is found experimentally for a reaction
for which absolute entropies are known for all but one species, then the
absolute entropy for this additional species can be calculated. A large number
of the entries in the tables of absolute entropies have been obtained in this
way. As an illustration, the absolute entropies for $H_2(g)$ and $C_2H_4(g)$ are 31.21
and 52.45 cal K^{-1} mol^{-1}, respectively, at 298 K. The experimental ΔS_{298}^0 was
found to be -28.8 cal K^{-1} mol^{-1} for reaction (7-45); hence

$$S_{298}^0[C_2H_6(g)] = -28.8 + 31.21 + 52.45 = 54.86 \text{ cal K}^{-1} \text{ mol}^{-1}$$

7-CN-1 Units and equilibrium constants

A minor point has to do with the use of unit
designations for equilibrium constants. It is
mentioned in Section 7-2 that K_P is usually given

in atmosphere units and K_C in moles per liter
units. Equation (7-12) relates ln K_P to ΔG^0, and
there appears to be a difficulty since it is mathematical nonsense to take the logarithm of a
quantity having dimensions. One is, in effect,
taking the base e to a power expressed in at-

mospheres or in moles per liter, which is meaningless; all exponents *must* be dimensionless numbers. Taking the case of a K_P as a specific, the problem traces back to the use of Eq. (7-9) rather than Eq. (7-8). Strictly speaking, K_P in Eq. (7-12) should be written

$$K_P = \frac{P_M^m P_N^n \cdots}{P_A^a P_B^b \cdots} (P_0)^{-\Delta n} \qquad (7\text{-}48)$$

where Δn is the moles of products minus the moles of reactants. That is, each pressure should be divided by P_0. However, since the reference pressure P_0 is unity, it is unnecessarily cumbersome to show the $(P_0)^{-\Delta n}$ term in an equilibrium constant expression. The usual practice is to omit it, but to specify the units of the first part of the expression:

$$K_P = \frac{P_M^m P_N^n \cdots}{P_A^a P_B^b \cdots} (\text{atm})^{\Delta n} \qquad (7\text{-}49)$$

The same type of procedure is used for a K_C.

7-CN-2   Chemical equilibrium and the second law of thermodynamics

A. Free energy and
   equilibrium

It is concluded in Section 6-7 that for a system at constant temperature and pressure the free energy will be a minimum at equilibrium. There is an aspect to this statement that may seem puzzling in the present context. For the general reaction (7-10), the quantity Q in Eq. (7-11) would vary from infinity to zero and $\ln Q$ from plus infinity to minus infinity as one goes from reactants only to products only. ΔG in Eq. (7-11) must thus vary monotonically with degree of reaction. Where, then, is the free energy minimum?

The answer to the above question is that the minimum occurs in the *total* free energy of the system as we vary the degree of reaction. Let ξ be the degree of reaction, which will vary from zero to one. The total free energy G_{tot} for the general reaction (7-10) is just the sum of the products (moles) (free energy per mole) for each species. Thus

$$
\begin{aligned}
G_{\text{tot}} = {} & a(1 - \xi)G_A^0 + b(1 - \xi)G_B^0 + \cdots \\
& + m\xi G_M^0 + n\xi G_N^0 + \cdots \\
& + a(1 - \xi)RT \ln P_A + b(1 - \xi)RT \ln P_B \\
& + \cdots + m\xi RT \ln P_M \\
& + n\xi RT \ln P_N + \cdots \qquad (7\text{-}50)
\end{aligned}
$$

By grouping terms, we obtain

$$
\begin{aligned}
G_{\text{tot}} = {} & aG_A^0 + bG_B^0 + \cdots + \xi \Delta G^0 \\
& + (1 - \xi)RT \ln Q_R + \xi RT \ln Q_P \\
& \qquad\qquad\qquad\qquad\qquad (7\text{-}51)
\end{aligned}
$$

where Q_R is the denominator of the Q expression, and Q_P is the numerator. It is this G_{tot} that goes through a minimum at equilibrium. If, for example, we differentiate with respect to ξ,

$$
\begin{aligned}
\frac{dG_{\text{tot}}}{d\xi} = {} & \Delta G^0 - RT \ln Q_R \\
& + RT \ln Q_P = \Delta G^0 + RT \ln Q \quad (7\text{-}52)
\end{aligned}
$$

On setting $dG_{\text{tot}}/d\xi = 0$, we regain Eq. (7-12).

EXAMPLE   As a simple illustration, we take the reaction to be $A \rightleftharpoons M$, with $P_A^0 = 1$ atm and $\Delta G^0 = -1$ kcal at 25°C. We find from Eq. (7-12) that $K_P = \exp[-(-1000)/(1.987)(298.15)] = 5.41$, whence $P_{A,\text{equil}} = 0.156$ atm (and $P_{M,\text{equil}} = 0.844$ atm).
Equation (7-49) reduces to

$$G_{\text{tot}} = G_A^0 - 1000\xi + (1 - \xi)(592) \ln P_A + \xi (592) \ln P_M \qquad (7\text{-}53)$$

Substitution of various ξ values gives

ξ	0	0.2	0.4	0.6	0.8	0.844	0.9	1
$(G_{\text{tot}} - G_A^0)$	0	−497	−798	−998	−1096	−1101	−1092	−1000

The results are plotted in Fig. 7-3(a), and there is indeed a minimum at $\xi = 0.844$, corresponding to $P_{B,equil} = 0.844$ atm.

B. Entropy and
   equilibrium

For the general case of a reacting system that is not at equilibrium, we have

$$\Delta G = \Delta H - T\Delta S \qquad (7\text{-}54)$$

and now consider how this relates to the dictum in Section 6-7 that entropy reaches a maximum at equilibrium for an isolated system (note Fig. 6-12a). Our reaction is at some constant temperature T, so there must be an associated heat reservoir at T to maintain this temperature. Thus if ΔH is negative, the heat reservoir will absorb some increment of evolved heat δq when the degree of reaction increases by $d\xi$, and conversely if ΔH is positive. The important

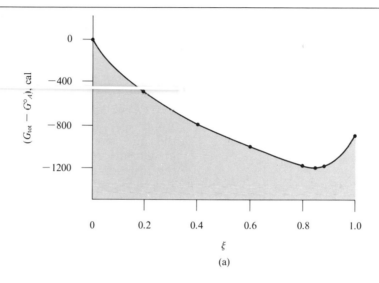

(a)

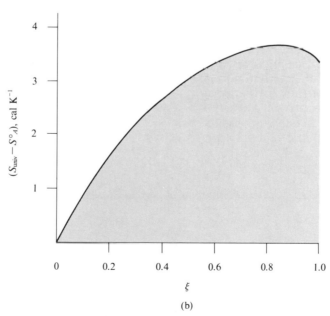

(b)

FIGURE 7-3
Variation of (a) $(G_{tot} - G_A^0)$ and (b) $(S_{univ.} - S_A^0)$ with degree of reaction (see examples in text).

EXAMPLE We can extend the preceding example involving the reaction A $\rightleftharpoons$ M with $\Delta G^0 = -1$ kcal at 25°C. We add the assumption that $\Delta H^0 = -2$ kcal and hence that $\Delta S^0 = -3.36$ cal K^{-1}. Equation (7-56) becomes

$$(S_{univ.} - S_A^0) = 3.35\ \xi - (1 - \xi)R \ln P_A - \xi R \ln P_M \tag{7-57}$$

with $P_A = (1 - \xi)$ atm and $P_B = \xi$ atm. The resulting values of $(S_{univ.} - S_A^0)$ vs. ξ are plotted in Fig. 7-3(b); we see that $S_{univ.}$ indeed goes through a maximum at the equilibrium point.

---

point is that it is the reaction system *plus* the heat reservoir that is our isolated system, that is, our "universe."

The entropy of the reaction system plus the heat reservoir, $S_{univ.}$, is given by

$$S_{univ.} = a(1 - \xi)S_A + b(1 - \xi)S_B + \cdots \\ + m\xi S_M + n\xi S_N + \cdots - \xi\Delta H/T \tag{7-55}$$

Equation (6-26) allows us to replace S by $S^0 - R \ln P$ for each species, to obtain

$$S_{univ.} = aS_A^0 + bS_B^0 + \cdots + \xi\Delta S^0 \\ - (1 - \xi)R \ln Q_R - \xi R \ln Q_P - \xi\Delta H/T \tag{7-56}$$

C. A second law triumph

Equations (7-30) and (7-31) constitute one of the major triumphs of the second law of thermodynamics. The connection between the temperature dependence of an equilibrium constant and the enthalpy change of a reaction is not otherwise deducible and provides a means for obtaining a calorimetric quantity, ΔH^0, from equilibrium constant data, that involves only measurements of equilibrium concentrations. The equations provide a means for verifying the second law, since the ΔH^0 of a reaction obtained using them should be the same as the directly determined calorimetric value. A num-

ber of such checks have been made, and three examples are given in Table 7-3. Since $\Delta n = 0$ for the reactions in question, $\Delta H^0 = \Delta E^0$. It is quite likely that the calculated ΔE^0 values are more accurate than the directly measured ones, but the two agree within experimental error.

D. The principle of microscopic reversibility

Another application of the second law yields the conclusion that a catalyst cannot affect the equilibrium constant for a reaction. Consider the situation illustrated in Fig. 7-4. A reaction A + B to give products M + N ordinarily proceeds by some direct path. Alternatively, it may occur with the aid of a catalyst. Thus the reaction $H_2 + I_2 = 2HI$ was long thought to be a direct reaction; it may also be catalyzed by platinum metal. Suppose that, as indicated in the figure, the catalyst affects only the forward reaction. In its presence, the sum of the forward rates would clearly be larger than otherwise, while the backward rate would be unchanged. The position of equilibrium would therefore shift to the right, by the law of mass action. If we suppose further that the reaction produces heat q when it occurs, then a violation of the second law would be possible. We first allow equilibrium to be reached without the catalyst (Pt, in the case of the formation of HI),

---

TABLE 7-3

Comparison of second law and calorimetric heats of reaction[a]

Reaction	$K_{P,298}$	ΔE^0 (kcal) Second law	ΔE^0 (kcal) Calorimetric
$2HCl = H_2 + Cl_2$	5.50×10^{-34}	43.96	44.0
$2HBr = H_2 + Br_2$	1.05×10^{-19}	24.38	24.2
$2HI = H_2 + I_2$	5.01×10^{-4}	2.95	2.94

[a]Adapted from E.A. Moelwyn-Hughes, "Physical Chemistry," 2nd revised ed., p. 994. Pergamon, Oxford, 1961.

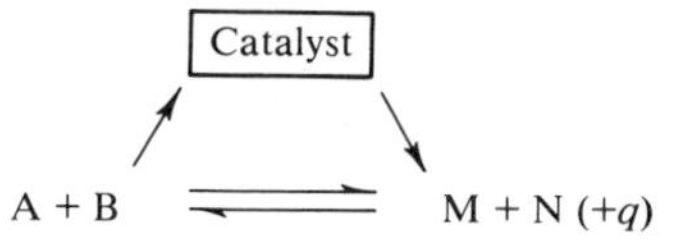

FIGURE 7-4
Illustration of the principle of microscopic reversibility.

and then add the catalyst, and heat δq is produced as the equilibrium is shifted. This heat is used to run a machine, and thus do work, cooling the system back to its original temperature in the process. We then remove the catalyst and the equilibrium shifts back. Heat δq is now extracted from the surroundings, which must warm the system back to the ambient temperature. A cycle has therefore been completed for which the net effect has been the isothermal conversion of heat energy into work, and a perpetual motion machine of the second kind has been found.

We conclude that the supposed situation is impossible and that the catalyst must accelerate the forward and backward reactions equally. A catalyst is simply a chemical intermediate that is regenerated and hence not consumed in the reaction; it is in this respect no different from any reaction intermediate through which the reaction can proceed. We also conclude that for any reaction path the forward and backward rates must be the same at equilibrium. Further, if a reaction can occur by two or more different paths, catalyzed or not, at equilibrium the molecular traffic must be balanced for each path separately. This general conclusion is known as the *principle of microscopic reversibility.* It will be of use in Chapter 14 (on chemical kinetics).

SPECIAL TOPICS

7-ST-1   Effect of pressure on chemical equilibria involving gases

There are two quite different ways in which the total pressure of a mixture of gases affects the equilibrium. The first was discussed in Section 7-3B and involves the Le Châtelier principle. The second, the subject here, involves nonideal behavior. The material so far has been developed on the basis that gases are ideal; the assumption is a good one for low-boiling gases at pressures around 1 atm. Otherwise rather serious errors can develop.

It is shown in Section 6-ST-2 that a quantity known as the fugacity f functions as a thermodynamic pressure. By this is meant that all thermodynamic equations calling for the partial pressure P_i of an ideal gas remain exact for nonideal gases if f_i is used instead. Equation (7-9) then reads

$$G_i = G_i^0 + RT \ln f_i \qquad (7\text{-}58)$$

and on carrying through the subsequent derivation, Eq. (7-12) becomes

$$\Delta G^0 = -RT \ln K_f \qquad (7\text{-}59)$$

where K_f has the same form as K_P, but with fugacities instead of partial pressures. Fugacity is so defined that $f_i \to P_i$ as $P_i \to 0$. It follows that $K_i \to K_P$ as $P \to 0$. If, further, we adopt the convenience of writing $f_i = \gamma_i P_i$ (see Section 6-ST-2), where γ_i is the activity coefficient, we have

$$
\begin{aligned}
K_f &= \frac{f_M{}^m f_N{}^n \cdots}{f_A{}^a f_B{}^b \cdots} \\[1em]
&= \frac{P_M{}^m P_N{}^n \cdots}{P_A{}^a P_B{}^b \cdots} \frac{\gamma_M{}^m \gamma_N{}^n \cdots}{\gamma_A{}^a \gamma_B{}^b \cdots} = K_P K_\gamma
\end{aligned}
\qquad (7\text{-}60)
$$

Since $\gamma_i \to 1$ as $P1_i \to 0$, it follows that $K_\gamma \to 1$ as $P \to 0$.

The procedure is to evaluate γ for each species, calculate K_γ, and then calculate K_f. Alternatively, if K_P is known at low pressures, so that $K_f \simeq K_P$, one may determine K_P at high pressures by estimating K_γ and using Eq. (7-60). Certain approximations are tolerated at this point. First, the law of corresponding states will begin to fail at very high pressures, as the individual shapes of molecules begin to be im-

EXAMPLE We can continue the example of Section 7-3 on the equilibrium, $3H_2 + N_2 = 2NH_3$, for which K_P was given as 2.49×10^{-5} at 500°C. This is actually the K_f value (as obtained from Tables 6-4 and 6-5). It was calculated that a pressure of 686 atm was required if the equilibrium mixture was to contain 40% NH_3. We obtain a more correct answer as follows. The various values of P_r, T_r, and γ are summarized in Table 7-4 and we obtain $K_\gamma = (0.90)^2/(1.29)^3(1.35) = 0.280$. At this total pressure, then, the value of K_P that should be used is $2.49 \times 10^{-5}/0.280 = 8.89 \times 10^{-5}$ and the equilibrium percent of ammonia should be about 50% instead of 40%.

portant. Second, it is usually assumed that in a mixture of gases the fugacity of any one component is determined by the reduced pressure of that component, based on the total gas pressure. Again at high pressures this assumption will lead to some error. The procedure is adequate, however, for most purposes.

7-ST-2 Application of statistical thermodynamics to chemical equilibrium

The various expressions for obtaining the translational, rotational, and vibrational contributions to the enthalpy, free energy, and entropy of an ideal gas are developed in Section 6-12. Their application to the case of $N_2(g)$ at 25°C is illustrated in Section 6-12 and the statistical thermodynamic calculation of the enthalpy change for the reaction $3H_2 + N_2 = 2NH_3$ is carried out in detail in Section 5-ST-3.

An important point is that statistical thermodynamics can give the absolute entropy of a substance if the translational, rotational, and vibrational partition functions are known but cannot give the absolute energy, enthalpy, or free energy. The reason is that while the third law allows us to set the entropy equal to a zero at 0 K, we do not know the energy at 0 K. As a result, the statistical thermodynamic calculations yield, for some temperature T, S^0,

$(H^0 - E_0)$ or $(H^0 - H_0^0)$, and $(G^0 - E_0)$ or $(G^0 - H_0^0)$. The subscript zero denotes values at 0 K, and enthalpies and free energies can only be obtained relative to E_0 or to H_0^0. The differences $H^0 - H_0^0$ and $(G^0 - H_0^0)/T$ are known as the enthalpy and free energy functions, respectively.

The consequence is that if all the partition functions are available, statistical thermodynamics can give ΔS^0 for a reaction and quantities such as $\Delta H^0 - H_0^0$ and $\Delta G^0 - \Delta H_0^0$. It is therefore necessary to have an experimental value of ΔH^0 for some one temperature; this allows ΔH_0^0 to be calculated, and hence ΔH^0 and ΔG^0, at any other temperature.

Because of their natural relationship to the statistical approach, the enthalpy and free energy functions are now often tabulated instead of free energies of formation, as in Tables 6-4 and 6-5. A further advantage is that the quantity $(G^0 - H_0^0)/T$ varies only slowly with temperature. It is therefore possible to tabulate it for rather widely spaced temperature intervals, and Tables 6-4 and 6-5 allow calculation of ΔG^0 and hence of K_P for a reaction over a wide range of temperature without the awkwardness of Eq. (7-35), as detailed in Section 6-ST-3. Yet another way of applying statistical thermodynamics to the formulation of equilibrium constants is given in Section 14-ST-1.

An alternative and often useful approach is as follows. Following Eq. (6-86), we can write the standard free energy of a substance as

$$G^0 = E_0 + G_T^0 \qquad (7-61)$$

TABLE 7-4	P_c(atm)	P_r	T_c (K)	T_r	γ
NH_3	112.2	6.11	405.5	1.91	0.90
N_2	33.5	20.5	126.0	6.14	1.35
H_2	12.8	53.6	33.2	23.3	1.29

that is, as the sum of its energy at absolute zero and of the free energy developed on raising the temperature to T. For an ideal gas the quantity G_T^0 is the sum of G_{trans}^0, G_{rot}, and G_{vib} as given, for example, by Eqs. (6-99), (6-109), and (6-111).

We next define a quantity **q** as

$$q = \exp\left(-\frac{G_T^0}{RT}\right) \tag{7-62}$$

Equation (7-12) may be written as

$$K = \exp\left(-\frac{\Delta G^0}{RT}\right) \tag{7-63}$$

and on using Eqs. (7-61) and (7-62), we obtain

$$K = K_q \exp\left(-\frac{\Delta E_0}{RT}\right) \tag{7-64}$$

where K_q is the product and quotient of the **q**'s for the reaction products and reactants; thus for a reaction A + B = C, $K_q = q_C/q_A q_B$. We can think of **q** as a statistical thermodynamic concentration (sometimes called the *rational activity*—see Section 9-CN-3).

If the species are all ideal gases, then the expression for **q** becomes rather simple. Thus combination of Eqs. (4-83) and (6-99) gives

$$q_{trans} = \left(\frac{2\pi mkT}{h^2}\right)^{3/2} \frac{1}{N_A} \tag{7-65}$$

provided the standard state is taken to be 1 molecule cm^{-3}. Equation (6-109) yields

$$q_{rot} = \frac{8\pi^2 IkT}{h^2} = \frac{T}{\sigma\theta_{rot}}$$

for a linear molecule and

$$q_{rot} = \frac{\pi^{1/2}}{\sigma}\left[\frac{8\pi^2(I_x I_y I_z)^{1/3} kT}{h^2}\right]^{3/2} \tag{7-67}$$

for a nonlinear molecule, where I_x, I_y, and I_z are the three principal moments of inertia and σ is a degeneracy factor defined as the number of equivalent ways of orienting the molecule in space. Thus $\sigma_{H_2} = 2$, $\sigma_{HI} = 1$, $\sigma_{H_2O} = 2$, $\sigma_{NH_3} = 3$, and $\sigma_{benzene} = 12$.

Finally, Eq. (6-111) gives

$$q_{vib} = \Pi\left[1 - \exp\left(-\frac{\theta_{vib}}{T}\right)\right]^{-1} \tag{7-68}$$

where the product is over all vibrational modes, the zero point energies being included in E_0. The total **q** is then the product $q_{trans}q_{rot}q_{vib}$.

If K is known for a gaseous equilibrium at some one temperature, a calculation of ΔE_0 is in principle possible from spectroscopic values of θ_{rot} and θ_{vib}. One may then calculate K for any other temperature.

GENERAL REFERENCES

BLINDER, S.M. (1969). "Advanced Physical Chemistry." Macmillan, New York.
LEWIS, G.N., AND RANDALL, M. (1961). "Themodynamics," 2nd ed (revised by K.S. Pitzer and L. Brewer). McGraw-Hill, New York.

CITED REFERENCES

FENTON, T.M., AND GARNER, W.E. (1930). *J. Chem. Soc.* **1930**, 694.
HOLLAND, C. (1913). *Z. Elektrochem.*, **18**, 234.

EXERCISES

ΔG_{298}^0 is -7.860 kcal for the reaction $C_2H_4 + H_2 = C_2H_6$, and is -33.718 kcal for the reaction $C_2H_2 + H_2 = C_2H_4$. Calculate K_P for the disproportionation reaction $C_2H_2 + C_2H_6 = 2\ C_2H_4$ at 25°C. All species are gaseous.   **7-1**

Ans. 1.047×10^7.

Referring to Exercise 7-1, what is K_P at 25°C for the reaction (a) $\frac{1}{2} C_2H_2 + \frac{1}{2} C_2H_6 = C_2H_4$ and (b) $C_2H_4 = \frac{1}{2} C_2H_2 + \frac{1}{2} C_2H_6$?   **7-2**

Ans. (a) 3.24×10^3, (b) 3.09×10^{-4}.

K_P is 350 atm^{-1} at 350°C for the reaction $CO + Cl_2 = COCl_2$ (phosgene). An initial pressure of 1×10^4 Pa of CO and 1×10^3 Pa of Cl_2 is introduced into a flask at 350°C. (a) Calculate K_P in Pa^{-1},   **7-3** that is, for a standard state pressure of 1 Pa, and (b) the percent conversion of the Cl_2 to phosgene at equilibrium.

Ans.(a) 3.45×10^{-3} Pa, (b) 96.9%.

7-4   A flask initially contains 1 mole of HI and then equilibrates according to the reaction $2HI = H_2 + I_2$. The equilibrium degree of this dissociation is 0.220 at 45°C and 1 atm total pressure. Calculate K_P. Calculate also the degree of dissociation if the equilibrium mixture is allowed to expand until the pressure falls to 0.1 atm.

Ans. $K_P = 0.0200$; $\alpha = 0.220$ (unchanged).

7-5   Given the reaction $CO_2(g) + H_2(g) = CO(g) + H_2O(g)$, calculate K_P, K_n, and K_C if after the mixing of 1 mole of CO_2 and 2 moles of H_2, the equilibrium mixture at 0.1 atm and 25°C contains 1.80% by volume of CO.

Ans. $K_P = K_C = K_n = 1.59 \times 10^{-3}$.

7-6   The amount 24.26 g of PCl_5 is introduced into a 10 liter flask, which is brought to 250°C. When the equilibrium $PCl_5(g) = PCl_3(g) + Cl_2(g)$ is established, the pressure is found to be 0.9035 atm. Calculate K_P and K_C for the equilibrium.

Ans. $K_P = 1.689$ atm, $K_C = 3.936 \times 10^{-3}$ M.

7-7 K_P changes by 1.5% per degree (around 100°C) for a certain reaction. Calculate ΔH^0.

Ans. 4.15 kcal.

7-8 K_P for the reaction of Exercise 7-6 is 5.28×10^6 at 25°C. Use this datum and the answer to the exercise to calculate ΔH_0.

Ans. -20.60 kcal.

7-9   Calculate ΔS^0_{298} and K_P at 25°C and at 200°C for the reaction $2HCl(g) + \frac{1}{2}O_2(g) = H_2O(g) + Cl_2(g)$. Use data from Tables 5-2 and 7-2.

Ans. $\Delta S^0_{298} = -5.21$ cal K^{-1}, $K_P = 7.39 \times 10^8$ atm$^{-1/2}$ and 1.47×10^5 atm$^{-1/2}$.

7-10   Verify the plot in Fig. 7-1 of the data in Table 7-1 and the calculated value of ΔH^0.

7-11   Use Tables 5-2 and 7-2 to find a reaction for which $K_P < 1$ at 25°C and for which ΔG^0 is unfavorable because (a) of an unfavorable ΔH^0 and (b) an unfavorable ΔS^0.

Ans. (a) Reverse of the reaction of Exercise 7-9, for example.

7-12   Calculate ΔG^0 and K_P at 25°C for the reaction $C(graphite) + 2H_2 = CH_4$.

Ans. $\Delta G^0 = -12,140$ cal, $K_P = 7.94 \times 10^8$ atm^{-1}.

7-13   The standard free energy of formation of $Ag_2O(s)$ at 25°C is -2.59 kcal mol^{-1}, and $\Delta H^0_{f,298}$ is -7.31 kcal mol^{-1}. Calculate (a) the dissociation pressure of Ag_2O, that is, the pressure of oxygen in equilibrium with Ag_2O, at 25°C, and (b) the temperature at which this dissociation pressure would be 1 atm.

Ans. (a) 1.60×10^{-4} atm, (b) 462 K.

7-14   Calculate K_P at 25°C for the reaction $2C(s) + 3H_2(g) = C_2H_6(g)$. How many moles of hydrogen must be added to a 5-liter flask containing 12 mg of carbon at 25°C in order to just convert all of the carbon to C_2H_6?

Ans. $K_P = 5.78 \times 10^5$ atm^{-2}; 1.83×10^{-3} mole.

PROBLEMS

7-15 ΔG^0_{298} is -15.89 kcal for the reaction $N_2O + 2O_2 = N_2O_5$, and is 21.18 kcal for the reaction $N_2O_5 = 2NO_2 + \frac{1}{2}O_2$. Calculate ΔG^0_{298} for the reaction $2NO_2 + N_2O = 4NO + \frac{1}{2}O_2$, and K_P (all species are gaseous). Additional data from Table 7-2 may be used.

The standard enthalpies of formation and absolute entropies at 25°C are -7.590 kcal mol^{-1} and 81.81 cal K^{-1} mol^{-1} and -18.46 kcal mol^{-1} and 70.00 cal K^{-1} mol^{-1} for *trans*-2-pentene and cyclopentane, respectively. Calculate the equilibrium constant for the conversion of the former to the latter compound at 25°C. **7-16**

Calculate the equilibrium constant for the reaction: 2 glycine(*aq*) = glycylglycine(*aq*) + H$_2$O at 25°C. **7-17**

Let ρ_r be the density of an equilibrium mixture of N$_2$O$_4$ and NO$_2$ divided by the density that would be observed at the same temperature and pressure if the N$_2$O$_4$ were not dissociated. Derive an expression for K_P involving only ρ_r and the equilibrium total pressure P. Find the degree of dissociation α and K_P for an equilibrium mixture if P is 500 Torr at 50°C (use Table 5-2). **7-18**

Determine the ratio of initial partial pressures of H$_2$ and N$_2$ that will give the maximum yield of NH$_3$ in the reaction 3H$_2$(g) + N$_2$(g) = 2NH$_3$(g), assuming a flask of constant volume and temperature. **7-19**

A 2:1 mixture of SO$_2$ to O$_2$ is placed in a flask at a certain temperature. At equilibrium for the reaction 2SO$_2$(g) + O$_2$(g) = 2SO$_3$(g), there is 10 mole % SO$_3$ present and the total pressure is 50 atm. Calculate K_P. **7-20**

Determine the ratio of initial partial pressures of C$_2$H$_2$ to H$_2$ that will give a maximum yield of C$_2$H$_6$ in the reaction C$_2$H$_2$(g) + 2H$_2$(g) = C$_2$H$_6$(g), assuming a flask of constant volume and temperature. **7-21**

An equimolar mixture of C$_2$H$_2$(g) and H$_2$(g) is introduced into a flask at 1250 K, with a catalyst for the reaction C$_2$H$_2$(g) + 2H$_2$(g) = C$_2$H$_6$(g). The equilibrium pressure and gas density are found to be 0.592 atm and 0.1370 g liter^{-1}, respectively. Calculate K_P. **7-22**

For the reactions (a) H$_2$ = 2H, (b) Cl$_2$ = 2Cl, (c) 2HCl = H$_2$ + Cl$_2$, the values of log K_P at 2000 K are (a) -5.509, (b) 0.245, and (c) -5.560 (pressures in atm). By successive approximations find the equilibrium pressures accurate to one or two places (that is, to about 10%) for HCl, H$_2$, Cl$_2$, H, and Cl on heating HCl(g) to 2000 K at the total pressure of 1 atm. **7-23**

A mixture of CO$_2$ and CF$_4$ was brought to equilibrium in a flask at 1000°C, and then quenched. Analysis showed that the equilibrium mixture was 40 mole % CF$_4$ and 20 mole % COF$_2$. Calculate K_P for the reaction 2COF$_2$(g) = CO$_2$(g) + CF$_4$(g). **7-24**

Derive an equation analogous to Eq. (7-30) but relating $d(\ln K_C)/dT$ to ΔH^0. **7-25**

Derive an expression analogous to Eq. (7-32) but for the case of a ΔH^0 varying with temperature, $\Delta H^0 = \Delta H_0^0 + bT$. **7-26**

Use data from Tables 5-2, 5-4, and 7-2 to calculate K_P for the reaction 3H$_2$(g) + N$_2$(g) = 2NH$_3$(g) at 1000°C. That is, include the variation of ΔH^0 with temperature. **7-27**

Given the following C_P^0 values (cal °C^{-1} mol^{-1}): **7-28**

   H$_2$(g): 6.85 + 0.00028T + 0.00000022T^2,

   CH$_4$(g): 4.38 + 0.01417T,

C(graphite): 1.22 + 0.00489T $-$ 0.000001111T^2.

For the reaction

   C(graphite) + 2H$_2$(g) = CH$_4$(g)

($\Delta H_{298}^0 = -18{,}062$ cal, $\Delta G_{298}^0 = -11{,}994$ cal), calculate (a) ΔC_P^0, (b) the equation giving ΔH^0 as a function of temperature, (c) ΔH^0 at 700°C, (d) the equation giving ΔG^0 as a function of temperature, and (e) ΔG^0 and K_P for the reaction at 700°C.

Calculate ΔG_{298}^0 for the reaction CO$_2$(g) + H$_2$(g) = CO(g) + H$_2$O(g). Use Table 5-4 to express the temperature dependence of ΔG^0 as a polynomial in temperature and plot ΔG^0 versus T from 25°C to 1500°C. Calculate and similarly plot ΔH^0 and ΔS^0 for the reaction. **7-29***

7-30 A study is made of the equilibrium $SeCl_4(s) = SeCl_2(g) + Cl_2(g)$. The total pressure of products in equilibrium with $SeCl_4(s)$ is found to be:

T, °C	120	129.5	140	150	161	170.5	175.5	180.5
P, kPa	3.19	5.33	8.73	15.65	27.12	41.41	51.94	64.26

Make the appropriate graph of the data and find ΔH^0; also, find K_P at 25°C.

7-31 Solid $NaHSO_4$ is transformed on heating to $Na_2S_2O_7$. The equilibrium pressure over the salts is 2.15 Torr at 140°C and 17.84 Torr at 180°C. What amount of heat must be added to 1 mole of $NaHSO_4$ to convert it all to $Na_2S_2O_7$ at constant temperature?

7-32 Ammonium carbamate dissociates completely in the vapor phase as shown by the equation $NH_4CO_2NH_2(s) = 2NH_3(g) + CO_2(g)$, and at 25°C the dissociation pressure at equilibrium is 0.117 atm. The dissociation pressure at 25°C for the equilibrium $LiCl \cdot 3NH_3(s) = LiCl \cdot NH_3(s) + 2NH_3(g)$ is 0.168 atm.

(a) Neglecting the volume of the solid phases in comparison with the volume of the vapor phase calculate the final total pressure when equilibrium is reached in a 24.4-liter vessel at 25°C when it initially contains 0.050 mole of $CO_2(g)$ and 0.20 mole of $LiCl \cdot 3NH_3(s)$.

(b) Calculate the number of moles of each solid phase present at equilibrium.

(c) Calculate the equilibrium total pressure at 25°C in a 24.4-liter vessel containing initially 0.050 mole $CO_2(g)$ and 0.10 mole of $LiCl \cdot 3NH_3(s)$.

7-33 $Ag_2O(s)$ is in equilibrium with $Ag(s)$ and $O_2(g)$ at 445°C. At this temperature the oxygen pressure is 207 atm. Write the equation for the reaction and obtain the corresponding K_P. Sketch a phase map analogous to that of Fig. 7-2(a).

7-34 Ammonium iodide (solid) dissociates into NH_3 and HI; in a particular experiment excess $NH_4I(s)$ is introduced into an evacuated flask and at 400°C the equilibrium pressure of the gases is 705 Torr. Eventually the HI dissociates into H_2 and I_2, its degree of dissociation being 21% at this temperature. Calculate the final equilibrium pressure above the solid ammonium iodide.

7-35* The following data are reported for the reaction $SO_2 + \frac{1}{2}O_2 = SO_3$.

t(°C)	K_p(atm$^{-1/2}$)	t(°C)	K_p(atm$^{-1/2}$)
528	31.3	727	1.86
579	13.8	789	0.96
627	6.55	832	0.63
680	3.24	897	0.36

Obtain the best least squares slope for the plot of $\ln K_P$ versus $1/T$ and calculate ΔH^0 for the reaction and its standard deviation.

SPECIAL TOPICS PROBLEMS

7-36 Calculate the plot corresponding to Fig. 7-3(b), but for $\Delta G^0 = -1$ kcal at 25°C and $\Delta H^0 = -5$ kcal.

7-37 Calculate K_f for the reaction $3H_2(g) + N_2(g) = 2NH_3(g)$ at 1200°C, using Tables 6-4 and 6-5; then evaluate K_γ if the pressure is 900 atm, and obtain the K_P for this condition.

7-38 Calculate K_f for the equilibrium $C_2H_2(g) + C_2H_6(g) = 2C_2H_4(g)$ at 500 K (use Tables 6-4 and 6-5, for example). Then calculate K_γ and K_P if the total pressure is 250 atm. Critical constants can be obtained from Table 1-4; also, T_c and P_c are 32°C and 48.8 atm, respectively, for ethane gas.

7-39 Calculate K_P for $H_2 + I_2 = 2HI$ from free energies of formation and K_q [Eq. (7-64)] from the spectroscopic constants for the molecules. Evaluate ΔE_0 and calculate K_P at 2000°C.

Calculate K_P for $CO + \frac{1}{2}O_2 = CO_2$ at 25°C from free energies of formation and find the K_q [Eq. (7-64)] from the spectroscopic constants for the molecules. Evaluate ΔE_0 for the reaction and calculate K_P at 1000°C. The rotational constant B_e is 0.3906 cm^{-1} for CO_2, where $B_e = k\theta_{rot}/hc$, and the fundamental vibrational wave numbers are $\nu_1 = 1388.3$ cm^{-1}, $\nu_2 = 667.3$ cm^{-1} (twice), and $\nu_3 = 2349.3$ cm^{-1}.

7-40

Derive Eq. (7-65); calculate q_{trans} for $^{35}Cl_2$ (note Table 6-1).

7-41

Calculate the statistical mechanical value of ΔG^0 for the reaction $^{35}Cl_2 = 2^{35}Cl$ at 1500°C. Take ΔE_0 to be 58 kcal.

7-42

chapter 8
Liquids and solids

8-1   Introduction

There is much emphasis on gases, especially ideal gases, in the preceding chapters. After all, the ideal gas is a universal state of matter, approached by all substances in the limit of low pressure. The ideal gas can be treated exactly by kinetic molecular theory; it provides our thermodynamic temperature scale; its thermodynamic properties are simple and its statistical thermodynamic description is relatively easy. In brief, the gaseous state is of fundamental importance theoretically and yet is fairly simple to deal with.

We turn, in this chapter, to some aspects of the physical chemistry of liquids and solids, that is, of the *condensed state*. As this last name suggests, liquids and solids are relatively dense. The molar volume of a gas at STP is 22.4 liters in contrast to molar volumes of liquids and solids, which are typically in the range of a few tens or hundreds of cm^3. Roughly speaking, the condensed state is a thousand times more dense than the gaseous state, which means that individual molecules are ten times closer together. Intermolecular forces now dominate the molecular picture, and these are discussed in some detail in Section 8-ST-1. While general thermodynamic relationships apply, of course, specific properties of liquids and solids vary greatly. There is no analogue to the ideal gas law as a general equation of state.

Another distinguishing feature of the condensed state is that there can be a boundary or surface. A solid or liquid can co-exist with its vapor, and the next section takes up the very important matter of vapor pressure. Also, part of this chapter is devoted to *surface chemistry*, a major field of physical chemistry. We learn about surface tension and how to include it in thermodynamic relationships.

The impression given up to this point is that liquids and solids form a class and can be treated alike. This conclusion is reasonably valid with respect to macroscopic properties, both thermodynamic ones and transport properties such as viscosity and diffusion. That is, the formal treatments are essentially the same for either condensed state. Liquids and solids differ greatly, however, in the matter of structure. There is not much structural ordering in the typical liquid, and in this respect it behaves like a very dense gas. Witness the ability of the van der Waals equation to deal with properties of liquids (Section 1-8). At the other extreme, crystalline solids have long range order; moreover, their properties may not be isotropic, that is, uniform with respect to direction. Some discussion of the structure of liquids is given in Section 8-CN-1.

8-2 The vapor pressure of liquids and solids

Perhaps the most important thermodynamic leverage on the properties of either a liquid or a solid substance is that obtained through its vapor pressure. The reason is that at equilibrium the molar free energy of the condensed phase must be the same as that of the vapor phase—otherwise, by the free energy criterion for equilibrium, Eq. (6-47), spontaneous evaporation or condensation should occur. The vapor phase will generally be nearly ideal as a gas, and will be so treated here; this means that the thermodynamics of an ideal gas can be related to that of a solid or a liquid.

We can write Eq. (6-48) for each phase separately. Thus, for the gas phase we have

$$dG_g = -S_g \, dT + V_g \, dP \tag{8-1}$$

for any small change in temperature and pressure. Similarly, for the condensed phase, solid or liquid, but for the moment designated as liquid we have

$$dG_l = -S_l \, dT + V_l \, dP \tag{8-2}$$

where P must be the vapor pressure since no other gas is present in the vapor phase. If the two phases are in equilibrium, G_g must equal G_l, and if they are to remain in equilibrium after a small change in temperature, then for this change dG_g must equal dG_l. This means that dP and dT must so change that

$$-S_g \, dT + V_g \, dP = -S_l \, dT + V_l \, dP \tag{8-3}$$

or

$$\frac{dP}{dT} = \frac{S_g - S_l}{V_g - V_l} = \frac{\Delta S}{\Delta V} \tag{8-4}$$

where ΔS and ΔV refer to the process

$$\text{liquid}(P, T) = \text{vapor}(P, T) \tag{8-5}$$

This is a constant-temperature process and is reversible, so that $\Delta S = q/T$; it is also at constant pressure, so that $q = q_P = \Delta H$. Equation (8-4) can therefore be written in the alternative form

$$\frac{dP}{dT} = \frac{\Delta H}{T \, \Delta V} \tag{8-6}$$

Equation (8-6) is known as the *Clapeyron equation*. On looking back over the derivation, we see that is applies to any phase transition. This includes not only a solid–vapor equilibrium, but also that between two condensed phases. Equation (8-6) is thus valid for the general process:

substance in phase $\alpha(P, T)$ = substance in phase $\beta(P, T)$ (8-7)

where α and β may be any two coexisting phases. Of course, if α and β are both condensed phases, then no vapor is present, and the pressure is whatever mechanical pressure is established. The most important example of this situation is that of the equilibrium between a solid and a liquid, discussed in Section 8-4.

Where β is a vapor phase we customarily make two approximations. The first is to neglect V_l as compared to V_g—as noted in the Introduction, for vapor pressures around 1 atm, the molar volume of the vapor will be a thousand times or so that of the liquid. We then further assume the vapor to be ideal and replace V_g by RT/P. Equation (8-6) then becomes

$$\frac{dP^\circ}{dT} = \frac{\Delta H_v}{T(RT/P^\circ)}$$

or

$$\frac{d(\ln P^\circ)}{dT} = \frac{\Delta H_v}{RT^2} \tag{8-8}$$

The "o" superscript to P identifies it as an equilibrium vapor pressure (we continue to use superscript zero to designate a standard state), and ΔH is now specifically designated as an enthalpy of vaporization. Equation (8-8) is known as the *Clausius–Clapeyron equation*. It is of sufficient importance that its behavior should be examined in detail.

Integration, assuming that ΔH_v does not vary with temperature, gives

$$\ln P^\circ = A - \frac{\Delta H_v}{RT} \tag{8-9}$$

or, if performed between the limits of T_1 and T_2,

$$\ln \frac{P_2^\circ}{P_1^\circ} = \frac{\Delta H_v}{R} \left(\frac{1}{T_1} - \frac{1}{T_2} \right) \tag{8-10}$$

Equation (8-9) leads us to expect that a plot of $\ln P^\circ$ versus $1/T$ should be a straight line. Moreover, the slope of the line should give $-\Delta H_v/R$. This expectation is fairly accurately met, provided that the equilibrium vapor density is not too high. This means that the liquid should be well below its critical temperature; there is usually no problem with solids.

Some experimental vapor pressures are plotted in Fig. 8-1(a) as a function of temperature and again in Fig. 8-1(b) in the form of a semilogarithmic graph of P° versus $1/T$. Note that Eq. (8-9) is rather well obeyed, not only for the liquids but also for the solids. The normal boiling point may be read off the graphs as well since it is, by definition, the temperature at which the vapor pressure is 1 atm.

The normal boiling point of water is, of course, 100°C. Its vapor pressure at 80°C is 0.4672 **EXAMPLE**
atm. On rearrangement of Eq. (8-10), we have

$$\Delta H_v = \frac{RT_2 T_1 \ln(P_2^\circ/P_1^\circ)}{\Delta T} \tag{8-11}$$

FIGURE 8-1
Temperature dependence
of some liquid vapor
pressures.

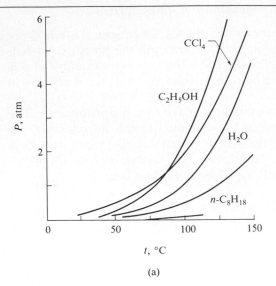

(a)

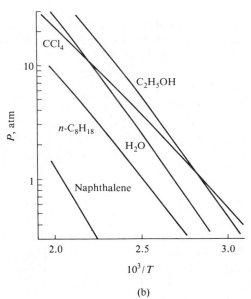

(b)

where $\Delta T = T_2 - T_1$. Then, since $\ln(P_2°/P_1°) = 0.7610$, we have

$$\Delta H_v = \frac{(1.987)(373.15)(353.15)(0.7610)}{20}$$

$$= 9960 \text{ cal mol}^{-1}$$

Alternatively, the graph for water in Fig. 8-1(b) is, around 100°C, a straight line for which $1/T = 2.456 \times 10^{-3}$ at $P° = 3$ atm and 2.919×10^{-3} at $P° = 0.3$ atm. The slope of the line is then

$$\text{slope} = \frac{(\ln 3 - \ln 0.3)}{(2.456 - 2.919)10^{-3}} = -4.974 \times 10^{-3}$$

By Eq. (8-9), ΔH_v is

$$\Delta H_v = -(R)(\text{slope}) = -(1.987)(-4.974 \times 10^3) = 9880 \text{ cal}$$

This result is lower than the first one mainly because the ln P versus $1/T$ plot is slightly curved, and we have in effect taken the slope of the straight line between two fairly well-separated points. This curvature is discussed further below.

Liquid or solid vapor pressure is just that—the equilibrium pressure of vapor in the presence of the condensed phase. In the experimental procedure, however, one must make sure that the measured pressure does not include that of air or any other foreign gas. A simple way of doing this is by means of the *isotensiscope*, illustrated in Fig. 8-2(a). A sample of the liquid is vaporized to form, by condensation, a manometer of liquid in the adjacent U-tube and, in the process, to sweep out any foreign gases. One then adjusts the level of the mercury manometer outside so that the levels of the inside manometer are the same and reads the vapor pressure on the mercury manometer. One may, alternatively, bubble a known amount of inert gas through a known weight of liquid and determine the weight loss of the liquid, as shown in Fig. 8-2(b). The exiting gas is saturated with respect to the liquid vapor, whose partial pressure is therefore $P°$. By Dalton's law, the ratio $P°/P$, where P is the atmospheric pressure, must equal the mole fraction of vapor present:

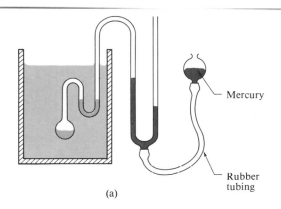

Mercury

Rubber tubing

(a)

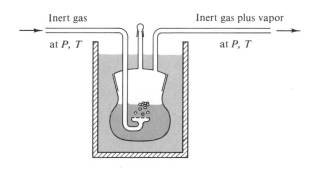

Inert gas
at P, T

Inert gas plus vapor
at P, T

(b)

FIGURE 8-2
Vapor pressure determination. (a) The isotensiscope. (b) The evaporation method.

$$\frac{n_v}{n_A + n_v} = \frac{P^\circ}{P} \tag{8-12}$$

The number of moles of inert gas n_A is known and the number of moles of liquid vaporized n_v is given by the weight loss. Thus P° can be calculated from Eq. (8-12).

EXAMPLE   In an experiment to determine the vapor pressure of water, 20 liters of air are bubbled through liquid water, resulting in 353.6 mg of water evaporated. Ambient temperature (of the air and of the water) is 20°C and the barometric pressure reading is 750 mm Hg. The number of moles of water vaporized is $0.3536/18 = 1.9642 \times 10^{-2}$, and the number of moles of air used is $(750/760)(20)/(0.082057)(293.15) = 0.8205$. Substitution into Eq. (8-12) gives $P^\circ/P = 2.338 \times 10^{-2}$, whence $P^\circ = 17.53$ mm Hg.

One of the assumptions that we made in obtaining the integrated forms, Eqs. (8-9) and (8-10), was that ΔH_v does not vary with temperature. Figure 8-1(b) shows that this is a fairly good assumption. A closer look does show, however, that some curvature does set in at the higher pressures. Some curvature is expected at all temperatures; we know from Eq. (5-28) that $[\partial(\Delta H_v)/\partial T]_P = \Delta C_P$. For water ΔC_P is about -10 cal K^{-1} mol^{-1}, so ΔH_v for water should decrease by 10 cal K^{-1}—not a very large effect.

The situation as one approaches the critical point is more serious. At the critical temperature the two phases have become the same and ΔH_v *must* approach zero. This decrease toward zero is not strong, however, until the vapor density is several percent of that of the liquid or until perhaps eight-

FIGURE 8-3
(a) Enthalpy of vaporization of CO_2 as a function of temperature and (b) variation of the separate liquid and vapor enthalpies with temperature.

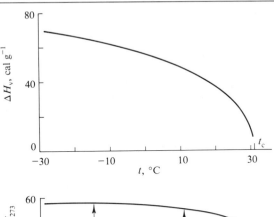

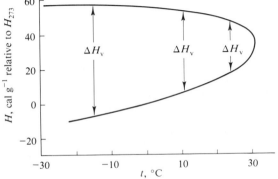

tenths of the critical temperature has been reached. This is illustrated in Fig. 8-3(b) for CO_2, which shows the liquid and vapor enthalpies (relative to that of the liquid at 0°C) as functions of temperature. The difference between the two enthalpies gives the heat of vaporization at each temperature. ΔH_v is dropping fairly rapidly even at -30°C, which is about $0.8\,T_c$; the vapor density at this point is about 4% of that of liquid CO_2. For this region, then, one must go back to Eq. (8-6) in treating the data; in fact the fugacity or thermodynamic pressure of the gas should be used in place of P (see Section 6-ST-2) since gaseous CO_2 has become appreciably nonideal.

8-3   Enthalpy and entropy of vaporization; Trouton's rule

The Clapeyron equation, the Clausius–Clapeyron equation, and the van't Hoff equation, which apply to chemical equilibrium (Section 7-4), belong to a family that marks one of the major achievements of thermodynamics. They all rest on the combined first and second laws of thermodynamics and have in common that the temperature dependence of an intensive quantity, vapor pressure in the present case, gives a thermochemical property, ΔH. That is, from vapor pressure measurements alone one is able to calculate the calorimetric heat of vaporization. In the case of the van't Hoff equation the temperature dependence of the equilibrium constant gives the $\Delta H°$ of reaction. The relationships are phenomenological—no model or other picture of molecular properties is needed. Except for deliberately introduced approximations, as in the Clausius–Clapeyron equation, they are as exact and as valid as the laws of thermodynamics themselves; and they can be verified by doing the calorimetry. These equations have become commonplace in physical chemistry, but we should not therefore become inured to their remarkable power. One practical consequence is that much laborious calorimetric work is made unnecessary; many thermochemical quantities are determined indirectly through these relationships.

With this preamble, let us look at some actual enthalpies of vaporization, assembled in Table 8-1, and mostly calculated from vapor pressure data. Enthalpies of vaporization are usually reported at the normal boiling point of the liquid, as in the table. This is partly a matter of convenience and of convention. Also, however, the normal boiling temperatures represent approximately corresponding states (Section 1-9); the normal boiling temperature of a liquid is often about two-thirds its critical temperature. In a related series of compounds, such as the n-alkanes, ΔH_v increases regularly with increasing molecular weight; otherwise, however, there is little consistency in this respect. Thus oxygen although it has a higher molecular weight than water, has a much lower enthalpy of vaporization. Iodine is heavier than most metals and yet the latter have far higher ΔH_v values. This matter is important because it bears directly on the size and nature of intermolecular forces. We have many indications that these forces are short-range. Van der Waals forces, responsible for the nonideality of vapors, fall off with the inverse sixth power of the intermolecular distance [Section 1-8 and Eq. (1-62)]. Other indications of the short-range nature of intermolecular forces come from surface chemistry. Thus the surface tension of a liquid is established within a surface layer only a few molecular diameters in depth.

The conclusion that intermolecular forces are short-range means that, to a

TABLE 8-1
Enthalpies and entropies of vaporization and fusion[a]

Substance	t_{nbp} (°C)	Liquid ⇌ Vapor ΔH_v^0 (kcal mol^{-1})	Liquid ⇌ Vapor ΔS_v^0 (cal K^{-1} mol^{-1})	t_f (°C)	Solid ⇌ Liquid ΔH_f^0 (kcal mol^{-1})	Solid ⇌ Liquid ΔS_f^0 (cal K^{-1} mol^{-1})
He	−268.944	0.020	4.7	−269.7	0.005	1.5
H_2	−252.77	0.216	10.6	−259.20	0.028	2.0
N_2	−195.82	1.333	17.24	−210.01	0.172	2.72
O_2	−182.97	1.630	18.07	−218.76	0.106	1.95
CH_4	−161.49	1.955	17.51	−82.48	0.225	2.48
C_2H_6	−88.63	3.517	19.06	−183.27	0.683	7.60
HCl	−85.05	3.86	20.5	−114.2	0.476	2.99
Cl_2	−34.06	4.878	20.40	−101.0	1.531	8.89
NH_3	−33.43	5.581	23.28	−77.76	1.351	6.914
SO_2	−10.02	5.955	22.63	−75.48	1.769	8.95
n-C_4H_{10}	−0.50	5.353	19.63	−138.350	1.114	8.263
CH_3OH	64.7	8.43	24.95	−97.90	0.757	4.32
CCl_4	76.7	7.17	20.5	−22.9	0.60	2.4
C_2H_5OH	78.5	9.22	26.22	−114.6	1.200	7.57
C_6H_6	80.10	7.353	20.81	5.533	2.531	8.436
H_2O	100.00	9.7171	26.040	0.000	1.4363	5.2581
$Fe(CO)_5$	105	8.9	23.5	−21	3.25	12.9
CH_3COOH	118.3	5.82	14.8	16.61	2.80	9.66
Hg	356.57	13.89	22.06	−38.87	0.557	2.37
Cs	690	16.32	16.95	28.7	0.50	1.6
Zn	907	27.43	23.24	419.5	1.595	2.303
NaCl	1465	40.8	23.8	808	6.5	6.3
Pb	1750	43.0	21.3	327.4	1.22	2.03
Ag	2193	60.72	24.62	960.8	2.70	2.19
Graphite[b]	4347	170.9	—	—	—	—

[a]From F.A. Rossini et al., Tables of Selected Values of Chemical Thermodynamic Quantities. Nat. Bur. Std. Circ. No. 500. U.S. Govt. Printing Office, Washington, D.C., 1959.

[b]Sublimes.

first approximation, the vaporization process can be viewed as shown in Fig. 8-4. Figure 8-4(a) illustrates schematically an attractive potential between two molecules eventually overridden by a short-range repulsion. The maximum attractive potential ϕ_0 occurs at the equilibrium separation r_0. If in the liquid a molecule has n nearest neighbors, then, to a first approximation, its total interaction with them is $n\phi$. As illustrated in Fig. 8-4(b), the energy of vaporization should then be $n\phi/2$, the factor of two entering because ϕ is a shared energy between two molecules.

The wide variation in enthalpies and hence in energies of vaporization then implies a wide variation in ϕ between different kinds of atoms and molecules. Intermolecular forces in fact fall into several main categories. The potential ϕ may arise from direct chemical bonding, as in diamond or graphite; it may reflect the metallic bond, as in metals; or it may result largely from direct Coulomb's law interactions, as in the alkali halide crystals. In many cases none of these types of interaction can be present, and we observe a weaker but very general attraction between molecules. These secondary forces are responsible for condensation to molecular liquids and crystals, as in the case

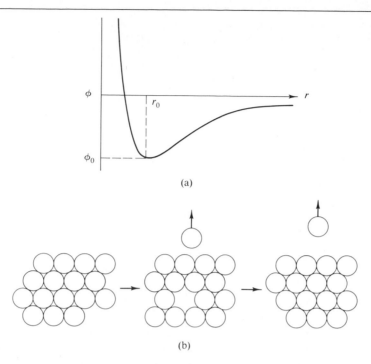

FIGURE 8-4
(a) Qualitative potential function for the interaction between two molecules having a van der Waals type of mutual attraction. (b) Schematic process for estimating energies of vaporization.

of the rare gases, nitrogen, and the hydrocarbons. Such interactions have come to be known as *van der Waals forces* because of their representation in the a constant of that equation. Finally, molecules having a somewhat acidic proton, such as water and alcohols, can hydrogen-bond. Thus in ice each oxygen has four hydrogens around it, two closer and two farther away; if stretched out, a single chain of water molecules would look like this:

The actual crystal of ice has a three-dimensional network of such hydrogen bonds as shown in Fig. 8-5, and liquid water is confidently thought to have local regions of hydrogen-bonded clusters. The structure of water is discussed further in Section 8-CN-1, and the subject of van der Waals forces is taken up in Section 8-ST-1. It is sufficient for the present to have emphasized the molecular significance of enthalpies of vaporization.

Returning to Table 8-1, we observe one very great regularity, namely that enthalpies of vaporization increase almost linearly with T_{nbp}. The result is that the entropy of vaporization is nearly constant. With the principal exceptions of He and H_2, the values lie between 17 and 26 cal K^{-1} mol^{-1} over a range of boiling points from $-195°C$ to over $2000°C$. A good average value for most ordinary liquids is about 21; this consistency of behavior was noted as early as 1884, in what is known as *Trouton's rule:*

FIGURE 8-5
Hydrogen-bonded clusters in liquid water. [From A. Nemethy and H. A. Scheraga, J. Chem. Phys. **36**, 3382 (1962).]

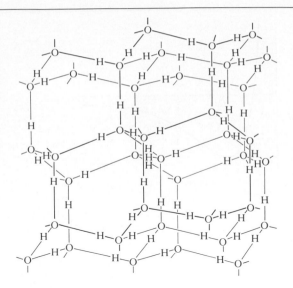

$$\frac{\Delta H_v}{T_{nbp}} \simeq 21 \text{ cal K}^{-1} \text{ mol}^{-1} \quad \text{or} \quad 88 \text{ J K}^{-1} \text{ mol}^{-1} \tag{8-13}$$

Trouton's rule may be rationalized on a very simple basis, and one which has suggested a useful model for liquids. Most liquids are somewhat expanded in comparison to their solid phase, usually by about 10% (an exception, water, is explained as due to the partial collapse of the very open ice crystal structure on melting). This expansion provides an extra or free volume V_{free} for the molecules to move around in; it amounts to perhaps 3 cm^3 mol^{-1} for an ordinary liquid. If we assume that the liquid is merely a highly compressed gas whose effective volume, or $V - b$ term in the van der Waals equation, is about 3 cm^3 mol^{-1} and further assume that Q_{int} is the same for both vapor and liquid (that is, that the rotational and vibrational partition functions are the same), then ΔS_v is assigned entirely to the difference in translational entropy between liquid and vapor. By the Sackur-Tetrode equation, Eq. (6-102), $S_{trans} = $ constant $+ R \ln V$, so $\Delta S_{v(trans)} = R \ln(V_g/V_{free})$. The molar vapor volume for a liquid boiling around 80°C is about 30,000 cm^3 mol^{-1} and so we compute $\Delta S_{v(trans)}$ to be about $R \ln(10^4)$ or about 18 cal K^{-1} mol^{-1}.

This is too simple to be more than suggestive of an approach to the statistical thermodynamics of liquids. One of the more serious treatments assumes each molecule to be in a cell formed by its near neighbors, that is, a cage. Its potential function along a cross section might look as in Fig. 8-6(b), as compared to that for an atom in a crystalline solid, illustrated in Fig. 8-6(c). The model allows estimations of energy levels and hence of the partition function. An alternative has been to view the free volume as present in the form of actual pockets or holes, each of about molecular size. The liquid can then be treated as an intimate mixture of condensed, even crystalline, phase and rapidly moving holes. In fact, the law of the rectilinear diameter (Section 1-CN-3) suggests that if molecular size holes are present, the concentration of holes in the liquid is always about equal to the concentration of molecules in the equilibrium vapor. In this way the average density of the two phases

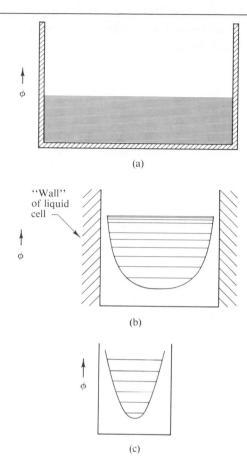

FIGURE 8-6
Potential function and corresponding energy levels for (a) a gas, with extremely closely spaced translational energy levels, (b) a liquid with each molecule in a liquid cell or cage, with closely spaced energy levels corresponding to vibrations in a relatively flat potential well, and (c) a solid with each molecule in a crystalline lattice with rather widely spaced vibrational levels corresponding to harmonic oscillation.

remains nearly independent of temperature. The "hole" model has been useful in the treatment of rate processes, such as viscous flow and diffusion (see Section 8-ST-3).

To return to the data of Table 8-1, we see that water is unusual in having both a larger heat of vaporization and a larger entropy of vaporization than other liquids of similar molecular weight. We regard these differences as reflections of a relatively high degree of structure in liquid water as a result of hydrogen bonding. Hydrogen bonds have a bond energy of 5–7 kcal mol^{-1} and the breaking of such bonds on vaporization gives water a much higher heat of vaporization than would be expected were only ordinary van der Waals interactions present. (Compare, for example, water with methane). The entropy of vaporization is high because of structure in the liquid. Structure reduces the number of ways in which a system can have energy, and hence reduces its thermodynamic probability and therefore its entropy. The entropy of liquid water is thus abnormally low, so ΔS_v is high.

Other molecules capable of hydrogen bonding generally show the same type of behavior; the Trouton constants are large for methanol and ethanol, for example. Liquid ammonia also is classified as hydrogen-bonded, as is acetic acid. In this last case, however, an unusually low entropy of vaporization is observed because the vapor contains a high percentage of dimers.

In fact, it is the study of the dissociation of dimeric acetic acid vapor that provides one estimate of hydrogen bond energies. The dimer is believed to have the structure

$$H_3C-C\underset{O---H-O}{\overset{O-H---O}{<}}C-CH_3$$

and its heat of dissociation of 14 kcal mol^{-1} gives the value of 7 kcal mol^{-1} per hydrogen bond quoted earlier. A better average value for an O—H——O type of hydrogen bond would be about 5 kcal mol^{-1}, as estimated for water and various alcohols.

8-4 Additional thermodynamic properties of liquids and solids

There are, of course, fairly extensive tabulations of the free energies, enthalpies, and entropies of various liquids. A selection of these may be found in the tables of Chapter 5 and Table 7-2. More complete listings are given by the National Bureau of Standards, Circular No. 500 (Rossini et al., 1959). The remaining thermodynamic properties of interest consist of coefficients such as those of thermal expansion and compressibility and of heat capacities. A sampling of these quantities is given in Table 8-2.

Coefficients of compressibility and of thermal expansion are small for liquids, of the order of 0.01–0.1% volume change per atmosphere or per degree. The two coefficients are related to the heat capacity difference $C_P - C_V$ as discussed in Section 6-8 and given by Eq. (6-67):

$$C_P - C_V = \frac{\alpha^2 VT}{\beta}$$

The difference can be surprisingly large. The explanation is, clearly, not in terms of the work of expansion against atmospheric pressure, as with gases, but rather in terms of expansion against the internal pressure, P_{int} (see Section 4-7). From the definition of P_{int} it follows that

$$C_P - C_V = \alpha V(P + P_{int}) \tag{8-14}$$

or

TABLE 8-2
Some thermodynamic coefficients at 25°C[a]

Liquid	$10^3\alpha$ (K^{-1})	$10^5\beta$ (atm^{-1})	V (cm^3 mol^{-1})	P_{int} (atm)	Heat capacity (cal K^{-1} mol^{-1}) $C_P(l)$	$C_V(l)$	$C_V(g)$
H_2O	0.48	4.6	18	3,110	18	17.3	6.0
CCl_4	1.24	10.6	97	3,486	31.5	21.4	18.7
CS_2	1.22	9.5	60	3,829	18.1	11.3	10
C_6H_6	1.24	9.7	89	3,811	32.1	22	22
Hg(l)	0.18	0.34	14.8	15,800	6.65	5.6	3.0
Zn(s)	0.0893	0.15	7.1	—	6	5.7	—

[a] $\alpha = (1/V)(\partial V/\partial T)_p$; $\beta = -(1/V)(\partial V/\partial P)^T$.

$$P + P_{\text{int}} = \frac{\alpha T}{\beta} \tag{8-15}$$

The table includes values for P_{int} and these are indeed far larger than P (which is 1 atm). Notice, however, that $C_P - C_V$ is relatively small for a solid, as illustrated by the listing for zinc metal.

One ordinarily measures C_P directly; this gets around the experimental task of confining a liquid to a fixed volume while measuring its heat capacity. Water, as usual, is unusual. The density of water goes through a maximum at 4°C, and α is therefore zero at this temperature, and is still unusually small at 25°C. The immediate consequence is that the difference $C_P - C_V$ is small, and P_{int} likewise is smaller than one would expect. All this is yet another reflection of the structural changes that are occurring in liquid water as it is warmed from 0°C (see Section 8-CN-1).

It is instructive to compare the values of $C_V(l)$ and $C_V(g)$. The increase from 3.0 to 5.6 cal K^{-1} mol^{-1} on going from Hg(g) to Hg(l) strongly suggests that the $3N_A$ degrees of freedom per mole of mercury atoms are present in the liquid as vibrations rather than as translations. That is, the heat capacity of the liquid is close to $3R$ in value, rather than to $\frac{3}{2}R$ (see Section 4-9). The same is true for water; the value of $C_V(l)$ of about 17 cal K^{-1} mol^{-1} is close to that for $9N_A$ degrees of vibrational freedom per mole, or $9R$ in heat capacity. Again, it appears that water molecules have exchanged translational motion for vibrations in going from the gas to the liquid phase. Moreover, the fact that the heat capacity is close to the equipartition value must mean that the vibrational energy states are close together.

8-5   Liquid–solid and solid–solid equilibria. Phase maps

It was noted that Eq. (8-6) applies to any phase equilibrium. We can apply it to the process

$$\text{solid}(P, T) = \text{liquid}(P, T) \tag{8-16}$$

The equation now reads

$$\frac{dP}{dT} = \frac{\Delta H_f}{T_f \Delta V_f} \tag{8-17}$$

where, in the absence of an equilibrium vapor phase, P is the pressure applied, either mechanically or by means of an inert gas, ΔV_f is the molar volume difference $V_l - V_s$; and ΔH_f is the enthalpy of fusion. Equation (8-17) is not amenable to much simplification, although for a small range of pressure ΔV_f will be approximately constant. The equation then represents a straight-line plot of P versus T_f. This is illustrated in Fig. 8-7(a) for the important case of the ice–water equilibrium. Here ΔV_f is negative; the molar volume of water is 18.02 cm^3 mol^{-1} and that of ice is 19.63 cm^3 mol^{-1} at 0°C. From Table 8-1, $\Delta H_f = 1436$ cal mol^{-1}, so we have

$$\frac{dP}{dT} = \frac{(1436)(82.06/1.987)}{(273.2)(18.02 - 19.63)} = -135 \text{ atm K}^{-1}$$

Thus around 0°C the freezing point of ice decreases by 0.74°C per 100 atm pressure. To obtain a more exact result, we would have to know ΔH_f and ΔV_f

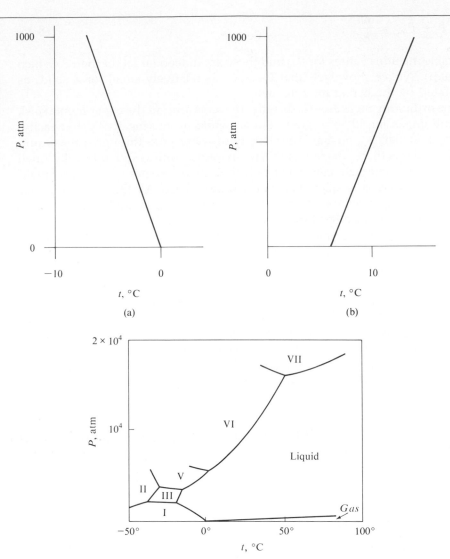

FIGURE 8-7 Variation of melting point with pressure for (a) water and (b) benzene. (c) Phase map or phase diagram for water, including the various high pressure forms.

as functions of temperature and of pressure. We will not concern ourselves with this level of complexity.

Water, as usual, is an exception. For most liquids ΔV_f is positive and, as illustrated in Fig. 8-7(b), the slope of the plot of P versus T_f is therefore also positive. The usual case is thus that melting point increases with increasing pressure.

A number of enthalpies and entropies of fusion are also given in Table 8-1. No particular regularities are present—there is no equivalent to Trouton's rule in the case of entropy of fusion. This quantity is now very dependent on the degree of structural loosening that occurs on melting. As indicated in Fig. 8-6, the potential function for the vibration of a molecule in a liquid is weaker than that in a solid; the energy level spacings are closer, the partition function is larger, and hence so is the entropy. The effect is too sensitive to details of structure to allow much generalization, although a simple monatomic liquid and its solid, such as argon, can be treated fairly well, and so can small, nonhydrogen-bonded molecules.

Returning to Figs. 8-7(a),(b), we see that a rather large pressure is needed to produce a significant change in a melting point. The field of high-pressure chemistry is an interesting one, and much of the pioneering work in this area was done by Bridgman and his co-workers (see Bridgman, 1949). Values as high as 400,000 atm have been reached using special construction materials, hydraulic pressures, and other devices. Parenthetically, the very measurement of such large pressures becomes difficult.

High-pressure chemistry leads to more than just simple extensions of diagrams such as Figs. 8-7(a),(b). Reactions become feasible that do not occur ordinarily, such as the conversion of graphite into diamonds (small ones) and the formation of new types of crystal phases. The water system now looks as shown in Fig. 8-7(c). At around 2000 atm, the ordinary hexagonal structure of ice rearranges, that is, undergoes a phase transition, to a different and denser crystal structure known as ice III. The freezing point curve now slopes to the right. Further phase transitions take place at successively higher pressures to give ice V, ice VI, and, finally, ice VII. Equation (8-6) applies also to transitions between two solid phases and governs the lines showing the effect of pressure on the transition temperature between any two of the solid forms of ice. The figure has taken on the appearance of a phase map. At a temperature below a two-phase equilibrium line only the lower-temperature phase is present; conversely, at a temperature to the right of the line only the higher-temperature phase exists. The various marked regions thus each consist of a single phase only and the boundaries are lines of two-phase equilibrium.

Other strange things can happen under high pressure. Oxygen, for example, becomes a red solid at room temperature!

An interesting current story is that of the development of the phase diagram for C(graphite), C(liquid), and diamond. The key to the synthesis of diamond from graphite, incidentally, has been the inclusion of catalysts, along with the use of high temperature and pressure. Figure 8-8 shows an example of a modern piece of high-pressure equipment. Such devices can be compact. A diamond anvil unit capable of producing 10^5 atm pressure can, for example, be held in the hand.

We return to the low-pressure region to include the liquid–vapor equilibrium line in the phase map for water, shown schematically in Fig. 8-9. Included as well is the ice–vapor equilibrium line, or the plot of the sublimation pressure of ice versus temperature to which the Clausius–Clapeyron equation (8-8) applies. The pressure scale of a diagram of this sort has a dual significance. In the case of the solid–vapor and liquid–vapor lines, pressure is the vapor pressure of the phase in question. For the solid–liquid line, however, no vapor phase is present, and pressure is now mechanically applied pressure.

Note that the solid–vapor line is steeper than the liquid–vapor line; this is because the enthalpy of sublimation is greater than that of vaporization of the liquid. This must always be so:

$$\text{(a) solid} = \text{liquid,} \qquad \Delta H_f$$

$$\text{(b) liquid} = \text{vapor,} \qquad \underline{\Delta H_v} \qquad\qquad\qquad (8\text{-}18)$$

$$\text{(c)} = \text{(a)} + \text{(b) solid} = \text{vapor} \qquad \Delta H_S = \Delta H_f + \Delta H_v$$

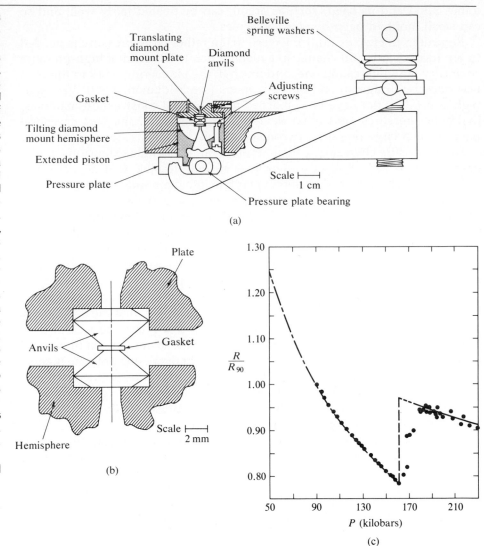

FIGURE 8-8
(a) A modern high-pressure cell. A screw-driven lever applies pressure to a piston, which in turn applies pressure to two diamond anvils. (b) Enlarged view of the diamond anvils. The material to be studied is placed between the anvils and is contained by a metal gasket. Diamond anvils are useful for optical studies since they are transparent. [From G. J. Piermarini and S. Block, Rev. Sci. Instr. **46,** 973 (1975).] (c) Data showing a high-pressure phase change in Pb; the ordinate is resistance relative to resistance at 90 kbar. Data of this type may be obtained by using electrode disks, in this case AgCl, on either side of the sample. Since no optical measurements were involved, carbaloy rather than diamond anvils were used. [From A. S. Balchan and H. G. Drickamer, Rev. Sci. Instr. **32,** 308 (1961).]

ΔH_S must be the sum of ΔH_f and ΔH_v, and since both are positive, the conclusion follows that $\Delta H_S > \Delta H_v$. One consequence is that the solid–vapor and liquid–vapor lines must cross in the manner shown. In the case of water, they do so at 0.0099°C and at the common vapor pressure of 4.579 Torr. This crossing is known as the *triple point* since three phases are in equilibrium: solid, liquid, and vapor. A further consequence is that the solid–liquid equilibrium line must also cross the other two at the triple point. The general rule is that any triple point marks the intersection of the lines representing the three ways of taking two phases at a time. Figure.8-7(c), for example, shows several triple points.

The subject of phase equilibria and phase maps, commonly called *phase diagrams,* is taken up in Chapter 11. However, one other often encountered phase diagram is that for sulfur, shown in Fig. 8-10. The pressure scale is

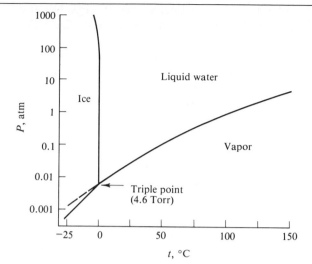

FIGURE 8-9
Phase diagram for water in the low-pressure region, showing the ice I–liquid–vapor triple point at 4.6 Torr.

somewhat schematic, so as to make the diagram more easily displayed. Below 95°C rhombic sulfur is the stable crystal modification and line *ab* gives the sublimation pressure for this form. Transformation to monoclinic sulfur occurs at 95°C and line *bc* gives its sublimation pressure: point *b* is then the rhombic–monoclinic–vapor triple point. Monoclinic sulfur melts at about 125°C and the vapor pressure of liquid sulfur is given by segment *cd*. A second triple point, at *c*, marks the monoclinic–liquid–vapor point. Rhombic sulfur is denser than monoclinic sulfur so the two-phase equilibrium line *be* slopes to the right; and monoclinic sulfur is denser than the liquid, so that line *ce* also slopes to the right. The relative slopes are such, however, that the two lines intersect at point *e*, a third triple point, at which one has a rhombic–monoclinic–liquid equilibrium.

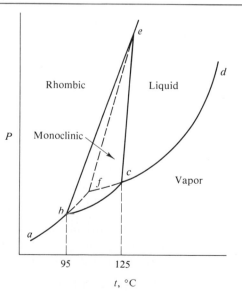

FIGURE 8-10
Phase diagram for sulfur. Dashed lines represent metastable equilibria.

An additional aspect shown in Figs. 8-9 and 8-10 is *metastable* phase equilibrium. Thus liquid water may be cooled below its freezing point (to as low as $-40°C$); the liquid–vapor line in this region represents a *metastable equilibrium* since the liquid is unstable toward freezing. Such lines are shown in the figures as dashed lines. In the water diagram the extension of the liquid–vapor line is possible; a solid cannot be metastable toward melting, however, nor can a liquid or a solid be metastable toward establishing an equilibrium vapor pressure. However, one crystal modification can be metastable with respect to another. Thus rhombic sulfur can be heated above 95°C without immediately converting to the more stable monoclinic form; its metastable vapor pressure curve is given by line *bf* in Fig. 8-10. Liquid sulfur can be supercooled along line *cf*. The diagram therefore shows a fourth, metastable triple point for the rhombic–liquid–vapor equilibrium. This completes the possibilities; that is, four phases can show a maximum of six two-phase equilibrium lines and four triple points. A triple point may be entirely hypothetical, however, and not just metastable. For example, if rhombic sulfur happened to be less dense than monoclinic sulfur, line *be* of Fig. 8-10 would then slope to the left, and the rhombic–monoclinic–liquid triple point would be a hypothetical or totally unstable one lying somewhere in the vapor phase region.

8-6   The free energy of a liquid and its vapor

Equation (8-1) gives G_g as a function of T and P; values of $(G_g - E_0)$ (E_0 being the internal energy at 0 K) can be obtained, for example, from the spectroscopic constants for the gas, using the appropriate partition functions, as discussed in the preceding chapter. The graphical display of G_g would look approximately as shown in Fig. 8-11; G_g decreases with increasing temperature and increases with increasing pressure. The function G_l for the liquid is not easily evaluated theoretically but should have the qualitative appearance shown in the figure; the pressure dependence should be much less than for G_g, for example, and at low temperatures the surface must lie below that for G_g since the liquid is then the stable phase. We also know that the line of intersection of the two surfaces must correspond to the P–T line of liquid–vapor equilibrium. A third surface for G_S also exists; its intersection with the other two would give the solid–liquid and solid–vapor equilibrium lines.

Along the liquid–vapor line, then, $G_g = G_l$, and a convenient way of describing G_l is in terms of the free energy of the vapor. We ordinarily deal with vapor pressures not much over 1 atm and will therefore make the assumption that the vapor behaves ideally. The free energy of a gas is given by Eq. (6-54), which becomes, in the present context,

$$dG_g = RT\, d(\ln P) \qquad \text{(ideal gas)} \tag{8-19}$$

or

$$G_g = G_g^0 + RT \ln P \qquad \text{(ideal gas)} \tag{8-20}$$

for a change in pressure at constant temperature. Here G_g^0 is the free energy of the vapor or gas phase at the temperature in question and 1 atm pressure, and is thus the standard-state free energy. It follows that

$$G_l^0 = G_g^0 + RT \ln P^\circ \tag{8-21}$$

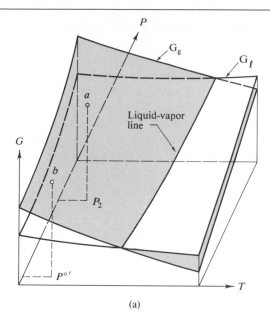

FIGURE 8-11
(a) Plot of typical G(T,P) function for a liquid and for its vapor. The crossing of the two surfaces marks the line of liquid–vapor equilibrium, and hence gives the vapor pressure versus temperature curve. (b) Isobaric cross-section; G versus T at 1 atm. (c) Isothermal cross-section; G versus P at the normal boiling point.

(a)

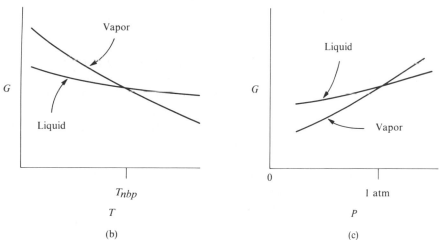

(b)

(c)

where G_l^0 is the free energy of the liquid in its standard state and P° is the liquid vapor pressure. Strictly speaking, G_l^0 refers to the liquid under its own vapor pressure, but the small difference between this value and that for the usual standard state of 1 atm pressure can be neglected here.

An important application of Eq. (8-21) is the following. We see from Eq. (8-2) that the free energy of a liquid increases with pressure at constant temperature, and integration gives

$$G_{l,2} = G_l^0 + \int_{P^\circ}^{P} V_l \, dP \simeq G_l^0 + V_l \Delta P \tag{8-22}$$

where P° is the vapor pressure of the liquid and P is some imposed mechanical pressure. If this higher pressure is established by using an inert gas to compress the liquid, there will still be a gas phase and the liquid will still have an equilibrium vapor pressure $P^{\circ\prime}$. This vapor pressure $P^{\circ\prime}$ corresponds to a vapor free energy of $G_g^0 + RT \ln P^{\circ\prime}$, so we have

$$G_{l,2} = G_g^0 + RT \ln P^{\circ\prime} \tag{8-23}$$

On combining Eqs. (8-21)–(8-23), we obtain

$$RT \ln \frac{P^{\circ\prime}}{P^\circ} = \int_{P^\circ}^{P} V_l \, dP \simeq V_l \Delta P \tag{8-24}$$

Returning to Fig. 8-11, $G_{l,2}$ corresponds to point a located on the free energy surface for the liquid at temperature T and mechanical pressure P_2. The vapor having the same free energy at T is in the state given by point b, or at vapor or gas pressure $P^{\circ\prime}$.

A numerical illustration is as follows. Water has a molar volume of 18 cm³ mol⁻¹ and if we neglect its compressibility, then for 100 atm pressure, the integral of Eq. (8-24) is just $(18)(100) = 1800$ cm³ atm. At 25°C, $RT = (82.06)(298.2) = 24{,}470$ cm³ atm and $\ln(P^{\circ\prime}/P^\circ)$ becomes $(1800)/(24{,}470) = 0.0736$ and $P^{\circ\prime}/P^\circ = 1.076$. The vapor pressure is thus increased 7.6% by the application of 100 atm pressure.

8-7 The surface tension of liquids. Surface tension as a thermodynamic quantity

We have so far considered only bulk properties, that is, properties of a portion of matter which, if extensive, are proportional to the amount of substance or which, if intensive, apply to the interior of the phase. We now recognize that energy resides in the interface between two phases and, more specifically, that work is required to form or to extend an interface. For the present, the discussion is limited to the liquid–vapor interface of pure liquids. The treatment of solutions is deferred to Section 9-ST-2.

The requirement of work to increase a liquid–vapor surface is most easily demonstrated experimentally by means of soap films, since these persist long enough to be handled. As illustrated in Fig. 8-12(a), if a soap film is formed in a frame one side of which can move freely, one finds that the film will spontaneously contract, and that to prevent contraction, a force f must be applied. If f is decreased slightly, contraction occurs, and if f is increased, the film extends. The process is then a reversible one, and the work involved is reversible, constant-temperature and constant-pressure work. It therefore corresponds to a free energy quantity.

For a small displacement dx of the movable side of the frame the work done by the film is

$$w = -f \, dx = -\gamma l \, dx \tag{8-25}$$

where γ is the force per unit length exerted by the film and is called the *surface tension*. The units of surface tension are evidently force per unit length, or, in the cgs system, dyne per centimeter. Since $l \, dx$ is the change in area of the film $d\mathscr{A}$, a less geometrically specific form of Eq. (8-25) is

$$w = -\gamma \, d\mathscr{A} \tag{8-26}$$

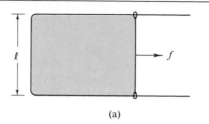

(a)

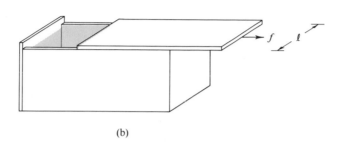

(b)

FIGURE 8-12
Work of extending a
surface: (a) Extending a
soap film held in a wire
frame by moving a side.
(b) Extending the surface
of a liquid by sliding back
a cover of inert material.

Evidently, γ can also be expressed in units of energy per unit area, or as erg per square centimeter. The two sets of units are equivalent, and surface tension may be stated in either way. The experiment with a pure liquid, rather than a soap solution, is the hypothetical one shown in Fig. 8-12(b). We imagine a container filled with the liquid and at a given T and P. There is a movable lid that can be slid back and forth to change the amount of liquid surface. (We stipulate that there is no interaction between the liquid and the material of the lid.) The force on the lid is then γl as before.

The free energy change associated with a change in surface area can be added to the usual statement of the combined first and second laws [Eq. (6-48)] to give

$$dG = -S\,dT + V\,dP + \gamma\,d\mathscr{A} \tag{8-27}$$

Thus

$$\gamma = \left(\frac{\partial G}{\partial \mathscr{A}}\right)_{T,P} = G^{S} \tag{8-28}$$

where G^{S} denotes the surface free energy per unit area. Further, since the process of Fig. 8-12(b) is a reversible one, the q associated with it is a q_{rev}, and we have

$$dq = T\,dS = TS^{S}\,d\mathscr{A} \tag{8-29}$$

where S^{S} is the surface entropy, also per unit area. Other thermodynamic relationships apply. By Eq. (6-50), $(\partial G/\partial T)_{P} = -S$, and likewise,

$$\left(\frac{\partial G^{S}}{\partial T}\right)_{P} = -S^{S} \tag{8-30}$$

or

$$\left(\frac{\partial \gamma}{\partial T}\right)_P = -S^s$$

Also

$$H^s = G^s + TS^s \tag{8-31}$$

Often, and as a good approximation, H^s and the surface energy E^s are not distinguished, so Eq. (8-31) is usually given in the form

$$E^s = G^s + TS^s \tag{8-32}$$

or

$$E^s = \gamma - T\frac{d\gamma}{dT} \tag{8-33}$$

(constancy of pressure being assumed for $d\gamma/dT$).

The thermodynamics of curved interfaces constitutes a very important topic, one aspect of which is developed here, namely the relationship between the mechanical pressure across an interface and its curvature. To illustrate the physical reality of the effect, we again use a soap film, now in the form of a soap bubble. As shown in Fig. 8-13(a), a manometer is attached to the soap bubble pipe, and after the bubble has been blown, the stem of the pipe is closed off. The manometer will now show a pressure difference. For an actual soap bubble this is about 1 mm H_2O for a bubble of 2 mm radius. The physical explanation for the higher pressure inside the bubble is that the film tends spontaneously to decrease its area and hence its total surface free energy but, in doing so, shrinks the bubble to the point that the increased pressure prevents further change.

The relevant equation may be derived as follows. Consider a bubble of

---

FIGURE 8-13
(a) The pressure inside a soap bubble is larger than that outside. (b) Derivation of the Laplace equation for a spherical surface.

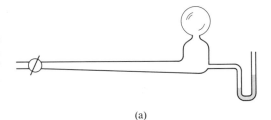

(a)

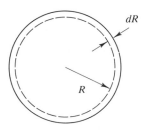

(b)

radius R as shown in Fig. 8-13(b). Its total surface free energy is $4\pi R^2\gamma$ and if the radius were to decrease by dR, then the change in surface free energy would be $8\pi R\gamma\, dR$. Since shrinking decreases the total surface free energy, at equilibrium the tendency to shrink must be balanced by a pressure difference across the film $\Delta\mathbf{P}^*$ such that the work against this pressure difference $(\Delta\mathbf{P})4\pi R^2\, dR$ is just equal to the decrease in surface free energy. Thus we have

$$(\Delta\mathbf{P})4\pi R^2\, dR = 8\pi R\gamma\, dR$$

or

$$\Delta\mathbf{P} = \frac{2\gamma}{R} \tag{8-34}$$

Equation (8-34) is the *Laplace equation* for a spherical interface. An important conclusion is that the smaller the bubble, that is, the smaller the value of R, the larger is $\Delta\mathbf{P}$. Conversely, $\Delta\mathbf{P}$ goes to zero in the limit of $R \rightarrow \infty$, which corresponds to the case of a plane interface.

Parenthetically, common usage defines γ as the surface tension for one interface. Because of this it would be better to use 2γ instead of γ in relationships such as Eq. (8-34) when they are applied to soap or other films.

In summary, the two most important relationships developed so far are Eq. (8-33) for surface energy, $E^S = \gamma - T(d\gamma/dT)$, and the Laplace equation (8-34), $\Delta\mathbf{P} = 2\gamma/R$. The latter is one of the four principal special equations of surface chemistry, the other three being the Kelvin equation (Section 8-9), the Gibbs equation (Section 9-ST-2), and Young's equation (see Adamson, 1982).

8-8 Measurement of surface tension

We do not usually devote much space to experimental procedures, but the various ways of measuring surface tension are so closely related to basic surface phenomena that discussion of the former illustrates the latter.

A. Capillary rise

If a small-bore tube is partially immersed in a liquid as shown in Fig. 8-14(a), one usually observes that the liquid rises to some height h above the level of the general liquid surface. This phenomenon may be treated by means of the Laplace equation. If, as shown in more detail in Fig. 8-14(b), the liquid wets the wall of the capillary tube, then if the radius of the tube r is small, the meniscus will be hemispherical in shape. Its radius of curvature is then equal to r, that is, $R = r$, and Eq. (8-34) becomes

$$\Delta\mathbf{P} = \frac{2\gamma}{r} \tag{8-35}$$

If, as further illustrated in the figure, $\mathbf{P}_1$ is the general atmospheric pressure, then the pressure just under the meniscus must be $\mathbf{P}_1 - (2\gamma/r)$. The $\Delta\mathbf{P}$ in

---

*We use boldface $\mathbf{P}$ to distinguish mechanical pressure from vapor pressure P.

FIGURE 8-14
(a) The capillary rise
phenomenon. (b) Detail
for a case of a spherical
meniscus.

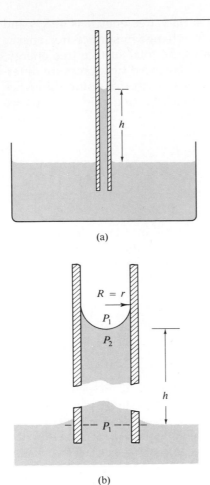

(a)

(b)

the Laplace equation is always such that the higher pressure is on the concave side of the curved surface. The pressure P_1 is also the pressure on the flat portion of the liquid surface outside the capillary tube, and hence the pressure at that liquid level inside the capillary. Then $P_1 - P_2$ must correspond to the change in hydrostatic pressure of the liquid, $\rho g h$. We can thus write

$$\rho g h = \frac{2\gamma}{r} \tag{8-36}$$

or

$$h = \frac{2\gamma}{\rho g r} \tag{8-37}$$

(Strictly speaking, one should use $\Delta\rho$, the difference in density between the liquid and vapor phases.) The quantity $2\gamma/\rho g$ is a property of the liquid alone and Eq. (8-37) may be put in the form

$$a^2 = rh \tag{8-38}$$

where a is called the *capillary constant*. Equation (8-38) tells us that the plot of capillary rise h versus r is a hyperbola. Equation (8-38) is exact only in the limit of small r; otherwise corrections are needed.

A more general case is that in which the liquid meets the capillary wall at some angle θ as illustrated in Fig. 8-15. Without going through the details of the geometry, we can say that the radius of curvature of the meniscus R is now given by $R = r/\cos \theta$ (assuming the meniscus is still spherical in shape) and Eq. (8-37) becomes

$$h = \frac{2\gamma \cos \theta}{\rho g r} \qquad (8\text{-}39)$$

A liquid making an angle of 90° with the wall would show no capillary rise at all. Mercury, whose contact angle against glass is about 140°, shows a capillary *depression*. That is, $\cos \theta$ is now negative and so is h.

The following is an alternative approach, but one that is equivalent to the preceding in the first approximation. The spontaneous contractile tendency of a liquid surface acts as though there were a surface force γ per unit length. In the capillary rise situation, the total such force is $2\pi r\gamma$ (or, more generally, the vertical component would be $2\pi r\gamma \cos \theta$). This force is balanced by the force exerted by gravity on the column of liquid which is supported, or $(\pi r^2)(\rho g h)$. On equating these two forces, we have Eq. (8-36) [or(8-39)].

The capillary rise phenomenon is not only the basis for an absolute and accurate means of measuring surface tension but is also one of its major manifestations. The phenomenon accounts for the general tendency of wetting liquids to enter pores and fine cracks. The absorption of vapors by porous solids to fill their capillary channels and the displacement of oil by gas or water in petroleum formations constitute specific examples of capillarity effects. The waterproofing of fabrics involves a direct application of Eq. (8-39). Fabrics are porous materials, the spaces between fibers amounting to small capillary tubes. As shown in Fig. 8-16(a), coating the fibers with material on which water has a contact angle of greater than 90° provides a Laplace pressure which opposes the entry or "rise" of water into the fabric. The material is *not* waterproof; once the Laplace pressure is exceeded (as, for example, by directing a jet of water onto the fabric or merely by rubbing of the wet fabric),

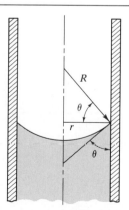

FIGURE 8-15
Capillary rise in the case of a nonwetting liquid. Detail shows the meniscus profile assuming a spherical shape.

FIGURE 8-16
Mechanism of water
repellancy.

Water

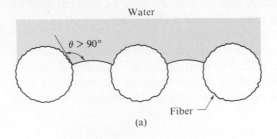

$\theta > 90°$

Fiber

(a)

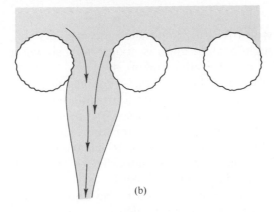

(b)

then penetration occurs, and water will seep freely through the filled channels
as illustrated in Fig. 8-16(b).

B. Drop weight
method

It was pointed out earlier that one can assign a force γ per unit length of a
liquid surface as giving the maximum pull the surface can support. In the
case of a drop of liquid which is formed at the tip of a tube and allowed to
grow by delivering liquid slowly through the tube, a size is reached such that
surface tension can no longer support the weight of the drop, and it falls.
The approximate sequence of shapes is shown in Fig. 8-17. To a first approximation, we write

$$W_{\text{ideal}} = 2\pi r \gamma \tag{8-40}$$

FIGURE 8-17
Sequence of shapes for a
drop that detaches from a
tip. [From A. W. Adamson,
"The Physical Chemistry of
Surfaces," 4th ed.
Copyright 1982, Wiley
(Interscience), New York.
Used with permission of
John Wiley & Sons, Inc.]

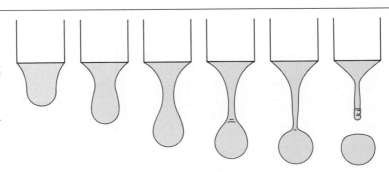

where W_{ideal} denotes the weight of the drop that should fall and r is the radius of the tube (outside the radius if the liquid wets the tube). Equation (8-40) is known as *Tate's law*.

As also illustrated in the figure, the detached drop leaves behind a considerable residue of liquid, and the actual weight of a drop is given by

$$W_{\text{act}} = W_{\text{ideal}}f \tag{8-41}$$

where f is a correction factor which can be expressed either as a function of r/a, where a is the square root of the capillary constant, or of $r/V^{1/3}$, where V is the volume of the drop. Table 8-3 gives some values of f; a more complete table is given in the reference cited therein. Note that the correction is considerable. The method, using the correction table, is accurate and convenient, however. In practice, one forms the drops within a closed space to avoid evaporation and in sufficient number that the weight per drop can be determined accurately. It is only necessary to form each drop slowly during the final stages of its growth.

A set of methods that are somewhat related in treatment to the drop weight method have in common that one determines the maximum pull to detach an object from the surface. If the perimeter of the object is p, then the maximum weight of meniscus that the surface tension can support is just

$$W_{\text{ideal}} = p\gamma \tag{8-42}$$

The so-called *du Nouy tensiometer* uses a wire ring, usually made of platinum, and p is then $4\pi R$, where R is the radius of the ring, and one is allowing for both the inside and the outside meniscus. As with the drop weight method, rather large correction factors are involved, and these have been tabulated (see Adamson, 1982).

If, instead of a ring, one uses a thin slide, either an actual microscope slide or a metal one of similar size, then it turns out that Eq. (8-42) is quite accurate and may be applied directly. The method is now known as the *Wilhelmy slide* method and is shown schematically in Fig. 8-18. Alternatively, the slide need not be detached but need merely be brought into contact with the liquid and then raised until the end is just slightly above the liquid surface. The extra weight is again the meniscus weight and is given by Eq. (8-42). The Wilhelmy method is one of those most widely used today. It is especially valuable in studying the surface tension of solutions and of films of insoluble substances.

C. Detachment methods

$r/V^{1/3}$	f	$r/V^{1/3}$	f
0.30	0.7256	0.80	0.6000
0.40	0.6828	0.90	0.5998
0.50	0.6515	1.00	0.6098
0.60	0.6250	1.10	0.6280
0.70	0.6093	1.20	0.6535

TABLE 8-3
Correction factors for the drop weight method[a]

[a]See A.W. Adamson, "Physical Chemistry of Surfaces," 4th ed. Wiley (Interscience), New York, 1982.

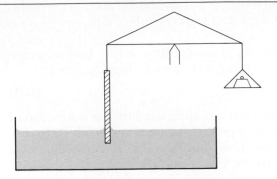

FIGURE 8-18
The Wilhelmy slide
method. A thin plate is
suspended from a balance.

8-9   Results of surface tension measurements

Some representative surface tension values are assembled in Table 8-4. Notice the very wide range in values, from 0.308 dyn cm^{-1} for liquid helium to 1880 dyn cm^{-1} for molten iron. An important datum is the value for water, 72.88 dyn cm^{-1} at 20°C; other common solvents have surface tensions around 20–30 dyn cm^{-1}. Some values for the interfacial tension between two liquids are included. The SI unit of surface tension is newton per meter (N m^{-1}); 1 N m^{-1} = 10^3 dyn cm^{-1}.

TABLE 8-4
Surface tension of liquids

Liquid	Temperature (°C)	γ (dyn cm^{-1})	$d\gamma/dT$
He	-270.7	0.308	-0.07
N$_2$	-198	9.41	-0.23
Perfluoroheptane	20	13.19	—
n-Heptane	20	20.14	—
n-Octane	20	21.62	-0.10
Ethanol	20	22.39	-0.086
CCl$_4$	20	26.43	—
Benzene	20	28.88	-0.13
Water	20	72.88	-0.15
KClO$_3$	368	81	—
NaNO$_3$	308	116.6	—
Ba(NO$_3$)$_2$	595	134.8	—
Sodium	97	198	-0.10
Mercury	20	486.5	-0.20
Silver	1100	879	-0.184
Iron	1535	1880	-0.43

Liquid–Liquid Interfaces, 20°C			
Liquids	γ (dyn cm^{-1})	Liquids	γ (dyn cm^{-1})
n-Butanol–water	1.8	Water–mercury	415
n-Heptanoic acid–water	7.0	Ethanol–mercury	389
Benzene–water	35.0	n-Heptane–mercury	378
n-Heptane–water	50.2	Benzene–mercury	357

One may use the table to illustrate characteristic values in the various methods for determining surface tension. For example, the constant a^2 is 0.15 cm^2 for water, about 0.05 cm^2 for a typical organic liquid, and around 0.5 for a molten metal. Equation (8-38) then allows an easy estimate of the capillary rise. Thus, for water in a 0.1-cm-diameter capillary, $h = 0.15/0.05 = 3.0$ cm and for a 0.01-cm-diameter capillary, $h = 30$ cm.

Had the drop weight method been used, an average weight of about 0.085 g would have been found using a tip of 0.3 cm outside radius. The quantity $r/V^{1/3}$ is 0.68 and from Table 8-3, $f = 0.61$. Thus

$$\gamma = \frac{(0.085)(981)}{(2\pi)(0.3)(0.61)} = 72.5 \text{ dyn cm}^{-1}$$

Finally, for a Wilhelmy slide having a typical width of 3 cm, the meniscus weight for water at 20°C would be

$$W = (2)(3)(72.88) = 437 \text{ dyn or } 0.446 \text{ g}$$

We assume the slide to be thin enough that the edges do not contribute appreciably to the perimeter. An edge effect does exist and usually is allowed for by using slides of varying widths.

The table lists values for $d\gamma/dT$ for several liquids. Surface tension decreases with temperature, often almost linearly, and a relationship due to Eötvös gives

$$\gamma V^{2/3} = k(T_c - T) \tag{8-43}$$

Another, due to Guggenheim, is of the form

$$\gamma = \gamma^\circ \left(1 - \frac{T}{T_c}\right)^n \tag{8-44}$$

where $n = 11/9$ for organic liquids. Both equations correctly predict that the surface tension should go to zero at the critical temperature T_c and the former, in combination with Eq. (8-33), requires that the surface energy E^s should be independent of temperature if V is constant. E^s does remain nearly constant until fairly close to the critical temperature, and then drops rapidly toward zero. The situation in this respect is similar to that for enthalpies of vaporization.

8-10   The Kelvin equation. Nucleation

According to the Laplace equation (8-34) a small drop of liquid should be under an excess mechanical pressure $\Delta P = 2\gamma/R$. We can combine this result with Eq. (8-24), which gives the effect of mechanical pressure on the vapor pressure of a liquid. If V_l is assumed to be constant, we get

$$RT \ln\frac{P^{\circ\prime}}{P^\circ} = V_l\Delta P = \frac{2\gamma V_l}{R}$$

or

$$RT \ln\frac{P^{\circ\prime}}{P^\circ} = \frac{2\gamma V_l}{r} \tag{8-45}$$

where r is the radius of the drop. Equation (8-45) is known as the *Kelvin equation* and, as mentioned earlier, is one of the four principal equations of surface chemistry.

The effect predicted by Eq. (8-45) is not very large until quite small values of r are reached. Thus for water at 25°C, $P^{\circ\prime}/P^{\circ}$ is about 1.001 if $r = 10^{-4}$ cm and 1.11 if $r = 10^{-6}$ cm. The equation has been verified experimentally for water, dibutyl phthalate, mercury, and other liquids, down to radii of about 10^{-5} cm (by using aerosols). For the inverse case of a vapor or gas bubble in a liquid, $P^{\circ\prime}$ is less than P°. There is also an effect of curvature on the *enthalpy* of vaporization (see Adamson and Manes, 1983).

The increase in vapor pressure called for by the Kelvin equation is not large enough to be of any concern in the case of macroscopic systems. The phenomenon is central, however, to the mechanism of vapor condensation. If a vapor is cooled or compressed to a pressure equal to the vapor pressure of the bulk liquid, then condensation should occur. The difficulty is that the first few molecules condensing can only form a minute drop and the vapor pressure of such a drop will be much higher than the regular vapor pressure. In practice, condensation may occur on the walls of the container or on dust particles, but in clean systems, large degrees of supersaturation are possible. In the case of water, for example, the vapor at around 0°C may be compressed to about five times the equilibrium vapor pressure before spontaneous homogeneous condensation sets in. Similar results have been found for many other vapors.

The crucial step in the condensation process is the formation of droplets big enough that their further growth by condensation of vapor onto them is rapid. Such droplets are called *nuclei* and their formation is called *nucleation*. The theory of homogeneous nucleation calculates the excess surface free energy of a drop as a function of its size. The free energy necessary to form a given drop is then determined by size and by the ratio of the actual pressure of the vapor P to P°. For each value of P there is a certain critical drop size; drops larger than this decrease in free energy with increase in size and should grow spontaneously. One next obtains the concentration of drops of this critical size, using the Boltzmann equation, and their rate of growth, using the surface collision frequency equation (2-59). The final conclusion is that only when P has reached a certain critical ratio with respect to P° will the rate of condensation suddenly become rapid. Nucleation theory has been fairly successful; predicted and observed values of P/P° agree well.

The same analysis and, with modifications, the same equations apply to the supercooling of a liquid, where the critical nuclei are now incipient crystals, and to the precipitation of a solute from solution. As an example of this last, aqueous $S_2O_3^{2-}$ slowly decomposes in acid solution to give elemental sulfur. At first the sulfur produced simply builds up in solution, but at a very sharp critical concentration nucleation begins, and thereafter all further sulfur produced adds to the existing nuclei. The result is a dispersion of sulfur which consists of spherical particles of virtually uniform size. Such monodisperse sulfur sols now show beautiful light scattering effects—incident white light is scattered as a rainbow of color. The same effect occurs, incidentally, with natural opals. These consist of close-packed spheres of silica, originally slowly deposited from solution, which are monodisperse. The flashes of color seen in an opal are again due to interferences among light rays scattered from the arrays of spheres.

8-11   Viscosity of liquids

The viscosity of liquids is considered briefly here as an example of an important nonthermodynamic property. The subject of viscosity was treated in an introductory way in Section 2-9 with special emphasis on the kinetic molecular theory of gas viscosities. Perhaps the most common manifestation of viscosity in the laboratory and in engineering is in the flow of a fluid through a pipe or tube. The basic observation is that the volume rate of flow is proportional to the pressure drop, provided the flow is streamline, as illustrated in Fig. 2-9. The actual equation, known as the *Poiseuille equation*, was also given earlier [Eq. (2-79)],

$$\eta = \frac{\pi(\mathbf{P}_1 - \mathbf{P}_2)r^4 t}{8Vl}$$

where η is the viscosity (in poise, abbreviated P), $\mathbf{P}_1 - \mathbf{P}_2$ is the pressure drop (in dynes per square centimeter), r is the radius of the tube (in centimeters), V is the volume flowing (in cubic centimeters), l is the length of the tube (in centimeters), and t is time (in seconds). Equation (2-79) is derived in Section 8-ST-2, and only its application is considered here.

Viscosities may be measured experimentally by the direct application of Eq. (2-79). Some liquids have viscosities that vary greatly with the rate of shear, and measurements on them should be made with a carefully controlled flow rate. Often a coiled capillary tube is used.

Calculate the flow rate of a liquid of 30 P through a 1 m capillary of 0.1 mm radius if the pressure drop is 10 atm. We solve Eq. (2-79) to obtain   **EXAMPLE**

$$\frac{V}{t} = \frac{(\pi)(10)(1.013 \times 10^6)(0.01)^4}{(8)(100)(30)} = 1.33 \times 10^{-5} \text{ cm}^3 \text{ s}^{-1}$$

In SI we use $\mathbf{P}$ as 1.013×10^6 Pa, r as 1×10^{-4} m, and η as 3 to obtain 1.33×10^{-11} $m^3 s^{-1}$.

In the case of the usual, well-behaved liquid it is much more convenient to use an *Ostwald viscometer*, illustrated in Fig. 8-19. A definite volume of liquid is introduced into bulb B and the liquid is sucked up into bulb A and past the mark a; the volume is such that bulb B is not quite emptied. The liquid is then allowed to flow back and the time for the meniscus to pass from a to b determined. The theory of the viscometer is essentially that of Eq. (2-79), except that the pressure drop changes with time, as the liquid level drops, and hence so does the flow rate. The average pressure drop is proportional to the density ρ of the liquid, however, and the other terms of Eq. (2-79) are constant. We can thus write

$$\eta = \mathbf{k}\rho t \tag{8-46}$$

where $\mathbf{k}$ is a constant for a given viscometer and is determined by measurement of the time of flow for a liquid of known viscosity.

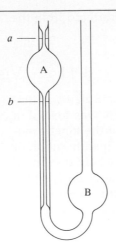

FIGURE 8-19
An Ostwald viscometer.

A selection of viscosity data is given in Table 8-5. Ordinary solvent liquids have viscosities of a few centipoise. Oils, such as heavy lubricating oils, lie in the range of a few poise; glycerin has a viscosity of about 10 P at 25°C. Viscosities are generally very temperature-dependent, as also illustrated by the data of the table. The fluidity φ, or reciprocal of the viscosity, usually varies nearly exponentially with temperature:

$$\phi = A \exp\left(-\frac{E^*}{RT}\right) \tag{8-47}$$

As the form of Eq. (8-47) suggests, the temperature dependence is attributed to the presence of a characteristic energy for flow to occur, and values for E^* are included in the table. These are around 2 kcal mol^{-1}, except for water, whose value is unusually high, presumably because of the structure-breaking that must occur in flow (see both Commentary and Notes and Special Topics sections).

TABLE 8-5
Viscosity of liquids[a]

Liquid	Viscosity (cP)			E^* (cal mol^{-1})
	0°C	20°C	40°C	
H_2O	1.792	1.005	0.656	4300
n-Octane	0.7060	0.5419	0.4328	3000
Ethanol	1.773	1.200	0.834	3200
Benzene	0.912	0.652	0.503	2540
CCl_4	1.329	0.969	0.739	2600
Mercury	1.684	1.547	1.483	600

[a]From International Critical Tables, Vol. 7. McGraw-Hill, New York, 1930.

COMMENTARY AND NOTES

8-CN-1   The structure of liquids

Do liquids have structure, that is, do molecules in a liquid show definite spacial relationships between each other? The answer depends somewhat on the viewpoint. Solids do have structure, and a crystalline solid has a regular, repeating structure. Attractive forces, either van der Waals or chemical bonding in nature, have brought the molecules together to a point where attraction is balanced by the repulsion due to overlapping electronic clouds; the separation is essentially that of the minimum in Fig. 8-4(a). The typical liquid is expanded about 10% relative to its solid, however, so there is some free space and this allows the molecules to assume a variety of loose-knit structures. The melting of a solid costs energy; molecules have moved somewhat apart, against their mutual attraction. They have more entropy, however, because of the greater variety of arrangements possible. The loose structure in a liquid allows molecules to move around, although by low energy vibrations rather than by actual free translational motions, and this means that the local structure around a given molecule fluctuates with time. The time scale is that of a few to a few hundred vibration times, or around 10^{-12} s.

The situation is the following. Properties that depend on the average behavior over intervals long compared to 10^{-12} s will show a liquid to be isotropic and tend to equate it to a dense gas. The picture is one of molecules in random motion, although strongly experiencing each other's potential fields. On the other hand, x-ray diffraction studies suggest a different emphasis. Although the diffraction patterns are very diffuse, they are not as diffuse as they would be if intermolecular distances were entirely random. The x-ray findings may be reported as the density, relative to the average density, of molecules at a given distance from some particular molecule. Two such results are illustrated in Fig. 8-20. In the case of liquid potassium it is probable that a potassium atom will have a neighbor at a distance of about 4.5 Å; the probability of finding one at twice this

distance is larger than for a random distribution, but beyond this the liquid appears isotropic. The result for mercury is similar, the nearest-neighbor spacing being about 3 Å. In the case of liquid water the nearest neighbors are at about 2.9 Å. The general picture of a liquid is thus one of a semicrystalline local order, with each molecule tending to have a certain number and geometry of nearest neighbors. The structure fluctuates with time, however, and is not perfectly regular. The consequence is that any regularity around a given molecule has essentially vanished by the third or fourth molecular diameter distance away.

This local crystallinity or organization is often called the liquid or *solvent cage*. Individual molecules are regarded as vibrating within their nearest-neighbor confines somewhat as though they were in a crystal, but only for a relatively few vibrational periods. Fluctuations break the cage and establish a new one. This structural picture of a liquid becomes very important in dealings with dynamic processes—viscous flow, diffusion, and rates of chemical reaction. It will be invoked in Chapter 15 (on chemical kinetics).

Water

Water has been described rather frequently in this chapter as "unusual." It does not obey Trouton's rule well, having too high an enthalpy and entropy of vaporization. The liquid is denser than the solid and shows a maximum density at 4°C, so that between 0°C and 4°C the density increases with temperature. Its heat capacity is abnormally large and suggests that all 9 N_A degrees of freedom per mole are involved as weak vibrations. The surface tension of water is unusually high for so small a molecular species. The x-ray scattering data give radial distributions (as in Fig. 8-20) suggesting about 4.5 nearest neighbors, as compared to 4.0 in the ice structure.

Water is, after all, the most important chemical in our lives, both biologically and geologically. The problem of the structure of water has thus presented a scientific challenge not

FIGURE 8-20
Radial distribution functions. (a) Liquid potassium. [From C. D. Thomas and N. S. Gingrich, J. Chem. Phys. **6**, 412 (1938).] (b) Liquid mercury. [See C. N. J. Wagner, H. Ocken and M. L. Jashi, Z. Naturforsch, **20a**, 325 (1965).]

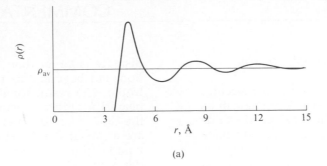

(a)

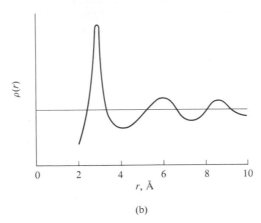

(b)

only of great intrinsic interest but also of great general significance. Ice has a very open structure, and the increase in density on melting was at first explained as just due to the breakdown of the ice into individual water molecules, with perhaps small hydrogen-bonded units present. This picture could not account for many of the properties, including the high critical temperature of water.

Next, a number of models were proposed which involved extensive structuring different from that of ice. L. Pauling, for example, suggested cages of hydrogen-bonded water molecules with an additional molecule in the center. Structures of this type have seemed too crystalline and rigid. Current theory is along the following lines. H. S. Frank and co-workers suggested that hydrogen bonding might be somewhat cooperative, that is, the formation of a second and a third hydrogen bond should be easier, based on dipole moment considerations, than that of the first bond. The effect would be that once a few molecules had hydrogen-bonded, a rather large cluster would

grow easily. A later fluctuation would cause the cluster break up. The idea is one of *"flickering clusters."* Dielectric studies on water show that molecules are able to respond to an alternating electric field up to very high frequencies, about 10^{10} s^{-1}. This means that any structured region must be very short-lived—about 10^{-10} s, or about 10^3 vibrations; hence the term "flickering clusters." Judging from the characteristic energy for the change in fluidity with temperature and similar energy values obtained from dielectric relaxation and from sound absorption, it would appear that a cluster lasts until fluctuation in the local energy density gives it 4–5 kcal of extra energy per molecule; it then breaks up while, on the average, another cluster is forming elsewhere in the liquid. A schematic illustration of this model is given in Fig. 8-21.

The cluster model was further developed by Nemethy and Scheraga (1962) in terms of a statistical thermodynamic calculation. Energies were assigned according to the average number of hydrogen bonds per oxygen, which

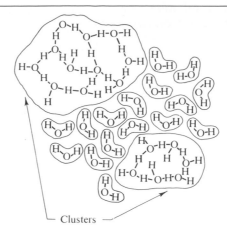

FIGURE 8-21
The "flickering cluster" model for liquid water. The structured and unstructured regions come and go quickly. [From A. Nemethy and H. A. Scheraga, J. Chem. Phys. **36**, 3382 (1962).]

Clusters

ranged from zero for uncoordinated molecules to a maximum of four. A product of partition functions was then set up, with appropriate expressions for vibration, rotation, and translation of each type of cluster. The result led to moderately good agreement with a number of properties of water, and this type of picture is generally accepted as correct in principle if not in detail. It leads to rather large clusters—ones with 50–100 water molecules are fairly probable. (See also Benson, 1978).

The thermodynamic and other properties of water are thus those of a highly structured medium. A related consequence is that water as a solvent behaves unusually. It appears, for example, that small ions order their immediately neighboring water molecules into a local structure which is not compatible with the general liquid structure, so that there is a nonstructured transition zone between the hydrated ion and the solvent. Nonpolar solutes such as hy-

drocarbons show a peculiar behavior. There is a small, negative ΔH of solution, strongly counterbalanced by a very large negative entropy change. An early idea, from H. S. Frank and M. J. Evans, was that such solutes induced a local structuring of the water—the picture was, in 1945, known as the "*iceberg*" model. A more recent, although similar, interpretation is that nonpolar molecules prefer the interstices between clusters and in this way promote structure and hence decrease the entropy of water.

Some mention should be made of "polywater." A storm of interest developed during the late 1960s in what appeared to be a new form of liquid water, formed by condensing the vapor into small capillaries. Eventually it was found that the new properties being observed were due to impurities. For an interesting account of the rise and fall of polywater or "anomalous" water, see Franks (1981).

SPECIAL TOPICS

8-ST-1 Intermolecular forces

It is noted in Section 8-3 that several distinct types of intermolecular and interatomic forces are known. Here we take up those that are loosely described as *van der Waals forces*, that is, interactions that are not due to actual chemical bond formation nor to simple electrostatic

forces as in an ionic crystal. The hydrogen bond is also excluded. What remains are the attractive forces that neutral, chemically saturated molecules experience.

These may be divided into the following categories: (a) dipole–dipole, (b) dipole–induced dipole and ion–induced dipole, and (c) induced dipole–induced dipole (dispersion).

It is best to start with Coulomb's law, according to which the force between two point charges is given by*

$$f = \frac{q_1 q_2}{r^2} \tag{8-48}$$

The potential energy of interaction, $\phi = \int f\, dx$, is given by

$$\phi = \frac{q_1 q_2}{r} \tag{8-49}$$

The potential V of a charge q is defined as

$$V = \frac{q}{r} \tag{8-50}$$

with the meaning that unit opposite charge q_0 will experience a potential energy $-qq_0/r$ at a separation r. The sign of V is negative if that of q is negative.

We next consider a molecule having a dipole moment μ (see Section 3-4), that is, one in which charges q^+ and q^- are separated by a distance l, giving a dipole moment $\mu = ql$. A dipole experiences no net interaction with a uniform potential since the two charges are affected equally and oppositely. If there is a gradient of the potential, or a field, defined as $\mathbf{F} = dV/dr$, then there is a net effect. As illustrated in Fig. 8-22(a), the potential energy of a dipole in a field is

$$\phi(\mu) = -Vq + \left(V - \frac{dV}{dr}l\right)q = -\mu \mathbf{F} \tag{8-51}$$

Conversely, a dipole produces a potential. Thus a test charge q_0 experiences the potential energy [Fig. 8-22(b)]

$$\phi(q_0) = -\frac{qq_0}{r} + \frac{qq_0}{r+l}$$

if l is small compared to r, then

$$\phi(q_0) = -\frac{qq_0}{r} + \frac{qq_0}{r}\left(1 - \frac{l}{r}\right)$$
$$= -\left(\frac{\mu}{r^2}\right)q_0 \tag{8-52}$$

*Note that we are using the traditional cgs-esu system. It is the natural one for dealing with electrostatic interactions; the alternative SI system is discussed in Section 3-CN-1.

or the potential of a dipole is

$$V(\mu) = \frac{\mu}{r^2} \tag{8-53}$$

and its field is

$$\mathbf{F}(\mu) = \frac{2\mu}{r^3} \tag{8-54}$$

(Potentials and fields are reported with a positive sign; the interaction energy of a charge or a dipole is minus if attractive and positive if repulsive.)

Two dipoles then interact each with the field of the other to give

$$\phi(\mu, \mu) = -\mu\frac{2\mu}{r^3} = -\frac{2\mu^2}{r^3} \tag{8-55}$$

This is the interaction for dipoles end-on, as shown in Fig. 8-22(c). In a liquid, thermal agitation tends to make the relative orientations random, while the interaction energy acts to favor alignment. The analysis is similar to that of Eq. (3-31) and leads to the result [due to W. H. Keesom in 1912]:

$$\phi(\mu, \mu)_{av} = -\frac{2\mu^4}{3kTr^6} \tag{8-56}$$

This *orientation* attraction thus varies inversely as the sixth power of the distance between the dipoles. Remember, however, that the derivation has assumed separations large compared to l.

A further type of interaction is that in which a field induces a dipole moment in a polarizable atom or molecule. We have from Eq. (3-19) that the induced moment is proportional to $\mathbf{F}$:

$$\mu_{ind} = \alpha \mathbf{F}$$

or, by Eq. (8-51),

$$\phi(\alpha, \mathbf{F}) = -(\mu_{ind})(\mathbf{F}) = -\tfrac{1}{2}\alpha \mathbf{F}^2 \tag{8-57}$$

(the factor $\tfrac{1}{2}$ enters because, strictly speaking, we must integrate $\int_0^{\mathbf{F}} \mu_{ind}\, d\mathbf{F}$). The induced dipole is instantaneous (as compared to molecular motions) and the potential energy between a dipole and a polarizable species is therefore independent of temperature:

$$\phi(\mu, \alpha) = -\frac{1}{2}\alpha\left(\frac{2\mu}{r^3}\right)^2$$
$$= -\frac{2\alpha\mu^2}{r^6} \tag{8-58}$$

FIGURE 8-22

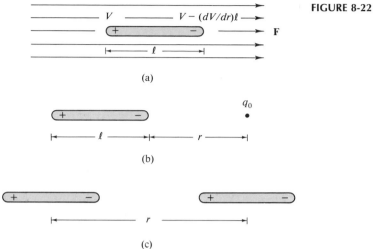

(a)

(b)

(c)

The interaction must be averaged over all orientations of the dipole; one then obtains

$$\phi(\mu, \alpha) = -\frac{\alpha\mu^2}{r^6} \tag{8-59}$$

This is known as the *induced* attraction, worked out by P. Debye in 1920.

Both types of attraction show an inverse-sixth-power dependence on distance, as is found experimentally, for example, in the a/V^2 term of the van der Waals equation. However, van der Waals interactions are not as sensitive to temperature as the orientation attraction calls for and also exist between molecules having no dipole moments, such as N_2 and other homopolar diatomic molecules, symmetric ones such as methane, and of course, rare-gas atoms. Yet another type of attraction must be present, and its source was discovered by F. London in 1930. The London effect is known as the dispersion attraction, and the derivation, while basically a quantum mechanical one, can be explained as follows.

The energy of an atom 1 in a field is, by Eq. (8-57),

$$\phi(\alpha_1, \mathbf{F}) = -\tfrac{1}{2}\alpha_1\mathbf{F}^2$$

The source of the field is attributed to the average dipole moment (root mean square average) for the oscillating electron–nucleus system of the second atom, or

$$\mathbf{F}_2 = \frac{2\overline{\mu}_2}{r^3}$$

Now the polarizability of an atom can be expressed as a sum over all excited states of the transition moment, *ex* (charge times displacement) squared, divided by the energy. If approximated by the largest term, then for atom 2 we have

$$\alpha_2 = \frac{\overline{(ex)^2}}{h\nu_0} \tag{8-60}$$

where $h\nu_0$ is the ionization energy. The quantity $\overline{(ex)^2}$ is now identified with μ_2^2, so we have

$$\phi(\alpha, \alpha) = -\frac{1}{2}\alpha_1\left(\frac{2\overline{\mu}_2}{r^3}\right)^2$$

$$= -\frac{2\alpha_1\alpha_2 h\nu_0}{r^6} = -\frac{2\alpha^2 h\nu_0}{r^6}$$

A more accurate derivation gives for the *dispersion* (or induced dipole–induced dipole) attraction

$$\phi(\alpha, \alpha) = -\frac{3\alpha^2 h\nu_0}{4r^6} \tag{8-61}$$

The situation to this point is summarized in Tables 8-6 and 8-7; in Table 8-7 expressions in boxes are for the three types of net attractive interactions between neutral molecules. They all vary with the inverse sixth power of the distance between molecules and all are approximate, particularly in assuming that the distance between molecules is large compared to the molecular size. Nonetheless, it is instructive to compare their relative magnitudes. The

TABLE 8-6 Electrical interactions	Produces:		Has an energy of interaction with:	
A molecule with:	Potential V	Field F	A potential V	A field of F
Charge q	$\lvert q/r \rvert$	$\lvert q/r^2 \rvert$	$\phi = qV$	—
Dipole μ	$\lvert \mu/r^2 \rvert$	$\lvert 2\mu/r^3 \rvert$	—	$\phi = -\mu\mathbf{F}$
Polarizability α	—	—	—	$\phi = -\tfrac{1}{2}\alpha\mathbf{F}^2$

calculated energies will be in ergs if dipole moments are given in units of esu centimeter, polarizabilities in cubic centimeter and ionization energy $h v_0$ in ergs. Table 8-8 summarizes some calculations. The dispersion attraction is large for all types of molecules, the induced attraction is relatively small for all types, and the orientation attraction can be dominant for very polar molecules. The numbers in the table are for ϕr^6 and actual energies can be obtained on insertion of a value for the intermolecular distance. For example, if we set $r = 4$ Å, then ϕ for argon becomes $48 \times 10^{-60}/4.1 \times 10^{-45}$ or 1.17×10^{-14} erg molecule^{-1} or 170 cal mol^{-1}. If the liquid is close-packed, each atom has 12 nearest neighbors and the energy of vaporization should be $12\phi/2$, or about 1000 cal mol^{-1}. The actual value is 1600 cal mol^{-1}. The calculation for water leads to about 5 kcal mol^{-1} for ΔE_v, and the much higher observed value of 10 kcal mol^{-1} is explained in terms of hydrogen bonding.

8-ST-2   The viscosity of liquids

Before we take up some further aspects of the viscosity of liquids, the derivation of the Poiseuille equation (2-79) is in order. The situa-

tion is illustrated in Fig. 8-23(a), which shows the flow lines for a liquid undergoing streamline flow down a circular tube. An inner cylinder of radius r has an area of $2\pi r$ per unit length of tube, or a total area of $2\pi rl$. It therefore experiences a frictional drag or force, according to the defining equation for viscosity, Eq. (2-78):

$$f = \eta(2\pi rl)\frac{dv}{dr} \tag{8-62}$$

This force is balanced by the pressure drop acting on the area πr^2, so we have

$$(\mathbf{P}_1 - \mathbf{P}_2)(\pi r^2) = \eta(2\pi rl)\frac{dv}{dr}$$

or

$$dv = \frac{\mathbf{P}_1 - \mathbf{P}_2}{2\eta l}r\,dr \tag{8-63}$$

Integration now gives the velocity profile. We take v to be zero at the wall, and so obtain

$$v_r = \frac{(\mathbf{P}_1 - \mathbf{P}_2)(r_0{}^2 - r^2)}{4\eta l} \tag{8-64}$$

where v_r is the velocity at radius r, and r_0 is the radius of the tube. The volume flow of an annulus of thickness dr is $v_r(2\pi r)\,dr$ and the total volume flow through the pipe is obtained by integration [Eq. (2-79)]

TABLE 8-7 Interaction energies ϕ	Ion q	Dipole μ	Polarizable molecule α
Ion q	$\lvert q^2/r \rvert$	$\lvert q\mu/r^2 \rvert$	$-\tfrac{1}{2}\alpha q^2/r^4$
Dipole μ	—	$\lvert 2\mu^2/r^3 \rvert$	$-2\alpha\mu^2/r^6$
		$\boxed{\phi_{av} = -2\mu^4/3kTr^6}$	$\boxed{\phi_{av} = -\alpha\mu^2/r^6}$
Polarizable molecule α	—	—	$\boxed{-\tfrac{3}{4}\alpha^2 h v_0/r^6}$

Molecule	$10^{18}\mu$ (esu cm)	$10^{24}\alpha$ (cm³)	$h\nu_0$ (eV)[b]	$10^{60}\phi r^6$ (erg cm⁶)		
				$\mu-\mu$[c]	$\mu-\alpha$	$\alpha-\alpha$
He	0	0.21	24.7	0	0	1.2
Ar	0	1.64	15.8	0	0	48
CO	0.12	1.99	14.3	0.0034	0.057	67.5
HCl	1.03	2.63	13.7	18.6	5.4	105
NH₃	1.5	2.21	16	84	10	93
H₂O	1.84	1.48	18	190	10	47

TABLE 8-8
Contributions to the interaction energies between neutral molecules[a]

[a]Adapted from J.O. Hirschfelder, C.F. Curtiss, and R.B. Bird, "Molecular Theory of Gases and Liquids," corrected ed., p. 988. Wiley, New York, 1964. See also N.J. Bridge and A.D. Buckingham, *Proc. Roy. Soc.*, A295 (1966), and T.M. Miller and B. Bederson, *Adv. At. Mol. Phys.*, **13**, 1 (1977).

[b]One electron-volt (eV) corresponds to 1.602×10^{-12} erg (or 23 kcal mol⁻¹).

[c]Calculated for 20°C.

$$\frac{V}{t} = \frac{\pi(P_1 - P_2)r_0^4}{8\eta l}$$

which is the desired equation.

We turn now to a different aspect. An important theory of liquid viscosity is due to H. Eyring. The liquid is viewed as being somewhat structured, as shown in Fig. 8-24, so that in order for a molecule to pass into a nearby hole, it must get through a bottleneck or energy barrier. It must, in other words, escape from the solvent cage spoken of earlier. Were it not for this barrier, the rate of jumping through would be about that corresponding to a vibration frequency—for a vibration of energy $h\nu_0$ equal to kT, the frequency ν_0 is kT/h, and we take this to be the natural jump frequency. The

presence of the barrier reduces the probability of the jump by the Boltzmann factor $e^{-\epsilon^*/kT}$, where ϵ^* is the energy required. We now have

$$\text{rate of jumping} = \frac{kT}{h}e^{-\epsilon^*/kT} \qquad (8\text{-}65)$$

In the absence of any stress, jumps occur equally frequently in either direction, but if a shear force is applied, then the effect is to make jumps easier one way than the other. If the force is f per unit area and the effective area of a molecule is σ, then the theory assumes that the energy barrier gains or loses an increment of energy equal to $f\sigma$ times the distance over which the force acts. This last is taken to be $\frac{1}{2}\lambda$, where λ is the jump distance. Thus we have

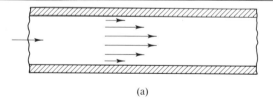

(a)

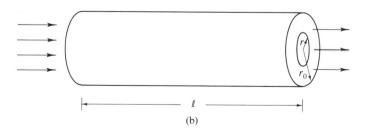

(b)

FIGURE 8-23
Streamline flow down a cylindrical tube:
(a) velocity profile,
(b) derivation of Eqs. (8-64) and (2-79).

FIGURE 8-24
Theoretical model for
viscous flow. (From J. O.
Hirschfelder, C. F. Curtiss,
and R. B. Bird, "Molecular
Theory of Gases and
Liquids," corrected ed.
Copyright 1964, Wiley,
New York. Used by
permission of John Wiley
& Sons, Inc.)

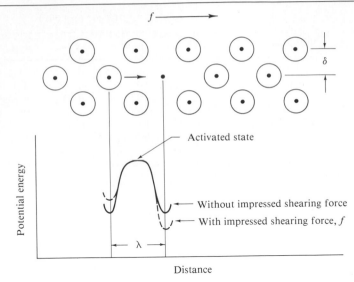

rate of forward jumps: $\dfrac{kT}{h}e^{-(\epsilon^* - \frac{1}{2}f\sigma\lambda)/kT}$

rate of backward jumps: $\dfrac{kT}{h}e^{-(\epsilon^* + \frac{1}{2}f\sigma\lambda)/kT}$

The net rate of jumping is the difference between the forward and backward rates:

net jump rate: $\dfrac{kT}{h}e^{-\epsilon^*/kT}[e^{f\sigma\lambda/2kT} - e^{-f\sigma\lambda/2kT}]$

The quantity $f\sigma\lambda/2kT$ is usually small compared to unity, and so the exponentials are customarily expanded and only the first terms taken:

net jump rate: $\dfrac{kT}{h}e^{-\epsilon^*/kT}\dfrac{f\sigma\lambda}{kT}$

The net flow velocity in the layer is the jump rate times λ, so we have

$$v = \dfrac{kT}{h}e^{-\epsilon^*/kT}\dfrac{f\sigma\lambda}{kT}\lambda$$

Since f is the relative force between adjacent layers, v is therefore the relative layer velocity. It follows that the velocity gradient dv/dx is just v/δ. From the definition of viscosity, Eq. (2-78), we have

$$\eta = \dfrac{\text{force/unit area}}{dv/dx}$$

so

$$\eta = \dfrac{f/\sigma}{v/\delta} = \dfrac{\delta}{\sigma^2\lambda^2}he^{\epsilon^*/kT} \tag{8-66}$$

or

$$\eta \simeq \mathbf{n}he^{\delta^*/kT} \tag{8-67}$$

where $\mathbf{n}$ denotes the concentration in molecules per cubic centimeter.

Equation (8-67) is now in the same form as Eq. (8-47), where A of the latter equation is equal to $1/\mathbf{n}h$ according to the Eyring theory. The preceding discussion is not complete in that the theory adds, in effect, that the jump frequency may be $(kT/h)A'e^{-(\epsilon^*/kT)}$, where A' is an added factor. Equation (8-67) does not fare too badly, however. As an example, $\epsilon^* = 2600$ cal mol^{-1} for CCl_4 (from Table 8-5) and $\mathbf{n} = 6.02 \times 10^{23}/97$. Insertion of these numbers into the equation gives $\eta = 0.51$ (centipoise) at 0°C, as compared to the observed value of 1.329.

An interesting point is that the energy for the jump process looks rather similar to that required for vaporization. One finds in fact that the characteristic energy for viscosity tends to be about one-third the energy of vaporization. The values are also not very different from the surface energies per mole. There are a number of interrelations among the various properties of liquids whose general explanation can be given now but whose rigorous treatment is yet to come.

ADAMSON, A. W. (1982). "The Physical Chemistry of Surfaces," 4th ed. Wiley (Interscience), New York.

HILDEBRAND, J. H., AND SCOTT, R. L. (1950). "The Solubility of Non-Electrolytes," 3rd ed. Van Nostrand-Reinhold, Princeton, New Jersey.

ADAMSON, A. W. (1982). "The Physical Chemistry of Surfaces," 4th ed. Wiley (Interscience), New York.

ADAMSON, A. W. AND MANES, M. (1984). *J. Chem. Ed.*, **61**, 590.

BENSON, S. W. (1978). *J. Amer. Chem. Soc.* **100,** 5640.

BRIDGMAN, P. W. (1949)." The Physics of High Pressures. Bell, London.

BUNDY, F. P., STRONG, H. M., and WENTORF, R. H. JR. (1973). *Chemistry and Physics of Carbon* **10**, 213.

FRANKS, F. (1981). "Polywater." The MIT Press, Cambridge, Mass.

NEMETHY, G., AND SCHERAGA, H. A. (1962). *J. Chem. Phys.* **36**, 3382.

ROSSINI, E. A., et al. (1959). Tables of Selected Values of Chemical Thermodynamic Quantities. Nat. Bur. Std. Circ. No. 500. Also, Evans W. H., Parker, V. B., Halow, I., Bailey, M., and Schumm, R. H., eds., Nat. Bur. Std. Tech. Note 270-3, 1968.

EXERCISES

The normal boiling point of methanol is 64.7°C and its heat of vaporization is 262.8 cal g^{-1}. Calculate its vapor pressure at 25°C.

8-1

Ans. 0.188 atm.

Calculate the boiling point of methanol under 2 atm pressure; use data from Exercise 8-1.

8-2

Ans. 84.5°C.

The vapor pressure of liquid nitrogen is 1 Torr at -226.1°C and 40 Torr at -214.0°C. Calculate the heat of vaporization of nitrogen and its normal boiling point.

8-3

Ans. 1686 cal mol^{-1}, 74.4 K (the actual boiling point is 77.3 K; why is the calculated answer so far off?).

Calculate the weight loss due to evaporation that should occur if ten liters of dry air are bubbled through liquid methanol at 25°C. The vapor pressure of methanol is 143 Torr at this temperature. The air volume is measured at 1 atm and 25°C, and atmospheric pressure is 750 Torr.

8-4

Ans. 3.09 g.

Estimate the vapor pressure at 25°C of a liquid which obeys Trouton's rule and whose normal boiling point is 64.7°C.

8-5

*Ans.*0.245 atm.

Calculate the vapor pressure at infinite temperature of a liquid which obeys Trouton's rule and whose heat of vaporization does not change with temperature.

8-6

Ans. 3.89 × 10^4 atm.

Calculate the freezing point of acetic acid under 1000 atm pressure. Use Table 8-1; also, the densities of liquid and solid acetic acid may be taken to be 1.049 and 1.10 g cm^{-3}, respectively.

8-7

Ans. 6.65°C.

Calculate ΔG for the process

8-8

Benzene (*l*, 80.1°C, 1 atm) = Benzene (*g*, 80.1°C, 0.1 atm)

The normal boiling point is 80.1°C.

Ans. -1620 cal mol^{-1}.

8-9   The melting point of a substance increases by 0.1500°C per 1500 atm applied pressure. The liquid and solid densities are 1.550 and 1.650 g cm^{-3}, respectively. Calculate the entropy of fusion.

Ans. 9.47 cal K^{-1} g^{-1}.

8-10   Liquid ethanol has a vapor pressure of 400 mm Hg at 63.5°C and its density is 0.789 g cm^{-3}. Calculate the free energy change and the increase in vapor pressure if the liquid is placed under 750 atm mechanical pressure.

Ans. 1060 cal mol^{-1}, 1550 Torr.

8-11   Calculate ΔP for a soap bubble of 0.5 cm radius. The soap solution used has a surface tension of 4 dyne cm^{-1}.

Ans. 0.0240 Torr.

8-12   Obtain the value for the capillary constant a for benzene at 20°C and the height of capillary rise for benzene in a 0.1-mm diameter capillary.

Ans. $a = 0.259$ cm, $h = 13.4$ cm.

8-13   Find the ideal and the actually observed drop weight in the case of ethanol at 20°C and a dropping tip of 2 mm diameter. The density of ethanol is 0.789 g cm^{-3}.

Ans. 14.36 mg, 9.93 mg.

8-14   What pull should be required to detach a 3-cm wide Wilhelmy slide from the surface of benzene?

Ans. 173 dyn or 177 mg.

8-15   Calculate E^s for liquid sodium at 97°C.

Ans. 235 erg cm^{-2}.

8-16   Calculate the percentage increase over the normal value of the vapor pressure for an aqueous fog at 20°C consisting of 0.5 μm droplets.

Ans. 0.21%.

8-17   How many m^3 per day of mercury should flow though a 1×10^{-4} m diameter, 0.15 m long tube at 40°C if the pressure drop is 5×10^5 Pa?

Ans. 4.77×10^{-4} m^3 per day.

PROBLEMS

8-18   Show that, according to Trouton's rule, all liquids should have the same hypothetical vapor pressure at infinite temperature. Assume that heats of vaporization do not vary with T.

8-19   Derive Eq. (8-8) from Eq. (8-6), assuming that the vapor obeys the equation of state $P(V_g - V_l) = RT$.

8-20   It is sometimes necessary to correct a boiling point to some ambient barometric pressure that is not exactly 760 mm Hg. Show that the fractional change in boiling point, $\Delta T/T_b$, is just 0.095 times the fractional deviation of pressure from 760 mm Hg. Assume Trouton's rule.

8-21   When 15 liters of nitrogen measured at 25°C and 760 mm Hg were bubbled through a saturator containing *tert*-butyl alcohol at 39.8°C, 6.782 g of the alcohol were evaporated. Calculate its vapor pressure. (The evaporator is thermostatted at 39.8°C.)

8-22   Ten liters of dry nitrogen at 25°C and 750 mm Hg pressure are bubbled through neat chlorobenzene, thereby becoming saturated with its vapor. The mixed gases are collected over water at 25°C, and found to have a volume of 10.587 liters at a pressure of 740 mm Hg. Calculate the vapor pressure of chlorobenzene at 25°C.

The vapor pressure of carbonyl selenide (COSe) is:

t, °C	-22.9	-31.3	-36.7	-44.3	-52.6
P, mm Hg	720.9	498.7	387.5	267.7	172.7

Calculate the ΔH_v and the value of the Trouton constant for this substance.

8-23

The vapor pressure of antimony is given here for various temperatures. Make a semilogarithmic plot of P versus $1/T$ and calculate the heat of vaporization, the normal boiling point, and the value for the Trouton constant.

P, Torr	50	100	200	300	400	500	600
t, °C	1070	1137	1190	1225	1252	1280	1300

8-24

Determine the best least squares straight line for the plot of Problem 8-24 and the standard deviation in the calculated heat of vaporization.

8-25*

Equation (8-9) was obtained with the assumption that the heat of vaporization does not change with temperature. Derive the corresponding equation assuming that there is a constant ΔC_P for the vaporization process. Would a plot of $\ln P$ versus $1/T$ still be linear? If not, would the slope at any point give the correct ΔH_v for that temperature?

8-26

The normal boiling point of toluene is 110.5°C and its vapor pressure is 1.333×10^4 Pa at 51.9°C. Calculate ΔH_v.

8-27

The normal boiling point of diethyl ether is 34.6°C and ΔH_v is 6219 cal mol^{-1}. The liquid density at this temperature is 0.7135 g cm^{-3}. (a) Calculate the value of dP/dT by exact and by approximate methods. (b) If diethyl ether is to be distilled at 20°C, to what value must the pressure be reduced?

8-28

The coefficient of compressibility of liquid acetic acid is 9.08×10^{-11} cm^2 dyn^{-1} and its coefficient of thermal expansion is 1.06×10^{-3} K^{-1}, both at 20°C. Calculate the internal pressure of acetic acid, and the difference $C_P - C_V$. Compare your value of P_{int} with that obtained from the van der Waals constant a for acetic acid [note Eq. (4-37)].

8-29

A compound of molecular weight 74 g mol^{-1} has a liquid density of 0.78 g cm^{-3}. Its melting point is 15°C and the enthalpy of fusion is 1250 cal mol^{-1}. If the melting point increases by 0.01°C per atm of applied pressure, what is the density of the solid?

8-30

The vapor pressure of solid CO_2 is 3073 and 1486 mm Hg at -60°C and -70°C, respectively, and that of the liquid is 7545 and 5128 mm Hg at -40°C and -50°C, respectively. (a) Find the triple point temperature and pressure and (b) calculate the heat of fusion of CO_2 using the above data.

8-31

Eighteen grams of water supercooled to -10°C are placed in a bomb, completely filling it. The bomb is placed in a thermostat at -10°C. Eventually ice forms and the pressure increases until equilibrium is established. Calculate the number of moles of ice present at equilibrium and the final pressure if the heat of fusion of water is 80 cal g^{-1} (independent of T) and the compressibility of liquid water is $\beta = -(1/V)(\partial V/\partial P)_T = 5 \times 10^{-5}$ atm^{-1} (independent of T). Assume ice to be incompressible; molar volumes are water, 18 cm^3 and ice, 20 cm^3 (independent of T).

8-32

Naphthalene melts at 80.2°C with a heat of fusion of 35.6 cal g^{-1}, and the heat of vaporization is 75.5 cal g^{-1} at the normal boiling point of 217.9°C. The density of the solid is 1.15 g cm^{-3} and that of the liquid may be taken to be 10% less. Construct the phase diagram for naphthalene along the lines of Fig. 8-9; calculate the vapor pressure at the triple point.

8-33

Carbon tetrabromide forms three solid phases. Phase II changes to I at 50°C and 1 atm; I melts at 92°C with an increase in volume; the liquid boils at 190°C. The triple point for I, II, and III is at 115°C and 1000 atm, and there are two phases at 2000 atm and 135°C and at 2000 atm and 200°C.

(a) Draw the phase diagram and letter the phase regions.
(b) Draw a curve showing how pressure changes with volume at 120°C for a pressure increase from 1 to 2000 atm.

8-34

8-35   Construct a semiquantitative phase diagram for CO_2. Use data from Problem 8-31 and Table 1-4; also, solid CO_2 is more dense than the liquid. Why does one not ordinarily see liquid CO_2?

8-36   Calculate the vapor pressure of water at 25°C if the water is present in a tank of compressed nitrogen gas at 2000 lb in^{-2} pressure (state your assumptions).

8-37   Use data from Table 7-2 plus the information that the densities of diamond and of graphite are 3.51 and 2.25 g cm^{-3}, respectively, to calculate the pressure at which one might expect the two forms to be in equilibrium.

8-38   Derive the equation for the height of capillary rise between two parallel glass plates d centimeters apart. Assume the liquid wets the glass and that end effects may be neglected, that is, the plates are very long, and only the rise in the middle is being considered. (See accompanying diagram.)

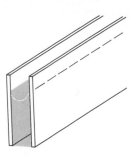

8-39   The surface tension of a liquid is determined by the drop weight method. Using a tip whose outside diameter is 6×10^{-4} m and whose inside diameter is 2×10^{-5} m, it is found that the weight of 20 drops is 8×10^{-4} kg. The density of the liquid is 950 kg m^{-3}, and it wets the tip. Using the appropriate correction factor, calculate the surface tension of the liquid.

8-40   For the case of zero contact angle, a more accurate capillary rise equation than that given is one which takes into account the weight of the meniscus: $\gamma = \frac{1}{2}\rho\, gr(h + \frac{1}{3}\, r)$. Derive this equation. A particular liquid (which wets the capillary) shows a capillary rise of 0.627 cm in a capillary of 1 mm radius. Calculate the surface tension of the liquid using both the simple equation and the above one.

8-41   Derive from Eq. (8-44) expressions for S^s and E^s. Assuming that $n = 11/9$, calculate γ_0 for water and S^s and E^s at 20°C.

8-42   The surface tension of Hg is 471 dyne cm^{-1} at 24.5°C. In a series of measurements the following drop weight data were obtained (diameter of tip in cm, weight of drop in g): (0.04293, 0.05443), (0.07651, 0.09093), (0.10872, 0.12331). Calculate the corresponding f and $r/V^{1/3}$ values to fill out Table 8-3.

8-43   The contact angle of water against Teflon (polytetrafluoroethylene) is 107° at 20°C. Calculate the value of h in the capillary rise equation for the case of a 0.2 mm diameter Teflon capillary tube dipping into water at 20°C.

8-44   Referring to Fig. 8-16, assume that the holes in the fabric are circular, of diameter 0.3 mm, and that water makes a contact angle of 115°C with the material of the fabric. How deep a layer of water will the fabric hold before it runs through (as shown in Fig. 8-16(b))?

8-45   Calculate the vapor pressure of water when present in a capillary of 1 μm radius (zero contact angle). Express your result as percent change from the normal value at 25°C.

8-46   Derive the equation: $\left(\dfrac{\partial\gamma}{\partial P}\right)_{\mathscr{A},T} = \left(\dfrac{\partial V}{\partial \mathscr{A}}\right)_{P,T}$.

8-47   Water at 20°C and under 100 atm pressure flows through a capillary tube of 2 μm radius at the rate of 0.0226 cm^3 hr^{-1}. A certain aqueous polymer solution has a viscosity of 2.00 cP at 20°C and its flow rate under the same conditions is 0.01120 cm^3 hr^{-1}. The effective radius of the tube is now less because of an adsorbed layer of polymer. Calculate the thickness of the polymer layer.

SPECIAL TOPICS PROBLEMS

Calculate (a) the interaction energy between an argon atom and a point electronic charge 5 Å away and (b) the interaction energy between two argon atoms separated by 5 Å. Explain in each case whether the interaction is attractive or repulsive. (Remember Table 3-1.) **8-48**

Make the calculations needed to add a line to Table 8-8 for HI. The ionization potential for HI is 12.8 eV. **8-49**

Consider two HCl molecules oriented so that their bond axes are collinear: H—Cl · · · Cl—H. Calculate the dipole-dipole interaction energy (including sign) when the centers of mass are (a) 5 Å apart and (b) 3 Å apart. Make the calculation using Eq. (8-55) and also the more accurate calculation in which the assumption used in obtaining Eq. (8-52) is *not* made. The bond length of HCl is 1.275 Å; assume the molecules actually to be $H^{35}Cl$. Report your calculated energies relative to kT at 25°C. **8-50**

Derive the equation corresponding to Eq. (8-64) but for the case of flow between parallel plates separated by distance r and of length l. The flow, V, is now per centimeter width. **8-51**

Look up the temperature dependencies and calculate E^* or ϵ^* for water and for hexane. Calculate the ratio of the theoretical to the experimental viscosity at 25°C, where the theoretical value is from Eq. (8-67). **8-52**

chapter 9
Solutions of nonelectrolytes

9-1   Introduction

We introduce, with this chapter, the physical chemistry of systems for which composition is a state variable. A solution is a mixture at the molecular level of two or more chemical species; it may be gaseous, liquid, or solid. If clusters of molecules are present, the situation becomes more complex. If the clusters are of the order of 100 Å to a few thousand Å or around 10^{-4} cm in size, the system is colloidal in nature. If they reach to 10^{-4} to 10^{-3} cm, we speak of the mixture as a suspension, an emulsion, or an aerosol. Beyond this, we simply have a mechanical mixture of two or more bulk phases.

There are no sharp natural boundaries in this sequence, but there are practical ones. Most physical chemists have concentrated on the extremes, that is, on molecular solutions or on systems having well-defined bulk phases which themselves may be solutions. On the other hand, much of the biological and physical world involves mixtures of the colloidal or intermediate type of dispersivity. The physical chemistry of these last systems is difficult, and its introduction is reserved for Chapter 21. We confine ourselves here to the simpler case of molecular solutions.

A solution or mixture of gases presents little problem, at least at the level of complexity of this text. Unless very dense, gases are always fully miscible and, in the usual laboratory pressure range of around 1 atm, form essentially ideal solutions. Dalton's law of partial pressures [Eq. (1–18)] is well obeyed. We will consider the entropy and free energy of formation of gaseous mixtures in Section 9-4.

Solid solutions, that is, molecular dispersions of two or more species in a solid phase, are quite common. Alloys are one example; also, many ionic crystals are able to substitute one type of ion for another (of the same charge) almost randomly within their crystal lattices. Solid solutions are more difficult to study experimentally, however, and are less studied than liquid ones. Their behavior is also more subject to eccentricities. Most of our data are from, and most of our common experience is with, liquid solutions. For these reasons, the material that follows refers mainly to liquid solutions. It should be remembered, however, that the formal thermodynamics is the same for both types of solutions.

It is now desirable to define the term *composition* more precisely. A phase may consist of a number of molecular species and yet still qualify thermo-dynamically as a pure substance. Liquid water, for example, contains not only a large assortment of transient clusters (see Section 8-CN-1), but also definite concentrations of hydrogen and hydroxide ions. These are all in equilibrium with each other, however, and their relative proportions are not subject to arbitrary change. Composition with respect to these species is not an inde-pendent variable; once we fix the temperature and pressure of a sample of water, we automatically also fix the various equilibrium constants and hence compositions. It is therefore not necessarily the number of species present that serves to characterize a solution. Nor is it necessarily the number of *constituents*. By constituents, we mean chemical species that we can physically measure out in making up a solution. The term *formal* composition denotes the composition calculated in terms of what is weighed out and mixed. We may, for example, prepare a mixture of hydrogen, nitrogen, and ammonia. In the presence of a catalyst, these would be in equilibrium, and it would be sufficient to specify the formal composition with respect to only two of the three species. In the absence of the catalyst, however, all three compositions are independently variable, and the formal composition with respect to all three would have to be specified.

We meet this type of complication by using the term *component*. The number of components of a solution (or of any mixture) is the least number of in-dependently variable chemical species required to define the composition of the solution (and of all phases present, if there is more than one). Hydrogen plus nitrogen plus ammonia plus catalyst is a two-component system; without the catalyst, it becomes a three-component system. Ordinary water is a one-component system; water plus solute is a two-component system, and so forth.

As to compositions themselves, various measures are used. A common one for this chapter will be the mole fraction, denoted by x_i for a liquid solution and, for clarity, by y_i for a gaseous solution. Mole fractions are strictly additive. That is, the total number of moles of a solution is just the sum of the numbers of moles of each species, and the mole fraction composition of a solution formed when two others are mixed is strictly obtainable on this basis. Volume fractions ϕ_i are sometimes used, but the volume of a solution is not in general equal to the sum of the volumes of the constituents mixed, and care must be taken in the definition of the exact experimental basis for a volume fraction.

Other measures of composition that are commonly used are the following. *Molality*, m, denotes a kind of mole fraction that is well suited to aqueous solutions. It means the number of moles of the solute component per 1000 g of the major or solvent component. Alternatively, it is convenient both for the laboratory chemist and in certain theoretical treatments to use concentra-tion in terms of volume of solution. The term *molarity*, M, will be used to denote moles per liter of solution (not of solvent alone), and n to denote molecules per cm^3 of solution. *Formality*, f, is useful from the analytical point of view; f is the number of formula weights of a substance used in making up a liter of solution.

Most of this chapter deals with the vapor pressure of solutions and the related matter of boiling point diagrams. We discuss both ideal and nonideal solutions, and how to treat them thermodynamically. The approach is from the phenomenological or classical point of view, although a bit of statistical thermodynamics is described in Section 9-CN-2. Some additional properties of solutions are covered in Section 9-CN-1, and the surface tension of solutions is developed in Section 9-ST-2.

9-2   The vapor pressure of solutions. Raoult's and Henry's laws

The vapor pressure and vapor composition in equilibrium with a solution of volatile substances constitute important and very useful information. The practical usefulness lies in the application of distillation processes, and the physical chemical importance, in the provision of a means of studying the thermodynamic properties of liquids and of testing models for the structure of liquids. One function of this section is therefore to introduce characteristic data and behavior as a preliminary to the thermodynamic treatments.

Vapor pressure data are customarily displayed in the form of vapor pressure–composition diagrams for some particular temperature, as illustrated schematically in Fig. 9-1. Here P_{tot} is the total combined vapor pressure above varying compositions of a solution of liquids A and B. The liquids are, in this example, taken to be fairly similar, and the P_{tot} curve, although not linear, decreases steadily from $P_A°$, the vapor pressure of pure A, to $P_B°$, that of pure B, as x_B is varied from 0 to 1. The vapor is a mixture of gaseous A and B, and the variations of the partial pressures P_A and P_B are also shown. We assume that no other gases are present and that the vapors are ideal, so that

A. Vapor pressure diagrams

$$P_{tot} = P_A + P_B \tag{9-1}$$

If A and B are very similar, the limiting case being one of two substances differing only in isotopic content, then the vapor pressure diagram takes on an especially simple appearance. A good example is provided by the benzene-toluene system, shown in Fig. 9-2. The values of P_{tot}, P_b, and P_t are now given by nearly straight lines. Thus we have

$$P_t = P_t°x_t \qquad P_b = P_b°(1 - x_t) = P_b°x_b \tag{9-2}$$

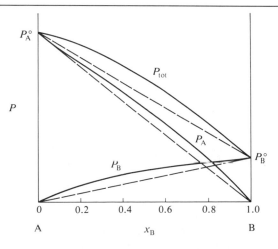

FIGURE 9-1
Vapor pressure—composition diagram. The dashed lines show the ideal behavior.

FIGURE 9-2
The benzene–toluene system at 20°C: (a) vapor pressure–liquid composition diagram; (b) liquid and vapor composition diagram.

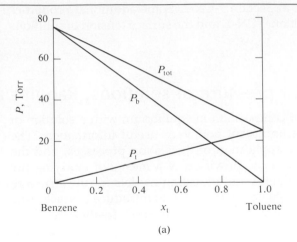

(a)

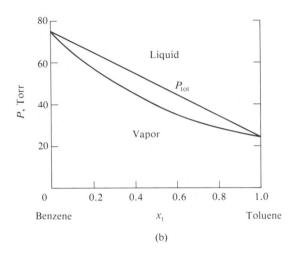

(b)

where x_t and x_b denote the mole fractions of toluene and of benzene, respectively, and the degree superscript indicates a pure phase. Then P_{tot} is simply

$$P_{tot} = P_t + P_b = P_b^° + (P_t^° - P_b^°)\, x_t \qquad (9\text{-}3)$$

which is the equation of the straight line connecting $P_b^°$ with $P_t^°$. A solution with this behavior is called an ideal solution, and the general form corresponding to Eqs. (9-2) is called *Raoult's law* (1884):

$$P_i = P_i^° x_i \qquad (9\text{-}4)$$

For simplicity, we will largely restrict the discussion to two-component systems, for which the Raoult's law statements are

$$P_1 = P_1^° x_1 \qquad P_2 = P_2^° x_2 \qquad (9\text{-}5)$$

with the corollary that

$$P_{tot} = P_1^° + (P_2^° - P_1^°)\, x_2 \qquad (9\text{-}6)$$

Raoult's law corresponds to a particularly simple picture of a solution—essentially one in which the components are distinguishable, but just barely, so that their physical properties are virtually identical. The situation is rather analogous to that of the ideal gas; the ideal gas law also corresponds to a particular, very simple picture. The ideal gas law is, moreover, the limiting law for all real gases, and, in this respect, is not hypothetical or approximate at all. The same is believed to be true for Raoult's law. Experimental evidence suggests that Raoult's law is the limiting law for all solutions, approached by each component as its mole fraction approaches unity. That is, as a limiting law statement, Eq. (9-4) reads

$$\lim_{x_i \to 1} P_i = P_i^\circ x_i \tag{9-7}$$

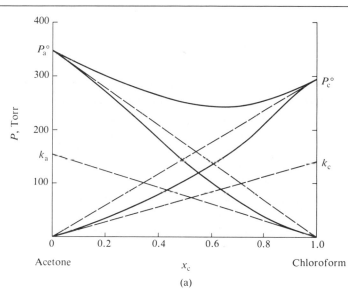

(a)

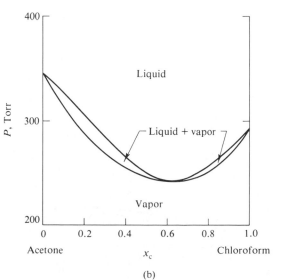

(b)

FIGURE 9-3
The acetone–chloroform system at 35°C, showing negative deviation from ideality; (a) vapor pressure–liquid composition diagram; (b) liquid and vapor composition diagram.

Notice that the curves for P_A and P_B are drawn in Fig. 9-1 so that they approach the Raoult's law line as x_A and x_B approach unity. The acetone–chloroform system shown in Fig. 9-3 provides a specific illustration. In this case, the partial pressure curves lie below the Raoult's law lines, whereas in Fig. 9-1, they lie above the Raoult's law lines. We speak of the first situation as one of negative deviation, and the second, as one of positive deviation (from ideality).

Raoult's law as an *ideal* law is easy to understand theoretically. It is the expected behavior if there is complete uniformity of intermolecular forces, just as the ideal gas law is the expected behavior in the complete absence of intermolecular forces. Raoult's law as a *limiting* law, Eq. (9-7), is essentially a statement—an important one—of common observation.

Acceptance of Raoult's limiting law provides a basis for the understanding of a second limiting law, *Henry's law*. Henry's law states that the partial pressure of a component becomes proportional to its mole fraction in the limit of zero concentration:

$$\lim_{x_i \to 0} P_i = k_i x_i \tag{9-8}$$

This is illustrated in Fig. 9-4, in which the vapor pressure curves of Fig. 9-1 are shown approaching the limiting slopes k_A and k_B; these slopes define straight lines whose intercepts are k_A and k_B. Under the limiting Henry's law condition each molecule of component A has become surrounded by B molecules. The environment is thus not one of pure A, but another environment determined by the nature of the A–B interactions. We can regard k_A as the hypothetical vapor pressure that pure A would exert if the molecules all had this different environment, and similarly, k_B as the hypothetical vapor pressure that pure B would exert in an environment consisting of A molecules.

Like Raoult's law, Henry's law applies as a limiting law to systems of both positive and negative deviation from ideality. The partial pressure curves in

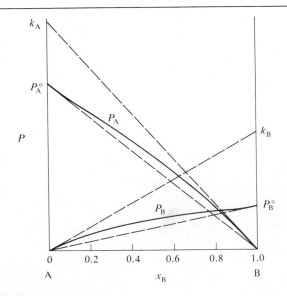

FIGURE 9-4
System showing positive deviation from ideality; illustration of Henry's and Raoult's laws.

Figs. 9-3 and 9-4 obey both limiting laws. There is a thermodynamic requirement (see Section 9-5B) that if a system shows positive (negative) deviation from ideality in the Raoult's law region, it must also show positive (negative) deviation in the Henry's law region. This condition is also illustrated in Fig. 9-3.

Henry's law is approximately valid for any solute in a dilute solution, and a particular application is to the solubility of gases in liquids. As an approximate law, Eq. (9-8) becomes

B. Solubility of gases

$$P_2 = k_2 x_2 \tag{9-9}$$

where species 2 in a two-component system will, by convention, be taken to refer to the solute. Values of k_2 are around 10^4 to 10^5 atm for common inert gases (such as N_2, O_2, CO, and CH_4) in water.

Water saturated with air at 25°C is 2.61×10^{-4} M in oxygen, and x_{O_2} is therefore $(2.61 \times 10^{-4})/(55.55 + 2.61 \times 10^{-4}) = 4.70 \times 10^{-6}$. Since air is only 20% oxygen, P_{O_2} is 0.2 atm, and substitution into Eq. (9-9) gives $k_{O_2} = (0.2)/(4.70 \times 10^{-6}) = 0.246 \times 10^5$ atm.

EXAMPLE

Some of the literature report Henry's law constants for gases in terms of the volume of gas dissolved, measured at the temperature and pressure in question, per unit volume of solvent. The Henry's law constant for oxygen becomes, on this basis, $(1.30 \times 10^{-3})(0.082)(298)/1.0$ or 0.032 liter of O_2 per liter of water.

A vapor pressure diagram also contains the information to give the composition of the vapor in equilibrium with a given composition of solution. Thus, for the system of Fig. 9-1, we have

C. Vapor composition diagrams

$$y_A = \frac{P_A}{P_{tot}} = \frac{P_A}{P_A + P_B} \qquad y_B = \frac{P_B}{P_{tot}} = \frac{P_B}{P_A + P_B} \tag{9-10}$$

where y_A and y_B denote the mole fractions of A and B in the vapor, respectively. A conventional way of supplying this information is to plot the vapor composition corresponding to each value of P_{tot}, along with P_{tot} versus liquid composition, as shown in Fig. 9-5. For example, for a liquid of composition x_1, P_{tot} has the value P_1, and the solution is in equilibrium with vapor of composition y_1. The corresponding vapor composition plots are included in Figs. 9-2(b) and 9-3(b).

Vapor composition diagrams are in effect phase maps or phase diagrams. If the system is contained in a piston and cylinder arrangement thermostated to the given temperature, then from Fig. 9-5, liquid of composition x_1 cannot vaporize if the pressure is greater than P_1; the system will consist of liquid phase only. The same will be true for any composition and pressure defining a point lying above the liquid line. The upper region of the diagram is one of liquid phase only. Similarly, a system of composition x_1 at a pressure less than P_3 will consist of vapor phase only. The lower region of the diagram must be one of vapor phase only. The liquid and vapor composition lines thus mark the boundaries of the liquid-only and vapor-only regions, respec-

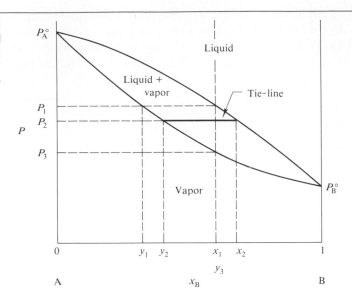

FIGURE 9-5
Use of vapor pressure and vapor composition diagrams—the lever principle.

tively. Finally, a system whose overall composition and pressure locate a point between the two lines consist of a mixture of phases.

A diagram such as that of Fig. 9-5 allows a complete tracing of the sequence of events as the pressure of a system is changed at contant temperature. Suppose, for example, that a system of composition x_1 is initially under some high pressure. As the pressure is reduced vaporization will begin at P_1, producing vapor of composition y_1. With further decrease in pressure more and more vaporization must occur, and since the vapor is richer in A than is the liquid, the latter must move to the right in composition. When the pressure has reached P_2, liquid of composition x_2 is now in equilibrium with vapor of composition y_2. Finally, when the pressure has been reduced to P_3 all the liquid will be vaporized and further reduction in pressure will merely expand the mixed vapors.

Since the entire system is a closed one, the vapor and liquid phases are always of some uniform relative composition, and their relative amounts can be calculated by material balance. For example, if the system consists of n moles total, then at any point

$$n = n_v + n_l \tag{9-11}$$

where n_l and n_v are the number of moles of liquid phase and of vapor phase, respectively. Thus for a system of overall composition x_1 and at pressure P_2 the material balance in B is

$$nx_1 = n_v y_2 + n_l x_2 = n_v y_2 + (n - n_v)x_2 \tag{9-12}$$

Equation (9-12) rearranges to give

$$\frac{n_v}{n} = \frac{x_2 - x_1}{x_2 - y_2} \tag{9-13}$$

Equation (9-13) can be given a very simple and useful graphical interpretation. The horizontal line connecting the points y_2 and x_2 is known as a *tie-*

line. In general a tie-line connects the compositions of equilibrium phases. The difference $x_2 - y_2$ corresponds to the length of the tie-line at P_2 and the difference $x_2 - x_1$, the length of the right-hand section of the line. Alternatively, if the tie-line is regarded as a balance pivoted at the point x_1, then weights proportional to n_v and n_l will just balance if placed at the y_2 and x_2 ends, respectively. Equation (9-13) with its associated graphical interpretation is known as the *lever principle.*

The acetone–chloroform system of Fig. 9-3 shows a minimum in P_{tot}. The physical interpretation is along the lines of Fig. 8-4, where for a pure liquid the energy of vaporization was attributed to $n\phi/2$, where n is the number of nearest neighbors and ϕ is the interaction energy. A negative deviation then suggests that ϕ_{AB} is greater than ϕ_{AA} or ϕ_{BB}, so that the ease of vaporization is reduced if A and B molecules mutually surround each other.

> **D.** Maximum and minimum vapor pressure diagrams

Such an increase in interaction energy between unlike molecules would, as an extreme, lead to the formation of an actual compound. In the case of Fig. 9-3 the appearance is more than of a tendency toward association.

The extreme case, in terms of this picture, would be one in which a very stable AB compound actually formed, as illustrated in Fig. 9-6. Systems in which the overall composition x_B is less than 0.5 consist of a solution of A and AB, those of composition greater than 0.5 consist of a solution of B and AB. These two solutions are shown as ideal but need not be. Note that there is a discontinuity in the slope of the P_{tot} line at $x_B = 0.5$. In the acetone–chloroform system, however, the P_{tot} line is rounded at the minimum— an indication that no very stable AB complex forms. One may, in fact, estimate the dissociation constant of such a complex from the degree of curvature around this minimum.

Deviations from ideality may, of course, be positive. This is illustrated in Fig. 9-7 for the system benzene–ethanol. The physical argument is now reversed; we conclude that ϕ_{AB} is less than ϕ_{AA} or ϕ_{BB}. The extreme of this situation is that in which the two liquids have limited solubility in each other. This means that two phases α and β different composition can coexist, and that therefore

$$P_A{}^\alpha = P_A{}^\beta = P_A{}^{\alpha\beta} \qquad P_A^\alpha = P_B{}^\beta = P_B{}^{\alpha\beta} \qquad P = P_A + P_B \qquad (9\text{-}14)$$

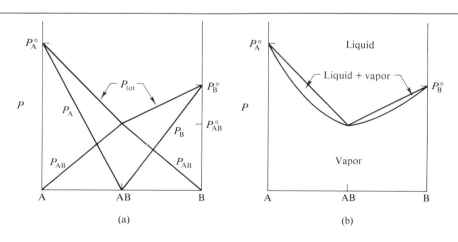

FIGURE 9-6
Formation of a stable compound AB, but with the solutions A + AB and B + AB ideal.

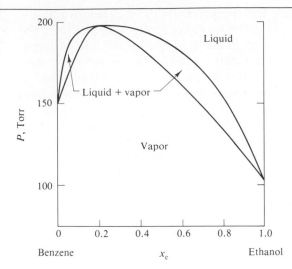

where the superscript α or β refers to a quantity for a single phase and the superscript $\alpha\beta$ stands for a quantity when both phases are present.

The limiting situation of complete immiscibility is shown in Fig. 9-8. The possible types of sequences are those for the systems x_2 and x_1. All systems now consist of two liquid phases initially, and when the pressure is reduced to P^{AB} vapor phase of composition y^{AB} forms and continues to do so until one liquid or the other is gone. The remaining liquid then continues to vaporize along the appropriate vapor composition line. The composition y^{AB} is in this case given by

$$y^{AB} = \frac{P_B{}^\circ}{P_A{}^\circ + P_B{}^\circ} = \frac{P_B{}^\circ}{P^{AB}} \tag{9-15}$$

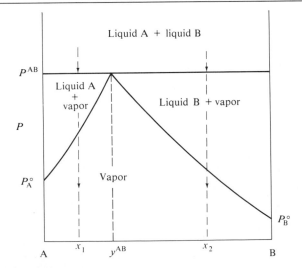

A fairly simple treatment developed by M. Margules in 1895 is still a very useful one. The difference between ϕ_{AB} and $\frac{1}{2}(\phi_{AA} + \phi_{BB})$ can be regarded as an energy term that enters as a Boltzmann factor modifying P_A over its ideal value. This energy difference should be approximately proportional to $x_B{}^2$ on the basis of arguments about the proportion of A–A and A–B interactions, and one writes

$$P_A = P_A°x_A \exp(\alpha x_B{}^2) \tag{9-16}$$

where α is a characteristic constant (and is temperature-dependent). Since the A–B interaction is a mutual one, a similar equation applies to P_B:

$$P_B = P_B° \, x_B \exp(\alpha x_A{}^2) \tag{9-17}$$

where α is the same constant as in Eq. (9-16). Notice the Eqs. (9-16) and (9-17) give Raoult's law as a limiting law, and reduce to Raoult's law for all compositions if $\alpha = 0$.

The model also provides a relationship between the Henry's law constants k_A and k_B [Eq. (9-8)]. Thus from Eq. (9-16) we have

$$\frac{dP_A}{dx_A} = P_A°[\exp(\alpha x_B{}^2)](1 - 2\alpha x_A x_B) \tag{9-18}$$

and in the limit when $x_A \rightarrow 0$

$$k_A = P_A°e^\alpha \tag{9-19}$$

The situation is symmetric, and so

$$k_B = P_B°e^\alpha \tag{9-20}$$

E. A model for nonideal solutions

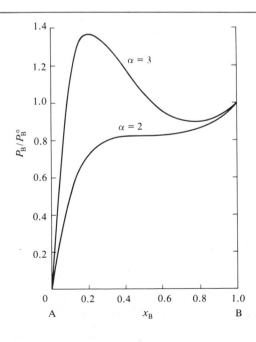

FIGURE 9-9
Plot of P_B according to the Margules equation (9-17). (The plot of P_A is similar, but increases from right to left, of course.)

Thus the two Henry's law constants are predicted to be in the ratio of the $P°$ values. This rule is obeyed reasonably well except for strongly associated liquids (α very negative).

In the case of positive deviation α is positive, and if sufficiently so, the curve calculated fromm Eq. (9-17) will show a maximum and a minimum, as illustrated in Fig. 9-9. The situation is reminiscent of that with respect to the van der Waals equation; the conclusion is that the experimental vapor pressure curve must show an equivalent horizontal portion and that the system is one of partial miscibility. The "critical temperature" is such that $\alpha = 2$. For this value of α the system just fails to show a miscibility gap.

9-3 The thermodynamics of multicomponent systems

Some aspects of the more formal thermodynamics of solutions, gaseous, liquid, or solid, must now be taken up. An important result will be the introduction of a new quantity, the chemical potential, which plays much the same role for solutions as G does for pure substances. The criterion for phase equilibrium is then expanded to the case of phases that are solutions, and an important new relationship, the *Gibbs–Duhem equation*, is introduced. Further developments are given in the Special Topics section, including the thermodynamic treatment of the surface tension of solutions. The principal new concept to be understood is that of *partial molal quantities*. Thermodynamics has become more complicated, but unavoidably so.

A. The chemical potential

The various thermodynamic functions must now include the amount of each component as a variable. That is, we consider what is called an *open system*, or one which may gain or lose chemical species. Thus the total energy $\mathbf{E}$ is a function of $\mathbf{S}$, v, and now n_i, where n_i denotes the number of moles of the ith species, and sans serif ($\mathbf{E}$, $\mathbf{S}$, and so on) denotes extensive quantities not on a per mole basis. The total differential for $\mathbf{E}$ is

$$d\mathbf{E} = \left(\frac{\partial \mathbf{E}}{\partial \mathbf{S}}\right)_{v,n_i} d\mathbf{S} + \left(\frac{\partial \mathbf{E}}{\partial v}\right)_{\mathbf{S},n_i} dv + \left(\frac{\partial \mathbf{E}}{\partial n_1}\right)_{\mathbf{S},v,n_i \neq n_i} dn_1$$
$$+ \left(\frac{\partial \mathbf{E}}{\partial n_2}\right)_{\mathbf{S},v,n_i \neq n_2} dn_2 + \cdots \quad (9\text{-}21)$$

Comparison with Eq. (6-17), to which Eq. (9-21) should reduce if the dn_i are zero, gives*

$$d\mathbf{E} = T\,d\mathbf{S} - P\,dv + \sum_i \left(\frac{\partial \mathbf{E}}{\partial n_i}\right)_{\mathbf{S},v,n_j \neq n_i} dn_i \quad (9\text{-}22)$$

Similarly, free energy is now a function of T, P, and also n_i, so we have

$$d\mathbf{G} = \left(\frac{\partial \mathbf{G}}{\partial T}\right)_{P,n_i} dT + \left(\frac{\partial \mathbf{G}}{\partial P}\right)_{T,n_i} dP + \sum_i \left(\frac{\partial \mathbf{G}}{\partial n_i}\right)_{T,P} dn_i \quad (9\text{-}23)$$

Comparison with Eq. (6-48) identifies the first two coefficients, so that

---

*To simplify the appearance of equations, the reminder $n_j \neq n_i$ will be taken for granted in derivatives such as $(\partial \mathbf{G}/\partial n_i)_{\mathbf{S},v,n_j \neq n_i}$.

$$dG = -S\, dT + v\, dP + \sum_i \left(\frac{\partial G}{\partial n_i}\right)_{T,P} dn_i \qquad (9\text{-}24)$$

Alternatively, however, we have

$$dG = dH - d(TS) = dE + P\, dv + v\, dP - T\, dS - S\, dT \qquad (9\text{-}25)$$

so, in combination with Eq. (9-22), it must also be true that

$$dG = -S\, dT + v\, dP + \sum_i \left(\frac{\partial E}{\partial n_i}\right)_{S,v} dn_i \qquad (9\text{-}26)$$

Thus we define

$$\mu_i = \left(\frac{\partial E}{\partial n_i}\right)_{S,v} = \left(\frac{\partial G}{\partial n_i}\right)_{T,P} \qquad (9\text{-}27)$$

where μ_i is a new quantity called the *chemical potential*.

Equations (9-22) and (9-26) can now be written

$$dE = T\, dS - P\, dv + \sum_i \mu_i\, dn_i \qquad (9\text{-}28)$$

$$dG = -S\, dT + v\, dP + \sum_i \mu_i\, dn_i \qquad (9\text{-}29)$$

The second of these equations is perhaps the more useful since it identifies the chemical potential μ_i as the free energy change of a system per mole of added component i, with temperature, pressure, and the other mole quantities kept constant. The chemical potential is a coefficient (like heat capacity) and we are really talking about the change dG for a small increment dn_i. The added amount of the ith component should not be sufficient to change the composition of the system appreciably since μ_i will depend on composition as well as on temperature and pressure.

We are dealing with equilibrium systems, and hence with systems for which no spontaneous change in temperature or pressure occurs. Equation (9-29) then reduces to

B. Criterion for phase equilibrium

$$dG = \sum_i \mu_i\, dn_i \qquad (9\text{-}30)$$

If the system consists of a single phase and is chemically isolated, that is, if no chemical species can enter or leave, then the criterion for equilibrium given by Eq. (6-47) applies, so we have

$$dG = 0 \qquad \sum_i \mu_i\, dn_i = 0 \qquad (9\text{-}31)$$

Alternatively, the system might consist of two (or more) phases in equilibrium. For each phase there will be an equation of the form of Eq. (9-30):

$$dG^\alpha = \sum_i \mu_i^\alpha dn_i^\alpha \qquad dG^\beta = \sum_i \mu_i^\beta\, dn_i^\beta \qquad (9\text{-}32)$$

and so forth, where

$$dG = dG^\alpha + dG^\beta + \cdots \qquad (9\text{-}33)$$

If the entire set of phases is in equilibrium and constitutes a chemically closed system overall, then again $d\mathbf{G}$ is zero, and we now have

$$0 = \sum_i \mu_i^\alpha \, dn_i^\alpha + \sum_i \mu_i^\beta dn_i^\beta + \cdots \tag{9-34}$$

Suppose that some process occurs in this equilibrium system of phases whereby dn_i moles of the ith species is transferred from phase α to phase β; all other dn quantities are zero. For this process it follows that

$$\mu_i^\alpha \, dn_i^\alpha + \mu_i^\beta \, dn_i^\beta = 0 \tag{9-35}$$

Since $dn_i^\alpha = -dn_i^\beta$ (the total number of moles of the ith species remains the same in the overall system), we have

$$\mu_i^\alpha = \mu_i^\beta \tag{9-36}$$

The very important conclusion is that for phase equilibrium between solutions the chemical potential of each species must be the same in every phase in which it is present. Equation (9-36) is a more general statement of the equilibrium condition for a pure substance, $G^\alpha = G^\beta$.

C. The Gibbs–Duhem equation

A very useful relationship may be obtained from Eq. (9-28). Since our applications will be restricted to two-component systems, we will make the derivation on that basis. The differentials of Eq. (9-28) are all of the form

(intensive property) $\times \, d$(extensive property).

It is permissible to integrate such equations, keeping the intensive properties constant. Thus Eq. (9-28) becomes

$$\int d\mathbf{E} = T \int d\mathbf{S} - P \int dv + \sum_i \mu_i \int dn_i \tag{9-37}$$

If the integration limits are from 0 to $\mathbf{E}$, 0 to $\mathbf{S}$, 0 to v, and 0 to n_i, respectively, we obtain

$$\mathbf{E} = T\mathbf{S} - Pv + \sum_i \mu_i \, n_i \tag{9-38}$$

The physical process can be imagined as one whereby we introduce additional amounts of the components, keeping the temperature, pressure, and composition (and hence the μ_i) constant.

Equation (9-38) is of general validity, and it follows that

$$d\mathbf{E} = T \, d\mathbf{S} + \mathbf{S} \, dT - \mathbf{P} \, dv - v \, d\mathbf{P} + \sum_i n_i d\mu_i + \sum_i \mu_i dn_i \tag{9-39}$$

Comparison with Eq. (9-28) leads to the important auxiliary relationship

$$\mathbf{S} \, dT - v \, d\mathbf{P} + \sum_i n_i d\mu_i = 0 \tag{9-40}$$

This reduces to

$$\sum_i n_i d\mu_i = 0 \tag{9-41}$$

for the case of T and P constant. [Note that Eq. (9-41) could have been obtained directly from Eq. (9-30).] We can divide by the total number of moles to obtain an alternative form of Eq. (9-41):

$$\sum_i x_i d\mu_i = 0 \tag{9-42}$$

Equations (9-41) and (9-42) are known as *Gibbs–Duhem equations.* Their great usefulness is in relating chemical potential changes in a system. The application of Eq. (9-42) will be illustrated in Section 9-5, for the case of a two-component system. For this case, we have

$$n_1 \, d\mu_1 + n_2 \, d\mu_2 = 0 \qquad x_1 \, d\mu_1 + x_2 \, d\mu_2 = 0 \tag{9-43}$$

The chemical potential is one of a family of partial molal quantities. If, in general, we have some extensive property $\mathscr{P}$, then for a solution $\overline{\mathscr{P}}_i$ is the partial molal property for the ith component:

D. Partial molal quantities

$$\overline{\mathscr{P}}_i = \left(\frac{\partial \mathscr{P}}{\partial n_i}\right)_{T,P} \tag{9-44}$$

[The condition n_j constant, $n_j \neq n_i$ is also made; see the footnote following Eq. (9-21).] Specific illustrations of Eq. (9-44) are:

$$\overline{V}_i = \left(\frac{\partial v}{\partial n_i}\right)_{T,P} \tag{9-45}$$

$$\overline{H}_i = \left(\frac{\partial \mathbf{H}}{\partial n_i}\right)_{T,P} \tag{9-46}$$

and

$$\overline{S}_i = \left(\frac{\partial \mathbf{S}}{\partial n_i}\right)_{T,P} \tag{9-47}$$

Further, the various thermodynamic coefficients that were derived for a single substance retain the same form for the ith component of a solution if the corresponding partial molal quantities are used. As useful examples, we have

$$\left(\frac{\partial \mu_i}{\partial P}\right)_T = \frac{\partial}{\partial P}\left(\frac{\partial \mathbf{G}}{\partial n_i}\right)_{P,T} = \frac{\partial}{\partial n_i}\left(\frac{\partial \mathbf{G}}{\partial P}\right)_T = \left(\frac{\partial v}{\partial n_i}\right)_{P,T} = \overline{V}_i \tag{9-48}$$

and, similarly,

$$\left(\frac{\partial \mu_i}{\partial T}\right)_P = -\overline{S}_i \tag{9-49}$$

Equation (9-48) may be applied to a mixture of ideal gases. By Eq. (9-45),

$$\overline{V}_i = \left(\frac{\partial v}{\partial n_i}\right)_{T,P} = \frac{RT}{P}\left(\frac{\partial n}{\partial n_i}\right)_{T,P} = \frac{RT}{P}$$

Then

$$\left(\frac{\partial \mu_i}{\partial P}\right)_T = \overline{V}_i = \frac{RT}{P}$$

and

$$d\mu_i = RT \frac{dP}{P} = RT \frac{dP_i}{P_i} = RT \, d \ln P_i \qquad \text{(ideal gas)} \tag{9-50}$$

Note that we can replace dP/P by dP_i/P_i since $dP_i = x_i dP$. We thus have

$$\mu_i(g) = \mu_i^\circ(g) + RT \ln P_i \qquad \text{(ideal gas)} \tag{9-51}$$

where $\mu_i^\circ(g)$ is the chemical potential of the gas in its standard state, ordinarily taken to be 1 atm. (The same equation applies for a nonideal gas with fugacity f_i replacing P_i—see Section 6-ST-2.) Since $\mu_i^\circ(g)$ refers to pure component i, it could just as well have been written $G_i^\circ(g)$; it seems preferable, however, to keep the notation symmetric. [See Robinson (1964) for a discussion of the preceding derivation.]

The derivation of the Gibbs–Duhem equation [Eq. (9-42)] is a specific example of a more general procedure. If the independent variables of some function $y = f(u,v,w,\dots)$ are extensive quantities, that is, ones which increase in proportion to the amount of the system, then by a theorem due to Euler, it must be true that

$$u\left(\frac{\partial f}{\partial u}\right)_{v,w,\dots} + v\left(\frac{\partial f}{\partial v}\right)_{u,w,\dots} + w\left(\frac{\partial f}{\partial w}\right)_{u,v,\dots} + \cdots = f \tag{9-52}$$

This theorem was invoked implicitly in integrating Eq. (9-28), with $E = f(S, v, n_i)$, to obtain Eq. (9-38).

In the case of partial molal quantities we restrict ourselves to a system at constant temperature and pressure, so that the amounts n_i of the various components are the only variables. Thus in the case of a two-component system, for some property $\mathscr{P}$, we have

$$d\mathscr{P} = \left(\frac{\partial \mathscr{P}}{\partial n_1}\right)_{n_2} dn_1 + \left(\frac{\partial \mathscr{P}}{\partial n_2}\right)_{n_1} dn_2 = \overline{\mathscr{P}}_1 \, dn_1 + \overline{\mathscr{P}}_2 \, dn_2 \tag{9-52}$$

where $\overline{\mathscr{P}}_1$ and $\overline{\mathscr{P}}_2$ are the partial molal values. Then, by Euler's theorem,

$$\mathscr{P} = n_1 \overline{\mathscr{P}}_1 + n_2 \overline{\mathscr{P}}_2 \tag{9-53}$$

Differentiation and comparison with Eq. (9-52) gives

$$n_1 \, d\overline{\mathscr{P}}_1 + n_2 d\overline{\mathscr{P}}_2 = 0 \tag{9-54}$$

If $\mathscr{P} = \mathbf{G}$, then

$$\mathbf{G} = n_1 \mu_1 + n_2 \mu_2 \tag{9-55}$$

and

$$n_1 \, d\mu_1 + n_2 \, d\mu_2 = 0 \qquad x_1 \, d\mu_1 + n_2 \, d\mu_2 = 0 \tag{9-43}$$

If $\mathscr{P} = v$, we have

$$v = n_1 \overline{V}_1 + n_2 \overline{V}_2 \tag{9-56}$$

$$n_1 \, d\overline{V}_1 + n_2 \, d\overline{V}_2 = 0 \qquad x_1 d\overline{V}_1 + x_2 d\overline{V}_2 = 0 \tag{9-57}$$

and if $\mathscr{P} = \mathbf{H}$, then

$$\mathbf{H} = n_1 \overline{H}_1 + n_2 \overline{H}_2 \tag{9-58}$$

and

$$n_1 d\overline{H}_1 + n_2 d\overline{H}_2 = 0 \tag{9-59}$$

There are some useful special procedures for obtaining partial molal quantities from experimental data, which can be illustrated easily for volume. We define the average molar volume as

$$V_{av} = \frac{v}{n_1 + n_2} \tag{9-60}$$

Then

$$\overline{V}_1 = \left(\frac{\partial v}{\partial n_1}\right)_{n_2} = V_{av} + (n_1 + n_2)\left(\frac{\partial V_{av}}{\partial x_2}\right)_{n_2}\left(\frac{\partial x_2}{\partial n_1}\right)_{n_2}$$

Since

$$\left(\frac{\partial x_2}{\partial n_1}\right)_{n_2} = -\frac{n_2}{(n_1 + n_2)^2}$$

it follows that

$$\overline{V}_1 = V_{av} - x_2 \frac{dV_{av}}{dx_2} \tag{9-61}$$

Equation (9-61) has a simple geometric meaning. If V_{av} is plotted against mole fraction, then dV_{av}/dx_2 is the slope of the tangent at composition x_2, and the intercept of the tangent at $x_2 = 0$ then gives $\overline{V}_1$. Since the equation is symmetric, the intercept at $x_2 = 1$ gives $\overline{V}_2$. The situation is illustrated in Fig. 9-10 for the acetone–chloroform system. Also, of course, Eq. (9-57) may be used for the calculation of $\overline{V}_2$ if $\overline{V}_1$ is known as a function of composition.

The volume change when the pure components are mixed is

$$\Delta V_M = V_{av} - x_1 V_1^\circ - x_2 V_2^\circ \tag{9-62}$$

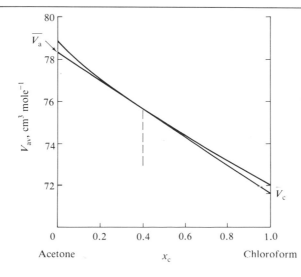

FIGURE 9-10
Variation of V_{av} with composition for the acetone–chloroform system at 25°C. The intercepts of the tangent give $\overline{V}_a$ and $\overline{V}_c$ for that composition.

where V_1° and V_2° are the molar volumes of the pure species. It follows from Eq. (9-56) that

$$V_{av} = x_1 \overline{V}_1 + x_2 \overline{V}_2 \qquad (9\text{-}63)$$

and Eq. (9-62) can alternatively be written

$$\Delta V_M = x_1(\overline{V}_1 - V_1^{\circ}) + x_2(\overline{V}_2 - V_2^{\circ}) \qquad (9\text{-}64)$$

Enthalpies are treated somewhat similarly, but a complication is that, unlike volumes, absolute enthalpies are not known. The subject is placed in Section 9-ST-1 because of this added complexity.

9-4   Ideal gas mixtures

The chemical potential of a component of an ideal gas mixture is given by Eq. (9-51),

$$\mu_i(g) = \mu_i^{\circ}(g) + RT \ln P_i$$

We can apply this equation to calculate the free energy and entropy change for the isothermal process

$$n_1(\text{gas 1 at } P) + n_2(\text{gas 2 at } P) = \text{mixture (at } P, \text{ with } P_1$$
$$= x_1 P \text{ and } P_2 = x_2 P) \qquad (9\text{-}65)$$

The process, physically, corresponds to the procedure shown in Fig. 9-11. For pure gases, $\mathbf{G}_1 = n_1[\mu_1^{\circ}(g) + RT \ln P]$ and $\mathbf{G}_2 = n_2[\mu_2^{\circ}(g) + RT \ln P]$. The free energy of the mixture is, by Eq. (9-55),

$$\mathbf{G}_{mix} = n_1[\mu_1^{\circ}(g) + RT \ln P_1] + n_2[\mu_2^{\circ}(g) + RT \ln P_2]$$

or, since $P_1 = x_1 P$ and $P_2 = x_2 P$, substitution gives

$$\mathbf{G}_{mix} = n_1[\mu_1^{\circ}(g) + RT \ln P + RT \ln x_1]$$
$$+ n_2[\mu_2^{\circ}(g) + RT \ln P + RT \ln x_2]$$

The free energy change for process (9-65) is $\mathbf{G}_{mix} - \mathbf{G}_1 - \mathbf{G}_2$ or

$$\Delta\mathbf{G}_M = n_1 RT \ln x_1 + n_2 RT \ln x_2 \qquad \text{(ideal gas)} \qquad (9\text{-}66)$$

The free energy of mixing per mole of mixture is

$$\Delta G_M = x_1 RT \ln x_1 + x_2 RT \ln x_2 \qquad \text{(ideal gas)} \qquad (9\text{-}67)$$

Since $\Delta S = -[\partial(\Delta G)/\partial T]_P$, the entropy of mixing per mole of solution is

$$\Delta S_M = -(x_1 R \ln x_1 + x_2 R \ln x_2) \qquad \text{(ideal gas)} \qquad (9\text{-}68)$$

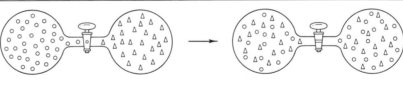

FIGURE 9-11
The mixing of two gases.

Gas 1
at P, T

Gas 2
at P, T

Mixture at P, T

Notice that the free energy of mixing is independent of P and that the entropy of mixing is independent of both P and T. The latter could have been obtained on the basis of the probability arguments of Section 6-9A (see also Section 9-CN-3). Finally, since

$$\Delta G_M = \Delta H_M - T\Delta S_M \tag{9-69}$$

it follows that for ideal gases

$$\Delta H_M = 0 \tag{9-70}$$

9-5   Ideal and nonideal solutions. Activities and activity coefficients

We can apply the criterion for equilibrium to the case of a solution and its vapor. The treatment will be in terms of a liquid solution, but it is equally applicable to solid solutions. The vapor phase is assumed, as usual, to be ideal, and for simplicity we consider only a two-component solution. The requirement that the chemical potential of a component be the same in two phases that are in equilibrium may be combined with Eq. (9-51) to give

$$\mu_1(l) = \mu_1(g) = \mu_1°(g) + RT \ln P_1 \tag{9-71}$$

In the case of a pure liquid $\mu_1(l)$ becomes $\mu_1°(l)$ [or just $G_1°(l)$] and Eq. (9-71) reduces to

$$\mu_1°(l) = \mu°_1(g) + RT \ln P_1° \tag{9-72}$$

Alternatively, we may add and subtract $RT \ln P_1°$ on the right-hand side of Eq. (9-71) to obtain

$$\mu_1(l) = [\mu_1°(g) + RT \ln P_1°] + RT \ln \frac{P_1}{P_1°}$$

or

$$\mu_1(l) = \mu_1°(l) + RT \ln \frac{P_1}{P_1°} \tag{9-73}$$

We thus have two ways of expressing the chemical potential of component one. Equation (9-71) does so in terms of $\mu_1°(g)$ and P_1, while Eq. (9-73) does so in terms of $\mu_1°(l)$ and $P_1/P_1°$. In the first case the standard or reference state is the vapor at unit pressure, 1 atm, and in the second case it is the pure liquid. The equations are symmetric with respect to the components and so a parallel set of relationships applies for component 2:

$$\mu_1(l) = \mu_2°(g) + RT \ln P_2 \tag{9-74}$$

$$\mu_2(l) = \mu_2°(l) + RT \ln \frac{P_2}{P_2°} \tag{9-75}$$

A. Ideal solutions

Equations (9-73) and (9-75) take on a very simple form if Raoult's law is obeyed, since then $P_1/P_2° = x_1$ and $P_2/P_2° = x_2$. Thus

$$\mu_1(l) = \mu_1°(l) + RT \ln x_1 \qquad \text{(ideal solution)} \tag{9-76}$$

$$\mu_2(l) = \mu_2°(l) + RT \ln x_2 \qquad \text{(ideal solution)} \tag{9-77}$$

We can use Eq. (9-55) to obtain the total free energy of the solution:

$$\mathbf{G} = n_1\mu_1°(l) + n_2\mu_2°(l) + n_1RT \ln x_1$$
$$+ n_2RT \ln x_2 \qquad \text{(ideal solution)} \qquad (9\text{-}78)$$

If we now consider the process of preparing the solution by mixing the pure liquids,

$$n_1(\text{component 1}) + n_2(\text{component 2}) = \text{solution} \qquad (9\text{-}79)$$

the corresponding free energy change is

$$\Delta\mathbf{G}_M = n_1RT \ln x_1 + n_2RT \ln x_2 \qquad \text{(ideal solution)} \qquad (9\text{-}80)$$

or, per mole of solution,

$$\Delta G_M = x_1RT \ln x_1 + x_2RT \ln x_2 \qquad \text{(ideal solution)} \qquad (9\text{-}81)$$

We can also obtain the entropy of mixing. From Eqs. (9-49) and (9-76),

$$-\overline{S}_1(l) = -S_1°(l) + R \ln x_1 \qquad \text{(ideal solution)} \qquad (9\text{-}82)$$

and similarly for component 2. The total entropy of the solution is then

$$\mathbf{S} = n_1S_1°(l) + n_2S_2°(l) - (n_1R \ln x_1 + n_2R \ln x_2) \qquad (9\text{-}83)$$

and, for the mixing process, per mole of solution,

$$\Delta S_M = -(x_1R \ln x_1 + x_2R \ln x_2) \qquad \text{(ideal solution)} \qquad (9\text{-}84)$$

Note that Eqs. (9-81) and (9-84) are identical to Eqs. (9-67) and (9-68) for the mixing of ideal gases (see Commentary and Notes section).

Since $\Delta G = \Delta H - T\,\Delta S$ for a constant-temperature process, it follows from Eqs. (9-81) and (9-84) that

$$\Delta H_M = 0 \qquad (9\text{-}85)$$

The heat of mixing for an ideal solution (and for ideal gases) is zero.

B. Nonideal solutions

Equations (9-73) and (9-75) could be used for nonideal solutions, but it would be inconvenient always to have to refer to vapor pressures. A more general form, preferably analogous to Eqs. (9-76) and (9-77) for ideal solutions, would be very advantageous. We obtain this form by introducing the effective or thermodynamic concentration, called the *activity*. Activity or effective mole fraction a is defined so as to retain the form of Raoult's law:

$$P_i = a_iP_i° \qquad (9\text{-}86)$$

Equation (9-73) becomes

$$\mu_1(l) = \mu_1°(l) + RT \ln a_1 \qquad (9\text{-}87)$$

and similarly for component 2. If we take Raoult's law as the limiting law for all solutions, as $x_1 \rightarrow 1$, a_1 must approach x_1. We may retain the Raoult's law form even more explicitly by using the term coefficient γ_i, defined as the factor by which a_i deviates from x_i:

$$a_i = \gamma_ix_i \qquad (9\text{-}88)$$

Since $a_i \rightarrow x_i$, $\gamma_i \rightarrow 1$ as $x_i \rightarrow 1$. Equation (9-87) becomes

$$\mu_1(l) = \mu_1°(l) + RT \ln x_1 + RT \ln \gamma_1 \tag{9-89}$$

The equation for the free energy of mixing, corresponding to the process of Eq. (9-79), can be put in a form that allows ΔG_M to be expressed as the sum of an ideal and a nonideal contribution. Equation (9-80) becomes

$$\Delta G_M = x_1 RT \ln a_1 + x_2 RT \ln a_2 \tag{9-90}$$

or

$$\Delta G_M = x_1 RT \ln x_1 + x_2 RT \ln x_2 + x_1 RT \ln \gamma_1 + x_2 RT \ln \gamma_2$$

Alternatively,

$$\Delta G_E = \Delta G_M - \Delta G_{M(ideal)} = x_1 RT \ln \gamma_1 + x_2 RT \ln \gamma_2 \tag{9-91}$$

The difference $\Delta G_M - \Delta G_{M(ideal)}$ is known as the *excess free energy of mixing* ΔG_E.

Similarly,

$$\Delta S_E = \Delta S_M - \Delta S_{M(ideal)} \tag{9-92}$$

The evaluation of ΔS_E involves the change in activity coefficients with temperature or, alternatively, a measurement of ΔH_M (see Special Topics section). Also, $\Delta G_E = \Delta H_E - T \Delta S_E$, where

$$\Delta H_E = \Delta H_M \tag{9-93}$$

since $\Delta H_{M(ideal)}$ is zero.

Note that if the vapor pressure shows a positive deviation from ideality, then a_1 and a_2 will be greater than the corresponding mole fractions and the γ's will be greater than unity. Conversely, if the deviation is negative, the γ's will be less than unity. The Gibbs–Duhem equation provides some important conclusions in this respect. If we evaluate $d\mu_1(l)$ from Eq. (9-87), then, by Eq. (9-43),

$$x_1 \, d(\ln a_1) + x_2 \, d(\ln a_2) = 0 \tag{9-94}$$

or

$$\int d(\ln a_2) = - \int \frac{x_1}{x_2} \, d(\ln a_1) \tag{9-95}$$

Thus if the activities of component 1 are known for a range of concentrations, integration of Eq. (9-95) allows a calculation of the change in activity of component 2 (see Sections 9-5C and 9-ST-1).

Some useful qualitative conclusions are possible. An alternative form of Eq. (9-95) is

$$\ln(a_1/x_1) = - \int_0^{x_2} (x_2/x_1) \, d(\ln(a_2/x_2)) \tag{9-96}$$

(The derivation is straightforward and is left as an exercise.) If component 2 of the system shows a positive deviation from ideality, $a_2/x_2 > 1$, then $\ln(a_2/x_2)$ is a positive number at $x_2 = 0$ and decreases to zero as x_2 approaches unity. The integral is therefore negative, and the right-hand side of Eq. (9-96) is thus positive, which means that $a_1/x_1 > 1$ or that component 1 also shows positive deviation from ideality. A parallel argument demonstrates that if component 2 shows negative deviation from ideality, then so will component 1.

Yet another form of the Gibbs–Duhem equation is

$$\frac{d(\ln a_1)}{d(\ln x_1)} = \frac{d(\ln a_2)}{d(\ln x_2)} \qquad (9\text{-}97)$$

Suppose that in dilute solution the solute obeys Henry's law, that is, $a_2 = k_2 x_2$. It follows that $d(\ln a_2)/d(\ln x_2) = 1$ and hence that $d(\ln a_1)/d(\ln x_1) = 1$ or that $a_1 = (\text{constant})(x_1)$. Since a_1 must equal x_1 as $x_2 \to 0$, the constant must be unity. Thus if the solute obeys Henry's law as $x_2 \to 0$, the solvent must obey Raoult's law in this limit.

C. Calculation of activities and activity coefficients

The preceding material is sufficiently complicated that we now offer a detailed numerical example to help clarify just how the various definitions and procedures are implemented. The data of Fig. 9-3 for the acetone–chloroform system will be used. Values for the two partial pressures, interpolated from the original data, are given in Table 9-1. There are a number of regularities and interrelations to notice. First, the activity coefficients are given either by $\gamma_c = a_c/x_c$ and $\gamma_a = a_a/x_a$ (c = chloroform and a = acetone) or by $\gamma_c = P_c/P_{c(\text{ideal})}$ and $\gamma_a = P_a/P_{a(\text{ideal})}$, where P_{ideal} is the Raoult's law value for the partial pressure. The two calculations are equivalent.

Next, the plot of the activity coefficients given in Fig. 9-12 shows that $\gamma_c \to 1$ as $x_c \to 1$, and γ_c approaches the limiting value of 0.485 as x_c approaches zero. This limiting value is just the ratio of the Henry's law limiting slope to $P_c°$; that is, from Fig. 9-3(a), $k_c = 142$ Torr and $P_c° = 293$ Torr, so that the ratio is $142/293 = 0.485$. Similarly, $\gamma_a \to 1$ as $x_a \to 1$, and γ_a approaches the limiting value of 0.449 as $x_a \to 0$; here the ratio is $155/345$, where $k_a = 155$ Torr and $P_a° = 345$ Torr.

The data serve to test the Margules model (Section 9-2E). By Eqs. (9-19) and (9-20), k_c/k_a should be equal to $P_c°/P_a°$ or $293/345 = 0.85$; the actual ratio is $142/155$ or 0.92. Alternatively, the model predicts that the limiting value of γ_c as $x_c \to 0$ should be the same as the limiting value of γ_a as $x_a \to 0$, and equal to e^α. The respective limiting values are actually 0.485 and 0.449, or about 8% different. The model is thus approximate in this case, but still is not too bad, considering its simplicity.

The data may also be used to illustrate the application of Eq. (9-95). Figure 9-13 shows the plot of x_a/x_c versus $\log a_a$. The shaded area corresponds to the integral between $x_c = 0.9$ and $x_c = 0.3$:

TABLE 9-1
The acetone (a)–chloroform (c) system at 35°C[a]

| | Vapor Pressure (Torr) | | | | | | | |
| | Observed | | Raoult's Law | | Activity[b] | | Activity Coefficient[c] | |
x_c	P_c	P_a	P_c	P_a	a_c	a_a	γ_c	γ_a
0.00	0	345	0	345	0	(1.00)	(0.485)	(1.00)
0.10	16.0	310	29.3	311	0.0546	0.899	0.546	0.998
0.20	35	270	59	276	0.119	0.783	0.597	0.978
0.30	57	227	88	242	0.195	0.658	0.648	0.940
0.40	82	185	117	207	0.280	0.536	0.700	0.894
0.50	112	140	147	173	0.382	0.406	0.765	0.812
0.60	142	102	176	138	0.485	0.296	0.808	0.739
0.70	180	65	205	104	0.614	0.188	0.878	0.628
0.80	219	37	234	69	0.747	1.107	0.934	0.536
0.90	257	16.5	264	34.5	0.877	0.048	0.975	0.478
1.00	293	0	293	0	(1.00)	0	(1.00)	(0.449)

[a]Data from the "International Critical Tables," Vol. 3. McGraw-Hill, New York, 1928.
[b]$a_c = P_c/P_c°$, $a_a = P_a/P_a°$.
[c]$\gamma_c = a_c/x_c$, $\gamma_a = a_a/x_a$.

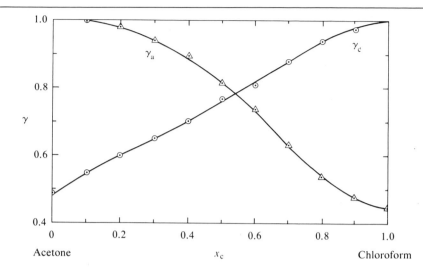

FIGURE 9-12
Activity coefficient plot for the acetone–chloroform system at 35°C.

$$\log a_{c(x_c=0.9)} - \log a_{c(x_c=0.3)} = \int_{x_c=0.3}^{x_c=0.9} \frac{x_a}{x_c} \, d(\log a_a) \tag{9-98}$$

The area is approximately -0.66, so

$$\log a_{c(x_c=0.3)} = \log a_{c(x_c=0.9)} - 0.66 = \log(0.877) - 0.66 \tag{9-99}$$

whence $a_c = 0.192$, in good agreement with the observed value of 0.195.

Finally, we can calculate the free energy and excess free energy of mixing, using Eqs. (9-90) and (9-91). We have

$$\Delta G_E = (1.987)(308)(x_c \ln \gamma_c + x_a \ln \gamma_a)$$
$$= 612(x_c \ln \gamma_c + x_a \ln \gamma_a) \tag{9-100}$$

The calculated values for ΔG_E are plotted in Fig. 9-14 for 25°C. Here ΔG_E is negative and goes through a minimum at about $x_c = 0.6$; it is zero, of course, for either pure liquid.

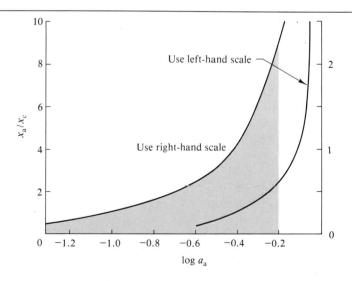

FIGURE 9-13
Application of the Gibbs–Duhem equation to the acetone–chloroform system at 35°C.

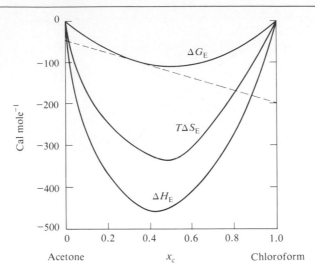

FIGURE 9-14
The acetone–chloroform system at 25°C. (See Section 9-ST-1 for explanation of the dashed line.) [Data from I. Prigogine and R. Defay, "Chemical Thermodynamics" (D. H. Everett, translator). Longmans, Green, New York, 1954.]

We must know the temperature dependence of ΔG_E in order to obtain ΔS_M and hence ΔS_E or, alternatively, calorimetric heat of mixing data so as to obtain ΔH_E. These quantities have been obtained, and are included in the figure. Notice that $T\,\Delta S_E$ and ΔH_E are both relatively large but partially cancel to give a much smaller ΔG_E. This often happens.

9-6   The temperature dependence of vapor pressures

We proceed in the next section to discuss boiling point diagrams, and it is especially in this connection that the temperature dependence of solution vapor pressures is important. We can obtain a relationship analogous to the Clausius–Clapeyron equation by proceding as follows. Differentiation of the equation

$$\mu_i(l) = \mu_i^\circ(g) + RT \ln P_i \qquad \text{[Eq. (9-71)]}$$

with respect to temperature, and use of Eq. (9-49), gives

$$-\overline{S}_i(l) = -S_i^\circ(g) + R \ln P_i + RT \frac{d(\ln P_i)}{dT}$$

The term $R \ln P_i$ is replaced by $[\mu_i(l) - \mu_i^\circ(g)]/T$ to give

$$RT \frac{d(\ln P_i)}{dT} = \frac{[\mu_i^\circ(g) + TS_i^\circ(g)]}{T} - \frac{[\mu_i(l) + T\overline{S}_i(l)]}{T} \qquad (9\text{-}101)$$

The defining equation for G,

$$G = H - TS \qquad \text{[Eq. (6-36)]}$$

becomes, for a component of a solution,

$$\mu_i = \overline{H}_i - T\,\overline{S}_i \qquad (9\text{-}102)$$

[obtained by differentiating Eq. (6-36) with respect to dn_i at constant T and P]. The terms in brackets in Eq. (9-101) may next be replaced by the corresponding enthalpies to give, on rearrangement,

$$\frac{d(\ln P_i)}{dT} = \frac{\Delta \overline{H}_{v,i}}{RT^2} \tag{9-103}$$

where

$$\Delta \overline{H}_{v,i} = H_i^\circ(g) - \overline{H}_i(l) \tag{9-104}$$

and is the partial molal heat of vaporization for the ith component. In the case of pure liquid $\overline{H}_i(l)$ becomes $H_i^\circ(l)$ and Eq. (9–103) reduces to the Clausius–Clapeyron equation. The same is true for an ideal solution, since ΔH_M is zero.

The ideal solution form of Eq. (9-103) may be developed more explicitly. First, for a pure liquid the Clausius–Clapeyron equation can be written

$$P_i^\circ = \exp\left[\frac{\Delta H_{V,i}^\circ}{R}\left(\frac{1}{T_{b,i}^\circ} - \frac{1}{T}\right)\right] \tag{9-105}$$

where $T_{b,i}^\circ$ is the normal boiling point of liquid i; Eq. (9-105) follows from Eq. (8-10) when we set the vapor pressure equal to 1 atm at T_b°. For an ideal solution, $P_i = x_i P_i^\circ$, so Eq. (9-105) becomes

$$P_i = x_i \exp\left[\frac{\Delta H_{V,i}^\circ}{R}\left(\frac{1}{T_{b,i}^\circ} - \frac{1}{T}\right)\right] \tag{9-106}$$

Equations (9-104) and (9-106) apply equally well to the vapor pressure of the ith component of a *solid* solution, ideal in the case of Eq. (9-106). The enthalpy quantities are then those for sublimation, of course.

9-7 Boiling point diagrams

A. General appearance

The material of Section 9-2 is now extended to show the various types of boiling point diagrams that one finds for a solution of two volatile liquids. It is first necessary to consider how the total vapor pressure of the solution should vary with composition and temperature. This is most easily done for the case of an ideal solution, for which the two partial pressures are, from Eq. (9-106).

$$P_A = x_A \exp\left[\frac{\Delta H_{V,A}^\circ}{R}\left(\frac{1}{T_{b,A}^\circ} - \frac{1}{T}\right)\right] \tag{9-107}$$

and

$$P_B = x_B \exp\left[\frac{\Delta H_{V,B}^\circ}{R}\left(\frac{1}{T_{b,B}^\circ} - \frac{1}{T}\right)\right] \tag{9-108}$$

The total pressure P is just

$$P = P_A + P_B \tag{9-109}$$

Substitution of the expressions for P_A and P_B into Eq. (9-109) gives the equation for the variation of P with composition and temperature.

The general appearance of this function is shown in Fig. 9-15; it is assumed that component A is the one with the lower boiling point.

The upper surface gives the total vapor pressure and the lower one the vapor composition for solutions of a given composition and temperature. The front of the projection corresponds to a cross section at constant temperature,

FIGURE 9-15
Variation of P with
temperature and
composition for an ideal
solution.

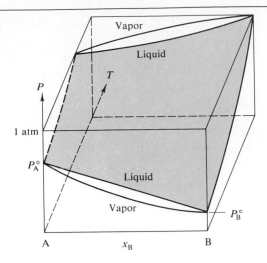

FIGURE 9-15
Variation of P with
temperature and
composition for an ideal
solution.

FIGURE 9-16
Two cross sections of Fig.
9-15. (a) At constant T,
giving the vapor pressure
and vapor composition
diagram. (b) At constant
P = 1 atm, giving the
boiling point and vapor
composition diagram.

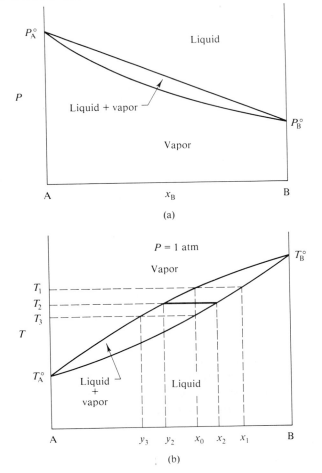

and thus constitutes the vapor pressure–composition diagram for that temperature, as shown in Fig. 9-16(a). The top surface in Fig. 9-15 corresponds to a cross section at constant pressure and therefore to the boiling point diagram for that pressure, as shown in Fig. 9-16(b). Notice that the liquid composition line is now curved and lies below rather than above the vapor composition line.

The normal boiling point diagram is given by a cross section at $P = 1$ atm. We can obtain the boiling point versus composition line analytically by setting $P = 1$ in Eq. (9-109):

$$1 = x_A \left\{ \exp\left[\frac{\Delta H^\circ_{V,A}}{R} \left(\frac{1}{T^\circ_{b,A}} - \frac{1}{T_b} \right) \right] \right\}$$
$$+ x_B \left\{ \exp\left[\frac{\Delta H^\circ_{V,B}}{R} \left(\frac{1}{T^\circ_{b,B}} - \frac{1}{T_b} \right) \right] \right\} \quad (9\text{-}110)$$

Equation (9-110) reduces to two variables since $x_A + x_B = 1$. Since it is transcendental, it is best solved by picking successive choices for T_b and solving for corresponding x_A or x_B. The resulting plot is shown in Fig. 9-16(b). The vapor line gives the compositions of vapor in equilibrium with boiling solutions and is calculated from the corresponding P_A and P_B values:

$$y_A = \frac{P_A}{P} \qquad y_B = \frac{P_B}{P} \quad (9\text{-}111)$$

P, the total vapor pressure, is one atmosphere in the case of a normal boiling point calculation.

The boiling point diagram is again a phase map. Referring to 9-16(b), we see that if a system of composition x_0 is contained in a cylinder with a piston arranged so that the pressure is always 1 atm, the system consists entirely of vapor if $T > T_1$. On cooling to T_1, liquid of composition x_1 begins to condense out and by temperature T_2 the system consists of liquid of composition x_2 and vapor of composition y_2. The relative amounts are given on application of the lever principle to the tie-line at T_2. At T_3 the last vapor, of composition y_3, has condensed, and below T_3 the system is entirely liquid.

Figure 9-16 illustrates another point, namely, that the boiling point diagram is (roughly) similar in appearance to that of the vapor pressure diagram turned upside down: The higher vapor pressure liquid is the lower boiling one, and the relative positions of the phase regions are reversed. A similar situation holds for nonideal systems as shown in Fig. 9-17. Positive deviation, leading to a maximum in the vapor pressure diagram, will usually give a minimum boiling system as in Fig. 9-17(a), whereas a negatively deviating system with a minimum in the vapor pressure diagram usually shows maximum boiling behavior, as in Fig. 9-17(b). Compare with Figs. 9-7 and 9-3.

B. Distillation

If a boiling system is arranged so that the vapors are continuously removed rather than being contained as in the cylinder and piston arrangement, a somewhat different sequence of events occurs. Referring to the case of Fig. 9-16(b), we see that liquid of composition x_0 would first boil at T_3, producing vapor of composition y_3. Since the vapor is richer in A than is the liquid, and is steadily being removed, the liquid composition progressively becomes richer in B, passing compositions x_2 and x_1, respectively. Unlike the situation with

FIGURE 9-17
Vapor pressure and boiling point diagrams: (a) Positive deviation from Raoult's law, giving a minimum boiling diagram. (b) Negative deviation from Raoult's law, giving a maximum boiling diagram.

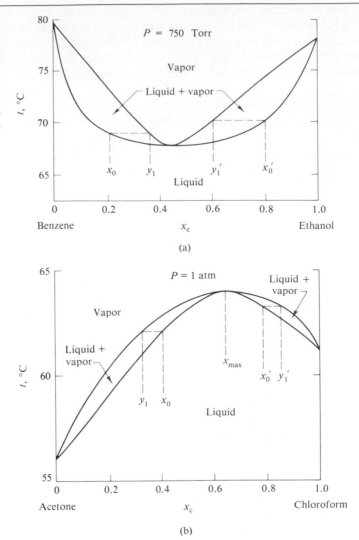

the closed system, however, liquid remains when T_1 is reached. This is because the overall vapor composition is not x_0, but rather the average of the compositions of the succession of vapors produced, ranging from y_3 to x_0. For example, this average vapor composition might be about equal to y_2, in which case the relative amount of liquid remaining would be given by the lever $(y_2 - x_0)/(y_2 - x_1)$, or about 50%. Continued boiling would continue to shift the liquid composition to the right, and the last drop of liquid remaining would be essentially pure B.

A similar analysis applies to Fig. 9-17. Liquids of composition either x_0 or x_0' produce initial vapors of composition y_1 or γ_1'; in both cases the vapor composition is closer to the minimum boiling composition than is the liquid composition. As a consequence, continued boiling of system x_0 moves the liquid composition progressively toward pure benzene, and continued boiling of system x_0' moves it toward pure ethanol. If there is a maximum boiling point, the vapor compositions lie away from the maximum, as compared to the liquid composition, as shown in Fig. 9-17(b). The result is that continued boiling of either liquid x_0 or x_0' eventually produces liquid of composition x_{max}, the maximum boiling composition. At this point the liquid and vapor compositions are the same and continued boiling produces no further change. The system now behaves as though it were a pure liquid and is called an *azeotropic mixture*. The value of x_{max} depends on the pressure; Fig. 9-17(b) is, after all, merely one particular isobaric cross section of a general diagram of the type shown in Fig. 9-17. It is possible, however, to use this maximum boiling feature as a means of preparing a standard solution. An example is the hydrochloric acid–water system, for which the maximum boiling composition is 20.222% HCl at 760 Torr (but shifts to 20.360% HCl at 700 Torr). The standardization procedure consists simply in boiling a solution until no further change in boiling point occurs and recording the concentration appropriate to the ambient or barometric pressure.

Fractional distillation comprises a series of evaporation–condensation steps. It is helpful at this point to refer to a diagram of the type shown in Fig. 9-18, in which vapor composition y is plotted against liquid composition x. The case illustrated is that of a relatively ideal solution. Liquid of composition x_0 produces some vapor of composition y_1. If this vapor is condensed, the effect is to locate a new liquid composition x_1 on the diagonal. Liquid x_1 produces vapor y_2 and on its condensation, liquid x_2 results. The series of steps gives the number of operations needed to reach the final liquid composition x_3. This analysis assumes that only a small amount of each liquid is vaporized; in actual practice the fraction is appreciable and so the vapor compositions are always less enriched in the more volatile component than in the ideal situation. The detailed treatment of fractional distillation constitutes a major subject in chemical engineering and is beyond the scope of this text.

A special case in distillation is that of two immiscible liquids. A mixture of two such liquids will boil when their combined vapor pressure reaches 1 atm. We thus write the separate Clausius–Clapeyron equations for each pure liquid:

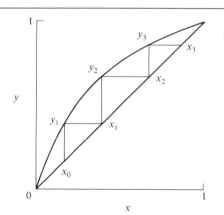

FIGURE 9-18
Plot of y versus x.

$$P_A^\circ = \exp\left[\frac{\Delta H_{v,A}^\circ}{R}\left(\frac{1}{T_{b,A}^\circ} - \frac{1}{T}\right)\right] \qquad (9\text{-}112)$$

$$P_B^\circ = \exp\left[\frac{\Delta H_{v,B}^\circ}{R}\left(\frac{1}{T_{b,B}^\circ} - \frac{1}{T}\right)\right] \qquad (9\text{-}113)$$

The normal boiling point of the mixture of liquid phases is given by

$$1 = \exp\left[\frac{\Delta H_{v,A}^\circ}{R}\left(\frac{1}{T_{b,A}^\circ} - \frac{1}{T_b}\right)\right] + \exp\left[\frac{\Delta H_{v,B}^\circ}{R}\left(\frac{1}{T_{b,B}^\circ} - \frac{1}{T_b}\right)\right] \qquad (9\text{-}114)$$

since we require that $P_A^\circ + P_B^\circ = 1$. The situation is illustrated in Fig. 9-19. Boiling of such a mixture produces vapor of composition

$$y_B = \frac{P_B^\circ}{P_A^\circ + P_B^\circ} \qquad (= P_B^\circ \text{ if } P = 1) \qquad (9\text{-}115)$$

where P_A° and P_B° are the vapor pressures of the pure liquids at T_b. On continued boiling, one or the other liquid phase will eventually disappear and the boiling point will then revert to that of the remaining liquid.

A procedure of this type is often known as a steam distillation, since a frequent application is the distillation of a mixture of water and an insoluble organic liquid or oil. The advantage is that the oil is thereby distilled at a much lower temperature than would otherwise be needed and with less danger of decomposition. It is also possible to obtain the molecular weight of the oil from Eq. (9-115). If component A is water, then P_A° is given by the measure T_b and P_B° is then the ambient or barometric pressure minus P_A° and y_B is given by Eq. (9-115). With y_B and the weight fraction of the distillate known, M_B can be calculated.

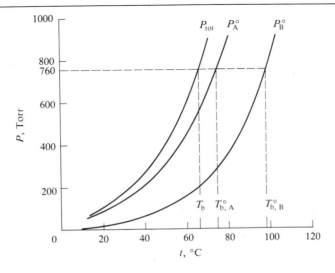

FIGURE 9-19
The case of two immiscible liquids.

EXAMPLE

Suppose that a mixture of an insoluble organic liquid and water boiled at 90.2°C under a pressure of 740.2 Torr. The vapor pressure of pure water is 530.1 Torr at this temperature. The condensed distillate is 71% by weight of the oil. Evidently $P_B°$ is 740.2 − 530.1 or 210.1 Torr; therefore y_B = 210.1/740.2 = 0.2838. Since

$$y_B = \frac{W_B/M_B}{(W_B/M_B) + (W_A/M_A)}$$

where W denotes weight of substance, or

$$\frac{W_B}{M_B} = \frac{y_B}{1 - y_B}\frac{W_A}{M_A}$$

then, per 100 g of distillate,

$$\frac{W_B}{M_B} = \frac{0.2838}{0.7162}\frac{29}{18.02} = 0.638$$

and M_2 = 71/0.638 = 111.2 g mol^{-1}.

9-8 Partial miscibility

The equilibrium between a liquid solution and a pure solid phase of one of the components is treated in Chapter 10 and that between liquid and solid solutions in Chapter 11. There remains the case of two partially miscible liquid phases. If liquid phases α and β are in equilibrium, then if the system is one of two components A and B, the condition for equilibrium is that

$$\mu_A{}^\alpha - \mu_A{}^\beta \quad \text{and} \quad \mu_B{}^\alpha = \mu_B{}^\beta \qquad (9\text{-}116)$$

As discussed in the Special Topics section, this means that a plot of the molar free energy of the solution, $\mathbf{G}/(n_A + n_B)$, versus composition shows a double minimum, and therefore so, too, does a plot of ΔG_{mix} versus composition.

The situation is one in which there is a limited solubility of B in A, giving A-rich solutions designated as α phase, and a limited solubility of A in B, giving B-rich solutions designated as β phase. The maximum solubilities may then be designated as S^α and S^β, where S^α is the composition of a solution saturated with respect to B and S^β is that of a solution saturated with respect to A. The compositions S^α and S^β are not very dependent on pressure; quite large pressures are needed to change the free energies of liquids appreciably. They are temperature-dependent, however, and this dependence is customarily shown in plots of solubility versus temperature at 1 atm pressure.

Figure 9-20 shows a schematic temperature–composition plot for two liquids A and B. The left-hand line gives the variation with temperature of S^α and the right-hand line the variation of S^β. The behavior illustrated is the very common one in which both solubilities increase with increasing temperature. There is therefore a temperature T_c at which they have become equal and above which the two liquids are completely miscible. The temperature T_c is known as a *consolute temperature*, in this case, an upper consolute temperature. The figure again has the properties of a phase map. Systems whose overall composition and temperature locate a point in the region between the S^α and S^β lines, such as system x_0 at T_1, will consist of two liquid

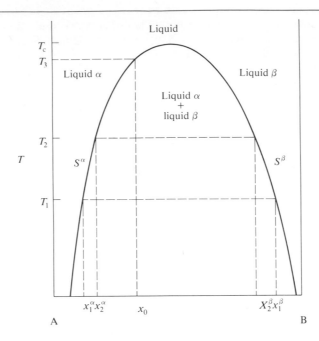

phases. The compositions are given by the ends of the tie-line as x_1^α and x_1^β, and the relative amounts present may be obtained by means of the lever principle. Thus we have

$$\frac{n^\alpha}{n^\alpha + n^\beta} = \frac{x_0 - x_1^\beta}{x_1^\alpha - x_1^\beta} \tag{9-117}$$

where n^α and n^β are, respectively, the number of moles of phase α of composition x_1^α and of phase β of composition x_1^β. When the system is warmed to T_2 the two phases are of composition x_2^α and x_2^β and application of the lever principle shows that the proportion of phase α has increased. At T_3 phase β disappears and the system consists of phase α of composition x_0.

One may, alternatively, make a horizontal traverse of the diagram. Thus addition of liquid B to pure liquid A at T_1 gives a phase of increasing mole fraction of B. When composition x_1^α is reached phase of composition x_1^β begins to appear, and further addition of B steadily increases the proportion of β phase. When the system composition reaches x_1^β no more α phase remains and continued addition of B now merely increases the concentration of B in the β phase.

The phenol–water system shows this type of behavior, as illustrated in Fig. 9-21(a). The upper consolute temperature in this case is about 70°C. It can also happen that the solubilities increase with decreasing temperature, as illustrated by the water–triethylamine system of Fig. 9-21(b). There is now a *lower consolute temperature*. Finally, both types of behavior may be shown, as in the case of the water–nicotine system of Fig. 9-21(c). The compositions are given in weight fraction in all these figures and application of the lever principle will therefore give the relative weights of the two phases that are present.

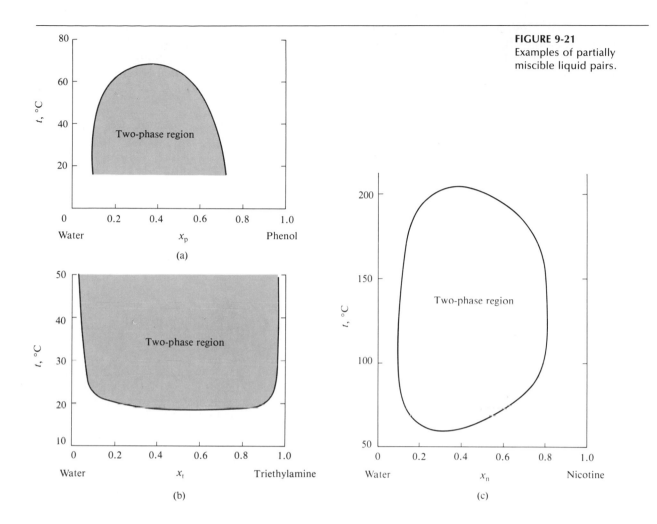

FIGURE 9-21
Examples of partially miscible liquid pairs.

COMMENTARY AND NOTES

9-CN-1 Ideal, regular, and athermal solutions. Some additional properties of solutions

Several of the attributes of an ideal solution have been introduced in the preceding sections. These, plus some further ones, are assembled here to provide an overall picture. The ideal solution obeys Raoult's law, so that

$$\mu_i(l) = \mu_i^\circ(l) + RT \ln x_i$$
[ideal solution, Eq. (9-76)]

The free energy of mixing to form an ideal solution is given, for the case of two components, by

$$\Delta G_M = x_1 RT \ln x_1 + x_2 RT \ln x_2$$
[ideal solution, Eq. (9-80)]

and the entropy of mixing by

$$\Delta S_M = -(x_1 R \ln x_1 + x_2 R \ln x_2)$$
[ideal solution, Eq. (9-84)]

The enthalpy of mixing is zero for an ideal solution, since $\Delta H_M = \Delta G_M + T\Delta S_M$. Also, $\overline{H}_i$

$= H^0_i$ for an ideal solution. It follows from their definitions (Section 9-5B) that ΔG_E and ΔS_E are zero. Differentiation of Eq. (9-76) with respect to P at constant T gives, with the use of Eq. 9-48, $\bar{V}_i = V_i^\circ$. The partial molal volume of an ideal solution component is thus the same as that of the pure component. Consequently $\Delta V_M = 0$.

Nonideal solutions show more complicated behavior, such as that illustrated in Fig. 9-14. Many show certain regularities, however, which allow some categorization. If, for example, Eqs. (9-16) and (9-17) are obeyed, then

$$\Delta G_E = x_1 x_2 \alpha RT \qquad (9\text{-}118)$$

Solutions obeying Eq. (9-118) have been called *"simple" solutions* by E. A. Guggenheim. One model for solutions concludes that α is just an interaction energy divided by RT. It follows that $d(\Delta G_E)/dT$ and hence the excess entropy of mixing should be zero, but not, of course, ΔH_M. This type of solution is known, after J. H. Hildebrand, as a *regular solution*. The implication is that the two components are randomly distributed in the solution as though it were ideal, although their interaction energies with themselves and with each other are different.

An *athermal solution* is one for which $\Delta H_M = 0$, although ΔS_E and ΔG_E are not necessarily zero. This situation may occur with components rather similar in chemical nature but very different in molecular size. A simple treatment on this basis leads to the equation for the free energy of mixing:

$$\Delta G_M = x_1 RT \ln \phi_1 + x_2 RT \ln \phi_2 \qquad (9\text{-}119)$$

where ϕ_1 and ϕ_2 are the volume fractions. There is some statistical thermodynamic basis for the supposition that for a solution to be ideal the molar volumes of the components should be about the same (see the next section) and Raoult's law may in fact be less general, even as a limiting law, than is customarily thought.

Although the vapor pressure of solutions is emphasized in this chapter, all of the various types of properties mentioned in previous chapters have been measured for solutions. As one example, the viscosity of a solution of similar substances will vary nearly linearly with composition; often a better straight line is obtained if the reciprocals or fluidities are used instead. Deviations from linearity in the plot of viscosity versus mole fraction tend to correlate with such deviations in the corresponding vapor pressure diagrams. The acetone–chloroform system, which shows a minimum in the vapor pressure diagram attributable to greater A–B than A–A or B–B types of interactions, has a maximum in the viscosity–composition plot. Diffusion has also been studied a good deal in binary liquid systems. There is a single mutual diffusion coefficient, but, in addition, self-diffusion coefficients may be obtained for each component separately by means of isotopic labeling. As with viscosity, there are a number of semiempirical models but no really satisfactory ones. The surface tension of solutions constitutes a large subject, aspects of which are discussed in Section 9-ST-2.

9-CN-2 Statistical thermodynamics of solutions

Needless to say, the statistical thermodynamic treatment of solutions is difficult and fragmentary in its achievements. One may, in principle, set up the partition functions, which now involve the chemical potential of a component as a weighting factor. Useful in this connection is what is called the *absolute* or *rational activity q*, given by

$$\mu_i = RT \ln q_i \qquad (9\text{-}120)$$

Like the chemical potential, the absolute activity of a species is the same in all equilibrium phases in which the species is present.

Certain simple entropy calculations can be made. One consideration is the following. It will be recalled that the translational partition function for a gas required the factor $1/N!$

$$\mathbf{Q}_{tot} = \frac{1}{N!}\, \mathbf{Q}^{N_A} \qquad [\text{Eq. (6-68)}]$$

In the case of a crystalline solid each molecule is restricted to its own volume V/N_A and the translational partition function becomes

$$\mathbf{Q}'_{tot} = \mathbf{Q}^{N_A}\left(\frac{V}{N_A}\right)^{N_A} \qquad (9\text{-}121)$$

Equation (9-121) follows from Eq. (4-84) if the volume is made V/N_A rather than v, the volume of the whole system. The difference between the two corresponding entropies is, by Eq. (6-

75), just $k \ln(\mathbf{Q}/\mathbf{Q}')$, since the energies for the two situations are the same. Thus

$$S(\text{gas}) - S(\text{lattice}) = k \ln\left(\frac{N_A^{N_A}}{N_A!}\right) = N_A k \tag{9-122}$$

[using Stirling's formula for $\ln(N_A!)$].

This entropy factor is known as the *communal entropy*, and is thought to develop by stages as a solid melts and the liquid is heated. The lattice model may be approximately applicable to the liquid, in other words. Although absolute calculations of thermodynamics quantities are most difficult, we can obtain the entropy of mixing. We assume the solution to consist of lattice sites, all equivalent, which may be occupied either by a molecule of component 1 or by a molecule of component 2. In addition to all other contributions there is now one which has to do with the ways in which N_1 molecules of species 1 and N_2 molecules of species 2 may be distributed, where $N_1 + N_2 = N_A$. The reasoning at this point is very similar to that of Secton 2-2. There are $N_A!$ ways in which N_A molecules can be arranged among the sites, but we assume that those of component 1 are indistinguishable among themselves, and likewise for those of component 2. We must then divide by the number of ways in which N_1 molecules can be arranged among themselves, and similarly for the N_2 molecules. The thermodynamic probability of the solution is

$$W = \frac{N_A!}{N_1! \, N_2!} \tag{9-123}$$

The relationship between entropy and thermodynamic probability [Eq. (6-77)] now gives for the mixing entropy of the solution

$$\Delta S_M = k \ln\left(\frac{N_A!}{N_1! \, N_2!}\right) \tag{9-124}$$

Use of Stirling's approximation for factorials and some straightforward algebraic maneuvering yields the final form

$$\Delta S_M = -(x_1 R \ln x_1 + x_2 R \ln x_2)$$
$$[\text{Eq. (9-84)}]$$

which is the same as that previously obtained. The uniform lattice concept does seem to imply that the two species should be of about the same molecular size for ideal solution behavior to hold, as mentioned in the preceding section. This is only an implication, however, since the lattice picture is not essential to obtain a correct expression for the entropy of mixing. Thus Eq. (9-84) also applies to the mixing of two ideal gases, although the basis of obtaining it is through the Sackur–Tetrode equation [Eq. (6-102)], which gives the volume dependence of the entropy of an ideal gas to be $R \ln V$. Then ΔS for the mixing process of Eq. (9-65) is, per mole,

$$\Delta S_M = x_1 R \ln \frac{V}{V_1} + x_2 R \ln \frac{V}{V_2} \tag{9-125}$$

where V_1 and V_2 are the initial volumes of the two gases and V is their common final volume. Since $V_1 = x_1 V$ and $V_2 = x_2 V$, Eq. (9-84) again results.

Thus two very different pictures have produced the same conclusion. One must be cautious in assuming that, simply because a particular model yields a correct equation, the model itself is therefore correct.

SPECIAL TOPICS

9-ST-1 More on enthalpies and free energies of mixing

The measurement and handling of parital molal volume quantities is discussed in Section 9-3D. Enthalpies are treated somewhat similarly, but a complication is that, unlike volumes, absolute enthalpies are not known. It is necessary, then, to deal entirely with heats of mixing. The enthalpy change for the process

$$n_1 \text{ (component 1)}$$
$$+ \, n_2 \text{ (component 2)} = \text{solution}$$

is called the *integral heat of solution*, as an alternative expression to the heat of mixing $\Delta \mathbf{H}_M$. For this process

$$\Delta\mathbf{H}_M = \mathbf{H}_{soln} - n_1 H_1^\circ - n_2 H_2^\circ \qquad (9\text{-}126)$$

where H_1° and H_2° are the enthalpies of the pure components. Alternatively, using Eq. (9-58), we obtain

$$\Delta\mathbf{H}_M = n_1(\overline{H}_1 - H_1^\circ) + n_2(\overline{H}_2 - H_2^\circ) \qquad (9\text{-}127)$$

or

$$\Delta\mathbf{H}_M = n_1\overline{Q}_1 + n_2\overline{Q}_2 \qquad (9\text{-}128)$$

where $\overline{Q}$ denotes the enthalpy relative to the pure component. According to Eq. (9-128), $\overline{Q}_2$ is given by

$$\overline{Q}_2 = \left[\frac{\partial(\Delta\mathbf{H}_M)}{\partial n_2}\right]_{n_1} \qquad (9\text{-}129)$$

and could be obtained experimentally from the slope of a plot of $\Delta\mathbf{H}_M$ versus n_2, from data on the heats of dissolution of various amounts of solute in a fixed amount of solvent. Therefore $\overline{Q}_2$ is called the *differential enthalpy of solution*. Equation (9-128) may alternatively be written in the form

$$\Delta H_M = \frac{\Delta\mathbf{H}_M}{n_1 + n_2} = x_1\overline{Q}_1 + x_2\overline{Q}_2 \qquad (9\text{-}130)$$

The same graphical procedure may now be applied as was used for obtaining partial molal volumes. Thus if a plot of ΔH_M versus mole fraction is constructed, then the tangent at a given composition will have intercepts at $x_1 = 1$ and $x_2 = 1$ of $\overline{Q}_1$ and $\overline{Q}_2$, respectively, as illustrated in Fig. 9-22.

Many of the results on heats of solution are for electrolytes or other solid solutes, and for such systems it is customary to polarize the treatment around the solute species, component 2. One refers heats of solution of solutes to the value $\overline{Q}_2^\circ$ for an infinitely dilute solution by introducing a quantity called the *relative enthalpy* of solution $\overline{L}$:

$$\overline{L}_2 = \overline{Q}_2 - \overline{Q}_2^\circ = \overline{H}_2 - \overline{H}_2^\circ \qquad (9\text{-}131)$$

where $\overline{H}_2^\circ$ is the partial molal enthalpy of the solute at infinite dilution and $\overline{Q}_2^\circ = \overline{H}_2^\circ - \overline{H}_2^\circ$. By definition $\overline{L}_2^\circ = 0$ Usually the pure liquid solvent is kept as the reference state for component 1, so we have

$$\overline{L}_1 = \overline{Q}_1 = \overline{H}_1 - \overline{H}_1^\circ \qquad (9\text{-}132)$$

Since $\overline{H}_2^\circ$ and H_1° are constants, insertion of the definitions for $\overline{L}_1$ and $\overline{L}_2$ into Eq. (9-59) gives

$$n_1\, d\overline{L}_1 + n_2\, d\overline{L}_2 = 0 \qquad (9\text{-}133)$$

The $\overline{L}$ quantities may thus be used in the same way as the $\overline{Q}$ quantities, or in general as ordinary partial molal quantities. We also have

$$\Delta\mathbf{H}_{M,x_2} - \Delta\mathbf{H}_M^\circ = L = n_1\overline{L}_1 + n_2\overline{L}_2 \qquad (9\text{-}134)$$

where $\Delta\mathbf{H}_{M,x_2}$ is the enthalpy of mixing of n_1 moles of solvent and n_2 moles of solute to give a solution of composition x_2, and $\Delta\mathbf{H}_M^\circ$ is the heat of solution of n_2 moles of solute to give an infinitely dilute solution.

One may obtain $\overline{L}_1$ or $\overline{Q}_1$ experimentally as suggested by the equation analogous to Eq. (9-129), that is, from the variation of $\Delta\mathbf{H}_M$ with n_1, as solvent is added to a fixed amount of solute. The alternative, and equivalent, measurement is that of the heat evolved on the

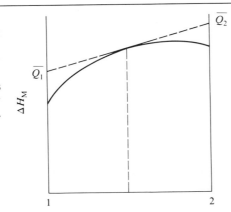

FIGURE 9-22
Variation of the molar heat of mixing ΔH_M with composition. The intercepts of the tangent give the differential heats of solution $\overline{Q}_1$ and $\overline{Q}_2$ for that composition.

$\bar{L}_1$ is 4.0 cal mol^{-1} and $\bar{L}_2$ is -248 cal mol^{-1} for a 1.11 m solution of sodium chloride.   **EXAMPLE**
The heat of solution at infinite dilution is, from Tables 5-2 and 5-3,

$$NaCl(s) + H_2O = NaCl(\text{infinitely dilute solution})$$

$$\Delta H = -97{,}302 - (-98{,}232) = 930 \quad \text{cal mol}^{-1}$$

The heat of solution to give a 1.11 m solution differs from this value by

$$L = n_1\bar{L}_1 + n_2\bar{L}_2 \qquad [\text{Eq. (9-134)}]$$

or by $(55.5)(4.0) + (1.11)(-248) = 222 - 275 = -53$ cal per 1.11 mole or by -48 cal mol^{-1}. The actual heat of solution is then $930 - 48 = 882$ cal mol^{-1}.

addition of a small amount of solvent to a solution of a given composition. This last is known as a *heat of dilution*. One may also obtain $\bar{L}_1$ by the graphical method of Fig. 9-22 or indirectly from $\bar{L}_2$ values by the integration of Eq. (9-133).

Heats of mixing or of solution are direct, calorimetrically determined quantities, and the preceding framework of relationships and definitions has been developed with this in mind. Free energies of mixing are determined indirectly, through vapor pressure measurements, but may still be treated in just the same way. The equation analogous to Eq. (9-130) is

$$\Delta G_M = x_1\mu_{1(\text{rel})} + x_2\mu_{2(\text{rel})} \qquad (9\text{-}135)$$

where $\mu_{1(\text{rel})} = \mu_1 - \mu_1^\circ$ and $\mu_{2(\text{rel})} = \mu_2 - \mu_2^\circ$. Equation (9-135) is the same as Eq. (9-90). Again, if ΔG_M is plotted against mole fraction, the tangent line at a given composition has intercepts at $x_1 = 1$ and $x_2 = 1$ corresponding to $\mu_{1(\text{rel})}$ and $\mu_{2(\text{rel})}$, that is, to $RT \ln a_1$ and $RT \ln a_2$, respectively. If the plot is of ΔG_E, the intercepts give $RT \ln \gamma_1$ and $RT \ln \gamma_2$, as indicated in Fig. 9-14 for $x_c = 0.4$.

Figure 9-23 is calculated from the Margules equations using an α of 2.5, and illustrates an important further point. Since there are two minima, there are two compositions, x^α and x^β, for which the a_1 and a_2 values are the same. The situation is one of partial miscibility and a system of overall composition lying between x^α and x^β will spontaneously separate into phases of those two compositions. The figure is symmetric because of the simplicity of the model; in most actual cases of two partially miscible liquids the two minima would not be symmetrically disposed.

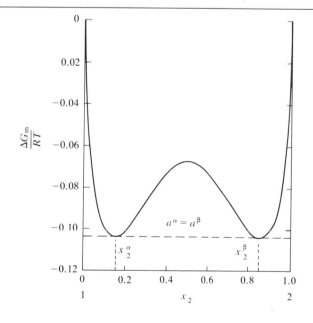

FIGURE 9-23
Free energy of mixing for a system obeying the Margules equations with an α of 2.5.

9-ST-2 The surface tension of solutions. The Gibbs equation

Solutions show quite a variety of surface tension behavior, and some typical categories of surface tension versus composition plots are shown in Fig. 9-24. In the case of similar liquids the surface tension plot is roughly symmetric relative to a straight line connecting the values of γ for the pure liquids, as exemplified by the data shown in Fig. 9-24(a). If the molecular areas σ of the two species are similar, then a simple treatment based on the energy $\gamma\sigma$ required to bring a molecule into the surface gives the equation

$$e^{-\gamma\sigma/kT} = x_1 e^{-\gamma_1\sigma/kT} + x_2 e^{-\gamma_2\sigma/kT} \qquad (9\text{-}136)$$

The surface tensions of the respective pure liquids are given by γ_1 and γ_2. Another form, derived for regular solutions (Section 9-CN-1), is

$$\gamma = \gamma_1 x_1 + \gamma_2 x_2 - \beta x_1 x_2 \qquad (9\text{-}137)$$

where β is an empirical constant related to the constant α of Eq. (9-16).

If the two liquids have rather different surface tensions, then the plot will look like that shown in Fig. 9-24(b) for the water–ethanol system. The surface tension drops rapidly over the first 10 or 20% of ethanol added and then approaches the value for pure ethanol more slowly. This type of behavior becomes accentuated in the case of a long-chain solutes having a polar end group, such as sodium lauryl sulfate. As illustrated in Fig. 9-24(c), there is a sharp drop in γ even for very dilute solutions. Such long-chain solutes include the common soaps and detergents and belong to the class of so-called surface-active agents (surfactants). As discussed further below, the thermodynamic implication is that the surfactant concentrates at the solution–air interface. With sodium lauryl sulfate, there is nearly a monolayer or complete film of the surfactant by 0.01 M concentration.

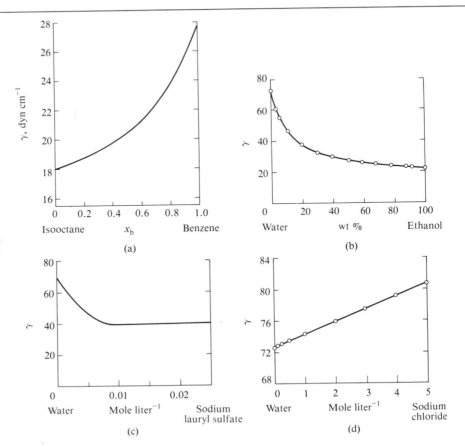

FIGURE 9-24 Surface tension–composition behavior: (a) similar liquids; (b) a high surface tension liquid and a low surface tension liquid; (c) an aqueous surfactant system; (d) an aqueous electrolyte system. [Part (a) from H. B. Evans, Jr. and H. L. Clever, J. Phys. Chem. **68,** 3433 (1964). Copyright 1964 by the American Chemical Society. Reprinted by permission of the copyright owner.]

A fourth category of system is that of an electrolyte solution such as aqueous sodium chloride, shown in Fig. 9-24(d). The surface tension increases somewhat with concentration. The thermodynamic implication is that the surface region is more dilute in electrolyte than the bulk solution, or that negative adsorption occurs at the interface.

As suggested above, an important application of thermodynamics is to the variation of surface tension of a solution with its composition. The following derivation is essentially that of J. W. Gibbs, and the result is known as the *Gibbs adsorption equation.* We wish to deal with surface thermodynamic quantities and we must somehow separate their contribution from those of the bulk phases that form the interface. We do this by locating an arbitrary dividing plane S–S roughly in the interfacial region as shown in Fig. 9-25. We then assign a total energy and entropy to bulk phase α assuming it to continue unchanged up to this dividing plane, and similarly for bulk phase β. The actual total energy and enthalpy of the system are then written as

$$\mathbf{E} = \mathbf{E}^{\alpha} + \mathbf{E}^{\beta} + \mathbf{E}^{S}$$
$$\mathbf{S} = \mathbf{S}^{\alpha} + \mathbf{S}^{\beta} + \mathbf{S}^{S} \qquad (9\text{-}138)$$

where $\mathbf{E}^{S}$ and $\mathbf{S}^{S}$ are now called the *surface excess energy* and *entropy.* Similarly, we have

$$n_1 = n_i^{\alpha} + n_i^{\beta} + n_i^{S} \qquad (9\text{-}139)$$

For a small, reversible change $d\mathbf{E}$ in the energy of the whole system,

$$dE = dE^{\alpha} + dE^{\beta} + dE^{S} = T\, dS^{\alpha}$$
$$- P\, dv + \sum_i \mu_i\, dn_i^{\alpha} + T\, dS^{\beta} - P\, dv^{\beta}$$
$$+ \sum_i \mu_i\, dn_i^{\beta} + T\, dS^{S}$$
$$+ \sum_i \mu_i\, dn_i^{S} + \gamma\, d\mathcal{A} \qquad (9\text{-}140)$$

where γ is surface tension and $\mathcal{A}$ is the area of the surface (the volume is entirely taken care of by $v^{\alpha} + v^{\beta}$). Equation (9-28) applies separately to phases α and β, so these terms all drop out, to leave

$$d\mathbf{E}^{S} = T\, dS^{S} + \sum_j \mu_i\, dn_i^{S} + \gamma\, d\mathcal{A}$$
$$(9\text{-}141)$$

The Euler theorem (Section 9-3D) may now be applied; that is, Eq. (9-141) may be integrated, with T, μ_i, and γ kept constant, to give

$$\mathbf{E}^{S} = T\mathbf{S}^{S} + \sum_i \mu_i\, n_i^{S} + \gamma\mathcal{A} \qquad (9\text{-}142)$$

Differentiation and comparison with the preceding equation gives

$$0 = \mathbf{S}^{S}\, dT + \sum_i n_i^{S}\, d\mu_i + \mathcal{A}\, d\gamma \quad (9\text{-}143)$$

For a two-component system at constant temperature,

$$n_1^{S}\, d\mu_1 + n_2^{S}\, d\mu_2 + \mathcal{A}\, d\gamma = 0 \quad (9\text{-}144)$$

It is convenient for us to divide through by the area $\mathcal{A}$ to obtain

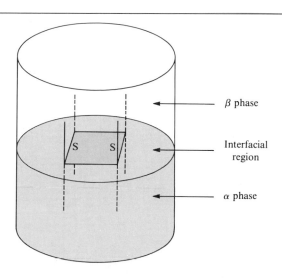

β phase

Interfacial region

α phase

FIGURE 9-25
Illustration of the location of a dividing plane S in the interfacial region between phases α and β. The dividing plane is parallel to the interface and of unit cross-section.

$$dγ = -Γ_1 dμ_1 - Γ_2 dμ_2 \qquad (9\text{-}145)$$

where $Γ_1$ and $Γ_2$ are the excess quantities per unit area.

The exact position of the dividing surface shown in Fig. 9-25 is not specified; clearly the values of $Γ$ will depend on this. We now specify the location to be such that $Γ_1 = 0$, so that Eq. (9-145) reduces to

$$Γ_2{}^1 = -\left(\frac{\partial γ}{\partial μ_2}\right)_T \qquad (9\text{-}146)$$

where the superscript is a reminder of the choice that has been made. The chemical potential may be expressed in terms of activity,

$$Γ_2{}^1 = -\frac{1}{RT}\frac{dγ}{d(\ln a_2)} \qquad (9\text{-}147)$$

Finally, in dilute solution the activity will be proportional to concentration, so that an approximate form is

$$Γ_2{}^1 = -\frac{C}{RT}\frac{dγ}{dC} \qquad (9\text{-}148)$$

The preceding are various forms of the Gibbs equation.

Figure 9-26 may help to explain the physical meaning of this conventional choice of location of the dividing surface. The figure shows schematically how the concentrations of solvent and of solute might vary across the interfacial region. The β phase is assumed to be vapor, so the concentrations in it are negligible. The surface excess is the difference between the amount actually present and that which would be present were the bulk phase to continue unchanged up to S–S so that the phase boundary became a step. The net shaded area for the solvent is then its surface excess, and S–S has been located so that this is zero. The surface excess of the solute is also given by its net shaded area and is positive in this example. An alternative, operational definition is as follows. If a sample of interface is taken, of 1 cm^2 area, and deep enough to include at least some bulk phase on either side, then $Γ_2{}^1$ is the (algebraic) excess of solute over the number of moles that would be present in a bulk region containing the same number of moles of solvent. A famous experiment made by J.W. McBain and co-workers in 1932 allowed $Γ_2{}^1$ to be measured directly. A fast-moving knife blade (called a *microtome*) scooped a 3.2 g sample of solution at 20°C from the surface of a trough having a surface area of 310 cm^2. The solution contained 5 g of phenol ($M = 94$ g mol^{-1}) per 1000 g of water, that is, it was 0.053 m. It was found, by means of an interferometer, that the sample contained 2.52×10^{-6} g more of phenol per gram of water than did the bulk solution. The value of $Γ_2{}^1$ is thus $(2.52 \times 10^{-6})(3.2)/(94)(310) = 2.77 \times 10^{-10}$ mol cm^{-2}.

The Gibbs equation allows an indirect calculation of $Γ_2{}^1$ from surface tension data. Continuing with the preceding example, the surface tensions of 0.05 m and 0.127 m solutions were 67.7 and 60.1 dyn cm^{-1}, respectively, at 20°C. A plot of $γ$ versus C gives a slope, $-dγ/dC$, of 100 dyn cm^{-1} M^{-1} and application of Eq. (9-148) yields

$$Γ_2{}^1 = \frac{(0.063)(100)}{(8.31 \times 10^7)(293)}$$
$$= 2.2 \times 10^{-10} \text{ mole cm}^{-2}$$

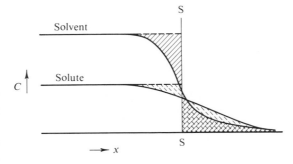

FIGURE 9-26
Illustration of the $Γ_2{}^1$ convention for the Gibbs equation. The dividing surface is located so that the shaded areas for the solvent curve balance. [From A.W. Adamson, "The Physical Chemistry of Surfaces," 3rd ed. Copyright 1976, Wiley (Interscience), New York. Used with permission of John Wiley & Sons, Inc.]

The two numbers agree fairly well—the microtome experiment is a very difficult one and was, in fact, a triumph of its day.

The value of Γ_2^1 obtained is, in one sense, a very small number. It corresponds, however, to about 80 Å^2 per molecule, or perhaps twice the value for a close-packed monolayer of molecules lying flat on the surface. The surface population is thus quite high.

As the example illustrates, if the surface tension of a solution decreases with increasing concentration, then Γ_2^1 is positive and the solute is concentrated at the interface. Conversely, as with electrolyte solutions, if $d\gamma/dC$ is positive, then Γ_2^1 is negative, meaning that the surface concentration of electrolyte is less than it is in solution. In the case of 1 M sodium chloride, the negative surface excess is equivalent to a surface layer of pure water about one molecule thick.

In sufficiently dilute solution, the surface tension will approach proportionality to concentration, that is,

$$\gamma = \gamma^\circ - bC \qquad (9\text{-}149)$$

where γ° is the surface tension of the pure solvent. The product $C\, d\gamma/dC$ is now just $-(\gamma^\circ - \gamma)$, so Eq. (9-148) reduces to

$$\Gamma_2^1 = \frac{\gamma^\circ - \gamma}{RT}$$

It is now convenient to introduce the quantity π, called the *surface* pressure, and defined as

$$\pi = \gamma^\circ - \gamma \qquad (9\text{-}150)$$

Also $\Gamma_2^1 = 1/\sigma$, where σ is the area per mole. With these substitutions Eq. (9-148) becomes

$$\pi\sigma = RT \qquad (9\text{-}151)$$

This is the equation of state of a two-dimensional ideal gas! The film pressure does in fact correspond to a two-dimensional pressure. As illustrated in Fig. 9-27, if a flexible membrane separates pure solvent from solution, the floating barrier will experience a force πl, where l is its length. Returning to the numerical example, the film pressure of a 0.05 m solution of phenol was given as 5.2 dyn cm^{-1}. Again, this seems like a small number. The force is being exerted by a monolayer, however, and so we obtain the equivalent three-dimensional pressure by dividing by the depth of a molecule. In this case, the result is about $5.2/(4 \times 10^{-8})$ or 1.3×10^8 dyn cm^{-2}, which corresponds to 130 atm. Thus the lateral compression on the monolayer is quite appreciable at the molecular level.

More generally, the Gibbs equation allows surface tension–concentration data to be translated into values of π versus σ and such plots often look much like the ones of P versus V for nonideal gas. A two-dimensional van der Waals equation may be used, for example,

$$\left(\pi + \frac{a}{\sigma^2}\right)(\sigma - b) = RT \qquad (9\text{-}152)$$

Such films are known as Gibbs monolayers, since they are normally studied through use of the Gibbs equation.

If the surfactant is quite insoluble, it may be spread directly onto the liquid surface, usually one of water. Now Γ_2^1 is known directly, as the amount placed on a known area of surface, and hence σ also is known. The film pressure is usually obtained from surface tension measurements, by the Wilhelmy slide method, although the force on a floating barrier may also be measured directly. The data are again usually reported as π versus σ plots.

Such plots may resemble those for bulk phases. As illustrated in Fig. 9-28, a film or *monolayer* of stearic acid has a very low compressibility; the π versus σ plot extrapolates to an area of about 22 Å^2 per molecule, corresponding to close-packing, and in general the

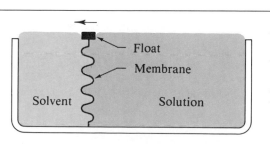

FIGURE 9-27
The PLAWM (Pockels–Langmuir–Adam–Wilson–McBain) trough—a means of measuring directly the film pressure π for a solution.

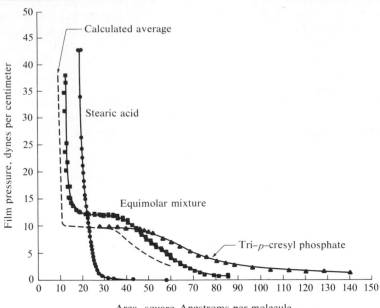

FIGURE 9-28

Isotherms of π versus σ for stearic acid, tri-p-cresyl phosphate, and an equimolar mixture [From H. E. Ries, Jr., and H. D. Cook, J. Colloid Sci. **9,** 535 (1954).]

Calculated average

Stearic acid

Equimolar mixture

Tri–p–cresyl phosphate

Film pressure, dynes per centimeter

Area, square Angstroms per molecule

film behaves as though it were a two-dimensional solid. On the other hand, the bulky tri-p-cresyl phosphate molecule forms a highly compressible film and one whose properties are like those of a low-density but viscous fluid. Note that nonideal two-dimensional mixtures are possible! The mixed film shows a π–σ behavior that departs significantly from that expected for an ideal solution. At low film pressures the tri-p-cresyl phosphate seems to dominate, whereas at high film pressures the mixed film behaves more like stearic acid.

GENERAL REFERENCES

General treatises cited in Chapter 1.

ADAMSON, A.W. (1982). "The Physical Chemistry of Surfaces," 4th ed. Wiley (Interscience), New York.

BERRY, R.S., RICE, S.A., AND ROSS, J. (1980). "Physical Chemistry." Wiley, New York.

HILDEBRAND, J.H., AND SCOTT, R.L. (1950). "The Solubility of Nonelectrolytes," 3rd ed. Van Nostrand–Reinhold, Princeton, New Jersey.

CITED REFERENCES

FRICKE, R., (1929). *Z. Elektrochem.* **35,** 631.

HILDEBRAND, J.H., AND SCOTT, R.L. (1950). "The Solubility of Nonelectrolytes," 3rd ed. Van Nostrand–Reinhold. Princeton, New Jersey.

McBAIN, J.W., AND HUMPHREYS, C.W. (1932). *J. Phys. Chem.* **36,** 300.

ROBINSON, P.J. (1964). *J. Chem. Ed.* **41,** 654.

EXERCISES

Take as exact numbers given to one significant figure.

9-1 Assume that toluene and *p*-xylene form ideal solutions. The normal boiling point of toluene is 110.6°C, and at this temperature the vapor pressure of *p*-xylene is 340 Torr. Calculate (a) the separate partial pressures and the total vapor pressure at 110.6°C of a solution of $x_t = 0.25$, and (b) the composition of that solution which would boil at 110.6°C if under the reduced pressure of 500 Torr.

Ans. (a) $P_t = 190$ Torr,
$P_x = 255$ Torr, $P_{tot} = 445$ Torr; (b) $x_t = 0.381$.

Calculate the composition of the vapor in equilibrium with each of the two solutions of Exercise 9-1.

9-2

Ans. $y_b = 0.427$, $y_b = 0.579$.

The vapor pressure of cyclohexane (ch) is 100 mm Hg at 25.5°C. A mixture of 0.4 mole of ch with 0.1 mole of n-hexane (h) has a total vapor pressure of 110 mm Hg at this temperature. Assuming ideal solution behavior, calculate (a) the vapor pressure of pure h at this temperature and (b) the composition of the vapor above the solution.

9-3

Ans. (a) $P_h = 150$ mm Hg, (b) $y_{ch} = 0.727$.

For what composition of solution at 25.5°C would the vapor phase be equimolar in the two components of Exercise 9-3?

9-4

Ans. $x_{ch} = 0.600$.

The molecular weight of substance B is 95 g mol^{-1}, and dissolving 0.35 g of B in 1.5 mole of a nonvolatile solvent A gives a solution with a vapor pressure of 3 Torr. Calculate the Henry's law constant for B dissolved in A.

9-5

Ans. 1224 Torr.

The Henry's law constant for Kr in water is 2.00×10^4 atm at 20°C. How many grams of Kr should dissolve in 1000 g of water at this temperature under pressure of 30 atm?

9-6

Ans. 6.99 g.

The Henry's law constant for N_2 in water is 8.71×10^9 Pa at 20°C. How many cm^3 of N_2, measured at 20°C and the pressure used, should dissolve in 1 cm^3 of water?

9-7

Ans. 0.0155 cm^3.

Two mole of toluene and 8 mole of benzene are introduced into a vessel at 20°C and the total vapor pressure is found to be 60 Torr. Using Fig. 9-2(b), estimate the number of moles of vapor formed and the compositions of the liquid and vapor phases present.

9-8

Ans. $x_t = 0.35$, $y_t = 0.18$, $n_v = 8.8$.

Calculate the vapor composition above a mixture of water and p-xylene, two essentially immiscible liquids, at 75.9°C. At this temperature the vapor pressures are 300 mm Hg and 100 mm Hg for water and p-xylene, respectively.

9-9

Ans. x_x eq 0.25.

Sketch the phase diagram for the system of Exercise 9-9 (the diagram analogous to Fig. 9-8). One tenth mole of water and 0.5 mole of p-xylene are introduced into an evacuated flask. On bringing the temperature to 75.9°C it is found that all of the water and 10% of the p-xylene has evaporated. Calculate the total pressure.

9-10

Ans. 300 mm Hg.

In a steam distillation of an insoluble oil the boiling point of the mixture is found to be 95°C and the distillate is found to contain 80% by weight of the oil. Atmospheric pressure is 755 Torr, and the vapor pressure of water at 95°C is 634 Torr. Calculate the molecular weight of the oil.

9-11

Ans. 377 g mol^{-1}.

Isopropyl alcohol (ipa) and benzene (b) form nonideal solutions. If x_{ipa} is 0.059, the partial pressure of ipa is 12.9 Torr at 25°C. The vapor pressures of the pure liquids are 44.0 and 94.4 Torr, respectively. Calculate k_{ipa} and α and estimate k_b.

9-12

Ans. $k_{ipa} - 269$ Torr, $\alpha = 1.81$, $k_b = 577$ Torr.

The partial molal volumes for water and ethanol at 20°C are 17 and 57 cm^3 mol^{-1}, respectively, for a solution of $x_e = 0.4$. Calculate the volume change on mixing sufficient ethanol with two moles of water to give this final composition.

9-13

Ans. -3.87 cm^3.

9-14   Calculate ΔG_M, ΔH_M, and ΔS_M for the process whereby O_2 and N_2, each at STP, are mixed to form 1 mole of air at STP. Assume ideal gas behavior; air is 20 mole % O_2.

Ans. $\Delta G_M = -272$ cal, $\Delta H_M = 0$, $\Delta S_M = 0.994$ cal K^{-1}.

9-15   Calculate ΔG_M, ΔH_M, and ΔS_M for the mixing of 1 mole of toluene with 3 moles of *p*-xylene at 25°C. Assume ideal solution behavior.

Ans. $\Delta G_M = -1330$ cal, $\Delta H_M = 0$, $\Delta S_M = 4.47$ cal K^{-1}.

9-16   Consider the solution described in Exercise 9-12. Calculate (a) the activity and the activity coefficient of ipa, taking pure ipa to be the standard state, and (b) the excess free energy of mixing of one mole of ipa with sufficient benzene to form a solution of the given composition.

Ans. (a) $a_{ipa} = 0.293$, $\gamma_{ipa} = 4.97$; (b) $\Delta G_E = 1006$ cal.

9-17   A certain amount of an ethanol–benzene solution of $x_e = 0.20$ is introduced into a flask; some of it vaporizes and the residual solution has a total vapor pressure of 750 Torr at 72.5°C. Find the compositions of the final solution and of the vapor phase in equilibrium with it at 72.5°C, and the percentage of original solution that vaporized.

Ans. x_e (final) $= 0.09$, $y_e = 0.205$, 95%.

9-18   An ethanol–benzene solution of $x_b = 0.2$ is boiled in an open flask until the boiling point rises to 75°C. Barometric pressure is 750 mm Hg. Estimate the number of moles of liquid remaining per mole originally present.

Ans. 0.4.

9-19   One hundred grams of a 60 mole % solution of phenol in water ($x_p = 0.60$) initially at 80°C is cooled. (a) At what temperature will the solution become turbid? (b) What are the amounts and compositions of the phases present at 40°C? (The abscissa scale of Fig. 9-21(a) is in mole fraction.)

Ans. (a) 55°C; (b) 95 g of phenol-rich phase with $x_p = 0.67$ and 5 g of water-rich phase with $x_p = 0.10$.

PROBLEMS

9-20   Derive a general equation that relates x_2 in a two-component solution to molality m.

9-21   Derive a general equation that relates x_2 in a two-component solution to molarity M.

9-22   The vapor pressures of ethylene bromide and propylene bromide are 172 and 127 Torr, respectively, at 80°C; the compounds form nearly ideal solutions. Thirty grams of ethylene bromide and 25 g of propylene bromide are equilibrated at 80°C and a total pressure of 150 Torr. Calculate the composition of the liquid phase and the moles of each compound in the vapor phase.

9-23   Suppose that *o*-, *m*-, and *p*-xylene form ideal three-component solutions. The respective vapor pressures at 60°C are 40, 50, and 50 mm Hg. We want to find all possible compositions of solutions such that $y_p = 0.60$. Show that the possible compositions obey an equation of the form $x_m = a + bx_o$ and evaluate the constants a and b.

9-24   Liquids A and B form an ideal solution. A solution of mole fraction $x_A = 0.4$ is treated as follows: (a) 0.30 mole is introduced into an evacuated vessel of volume such that, at 25°C, 15% of the liquid (mole %) evaporates. The final total pressure is 82.2 Torr at 25°C. (b) A portion of the equilibrium vapor is drawn off, condenses completely, and then found to have a total vapor pressure of 92.3 Torr at 25°C. Calculate the vapor pressures of pure liquids A and B at 25°C.

9-25   *n*-Hexane and *n*-heptane form ideal solutions, and their respective vapor pressures at 50°C are 53.3 and 18.7 kPa. A liquid mixture of the two, containing 0.02 mole of each, is introduced into an evacuated 2-liter flask at 50°C. Neglecting the volume of the liquid phase in comparison with 2 liters, calculate the equilibrium composition of the liquid that remains after vaporization has occurred.

The Henry's law constant for ammonia in water at 20°C is 688 Torr. For a 1 m ammonia solution at 20°C, calculate (a) the total vapor pressure and (b) the vapor composition. The vapor pressure of water at 20°C is 17.36 Torr.

9-26

The absorption coefficient of a gas is defined as the volume of gas that dissolves in one volume of water at a given temperature. The gas volume is measured at the same pressure P as is in equilibrium with the solution, but is reduced to 0°C. Calculate the Henry's law constant for SO_2 if the absorption coefficient is 32.8 at 25°C.

9-27

Demonstrate that the critical temperature for a system obeying the Margules equations is one for which $\alpha = 2$.

9-28

The pressure is steadily reduced over a 50 weight % solution of benzene–ethanol kept at 35°C. (a) At what pressure will the solution begin to boil, and what will be the composition of the first vapor? (b) At what pressure will the last drop of liquid evaporate, and will be its composition?

9-29

Liquids A and B form solutions that obey the Margules equation. P_A^0 and P_B^0 are 200 and 400 mm Hg, respectively, and k_A is 150 mm Hg. Calculate k_B and the activity coefficient of B as $x_B \rightarrow 0$.

9-30

A solution of 1 mole of NaOH in 4.559 moles of water has a vapor pressure of 4.474 Torr at 15°C, whereas the vapor pressure of pure water is 12.788 Torr at 15°C. What is (a) the activity of water in the solution (that is, the effective mole fraction) and (b) the difference between the chemical potential (that is, molar free energy) of the water in the solution and in pure water (Fricke, 1929!)?

9-31

The apparent molal volume ϕ_2 of a solute is defined as $\phi_2 = (V - n_1 V_1^0)/n_2$, where V is the volume of solution containing n_1 moles of solvent and n_2 moles of solute. In the case of aqueous NaCl solutions at 25°C, $\phi_2 = 16.6253 + 1.7738\, n_2^{1/2} + 0.1194\, n_2$. (a) Derive the expression for V in terms of molality m of NaCl. (b) Calculate $\bar{V}_1$ and $\bar{V}_2$ for an infinitely dilute solution and for a 0.5 m solution.

9-32

The apparent molal volumes (see Problem 9-32) of aqueous acetic acid at 5.5°C are:

9-33

ϕ_2, cm³	49.61	49.69	49.72	49.85	49.88	49.93	50.21	51.05
m	0.002	0.005	0.01	0.03	0.05	0.10	1.05	6.70

Make a suitable plot of the data and use the graphical method to obtain $\bar{V}_1$ and $\bar{V}_2$ for a 0.5 m solution of acetic acid.

The relative partial molal volume $\bar{V}_r$ may be defined as $\bar{V}_r = \bar{V} - V°$, where $V°$ is the molar volume of the pure species. The $\bar{V}_r$ values are -0.5 and -1.3 cm³ mol⁻¹ for ethyl iodide and ethyl acetate, respectively, for a solution of mole fraction 0.4 in ethyl acetate. The densities of the pure liquids are 1.933 and 0.901 g cm⁻³, respectively. Calculate the volume of 50 g of such a solution, and the volumes of the components separately.

9-34

Calculate the minimum work to "unmix" air, that is, to obtain 80 liters of pure nitrogen and 20 liters of pure oxygen, each at 25°C and 1 atm pressure, from 100 liters of air at this pressure and temperature. The ΔH_M may be assumed to be zero.

9-35

The following activity coefficient data are for the acetone–methanol system at 56°C:

9-36

x_1	0.280	0.400	0.600	0.676
γ_1	1.392	1.248	1.095	1.089
γ_2	1.057	1.113	1.246	1.352

Acetone is component 1 and methanol is component 2; $P_1° = 760$ mm Hg and $P_2° = 525$ mm Hg. (a) Calculate and plot the corresponding vapor pressure diagram. (b) Calculate and plot the partial molal free energies and the total excess free energy as functions of solution composition.

The International Critical Tables (1928) give the following data for the partial pressure of acetic acid above acetic acid–benzene solutions at 50°C:

9-37

x_a(%)	1.60	4.39	8.35	11.38	17.14	29.79
P(Torr)	3.63	7.25	11.51	14.2	18.4	24.8

x_a(%)	36.96	58.34	66.04	84.35	99.31
P(Torr)	28.7	36.3	40.2	50.7	54.7

Calculate and plot the activities and activity coefficients of acetic acid as a function of composition. Apply the method of Fig. 9-13 to obtain the activities and activity coefficients of benzene for several concentrations and plot these results as well.

9-38   A semi-empirical equation known as the van Laar equation gives

$$\ln\gamma_1 = (A_{12}b_1/RT)z_2^2 \qquad \ln\gamma_2 = (A_{12}b_2/RT)z_1^2 \qquad (9\text{-}153)$$

where

$$z_1 = b_1x_1/(b_1x_1 + b_2x_2) \qquad z_2 = b_2x_2/(b_1x_1 + b_2x_2)$$

The data for the carbon disulfide–isopentane system at 25°C can be fit with $b_2/b_1 = 1.94$ and $(A_{12}b_1/RT) = 0.58$, where carbon disulfide is component 1. Also, $P_1° = 380$ mm Hg and $P_2° = 760$ mm Hg. (a) Calculate ΔG_E for an equimolar mixture and (b) calculate and plot the vapor pressure diagram for the system, including the vapor composition curve.

9-39   Derive Eqs. (9-96) and (9-97).

9-40   Calculate ΔG_E for $x_c = 0.2$ and 0.5 in the acetone–chloroform system at 35°C.

9-41*   The vapor pressures of many liquids can be adequately represented by the equation $\log_{10}P$ (mm Hg) $= -0.05223a/T + b$. For n-hexane, $a = 31,679$ and $b = 7.724$; for n-heptane, $a = 37,358$ and $b = 8.2585$. Assuming the solutions to be ideal, calculate the total vapor pressure of solutions of n-hexane and n-heptane of composition 0.2, 0.4, 0.6, 0.8 (mole fraction) for the temperatures 75, 80, 85, and 90°C. Plot the results as P versus mole fraction, and also calculate and put on the same graph the composition of vapor in equilibrium with each of the solutions at each temperature. Construct from this information a plot of the normal boiling point versus composition. Give also the vapor composition line.

A small amount of the vapor in equilibrium at 85°C with solution of $x_{hex} = 0.5$ is condensed. What is the vapor pressure of the condensate at 85°C? What is the normal boiling point of the original solution? Of the condensate?

9-42*   Use Eq. (9-110) to calculate the normal boiling point diagram for the system n-heptane–n-decane. The respective normal boiling points of the two liquids are 98.4 and 174.1°C. You may assume Trouton's rule to hold. Include in the diagram the vapor composition line.

A solution of mole fraction 0.5 is boiled under 1 atm pressure until only 10 mole % of liquid remains. Determine the boiling point of this residual liquid.

9-43   One mole of acetone is added to 1 mole of a maximum boiling solution of acetone and chloroform. Determine the normal boiling point of the new solution, and the composition of the vapor in equilibrium with it.

9-44   The vapor pressure of camphor is given by the equation of Problem 9-41, with $a = 53,560$ and $b = 8.800$. Camphor and water are essentially immiscible. Find the boiling point of a mixture of camphor and water if the barometric pressure is 740 mm Hg. Calculate the number of grams of camphor that should steam distill at 740 mm Hg, per 1000 g of water.

9-45   A solution containing 20 mole % of phenol in water is cooled to 50°C and the phenol-rich layer which separates out is drawn off. This phenol-rich layer is then cooled to 30°C, and of the two layers present, the one richest in phenol is again drawn off. If 100 g of original solution were used, how many grams of the final phenol-rich layer are obtained and what is the composition of this layer?

9-46   (a) Triethylamine is added to 0.2 mole of water at 30°C until the solution just becomes turbid. How many moles are added? (b) Water is added to 0.3 mole of triethylamine at 30°C until the solution just becomes turbid. How many moles are added? (c) The solutions of (a) and (b) are combined. Give the compositions and amounts of the phases present.

An interesting relationship for ideal gas mixtures is

$$\left(\frac{\partial P_i}{\partial n_i}\right)_{T,P,n_j} = \frac{RT}{v}(1 - x_i) \tag{9-154}$$

Derive this equation. [It may be used in an alternative derivation of Eq. (9-50).]

9-47

SPECIAL TOPICS PROBLEMS

Show that $\overline{Q}_2 = \Delta H_M - x_1 [d(\Delta H_M)/dx_1]$.

9-48

For the process

$$\text{NaCl (solution, } m) \rightarrow \text{NaCl (infinite dilution)}$$

9-49

$-\Delta H$ has been found to vary as follows at 25°C:

m	0.2	1.0	2.0	3.0	4.0	5.0	6.12
$-\Delta H$(cal mol^{-1})	90	-23	-177	-304	-395	-453	-483

Calculate $\overline{L}_1$ and $\overline{L}_2$ for 2 m NaCl and for the saturated solution (6.12 m), and find the value for ΔH for the process

$$\text{NaCl}(s) + \text{water} \rightarrow \text{saturated solution}$$

(remember Table 5-3).

Plot the data of Table 9-1 in the form of Fig. 9-22 and calculate $\mu_{1(rel)}$ and $\mu_{2(rel)}$ for solutions of $x_1 = 0.2$ and $x_1 = 0.8$ (species 1 being acetone).

9-50

The surface tension of an aqueous solution varies with the concentration of solute according to the equation $\gamma = 72 - 350C$ provided that C is less than 0.05 M. Calculate the value of k in the equation $\Gamma_2^1 = kC$. The temperature is 25°C.

9-51

Find the expression for γ as a function of C if $\Gamma = aC/(1 + bC)$, where a and b are constants.

9-52

A 2% by weight aqueous surfactant solution has a surface tension of 69.0 dyne cm^{-1} at 20°C. (a) Calculate σ, the area of surface containing one molecule of surfactant. State any assumptions that must be made. (b) The additional information is now supplied that a 2.2% solution has a surface tension of 68.8 dyne cm^{-1}. If the surface adsorbed film obeys the equation of state $\pi(\sigma - \sigma_0) = kT$, calculate from the combined data a value of σ_0, the actual area of a molecule.

9-53

chapter 10
Dilute solutions
of nonelectrolytes.
Colligative properties

The material that follows has been separated from the preceding chapter primarily because a somewhat special emphasis is involved. A very important situation is that in which two phases are in equilibrium, one of which is a solution and the other of which is a *pure phase* of one of the components. The description is a general one of a system exhibiting a *colligative property* (from the Latin *colligatus* or collected together). The common feature of colligative properties is that, to a first approximation, the observed behavior depends on the mole fraction composition of the solution and on the physical properties of the component which is present in both phases but not on the nature of the second component. The former will be defined as the solvent and the latter as the solute.

This common feature is an exactly observed one if the solution phase is ideal in its behavior. What is actually required is that the activity coefficient of the solvent be independent of the chemical nature of the solute; this condition will ordinarily be approached as $x_1 \rightarrow 1$, since Raoult's law is the limiting law for solutions. Alternatively stated, it is desirable that the solution phase be dilute with respect to the solute. There is therefore considerable emphasis in this chapter on the behavior of dilute solutions.

The most important types of phase equilibria which give rise to colligative phenomena are given in Table 10-1. Other combinations are possible, of course, but are not encountered very often and are therefore ignored here.

One of the important applications of colligative phenomena is to the determination of the molecular weight of a solute present in dilute solution. There are a number of other methods for such a determination, and these are reviewed in Section 10-8 so as to provide a general picture of this aspect of physical chemistry.

Colligative property measurements on nonideal solutions give the thermodynamic activity of the solvent and, indirectly, the activity and activity coefficient of the solute. This constitutes a second major application of colligative phenomena and is described in some detail in the Special Topics section. Although appropriate to this chapter, chemical equilibria involving nonelectrolytes are more conveniently discussed in Chapter 12 along with equilibria involving electrolytes.

TABLE 10-1 Colligative property types of phase equilibria	Phases in equilibrium	Restriction or condition	Name of resulting colligative property
	Liquid solution $\leftrightarrows$ vapor	Solute must be nonvolatile	Vapor pressure lowering Boiling point elevation
	Solid $\leftrightarrows$ liquid solution	Solid phase must consist of pure solvent	Freezing point depression
	Liquid solution $\leftrightarrows$ liquid solvent	Semipermeable membrane prevents the solute from entering the pure solvent phase	Osmotic pressure

10-1   Vapor pressure lowering

The general statement for the vapor pressure of the solvent component of a liquid solution is

$$P_1 = a_1 P_1^\circ \qquad \text{[Eq. (9-86)]}$$

or

$$\frac{\Delta P_1}{P_1^\circ} = 1 - a_1 \qquad (10\text{-}1)$$

where ΔP_1 is the vapor pressure lowering of the solvent, $P_1^\circ - P_1$. If the solute is nonvolatile, then P is the total vapor pressure above the solution and ΔP_1 is the total vapor pressure lowering ΔP.

If the solution is ideal, then Raoult's law, Eq. (9-4), holds, which means that $a_1 = x_1$. For a one-component system Eq. (10-1) becomes

$$\frac{\Delta P}{P_1^\circ} = x_2 \qquad (10\text{-}2)$$

still assuming the solute to be nonvolatile. Thus a measured behavior of the solvent has given the mole fraction of the solute, independent of the chemical nature of the latter. As will be illustrated later, if the weight composition of the solution is known, then the value of x_2 can be used for calculation of the molecular weight of the solute.

10-2   Boiling point elevation

We can use the procedures of Section 9-6 to relate the change in normal boiling point of a solution to the solvent activity if the solute is nonvolatile. The situation is that since $P = 1$ atm, $\mu_1(g) = \mu_1^\circ(g)$, and for phase equilibrium $\mu_1(l)$ is therefore equal to $\mu_1^\circ(g)$. Thus

$$\mu_1^\circ(g) = \mu_1(l) = \mu_1^\circ(l) + RT \ln a_1 \qquad (10\text{-}3)$$

[from the definition of a_1, Eq. (9-87).

Differentiation with respect to temperature gives

$$-S_1^\circ(g) = -S_1^\circ(l) + R \ln a_1 + RT \frac{d(\ln a_1)}{dT} \tag{10-4}$$

and on replacement of $R \ln a_1$ by $[\mu_1^\circ(g) - \mu_1^\circ(l)]/T$ and following the procedure of Section 9-6,

$$\frac{d(\ln a_1)}{dT} = \frac{H_1^\circ(l) - H_1^\circ(g)}{RT^2} = -\frac{\Delta H_{v,1}^\circ}{RT^2} \tag{10-5}$$

where $\Delta H_{v,1}^\circ$ is the standard heat of vaporization of the pure solvent. Equation (10-5) may be integrated between limits, with $a_1 = 1$ at $T = T_{b,1}^\circ$, the normal boiling point of the pure solvent, assuming that ΔH_v is independent of T:

$$\ln a_1 = \frac{\Delta H_{v,1}^\circ}{R} \left(\frac{1}{T_b} - \frac{1}{T_{b,1}^\circ} \right) \tag{10-6}$$

with T_b denoting the normal boiling point of a solution of solvent activity a_1.

We next consider the solution to be ideal. Equation (10-6) becomes

$$\ln x_1 = \frac{\Delta H_{v,1}^\circ}{R} \left(\frac{1}{T_b} - \frac{1}{T_{b,1}^\circ} \right) \tag{10-7}$$

Again, a measurement of a property due to the solvent has yielded the mole fraction composition of the solution: x_2 may, of course, be calculated from $\ln x_1$.

It was noted in the opening remarks that we obtain the condition of ideality in practice by using a dilute solution. Equation (10-7) undergoes considerable simplification under this condition. First, $\ln x_1$ may be expanded and only the first term of the series used, so that $\ln x_1 \simeq -(1 - x_1) = -x_2$. For example, if $x_2 < 0.01$, the error in the approximation is less than 1%. With this substitution, and after rearrangement, Eq. (10-7) becomes

$$\Delta T_b = \frac{RT_{b,1}^\circ T_b}{\Delta H_{v,1}^\circ} x_2 \tag{10-8}$$

where ΔT_b is the normal boiling point elevation of the solution, $T_b - T_{b,1}^\circ$.

The expansion of the $\ln x_1$ term has already limited the maximum value of x_2 to be tolerated, and no significant further error is introduced if T_b is approximated by $T_{b,1}^\circ$ in the numerator of Eq. (10-8). Also, we have

$$x_2 = \frac{n_2}{n_2 + n_1} = \frac{m}{m + (1000/M_1)} \tag{10-9}$$

where m is the molality of the solute and M_1 is the molecular weight of the solvent. The error due to neglecting m in the denominator of Eq. (10-9) is again not important. With these further approximations Eq. (10-8) becomes

$$\Delta T_b = \left[\frac{R(T_{b,1}^\circ)^2 M_1}{1000 \, \Delta H_{v,1}^\circ} \right] m = K_b m \tag{10-10}$$

where K_b is called the *boiling point elevation constant*. K_b depends only on properties of the pure solvent. The value for water is 0.514; further values are given in Table 10-2 in Section 10-4.

TABLE 10-2
Boiling point elevation
and freezing point
depression constants[a]

Solvent	$t_b(°C)$	K_b	$t_f(°C)$	K_f
Ethyl ether	34.4	2.11	—	—
Chloroform	61.2	3.63	—	—
Ethanol	78.3	1.22	—	—
Benzene	80.2	2.53	5.6	5.12
Water	100.0	0.514	0.00	1.855
Acetic acid	118	3.07	17	3.90
Bromobenzene	155.8	6.20	—	—
Naphthalene	—	—	80.2	6.8
Camphor	—	—	178	40

[a]See Table 8-1 for some values of $\Delta H_v^°$ and $\Delta H_f^°$.

10-3   Freezing point depression

The equilibrium between a solid phase consisting of pure solvent and a liquid solution is treated as follows. The condition of equilibrium is that $\mu_1^°(s) = \mu_1(l)$. Then, by Eq. (9-87), we have

$$\mu_1^°(s) = \mu_1^°(l) + RT \ln a_1 \tag{10-11}$$

The rest is analogous to the preceding derivation. Differentiation with respect to temperature yields

$$-S_1^°(s) = -S_1^°(l) + R \ln a_1 + RT \frac{d(\ln a_1)}{dT} \tag{10-12}$$

and, on replacement of R in a_1 by $[\mu_1^°(s) - \mu_1^°(l)]/T$ and rearrangement, we have

$$\frac{d(\ln a_1)}{dT} = \frac{H_1^°(l) - H_1^°(s)}{RT^2} = \frac{\Delta H_{f,1}^°}{RT^2} \tag{10-13}$$

Equation (10-13) is then integrated between limits, with $a_1 = 1$ at $T_{f,1}$, the melting (or freezing) point of the pure solvent, to give

$$\ln a_1 = \frac{\Delta H_{f,1}^°}{R} \left(\frac{1}{T_{f,1}^°} - \frac{1}{T_f} \right) \tag{10-14}$$

where T_f is the freezing point of the solution of solvent activity a_1. Notice the parallel that is developing between the boiling point elevation and freezing point depression effects.

If the solution is ideal, Eq. (10-14) becomes

$$\ln x_1 = \frac{\Delta H_{f,1}^°}{R} \left(\frac{1}{T_{f,1}^°} - \frac{1}{T_f} \right) \tag{10-15}$$

Again, the most valid application will be to dilute solutions, for which $\ln x_1$ may be approximated by $-x_2$, so that

$$\Delta T_f = \frac{RT_{f,1}^° T_f}{\Delta H_{f,1}^°} x_2 \tag{10-16}$$

and, by setting $T_f = T_{f,1}^°$ and neglecting m in the denominator of Eq. (10-9), we obtain

$$\Delta T_f = \left[\frac{R(T_{f,1}^\circ)^2 M_1}{1000 \, \Delta H_{f,1}^\circ} \right] m = K_f m \tag{10-17}$$

K_f is called the *freezing point depression constant;* its value depends only on properties of the solvent. For water, $K_f = 1.86$. Further values are given in Table 10-2.

We have just treated the phase equilibrium

pure solid (1) = solution of (1) and (2) (10-18)

in terms of a freezing point depression of solvent species (1). The laboratory appearance of such an equilibrium mixture, however, is simply one of a suspension of a solid in a solution, and the system could just as well be described as a saturated solution of solid species (1) in species (2) as *solvent.* When viewed according to this alternative perspective, Eqs. (10-14) and (10-15) give the *solubility* of component (1), now thought of as the *solute.* T_f in the above equations is no longer described as a freezing point, but just as T, the temperature at which the solubility is measured.

The freezing point of a 0.25 mole fraction solution of, say, ethanol in acetic acid should **EXAMPLE** be approximately 0°C (assuming ideal solution behavior). Alternatively, this same mole fraction denotes the solubility of solid acetic acid in ethanol at 0°C.

Equation (10-15) in particular provides some useful qualitative rules on solubility. (a) The solubility of a compound should increase as temperature increases, and (b) the solubility of a series of compounds should decrease as their melting point increases. This last rule is fairly well obeyed provided that the heats of fusion do not vary much along the series. It should be remembered, however, that in solubility situations, x_1 may be fairly small, and Eq. (10-15) can be seriously in error due to approximating a_1 by x_1.

10-4   Summary of the first three colligative properties

The preceding two derivations were made on a somewhat formal thermodynamic basis and it is possible to give a physical explanation of all of the effects so far discussed. This is done in Fig. 10-1, which shows schematic vapor pressure plots for pure solid and pure liquid solvent and for a dilute solution of a nonvolatile solute. The vapor pressure lowering follows directly. The boiling point elevation is the difference between the two temperatures at which the $P = 1$ atm line crosses the $P_1^\circ(l)$ and $P_1(l)$ curves.

As noted in Section 8-5, at the melting point of a pure substance the solid and liquid vapor pressures must be the same. Therefore $T_{f,1}^\circ$ is the temperature of crossing of the $P_1^\circ(s)$ and $P_1^\circ(l)$ curves. Since the solute is not present in the solid phase, the vapor pressure of the solid is not altered by the presence of solute, and T_f must therefore be the temperature of crossing of the $P_1^\circ(s)$ and $P_1(l)$ curves.* One objection to this last analysis is that it is not strictly necessary that the solute be nonvolatile in the freezing point depression effect. It is only necessary that it be insoluble in the solid solvent phase.

---

*Strictly speaking, T_f° and T_f in Fig. 10-1 are the freezing points for the system under its *own vapor pressure.* Correction to 1 atm pressure would be small, however.

FIGURE 10-1
Explanation of boiling point elevation and freezing point depression effects in terms of vapor pressure changes. The solute is assumed to be nonvolatile.

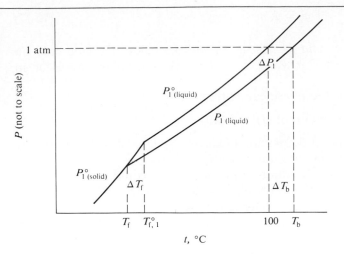

The boiling point elevation and freezing point depression effects are the more commonly used of the three colligative properties, and some of the experimental aspects are as follows. Figure 10-2 shows a Cottrell boiling point apparatus, the main feature of which is the tubular yoke around the bulb of the thermometer. The purpose is to bathe the bulb in boiling solution; were vapor simply allowed to condense onto the bulb, the temperature registered

FIGURE 10-2
Cottrell boiling point apparatus.

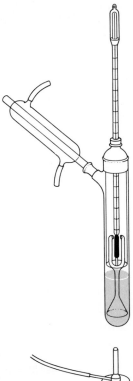

would tend to be that of the boiling point of the solvent, since the vapor consists of pure solvent.

Freezing point depressions may be obtained with very simple equipment, often consisting merely of a Dewar flask equipped with a Beckmann thermometer and a stirrer. The freezing point of the pure solvent is first determined, as, for example, by using a mixture of ice and water if the system is aqueous or in general by gradual cooling of the solution until freezing sets in. The experiment is repeated with the solution. Liquids tend to supercool, and it is necessary to be sure that enough finely divided solid solvent is present to ensure equilibrium.

It is also essential to know the composition of the equilibrium solution that corresponds to the temperature measurement. This is not difficult in the case of boiling elevation; if the boiling solution is under reflux, its composition is essentially the same as that initially. In the freezing point depression experiment, however, there is apt to be appreciable freezing or melting before the final temperature reading, and a sample of the solution should be withdrawn for analysis at that point.

The vapor pressure lowering effect, while less frequently used, is actually quite important for very precise work. Rather than attempting direct vapor pressure measurements, however, one usually compares the unknown with a known solution by allowing the two solutions to come to vapor pressure equilibrium at a known temperature. For example, an open beaker of each solution might be placed in an otherwise empty desiccator, which is then thermostated. Solution A contains a known nonvolatile solute and solution B contains the nonvolatile one being studied. The solvent vapor pressures above the two solutions will initially not be the same. Suppose that at first $P_{1,A}$ is greater than $P_{1,B}$. Then solvent will distill from solution A to solution B, concentrating the former and diluting the latter. Eventually *isopiestic equilibrium* is reached, that is $P_{1,A} = P_{1,B}$. The solutions must now have identical values for solvent activity a_1. Analysis of solution A then determines the actual value of $a_{1,A}$. Since the solutions are in equilibrium, $a_{1,A} = a_{1,B}$. If solution B is ideal or is dilute enough that Raoult's law has been reached as a limiting law, then $a_{1,B} = x_{1,B}$.

The isopiestic method has an advantage over the other two in being an isothermal one, so that corrections for changes in solvent properties with temperature are not needed. The isopiestic method is tedious, however, and is only used when very accurate results are wanted.

Representative values of K_b and K_f are given in Table 10-2; some are quite large, and such solvents are often used in qualitative molecular weight determinations because of the ease of measuring ΔT_b or ΔT_f. It should be pointed out that both constants, but especially K_b, depend on the ambient pressure. This is because of the $(T_{b,1}^\circ)^2$ factor in Eq. (10-10), which is pressure dependent. Thus K_b for benzene changes by 0.025% per Torr difference between 760 Torr and the actual barometric pressure.

The following illustrates the three effects. Suppose that an aqueous solution of an organic compound is in isopiestic equilibrium with 0.1 *m* sucrose at 25°C. What is the vapor pressure of the solution if that of pure water is 23.76 Torr at 25°C? What would be the boiling point elevation and freezing point depression? If the solution contains 2.50 wt % of the compound, what is its molecular weight?   **EXAMPLE**

To answer the first question, we need the mole fraction of the sucrose solution: $x_2 = 0.1/[0.1 + (1000/18)] = 0.00180$. The vapor pressure lowering is then $(0.0018)(23.76) = 0.0428$ Torr. The solution of the compound must have the same value for a_1 as the sucrose solution and is dilute enough that we can use the dilute solution approximation; the molality of the two solutions is therefore also the same. The boiling point elevation is then $(0.514)(0.1) = 0.0514°C$, and the freezing point depression is $(1.855)(0.1) = 0.1855°C$.

We have 25 g of compound per 975 g of water or 25.6 g per 1000 g of water. Since the solution is 0.1 m, the molecular weight must be 256.

10-5   Osmotic equilibrium

An osmotic equilibrium is one between a solution and pure solvent, both liquid. It is necessary to have a barrier between the liquids to separate them. The barrier must be impermeable to the solute, yet for equilibrium it must be permeable to the solvent. Such barriers or membranes (since they usually are thin) are called *semipermeable membranes*. The solvent activity is different in the two liquids, being lower in the solution than in the pure solvent; to bring the system to equilibrium, it is necessary to have the solution under mechanical pressure. The effect of such pressure on a pure liquid was shown in Section 8-5 to increase its vapor pressure or activity, and the same is true for a solution. The pressures involved can be more than atmospheric, and a semipermeable "membrane" usually must be either well supported or itself quite strong.

A schematic osmotic pressure apparatus is shown in Fig. 10-3. In the form pictured, solvent passes through the semipermeable membrane until sufficient hydrostatic head develops on the solution side for the activity of water in the solution to be equal to that of the pure solvent. Alternatively, the solution may be placed under sufficient mechanical pressure to just prevent net flow of solvent. A simple demonstration of osmotic pressure is described in Fig. 10-4.

Osmotic equilibrium is independent of how the membrane acts so long as

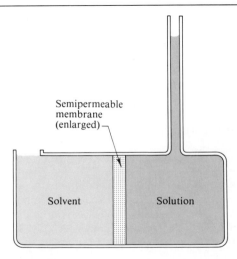

FIGURE 10-3
Schematic osmotic pressure apparatus.

Semipermeable membrane (enlarged)

Solvent

Solution

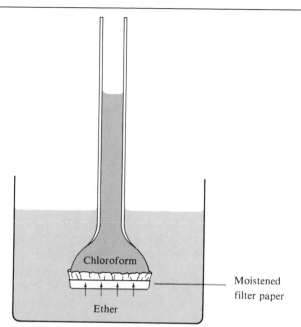

FIGURE 10-4
Filter paper moistened with water and acting as a "leaky" membrane between chloroform (in the tube) and ether. Chloroform is quite insoluble in water and the wet filter paper acts as a barrier to its passage. Ether, however, is sufficiently soluble in water to diffuse through the "membrane."

Chloroform

Moistened filter paper

Ether

it is in fact permeable only to the solvent. The condition for equilibrium is that $\mu_1{}^\cap(l) = \mu_1(l)$, where $\mu_1(l)$ is the chemical potential of the solvent in the solution. For a solution not under pressure,

$$\mu_1(l) = \mu_1{}^\circ(l) + RT \ln a_1 \qquad [\text{Eq. (9-87)}]$$

and the effect of pressure is, according to Eq. (8-24), to increase $\mu_1(l)$ by the integral of $\overline{V}_1 dP$:

$$\mu_1(l) = \mu_1{}^\circ(l) + RT \ln a_1 + \int_{P_1{}^\circ}^{\Pi + P_1{}^\circ} \overline{V}_1 dP \qquad (10\text{-}19)$$

At equilibrium the free energy of the solvent must be the same on both sides of the diaphragm, or

$$\mu_1{}^\circ(l) = \mu_1(l) + RT \ln a_1 + \int_{P_1{}^\circ}^{\Pi + P_1{}^\circ} \overline{V}_1 dP$$

or

$$-RT \ln a_1 = \int_{P_1{}^\circ}^{\Pi + P_1{}^\circ} \overline{V}_1 dP \qquad (10\text{-}20)$$

The required mechanical pressure Π to produce equilibrium is known as the *osmotic pressure*. Ordinarily Π is large compared to $P_1{}^\circ$ and the change in $\overline{V}_1$ with pressure can be neglected, so Eq. (10-20) simplifies to

$$-RT \ln a_1 = \Pi \overline{V}_1 \qquad (10\text{-}21)$$

If the solution is ideal, then $a_1 = x_1$ and

$$-RT \ln x_1 = \Pi \overline{V}_1 \qquad (10\text{-}22)$$

If the solution has approached ideality by also being dilute, then $\ln x_1$ may be approximated by $-x_2$, as before. At the same level of approximation $x_2 \simeq n_2/n_1$, and so Eq. (10-22) takes the very simple form

$$RT \frac{n_2}{n_1} = \Pi \overline{V}_1{}^\circ$$

In the case of a dilute solution $n_1 \overline{V}_1{}^\circ$ is essentially the volume of the solution, so

$$\Pi V = n_2 RT \qquad \text{or} \qquad \Pi = CRT \qquad\qquad (10\text{-}23)$$

The interesting final result is that in dilute solution osmotic pressure obeys the ideal gas law. The final approximation is unnecessarily severe, and a very useful dilute solution form is one in which C is the number of moles of solute per unit volume of solvent, rather than per unit volume of solution.

Van't Hoff proposed a simple physical explanation of Eq. (10-23), illustrated in Fig. 10-5. Since solvent molecules can pass freely through the membrane, they are "invisible" to it. The solute molecules are confined by the membrane, however, and exert the same pressure on it as would an ideal gas at that concentration and pressure. This "bombardment" picture turns out not to be very useful, in that attempts to pursue it in detail have not been fruitful. The modern tendency is to avoid such mechanical models.

EXAMPLE The example of the preceding section may be extended to the calculation of the osmotic pressure of the solutions. The osmotic pressure Π must be the same for the sucrose solution and that of the unknown compound at 25°C since they are in isopiestic equilibrium. Then $C = 0.1$ mole liter^{-1} of solvent and $\Pi = (0.1)(0.0821)(298) = 2.45$ atm. Note how large the osmotic pressure is compared to the other colligative effects.

Some data due to Berkeley and Hartley in 1906 on the osmotic pressure of sucrose solutions are given in Table 10-3. The simple formula (10-23) fails badly by about 1 m sucrose concentration, and even the more exact form, Eq. (10-22), is inadequate; it would predict an osmotic pressure of 92 atm for the most concentrated solution, instead of the observed 134.7 atm. We have confidence in the thermodynamics, and evidently the solution must not be ideal, as assumed in writing Eq. (10-22). It turns out, however, that the quite

FIGURE 10-5
Simple physical illustration of Eq. (10-23).

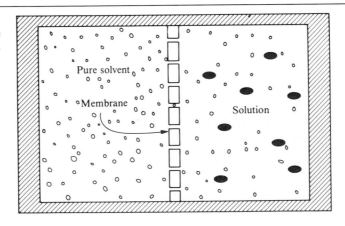

| Concentration (grams per liter of solution) | $(n_2/n_1) \times 10^4$ | Osmotic pressure (atm) | | | TABLE 10-3 |
		Observed	Equation (10-23)	Modified Eq. (10-22)[b]	Osmotic pressure of aqueous sucrose at 0°C[a]
2.02	1.064	0.134	0.132	0.133	
10.0	5.294	0.66	0.655	0.661	
45.0	24.22	2.97	2.947	3.056	
300	193.6	26.8	19.65	26.99	
750	737.2	134.7	49.14	154.5	

[a]Source: E. A. Moelwyn-Hughes, "Physical Chemistry," p. 803. Pergamon, Oxford, 1961.
[b]See text.

reasonable agreement shown by the last column of Table 10-3 results if one supposes that each sucrose molecule binds six molecules of water. The actual number of moles of solvent present is thus reduced by six per mole of sucrose present, and x_2 is correspondingly increased. This example is cited to illustrate one way in which a formal nonideality may be accounted for.

10-6   Activities and activity coefficients for dilute solutions

The use of activities and activity coefficients for solutions is discussed in Section 9-5. It will be recalled that activity a is defined as the effective mole fraction such as to keep the form of Raoult's law:

$$P_i = a_i P_i^\circ \quad \text{[Eq. (9-86)]}$$

The activity coefficient γ is defined as the factor whereby a differs from x,

$$a_i = \gamma_i x_i \quad \text{[Eq. (9-88)]}$$

so that the chemical potential for the ith species is

$$\mu_i(l) = \mu_i^\circ(l) + RT \ln a_i \quad \text{[Eq. (9-87)]}$$

Since Raoult's law is approached by all solutions as $x_i \to 1$, it follows that in this limit $a_i \to x_i$ and $\gamma_i \to 1$.

This convention is fine for solutions of two liquids, whose compositions can range from pure component 1 to pure component 2. It is, however, very awkward for dilute solutions, and an alternative set of definitions has become customary for the solute. That for the solvent remains the same, but, since Henry's law is the limiting law for the solute, we now define an activity a_2' such that Henry's law is obeyed,

$$P_2 = a_2' k_2 \tag{10-24}$$

The corresponding activity coefficient γ_2' is given by

$$a_2' = \gamma_2' x_2 \tag{10-25}$$

These conventions are to be applied only to the solute, and Eqs. (10-24) and (10-25) have been written on this basis. Since Henry's law is approached as $x_2 \to 0$, then $a_2' \to x_2$ and $\gamma_2' \to 1$ at this limit.

In the case of aqueous solutions, it is customary to express the concentration of solute as molality m rather than as x_2. Henry's law may still be used as a limiting law since by Eq. (10-9), m and x_2 become proportional as $x_2 \to 0$, and in this limit

$$P_2 = mk_m \tag{10-26}$$

Activity is now defined by

$$P_2 = a_m k_m \tag{10-27}$$

and activity coefficient as

$$a_m = \gamma_m m \tag{10-28}$$

As x_2, and hence m, approaches zero, $a_m \to m$ and $\gamma_m \to 1$. Finally, x_2 or m becomes proportional to C, the concentration in moles per liter of solution, as the concentration approaches zero. Henry's law becomes $P_2 = Ck_C$, and a concentration-based activity a_C may be defined as

$$P_2 = a_C k_C \tag{10-29}$$

The corresponding activity coefficient γ_C is

$$a_C = \gamma_C C \tag{10-30}$$

Again, as x_2, and hence C, approaches zero, $a_C \to C$ and $\gamma_C \to 1$.

The subject is discussed in more detail in the Special Topics section, but it turns out that the effect of using the Henry's law conventions is to change the reference or standard state, and hence the value of $\mu_i^\circ(l)$ of Eq. (9-87). All equations involving only changes in activity remain the same, but in using the Gibbs–Duhem equation, as in Section 9-5, the first integration limit is $x_2 = 0$ rather than $x_2 = 1$ since it is now the former condition for which $\gamma = 1$ and hence $\ln \gamma = 0$. Alternatively, γ_2' differs from γ_2 by a constant factor. However, mole fraction, molality, and concentration cease to be proportional in concentrated solutions, and as a consequence the three types of actvitiy coefficient differ. The general relationship among them is

$$\gamma_2' = \gamma_m(1 + 0.001mM_1) = \gamma_C \frac{\rho + 0.001C(M_1 - M_2)}{\rho^\circ} \tag{10-31}$$

where M_2 is the solute molecular weight, ρ is the density of the solution, and ρ° is the density of the pure solvent.

10-7 Colligative properties and deviations from ideality

The equations for the four colligative properties are summarized in Table 10-4. They group into three categories: (1) the forms that are strictly valid (except for the assumed constancy of heats of vaporization and freezing, and of activity coefficients with temperature), (2) those that assume an ideal solution, and (3) those that assume an ideal and dilute solution. A number of systems have been studied very precisely, and there seems to be no question that Raoult's law is a valid limiting law. Also, heats of fusion from K_f values agree well with those from direct calorimetric determinations.

Although these affirmations are important, it is also true that deviations from the simple behavior are common. We discount immediately cases in which the dilute solution forms have been misused in that the conditions for

		Equation assuming	
			Ideal and Dilute Solution
Effect	Exact equation	Ideal Solution	
Vapor pressure lowering	$(\Delta P)/P_1{}^\circ = 1 - a_1$	$(\Delta P)/P_1{}^\circ = x_2$	—
Boiling point elevation	$\ln a_1 = (\Delta H_{v,1}^\circ/R)$ $\times [(1/T_b) - (1/T_{b,1}^\circ)]$	$\ln x_1 = (\Delta H_{v,1}^\circ/R)$ $\times [(1/T_b) - (1/T_{b,1}^\circ)]$	$\Delta T_b = K_b m$
Freezing point depression	$\ln a_1 = (\Delta H_{f,1}^\circ/R)$ $\times [(1/T^\circ{}_{f,1}) - (1/T_f)]$	$\ln x_1 = (\Delta H_{f,1}^\circ/R)$ $\times [(1/T^\circ{}_{f,1}) - (1/T_f)]$	$\Delta T_f = K_f m$
Osmotic pressure effect	$-RT \ln a_1 = \Pi \overline{V}_1$	$-RT \ln x_1 = \Pi \overline{V}_1$	$\Pi = CRT$

TABLE 10-4
Colligative property equations

[a]The last three of these equations are not fully exact. Assumptions such as constancy of heat vaporization or heat of freezing have been made in integration over ΔT_b or ΔT_f. In the case of osmotic pressure, other assumptions, noted in the text, have been made. These are not drastic assumptions, and in this text the equations given are regarded as thermodynamically exact.

the mathematical approximations have not been met. There remain several explanations. One is that the solute is either associated or dissociated to some extent, so that its apparent molecular weight differs from its formula weight. Remember that the colligative property effects are determined by x_2, the mole fraction of solute. This mole fraction is determined by the actual species present. As one example, the molecular weight of benzoic acid is just the formula weight of 122.1 in acetone solution, but in benzene, the apparent molecular weight is 242. That is, the colligative property measurement reports half as many solute molecules as expected from the amount weighed out in making up the solution. The explanation is that benzoic acid is largely dimerized in benzene solution.

Alternatively, if the freezing point depression is determined for an aqueous sodium chloride solution, the result gives an apparent molecular weight that is about half the formula weight. In this case we have a strong electrolyte which has dissociated into two ions, each of which acts as a solute species with respect to a colligative property. Van't Hoff, who contributed much to the physical chemistry of the osmotic pressure effect, defined a factor i as

$$\Pi v = i n_2 RT \qquad (10\text{-}32)$$

where the simple meaning of i is that it gives the average number of moles of particles produced per formula weight of solute. Thus i would be $\frac{1}{2}$ for benzoic acid in benzene and 2 for aqueous sodium chloride. The approach, while oversimplified, emphasizes the point that the colligative property measurement gives an average molecular weight. One may thus use the measurement to calculate the degree of association or, for a weak electrolyte, the degree of dissociation. In the case of polymer solutions, where a wide range of molecular weight species is present, the average molecular weight that results is known as a number average (see Section 10-8E).

Another source of deviation from ideal behavior is illustrated by the data of Table 10-3. It appears, from this as well as from other evidence, that sucrose tightly binds about six molecules of water. The situation is essentially one of compound formation between solute and solvent, and the effect on a colligative property measurement is that the effective value of x_1 is reduced while that of x_2 is increased. A fair measure of the departure from ideal behavior of electrolyte solutions can be accounted for if each ion, especially the positive ion, is assumed to bind a certain number of water molecules.

The resemblance of the ideal osmotic pressure equation, Eq. (10-23), to the ideal gas law was noted in Section 10-5. The resemblance extends further. Just as nonideal gases can be represented by a virial equation (Eq. 1-28), so can the osmotic pressure of a nonideal solution,

$$c\frac{\Pi}{RT} = \frac{1}{M} + B(T)c \cdots \tag{10-33}$$

where c is concentration in grams per cubic centimeter.

Finally, departures from ideality may always be treated in a nonspecific way by calculating a_1 from use of the exact equation. The solute activity, a_2, and activity coefficient, γ_2, may then be obtained by the methods of Section 10-ST-1.

10-8   Other methods of molecular weight determination

One of the important uses of colligative property measurements is to obtain the molecular weight of a solute. This is particularly true for new substances being characterized for the first time, including biological species such as proteins or nucleic acids. Polymer chemistry is also a highly developed field, and a central piece of information is again the molecular weight of a particular preparation. In this case there will usually be a mixture of various molecular weights and one seeks to determine the average value (as will be discussed later).

There are several other means whereby either the size or the molecular weight of a solute may be determined, and it is appropriate to include these in the present chapter.

A. Sedimentation velocity

It is possible to estimate the size of a molecule by the speed with which it moves through a solution under the influence of some force. One such force is that due to gravity. A particle of mass m_2 experiences a downward pull of $m_2 g$, opposed in solution by the buoyancy of the solvent, where g is acceleration due to gravity. In effect, the particle displaces volume m_2/ρ_2, corresponding to a mass of solvent equal to $(m_2/\rho_2)\,\rho_1$, where ρ_2 and ρ_1 are the densities of the particle and of the solvent, respectively. The net force due to gravity is then

$$m_2 g\left(1 - \frac{\rho_1}{\rho_2}\right) = m_2 g(1 - \bar{v}_2\rho_1) \tag{10-34}$$

where $\bar{v}_2$ is the partial specific volume of the particle.

The particle will fall through the solution if $\rho_2 > \rho_1$, but as it accelerates a viscous drag develops which is proportional to the velocity, or equal to $f\,v$, where v is the velocity and f is called the *friction coefficient*. A limiting or terminal velocity is reached when the force due to gravity and that of the viscous drag have become equal, or when

$$v = \frac{m_2 g(1 - \bar{v}_2 \rho_1)}{f} \qquad (10\text{-}35)$$

In liquids of ordinary viscosity, this terminal velocity is reached very quickly, and the experimental observation is that the particle immediately assumes a constant speed of fall.

The friction coefficient depends in a complicated way on the shape of the particle and on the viscosity of the solvent. However, if the particle is spherical, a simple formula due to Stokes applies:

$$f = 6\pi\eta r \qquad (10\text{-}36)$$

where η is the viscosity of the medium and r is the radius of the particle. The velocity of the fall is then

$$v = \frac{m_2 g}{6\pi\eta r}(1 - \bar{v}_2 \rho_1) \qquad (10\text{-}37)$$

The particle is to be a solute molecule, although perhaps a large one such as a protein, and its rate of fall due to gravity will be extremely small—so small, in fact, that, as will be seen, it is quickly nullified by back-diffusion. It is therefore ordinarily necessary to increase the sedimentation force by using a centrifuge. Acceleration due to gravity is replaced by $\omega^2 x$, where ω is the angular velocity (in radians per second) and x is the distance from the center of rotation. Equation (10-35) becomes

$$\frac{dx}{dt} = \frac{m_2 \omega^2 x(1 - \bar{v}_2 \rho_1)}{f} \qquad (10\text{-}38)$$

or

$$s = \frac{m_2}{f}(1 - \bar{v}_2 \rho_1) \qquad s = \frac{dx/dt}{\omega^2 x} \qquad (10\text{-}39)$$

where s is called the *sedimentation coefficient* and is characteristic of a given solute and solvent.

For a laboratory centrifuge, x might be about 10 cm and the speed of rotation 50 rps. The acceleration $\omega^2 x$ is then $(2\pi)^2(50)^2(10)$ or about 1×10^6 cm s^{-2} or $10^3 g$. A protein molecule of molecular weight 100,000, or particle weight 1.66×10^{-19} g, and of specific volume 0.80 cm^3 g^{-1} would then sediment with a rate of $(1.66 \times 10^{-19})(1 \times 10^6)(1 - 0.8)/f$, or about $3.3 \times 10^{-14}/f$, in cm s^{-1}. If the particle is spherical, we have

$$\tfrac{4}{3}\pi r^3 = (1.66 \times 10^{-19})(0.8) \simeq 1.3 \times 10^{-19}$$

from which r is found to be 3.2×10^{-7} cm. The viscosity of water is about 0.89 cP and, by Stokes' law, $f = 6\pi(0.0089)(3.14 \times 10^{-7}) = 5.4 \times 10^{-8}$ cm s^{-2}. The velocity of sedimentation of the protein molecules would then be $(3.3 \times 10^{-14})/(5.4 \times 10^{-8})$, or about 6.1×10^{-7} cm s^{-1}, a rather small number.

EXAMPLE

The calculation illustrates the difficulty in carrying out sedimentation experiments with ordinary centrifuges. Thé Svedberg pioneered in the 1920s the development of very high-speed centrifuges or *ultracentrifuges*. Forces as high as 300,000g have been obtained. The sedimentation rate would be 300 times larger than in the preceding example, or about 2×10^{-4} cm s^{-1}, an easily

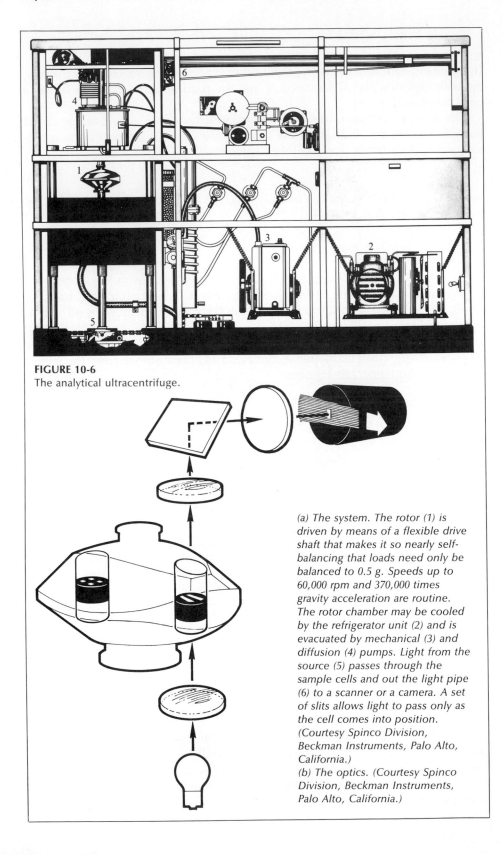

FIGURE 10-6
The analytical ultracentrifuge.

(a) The system. The rotor (1) is driven by means of a flexible drive shaft that makes it so nearly self-balancing that loads need only be balanced to 0.5 g. Speeds up to 60,000 rpm and 370,000 times gravity acceleration are routine. The rotor chamber may be cooled by the refrigerator unit (2) and is evacuated by mechanical (3) and diffusion (4) pumps. Light from the source (5) passes through the sample cells and out the light pipe (6) to a scanner or a camera. A set of slits allows light to pass only as the cell comes into position. (Courtesy Spinco Division, Beckman Instruments, Palo Alto, California.)
(b) The optics. (Courtesy Spinco Division, Beckman Instruments, Palo Alto, California.)

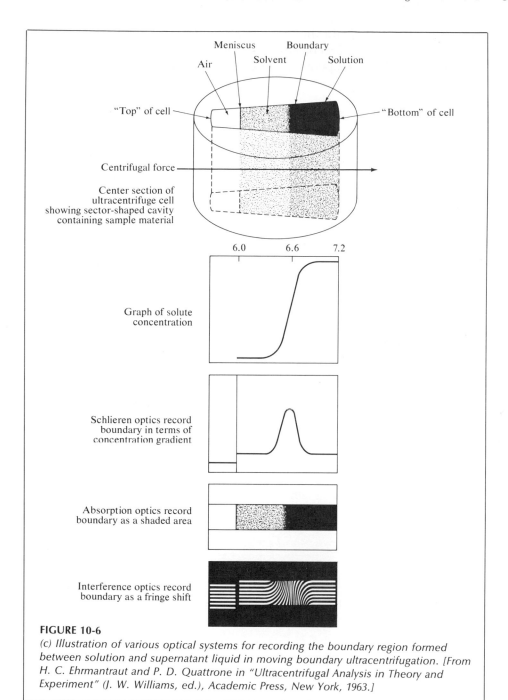

Air

Meniscus   Boundary

Solvent   Solution

"Top" of cell

"Bottom" of cell

Centrifugal force

Center section of
ultracentrifuge cell
showing sector-shaped cavity
containing sample material

6.0   6.6   7.2

Graph of solute
concentration

Schlieren optics record
boundary in terms of
concentration gradient

Absorption optics record
boundary as a shaded area

Interference optics record
boundary as a fringe shift

FIGURE 10-6

*(c) Illustration of various optical systems for recording the boundary region formed
between solution and supernatant liquid in moving boundary ultracentrifugation. [From
H. C. Ehrmantraut and P. D. Quattrone in "Ultracentrifugal Analysis in Theory and
Experiment" (J. W. Williams, ed.), Academic Press, New York, 1963.]*

measurable rate. Centrifuges of this type may be air-suspended and air-dri-
ven. More elegantly, the rotor can be suspended by a magnetic field and
driven by induction. In the latter version, the rotor chamber is evacuated so
that friction is virtually absent. The cell containing the solution is placed in
the rotor and can be viewed through windows. Stroboscopic pictures of the
boundary formed by sedimenting material can be obtained with the use of

synchronous lighting, and the sedimentation velocity thus determined. The molecular weight may then be found, essentially by reversal of the preceding calculation. One contemporary type of ultracentrifuge is described in Fig. 10-6.

B. Diffusion   A second type of kinetic approach which yields information about molecular size is that of diffusion studies. The *coefficient of diffusion* is defined phenomenologically by Fick's law,

$$J = -\mathcal{D}\frac{d\mathbf{n}}{dx} \qquad \text{[Eq. (2-87)]}$$

where $\mathcal{D}$ is the diffusion coefficient and J is the molecular flux along the concentration gradient, in molecules per square centimeter per second. Diffusion occurs because molecules in solution drift around under random kinetic motion and more move from a high-concentration region than from a low-concentration region. It is purely as a statistical effect that net flow occurs down a concentration gradient. The flow may be treated as a net bias in the otherwise random motion, whereby molecules appear to exhibit a net velocity v in the direction of the concentration gradient. The flux J can therefore be represented as

$$J = \mathbf{n}v \tag{10-40}$$

We now invoke a force to produce the velocity v, namely the gradient of chemical potential $-d\mu_2/dx$ or $-kT\, d(\ln a)/dx$. The velocity is then

$$v = -\frac{kT\, d(\ln a)/dx}{f} \tag{10-41}$$

and combination of Eqs. (2-87), (10-40), and (10-41) gives

$$-\frac{kT\mathbf{n}\, d(\ln a)/dx}{f} = -\mathcal{D}\frac{d\mathbf{n}}{dx}$$

or

$$\mathcal{D} = \frac{kT}{f}\frac{d(\ln a)}{d(\ln \mathbf{n})} \tag{10-42}$$

It is convenient to use the convention of a_C, so that $d(\ln a_C) = d(\ln C) + d(\ln \gamma_C)$ (see Section 10-6) and since $d(\ln \mathbf{n}) = d(\ln C)$ ($\mathbf{n}$ and C are proportional), Eq. (10-42) reduces to

$$\mathcal{D} = \frac{kT}{f}\left(1 + \frac{d(\ln \gamma_C)}{d(\ln C)}\right) \tag{10-43}$$

In the case of dilute nonelectrolyte solutions γ_C will approach unity, and Eq. (10-43) is often used in the limiting form

$$\mathcal{D} = \frac{kT}{f} \tag{10-44}$$

Equation (10-43) was obtained by Einstein in 1905 (see Section 16-CN-1 for a historical sketch).

The formalism of the chemical potential gradient as a driving force in diffusion is just that—a formalism. It is a very convenient one, however, and more elaborate analyses based on statistics of individual molecular motions give the same result.

As in the case of sedimentation rates, one may consider the diffusing molecule to be spherical and apply Stokes' law. The diffusion coefficient becomes

$$\mathcal{D} = \frac{kT}{6\pi\eta r} \tag{10-45}$$

It is thus possible to obtain a molecular radius from diffusion data. Small molecules have diffusion coefficients in water of about 10^{-5} cm^2 s^{-1} at room temperature and hence corresponding r values of about $(1.37 \times 10^{-16})(298)/(6\pi)(0.01)(10^{-5}) = 2.2 \times 10^{-8}$ cm, or of a few angstroms. The values obtained agree fairly well with estimates from van der Waals constants or from crystallographic determinations of molecular sizes. In general, even though the exact form of Stokes' law may not be correct, Eq. (10-45) is often obeyed to the extent that $\mathcal{D}\eta$ is a constant; the observation is known as *Walden's rule*.

It is difficult to measure diffusion coefficients smaller than about 10^{-7} cm^2 s^{-1}, so that the method begins to fail for molecular weights above 10^6 (see Table 10-5). In summary, diffusion coefficients provide a measure of the hydrodynamic radius of a molecule and thereby information about its molecular weight.

We may now combine the two preceding subsections to treat sedimentation equilibrium. The dynamic picture is that a molecule under the influence of gravity or a centrifugal field will move in the direction of the force, so that a concentration gradient develops. Eventually the back-diffusion rate becomes equal to the sedimentation rate. That is, v in Eq. (10-35) and v in Eq. (10-41) may be equated to give

C. Sedimentation equilibrium

$$m_2 g(1 - \bar{v}_2\rho_1) = kT\frac{d(\ln a)}{dx} \tag{10-46}$$

The friction coefficient is the same for both processes, and therefore cancels out. Rearrangement gives

$$\frac{d(\ln a)}{dx} = \frac{m_2 g}{kT}(1 - \bar{v}_2\rho_1) \tag{10-47}$$

Substance	Molecular weight (g mol^{-1})	$10^5 \times \mathcal{D}$ (cm^2 s^{-1})	$10^{13} \times s$ (s)	$\bar{v}_2$ (cm^3 g^{-1})
Water	18	2.14[a]	—	—
Urea	60	1.18 (20°C)	—	—
Sucrose	342	0.52	—	—
Ovalbumin	44,000	0.078	3.6	0.75
Hemoglobin (man)	63,000	0.069	4.46	0.75
Urease	490,000	0.034	18.6	0.73
Tobacco mosaic virus	40×10^6	0.0053	185	0.72

TABLE 10-5
Diffusion and sedimentation coefficients (in water at 225°C)

[a]This is a self-diffusion coefficient obtained by isotopic labeling.

or, in dilute solution

$$\frac{d(\ln C)}{dx} = \frac{m_2 g}{kT}(1 - \bar{v}_2 \rho_1) \tag{10-48}$$

Except for the necessary buoyancy correction, Eq. (10-48) is the same as the barometric equation (2-27); if integrated, with $dx = -dh$, it yields

$$C = C_0 \exp\left(-\frac{m_2 g' h}{kT}\right)$$

where, to bring out the comparison, g has been replaced by the effective force due to gravity g', where $g' = g(1 - \bar{v}_2 \rho_1)$

Alternatively, Eq. (10-44) may be used to eliminate f in Eq. (10-39), to give

$$s = \frac{m_2 \mathcal{D}}{kT}(1 - \bar{v}_2 \rho_1) \tag{10-49}$$

or

$$M_2 = \frac{RTs}{\mathcal{D}(1 - \bar{v}_2 \rho_1)} \tag{10-50}$$

Thus if both the sedimentation and diffusion coefficients are known, the molecular weight may be determined without any assumptions as to the size or shape of the molecule.

To return to sedimentation equilibrium, we see that if g in Eq. (10-48) is replaced by $\omega^2 x$ and the equation is integrated between the limits C_1 at x_1 and C_2 at x_2, the result is

$$M = \frac{2RT \ln(C_2/C_1)}{\omega^2(x_2^2 - x_1^2)(1 - \bar{v}_2 \rho_1)} \tag{10-51}$$

Sedimentation equilibrium studies using ultracentrifuges have provided a great deal of information about the molecular weights of large molecules such as proteins, other biological substances, and polymers. The method is tedious in that considerable time is required to attain equilibrium. A modern variant involves the use of a gradient in the density of the medium. The effect is greatly to accentuate the equilibrium concentration distribution, and molecular weights as low as about 300 can be determined by this means.

EXAMPLE The numerical example of Section 10-8A may be extended to Eq. (10-51). We take ω^2 to be 1×10^5, the molecular weight to be 100,000, and $\bar{v}_2$ to be 0.80, and assume that x_2 and x_1 are 4 and 3 cm, respectively. Then

$$\ln \frac{C_2}{C_1} = \frac{(100,000)(1 \times 10^5)(4^2 - 3^2)(1 - 0.80)}{(2)(8.3 \times 10^7)(298)} \simeq 0.28$$

Some values for diffusion and sedimentation coefficients are given in Table 10-5.

D. Light scattering   The subject of light scattering by small particles is a complicated one and will be treated here only in a superficial way. In terms of the electromagnetic theory of light, incident radiation induces an oscillating dipole in a molecule or particle, which in turn reradiates. The intensity of scattered light therefore depends on the polarizability α; it is also a function of angle, being a maximum in the direction of the incident beam.

The detailed analysis was begun by Lord Rayleigh in 1900, extended by G. Mie around 1908, and further extended by P. Debye in 1947. The intensity of unpolarized light scattered by a single molecule is given by

$$\frac{i}{I_0} = \frac{16\pi^4\alpha^2(1 + \cos^2\theta)}{\lambda^4 r^2} \tag{10-52}$$

where, as illustrated in Fig. 10-7, I_0 is the incident light intensity per square centimeter, i is the intensity per square centimeter observed at scattering angle θ, α is the polarizability of the scattering molecule, λ is the wavelength of light, and r is the distance from the scattering particle. Some of the important features of Eq. (10-52) are as follows:

1. The intensity depends inversely on λ^4 and is therefore much stronger for blue than for red light. Small-molecule scattering in our atmosphere brings blue light to us from the sky and reddens direct sunlight, especially at sunset when the distance traveled through the atmosphere is larger than at other times.
2. The minimum scattering occurs at $90°$. The qualitative reason is that the induced dipoles are oriented perpendicularly to the incident light and reradiate most strongly in the same direction.
3. The dependence on the polarizability has been accounted for.

In the case of a dilute solution or suspension of scattering molecules or particles, a useful modification of Eq. (10-52) is the following. Assuming additivity of molar refractions, (Eq. (3-15) may be manipulated to give $n^2 - n_0^2 = 4\pi N_0 \alpha c/M$. Here, n is the index of refraction of the solution and n_0 that of the solvent; c is the concentration in grams per cubic centimeter of scattering molecules of polarizability α and molecular weight M. For a dilute solution, $n^2 - n_0^2 = (n + n_0)(n - n_0) \simeq 2n_0 c\, dn/dc$ and we obtain

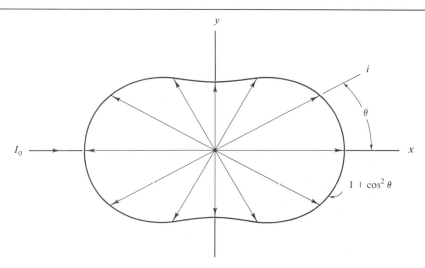

FIGURE 10-7
Intensity envelope for scattered light. The envelope is a figure of revolution about the axis defined by the incident light.

$$\frac{i}{I_0} = \frac{4\pi^2 M^2 (1 + \cos^2 \theta)\, n_2^0 (dn/dc)^2}{N_A \lambda^4 r^2} \tag{10-53}$$

Since the number of particles per cubic centimeter is cN_A/M, the total scattering is

$$\frac{i}{I_0} = \frac{4\pi^2 (1 + \cos^2 \theta)\, n_2^0 (dn/dc)^2}{N_A \lambda^4 r^2}\, Mc \tag{10-54}$$

We next integrate the scattered intensity over all angles, to obtain the total proportion of light removed from the incident beam. The removal of light by scattering produces the same mathematical law as for the absorption of light (see Section 3-2),

$$I = I_0 e^{-\tau x} \tag{10-55}$$

where I is the intensity of unscattered light and τ is called the *turbidity*. The proportion of light that is scattered is then $(I_0 - I)/I_0$ and for small amounts of scattering this is equal to τ for a 1-cm path length. The result of the integration over all angles and the introduction of τ is

$$\tau = \frac{32\pi^3 n_0^2 (dn/dc)^2}{3 N_A \lambda^4}\, Mc = HMc \tag{10-56}$$

where

$$H = \frac{32\pi^3 n_0^2 (dn/dc)^2}{3 N_A \lambda^4} \tag{10-57}$$

and therefore

$$M = \frac{\tau}{Hc} \tag{10-58}$$

Actual practice is somewhat more complicated. Solution nonidealities cause deviation from Eq. (10-58), so the quantity Hc/τ is usually plotted against c and extrapolated to zero concentration. Further, the reduction in light intensity of the direct beam is rather small, so it is difficult to measure $I_0 - I$ accurately. A better procedure is to determine i at some definite angle, most commonly 90°, and use Eq. (10-52) to relate the scattering intensity at that angle to the total scattering corresponding to τ. A further complication is that if the particles are comparable in size to the wavelength of the light used, there will be interference effects arising from light scattered from different parts of the same particle. If this is a problem, then i/I_0 is measured at various angles and the calculated H values are extrapolated to $\theta = 0$, where such interference vanishes. This is done for each of several concentrations and the set of extrapolated points lies on a line which can in turn be extrapolated to zero concentration. A clever way of making the two extrapolations (to zero c and to zero θ) is due to Zimm (1948).

Light scattering can be used for any size molecule; one may, for example, observe scattering from a pure liquid due just to the fluctuations in density that occur, and the scattering by the atmosphere has already been mentioned.

Equation (10-56) states, however, that for a given concentration in grams per cubic centimeter the larger the molecular weight, the greater the scattering. The method is then most sensitive for macromolecules and has been widely used in the study of biological and synthetic polymers.

10-9 Types of molecular weight averages

It is particularly true for polymer solutions (discussed further in Chapter 20) that a preparation will consist of a range of molecular weights. A molecular weight determination by one of the methods of this chapter will give an average value but, it turns out, not always the same average as that given by some other method.

There are two principle ways in which one may average the molecular weight of a collection of solute molecules. The first is the intuitive one of simply dividing the total weight by the total number of particles. This is known as the *number average molecular weight* M_n and is defined formally as follows:

$$M_n = \frac{n_1M_1 + n_2M_2 + \cdots}{n_1 + n_2 + \cdots} = \frac{\sum_i n_iM_i}{\sum_i n_i} = \sum_i x_iM_i \qquad (10\text{-}59)$$

Thus a sample consisting of equal numbers of molecules of molecular weight 100, 1000, and 10,000 g mol^{-1} would have a number average molecular weight of

$$M_n = \frac{\frac{1}{3}(100) + \frac{1}{3}(1000) + \frac{1}{3}(10,000)}{1} = 3700$$

In this case each molecular weight is weighted by the number of particles involved. An alternative weighting is by the weight of particles of a given molecular weight (or essentially by their size):

$$M_w = \frac{w_1M_1 + w_2M_2 + \cdots}{w_1 + w_2 + \cdots} = \frac{\sum_i w_iM_i}{\sum_i w_i} = \frac{\sum_i n_iM_i^2}{w_{\text{tot}}} \qquad (10\text{-}60)$$

M_w is known as the *weight average molecular weight*. Referring to the example just given, since there are equal numbers of each kind of molecule,

$$M_w = \frac{(100)^2 + (1000)^2 + (10,000)^2}{100 + 1000 + 10,000} = 9100$$

In this case, as in general, the weight average molecular weight is greater than the number average molecular weight. The ratio of the two gives a measure of the total spread of molecular weights in the sample.

The colligative property measurements essentially count the number of particles and one obtains the molecular weight by dividing the weight of material present by this number. The result is therefore a number average molecular weight. Sedimentation experiments weight each particle according to its mass and therefore give a weight average molecular weight. The same is true of light scattering measurements.

COMMENTARY AND NOTES

10-CN-1   Water desalination

The material of this chapter has some bearing on the important subject of water desalination. This is at present accomplished mainly by distillation, and large-scale installations are found in the southwestern United States, in parts of South America, and in the Middle East. It seems inevitable that the coastal regions of the world will draw increasingly on this method as a source of both agricultural and potable water. The current prevalence of distillation may be largely a reflection of the fact that engineering is most highly developed for this type of process. Other methods may eventually dominate, and many are now under active investigation.

Sea water contains about 35,000 ppm of dissolved salts, mostly sodium chloride, with magnesium, calcium, and potassium following in order of importance; it is roughly equivalent to a 0.7 M sodium chloride solution, and the vapor pressure of sea water at 25°C averages about 0.78 Torr lower than that of pure water. The basic desalination process is then

sea water (25°C, P_1

$$= 22.98 \text{ Torr})$$ (10-61)
$$= \text{pure water (25°C, } P_1^\circ$$
$$= 23.76 \text{ Torr})$$

This supposes that the pure water is obtained from so large an amount of sea water that the concentration of the latter is not materially affected. The free energy change is

$$\Delta G = RT \ln \frac{P_1^\circ}{P_1} = 592 \ln \frac{23.76}{22.98}$$
$$= 19.8 \text{ cal mol}^{-1}$$

This is the reversible work required and is therefore the minimum possible. The free energy change ΔG would be somewhat larger for a process in which the products consisted of more highly concentrated sea water plus pure water. In either case, however, the minimum energy required is far less than the approximately 10,000 cal mol^{-1} needed to vaporize the water directly. The relatively high efficiency of contemporary distillation plants is a result of using very efficient multistep evaporation as well as heat exchangers. In effect, the 10,000 cal mol^{-1} needed for evaporation is largely recovered from the heat liberated on condensation. The heat "turnover" is very large, however, and it has seemed that some process whose direct energy requirement was closer to the minimum could be efficient without the large capital investment required by the complex engineering of distillation plants.

One proposal has been to produce pure water by partially freezing sea water and then separating out the ice. The heat investment in the initial step is now only about 1400 cal mol^{-1}. However, the refrigeration plant needed introduces a new fundamental inefficiency. More important, perhaps, is the finding that the ice so produced tends to occlude considerable salt so that more than one cycle of purification is necessary.

Osmotic devices have attracted much interest. If sea water is separated from fresh water by a semipermeable membrane, then application of mechanical pressure exceeding the osmotic pressure will cause water to pass to the pure water side. The process is one of *reverse osmosis*. The pressure could be applied by means of pumps. Conceivably a pipe could be lowered into the ocean to a depth such that the hydrostatic head exceeded the osmotic pressure. A semipermeable membrane covering the end of the pipe would pass fresh water. In effect, one would have a fresh water well for which the only work required would be that needed to pump the water to the surface. In fact, not even that work need be required [Levenspiel and de Nevers (1974)]!

A problem has been to find membranes that will not pass electrolytes but will pass water rapidly. The osmotic pressure of sea water is some thirty atmospheres, and either a membrane must be very strong or several stages would be needed. Some progress has been made in the use of electrical energy to do the osmotic work with ion-selective membranes in electrolyte cells. As illustrated in Fig. 10-8, cations may pass out of the central compartment at the right, but anions cannot enter; and anions may pass out at the left, but cations cannot enter. The efficiency increases in proportion to the

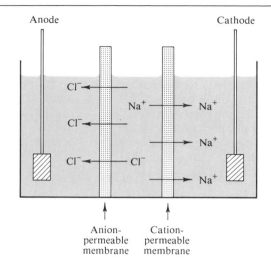

FIGURE 10-8
Electrolytic method for water desalination.

number of such units in parallel but otherwise decreases in proportion to the salt concentration. Cells of this type are being used for the desalination of very hard well water (in Israel) but do not appear so far to be economic for the treatment of sea water. Some recent developments with ion-exchange and clay-type membranes have given encouraging results in an ordinary reverse osmotic process but only in small-scale tests.

It should be pointed out that the problem is heightened by the economic requirements. It is necessary to produce water at a cost of no more than about 35 cents per 1000 gal. This is a difficult accomplishment for what amounts to a chemical processing plant. Efficiency must be very high if power costs alone are not to

exceed this cost figure, even with no allowance for capital investment. Large amounts of low-cost power are thus a vital ingredient—hence the emphasis on coupling desalination plants with nuclear power generators.

10-CN-2 Osmotic pressure

There are many interesting sidelights to colligative properties. The osmotic effect was probably the first studied because of its great importance in biological systems. As just one example, plant cells consist of a relatively rigid cellulose wall enclosing a bag or *vacuole* filled with sap (salts, sugars, coloring matter). The

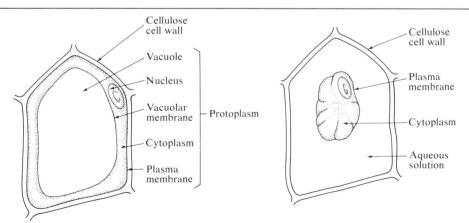

FIGURE 10-9
Left: turgid plant cell; right: plasmolyzed cell. (From S. Gladstone, "Textbook of Physical Chemistry," 2nd ed. D. Van Nostrand, New York, 1946. By permission.)

bag is lined by the vacuolar membrane, and the space between it and the cellulose wall contains the cell cytoplasm and nucleus. The cytoplasm is lined by a second membrane, the plasma membrane, as shown in Fig. 10-9. The cellulose wall is relatively permeable to both water and solutes, but the two membranes will not pass large molecules such as sugars. In the normal cell, water has passed through the membranes to set up an osmotic pressure so that the vacuole and cytoplasm press against the cellulose wall. The cell is rigid or "turgid."

If now the cell is immersed in a salt or sugar solution of higher osmotic pressure, water leaves, the cytoplasm and vacuole shrink, and the cell becomes flaccid. Conversely, if the cell is placed in pure water, the osmotic pressure developed inside the cell may become large enough to burst the cellulose wall. Such effects are called *plasmolysis*. A solution that neither shrinks nor expands the cell is said to be *isotonic* with it. Osmotic effects can play an important role in the treatment of foods, as in salting or pickling.

SPECIAL TOPICS

10-ST-1   Dilute solution conventions for activities and activity coefficients

The concepts of activity and activity coefficient are introduced in Section 9-5B. Activity is defined by Eq. (9-86) as the effective mole fraction required to retain the form of Raoult's law and therefore that of the general thermodynamic treatment of ideal solutions. The effect of deviation from ideality is extracted yet more clearly by means of an activity coefficient, or the factor whereby the activity deviates from mole fraction. Since Raoult's law is obeyed as a limiting law for all solutions, it follows that $a_i \to 1$ and $\gamma_i \to 1$ as $x_i \to 1$. Thus by Eq. (9-87),

$$\mu_1(l) = \mu_1°(l) + RT \ln a_1$$

and $\mu_1(l) \to \mu_1°(l)$ as $x_1 \to 1$. The standard or reference state, for which $\mu_1°(l)$ is the chemical potential, is therefore the pure component 1. The same equations but with the subscript 2 apply to component 2.

The second important equation of the preceding chapter is the Gibbs–Duhem relationship, which allows the calculation of the activity of one component of a solution if that of the other component is known over a range of compositions,

$$x_1 d(\ln a_1) + x_2 d(\ln a_2) = 0 \quad \text{[Eq. (9-94)]}$$

or

$$d(\ln a_2) = -\frac{x_1}{x_2} d(\ln a_1) \quad \text{[Eq. (9-95)]}$$

An example of the use of Eq. (9-95) is given in Section 9-5C.

It is now desirable to extend the applications of the Gibbs–Duhem equation to a two-component system. Since $a_1 = \gamma_1 x_1$,

$$x_1 d(\ln a_1) = x_1[d(\ln x_1) + d(\ln \gamma_1)] = dx_1 + x_1 d(\ln \gamma_1) \quad (10\text{-}62)$$

Similarly,

$$x_2 d(\ln a_2) = dx_2 + x_2 d(\ln \gamma_2) \quad (10\text{-}63)$$

Since $dx_1 = -dx_2$, substitution into Eq. (9-94) yields

$$x_1 d(\ln \gamma_1) = -x_2 d(\ln \gamma_2) \quad (10\text{-}64)$$

or

$$\ln \gamma_2 = - \int_{x_2 = 1}^{x_2} \frac{x_1}{x_2} d(\ln \gamma_1) \quad (10\text{-}65)$$

where the integration limits are from $x_2 = 1$ (and hence $\gamma_2 = 1$ and $\ln \gamma_2 = 0$) to some specific mole fraction x_2.

Returning to Eq. (9-94), we may divide the first term by dx_1 and the second term by $-dx_2$ (which is the same as dx_1) to get

$$\frac{d(\ln a_1)}{d(\ln x_1)} = \frac{d(\ln a_2)}{d(\ln x_2)} \quad \text{[Eq. (9-97)]}$$

This is known as the *Duhem–Margules or Gibbs–Duhem equation*. It tells us, for example, that if component 1 obeys Raoult's law, so that $d(\ln a_1)/d(\ln x_1) = 1$, then it must also be true that $d(\ln a_2)/d(\ln x_2) = 1$. This condition is only required to hold in differential form, and on integration we find

$$\ln a_2 = \ln x_2 + \ln k_2'$$

or

$$a_2 = k_2' x_2 \qquad (10\text{-}66)$$

Since a_2 is defined as equal to $P_2/P_2°$, Eq. (10-66) can be written as

$$P_2 = k_2 x_2 \qquad [\text{Eq. (9-9)}]$$

which is a statement of Henry's law. The conclusion is that, in the case of a nonideal solution, as component 1 approaches Raoult's law behavior $(x_1 \to 1)$, component 2 must approach Henry's law behavior $(x_2 \to 0)$. Thus the validity of Raoult's law as $x_1 \to 1$ implies that of the Henry's law limiting behavior as $x_2 \to 0$.

The preceding framework of thermodynamic treatment is very useful in the case of solutions which can range in composition from pure component 1 to pure component 2 with both components volatile. More often, however, the solute is not a liquid in the pure state and is not volatile. Physical chemical studies are then restricted to the range of composition from $x_2 = 0$ up to the solubility of the solute, and therefore often to relatively dilute solutions. It has therefore been necessary to devise an alternative framework, and this has been done with Henry's law as a reference rather than Raoult's law. The procedure follows.

We start with Eq. (9-71), written for solute species (2),

$$\mu_2(l) = \mu_2(g)$$
$$= \mu_2°(g) + RT \ln P_2$$

but this time add and subtract $RT \ln k_2$ on the right-hand side to obtain

$$\mu_2(l) = \mu_2°(g) + RT \ln k_2$$
$$+ RT \ln \frac{P_2}{k_2} \qquad (10\text{-}67)$$

or

$$\mu_2(l) = \mu_2°' + RT \ln \frac{P_2}{k_2} \qquad (10\text{-}68)$$

[we do not write $\mu_2°'(l)$ since the pure solute is not necessarily a liquid], where

$$\mu_2°' = \mu_2°(g) + RT \ln k$$
$$= \mu_2°(l) + RT \ln \frac{k_2}{P_2°} \qquad (10\text{-}69)$$

A new activity a_2' is now defined as

$$P_2 = a_2' k_2 \qquad [\text{Eq. (10-24)}]$$

so that Eq. (10-68) becomes

$$\mu_2(l) = \mu_2°' + RT \ln a_2' \qquad (10\text{-}70)$$

Also,

$$a_2' = x_2 \gamma_2' \qquad [\text{Eq. (10-24)}]$$

so, alternatively,

$$\mu_2(l) = \mu_2°' + RT \ln x_2$$
$$+ RT \ln \gamma_2' \qquad (10\text{-}71)$$

Equations (10-70) and (10-71) have the same form as do Eqs. (9-88) and (9-89), but there has been a change in the standard or reference state, as given by Eq. (10-69). Henry's law is the limiting reference condition and $a_2' \to x_2$ and $\gamma_2' \to 1$ as $x_2 \to 0$. Alternatively, a_2' is now defined as P_2/k_2, whereas a_2 was defined as $P_2/P_2°$. A numerical illustration will be given later.

Mole fraction composition units are convenient for relatively symmetric systems, but where the emphasis is on dilute solutions and one of the components is always thought of as the solute it is more customary to use molalities. This is particularly true for electrolyte solutions. Molality and mole fraction are related by Eq. (10-9), which reduces to

$$x_2 = \frac{m M_1}{1000} \qquad (10\text{-}72)$$

in the limit of infinite dilution. Henry's law may now be written

$$P_2 = \frac{k_2 M_1}{1000} m = m k_m \qquad (10\text{-}73)$$

We proceed as before, but now adding and subtracting $RT \ln k_m$ on the right-hand side of Eq. (9-71), to obtain

$$\mu_2(l) = \mu_2°(g) + RT \ln k_m$$
$$+ RT \ln \frac{P_2}{k_m} \qquad (10\text{-}74)$$

or

$$\mu_2(l) = \mu_m° + RT \ln \frac{P_2}{k_m} \qquad (10\text{-}75)$$

where

$$\mu_m{}^\circ = \mu_2^{\circ\prime} + RT \ln \frac{M_1}{1000} \qquad (10\text{-}76)$$

A new activity a_m is then defined as

$$P_2 = a_m\left(\frac{M_1}{1000}\right)k_2$$

$$= a_m k_m \qquad [\text{Eq. (10-26)}]$$

so that

$$\mu_2(l) = \mu_m{}^\circ + RT \ln a_m \qquad (10\text{-}77)$$

Again, if

$$a_m = \gamma_m m \qquad [\text{Eq. (10-28)}]$$

then

$$\mu_2(l) = \mu_m{}^\circ + RT \ln m$$
$$+ RT \ln \gamma_m \qquad (10\text{-}78)$$

The form is the same as before, with the standard state given by Eq. (10-76). The reference condition is again that of infinite dilution, since as $m \to 0$, $a_m \to m$ and $\gamma_m \to 1$.

We must repeat the process once more. It is occasionally necessary to deal with molarity rather than molality concentrations in connection with nonideal solutions. Mole fraction and molarity C are related:

$$C = \frac{1000\rho n_2}{n_1 M_1 + n_2 M_2}$$

$$= \frac{1000\rho x_2}{M_1 + x_2(M_2 - M_1)} \qquad (10\text{-}79)$$

where ρ is the density of the solution. The limiting relationship at infinite dilution is

$$x_2 = \frac{M_1 C}{1000\rho_0} \qquad (10\text{-}80)$$

where ρ_0 is the density of the pure solvent. The limiting Henry's law form becomes

$$P_2 = C k_c = C\left(\frac{M_1}{1000\rho_0}\right)k_2 \qquad (10\text{-}81)$$

We add and subtract the quantity $RT \ln k_C$ on the right-hand side of Eq. (9-71) to give

$$\mu_2(l) = \mu_2^\circ(g) + RT \ln k_C$$
$$+ RT \ln \frac{P_2}{k_C} \qquad (10\text{-}82)$$

or

$$\mu_2(l) = \mu_C{}^\circ + RT \ln \frac{P_2}{k_C} \qquad (10\text{-}83)$$

where

$$\mu_C{}^\circ = \mu_2^{\circ\prime} + RT \ln \frac{M_1}{1000\rho_0} \qquad (10\text{-}84)$$

A new activity a_C is defined as

$$P_2 = a_C k_C = a_C\left(\frac{M_1}{1000\rho_0}\right)k_2 \qquad (10\text{-}85)$$

so we have

$$\mu_2(l) = \mu_C{}^\circ + RT \ln a_C \qquad (10\text{-}86)$$

The corresponding activity coefficient is

$$a_C = \gamma_C C \qquad (10\text{-}87)$$

so that

$$\mu_2(l) = \mu_C{}^\circ + RT \ln C$$
$$+ RT \ln \gamma_C \qquad (10\text{-}88)$$

The three conventions give the following alternative expressions for $\mu_2(l)$:

$$\mu_2(l) = \mu_2^\circ + RT \ln a_2$$
$$= \mu_2^{\circ\prime} + RT \ln a_2{}'$$
$$= \mu_m{}^\circ + RT \ln a_m$$
$$= \mu_C{}^\circ + RT \ln a_C \qquad (10\text{-}89)$$

Since the standard-state values are constants, we have

$$d\mu_2(l) = d \ln a_2 = d \ln a_2{}'$$
$$= d \ln a_m = d \ln a_C \qquad (10\text{-}90)$$

and substitution into the Gibbs–Duhem equation gives the same form, Eq. (9-95), in all three cases. The subsequent steps leading to Eq. (10-65) are exactly the same as for $\gamma_2{}'$, so that we have

$$\ln \gamma_2{}' = -\int_{x_2=0}^{x_2} \frac{x_1}{x_2} \, d(\ln \gamma_1) \qquad (10\text{-}91)$$

The integration limit, however, is from $x_2 = 0$ (and hence $\gamma_2{}' = 1$ and $\ln \gamma_2{}' = 0$) to x_2. In the case of γ_m the convenient cancellation of the $d(\ln x_1)$ and $d(\ln x_2)$ terms does not occur. However, since $x_2/x_1 = m/(1000/M_1)$ and

$$d(\ln a_m) = d(\ln \gamma_m) + d(\ln m),$$

one obtains

$$\ln \gamma_m = -\int_{m=0}^{m} \left[\frac{1000}{mM_1} d(\ln a_1) - d(\ln m) \right] \quad (10\text{-}92)$$

In actual practice, we make some further algebraic manipulations to facilitate the evaluation of the integral, but the point is that the Gibbs–Duhem equation still allows an evaluation of γ_m if the activity of the solvent is known over a range of concentration. Notice, incidentally, that the Raoult's law standard state is retained for the solvent; that is $\gamma_1 \rightarrow 1$ as x_2 or m approaches zero. The computation of γ_c values follows a similar procedure.

The complications of the preceding methods for changing standard state are regrettable but cannot be avoided. However, the operations are not difficult. The example of Section 9-5C may be extended to the calculation of a_2' and γ_2', with chloroform assumed to be the solute. Since k_C was found to be 142 Torr, $a_C' = P_C/142$ and $\gamma_C'/x_2/x_2 = P_C/142x_2$.* The values of γ_C' differ from those of γ_C by the constant factor of $1/0.485$, reflecting the fact that only a change in the choice of reference state is involved. The acetone–chloroform system is a symmetric one, and a set of a_a' and γ_a' values could also be calculated. This is left as an exercise. Values of γ_m and γ_C are not given, but manipulation of the preceding equations gives

$$\gamma_2' = \gamma_m(1 + 0.001mM_1)$$

$$= \gamma_C \frac{p + 0.001C(M_1 - M_2)}{\rho^\circ}$$

$$[\text{Eq. (10-31)}]$$

*Subscript c denotes chloroform and subscript a denotes acetone.

The preceding example was for a system of two volatile liquids. The values of a_2' and hence of γ_2' could be calculated from those of P_2. It is not necessary, however, that P_2 be measured or even that the solute be volatile. Equations (10-91) and (10-92) allow the calculation of γ_2' and γ_m through an integration of the Gibbs–Duhem equation. This may be done if the activity of the solvent is known over a range of concentration, as illustrated in Section 9-5C by the analogous calculation of a_2 from the data for a_1. As summarized in Table 10-4 the exact equations for the colligative phenomena all involve a_1 or $\ln a_1$. The precise use of these equations involves obtaining a_1 values over a range of concentration and then calculating from them the activity or the activity coefficient of the solute for various concentrations. Depending on the way in which the Gibbs–Duhem integration is set up, one obtains γ_2, γ_2', γ_m, or γ_C.

One final comment. The three Henry's law reference procedures introduce a slightly paradoxical situation. The reference condition is that of infinite dilution; it is at infinite dilution that γ', γ_m, and γ_C approach unity, so that this becomes the integration limit for relationships such as Eqs. (10-91) and (10-92). On the other hand, $\mu_2(l)$ in Eqs. (10-70) and (10-77) equals $\mu_2^\circ{}'$, μ_m°, or μ_C° when the corresponding activities are *unity* (not zero!). In the acetone–chloroform system, for example, a_2' is unity at $x_C = 0.60$. We speak of $\mu_2^\circ{}'$, μ_m°, and μ_C° as, respectively, the *hypothetical unit mole fraction, unit molality,* and *unit molality* standard states. The reason for the term hypothetical is that there is no condition in which the solute has both unit activity and unit activity coefficient. By contrast, in the Raoult's law system, the pure liquid does meet this condition. Hypothetical standard states are perfectly definite ones; the activities and activity coefficients are given by operational procedures. Their lack of simple physical meaning is therefore no handicap to their use.

GLASSTONE, S., LAIDLER, K.J., and EYRING, H. (1941). "The Theory of Rate Processes." McGraw-Hill, New York.

HILDEBRAND, J.H., AND SCOTT, R.L. (1950). "The Solubility of Nonelectrolytes," 3rd ed. Van Nostrand–Reinhold, Princeton, New Jersey.

HIRSCHFELDER, J.O., CURTIS, C.F., AND BIRD, R.B. (1964). "Molecular Theory of Gases and Liquids," corrected ed. Wiley, New York.

LEWIS, G.N., AND RANDALL, M. (1961). "Thermodynamics," 2nd ed. (revised by K. S. Pitzer and L. Brewer). McGraw-Hill, New York.

MOELWYN-HUGHES, E.A. (1961). "Physical Chemistry," 2nd revised ed. Pergamon, Oxford.

GENERAL REFERENCES

CITED
REFERENCES

BRØNSTED, J.N. (1906). *Z. Phys. Chem.* **55**, 371.
LEVENSPIEL, O., AND DE NEVERS, N. (1974). *Science* **183**, 157.
SCHRÖDER, I. (1893), *Z. Phys. Chem. (Leipzig)* **11**, 457.
SIMONS, J.H. AND POWELL, M.G. (1945). *J. Amer. Chem. Soc.* **67**, 77.
WALL, F.T., AND ROSE, P.E. (1941). *J. Amer. Chem. Soc.* **63**, 3302.
ZIMM, B.H. (1948). *J. Chem. Phys.* **16**, 1093.

EXERCISES

Take as exact numbers given to one significant figure.

10-1   Calculate the vapor pressure lowering of a 0.5 m aqueous solution of the sugar mannose at 20°C. The vapor pressure of water at this temperature is 17.535 mm Hg.

Ans. 0.156 mm Hg.

10-2   A 2.00 weight % solution of a nonvolatile compound in ethanol produces a vapor pressure lowering of 0.45%. Calculate its molecular weight.

Ans. 79.6 g mol^{-1}.

10-3   Calculate the boiling point elevation for a 5 m aqueous sugar solution using (a) Eq. (10-7) and (b) Eq. (10-10).

Ans. (a) 2.57°C, (b) 2.47°C.

10-4   (a) Calculate K_b for ethanol (look up the necessary data). (b) The boiling point elevation is 0.180°C for a solution of 10 g of a nonvolatile solute in 200 g of ethanol. Calculate the molecular weight of the solute.

Ans. (a) 1.226°C m^{-1}, (b) 341 g mol^{-1}.

10-5   Calculate the freezing point depression for the solution of Exercise 10-3 by Eq. (10-15) and by Eq. (10-17).

Ans. 8.61°C, 9.28°C.

10-6   Ethylene glycol, CH_2OHCH_2OH, is used as an antifreeze. Assuming ideal solutions, what weight per cent of ethylene glycol should be present in a car's radiator to protect it down to 0°F?

Ans. 41%

10-7   The vapor pressure lowering of a 1 m solution of KCl at 100°C is 22.3 mm Hg. What is the van't Hoff i factor?

Ans. 1.66.

10-8   It is noted in Section 10-CN-1 that sea water corresponds to about 0.7 m NaCl. Assuming the NaCl to be fully dissociated and the solution ideal, calculate (a) the vapor pressure lowering at 25°C, (b) the boiling point elevation, (c) the freezing point depression, and (d) the osmotic pressure at 25°C.

Ans. (a) 0.296 mm Hg, (b) 0.710°C, (c) 2.55°C, (d) 33.8 atm.

10-9   Calculate the diffusion coefficient in water at 25°C of a protein of molecular weight 250,000, assumed to be spherical and of density 1.345 g cm^{-3}.

Ans. 5.83 × 10^{-7} cm^2 s^{-1}.

10-10   How fast should an aqueous suspension of colloidal gold settle at 25°C if the particles are 0.5 μm (10^{-4} cm) in diameter and the density of gold is 19.3 g cm^{-3}? Assume the particles to be spherical.

Ans. 2.79 × 10^{-4} s^{-1}.

10-11   How fast should the suspension of Exercise 10-10 settle in a centrifuge operating at 50 rps and having a radius of 5 cm?

Ans. 0.104 cm s^{-1}.

The protein of Exercise 10-9 is studied by centrifugation. What should be the ratio of concentrations between $x = 5$ and $x = 5.5$ cm if the speed of the centrifuge is 50 rps?   **10-12**

Ans. 0.512.

Estimate the diffusion coefficient of ovalbumin in water at 98.6°F. You may use data from Table 10-5 and Walden's rule.   **10-13**

Ans. 1.00×10^{-6} cm^2 s^{-1}.

A polymer consists of the following molecular weight fractions: $M = 2 \times 10^3$, 20%; $M = 1 \times 10^4$, 60%; $M = 3 \times 10^4$, 20% (percentages are by weight). Calculate the number and the weight average molecular weights.   **10-14**

Ans. $M_n = 6.0 \times 10^3$ g mol^{-1}, $M_w = 1.24 \times 10^4$ g mol^{-1}.

PROBLEMS

Calculate the boiling point elevation constant for water that would be found in Denver, Colorado, where the barometric pressure is 650 mm Hg.   **10-15**

An isopiestic measurement shows that an aqueous sugar solution with 5% by weight of an unknown sugar has the same vapor pressure as a 1% aqueous sodium chloride solution. Calculate the molecular weight of the sugar, assuming ideal solution behavior.   **10-16**

Equation (10-6) was obtained by an integration in which it was assumed that ΔH_v° is constant. Obtain a more accurate equation with the assumption that $\Delta H_v = \Delta H_v^\circ + \Delta C_P y$, where $y = T_b^\circ - T_b$.   **10-17**

The triple point of water, that is, the temperature of ice–liquid water–water vapor equilibrium, differs from 0°C because of two effects. The standard freezing point of 0°C is for water under 1 atm of air pressure rather than the 4.6 mm Hg of water vapor pressure. Also, water under 1 atm of air has about 5.7×10^{-4} m N$_2$ and 2.6×10^{-4} m O$_2$ dissolved in it at 0°C. Calculate the triple point of water.   **10-18**

Calculate the freezing point depression of water for various concentrations of ethylene glycol (antifreeze) and, similarly, the freezing point depression of ethylene glycol for various concentrations of water. Plot the two curves as T_f versus mole fraction and find the composition and temperature of the eutectic (see Section 11-3A). Assume ideal solutions; the freezing point of ethylene glycol is -11°C and its heat of fusion is 42 cal g^{-1}. Put on the same graph some actual freezing point data taken from a handbook, and comment.   **10-19**

Calculate the solubility of anthracene in benzene at 25°C. The melting point of anthracene is 217°C and its heat of fusion is 38.7 cal g^{-1}.   **10-20**

The vapor pressure of an aqueous solution of a nonvolatile solute is 4.550 Torr at 0°C, at which temperature the vapor pressure of water is 4.579 Torr. Calculate the freezing point of the solution. State all the assumptions and approximations involved in your calculation.   **10-21**

The freezing point of CCl$_4$ is lowered by 5.97°C if 66.83 g of VCl$_4$ is present per 1000 g of solvent. Find the equilibrium constant for the reaction V$_2$Cl$_8$ = 2VCl$_4$ at the temperature of the freezing solution. Carbon tetrachloride melts at -22.9°C and its heat of fusion is 640 cal mol^{-1}. Assume ideal solution behavior. [See Simons and Powell (1945).]   **10-22**

The freezing point lowering of a solution containing 378 mg of acetic acid in 50 g of benzene is 0.16°C. Calculate the molecular weight of acetic acid in the solvent.   **10-23**

Camphor is sometimes used as a solvent for approximate molecular weight determinations because of its large K_f value, 40. What molecular weight would be found for phenolphthalein if 10 mg dissolved in 150 mg of camphor showed an 8.71°C lowering of the freezing point?   **10-24**

10-25 Calculate the solubility of phenanthrene in benzene at 25°C (in mole %) if its heat of fusion is 4450 cal mol^{-1}. The melting point is 100°C and the solution is to be assumed ideal.

10-26 It is found that 6.6 g of benzoic acid will dissolve in 0.1 mole of a certain solvent at 60°C. The melting point of benzoic acid is 122°C. Calculate its heat of fusion.

10-27 A problem to which considerable attention is being devoted is the economic conversion of sea water to potable water. Sea water may be considered as an approximately 1 m solution of NaCl, and two typical processes might be (each at 25°C):

(a) H_2O (in infinite amount of sea water) $=$ H_2O (pure)
(b) NaCl $+$ 55.5H_2O $=$ NaCl $+$ 18H_2O $+$ 37.5 H_2O (pure).
  (as a solution)   (as a solution)

Calculate the minimum work for each process. The vapor pressure of sea water is 0.78 Torr lower than that of pure water (at 25°C), and that of concentrated NaCl solution in (b) is 2.5 Torr lower than that of pure water. Calculate the cost per 1000 gal water (fresh) produced by each process, assuming 0.5 ¢ per kW hr. The approximate cost of water (wholesale) in this area is 15 ¢ per 1000 gal.

10-28 The osmotic pressure at 25°C of a series of solutions of a rubber type polymer was measured in toluene as solvent, and expressed as height h of solution. The solution is taken to have the same density as the pure solvent, namely 0.8566 g cm^{-3}. The concentration c is given as g cm^{-3} and the values of (c,h) are as follows: $(3 \times 10^{-3}, 2.19)$, $(1 \times 10^{-3}, 0.6)$, $(3 \times 10^{-4}, 0.167)$, where h is in cm. Calculate the approximate molecular weight of the polymer, and, by an extrapolation method, a presumably accurate value.

10-29 An improved form of Eq. (10-23) is $\Pi (V - b) = RT$, where V is the volume of solution per mole of solute and b can be regarded as the actual molar volume of the solute. Obtain Eq. (10-33) from the above (make and explain a suitable approximation).

10-30 Calculate the ratio $\Delta P/\Pi$ for a dilute aqueous solution at 25°C, where ΔP is the vapor pressure lowering. The vapor pressure of water at 25°C is 23.756 mm Hg and its density is 0.99704 g cm^{-3}.

10-31 Referring to Section 10-CN-1, one end of a pipe is fitted with a membrane permeable only to water; the pipe is to be lowered into sea water of temperature 4°C. (a) To what depth h should the pipe be lowered for fresh water just to appear at the bottom of the pipe? (b) Recognizing that the density of sea water is 1.02 g cm^{-3}, what depth of fresh water would there be in the pipe if it were lowered to the depth h of part (a)?

10-32 Several ways of determining Avogadro's number are implicit in the material of this chapter: Describe specific illustrative experiments.

10-33 The boundary between solvent and a solution of a species of hemoglobin moved 0.074 cm in 40 min under centrifugation at 4×10^4 rpm, the boundary being 4.5 cm from the axis of rotation. The diffusion coefficient of this hemoglobin is 7.0×10^{-7} cm^2 s^{-1} and its specific volume is 0.76 cm^3 g^{-1}. The measurements are at 35°C. Calculate the molecular weight of the protein.

10-34 The molecular weight of a particular DNA is found to be 4.00×10^6 g mol^{-1} and its sedimentation coefficient is 12.0×10^{-13} in cgs units. The product $\bar{v}_2\rho$ is 0.52 (in water at 25°C). Calculate the diffusion coefficient for this DNA.

10-35 A dilute aqueous suspension contains 100 mg of particles per liter, of which 20 mg consist of uniform particles having a radius of 4 μ and 80 mg consist of uniform particles having a radius of 1.5 μ. The density of the particles is 1.80 g cm^{-3} and the temperature is 25°C. The height of the column of suspension is 40 cm, but the solution-suspension boundary moves downward, of course, as the particles settle. Calculate (a) the time it should take for the larger particle size fraction to settle out completely, (b) the fraction by weight of the suspension that settles out in this time, and (c) the composition of the sediment.

10-36 The molecular weight of an albumin protein is determined by the sedimentation equilibrium method at a speed of 140 rps. The protein has a density of 1.35 g cm^{-3} and that of the aqueous medium is 1.00 g cm^{-3}. The equilibrium concentration gradient is such that at 25°C the solution contains 0.65

wt% of protein at a distance of 4.30 cm from the axis of rotation and 1.300 wt% at a distance of 4.60 cm. Calculate the molecular weight.

Explain under what circumstances the number average and the weight average molecular weight of a collection of colloidal particles will be the same. **10-37**

A solution of polymer contains equal numbers of polymer molecules of molecular weights 3×10^4 and 4.5×10^4 g mol^{-1}. Calculate the number and weight average molecular weights. Repeat the calculation for the case of a solution containing equal *weights* of the two sizes of molecules. **10-38**

A solution is known to contain equal molar concentrations of two proteins, of molecular weights, M_1 and M_2. From a colligative property measurement, M_n is found to be 2×10^5 g mol^{-1}, and from viscosity measurements M_w is found to be 3.5×10^5 g mol^{-1}. Calculate M_1 and M_2. **10-39**

Calculate the molecular weight from the following light scattering data for an aqueous solution of a substance if the following turbidities are obtained (for 426 nm and 25°C; $dn/dc = 0.105$): **10-40**

Concentration (g cm^{-2})	0.080	0.070	0.057	0.050	0.038	0.026	0.015	0.008
$10^3\tau$ (cm^{-1})	0.700	0.633	0.544	0.492	0.394	0.286	0.174	0.096

Calculate the turbidity of a 0.02 *M* solution of sucrose in water at 20°C, assuming the solution to be ideal. The refractive indices of 0.5, 1.0, 1.5, and 2.0 wt% sucrose solutions at 20°C are 1.3337, 1.3344, 1.3351, and 1.3359, respectively. Assume a wavelength of 589 nm. **10-41**

SPECIAL TOPICS PROBLEMS

Derive Eq. (10-31). **10-42**

Calculate a_c' and γ_c' from the data of Table 9-1 for each of the compositions and plot the results versus composition. **10-43**

Calculate γ_C and γ_m for acetone from the data of Table 9-1 and plot the results versus composition. **10-44**

chapter 11
Heterogeneous equilibrium

Equilibria involving two or more phases—heterogeneous equilibria—have been discussed at various points in preceding chapters. The function of the present chapter is to organize examples of such systems according to a more formal framework and to extend the complexity and variety of systems considered. The thermodynamic criterion for equilibrium finds a more general form known as the *Gibbs phase rule*. There will be considerable emphasis on graphical methods for representing phase equilibria. Graphs showing pressure–temperature–composition domains for phases were called phase maps in preceding chapters; the more common term of *phase diagram* will now be used.

11-1   The Gibbs phase rule

The criterion for phase equilibrium was developed in Section 9-3B, where it was concluded that the chemical potential of each component must be the same in all phases. Thus for two phases in equilibrium the condition for the ith component is

$$\mu_i^\alpha = \mu_i^\beta = \cdots \qquad \text{[Eq. (9-36)]}$$

where α, β, and so on denote the various phases in equilibrium. The effect of this condition is to reduce the number of independent variables that are needed to specify the state of the system.

Consider first the case of a pure substance. If it is present as a single phase α, then we must specify pressure and temperature to fix the state of the substance. That is, we write

$$\mu^\alpha = f^\alpha(P, T) \tag{11-1}$$

and phase α will have a region of existence over some range of P and T, as is illustrated in Fig. 8-11. The same is true for some second phase β:

$$\mu^\beta = f^\beta(P, T) \tag{11-2}$$

(It would be perfectly acceptable to use G^α and G^β instead of μ^α and μ^β, since composition is not a variable, but it seems better to keep a uniform nomenclature.)

Equilibrium between the two phases corresponds to a line of crossing of the surfaces generated by f^α and f^β in the three-dimensional plots of μ versus P and T. In effect we solve Eqs. (11-1) and (11-2) simultaneously to obtain as the condition of phase equilibrium

$$f^\alpha(P, T) = f^\beta(P, T) \qquad \text{or} \qquad P = \phi^{\alpha\beta}(T) \qquad (11\text{-}3)$$

There is now only one *degree of freedom;* that is, one independently variable intensive quantity determines the state of the system having two phases present.

If a third phase γ is possible, then for α–γ equilibrium we have

$$f^\alpha(P, T) = f^\gamma(P, T) \qquad \text{or} \qquad P = \phi^{\alpha\gamma}(T) \qquad (11\text{-}4)$$

We now ask that all three phases be in equilibrium, that is, coexist at the same P and T. We obtain the condition by solving Eqs. (11-3) and (11-4) simultaneously. With two unknowns (P and T) and two equations, the solution gives unique values for pressure and temperature. The number of degrees of freedom is now zero.

The graphical analogue of the above algebra may be helpful. We have seen that Eq. (11-3) corresponds to the P versus T line formed by the intersection of the two surfaces generated by Eqs. (11-1) and (11-2). We now imagine that a third surface, generated by the equation

$$\mu^\gamma = f^\gamma(P, T) \qquad (11\text{-}5)$$

is added to Fig. 8-11. This third surface will cut the $P = \phi^{\alpha\beta}(T)$ line at a *point.* At this unique point $\mu^\alpha = \mu^\beta = \mu^\gamma$, and the three phases are in equilibrium. Figure 9-9 provides a specific illustration. The liquid–vapor two-phase line crosses the solid–vapor two-phase line at the triple point (for water in this case). Note that the third two-phase line, that for solid–liquid equilibrium, *must* also go through this point (why?).

A system consisting of a single substance is called a *one-component* system, meaning that the composition of each phase present is fixed at any given P and T. A system for which the relative amounts of two substances must be specified for composition to be fixed is called a *two-component* system. Some complexities are discussed in the next section; the important point at the moment is that it is the *relative* amounts of the two components that suffices to fix composition. Knowing the total amounts of each component would tell us the *size* of the system, but this is not relevent to phase equilibria. We are dealing with equality of chemical potentials, and chemical potential is an *intrinsic* property, determined by composition (and P and T). Composition can be expressed in various ways, as discussed in Section 9-1, but a convenient choice here is mole fraction x. Note that in a two-component system only x_1 is needed since, by definition, $x_2 = 1 - x_1$.

The statements of chemical potential for phase α in a two-component system are thus

$$\mu_1{}^\alpha = f_1{}^\alpha(P, T, x_1{}^\alpha) \qquad \text{and} \qquad \mu_2{}^\alpha = f_2{}^\alpha(P, T, x_1{}^\alpha) \qquad (11\text{-}6)$$

There are three degrees of freedom in this case; we must specify, P, T, and x_1 (or x_2) in order to define the state of the system.

If two phases α and β are in equilibrium, then

$$\mu_1{}^\alpha = \mu_1{}^\beta \qquad \text{and} \qquad \mu_2{}^\alpha = \mu_2{}^\beta \qquad (11\text{-}7)$$

or

$$f_1^\alpha(P, T, x_1^\alpha) = f_1^\beta(P, T, x_1^\beta) \quad \text{and} \quad f_2^\alpha(P, T, x_1^\alpha) = f_2^\beta(P, T, x_2^\beta)$$

$$(11\text{-}8)$$

There are now four variables, P, T, x_1^α, and x_1^β, and two equations, so that two of the variables are no longer independent; there are two degrees of freedom. In the case of an equilibrium between a solution and its vapor, we customarily choose T and $x_1(l)$ as the independent variables. If these are specified, then both the total vapor pressure and the vapor composition are determined.

The general case is developed as follows. If a system has C components, then the number of variables required to define its state will be two (P and T), plus $(C - 1)$. That is, C mole fractions are involved, one of which is immediately eliminated by the condition that their sum must be unity. Since $(C - 1)$ composition variables are needed for *each* phase, the total number of variables is

$$\text{number of variables} = \mathscr{P}(C - 1) + 2 \tag{11-9}$$

where $\mathscr{P}$ denotes the number of phases present. The matter of how many phases there are usually presents no difficulty, but there can be subtleties, as discussed in the next section.

The condition for equilibrium generates the equations

$$\mu_1^\alpha = \mu_1^\beta = \mu_1^\gamma = \mu_1^\delta = \cdots$$

$$\mu_2^\alpha = \mu_2^\beta = \mu_2^\gamma = \mu_2^\delta = \cdots \tag{11-10}$$

$$\vdots$$

or $\mathscr{P} - 1$ equations for each component. The total number of equations is

$$\text{number of equations} = C(\mathscr{P} - 1) \tag{11-11}$$

The number of degrees of freedom is the difference between Eqs. (11-9) and (11-11),

$$F = \mathscr{P}(C - 1) + 2 - C(\mathscr{P} - 1)$$

or

$$F + \mathscr{P} = C + 2 \tag{11-12}$$

Equation (11-12) is known as the *Gibbs phase rule*.

The phase rule is useful in several ways. It tells us the maximum possible number of equilibrium phases that can coexist. In the more complex cases, it helps to determine the number of components present and whether the phases present are well defined.

11-2   Some further explanation of the terms component and phase

We can summarize the definitions of the preceding section as follows:

Number of components: the minimum number of independently variable constituents by means of which the composition of each phase can be expressed.

Number of phases: the number of different kinds of states of matter present.

Number of degrees of freedom: the number of independently variable intensive quantities whose values must be fixed to determine the state of the system.

The subject is full of subtleties, however, and we now take a more careful look at what is meant by the terms component and phase.

An introduction to the concept of a component is given in Section 9-1. A given phase may contain a variety of molecular species, but many of these will be in chemical equilibrium with each other and cannot therefore be varied independently. Pure liquid water has monomers, dimers and so on, and clusters, as well as H^+ and OH^- ions; it is a one-component system because these species are all in equilibrium with each other. Water plus alcohol has water–alcohol clusters and so on, but all compositions are fixed if the proportions of water-substance and alcohol-substance are given. The system is a two-component one.

A complication develops in the case of electrolyte solutions. Aqueous sodium chloride is a two-component system; sodium and chloride ions cannot be varied independently because the system must remain electrically neutral. An aqueous mixture of NaCl and KNO_3 might be thought to constitute a three-component system. It is actually a special case of a four-component system. The solution contains Na^+, K^+, Cl^-, and NO_3^- ions; the four compositions are reduced by the electroneutrality requirement to three independent ones, plus water, for a total of four. Alternatively, the mole fraction of water, that of total salt, the Na^+/K^+ ratio, and the Cl^-/NO_3^- ratio define all possible compositions.

A second kind of problem is that the number of components may depend on whether or not a given chemical reaction is rapid within the time scale of the phase equilibrium studies. As mentioned in Section 9-1, the system hydrogen–nitrogen–ammonia consists of either two or three components according to whether or not chemical equilibrium is reached rapidly. In dealing with freezing point diagrams, we took up briefly cases of compound formation. If such formation is very slow, so that the compound is not in equilibrium with the reactants that form it, then the compound must be treated as an additional component.

Another kind of complexity can arise if more than one phase is present. Consider first an aqueous solution of sodium acetate, NaAc. This is a two-component system. Although H^+, OH^-, Na^+, and Ac^- ions are present, along with some acetic acid, HAc, from the hydrolysis of the NaAc, all these species are in equilibrium with each other and their composition is fixed if we specify the relative amounts of water and of NaAc used in making up the system (at a given P and T). Suppose now that some benzene is added, to form a second phase. It might be thought that the system is now a three-component one, but this is not correct! The difficulty is that HAc is fairly soluble in benzene but NaAc is virtually insoluble. As illustrated in Fig. 11-1, the consequence is that benzene layer will contain essentially only acetic acid, and the aqueous layer will consist mainly of a solution of NaAc and NaOH. The composition of neither layer can be described purely in terms of mole fractions of water, benzene, and NaAc. One must in addition specify *either* a mole fraction of HAc or a mole fraction of NaOH (in the latter case, $x_{HAc} = x_{NaAc} +$

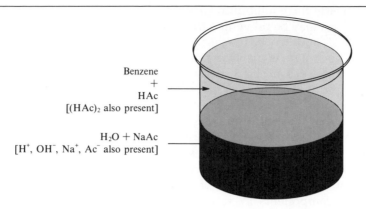

FIGURE 11-1
The addition of benzene to the two-component aqueous NaAc solution produces a four- rather than a three-component system. The composition of each phase is determined by the relative amounts of water, benzene, NaAc, and HAc or NaOH present.

Benzene
+
HAc
[(HAc)$_2$ also present]

H$_2$O + NaAc
[H$^+$, OH$^-$, Na$^+$, Ac$^-$ also present]

$x_{H_2O} - x_{NaOH}$). There is thus an additional composition variable and the system is a *four*-component one.

This last example helps to explain why the term "each phase" appears in the above definition of number of components. The reader can perhaps appreciate at this point why the phase rule often helps one decide just how may components are effectively present in a given system. We have mentioned the preceding complications in order to give depth to the definition of component. The actual systems described in this chapter present no such problems.

We next consider the definition of the term phase. A phase is first of all a homogeneous portion of matter, by which we mean that its time-average properties do not vary discontinuously from one spot to another. A phase region must be large enough that random, thermal fluctuations are small; otherwise the operational definition would become impossible to apply. Adjacent phase regions are marked by an interface or discontinuity in properties. The derivation of the phase rule is based on the further requirement that the chemical potentials of all components present are determined by the variables P, T, and $(C - 1)$ compositions. Neither these potentials nor any other molar thermodynamic properties ordinarily depend appreciably on the state of subdivision of the phase. Thus H$_2$O(l) and one piece of ice is a two-phase system; likewise H$_2$O(l) and two pieces of ice, and so on. The chemical potential of each piece of ice is the same and is independent of how many pieces are present.

A problem develops if the particles of a phase are so small that their chemical potential does depend appreciably on size owing to the surface energy contribution. The Kelvin equation (8-45) gives the free energy of a small particle as a function of T and of particle radius r. Ideally, particles of different r would not be in equilibrium. The two pieces of ice in the preceding example should eventually become a single piece. The mechanism would probably be through the dissolving of the smaller piece (its free energy being greater because of the smaller r) and corresponding growth of the larger piece.

This matter is not a trivial one. A precipitate of AgCl, for example, initially consists of a dispersion of particle sizes. Is, now, the measured solubility that of the smaller or of the larger crystals? (It appears to be that of the smaller ones.) Such precipitates will usually age or equilibrate to what appears to be nearly the equilibrium solubility. In colloidal suspensions, however, the particles may be prevented from merging by interparticle repulsions (see Section 21-1) and may be too insoluble to age by a dissolution–reprecipitation process. The system is then metastable in this respect, even though a range of particle size is present. How many phases are present? In the case of aqueous colloidal electrolytes, there is a concentration above which aggregates of 50 to 100 monomer units, called *micelles*, form. This critical micelle concentration is not as sharply defined as is a solubility limit, but almost so. Do micelles represent an new phase? (Actually, no.)

Questions such as these arise in the study of phase equilibria, and in difficult situations the phase rule itself may be used as the criterion for establishing the number of phases present. That is, if one knows the number of components in a system and can determine the number of degrees of freedom one defines the number of phases present.

In conclusion, it should be noted that Eq. (11-12) is for the special (but usual) case in which P and T are the only noncomposition variables that determine chemical potential. For example, should variations in electric, magnetic, or gravitational field be important, the phase rule would take the form $F + \mathcal{P} = C + n$ where $n > 2$.

11-3 One-component systems

One-components systems are summarized in Chapter 8. The phase rule reduces to $F + \mathcal{P} = 3$. We may have single phases existing over a region of P and T, two-phase equilibria governed by the Clapeyron (or Clausius–Clapeyron) equation and thus defined by a P–T line, and three-phase equilibria characterized by a triple point, or unique values of P and T. If n phases are known, there will be $n(n - 1)/2$ lines of two-phase equilibria; this is the number of distinguishable ways of picking two out of n objects. Also, there will be $n(n - 1)(n - 2)/3!$ possible triple points.

From the graphical point of view, a three-dimensional model is needed; one must show $\mu = f(P, T)$ as in Fig. 8-11, or the equation of state of each phase $V = g(P, T)$, as illustrated in Fig. 1-8. One may take various cross sections of the three-dimensional plot, usually either isothermal or isobaric ones. The surfaces of state for gaseous, liquid, and solid water intersect to give gas–liquid, gas–solid, and liquid–solid lines, which in turn intersect at the triple point.

Two-phase equilibria can, of course, be shown as P versus T plots, as in Fig. 8-1 for vapor pressures. A set of two-phase equilibrium lines constitutes a phase diagram, as in Figs. 8-8(c) and 8-10. Two additional systems are shown in Fig. 11-2. That for phosphorus (a) illustrates a case where, although two solid forms are known, only one is ever stable. White phosphorus may be prepared chemically but is always unstable with respect to red phosphorus. The ammonium nitrate system (b) is one in which a rather long succession of crystal modifications occur, each with a region of stability.

11-4 Two-component systems

The phase rule for a two-component system is given by $F + \mathcal{P} = 4$. A four-dimensional plot would be needed to show the state of a two-component system, that is, to plot Eqs. (11-6) or to show $V = g(P, T, x_1)$. One must

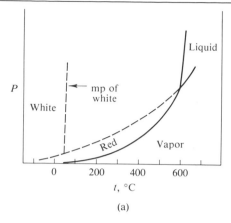

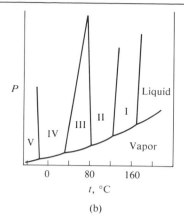

FIGURE 11-2
(a) Phase diagram for phosphorus. (b) Phase diagram for NH₄NO₃ [see R.G. Early and T.M. Lowry, J., Chem. Soc. 1387 (1919)].

therefore proceed immediately to one or another cross section, or use tabulations. Our interest, however, is in phase equilibria. The equilibrium between two phases can be shown as a P–T-composition plot or three-dimensional model, as in Fig. 9-15. These are awkward to use, so in actual practice one customarily takes either isothermal or isobaric cross sections. The vapor pressure-composition diagrams of Chapter 9 are examples of the former, and the boiling point diagrams are examples of the latter. Figure 7-2 similarly shows isothermal and isobaric cross sections for the CaO–CO₂ system and Fig. 9-21 shows an isobaric cross section depicting the equilibrium between two partially miscible liquids. The freezing point depression equation (10-15) gives the line of equilibrium between a solution and a pure solid for a constant pressure of 1 atm.

These examples represent isolated portions of various phase diagrams, and the material that follows will assemble these various portions into a more complete picture. Boiling point diagrams are covered in Section 9-7, and the emphasis here will be on freezing point and solubility diagrams. Solid phases will be taken to be immiscible; some cases of partial miscibility are considered in Section 11-ST-1.

We consider the simplest case first, namely, that in which the miscibility gap essentially coincides with the A and B sides of the diagram, that is, the case of complete immiscibility. If the liquid solution is ideal, then the solution–solid equilibrium line is given by the freezing point depression equation (10-15). This equation is plotted in Fig. 11-3, assuming first that pure solid A separates out; line ab results. A system of this type is symmetric, and the freezing point of liquid B will similarly be lowered by the presence of dissolved A. This second freezing point depression curve is shown by line cd in the figure. At the point of crossing of these two lines solution of composition x_E is simultaneously in equilibrium with pure solid A and pure solid B. On attempted cooling, the system can pursue neither the dashed extension of ab nor that of cd without departing from equilibrium with respect to either solid B or solid A. The consequence is that the temperature must remain invariant at T_E so long as the three phases are present. This invariance is demanded

A. Freezing (or melting) point diagrams

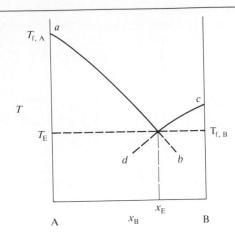

FIGURE 11-3
Freezing point diagram for an ideal solution.

by the phase rule—the one degree of freedom having been used in setting the pressure at 1 atm.

The completed phase diagram is shown in Fig. 11-4; it is called a *eutectic diagram*. The temperature T_E is the *eutectic temperature* and x_E the *eutectic composition*. The phase map aspect may be developed as follows. A solution (or melt, if the system is a high-temperature one) of composition x_1 will begin to freeze at T_1. With further cooling, solid A appears, and at some temperature T_2 the system consists of solid A and liquid of composition x_2. The proportions are given by the lever principle (Section 9-2C). The lengths l_1 and l_2 of the two sections of the tie-line at T_2 are in the same proportion as are the amounts of solution and of solid:

$$l_1/l_2 = \text{(amount of solution)/(amount of solid)}$$

If composition is plotted as mole fraction, the ratio will be of the number of moles of solution to the number of moles of solid; and if composition is in weight fraction or percent, the ratio will be that of weight of solution to weight of solid.

At T_E the solution has reached composition x_E, and further withdrawal of

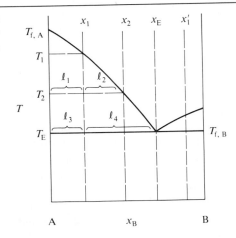

FIGURE 11-4
Application of the lever principle.

heat leads to the simultaneous freezing out of solid A and solid B. Since the solution cannot change composition, the proportion of A and B in the mixed solids must be given by x_E. The eutectic solid is thus one of definite composition, and perhaps for this reason it has sometimes been misnamed a *eutectic compound*. It is not a compound; its composition varies with pressure, for example. In the case of metal systems especially the superficial appearance of the eutectic mixture may be that of a single phase, but microscopic examination shows a mechanical mixture of small crystals of the two pure phases.

Eventually the liquid phase disappears and the system consists of solid A and solid B in overall proportion corresponding to x_1.

Cooling of a liquid of composition x_1' gives the same general sequence of changes. The first phase to separate out is now pure solid B, however. At T_E both A and B freeze out together, and when the liquid phase has disappeared one has the mixed solid phases of overall composition x_1'.

The preceding system is called a simple eutectic one, for reasons that are made especially apparent in the Special Topics section; freezing point diagrams can take on a rather complicated appearance. Simple eutectic, as well as the more complex systems, may be studied by what is known as the method of *thermal analysis*. The experiment consists in placing the liquid solution or melt in a container from which heat is withdrawn at a steady rate. That is, dq/dt is kept approximately constant. The liquid has a certain heat capacity $C_P(l)$ and the rate of change of its temperature with time is

B. Cooling curves for a simple eutectic system

$$\frac{dT}{dt} = \frac{1}{C_P(l)} \frac{dq}{dt} \tag{11-13}$$

(dq, and hence dT/dt, is negative since the system is losing heat). If we suppose the solution to be of composition x_1 in Fig. 11-4, then at T_1 solid A begins to freeze out. This means that its latent heat of freezing $\Delta H_{f,A}^\circ$ is liberated. The shape of the line ab [Fig. 11-3] determines dn_A/dT, that is, the number of moles of A appearing per unit temperature drop. Due to this effect, heat is supplied to the system at the rate

$$\frac{dq_f}{dT} = -\Delta H_{f,A}^\circ \frac{dn_A}{dT} \tag{11-14}$$

The consequence is that the heat capacity of the system is increased by an amount $C_{P,f}$, and the net heat capacity becomes

$$C_P(\text{net}) = C_P(l)n_L + \Delta H_{f,A}^\circ \frac{dn_A}{dT} + C_P(A)n_A \tag{11-15}$$

where n_L and n_A are the number of moles of liquid and of solid A present respectively. The rate of cooling is therefore reduced.

A schematic cooling curve for composition x_1 is shown in Fig. 11-5. The slope of the first section is given by Eq. (11-13). At T_1 (see Fig. 11-4) the slope changes to one determined by $C_P(\text{net})$; the cooling line is actually curved since $C_P(\text{net})$ changes with the relative amounts of liquid and solid phase. At T_E the temperature remains constant; the removal of heat now goes to freezing out A and B together.

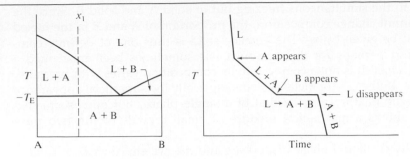

At the end of the halt, liquid phase has disappeared, and cooling resumes, with dT/dt given by Eq. (11-13) but with a heat capacity corresponding to that of the mixture of solids:

$$C_P(A + B) = n[x_1 C_B + (1 - x_1) C_A] \tag{11-16}$$

where n denotes the total number of moles of the system. Experimental cooling curves may show small temperature minima at breaks and halts, due to supercooling.

It helps to define exactly what is happening along each section of a cooling curve if the phase or phases present are indicated as is done in Fig. 11-5. A halt corresponds to a so-called phase reaction, that is, to the physical process of interconversion of phases. In the present example, the phase reaction is $L \rightarrow A + B$. One should also show at each break what phase or phases are appearing or disappearing.

Figure 11-6 shows a series of cooling curves for various initial compositions of the system of Fig. 11-4. Notice that the locus of the temperatures of the first break defines the freezing point lines [ab and cd in Fig. 11-3]. The unique halt temperature identifies the system as one having a single eutectic, and the cooling curve of composition labeled x_3 identifies the eutectic composition since preliminary separation of neither A nor B alone occurs.

The relative lengths of the various sections of the cooling curves should also be consistent with the phase diagram. Suppose, for simplicity, that there is always one mole total of system. The length of the halt for the cooling of pure liquid A is then $\Delta H_{f,A}^\circ/(dq/dt)$ and that for the cooling of pure liquid B is $\Delta H_{f,B}^\circ/(dq/dt)$. For intermediate compositions the length of the eutectic halt

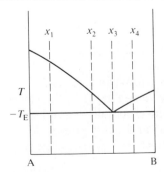

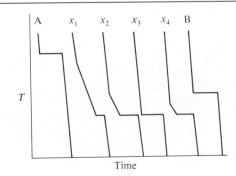

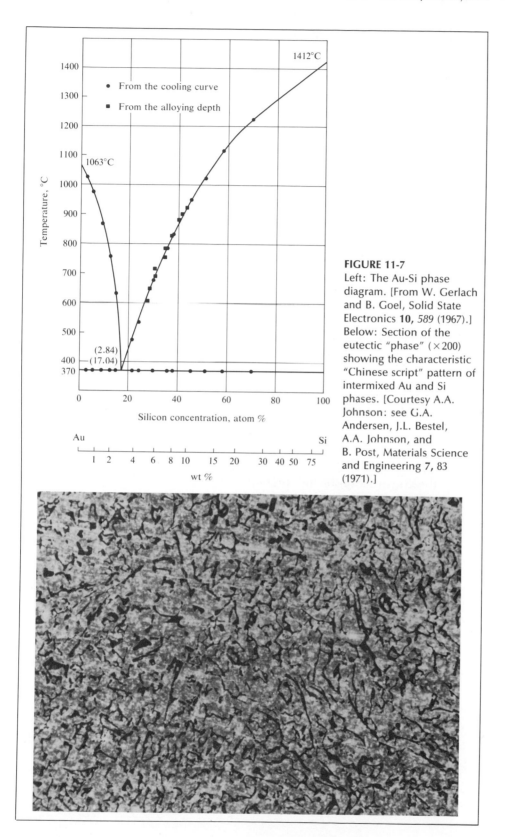

FIGURE 11-7
Left: The Au-Si phase diagram. [From W. Gerlach and B. Goel, Solid State Electronics **10**, *589* (1967).] Below: Section of the eutectic "phase" (×200) showing the characteristic "Chinese script" pattern of intermixed Au and Si phases. [Courtesy A.A. Johnson: see G.A. Andersen, J.L. Bestel, A.A. Johnson, and B. Post, Materials Science and Engineering **7**, 83 (1971).]

is proportional to n_E, the amount of eutectic solution that must freeze out. Returning to Fig. 11-4, we see that $n_E = l_3/(l_3 + l_4)$. Clearly, the closer the initial composition is to x_E, the greater will be this proportion, and the longer will be the halt in the cooling curve. An alternative way for one to locate the eutectic composition is then to plot the length of halt against initial composition and extrapolate both sections to find the maximum. One thus avoids having to hit the composition x_E exactly in a series of thermal analysis experiments.

Simple eutectic behavior is illustrated by the Au–Si system, shown in Fig. 11-7. The gold–silicon junction is an important one in the semiconductor industry and the composition and morphology of the eutectic "phase" have been much studied.

C. Compound formation

It happens quite often that the two components form one or more solid compounds—AB, A_2B, AB_2, and so on. If these are stable, they show a normal melting point, or are said to melt congruently. The effect is simply to divide the phase diagram into as many separate sections. This is illustrated in Fig. 11-8 for the case of a compound AB_2. The two eutectics have no necessary connection with each other, nor, in fact, do any aspects of the two portions of the diagram. Each may be treated separately.

What sometimes happens, however, is that a compound decomposes to give liquid and one of the other solid phases. Figure 11-8 would now take on the appearence of Fig. 11-9, where compound AB_2 decomposes at T_d to give liquid of composition x_d and solid B. Such a decomposition is called an *incongruent melting*, meaning that the liquid is not of the same composition as the solid and avoiding the implication of irreversibility suggested by the term decomposition. It is helpful at this point to label the various phase regions. This not only clarifies the geography of the phase map but is of direct use in the construction or interpretation of cooling curves. The general rule is that a break occurs when, on cooling, the system composition line passes from one type of phase region to another, such as on cooling along the x_1

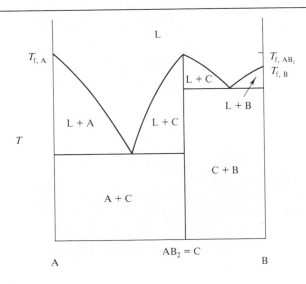

FIGURE 11-8
The case of compound formation.

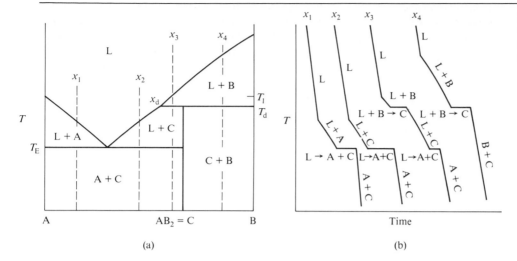

FIGURE 11-9
(a) Unstable compound formation. (b) Cooling curves.

line shown in Fig. 11-9. A halt must occur when a line of three phase equilibrium is reached, and cooling must resume when a two-phase or a one-phase region is entered.

Systems of composition x_1 and x_2 show normal eutectic cooling curves, but the behavior of the curve for x_3 is more complex. Solid B begins to separate out at T_1 and at T_d the solution has reached composition x_d and is now in equilibrium with both solid B and compound C ($= AB_2$). There must therefore be a halt in the cooling curve until at least one of the three phases disappears.

The phase reaction that occurs at T_d may be determined as follows. Just above T_d the phases present are L and B; just below T_d they are L and C. Evidently B disappears and C appears. However, B cannot by itself produce C, and the phase reaction is evidently L + B → C. We may confirm this by noting that for composition x_3, use of the lever principle shows that the proportion of L present decreases when the three-phase line is traversed.

After the halt at T_d, the system x_3 consists of L + C, and further cooling freezes out more C until T_E is reached. At this point the liquid is at x_E and the rest of the cooling curve is the same as for the simple eutectic case.

Finally, a system of composition x_4 traverses the T_d line on cooling but then enters the C + B phase region. The phase reaction is always the same anywhere along a three-phase line, but what now happens is that as the process L + B → C occurs it is the liquid phase which is used up first.

D. Rules of construction

Although not important for the relatively simple diagrams considered here, there are some rules that assist in the construction and labeling of more complicated diagrams. A two-component phase diagram consists of one-phase and two-phase regions and lines of three-phase equilibria. The latter always connect the three compositions of the three phases that are in equilibrium and must, of course, be exactly horizontal lines. There will always be two end phases and one of intermediate composition, and a first rule is that the end phases are always present both above and below the temperature of the three-phase line. The phase of intermediate composition may exist only above the line, as in eutectic diagrams, or only below the line, as with C in Fig.

11-9. In this case the line is called a *peritectic line*. Alternatively, a line of three-phase equilibrium is a boundary for three two-phase regions. There are two such regions above a eutectic line and one below it; conversely, there is one such region above a peritectic line and two regions below it.

A second rule is that a horizontal traverse must encounter alternate one-phase and two-phase regions. In applying the rule to a diagram such as that of Fig. 11-9, we regard pure A, pure C, and pure B as very narrow one-phase regions. The reader is encouraged to test the various two-component diagrams of this chapter against these rules.

11-5   Sodium sulfate-water and other systems

A specific system that both illustrates and extends a number of the points discussed in the preceding section is that for sodium sulfate and water. The usual phase diagram is shown in Fig. 11-10 and displays the following features. The lower left curve *ab* is the ordinary freezing point depression curve for aqueous sodium sulfate solutions. Curve *bc* is the solubility curve for $Na_2SO_4 \cdot 10H_2O$, showing a normal increasing solubility with increasing temperature. (The curve can be viewed alternatively as the freezing point depression curve of $Na_2SO_4 \cdot 10H_2O$, as discussed in Section 10-3.) This region of the diagram then shows a simple eutectic behavior, T_E being at $-1.3°C$, and the eutectic solution being 0.33 *m* in sodium sulfate. As an item of incidental information, a eutectic involving ice as one of the solid phases is often called a *cryohydric point*.

The decahydrate decomposes at 32.4°C to give the anhydrous salt whose solubility curve is given by *cd*. The situation is that of unstable compound formation and the three-phase line at 32.4°C is therefore of the peritectic type. The diagram is schematic in that both the temperature and the composition

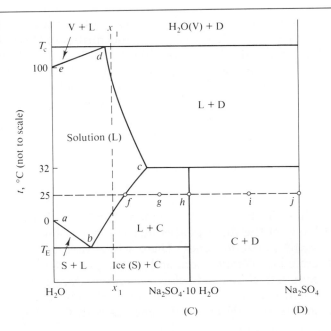

FIGURE 11-10
The H_2O–Na_2SO_4 system
for 1 atm pressure.

scales have been somewhat distorted so as to display the various phase regions more clearly. These are labeled, however, and a cooling curve, such as for the system of composition x_1, leads to the sequence of phase changes determined by the regions traversed; it will show the usual eutectic halt at T_E.

In a system of this type one of the components is relatively volatile, and one may consider an evaporation sequence. For example, if a dilute sodium sulfate solution is evaporated at 25°C the system will traverse the horizontal line shown in Fig. 11-10. Saturation with respect to $Na_2SO_4 \cdot 10H_2O$ occurs at system composition f, and continued evaporation deposits solid $Na_2SO_4 \cdot 10H_2O$ in increasing amounts. When the overall composition has reached point g, solution f and solid decahydrate are present in amounts given by the lever principle. At system composition h, solution has disappeared and only $Na_2SO_4 \cdot 10H_2O$ is present. Further evaporation—or removal of water—begins to transform the solid into anhydrous salt; at system composition i there would be about equal parts of $Na_2SO_4 \cdot 10H_2O$ and Na_2SO_4. At point j, of course, only the anhydrous salt is present.

We now turn to the higher-temperature region of the diagram. Pure water is in equilibrium with vapor at 100°C, this being the diagram for 1 atm pressure, and there is a boiling point elevation with increasing salt concentration, given by line de. At the intersection of lines de and cd, solution, vapor, and Na_2SO_4 are in equilibrium, and above T_c the solution has evaporated to give water vapor and solid Na_2SO_4.

The boiling point and therefore also T_c decrease with decreasing pressure. At a sufficiently low pressure, the phase diagram takes on the appearance shown in Figure 11-11(a). The decahydrate now decomposes at T_c' to give water vapor and the anhydrous salt. At one particular pressure $T_c = T_c'$, and at this temperature four phases, V, L, C, and D, can be in equilibrium, as shown in Fig. 11-11(b). This is the quadruple point for the system.

There is a large variety of phase diagrams of the salt hydrate type. Another common one is for the $H_2O–FeCl_3$ system, shown in Fig. 11-12. A succession of hydrates exists, each melting congruently. An interesting exercise is to trace out the series of changes occurring on evaporation of a $FeCl_3$ solution at a temperature just below the melting point of the heptahydrate (see Problem

---

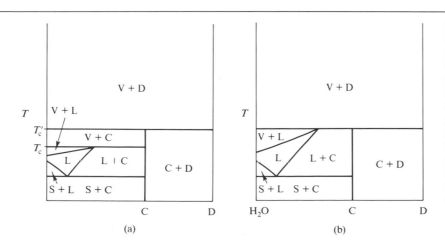

(a)   (b)

FIGURE 11-11
Effect of reduction of pressure on the phase diagram for the $H_2O–Na_2SO_4$ system. In (a) the pressure is low enough that the boiling point T_c is well below the temperature of point (c) in Fig. 11-10. In (b), further reduction of pressure has brought T_c and T_c' into coincidence. This is the quadruple point pressure.

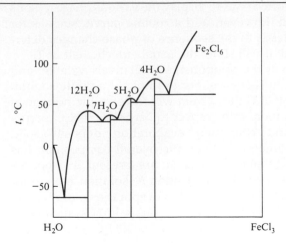

FIGURE 11-12
The H_2O–$FeCl_3$ system.

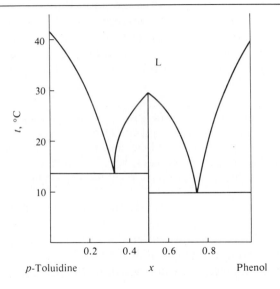

FIGURE 11-13
The p-toluidine–phenol
system.

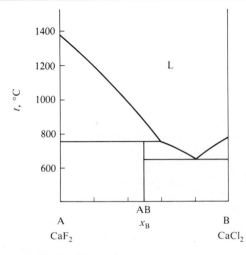

FIGURE 11-14
The CaF_2–$CaCl_2$ system,
showing unstable
$CaF_2 \cdot CaCl_2$ compound
formation.

11-19). A very simple example of 1:1 compound formation involving organic compounds is provided by the *p*-toluidine–phenol system shown in Fig. 11-13; the CaF_2–$CaCl_2$ system shown in Fig. 11-14 illustrates unstable compound formation.

11-6   Three-component systems

Some formidable problems of representation develop with three-component systems. Graphing of the equation of state $V = g(P, T, x_1, x_2)$ now requires five-dimensional space! The requirement is reduced to a four-dimensional one if two-phase equilibria are to be shown over a region of existence. Fortunately, or perhaps because of this difficulty, most studies of three-component systems have been carried out with only condensed phases present so that pressure is not an important variable, and isobaric diagrams supply the important information. These are three-dimensional, and their use is awkward but feasible. The chemical engineer who deals with the distillation of multicomponent systems has a problem that we will not consider here.

In an isobaric phase diagram two coordinates are used to fix composition and a third is used to fix the temperature. There are various ways in which the composition of a ternary or three-component system may be plotted, the choice being made in terms of the type of system; various methods are discussed by MacCarthy (1983). A very convenient one makes use of the properties of an equilateral triangle. As illustrated in Fig. 11-15, one of these properties is that the sum of the perpendicular distances to an interior point, $ad + bd + cd$, is always equal to the altitude of the triangle h. Triangular graph paper is therefore ruled with three sets of lines, each set parallel to one of the bases. A point such as d may then be read as having the coordinates 50% h_A, 20% h_B, and 30% h_C. The coordinate system is then used to express composition. For example, point d would be the pivot or balance point for the triangle if weights in the proportion 50:20:30 were hung from the A, B,

A. Graphical methods

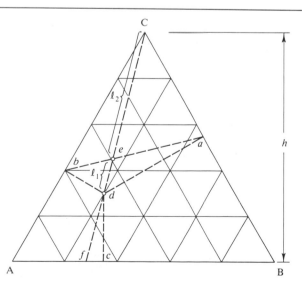

FIGURE 11-15
Triangular coordinate paper.

and C corners, respectively. Point d thus corresponds to a system of composition 50% A, 20% B, and 30% C.

A valuable feature of the triangular graph is that addition of one of the components will cause the system composition to move along the line drawn between it and that corner. Thus addition of C to a mixture of composition d will produce the succession of new compositions along the line dC. The lever principle applies. Thus at point e the ratio of the amount of added C to the amount of original mixture is l_1/l_2.

If C is removed from mixture d (by evaporation or freezing), the composition of the remaining mixture will move toward f, still on the dC line. It is not necessary that the added or removed material be pure component. The line connecting any two compositions is the locus of all mixtures of such compositions. For example, point e happens to be on the line connecting points a and b. Addition of mixture a to mixture e would move the system composition along the aeb line, toward a. Alternatively, mixture e could be made up by the combination of a and b in the proportions given by the lever principle.

B. The simple ternary
eutectic diagram

Isobaric diagrams may be constructed with a coordinate system based on the equilateral triangular prism, with temperature measured along the prism axis. Each face of the prism constitutes a two-component system; the three faces thus show the A–B, B–C, and A–C phase diagrams. This is illustrated in Fig. 11-16 for a simple eutectic system, that is, one in which A, B, and C are fully miscible in the liquid phase but are essentially immiscible in the solid state.

Examination of the figure shows, for example, that the freezing point T_{AC} of the binary eutectic solution ac is further lowered on addition of B to the system. The composition of the series of freezing A–C eutectic solutions moves along the line ac-abc. There is a corresponding depression of the bc eutectic by the addition of A and of the ab eutectic by the addition of C. The three freezing point depression curves for the three binary eutectic solutions meet at point abc at T_{ABC}. This is the quadruple point of the diagram; at T_{ABC} solids A, B, and C are in equilibrium with solution of composition abc.

The sequence of events on cooling of a solution of composition (1) is also shown. At temperature T_1, or point e on the diagram, C begins to freeze out, and the solution composition moves along the dashed line ef. When it reaches the ac-abc line and temperature T_3, solid A begins to freeze out as well, and with further cooling A and C freeze out together while the solution composition moves toward abc. At T_{ABC} the solution is of composition abc and is now also saturated with respect to B. The system is invariant at this point and the temperature remains constant while A, B, and C freeze out together. Cooling resumes when the solution has disappeared.

It is necessary to work with isothermal sections of the diagram if the analysis is to be more accurate. A succession of these is shown in Fig. 11-17; the temperatures correspond to those marked in Fig. 11-16. The section at T_1, Fig. 11-17(a), cuts the three "cloverleaves" of the solid model, the curved lines marking the intersection of the T_1 plane with each surface and hence giving the compositions of solutions saturated with respect to A, to B, and to C. The temperature T_1 is such that a system of composition (1) is just on the solubility curve for solutions saturated with respect to C.

The section at T_2, Fig. 11-17(b), is lower down, and so the three solubility

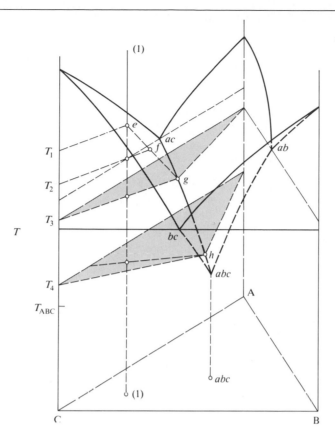

FIGURE 11-16
A simple ternary eutectic
system.

curves have moved inward or in the direction of decreased solubility. Point
(1) now lies within the pie-shaped region at the C corner. The system consists
of solid C and solution of composition f in the ratio l_1/l_2. The pie-shaped
regions are thus two-phase regions, as marked in the figure.

Figure 11-17(c), for T_3, shows that the system is now in equilibrium with
solution g, which is on the tie-line to the A corner as well as on that to the
C corner; the solution is therefore saturated with respect to both solids. At
T_4 the system composition lies within the triangular region AhC; the system
consists of solids A and C and solution h. The other two triangular regions
that have also appeared consist, respectively, of A, B, and solution j, and of
B, C, and solution i. We will return later to show how the proportion of each
phase can be determined. Continuing to the final diagram of the sequence,
at temperature T_{ABC} the three triangular regions have just joined to leave a
vestigial point marking the composition of the ternary eutectic solution abc.
The cross section at a yet lower temperature would show no features at all—
the system consists of solids A, B, and C in amounts given by the overall
composition. This is shown in Fig. 11-17(e).

We return now to Figure 11-17(d), the triangular region of which is reproduced in larger form in Fig. 11-18. To repeat, system of composition (1) is
present as solid A, solid C, and solution of composition h. The relative amounts
of these three phases are such that weights corresponding to them would, if
placed at the A, C, and h corners, make point (1) a balance point for the

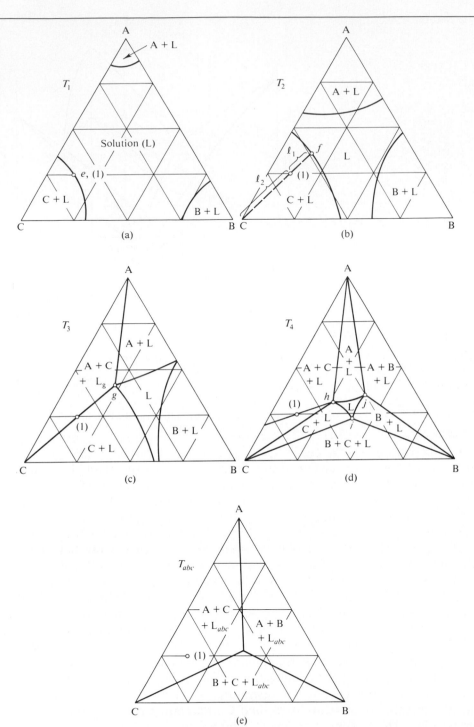

FIGURE 11-17
Cross sections taken at
successively lower
temperatures (see Fig.
11-16).

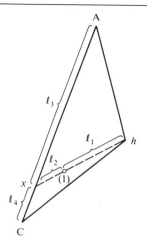

FIGURE 11-18
Application of the lever principle.

triangle. The equivalent graphical procedure is as follows. We first divide the system into solution h and mixed solids A and C. The relative amounts are given as l_2/l_1 (and would be about 1:2 in this case). If the diagram were on a weight percent basis, this means that 100 g of the system would consist of 33.3 g of solution h and 66.6 g of mixed solids A and C.

The proportion of A to C in the mixed solids is next given by the lever AxC; $A/C = l_4/l_3$, or about one-fifth in this example. The 66.6 g therefore is made up of 11.1 g of A and 55.5 g of C.

There are a fair number of simple three-component eutectic systems known. A specific example is the Bi–Pb–Sn system shown in Fig. 11-19. Other types of example are provided by trios of salts such as KCl–NaCl–CdCl$_2$ or of organic compounds such as o-, m-, and p-dichlorobenzene.

C. Ternary eutectic systems

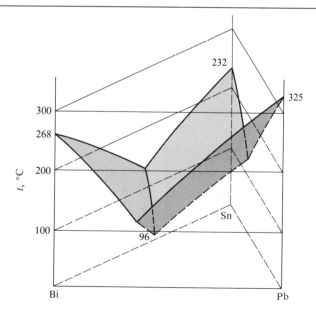

FIGURE 11-19
The *Bi–Pb–Sn* system.

Often the solid phases will show some mutual solubility. Also, various compounds may be formed as well. The complete, three-dimensional phase diagrams for such systems become rather complicated; the simple treatment of particular isothermal sections is discussed briefly in the Special Topics section.

11-7 Three-component solubility diagrams

It is pointed out in Section 10-3 that the terms freezing point and solubility merely represent different emphases of the same phenomenon. Consider the system H_2O–NaCl–KCl. The respective melting points are 0°C, 1413°C, and 1500°C and the solids are mutually insoluble, so that the phase diagram is of the simple eutectic type but is highly distorted because of the large difference between the melting point for water and those for the salts. The diagram is sketched in Fig. 11-20 and the isothermal cross section at 25°C is shown in Fig. 11-21. The context is such that we prefer to call line *ab* the solubility curve for NaCl in mixed NaCl–KCl solutions rather than the freezing composition curve. Similarly, line *bc* gives the solubility curve for KCl in mixed NaCl–KCl solutions.

As in the case of the H_2O–Na_2SO_4 system, one may discuss evaporation sequences. Evaporation of a solution of composition (1) (Fig. 11-21) means that water is removed from the system, and the system composition therefore moves along the line drawn from the H_2O corner through point (1). When the system reaches composition *d* it is now saturated with respect to KCl, and further evaporation precipitates this salt. The solution composition moves along the *cb* line toward *b*, and at system composition *e* solution of composition

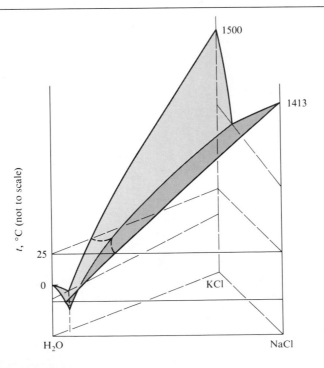

FIGURE 11-20
The *H_2O–KCl–NaCl* system.

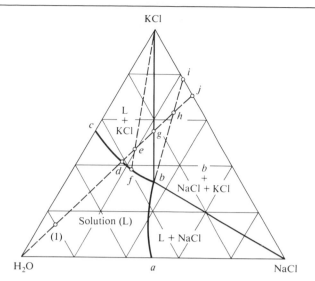

FIGURE 11-21
Isothermal cross section (near 25°C) for the H_2O–KCl–$NaCl$ system. Illustration of an evaporation sequence.

f and solid KCl are present. The relative amounts are given, as usual, by the lever principle. At system composition *g* the solution is at *b* and has just become saturated with respect to NaCl as well. Since temperature as well as pressure is kept constant, the system has no further variance, and continued evaporation changes the proportions but not the compositions of the three phases present. For system composition *h* the lever *bhi* gives the relative amount of solution *b* and mixed solids KCl and NaCl. The lever KCl–*i*–NaCl then gives the proportion of KCl to NaCl in the mixed solids.

Another example is as follows. It will be recalled that compound formation occurred in the H_2O–Na_2SO_4 binary system; below 32°C the solid in equilibrium with saturated solution was $Na_2SO_4 \cdot 10H_2O$. It is of interest therefore to consider the ternary system H_2O-Na_2SO_4–$NaCl$. The 25°C cross section is shown in Fig. 11-22. The line *cd* is for solutions saturated with respect to NaCl. Something interesting has happened, however, along the section *bc*. This is now a solubility curve for anhydrous sodium sulfate. This equilibrium, not previously possible at 25°C, now occurs. The physical explanation is that addition of NaCl has reduced the activity of water in the solution sufficiently to make Na_2SO_4 the preferred solid phase. Further, at point *b* the solution is in equilibrium with both Na_2SO_4 and $Na_2SO_4 \cdot 10H_2O$, and water vapor pressure above this solution must be that of Fig. 11-11(a), since these two solids are in equilibrium with water vapor at this pressure and $T_c' = 25°C$.

The topology of the H_2O–Na_2SO_4–$NaCl$ diagram is such as to show the phenomenon of *retrograde solubility*. Evaporation of a solution of composition (1) leads first to precipitation of $Na_2SO_4 \cdot 10H_2O$, the composition moving toward *b*. When the solution composition reaches *b*, Na_2SO_4 is ready to precipitate, and further evaporation converts $Na_2SO_4 \cdot 10H_2O$ into Na_2SO_4 (plus solution). The decahydrate therefore *disappears* with further evaporation and the solution then moves toward composition *c*. From this point on NaCl and Na_2SO_4 precipitate out together until solution *b* has dried up.

A large number of salt solubility diagrams have been studied, and a con-

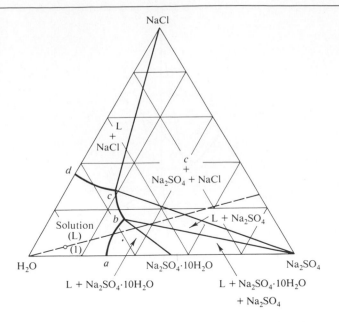

FIGURE 11-22
The H_2O–$NaCl$–Na_2SO_4
system (near 25°C).
Illustration of hydrate
formation.

cluding example illustrates a case of double salt formation, shown in Fig.
11-23. Here, the compound $Li_2SO_4 \cdot (NH_4)_2SO_4$ forms.

The more symmetric ternary eutectic systems are of great general importance in metallurgy. Salt solubility systems are similarly central to the understanding of the evaporative recovery of salts from brines. The evaporation
of brines, either from sea water or from dissolved natural salt deposits, constitutes our major source of most of such minerals.

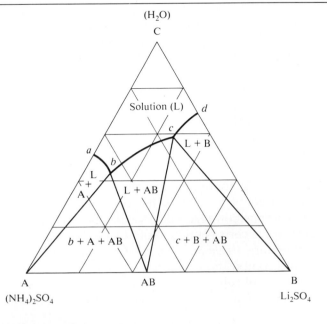

FIGURE 11-23
The
H_2O–Li_2SO_4–$(NH_4)_2SO_4$
system (near 25°C),
illustrating double salt
formation.

SPECIAL TOPICS

11-ST-1 Two-component freezing point diagrams. Partial miscibility

The freezing point diagrams so far considered have all been ones for systems in which solid phases were entirely immiscible. The other extreme is that of complete miscibility in both liquid and solid phases. The appearance would be that of Fig. 11-24(a). We next imagine a progressive change in the properties of A and of B such that the miscibility in the solid state gradually decreases and the miscibility gap moves upward. In Fig. 11-24(b) the miscibility gap has come close to the melting points of A

and B, and the incipient immiscibility is reflected in the minimum of the solid solution–liquid solution composition curves.

In Fig. 11-24(c) the miscibility gap has impacted the freezing point curves and the system is now the eutectic type. Solution of composition x_E is in equilibrium at T_E with two solid phases. These are not pure A and B, but solid solutions of composition z^α and z^β. The letter z will be used to denote compositions of solid solutions and the Greek superscripts to indicate the type of phase. Thus an α phase is one that is rich in component A and a β phase one that is rich in component B. Further diminution of the degree of miscibility leads to Fig. 11-24(d), essentially the same as Fig. 11-3.

Figure 11-3 also illustrates a succession of boiling point diagrams for progressively less and less miscible liquids. We need only to change the labeling; L is replaced by vapor phase V, and solid solutions α and β by liquid solutions L_α and L_β (note Fig. 9-20).

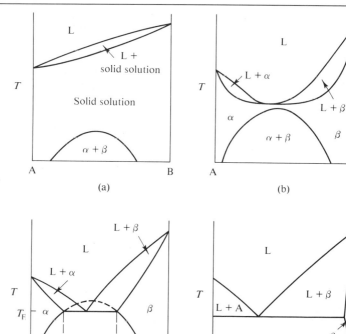

FIGURE 11-24
Progression of appearance of a freezing point diagram with increasing degree of immiscibility of the solid phases.

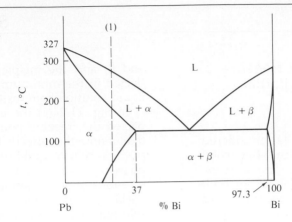

FIGURE 11-25
The *Pb–Bi* system.

The Pb–Bi system is of the type of Fig. 11-24(c), as shown in Fig. 11-25. The cooling of a melt of composition (1) leads to a break at about 275°C as α phase begins to freeze out, and liquid and α phases would then vary in composition along their respective curves, with the latter increasing in amount. At about 175°C the system is entirely α phase [of composition (1)] and the cooling rate increases. At about 50°C the system composition line crosses the solubility curve for α phase, and β phase begins to form. There is a break in the cooling curve at this point, but not a marked one since the heat of separation into the two solid solutions is probably not large. Cooling curves for compositions between 37 and 97% Bi are similar to those for a simple eutectic, except that the solid phases are α and β rather than the pure components.

The sequence of Fig. 11-24 is not the only possible one. If the melting points of the two components are fairly different and the mis-

cibility gap is narrow, then the sequence in Fig. 11-26 may result. The diagram 11-26(b) is now of the peritectic type. At T_P liquid of composition x_p is in equilibrium with solid solutions z^α and z^β; notice that the liquid composition lies outside of those of the two solid solutions rather than between them.

Iron and gold form a peritectic system, shown in Fig. 11-27, and some representative cooling curves may be considered. That for a system of composition (1) is analogous to the cooling curve for Fig. 11-25 and need not be explained further. A system of composition (2) will show a break at about 1400°C, when α phase first forms. With continued cooling, liquid and α phase shift in composition along their respective lines, α phase increasing in amount. At 1170°C, or T_P, the solution is now also in equilibrium with β phase, and there is a halt in the cooling curve while α and β phases (30% Au and 65% Au, respectively) crystallize out. The phase reaction is that of a peritectic system (see

FIGURE 11-26
Development of peritectic type of freezing point diagram.

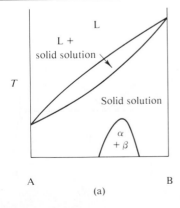

(a)

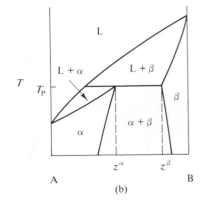

(b)

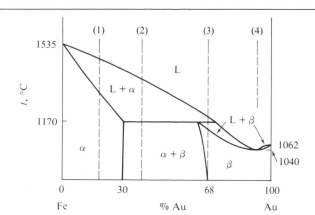

FIGURE 11-27
The *Fe–Au* system.

Section 11-3C), or L + α → β. In this case L phase is used up first, and at the end of the halt the system consists of α and β phases only.

The cooling curve for a system of composition (3) is similar to that for (2) up to the halt. Phase α is now used up first in the phase reaction, however, so L and β phase remain at the end of the halt. With further cooling, L and β phase compositions shift along their respective curves, with β phase increasing in proportion, and at about 1150°C only β phase remains. Compositions to the right of the three-phase line give cooling curves similar to (1), but with β phase forming rather than α phase. Note that there is a slight minimum in the freezing point curve at (4), or 94% Au. A liquid of this composition freezes to β phase of the same composition, and the cooling curve has

the same appearance as that for a pure substance.

Partial miscibility may occur in the liquid region to give a phase diagram such as illustrated in Fig. 11-28. The diagram is labeled, and one can work out the various cooling curves by following a system composition line through the various phase regions. A system of composition (1) would, for example, show one halt at T_1 while L^α phase converted to L^β and α phase, and a second halt at T_E while L^β phase converted to α and β phases.

Finally, no discussion of phase diagrams seems complete without a mention of the iron–carbon system. The diagram, shown in Fig. 11-29, illustrates yet another kind of partial miscibility—that between two different crystalline modifications of a solid phase. The sta-

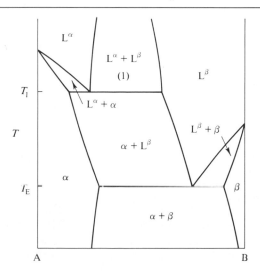

FIGURE 11-28
Partial miscibility in the liquid phases as well as in the solid phases.

FIGURE 11-29
The *Fe–C* system.

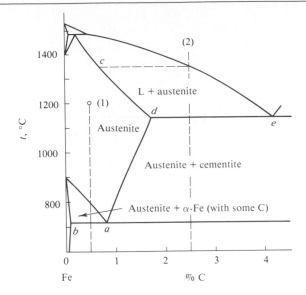

ble crystalline form of pure iron below 910°C is body-centered cubic and is called α-iron. At 910°C α-iron changes to a face-centered type of crystal lattice called γ-iron and then, at 1401°C, there is a reversion back to a body-centered type of structure, now called δ-iron. The melting point of iron is 1535°C.

We explore the diagram by first dissolving some carbon in γ iron at about 1200°C, to give a solid solution called *austenite* of composition (1). On cooling to about 800°C α-iron containing some dissolved carbon begins to separate out. A eutectic-type three-phase line is reached at 700°C, at which point austenite phase of composition *a* is in equilibrium with α phase of composition *b* and Fe_3C (called *cementite*). Below 700°C the system consists of α-iron (with some dissolved carbon) and cementite.

This sequence corresponds to the heat treatment of *steel*. Iron with less than about 2% carbon can be heated to the austenite single-phase region. It is then easily rolled or otherwise formed. On cooling, the separation into α-iron and cementite occurs, and the extreme hardness of cementite gives steel its strength. The rate of cooling affects the particle size of the two-phase mixture and hence the mechanical properties of the steel.

Consider next a system of composition (2) corresponding to about 2.5% C. On cooling of a molten iron–carbon solution, austenite phase of composition *c* begins to form, and with further cooling the liquid and solid solutions move

in composition toward *e* and *d*, respectively, the proportion of the latter type of phase increasing. At 1125°C the eutectic is reached, and the system is now in equilibrium with Fe_3C; during the halt austenite phase of composition *d* and Fe_3C crystallize out together. Below 1125°C the system consists of Fe_3C and austenite phase of composition moving along the *da* line. At 700°C the α-iron–austenite–Fe_3C three-phase line is reached, and the further changes are as described earlier. An iron of this composition, unlike a steel, does not become a single phase until the melting point is reached. Such iron is called *cast iron*—it is not malleable when hot, but has valuable corrosion-resistant properties. If it is of composition close to the eutectic, its melting point is low enough to allow a fairly easy casting procedure.

The remaining small region in the upper left is of no great interest here. One can trace the details of its phase behavior by following system composition lines that pass through various portions of the region.

11-ST-2   Partial miscibility in three-component systems

The water–phenol system is shown in Fig. 9-21(a); there is an upper consolute temperature of about 70°C and around room temperature the miscibility is rather limited. Acetone is fully

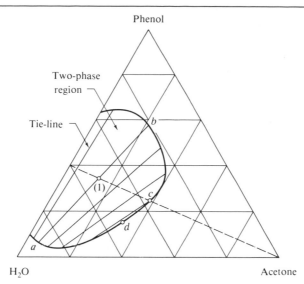

FIGURE 11-30
The three-component
miscibility gap. The
H_2O–acetone–phenol
system.

miscible with water and with phenol separately and when it is added to a water–phenol two-phase system the effect is to increase the mutual solubility. The solubility behavior is shown in Fig. 11-30 for a temperature of about 30°C. The tie-lines connect pairs of mutually saturated solutions. Thus a system of overall composition (1) consists of liquid phases of compositions a and b and in amounts given by the lever principle. Addition of acetone moves the system composition along the line drawn from point (1) to the acetone corner, the compositions of the two liquid phases being given by

the ends of each successive tie-line. When the system composition reaches c only a single phase is present and further addition of acetone merely dilutes the liquid solution. The tie-lines themselves reach a vanishing point at d called the *critical point*.

The whole region of partial miscibility diminishes in size with increasing temperature, and a set of isothermal cross sections generates a helmet-shaped two-phase region in the three-dimensional phase diagram, as illustrated in Fig. 11-31. The locus of successive critical points goes through the peak of the helmet (at 92°C,

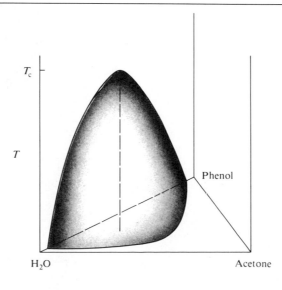

FIGURE 11-31
The three-component
miscibility gap in three
dimensions.

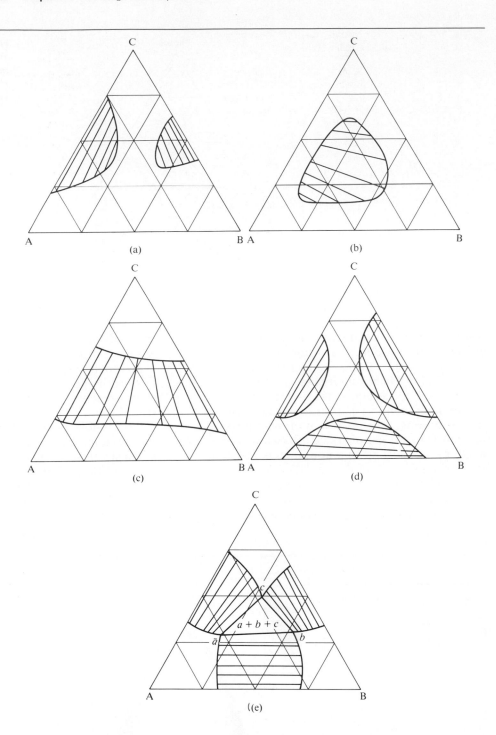

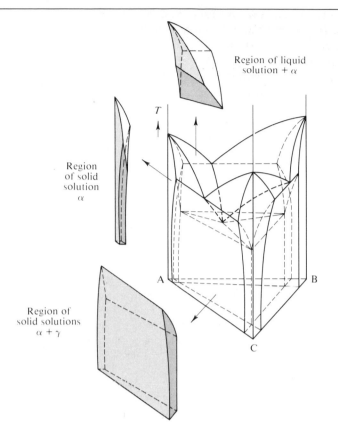

FIGURE 11-33
Three-component eutectic system with partial miscibility in the solid phases.

Region of liquid solution + α

Region of solid solution α

Region of solid solutions α + γ

59% water and 29% phenol), which is the *critical temperature* of the ternary system. Various cooling or dilution sequences may be traced out in the usual way by following the system composition line and, where it is in the two-phase region, using the ends of the tie-line through it to determine the compositions of the liquid phases present.

It must be remembered that while a tie-line is an isothermal line, there is no requirement that a set of lines such as for Fig. 11-30 be parallel. The ends of each successive line must be determined by individual experiment, and a set of tie-lines must be shown if the diagram is to be usable.

A number of types of miscibility gaps are possible for a ternary system, as illustrated in Fig. 11-32. Figure 11-32(d) represents the case of partial miscibility in all three binary systems and Fig. 11-32(e) shows the result of a further decrease in miscibility, such as by lowering of the temperature. The three miscibility gaps have now emerged to frame a central triangular region of three phase equilibrium.

Three-component freezing point diagrams with partial miscibility in the solid phases add partial miscibility regions to Fig. 11-16. A perspective drawing of such a model is shown in Fig. 11-33; the reader may enjoy working out the various phase regions. As will all models based on the prismatic coordinate scheme, each side shows a two-component phase diagram.

MARSH, J. S. (1935). "Principles of Phase Diagrams." McGraw-Hill, New York.

MASING, G. (1944). "Ternary Systems" (translated by B. A. Rogers). Dover, New York.

RICCI, J. E. (1951). "The Phase Rule and Heterogeneous Equilibrium." Van Nostrand–Reinhold, Princeton, New Jersey.

WETMORE, F. E. W., AND LeROY, D. J. (1951). "Principles of Phase Equilibria." McGraw-Hill, New York.

GENERAL REFERENCES

CITED
REFERENCES

Lightfoot, W. J., and Prutton, C. F. (1947). *J. Amer. Chem. Soc.* **69**, 2098.
MacCarthy, P. (1983). *J. Chem. Ed.* **60**, 922.

EXERCISES

11-1 Explain how many degrees of freedom are present for the following systems: (a) ice plus an aqueous solution of ethylene glycol; (b) a layer of water saturated with benzene plus a layer of benzene saturated with water; (c) system (b) with vapor phase present; (d) a solution of NaOH plus equilibrium vapor; (e) a solution of NaOH and KOH plus equilibrium vapor.

Ans. (a) 2; (b) 2; (c) 1; (d) 3; (e) 2.

11-2 Explain how many degrees of freedom are present for a solution of H_2O and D_2O (and hence also containing an equilibrium amount of HDO) and equilibrium vapor.

Ans. 2.

11-3 A system contains equilibrium amounts of $CaCO_3(s)$, $CaO(s)$, and $CO_2(g)$ at a given temperature. Explain how many degrees of freedom there are.

Ans. 0.

11-4 In Fig. 11-4, x_1' is about 90% B and x_E is about 70% B. A melt of composition x_1' is cooled to just above T_E. Calculate the percentage of the system that has frozen out at this point. If the system contained 1 mole total, how many moles of liquid are present (explain)?

Ans. 2/3; 1/3 mole.

11-5 As a continuation of Exercise 11-4, how long should the halt at T_E be if the heats of fusion of A and B are 3 and 5 kcal mol^{-1}, respectively, and the rate of removal of heat is 400 cal min^{-1}? Show your work.

Ans. 3.67 min.

11-6 The break and halt temperatures for the cooling curves of molten solutions of metals A and B are given in the following listing. Construct a phase diagram consistent with these data and label the phase regions. Give the probable formula(s) of any compound(s). The data are given as (mole % A, °C of break in cooling curve, °C of halt in cooling curve): (100, —, 950); (90, 800, 500); (80, 600, 500); (70, 770, 500); (60, 750, 300); (50, 650, 300); (40, 510, 300); (30, —, 300); (20, 400, 300); (10, 520, 300); (0, —, 600).

Ans. The compound A_2B is indicated.

11-7 The decomposition temperature of $FeCl_3 \cdot 12H_2O$ is 45°C. Explain what sequence of solid phases should be seen as a dilute solution of $FeCl_3$ is evaporated at 46°C.

Ans. $FeCl_3 \cdot 5H_2O$, $FeCl_3 \cdot 4H_2O$, Fe_2Cl_6.

11-8 Two g of water, 5 g of $Na_2SO_4 \cdot 10H_2O$, and 20 g of Na_2SO_4 are equilibrated at 25° C. Describe what phase or phases should be present.

Ans. Solution of composition *f* in Fig. 11-10 plus $Na_2SO_4 \cdot 10H_2O$.

11-9 Assume that Fig. 11-21 is scaled in weight percent. One hundred grams of solution of composition (1) is evaporated at 25°C until KCl first begins to precipitate. (a) How many grams of solution remain at this point? The system is then evaporated until NaCl begins to precipitate out. (b) How many grams of solution remain at this point? (c) Give the amounts and compositions of the phases present when 95% of the original water has been evaporated.

Ans. (a) 35 g; (b) 19 g; (c) 9 g or solution of composition (b), 3.0 g of NaCl and 12 g of KCl.

PROBLEMS

The general instruction for the following problems is to draw the simplest phase diagram consistent with the information given and to label the various phase regions as to the number and nature of the phases present. The instructor may ask for various cooling curves to be sketched and labeled. Some needed information is:

Melting points(°C)					
Al	658	Pb	327	Ni	1452
Bi	273	Mg	651	Na	98
Ca	810	Hg	−39	Tl	303
Co	1480	Mo	2535	K	62

11-10 Answer the question "why" at the end of the paragraph containing Eq. (11-15).

11-11 Explain at what temperature the evaporation of aqueous ferric chloride should lead to two successive solidifications and liquifactions of the system.

11-12 The break and halt temperatures for the cooling curves of molten solutions of metals A and B are given in the following listing. Construct a phase diagram consistent with the data and label the phase regions. Give the probable formula(s) of any compound(s). The data are given as (mole % A, °C of break in cooling curve, °C of halt in cooling curve, °C of any second halt in cooling curve): (100, —, 700); (90, 640, 450, —); (80, 550, 450, —); (70, —, 400, —); (60, 580, 400, —): (50, 670, 650, 400); (40, 750, 650, 400); (30, 820, 650, —); (20, 880, 650, —); (10, 920, 650, —); (0, —, 900, —).

11-13 Two hundred grams of a 10% by weight melt of silicon in gold is cooled to 600°C and the melt that is present at this point is separated from the solid that has formed; the melt is now cooled to just above the eutectic temperature. Show what phases are present and calculate the amount of each phase and its composition.

11-14 Mercury and lead dissolve in all proportions in the liquid state and they form no compounds. A liquid phase is in equilibrium at −40°C with two crystalline phases containing 35% and 100% Hg, respectively.

11-15 Bismuth and lead form no compounds; solid solutions containing 1% and 63% lead are in equilibrium with a liquid containing 43% lead at 125°C.

11-16 Potassium and sodium form the compound Na_2K, which decomposes at 7°C into Na(s) and a melt containing 45% Na. At −10°C, melt containing 20% Na is in equilibrium with K(s) and $Na_2K(s)$. Na and K are completely miscible in the liquid state.

11-17 CuCl and $FeCl_3$ melt at 422 and 282°C, respectively, and are miscible in the liquid state. A compound containing 62% $FeCl_3$ by weight melts at 310°C; there are eutectics at 295 and 255°C.

11-18 Calcium and sodium mix in all proportions in the liquid state above 1150°C, the mutual solubilities are 33% Na and 82% Na at 1000°C, liquids of 14 and 93% Na are in equilibrium with solid calcium at 710°C; the eutectic temperature is 97.5°C and no compounds or solid solutions form.

11-19 Liquid and solid phases of the composition Hg_5Tl_2 are in equilibrium at 14°C; the phases at 0°C are a liquid of 40%Tl and solid solutions of 32% and 84% Tl; the phases at −59°C are Hg, a liquid of 8% Tl, and a solution of 22% Tl.

11-20 Magnesium and nickel form a compound $MgNi_2$ that melts at 1145°C and a compound Mg_2Ni that decomposes at 770°C into a liquid containing 30 mole% nickel and the other compound. The eutectics are at 23% nickel and 510°C and 89% nickel and 1080°C.

FIGURE 11-34
Set of cooling curves for
Problem 11-25.

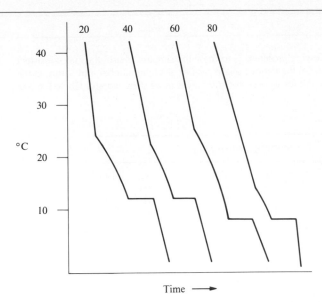

11-21 Aluminum and cobalt form three compounds, of which AlCo melts at 1630°C, Al_5Co_2 decomposes at 1170°C, and Al_4Co decomposes at 945°C. AlCo and Co form an incomplete series of solid solutions, with solid containing 84 mole% Co, liquid containing 89 mole% Co, and solid containing 92 mole% Co in equilibrium at 1375°C.

11-22 An aqueous solution just saturated with respect to Na_2SO_4 at 80°C is cooled to −10°C. Describe the sequence of phase changes and sketch a semiquantitative cooling curve.

11-23 Draw a semiquantitative set of cooling curves for the compositions 0%, 25%, 50%, 75%, and 100% in the CaF_2–$CaCl_2$ system (Fig. 11-14).

11-24 The Au-Si diagram is shown in Fig. 11-7. Make a quantitative calculation of the cooling curve for a melt containing 40 mole % Si. Take the heats of fusion of Au and of Si both to be 16 cal g^{-1} and the heat capacity of molten solutions to be 6 cal K^{-1} mol^{-1}. The rate of heat removal is constant at 100 cal s^{-1} and there is 1 g of initial melt.

11-25 Figure 11-34 shows a set of cooling curves for a pair of organic compounds, A and B, which are miscible in the liquid state. The melting points of A and B are 42°C and 39°C, respectively, and the numbers in the figure indicate mole percent of B in the melt being cooled.

11-26 Show that the line *Cedf* in Fig. 11-15 is the locus of compositions for which the ratio of A to B is constant.

11-27 The following data refer to the system K_2SO_4–$(NH_4)_2SO_4$–H_2O at 30°C. In each line of data are given the compositions of a saturated solution and of a moist solid (solid solutions are formed) (in weight per cent).

Draw the triangular diagram and label all phase regions for the K_2SO_4–$(NH_4)_2SO_4$–H_2O system. A solution containing 95 g of water, 5 g of K_2SO_4, and 10 g of $(NH_4)_2SO_4$ is evaporated isothermally at 30°C. (a) What is the composition of the solution when a precipitate first appears and what is the composition of the precipitate? (b) What is the composition of the last solution to exist before the system solidifies? Give the compositions and amounts of the solid(s) present. (c) When the system reaches the point of containing 40% by weight of water what phases are present and in what amounts? Show your calculations and indicate them graphically on your diagram insofar as possible.

Saturated solution		Solid	
K_2SO_4	$(NH_4)_2SO_4$	K_2SO_4	$(NH_4)_2SO_4$
0	44.2	[only $(NH_4)_2SO_4$]	
1.2	42.7	1.6	90.3
2.4	40.9	22.1	69.3
4.1	37.8	33.9	57.8
4.9	33.5	50.8	42.3
6.4	31.0	62.2	30.0
9.1	18.5	84.0	12.7
10.7	8.4	93.2	3.1
11.2	0	(only K_2SO_4)	

11-28 The following data were obtained by W.J. Lightfoot and C.F. Prutton (1947) on the system $CaCl_2$–KCl–H_2O at 75°C:

Saturated solution (wt%)		
$CaCl_2$	KCl	Solid Phase
0	33.16	KCl
11.73	21.62	KCl
18.27	16.00	KCl
28.47	9.62	KCl
37.65	6.77	KCl
47.65	8.43	KCl
50.19	10.32	KCl and $2KCl \cdot CaCl_2 \cdot 2H_2O$
50.92	9.36	$2KCl \cdot CaCl_2 \cdot 2H_2O$
53.85	6.21	$2KCl \cdot CaCl_2 \cdot 2H_2O$
56.33	4.51	$2KCl \cdot CaCl_2 \cdot 2H_2O$
57.62	3.60	$2KCl \cdot CaCl_2 \cdot 2H_2O$ and $CaCl_2 \cdot 2H_2O$
57.77	2.56	$CaCl_2 \cdot 2H_2O$
58.58	0	$CaCl_2 \cdot 2H_2O$

Construct the triangular plot for this system and label all phase regions. Describe what would happen on isothermal evaporation of a solution initially containing 100 g of water and 5 g each of $CaCl_2$ and KCl. How many grams of water have been evaporated when half the weight of the system is in solids?

11-29 Figure 11-35 shows a series of binary phase diagrams, real or hypothetical. They are melting point diagrams in the sense that the upper regions contain only liquid phases. State the *number* and *type* of phases present in each *area* and label each *line* of three-phase equilibrium.

11-30 Figure 11-36 shows two isothermal, isobaric ternary diagrams. All phases present are either liquid or solid. State the number and type of phases present in each area.

11-31 Metals A and B form ideal solutions and are completely immiscible in the solid phase. Calculate the simple eutectic diagram for this system using the freezing point depression equation of Chapter 10; the melting points are 500°C and 800°C, respectively, the pure metals both have 4.0 cal K^{-1} mol^{-1} entropy of fusion, and both obey the Dulong and Petit heat capacity relationship. Calculate and plot the actual cooling curves for solutions containing 100% A, 80% A, and of the eutectic composition, assuming 0.1 mole of solution is used in each case and that the rate of heat removal is 200 cal min^{-1}.

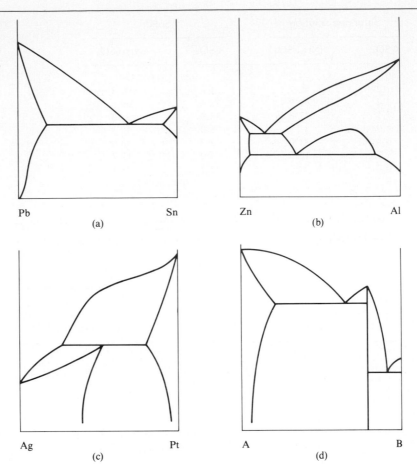

FIGURE 11-35
Some assorted binary phase diagrams. See Problem 11-29.

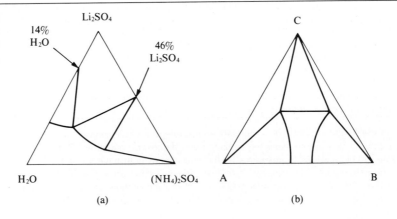

FIGURE 11-36
Two isothermal, isobaric ternary diagrams. See Problem 11-30.

SPECIAL TOPICS PROBLEMS

The following cooling curve data have been obtained for melts of metals A and B (mole % A, °C of 1st break, °C of second break, °C of halt): (100, —, —, 750); (90, 725, 625, —); (80, 690, 475, —); (70, 670, —, 475); (60, 625, —, 475); (50, 560, —, 475); (40, 525, 430, 475); (30, 475, 410, —); (20, 450, 380, —); (10, 420, 370, —); (0, —, —, 360).
11-32

Sketch and label a set of cooling curves for the system of Fig. 11-26(b); these should be for 10, 20, 30, 50, 70, and 95% B.
11-33

Sketch and label the set of cooling curves for the marked compositions in Fig. 11-27.
11-34

Describe the sequence of events when a 50/50 mixture of phenol and acetone is added to a system of composition (1) in Fig. 11-30. Assuming the diagram is on a weight basis, how many grams of acetone, water, and/or phenol must be added to 100 g of system to composition *d* to obtain a system of composition *b*?
11-35

For each of the systems of Fig. 11-31, describe the sequence of changes if A is added to a 1:1 mixture of B and C.
11-36

Sketch and label the cooling curve to be expected for a system of composition labeled (1) in Fig. 11-33.
11-37

chapter 12
Solutions
of electrolytes

12-1   Introduction

The historical development of electrochemistry constitutes one of the more interesting stories of science. It begins, for us, with observations made in 1600 by W. Gilbert, physician to Queen Elizabeth, on the ability of amber to attract pith or other light objects when rubbed with a piece of fur. Gilbert coined the word *electric* (from the Greek for amber) in describing such behavior. Benjamin Franklin became interested in the subject and suggested, around 1750, that the different behavior of electrostatically charged glass and amber was not due to two kinds of electricity, but rather to an excess or deficiency of an electric "fluid." Thus began the subject of electrostatics, which culminated around 1890 with the identification of the electron as the unit of electricity.

Another chain of events led to the discovery of a second manifestation of electricity. The story traces back to 1678 when Swammerdam demonstrated before the Grand Duke of Tuscany that a frog's leg resting on a copper support would twitch when touched with a silver wire connected to the copper. We are more familiar with the experiments of L. Galvani in 1790, who observed that a frog's leg (again) would twitch when connected to a static electricity generator or when the nerve was merely touched with a metal strip which was connected to the end of the leg. The terms *galvanic electricity* and *galvanometer* honor this discovery. By 1800 A. Volta succeeded in producing visible sparks from a stack of alternate silver and zinc plates. This was the first battery, then called a *voltaic pile* or a *galvanic cell*. The first definite experiment in electrochemistry appears to have been made by W. Nicholson and A. Carlisle in 1800, who used a galvanic cell to electrolyze water. By 1807 H. Davy had isolated sodium and potassium by the electrolysis of their hydroxides.

The foundation of modern electrochemistry was laid by M. Faraday, working around 1830 at the Royal Institution. He showed that a given quantity of electricity produced a fixed amount of electrolysis and formulated the following now well-known laws.*

---

*Faraday's law is not always experimentally obvious; for a puzzling discrepancy see Palit (1975).

1. The amount of chemical decomposition produced by a current is proportional to the quantity of electricity passed.
2. The amounts of different substances deposited or dissolved by a given quantity of electricity are proportional to their chemical equivalent weights.

Faraday's contribution is of an importance comparable to that of Joule in finding the mechanical equivalent of heat. In modern language, Faraday determined the electrochemical equivalence, or the amount of electricity corresponding to one mole of electrons. This equivalence, which we call the *Faraday constant* $\mathscr{F}$, is

$$\mathscr{F} = N_A e = (6.02205 \times 10^{23})(1.60219 \times 10^{-19}) = 96,484.7 \text{ C mol}^{-1}$$

where N_A is Avogadro's number and e is the charge on the electron; C is the abbreviation for the coulomb. Our name for the natural unit of charge, the *electron*, was proposed, incidentally, by G. Stoney in 1874.

A great number of interesting and perceptive experiments have been passed over in this account. Much of our basic nomenclature originated during Faraday's time: *anode* and *cathode* for the positive and negative pole, respectively, of a battery; *ion* (Greek for wanderer) for the carrier of electricity in solution; *ampere* for the unit of current; and *ohm* for the measure of electrical resistance. The latter two terms are, of course, in recognition of the pioneer investigations of A.M. Ampère and G.S. Ohm.

The mechanism whereby current is transported through solutions of salts took nearly a century to be understood after the first observations of the phenomenon by Davy and others. An early idea was that of T. von Grotthus, who suggested in 1805 that an electrolyte consisted of polar molecules which lined up in an electric field and, by exchanging ends, passed electricity down a chain. He was trying to explain how electricity could go through a solution and yet cause electrolysis only at the electrodes. As illustrated in Fig. 12-1, his thought was the application of a potential across the electrodes caused the polar molecules to break and reform in such a way as to leave a negative piece of one at the anode and a positive piece of another at the cathode. The idea was clever, and will be invoked in more modern language in explanation of the motion of hydrogen ions.

As originally stated, the Grotthus mechanism was untenable on various

FIGURE 12-1
Illustration of the original
Grotthus mechanism.

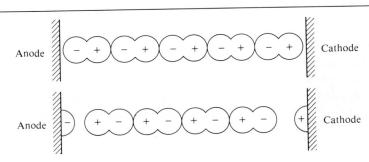

grounds: Electrolyte "molecules" are not that close together in solution; why should even the weakest potential be able to cause strong molecules to break up; and so on. The hypothesis was largely demolished in 1857 by R. Clausius. Clausius proposed instead that the positive and negative parts of the electrolyte molecule were always present to some extent as fragments or ions, which then carried the current. We would call his proposal one of partial dissociation. It was a compromise in the sense that it was very difficult to accept that a molecule could break up or dissociate, and Clausius theorized that only a small fraction of the molecules actually did so.

We come now to the last quarter of the nineteenth century. J. van't Hoff and his group made colligative property measurements on sugar solutions and then on aqueous electrolytes. They reported large i factors [Eq. (10-32)] in the latter case. Thus i was close to two for NaCl solutions. S. Arrhenius drew heavily on van't Hoff's work in proposing the theory of electrolytic dissociation in 1883. The theory amounted to an assertion that electrolytes were mostly and not just partially dissociated into ions. During this last period the giants of electrochemistry were Arrhenius, F. Kohlrausch, W. Ostwald, and van't Hoff. Kohlrausch and his school carried out a monumental number of experiments on the conductance of electrolyte solutions, establishing that they obeyed Ohm's law. Thus the terms *specific conductivity* and *equivalent conductivity* were defined, and the major rules governing the variation of conductance with concentration were formulated. Ostwald did much to clarify the behavior of what we now call *weak electrolytes,* or ones that behave essentially as Clausius had proposed much earlier.

The basic framework of electrochemistry was thus in place by the turn of the century. The early 1900s were spent in more and more precise studies of electrolyte solutions. The next major advances were made in the 1920s, in the treatment of the forces between ions, which determined the quantitative aspects of their motion through a solvent and their thermodynamic properties. A major theory by P. Debye and E. Hückel led, in 1923, to an enduring picture of dilute electrolyte solutions. Each ion tends to have around it an excess concentration of oppositely charged ions, which form a statistical or diffuse *atmosphere.* This atmosphere contributes to the thermodynamic chemical potential of the ion and hence to its activity coefficient. At about the same time, J. Brønsted and N. Bjerrum (in Copenhagen) made lasting contributions both to the understanding of acid and base strengths and in establishing that ions could form *ion pairs* in more concentrated solutions.

Late in the same decade, L. Onsager extended the Debye–Hückel theory to treat dynamic effects such as conductance and diffusion. Later, with I. Prigogine, he was a leader in developing the general thermodynamic treatment of irreversible processes. Still later, R. Tolman and J. Kirkwood pioneered the development of the statistical mechanical theory of solutions—a task yet to be finished.

We begin the subject of electrochemistry with the important new property of electrolyte solutions—that of *conductance.* The separate behavior of each kind of ion is then discussed in terms of ionic mobilities and of transference numbers. The chapter moves on to a presentation of the Debye–Hückel treatment of the nonideality of electrolyte solutions and concludes with a study of ionic equilibria.

12-2   Conductivity—experimental definitions and procedures

A. Defining equations

Electrical units are discussed in Section 3-CN-1, with emphasis on how electrostatic and SI units relate to Coulomb's law. The situation is less complicated here. Traditional electrochemistry makes use of the volt (V), ampere (A), coulomb (C), and ohm (Ω)—all accepted in the SI recommendations. There is still room for confusion in that various sources may report values in cm or in m units, and concentrations in equivalents or in moles per unit volume, unit volume being either 1 cm³, 1 liter (or 1 dm³), or 1 m³! Some clarification is attempted in what follows.

The ampere is defined in terms of Ampere's law (of course):

$$f = 2\delta\frac{I^2}{d} \tag{12-1}$$

Equation (12-1) gives the force f per unit length between two infinitely long, parallel wires separated by distance d and carrying current I. In the SI system, f will be in newtons if I is in amperes and d is in meters, but with the constant δ equal to 1×10^{-7} (exactly).* The coulomb is the quantity of electricity q corresponding to a current of 1 A flowing for 1 s, or in general, $q = It$ (or $\int I\,dt$). The unit of potential is the volt; it requires 1 J (joule) of energy to transport 1 C of charge across 1 V potential difference. Finally, resistance R is defined in terms of Ohm's law,

$$V = IR \tag{12-2}$$

A current of 1 A flowing through a resistance of one ohm produces a voltage drop of 1 V. The resistance R is a function of temperature and, in the case of electrolyte solutions, of concentration.

Ohm's law can be thought of both as an ideal law and as the limiting law for small V and I. It is well obeyed by all substances, provided the energy dissipated does not result in appreciable local heating. In the case of electrolyte solutions Ohm's law begins to fail at high voltages because ionic velocities become large enough that the distortion of the diffuse ion atmosphere around each ion ceases to be proportional to its velocity. In the case of metals current is carried by electrons, and there is no problem in this last respect. The same is true for semiconductors, which differ from metals mainly in that the concentration of conduction electrons is small and increases exponentially with temperature so that the resistance is very temperature-dependent.

The resistance of a substance is proportional to its thickness l and inversely proportional to the cross-sectional area $\mathcal{A}$. One therefore generally reports a specific resistance or *resistivity* ρ defined by

$$R = \frac{l}{\mathcal{A}}\rho \tag{12-3}$$

The resistivity ρ is given in ohm centimeter in the cgs system and in ohm meter in the SI system. Although resistance is the measured quantity, its

---

*In the older emu system of units, $\delta = 1$, with the force in dynes, distances in cm, and current in abamperes, 1 abampere = 10 amperes. Note that in SI, δ is just $\mu_0/4\pi$ (Section 3-CN-1).

reciprocal, the *conductance*, is more useful in dealing with electrolyte solutions. Conductance L is defined as

$$L = \frac{1}{R} \tag{12-4}$$

The SI unit of conductance is the siemens, $1\text{ S} = 1\ \Omega^{-1}$. *Specific conductance*, or *conductivity*, κ, is the reciprocal of ρ,

$$\kappa = \frac{1}{\rho} = \frac{l}{\mathscr{A}}\frac{1}{R} \tag{12-5}$$

As will be seen later, the ratio $l/\mathscr{A}$ is usually treated as an apparatus or cell constant and is given the symbol k. Thus

$$R = k\rho \tag{12-6}$$

$$\kappa = kL \tag{12-7}$$

An electrolyte solution conducts electricity by several paths. Each ion contributes, including those from the self-ionization of the solvent. The situation is therefore one of resistances in parallel so that the total resistance obeys the law

$$\frac{1}{R_{\text{obs}}} = \frac{1}{R_1} + \frac{1}{R_2} + \frac{1}{R_3} + \cdots$$

Conductances are thus additive, that is,

$$L_{\text{obs}} = L_1 + L_2 + L_3 + \cdots \tag{12-8}$$

hence their great utility in this situation. One is ordinarily interested just in the contribution to the observed conductance by the electrolyte and therefore subtracts out that due to the medium L_0:

$$L = L_{\text{obs}} - L_0 \tag{12-9}$$

The same relation will, of course, apply to specific conductivities:

$$\kappa = \kappa_{\text{obs}} - \kappa_0 \tag{12-10}$$

One of the first quantitative observations was that the net specific conductivity of a solution is approximately proportional to the electrolyte concentration. It should be exactly so if each ion were a completely independent agent since each would then make its separate, additive contribution to κ. A very useful quantity is therefore the *molar conductivity*, Λ_m, which is the value of κ contributed by 1 mole of electrolyte. We derive the relationship between Λ_m and κ as follows. It is convenient to use length in cm rather than in meters, and we consider a portion of the electrolyte solution that is 1 cm deep and of area such that the volume contains one mole of electrolyte. The volume (in cm^3) required to contain one mole of electrolyte is $1000/C$, where C is concentration in mol liter^{-1} (or mol dm^{-3}, which is equivalent). It follows that $\mathscr{A}$ in Eq. (12-5) is $1000/C$ in cm^2 (and l is 1 cm). Since Λ_m is just the value of $1/R$ under the above conditions, we have

$$\Lambda_m\ (= 1/R \text{ for 1 mole}) = \kappa\,\frac{\mathscr{A}}{l} \tag{12-11}$$

or

$$\Lambda_m = \frac{1000}{C} \kappa \tag{12-12}$$

In using Eq. (12-12), κ should be in units of ohm^{-1} cm^{-1}, and Λ_m will then be in units of cm^2 mol^{-1} ohm^{-1}.*

An alternative treatment of electrolyte conductivity uses C_e, where C_e is in *equivalents* per liter, to obtain Λ_e, the *equivalent conductivity* (see K.S. Pitzer and L. Brewer, 1961). Equation (12-12) now reads

$$\Lambda_e = \frac{1000}{C_e} \tag{12-14}$$

There are pros and cons as to which system is the more convenient, but interconversion is easy. The number of equivalents per mole of an electrolyte is just the number of moles of positive (or negative) ionic charge per mole of substance. There is one equivalent per mole of NaCl, for example, but two equivalents per mole of Na_2SO_4. It follows that $\Lambda_{m,NaCl} = \Lambda_{e,NaCl}$ but $\Lambda_{m,Na2SO4} = 2\Lambda_{e,Na2SO4}$. In general, if the formula for an electrolyte shows ν_+ positive ions of charge number z_+ and ν_- negative ions of charge number z_-, then $\nu_+ z_+ = \nu_- z_-$ (electroneutrality), and $\Lambda_m = (\nu_+ z_+)\Lambda_e = (\nu_- z_-)\Lambda_e$.

The various defining equations are summarized in Table 12-1.

B. Measurement of conductance

The actual quantity usually measured for an electrolyte solution is its electrical resistance R. This is usually done by means of a Wheatstone bridge, the basic scheme of which is shown in Fig. 12-2. In its simplest form, the galvanometer G shows no current flow when $R_x/R_1 = R_2/R_3$. In practice, it is very important first of all that the electrodes of the cell used have identical electrode potentials; otherwise a voltage change across the cell will be present in addition to the voltage drop due to Ohm's law. This requirement is usually met by the use of identical reversible electrodes, such as platinized platinum, and an alternating source of potential. This last avoids having any appreciable net electrolysis at the electrodes and hence buildup of electrolysis products. The use of the ac bridge introduces another problem, however, namely that due to the capacitance of the cell. There must now be a means for balancing out capacity differences across the two legs of the bridge, or otherwise the ac galvanometer (often an oscilloscope) will not register a point of zero current. The reader is referred to experimental texts for further details.

The typical measurement, then, is of R_x, the resistance of the cell, first when filled with the solvent medium and then when filled with the electrolyte

---

*In the SI system, Eq. (12-12) becomes

$$\Lambda_m = \kappa/C \tag{12-13}$$

Remember, however, that κ (SI) is in ohm^{-1} m^{-1} and is thus equal to 100 κ (cgs), and that C is now in mol m^{-3}. The consequence is that Λ_m (SI) = 10^{-4} Λ_m (cgs). It is the inconvenient size of the SI unit of molar conductivity that prompts the use of the more conventional definition of Λ_m.

| Quantity | Symbol | Units | | Defining relationship | Equation number |
		Cgs	SI		
Resistance	R	ohm	ohm	$V = IR$[a]	(12-2)
Resistivity	ρ	ohm cm	ohm m	$\rho = \dfrac{\mathscr{A}}{l} R$	(12-3)
Conductance	L	ohm⁻¹ (or mho)	ohm⁻¹	$L = \dfrac{1}{R}$	(12-4)
Conductivity	κ	ohm⁻¹ cm⁻¹	ohm⁻¹ m⁻¹	$\kappa = \dfrac{1}{\rho} = \dfrac{1}{\mathscr{A}R}$	(12-5)
Cell constant	k	cm⁻¹	m⁻¹	$k = \dfrac{1}{\mathscr{A}}$	(12-6)
Molar conductivity	Λ_m	cm² mol⁻¹ ohm⁻¹	m² mol⁻¹ ohm⁻¹	$\Lambda_m = \dfrac{1000}{C}$	(12-12)
Equivalent conductivity	Λ_e	cm² equiv⁻¹ ohm⁻¹	m² equiv⁻¹ ohm⁻¹	$\Lambda_e = \dfrac{1000}{C_e} \kappa$	(12-14)

TABLE 12-1
Definitions of electrical quantities

[a]Here V is the potential difference and I the current between electrodes of effective area $\mathscr{A}$ and separation l. Further, C is concentration in mol liter⁻¹ and C_e is concentration in equivalents liter⁻¹.

solution. Application of Eq. (12-9) then gives the net conductance L due to the solute.

The second experimental problem is the reduction of L to a conductivity κ. The defining equation (12-5) is not easily applied in practice since l is not exactly the distance between the electrodes and $\mathscr{A}$ is not exactly the area of the electrodes. The reason is that ions lying outside of the cylinder defined by the two electrodes still experience potential gradients and contribute to the conductivity. Extensive and careful studies have allowed for this contribution, with the result that accurate specific conductivities are known for several standard electrolyte solutions. One such standard is a solution containing 0.74526 g KCl per 1000 g of solution, whose conductivity at 25°C is 0.0014088 ohm⁻¹ cm⁻¹. [See Daniels *et al.* (1956) for further details.] Alternatively, the conductivities of 0.1 M and 0.01 M KCl are 0.012886 and 0.0014114 ohm⁻¹ cm⁻¹ at 25°C, respectively.

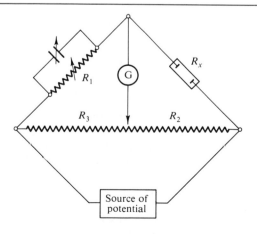

FIGURE 12-2
Schematic diagram of a Wheatstone bridge.

The availability of precise, absolute conductivities allows the problem of determining the l and $\mathcal{A}$ values for a cell to be bypassed. One measures the resistance of the cell when filled with a standard solution and calculates the cell constant k by means of Eq. (12-6). The same cell constant applies to a measurement with any other solution (in very precise work, the standard and unknown solutions should have about the same resistance). One then uses k to calculate ρ or κ from the measured resistance R or conductance L.

Resistances can be measured with great precision, and conductance determinations can therefore be extended to very dilute solutions. At this point, however, one must take major precautions to rinse the cell clean of any impurities adsorbed on the wall or on the electrodes and to free the solvent of all traces of electrolytes. Kohlrausch and co-workers, for example, were able to obtain water virtually free of ionic impurities and found a residual conductivity of 6.0×10^{-8} ohm^{-1} cm^{-1} at 25°C. As discussed in Section 12-4A, this residual value is due to the $\sim 10^{-7}$ mol liter^{-1} of H^+ and OH^- ions from the dissociation of the water itself. This concentration then represents the upper limit of dilution possible with an aqueous electrolyte before its contribution to the conductivity of the solution is swamped by that of the solvent water.

12-3   Results of conductance measurements

The decades preceding and following 1900 were ones in which the conductance of aqueous salt solutions was studied in great detail. Most soluble mineral salts, such as NaCl or KNO_3, show a pattern of behavior that led Arrhenius to propose his theory of electrolytic dissociation, namely that salts are largely dissociated in aqueous solution and fully dissociated at extreme dilution. We agree with this interpretation today and call such substances *strong electrolytes*.

Other salts, and especially some acids and bases, behave quite differently from strong electrolytes. It is necessary to suppose that only a small degree of dissociation takes place; such substances are known as *weak electrolytes*. We take up these two categories of electrolytes in turn and then consider some of the more modern treatments of their behavior.

A. Strong electrolytes   It is typical of a strong electrolyte solution that its conductivity is proportional to its concentration, to a first approximation. This behavior is illustrated in Fig. 12-3(a), and its observation led to the introduction of equivalent conductivity Λ_e as a useful quantity. Nowadays, the closely related molar conductivity, Λ_m, is coming into more common use. According to Eqs. (12-12) and (12-14), κ should be a linear function of concentration if Λ_m or Λ_e is an intrinsic property of the electrolyte, independent of its concentration. The linearity law is not exact, however; as shown in Fig. 12-3(b), Λ for strong electrolytes decreases noticeably with concentration. To allow better comparison between electrolytes, the quantity plotted in the figure is Λ_e. Λ_m and Λ_e are the same for uni-univalent electrolytes, of course, but Λ_m for K_2SO_4 is twice Λ_e, and the ratio is 4:1 for $K_4Fe(CN)_6$.

It was found very early that the decrease in Λ with concentration is closely linear in $\sqrt{C}$, that is

$$\Lambda_m = \Lambda_m^\circ - A\sqrt{C} \tag{12-15}$$

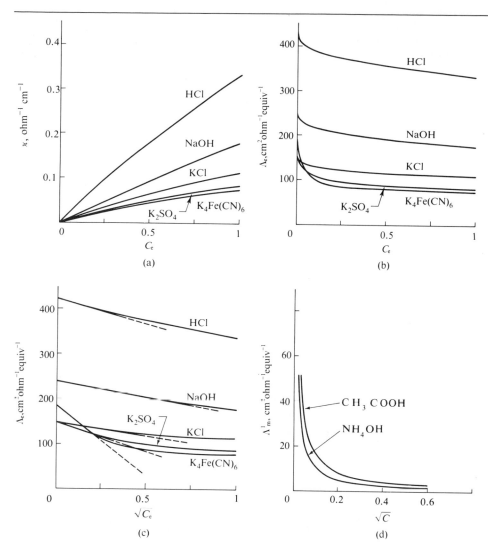

FIGURE 12-3
(a) Conductivity is approximately proportional to C_e (and hence to C since $C_e = \nu_+ z_+ C$). (b) Molar conductivity decreases with increasing C and (c) is nearly linear in $\sqrt{C}$. Note that to facilitate comparison between electrolytes, Λ_e rather than Λ_m values are plotted, where $\Lambda_m = \nu_+ z_+ \Lambda_e$. (d) Molar conductivity decreases strongly with increasing C in the case of a weak electrolyte. All data are for 25°C.

Figure 12-3(c) shows that Eq. (12-15) is well obeyed up to about 0.1 M for uni-univalent concentration. Deviations set in much earlier, however, with more highly charged salts.

Equation (12-15) has been explained theoretically as a consequence of in-terionic attractions and repulsions, that is, as a result of long-range Coulomb forces between ions. The effect is, in a sense, an incidental one since it arises from the charge on ions rather than from more specific chemical properties. Application of the same type of theory to electrolyte nonideality is discussed in Section 12-8.

An important practical use of Eq. (12-15) is the extrapolation of data to obtain Λ_m° (or Λ_e°). Extrapolations of this type constitute an important technique in physical chemistry; they allow the determination of the value of a physical property under experimentally inaccessible conditions. (This is done in Fig. 1-3, to obtain the molecular weight of a real gas.) The present extrapolation gives us Λ_m°, which is the value of Λ_m when the ions are so far apart that

TABLE 12-2
Some molar
conductivities for aqueous
electrolytes at 25°C[a]

Electrolyte	$\Lambda_m^{\circ\,b}$ (cm² mol⁻¹ ohm⁻¹)	Electrolyte	$\Lambda_m^{\circ\,b}$ (cm² mol⁻¹ ohm⁻¹)
HCl	426.16	CaCl₂	271.68
LiCl	115.03	Ca(NO₃)₂	381.88
NaCl	126.45	BaCl₂	279.96
KCl	149.86	Na₂SO₄	259.8
NH₄Cl	149.7	Na₂C₂O₄	248
KBr	151.9	CuSO₄	267.2
NaNO₃	121.55	ZnSO₄	265.6
KNO₃	144.96	LaCl₃	437.4
AgNO₃	133.36	K₄Fe(CN)₆	738.0
NaOOCCH₃	91.0	NaOH	248.1
MgCl₂	258.8		

[a]Source: H. S. Harned and B. B. Owen, "The Physical Chemistry of Electrolyte Solutions," 3rd ed. Van Nostrand–Reinhold, Princeton, New Jersey, 1958.

[b]To convert to SI units, multiply by 10^{-4}; thus $\Lambda^{\circ}_{m,HCl} = 426.16 \times 10^{-4}$ m² mol⁻¹ ohm⁻¹.

there is no electrostatic interaction between them. We call Λ_m° the limiting value of Λ_m at infinite dilution. Table 12-2 gives Λ_m° values for a selection of electrolytes.

An important point is that Λ_m° for an electrolyte should just be the sum of the separate contributions of the ions, since at infinite dilution all interionic attractions and repulsions have vanished. We can therefore write

$$\Lambda_m^{\circ} = \nu_+\lambda_+^{\circ} + \nu_-\lambda_-^{\circ} \tag{12-16}$$

where λ_+° and λ_-° are the limiting molar conductivities of the individual ions. The experimental determination of ionic molar conductivities is discussed in Section 12-6, but we can point out here an important application of the principle of Eq. (12-16). Suppose, for example, that we have Λ_m° values for KCl, NaNO₃, and KNO₃. Equation (12-16) allows us to calculate Λ_m° for NaCl. Thus

$$\Lambda_{m,NaCl}^{\circ} = \Lambda_{m,Na^+}^{\circ} + \Lambda_{m,\,Cl^-}^{\circ}$$

$$= \Lambda_{m,\,K^+}^{\circ} + \Lambda_{m,\,Cl^-}^{\circ} + \Lambda_{m,\,Na^+}^{\circ} + \Lambda_{m,\,NO_3^-}^{\circ}$$

$$- \Lambda_{m,\,K^+}^{\circ} - \Lambda_{m,\,NO_3^-}^{\circ} \tag{12-17}$$

or

$$\Lambda_{m,\,NaCl}^{\circ} = \Lambda_{m,\,KCl}^{\circ} + \Lambda_{m,\,NaNO_3}^{\circ} - \Lambda_{m,\,KNO_3}^{\circ} \tag{12-18}$$

On putting in the actual numbers,

$$\Lambda_{m,NaCl}^{\circ} = 149.86 + 121.55 - 144.96 = 126.45$$

which agrees with the value obtained by direct measurement, given in Table 12-2.

We can extend the above illustration to a case involving polyvalent ions. Suppose we want Λ_m° for $Na_3Co(NH_3)_6$ but have a value only for $K_3Co(NH_3)_6$. We can write

$$\Lambda_{m,\ Na_3Co(NH_3)_6}^\circ = \Lambda_{m,\ K_3Co(NH_3)_6}^\circ + 3\Lambda_{m,\ NaCl}^\circ - 3\Lambda_{m,KCl}^\circ \tag{12-19}$$

On inserting the Λ_m° values, we get $\Lambda_{m,Na_3Co(NH_3)_6}^\circ = 526.5 + (3)(126.4) - (3)(149.9) = 456.0 \text{ cm}^2 \text{ mol}^{-1} \text{ ohm}^{-1}$.

EXAMPLE

B. Weak electrolytes

One of the early problems was that not all electrolytes behaved as illustrated in Fig. 12-3(a–c). Thus, unlike HCl and NaOH, acetic acid and aqueous ammonia give the results shown in Fig. 12-3(d). In this case, however, the behavior was explainable on the basis of a dissociation equilibrium. For a binary electrolyte AB whose formal concentration is C we write

$$\begin{array}{ccc} AB & = A^+ & + B^- \\ (1 - \alpha)C & \alpha C & \alpha C \end{array}$$

where α is the degree of dissociation. The equilibrium constant is

$$K = \frac{(\alpha C)^2}{(1 - \alpha)C} = \frac{\alpha^2 C}{1 - \alpha} \tag{12-20}$$

If one measures the conductivity of the weak electrolyte solution and applies Eq. (12-12), an apparent molar conductivity Λ_m' results because only part of the weak electrolyte is actually present as ions; it is the Λ_m' values that are shown in Fig. 12-3(d). Thus for a weak electrolyte,

$$\Lambda_m' = \frac{1000}{C} \kappa \tag{12-21}$$

The true molar conductivity of the ions present is given by

$$\Lambda_m = \frac{1000}{C^*} \kappa \tag{12-22}$$

where $C^* = \alpha C$ and is the concentration of the dissociated portion. The degree of dissociation α is then

$$\frac{\Lambda_m'}{\Lambda_m} = \frac{C^*}{C} = \alpha \tag{12-23}$$

Substitution of this result into Eq. (12-20) gives

$$K = \frac{(\Lambda_m')^2\ C}{\Lambda_m(\Lambda_m - \Lambda_m')} \tag{12-24}$$

If interionic attraction effects can be neglected, as is often the case for weak electrolyte solutions if the actual ionic or dissociated concentration is low, Λ_m can be replaced by Λ_m° and Eqs. (12-23) and (12-24) become

$$\frac{\Lambda_m'}{\Lambda_m^\circ} = \alpha \tag{12-25}$$

and

$$K = \frac{(\Lambda'_m)^2 C}{\Lambda^\circ_m (\Lambda^\circ_m - \Lambda'_m)} \tag{12-26}$$

Equation (12-26) is known as the *Ostwald dilution law*, after W. Ostwald, who proposed it. It allows a determination of the dissociation constant of a weak electrolyte from conductance data (note Problem 12-34).

C. General treatment
of electrolytes

The great success of Eq. (12-26) in accounting for the behavior of weak electrolytes such as acetic acid led Ostwald, Arrhenius, and others to apply the same equation to salts such as KCl and to acids such as HCl. The results were confusing. Although the K values computed on the basis of Eq. (12-26) were reasonably constant for a weak electrolyte, they would vary 10- or 20-fold when so computed for strong electrolytes over a range of concentration.

It was eventually realized that two different effects are present. The predominant one in the case of a weak electrolyte is that of its partial dissociation into ions. Superimposed on this, however, is the effect of interionic attraction, which causes the value of Λ_m (as distinct from Λ'_m) to vary with concentration. If the electrolyte is largely or fully dissociated, then the interionic attraction aspect becomes the dominant one. This is the situation with strong electrolytes.

Neither effect can be neglected for any electrolyte. In the case of a weak electrolyte the correct equation to use is Eq. (12-24), where Λ_m must have the value corresponding to the actual ionic environment that is present. That is, the dissociation produces an ion concentration C^*, which in turn affects Λ_m according to Eq. (12-15). The constant A of this equation,

$$\Lambda_m = \Lambda^\circ_m - A \sqrt{C^*} \qquad \text{[Eq. (12-15)]}$$

has been evaluated theoretically, and is given by

$$A = 0.2289 \, \Lambda^\circ_m + 60.19 \tag{12-27}$$

Equation (12-27) is valid for aqueous univalent ions at 25°C at concentrations below about 0.1 M.

12-4   Some sample calculations

A. Calculation of
dissociation
constants

The first two examples are ones in which interionic attraction effects are neglected. That is, we assume that Λ°_m gives the equivalent conductivity of the ions. We obtain the following resistances measured at 25°C: 1×10^6, 24.96, and 1982 ohms, for solvent water, 0.1 M KCl, and 0.01 M acetic acid, respectively.

We wish to calculate the dissociation constant of acetic acid. It is first necessary to obtain the cell constant. The net conductance due to the KCl is

$$\frac{1}{24.96} - \frac{1}{10^6} = 0.04006$$

The conductivity of 0.1 M KCl is known to be 0.012886 ohm^{-1} cm^{-1}, and use of Eq. (12-6) gives the cell constant:

$$k = \frac{\kappa}{L} = \frac{0.012886}{0.04006} = 0.3217 \text{ cm}^{-1}$$

The net conductance of the acetic acid is

$$L = \frac{1}{1982} - \frac{1}{10^6} = 5.035 \times 10^{-4}$$

from which we find

$$\kappa = (5.035 \times 10^{-4})(0.3217) = 1.620 \times 10^{-4}$$

The apparent molar conductivity is therefore

$$\Lambda_m' = \frac{1000}{0.01} 1.620 \times 10^{-4} = 16.20 \text{ cm}^2 \text{ mol}^{-1} \text{ ohm}^{-1}$$

We may obtain Λ_m° for the ions of acetic acid from the additivity rule, using the values 91.0, 426.2, and 126.5 for the Λ_m° values of sodium acetate NaAc, HCl, and NaCl, respectively:

$$\Lambda_{m,HAc}^\circ = \Lambda_{m,NaAc}^\circ + \Lambda_{m,HCl}^\circ - \Lambda_{m,NaCl}^\circ$$

$$= 91.0 + 426.2 - 126.5 = 390.7$$

The degree of dissociation is then

$$\alpha = \frac{16.20}{390.7} = 0.0415$$

and the dissociation constant is

$$K = \frac{(0.0415)^2(0.01)}{1 - 0.0415} = 1.797 \times 10^{-5}$$

The resistance of 1×10^6 quoted for the solvent water was deliberately picked as a somewhat low value. It corresponds to a specific conductivity of

$$\kappa = (10^{-6})(0.3217) = 3.217 \times 10^{-7} \text{ ohm}^{-1} \text{ cm}^{-1}$$

The value for pure water was quoted in Section 12-2B as 6.0×10^{8} ohm ^{1} cm^{-1}, and the difference, 2.62×10^{-7}, is therefore to be attributed to impurities. Taking NaCl as a typical impurity, we get the corresponding concentration from Eq. (12-12) using $\Lambda_m^\circ = 126$:

$$C = \frac{1000}{\Lambda}\kappa = \frac{1000}{126}(2.62 \times 10^{-7}) = 2.08 \times 10^{-6} M$$

This concentration of ionic impurity is not likely to affect the dissociation of the 0.01 M acetic acid, so the slightly impure water used is acceptable for the particular experiment.

A value of 6.0×10^{-8} ohm^{-1} cm^{-1} for the specific conductivity of pure water at 25°C was obtained by Kohlrausch and co-workers in 1894; the measurement allows a calculation of the dissociation constant for water itself. We write

$$H_2O = H^+ + OH^- \qquad K_w = (H^+)(OH^-) \qquad (12\text{-}28)$$

The concentration of water does not appear in K_w since we take pure liquid water to be the standard state for water substance, and hence of unit activity.

The limiting equivalent conductivity for the electrolyte H^+, OH^- may again be calculated from the additivity rule:

$$\Lambda^{\circ}_{m,H^+,OH^-} = \Lambda^{\circ}_{m,HCl} + \Lambda^{\circ}_{m,NaOH} - \Lambda^{\circ}_{m,NaCl}$$

$$= 426.16 + 248.1 - 126.45 = 547.81$$

Equation (12-12) then gives the concentration of ions present:

$$C = \frac{1000}{547.81}(6.0 \times 10^{-8}) = 1.10 \times 10^{-7} \, M$$

The value of K_w at 25°C is therefore

$$K_w = (1.10 \times 10^{-7})^2 = 1.20 \times 10^{-14}$$

It is very difficult to remove the last traces of ionic impurities from water and this value for K_w is a little high; a better (modern) value is

$$K_w = 1.01 \times 10^{-14} \tag{12-29}$$

B. Correction for the interionic attraction effect

We can improve the preceding calculation of the dissociation constant for acetic acid if we correct for the error introduced by using $\Lambda^{\circ}_{m,HAc}$ rather than Λ_{m,H^+,Ac^-} in determining it (we write Λ_{m,H^+,Ac^-} as a reminder that this is the actual equivalent conductivity of the ions H^+ and Ac^-). The value of 0.0415 for α is now regarded as a first approximation, giving an ion concentration

$$C^* = (0.0415)(0.01) = 4.15 \times 10^{-4} \, M$$

The constant A of Eq. (12-15) is as given by Eq. (12-27),

$$A = (0.2289)(390.7) + 60.19 = 149.62$$

Equation (12-15) then gives

$$\Lambda_{m,H^+,Ac^-} = 390.7 - (149.62)(4.15 \times 10^{-4})^{1/2} = 387.7$$

The second approximation to α is therefore

$$\alpha = \frac{16.20}{387.7} = 0.0418$$

The change is small enough in this case that another round of approximation is not needed (but may be carried out by the reader as an exercise). The final value for K is now

$$K = \frac{(0.0418)^2(0.01)}{1 - 0.0418} = 1.82 \times 10^{-5} \tag{12-30}$$

C. Calculation of solubility products

A useful application of conductivity measurements is to the calculation of the solubility of slightly soluble salts. Consider the following set of data: The resistance of a cell is 227,000 ohms when filled with water and is 21,370 ohms when filled with saturated calcium oxalate solution. The cell constant is 0.25 cm^{-1}.

We calculate first the net conductivity of the calcium oxalate. The conductivity of the water is, by Eq. (12-5), $0.25/227,000 = 1.10 \times 10^{-6}$ (the water is again somewhat impure), and that of the saturated solution is 0.25/ 21,370 = 1.170 × 10⁻⁵. The net value is 1.060 × 10⁻⁵ ohm⁻¹ cm⁻¹. The value of Λ_m° for Ca^{2+} and oxalate, Ox^{2-}, ions is obtained through use of the additivity rule:

$$\Lambda_{m,\ CaOx}^\circ = \Lambda_{m,\ Ca(NO_3)_2}^\circ + \Lambda_{m,\ Na_2Ox}^\circ - 2\Lambda_{m,\ NaNO_3}^\circ$$

$$= 261.88 + 248 - (2)(121.55) = 267$$

The concentration of Ca^{2+} and Ox^{2-} ions is then

$$C = \frac{1000}{267}\,1.060 \times 10^{-5} = 3.97 \times 10^{-5}\ M$$

The solubility product is therefore $K_{sp} = (Ca^{2+})(Ox^{2-}) = 1.58 \times 10^{-9}$. As a reminder, the solubility equilibrium is

$$CaOx(s) = Ca^{2+} + Ox^{2-}$$

but the concentration of $CaOx(s)$ does not appear in the K_{sp} expression since solid calcium oxalate is in its standard state and hence has unit activity (note Section 7-6).

12-5   Ionic mobilities

The conductance of an electrolyte solution is understood to be due to the motion of ions through the solution as a result of the applied potential. Positive ions move toward the electrode which is negatively charged in solution, or the cathode, and negative ions move toward the electrode which is positively charged in solution, or the anode. The current due to each ion is given by the product of the velocity of the ion and its charge. We consider first the quantitative aspect of this motion of ions under the influence of an applied potential, and then methods for measuring ion velocities.

A. Defining equations

The motion of a particle in solution when subjected to some force f is discussed in Section 10-8A. To review the situation, we recall that a limiting velocity v is quickly reached such that

$$v = \frac{f}{\mathit{f}} \tag{12-31}$$

where f is the friction coefficient. In the case of spherical particles, recall that a relationship due to Stokes gives $\mathit{f} = 6\pi\eta r$, where η is the viscosity of the medium and r is the particle radius. It has become customary when dealing with ions to use the reciprocal of f, called the *mobility* ω. Equation (12-31) thus becomes

$$v = \omega f \tag{12-32}$$

We want now to consider the case of an ion in solution which experiences a force due to the imposed potential difference between the electrodes. The potential energy ϕ of a charge q in a potential V is, by the definition of potential, just qV (and will be in joules if q is in coulombs and V in volts).

The force acting on a particle of this charge is then $d\phi/dx$ or $q\mathbf{F}$, where $\mathbf{F}$ is the field and is equal to dV/dx. The force in dynes is then

$$f = 10^7 \mathbf{F}ze \tag{12-33}$$

where $\mathbf{F}$ is in volts per centimeter, z is the charge number of the ion, and e is the electronic charge in coulombs, $e = 1.602 \times 10^{-19}$ C.*

The velocity of the ith ion is

$$v_i = 10^7 \mathbf{F}z_i\omega_i e \tag{12-34}$$

or

$$v_i = u_i\mathbf{F} \tag{12-35}$$

where u_i is called the *electrochemical mobility*, usually expressed as (cm sec^{-1})/(V cm^{-1}) or cm^2 V^{-1} sec^{-1}, and

$$u_i = 10^7 z_i\,\omega_i\,e \tag{12-36}$$

We now want to determine the molar conductivity of this ion. According to the definition of molar conductivity, the electrodes are to be 1 cm apart and of area such that the enclosed volume contains 1 mole. The conductance so measured is then the molar conductivity. In the present case the current carried will be

$$I_i = v_iz_i\mathscr{F} \tag{12-37}$$

since z times Faraday's number gives the coulombs of charge carried by the ith ion. The potential difference is just $\mathbf{F}$, since the electrodes are 1 cm apart, and we find from Ohm's law

$$\lambda_i = \frac{1}{R_i} = \frac{I_i}{\mathbf{F}} = \frac{v_iz_i\mathscr{F}}{\mathbf{F}} = \frac{u_i\mathbf{F}z_i\mathscr{F}}{\mathbf{F}}$$

or

$$\lambda_i = z_iu_i\mathscr{F} \tag{12-38}$$

We can combine Eqs. (12-38) and (12-36) with ω_i replaced by the Stokes' law expression $1/6\pi\eta r_i$, to obtain

$$\lambda_i = \frac{10^7\,z_i^2\,e\,\mathscr{F}}{6\pi\eta r_i} \tag{12-39}$$

Two qualitative conclusions follow. First, for similar size ions the ratio λ_i/z_i^2 should be approximately constant; and second, the product $\lambda_i\eta$ should be approximately constant for a given ion if one changes the viscosity of the medium. This last is another illustration of *Walden's rule* (note Eq. (10-45) and following discussion).

The molar conductivity for an electrolyte is the sum of the contributions of the separate ions; for an electrolyte with only two kinds of ions, Eq. (12-16) may be generalized to

$$\Lambda_m = \nu_+\lambda_+ + \nu_-\lambda_- \tag{12-40}$$

---

*Equation (12-33) becomes $f = \mathbf{F}ze$ in SI units, with f in newtons and $\mathbf{F}$ in V m^{-1}. Alternatively, if e is given in esu, Eq. (12-33) becomes $f = \mathbf{F}ze/300$ with f in dynes and $\mathbf{F}$ in V cm^{-1}.

Since Λ_m can be found from conductance measurements, if λ_+ can be found by some independent means, then λ_- may be calculated, and vice versa. As discussed below, one way of doing this is to measure the velocity of an ion in a known field.

The λ value for Ca^{2+} is 119 cm^2 mol^{-1} ohm^{-1}. How fast should Ca^{2+} ions move under a field of 2 V cm^{-1}? We use Eq. (12-38) to find $u_{Ca^{2+}} = (119)/(2)(96485) = 6.17 \times 10^{-4}$ cm^2 V^{-1} s^{-1}, and Eq. (12-35) to conclude that $v_{Ca^{2+}} = (6.17 \times 10^{-4})(2) = 1.23 \times 10^{-3}$ cm s^{-1}. Such a velocity is far less than the kinetic energy velocity, around 10^4 cm s^{-1}, and the effect of the imposed electric field is evidently to give a very minor bias to the random thermal motion of the ion in solution.

We can next ask for the effective radius of the Ca^{2+} ion, assuming that the λ value is for water at 25°C and that Stokes' law applies. We use Eq. (12-39):

$$r_{Ca^{2+}} = \frac{(10^7)(2)^2(1.6022 \times 10^{-19})(96485)}{(6)(3.142)(119)(0.008937)} = 3.08 \times 10^{-8} \text{ cm}$$

As discussed further below, this is an *effective* radius—one which includes the hydration shell around each ion.

The calculations may be repeated in SI units. Remember that $\lambda(SI) = 10^{-4} \lambda(cgs)$, $\eta(SI) = 0.1 \eta(cgs)$, **F** is in V m^{-1}, and the factor of 10^7 disappears from Eq. (12-39).

EXAMPLE

The rate of motion of a given kind of ion under an imposed electric field may be observed directly if a boundary is present between two kinds of electrolytes. The method, called the *moving boundary method*, is illustrated schematically in Fig. 12-4. A vertical tube is layered with a solution of MX' (sodium acetate in this example) on top of one of MX (sodium chloride), which in turn is on top of one of M'X (lithium chloride). The concentrations are such that the solutions decrease in density upward, so that convective mixing will not occur. The applied potential is such as to cause positive and negative ions to move in the directions shown. The lower boundary is one between M'X and MX, and hence between two types of positive ions, while the upper boundary is

B. Measurement of ionic mobilities

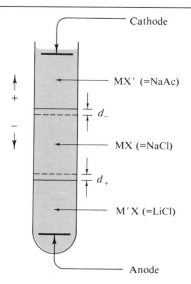

FIGURE 12-4
Illustration of the moving boundary method.

between MX and MX', or between two types of negative ions. The positions of these boundaries can be measured optically (and are visible to the eye) because of the change in refractive index.

A further requirement is that the mobility of M be greater than that of M' and that the mobility of X be greater than that of X'. In terms of the particular example, $u_{Na^+} > u_{Li^+}$ and $u_{Cl^-} > u_{Ac^-}$. If this condition is met, then M' cations will not overtake the M cations but will lag behind slightly until enough extra potential drop occurs locally to bring them just up to the velocity of the M cations. The boundary between MX and M'X then remains sharp and its motion is determined by the velocity of the cations M. Similarly, the upper boundary remains sharp and its motion is determined by that of the anions X.

On application of a potential across the electrodes, after a time t the lower boundary has moved up a distance d_+ and the upper one down a distance d_-. The velocities of the M and X ions are then proportional to d_+ and d_-, respectively, and since the ions experience the same potential gradient, the velocities are in turn proportional to the respective mobilities, and, by Eq. (12-38), to the respective λ values. Thus

$$\frac{\lambda_M}{\lambda_X} = \frac{d_+}{d_-} \tag{12-41}$$

By Eq. (12-40),

$$\Lambda_{m,\ MX} = \lambda_M + \lambda_X = \lambda_X\left(\frac{d_+}{d_-} + 1\right)$$

or

$$\lambda_X = \Lambda_{m,\ MX}\left(\frac{d_-}{d_+ + d_-}\right)$$

and

$$\lambda_M = \Lambda_{m,\ MX}\left(\frac{d_+}{d_+ + d_-}\right) \tag{12-42}$$

One can therefore determine λ_X and λ_M from the motions of the two boundaries if $\Lambda_{m,\ MX}$ is known.

The restrictions on this type of procedure are somewhat severe, and act to limit the number of systems that can be studied. An alternative is to deal with just one boundary, say the one between MX and M'X. The motion of this boundary through distance d_+ sweeps out a volume $d_+\mathcal{A}$, where $\mathcal{A}$ is the cross-sectional area of the tube. If the concentration of MX is C moles per liter, then the total quantity of electricity carried by the M cations is

$$q_M = \frac{C_{MX}}{1000}(d_+\mathcal{A})\mathcal{F} \tag{12-43}$$

Had the motion of the upper boundary been observed, then, similarly,

$$q_X = \frac{C_{MX}}{1000}(d_-\mathcal{A})\mathcal{F} \tag{12-44}$$

It follows that

$$\lambda_M = \Lambda_{m,\,MX} \frac{q_M}{q_M + q_X} \tag{12-45}$$

since q_M and q_X are proportional to d_+ and d_-, respectively. However, $q_M + q_X = q_{tot} = It$, where I is the current through the system and t is the elapsed time. Equation (12-45) can therefore be written

$$\lambda_M = \Lambda_{m,\,MX} \frac{(C_{MX}/1000)(d_+ \mathscr{A})\mathscr{F}}{It} \tag{12-46}$$

It is thus possible to determine the ratio $\lambda_M/\Lambda_{m,\,MX}$ from measurements of the motion of just one of the two boundaries. This ratio is known as the *transference number* for the ion M in salt MX (see Section 12-6).

EXAMPLE

A simple version of the moving boundary method is that illustrated in Fig. 12-5. The upper compartment contains 2×10^{-3} M $AgNO_3$ and the lower one contains a slightly denser solution of $NaNO_3$. After passage of 1.5 mA for 10 min, the boundary between the two solutions is found to have moved 0.619 cm. The area of the cell is 3.5 cm². Calculate the transference number of Ag^+. Since Ag^+ is the faster moving of the two cations, the boundary remains sharp and its motion is due to the migration of Ag^+ ions. Substitution into Eq. (12-46) gives

$$\frac{\lambda_{Ag^+}}{\Lambda_{m,AgNO_3}} = \frac{(2 \times 10^{-3}/1000)(0.619)(3.5)(96485)}{(1.5 \times 10^{-3})(10)(60)} = 0.465$$

Ionic mobilities may be determined by the preceding method or indirectly, through the measurement of transference numbers, discussed in the next section. Some representative values are given in Table 12-3, extrapolated to infinite dilution. As suggested in the numerical example of Section 12-5A, the values are of the order of 10^{-4} cm² V⁻¹ s⁻¹, and ions in solution do not actually move very rapidly under ordinary voltage gradients.

The example cited also illustrates the use of the Stokes' expression for

C. Discussion of results

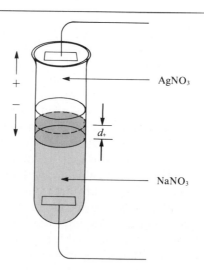

FIGURE 12-5
Moving boundary method using a single boundary. See Example.

TABLE 12-3
Mobilities of aqueous ions at infinite dilution at 25°C

Cation	Mobility (cm² V⁻¹ s⁻¹ × 10⁴)	Anion	Mobility (cm² V⁻¹ s⁻¹ × 10⁴)
H^+	36.30	OH^-	20.50
Li^+	4.01	F^-	5.70
Na^+	5.19	Cl^-	7.91
K^+	7.62	Br^-	8.13
Rb^+	7.92	I^-	7.95
Ag^+	6.41	NO_3^-	7.40
Cs^+	7.96	CH_3COO^-	4.23
NH_4^+	7.60	CO_3^{2-}	7.46
Ca^{2+}	6.16	SO_4^{2-}	8.27
La^{3+}	7.21		

mobility to calculate apparent or *hydrodynamic* ionic radii. As another case, the corresponding calculation gives r_{Na^+} as 1.83 Å. The radius of the Na^+ ion in a crystal of NaCl is found to be only 0.95 Å, and to the extent that Stokes' law is accurate, it appears that the Na^+ ion is larger in solution than is the bare ion. The accepted explanation is that the ion strongly binds a sphere of water molecules around it, so that the size of the *hydrodynamic* unit is increased. This interpretation is supported by the fact that the corresponding calculation for a large ion such as $N(CH_3)_4^+$ gives essentially the crystallographic radius. Tetramethylammonium ion is large and nonpolar, so that its interaction with water molecules is not expected to be so strong as to bind them to it as a kinetic unit.

This example illustrates that quite reasonable ionic radii result from calculations based on mobility measurements, provided that ions are regarded as capable of binding water molecules to form a hydrodynamic unit. Table 12-3 gives the order of increasing mobility of the alkali metal cations as

$$Li^+ < Na^+ < K^+ < Rb^+ < Cs^+$$

Since by Eq. (12-39) mobility should vary inversely with the ion radius, the series is just the *opposite* of what might be expected in terms of the actual ion sizes—Li^+ ion should be the smallest and the most mobile of the series. The qualitative explanation of this reversal is that the binding of water molecules by an ion is strongly affected by the charge density at its surface, and therefore varies along the series. The potential of a point charge q is q/r at a distance r (as discussed in Section 8-ST-1), and is consequently larger at the surface of a Li^+ ion than at the surface of a Cs^+ ion, to consider the extremes. Evidently, aqueous Li^+ ions carry a number of water molecules with them, tightly held by the strong local field, while the surface potential of a Cs^+ ion is small enough that it appears to carry few if any water molecules with it. The mobility series for the alkali metal ions thus provides one of the strong evidences for the hydration of ions in solution.

Theoretical calculations have been made of the distribution of water molecules around an ion. A recent example (Impay, Madden, and McDonald, 1983) involved some molecular dynamic calculations. A "system" was chosen, consisting of the ion in question and of a number of water molecules, and a theoretical potential function was chosen to describe the interaction between the dipolar water molecule and the ion. A temperature was chosen and the molecules given an appropriate initial kinetic energy. The motions or trajectories of all of the molecules were then calculated as a function of time until further change in average molecular positions could be neglected. Figure 12-6(a) shows the results for K^+ ion in water at 274 K, in the form of radial distributions (note Section 8-CN-1) for ion-oxygen and ion-hydrogen distances. Notice the sharp maxima (at 2.76 Å and 3.35 Å for the ion-oxygen and ion-hydrogen distances, respectively) and the secondary maxima, corresponding to the second solvation shell. The area under the first peak is taken to give the *coordination number* or the average number of water molecules in the first coordination sphere. The values were 7.5 for K^+, 5.3 for Li^+, and 7.2 for Cl^-.

According to the above numbers, the hydrated Li^+ ion is *smaller* than the hydrated K^+ ion, in seeming contradiction to the mobility results. What is needed, however, is the *dynamic hydration number,* essentially the number of water molecules that follow the ion as it makes little diffusional jumps. The molecular dynamic calculations gave this last number to be about 2.9 for K^+, 4.6 for Li^+, and 2.6 for Cl^- ions. Comparison of the coordination and dynamic hydration numbers confirms the previous qualitative conclusion; Li^+ ion hangs on to almost all of its waters of hydration as it moves through the solution, while for K^+ ion, the waters are more loosely held and only about half actually move with the ion.

The positions of the peaks in the ion-oxygen and ion-hydrogen radial distribution functions allow a calculation of the probable orientation of a water molecule around an ion. As shown in Fig 12-6(b), the water molecule is oriented with the oxygen closest to the cation (as might be expected) and with the plane of the molecule at an angle to the ion-oxygen line. In the case of K^+, this angle θ was found to be 36°.

In the case of di- and trivalent ions, the rate of exchange between coordinated and free water is slow enough to be followed as a chemical reaction. The coordination chemist thus regards an aqueous cation in much the same light as any other complex ion; the ion holds water more or less tightly in its

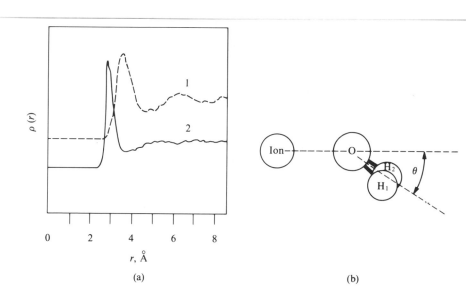

(a)   (b)

FIGURE 12-6
Results of molecular dynamics calculations for aqueous K^+ ion. (a) Radial distribution functions for (1) ion-hydrogen and (2) ion-oxygen distances. The "system" consisted of K^+ ion and 125 water molecules, at 274 K. (b) Structure of the cation-water complex. [From R.W. Impay, P. A. Madden, and I.R. McDonald, J. Phys. Chem., **87**, 5071 (1983).]

coordination sphere, and this water must leave if some other group or ligand is to be coordinated in its place.

The case of the H^+ ion is a very interesting one. The simple ion, is, of course, just a proton, and its radius on this basis would be about 10^{-13} cm. We know that this is not the situation and that H^+ is coordinated to a water molecule, so that it is more appropriately written as H_3O^+. However, as discussed in Section 8-CN-1, liquid water is a very complex mixture of individual molecules and clusters of hydrogen-bonded molecules. These last were described as "flickering clusters" since they form, break up, and reform elsewhere very rapidly. It is therefore difficult to assign H^+ to any one single unit and the formulation H_3O^+ is used here in the same way that H_2O is used to describe the molecular unit of liquid water.

The formula H_3O^+ is, however, the minimum size of the positively charged unit, and the observed mobility for H^+ ion is much too large to be explainable in terms of this size [note Eq. (12-48)]. It seems very likely that a variation of the Grotthus mechanism of Fig. 12-1 is operative in this case. As an illustration of the proposed mechanism, one of the clusters of Fig. 8-21 may be stretched out in linear form and imagined to carry an extra proton. This is illustrated in Fig. 12-7, which shows the proton at the left end of the chain. The long spacings represent hydrogen bonds, and the short ones, regular bonds. Hydrogen bonds are mobile in that the hydrogen atom can easily jump from a O—H——O to a O——H—O configuration. Such a shift down the chain has the effect of putting the extra proton on the right-hand oxygen in the figure. In this case, a unit of positive charge has moved four molecular distances without any appreciable molecular motion being required. The effect would be to give the proton a large apparent mobility although it is charge rather than actual hydrogen atoms that is doing the moving.

The process as just described is not quite complete, however, since after the charge transfer each water molecule has, in effect, been rotated 180°. Examination of Fig. 12-7 makes this apparent. For the chain to be ready to transport another unit of charge, the molecules must rotate back to their original configuration. It may be that either the rotation time for a water molecule or the rate of formation and disruption of clusters determines the speed at which charge can move through the solution. M. Eigen, from his studies of very fast reactions in aqueous solutions, has been led to suggest that, rather than the proton being associated randomly with molecules and clusters, a special unit is the most stable one:

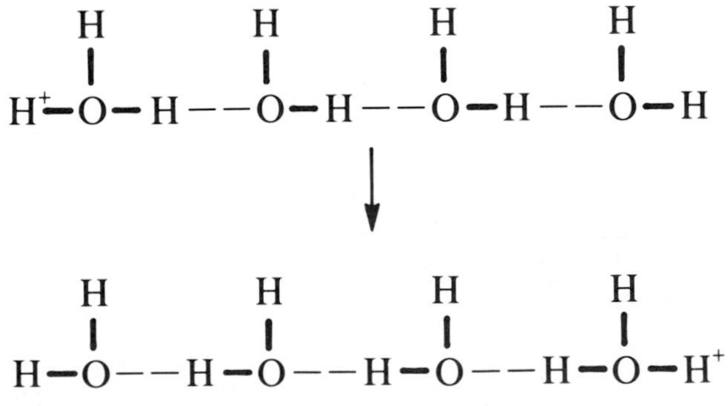

FIGURE 12-7
Illustration of the possible mechanism of conduction in water.

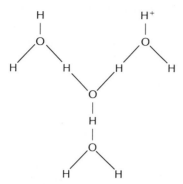

If so, the rate of reformation of $H_9O_4^+$ may be the slow step that determines the mobility of protons in water. The subject is as complicated as the structure of water itself.

The mobility of the OH^- ion is also unusually high, and mechanisms have been proposed that are similar to those for proton migration.

A special case of the moving boundary method should be discussed briefly. It has been very highly developed for use in the study of proteins and other large molecules, and the process is now called *electrophoresis*. As an example, a boundary is formed between two buffer solutions, one of which contains protein. The boundary is observed to move on application of a potential between the electrodes and one thus obtains the rate of motion of the protein and, from the potential gradient, a value for its electrochemical mobility. The temperature is often set at about 4°C or at the point of maximum density of the solution. In this way convection currents due to temperature fluctuations are largely eliminated. Alternatively, a gel matrix may be used to eliminate convection. Even better, the method would perform ideally in a space laboratory, in a gravity-free situation!

Electrophoresis measurements serve to distinguish protein molecules qualitatively, since each shows a characteristic rate of motion. If a mixture is initially present as a thin band, separation into several distinct zones will occur. The method may thus be used for the analysis of a protein or other mixture. On a more quantitative level the determination of the mobility u allows us to estimate the charge z that is carried. If the size of the protein molecule is known, then Eq. (12-48) may be used for the calculation of z. The more elegant approach is to determine the friction coefficient f by means of sedimentation or diffusion measurements (see Section 10-7) and to write the more general expression

$$u_i = \frac{10^7 z_i e}{f} \qquad (12\text{-}47)$$

The charge may then be computed without making any assumptions about molecular size or shape.

It should be noted that in the case of amino acids and proteins, the charge on a given molecule fluctuates with time as it adds or loses protons in rapid acid–base equilibria (note Section 12-CN-2). The electrophoretic velocity depends on the average charge, and is therefore very pH-dependent.

Electrophoretic measurements may be extended to particles large enough

D. Electrophoresis

to be visible under a microscope, that is, to colloidal systems. For reasons that are not discussed until Section 21-ST-1, the equipment is now called a *zetameter*, but the method is essentially the same as that just described. One applies a known potential gradient and observes the slow motion of the colloidal particle along the gradient. We can then calculate the mobility from the rate of such motion and the field and use Eq. (12-48) to obtain the total charge on the particle.

$$u_i = \frac{10^7 z_i e}{6\pi\eta r}. \tag{12-48}$$

12-6   Transference numbers—ionic equivalent conductivities

The direct measurement of ionic mobilities is discussed in the preceding section. An alternative approach is to determine the fraction of current carried by a particular ion, rather than its actual velocity.

A. Defining equations

The *transference* (or *transport*) *number* of an ion in an electrolyte solution is defined as the fraction of the total current which that ion carries when a potential gradient is applied. The transference number for the *i*th ion, t_i, is therefore given by the ratio of the conductivity κ_i contributed by that ion to the total conductivity of the solution, $\Sigma_i \kappa_i$. Equation (12-12) may be written for a particular ionic species,

$$\lambda_i = \frac{1000}{C_i} \kappa_i \tag{12-49}$$

and it follows that

$$t_i = \frac{\lambda_i C_i}{\Sigma_i \lambda_i C_i} \tag{12-50}$$

Equation (12-50) is a general one, valid for a mixture of electrolytes. For a single electrolyte, $C_+ = \nu_+ C$ and $C_- = \nu_- C$, so that Eq. (12-50) reduces to

$$t_+ = \frac{\nu_+ \lambda_+}{\Lambda_m} \qquad t_- = \frac{\nu_- \lambda_-}{\Lambda_m} \tag{12-51}$$

with the help of Eq. (12-40).

Some representative transference numbers are given in Table 12-4. Note

TABLE 12-4	Concentration C	Cation transference number					
Cation transference numbers in aqueous solutions at 25°C[a]		HCl	NaCl	KCl	Na₂SO₄	K₂SO₄	LaCl₃
	0.01	0.8251	0.3918	0.4902	0.3848	0.4829	0.4625
	0.02	0.8266	0.3902	0.4901	0.3836	0.4848	0.4576
	0.05	0.8292	0.3876	0.4899	0.3829	0.4870	0.4482
	0.10	0.8314	0.3854	0.4898	0.3828	0.4890	0.4375
	0.20	0.8337	0.3821	0.4894	0.3828	0.4891	0.4233

[a]As determined by the moving boundary method. See H. S. Harned and B. B. Owen, "The Physical Chemistry of Electrolyte Solutions," 3rd ed. Van Nostrand–Reinhold, Princeton, New Jersey, 1958.

that these vary with concentration. The reason is that λ_+ and λ_- vary differently with concentration.

EXAMPLE

As a first exercise, we can calculate λ_{Na^+} in 0.001 M Na_2SO_4 if $\lambda_{SO_4^{2-}}$ is 159.6 cm^2 mol^{-1} ohm^{-1} and t_{Na^+} is 0.3857. It follows from Eq. (12-51) that $\Lambda_{m,\ Na_2SO_4}$ = 159.6/(1 − 0.3857) = 259.8 cm^2 mol^{-1} ohm^{-1}. Equation (12-40) then gives λ_{Na^+} = (259.8 − 159.6)/2 = 50.1 cm^2 mol^{-1} ohm^{-1}.

We next suppose that the solution is also 3 × 10^{-4} M in $La_2(SO_4)_3$, with $\lambda_{La^{3+}}$ = 208.8 cm^2 mol^{-1} ohm^{-1}, and ask for t_{Na^+} in this mixed electrolyte solution. The total SO_4^{2-} concentration is 1.9 × 10^{-3} M and the La^{3+} concentration is 6 × 10^{-4} M. Application of Eq. (12-50) gives

$$t_{Na^+} = \frac{(2 \times 10^{-3})(50.1)}{(2 \times 10^{-3})(50.1) + (6 \times 10^{-4})(208.8) + (1.9 \times 10^{-3})(159.6)}$$

$$= 0.1895$$

Note: the data used are those for infinite dilution; the solutions are not actually sufficiently dilute for this use to be very accurate. Not only will the actual molar conductivities be different from the limiting values, but those for Na$^+$ and SO$_4^{2-}$ ions will be different for the mixed solution than for 0.001 M Na$_2$SO$_4$ alone, because of the different ionic environment.

The fraction of current carried by each ion of an electrolyte, and hence the transference numbers, may be determined directly with a special type of electrolysis cell. The procedure is known as the *Hittorf method* (introduced in 1853) and a Hittorf cell is sketched in Fig. 12-8. The cell is constructed so that the three compartments may be isolated from each other and drained separately. The basis of the method is that the change in the amount of electrolyte in either end compartment depends both on the electrolysis reaction and on the number of ions that have migrated in or out in the process of carrying current.

The cell shown in the figure is filled with silver nitrate solution and has silver electrodes. The general situation is that since the mobilities of Ag$^+$ and NO$_3^-$ ions are about equal, each carries about half the current. Thus for each

B. Transference numbers by the Hittorf method

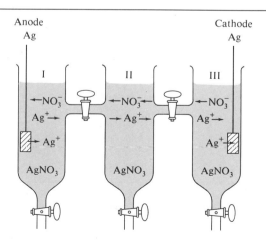

FIGURE 12-8
Schematic drawing of a Hittorf transference cell.

faraday of electricity put through the cell, $\frac{1}{2}$ equiv of Ag^+ ions will pass from left to right across the dividing line I–II between cells I and II and $\frac{1}{2}$ equiv of NO_3^- ions will pass across the line I–II from right to left. Oxidation occurs at the anode, so that, per faraday, 1 equiv of Ag^+ ions is delivered into cell I by the silver electrode. Between the gain by electrolysis and the loss by migration, there is a net gain of $\frac{1}{2}$ equiv of Ag^+ ions in cell I. Since the NO_3^- ion is not involved in the electrode reaction, the gain of $\frac{1}{2}$ equiv by migration is net. Cell I will thus show an overall gain of $\frac{1}{2}$ equiv of silver nitrate per faraday.

The details of the analysis of the Hittorf method are given in the Special Topics section. It is sufficient here to note that analysis of the changes in amounts present in the electrode compartments of a Hittorf cell allows the calculation of the fraction of current carried by each ion, and hence the determination of its transference number.

C. Ionic molar conductivities

There are two principal methods for obtaining ionic molar conductivities. The first is by direct measurement of the ionic mobility, as in the moving boundary experiment, and the second is through a determination of the cation or anion transference number for an electrolyte of known equivalent conductivity, as by means of the Hittorf method.

As in the case of the molar conductivity of an electrolyte, the usual quantity tabulated is the molar conductivity of the ion at infinite dilution. This value is characteristic of the isolated ion free of long-range interionic attraction effects. A number of such values are given in Table 12-5. These are, of course, parallel to the mobilities in Table 12-3, being related by Eq. (12-38). The same general comments apply here as were made in Section 12-5C.

TABLE 12-5
Molar conductivities of aqueous ions at 25°C[a]

Cation	Molar conductivity[b] $(cm^2\ mol^{-1}\ ohm^{-1})$	Anion	Molar conductivity[b] $(cm^2\ mol^{-1}\ ohm^{-1})$
H^+	349.8	OH^-	198.3
Li^+	38.68	F^-	55.4
Na^+	50.10	Cl^-	76.35
K^+	73.50	Br^-	78.14
Rb^+	77.81	I^-	76.84
Cs^+	77.26	NO_3^-	71.46
Ag^+	61.90	ClO_4^-	67.36
NH_4^+	73.55	HCO_3^-	44.50
Be^{2+}	90	CO_3^{2-}	138.6
Mg^{2+}	106.1	SO_4^{2-}	160.0
Ca^{2+}	118.9	$HCOO^-$	54.59
Ba^{2+}	127.3	$HC_2O_4^-$	40.2
La^{3+}	209	$C_2O_4^{2-}$	148.3
$Co(NH_3)_6^{3+}$	306	CH_3COO^-	40.90

[a]See R. A. Robinson and R. H. Stokes, "Electrolyte Solutions," 2nd ed. (rev.), Butterworths, London (1970).
[b]To convert to SI units, multiply by 10^{-4}. Thus $\lambda_{Na^+} = 50.10\ cm^2\ mol^{-1} = 50.10 \times 10^{-4}\ m^2\ mol^{-1}\ ohm^{-1}$.

12-7   Activities and activity coefficients of electrolytes

An electrolyte, like any other solute, tends to give nonideal solutions, approaching Henry's law behavior at infinite dilution. We know that at high dilution the positive and negative ions act independently. The colligative property effects report, for example, the number of particles expected from the complete dissociation of the electrolyte. On the other hand, it is not possible to vary a single ion concentration, keeping everything else constant. An attempt to do so would immediately result in the solution acquiring an enormous electrostatic charge. An excess of even 10^{-10} mol liter^{-1} of one kind of ion over another would result in a static charging of the solution to a potential of about 10^6 V! In other words, we cannot prepare a solution containing only one kind of ion and therefore cannot determine individual ion activities or activity coefficients; we can only observe a mean value for the positive and negative ions present.

A. Introductory comments

The situation is illuminated if we consider the case of a solubility product equilibrium. We write for a slightly soluble salt MX the solubility equilibrium

$$MX(s) = MX(\text{solution}) = M^+ + X^- \tag{12-52}$$

The thermodynamic criterion for equilibrium is satisfied if we write

$$\mu_{MX}(s) = \mu_{MX}(\text{solution})$$

as required by Eq. (9-36). We can introduce the activity of the electrolyte species:

$$\mu_{MX}(\text{solution}) = \mu_{MX}^\circ(\text{solution}) + RT \ln a_{MX} \tag{12-53}$$

The activity a_{MX} corresponds to the solute activity a_2 in the equations of Chapter 10. It can be obtained, for example, by application of the Gibbs–Duhem integration procedure to solvent activities as determined from colligative property measurements. As with solutes generally, we use a Henry's law standard state, usually the one based on molality as a concentration unit (see Section 10-6).

The treatment up to this point is rather unsatisfactory, however. It does not tell us that the electrolyte is dissociated or how a_{MX} is apt to be influenced by the presence of a common ion in the solution. Returning to the solubility equilibrium, the normal way of writing the equilibrium constant for process (12-52) is in terms of a solubility product:

$$K_{SP} = (M^+)(X^-) \tag{12-54}$$

We know that the constant K_{sp} is well–behaved in dilute solutions. For example, while we cannot avoid having essentially equal numbers of positive and negative ions present in any solution, we can, by using mixed electrolytes, vary the concentrations of specific kinds of ions, such as M^+ or X^-, independently. If S denotes the solubility of MX(s), then in pure water we have

$$K_{sp} = S^2 \tag{12-55}$$

If added X^- ion is present, as in the form of a concentration C of NaX, then Eq. (12-54) becomes

$$K_{sp} = (S)(S + C) \tag{12-56}$$

We now have the common ion effect whereby added X^- ion depresses the solubility of MX(s). Since the solution is still in equilibrium with the solid, a_{MX} must not have changed, even though the individual values of (M^+) and (X^-) are now quite different from before. Thus observation tells us that in dilute solution

$$a_{MX} = (M^+)(X^-) \tag{12-57}$$

Use of Eq. (12-57) allows a more realistic treatment of colligative effects in dilute solution. The Gibbs–Duhem integration gives $\ln a_{MX}$, and for a single electrolyte solute MX, $(M^+) = (X^-) = m$. We then have $\ln a_{MX} = \ln m^2 = 2 \ln m$, and the experimentally observed factor of 2 has now appeared.

Consider, for example, the osmotic pressure effect. The basic equation is

$$-\ln a_1 = \frac{V_1}{RT} \Pi \qquad \text{[Eq. (10-21)]}$$

and this is to be used with the Gibbs–Duhem relation

$$x_1 \, d(\ln a_1) + x_2 \, d(\ln a_2) = 0 \qquad \text{[Eq. (9-94)]}$$

where x_1 and x_2 must denote the mole fraction of components 1 (solvent) and 2 (salt MX). It is simpler for the present illustration to use mole fraction rather than molality for the salt concentration, and so we write $a_{MX} = (x_M^+)(x_X^-) = x_2^2$ (rather than $a_{MX} = m^2$). This amounts to using the first of the Henry's law conventions of Section 10-6. Since $d(\ln a_{MX}) = 2 \, d(\ln x_2)$, combination of Eqs. (10-21) and (9-94) gives

$$2x_2 \, d(\ln x_2) = -x_1 \, d(\ln a_1) = \frac{V_1}{RT} x_1 \, d\Pi \tag{12-58}$$

or

$$\frac{V_1}{RT} \Pi = 2 \int_0^{x_2} \frac{x_2}{x_1} \, d(\ln x_2) = 2 \int_0^{x_2} \frac{dx_2}{x_1}$$

The integral of dx_2/x_1 is $-\ln(1 - x_2)$ and since the solution is to be dilute, this becomes just x_2. The final result is

$$\frac{V_1}{RT} \Pi = 2x_2 \tag{12-59}$$

Thus the osmotic pressure is predicted to be twice the value expected just from the mole fraction of the salt, or the van't Hoff i factor comes out equal to 2, as observed.

B. Defining equations for activity and activity coefficient

The preceding analysis was presented to show that the activity of an electrolyte is equal to the product of the ion concentrations in very dilute solutions. For the more general case of a nonideal solution we therefore write

$$a_2 = a_+ a_- \tag{12-60}$$

where a_2 denotes the electrolyte activity and a_+ and a_- the individual ion activities, which become equal to m_+ and m_-, respectively, in the limit of infinite dilution. We then further define the activity coefficients γ_+ and γ_-:

$$a_+ = \gamma_+ m_+ \quad \text{and} \quad a_- = \gamma_- m_- \tag{12-61}$$

Since a_2 involves the square of a concentration, it is convenient to define a new activity $a_\pm$ as the square root of a_2:

$$a_2 = a_\pm{}^2 \tag{12-62}$$

The activity $a_\pm$ is known as the *mean activity*. Similarly, $\gamma_\pm$ is called the *mean activity coefficient*:

$$\gamma_\pm{}^2 = \gamma_+ \gamma_- \tag{12-63}$$

and $m_\pm$ the *mean molality*:

$$m_\pm{}^2 = m_+ m_- \tag{12-64}$$

Then

$$a_\pm = \gamma_\pm m_\pm \tag{12-65}$$

This set of definitions is for the specific case of a 1–1 electrolyte, that is, for an electrolyte which produces 1 mole of ions of each kind per formula weight. The general treatment is suggested by consideration of the solubility product for a salt $M_{\nu_+} X_{\nu_-}$:

$$M_{\nu_+} X_{\nu_-}(s) = \nu_+ M^{z+} + \nu_- X^{z-} \qquad K_{sp} = (M^{z+})^{\nu_+}(X^{z-})^{\nu_-} \tag{12-66}$$

where z_+ and z_- are the respective ion charges and ν_+ and ν_- are the numbers of ions of each type. Electroneutrality requires that

$$\nu_+ z_+ = \nu_- z_-. \tag{12-67}$$

We want to define the activity of this general electrolyte in terms of individual ion activities such that in the limit of infinite dilution we obtain the expression on the right-hand side of Eq. (12-66). The definitions are then

$$a_2 = a_+{}^{\nu_+} a_-{}^{\nu_-} \tag{12-68}$$

$$a_\pm{}^\nu = a_+{}^{\nu_+} a_-{}^{\nu_-} \qquad \text{where } \nu = \nu_+ + \nu_- \tag{12-69}$$

$$\alpha_+ = \gamma_+ m_+ \qquad a_- = \gamma_- m_- \qquad \text{[Eq. (12-61)]}$$

$$\gamma_\pm{}^\nu = \gamma_+{}^{\nu_+} \gamma^{\nu_-} \tag{12-70}$$

$$m_\pm{}^\nu = m_+{}^{\nu_+} m_-{}^{\nu_-} \tag{12-71}$$

and, as before,

$$a_\pm = \gamma_\pm m_\pm \qquad \text{[Eq. (12-65)]}$$

The complications introduced by these definitions are regrettable. They develop naturally, however, when we deal with nonideal electrolyte solutions.

C. Activity coefficients from solubility measurements

The preceding material allows us to write the thermodynamic equilibrium constant for the solubility equilibrium of an electrolyte. Thus for AgCl we have

$$AgCl(s) = Ag^+ + Cl^- \tag{12-72}$$

$$K_{th} = a_{Ag^+}a_{Cl^-} = (Ag^+)(Cl^-)\,\gamma_{Ag^+}\gamma_{Cl^-} \tag{12-73}$$

or

$$K_{th} = K_{sp}\gamma_\pm^{2} \tag{12-74}$$

where K_{th} is the thermodynamic solubility product and K_{sp} is the usual form in which concentrations are used. The solubility of AgCl in water is about 10^{-5} m and it seems likely that $\gamma_\pm$ will be unity for so low a concentration. We can investigate the situation by adding some neutral or noncommon-ion electrolyte. We find experimentally that on doing so the solubility of AgCl increases, and hence so does its K_{sp} value. Evidently $\gamma_\pm$ is varying with the concentration of added neutral electrolyte, since K_{th} must remain a constant.

A useful way of graphing such data follows if we write Eq. (12-74) in the form

$$\log K_{sp} = \log K_{th} - 2 \log \gamma_\pm \tag{12-75}$$

We then plot $\log K_{sp}$ against some function of concentration. Figure 12-9 gives the results of measurements that show the effect of added KNO_3 and $MgSO_4$ on the solubility and hence the K_{sp} for AgCl. As with equivalent conductivity, the experimental observation is that a plot is nearly linear if the square root of the added electrolyte concentration is used. The theoretical explanation is given in the next section, as is the definition of ionic strength, I.

Since we are using the Henry's law convention for activities and activity coefficients, $\gamma_\pm$ approaches unity at infinite dilution. The intercept of Fig. 12-9 therefore gives $\log K_{th} = -9.790$, or $K_{th} = 1.62 \times 10^{-10}$. We may then calculate $\gamma_\pm$ for the ions Ag^+ and Cl^- in the presence of added KNO_3 by

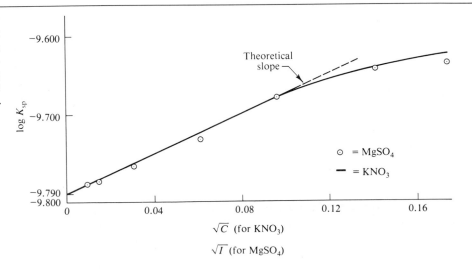

FIGURE 12-9
Variation of K_{sp} for AgCl at 25°C with increasing KNO_3 concentration (plotted as $\sqrt{m}$) and increasing $MgSO_4$ concentration (circles, plotted as $\sqrt{I}$).

inserting K_{th} and the measured K_{sp} into Eq. (12-75). Thus log $K_{sp} = -9.645$ at $\sqrt{m} = 0.175$ or $m = 0.0306$; then log $\gamma_{\pm} = [(-9.790) - (-9.645)]/2 = -0.0725$, whence $\gamma_{\pm} = 0.846$. The data allow a tabulation of $\gamma_{\pm}$ versus concentration of added KNO_3 (strictly speaking, m includes the contribution of the dissolved AgCl to the total salt concentration).

Studies such as the preceding were carried out for a number of systems, especially by V. LaMer and co-workers, and these confirmed an earlier observation by G. Lewis that the activity coefficient of an electrolyte depends mainly on the total concentration of ions and only secondarily on their specific chemical natures. Thus for concentrations below about 0.05 m, the data of Fig. 12-9 would be essentially the same if $NaNO_3$, $KClO_4$, and so on were used instead of KNO_3. The results, incidentally, would also have been the same if NaCl were the added electrolyte. Although there is now a common-ion effect, depressing the solubility of the AgCl, the experimental K_{sp} values will still show the same variation with concentration of total electrolyte present.

D. The ionic strength principle

Empirical observation showed, however, that if other than uni-univalent salts were used, there was an increased effect on activity coefficients. It was found that all types of electrolytes can be put on a common basis by expression of the ionic concentration in terms of a quantity called the ionic strength I, where

$$I = \tfrac{1}{2} \sum_i m_i z_i^2 \tag{12-76}$$

In the case of a uni-univalent electrolyte, $I = m$, but for, say K_2SO_4, $I = \frac{1}{2}(2m + 4m) = 3m$, while for $MgSO_4$, $I = \frac{1}{2}(4m + 4m) = 4m$. The increase in solubility of a salt in the presence of any electrolyte is nearly the same if the results are plotted in terms of I rather than m. The circles in Fig. 12-9 are the points obtained with $MgSO_4$ as the neutral salt and with the abscissa reading $\sqrt{I}$ rather than $\sqrt{m}$.

The discovery of the *ionic strength principle* was a major step toward the understanding of nonideality effects in dilute electrolyte solutions. The principle constituted strong evidence that it was the charges on ions and not their particular chemical natures that determined activity coefficients. It paved the way for the development of the interionic attraction theory (described in Section 12-8).

The change in solubility of a salt with ionic strength provides one means for determination of the activity coefficient of the salt in the presence of some other electrolyte. The method is not applicable to soluble salts, and, moreover, necessarily gives activity coefficients in mixed electrolyte solutions. The activity coefficient of an electrolyte in solutions containing no other ionic species is of more general importance.

E. Methods of determining activity coefficients

As indicated in Section 12-7A, a_2, the solute activity, may be obtained from colligative property measurements through use of the Gibbs–Duhem equation. Equation (12-58) would be written with $d(\ln a_2)$ rather than $d(\ln x_2)$, and then manipulated into the form of Eq. (9-95). The detailed procedures for simplifying the handling of this integration may be found in more specialized texts [for example, Lewis and Randall (1961)]. Much of our activity coefficient data comes from measurements of the emf of electrochemical cells. This approach is discussed in detail in Chapter 13 and will not be reviewed here.

Activity coefficients may also be obtained from a study of ionic equilibrium. As a specific example, Table 12-6 gives the concentration equilibrium constant for the dissociation of acetic acid as determined from conductivity measurements at various concentrations. The concentration equilibrium constant is:

$$HAc = H^+ + Ac^- \qquad K = \frac{\alpha^2 C}{1 - \alpha} \qquad (12\text{-}77)$$

and α in the table has been corrected for the ion atmosphere effect on Λ, as described in Section 12-3C. The residual variation in K is attributed to nonideality, and Eq. (12-77) is written

$$K_{th} = K\frac{\gamma_{H^+}\gamma_{Ac^-}}{\gamma_{HAc}} = KK_\gamma \qquad (12\text{-}78)$$

The solutions are dilute enough that, as a nonelectrolyte, HAc is probably in the Henry's law region of behavior so $\gamma_{HAc} = 1$. Therefore $K_\gamma = \gamma_{\pm}^2$, and Eq. (12-78) may be plotted according to the form

$$\log K = \log K_{th} - 2 \log \gamma_\pm \qquad (12\text{-}79)$$

The ionic strength derives entirely from the H^+ and Ac^- ions, whose concentration is αC, and the plot of $\log K$ versus $(\alpha C)^{1/2}$ is shown in Fig. 12-10. Extrapolation to zero αC gives $\log(10^5 K_{th}) = 0.2440$, or $K_{th} = 1.754 \times 10^{-5}$. Insertion of this value into Eq. (12-78) allows K_γ, and hence $\gamma_\pm$, to be calculated for each concentration. Thus for 0.050 m HAc, $\gamma_\pm^2 = 1.754 \times 10^{-5}/1.849 \times 10^{-5} = 0.949$, or $\gamma_\pm = 0.97$.

The same approach would be used in the case of an equilibrium not involving ionic species. Thus for the esterification reaction in aqueous solution,

$$CH_3COOH + C_2H_5OH = CH_3COOC_2H_5 + H_2O$$

one again writes

$$K_{th} = \frac{(CH_3COOC_2H_5)(H_2O)}{(CH_3COOH)C_2H_5OH} \frac{\gamma_{CH_3COOC_2H_5}\gamma_{H_2O}}{\gamma_{CH_3COOH}\gamma_{C_2H_5OH}} = KK_\gamma$$

and extrapolates the measured values of K to infinite dilution. Pure water is taken to be in its standard state so that γ_{H_2O} approaches unity at infinite

TABLE 12-6	$C \times 10^3$ (M)	$\alpha C \times 10^4$	$K \times 10^5$ (M)
Dissociation constant of acetic acid at 25°C[a]	0.028014	0.151	1.768
	0.2184	0.540	1.781
	2.4140	2.00	1.809
	9.8421	4.12	1.834
	20.000	5.96	1.840
	50.000	9.50	1.849

[a]Adapted from S. Glasstone, "Textbook of Physical Chemistry," p. 955. Van Nostrand–Reinhold, Princeton, New Jersey, 1946.

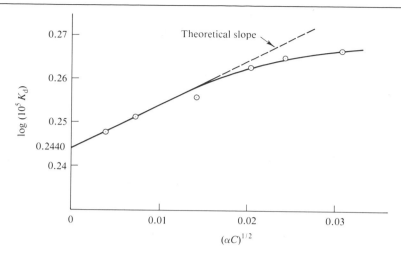

FIGURE 12-10
Variation of K_d for acetic acid with concentration at 25°C.

dilution of the other species, and the extrapolated value of K is then equal to K_{th}. We are primarily concerned with rather dilute solutions in this chapter, and activity coefficients for nonionic species are close to unity. It is only in the case of ions that the long-range Coulomb interactions lead to nonideal behavior even in quite dilute solutions ($C < 0.1m$).

As noted earlier, a number of methods may be used for the determination of the activity coefficient of an electrolyte. The results have been collected and tables of standard activity coefficients are available. A selection of such results is given in Table 12-7 and the mean activity coefficients of some typical electrolytes are plotted against concentration in Fig. 12-11.

The points to notice are the following. First, of course, even 0.1 m solutions can be drastically nonideal in their behavior. Second, electrolytes of a given charge type tend to show similar variations of $\gamma_\pm$ with concentration at first

F. Activity coefficients

Electrolyte	Mean molal activity coefficient								
	0.001 m	0.005 m	0.01 m	0.05 m	0.1 m	0.5 m	1.0 m	2.0 m	4.0 m
HCl	0.965	0.928	0.904	0.830	0.796	0.757	0.809	1.009	1.762
NaCl	0.966	0.929	0.904	0.823	0.778	0.682	0.658	0.671	—
KCl	0.965	0.927	0.901	0.815	0.769	0.650	0.605	0.575	0.582
HNO$_3$	0.965	0.927	0.902	0.823	0.785	0.715	0.720	0.783	0.982
KNO$_3$	—	—	—	—	0.733	0.542	0.548	0.481	—
AgNO$_3$	—	0.92	0.90	0.79	0.72	0.51	0.40	0.28	—
NaOH	—	—	—	0.82	—	0.69	0.68	0.70	0.89
H$_2$SO$_4$	0.830	0.639	0.544	0.340	0.265	0.154	0.130	0.124	0.171
K$_2$SO$_4$	0.89	0.78	0.71	0.52	0.43	—	—	—	—
BaCl$_2$	0.88	0.77	0.72	0.56	0.49	0.39	0.39	—	—
CuSO$_4$	0.74	0.53	0.41	0.21	0.16	0.068	0.047	—	—
K$_4$Fe(CN)$_6$	—	—	—	0.19	0.14	0.067	—	—	—

TABLE 12-7
Mean activity coefficients for aqueous electrolytes at 25°C[a]

[a]Adapted from W. M. Latimer, "The Oxidation States of the Elements and Their Potentials in Aqueous Solutions," 2nd ed. Prentice-Hall, Englewood Cliffs, New Jersey, 1952, and H. S. Harned and B. B. Owen, "The Physical Chemistry of Electrolytic Solutions," 3rd ed. Van Nostrand–Reinhold, Princeton, New Jersey, 1958.

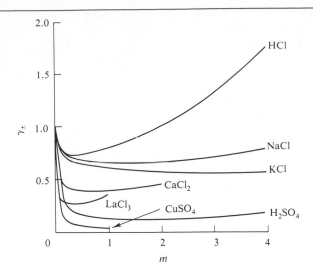

but deviate from each other at higher concentrations. Third, the higher the product z_+z_-, the earlier the deviations from ideality set in. The differences within a given family of the same z_+z_- product appear to be due to several causes. Extensive hydration of the ions ties up the solvent water and makes the true mole fraction higher than the apparent one; a similar explanation was offered for the behavior of aqueous sucrose solutions (note Table 10-3). This may be a major reason for the rather high activity coefficients of electrolytes such as HCl in concentrated solutions. A second factor is that many electrolytes are not fully dissociated. Although a chemical bond is not ordinarily expected to form in the case of electrolytes involving a rare gas type of ion such as K^+, ions may associate strongly as an ion pair. Salts such as $CuSO_4$, however, may actually form a coordinate bond in their association. Acids such as H_2SO_4 are relatively weak, and are not fully dissociated except in dilute solution. If association occurs for any of these reasons, the result is generally to lower the activity coefficient of the electrolyte.

A less chemically specific effect is that the actual size of the ion becomes important in concentrated solutions. The result is an increase in activity coefficient over what it would otherwise be.

These various explanations are important in the sense that they provide some rationale for an otherwise bewildering variety of behavior. They are discussed in somewhat more detail in the Commentary and Notes section. On the other hand, an activity coefficient is a *phenomenological* quantity; when multiplied by the mean molality, it gives the mean activity of the electrolyte and therefore its chemical potential, regardless of explanation.

12-8   The Debye–Hückel theory

Repeated remarks have been made throughout this chapter about the effect of the long-range Coulomb interactions between ions on both their transport behavior or mobilities and their activity coefficients. The theoretical approach to the former effect is mentioned further in Section 12-ST-1. The same interactions that affect ion mobilities also affect their activity coefficients, and the treatment of this second effect is important enough to cover here.

The theory of Debye and Hückel (1923) constituted a major breakthrough. Earlier observations, especially by G. Lewis, had shown that the activity coefficients of electrolytes were, in dilute solutions, determined more by ionic strength than by any specific chemical property. The implication that nonideality effects are due mainly to long-range Coulomb interactions between ions was accepted, but the problem of treating a collection of ions seemed insurmountable. The normal statistical mechanical approach would require one somehow to list the energy states of each ion while recognizing that its energy was affected by the location of all other ions.

The path that Debye and Hückel found was through the assumption that there exists some average potential around each ion and that each potential field is independent, that is, one does not perturb another. Further assumptions were that the electrolyte is completely dissociated and, at first, that the ions are of negligible size. It was also assumed that electrical interactions are solely responsible for deviations from nonideality. The treatment is therefore limited to rather dilute solutions.

We proceed as follows, taking for simplicity the electrolyte to be of the uni-univalent type. It is assumed that there is some average potential ψ which is a function of distance r from any particular ion. The potential energy of an ion in this potential is $e\psi$, where e is the electronic charge. Since we are dealing with Coulomb's law, ψ will be in esu units and likewise e. The probability of finding a positive ion in a region of potential ψ around a particular ion of like charge is given by the Boltzmann principle:

$$\mathbf{n}_+ = \mathbf{n}e^{-e\psi/kT} \tag{12-80}$$

where $\mathbf{n}$ is the average concentration in molecules per cubic centimeter. Similarly

$$\mathbf{n}_- = \mathbf{n}e^{e\psi/kT} \tag{12-81}$$

The net charge density ρ is then

$$\rho = (\mathbf{n}_+ - \mathbf{n}_-)e = \mathbf{n}e(e^{-e\psi/kT} - e^{e\psi/kT}) \tag{12-82}$$

The next major assumption, without which further progress would have stopped, was that a theorem from electrostatics, known as the Poisson equation, could be used. This states that the rate of change or the divergence of the gradient of the electrostatic potential at a given point is proportional to the charge density at that point. The equation is valid for a continuous medium of uniform dielectric constant D and the equation for spherical coordinates is

$$\nabla^2\psi = \frac{1}{r^2}\frac{d}{dr}\left(r^2\frac{d\psi}{dr}\right) = -\frac{4\pi\rho}{D} \tag{12-83}$$

where ∇^2 is known as the Laplacian operator, *del squared*. In rectangular coordinates, ∇^2 is

$$\nabla^2\psi = \frac{\partial^2\psi}{\partial x^2} + \frac{\partial^2\psi}{\partial y^2} + \frac{\partial^2\psi}{\partial z^2} \tag{12-84}$$

but the polar coordinate system is useful here since we assume ψ to be a function of r only, that is, to be spherically symmetric. We require here only that part of ∇^2 involving r (see Section 16-7 for the complete expression of ∇^2 in polar coordinates).

Equation (12-83) is written for ψ and ρ in esu; if SI units are to be used, we have

$$\nabla^2\psi = -\frac{\rho}{\epsilon^\circ D} \tag{12-85}$$

where ϵ° is the "permittivity of vacuum" (see Section 3-CN-1). Since we are dealing with a derivation in electrostatics it seems easier to use the esu system.

We now combine Eqs. (12-82) and (12-83) to obtain what has come to be known as the *Poisson–Boltzmann equation*:

$$\nabla^2\psi = -\frac{4\pi ne}{D}\left(e^{-e\psi/kT} - e^{e\psi/kT}\right) \tag{12-86}$$

The assumptions implicit up to this point (such as the independence of the potential fields around each ion) require that the electrostatic effect not be a large one, that is, that $e\psi/kT$ be a small number. We therefore proceed to expand the exponentials, keeping only the first term, to obtain

$$\nabla^2\psi = \left(\frac{8\pi ne^2}{DkT}\right)\psi \tag{12-87}$$

The collection of quantities multiplying ψ on the right-hand side of Eq. (12-87) is assembled into a single parameter κ, defined as

$$\kappa^2 = \frac{4\pi e^2}{DkT}\sum_i z_i^2 \mathbf{n}_i \tag{12-88}$$

Equation (12-88) applies to the general case of a collection of ions of charges z_i and reduces to $\kappa^2 = 8\pi ne^2/DkT$ for a uni-univalent electrolyte. Equation (12-87) then becomes

$$\nabla^2\psi = \kappa^2\psi \tag{12-89}$$

The solution to Eq. (12-89) is

$$\psi(r) = \frac{ze}{Dr}e^{-\kappa r} \tag{12-90}$$

This may be verified by insertion of the expression for $\psi(r)$ back into Eq. (12-89). Thus

$$\frac{d\psi}{dr} = -\frac{ze}{Dr^2}e^{-\kappa r} - \frac{ze\kappa}{Dr}e^{-\kappa r}$$

$$r^2\frac{d\psi}{dr} = -\frac{ze}{D}e^{-\kappa r} - \frac{ze\kappa r}{D}e^{-\kappa r}$$

$$\frac{d}{dr}\left(r^2\frac{d\psi}{dr}\right) = \frac{ze\kappa}{D}e^{-\kappa r} - \frac{ze\kappa}{D}e^{-\kappa r} + \frac{ze\kappa^2 r}{D}e^{-\kappa r}$$

$$= \frac{ze\kappa^2 r}{D}e^{-\kappa r}$$

$$\frac{1}{r^2}\frac{d}{dr}\left(r^2\frac{d\psi}{dr}\right) = \frac{ze\kappa^2}{Dr}e^{-\kappa r} = \kappa^2\psi \qquad \text{Q.E.D.}$$

We again expand the exponential, keeping only the first terms, to get

$$\psi(r) = \frac{ze}{Dr} - \frac{ze}{D}\kappa \qquad (12\text{-}91)$$

The first term on the right is just the potential due to the charge on the ion itself, and we are interested only in the second term, $-ze\kappa/D$, which gives the alteration in the potential due to the distribution of other ions around the given ion. Notice that this second term corresponds to the potential of a charge $-ze$ as observed at a distance $1/\kappa$. The quantity $1/\kappa$ has the dimensions of distance (centimeters in the cgs system) and has come to be known as the *effective* or *equivalent radius of the ion atmosphere*. The simple physical picture, illustrated in Fig. 12-12, is thus one of an ion of charge ze having a statistical excess of ions of opposite charge around it, the excess amounting to just $-ze$ and behaving as though it were located on a spherical shell of radius $1/\kappa$.

It is next necessary to find the free energy associated with this extra potential originating from the ion atmosphere. This is done by calculating the reversible work needed to form the atmosphere: We integrate the product of potential times charge as the ion is allowed to build up its charge from zero to the full value:

$$\text{electrical work} = \int_0^{ze} \psi\text{atm}\, d(ze) = \int_0^{ze} \left(-\frac{ze\kappa}{D}\right) d(ze)$$

or

$$w_{el} = -\frac{1}{2}\frac{\kappa}{D}(ze)^2$$

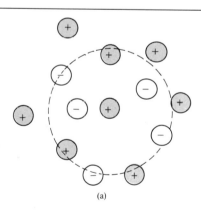

(a)

(b)

FIGURE 12-12
(a) A distribution of ions around a given positive ion. The dashed circle is drawn to indicate that there will tend to be an excess of negative over positive ions in the vicinity of a given positive ion. (b) In Debye–Hückel theory this excess negative charge acts as though it were located on a spherical shell of radius $1/\kappa$.

This electrical work then contributes to the chemical potential μ_i of the ith ion:

$$\mu_i = \mu_i^\circ + kT \ln a_i + w_{el}$$

Since the whole derivation is for a very dilute solution, it seems safe to assume $a_i = m_i$, that is, that the ion obeys Henry's law *apart from the electrical contribution*. This last is the source of the observed deviation from ideality, reported experimentally in terms of an activity coefficient. The conclusion is then

$$w_{el} = kT \ln \gamma_i = -\frac{z_i^2 e^2 \kappa}{2D}$$

or

$$\ln \gamma_i = -\frac{z_i^2 e^2 \kappa}{2DkT} \tag{12-92}$$

We next need to find the mean activity coefficient for the case of an electrolyte having just two kinds of ions. From Eq. (12-70) we have

$$\ln \gamma_\pm = \frac{\nu_+}{\nu} \ln \gamma_+ + \frac{\nu_-}{\nu} \ln \gamma_- \tag{12-93}$$

Algebraic manipulation of Eqs. (12-93) and (12-67) gives

$$\ln \gamma_\pm = -|z_+ z_-| \frac{e^2 \kappa}{2DkT} \tag{12-94}$$

Finally, κ can be related to the ionic strength I [Eq. (12-76)] since $\mathbf{n}_i = (C_i/1000)N_A$ and in dilute solution $C_i = m_i \rho_0$, where ρ_0 is the solvent density. Equation (12-88) becomes

$$\kappa^2 = \frac{8\pi e^2 N_A^2 \rho_0}{1000 DRT} I \tag{12-95}$$

and so Eq. (12-94) becomes

$$\ln \gamma_\pm = -|z_+ z_-| \frac{e^3 N_A^2}{DRT} \left(\frac{2\pi \rho_0}{1000 DRT}\right)^{1/2} I^{1/2}$$

$$= -A|z_+ z_-| I^{1/2} \tag{12-96}$$

Insertion of the values for the general constants (e in esu and R in erg mol^{-1} K^{-1}) gives

$$\ln \gamma_\pm = -4.198 \times 10^6 |z_+ z_-| \left(\frac{\rho_0}{D^3 T^3}\right)^{1/2} I^{1/2} \tag{12-97}$$

For water at 25°C, $D = 78.54$ and Eq. (12-97) reduces to*

$$\ln \gamma_\pm = -1.172 |z_+ z_-| I^{1/2} \tag{12-98}$$

---

*As noted earlier, the derivation is made in the cgs/esu system, as the more natural one to use. The equations are the same in SI except that D is everywhere multiplied by $4\pi\varepsilon_0$ (see Section 3-CN-1 and, for more detail, Section 8-ST-1) and the factor 1000 disappears from the denominators of Eqs. (12-95) and (12-96). Ionic strength remains in m (mol kg^{-1} of solvent) but ρ_0 is now in kg m^{-3} and $1/\kappa$, the ion atmosphere radius, in meters.

Equation (12-98) is often called the *limiting law* of interionic attraction theory, since it approaches exact validity only in the limit of concentration approaching zero.

The Debye-Hückel treatment leads, first of all, to a theoretical explanation of the empirical observation that the activity coefficient of an electrolyte is determined by the ionic strength I of the medium. The quantitative predictions, as for example, from Eq. (12-98), have been well verified for uni-univalent electrolytes. The theoretical slopes are included in Figs. 12-9 and 12-10, and the experimental points approach agreement at low concentrations. One may make a more direct check by plotting the activity coefficient data of Table 12-7 according to Eq. (12-98). This is done in Fig. 12-13, and it is apparent that the uni-univalent electrolytes approach agreement with theory. It is less clear whether accurate agreement occurs in the case of the higher charge types, although it is assumed that it does in sufficiently dilute solution. See the next section for further discussion.

EXAMPLE

Calculate the ion atmosphere radius and the mean activity coefficient for 0.1 m Na$_2$SO$_4$ at 25°C. First, by Eq. (12-76), $I = \frac{1}{2}[(0.2)(1)^2 + (0.1)(2)^2] = 0.3$. We use Eq. (12-95) to obtain

$$\kappa^2 = \frac{(8)(\pi)(4.803 \times 10^{-10})^2(6.022 \times 10^{23})^2\,(1)(0.3)}{(1000)(78.54)(8.3144 \times 10^7)(298.15)} = 3.240 \times 10^{14}\ \text{cm}^{-2}$$

whence $1/\kappa = 5.56$ Å. For the same calculation in SI, we use $4\pi\epsilon_0 D$ in place of D and omit the 1000 in the denominator, to obtain

$$\kappa^2 = \frac{(2)(1.6022 \times 10^{-19})^2(6.022 \times 10^{23})^2(1 \times 10^3)(0.3)}{(8.8542 \times 10^{-12})(78.54)(8.3144)(298.15)} = 3.240 \times 10^{18}\ \text{m}^{-2}$$

or $1/\kappa = 5.56 \times 10^{-10}$ m. We can obtain the activity coefficient directly from Eq. (12-98): $\ln \gamma_\pm = -1.172\,(1 \times 2)(0.3)^{1/2} = -1.2839$ and $\gamma_\pm = 0.277$. (The experimental value is 0.452.)

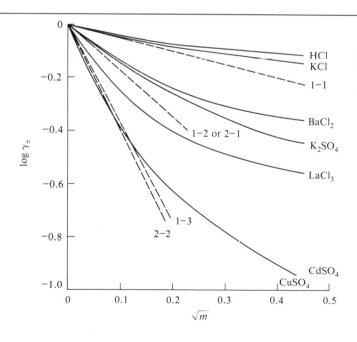

FIGURE 12-13
Plot of activity coefficient data at 25°C so as to test the Debye–Hückel theory. Dashed lines give the various limiting law slopes.

12-9 Activity coefficients for other than dilute aqueous solutions

It is noted at the end of the preceding section that the activity coefficients of uni-univalent ions do approach the Debye–Hückel limiting law in dilute solutions, and that the values for other valence types are assumed to do so at extreme dilution. In effect, however, the simple Debye–Hückel treatment is valid only for slightly "contaminated" water, and a great deal of effort has been devoted to the physical chemical treatment of more concentrated aqueous solutions and of nonaqueous solutions. The first improvement was with the recognition of the finite size of ions, based on the assumption that there is some distance a of closest approach; the effect is to modify ψ in such a way as to lead to the following equation:

$$\ln \gamma_i = -\frac{z^2 e^2}{2DkT} \frac{\kappa}{1 + \kappa a} \tag{12-99}$$

or

$$\ln \gamma_\pm = -\frac{AI^{1/2}}{1 + BI^{1/2}} \tag{12-100}$$

where A is as defined in Eq. (12-96) and B is given by

$$B = \frac{35.56}{(DT)^{1/2}} \mathring{a} \tag{12-101}$$

where $\mathring{a}$ is now in Ångstroms. For water at 25°C, $B = 0.232\mathring{a}$. Equation (12-100) fits activity coefficient data for uni-univalent electrolytes, up to perhaps 0.1 m, with reasonable values for $\mathring{a}$.

The next efforts were in various different directions. V. LaMer and others attempted to solve the Poisson–Boltzmann equation (12-86) more exactly, but encountered great mathematical difficulties. Although major departures from the simple theory were shown to be expected, especially for asymmetric electrolytes, such as those of the 1–2 type, the extended terms that they obtained are hard to use and are not very satisfactory. A semiempirical equation due to Brønsted suggests that at higher concentrations log $\gamma_\pm$ should be proportionnal to m, and a commonly used form is

$$\ln \gamma_\pm = -\frac{AI^{1/2}}{1 + BI^{1/2}} + Cm \tag{12-102}$$

B and C may be treated as empirical constants, but may also be related semi-empirically to parameters such as the hydration number of the cation, that is, the number of bound water molecules (Robinson and Stokes, 1970).

A somewhat different and stimulating approach was that of N. Bjerrum, also of the Copenhagen school, who pointed out that in terms of the Debye–Hückel model, the probability of finding an ion of opposite charge near a given ion went through a minimum at a characteristic distance q:

$$q = \frac{e^2 z_1 z_2}{DkT} \tag{12-103}$$

For a 1–1 electrolyte at 25°C, q is about 3.5 Å and Bjerrum suggested that ions closer than this distance should be regarded as ion pairs. The existence of a distance q arises as somewhat of an accident, however; $\psi(r)$ decreases steadily outward, but the volume of space increases with r^3 and it is the combination of these opposite factors that leads to q as the radius of the spherical shell having a minimum probability of finding a counter ion. The approach, however, established a pattern of thought that has been very useful. See the further discussion in Section 12-CN-1.

12-10   Ionic equilibria

There are two somewhat separate aspects of the treatment of ionic equilibria. The first, introduced in Section 12-7, is that of the determination of the true or thermodynamic equilibrium constant. Recall that the basic procedure is the extrapolation of a concentration equilibrium constant, K_{sp} in the case of solubility and K in the case of homogeneous equilibrium, to infinite dilution. To review this aspect briefly, we write for each species $\mu_i = \mu_i^\circ + RT \ln a_i$, and, following the derivation of Eq. (7-11), obtain

$$\Delta G = (m\mu_M + n\mu_N + \cdots) - (a\mu_A + b\mu_B + \cdots)$$

or

$$\Delta G = \Delta G^0 + RT \ln Q_{th}, \qquad Q_{th} = \frac{a_M{}^m a_N{}^n \cdots}{a_A{}^a a_B{}^b \cdots} \tag{12-104}$$

At equilibrium $\Delta G = 0$, and Q_{th} becomes the thermodynamic equilibrium constant K_{th}:

$$\Delta G^0 = -RT \ln K_{th} \tag{12-105}$$

It is convenient for us to write each activity as a product of concentration and activity coefficient, to obtain

$$K_{th} = \frac{(M)^m(N)^n \cdots}{(A)^a(B)^b \cdots} \frac{\gamma_M{}^m \gamma_N{}^n \cdots}{\gamma_A{}^a \gamma_B{}^b} = KK_\gamma$$

or

$$K = \frac{K_{th}}{K_\gamma} \tag{12-106}$$

where K is the concentration equilibrium constant, using mole fraction, molality, or molarity, and K_γ is the activity coefficient constant using the corresponding activity coefficients.

Solution equilibria are usually studied in fairly dilute solutions, and one commonly assumes the nonionic species obey Henry's law, that is, have unit activity coefficients. However, we have seen that electrolytes show serious departures from ideality even at high dilution and quite appreciable error can result if we treat ionic equilibria without considering K_γ. Several procedures are possible. First, as already discussed, one may extrapolate K values to infinite dilution so as to obtain K_{th}, and then back calculate the K_γ value for each equilibrium condition. One may estimate K_γ by evaluating each ion activity coefficient from the Debye–Hückel equation (12-97) or (12-98). One may, alternatively, use an experimental mean activity coefficient from some independent source. Thus if K_{th} is for a solubility equilibrium, then, as discussed in Section 12-7, K_γ will correspond to the mean activity coefficient of the electrolyte raised to the power ν, and standard tables of activity coefficients, such as Table 12-7, can be used. In the case of an equilibrium such as

$$Ac^- + H_2O = HAc + OH^-$$

where ions of more than one electrolyte are involved, it is always possible to give K in terms of mean activity coefficient ratios. Thus in this case, we have

$$K_\gamma = \frac{\gamma_{OH^-}}{\gamma_{Ac^-}} \tag{12-107}$$

and if the cation present in the solution is Na^+, multiplication of the numerator and denominator of Eq. (12-107) gives

$$K_\gamma = \frac{\gamma^2_{\pm,NaOH}}{\gamma^2_{\pm,NaAc}} \tag{12-108}$$

It is usually acceptable to take activity coefficients in mixed electrolytes, $\gamma_{\pm,NaOH}$ and $\gamma_{\pm,NaAc}$ in this case, to be the same as for the pure electrolyte species at the same overall ionic strength, although specific effects would have to be allowed for in accurate work.

A remaining device that is employed is to examine the equilibrium in the presence of a large excess of nonparticipating electrolyte, often $NaClO_4$. The nature of the solvent medium is in this way made essentially independent of the concentrations of the species and K_{th}/K_γ is therefore constant. The use of a swamping electrolyte is very common, for example, in the study of reaction kinetics involving ionic species.

The second aspect of the treatment of ionic equilibria, and the principal subject of this section, is that of the algebraic relationships and manipulations that are useful. These are for the most part standard, and the discussion here will be cursory except for some special cases. Table 12-8 gives some representative ionic equilibrium constants, and we now proceed to consider the use of solubility products in calculating solubilities under various circumstances, and then the treatment of simple acids and bases. A more detailed approach to acid–base equilibria is given in the Special Topics section.

A. Solubility equilibria   The solubility product expression for a 1–1 electrolyte is

$$MX(s) = M^{z+} + X^{z-} \qquad K_{sp} = (M^{z+})(M^{z-})$$

and, for the general case,

$$M_{\nu_+} X_{\nu_-}(s) = \nu_+ M^{z+} + \nu_- X^{z-}$$
$$K_{sp} = (M^{z+})^{\nu_+} (X^{z-})^{\nu_-} \qquad [\text{Eq. (12-66)}]$$

The solid salt is taken to be in its standard state and hence to have unit activity, but we must remember that for K_{sp} to apply, the solid must in fact be in equilibrium with the solution.

The solubility S of an electrolyte is defined as the number of gram formula weights that dissolve in the particular medium. If there is no added common ion, then $S = (M^{z+})/\nu_+ = (X^{z-})/\nu_-$ so in the general case

$$K_{sp} = (\nu_+)^{\nu_+} (\nu_-)^{\nu_-} S^\nu \tag{12-109}$$

Solubility products

SALT	K_{sp}	SALT	K_{sp}
CuCl	1.0×10^{-6b}	$Ag(C_2H_3O_2)$	1.8×10^{-3b}
$PbCl_2$	1.63×10^{-5c}	AgCl	1.56×10^{-10}
$PbCrO_4$	1.8×10^{-14b}	AgBr	7.7×10^{-13}
$Mg(C_2O_4)$	8.6×10^{-5b}	AgI	1.5×10^{-16}
$Ca(IO_3)_2 \cdot 6H_2O$	64.4×10^{-8b}	Ag_2CrO_4	9×10^{-12}
Hg_2Cl_2	1.7×10^{-18c}	TlCl	1.87×10^{-4c}

Dissociation of simple acids and bases

ACID	K_a	BASE	K_b
H_2O	1.002×10^{-14}	NH_4OH	1.77×10^{-5}
CH_3COOH	1.76×10^{-5}	$C_6H_5NH_3OH$	4.27×10^{-10}
$CH_2ClCOOH$	1.40×10^{-3}	$C_2H_5NH_3OH$	5.62×10^{-4}
CCl_3COOH	0.2	pyridine	1.78×10^{-9}
C_6H_5COOH	6.46×10^{-5}	hydrazine	3.00×10^{-6}
HCN	4.93×10^{-10}		

Dissociation of polybasic acids

ACID	K_1	K_2	K_3
H_2CO_3	4.30×10^{-7}	5.61×10^{-11}	—
$H_2C_2O_4$	6.5×10^{-2}	6.1×10^{-5}	—
$C_4H_4O_4$ (succinic acid)	6.5×10^{-5}	2.7×10^{-6}	—
H_2S	9.1×10^{-8b}	1.1×10^{-12b}	—
H_2SO_4	Strong	0.012	—
H_2SO_3	0.0154^b	1.02×10^{-7b}	—
H_3PO_4	7.52×10^{-3}	6.23×10^{-8}	2.2×10^{-13}
H_3AsO_4	5.62×10^{-3b}	1.70×10^{-7b}	3.95×10^{-12b}

TABLE 12-8

Ionic equilibrium constants[a]

[a] Aqueous solution at 25°C. Concentration constants (not K_{th}) unless otherwise indicated.
[b] 18°C.
[c] Thermodynamic constant, K_{th}.

Sources: Various, including "Handbook of Chemistry and Physics," 57th ed., CRC Press, Cleveland, 1976–77; K.S. Pitzer and L. Brewer, "Thermodynamics", 2nd ed., McGraw-Hill, New York, 1961.

If, however, an electrolyte is present which furnishes a common ion, say X^{z-}, then the solubility is given by $(M^{z+})/\nu_+$. The expression for K_{sp} becomes

$$K_{sp} = (\nu_+ S)^{\nu_+}(C + \nu_- S)^{\nu_-} \tag{12-110}$$

where C is the concentration of added X^{z-}.

In the case of a saturated solution of a slightly soluble salt, the concentration may be low enough that $K_{sp} \approx K_{th}$, that is, K_γ may be essentially unity. In general, however, it is K_{th} that is the true constant and one should always be aware that activity coefficient corrections may not be negligible.

Calculate the solubility of Ag_2CrO_4 at 25°C in water and in 0.10 m Na_2CrO_4. Table 12-8 gives the K_{sp} as 9×10^{-12}, so for the solubility in water we have $9 \times 10^{-12} = (2)^2(1)S^3$, whence $S = 1.3 \times 10^{-4}$ m. If the solvent is 0.10 m Na_2CrO_4, then

$$9 \times 10^{-12} = (2S)^2 (0.10 + S).$$

EXAMPLE

The resulting cubic equation can be solved by successive approximations:

$$S_1 = \left[\frac{9 \times 10^{-12}}{(4)(0.10)} \right]^{1/2} = 4.7 \times 10^{-6}$$

The first approximation, S_1, gives a result that is small compared to 0.10, and is therefore adequate. The serious error lies in the neglect of K_γ, that is, the $(\gamma_{Ag^+})^2$ $(\gamma_{CrO_4^{2-}})$ factor, which is the same as $(\gamma_{\pm, Ag_2CrO_4})^3$. We don't have the actual activity coefficient data but can estimate the effect as follows. We use Eq. (12-78) to estimate $\gamma_{\pm, Ag_2CrO_4}$ for a 1.3×10^{-4} m solution (the solubility in water). The ionic strength is $I = \frac{1}{2}[(1)^2(2)(1.3 \times 10^{-4}) + (2)^2 (1.3 \times 10^{-4}] = 3.9 \times 10^{-4}$, so that $\ln \gamma_\pm = -(1.172)(1)(2)(3.9 \times 10^{-4})^{1/2} = -0.0463$ and $\gamma_\pm = 0.955$. It follows that $K_{th} = K_{sp}K_\gamma = (9 \times 10^{-12})(0.955)^3 = 7.83 \times 10^{-12}$. Even at this low concentration, a 2–1 electrolyte probably deviates from the limiting law, so our value of K_{th} is somewhat approximate.

We turn next to the 0.10 m Na_2CrO_4 solution. We need the activity coefficient of Ag_2CrO_4 in this solution, which is not available, and we make the approximation that $\gamma_\pm$ for Ag_2CrO_4 will be about that for Na_2CrO_4, but this is also not available. As a reasonable guess, however, we can use $\gamma_\pm$ for Na_2SO_4, a salt of the same charge type, which is 0.452 for a 0.1 m solution (the amount of dissolved Ag_2CrO_4 adds negligibly to the ionic strength). Our new K_{sp} value is K_{th}/K_γ or $(7.83 \times 10^{-12})/(0.452)^3 = 8.48 \times 10^{-11}$. The corrected solubility is therefore

$$S_1 = \left[\frac{8.48 \times 10^{-11}}{(4)(0.10)} \right]^{1/2} = 1.46 \times 10^{-5}$$

The activity coefficient correction thus changes the calculated solubility about threefold! This example illustrates the difficulty in making accurate calculations of ionic equilibria in solutions of appreciable ionic strength.

Another type of complication is that the dissolved salt may not be fully dissociated. For example, appreciable amounts of CoOx (where Ox denotes oxalate ion) are present in solution as the undissociated molecule. If the solution is saturated with respect to CoOx(s), the concentration S_0 of undissociated CoOx in solution is a constant. The observed solubility S is then $S_0 + (Co^{2+})$, and if Na_2Ox is added, S decreases due to the common ion effect. With further addition of Na_2Ox, however, the complex $(CoOx)_2^{2-}$ begins to form and the solubility, that is, total dissolved CoOx(s), increases. Other cases of complex formation with slightly soluble salts include the well-known example of $Ag(CN_2^-)$, as well as that of $AgCl_2^-$ and other silver halide complexes.

Finally, if the anion of the slightly soluble salt is one of a weak acid, then the solubility will depend on the pH of the solution. Thus in the case of silver acetate the following simultaneous equilibria would hold:

$$AgAc(s) = Ag^+ + Ac^- \qquad K_{sp} = 1.8 \times 10^{-3},$$

$$HAc = H^+ + Ac^- \qquad K = 1.76 \times 10^{-5}$$

If the pH is known, this determines the degree of dissociation α of the acid, and

$$K_{sp} = (Ag^+)(Ac^-) = S(\alpha S)$$

In this case the total acetic acid substance is given by the solubility S, but the actual Ac^- concentration is only αS.

B. Equilibrium across a semipermeable membrane

An important type of ionic equilibrium is that across a membrane which is permeable only to certain of the ions present. The situation is known as one of *Donnan equilibrium.* As an example, consider the arrangement shown in

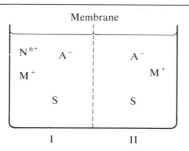

FIGURE 12-14
Donnan equilibrium.

Fig. 12-14. The membrane is permeable to solvent and to M^+ and X^- ions, but *not to* N^+ ions.

The condition for ionic equilibrium is that the activity a_2 [see Eq. (12-58)] be the same on both sides of the membrane for that electrolyte to which it is permeable, in this case M^+, X^-. Thus

$$a_{M^+}^{I} \, a_{X^-}^{I} = a_{M^+}^{II} \, a_{X^-}^{II} \tag{12-111}$$

If the solutions are dilute, activities may be replaced by concentrations. Also, electroneutrality requires that $(M^+)^I = (X^-)^I = C$ and $(M^+)^{II} + (N^+)^{II} = (X^-)^{II}$. Equation (12-111) thus reduces to

$$C^2 = (M^+)^{II} [(M^+)^{II} + (N^+)^{II}] \tag{12-112}$$

Suppose that $C = 0.01 \, m$ and that $(N^+) = 0.1 \, m$. We have $(0.01)^2 = (M^+)^{II} [(M^+)^{II} + 0.1]$, whence $(M^+)^{II} = 9.9 \times 10^{-4} \, m$. **EXAMPLE**

The example illustrates that the Donnan effect acts to exclude M^+ from side II.

The physical basis for the exclusion effect is that a potential difference, the *Donnan potential* ϕ, makes side II positive relative to side I. It can be shown that

$$\phi = \frac{RT}{\mathscr{F}} \ln \frac{(M^+)^I}{(M^+)^{II}} \tag{12-113}$$

where $\mathscr{F}$ is Faraday's number [see Eq. (13-12)]. In the example, ϕ is 0.0594 V for 25°C.

Donnan potentials are important in biology. For example, nerve cells or axons appear to be permeable to K^+ but not to Na^+ ions and application of Eq. (12-113) gives about the observed potential across the resting cell membrane. Reduction of this potential by more than a certain threshold amount makes the membrane permeable to Na^+ ions and a wave of local depolarization races along from cell to cell. Such nerve impulses travel at some 100 ft s^{-1}. The contraction of muscle cells also involves changes in cell membrane potential and in ratio of permeability to Na^+ versus K^+ ions.

The treatment of dissociation equilibria involving weak acids and bases will be limited here to the cases of a simple acid HA and a simple base BOH. Even so, the general solutions can involve rather complex algebraic manip- **C.** Weak acids and bases ulations. A powerful alternative approach is given in the Special Topics sections.

We consider first the weak monobasic acid HA:

$$HA = H^+ + A^- \qquad K_a = \frac{(H^+)(A^-)}{(HA)} \qquad (12\text{-}114)$$

The general situation is one of a solution prepared by dissolving amounts of HA and of the salt MA so as to give the formalities f_{HA} and f_{MA}. The cation M^+ is assumed not to hydrolyze. With the solvent taken to be water, we have*

$$H_2O = H^+ + OH^- \qquad K_W = (H^+)(OH^-) \qquad (12\text{-}115)$$

The solution must be electrically neutral, and so

$$(M^+) + (H^+) = (A^-) + (OH^-) \qquad (12\text{-}116)$$

where $(M^+) = f_{MA}$. Finally, by material balance

$$(HA) + (A^-) = f_{HA} + f_{MA} \qquad (12\text{-}117)$$

Note that (HA) and (A^-) denote the actual concentrations of these species, whereas f_{HA} and f_{MA} are the amounts weighed out per 1000 g of water when the solution is made up.

The preceding four equations must be solved simultaneously, and it is helpful to reduce them as follows. From Eq. (12-116) we have

$$(A^-) = f_{MA} + (H^+) - (OH^-)$$

and insertion of this result into Eq. (12-117) gives

$$(HA) = f_{HA} - (H^+) + (OH^-)$$

Equation (12-114) then becomes

$$K_a = \frac{(H^+)[f_{MA} + (H^+) + (OH^-)]}{f_{HA} - (H^+) + (OH^-)} \qquad (12\text{-}118)$$

The simultaneous solution of Eqs. (12-115) and (12-118) then gives (H^+) and (OH^-) for any formal composition.

Equation (12-118) simplifies considerably under various special conditions.

Case 1. $f_{MA} = 0$. Then

$$K_a = \frac{(H^+)[(H^+) - (OH^-)]}{f_{HA} - (H^+) + (OH^-)} \qquad (12\text{-}119)$$

If $(OH) \ll (H^+)$, then $K_a = (H^+)^2/[f_{HA} - (H^+)]$. This condition holds if $(H^+)^2 > 10^{-12}$ and hence if $K_a f_{HA} > 10^{-12}$. If also $f_{HA} \gg (H^+)$, then $K_a = (H^+)^2/f_{HA}$. This condition holds if $f_{HA} > 100(H^+)$ and hence if $f_{HA} > 10^4 K_a$.

As examples, for 0.1 m HAc, $K_a f_{HA} = 1.76 \times 10^{-6}$ and is much greater than 10^{-12}. Since f_{HA} is about $10^4 K_a$, the last approximation can just be used (to 1% error) and $(H^+)^2 = 1.76 \times 10^{-6}$, $(H^+) = 1.32 \times 10^{-3}$. However, if f_{HA} were 10^{-8} m, then $K_a f_{HA}$ would be 1.76×10^{-13}, and the full equation (12-118) would be needed.

*It is conventional to take a_{H_2O} to be unity. For comparison with other acids, however, one sometimes uses $K'_w = (H^+)(OH^-)/(H_2O)$ where $C_{H_2O} = 55.5$ M at 25°C.

Case 2. $f_{HA} = 0$. Then

$$K_a = \frac{(H^+)[f_{MA} + (H^+) - (OH^-)]}{-(H^+) + (OH^-)}$$

If we divide K_a by K_w to give K_h, the hydrolysis constant, we obtain

$$K_h = \frac{K_w}{K_a} = \frac{(OH^-)[(OH^-) - (H^+)]}{f_{MA} + (H^+) - (OH^-)} \qquad (12\text{-}120)$$

This is of the same form as Eq. (12-119), and the same two types of approximation follow:

$$\text{If } K_h f_{MA} > 10^{-12}, \qquad \text{then } K_h = \frac{(OH^-)^2}{f_{MA} - (OH^-)}$$

$$\text{If } f_{MA} > 10^4 K_h, \qquad \text{then } K_h = \frac{(OH^-)^2}{f_{MA}}$$

For example, $K_h = 1.00 \times 10^{-14}/1.76 \times 10^{-5} = 5.68 \times 10^{-10}$. For 0.1 m NaAc, $K_h f_{MA} = 5.68 \times 10^{-11}$, or more than 10^{-12}, and $f_{MA} = 0.1$, or more than $10^4 K_h$. The simplest form may then be used and $(OH^-)^2 = (0.1)(6.58 \times 10^{-10})$, $(OH^-) = 7.5 \times 10^{-6}$, $(H^+) = 1.3 \times 10^{-9}$.

Case 3. If f_{MA} and f_{HA} are each greater than (H^+) or (OH^-), then Eq. (12-118) reduces to

$$K_a = \frac{(H^+) f_{MA}}{f_{HA}} \qquad (12\text{-}121)$$

The solution is now said to be *buffered*. That is, in order to change (H^+) appreciably, we must add sufficient acid or base to change f_{MA} or f_{HA} appreciably. Thus in a solution 0.1 f in HAc and 0.1 f in NaAc, (H^+) will be 1.76×10^{-5}. Addition of 0.01 f HCl changes f_{HAc} to 0.11 and f_{NaAc} to 0.09, and hence (H^+) only changes to 2.1×10^{-5}.

A parallel set of relationships holds for the weak base BOH:

$$BOH = B^+ + OH^- \qquad K_b = \frac{(B^+)(OH^-)}{(BOH)}$$

The analog of Eq. (12-118) is

$$K_b = \frac{(OH^-)[f_{BX} + (OH^-) - (H^+)]}{f_{BOH} - (OH^-) + (H^+)} \qquad (12\text{-}122)$$

where X_- is a nonhydrolyzing anion. The various special cases are similarly analogous to those for the weak acid.

We can find the equilibrium concentrations in a solution containing HA, MA, BOH, and BX, by solving Eqs. (12-115), (12-118), and (12-122) simultaneously. If all of the concentrations (HA), (A$^-$), (BOH), and (B$^+$) are much larger than that of (H$^+$) or (OH$^-$), then these ions may be ignored in the charge balance, and the simultaneous equations to be solved are

$$BOH + HA = B^+ + A^- + H_2O \qquad K_{ab} = \frac{(B^+)(A^-)}{(BOH)(HA)} = \frac{K_a K_b}{K_w}$$

$$(B^+) + f_{MA} = (A^-) + f_{BX} \qquad f_{HA} + f_{MA} = (HA) + (A^-)$$

$$f_{BOH} + f_{BX} = (BOH) + (B^+)$$

D. Titration curves

An acid–base titration consists of adding successive amounts of a base to a solution of an acid, or of an acid to a solution of a base, and noting how the hydrogen ion concentration [or $pH = -\log(H^+)$] varies. We will consider only the first situation here.

The calculation of a titration curve is clarified if one recognizes that a solution which initially has a formality f_{HA} in acid is converted into one having formalities f'_{HA} and f'_{MA} by addition of a strong base, where f'_{MA} is the formality of the added base and $f'_{HA} + f'_{MA} = f_{HA}$. As a specific example, consider a solution that initially is 0.1 f in acetic acid. Addition of sodium hydroxide sufficient to make the solution 0.05 f in NaOH must yield an equilibrium mixture identical to that which would be obtained if the solution were 0.05 f in HAc and 0.05 f in NaAc. In effect, the formal composition of the solution may be expressed in two alternative ways:

$$\left.\begin{array}{l} 0.1 \ f \text{ in HAc} \\ 0.05 \ f \text{ in NaOH} \end{array}\right\} \sim \left\{\begin{array}{l} 0.05 \ f \text{ in HAc} \\ 0.05 \ f \text{ in NaAc} \end{array}\right.$$

Both statements specify the same amount of acetic acid substance and sodium ion.

The general procedure for finding such alternative ways of expressing a formal composition is to consider species that would tend to react and then to suppose that the reaction goes to completion. The new formal composition is expressed in terms of the products. Thus HAc and NaOH tend to react:

$$HAc + Na^+ + OH^- = Na^+ + Ac^- + H_2O \tag{12-123}$$

The alternative expression of formal composition follows if we suppose that the neutralization of the 0.1 m HAc by 0.05 m NaOH goes to completion, giving 0.05 m NaAc and residual 0.05 m HAc. The reaction does not in fact go entirely to completion, but the formal composition can be given as 0.05 f NaAc and 0.05 f HAc.

With this preamble, the calculation of a titration curve reduces to the calculation of (H^+) from Eq. (12-118) for various ratios of f_{MA} to f_{HA}, with their total kept constant. The equivalence point is that for which $f_{HA} = 0$. The result of such a calculation for a 0.002 m acetic acid solution is shown in Fig. 12-15, where F_N is the degree of neutralization. (It has been assumed that the

---

FIGURE 12-15
Titration curve for acetic acid using a strong base.

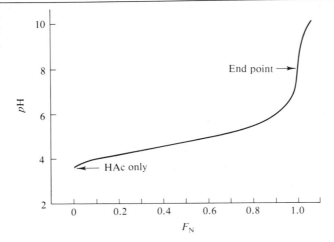

added sodium hydroxide solution is sufficiently concentrated that dilution effects can be neglected.)

Calculations of this type can be quite tedious, and as in the preceding subsection, the reader is reminded that a more general and powerful approach is given in the Special Topics section.

A titration may be followed conductimetrically. In a neutralization reaction such as that of Eq. (12-123), each portion of NaOH added converts an equivalent amount of HAc into $Na^+ + Ac^-$. The conductance of the solution therefore increases since a weak electrolyte is being replaced by a strong one. At the end point, further addition of NaOH adds Na^+ and OH^- ions to those of the NaAc, and the conductance increases more rapidly, as illustrated in Fig. 12-16. If the acid being titrated is a strong acid, then the neutralization converts the ions $H^+ + A^-$ into $Na^+ + A^-$ or, in effect, substitutes sodium ion for hydrogen ion. Since the molar conductivity of sodium ions is much less than that of hydrogen ions, the result is that the conductance drops during the titration until the end point is reached, and then rises as before.

Suppose that 0.1 m HCl is being titrated with concentrated NaOH (we therefore neglect volume changes) and that the conductance is followed using an immersion-type conductivity cell of cell constant 0.2. The initial conductance is $L_0 = (C\Lambda)/1000k = (0.1)(426)/(1000)(0.2) = 0.213$ ohm^{-1}. When enough NaOH has been added to make the solution 0.05 f in NaOH or, alternatively, 0.05 m in NaCl and 0.05 m in residual HCl, the conductance L_1 is

$$L_1 = \frac{(0.05)(426) + (0.05)(126)}{(1000)(0.2)} = 0.138 \text{ ohm}^{-1}$$

At the end point the formality of the added NaOH is 0.1, and the solution consists of just 0.1 m NaCl. The conductance $L_2 = (0.1)(126)/(1000)(0.2) = 0.063$ ohm^{-1}. When the end point is overshot with 0.15 f added NaOH, the system is 0.1 m in NaCl and 0.05 m in NaOH, and L_3 is now

$$L_3 = \frac{(0.1)(126) + (0.05)(248)}{(1000)(0.2)} = 0.125 \text{ ohm}^{-1}$$

Thus the series of L values 0.213, 0.138, 0.063, and 0.125 goes through a minimum at the end point.

EXAMPLE

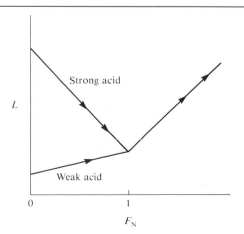

L

Strong acid

Weak acid

0   1

F_N

FIGURE 12-16
Conductimetric titration curves for a strong acid (upper left line) and a weak acid (lower left line).

COMMENTARY AND NOTES

12-CN-1   Electrolytic dissociation

A brief further discussion of the theory of electrolytic dissociation seems worthwhile. This is something we now take for granted, yet the reasons for the original resistance to Arrhenius' proposal were serious ones and should be appreciated. It must be remembered that in 1890 the electron as such had not yet been discovered; it would be many years before x-ray crystallography revealed the crystal structures of solids; and, of course, wave mechanics was not imagined. Mineral solids, such as NaCl, were known as hard, high-melting materials, similar in these respects to metals, and evidently held together by strong chemical forces. It seemed incredible that the mere dissolving of such a salt in water would break it up into fragments. The heat of solution of most salts is relatively small, and so it appeared that little play of energy was present; the most natural supposition was that solutions of mineral salts consisted of the molecular units.

Two types of observations led Arrhenius to attack this very reasonable picture. The first was the difficulty in accounting for Ohm's law being obeyed by electrolyte solutions. The original Grotthus idea of salt molecules exchanging polar parts (Fig. 12-1) implied that even the most minute applied potential would induce a chain of bond breaking and making. This was contrary to experimental evidence, which strongly indicated that electrical charge carriers were ever-present, waiting to carry current, and did not have to be created by an applied potential. The second type of observation was that of the colligative behavior of electrolyte solutions. Van't Hoff, in particular, had found it necessary to introduce his (now famous) i factor into the osmotic equation [Eq. (10-32)]. The experimental result is that $i \approx 2$ for aqueous NaCl and this was entirely mysterious until electrolytic dissociation was proposed. The early situation was clouded by what we now consider to be a combination of interionic attraction and partial dissociation effects, the consequence of which is that i factors are not found to be simple integers except at extreme dilution. Nonetheless, it was the combined impact of colligative property and conductance measurements that established the theory of electrolytic dissociation.

As noted in Section 12-3, it is now useful to class electrolytes as strong or as weak. Weak electrolytes obey the Ostwald dilution law well, and their behavior is thus determined primarily by a dissociation equilibrium. Strong electrolytes, however, do not obey a dissociation equilibrium constant at all well, and it was not until the development of interionic attraction theory that they could be understood.

We can also explain how it is possible that strongly bonded salt crystals can so readily dissociate in water solution. X-ray crystallography tells us that crystalline NaCl consists of a closely packed array of alternate Na^+ and Cl^- ions and not of molecular Na—Cl units (Section 19-3). The strong forces that hold the crystal together are primarily Coulomb attractions between unlike charged ions. The Coulomb attractive potential between one Na^+ and an adjacent ion is

$$\phi = \frac{e^2}{r} \qquad (12\text{-}124)$$

where r is the distance between the centers of the ions. The picture is essentially that of Fig. 8-4, and a summation of all of the mutual Coulomb interactions gives the total crystal or so-called lattice energy. Such calculations (see Section 20-ST-3) agree well with the observed heats of sublimation, 186 kcal mol^{-1} in the case of NaCl.

Equation (12-124) applies if the medium in which the ions are present is vacuum; otherwise the dielectric constant enters in the denominator. Water has a dielectric constant of about 80, so the energy to dissociate NaCl would be reduced to about 1.5 kcal mol^{-1}. Thus by inserting itself between the ions, water reduces their mutual attraction to a point where the gain in entropy on dissolution can make the process occur. The 125 kcal mol^{-1} or other analogous quantity has not disappeared. This much energy is still required to take the aqueous Na^+ and Cl^- ions into the gas phase. In other words, the strong mutual binding in the crystal has, in solution, been replaced by a strong binding to water. We speak of this binding as the *hydration energy* and ionic crystals may be said to dissolve in water because the hydration energy nearly balances the lattice energy. It is interesting, for example, that the heat of solution of an anhydrous salt is generally more negative (more heat is evolved) than that of a corresponding hydrate. Thus the heat of solution of Na_2SO_4 is -0.5 kcal mol^{-1}, while that of $Na_2SO_4 \cdot 10H_2O$ is $+19$ kcal mol^{-1}; the presence of water of hydration in the crystal has noticeably diminished the amount of hydration energy available when the crystal dissolves.

The Coulomb forces between ions in aqueous solution, although greatly reduced from those between ions in the crystal, are still important; they give rise to the interionic attraction effects in conductance and in the Debye–Hückel treatment of nonideality. It seems certain that ion pairs and possibly ion clusters form in moderately concentrated solution; it is noted in Section 12-9 that N. Bjerrum used the concept of ion pair formation as an aid in explaining deviations from the Debye–Hückel limiting law.

At this point a distinction should be made between ion pairing and chemical association. Acetic acid is highly associated, but as a result of chemical bond formation between H^+ and Ac^-. Transition metal ions form coordinate bonds with anionic ligands; the case of CoOx and $Co(Ox)_2^{2-}$ mentioned in Section 12-10A is but one example. As just one further example, the Fe^{3+} ion forms the complex FeX^{2+}, where X may be a halogen, pseudohalogen, or hydroxide ion. In fact, transition metal ions have a solvation shell of coordinated water, and the formation of FeX^{2+} would more properly be written

$$Fe(H_2O)_6^{3+} + X^- = Fe(H_2O)_5X^{2+} + H_2O$$

Many of these ligand substitution reactions are in rapid equilibrium, and the apparent nonideality of, say, aqueous $FeCl_3$ is largely due to complex formation. On the other hand, an ion such as Na^+ does not have much ability to form coordinate bonds, and the attraction between Na^+ and Cl^- is, as noted in the preceding section, mainly of Coulombic origin. We speak of *ion association* (or ion pairing), then, when we wish to think of the unit as consisting of the intact ions. The distinction between ion pairing

and coordinative association is not always easy to make. The association which occurs with, say, Zn^{2+} and SO_4^{2-} may be partly Coulombic and partly coordinative—Zn^{2+} does, after all, form well-known complex ions. In the case of $Co(NH_3)_6^{3+}$, however, the coordination sphere is saturated, and the association constant of 1000 observed between this ion and SO_4^{2-} is therefore spoken of confidently as due to ion pairing.

To return to the matter of activity coefficients, there is no question that ion pairing does occur in more concentrated solutions, and even in dilute ones if the ions are highly charged. R. M. Fuoss and C. A. Kraus have estimated ion pair dissociation constants for a number of electrolytes from conductivity measurements. For uni-univalent electrolytes values are of the order of unity and for di-divalent ones they are around 0.01 or less. Thus the estimated dissociation constants are about 1.4 for KNO_3, 0.15 for KSO_4^-, and 5.3×10^{-3} for $ZnSO_4$ (aqueous solutions at 25°C). The conclusion is that the Debye–Hückel theory must be used with great caution for other than dilute solutions of uni-univalent electrolytes. Semiempirical equations such as Eq. (12-102) are very helpful, but it must be remembered that the constants B and C of the equation depend on the total ionic makeup of a solution and not just on the particular electrolyte in question.

Ion pairing, important in aqueous solutions, becomes a dominant feature in nonaqueous systems. As the dielectric constant of a medium is reduced, the Coulomb forces between ions increase, and the degree of dissociation of an electrolyte drops dramatically. As an example, the dissociation constant of KI is only about 2×10^{-4} in pyridine as solvent. There are a few nonaqueous solvents, of fairly high dielectric constant, in which electrolytes having large ions may be moderately dissociated. The salt $K^+[Cr(NH_3)_2(NCS)_4]^-$ is fully dissociated in nitromethane, and similar complex ions are moderately dissociated in dipolar aprotic solvents such as dimethylsulfoxide, dimethylformamide, and the like. More than dielectric constant is involved, however. Formamide has a dielectric constant of 109, but is not generally a better solvent for electrolytes than water (dielectric constant 78). Hydrogen bonding ability may also be important; it is present in water but not in *aprotic* solvents such as formamide.

12-CN-2   Acids and bases

A. Aqueous solutions

There are many interesting regularities in the dissociation constants of weak acids and bases. The following are a few examples that draw on the data of Table 12-8. Carboxylic acids tend to have K_a values of around 10^{-5}, except that as the carbon atom adjacent to the carboxyl group is substituted with electronegative groups, such as halogens, the acidity increases. For CCl_3COOH, $K_a = 0.13$, for example. This appears to be an inductive effect by which the carboxyl carbon atom is made more positive and the O—H bond is electrostatically weakened. A large number of acids of this type have been studied, and L. P. Hammett has used the data to characterize substituents. Thus in the case of substituted benzoic acids a substituent is assigned a parameter σ, defined as $\sigma = pK_0 - pK_s$, where $pK_0 = -\log K_0$, with K_0 the dissociation constant of benzoic acid, and $pK_s = -\log K_s$, with K_s the dissociation constant of the substituted acid. The σ values are then found to correlate with the behavior of the compounds in other equilibria or in their reaction kinetics.

The electrostatic effect appears to be very important in determining the pK values for oxyacids. The approximately 10^5-fold reduction in successive K values for H_3PO_4 is one example. The three protons should be essentially equivalent, except for the increased electrostatic work of dissociation in the series H_3PO_4, $H_2PO_4^-$, HPO_4^{2-}. A useful empirical rule is that for an oxyacid of the general formula $MO_m(OH)_n$,

pK_1 is approximately $7 - 5m$, pK_2 is $12 - 5m$, and so on. Thus for H_3PO_4, pK_1 should be $7 - 5 = 2$, pK_2 should be about $12 - 5 = 7$, and so on; for H_2SO_4, pK_1 is predicted to be -3, in agreement with the observation that the first dissociation is that of a strong acid, while pK_2 should be 2, again about as observed.

Another type of situation is that in which two acid functions are sufficiently separated in a molecule that they should be essentially free of the preceding electrostatic effects. An example is $HOOC—(CH_2)_n—COOH$, where n is two or more. It might be supposed that K_1 and K_2 should be the same since the two groups are independent. A statistical factor remains, however. If the acid is represented by H_1H_2A, there are two ways for the first-stage dissociation to occur, namely to give H_1A^- or H_2A^-. There is only one choice of hydrogen to dissociate in the second stage, but the association reaction $H^+ + A^{2-}$ can occur on either position. Thus K_1 is enhanced and K_2 is diminished by this effect. In the case of succinic acid, $HOOC—(CH_2)_2—COOH$, $K_1 = 6.4 \times 10^{-5}$, or somewhat more than twice K_a for acetic acid, and K_2 is 2.7×10^{-6}, or about half K_a for acetic acid.

A related behavior is that of amino acids, typifying electrolytes having separated weakly acidic and weakly basic groups. An amino acid is sometimes shown in the form $H_2N—R—COOH$, but it appears certain that internal proton transfer is largely complete, so that $^+H_3N—R—COO^-$ is the actual species in aqueous solution. This last is known as a *zwitterion*. The behavior of an amino acid is then represented in terms of the two stages of dissociation of a dibasic acid:

$$^+H_3N—R—COOH = H^+ + {}^+H_3N—R—COO^- \qquad K_1 = \frac{(H^+)(A^\pm)}{(A^+)}$$
$$\underset{A^+}{} \qquad\qquad\qquad \underset{A^\pm}{}$$

$$^+H_3N—R—COO^- = H^+ + H_2N—R—COO^- \qquad K_2 = \frac{(H^+)(A^-)}{(A^\pm)}$$
$$\underset{A^\pm}{} \qquad\qquad\qquad \underset{A^-}{}$$

In the case of glycine, H_2NCH_2COOH, $K_1 = 4.5 \times 10^{-3}$ and $K_2 = 2.24 \times 10^{-12}$ at 25°C.

The *isoelectric* point is a state of special importance for an amino acid. This is the pH value such that $(A^+) = (A^-)$. Acid–base equilibria are rapid, which means that an individual amino acid molecule rapidly samples all its possible states of dissociation and at the isoelectric point it therefore spends equal times as A^+ and as A^-. The effect is that the amino acid behaves as though it were electrically neutral, even

though it is still a conducting electrolyte; it displays essentially no net motion in an electrophoresis experiment, for example. Analysis of the preceding equilibrium relationships shows that at the isoelectric point $(H^+) = (K_1K_2)^{1/2}$. It can also be shown that at this pH the total degree of ionization of the amino acid is at a minimum. For this reason many of the physical properties of an amino acid solution exhibit maxima or minima at the isoelectric point.

B. Brønsted treatment of acids and bases

The phenomenology of electrolyte behavior in aqueous solutions made it natural for Arrhenius and Ostwald to define an acid as a substance furnishing H^+ ions and a base as one furnishing OH^- ions. Neutralization then consisted of the reaction of these ions to give water. This formalism is an adequate and functioning one for the treatment of acid–base equilibria in water solution, although it is somewhat misleading chemically. The H^+ ion exists in water at least as H_3O^+, if not in more complex forms, and the dissociation of a weak acid is more correctly written as

$$HA + H_2O = A^- + H_3O^+ \qquad (12\text{-}125)$$

than as $HA = H^+ + A^-$. Similarly, a weak base may produce OH^- ions by the reaction

$$H_2O + B = OH^- + HB^+ \qquad (12\text{-}126)$$

(as, for example, with $B = NH_3$).

Brønsted and Lowry illuminated the chemistry of weak acids and bases by recognizing (in 1923) that the degree of a reaction such as those given by Eqs. (12-125) and (12-126) must depend on the natures of both reactants and that the reactions are really symmetric. They introduced the more general definitions that an *acid* is a *proton donor* and a *base* is a *proton acceptor*. Equations (12-125) and (12-126) now fall into the common form

acid + base = conjugate base + conjugate acid

$$HA + H_2O = A^- + H_3O^+$$

$$H_2O + B = OH^- + HB$$

An acid, on yielding a proton, becomes its conjugate base, and a base, on accepting a proton, becomes its conjugate acid.

The Brønsted picture has been useful in two major ways. First, it emphasizes that the anion of a weak acid is a proton acceptor, or a base, as well as OH^- ion and may be capable of reacting directly with a proton donor without going through the route of accepting an H^+ ion released by it. There are a number of cases of acid- or base-catalyzed reactions in which direct reaction evidently occurs. The rate of reaction in such cases is found to depend on the specific acid or base present and not just on the pH of the solution.

The second important feature of the Brønsted picture is that it relates aqueous to nonaqueous systems. Other solvents can now be seen in striking analogy to water. Liquid ammonia, for example, autoionizes to give NH_4^+ and NH_2^- ions, in analogy to the H_3O^+ and OH^- ions of water. Acetic acid in liquid ammonia solution than dissociates according to the reaction

$$HAc + NH_3 = Ac^- + NH_4^+ \qquad (12\text{-}127)$$

This dissociation is virtually complete. That is, acetic acid is a strong acid in liquid ammonia solvent, and the reason is clearly that NH_3 is a much better proton acceptor than is H_2O. Acids that are too weak to be studied in water solution can be observed easily in liquid ammonia.

Pure acetic acid may be used as a solvent, and the autoionization reaction gives $CH_3CO_2H_2^+$ and $CH_3CO_2^-$ ions. Then HCl in acetic acid solution dissociates according to the reaction

$$HCl + HAc = Cl^- + H_2Ac^+ \qquad (12\text{-}128)$$

However, HAc is a weak base and this dissociation is not complete. That is, HCl behaves as a weak acid in acetic acid solvent. In this solvent, the series of acid strengths

$$HClO_4 > HBr > H_2SO_4 > HCl > HNO_3$$

is established. All of these acids are so completely dissociated in water that distinctions among them cannot be made and it is only by balancing their proton-donating ability against a rather poor acceptor that the real differences become apparent.

Other acid–base systems have some value. In molten oxide mixtures an acid and a base may be defined as an acceptor and a donor, respectively, of oxide ion. In the reaction $BaO + CO_2 = BaCO_3$, BaO is then the base and CO_2 the acid. G. N. Lewis proposed the rather general definitions of an *acid* as an *electron pair acceptor* and a *base* as an *electron pair donor*. The designations are consistent with the Brønsted scheme in that a Brønsted base has an unshared pair of electrons capable of bonding with H^+. The Lewis definition also allows the reaction $BF_3 + (CH_3)_3N: = (CH_3)_3N:BF_3$ to be characterized as an acid–base one, BF_3 being the acid. The Lewis formalism does not provide the symmetry of the Brønsted formalism, however, and really is more a notation indicating that an important type of chemical bond is one in which the electron pair derives from one of the associating molecules. Such a bond is also often called a *coordinate bond*, and in the Lewis sense coordination chemistry is a study of acid–base reactions. J. Chatt and R.G. Pearson have subdivided Lewis acids and bases into

categories A and B, alternatively known as "hard" and "soft," depending roughly on their charges and polarizabilities. This author feels that it is more profitable to study the coordinate bond as it is, rather than in terms of often hypothetical acid plus base reactions. Thus in the case of $Cr(CO)_6$, Cr is supposedly the acid and CO the base, but this approach does not seem very illuminating of the actual wave-mechanical description of the Cr—CO bond.

SPECIAL TOPICS

12-ST-1   Ionic diffusion coefficients

It is shown in Section 10-7B that the diffusion coefficient for a particle in solution is given by

$$\mathcal{D} = \frac{kT}{f} \qquad \text{[Eq. (10-44)]}$$

where f is the friction coefficient. The same treatment applies to individual ions, and writing $\omega = 1/f$, Eq. (10-44) becomes

$$\mathcal{D}_i = \omega_i kT \qquad (12\text{-}129)$$

It is now possible to relate $\mathcal{D}_i$ and electrochemical mobility u_i, using Eq. (12-36):

$$\mathcal{D}_i = 10^{-7} \frac{u_i kT}{z_i e}$$

or, using Eq. (12-38),

$$\mathcal{D}_i = 10^{-7} \frac{\lambda_i kT}{z_i^2 e \mathcal{F}} = \frac{\lambda_i RT}{z_i^2 \mathcal{F}^2} \qquad (12\text{-}130)$$

where R is now in joules per degree Kelvin per mole.

Equation (12-130) gives the diffusion coefficient for a single ion, as might be observed in a self-diffusion experiment. Ordinarily, however, one studies the diffusion of an electrolyte as a whole, and the electroneutrality requirement enforces the same diffusional velocity on both ions. The intrinsically slower diffusing type of ion will lag behind the faster one until a local potential (called the *diffusion potential*) builds up. This potential ϕ acts to retard the faster ions and speed up the slower ones, and Eq. (2-87) for the diffusional flux therefore becomes

$$J_1 = -\mathcal{D}_1 \frac{dC_1}{dx} + \phi u_1 C \qquad (12\text{-}131)$$

$$J_2 = -\mathcal{D}_2 \frac{dC_2}{dx} - \phi u_2 C \qquad (12\text{-}132)$$

where 1 and 2 denote the two kinds of ion present and C is the ion concentration assuming a 1–1 electrolyte. In other words, ϕu gives the increment in velocity due to the local field and $\phi u C$ gives the corresponding increment to the molecular flow across a unit area. We require that $J_1 = J_2$, that is, that no excess of one ion build up over the other during the diffusion, and on equating (12-131) and (12-132), we obtain

$$\phi = \frac{(\mathcal{D}_1 - \mathcal{D}_2)\, dC/dx}{(u_1 + u_2)C}$$

Substitution of this result into Eq. (12-131) [or (12-132)] gives

$$J_1 = -\left(-\mathcal{D}_1 + \frac{\mathcal{D}_1 - \mathcal{D}_2}{u_1 + u_2} u_1 \right) \frac{dC}{dx} \qquad (12\text{-}133)$$

Since u_i is proportional to $\mathcal{D}_i$ and J_1 is now the flux for the diffusion of the electrolyte as a whole, we have

$$J_1 = -\mathcal{D} \frac{dC}{dx}$$

Equation (12-133) reduces to

$$\mathcal{D} = \frac{2\mathcal{D}_1 \mathcal{D}_2}{\mathcal{D}_1 + \mathcal{D}_2} = \frac{2RT}{\mathcal{F}^2} \frac{\lambda_1 \lambda_2}{\Lambda} \qquad (12\text{-}134)$$

Equation (12-134) is known as the *Nernst diffusion equation*; it permits the calculation of the diffusion coefficient for a 1–1 electrolyte if the molar conductivities are known.

The more complete form of Eq. (12-134) allows for nonideality in the same manner as does Eq. (10-43):

$$\mathscr{D} = \frac{2RT}{\mathscr{F}^2} \frac{\lambda_1\lambda_2}{\Lambda} \left[1 + \frac{d(\ln \gamma c)}{d(\ln C)} \right] \quad (12\text{-}135)$$

Strictly speaking, interionic attractions affect the equivalent conductivities differently from the diffusional mobilities. In the former case the two kinds of ion are moving in opposite directions, whereas in diffusion they are moving in the same direction. In brief, two kinds of effects are recognized in conductance. The motion of a given ion in a direction opposite to its atmosphere produces an electrical drag or *relaxation effect,* given by the first term of Eq. (12-27). In addition, the ion atmosphere entrains solvent, so that each ion is in effect swimming in a countercurrent of moving solvent. This is known as the *electrophoretic effect* and is given by the second term of Eq. (12-27). In diffusion, there is no relaxation effect, but there is still an electrophoretic one. The theories of these effects are approximate, and the treatment of the electrophoretic effect is open to certain objections, particularly for other than uni-univalent electrolytes. Also, the derivations are intimately tied to the Debye–Hückel assumptions, and show these approximations. [See Harned and Owen (1950) for further details.]

12-ST-2 The Hittorf method

The analysis of the Hittorf method for the determination of transference numbers was greatly

abridged in Section 12-6, and the more detailed presentation appears here.

We consider first the system shown in Fig. 12-8, for which a qualitative analysis was given. The more exact bookkeeping is as follows, based on 1 $\mathscr{F}$ of electricity passing through the cell. The overall change per faraday is that the anode compartment gains t_- equiv of $AgNO_3$ and the cathode compartment loses t_- equiv. The middle compartment should not change in content since the gains and losses due to the migration of ions past the dividing lines I–II and II–III exactly compensate.

This analysis applies to any cell containing a single electrolyte and having electrodes that generate or consume the cation of the electrolyte. The cell of Fig. 12-8 might have been filled with $CuSO_4$ solution, for example, and have copper electrodes. We would conclude that, per faraday, the anode compartment would gain t_- equiv of $CuSO_4$ and the cathode compartment would lose t_- equiv.

Alternatively, the electrodes might be reversible to the anion, that is, the electrode reaction might produce and consume anion. This would be the case if, for example, the cell were filled with NaCl solution, and the electrodes were silver coated with silver chloride. The bookkeeping is now as follows, on page 486. The opposite working of the electrode reaction thus has the effect of making the anode compartment undergo a net loss of t_+ equiv of NaCl and the cathode compartment undergo a net gain of t_+ equiv of NaCl, in contrast to the preceding example.

It is not necessary, of course, that the electrode reaction involve either of the ions of the electrolyte. If platinum electrodes were used in

---

Cell I. Anode compartment
  Anode reaction: $Ag = Ag^+ + e^-$: gain of 1 equiv of Ag^+
  migration of Ag^+: loss of t_+ equiv of Ag^+
  migration of NO_3^-: gain of t_- equiv of NO_3^-.
Net change: Gain of $1 - t_+$ or t_- equiv of Ag^+, gain of t_- equiv of NO_3^-; or
    gain of t_- equiv of $AgNO_3$.
Cell III. Cathode compartment
  Cathode reaction: $Ag^+ + e^- = Ag$: loss of 1 equiv of Ag^+
  migration of Ag^+: gain of t_+ equiv of Ag^+
  migration of NO_3^-: loss of t_- equiv of NO_3^-.
Net change: Loss of $1 - t_+$ or t_- equiv of Ag^+, loss of t_- equiv of NO_3^-; or
    loss of t_- equiv of $AgNO_3$.

Cell I.   Anode compartment
    Anode reaction: $Ag + Cl^- = AgCl + e^-$:   loss of 1 equiv of Cl^-
    migration of Na^+:   loss of t_+ equiv of Na^+
    migration of Cl^-:   gain of t_- equiv of Cl^-.
Net change: Loss of t_+ equiv of Na^+, loss of $1 - t_-$ or t_+ equiv of Cl^-; or loss of t_+
               equiv of NaCl

Cell III.   Cathode compartment
    Cathode reaction: $AgCl + e^- = Ag + Cl^-$:   gain of 1 equiv of Cl^-
    migration of Na^+:   gain of t_+ equiv of Na^+
    migration of Cl^-:   loss of t_- equiv of Cl^-.
Net change: Gain of t_+ equiv of Na^-, gain of $1 - t_-$ or t_+ equiv of Cl^-; or gain of t_+
               equiv of NaCl.

---

the example just given, the analysis would be as follows below:

This last example illustrates some further points. First, the net change can be expressed either in terms of gains and losses of individual types of ions or in terms of gains and losses of complete electrolytes; the alternative statements are seen on examination to be entirely equivalent. The purpose of the middle compartment now becomes apparent. The analysis assumes that only Na^+ and Cl^- ions carry current past the I–II and II–III dividing lines. The electrode reactions are producing H^+ and OH^- ions, however, and if mixing occurs in compartments I and III during the electrolysis, then these electrode products will carry part of the current between compartments. The gains and losses of Na^+ and Cl^- ions would then be less than expected, and the overall analysis would be in error. The test of whether electrolysis products reached compartment II would, in this case, be whether the pH of II changed or not.

Notice that all of these analyses are couched in terms of amounts gained or lost during electrolysis. The concentrations present do not enter directly. The transference numbers are themselves concentration-dependent, however, since λ_+ and λ_- will in general change differently with concentration. It is desirable for this reason that no great change in composition occur during the electrolysis. Ordinarily, then, the amount of electricity passed through the cell will be some small fraction of a faraday.

It is, of course, necessary to measure the quantity of electricity involved in a Hittorf experiment so that the results can be put on a per faraday basis. It is not necessary, however, to know the applied potential or the actual current—only the total quantity of electricity. One usually obtains this by determining the amount of an electrode reaction that has occurred. Thus in the first example, the loss in weight of the silver anode gives the number of faradays used. It would be possible in the third example to

---

Cell I.   Anode compartment
    Anode reaction: $\frac{1}{2}H_2O = \frac{1}{4}O_2 + H^+ + e^-$:   gain of 1 equiv of H^+
    migration of Na^+:   loss of t_+ equiv of Na^+
    migration of Cl^-:   gain of t_- equiv of Cl^-.
Net change: Gain of 1 equiv of H^+, loss of t_+ equiv of Na^+, gain of t_- equiv of Cl^-;
               or gain of 1 equiv of HCl, loss of t_+ equiv of NaCl.

Cell III.   Cathode compartment
    Cathode reaction: $e^- + H_2O = \frac{1}{2}H_2 + OH^-$:   gain of 1 equiv of OH^-
    migration of Na^+:   gain of t_+ equiv of Na^+
    migration of Cl^-:   loss of t_- equiv of Cl^-.
Net change: Gain of 1 equiv of OH^-, gain of t_+ equiv of Na^+, loss of t_- equiv of
               Cl^-; or gain of 1 equiv of NaOH, loss of t_- equiv of NaCl.

---

titrate the solution in the anode compartment to find the amount of H^+ ion produced. However, it is generally more convenient to have a second cell in series with the Hittorf cell, which functions as a *coulometer*. This might consist of a silver anode dipping into silver nitrate solution and a platinum cathode. Either the loss in weight of the anode or the amount of silver deposited on the cathode would show the number of faradays that had passed through the cell. A numerical example is as follows.

12-ST-3   Treatment of complex ionic equilibria

The treatment of the dissociation of a weak acid or a weak base is developed along standard lines in Section 12-10 but stops at the point of simple acids and bases, HA and BOH. The solution of the simultaneous equations for (H^+) or (OH^-) becomes very unwieldy if more than one stage of dissociation can occur or if there is a mixture of weak electrolytes. An alternative procedure becomes much more practical.

We can construct an illustrative problem using the cell in which an NaCl solution was electrolyzed with platinum electrodes. Suppose that the initial NaCl solution is 0.1000 m, and thus contains 5.844 g of NaCl per 1000 g of water. The electrolysis is carried out until 1.92 g of silver have been deposited in a coulometer in series with the cell, at which point the anode compartment is drained and then rinsed with some of the *original* NaCl solution, so that a total of 301.35 g of solution results which is found to contain 0.02315 equiv of NaCl. Calculate t_+ and t_-.

    EXAMPLE

First, 1.92 g of silver corresponds to $1.92/107.87 = 0.01780$ equiv so that 0.01780 $\mathscr{F}$ of electricity passed through the cell. At the end of the experiment, the 0.02315 equiv of NaCl present in the anode solution plus rinse corresponds to 1.35 g, and so the amount of water in the solution was $301.3 - 1.35 = 300.0$ g. Now the amount of NaCl associated originally with 300 g of water is $(0.1)(300)/1000 = 0.03000$ equiv. There has been a loss of $0.03000 - 0.02315 = 0.00685$ equiv of NaCl.

This loss becomes, per faraday, $0.00685/0.01780 = 0.385$. From the earlier bookkeeping analysis, the loss should be t_+ equiv of NaCl per faraday. The transference number of Na^+ is therefore 0.385 and that of Cl^- is 0.615. The equivalent conductivity of 0.1 m NaCl is 106.74, so that $\lambda_{Na^+} = (0.385)(106.74) = 41.09$ and $\lambda_{Cl^-} = (0.615)(106.74) = 65.65$.

The use of a rinse of original solution is a characteristic procedure in a transference experiment. The purpose of the rinsing, of course, is to displace all of the electrolyzed solution, and the use of original solution for the rinse simplifies the ensuing calculation. We are not interested in concentrations and need only to compare the amount of NaCl present in electrolyzed solution plus rinse with the amount originally associated with the quantity of water in the combined solutions. It is thus not necessary to know exactly how much solution was originally present in the anode compartment.

An alternative but less accurate procedure would have been to fill the anode compartment with a known volume of the 0.1 m solution and to determine the change in concentration resulting from the electrolysis. For example, had the anode compartment originally contained 200 cm^3 of solution, or approximately 0.02 equiv of NaCl, then at the end of the electrolysis the amount remaining would have been $0.02 - 0.00685 = 0.01315$ equiv, corresponding to a 0.0657 m NaCl solution. The solution would also have contained 0.01780 equiv of HCl, or 0.0890 m HCl.

When we calculate transference numbers by the Hittorf method we implicitly assume that the solvent water does not migrate during the electrolysis. The water of hydration of the ions should move with them, however, and Washburn suggested the ingenious experiment of adding a sugar to the solutions in a Hittorf cell and referring the calculations to the sugar as the nonmigrating species. The result showed that several molecules of water do in fact migrate with each ion.

A. Acid HA or base BOH

We first illustrate the method for a monobasic acid HA before proceeding to more complex situations. The material balance equation (12-117) is now written in the form

$$f_{\text{tot}} = f_{\text{HA}} + f_{\text{MA}} = (HA) + \frac{(HA)K_a}{(H^+)}$$

using Eq. (12-114) to eliminate (A^-). Rearrangement gives

$$F_{\text{HA}} = \frac{1}{[K_a/(H^+)] + 1} \qquad (12\text{-}136)$$

where $F_{\text{HA}} = (HA)/f_{\text{tot}}$ and is the fraction of the total acid-substance which is present as undissociated acid HA. Alternatively, we may eliminate (HA) to obtain

$$F_{A^-} = \frac{1}{1 + [(H^+)/K_a]} \qquad (12\text{-}137)$$

Equations (12-136) and (12-137) are usually displayed as a plot of $\log F$ versus pH. Such a plot is called a *logarithmic diagram*. That for acetic acid is shown in Fig. 12-17. Notice that if the pH is much less than pK_a, Eq. (12-137) reduces to

$$\log F_{A^-} = pH - pK_a$$

and so the plot of $\log F_{A^-}$ versus pH becomes a straight line of slope $+1$ and intercept $-pK_a$

in the region of high acidity. Similarly, if $pH \gg pK_a$, then the plot of $\log F_{\text{HA}}$ versus pH becomes a straight line of slope -1 and intercept pK_a.

The logarithmic diagram for a weak electrolyte resembles a phase diagram. It provides a map of the pH domain of each species so that one can tell at a glance the general makeup of a solution at any given pH. The diagram is used quantitatively as follows. The charge balance equation

$$(M^+) + (H^+) = (A^-) + (OH^-)$$
$$[\text{Eq. (12-116)}]$$

can be put in the form

$$(M^+) + (H^+) = F_{A^-}f_{\text{tot}} + (OH^-) \qquad (12\text{-}138)$$

One knows f_{HA} and f_{MA} and hence f_{tot} and (M^+), and Eq. (12-138) is solved by successive approximations. A first guess at (H^+) allows F_{A^-} to be read off the logarithmic diagram, and substitution into the equation will indicate whether the value is too high or too low. This guides a second choice of (H^+), and so on, until a self-consistent answer is reached.

The treatment of a simple weak base BOH is analogous to that just given:

$$F_{\text{BOH}} = \frac{1}{[K_b/(OH^-)] + 1}$$

$$\qquad (12\text{-}139)$$

$$F_{B^+} = \frac{1}{1 + [(OH^-)/K_b]}$$

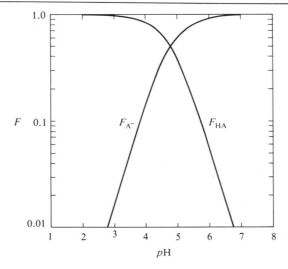

FIGURE 12-17
Logarithmic diagram for acetic acid at 25°C.

We can calculate the pH of a solution which is 0.002 f in HAc and 0.001 f in NaAc;   **EXAMPLE**
f_{tot} is then 0.003 and Eq. (12-138) becomes

$$0.001 + (H^+) = F_{A^-}(0.003) + (OH^-)$$

or

$$F_{A^-} = 0.333 + \frac{(H^+) - (OH^-)}{0.003}$$

We can guess immediately a pH of 4.45, at which $F_{A^-} = \frac{1}{3}$. The test of the equation is then

$$0.333 \overset{?}{=} 0.333 + \frac{3.54 \times 10^{-5}}{0.003} = 0.345$$

F_{A^-} is slightly too small, and we next try a pH of 4.46, $F_{A^-} = 0.345$. The test is now

$$0.345 \overset{?}{=} 0.333 + \frac{3.45 \times 10^{-5}}{0.003} = 0.345$$

The desired pH is then close to 4.46.

One may also calculate a titration curve. In Eq. (12-138), f_{tot} is now the initial formality of the acid, and (M^+) gives the concentration of added strong base. One may now insert successive choices for (H^+) and calculate (M^+) for each. The fraction of neutralization F_N is just

$$F_N = \frac{(M^+)}{f_{tot}}$$

Several points on the titration curve for 0.002 f HAc are summarized thus. Equation (12-138) becomes $(M^+) + (H^+) = 0.002 \, f_{A^-} + (OH^-)$, or $F_N = f_{A^-} - \{[(H^+) - (OH^-)]/0.002\}$; some sample calculations are given in Table 12-9. The calculation illustrates several items. The first two values for F_N are negative, meaning that excess strong acid must be present. The third line gives the actual starting point of the titration; the relatively rapid change in F_N between pH 4 and 6 characterizes the buffer region, and the asymptotic approach to $F_N = 1$ marks the endpoint. The last entry shows that 5.3% excess base has been added. The curve is plotted in Fig. 12-14.

pH	(H^+)	(OH^-)	f_{A^-}	$[(H^+) - (OH^-)]/0.002$	F_N
2	0.01	—	1.7×10^{-3}	5	-5
3	0.001	—	1.7×10^{-2}	0.5	-0.483
3.75	1.78×10^{-4}	—	0.089	0.089	0
4	1×10^{-4}	—	0.149	0.05	0.099
4.76	1.75×10^{-5}	—	0.500	0.009	0.491
6	1×10^{-6}	1.05×10^{-8}	0.9461	—	0.9461
8	1×10^{-8}	1.05×10^{-6}	$1 - (6 \times 10^{-4})$	—	0.9994
10	1×10^{-10}	1.05×10^{-4}	1.000	-0.053	1.053

TABLE 12-9

Calculation of a titration curve using the logarithmic diagram method

The logarithmic diagram is still a plot of log F versus pH, and the associated charge balance equation is

$$F_{B^+} f_{tot} + (H^+) = (X^-) + (OH^-) \qquad (12\text{-}140)$$

where $f_{tot} = f_{BOH} + f_{BX}$, where X is a non-hydrolyzing anion. The plots for ammonium hydroxide are included in Fig. 12-20.

Logarithmic diagrams may be used for any mixture of weak electrolytes. Thus, for a mixture which contains both a weak acid and a weak base, the charge balance equation becomes

$$(M^+) + F_{B^+} f_{(BOH+BX)} + (H^+)$$
$$= (X^-) + F_{A^-} f_{(HA+MA)} + (OH^-) \qquad (12\text{-}141)$$

and the correct pH for a given mixture is found by trial-and-error solution. As an example, a solution which is $0.002\ f$ in HAc, $0.001\ f$ in NaAc, $0.003\ f$ in NH_4OH, and $0.005\ f$ in NH_4Cl would have a pH such that the equation

$$0.001 + 0.008F_{NH_4^+} + (H^+)$$
$$= 0.005 + 0.003F_{Ac^-} + (OH^-)$$

is obeyed. In this case, (H^+) and (OH^-) will be negligible compared to the other terms, and so

$$F_{NH_4}^+ = 0.5 + 0.375F_{Ac}$$

A few successive choices of pH should serve to locate the value such that $F_{NH_4^+}$ and F_{Ac^-} as read off the logarithmic diagrams satisfy the equation.

B. Successful dissociations of a weak acid

The logarithmic diagram approach offers little advantage in speed over the straight attack in the case of a simple acid or base. If, however, there are multiple stages of dissociation, the logarithmic diagram method becomes the one of choice. It is not highly precise, being graphical, but this is not a matter of great importance since the overriding uncertainty will be in the values of the K_γ's, that is, in the size of the activity coefficient corrections.

In the case of a dibasic acid H_2A the equilibrium constants are

$$K_1 = \frac{(H^+)(HA^-)}{(H_2A)} \qquad \text{and} \qquad (12\text{-}142)$$

$$K_2 = \frac{(H^+)(A^{2-})}{(HA^-)}$$

and substitution into the material balance statement $f_{tot} = (H_2A) + (HA^-) + (A^{2-})$ gives

$$F_{H_2A} = \frac{1}{1 + [K_1/(H^+)] + [K_1K_2/(H^+)^2]}$$
$$(12\text{-}143)$$

$$F_{HA} = \frac{1}{[(H^+)/K_1] + 1 + [K_2/(H^+)]}$$
$$(12\text{-}144)$$

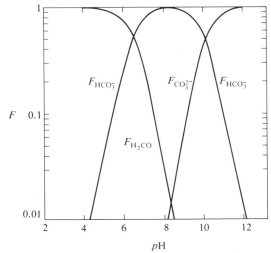

FIGURE 12-18
Logarithmic diagram for H_2CO_3 at 25°C.

$F_{HCO_3^-}$ $F_{CO_3^{2-}}$ $F_{HCO_3^-}$

F_{H_2CO}

pH

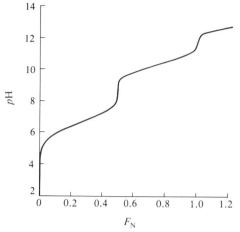

FIGURE 12-19
Titration curve for H_2CO_3.

$$F_{A^{2-}} = \frac{1}{[(H^+)^2/K_1K_2] + [(H^+)/K_2] + 1}$$

$$(12\text{-}145)$$

The logarithmic diagram for H_2CO_3 is shown in Fig. 12-18. Its use follows the same scheme as before, the charge balance equation now being

$$(M^+) + (H^+) = F_{HA^-} f_{tot}$$
$$+ 2F_{A^{2-}} f_{tot} + (OH^-) \quad (12\text{-}146)$$

Figure 12-19 shows the titration curve for 0.1 M H_2CO_3 calculated by the same procedure as before. That is, each assumed pH provides a value for (M^+) through Eq. (12-146) and hence for the degree of neutralization.

The treatment for a tribasic acid H_3A yields the equation

$$F_{H_3A} =$$

$$\frac{1}{1 + [K_1/(H^+)] + [K_1K_2/(H^+)^2] + [K_1K_2K_3/(H^+)^3]}$$

$$(12\text{-}147)$$

and so on. The logarithmic diagram for phosphoric acid is given in Fig. 12-20. Phosphate buffers are much used and the figure allows the calculation of the pH of any mixture of phosphoric acid and its various salts. The dashed curves are for ammonium hydroxide, so the combined plots can be used for mixtures that include ammonium salts as well as phosphates.

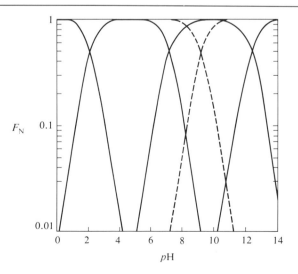

FIGURE 12-20
Logarithmic diagrams for H_3PO_4 (full lines) and NH_4OH (dashed lines).

C. Complex ions

The logarithmic diagram approach is not limited to weak acids and bases but may be applied to any set of successive dissociations. For example, Co^{2+} forms a succession of ammine complexes:

$$H_2O + Co(NH_3)_6^{2+} = Co(NH_3)_5(H_2O)^{2+} + NH_3 \qquad pK_6 = -0.74$$

$$H_2O + Co(NH_3)_5(H_2O)^{2+} = Co(NH_3)_4(H_2O)_2^{2+} + NH_3 \qquad pK_5 = 0.06$$

and so on. The values of pK_1, pK_2, pK_3, and pK_4 are 1.99, 1.51, 0.93, and 0.64, respectively. Note that the reactions have been written as interchanges of water for ammonia in the coordination sphere, in recognition of our belief that Co^{2+} is octahedrally coordinated.

Application of the standard procedure leads to

$$F_{Co(NH_3)_6^{3+}} = \cfrac{1}{1 + [K_6/(NH_3)] + [K_5 K_6/(NH_3)^2] + [K_4 K_5 K_6/(NH_3)^3] + \cdots}$$

and similarly for the other species. The logarithmic diagram now consists of plots of the log F's against $p(NH_3)$, that is, against $-\log(NH_3)$. Application of the charge balance equation then allows the calculation of the composition of any mixture of ammonia and a cobalt salt.

GENERAL
REFERENCES

HARNED, H. S., AND OWEN, B. B. (1950). "The Physical Chemistry of Electrolyte Solutions." Van Nostrand–Reinhold, Princeton, New Jersey.

BOCKRIS, J. O'M., AND CONWAY, B. E. (eds.) (1954). "Modern Aspects of Electro-Chemistry." Butterworths, London and Washington, D.C.

LEWIS, G. N., AND RANDALL, M. (1961). "Thermodynamics," 2nd ed. (revised by K.S. Pitzer and L. Brewer). McGraw-Hill, New York.

ROBINSON, R. A., AND STOKES, R. H. (1970). "Electrolyte Solutions," 2nd ed. (rev.). Academic Press, New York.

MacINNES, D. A. (1939). "The Principles of Electrochemistry." Van Nostrand–Reinhold, Princeton, New Jersey.

CITED
REFERENCES

IMPAY, R. W., MADDEN, P. A., AND McDONALD, I. R. (1983). *J. Phys. Chem.* **87**, 5071.

PALIT, S. R. (1975). *Chemistry* **48**, 16.

PITZER, K. S., AND BREWER, L. (1961). "Thermodynamics," 2nd ed. McGraw-Hill, New York.

ROBINSON, R. A., AND STOKES, R. H. (1970). "Electrolyte Solutions," 2nd ed. (rev.). Academic Press, New York.

EXERCISES

Neglect interionic attraction effects in the following Exercises and Problems unless specifically directed otherwise. Take as exact numbers given to one significant figure.

12-1   The measured resistance of a 0.01 M solution of KCl is 1400 ohms in a cell whose path length can be taken to be 8 cm. Calculate (a) the conductivity of the solution, (b) its conductance, (c) the effective area of the electrodes, and (d) the cell constant.

Ans. (a) 1.499×10^{-3} ohm^{-1} cm^{-1},
(b) 7.143×10^{-4} ohm^{-1}, (c) 3.812 cm^2, (d) 2.100 cm^{-1}.

12-2   The resistance of electrolyte solution A is 40 ohms in a given cell and that of electrolyte solution B is 80 ohms in the same cell. Equal volumes of solutions A and B are mixed. Calculate the resistance of this mixture, again in the same cell.

Ans. 53.33 ohms.

Calculate the equivalent conductivity, Λ_e, from the Λ_m for (a) $CaCl_2$, (b) $ZnSO_4$, (c) $LaCl_3$, and (d) $K_4Fe(CN)_6$.

Ans. (a) 135.84, (b) 132.8, (c) 145.8, (d) 184.5 $(cm^2\ equiv^{-1}\ ohm^{-1})$.

12-3

The conductivity of a saturated solution of AgBr in water at 25°C is $1.880 \times 10^{-7}\ ohm^{-1}\ cm^{-1}$, and the resistivity of the water used is $1.538 \times 10^7\ ohm\ cm$. Calculate $\Lambda_{m,\ AgBr}$ from data in Table 12-2 and thence K_{sp} for AgBr.

Ans. 140.30 $cm^2\ mol^{-1}\ ohm^{-1}$, $7.68 \times 10^{-13}\ M^2$.

12-4

Calculate Λ_m for (a) $Li_4Fe(CN)_6$, (b) $La(NO_3)_3$, and (c) the double salt $K(NH_4)SO_4$.

Ans. (a) 598.68, (b) 422.70, (c) 306.46 $(cm^2\ mol^{-1}\ ohm^{-1})$.

12-5

Condosity, T, is defined as the concentration of a NaCl solution having the same conductivity as a given solution. Calculate T for a 0.02 M solution of Na_2SO_4.

Ans. 0.0411 M.

12-6

Calculate the resistance of a 0.02 M solution of $LaCl_3$ in a conductivity cell having a cell constant of 1.75 cm^{-1}.

Ans. 200 ohm.

12-7

The solubility product for silver chromate is $9 \times 10^{-13}\ M^3$ at 25°C. Calculate the conductivity of a saturated solution at this temperature. The molar conductivity of CrO_4^{2-} may be taken to be the same as that of SO_4^{2-}.

Ans. $3.72 \times 10^{-5}\ ohm^{-1}\ cm^{-1}$.

12-8

The dissociation constant for NH_4OH is 1.77×10^{-5} at 25°C. Calculate Λ_m' and the conductivity of a 0.015 M solution.

Ans. 9.19 $cm^2\ mol^{-1}\ ohm^{-1}$, $1.38 \times 10^{-4}\ ohm^{-1}\ cm^{-1}$.

12-9

It is found that for a certain weak acid, HA, at 25°C, Λ_m' is 84.00 $cm^2\ mol^{-1}\ ohm^{-1}$ at a concentration of $4 \times 10^{-4}\ M$, and 42.00 $cm^2\ mol^{-1}\ ohm^{-1}$ at $1.8 \times 10^{-2}\ M$. Calculate Λ_m° and K_a.

Ans. 420 $cm^2\ mol^{-1}\ ohm^{-1}$, $2.00 \times 10^{-4}\ M^2$.

12-10

Calculate the molar conductivity of 0.05 M HCl at 25°C, allowing for interionic attraction effects.

Ans. 391 $cm^2\ mol^{-1}\ ohm^{-1}$ (the experimental value is 399.32).

12-11

How long should it take for a Na^+ aqueous ion in a field of 2 V cm^{-1} to move a distance equal to its Stokes' law or hydrated radius? Show your work.

Ans. 1.76×10^{-5} s (if 25°C).

12-12

The molar conductivity of a cation M^+ is determined at 25°C by means of a moving boundary experiment. A 0.005 M solution of MCl is used, and after 30 min of passing a current of 0.2 mA the cation boundary has moved 0.208 cm; the area of the column of solution is 1.50 cm^2. Calculate λ_{M^+} (Λ_{MCl} = 132 $cm^2\ mol^{-1}\ ohm^{-1}$.)

Ans. 55.2 $cm^2\ mol^{-1}\ ohm^{-1}$.

12-13

Estimate, with explanation, the molar conductivity of an aqueous M^{2+} ion at 50°C if its hydrated radius is taken to be the same as that of Na^+ ion.

Ans. 326 $cm^2\ mol^{-1}\ ohm^{-1}$.

12-14

Calculate the transference number at 25°C for Cl^- ion in 0.1 M KCl, and in a solution that is also 0.05 M in $NaNO_3$.

Ans. 0.510, 0.362.

12-15

The solubility of $Ba(IO_3)_2$ in water at 25°C is $4.52 \times 10^{-4}\ M$, and the K_{sp} is $6 \times 10^{-10}\ M^3$ in a certain concentration of aqueous KNO_3. Calculate (a) the solubility product in water, (b) the mean molarity

12-16

of the saturated solution, (c) the solubility in the KNO_3 solution, and (d) the mean activity coefficient for the electrolyte Ba^{2+}, $2\ IO_3^-$ in the KNO_3 solution.

Ans. (a) $3.69 \times 10^{-10}\ M^3$, (b) $7.18 \times 10^{-4}\ M$,
(c) $5.31 \times 10^{-4}\ M$, (d) 0.85 (assuming the answer to (a) is also K_{th}).

12-17   If the aqueous KNO_3 solution of Exercise 12-16 is 0.005 m, in what concentration of $La(NO_3)_3$ would the K_{sp} also be $6 \times 10^{-10}\ M^3$?

Ans. $8.33 \times 10^{-4}\ m$.

12-18   Calculate K_{th} for $Ba(IO_3)_2$ from the data of Exercise 12-16, and the actual concentration of the KNO_3 solution. Assume limiting law behavior.

Ans. 2.85×10^{-10}, 0.10 m.

12-19   Calculate $\gamma_\pm$ for 0.05 m acetic acid at 25°C, using Table 12-6.

Ans. 0.926.

12-20   Assuming the ionic strength principle, estimate $\gamma_\pm$ for 0.0333 m $Cu(NO_3)_2$ using the data of Table 12-7 (and assuming that the principle is obeyed by each ion separately).

Ans. 0.50.

12-21   Calculate $\gamma_{Ba^{2+}}$ and γ_{Cl^-} (separately) at 25°C in 0.01 m $BaCl_2$, using the Debye–Hückel theory, and compare the resulting $\gamma_\pm$ with the value in Table 12-7.

Ans. 0.444, 0.816, 0.666 (vs. 0.72).

12-22   What is the Debye–Hückel value for $\gamma_\pm$ of 0.5 m NaCl at 50°C?

Ans. 0.42.

12-23   The solubility product of Ag_2CrO_4 is 9×10^{-12} at 25°C. Calculate the solubility of this salt in 0.1 m silver nitrate, recognizing nonideality.

Ans. $8.32 \times 10^{-9}\ m$.

12-24   Calculate the pH and the degree of dissociation of 0.1 m dichloroacetic acid in 0.05 m HCl at 25°C. K_a is $3.32 \times 10^{-2}\ M^2$.

Ans. pH = 0.332, α = 0.416.

12-25   Repeat the calculation of the pH of the solution of Exercise 12-24 if 0.07 mol liter^{-1} of NaOH were added to it.

Ans. 0.877.

SOME EXERCISES IN SI UNITS

12-26   Calculate the force in newtons per meter between two wires 0.1 meter apart and carrying a current of 10 A.

Ans. 2×10^{-4} N/m.

12-27   The molar conductivity of Na_2SO_4 is 1.299×10^{-2} m^2 mol^{-1} ohm^{-1}. Calculate the resistance of a 0.005 M solution in a cell whose cell constant is 150 m^{-1}.

Ans. 1.155×10^3 ohm.

12-28   The SI unit of conductance is the siemens, S. Show what the units of S are (in kg, m, s, A).

Ans. A^2 kg^{-1} m^{-2} s^3.

12-29   Repeat in SI units the calculations of the example following Eq. (12-40).

Calculate the conductivity of a saturated solution of AgCl. Assume the water used has a conductivity of 70×10^{-6} ohm^{-1} m^{-1}.

12-30

Ans. 2.42×10^{-4} ohm^{-1} m^{-1}.

Calculate the ion atmosphere radius for a 0.02 m solution of a 1–1 electrolyte.

12-31

Ans. 2.155 nm.

PROBLEMS

The following data apply to aqueous solutions of potassium bromate at 25°C:

12-32

Concentration ($M \times 10^3$)	0.5443	0.8370	1.9640	3.2819
κ (ohm^{-1} cm^{-1}), $\times 10^3$	0.0693	0.1061	0.2462	0.4076

Find the limiting molar conductivity of potassium bromate.

A conductivity cell has a resistance of 3736 ohm when filled with 0.028 m H$_2$S solution and of 59.0 ohm when filled with 0.0100 m KCl. The measurements are made at 18°C, at which temperature the molar conductivity for HS$^-$ is 62 cm^2 mol^{-1} ohm^{-1}. Calculate K_a for H$_2$S (relevant data are available in the text, but should be corrected to 18°C with the use of Walden's rule).

12-33

Equation (12-26) can be put in the form $1/\Lambda'_m = a + b\Lambda'_m C$. Show what the constants a and b are. The following data are obtained for aqueous solutions of propionic acid at 25°C:

12-34

Concentration ($M \times 10^3$)	0.5669	0.8712	1.8650	4.8026
Λ'_m (cm^2 mol^{-1} ohm^{-1})	55.32	45.348	31.657	20.099

Use the linear form of Eq. (12-26) to find Λ°_m for propionic acid and its acid dissociation constant. Optional: Evaluate a and b by a least squares method.

The conductivity of a saturated solution of silver iodate at 80°C is 1.25×10^{-5} ohm^{-1} cm^{-1}; that of the water used is 1.4×10^{-6} ohm^{-1} cm^{-1} and the molar conductivity of silver iodate is 89 cm^2 mol^{-1} ohm^{-1}. Calculate the solubility product for silver iodate.

12-35

Calculate the mobilities of H$^+$ and Cl$^-$ ions at 0°C using Walden's rule, and the velocity with which each ion should move in a moving boundary experiment if the field is 0.1 V cm^{-1}. Calculate also the current if the solution is 0.02 M and the cross section of the tube is 0.5 cm^2.

12-36

The molar conductivity of tetramethylammonium bromide is 123.06 cm^2 mol^{-1} ohm^{-1} at 25°C. Use data from Table 12-2 to calculate the molar conductivity of tetramethylammonium chloride.

12-37

The molar conductivity of samarium chloride, SmCl$_3$, is 434.8 cm^2 mol^{-1} ohm^{-1} at 25°C. Use data from Table 12-2 to calculate the molar conductivity of samarium nitrate.

12-38

The molar conductivities at 25°C are 621.6, 459.9, and 1987.2 cm^2 mol^{-1} ohm^{-1} for Sr$_2$(P$_2$O$_7$), La(P$_3$O$_9$), and La$_4$(P$_2$O$_7$)$_3$, respectively. Calculate the molar conductivity for Sr$_3$(P$_3$O$_9$)$_2$, assuming that all the salts are soluble and fully dissociated. Optional: Make the calculation using equivalent conductivities, obtaining the equivalent conductivity of Sr$_3$(P$_3$O$_9$)$_2$.

12-39

One hundred cubic centimeters of 0.10 N sodium acetate solution is titrated with 0.1 N HCl solution. Calculate the conductivity of the resulting solution when 90, 99, 101, and 110 cm^3 of the HCl solution has been added. Bear in mind that acetic acid is only slightly ionized in the presence of HCl and NaAc; in these calculations neglect the variation of molar conductivity with concentration and use the value for infinite dilution. Your answers need be correct to only 1%. Make a semiquantitative plot of your calculated specific conductivities (as ordinate) versus the volume of HCl solution added.

12-40

12-41   If pure water has a conductivity of 4.5×10^{-8} ohm^{-1} cm^{-1} at 20°C, calculate the conductivity of a saturated solution of CO_2 in water at 20°C if the CO_2 pressure is maintained at 20 Torr and the equilibrium constant for the reaction $H_2O(l) + CO_2(aq) = HCO_3^- + H^+$ is 4.35×10^{-7}. The solubility of CO_2 in water follows Henry's law with a constant of 0.03353 mol $liter^{-1}$ atm^{-1}.

12-42   Calculate the molar conductivity for (a) Sm^{3+} and (b) $P_3O_9^{3-}$ if the cation transference numbers at 25°C are 0.473 for $SmCl_3$ and 0.416 for $Sr_3(P_3O_9)_2$. The necessary data are given in Problems 12-38 and 12-39.

12-43   For an incompletely dissociated electrolyte the Onsager equation is

$$\Lambda_m = \alpha[\Lambda_m^\circ - (A + B\Lambda_m^\circ)(\alpha C)^{1/2}] = \alpha\Lambda_m'$$

where $A = 82.4/\eta(DT)^{1/2}$ $B = 8.2 \times 10^5/(DT)^{3/2}$ D is the dielectric constant, and η is the viscosity. The molar conductivity of dichloroacetic acid in 0.03 M solution is 273 cm^2 mol^{-1} ohm^{-1}. For dichloroacetic acid, $\Lambda_m^\circ = 388.5$. Calculate the degree of dissociation α from the Onsager equation. [*Hint:* use a procedure of successive approximations, starting with $\alpha = \Lambda_m/\Lambda_m^\circ$] Assume 25°C.

12-44   In a transport experiment in 0.02 M NaCl solution at 25°C, using the moving boundary method, Longsworth found the boundary between NaCl and $CdCl_2$ solutions to move 6.0 cm in 2070 s with a current of 0.00160 A. The cross section of the tube was 0.12 cm^2. Calculate t_+.

12-45   Rewrite Eqs. (12-51) for the case of λ and Λ denoting equivalent conductivities.

12-46   The mobility of tetramethylammonium ion is 4.66×10^{-4} cm^2 V^{-1} s^{-1}. Calculate the hydrated ion radius. The hydrated ion radius for tetraethylammonium is 38% larger than that for tetramethylammonium ion. Calculate the molar conductivity of tetraethylammonium ion.

12-47   A solution of LiCl was layered over a 33.3×10^{-3} M solution of $GdCl_3$. Because Gd^{3+} has the higher mobility, the boundary remained sharp; after passing a current of 5.6 mA for 1 h 7 m, the boundary moved a distance down the tube corresponding to a volume of 1.00 cm^3. Calculate t_+ for $GdCl_3$.

12-48   A 1×10^{-4} M solution of $NaIO_3$ gives a resistance of 1.6411×10^5 ohm in a cell of cell constant 1.5 cm^{-1}. On saturating the solution with $Ca(IO_3)_2 \cdot 6H_2O$, the resistance dropped to 1.2740×10^4 ohm. The water used had a conductivity of 8.0×10^{-8} ohm^{-1} cm^{-1}. Calculate (a) the molar conductivity of $NaIO_3$ and (b) the solubility product of $Ca(IO_3)_2 \cdot 6H_2O$. You may use data from Table 12-5.

12-49   Derive Eq. (12-98) from Eq. (12-92).

12-50   Derive the equation

$$(1/v) \ln(K_{sp}/K_{sp}^\circ) = \ln \gamma_\pm^\circ - \ln \gamma_\pm$$

where K_{sp}° is the solubility of a salt in water, K_{sp} is the solubility in the presence of added neutral electrolyte, $\gamma_\pm^\circ$ is the mean activity coefficient in the water-saturated solution and $\gamma_\pm$ is the value for the slightly soluble salt in the presence of the added electrolyte.

   The solubility of a certain 1:3 salt is 1.50×10^{-4} M in water at 25°C, and is 2.32 times larger when enough 1:1 neutral electrolyte (no common ion) is added to make the ionic strength 0.3. Calculate (a) the ratio $(\gamma_\pm/\gamma_\pm^\circ)$ and (b) what the increase in solubility should be were the limiting law obeyed. [The salt was $[Co(NH_3)_6][Co(NH_3)_2(NO_2)_2(C_2O_4)]_3$ and thus consisted of a large cation and a large anion.]

12-51   The acid dissociation constant of acetic acid has been measured at 25°C as a function of ionic strength (I was increased by adding NaCl). Given the following data:

$K_a \times 10^5$	1.754	2.292	2.622	3.158	2.475
I	0	0.02	0.06	1.01	2.01

Calculate $\gamma_\pm$ for acetic acid at each of the ionic strengths and plot the values against $\sqrt{I}$.

12-52   Equation (12-100) can be put in the form $(1/\ln\gamma_\pm) = a + b/\sqrt{I}$ Derive this form and make a plot of the activity coefficient data for KCl from Table 12-7 to obtain a value for $\overset{\circ}{a}$. Optional: Make a least squares fit of the data up to 1 m KCl to obtain your values of a and b.

Calculate the concentration of (a) a 1:1 electrolyte, (b) a 1:2 electrolyte, and (c) a 2:2 electrolyte such that in each case the ion atmosphere radius is 10 Å.　**12-53**

The solubility of lead chloride is 0.005 mol liter^{-1} at 25°C. Calculate the activity product (that is, make correction for activity coefficients). Calculate also the solubility of $PbCl_2$ in 0.02 m sodium nitrate solution.　**12-54**

Barney et al. obtained data on the solubility of cobaltous oxalate (CoOx) as a function of added potassium oxalate concentration. The solubility is at first depressed due to the common ion effect, then increased, due to the complex formation:　**12-55**

$$Co^{2+} + 2Ox^{2-} \overset{k}{=} Co(Ox)_2^{2-}$$

The minimum in solubility occurs at $S = 9.2 \times 10^{-5}$ and $(K_2Ox) = 4.5 \times 10^{-4}$ M; the solubility of CoOx in pure water is 1.44×10^{-4} M. Derive the equation whereby K can be calculated from the solubility minimum, and calculate K.

At 25°C, the pK for the dissociation of $Ag(NH_3)_2^+$ into its components is 7.22. For AgCl, AgBr, and AgI, pK_{sp} is 9.77, 12.48, and 16.07, respectively. Calculate how many milligrams of these salts are dissolved by 1 liter of 1.0 m NH_3. Neglect $Ag(NH_3)^+$, NH_4^+, and OH^-.　**12-56**

Ion-exchange particles exhibit the Donnan effect. Here, the porous particle corresponds to the left side of Fig. 12-13 in the case of an anion exchanger. That is, N^+ denotes the positively charged exchange sites that are attached to the ion-exchange polymer matrix. Suppose that the concentration of exchange sites is 2.5 M for a certain ion exchanger. Calculate the interior concentration of Na^+ and of Cl^- when the exchanger is immersed in 0.01 M NaCl solution.　**12-57**

Calculate the concentration of H_2S, HS^-, S^{2-}, and H^+ in a 0.200 M NaHS solution (to an accuracy of about 1%). [Note: The solution to this problem can be much simplified by making judicious approximations at each stage.]　**12-58**

Calculate the concentration of each ion present in a solution made up of 0.01 mole of acetic acid and 0.005 mole of chloroacetic acid in one liter volume.　**12-59**

Solid NaOH is added to 1 liter of a 0.1 f solution of H_2SO_3 until the pH rises to 6. Calculate (a) the initial pH and (b) the number of moles of NaOH and the concentrations of all species present in the final solution.　**12-60**

Leucylglycine is an amino acid that dissociates into both hydrogen and hydroxide ions. At 25°C the constant for the dissociation into anions and hydrogen ions is 1.51×10^{-8}. The apparent dissociation constant for the dissociation into cations and hydroxyl ions is 3.02×10^{-11}. Calculate the pH at which the degree of dissociation into hydrogen ions and hydroxyl ions is the same.　**12-61**

How many millimoles of acid HA should be added to 100 cm^3 of a 0.02 M solution of base BOH in order to give a solution of pH 9? pK_a is 5 for HA and pK_b is also 5.　**12-62**

One hundred cm^3 of a 0.1 f solution of H_2SO_4 is mixed with an equal volume of 0.025 f NaOH. Calculate the pH of the resulting solution.　**12-63**

SPECIAL TOPICS PROBLEMS

Calculate the limiting value for the diffusion coefficient of aqueous KNO_3 at 25°C.　**12-64**

Derive the equation　**12-65**

$$\Lambda_m^\circ = \frac{\mathscr{F}^2}{RT}(\nu_+ z_+^2 \mathscr{D}_+^\circ + \nu_- z_-^2 \mathscr{D}_-^\circ)$$

The symbol $^\circ$ denotes limiting value.

12-66 Calculate $\mathcal{D}$ for 0.1 m KCl at 25°C, assuming that the Debye–Hückel limiting law applies for activity coefficients and neglecting other corrections. Take $\gamma_C = \gamma_{\pm}$ and $C = m$ in Eq. (12-135). Also, $t_+ = 0.4898$ and $\Lambda_m = 128.96$ cm^2 mol^{-1} ohm^{-1} for 0.1 m KCl.

12-67 A solution containing 0.14941 wt% KCl is electrolyzed in a Hittorf cell at 25°C with silver/silver chloride electrodes. After the electrolysis, which deposits 0.16024 g of silver in a silver nitrate coulometer, one of the end cells is analyzed and found to contain 120.99 g of solution of 0.19404 wt% KCl. Calculate the transference number for K$^+$.

12-68 A 0.01 m solution of HCl is electrolyzed in a Hittorf cell with a Pt/H$_2$ anode and a silver/silver chloride cathode. Each compartment contains 50 cm^3 of solution. A current of 2 mA is passed through the cell for 1 hr. Calculate the final concentrations of the various species present in the anode and cathode compartments.

12-69 Find the pH of a solution that is 0.05 f in ammonia and 0.1 f in KHCO$_3$, using Figures 12-17 and 12-19.

12-70* The acid dissociation constants for monochloroacetic acid, HM, and dichloroacetic acid, HD, are 1.55×10^{-3} M^2 and 3.32×10^{-2} M^2, respectively. (a) Make a semilogarithmic plot of the fractions of HM, M$^-$, HD, and D$^-$ vs. pH. (b) Calculate, using this plot, the pH of a solution that is 0.02 f in HD and 0.03 f in HM. (c) Calculate the pH if the solution in (b) is made 0.02 f in NaOH, and if the solution is made 0.03 f in NaOH.

12-71 Calculate the pH of a solution that is 0.02 f in Na$_2$HPO$_4$.

12-72* Ethylenediaminetetraacetic acid is a tetrabasic acid H$_4$Y. The successive pK's are 2.00, 2.67, 6.16, and 10.26. It forms a stable complex with Co ion, Co^{2+} + Y^{4-} = CoY^{2-}.

For this complex formation $K = 10^{16}$; Ba^{2+} forms a similar complex with $K = 10^7$. Both CoY^{2-} and BaY^{2-} are the anions of strong acids, that is, one does not get H$_2$CoY and so on in acid solution (although the solids are known). Calculate and plot the logarithmic diagram for the various forms of ethylenediaminetetraacetic acid for the pH range from 0 to 7, and the similar diagram for the distribution of cobalt and barium between Co^{2+} and Ba^{2+} and CoY^{2-} and BaY^{2-} as a function of pC, where C is the concentration of Y^{4-}. Calculate the percentage of complexing of Co^{2+} and Ba^{2+} in solutions of (a) 0.01 f Co(NO$_3$)$_2$, 0.01 f Ba(NO$_3$)$_2$, 0.1 f Na$_2$H$_2$Y; (b) the same as (a) but also 0.1 f in NaOH. Also give the pH's of these solutions.

12-73 Derive Eq. (12-113). Calculate the osmotic pressure at 25°C if $C = 0.01$ m and $(N^+) = 0.1$ m, neglecting activity coefficient effects.

chapter 13
Electrochemical cells

The preceding chapter dealt primarily with the physical chemistry of electrolyte solutions; we now concern ourselves with the overall chemical process that occurs when electricity is passed through a conducting solution. The emphasis will be on the work associated with this overall change, as measured by the reversible cell potential. Since reversible work at constant temperature and pressure corresponds to a free energy change, we will thus be able to bring the emf of cells into the general scheme of thermodynamics. The chapter concludes with a discussion of irreversible electrode processes, that is, with the physical chemistry of the approach of ions to, and their reaction at, the surface of an electrode.

13-1   Definitions and fundamental relationships

A. Cell conventions

An electrochemical cell has, as essential features, a current-carrying solution and two electrodes at which oxidation and reduction processes occur as current flows. Figure 13-1 gives a schematic of a fairly typical cell for this chapter; we have hydrogen and silver–silver chloride electrodes dipping into an aqueous solution of HCl. The hydrogen electrode, incidentally, typically consists of a platinized platinum metal surface arranged so that hydrogen gas bubbles past as it dips partly into the solution, the object being to provide the most intimate possible gas–solution–metal contact. Platinized platinum is merely platinum metal on which additional, very finely divided platinum has been deposited electrolytically; the result is a high area, catalytically active surface. A silver–silver chloride electrode consists of silver on which a fine-grained, adherent deposit of silver chloride has been placed, again electrolytically. The terminals of the cell might be connected to a motor, so as to provide electrochemical energy, or, in the laboratory, to a potentiometer circuit (see Section 13-2), so that the potential difference could be measured.

It is awkward to describe cells in a pictorial manner, and the conventional representation of the cell of Fig. 13-1 is

$$Pt/H_2(P \text{ atm})/HCl(m)/AgCl/Ag \qquad (13\text{-}1)$$

FIGURE 13-1
Schematic diagram of an
electrochemical cell. The
cell reaction is given by
Eq. (13-2).

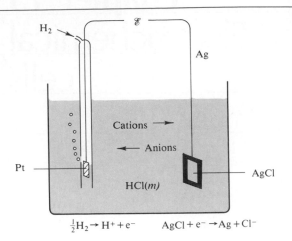

Equation (13-1) is known as a *cell diagram.* The rule is that one writes in order each successive phase that makes up the electrical circuit of the cell, using a diagonal bar to separate phases. One should in general specify not only the temperature of the cell, but also the composition of each condensed phase and the partial pressure of any gaseous one; we assume the general mechanical pressure to be 1 atm.

A potentiometric measurement on a cell will report a potential difference between the electrode terminals, and a second convention is needed to specify the sign of this cell potential $\mathscr{E}$. There is some variation in practice and consequent ambiguity, and the least confusing way of stating the convention used here seems to be the following. We first define the cell reaction as the chemical change that occurs per faraday of electricity passed through the cell in the direction such that oxidation occurs at the left-hand electrode of the cell diagram. This electrode will be called the *anode*; it is also the electrode toward which anions migrate in the cell solution as they carry current. The right-hand electrode of the cell diagram is then the one at which reduction occurs, and it is called the *cathode*; it is also the electrode toward which cations migrate as they carry current. The cell reaction corresponding to (13-1) is

$$\text{anode} \qquad \tfrac{1}{2}H_2(1\ \text{atm}) = H^+(m) + e^-$$

$$\text{cathode} \qquad AgCl + e^- = Ag + Cl^-(m) \tag{13-2}$$

$$\text{net} \qquad \tfrac{1}{2}H_2(1\ \text{atm}) + AgCl = Ag + H^+(m) + Cl^-(m)$$

Since the potential is a function of concentration as well as of the chemical species involved, statements of cell reactions should include the concentration or partial pressure of each substance.

The sign of $\mathscr{E}$ is now defined to be positive if the cell reaction occurs spontaneously in the direction written and to be negative if the reverse direction of reaction is the spontaneous one. In this particular example $\mathscr{E}$ would be positive. That is, silver chloride is spontaneously reduced by hydrogen gas. Although the mechanical arrangement of the cell is such that the direct chemical reaction is prevented from occurring, the process will take place spontaneously when the electrode terminals are connected.

This last is an important feature of electrochemical cells. The cell reaction has to be spontaneous in one direction or the other, and the cell must always be so designed that the direct chemical reaction is physically prevented from occurring. In the case of the cell of Eq. (13-1) the reactants H_2 and AgCl are isolated at the separate electrodes. Another way in which direct chemical reaction is prevented is illustrated by the *Daniell cell* shown in Fig. 13-2; this consists of a zinc anode dipping into $ZnSO_4$ solution and a copper cathode dipping into $CuSO_4$ solution. The two solutions are separated by a porous diaphragm which allows electrical contact but prevents gross mixing. The cell diagram is then

$$Zn/ZnSO_4(m_1) \; / \; CuSO_4(m_2)/Cu \tag{13-3}$$

where the dashed diagonal conventionally is used to indicate two miscible phases that are physically prevented from mixing. This situation will later be referred to as one of a *liquid junction*. The cell reaction is

anode $\qquad\qquad \frac{1}{2}Zn = \frac{1}{2}Zn^{2+}(m_1) + e^-$

cathode $\qquad \frac{1}{2}Cu^{2+}(m_2) + e^- = \frac{1}{2}Cu \tag{13-4}$

net $\qquad \frac{1}{2}Zn + \frac{1}{2}Cu^{2+}(m_2) = \frac{1}{2}Zn^{2+}(m_1) + \frac{1}{2}Cu$

Again, the cell has been written in such a way that the cell reaction occurs spontaneously—if placed directly into the $CuSO_4$ solution, the zinc electrode would react as shown—and so the measured emf of this cell would be reported as positive.

To return to the matter of sign convention, we note that the usual source of confusion is in the plus and minus markings of electrodes. The cells of Figs. 13-1 and 13-2 are drawn so that in spontaneous action the left-hand electrode is the anode. This means that the anode bears a positive charge relative to the cathode *at the solution end* and a negative charge relative to the cathode *at the exposed terminals*. It is the terminals that are marked, hence it is the spontaneously operating anode of a cell that bears the negative sign. To repeat, when we write a cell reaction the various signs and directions of

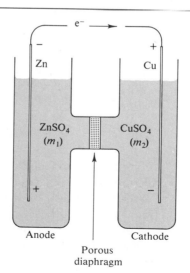

FIGURE 13-2
The Daniell cell.

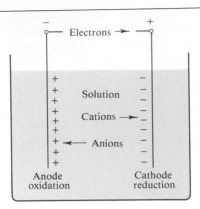

flow are taken to be as shown in Fig. 13-3. If this is the spontaneous direction of flow, the $\mathscr{E}$ is reported as a positive number.

A further feature of electrochemical cells that are used in precise measurements is that they are *reversible.* That is, the cell reaction must take place readily in either direction. If the cell is short-circuited, the reaction should proceed in its spontaneous direction; and if an external potential is applied which overrides the natural cell potential, then the reaction should just as readily proceed in the opposite direction. The reversible electrochemical cell is thus one which may be held in a state of dynamic balance by application of an external counterpotential just equal to $\mathscr{E}$. For example, in the case of the Daniell cell $\mathscr{E}$ is about 1.1 V if $m_1 = m_2 = 1$. External application of an opposing 1.1 V will just prevent reaction from occurring; a slightly smaller opposing potential will allow the cell reaction to occur as written, and a slightly larger opposing potential will make the reaction go in the opposite direction.

The customary way of determining $\mathscr{E}$ for a cell is, in fact, to find that opposing emf which puts the cell in balance. The procedure thus defines $\mathscr{E}$ as the reversible emf of the cell. If current is allowed to flow through the cell under conditions such that applied potential remains essentially equal to $\mathscr{E}$, then the work done is a reversible work. From the definition of potential, work in joules is given by qV, where q is the amount of charge carried through potential difference V. The reversible work for an electrochemical cell is then $n\mathscr{F}\mathscr{E}$, where n is the number of faradays passed through the cell.

For the Daniell cell as written in Eq. (13-4), the reversible work is $(96,485)(1.1) = 1.06 \times 10^5$ J ($m_1 = m_2 = 1$, and 25°C). Had the reaction been written for $2\mathscr{F}$,

$$Zn + Cu^{2+}(m_2) = Zn^{2+}(m_1) + Cu \tag{13-5}$$

the reversible work would be 2.12×10^5 J, and had the reaction been written in the opposite direction, $\mathscr{E}$ would be reported as -1.1 V and the corresponding reversible work would be -1.06×10^5 J. Thus the sign of $\mathscr{E}$ and both the sign and magnitude of the reversible work depend on how the cell reaction is written. A useful conversion factor is that for $n = 1$ an emf of 1 V $= 23.06$ kcal. This unit is sometimes called the *electron-volt:* 1 eV $= 23.06$ kcal mol^{-1}.

The reversible emf that is measured for a cell gives the reversible work associated with the cell reaction. Since this is reversible work at constant temperature and pressure, it is therefore the free energy change and, by Eq. (6-39), the sign convention is such that we must write

$$\Delta G = -n\mathcal{F}\mathcal{E} \qquad (13\text{-}6)$$

that is, a positive cell emf corresponds to a spontaneous cell reaction and hence to a negative free energy change.

We may now rewrite several important equations of Section 7-4 in terms of emf's. Thus Eqs. (7-24) and (7-25) become

$$\Delta S = n\mathcal{F}\left(\frac{\partial \mathcal{E}}{\partial T}\right)_P \qquad (13\text{-}7)$$

$$\Delta S^0 = n\mathcal{F}\left(\frac{\partial \mathcal{E}^0}{\partial T}\right)_P \qquad (13\text{-}8)$$

and, since $G = H - TS$ by definition, for a constant-temperature process we have

$$\Delta H = \Delta G + T\Delta S \qquad (13\text{-}9)$$

$$\Delta H = -n\mathcal{F}\left[\mathcal{E} - T\left(\frac{\partial \mathcal{E}}{\partial T}\right)_P\right] \qquad (13\text{-}10)$$

$$\Delta H^0 = -n\mathcal{F}\left[\mathcal{E}^0 - T\left(\frac{\partial \mathcal{E}^0}{\partial T}\right)_P\right] \qquad (13\text{-}11)$$

Equation (13-10) or (13-11) is known as the *Gibbs–Helmholtz equation.*

The emf of the cell Cd/solution saturated with $CdCl_2 \cdot 2.5H_2O$/AgCl/Ag is 0.6753 V at 25°C, and $d\mathcal{E}/dT = -0.00065$ V K^{-1}. If we write the cell reaction as Cd + 2AgCl = $CdCl_2$(sat. soln.) + 2Ag, then $\Delta S = (2)(96,485)(-0.00065) = -125\,JK^{-1}$ or -29.9 cal K^{-1}, $\Delta H = -(2)(96,485)[0.6753 - (298.1)(-0.00065)] = -1.677 \times 10^5$ J or -40.08 kcal (as compared with -39.5 kcal from thermochemical measurements), and $\Delta G = -(2)(96,485)(0.6753) = -1.303 \times 10^5$ J or -31.14 kcal.

EXAMPLE

An interesting point is that the q for a reversibly operating cell is given by $T\Delta S$; q would be $-(125)(298.1)$ or -37.27 kJ in the example. In terms of Eq. (13-9) the measured q is given by ΔH when the reaction occurs directly, as in a thermochemical experiment; in the reversible cell the energy ΔG goes to do useful work, and the observed q is then determined by the entropy change. Thus the statement $\Delta H = \Delta G + T\Delta S$ amounts to saying:

(total energy change)
$$= \begin{pmatrix}\text{energy available} \\ \text{to do work}\end{pmatrix} + \begin{pmatrix}\text{energy not available} \\ \text{to do work}\end{pmatrix}$$

(in a constant pressure system).

C. The Nernst equation

A very important relationship is obtained as follows. For a general cell reaction

$$aA + bB + \cdots = mM + nN + \cdots$$

we have from Eq. (12-104) that

$$\Delta G = \Delta G^0 + RT \ln Q_{th} \qquad [\text{Eq. (12-104)}]$$

where, it will be remembered, Q_{th} has the same form as an equilibrium constant but contains the activities of the products and reactants as arbitrarily specified by the stated reaction. If the system is at equilibrium, however, $\Delta G = 0$, and we then obtain

$$\Delta G^0 = -RT \ln K_{th} \qquad [\text{Eq. (12-105)}]$$

where K_{th} is the thermodynamic equilibrium constant.

Combination of Eqs. (13-6) and (12-104) gives the Nernst equation:

$$\mathscr{E} = \mathscr{E}^0 - \frac{RT}{n\mathscr{F}} \ln Q_{th} \qquad (13\text{-}12)$$

Insertion of the numerical constants for 25°C yields

$$\mathscr{E} = \mathscr{E}^0 - \frac{0.02569}{n} \ln Q_{th} = \mathscr{E}^0 - \frac{0.05917}{n} \log Q_{th} \qquad (13\text{-}13)$$

Equation (13-12) is the central equation of electrochemistry. By means of it we can determine how the emf of a cell should vary with composition, and we can also determine $\mathscr{E}^0$ for a cell reaction, which in turn enables us to obtain activity coefficients for electrolytes.

EXAMPLE   The Nernst equation for the cell reaction of Eq. (13-2) is

$$\mathscr{E} = \mathscr{E}^0 - 0.02569 \ln \frac{a_{H^+}\, a_{Cl^-}}{P_{H_2}^{1/2}} \qquad (13\text{-}14)$$

Ag and AgCl are in their standard states and hence have unit activity. The hydrogen pressure will be 1 atm, and, for the moment, we neglect activity coefficient effects, so Eq. (13-14) reduces to

$$\mathscr{E} = \mathscr{E}^0 - 0.02569 \ln[(H^+)(Cl^-)] = \mathscr{E}^0 - 0.05139 \ln m \qquad (13\text{-}15)$$

The observed emf of this cell is 0.49844 at 25°C and $m = 0.005$ and we may use the Nernst equation to calculate $\mathscr{E}$ for some other concentration, say 0.01 m. Since $\mathscr{E}^0$ is a constant, it follows from Eq. (13-15) that

$$\mathscr{E}_{0.01m} = \mathscr{E}_{0.005m} - 0.05139 \ln \frac{0.01}{0.005} = 0.46282$$

The observed value is 0.46419, a difference we will shortly be attributing to the nonideality of aqueous HCl.

As a different kind of example, for the reaction of the Daniell cell, Eq. (13-4), we have

$$\mathscr{E} = \mathscr{E}^0 - 0.02569 \ln \frac{a_{Zn^{2+}}^{1/2}}{a_{Cu^{2+}}^{1/2}} = \mathscr{E}^0 - 0.01285 \ln \frac{a_{Zn^{2+}}}{a_{Cu^{2+}}}$$

If Zn^{2+} and Cu^{2+} are at unit activity (or, very roughly, 1 m), $\mathscr{E} = 1.10$ V at 25°C. This means that $\mathscr{E}^0$ is also 1.10 V, since the log term is zero. If an excess of zinc metal is placed in a solution of copper sulfate which is initially at unit activity, the direct spontaneous reaction will occur to form copper metal and zinc ion. Eventually the solution will consist of roughly 1 m or unit activity Zn^{2+} and some small equilibrium concentration of Cu^{2+}. We can use the Nernst equation to calculate this last. Since the final state is at equilibrium, ΔG and hence $\mathscr{E}$ must be zero, and so

$$\mathscr{E}^0 = 1.10 = 0.01285 \ln \left(\frac{a_{Zn^{2+}}}{a_{Cu^{2+}}}\right)_{equil}$$

$$= 0.01285 \ln \left(\frac{1}{a_{Cu^{2+}}}\right)_{equil}$$

The equilibrium (Cu^{2+}) is then $\exp(-1.10/0.01285)$, or about 10^{-37} $m(!)$.

13-2   Experimental procedures

A. The potentiometer

The reader is referred to experimental texts for details, but the principle employed in the measurement of emf's of cells should be described at least briefly. An elementary potentiometer arrangement is shown in Fig. 13-4. One sets up a closed circuit involving a working battery, usually a wet cell capable of delivering a reasonable amount of current without changing its voltage. The circuit contains a moderately high resistance, perhaps 1000 ohms, which may be in the form of a slide wire, or which may be tapped at close intervals. This resistance R then has an ohmic drop in potential IR across it. The electrochemical cell is connected as shown, with the electrodes in the same direction as for the working cell. One now moves the point of contact A until the galvanometer G shows no current flow. At this point the potential drop AB is the same as $\mathscr{E}$ for the cell.

The circuit is usually calibrated by means of a standard cell, or one of accurately known emf $\mathscr{E}_{ref}$. By determining point A' when the standard cell is in balance, one therefore knows the voltage drop $A'B$. The desired emf is then $\mathscr{E}_{ref}(R_{AB}/R_{A'B})$. In actual practice one adjusts subsidiary resistances in the circuit (not shown) in calibrating with the standard cell so that the tapped or slide wire resistance position A will read directly in volts.

Potentiometric measurements are among the most accurate of physical

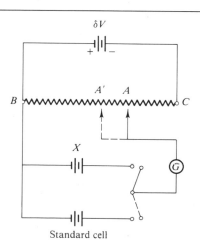

FIGURE 13-4
A potentiometer circuit.

chemistry. Cell potentials may be measured to about 10^{-5} V, the limiting accuracy usually being that of the voltage of the reference cell. If the unknown cell is one of very high resistance, then the sensitivity of the galvanometer may become the limiting factor. In the case of a pH meter (Section 13-8), for example, the glass electrode may have 10^5 ohms resistance, and one must use an electronic null-meter to detect the balance point.

The potentiometric method is a null method—when the circuit is balanced the galvanometer shows no current flow, and a slight shift in the position of the contact A to one side or the other causes a galvanometer deflection in one direction or the other. A barely detectable galvanometer deflection need correspond to no more than perhaps 10^{-12} A, so the change in condition needed to reverse the current flow is very small and the measured potential is essentially the reversible one. If a relatively large current, say 10^{-3} A, is drawn through a cell, then the potential drops as various irreversible processes occur, such as polarization, discussed in Section 13-10. The magnitude of the measured potential is therefore at a maximum at zero current.

B. Standard cells   The most widely accepted reference cell is the *Weston cell*, illustrated in Fig. 13-5. The anode consists of a layer of solid cadmium amalgam containing 12.5% cadmium, and the cathode consists of a pool of mercury layered with a thick paste of Hg_2SO_4. The solution is saturated with $CdSO_4 \cdot \frac{8}{3}H_2O$, with some excess crystals present on both sides, to maintain saturation. The cell diagram is

$$Pt/Cd(Hg)/\text{saturated } CdSO_4 \cdot \tfrac{8}{3}H_2O/Hg/Pt$$

and the corresponding cell reaction is

$$\tfrac{8}{3}H_2O + Cd(\text{amalgam}) + Hg_2SO_4(s)$$
$$= CdSO_4 \cdot \tfrac{8}{3}H_2O(s) + 2\,Hg \tag{13-16}$$

The emf at 20°C is 1.0186 V; the temperature dependence is small, -4.06×10^{-5} V K^{-1}.

FIGURE 13-5
The Weston cell.

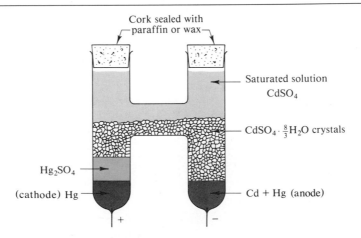

Cork sealed with paraffin or wax

Saturated solution CdSO$_4$

CdSO$_4 \cdot \frac{8}{3}$H$_2$O crystals

Hg$_2$SO$_4$

(cathode) Hg

Cd + Hg (anode)

An electrochemical cell consists essentially of the two parts defined by the two electrodes, and one often constructs a cell in which one electrode is that under investigation and the other is a conventional electrode of known properties. This last is called a *reference electrode*. A very common and easily constructed one is the *calomel reference electrode*, illustrated in Fig. 13-6(a). Platinum wire dips into a pool of mercury which is layered with a paste of Hg_2Cl_2, followed by a solution which is usually 0.1 N KCl, 1 N KCl, or saturated KCl. Electrolytic connection must be made to the rest of the cell, and this is done through a side arm in which the KCl solution has usually been stiffened with agar-agar or gelatine. This type of electrolytic connection is known as a *salt bridge* (see Section 13-ST-1).

A complete cell might then appear as Fig. 13-6(b). The cell diagram is

$$Pt/Hg/Hg_2Cl_2(s)/1\ N\ KCl\ \text{or saturated KCl} \not/ HCl(m)/H_2(1\ atm)/Pt$$

If the HCl is at unit activity, then at 25°C

$$\mathscr{E}_{298} = -0.2807\ \text{V (1 } N \text{ KCl) and } \mathscr{E}_{298}$$
$$= -0.2415\ \text{V (saturated KCl)}$$

The boundary between the KCl solution and that of the electrolyte of the second part of the cell is known as a *liquid junction*. Since ions are carrying the current in solution, passage of electricity means that ions move across the junction, just as in a transference experiment. In terms of this cell some K^+ ions must move from the KCl solution into the HCl one, and some Cl^- ions must move from a concentration m in the HCl to that in the KCl. Some net changes thus occur at the liquid junction, whose free energy requirement contributes to the emf of the cell as a whole. This contribution is known as the *junction potential*, and fortunately it is small if the bulk of the current is carried by oppositely charged ions of the same mobility. This is essentially the situation in the case of a KCl salt bridge; K^+ and Cl^- do have nearly the same mobility. The more detailed treatment of junction potentials is given in Section 13-ST-1, and it is sufficient here to note that for most purposes the junction potential for a saturated KCl (or a concentrated NH_4NO_3) salt bridge can be neglected. However, this effect does impair the accuracy of a cell involving a calomel reference electrode.

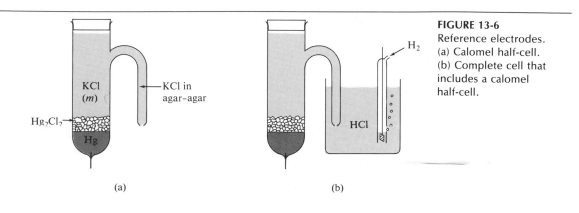

(a) (b)

FIGURE 13-6
Reference electrodes.
(a) Calomel half-cell.
(b) Complete cell that includes a calomel half-cell.

Another reference electrode is the hydrogen electrode itself. As illustrated in Fig. 13-6(b), hydrogen gas is bubbled past a Pt electrode; this last is *platinized*, that is, coated with a deposit of finely divided platinum. While capable of great accuracy, the hydrogen electrode is inconvenient to use, mainly because of the need to supply hydrogen gas at an accurately known pressure (why must the pressure be known?). Other reversible electrodes are the silver–silver ion and silver–silver chloride ones. The latter is illustrated in Fig. 13-1, and consists of a silver wire or piece of sheet silver on which an adherent deposit of silver chloride has been placed (by electrodeposition).

For accurate work a reference electrode should be a direct part of the cell, but it may be more convenient to connect the electrode being studied to the reference electrode via a salt bridge and to either neglect or try to estimate the value of the junction potential.

13-3   Determination of $\mathscr{E}^0$ values and activity coefficients

Equation (13-12) may be written in the form

$$\mathscr{E} = \mathscr{E}^0 - \frac{RT}{n\mathscr{F}} \ln Q - \frac{RT}{n\mathscr{F}} \ln Q_\gamma \tag{13-17}$$

where

$$Q_{th} = \frac{a_M{}^m a_N{}^n \cdots}{a_A{}^a a_A{}^b \cdots} = \frac{(M)^m (N)^n \cdots}{(A)^a (B)^b \cdots} \frac{\gamma_M{}^m \gamma_N{}^n \cdots}{\gamma_A{}^a \gamma_B{}^b \cdots} = Q Q_\gamma \tag{13-18}$$

We now write

$$\mathscr{E}' = \mathscr{E} + \frac{RT}{n\mathscr{F}} \ln Q = \mathscr{E}^0 - \frac{RT}{n\mathscr{F}} \ln Q_\gamma \tag{13-19}$$

The quantity $\mathscr{E}'$ is determined for a series of concentrations and plotted against concentration. At infinite dilution the activity coefficients and hence Q_γ approach unity, and $\ln Q_\gamma$ approaches zero; the extrapolated value of $\mathscr{E}'$ is thus equal to $\mathscr{E}^0$.

The procedure may be illustrated for the cell corresponding to reaction (13-2) at 25°C. We write Eq. (13-15) in the form

$$\mathscr{E} = \mathscr{E}^0 - 0.05139 \ln m - 0.05139 \ln \gamma_\pm$$

or

$$\mathscr{E}' = \mathscr{E} + 0.05139 \ln m = \mathscr{E}^0 - 0.05139 \ln \gamma_\pm \tag{13-20}$$

According to the Debye–Hückel limiting law, Eq. (12-97), $\ln \gamma_\pm$ should be proportional to $\sqrt{m}$, so $\mathscr{E}'$ is plotted against $\sqrt{m}$ as shown in Fig. 13-7. Extrapolation to zero concentration gives $\mathscr{E}^0_{298} = -0.22239$ V. Having determined the $\mathscr{E}^0$ for the cell reaction, one may then insert its value back into Eq. (13-20) and thus obtain $\gamma_\pm$ for HCl at each concentration.

This procedure illustrates how a number of very accurate values of $\mathscr{E}^0$ and of activity coefficients have been obtained. One may also, of course, estimate Q_γ either theoretically or from other activity coefficient data and thus calculate $\mathscr{E}^0$ from the measured $\mathscr{E}$.

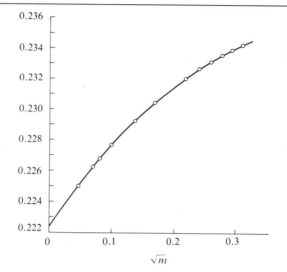

FIGURE 13-7
Determination of $\mathscr{E}^0_{298}$ for
the cell Pt/H$_2$(1 atm)/
HCl(m)/AgCl/Ag.

13-4   Additivity rules for emf's. Standard reduction potentials

Since the $\mathscr{E}$ or $\mathscr{E}^0$ for a cell reaction is just the free energy change per equivalent, emf's obey essentially the same additivity rules as do free energies. For example,

(a) $Cu^{2+} + H_2 \;=\; Cu \;+\; 2H^+$
(b) $Zn^{2+} + H_2 \;=\; Zn \;+\; 2H^+$

_____

(c) = (a) − (b) $Zn \;+\; Cu^{2+} = Zn^{2+} + Cu.$

We know that $\Delta G^0_{(c)} = \Delta G^0_{(a)} - \Delta G^0_{(b)}$, hence

$$-n\mathscr{F}\mathscr{E}^0_{(c)} \;=\; -n\mathscr{F}\mathscr{E}^0_{(a)} - (-n\mathscr{F}\mathscr{E}^0_{(b)}) \tag{13-21}$$

or $\mathscr{E}^0_{(c)} = \mathscr{E}^0_{(a)} - \mathscr{E}^0_{(b)}$. In general, if two cell reactions are added (subtracted) to produce a complete reaction, the resultant emf is the sum (difference) of those for the two reactions.

We make use of this attribute in much the same way as is done in formulating enthalpies and free energies of formation. First, all emf data are expressed as $\mathscr{E}^0$ relative to the hydrogen electrode as anode. Second, a cell reaction is expressed as a combination of two *half-cells*. For example, reaction (a) is broken down as

$Cu^{2+} \;+\; 2e^- \;=\; Cu$ $\mathscr{E}^0_{Cu^{2+}/Cu}$
$H_2 \;=\; 2H^+ \;+\; 2e^-$ $-\mathscr{E}^0_{H^{2+}/H_2}$

_____

$Cu^{2+} \;+\; H_2 \;=\; Cu + 2H^+$ $\mathscr{E}^0 = \mathscr{E}^0_{Cu^{2+}/Cu} - \mathscr{E}^0_{H^+/H_2}$

Thus any cell reaction may be written as the difference between two half-cell reduction reactions, and an $\mathscr{E}^0$ as the difference between two standard half-cell reduction potentials. $\mathscr{E}^0$ for reaction (c) above may therefore be written as

$$\mathscr{E}^0_{(c)} = (\mathscr{E}^0_{Cu^{2+}/Cu} - \mathscr{E}^0_{H^+/H_2}) - (\mathscr{E}^0_{Zn^{2+}/Zn} - \mathscr{E}^0_{H^+/H_2})$$
$$= (\mathscr{E}^0_{Cu^{2+}/Cu} - \mathscr{E}^0_{Zn^{2+}/Zn}) \quad (13\text{-}21)$$

Some indirect estimates suggest that the absolute standard half-cell potential for H^+/H_2 is small, but it is apparent that its actual value is immaterial in combining equations, since it cancels out. The next step, accordingly, is to make the convenient, arbitrary assignment that $\mathscr{E}^0_{H^+/H_2} = 0$ and to report the measured $\mathscr{E}^0$ values for reactions such as (a) and (b) above as the actual values for the half-cells. Thus $\mathscr{E}^0_{298} = 0.337$ V and -0.763 V for reactions (a) and (b), respectively, and we report that

$$\mathscr{E}^0_{Cu^{2+}/Cu} = 0.337 \text{ V} \qquad \text{and} \qquad \mathscr{E}^0_{Zn^{2+}/Zn} = -0.763 \text{ V}$$

Then $\mathscr{E}^0_{298}$ for reaction (c) is $0.337 - (-0.763) = 1.10$ V. This result is *independent* of the assumption regarding $\mathscr{E}^0_{H^+/H_2}$. Reaction (c) is the same as reaction (13-5), corresponding to the cell described by Eq. (13-3) and Fig. 13-2. Notice that the above calculation amounts to subtracting the reduction potential for the right-hand electrode from that for the left-hand one.

The general convention is that the emf of a cell is given by

$$\mathscr{E}^0_{cell} = \mathscr{E}^0_{right} - \mathscr{E}^0_{left} \qquad (13\text{-}22)$$

where $\mathscr{E}^0_{right}$ and $\mathscr{E}^0_{left}$ are the *standard reduction potentials* of the half-cells corresponding to the cathode and anode of the full cell, respectively.*

The general mass of emf data has been reduced by means of this formalism, and a number of standard half-cell reduction potentials are given in Table 13-1. Their use follows the example just given. Each value is actually the $\mathscr{E}^0$ for the cell whose cathode is the stated half-cell and whose anode is the standard hydrogen electrode.

A further aspect of the use of half-cell potentials is illustrated by the following example. Consider the reaction

$$\text{(c) } 2Hg + 2AgCl(s) = Hg_2Cl_2(s) + 2Ag \qquad (13\text{-}22)$$

which can be written as the difference between

$$\text{(a) } AgCl(s) + e^- = Ag + Cl^- \qquad \mathscr{E}^0_{298} = 0.2224$$

and

$$\text{(b) } Hg_2Cl_2(s) + 2e^- = 2Hg + 2Cl^- \qquad \mathscr{E}^0_{298} = 0.2676$$

The cell diagram corresponding to reaction (c) is

$$Hg/Hg_2Cl_2(s) \Big/ \begin{array}{c}\text{solution of electro-}\\ \text{lyte containing } Cl^-\end{array} \Big/ AgCl(s)/Ag \qquad (13\text{-}24)$$

and it follows from Eq. (13-22) that

---

*The older American convention has been to use *oxidation* rather than reduction potentials. There is no change in the sign convention for a cell, but now $\mathscr{E}^0_{cell} = \mathscr{E}^0_{left} - \mathscr{E}^0_{right}$ where the half-cell potentials are for oxidation (and opposite in sign to reduction potentials). The change to use of reduction potentials has been largely in obeisance to European practice.

TABLE 13-1
Standard reduction
potentials at 25°C[a]

Half-cell reaction	$\mathscr{E}^0_{298}$	Half-cell reaction	$\mathscr{E}^0_{298}$
$Li^+ + e^- = Li$	-3.045	$Cu^{2+} + 2e^- = Cu$	0.337
$K^+ + e^- = K$	-2.925	$Cu^+ + e^- = Cu$	0.521
$Ca^{2+} + 2e^- = Ca$	-2.87	$I_2(s) + 2e^- = 2I^-$	0.5355
$Na^+ + e^- = Na$	-2.714	$Ag_2SO_4 + 2e^- = 2Ag + SO_4^{2-}$	0.653
$Mg^{2+} + 2e^- = Mg$	-2.37	$2H^+ + O_2 + 2e^- = H_2O_2$	0.682
$Al^{3+} + 3e^- = Al$	-1.66	$Fe(cN)_6^{3-} + e^- = Fe(CN)_6^{4-}$	0.69
$Zn^{2+} + 2e^- = Zn$	-0.763	$Fe^{3+} + e^- = Fe^{2+}$	0.771
$Fe^{2+} + 2e^- = Fe$	-0.440	$Hg_2^{2+} + 2e^- = 2Hg$	0.789
$Cd^{2+} + 2e^- = Cd$	-0.403	$Ag^+ + e^- = Ag$	0.7991
$PbSO_4(s) + 2e^- = Pb + SO_4^{2-}$	-0.356	$2Hg^{2+} + 2e^- = Hg_2^{2+}$	0.920
$Tl^+ + e^- = Tl$	-0.3363	$Br_2(\ell) + 2e^- = 2Br^-$	1.0652
$Pb^{2+} + 2e^- = Pb$	-0.126	$H^+ + \frac{1}{4}O_2 + e^- = \frac{1}{2}H_2O$	1.23
$AgI(s) + e^- = Ag + I^-$	-0.156	$Tl^{3+} + 2e^- = Tl^+$	1.25
$Fe^{3+} + 3e^- = Fe$	-0.036	$Cl_2(g) ;+ 2e^- = 2\,Cl^-$	1.36
$2H^+ + 2e^- = H_2$	0.0000	$PbO_2 + SO_4^{2-} + 4H^+ + 2e^-$	
$AgBr(s) + e^- = Ag + Br^-$	0.0713	$\quad = PbSO_4 + 2H_2O$	1.685
$Cu^{2+} + e^- = Cu^+$	0.153	$F_2(g) + 2e^- = 2F^-$	2.87
$AgCl(s) + e^- = Ag + Cl^-$	0.22239		
Saturated calomel	0.242	*Basic solutions*	
$Hg_2Cl_2(s) + 2e^- = 2Hg + 2Cl^-$	0.2676	$SO_4^{2-} + H_2O + 2e^- = SO_3^{2-} + 2OH^-$	-0.93
Normal calomel	0.280	$2H_2O + 2e^- = H_2 + 2OH^-$	-0.8281
0.1 N calomel	0.3358	$Ni(OH)_2 + 2e^- = Ni + 2OH^-$	-0.72
		$H_2O + HO_2^- + 2e^- = 3OH^-$	0.88

[a] Largely adapted from G.N. Lewis and M. Randall, "Thermodynamics," 2nd ed. (revised by K.S. Pitzer and L. Brewer). McGraw-Hill, New York, 1961.

$$\mathscr{E}^0_{(c)} = 0.2224 - 0.2676 = -0.0452$$

The point of this example is that although we must multiply reaction (a) by 2 before combining it with reaction (b) in order to obtain reaction (c), we subtract the emf's directly. This is because an emf corresponds to the free energy change *per Faraday* [Eq. (13-6)], so that all emf's are on the same basis.

There is one situation where the additivity principle must be handled with care, namely in the combining of two half-cell reactions to give a third half-cell reaction. Each emf should be weighted by the number of faradays for which the half-cell reaction is written. As an example, consider the case

(a) $\quad Fe^{2+} + 2e^- = Fe \qquad\qquad \Delta G^0_{(a)} = -2\mathscr{F}\mathscr{E}^0_{(a)}$

(b) $\quad Fe^{3+} + e^- = Fe^{2+} \qquad\qquad \Delta G^0_{(b)} = -\mathscr{F}\mathscr{E}^0_{(b)}$

(c) $\quad Fe^{3+} + 3e^- = Fe \qquad\qquad \Delta G^0_{(c)} = -3\mathscr{F}\mathscr{E}^0_{(c)}$

Since $\Delta G^0_{(c)} = \Delta G^0_{(a)} + \Delta G^0_{(b)}$, it follows that $\mathscr{E}^0_{(c)} = (2\mathscr{E}^0_{(a)} + \mathscr{E}^0_{(b)})/3$. At 25°C, we get $\mathscr{E}^0_{(c)} = [(2)(-0.440) + (0.771)]/3 = -0.036$ (as confirmed in Table 13-1).

The general equation for the addition of half-cell reactions (a) and (b) to give half-cell (c) is

$$\mathscr{E}^0_{(c)} = \frac{n_{(a)}\,\mathscr{E}^0_{(a)} + n_{(b)}\,\mathscr{E}^0_{(b)}}{n_{(c)}} \tag{13-25}$$

where n denotes number of faradays. Note that if two half-cell reactions are combined to give a whole reaction, that is, one in which electrons have cancelled out, then $n_{(a)} = n_{(b)} = n_{(c)}$ and we revert to the simple situation of Eq. (13-21). Whenever one has any questions about combining cell emf's, it is best to start with the ΔG's since these are always additive.

13-5   Thermodynamic quantities for aqueous ions

The free energies of formation of aqueous ions are closely related to half-cell potentials. Consider, for example, the reaction

$$K(s) + H^+ = K^+(aq) + \tfrac{1}{2}H_2(g) \qquad \Delta G^0_{298} = -282.22 \text{ kJ} \qquad (13\text{-}26)$$

The customary convention (as might be expected) is to take the standard free energy of formation of $H^+(aq)$ to be zero, that is, to say that

$$\tfrac{1}{2}H_2(g) = H^+(aq) + e^- \qquad \Delta G^0 = 0 \qquad (13\text{-}27)$$

The standard free energy of formation of $K^+(aq)$ is then just ΔG^0 for Eq. (13-26),

$$K(s) = K^+(aq) + e^- \qquad \Delta G^0_{f,K^+} = -282.22 \text{ kJ} \qquad (13\text{-}28)$$

Recalling Eq. (13-6), $\mathscr{E}^0_{298}$ for Eq. (13-28) is $-(-282,220)/(96,485) = 2.925$ V; this is just the standard *oxidation* potential for the K/K^+ half-cell. The standard reduction potential is for the reverse of Eq. (13-28), or $\mathscr{E}^0_{K^+/K} = -2.925$ V (as confirmed in Table 13-1).

The above illustrates that the standard half-cell reduction potential for a metal ion to the metal may alternatively be stated in terms of a standard free energy of formation of that ion. For the general case of

$$M(s) = M^{n+} + ne^- \qquad (13\text{-}28)$$

where M is a metal,

$$\Delta G^0_{f,M^{n+}} = n\mathscr{F}\mathscr{E}^0_{M^{n+}/M} \qquad (13\text{-}29)$$

Corresponding relationships can be worked out to obtain the standard free energy of formation of a negative ion from its elements.

There are other ways of determining the free energy of formation of an aqueous electrolyte than from emf data. As an example, the solubility of KCl is 4.82 M at 25°C and the mean activity coefficient in the saturated solution is 0.588. The standard free energy of solution is given by the thermodynamic solubility product [see Eq. (12-106)]:

$$\Delta G^0_f(KCl,aq) - \Delta G^0_f(KCl,s) = -RT \ln K_{th} = -2RT \ln a_\pm \text{ (sat)}$$

$$= -(2)(8.3144)(298.15) \ln[(4.82)(0.588)]$$

$$= -5.164 \text{ kJ mol}^{-1}$$

Since we know $\Delta G^0_f(KCl,s)$ at 25°C (Table 7-2), we can get $\Delta G^0_f(KCl,aq)$ as -413.48 kJ mol^{-1}. This is the sum of the standard free energies of formation of K^+ and Cl^- ions, and if we subtract our previously determined value of -282.22 kJ mol^{-1} for K^+, we find $\Delta G^0_f(Cl^-,aq) = -131.26$ kJ mol^{-1}. In the case of AgCl, the corresponding calculation might be based on the thermodynamic solubility product as determined by extrapolation of the K_{sp} to zero ionic strength (as illustrated in Fig. 12-9). Having $\Delta G_f^0(AgCl, aq)$ and $\Delta G_f^0(Cl^-, aq)$, we can get the value for $\Delta G_f^0(Ag^+, aq)$.

The standard free energies of formation of a number of common ions have been worked out by procedures of these types, and a selection of values is given in Table 13-2.

Table 13-2 includes standard entropies of formation. These may be obtained by two routes. One way is to use Eq. (13-8), which requires a knowledge of the temperature dependence of the half-cell potential. Alternatively, the standard enthalpy of formation may be obtained from thermochemical data (see Table 5-3) or from the temperature dependence of a thermodynamic solubility product, and the relationship $\Delta G_f^0 = \Delta H_f^0 - T\Delta S_f^0$ used. Note that in Table 13-2 the added step of converting ΔS_f^0 values to absolute entropies has been taken.

All three of these thermodynamic quantities are obtained on the basis that their values are zero for $H^+(aq)$. It is worth re-emphasizing that this assumption cancels out when the values are combined to get a ΔG^0, ΔH^0, or ΔS^0 for a complete reaction, just as with half-cell potentials. Nor does it matter what the state of the electron is in equations such as Eq. (13-27), since the assumption likewise cancels out.

EXAMPLE

Calculate ΔG_f^0 ΔS_f^0 and ΔH_f^0 for $KClO_4(aq)$ at 25°C. We use Table 13-2 to find $\Delta G_f^0 = -282.3 + (-10.33) = -292.6$ kJ mol^{-1}. The entropies in Table 13-2 are absolute entropies, which means we have to go to Table 6-2 to get all the values needed for the formation reaction, $\frac{1}{2}Cl_2 + 2O_2 + K = KClO_4(aq)$. We get $\Delta S_f^0 = 102.5 + 180.7 - \frac{1}{2}(223.0) - (2)(205.1) - 63.6 = -302.1$ J K^{-1} mol^{-1}. Then ΔH_f^0 is $-292.6 + (298.15)(-302.1) = -382.7$ kJ mol^{-1}. This may be compared with the value of $-251.21 - 131.42 = -382.63$ from Table 5-3.

Ion	$\Delta\bar{S}^0$		$\Delta G_f^{0\,b}$		**TABLE 13-2**
	cal K^{-1} mol^{-1}	J K^{-1} mol^{-1}	kcal mol^{-1}	kJ mol^{-1}	Standard free energies of formation and absolute entropies of aqueous ions at 25°C[a]
H^+	(0.000)	(0.000)	(0.000)	(0.000)	
Li^+	3.4	14.2	−70.22	−298.3	
Na^+	14.4	60.2	−62.59	−261.87	
K^+	24.5	102.5	−67.46	−282.3	
Ag^+	17.67	73.9	18.43	77.1	
OH^-	−2.52	−10.5	−37.59	−157.3	
F^-	−2.3	−9.6	−66.08	−276.5	
Cl^-	13.2	55.2	−31.35	−131.2	
Br^-	19.29	80.7	−24.57	−102.8	
I^-	26.14	109.4	−12.35	−51.7	
ClO_4^-	43.2	180.7	−2.47	−10.33	
SO_4^{2-}	4.1	17.2	−177.34	−742.0	
CO_3^{2-}	−12.7	−53.1	−126.22	−528.1	

[a]Adapted from G.N. Lewis and M. Randall, "Thermodynamics," 2nd ed. (revised by K.S. Pitzer and L. Brewer). McGraw-Hill, New York, 1961.

[b]We are dealing with a species in solution and, strictly speaking, these are partial molal free energies or $\Delta\mu_f^0$ values.

13-6 Emf and chemical equilibria

A. Thermodynamic relationships

Combination of Eqs. (13-6) and (12-105), $\Delta G^0 = -RT \ln K_{th}$, gives an important relationship between the $\mathscr{E}^0$ for a cell and the equilibrium constant for the cell reaction:

$$\mathscr{E}^0 = \frac{RT}{n\mathscr{F}} \ln K_{th} \tag{13-30}$$

or, for 25°C

$$\log K_{th} = \frac{n}{0.05917} \mathscr{E}^0 \tag{13-31}$$

Thus every 59 mV in emf corresponds to one power of ten in K, if $n = 1$. Alternatively,

$$\ln K_{th} = \frac{n}{0.02569} \mathscr{E}^0 \tag{13-32}$$

B. Direct applications

A very direct and useful application of $\mathscr{E}^0$ values is to the treatment of oxidation–reduction equilibria in solution. The following examples illustrate typical situations.

EXAMPLE

To what extent will Zn reduce 0.01 m Fe^{2+} at 25°C? The reaction is

$$Zn + Fe^{2+} = Zn^{2+} + Fe$$

and, from Table 13-1, $\mathscr{E}^0_{298} = -0.440 - (-0.763) = 0.323$. Then $K_{th} = \exp[(2)(0.323)/(0.02569)] = 8.3 \times 10^{10}$, or

$$8.3 \times 10^{10} = \frac{a_{Zn^{2+}}}{a_{Fe^{2+}}}$$

The reaction will go virtually to completion, or, if activity coefficients are neglected, until (Zn^{2+}) is essentially 0.01 m; the equilibrium (Fe^{2+}) is then $0.01/8.3 \times 10^{10} = 1.2 \times 10^{-13}$.

EXAMPLE

To what extent should 0.01 m Hg_2^{2+} disproportionate into Hg and Hg^{2+} at 25°C? We combine the following half-cell reactions:

$$\text{(a)} \quad \tfrac{1}{2} Hg_2^{2+} + e^- = Hg(\ell) \qquad \mathscr{E}^0_{(a)} = 0.789$$

$$-[\text{(b)} \quad Hg^{2+} + e^- = \tfrac{1}{2} Hg_2^{2+} \qquad \mathscr{E}^0_{(b)} = 0.920]$$

$$\text{(c)} \quad Hg_2^{2+} = Hg(\ell) + Hg^{2+} \qquad \mathscr{E}^0_{(c)} = -0.131$$

$K_{th} = \exp(-0.131/0.02569) = 6.10 \times 10^{-3}$, so we have

$$6.10 \times 10^{-3} = \frac{(Hg^{2+})}{(Hg_2^{2+})} = \frac{(Hg^{2+})}{0.01 - (Hg^{2+})}$$

from which $(Hg^{2+}) = 6.06 \times 10^{-5}$. The calculation supposes that some $Hg(\ell)$ is formed or is present.

Several of the half-cell reactions of Table 13-1 are written as the reduction of a slightly soluble salt to give a metal and an aqueous anion. The potential for such a half-cell reaction may be combined with the one for the simple reduction of the metal ion to give the solubility product of the salt. If we have

C. Determination of solubility products

$$
\begin{array}{llll}
\text{(a)} & MX(s) + e^- = M + X^- & \mathscr{E}^0_{(a)} \\
-[\text{(b)} & M^+ + e^- = M & \mathscr{E}^0_{(b)}]
\end{array}
$$

$$
\begin{array}{llll}
\text{(c)} & MX(s) = M^+ + X^- & \mathscr{E}^0_{(c)} = \mathscr{E}^0_{(a)} - \mathscr{E}^0_{(b)}
\end{array}
$$

then, neglecting activity coefficients, $K_{sp} = \exp(\mathscr{E}^0_{(c)}/0.02569)$ at 25°C.

An alternative and sometimes very useful approach is the following. We write the Nernst equation for reaction (b):

$$\mathscr{E}_{(b)} = \mathscr{E}^0_{(b)} + 0.02569 \ln a_{M^+} \tag{13-33}$$

The potential at a metal–metal ion electrode must always reflect the chemical potential of that ion; in the presence of X^-, $MX(s)$ forms, which decreases a_{M^+} according to the solubility constant, $a_{M^+} = K_{th}/a_{X^-}$. The standard potential of reaction (a) must therefore be the same as the potential for the metal ion–metal electrode for that value of a_{M^+} which is present when a_{X^-} is unity. This value of a_{M^+} is just K_{th}, since $K_{th} = a_{M^+}a_{X^-}$. We thus have (for 25°C)

$$\mathscr{E}^0_{(a)} = \mathscr{E}^0_{(b)} - 0.02569 \ln(1/a_{M^+}, \, a_{X^-} = 1) = \mathscr{E}^0_{(b)} + 0.02569 \ln K_{th}$$

or

$$K_{th} = \frac{\mathscr{E}^0_{(a)} - \mathscr{E}^0_{(b)}}{0.02569} \tag{13-34}$$

which is the same result as before. This alternative viewpoint, although longer as presented here, becomes advantageous when one is dealing with more complicated situations.

Calculate the solubility product for Ag_2SO_4 if $\mathscr{E}^0_{298} = -0.627$ V for the cell

EXAMPLE

$$Ag/Ag_2SO_4/H_2SO_4(m)/H_2/Pt$$

Since $\mathscr{E}^0_{H^+/H_2}$ is zero by convention, the $\mathscr{E}^0$ for the cell is $-\mathscr{E}^0_{Ag_2SO_4/Ag}$ that is,

$$
\begin{array}{lll}
\text{anode reaction:} & 2Ag + SO_4^{2-} = Ag_2SO_4(s) + 2e^- & \mathscr{E}^0 = -\mathscr{E}^0_{Ag_2SO_4/Ag} \\
\text{cathode reaction:} & 2H^+ + 2e^- = H_2 & \mathscr{E}^0 = 0
\end{array}
$$

$$
\begin{array}{lll}
\text{cell reaction:} & 2Ag + SO_4^{2-} + 2H^+ = Ag_2SO_4(s) + H_2 & \mathscr{E}^0_{cell} = -\mathscr{E}^0_{Ag_2SO_4/Ag}
\end{array}
$$

If we combine the anode half reaction with

$$2Ag^+ + 2e^- = 2Ag \qquad \mathscr{E}^0_{Ag^+/Ag}$$

we obtain

$$2Ag^+ + SO_4^{2-} = Ag_2SO_4(s) \qquad \mathscr{E}^0 = \mathscr{E}^0_{Ag^+/Ag} - \mathscr{E}^0_{Ag_2SO_4/Ag}$$

or $\mathscr{E}^0 = 0.799 - 0.627 = 0.172$. It follows that

$$0.172 = (\mathscr{E}^0_{Ag^+/Ag} - \mathscr{E}^0_{Ag_2SO_4/Ag}) = \frac{0.02569}{2} \ln(1/K_{th}) \tag{13-35}$$

where K_{th} is the thermodynamic solubility product, or

$$K_{th} = \exp[-(2/0.02569)(0.172)] = 1.53 \times 10^{-6}$$

Application of the alternative procedure is as follows. From the point of view of emf, an equivalent statement of the cell reaction is

$$2Ag = 2Ag^+ + 2e^-$$

but with a_{Ag^+} at the value that occurs with $a_{SO_4^{2-}} = 1$. Therefore,

$$\mathscr{E}^0_{cell} = -0.627 = -\mathscr{E}^0_{Ag^+/Ag} - \frac{0.02569}{2} \ln a^2_{Ag^+}$$

$$= 0.799 - \frac{0.02569}{2} \ln K_{th},$$

which rearranges to Eq. (13-35).

D. Determination of dissociation constants

The alternative procedure described above is useful in dealing with cells involving a metal ion–metal couple where the metal ion is mostly present as a complex. If one knows the emf of the cell, one can calculate $a_{M^{z+}}$ where M^{z+} is the free or uncomplexed metal ion. This in turn allows the calculation of the equilibrium constant for complex formation.

EXAMPLE

The potential for the cell

$$Cu/0.02 \ f \ Cu(II) \ in \ 0.5 \ f \ NH_3/normal \ calomel \ electrode$$

is 0.26 V at 25°C. [We use formalities since the copper is largely present as the complex $Cu(NH_3)_4^{2+}$ and we wish merely to describe the overall makeup of the solution.] We treat the left-hand electrode as a Cu/Cu^{2+} electrode whose emf is determined by $a_{Cu^{2+}}$ in the solution: The emf of the cell is therefore written

$$\mathscr{E}_{cell} = \mathscr{E}_{ref} - \left[\mathscr{E}^0_{Cu^{2+}/Cu} - \frac{0.02569}{2} \ln(1/a_{Cu^{2+}}) \right]$$

or

$$0.26 = 0.280 - \left[0.337 + \frac{0.02569}{2} \ln a_{Cu^{2+}} \right]$$

whence $a_{Cu^{2+}} = 1.91 \times 10^{-11}$, and $(Cu^{2+}) \approx 1.91 \times 10^{-11} \ M$.

Virtually all of the Cu^{2+} is in the form of $Cu(NH_3)_4^{2+}$ so that $(NH_3) = 0.5 - (4)(0.02) = 0.42$, and we evaluate the equilibrium constant for complex formation as

$$K = \frac{(Cu(NH_3)_4^{2+})}{(Cu^{2+})(NH_3)^4} = \frac{0.02}{(1.91 \times 10^{-11})(0.42)^4} = 3.37 \times 10^{10}$$

(neglecting activity coefficients).

A number of equilibrium constants for the dissociation of complex ions have been determined in this way. We see in Section 13-9 that an analogous procedure may be applied to the determination of a_{H^+} in a solution.

13-7 Concentration cells

The term *concentration cell* is used to designate a cell whose net reaction involves only changes in concentrations of species (or of gas pressures) and no *net* oxidation or reduction. The $\mathscr{E}^0$ for such a cell must be zero since all species in the cell reaction are then to be at unit activity, in which case no net change at all accompanies the passage of electricity.

A very straightforward type of concentration cell is the following:

A. Electrode concentration cells

$$\text{Pt/H}_2(P_1)/\text{HCl } (m)/\text{H}_2(P_2)/\text{Pt}$$

for which the cell reaction is

anode	$H_2(P_1) = 2H+ (m) + 2e^-$
cathode	$2H^+(m) + 2e^- = H_2(P_2)$
net reaction	$H_2(P_1 = H_2 (P_2),$

with

$$\mathscr{E}_{298} = -\frac{0.02569}{2} \ln \frac{P_2}{P_1} \tag{13-36}$$

In the case of a metal electrode the metal may be present as an amalgam:

$$\text{Cd}(x_1, \text{ in Hg})/\text{CdSO}_4(m)/\text{Cd}(x_2, \text{ in Hg})$$

with

$$\mathscr{E}_{298} = -\frac{0.02569}{2} \ln \frac{a_{x_2}}{a_{x_1}} \tag{13-37}$$

Examination of Eqs. (13-36) and (13-37) shows that for every power of ten in the pressure or mole fraction ratio, there will be a contribution of ~30 mV.

In a simple electrolyte concentration cell liquid junctions are avoided by setting up two opposing cells that differ only in their electrolyte concentration. The following is an example:

B. Simple electrolyte concentration cells

$$\text{Ag/AgCl/HCl}(m_1)/\text{H}_2(1 \text{ atm})/\text{Pt} - \text{Pt/H}_2(1 \text{ atm})/\text{HCl}(m_2)/\text{AgCl/Ag},$$

first anode	$Ag + Cl^-(m_1) = AgCl + e^-$
first cathode	$H^+(m_1) + e^- = \frac{1}{2}H_2(1 \text{ atm})$
second anode	$\frac{1}{2}H_2(1 \text{ atm}) = H^+(m_2) + e^-$
second cathode	$AgCl + e^- = Ag + Cl^-(m_2)$
net reaction	$H^+(m_1) + Cl^-(m_1) = H^+(m_2) + Cl^-(m_2)$

Note that it is important to write each ionic species separately if the electrolyte is in fact treated as fully dissociated. The emf for this cell is

$$\mathscr{E}_{298} = -0.02569 \ln \frac{a_{H^+, \, m_2} a_{Cl^-, \, m_2}}{a_{H^+, \, m_1} a_{Cl^-, m_1}}$$

$$\mathscr{E}_{298} = -0.05138 \ln \frac{m_2}{m_1} - 0.05138 \ln \frac{\gamma_{\pm, m_2}}{\gamma_{\pm, m_1}}$$

Cells of this type may be used to obtain the ratio of activity coefficients of an electrolyte at two different concentrations.

Electrolyte concentration cells having a liquid junction are discussed in the Special Topics section.

13-8   Oxidation-reduction reactions

We consider here the situation in which both the oxidized and reduced forms of the half-cell couple are solution species. Examples are the couples Fe^{2+}/Fe^{3+} and Cu^+/Cu^{2+}. Both partners of the redox couple are in solution, and since they can be exposed only to one of the electrodes (or else the cell would be short-circuited), it is mandatory that the second electrode be connected by means of a salt bridge. Thus we have

$$Pt/Cu^+(m_1), \, Cu^{2+}(m_2)/\text{standard calomel electrode}$$

The potential of such a cell is given by the general form

$$\mathscr{E}_{298} = \mathscr{E}_{ref} - \left[\mathscr{E}^0_{anode} - \frac{0.02569}{n} \ln(a_{red}/a_{oxid}) \right] \tag{13-38}$$

where the anode reaction is

$$M(red) = M(oxid) + ne^-$$

and a_{red} and a_{oxid} refer to the activities of the reduced and the oxidized forms of the species of the redox couple. As mentioned in Section 13-2C, the presence of a liquid junction introduces some unavoidable inaccuracy in the measured emf; to a first approximation, however, the correction is a constant one as the ratio a_{red}/a_{oxid} is varied. (See Section 13-ST-1.)

Cells of this type have been used in the determination of the redox emf not only for metal ion couples but also for a variety of inorganic coordination compounds and for many organic systems. An example of the second type is the $Fe(CN)_6^{3-}/Fe(CN)_6^{4-}$ couple, for which $\mathscr{E}^0_{298} = 0.36$ V, and an example of an organic system is the quinone-hydroquinone couple

$$[O=C_6H_4=O] + 2H^+ + 2e^- = [HO-C_6H_4-OH] \qquad \mathscr{E}^0_{298} = 0.6994 \text{ V}$$

$$\tag{13-39}$$

It will be seen in the next section that the above couple is useful in pH determinations.

Returning to inorganic examples, we note that most elements exist in several oxidation states and that the potentials between these states are generally determined by cells of the type represented by Eq. (13-38). Table 13-3 summarizes a few of these emf relationships. The number between oxidation states is the standard reduction potential at 25°C. We assume acid solutions and use H^+ and H_2O as needed in writing the balanced half-cell reactions.

A cell corresponding to Eq. (13-38) may also be used to follow a redox titration. Let us say that ion M_1 is initially present in reduced form and that an oxidizing agent, ion M_2, is added progressively. The reaction

$Cr^{-0.91}$ $Cr^{2+-0.41}$ $Cr^{3+1.33}$ $Cr_2O_7^{2-}$	**TABLE 13-3**
$Mn^{-1.1}$ $Mn^{2+1.51}$ $Mn^{3+0.95}$ $MnO_2^{2.26}$ MnO_4^{2-} $^{0.564}$ MnO_4^-	Selected standard
$Fe^{-0.44}$ $Fe^{2+0.771}$ Fe^{3+}	reduction potentials for
$Co^{-0.277}$ $Co^{2+1.82}$ $Co^{3+>1.8}$ CoO_2^-	the elements in their
$Ni^{-0.250}$ $Ni^{2+1.78}$ NiO_2	various valence states[a]
$Cu^{0.521}$ $Cu^{+0.153}$ Cu^{2+}	
$Cl^{-1.3595}$ $Cl_2^{1.63}$ $HClO^{1.645}$ $HClO_2^{1.21}$ $ClO_3^{-1.19}ClO_4^-$	
$Br^{-1.07}$ $Br_2^{1.59}$ $HBrO^{1.49}$ BrO_3^-	
$I^{-0.535}$ $I_2^{1.45}$ $HIO^{1.14}$ $IO_3^{-(1.7)}$ H_5IO_6	

[a]Adapted from W. M. Latimer, "The Oxidation States of the Elements and Their Potentials in Aqueous Solution," 2nd ed. Prentice-Hall, Englewood Cliffs, New Jersey, 1952.

$$M_1(red) + M_2(oxid) = M_1(oxid) + M_2(red)$$

occurs. We assume that the reaction is rapid so that each stage of the titration the system is in in equilibrium, so that

$$K = \frac{a_{M_1}(oxid)a_{M_2}(red)}{a_{M_1}(red)a_{M_2}(oxid)} \qquad (13\text{-}40)$$

This means that the ratio $[a_{M_1}(red)/a_{M_1}(oxid)]/[a_{M_2}(red)/a_{M_2}(oxid)]$ is a constant.

The detailed treatment is most easily explained in terms of a specific example. Consider the reaction of Cu^+ with Fe^{3+}:

$$Cu^+ + Fe^{3+} = Cu^{2+} + Fe^{2+} \qquad (13\text{-}41)$$

The cell would be Pt/Cu^+(plus Cu^{2+}, Fe^{3+}, Fe^{2+} as the titration proceeds)/calomel reference electrode.

That is, one starts with the left-hand cell compartment containing only Cu^+ and measures the emf as successive amounts of Fe^{3+} are added.

The cell potential is $\mathscr{E}_{cell} = \mathscr{E}_{ref} - \mathscr{E}_{Pt}$, where $\mathscr{E}_{Pt}$ is the half-cell potential at the platinum electrode. There can only be one such potential, so it must be true that

$$\mathscr{E}_{Pt} = \mathscr{E}_{Fe^{3+}/Fe^{2+}} = \mathscr{E}_{Cu^{2+}/Cu^+} \qquad (13\text{-}42)$$

That is, the two redox couples cannot be establishing two *different* potentials at the platinum electrode! Equation (13-42) must apply at all stages of the titration. Thus, for 25°C,

$$\mathscr{E}_{Pt} = \mathscr{E}^0_{Cu^{2+}/Cu^+} - 0.02569 \ln \frac{(Cu^+)}{(Cu^{2+})}$$

$$= \mathscr{E}^0_{Fe^{3+}/Fe^{2+}} - 0.02569 \ln \frac{(Fe^{2+})}{(Fe^{3+})}$$

(We have neglected activity coefficients.) Equation (13-43) rearranges to

$$\mathscr{E}^0_{Cu^{2+}/Cu^+} - \mathscr{E}^0_{Fe^{3+}/Fe^{2+}} = -0.02569 \ln \frac{(Fe^{2+})(Cu^{2+})}{(Fe^{3+})(Cu^+)} = -0.618$$

Equation (13-44) defines the equilibrium constant for Eq. (13-41) and may be solved for each stage of the titration. One inserts the equilibrium ratio of $(Cu^+)/(Cu^{2+})$ or that of $(Fe^{2+})/(Fe^{3+})$ in Eq. (13-43) to obtain $\mathscr{E}_{Pt}$ and thence $\mathscr{E}_{cell}$. Equation (13-44) tells us that the equilibrium constant is large, which makes the calculation relatively easy. For example, if at a certain stage in the titration we have $(Cu_{tot}) = 0.01$ f and have added enough Fe^{3+}

to make the solution 0.0025 f in iron, then $(Cu^{2+})/(Cu^+) = (.0025/0.0075)$ to a good approximation.

At the endpoint, $(Fe_{tot}) = (Cu_{tot})$, and since $(Fe^{2+}) = (Cu^{2+})$ by the stoichiometry of Eq. (13-41), it follows that $(Cu^+) = (Fe^{3+})$. Equation (13-44) reduces to

$$\mathscr{E}^0_{Cu^{2+}/Cu^+} - \mathscr{E}^0_{Fe^{3+}/Fe^{2+}} = -(0.02569)(2)\ln\left[\frac{(Fe^{2+})}{(Fe^{3+})}\right] \quad \text{endpoint}$$

Insertion of this relationship into the right-hand side of Eq. (13-43) to eliminate the ln term leads to

$$\mathscr{E}_{Pt,\ endpoint} = \tfrac{1}{2}\left(\mathscr{E}^0_{Cu^{2+}/Cu^+} + \mathscr{E}^0_{Fe^{3+}/Fe^{2+}}\right) \qquad (13\text{-}45)$$

Equation (13-45) gives the value of $\mathscr{E}_{Pt}$ and hence of $\mathscr{E}_{cell}$ at the endpoint. The plot of this particular titration is shown in Fig. 13-8, where F is the degree of advancement of the titration.

13-9   Determination of pH

The term pH was coined by S. Sorensen in 1909 to mean $-\log(H^+)$, and we have come to use the symbol $p(X)$ as an operator meaning $-\log X$, where X may be a concentration, an activity, or an equilibrium constant; pK_a means $-\log K_a$, for example. A potential measurement always reflects the activity of the species present, and only at infinite dilution can activity be equated with concentration. Potentiometric methods for pH determination therefore measure some type of hydrogen ion activity, although it turns out that the exact nature of what is measured depends on the cell that is used. The modern procedure is to define pH as essentially

$$pH = -\log a_{H^+} \qquad (13\text{-}46)$$

but with the exact meaning of a_{H^+} determined by the cell used.

Consider first a cell such as (13-1):

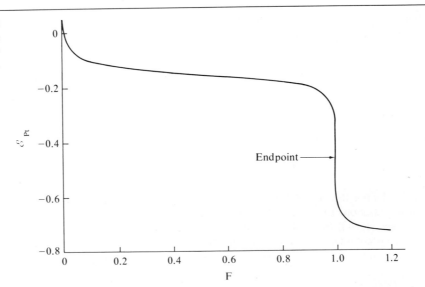

FIGURE 13-8
Potentiometric titration curve for $Cu^+ + Fe^{3+} = Cu^{2+} + Fe^{2+}$.

$$Pt/H_2(1\ atm)/HCl(m)/AgCl/Ag$$

The corresponding Nernst equation is

$$\mathcal{E} = \mathcal{E}^0 - 0.05139 \ln a_{\pm,HCl} \tag{13-47}$$

The procedure described in Section 13-3 allows the determination of $\mathcal{E}^0$, and so the cell measurements provide values for the mean activity of HCl in any solution, which might be thought to correspond to a_{H^+} Suppose that now we replace the electrolyte by $[HAc(m_1) + KCl(m_2)]$. The $\mathcal{E}^0$ remains the same, as does Eq. (13-47). However, the calculated mean activity for HCl will now depend on both m_1 and m_2. Thus such a cell is not suited for *p*H determination.

Next, consider the cell

$$Pt/H_2(1\ atm)/solution \ / \ calomel\ electrode$$

This emf is

$$\mathcal{E}_{298} = \mathcal{E}_{ref} - 0.02569 \ln(1/a_{H^+}) = \mathcal{E}_{ref} - 0.05916\ pH \tag{13-48}$$

and it appears that our goal is achieved. The problem is that $\mathcal{E}_{ref}$ contains the junction potential at the solution $/$ KCl interface (see Sections 13-2C and 13-ST-1) and is not rigorously measurable. What is actually done is to determine $\mathcal{E}_{ref}$ such that a_{H^+} corresponds to $(H^+)\ \gamma_\pm$ as observed, by independent means, to apply to various solutions of weak acids. In effect, it is assumed, not quite correctly, that $\gamma_{H^+} = \gamma_\pm$; it is also assumed, not quite correctly, that the junction potential incorporated in $\mathcal{E}_{ref}$ will not vary with the nature of the solution studied.

The result of all this is that the operational definition of *p*H is given by Eq. (13-48) with the $\mathcal{E}_{ref}$ values of Table 13-1 for the various calomel reference electrodes. The definition allows very precise *p*H measurements, but ones whose accuracy is subject to some (but probably not much) uncertainty so long as fairly dilute solutions are involved. The interpretation of *p*H in concentrated solutions or in nonaqueous solvents can be quite a problem.

The hydrogen electrode is a demanding one to use experimentally; the platinized platinum can be poisoned (lose its catalytic ability due to adsorption of solution components), so that the electrode fails to function well. An alternative *p*H-determining cell is that corresponding to Eq. (13-39):

$$Pt/solution\ with\ quinone\ (Q)\ plus\ hydroquinone\ (H_2Q)\ /$$
$$calomel\ electrode.$$

One convenience is that quinone forms a 1:1 compound with hydroquinone, *quinhydrone*, so that by dissolving the compound in the solution to be tested, one establishes equal concentrations of both species. With the added assumption that, being nonelectrolytes, their activity coefficients will be unity, the Nernst term,

$(RT/n\ \mathcal{F}) \ln(a_{H_2Q}/a_Q)$, drops out and one has for 25°C

$$\mathcal{E}_{298} = \mathcal{E}_{ref} - 0.6994 + 0.05916\ pH \tag{13-49}$$

(subject to the same reservations about liquid junctions as stated earlier).

The most widely used pH-determining cell is that known as the pH *meter* or *glass electrode*. The cell diagram is

glass membrane

$$\text{reference electrode in solution A} \,/\!/ \,/ \, \text{unknown solution} \,/ \, \text{calomel electrode}$$

(13-50)

where the anode may be either Ag/AgCl or a calomel electrode, but with the solution buffered at some constant pH. The glass membrane has the property of passing essentially only hydrogen ions, and, per faraday, the cell reaction is

$$H^+(\text{solution A}) = H^+(\text{unknown solution})$$

so

$$\mathscr{E}_{298} = \text{constant} + 0.05916\, pH \tag{13-51}$$

where the constant contains $\mathscr{E}^0$ for the cell, a constant Nernst term, and the junction potential. In practice, one calibrates the pH meter scale by measuring the emf when a known buffer solution is used in place of the unknown solution, and one finds that it gives almost the same pH values for other solutions as a hydrogen electrode does [see Dole (1941) and Bates (1954) for details].

Other ion-selective membranes have been developed that pass only one or another cation or anion, and an electrode incorporating such a membrane allows the determination of a pM or a pX, where M or X is the ion passed. The selectivity is not always rigorous, however, and one must test such electrodes carefully in the actual situation before accepting the results of their use.

13-10 Irreversible electrode processes

The material so far has dealt with reversible emf's, the experimental measurements being made under conditions of virtually no current flow through the cell. Practical applications of electrochemistry, except for standard reference cells, involve appreciable current flows and we now encounter a new set of phenomena which involve irreversible processes. One is that of overvoltage, important in electrodepositions and also in the study of the kinetics of electrode reactions. Another is polarography, discussed in Section 13-ST-2.

A. Electrodeposition

Electrodeposition is the production of an electrolysis product, usually a metal, on a preparative scale. The situation is illustrated in Fig. 13-9. When a potential is applied across the terminals of a cell very little happens until a critical voltage is reached, beyond which the current increases about linearly with potential. This is what occurs: Suppose that we have two platinum wires dipping in an HCl solution. The cell reaction and reversible potential are

anode $\qquad\qquad\qquad 2Cl^-(m) = Cl_2(P) + 2e^-$

cathode $\qquad 2H^+(m) + 2e^- \;\; = H_2(P)$

and

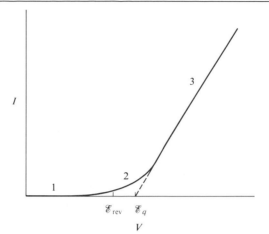

FIGURE 13-9
Variation of current I with
voltage V. 1–2: current
determined by diffusion of
electrode products away
from the electrode region;
2–3: current determined
by Ohm's law.

$$\mathscr{E} = \mathscr{E}^0_{H^+/H_2} - \mathscr{E}^0_{Cl_2/Cl^-} - \frac{RT}{2\mathscr{F}} \ln(a_{H_2}a_{Cl_2}) - \frac{RT}{2\mathscr{F}} \ln\left(\frac{1}{a^4_{\pm,HCl}}\right) \tag{13-52}$$

where $\mathscr{E}^0_{H^+/H_2} - \mathscr{E}^0_{Cl_2/Cl^-} = -1.395$ V at 25°C and a_{H_2} and a_{Cl_2} are the fugacities or, approximately, the gas pressures. In the present experiment, no gases are being supplied so that their pressures are essentially zero, making $\mathscr{E}$ a hypothetical large positive number. However, when the electrodes are dipped into the solution and connected, but with no potential applied, some fluctuation will decide one to be the anode and the other the cathode, and some minute amount of electrolysis will occur, forming a little dissolved hydrogen and chlorine next to the electrodes and bringing their local activities up to some low but nonzero value. The activity of the dissolved gases is far below that for 1 atm pressure and no bubbles can form, but the dissolved gases will diffuse steadily away from the electrodes into the solution. The result is that a small current will flow, and if a very sensitive probe could be used, each electrode would be observed to have some potential relative to the probe. The electrodes are said to be *polarized,* and the small current is called the *residual polarization current.*

If we now apply a small potential V between the electrodes, additional electrolysis will take place, building up a_{H_2} and a_{Cl_2} so that a back emf $\mathscr{E}_q$ develops in opposition to the applied potential difference. The increased hydrogen and chlorine activities lead to an increased diffusion rate away from the electrodes and a consequent increase in the very low steady-state current. We thus observe section 1–2 of the curve shown in Fig. 13-9. With continued increase in applied voltage a_{H_2} and a_{Cl_2} will become equal to the value of the gases at 1 atm, and actual bubble formation will occur. Then $\mathscr{E}_q$ is at its maximum value $\mathscr{E}_{q(max)}$ and cannot increase further, and beyond this point the increase in applied potential goes into an ohmic potential drop through the solution, that is, $V - \mathscr{E}_{q(max)} = IR$. The section 2–3 of the plot is thus linear.

This analysis indicates that extrapolation of the linear portion of the plot of Fig. 13-9 gives $\mathscr{E}_{q(max)}$; this is called the *decomposition potential.* Ideally, $\mathscr{E}_{q(max)}$ is equal to $\mathscr{E}_{rev}$ for the cell, but in practice it may be more positive, and the difference between the two is then called the *overvoltage.* In electrodeposition one is operating in this linear region, and although the reversible potential is the ideal decomposition potential, in practice one must know what the overvoltage will be.

Suppose that we wish to deposit cadmium from a 0.1 m solution of a cadmium salt at pH 7. The reversible potential for the reaction $Cd^{2+} + 2e^- = Cd$ will be, at 25°C and with activity coefficient effects neglected,

$$\mathscr{E}_{Cd} = -0.403 - \frac{0.02569}{2} \ln \frac{1}{0.1} = -0.432 \text{ V}$$

The value for hydrogen, that is, $\mathscr{E}$ for evolving hydrogen gas, is

$$\mathscr{E}_{H_2} = 0 - \frac{0.02569}{2} \ln \left(\frac{1}{10^{-7}} \right)^2 = -0.414 \text{ V}$$

One thus expects the evolution of hydrogen to occur first as potential is applied to the electrolysis cell and that, in fact, it would be virtually impossible to cause cadmium metal to deposit on the electrode. The actual situation is just the reverse because the overvoltage for hydrogen evolution from a cadmium surface is about 0.5 V; as a consequence, the cadmium deposits and no hydrogen evolution occurs.

Some representative hydrogen evolution overvoltages are as follows: platinized platinum, zero or small; smooth platinum, 0.09; silver, 0.15; copper, 0.23; lead, 0.64; zinc, 0.70; and mercury, 0.78 (for zero current). An interesting observation is that the hydrogen overvoltage parallels the heat of adsorption of atomic hydrogen on the metal. The overvoltage thus seems to be related to the energy required to produce surface hydrogen atoms as the intermediate to H_2 formation.

B. Theory of overvoltage

The excess of an applied potential over the reversible decomposition potential, or the overvoltage, is usually given the symbol η, and Fig. 13-9 could be redrawn so as to show a plot of I versus η. However, there are three general types of contribution to an observed overvoltage, only two of which are of fundamental interest. As mentioned earlier, the linear portion of Fig. 13-9 arises primarily from the Ohm's law drop in potential across the solution and is trivial from a theoretical point of view. Second, as active decomposition occurs, electrolysis products accumulate and reactants are depleted in the vicinity of the electrode; the effect is to change $\mathscr{E}_q$ from the reversible value calculated for the average compositions. This contribution to η is known as *concentration polarization,* and is of practical but not much theoretical interest (see Section 13-ST-2). We wish to concentrate

---

FIGURE 13-10
Apparatus for measuring overvoltages. [From A. W. Adamson, "Physical Chemistry of Surfaces," 4th ed. Copyright 1982, Wiley (Interscience), New York. Used by permission of John Wiley & Sons, Inc.]

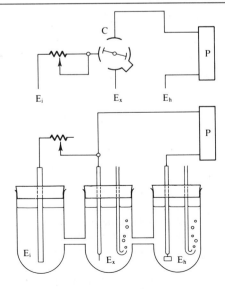

C-Commutator
P-Potentiometer

on that portion of η due to the intrinsic chemistry of the electrode process itself, often called *activation polarization*.

The experimental method is illustrated in Fig. 13-10. The desired current I is passed through the electrode to be studied $\mathscr{E}_x$ by means of the circuit $\mathscr{E}_1 - \mathscr{E}_x$ and the potential variation at $\mathscr{E}_x$ is measured by means of the separate potentiometer circuit $\mathscr{E}_x - \mathscr{E}_h$. Thus the ohmic contribution is essentially eliminated, and, with good stirring, most of the concentration polarization. The commutator device shown allows η to be measured either during current flow or immediately after cessation of flow; in this last case that portion of η due to activation polarization will often decay exponentially with time.

Activation polarization usually obeys an equation due to J. Tafel in 1905, known as the first law of electrode kinetics:

$$\eta = a + b \ln I \qquad (13\text{-}53)$$

A derivation of Eq. (13-53) is given in Section 13-ST-2, but the qualitative basis for it is that there is some slow step in the electrode process which involves an energy barrier, so that a Boltzmann factor $\exp(-\Delta G^*/RT)$ determines I. The applied potential acts to lower the level of this barrier, and hence the increased current is logarithmically related to η.

COMMENTARY AND NOTES

13-CN-1   Storage batteries

Electrochemical cells have great practical application. They are called *primary cells* if electrical energy is supplied on a one-time basis and *secondary cells* or *storage batteries* if an external source of energy is required to charge them.

One common type of storage battery is the lead–sulfuric acid cell for which the electrode reactions are

anode $Pb + H_2SO_4$

   $= PbSO_4(s) + 2H^+ + 2e^-$

cathode $PbO_2(s) + H_2SO_4 + 2H^+ + 2e^-$

   $= Pb + PbSO_4(s) + 2H_2O$

The net reaction is

$$PbO_2(s) + PbSO_4 + 2H_2SO_4$$
$$= 2PbSO_4(s) + 2H_2O \qquad (13\text{-}54)$$

The potential is about 2 V; most lead storage batteries consist of several cells in series. The traditional lead storage battery is not sealed; this is because some hydrogen and oxygen evolution may occur during charging, or even on standing. The newer sealed battery gets around the gasing problem in various ways. Lead-cal-cium grid alloys may be used to reduce gas evolution during self-discharge; the electrolyte may be immobilized as a gel or by absorption into microporous panels, to promote oxygen reduction at the lead electrode; and, while the battery is leakproof, there is a small, one-way vent to allow release of small amounts of hydrogen (Mahato and Laird, 1975).

The Leclanché dry cell has a potential of about 1.6 V and consists of a carbon electrode surrounded by manganese dioxide and graphite immersed in a starch paste containing zinc chloride and excess solid ammonium chloride. The electrode reactions of this cell are

anode $Zn + 2NH_3$

   $= Zn(NH_3)_2^{2+} + 2e^-$

cathode $2MnO_2 + 2NH_4^+ + 2e^-$

   $= Mn_2O_3 + H_2O + 2NH_3$

The lead storage battery loses a good deal of power under high loads and, of course, is very heavy. A more expensive but otherwise very attractive storage battery has been in use in Europe for many years, and, more recently, in the United States. This is the Jungner nickel–cadmium battery, which consists of nickel and cadmium electrodes and a KOH or other alkali electrolyte. The cell reactions are anode $Cd + 2OH^-$

$= Cd(OH)_2 + 2e^-$

cathode $Ni(OH)_3 + e^-$

$= Ni(OH)_2 + OH^-$

The net reaction is

$Cd + 2Ni(OH)_3$
$= Cd(OH)_2 + 2Ni(OH)_2$ (13-55)

Notice that the electrolyte serves merely as a vehicle for the transportation of OH^- ions. The Ni–Cd battery is rechargeable and can deliver very high currents without appreciable loss of power. An important use is in battery-powered tools and appliances.

An important type of primary cell is the fuel cell. A simple example is the cell

$$Pt/H_2(g)/electrolyte/O_2(g)/C$$

for which the net reaction is $H_2 + \frac{1}{2}O_2 = H_2O$. The power available tends to be limited by the relatively slow rate of reduction of O_2, but hydrogen/oxygen fuel cells have been used in the space program and may eventually find widespread use in permanent installations.

Much current research is being carried out on the development of suitable electrodes for the reversible combustion of hydrocarbons so as to obtain electrical energy directly from a reaction such as $CH_4 + 2O_2 = CO_2 + 2H_2O$. A methanol fuel cell using KOH as the electrolyte has found some commercial use.

The thermodynamic characteristics of some possible fuel cells are given in Table 13-4; the potentials are all about 1 V. The third column gives the reversible heat of the cell reaction, the $T\Delta S$ term; it is desirable that this heat be small so as to minimize the need for cooling the cell.

13-CN-2 Electrocapillarity. Absolute electrode potentials

A very interesting electrochemical effect is that of the change in interfacial tension with applied potential. Not many interfaces allow an accurate study of this effect, however, because an attempt to apply a potential difference often results in electrolysis or in electrolytic transport across the interface. An especially well adapted system is that of the mercury–electrolyte solution interface. This is highly polarizable, meaning that applied potentials result in little electrolysis in the absence of easily reducible ions, and, since mercury is a liquid, the interfacial tension can be observed by the methods of capillarity.

The typical experimental observation is illustrated in Fig. 13-11, which shows the mercury–solution interfacial tension to increase, go through a maximum, and then decrease as potential is applied. The experimental arrangement is that depicted in Fig. 13-12; a mercury reservoir terminates in a fine capillary tube and the position of the meniscus is viewed through a traveling microscope. Since the contact angle glass–mercury–solution is obtuse, there is a capillary depression [note Eq. (8-39)], so a positive head of mercury is needed to force the meniscus toward the end of the tapering capillary tube. The electrical connection is from the mercury through a source of potential to a calomel electrode and thence to the electrolyte solution. Measurement of the head required to maintain the meniscus yields the capillary depression and hence the surface tension at the mercury–solution interface. It is the variation of this surface tension with applied potential that is reported in Fig. 13-11; the abscissa is the potential relative to that at the maximum, called the *rational potential*.

TABLE 13-4 Possible fuel cell reactions	Reaction	$\mathscr{E}^0_{298}$ V	ΔH^0_{298}[a] kJ	q_{298}[a] kJ
	$H_2 + \frac{1}{2}O_2 = H_2O$	1.23	-285	-49
	$CH_4 + 2O_2 = CO_2 + 2H_2O$	1.06	-892	-73
	$CH_3OH(\ell) + \frac{3}{2}O_2 = CO_2 + 2H_2O$	1.21	-727	-24

[a]Per mole of fuel.

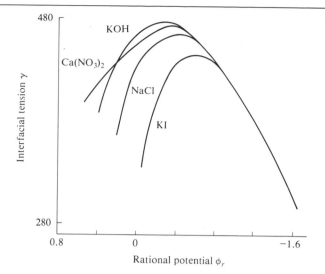

FIGURE 13-11
Electrocapillarity curves.
[From D.C. Grahame,
Chem. Rev. **41,** 441
(1947).© 1947, The
Williams and Wilkins Co.,
Baltimore, Maryland.]

The thermodynamic explanation of the electrocapillarity effect is that the derivative of surface tension with respect to potential gives the surface charge density σ:

$$\frac{\partial \gamma}{\partial V} = -\sigma. \tag{13-56}$$

Just as the adsorption of a surfactant at an interface lowers the interfacial tension, so does the concentration of charge at an interface. At the maximum the derivative in (13-56) is zero and so therefore must be the surface charge density; this suggests that the absolute potential difference across the interface is also zero. The voltage applied to reach this electrocapil-

larity maximum therefore just balances the natural potential difference between the phases. This applied potential difference is 0.48 V if the electrolyte is one not apt to interact with the mercury surface, such as potassium carbonate or sulfate, which implies that the absolute half-cell potential of the calomel electrode is 0.48 V (as compared to 0.28 V on the hydrogen scale).

The problem is that even though the charge density must be zero at the electrocapillarity maximum, there will still be adsorbed and polarized solvent and solute molecules, so that $\Delta\phi$, the galvanic potential difference across the interface, is not necessarily zero. Notice, for example, that the position of the electrocapil-

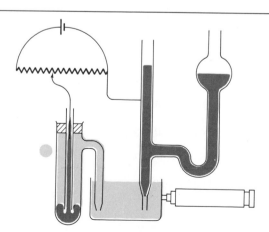

FIGURE 13-12
The Lippmann apparatus
for observing the
electrocapillary effect.
[From A.W. Adamson,
"Physical Chemistry of
Surfaces," 4th ed.
Copyright 1982, Wiley
(Interscience), New York.
Used with permission of
John Wiley & Sons, Inc.]

larity maximum shown in Fig. 13-11 varies with the nature of the electrolyte as the anion is changed; the same happens if the cation is varied or if the solvent medium is altered.

Various other attempts have been made to determine absolute half-cell potentials, but, as in this case, certain unprovable assumptions are always involved. It does seem likely, though, that the absolute values are not greatly different from those reported on the hydrogen scale.

SPECIAL TOPICS

13-ST-1 Liquid junctions

The existence and importance of the potential at a liquid junction have been noted in several places in this chapter. The detailed treatment is somewhat specialized, however, and has therefore been placed in this section. We consider first the potential of a cell having a liquid junction and then some aspects of the electrochemistry of the liquid junction itself.

A. Concentration cells
 with a liquid
 junction

The least complicated type of cell which has a liquid junction is a concentration cell since the electrode reactions are then the same, except for a difference in concentration of the electrolyte across the junction. The general cell diagram is

anode/solution at m_1 ⫶ solution at m_2/cathode

where the dashed line means that the two solutions are in electrolytic contact but are somehow prevented from mechanical mixing. This may be by means of a porous diaphragm, as in the Daniell cell of Fig. 13-2, or by a stiffening of one of the solutions at its point of contact with the other by agar-agar or gelatine, as is done in a salt bridge.

An example of a concentration cell with a liquid junction is the following:

$$\text{Pt/H}_2(1\text{ atm})/\text{HCl}(m_1) \; ⫶ \; \text{HCl}(m_2)/\text{H}_2(1\text{atm})/\text{Pt} \tag{13-57}$$

anode reaction $\frac{1}{2}\text{H}_2(1\text{ atm}) = \text{H}^+(m_1) + \text{e}^-$

cathode reaction $\text{H}^+(m_2) + \text{e}^- = \frac{1}{2}\text{H}_2(1\text{ atm})$

net reaction $\text{H}^+(m_2) = \text{H}^+(m_1)$

$$\tag{13-58}$$

The emf of a cell corresponds to the free energy change associated with the sum of *all* processes occurring per faraday of electricity passed through the cell. Current is carried by the ions in the solution, which means that t_+ equiv of hydrogen ion must cross the junction from left to right and t_- equiv of chloride ion must do so from right to left, where t denotes transference number (see Section 12-6). The changes that occur in addition to the electrode reactions are then

$$t_+\text{H}^+(m_1) = t_+\text{H}^+(m_2) \tag{13-59}$$
$$t_-\text{Cl}^-(m_2) = t_-\text{Cl}^-(m_1)$$

The sum of process (13-58) and processes (13-59) is

$$t_-\text{H}^+(m_2) + t_-\text{Cl}^-(m_2)$$
$$= t_-\text{H}^+(m_1) + t_-\text{Cl}^-(m_1) \tag{13-60}$$

bearing in mind that $t_+ + t_- = 1$. The corresponding Nernst equation is

$$\mathscr{E} = -\frac{RT}{\mathscr{F}} \ln \frac{a_{\text{H}^+,m_1}^{t_-} \, a_{\text{Cl}^-,m_1}^{t_-}}{a_{\text{H}^+,m_2}^{t_-} \, a_{\text{Cl}^-,m_2}^{t_-}}$$

or

$$\mathscr{E} = -2t_-\frac{RT}{\mathscr{F}} \ln\frac{a_{\pm,m_1}}{a_{\pm,m_2}} \tag{13-61}$$

If m_1 and m_2 are 0.1 and 0.01, respectively, then the $\gamma_\pm$ values are 0.796 and 0.904, respectively, and the t_- values are 0.1686 and 0.1749, respectively. The average value of t_- is 0.1717, and the emf of the cell at 25°C is calculated to be

$$\mathscr{E} = -(2)(0.1717)(0.02569) \ln\frac{(0.1)(0.796)}{(0.01)(0.904)}$$

$$= -0.1919$$

The overall cell reaction is the sum of processes (13-58) and (13-59), and the separate Nernst expressions for these are

$$\mathscr{E}_E = -\frac{RT}{\mathscr{F}} \ln\frac{a_{H^+,m_1}}{a_{H^+,m_2}} \tag{13-62}$$

and

$$\mathscr{E}_J = -t_+ \frac{RT}{\mathscr{F}} \ln\frac{a_{H^+,m_2}}{a_{H^+,m_1}}$$

$$- t_- \frac{RT}{\mathscr{F}} \ln\frac{a_{Cl^-,m_1}}{a_{Cl^-,m_2}} \tag{13-63}$$

Inspection confirms that the sum of Eqs. (13-62) and (13-63) gives Eq. (13-61); thus

$$\mathscr{E} = \mathscr{E}_E + \mathscr{E}_J$$

The emf for the cell can therefore be viewed as made up of two terms: that due to the electrode reactions $\mathscr{E}_E$, and the junction potential $\mathscr{E}_J$, which gives the work associated with the transport of ions across the liquid junction.

We can extend the preceding numerical example to the calculation of $\mathscr{E}_E$ and $\mathscr{E}_J$ only by making some assumption regarding individual ionic activity coefficients. If, for example, we assume that $\gamma_{H^+} = \gamma_{Cl^-} = \gamma_\pm$, then Eqs. (13-62) and (13-63) give $\mathscr{E}_E = -0.05588$ V and $\mathscr{E}_J = 0.03669$ V. The junction potential represents a major correction in this case.

In using Eq. (13-61), an average value of t_- for the two concentrations was used. It turns out that the correct average is not the simple arithmetic one, and the work done in transporting an ion across the junction actually involves the integral

$$RT \int_{m_1}^{m_2} t_i d(\ln a_i) \tag{13-64}$$

so that the corrected version of Eq. (13-45) is

$$\mathscr{E} = -2\frac{RT}{\mathscr{F}} \int_{m_1}^{m_2} t_- d(\ln a_\pm) \tag{13-65}$$

A study of cells of this type evidently permits the determination of transference numbers as a function of concentration.

The overall cell reaction will depend on whether the electrodes are reversible to the anion or to the cation. Thus for the cell

$$Ag/AgCl/HCl)(m_1) \mathop{/}\limits^{/}_{/} HCl(m_2)/AgCl/Ag$$

Eqs. (13-61) and (13-65) would have t_+ rather than t_-, and $\mathscr{E}_E$ would have the opposite sign. The numerical example would then give

$$\mathscr{E} = \mathscr{E}_E + \mathscr{E}_J \quad \text{or}$$
$$0.09257 = 0.05588 + 0.03669$$

The sign of the junction potential remains the same. Thus, although we analyze in detail each specific concentration cell to determine the correct overall cell reaction, the junction potential remains the same for any given junction.

B. Junction potentials

Equation (13-63) simplifies considerably if activity coefficient effects are neglected. We then obtain

$$\mathscr{E}_J = (t_+ - t_-)\frac{RT}{\mathscr{F}} \ln\frac{m_1}{m_2} \tag{13-66}$$

This form brings out the important point that $\mathscr{E}_J$ depends on $t_+ - t_-$. It is largely with this factor in mind that one expects the junction potential to be small in the case of a KCl bridge; t_+ for 1 m KCl is 0.488, and so $t_+ - t_-$ is only 0.024. The same situation applies in the case of NH_4NO_3, an electrolyte that is used in a salt bridge if it is not desirable that Cl^--ion-containing solution have contact with the electrolyte of the half-cell being studied.

If various electrolytes are present on either side of a liquid junction, then the total work of transporting ions across the junction per faraday, which gives $\mathscr{E}_J$, depends on weighted averages or integrals such as that of Eq. (13-64). By using concentrated KCl in the salt bridge, one ensures that the important t_i will be those of K^+ and Cl^-, so that $\mathscr{E}_J$ will not be very sensitive to the nature of the more dilute electrolyte solution of the half-cell into which the salt bridge dips. The analysis makes it clear, however, that although we can reduce a junction potential to a fairly small value, cells with liquid junctions retain a residual uncertainty in the interpretation of their emf's and one which is difficult to analyze exactly.

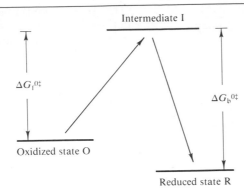

FIGURE 13-13
Mechanism for activation
polarization.

Intermediate I

$\Delta G_{f}^{0\ddagger}$

$\Delta G_{b}^{0\ddagger}$

Oxidized state O

Reduced state R

13-ST-2 Polarization at electrodes. Polarography*

Three types of polarization effects were noted in Section 13-10: ohmic, concentration, and activation. We now discuss these last two in more detail.

A. Activation polarization

The Tafel equation (13-53), which relates the activation polarization component of the overvoltage to the current I, may be accounted for in terms of the following analysis. Consider a general electrode reaction

$$e^- + O \text{ (oxidized state)}$$
$$= R \text{ (reduced state)}$$

where, as illustrated in Fig. 13-13, the oxidized state must pass through some high-energy intermediate I in the process of being reduced. We write an equation analogous to Eq. (8-65) for the rate of the forward process:

$$R_f = \left[\frac{kT}{h} \exp\left(\frac{-\Delta G_f^{0\ddagger}}{RT}\right)\right] (O) \qquad (13\text{-}67)$$

where $\Delta G_f^{0\ddagger}$ is the standard free energy change to go from O to I. It will be known in the next chapter as the free energy of activation. Similarly, the rate of the back reaction is

$$R_b = \left[\frac{kT}{h} \exp\left(\frac{-\Delta G_b^{0\ddagger}}{RT}\right)\right] (R) \qquad (13\text{-}68)$$

*An alternative positioning of this subject is as one of the special topics sections of Chapter 15 (on kinetics of reaction in solution).

The presence of a potential difference ϕ between the electrode and the solution contributes to $\Delta G_f^{0\ddagger}$ and we write this last as the sum of a chemical component and an electrical one:

$$\Delta G_f^{0\ddagger} = \Delta G_{f,\text{chem}}^{0\ddagger} - \alpha\mathcal{F}\phi \qquad (13\text{-}69)$$

which allows that only some of this potential difference is effective; this fraction is called the *transfer coefficient* α. The current due to the forward reaction is then

$$i_f = \left[\frac{kT}{h} \exp\left(\frac{-\Delta G_{f,\text{chem}}^{0\ddagger}}{RT}\right) \right.$$
$$\left. \exp\left(\frac{\alpha\mathcal{F}\phi}{RT}\right)\right] (O) \qquad (13\text{-}70)$$

where current is in faradays per square centimeter per second. At equilibrium there will be equal and opposite currents in the two directions, or $I_f^\circ = I_b^\circ = I^\circ$, where I° is known as the exchange current, and ϕ has the value ϕ° and corresponds to $\mathcal{E}^\circ$. Then

$$I^\circ = \left[\frac{kT}{h} \exp\left(\frac{-\Delta G_{f,\text{chem}}^{0\ddagger}}{RT}\right) \right.$$
$$\left. \exp\left(\frac{\alpha\mathcal{F}\phi^\circ}{RT}\right)\right] (O) \qquad (13\text{-}71)$$

Combination of Eqs. (13-70) and (13-71) gives

$$I_f = I^\circ \exp\frac{\alpha\mathcal{F}(\phi - \phi_0)}{RT}$$
$$= I^\circ \exp\frac{\alpha\mathcal{F}\eta}{RT} \qquad (13\text{-}72)$$

where η is the overvoltage. Equation (13-72) may be written in the form

$$\ln I = \ln I^\circ + \frac{\alpha\mathcal{F}\eta}{RT}$$

which rearranges to the Tafel equation.

In a stricter analysis, an equation analogous to Eq. (13-70) is written for I_b, with $I = I_f - I_b$; the effect is to alter the coefficient of η in Eq. (13-72) to give

$$I = I_0 \left[\exp\left(\frac{\alpha \mathscr{F} \eta}{RT} \right) - \exp\left(-\frac{(1 - \alpha)\mathscr{F}\eta}{RT} \right) \right] \quad (13\text{-}73)$$

As the equation suggests, the situation is symmetric; η may be positive or negative, so that the electrode reaction is driven either forward or backward. Figure 13-14 shows the general behavior of Eq. (13-73).

This treatment, although somewhat sketchy, is designed to indicate how the detailed study of overvoltage effects can lead to information about the intrinsic rate of the electrode reaction, through I°, and the energy of the reaction barrier. As an example, although the mechanism for the reduction of H^+ ion at a metal electrode is still not fully elucidated, the evidence suggests that the rate-determining step is probably the reaction of H_3O^+ with the metal surface to give adsorbed hydrogen atoms and water:

$$H_3O^+ + M + e^- = M - H + H_2O$$

B. Concentration polarization

The preceding analysis was based on the assumption that the rate-limiting step of the electrode process was some activated chemical reaction at the electrode surface. Another process must become rate-controlling at sufficiently high current densities, namely the rate of diffusion of reactant to and of product away from the electrode surface. If the reaction is one of the deposition of a metal from solution, only the rate of diffusion of the metal ion to the electrode is to be considered. Recalling Eq. (2-87) and the discussion of Section 10-8B, we can write

$$J = -\mathscr{D}\, \frac{dC}{dx} \quad (13\text{-}74)$$

where J is now in moles per square centimeter per second. If there is an excess of inert electrolyte in the solution so that the ion being reduced carries little of the current, the potential term in Eq. (12-131) will not be important and $\mathscr{D}$ will be a constant, equal to the ion diffusion coefficient as given by Eq. (12-129). Further, a reasonable approximation to the physical situation is that the stirring conditions leave a thin stagnant layer of thickness δ within which C varies linearly from C, the bulk solution con-

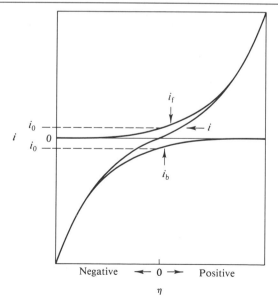

FIGURE 13-14
A general current versus voltage diagram.

centration, to C^S, the concentration at the electrode surface. We expect the linear variation of C with x since in a steady-state condition J is constant, and since $\mathcal{D}$ is constant, so must be dC/dx. Equation (13-74) then becomes

$$I_d = zJ = -\frac{z\mathcal{D}(C^S - C)}{\delta} \qquad (13\text{-}75)$$

where I_d is the diffusion current density in faradays per square centimeter per second and z is the valence number of the metal ion being discharged. Under the conditions of the preceding subsection it is assumed that $\mathcal{D}/\delta$ is large enough that $C^S = C$, and the observed current is determined by the rate of the chemical reduction reaction at the electrode, I_a. That is, under steady-state conditions there is no accumulation of material in the diffusion zone and $I_d = I_a$. As the overvoltage is increased, I_a increases by Eq. (13-73), and C^S drops enough for I_d to match the increase in I_a. The maximum possible diffusion rate $i_{d(max)}$ is reached when C^S has dropped to zero. Further increase in the overvoltage cannot increase the current any further, and the plot of I versus V must level off at $I_{d(max)}$, or at

$$I_{d(max)} = \frac{z\mathcal{D}C}{\delta} \qquad (13\text{-}76)$$

Thus the effect of concentration polarization is to give a maximum current which is proportional to the concentration of the ion being discharged.

If a mixture of potentially reducible ions is present, again with a supporting electrolyte (that is, an excess of inert electrolyte) present, then as the applied reducing potential is increased the first metal ion begins to deposit, reaches its I_{max}, and the second metal ion begins to deposit at some higher potential, and reaches its I_{max}, so that the plot of I versus V is as shown in Fig. 13-15. The characteristic deposition potential for each ion could be identified as in Fig. 13-9 but when one is operating in the diffusion-controlled region C^S and hence $\mathcal{E}_{rev}$ is changing with V so that one obtains a sinusoidal rather than a linear I versus V plot. Consequently, it is much more accurate to pick the potential at the half-way point of the step, or the half-wave potential $V_{1/2}$, as the characteristic one.

Since the overvoltage η is given by

$$\eta = \mathcal{E}_i - \mathcal{E}_{rev} = -\frac{RT}{z\mathcal{F}}\ln\frac{C^S}{C} \qquad (13\text{-}77)$$

where ohmic polarization is neglected, as are activity coefficient terms, we can solve for C^S/C. Also, combination of Eqs. (13-75) and (13-76) gives

$$I_d = I_{d(max)}\left(1 - \frac{C^S}{C}\right) \qquad (13\text{-}78)$$

The result is

$$\frac{I_d}{I_{d(max)}} = 1 - \exp\left(-\frac{z\mathcal{F}\eta}{RT}\right) \qquad (13\text{-}79)$$

Equation (13-79) gives a symmetric curve such as shown in Fig. 13-15, for which $V_{1/2}$ occurs at the inflection point.

FIGURE 13-15
Current versus voltage in the region of concentration polarization.

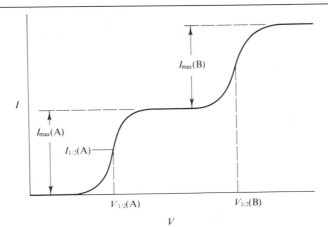

C. Polarography

The preceding analysis suggests that we could analyze for metal ions in solution by obtaining experimental I versus V curves, each limiting diffusion current giving the concentration of a particular ion, the ion being identified by its $V_{1/2}$. This procedure is indeed used. To maintain δ more constant than is possible by stirring the solution, a common practice is to use a rotating electrode. There is still a problem with accumulation of reduction products, and a very ingenious alternative procedure was devised by J. Heyrovsky in 1922.

The basic experimental features of the *Heyrovsky polarograph* are shown in Fig. 13-16. The cathode is a mercury drop that is steadily growing at the tip of a capillary immersed in the electrolyte solution, and the anode is a large pool of mercury. By having a tiny drop as the anode surface, concentration polarization effects can be made to develop at relatively small currents; the large area of the anode pool of mercury essentially eliminates concentration polarization at its surface. As each drop grows and falls the anode surface is steadily kept fresh and reproducible, and with a supporting electrolyte the same general analysis applies as for a stationary electrode.

A typical polarogram is shown in Fig. 13-17. The oscillations are a result of the successive appearance of new drops at intervals of a few seconds; currents may be only microamperes in order of magnitude. Each step is called a *polarographic wave* and is characterized by its half-wave potential $V_{1/2}$ and *diffusion current* $I_{d(\text{max})}$. The detailed algebraic analysis is complicated by the situation of an expanding cathode surface, and an equation derived by D. Ilkovic in 1938 will be given without the derivation:

$$I_{d(\text{max})} = 70.82 zw \left(\tfrac{2}{3}t\right)\left(\tfrac{1}{6}\mathscr{D}\right)\left(\tfrac{1}{2}C\right) \tag{13-80}$$

where z is the valence number of the ion, w is the flow of the mercury in grams per second, t is the drop time in seconds, and C is the concentration in moles per liter. Equations have also been derived for the half-wave potential, but this remains essentially an empirically determined quantity for each species.

Polarography is capable of measuring ion concentrations as low as 10^{-4} M and is a rapid as well as a sensitive analytical tool. The physical chemist uses it to study the chemistry of reduction (or of oxidation) processes. One can determine from the diffusion current whether the reduction occurs as a one- or a two-electron step. If complexing ions are added to the solution, both the reduction potentials and the formation constants of complex ions can be found. With a commutator to interrupt the applied voltage, chemical rate processes can be followed.

A modern variant of polarography uses a fixed (not a dropping mercury) electrode and current is recorded as the voltage is changed at a fixed rate. In *cyclic voltammetry* the voltage sweep is first in one direction and then in the other. Sweep speeds of up to 100 V s^{-1} may be used.

We can analyze what happens in terms of a fairly typical voltammogram for the reversible

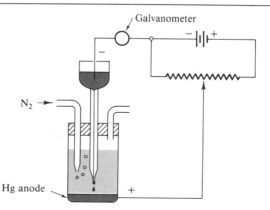

FIGURE 13-16
Schematic diagram of a polarographic cell.

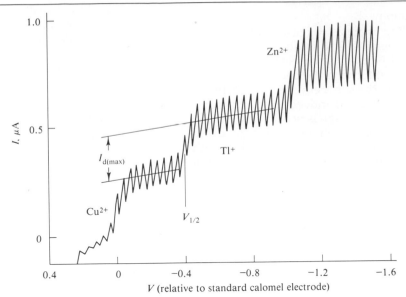

V (relative to standard calomel electrode)

couple $Fe(CN)_6^{3-}/Fe(CN)_6^{4-}$ The solution consists initially of $Fe(CN)_6^{3-}$ in a supporting electrolyte and, as shown in Fig. 13-18(a), the voltage applied to the working electrode is reduced linearly with time from an initial value V_a to the "switching" value V_b, at which point the voltage is made to increase linearly with time to the switching value V_c. Thereafter, the voltage cycles between V_b and V_c. As the voltage drops from V_a to V_b, the rate of reduction of the $Fe(CN)_6^{3-}$ and hence the current increases exponentially. The applicable equation is

$$I_{red} = I° \exp\left[-\frac{\alpha n \mathscr{F}}{RT}(\phi - \phi_0')\right] \quad (13\text{-}81)$$

Equation (13-81) is essentially the same as Eq. (13-72), except that ϕ_0' is semi-empirical in that it is the standard reduction potential combined with activity coefficient effects.

The actual voltammogram is shown in Fig. 13-18(b); section 1 shows the exponentially rising current. The electrolysis that is going on, however, depletes $Fe(CN)_6^{3-}$ near the cathode surface, and this depletion zone deepens with continuing current flow. The effect is to cause the current to reach a maximum value, at V_{pc} (peak cathodic current), after which the current I drops. In region 2 the current is limited by the increasingly slow rate of diffusional transport of $Fe(CN)_6^{3-}$ to the electrode surface. At

V_b the applied voltage starts increasing linearly, and when it reaches the oxidation potential of $Fe(CN)_6^{4-}$ the current flow increases exponentially in the opposite direction; the equation is analogous to Eq. (13-81) except that the transfer coefficient α is replaced by $(1 - \alpha)$.

Ideally, all of the $Fe(CN)_6^{4-}$ formed during the reduction section has remained close to the electrode and is available for reoxidation. Again, depletion occurs, and the anodic current peaks at V_{pa} and thereafter is limited by the rate of diffusional migration to the electrode. The cycle of Fig. 13-18(b) then repeats when the applied voltage reaches V_c and starts falling again.

One usually takes the average of V_{pc} and V_{pa} as the standard reduction potential for the redox couple (correcting for the emf of the reference electrode, often a calomel or a silver–silver chloride electrode). Also, for a reversible redox couple, the difference $(V_{pc} - V_{pa})$ should be about $0.059/n$ V at 25°C, which allows estimation of n.

Redox couples are not always reversible, of course. One of the many possible situations is illustrated in Fig. 13-19. Here, reversible reduction to the reduced form R occurs, but R can disappear by a slow chemical reaction to some nonoxidizable species C. The numbers in the figure give the ratio of the rate constant for the R → C reaction to the scan rate. If this is large, then, as illustrated by the curve labeled

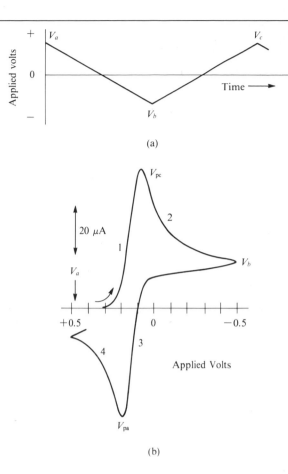

FIGURE 13-18
Voltammetry of 1 mM $K_3Fe(CN)_6$ in aqueous 0.1 M KCl, using a Pt working electrode and a Ag/AgCl reference electrode. (a) Scan cycle, at 100 mV s^{-1}. (b) Observed voltammogram. [From G.A. Mabbott, J. Chem. Ed., **60**, 697 (1983).]

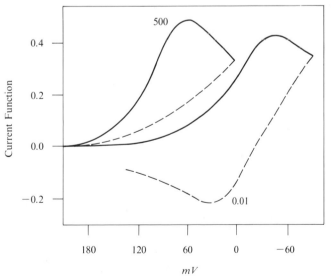

FIGURE 13-19
Cyclic voltammograms for a reversible reduction step followed by an irreversible chemical reaction to a redox-inert product. The numbers labeling the two curves are the ratios of the rate constant for the irreversible reaction to the scan rate. [From G.A. Mabbott, J. Chem. Ed., **60**, 697 (1983).]

500, most of the R formed in the reduction section has gone on to C, and the return peak is wiped out. However, if the scan rate is increased, then, as illustrated by the curve labeled 0.01, R is oxidized back to starting ma-

terial before much reaction to form C can occur, and a more nearly normal oxidation peak is seen. Studies of this type can give a great deal of information about electrochemical mechanisms and associated rate constants.

GENERAL REFERENCES

ADAMSON, A.W. (1982). "The Physical Chemistry of Surfaces," 4th ed. Wiley (Interscience), New York.

BARD, A.J., AND FAULKNER, L.R. (1980). "Electrochemical Methods, Fundamentals and Applications." Wiley, New York.

DANIELS, F., MATHEWS, J.H., WILLIAMS, J.W., BENDER, P., AND ALBERTY, R.A. (1956). "Experimental Physical Chemistry," 5th ed. McGraw-Hill, New York.

DOUGLAS, B.E., AND McDANIEL, D.H. (1965). "Concepts and Models of Inorganic Chemistry." Ginn (Blaisdell), Boston, Massachusetts.

KORTUM, G. (1965). "Treatise on Electrochemistry." Elsevier, Amsterdam.

LATIMER W.M. (1952). "The Oxidation States of the Elements and Their Potentials in Aqueous Solution," 2nd ed. Prentice-Hall, Englewood Cliffs, New Jersey.

LEWIS, G.N., AND RANDALL, M. (1961). "Thermodynamics," 2nd ed. (revised by K.S. Pitzer and L. Brewer), McGraw-Hill, New York.

CITED REFERENCES

BATES, R.G. (1954). "Electrometric pH Determinations." Wiley, New York.

DOLE, M. (1941). "The Glass Electrode." Wiley, New York.

HARNED, H.S., AND EHLERS, R.W. (1932). J. Amer. Chem. Soc. 54, 1350.

LATIMER, W.M. (1952). "The Oxidation States of the Elements and Their Potentials in Aqueous Solution," 2nd ed. Prentice-Hall, Englewood Cliffs, New Jersey.

LEWIS, G.N., AND RANDALL, M. (1961). "Thermodynamics," 2nd ed. (revised by K.S. Pitzer and L. Brewer). McGraw-Hill, New York.

MAHATO, B.K., AND LAIRD, E.C. (1975). "Power Sources," Vol. 5, D.H. Collins, ed. Academic Press, New York.

EXERCISES

Activity coefficient effects are to be neglected in Exercises and Problems unless specifically noted otherwise. Assume 25°C unless otherwise specified. Take as exact numbers given to one significant figure.

13-1 Give the right-hand and left-hand half-cell reactions, the overall cell reaction, and $\mathscr{E}^0$ for the following cells. (a) Pt/H$_2$(g)/HBr(aq)/AgBr(s)/Ag, (b) Hg/Hg$_2$Cl$_2$(s)/0.1 m KCl/AgCl(s)/Ag, (c) Cd/Cd(NO$_3$)$_2$(aq), Tl(NO$_3$)(aq)/Tl.

Ans. The $\mathscr{E}^0$'s are (a) 0.0713 V, (b) −0.0453 V, (c) 0.0667 V.

13-2 The emf is $\mathscr{E}_{298} = 0.2434$ V for the cell Pt/H$_2$(g)/HSCN(0.05 m)/Ag(SCN)(s)/Ag, where HSCN is a strong acid. Calculate $\mathscr{E}^0_{298}$.

Ans. 0.0895 V.

13-3 Calculate the value of $\mathscr{E}_{298}$ for the cell Pb/Pb(NO$_3$)$_2$(0.05 m), Cd(NO$_3$)$_2$(0.02 m)/Cd.

Ans. −0.289 V.

13-4 Write the cell diagram for a cell whose cell reaction is Pb(SO$_4$)(s) + 2Ag = Ag$_2$(SO$_4$)(s) + Pb, and find its $\mathscr{E}^0_{298}$ value.

Ans. −1.009 V.

13-5 $\mathscr{E}^0$ for the cell Pt/H$_2$(g)/HCl(aq)/AgCl(s)/Ag is 0.22234 V at 25°C and 0.21565 V at 35°C. Write the cell reaction and calculate for it ΔG^0_{298}, ΔH^0, and ΔS^0.

Ans. $\Delta G^0_{298} = -21.452$ kJ, $\Delta H^0 = -40.683$ kJ, $\Delta S^0 = -64.5$ J.

Calculate $\mathscr{E}^0$ and K at 25°C for the reaction $Fe + Cu^{2+}(aq) = Fe^{2+}(aq) + Cu$.   **13-6**

Ans. 0.593 V, 1.12×10^{20}.

Calculate K at 25°C for the reaction $\frac{1}{2}Pb + Tl^+(aq) = \frac{1}{2}Pb^{2+}(aq) + Tl$, and the ratio $(Tl^+)/(Pb^{2+})$ if   **13-7**
excess lead is added to a 0.1 m solution of $Tl(NO_3)$.

Ans. 2.79×10^{-4}, 1.29×10^8.

Use Tables 5-2 and 7-2 to obtain $\mathscr{E}^0_{298}$ and $d\mathscr{E}^0/dT$ for the fuel cell reaction $C_2H_6(g) + \frac{7}{2}O_2 = 2CO_2(g)$ **13-8**
$+ 3H_2O(\ell)$.

Ans. 1.086 V, -2.29×10^{-4} V K^{-1}.

Calculate the standard reduction potentials at 25°C for the half-cell reactions (a) $Cu^{2+} + e^- = Cu^+$ **13-9**
and (b) $Fe^{3+} + 3e^- = Fe$. Use the half-cell potentials for Cu^{2+}/Cu, Cu^+/Cu, Fe^{2+}/Fe, and Fe^{3+}/Fe^{2+}
and compare your answers with the values in Table 13-1.

Calculate the K_{th} for the solubility equilibrium of $Mg(C_2O_4)(s) = Mg^{2+}(aq) + C_2O_4^{2-}(aq)$ if $\mathscr{E}^0_{298}$ is   **13-10**
2.49 V for the half-cell $Mg(C_2O_4)(s)/Mg$. Table 13-1 is available.

Ans. 8.8×10^{-5}.

$\mathscr{E}^0_{298}$ is 1.12 V for the couple $Cu^{2+}/Cu(CN)_2^-$. Calculate the equilibrium constant for the dissociation   **13-11**
$Cu(CN)_2^- = Cu^+ + 2CN^-$ and the fraction of $Cu(CN)_2^-$ dissociated in a solution that is 0.01 f in
cuprous nitrate and 0.03 f in potassium cyanide.

Ans. 4.49×10^{-17}, 4.49×10^{-13}.

Calculate $\mathscr{E}_{298}$ for the following concentration cells. (a) $Pt/H_2(P = 1$ atm$)/HCl(0.1\ m)/H_2(P = 0.1$ **13-12**
atm$)/Pt$, (b) $Cd(Hg,x_{Cd} = 0.05)/Cd^{2+}(0.1\ m)/Cd\ (Hg,x_{Cd} = 0.01)$, (c) $Ag/AgCl(s)/HCl(m = 0.1)/H_2(0.1$
atm$)/Pt$—$Pt/H_2(1$ atm$)/HCl(m = 0.5)/AgCl(s)/Ag$.

Ans. (a) 0.296 V, (b) 0.0413 V, (c) -0.0531 V.

Recalculate the emf for cell (c) in Exercise 13-12, (a) using the appropriate activity coefficients for   **13-13**
aqueous HCl and (b) assuming ideal behavior but 45°C. (Note Table 12-7.)

Ans. (a) -0.0505 V, (b) -0.0567 V.

Calculate the standard free energy of formation of $Pb^{2+}(aq)$.   **13-14**

Ans. 243 kJ.

Calculate $d\mathscr{E}^0/dT$ for the half-cell $Na^+(aq) + e^- = Na(s)$, for around 25°C.   **13-15**

Ans. -9.54×10^{-5} V K^{-1}.

A pH meter reads 100 mV for a standard buffer of pH 3. Calculate a_{H^+} for an unknown solution for   **13-16**
which the meter reads 200 mV.

Ans. 2.04×10^{-5}.

PROBLEMS

Molecular weights are given in kg in the SI system; thus, the molecular weight of O_2 is 0.032 kg. In   **13-17**
going from the cgs to the SI system, state whether the numerical value of each of the following is
changed, and if it is, calculate the new value. (a) Avogadro's number. (b) Faraday's number. (c) $\mathscr{E}^0$
for the cell $Pt/H_2(1$ atm$)/HCl/AgCl/Ag$. (d) ΔG^0 for the cell reaction in (c). (e) $\mathscr{E}$ at 25°C for the cell
$Pt/H_2(1$ atm$)/HCl(0.001\ m)/AgCl/Ag$. What would your answers be had the molecular weight of O_2
been defined as 32 kg mol^{-1} (and m had been kg molecular weights per 1000 kg of solvent)?

Table 13-1 gives the potentials for the normal and 0.1 N calomel half-cells. Explain why the values   **13-18**
do not differ by exactly 0.05917 V.

13-19   The emf of the saturated Weston cell has been studied very carefully, and its dependence on temperature t (°C) is: $\mathscr{E} = 1.01864 - 4.06(t - 20) - 9.5 \times 10^{-7} (t - 20)^2$. Calculate $\mathscr{E}_{298}$ and ΔH and ΔS for the cell reaction of Eq. (13-16).

13-20   Write cell diagrams for cells having the following cell reactions. (a) $AgCl + Br^- = AgBr + Cl^-$, (b) $NaOH(aq, m_1) = NaOH(aq, m_2)$, and (c) $Zn + HgO = ZnO + Hg$.

13-21   The voltage of the cell $Ag/Ag_2SO_4(s)$/saturated solution of Ag_2SO_4 and $Hg_2SO_4/Hg_2SO_4/Hg$ is 0.140 V at 25°C and its temperature coefficient is 0.00015 V °C^{-1}.

   (a) Give the cell reaction.
   (b) Calculate the free energy change for the cell reaction.
   (c) Calculate the enthalpy change for the cell reaction.
   (d) Calculate the entropy change for the cell reaction.
   (e) Does the cell absorb or emit heat as the cell reaction occurs? Calculate the number of calories per mole of cell reaction.
   (f) One mole each of Hg, $Ag_2SO_4(s)$, $Hg_2SO_4(s)$, and some saturated solution of the two salts are mixed. What solid phases finally will be present and in what amounts?

13-22   Given that $\mathscr{E}^0_{298} = 0.3551$ V for $Ag(IO_3)/Ag$ and 0.7991 V for Ag^+/Ag, calculate the solubility product for $Ag(IO_3)$.

13-23   Write the cell diagram for a cell whose emf could be used to determine the solubility product for $PbSO_4$. What additional information would be needed besides the measured emf?

13-24   Calculate the percentage of mercury in the mercuric state in a solution of mercuric nitrate that is in equilibrium with liquid mercury.

13-25   R. Ogg found $\mathscr{E} = -0.029$ V at 25°C for the cell

$$Hg \Bigg/ \left(\begin{array}{c} \text{mercurous nitrate } (0.01\ N) \\ HNO_3 (0.1\ m) \end{array} \right) \Bigg/\!\!\Bigg/ \left(\begin{array}{c} \text{mercurous nitrate } (0.001\ N) \\ HNO_3\ (0.1\ m) \end{array} \right) \Bigg/ Hg$$

   What is the formula for mercurous ion? Show how your conclusion follows from this information.
   The actual dissociation constant for $Hg_2^{2+} = 2Hg^+$ is not known, but suppose that $\mathscr{E}$ for this cell is found to be -0.059 V when the higher and lower mercurous nitrate normalities are 2×10^{-4} and 3×10^{-6}, respectively. Calculate K_{diss} for Hg_2^{2+} from this data. (In both parts of the problem neglect any junction potential.)

13-26   Calculate the emf of the following cells (neglect any junction potentials). (a) $Pt/H_2(1\ atm)/H_2SO_4(0.5\ m)$, $H_2O_2\ (0.1\ m)/O_2(0.1\ atm)/C$, (b) $Pt/H_2(0.5\ atm)/HI(0.02\ m)/AgI(s)/Ag$, (c) $Pt/K_4Fe(CN)_6(0.02\ m)$, $K_3Fe(CN)_6(0.03\ m)$, $HClO_4(0.5\ m)$,$/HClO_4(0.5\ m)/H_2(1\ atm)/Pt$.

13-27   Is $H_2O_2(aq)$ stable? Calculate the equilibrium constant for $H_2O_2(aq) = H_2O + \frac{1}{2} O_2$ and the oxygen pressure in equilibrium with 0.01 m H_2O_2.

13-28   Calculate $\mathscr{E}^0_{298}$ for the half-cell $H_2O_2(aq)/H_2O$.

13-29   If lead and silver are assumed to be negligibly soluble in each other as solids, what must be the concentration of Ag^+ in a solution that is 0.2 m in Pb^{2+} in order that the two metals may be plated out together in an electrodeposition experiment? Assuming no overvoltage for hydrogen evolution, what is the lowest acceptable pH if metal deposition rather than hydrogen evolution is to occur? Assuming this pH, calculate the overall cell potential, given also that oxygen is formed at the anode at one atm pressure and at an overvoltage of 0.5 V.

13-30   It is desired to separate Cd^{2+} from Pb^{2+} by electrodeposition from a solution which is 0.1 m in each ion and at pH 2. Calculate the sequence in which Cd and Pb metals and H_2 are produced and the concentrations of the various species in solution when a new stage of electrolysis occurs. Include overvoltage effects in considering H_2 evolution.

13-31   The dissociation constant for water may be determined by means of emf measurements on the cell $Pt/H_2(1\ atm)/LiOH(m_1)$, $LiCl(m_2)/AgCl(s)/Ag$. Show that the following equation obtains at 25°C:

$(\mathscr{E} - 0.22239/0.05917 + \log(m_2/m_1) = -\log K_w - \log(\gamma_{Cl^-}/\gamma_{OH^-})$. The left-hand side is experimentally measurable, and we call it pK'. The following data are obtained (pK', m): (14.9915, 0.02), (14.9835, 0.04), (14.9755, 0.06), (14.9675, 0.08). Here $m = m_1 + m_2$. Plot pK' vs. m to obtain a value for K_w. Can you explain why the plot should in this case be against m rather than $\sqrt{m}$?

13-32 At 25°C, $\mathscr{E}$ for the cell Pt/H$_2$(1 atm)/HCl(m)/AgCl/Ag is as follows (Harned and Ehlers, 1932):

m	0.01002	0.01010	0.01031	0.04986	0.05005	0.09642
$\mathscr{E}$ (V)	0.46376	0.46331	0.46228	0.38582	0.38568	0.35393

m	0.09834	0.2030
$\mathscr{E}$ (V)	0.35316	0.31774

Calculate $\mathscr{E}^0$ for the cell by the extrapolation method, and then the activity coefficients for HCl at these concentrations and plot them against $\sqrt{m}$; compare the limiting slope with the value from theory.

13-33 The emf of the following cell has been measured at 25°C for various $m = m_1 = m_2$: Pt/H$_2$(1 atm)/ HClO$_4$(m_1) $\mathbin{/\!\!/}$ salt bridge $\mathbin{/\!\!/}$ HReO$_4$(m_2)/ReO$_3$(s)/Pt. The following data are obtained ($\mathscr{E}$, m): (0.5, 0.0540), (0.492, 0.0270), (0.468, 0.0068). Calculate the standard reduction potential for the half-cell ReO$_4^-$/ReO$_3$. The junction potential is neglected, with the extra justification of the probable similarity of activity coefficients and mobilities of perchlorate and perrhenate ions. Calculate also the mean activity coefficient of HReO$_4$ at each of the above concentrations.

13-34 At 25°C, $\mathscr{E}$ for the cell Pb/PbSO$_4$(s)/H$_2$SO$_4$(m)/H$_2$(1 atm)/Pt changes with the molality of the sulfuric acid as follows:

m	0.00100	0.00200	0.00500	0.0100	0.0200
$\mathscr{E}_{298}$ (V)	0.1017	0.1248	0.1533	0.1732	0.1922

By the extrapolation method, obtain $\mathscr{E}^0$ for this cell. From this value calculate the thermodynamic solubility product for PbSO$_4$.

13-35 The standard reduction potential for AgBr(s)/Ag varies with temperature as follows: $\mathscr{E}^0 = 0.07131 - 4.99 \times 10^{-4}(t - 25) - 3.45 \times 10^{-6}(t - 25)^2$, where t denotes °C. Calculate $\mathscr{E}^0_{308}$ and ΔH^0 and ΔS^0 at 25°C.

13-36 Given the standard reduction potentials $\mathscr{E}^0_{M^{3+}/M^+} = 1.25$ V and $\mathscr{E}^0_{M^+/M} = -0.336$ V, calculate $\mathscr{E}^0_{M^{3+}/M}$.

13-37 Given the standard reduction potentials at 25°C, $\mathscr{E}^0_{MnO_4^-/MnO_2} = 1.679$ V and $\mathscr{E}^0_{MnO_4^-/Mn^{2+}} = 1.491$ V, calculate $\mathscr{E}^0_{MnO_2/Mn^{2+}}$.

Note: the above potentials are for acid solution; that is, the half-cell reaction is balanced with H$^+$ and H$_2$O, and the standard potential is for a_{H^+} unity. Calculate $\mathscr{E}^0_{MnO_4/MnO_2}$ for basic solutions, that is, with the half-cell reaction balanced with OH$^-$ and H$_2$O, so that the standard potential is now for a_{OH^-} unity.

13-38 The standard reduction potential for basic solutions (see Problem 13-37) is $\mathscr{E}^0_{Cu_2O/Cu} = -0.361$ V. Calculate the solubility product for Cu$_2$O, that is, K for the dissolving of Cu$_2$O(s) to give Cu$^+$ and OH$^-$ ions.

13-39 At 25°C, $\mathscr{E}$ for the cell

Ag/AgCl/NaCl(m)/NaHg(amalgam) — NaHg(amalgam)/NaCl(0.100 m)/AgCl/Ag is as follows:

m	0.200	0.500	1.000	2.000	3.000	4.000
$\mathscr{E}_{298}$ (V)	0.03252	0.07584	0.10955	0.14627	0.17070	0.19036

The mean activity coefficient of 0.100 m NaCl is 0.773. Calculate the mean activity coefficients of NaCl in solutions of these concentrations and plot the values against the square root of m.

13-40   Calculate the standard free energy of formation of $SO_3^{2-}(aq)$ at 25°C.

13-41   Calculate the free energy of formation of $AgBr(s)$ using the solubility product for AgBr and standard free energies of formation of ions.

13-42   The pH of a solution may be found with the use of a hydrogen electrode and a calomel half-cell. Suppose that the pH of a certain buffer solution is measured at a high altitude such that the ambient pressure is 600 Torr but that the observer neglects this aspect and calculates the pH assuming the pressure of the hydrogen bubbling past the electrode to be 1 atm. If he reports a pH of 5.50, what is the correct pH of the solution?

13-43   Use the data of Table 13-3 to obtain $\mathcal{E}^0_{H_5IO_6/I_2}$ and $\mathcal{E}^0_{IO_3^-/I_2}$.

13-44 $\mathcal{E}^0_{PtCl_4^{2-}/Pt}$ is 0.73 V, while $\mathcal{E}^0_{Pt^{2+}/Pt}$ is estimated to be 1.20 V. Calculate K for the equilibrium $PtCl_4^{2-} = Pt^{2+} + 4Cl^-$.

13-45   A thallous (Tl^+) solution is titrated with ceric (Ce^{4+}) solution. The reaction is

$$Tl^+ + 2Ce^{4+} = Tl^{3+} + 2Ce^{3+}$$

Calculate the half-cell potential at 25°C for the reaction $Tl^{3+} + 2e^- = Tl^+$ *at the endpoint,* that is, the equivalence point of the titration. $\mathcal{E}^0_{Ce^{4+}/Ce^{3+}} = 1.60$ V [this is actually for a certain sulfuric acid concentration and the Ce(IV) is heavily complexed]. Also obtain the potentiometric titration curve for 0.1 m Tl^+ titrated with 0.1 m Ce^{4+}, assuming the reference cathode to be the normal calomel half-cell and neglecting junction potentials.

13-46   Use Tables 5-2 and 7-2 to obtain $\mathcal{E}^0_{298}$ and q_{298} for the possible fuel cell: $C(graphite) + O_2 = CO_2$.

SPECIAL TOPICS PROBLEMS

13-47   Calculate $\mathcal{E}$, $\mathcal{E}_E$, and $\mathcal{E}_J$ at 25°C for the cell $Cd(Hg\ amalgam)/CdCl_2(0.01\ m)\ /\!/\ CdCl_2(0.001\ m)/Cd(Hg\ amalgam)$. The transference number of Cd^{2+} in $CdSO_4(aq)$ is 0.38.

13-48   Calculate the emf's of concentration cells with transference for the following electrolytes in 0.01 m and 0.001 m solution, with appropriate silver chloride or mercurous sulfate electrodes: (a) $BaCl_2$; (b) $[Co(NH_3)_6]_2(SO_4)_3$.

13-49   Calculate the emf of the following cells with transference at 25°C:

(a) $Ag/AgCl/CaCl_2(0.01\ m)\ /\!/\ CaCl_2(0.001\ m)/AgCl/Ag$.
(b) $Hg/Hg_2SO_4/K_2SO_4(0.01\ m)\ /\!/\ K_2SO_4(0.001\ m)/Hg_2SO_4/Hg$.

13-50   Given the cell $Ag/Ag_2SO_4(s)/K_2SO_4\ (0.01\ m)\ /\!/\ K_2SO_4(0.02\ m)/Ag_2SO_4(s)/Ag$, (a) write the anode and cathode reactions per faraday, and the net cell reaction, and (b) calculate $\mathcal{E}_E$ and $\mathcal{E}_J$. Include activity coefficients (Table 12-7) and allow for variation of transference numbers with concentration (Table 12-4).

13-51   Calculate the plot of I versus ϕ for the case of $I_0 = 50$ mA, $\alpha = 0.4$, and $\mathcal{E}^0_{298} = 0.69$ V.

chapter 14
Kinetics of gas-phase reactions

The emphasis up to this point has been on the equilibrium properties of substances in pure or solution form. The rate processes considered have been restricted to transport phenomena such as diffusion and conductance. Chemical kinetics is a more complicated subject than that of chemical equilibrium because time is now a variable in the description of the state of a system and because there may be more than one reaction path whereby reactants become products. The theory is difficult because such paths involve molecules in energetic and otherwise unusual states; these states can rarely be studied by themselves independently, so theoretical models contain assumptions which cannot be corroborated in detail.

There are three major facets to the study of chemical kinetics. The first might be termed the experimental side but involves not only the measurement of reaction rates but also their reduction to what we will call a *rate law*. In chemical kinetics we retain the historic mass action rate law (Section 7-1) by expressing the rate R of a reaction at constant temperature as a function of the composition of the system. Thus if we have a reaction expressed as

$$aA + bB + cC + \ldots = mM + nN + \ldots.$$

the general expression for the rate law is:

$$R = -\frac{d(A)}{dt} = f[(A), (B), (C), \ldots] \tag{14-1}$$

The complete function can be complicated, but it is often of the form

$$R = k(A)^x(B)^y(C)^z \cdots \tag{14-2}$$

The form of Eq. (14-2) may come about naturally or by a deliberate choice of experimental conditions. As examples of this last situation, the complete rate law might be

$$R = \frac{k(A)(B)}{1 + k'(C)} \quad \text{or, as another example,} \quad R = k(A) + k'(A)(C)$$

If, however, a large excess of C were present, so that (C) was essentially constant, then the above rate laws would reduce to

$$R = (\text{constant})(A)(B) \qquad \text{or} \qquad R = (\text{constant})(A)$$

which is the form of Eq. (14-2). The role of species C in the rate law could be determined by carrying out additional experiments in which (C) was varied.

The constant k in Eq. (14-2) is called the *rate constant* (or *rate coefficient*), and, as noted above, an expression that gives R as a function of concentrations is known as the *rate law*. The first goal of the experimental stage of a kinetic study is to devise and carry out experiments that will establish the algebraic form of a mass action type of expression for R and to evaluate the rate constant or constants that it contains. The next experimental stage is that of determining how such constants vary with temperature and with the nature of the reaction medium.

The second facet of chemical kinetics is that of reaction mechanism. It turns out that with few exceptions the fundamental act of chemical change involves just one or two molecules at a time; that is, *elementary reactions* in kinetics are nearly always *unimolecular* or *bimolecular*. The overall reaction that is being studied may, however, occur through a series of such elementary steps, which then gives the experimental rate law some complicated form. The steps, typically, will involve transient chemical species or reaction intermediates whose presence is not always independently provable. The reaction between H_2 and Br_2, for example, is considered to occur through hydrogen and bromine atoms as intermediates, and steps in the overall reaction include $H + Br_2 = HBr + Br$.

It is this detailing of the intermediate bimolecular steps constituting the path for the overall reaction that is known as the *reaction mechanism*. One uses mainly the experimentally observed rate law, but in general all available information, to infer a reaction mechanism that is consistent with the kinetics of the system. The study of reaction mechanisms is the final goal of the more chemically oriented investigator; the field can be a very controversial one because it has happened that two or more mechanisms, each equally capable of explaining the data, have each received strong partisan support.

Third, the theoretical study of elementary reactions is now highly developed. It is concerned with the detail of how molecules approach each other, what their special requirements are for reaction—for example, with respect to energy—and with just how the crucial act of chemical bond making or breaking occurs. The two principal theoretical models of current importance differ considerably in flavor. If the elementary reaction is one of A with B to give products, and we suppose that state [AB]* is that in which the two reactants have come together to form a complex with the necessary energy and orientation for reaction, but have not yet fallen apart into products, we can write

$$A + B \underset{k_{-2}}{\overset{k_2}{\rightleftharpoons}} [AB]^* \overset{k_1}{\rightarrow} \text{products} \tag{14-3}$$

The first model, that of collision theory, deals with the rate at which A and B can form [AB]* and thus emphasizes k_2, and, in its simpler form, ignores k_{-2} so that [AB]* is considered always to go on to products. The second model, that of transition-state theory, treats [AB]* as being in equilibrium with A and

B, so that the reaction rate is determined by the [AB]* concentration and by the frequency or rate constant k_1.

The two models approach each other when pursued to a more sophisticated level, but each in its simple form has provided a framework or a language for qualitative explanations as to why a given reaction is fast or slow. Both are therefore presented.

Although chemical kinetics can be outlined as a unified general subject, as has been done here, when details are explored facets of the treatment of gaseous and solution systems tend to separate as distinguishable subfields. The three aspects just described develop different emphases and different bodies of empirical fact. For this reason the material has been separated into two chapters. We first take up the subject of rate laws and their algebraic manipulation and then that of some important mechanisms of gas-phase reactions. The two major theories are next presented, with emphasis on their application to gaseous systems. Chapter 15 covers some of the experimental techniques special to solution systems and then proceeds to a discussion of how both collision and transition-state theory are modified in the case of solutions.

14-2   Definition of reaction rate

There is room for ambiguity in dealing with a "reaction rate." Consider, for example, the reaction $H_2 + Br_2 \rightarrow 2HBr$. One can specify the rate in terms of dn_{H_2}/dt, dP_{H_2}/dt, or $d(H_2)/dt$, or similarly using Br_2 or HBr as the indicated species. Furthermore, $|d(HBr)/dt|$ will be twice $|d(H_2)/dt|$ because of the reaction stoichiometry.

It is ordinarily all right to use any of the above types of alternatives if one specifies clearly what is done. There is, however, a formal convention, recommended by the International Union of Pure and Applied Chemistry (IUPAC), which goes as follows. We write a general reaction

$$aA + bB = mM + nN \tag{14-4}$$

where the coefficients a, b, etc. are the smallest possible integers that give the correct stoichiometry. For later convenience we write Eq. (14-4) in the form

$$\sum_i v_i dX_i = 0 \tag{14-5}$$

where the v_i's are the above stoichiometric coefficients, but defined to be *negative* for reactants and *positive* for products, and the X_i's denote the reactant and product species. Since rate laws are generally expressed in terms of concentrations, as in Eq. (14-1), it is logical to express a reaction rate in terms of a rate of concentration change. We will use the formal definition

$$R = d(\xi/v)/dt = -\frac{1}{a}d(A)/dt = -\frac{1}{b}d(B)/dt = \frac{1}{m}d(M)/dt = \frac{1}{n}d(N)/dt$$

or, in general,

$$R = d(\xi/v)/dt = \frac{1}{v_i} d(X_i)/dt \tag{14-6}$$

The quantity ξ is the *advancement* of the reaction; thus $n_i = n_i^0 + v_i\,\xi$, where n_i^0 is the initial number of moles of the ith species; v is the total volume of the system. (The somewhat related idea of degree of reaction, was used in Section 7-CN-2).

We now have a definition of R based on the amount of overall reaction rather than on some one reactant or product. Note that R is given in terms of a rate of change of a concentration, ξ/v, with time, and that R is a positive number if the reaction is proceeding from left to right. Also, our discussions will be limited to reactions occurring under constant volume conditions, so that R can be written as $(1/v)d\xi/dt$; if this is not the case (note Problem 14-28), a dv/dt term must be recognized in the handling of a $d(X_i)/dt$ quantity.

The above formal definition of R is often unnecessarily cumbersome, and it is common to see R expressed merely as a particular $d(X_i)/dt$, where X_i is one of the reactants. The discussion does serve to emphasize, however, the need to recognize the reaction stoichiometry in relating one rate of concentration change to another.

14-3   Rate laws and simple mechanisms

The principal concern of this section is with the mathematical characteristics of specific functions $f[(A), (B), (C), \ldots]$, including their integrated forms. There is virtually no limit to the complexity of functions that may be encountered, and only the more important ones are taken up here. Also, we will consider only reactions that go to completion, deferring the more general case until Chapter 15.

A major category of rate laws is that for which R is given by an expression of the form of Eq. (14-2). We then speak of the *order* of the rate law as the sum of the exponents $(x + y + z \cdots)$, the important cases being those of first- and second-order reactions. Although the rate law bears no *a priori* relationship to the chemical equation for the overall reaction, the *stoichiometry* of this last must always be kept in mind. This point was noted in the preceding section and will be illustrated in Section 14-3B.

A. The first-order rate law Suppose that the forward rate R_f of a reaction is found to depend linearly on the concentration of some reactant A. If the overall reaction is

$$aA + \text{other reactants} \xrightarrow{R_f} \text{products} \tag{14-7}$$

then according to Eq. (14-6) we have

$$-\frac{1}{a}\frac{d(A)}{dt} = k(A) \tag{14-8}$$

If the reaction goes to completion, Eq. (14-8) provides the complete description of the rate behavior since R_b, the rate of the back reaction, is zero. Integration gives

$$\int \frac{d(A)}{(A)} = -ak\int dt$$

or

$$\ln(A) = -akt + \text{constant} \tag{14-9}$$

A plot of $\ln(A)$ vs. t gives a straight line of slope equal to $-ka$, as in Fig. 14-1(a), where the rate constant k has the dimension time^{-1} (usually s^{-1}).

It is usually convenient to evaluate the constant of integration by setting $(A) = (A)_0$ at $t = 0$, and Eq. (14-9) becomes, in exponential form,

$$(A) = (A)_0\, e^{-akt} \tag{14-10}$$

The graphical appearance is shown in Fig. 14-1(b). It is quite common in the chemical literature for the stoichiometry factor a to be combined with k and to write an equation such as Eq. (14-10) as

$$(A) = (A)_0\, e^{-k't} \tag{14-11}$$

where $k' = ak$; the units of k and k' are the same since a is a dimensionless quantity. Very often, a is unity, and $k = k'$.

A characteristic time is that at which $(A)/(A)_0 = \frac{1}{2}$; this is called the *half-life* of the reaction, and is very simply related to the rate constant:

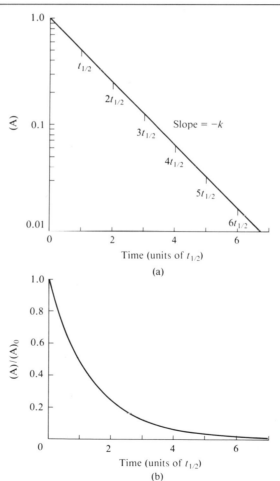

FIGURE 14-1
The first-order rate law. (a) Plotted according to Eq. (14-9). (b) $(A)/(A)_0$ as a function of time, showing the constance of $t_{1/2}$.

$$\frac{(A)}{(A)_0} = \tfrac{1}{2} = e^{-akt_{1/2}} \qquad \ln \tfrac{1}{2} = -akt_{1/2} \tag{14-12}$$

or

$$k't_{1/2} = akt_{1/2} = 0.6931 \tag{14-13}$$

Either k or $t_{1/2}$ defines the rate of a first-order reaction. Note that if $t = 2t_{1/2}$ the effect is to double the exponential in Eq. (14-12) so that $(A)/(A)_0 = (\tfrac{1}{2})^2 = \tfrac{1}{4}$. In general, if $t = nt_{1/2}$,

$$\frac{(A)}{(A)_0} = (\tfrac{1}{2})^n \tag{14-14}$$

This is illustrated in Fig. 14-1(b): $(A)/(A)_0 = \tfrac{1}{2}$ at $t_{1/2}$, $\tfrac{1}{4}$ at $2t_{1/2}$, $\tfrac{1}{8}$ at $3t_{1/2}$, and so on.

EXAMPLE   It takes 10 min for a certain first-order reaction to go 20% toward completion; the coefficient a in Eq. (14-7) is unity. What are k and $t_{1/2}$ and how long should it take for 75% reaction? By Eq. (14-10), $\ln(0.8) = -k(10)$, whence $k = 0.0223$ min^{-1} and $t_{1/2} = 0.06931/0.0223 = 31.1$ min. At 75% reaction, $(A)/(A)_0 = 0.25$ or $\tfrac{1}{4}$; the required time is two half-lives or 62.2 min.

There is nothing unique about the time for half reaction—any other fraction could be picked. Were we to speak of $t_{1/3}$, the analogue of Eq. (14-12) would be $\ln \tfrac{1}{3} = -akt_{1/3}$, and in general if t_r is the time for (A) to drop to the fraction of r of its initial value,

$$\frac{(A)}{(A)_0} = r = e^{-akt_r} \qquad \ln r = -akt_r \tag{14-15}$$

and

$$\frac{(A)}{(A)_0} = (r)^{n_r} \tag{14-16}$$

where n_r is the elapsed time expressed in multiples of t_r. As with half-life, an important characteristic of an exponential function is that if it takes time t_r for (A) to decrease by the factor r, then in a second such interval of time a further decrease by the same factor will occur, to give an overall decrease in (A) by a factor of r^2, and so on. A numerical illustration is given in Section 14-4.

It is sometimes useful to take $r = 1/e$, in which case t_r is given the special symbol τ and is known as the *mean lifetime* for the reaction since it is, in fact, the average time elapsed before a given molecule of A reacts. Substitution into Eq. (14-15) gives $1/e = e^{-ak\tau}$, or $-1 = -ak\tau$, whence

$$ak\tau = 1 \tag{14-17}$$

The quantity τ can be shown to be the mean lifetime as follows. For the moment let the mean lifetime be denoted by $\bar{t}$, which we define by weighting each value of t by the amount of A present:

$$\bar{t} = \frac{\int_{(A)=(A)_0}^{(A)=0} t\, d(A)}{\int_{(A)=(A)_0}^{(A)=0} d(A)} = \frac{[(A)_0/k]\int_{t=0}^{t=\infty} (akt)e^{-akt}d(akt)}{(A)_0}$$

$$\bar{t} = \frac{1}{ak}\,|\, -e^{-akt}(1 + akt)|_0^\infty = \frac{1}{ak}$$

Thus $\bar{t}$ is identical to τ.

Consider the case where the overall order of reaction in Eq. (14-2) is found to be two, that is $(x + y) = 2$, either because the rate is proportional to $(A)^2$ or because it is proportional to $(A)(B)$. In the first case we can use Eq. (14-7) as giving the overall process, and write the rate law as

$$R_f = -\frac{1}{a}\frac{d(A)}{dt} = k(A)^2 \tag{14-18}$$

We will again assume that the reaction goes to completion, so that R_b is zero. Note that the rate constant k now has the dimensions of $(\text{concentration})^{-1}$ $(\text{time})^{-1}$; time is usually given in seconds, and in the case of gases, concentration may be expressed in terms of partial pressure (the two are proportional for an ideal gas mixture at constant temperature). The rate constant would now be written in terms of $(\text{pressure})^{-1}(\text{s})^{-1}$, where pressure might be in atm, Pa, and so on.

The integration of Eq. (14-18) is straightforward:

$$-\int \frac{d(A)}{(A)^2} = ak \int dt \quad \text{or} \quad \frac{1}{(A)} = akt + \text{constant} \tag{14-19}$$

As illustrated in Fig. 14-2(a), a plot of $1/(A)$ versus t gives a straight line, with intercept $1/(A)_0$ and slope equal to ak, so that

$$\frac{1}{(A)} = \frac{1}{(A)_0} + akt \tag{14-20}$$

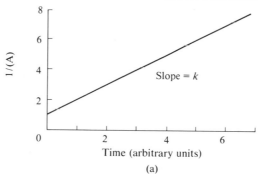

Slope = k

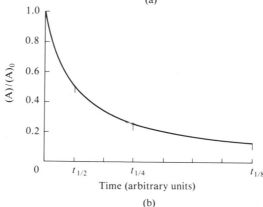

FIGURE 14-2
The second-order rate law. (a) Plotted according to Eq. (14-20). (b) $(A)/(A)_0$ as a function of time, showing the successive doubling of $t_{1/2}$.

Again, it will not be uncommon to find the stoichiometry factor a combined with the rate constant, and Eq. (14-20) written with k', where $k' = ak$. The concept of half-life may be used; we set $(A)/(A)_0 = \frac{1}{2}$ at $t_{1/2}$ in Eq. (14-20) to obtain

$$2 - 1 = (A)_0 ak t_{1/2} \qquad \text{or} \qquad t_{1/2} = \frac{1}{ak(A)_0} \tag{14-21}$$

Unlike the case of a first-order reaction, the time for successive reductions in (A) by a factor of one-half is *not* a constant, but doubles with each decrement. Thus $t_{1/4} = 3/ak(A)_0$ and $t_{1/8} = 7/ak(A)_0$, so that, as illustrated in Fig. 14-2(b), the first interval is $t_{1/2}$, the second $2t_{1/2}$, and so on.

If the reaction is second order, but with the rate proportional to (A)(B), the overall equation can be written as

$$aA + bB + \text{other reactants} \xrightarrow{R_f} \text{products} \tag{14-22}$$

and the rate law takes the form:

$$R_f = -\frac{1}{a}\frac{d(A)}{dt} = -\frac{1}{b}\frac{d(B)}{dt} = k(A)(B) \tag{14-23}$$

We again assume that there is no back reaction. It is convenient at this point to define x as the decrease in concentration of A at time t, or $x = (A) - (A)_0$. It follows that $(A) = (A)_0 - x$ and also that $(B) = (B)_0 - \frac{b}{a}x$; that is, for every mole of A reacting, $\frac{b}{a}$ moles of B react according to Eq. (14-22). Equation (14-23) becomes

$$\frac{dx}{dt} = ak(A)(B) = ak[(A)_0 - x][(B)_0 - \frac{b}{a}x] \tag{14-24}$$

or

$$\frac{dx}{[(A)_0 - x]\,[(B)_0 - \frac{b}{a}x]} = ak\,dt \tag{14-25}$$

On integration and setting $x = 0$ at $t = 0$, we obtain

$$\frac{1}{[b(A)_0 - a(B)_0]} \ln \frac{(B)_0[(A)_0 - x]}{(A)_0[(B)_0 - \frac{b}{a}x]} = kt \tag{14-26}$$

or

$$\frac{1}{[b(A)_0 - a(B)_0]} \ln \frac{(B)_0(A)}{(A)_0(B)} = kt \tag{14-27}$$

It is fairly common that $a = b = 1$, in which case Eq. (14-27) reduces to:

$$\ln \frac{(B)_0(A)}{(A)_0(B)} = [(A)_0 - (B)_0]kt \tag{14-28}$$

There are two special cases of interest. First, if $(A)_0 \ll (B)_0$, then very little B is consumed so that (B) is essentially constant and equal to $(B)_0$. The effect is that Eq. (14-23) reduces to the form

$$-\frac{d(A)}{dt} = [ak(B)_0](A) = k_{app}(A) \tag{14-29}$$

or to the same form as the first-order rate law, Eq. (14-8). Since k_{app} is not a pure rate constant but is the product of ak and the concentration $(B)_0$, it is customary to call Eq. (14-29) a *pseudo-first-order* rate law. This term is applied whenever a first-order rate expression has been obtained by virtue of some concentration or concentrations being held constant. The distinction is important to the mathematics of the situation, in that k_{app} will vary with $(B)_0$. Thus Eq. (14-29) gives the *time* rate law for a particular experiment, but not the *full concentration* rate law. The distinction can also be quite important in any theoretical interpretation of the rate constant.

The second special case is that for $b(A)_0 = a(B)_0$. Equation (14-27) becomes indeterminate, and it is necessary (as always in such a circumstance) to return to the original differential equation. Since $b(A)_0 = a(B)_0$, it follows that $b(A) = a(B)$ at all times (why?) and Eq. (14-23) reduces to

$$-\frac{d(A)}{dt} = ak(A)\left[\frac{b}{a}(A)\right] = -b(A)^2$$

or to the form of Eq. (14-18).

By choosing $(A)_0 \ll (B)_0$, we reduce the rate law to a first-order one, and by choosing $b(A)_0 = a(B)_0$, we reduce it to a much simpler mathematical form of a second-order rate law. Part of the art of experimental chemical kinetics lies in the designing of experimental conditions so that the mathematical complexity of the *time* rate law is reduced.

The rate law for the reaction $A + B \rightarrow$ products is $d(A)/dt = -k(A)(B)$ with $k = 0.02$ M^{-1} min^{-1}. What percent of A has reacted after 15 min if (a) $(A)_0 = 0.1$ M and $(B)_0 = 0.3$ M, (b) $(A)_0 = (B)_0 = 0.1$ M, and (c) if $(A)_0 = 0.001$ M and $(B)_0 = 0.3$ M? (a) By Eq. (14-28), $\ln[(0.3)(A)/(0.1)(B)] = (0.1 - 0.3)(0.02)(15) = -0.06$, whence $(A)/(B) = 0.3139$. Since $(A)/(B) = (0.1 - x)/(0.3 - x)$, we find $x - 8.49 \times 10^{-3}$ M, corresponding to 8.49% reaction of A. (b) We now use Eq. (14-20), which gives $1/(A) = 1/0.1 + (0.02)(15)$, whence $(A) = 0.09709$ M and the percent reacted is 2.91%. (c) The reaction is pseudo-first-order because of the great excess of B, so $\ln[(A)/(A)_0] = -k_{app}t = -(0.02)(0.3)(15) = -0.09$, whence $(A)/(A)_0 = 0.9139$, corresponding to 8.61% reaction. **EXAMPLE**

The reaction stoichiometry might be $2A + B \rightarrow$ products, but with the rate still first order in (A) and first order in (B). Eq. (14-23) now reads $-d(A)/dt = 2k(A)(B)$. The answers to parts (a), (b), and (c) above will now be different—see Problem 14-33. **NOTE**

C. The zero-order and other rate laws

The term "zero order" is applied to a reaction whose rate is independent of time, that is,

$$R_f = -\frac{1}{a}\frac{d(A)}{dt} = k \tag{14-30}$$

or

$$(A) = (A)_0 - akt \tag{14-31}$$

Such rate laws are more properly called pseudo-zero-order, since Eq. (14-31) gives the time dependence of (A) but cannot be the full description of the factors affecting the rate.

This point may be illustrated as follows. First, in photochemical reactions if the entire incident radiation is absorbed by the reacting species, then the rate of reaction will not depend on concentration and hence will be of zero order. However, the complete rate law is actually $R = I_{abs}\phi$, where I_{abs} is the intensity of the absorbed light and ϕ is an efficiency factor called the quantum yield for the reaction (see Section 18-4E); in the form written I_{abs} would be expressed as quanta of light per unit volume per second. Thus although the rate does not depend on (A), it does depend on I_{abs}; at a sufficiently low concentration of A, I_{abs} will become proportional to (A), and the rate law will revert to first order in (A).

Second, many examples of zero-order reactions occur in heterogeneous catalysis. The situation is one in which the reaction takes place on the surface of the catalyst so that $R = k'\theta\mathscr{A}$, where θ is the fraction of surface covered by the adsorbed reactant and $\mathscr{A}$ is the total catalyst surface area. If the concentration or pressure of A is large enough, $\theta = 1$, and the reaction is zero order; at sufficiently low pressures, however, θ becomes proportional to (A) and the reaction reverts to first order. Note that R will depend on the amount of catalyst, that is, on the area $\mathscr{A}$, as well. See the Special Topics section for further details.

Experimental rate laws may be of some general order n, the simplest case of which is

$$-\frac{1}{a}\frac{d(A)}{dt} = k(A)^n \tag{14-32}$$

Integration gives

$$\frac{1}{(A)^{n-1}} = \frac{1}{(A)_0^{n-1}} + (n-1)akt \tag{14-33}$$

provided that $n \neq 1$ Some examples are the interconversion of *ortho-* and *para-* hydrogen, for which $n = \frac{3}{2}$, the formation of phosgene ($CO + Cl_2 = COCl_2$), for which $n = \frac{5}{2}$, and a variety of reactions for which $n = 3$.

Various rate laws and their integrated forms are summarized in Table 14-1.

14-4 Experimental methods and rate law calculations

A. Use of additive properties

As stated earlier, an important goal of the experimentalist is that of establishing the rate law which governs a given reaction. The initial task is very similar to that in the study of chemical equilibrium—one needs some way of following the degree of action. As discussed in Section 7-3, one may quench or otherwise stop a reaction and then proceed at one's leisure to analyze chemically for reactants or products. However, very often some specific physical property of one or more of the species may be measured *in situ*, so that one may follow the reaction continuously as it proceeds. If one of the reactants

TABLE 14-1
Rate laws and their
integrated forms

Order	Rate Law $-\dfrac{1}{a}\dfrac{d(A)}{dt} =$	Type of reaction	Integrated form	Equation
0	k	aA + other reactants $\rightarrow$ products	$(A) = (A)_0 - akt$	14-30
$\frac{1}{2}$	$k(A)^{1/2}$	same	$(A)^{1/2} = (A)_0^{1/2} - \frac{1}{2}akt$	
1	$k(A)$	same	$\ln[(A)/(A)_0] = -akt$	14-9
			$(A)/(A)_0 = e^{-akt}$	14-10
2	$k(A)^2$	same	$\dfrac{1}{(A)} = \dfrac{1}{(A)_0} + akt$	14-20
	$k(A)(B)$	$aA + bB +$ other reactants $\rightarrow$ products	$\dfrac{1}{b(A)_0 - a(B)_0}\ln\dfrac{(B)_0(A)}{(A)_0(B)} = kt$	14-27
	$k(A)(B)$	$A + B +$ other reactants $\rightarrow$ products	$\ln\dfrac{(B)_0(A)}{(A)_0(B)} = [(A)_0 - (B)_0]kt$	14-28
n	$k(A)^n$	aA + other reactants $\rightarrow$ products	$\dfrac{1}{(A)^{n-1}} = \dfrac{1}{(A)_0^{n-1}}$ $+ (n - 1)akt$	14-33

has a characteristic light absorption, for example, the optical density of the solution may be monitored at a particular wavelength. In the case of a gas-phase reaction the change in the total pressure may be used as a measure of the degree of reaction (provided the number of moles of gaseous products differs from that of the reactants).

These last two quantities, optical density (or absorbance) and total pressure, are examples of additive properties $\mathscr{P}$ discussed in Section 3-1. There is a very useful relationship whereby $\mathscr{P}$ for a reacting mixture may be used to give the advancement ξ of the reaction. Consider the general reaction

$$aA + bB + \ldots = mM + nN + \ldots. \tag{14-34}$$

We suppose that A_0 moles of A, B_0 moles of B, and so on are present initially, but no products, so that

$$\mathscr{P}_0 = A_0\mathscr{P}_A + B_0\mathscr{P}_B + \cdots \tag{14-35}$$

After some time t we have

$$\mathscr{P}_t = [A_0 - a\xi]\mathscr{P}_A + [B_0 - b\xi]\mathscr{P}_B \\ + \cdots + m\xi\mathscr{P}_M + n\xi\mathscr{P}_N + \cdots \tag{14-36}$$

or, in the terminology of Section 14-2,

$$\mathscr{P}_t = \sum_i (X_{i,0} + \nu_i\xi)\mathscr{P}_i \tag{14-37}$$

Equation (14-36) rearranges to

$$\mathscr{P}_0 - \mathscr{P}_t = a\xi\mathscr{P}_A + b\xi\mathscr{P}_B + \cdots - m\xi\mathscr{P}_M - n\xi\mathscr{P}_N - \cdots \\ = -\sum_i \nu_i\xi\mathscr{P}_i \tag{14-38}$$

At infinite time, that is, when the reaction has proceeded to equilibrium (or, if the back reaction is negligible, to completion), ξ has the value ξ_∞ and

$$\mathscr{P}_\infty = [A_0 - a\xi_\infty]\mathscr{P}_A + [B_0 - b\xi_\infty]\mathscr{P}_B + \cdots + m\xi_\infty\mathscr{P}_M + n\xi_\infty\mathscr{P}_N \cdots$$

so that

$$\mathscr{P}_0 - \mathscr{P}_\infty = a\xi_\infty\mathscr{P}_A + b\xi_\infty\mathscr{P}_B + \cdots - m\xi_\infty\mathscr{P}_M - n\xi_\infty\mathscr{P}_N - \cdots \qquad (14\text{-}39)$$

and therefore

$$\frac{\mathscr{P}_0 - \mathscr{P}_t}{\mathscr{P}_0 - \mathscr{P}_\infty} = \frac{\xi}{\xi_\infty} \qquad (14\text{-}40)$$

the term $(a\mathscr{P}_A + b\mathscr{P}_B + \cdots - m\mathscr{P}_M - n\mathscr{P}_N - \cdots)$ being a common factor to both Eqs. (14-38) and (14-39). Alternatively, we have

$$\frac{\mathscr{P}_t - \mathscr{P}_\infty}{\mathscr{P}_0 - \mathscr{P}_\infty} = 1 - \frac{\xi}{\xi_\infty} = \frac{(A)_t - (A)_\infty}{(A)_0 - (A)_\infty} \qquad (14\text{-}41)$$

We see that the second part of Eq. (14-41) is correct by writing $(A)_t$ as $[A_0 - a\xi]/v$, $(A)_\infty$ as $[A_0 - a\xi_\infty]/v$, and $(A)_0$ as A_0/v, where v is the volume of the system, and observing that the result simplifies to $1 - (\xi/\xi_\infty)$. It might be noted that if the reactants are present initially in stoichiometric ratio and the reaction proceeds completely to products, then ξ/ξ_∞ is just the degree of advancement quantity used in Section 7-CN-1.

The application of Eq. (14-41) may be illustrated as follows. Paraldehyde, $(CH_3CHO)_3$, dissociates into acetaldehyde, the overall reaction being of the form

$$A = 3B$$

The change in total pressure with time at 260°C is given in Table 14-2 and application of Eq. (14-41), with $\mathscr{P} = P_{tot}$ (total pressure), then gives $P_A/P_{0,A}$, or the fraction of paraldehyde remaining ($P_{\infty,A}$ being zero in this case since the reaction goes to completion). The results are plotted in Fig. 14-3(a), according to Eq. (14-10), and the linearity of the plot indicates that the reaction is first order. In Fig. 14-3(b), the data are also plotted according to Eq. (14-20), which tests for possible obedience to the second-order rate law, but the points now fall on a curve. As indicated in the table itself, one could determine the first-order character directly by noting that $P_A/P_{0,A}$ drops by the constant factor 0.64 with each interval of 1 hr.

The rate law for the reaction is therefore

$$\frac{dP_A}{dt} = -kP_A$$

	Time (hr)	P_{tot} (Torr)	$P_A/P_{0,A}$
TABLE 14-2 Thermal dissociation of paraldehyde at 260° C	0	100	1.00
	1	173	0.64
	2	218	$0.41 = (0.64)^2$
	3	248	$0.26 = (0.64)^3$
	4	266	$0.17 = (0.64)^4$
	∞	300	0.00

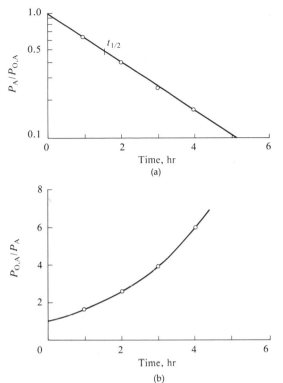

FIGURE 14-3
The thermal decomposition of paraldehyde at 260°C. (a) Data plotted according to the first-order rate law, Eq. (14-10) and (b) data plotted according to the second-order rate law, Eq. (14-20). In this case the solid line is merely a smoothing line drawn through the experimental points.

and there are several ways in which k might be obtained. The slope of the line in Fig. 14-3(a) gives $k = -[\ln(0.1) - \ln(1)]/5.2 = 0.44 \, \text{hr}^{-1}$. Alternatively, $t_{1/2}$ is read from the graph at 1.57 hr, and by Eq. (14-13), $k = 0.693/1.57 = 0.44 \, \text{hr}^{-1}$. A still different approach would be to take the 1 hr point, for which $k(1) = -\ln(.64)$; whence $k = 0.44 \, \text{hr}^{-1}$.

B. The isolation method

The preceding subsection dealt with one very common method of determining an experimental rate law—that of fitting data to the integrated form of a specific rate law. This procedure is accurate and very satisfactory if the reaction rate is some integral order in a single reactant, or a second-order reaction in two reactants. If the data do not fit one or another simple integrated form, the number of more complicated possibilities rises rapidly, and the approach becomes increasingly less definitive, that is, the various more complicated integrated forms may not actually differ enough to allow a clear experimental distinction among them. At this point it becomes very useful to fix one or another concentration so as to *isolate* the dependence of the rate on each species in turn. One usually does this using a large excess of first one reactant and then another.

An illustration may be constructed as follows. The reaction

$$2NO + H_2 = N_2O + H_2O \tag{14-42}$$

is known to obey the rate law

$$\frac{dP_{N_2O}}{dt} = kP_{NO}^2 P_{H_2} \tag{14-43}$$

with $k = 1.00 \times 10^{-7}$ Torr^{-2} s^{-1} at 820°C. Let us examine how this conclusion might have been arrived at by the isolation method. Suppose that first the reaction was studied with a large excess of NO: $P_{0,NO} = 600$ Torr and $P_{0,H_2} = 10$ Torr. P_{NO} will be virtually constant, and the rate law becomes pseudo-first-order:

$$-\frac{dP_{H_2}}{dt} = \frac{dP_{N_2O}}{dt} = [kP_{0,NO}^2]\, P_{H_2} = k_{app}P_{H_2}$$

where $k_{app} = (1.00 \times 10^{-7})(600)^2 = 0.036$ s^{-1}. The reaction would be found to obey Eq. (14-9) or Eq. (14-10), with $t_{1/2} = 0.693/0.036 = 19.3$ s, and the experimenter would therefore conclude that the rate law contains P_{H_2} as a term. We confirm this by observing that $t_{1/2}$ does not depend on P_{0,H_2} so long as NO is in large excess.

The next step would be to reverse the situation and use a large excess of $H_2 : P_{0',NO} = 10$ Torr and $P_{0,H_2} = 600$ Torr. The reaction would now be pseudo-second-order:

$$-\frac{1}{2}\frac{dP_{NO}}{dt} = \frac{dP_{N_2O}}{dt} = (kP_{0,H_2})\, P_{NO}^2 = k_{app}P_{NO}^2$$

or

$$\frac{dP_{NO}}{dt} = -k_{app}P_{NO}^2$$

with k_{app} now equal to $2(1.00 \times 10^{-7})(600) = 1.20 \times 10^{-4}$ Torr^{-1} s^{-1}. Note the presence of the stoichiometry factor of 2. The rate data would now obey Eq. (14-20), and would show $t_{1/2} = 1/(1.20 \times 10^{-4})(10) = 833$ s. The experimenter would conclude from this behavior that the rate law contains P_{NO}^2. Combination of the two results would thus yield the full rate law.

C. Use of initial rates

The procedures so far have made use of integrated rate laws. One must obtain data over a sufficient degree of reaction to establish agreement with one or another particular form. Considerable difficulties can develop with this procedure, however. The reaction may not be clear-cut, so that side reactions obscure the true course of the process being studied. Or the reaction may not proceed to completion, so that the term for the back reaction has mistakenly been omitted from the rate laws tested. It must be emphasized that quite different algebraic forms will often fit a given set of data equally well, especially if the results are of the usual accuracy of about 1%.

In either case, the difficulty is avoided if the reaction is studied during its initial stages only. The experimental problem is that the analytical procedure must now be one suited to determining small amounts of a product in large amounts of reactants. Conventional quenching techniques followed by specific product analysis are often best. A nonvolatile product might be selectively condensed out of the reaction mixture or separated from it by gas chromatography, for example. If a product has a distinctive absorption spectrum, then even small amounts may be measured *in situ*. General additive properties such as total pressure are not very sensitive to small degrees of reaction, however, and become difficult to use.

The measurement of an initial reaction rate provides no information in itself as to the form of the rate law; it is necessary to make several experiments in which the initial concentrations are varied. In terms of the previous example, $(dP_{N_2O}/dt)_{initial}$ would have been found to be 0.36 Torr s^{-1} with $P_{0,NO} = 600$ Torr and $P_{0,H_2} = 10$ Torr (at 820°C) and 0.72 Torr s^{-1} if P_{0,H_2} had been increased to 20 Torr. Thus, doubling (H$_2$) at constant (NO) doubled the initial rate, and on the assumption that the rate law was of the form

$$\frac{dP_{N_2O}}{dt} = kP_{NO}^x P_{H_2}^y$$

one would write

$$\frac{0.72}{0.36} = \frac{k(600)^x(20)^y}{k(600)^x(10)^y} = 2^y \quad \text{or} \quad 2 = 2^y, \, y = 1$$

The conclusion would thus again have been that the reaction was first order in H$_2$. Had $P_{0,NO}$ been 300 Torr and P_{0,H_2} 10 Torr, the initial rate would have been 0.090 Torr s^{-1} and the corresponding quotient of rate expressions would be

$$\frac{0.36}{0.090} = \frac{k(600)^x(10)^y}{k(300)^x(10)^y} = (2)^x \quad \text{or} \quad 4 = 2^x, \, x = 2$$

The reaction would thus have been shown to be second order in NO.

The analytical techniques so far mentioned require from a few seconds to carry out, as with optical density determinations, to minutes or hours, for chemical procedures. Many reactions are much faster than this, and a number of ingenious methods have been developed for such cases. For example, reaction times of as low as a few milliseconds can be studied with fast-mixing reactors. As illustrated in Fig. 14-4, the separate reactant gases are jetted into a chamber so designed that mixing is rapid and complete, and the mixed

D. Fast reaction techniques

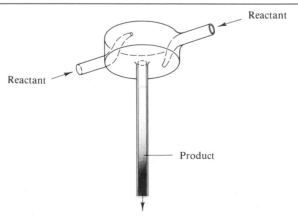

Reactant

Reactant

Product

FIGURE 14-4
Mixing cell for the study of fast reactions.

gases exit into a viewing chamber. One then determines optical density of the mixture by allowing a steady light beam to shine through the chamber and measuring its reduction in intensity. The average lifetime of the mixture can be estimated from the inlet flow rates and the volume of the mixing chamber, and each measurement is thus for one particular time of reaction. One varies the inflow rates to vary the reaction time.

A widely used variant, especially for pyrolysis reactions, is to use the mixing chamber itself as the reaction vessel, the exiting gases being quenched by rapid cooling. The time of reaction is the residence time of molecules in the chamber, which can be calculated from the flow rates and the volume of the chamber, and the system is essentially a steady-state one in which a constant (and usually small) degree of reaction occurs before the mixture exits to be quenched. Reactions such as the thermal decomposition of toluene have been studied by this means.

There are difficulties associated with either version of the flow method, mostly having to do with the accurate definition of temperature, gas pressure, and reaction time. There will be, for example, a pressure drop in the flow system, which means that reactant concentrations are changing, and hence the lifetime of any bimolecular or higher order reaction. In a pyrolysis reaction, where cool reactants enter a hot mixing chamber and the product mixture is quenched by cooling, it is not easy to calculate the effective average temperature or the time of reaction.

A somewhat related method is that of the *shock tube*. The reaction mixture is separated by means of a diaphragm from some inert gas which is at a high pressure. On rupture of the diaphragm, a shock wave passes down the reaction mixture, rapidly heating it by hundreds of degrees. As a result of the change in temperature, reaction occurs, and the rate of formation of products may be followed by rapid spectroscopic methods.

A very powerful method is that of *flash photolysis*. The reactive species is now produced by a short, intense flash of light. Light flashes containing perhaps 10^{-7} mole of light quanta can be made to occur within a few microseconds by triggering the discharge of high-voltage capacitors through a gas-filled tube; xenon at about 10 Torr is often used. As illustrated in Fig. 14-5, two or more flash tubes fire simultaneously to start the reaction in the reaction cell.

The ensuing reaction may be followed spectroscopically in one of two ways. As shown in the figure, monochromatic light from a source S passes through the length of the reaction cell, and its intensity is monitored by means of a photomultiplier tube connected to an oscilloscope. One obtains in this way the absorbance at a particular wavelength as a function of time. Successive experiments then allow the determination of absorbance versus time at other wavelengths.

Alternatively, there may be a second, smaller flash tube triggered so as to fire at a preset time following the main flash. The "white" light from this second flash passes down the length of the reaction cell and thence into a spectrograph, which allows the photographic recording of the entire absorption spectrum. The absorption spectrum of the system may now be obtained for various delay times.

Figure 14-5 shows also the type of data obtained for the system:

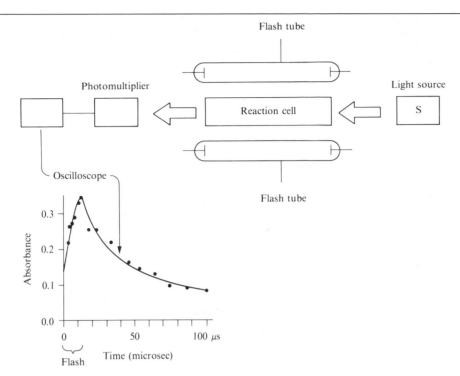

FIGURE 14-5
Schematic of a flash photolysis set-up. See text for description. The inset graph shows the results of an experiment in which azomethane, CH_3N_3, is flash photolyzed to produce methyl radicals, whose presence is monitored by their absorption at 216 nm. The concentration rises during the period of the flash, then decays by second-order kinetics due to the recombination reaction $2CH_3 \cdot \rightarrow C_2H_6$. In these particular experiments, the absorption spectrum of the reacting system was taken spectrographically at a series of delay times, that is, intervals of elapsed time after a flash. The absorbance at 216 nm was then measured on each of the series of photographic plates obtained from the spectrometer, and the results plotted against delay time. As noted in the text, the alternative approach would have been to obtain a continuous oscilloscope monitoring of the absorbance at 216 nm. [See N. Basco, D.G.L. James, and R.D. Suart, *Intern. J. Chem. Kin.*, **2**, 215 (1970).]

$$CH_3N_3 \xrightarrow{h\nu} CH_3\cdot + N_3 \cdot$$

$$2CH_3 \cdot \rightarrow C_2H_6 \quad \text{(and } 2N_3^{\cdot} \rightarrow 3N_2\text{)}$$

Note the rise in methyl radical concentration during the flash, and the subsequent second order decay.

The flash tube produces an intense flash, but over a period of some microseconds. The time scale may be shortened to nanoseconds and even picoseconds with the use of a pulsed laser (see Fig. 19-32). The photolysis is now at a particular wavelength characteristic of the laser, such as the 530 nm or 353 nm harmonic in the case of a Nd laser. A "white" light source for monitoring may be obtained, however, by splitting off part of the laser beam and focussing it to a point in some medium such as CCl_4; the intense electromagnetic field of the focussed laser light leads to multiphoton processes, the effect of which is to produce a brilliant "spark" of white light. Various delay arrangements allow the taking of a complete absorption spectrum of the reaction system at a specified time after the main laser pulse. Alternatively, a continuous monochromatic monitoring source may be used, as in the flash tube experiments, to obtain the variation in absorbance with time for a particular wavelength. This monochromatic monitoring source may be a second laser, such as a tunable dye laser.

One problem with gas phase flash or laser pulse photolysis as a means of initiating chemical reaction is that considerable transient heating may occur (amounting to tens or even hundreds of degrees). The complication can be largely eliminated, however, if sufficient inert gas is present in the reaction mixture to act as a thermal buffer.

14-5   Rate laws and reaction mechanisms

Once an experimental rate law has been determined, attention turns to the actual sequence of chemical steps that produces the reaction, that is, to the nature of the reaction path or mechanism. Such steps are known as *elementary chemical reactions,* as distinguished from the overall reaction. With two exceptions, discussed further in Sections 14-7 and 14-5E, we take *all* elementary reactions to be bimolecular, that is, to involve the reaction together of just two molecules. We proceed in this section to see how various overall rate laws can result.

A. Simple reactions   A *simple reaction* is defined here as one for which the overall process and the elementary reaction are the same. The simple reactions that have been studied fall into perhaps three categories, summarized in Table 14-3. *Association reactions,* as the name implies, involve the combining of two molecules to give a single product. *Exchange reactions* are ones in which an atom or a group is transferred from one molecule to another. A very large number of such reactions have been studied; for example, many organic molecules and radicals can exchange hydrogen or halogen atoms. *Decomposition reactions* may be ones in which two molecules of a species combine to then break up into simpler products. These are often exchange reactions.

TABLE 14-3
Some simple reactions

Type	Example
Association	$2NO_2 \rightarrow N_2O_4$
	$2CH_3 \rightarrow C_2H_6$
Exchange	$NO + ClNO_2 \rightarrow NOCl + NO_2$
	$NO_2 + O_3 \rightarrow NO_3 + O_2$
	$CO + Cl_2 \rightarrow COCl + Cl$
	$H + D_2 \rightarrow HD + D$
	$H + HCl \rightarrow H_2 + Cl$
	$CH_3 + H_2 \rightarrow CH_4 + H$
	$CH_3 + NH_3 \rightarrow CH_4 + NH_2$
Decomposition	$C_3H_7I \rightarrow C_3H_6 + HI$

The experimental observation is that the mass action law, in its historic form, applies to simple reactions. That is, the rate law corresponds to the overall chemical reaction. Thus for the exchange reaction

$$NO + ClNO_2 = NOCl + NO_2$$

the rate law is

$$R_f = -\frac{d(NO)}{dt} = k(NO)(ClNO_2)$$

A rather frequent situation is that in which a bimolecular reaction produces an intermediate I which in turn reacts with itself or with one of the original reactants. The reaction $2NO + H_2 \rightarrow N_2O + H_2O$ [Eq. (14-42)] is one example. The mechanism seems likely to be

B. Two-step mechanisms. Rules for obtaining a rate law from a mechanism

$$2NO \overset{K}{=} N_2O_2 \qquad \text{(rapid equilibrium)}$$

$$N_2O_2 + H_2 \overset{k}{\rightarrow} N_2O + H_2O \qquad \text{(slow step)}$$

The rules for constructing the rate law from a reaction mechanism are that, first, the mass action principle is applied to each elementary step, or to the slow step if all others are fast. This slow step determines the overall reaction rate, but it is then conventional to use the equilibrium constants and the stoichiometry of the other steps of the mechanism to express the rate law purely in terms of species that appear in the overall reaction.

The example provides an illustration of these rules. We first apply the mass action law to the slow step:

$$\frac{d(N_2O)}{dt} = -\frac{d(N_2O_2)}{dt} = k(N_2O_2)(H_2)$$

However, N_2O_2 does not appear in the overall equation (14-42), and we may eliminate it by using the equilibrium constant expression $K = (N_2O_2)/(NO)^2$ to get

$$\frac{d(N_2O)}{dt} = kK(NO)^2(H_2) \tag{14-44}$$

which is the observed rate law [Eq. (14-43)].

Note, however, that the alternative mechanism

$$NO + H_2 \overset{K}{=} NO \cdot H_2 \qquad \text{(rapid equilibrium)}$$
$$NO \cdot H_2 + NO \overset{k}{\rightarrow} N_2O + H_2O \qquad \text{(slow step)} \tag{14-45}$$

where $NO \cdot H_2$ is an association complex, equally well reproduces the experimental rate law (the reader might verify this statement). Thus even in this simple situation at least two alternative paths can be thought of, both chemically reasonable and both agreeing with the kinetic results. The similar type of reaction $2NO + X_2 = 2NOX$ is known, with $X_2 = O_2$, Cl_2, or Br_2, and the same ambiguity of mechanism is present. One of the applications of fast reaction techniques has been to the identification of reaction intermediates so as to allow a decision to be made between alternative mechanisms. It has not yet been determined, however, whether in the cases cited the intermediate is N_2O_2 or $NO \cdot X_2$.

A further point is that by either mechanism the experimental rate constant is seen to be a product of a true rate constant and the equilibrium constant for the precursor reaction. This type of situation presents a very real problem in the theoretical analysis of rate data; reaction rate theories deal with elementary reactions and can easily lead to erroneous conclusions if applied to composite rate constants.

C. The stationary-state hypothesis

The two-step mechanism just discussed is a special case of a more general situation. It was assumed that the first step consisted of a rapid equilibrium, but the more complete analysis would be as follows:

$$2NO \underset{k_{-1}}{\overset{k_1}{\rightleftharpoons}} N_2O_2$$

$$N_2O_2 + H_2 \overset{k_2}{\rightarrow} N_2O + H_2O$$

We now write the sum of the mass action rate expressions for *each* process whereby a given species should change in concentration with time:

$$\frac{1}{2}\frac{d(NO)}{dt} = -k_1(NO)^2 + k_{-1}(N_2O_2) \tag{14-46}$$

$$\frac{d(N_2O_2)}{dt} = k_1(NO)^2 - k_{-1}(N_2O_2) - k_2(N_2O_2)(H_2) \tag{14-47}$$

$$\frac{d(N_2O)}{dt} = k_2(N_2O_2)(H_2) \tag{14-48}$$

The set of three differential equations must now be solved simultaneously. Although this can be done in the present case, the mathematics of such situations rapidly becomes intractable, and a very useful approximation is usually made so as to simplify matters. If an intermediate I, in this case N_2O_2, is being produced and consumed in such a manner that its concentration never becomes appreciable, this means that the total reaction "traffic" is large compared to the amount of intermediate and hence that the latter rapidly attains a steady level of concentration which then slowly drops as the reactants are consumed. The further implication is that, in this case, $d(N_2O_2)/dt$ is small compared to $d(NO)/dt$ or $d(N_2O)/dt$; if (N_2O_2) is, say, $10^{-3}(NO)$, then $d(N_2O_2)/dt$ should be about $10^{-3}d(NO)/dt$, and hence 10^{-3} times each of the terms of Eq. (14-41). The approximation that is therefore made is to set $d(I)/dt$ equal to zero. This is known as the *steady-* or *stationary-state approximation*.

Application of the stationary-state approximation to the present example sets $d(N_2O_2)/dt = 0$ and allows Eq. (14-47) to be solved for (N_2O_2):

$$(N_2O_2) = \frac{k_1(NO)^2}{k_{-1} + k_2(H_2)} \tag{14-49}$$

Insertion of this result into Eq. (14-48) gives

$$\frac{d(N_2O)}{dt} = \frac{k_2k_1(NO)^2(H_2)}{k_{-1} + k_2(H_2)} \tag{14-50}$$

Notice that if k_{-1} is large compared to $k_2(H_2)$, then Eq. (14-50) reduces to Eq. (14-44), since $K = k_1/k_{-1}$. On the other hand, at very large (H_2), it should be possible to obtain the other limiting form:

$$\frac{d(N_2O)}{dt} = k_1(NO)^2 \tag{14-51}$$

Were the alternative mechanism, involving $NO \cdot H_2$ as intermediate, the correct one, then at large (NO), the limiting form should be

$$\frac{d(N_2O)}{dt} = k_1(NO)(H_2) \tag{14-52}$$

Investigators have attempted such studies as a means of distinguishing between the two mechanisms, but without success in that they could not reach sufficiently high pressures to observe departures from the normal rate law. However, success in this respect has been achieved in some related situations that will be discussed in Section 14-8.

We can examine the steady-state hypothesis in more detail by taking a hypothetical system. Suppose that the reaction sequence is

$$A \underset{k_{-1}}{\overset{k_1}{\rightleftharpoons}} I \overset{k_2}{\rightarrow} B \tag{14-53}$$

where I is an intermediate. According to the procedure of Section 14-5B, we would write $K_1 = k_1/k_{-1} = (I)/(A)$ and

$$\frac{d[(B)/(A)_0]}{dt} = k_2[(I)/(A)_0] = \left(\frac{k_1k_2}{k_{-1}}\right)\frac{(A)}{(A)_0} - k'_{app}\frac{(A)}{(A)_0} \tag{14-54}$$

However, according to the steady-state treatment, we write

$$\frac{d[(I)/(A)_0]}{dt} = k_1\frac{(A)}{(A)_0} - (k_{-1} + k_2)\frac{(I)}{(A)_0} \tag{14-55}$$

and we make the approximation that $d[(I)/(A)_0]/dt = 0$, so that

$$\frac{(I)}{(A)_0} \simeq \frac{k_1}{k_{-1} + k_2}\frac{(A)}{(A)_0} \tag{14-56}$$

and

$$\frac{d[(B)/(A)_0]}{dt} = k_2\frac{(I)}{(A)_0} = \frac{k_1k_2}{k_{-1} + k_2}\frac{(A)}{(A)_0} = k''_{app}\frac{(A)}{(A)_0} \tag{14-57}$$

Equations (14-53) can be solved exactly, however. The result is (Espenson, 1981):

$$\frac{(A)}{(A)_0} = \frac{k_1}{\lambda_2 - \lambda_3}\left[\frac{\lambda_2 - k_2}{\lambda_2}e^{-\lambda_2 t} - \frac{\lambda_3 - k_2}{\lambda_3}e^{-\lambda_3 t}\right] \tag{14-58}$$

$$\frac{(I)}{(A)_0} = \frac{k_1}{\lambda_2 - \lambda_3}[e^{-\lambda_3 t} - e^{-\lambda_2 t}] \tag{14-59}$$

and $(B)/(A)_0 = 1 - (A)/(A)_0 - (I)/(A)_0$. The decay constants λ_2 and λ_3 are given by $\lambda_2 = \frac{1}{2}(p + q)$ and $\lambda_3 = \frac{1}{2}(p - q)$ where $p = (k_1 + k_{-1} + k_2)$ and $q = (p^2 - 4k_1k_2)^{1/2}$. The rate quantities $d[(A)/(A)_0]/dt$ and $d[(I)/(A)_0]/dt$ can be obtained by differentiating Eqs. (14-58) and (14-59) with respect to time, and

$$\frac{d[(B)/(A)_0]}{dt} = -\frac{d[(A)/(A)_0]}{dt} - \frac{d[(I)/(A)_0]}{dt} \tag{14-60}$$

We take the specific case of $k_1 = 1$ min^{-1}, $k_{-1} = 10$ min^{-1}, and $k_2 = 20$ min^{-1}, so that $\lambda_2 = 30.341$ min^{-1} and $\lambda_3 = 0.65918$ min^{-1}. The results of the exact calculation are summarized in Table 14-4 and Fig. 14-6. The figure shows clearly that the concentration of the intermediate I is small compared to that of A and, except at the very beginning, also small compared to that of product B.

Turning to Table 14-4, the entries in columns 6 and 7 correspond to the two terms on the right-hand side of Eq. (14-55); their difference gives $d[(I)/(A)_0]/dt$, and it is seen that after about 0.25 min this difference is indeed small compared to either term separately, thus justifying the steady-state approximation. As a direct check, columns 3 and 8 agree to within a few percent, again after the first 0.1 to 0.2 min. Another point is that the steady-state approximation is *not* a good one at short times, as in the first row of data. At short times $(I)/(A)_0$ is rising toward a maximum value (note the change in sign in column 5), and it is only thereafter that the system settles down to near-steady-state behavior.

Finally, in the usual case of a reactive intermediate the ratios k_{-1}/k_1 and k_2/k_1 would be much larger than in the above example. In such a case $(I)/(A)_0$ would reach its maximum very quickly, its value would be smaller, and the steady-state approximation would be much better.

D. Chain reactions and explosions

A chain reaction is one in which some intermediate or intermediates are consumed and regenerated in a cycle of reactions the net result of which is to carry forward the overall reaction. Analysis of such systems can be quite complicated, and the following example will serve to illustrate both the type of mechanism encountered and a further application of the stationary-state hypothesis.

The reaction

$$H_2 + Br_2 = 2HBr \tag{14-61}$$

has been extensively studied from the time of M. Bodenstein around 1906. Unlike the situation with the seemingly analogous reaction $H_2 + I_2 = 2HI$, the experimental rate law is very complex:

$$\frac{d(HBr)}{dt} = \frac{k_t(H_2)(Br_2)^{1/2}}{1 + [k_i(HBr)/(Br_2)]} \tag{14-62}$$

where k_i has been called the inhibition constant and the term in the denominator reflects the inhibition of the reaction rate by the product HBr. The mechanism for this reaction seems now well established as the following:

TABLE 14-4
Solutions to eqs. (14-53) for the case of $k_1 = 1$ min^{-1}, $k_{-1} = 10$ min^{-1}, and $k_2 = 20$ min^{-1}

Time (min)	$(A)/(A)_0$	$(I)/(A)_0$	$(B)/(A)_0$	$\dfrac{d[(I)/(A)_0]}{dt}$	$k_1[(A)/(A)_0]$	$(k_{-1} + k_2)[(I)/(A)_0]$	$[(I)/(A)_0]_{ss}$[a]
0.05	0.95899	0.02520	0.01581	0.20274	0.95899	0.75600	0.03197
0.1	0.92601	0.02992	0.04407	0.02840	0.92601	0.89760	0.03087
0.25	0.83834	0.02856	0.13310	−0.01831	0.83834	0.85680	0.02794
0.5	0.71096	0.02423	0.26481	−0.01597	0.71096	0.72690	0.02370
1	0.51134	0.01743	0.47123	−0.01149	0.51134	0.52290	0.01704
2	0.26450	0.00901	0.72649	−0.00594	0.26450	0.27030	0.00882
4	0.07077	0.00241	0.92682	−0.00159	0.07077	0.07230	0.00236

[a] From Eq. (14-56).

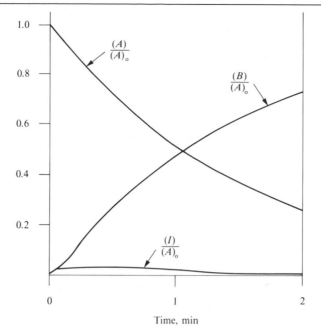

FIGURE 14-6
Solutions to Eqs. (14-53)
for the case of $k_1 = 1$
min^{-1}, $k_{-1} = 10$ min^{-1}, and
$k_2 = 20$ min^{-1}. Note that
$d[(I)/(A)_0]/dt$ is small
compared to $d[(A)/(A)_0]/dt$
and $d[(B)/(A)_0]/dt$ except at
short times.

$$Br_2 + M \xrightarrow{k_1} 2Br + M \qquad \text{(chain initiation)}$$

$$2Br + M \xrightarrow{k_2} Br_2 + M \qquad \text{(chain termination)} \qquad \text{(14-63)}$$

$$Br + H_2 \underset{k4}{\overset{k_3}{\rightleftharpoons}} HBr + H \qquad \text{(chain propagation)}$$

$$H + Br_2 \xrightarrow{k_5} HBr + Br \qquad \text{(chain propagation)}$$

(M is any gaseous species, and serves to supply the energy for the dissociation—see Section 14-5E). Note that the last two reactions together constitute a cycle, the net effect of which is to carry out the overall reaction [Eq. (14-61)]; this pair then constitutes the chain reaction.

The stationary-state assumption is now applied to the intermediates H and Br:

$$\frac{d(Br)}{dt} = 0 = 2k_1(Br_2)(M) - k_2(Br)^2(M) - k_3(Br)(H_2)$$
$$+ k_4(HBr)(H) + k_5(H)(Br_2) \qquad (14\text{-}64)$$

and

$$\frac{d(H)}{dt} - 0 = k_3(Br)(H_2) - k_4(HBr)(H) \quad k_5(H)(Br_2) \qquad (14\text{-}65)$$

or

$$(H) = \frac{k_3(Br)(H_2)}{k_4(HBr) + k_5(Br_2)} \qquad (14\text{-}66)$$

Also, addition of Eqs. (14-64) and (14-65) leads to

$$(Br) = \left[\frac{2k_1(Br_2)}{k_2}\right]^{1/2} = K_{1,2}^{1/2}(Br_2)^{1/2} \tag{14-67}$$

where $K_{1,2}$ is the equilibrium constant for the dissociation of Br_2 into atoms; notice that (M) has cancelled out.

The rate of production of HBr is given by

$$\frac{d(HBr)}{dt} = k_3(Br)(H_2) - k_4(HBr)(H) + k_5(H)(Br_2) \tag{14-68}$$

and replacement of (H) and (Br) by the appropriate expressions yields

$$\frac{d(HBr)}{dt} = \frac{2k_3K_{1,2}^{1/2}(H_2)(Br_2)^{1/2}}{1 + [k_4(HBr)/k_5(Br_2)]} \tag{14-69}$$

which is the same as the observed rate law [Eq. (14-62)].

The thermal decomposition of many gaseous organic molecules involves free radical chains. Usually, however, thermal decomposition or *pyrolysis* is complicated by the presence of a large variety of products due to the various types of decomposition reactions that can occur. It has therefore been useful to initiate such decompositions photochemically; the first step is then apt to be a simple one, and the whole reaction sequence can be followed at a low enough temperature that intermediates and reaction products do not themselves pyrolyze.

Perhaps the most studied system of this last type is that of the photodecomposition of acetone. Hundreds of publications have appeared on the subject, and it must suffice here to sketch the main conclusions. Irradiation of acetone vapor with light of around 254 nm wavelength results in its fragmentation to give CH_3CO, CH_3, and CO:

$$CH_3COCH_3 \xrightarrow{h\nu} \begin{cases} CH_3CO \cdot + \cdot CH_3 \\ CO + 2 \cdot CH_3 \end{cases}$$

The following reactions then appear to be important:

$$2 \cdot CH_3 \rightarrow C_2H_6 \qquad \text{(association)}$$

$$2 \cdot COCH_3 \rightarrow (CH_3CO)_2 \qquad \text{(association)}$$

$$\cdot CH_3 + \cdot COCH_3 \rightarrow CH_3COCH_3 \qquad \text{(recombination)}$$

$$\cdot CH_3 + \cdot COCH_3 \rightarrow CH_4 + CH_2 = CO \qquad \text{(H exchange)}$$

$$\cdot COCH_3 + M \rightarrow CO + \cdot CH_3 + M \qquad \text{(decomposition)}$$

$$\cdot CH_3 + \cdot CH_2COCH_3 \rightarrow CH_3CH_2COCH_3 \qquad \text{(association)}$$

$$\cdot CH_3 + CH_3COCH_3 \rightarrow CH_4 + \cdot CH_2COCH_3 \qquad \text{(H exchange)}$$

$$\cdot CH_2COCH_3 + M \rightarrow \cdot CH_3 + CH_2 = CO + M$$

(For clarity radicals are marked with an electron dot.) The products thus include CO, CH_4, C_2H_6, $C_2H_5COCH_3$, and $CH_2 = CO$, with the various indicated radicals as chain carriers. Notice that all but one of the elementary reactions are bimolecular. The last reaction is written as a unimolecular decomposition, but even here a second molecule is needed to supply the necessary energy for the reaction to take place; see Section 14-8.

Chain reactions can be very rapid; the chain carriers tend to be high-energy, fast-reacting species and can cause tens or hundreds of molecules to react in the chain carrying steps for each molecule reacting in the initiation step. It is not surprising that a chain reaction can occur in an explosive fashion. There are, however, two basic types of explosive situations. The first is known as a *thermal explosion*. As we see in the next section, reaction rates generally increase with increasing temperature, often dramatically. In a thermal explosion, the conditions are such that the heat produced in an exoergic reaction warms up the gas mixture, thus increasing the rate of reaction, and increasing the rate of heat generation. If the rate is such that heat is generated faster than it can escape by thermal conduction, convection, or radiation, the whole process builds on itself. The result is an explosion. The mathematics of a single-mechanism chain reaction indicates that there will be a critical pressure above which explosion can occur (once the temperature reaches a certain point) and below which no explosion is possible.

An explosion may occur because a reaction builds on itself autocatalytically, rather than just by self-heating. This can happen if *chain branching* occurs. In this mechanism the chain carrier or carriers produced in the initiation step can react in a way that multiplies their number. The chain sequence of Eq. (14-63) is *not* one of chain branching. In the chain carrying cycle, each Br atom produces a H atom and each H atom produces a Br atom. There is *no* change in the number of atoms. Consider, however, the mechanism of the reaction of H_2 with O_2:

$$\begin{aligned}
&\text{(1) } H_2 + \text{wall} \rightarrow H \text{ (on wall)} + H \\
&\text{(2) } H + O_2 \rightarrow HO + O \\
&\text{(3) } O + H_2 \rightarrow HO + H \\
&\text{(4) } HO + H_2 \rightarrow HOH + H \\
&\text{(5) } H + \text{wall} \rightarrow \text{chain termination}
\end{aligned} \qquad (14\text{-}70)$$

The initiation step, step 1, is one of several possibilities; another would be a slow direct reaction between H_2 and O_2 to produce H and HO_2. Steps 2 and 3 are said to be *branching* in that one radical produces *two* new ones, so that the effect is to multiply the number of chain-carrying intermediates present.

The consequence of such chain branching is that the reaction proceeds at an accelerating rate. If the pressure is low enough, there is still no explosion because the reactants get used up before an explosively fast rate is achieved; above a certain pressure (and hence reactant concentrations) rates are fast enough that the system can accelerate to an explosion.

The typical behavior of a chain reacting system is illustrated in Fig. 14-7. There is a first explosion limit, given by the *ab* line. At pressures and temperatures below this line, reaction is too slow for the system to accelerate to an explosion before the reactants are used up. If the pressure is increased, following the dashed line in the figure, the *bc* boundary is reached, above which *no* explosion occurs. The explanation is that the pressure and hence concentrations are such that termolecular reactions (see below) such as

$$H + O_2 + M \rightarrow HO_2 + M \qquad (14\text{-}71)$$

can occur, where M is a nonreacting molecule whose function is to carry off the energy of the reaction. The HO_2 molecule is not a good chain-carrying

FIGURE 14-7
Explosion limits of a stoichiometric mixture of $H_2 + O_2$. (From S. W. Benson, "The Foundations of Chemical Kinetics," Copyright 1960, McGraw-Hill, New York. Used with permission of McGraw-Hill Book Co.)

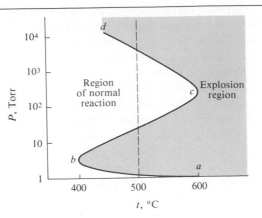

one and, moreover, can act as a chain terminater through reactions such as $H + HO_2 \rightarrow H_2 + O_2$. Mainly, however, the effect of the onset of termolecular reactions is to convert a chain branching to an ordinary chain reaction. It is the presence of the *bc* branch that distinguishes a branching from an ordinary chain reacting system.

Returning to Fig. 14-7, if line *cd* is crossed by further increase in pressure along the dashed line, the system again is explosive! This is usually just a region of thermal explosion, however, and no new chemistry need be occurring. Many oxidation reactions show the behavior of Fig. 14-7; examples include the oxidation of P_4, PH_3, and NH_3.

E. Termolecular reactions

All elementary reactions considered so far have been bimolecular or unimolecular, and even the so-called unimolecular processes are explained physically on the basis of activation by bimolecular collisions. The reason elementary reactions of a higher order are not usually important is simply that their frequency factor is expected to be small. The situation may be examined from either the point of view of collision theory or from that of transition-state theory.

Simple collision theory treats molecules as hard spheres so that the duration of a collision is so very short that the chance of a third molecule hitting a colliding pair is small. Thus we have a problem in defining exactly what is meant by a triple collision; roughly, we want the chance of three molecules being within a molecular diameter of each other. One way of estimating this is to assume that a collision between two molecules lasts about one vibrational period, or 10^{-13} s. The collision frequency Z_{12} for two molecules is about $10^{11}(A)(B)$, in liter mole^{-1} s^{-1}, and the concentration of a collision complex, $[AB]$, is then (rate of formation of $[AB]$) × (lifetime of $[AB]$) or about $10^{-13}Z_{12}$. The frequency of collisions between $[AB]$ and another molecule C would be about $10^{11}([AB])(C)$ or $(10^{11})(10^{-13})[10^{11}(A)(B)](C)$ or $10^9(A)(B)(C)$, as compared to $10^{11}(A)(B)$ for a bimolecular reaction. The rate constant for a termolecular reaction is thus expected to be about one hundredth of that for a bimolecular one, other factors being the same.

If all reactants are at unit concentration, 1 M, then a termolecular reaction would be significantly but not enormously slower than a bimolecular one, as an approximate rule. Most gas phase reactions are studied at relatively low pressures, however. Even 0.1 atm corresponds to a concentration of about 0.004 M. If all reactants are at 0.1 atm, the rate (not the rate constant) of a termolecular reaction would be $(0.004)(0.01)$ or 4×10^{-5} of that of a bimolecular one, or very much slower, other factors again being the same.

As a rule, termolecular reactions should thus be very much slower than bimolecular ones, and it is for this reason that the elementary reactions making up a reaction mechanism are ordinarily restricted to biomolecular ones. There is one type of reaction that can occur *only* as a termolecular one, however. If two *atoms* are to combine to give a molecule,

$$A + A = A_2$$

a third body must be present to carry off the energy of the A—A bond— otherwise the atoms can only fly apart again. The process must then be

$$A + A + M = A_2 + M$$

For example, in writing the mechanism for reaction (14-61), the step

$$2Br + M \rightarrow Br_2 + M \tag{14-72}$$

is involved. Further, by the principle of microscopic reversibility (see Section 7-CN-2D), the reverse process must occur by the same path:

$$Br_2 + M \rightarrow 2Br + M$$

The rates of a number of termolecular reactions have been measured. That for reaction (14-72), for example, is indeed given by $d(Br_2)/dt = 10^9(Br)^2(M)$. There is some further discussion of termolecular reactions in connection with Table 14-6.

14-6   Temperature dependence of rate constants

The preceding material has presented the customary procedure of expressing a reaction rate in terms of a rate law, or function of concentrations of species, and the concept of reaction mechanism, whereby the mass action principle is applied to the one or more elementary reactions that are responsible for the overall process. All dependences of a reaction rate other than on concentration are thus contained in the rate constant (or constants if the rate law is a complex one). We now examine the temperature dependence of reaction rates, that is, of rate constants.

A preliminary historical review seems appropriate at this point. The mass action principle developed during the period 1850–1890, beginning with the observation by L. Wilhelmy that the rate of inversion of cane sugar was proportional to the amount of unconverted sugar, the reaction being

$$H_2O + C_{12}H_{22}O_{11} \rightarrow C_6H_{12}O_6 \text{ (glucose)} + C_6H_{12}O_6 \text{ (fructose)}$$

Wilhelmy integrated the first-order rate equation and, in effect, developed much of the material of Section 14-3A. The first emphasis on chemical equilibrium as the result of a dynamic balance of equal forward and reverse rates came from C. Guldberg and P. Waage in 1867, who also clearly formulated the law of mass action. Later, van't Hoff added that the equilibrium constant should then be given by $K = k_f/k_b$. It was not until 1865, however, that the first second-order reaction was clearly defined experimentally, by Harcourt and Essen (1865, 1866, 1867), in a study of the reaction between permanganate and oxalate ions. Thus the subject of reaction kinetics evolved through the study of solution rather than gas-phase reactions, although we will see that it is for the latter that theory is best developed today.

The preceding developments set the stage for Arrhenius to observe in 1889 that rate constants showed much the same temperature dependence behavior as did equilibrium constants, namely,

$$\frac{d(\ln k)}{dT} = \frac{constant}{T^2}$$

By analogy with the second law equation, Eq. (7-30), for equilibrium constants, it was natural to write

$$\frac{d(\ln k)}{dT} = \frac{E^*}{RT^2} \tag{14-73}$$

where E^* represents some characteristic energy that must be added to the reactants for reaction to occur. We call E^* the *activation energy* of the reaction.

Equation (14-73) has the usual alternative forms. Integration gives

$$\ln k = constant - \frac{E^*}{RT} \tag{14-74}$$

so that a plot of $\ln k$ versus $1/T$ should give a straight line of slope $-E^*/R$.

An equivalent form is

$$k = A\, e^{-E^*/RT} \tag{14-75}$$

an integration between limits gives

$$\ln \frac{k_2}{k_1} = \frac{E^*}{R} \left(\frac{1}{T_1} - \frac{1}{T_2} \right) \tag{14-76}$$

These various forms are all known as the *Arrhenius equation* for rate constants; the constant A is called the *preexponential* or *frequency factor*.

As the equations imply, reaction rates increase as temperature increases; around 25°C, for example, a doubling of k with a 10°C rise in temperature corresponds to about 12 kcal mol^{-1} for E^*, and a quadrupling of the rate, to 24 kcal mol^{-1}, and so on.†

EXAMPLE E^* is 35 kcal mol^{-1} for a certain reaction. By what factor would k increase between 100°C and 110°C? By Eq. (14-76), $\ln(k_{383}/k_{373}) = [(35,000)/(1.987)][(10)/(383.15)(373.15)] = 1.23$. The factor is thus 3.43.

The Arrhenius equation is amazingly well obeyed by systems showing a rate law of the type of Eq. (14-2). The reaction $2HI \rightarrow H_2 + I_2$ obeys second-order kinetics over a wide range of conditions and some of Bodenstein's data (1894–1899) are plotted in Fig. 14-8 according to the form of Eq. (14-74). The values of k are given in Table 14-5 along with those calculated from the best-

---

†The factor by which k increases over a 10°C interval is called the *temperature coefficient* in older literature.

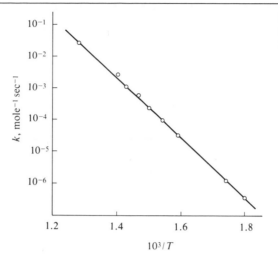

FIGURE 14-8
Arrhenius plot for the
reaction $2HI \rightarrow H_2 + I_2$.

fitting straight line in the Arrhenius plot. The agreement is within about 25%
over five orders of magnitude variation in k. This best straight line is given
by

$$k_f = 9.17 \times 10^{10}\, e^{-44,450/RT} = 9.17 \times 10^{10} e^{-22,370/T} \tag{14-77}$$

where k_f is in liters mol^{-1} s^{-1}.

Excellent though this agreement is, there is some systematic deviation of
the data from the straight line; a better fit is obtained by an equation of the
form

$$k_f = A'T^{1/2}\, e^{-E^*/RT} \tag{14-78}$$

or

$$k_f = 1.05 \times 10^9 T^{1/2}\, e^{-42,800/RT} \tag{14-79}$$

	k (liter mol^{-1} s^{-1})	
T (K)	Observed	Calculated from Arrhenius equation
556	3.52×10^{-7}	3.11×10^{-7}
575	1.22×10^{-6}	1.18×10^{-6}
629	3.02×10^{-5}	3.33×10^{-5}
647	8.59×10^{-5}	8.96×10^{-4}
666	2.19×10^{-4}	1.92×10^{-4}
683	5.12×10^{-4}	5.53×10^{-4}
700	1.16×10^{-3}	1.21×10^{-3}
716	2.50×10^{-3}	2.53×10^{-3}
781	3.95×10^{-2}	3.33×10^{-2}

TABLE 14-5
The reaction
$2HI \rightarrow H_2 + I_2{}^a$

[a]Adapted from E. A. Moelwyn-Hughes, "Physical Chemistry." Pergamon,
Oxford, 1961.

with E^* in kcal mol^{-1}. As will be seen in the next section, Eq. (14-79) is suggested on theoretical grounds to be better than the simple Arrhenius equation. An important point to note is that the slight improvement in fit has resulted in rather different A and E^* parameters; kineticists are not always properly sensitive to this situation.

The reverse reaction, $H_2 + I_2 \rightarrow 2HI$, has also been studied extensively, and the second-order rate constant k_b is given by

$$k_b = 3.30 \times 10^9 T^{1/2} e^{-38,900/RT} \tag{14-80}$$

where k_b is again in liter mol^{-1} s^{-1} and E^* in kcal mol^{-1}.

According to the mass action relationship,

$$K = \frac{(H_2)(I_2)}{(HI)^2} = \frac{k_f}{k_b} = 0.32 e^{-3900/RT} \tag{14-81}$$

That is, ΔH^0 for the reaction is 3900 kcal mol^{-1} according to the kinetic data. Table 5-2 gives 6.20 kcal mol^{-1} as the heat of formation of HI [from $I_2(s)$], so the corresponding thermochemical $\Delta H^0 = -(2)(6.20) + 14.88 = 2.48$ kcal [where 14.88 is the heat of sublimation of $I_2(s)$]. The agreement, 2.48 versus 3.90, is quite good, especially if the thermochemical ΔH is corrected by the (small) ΔC_P of reaction to the temperature region of the kinetic studies. The absolute values of K as determined directly and as found by the mass action equation also agree well.

Application of the van't Hoff and Arrhenius equations to the mass action expression $K = k_f/k_b$ thus leads to the conclusion that

$$\Delta E = E_f^* - E_b^* \tag{14-82}$$

as confirmed by the example. (The distinction between ΔH and ΔE is unimportant in the present case since $H_2 + I_2 = 2HI$ the two are equal.) The picture that emerges is one of a critical energy content of reacting species which is the same for either direction of reaction as illustrated in Fig. 14-9. This conclusion is in accord with the principle of microscopic reversibility (see Section 7-CN-2D). The nature of the common high-energy state will be the subject of discussion in Sections 14-7 and 14-8.

This presentation has left the implication that the $H_2 + I_2 \rightarrow 2HI$ reaction is a simple one, that is, one that proceeds by the single-step, bimolecular path indicated by the overall equation. The reaction has, in fact, been cited in

FIGURE 14-9
Relationships among E_f^*, E_b^*, and ΔE.

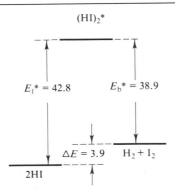

countless texts as the clearest and best understood example of a bimolecular reaction. It was a shock to many persons (including this writer) to learn that work by Sullivan (1967) has shown the actual mechanism probably to either

$$I_2 \overset{K_1}{=} 2I \tag{14-83}$$

$$H_2 + 2I \underset{k_{-2}}{\overset{k_2}{\rightleftharpoons}} 2HI \tag{14-84}$$

or the preequilibrium (14-83) followed by

$$H_2 + I \overset{K_3}{=} H_2I \tag{14-85}$$

$$H_2I + I \underset{k_{-4}}{\overset{k_4}{\rightleftharpoons}} 2HI \tag{14-86}$$

The first mechanism gives (after the procedure of Section 14-5B)

$$k_f = K_1 k_2 \qquad k_b = k_{-2}$$

and the second gives

$$k_f = K_1 K_3 k_4 \qquad k_b = k_{-4}$$

Notice that the *rate law* remains second order in both directions. Only the *interpretation* of k_f and of k_b has changed—the former now contains either one or two preequilibrium constants. None of the earlier mass action analysis is invalidated—k_f/k_b still gives the correct equilibrium constant—but the detailed picture of the reaction path has changed profoundly. The revised mechanisms were arrived at, incidentally, by the photochemical generation of iodine atoms. By this means, their concentration could be varied independently of that of I_2. The same series of investigations also indicated that above 600 K an increasing proportion of the reaction goes through a short chain mechanism similar to that for the reaction of hydrogen with bromine. Thus the period from Bodenstein (*ca.* 1895) to the present has barely sufficed to unravel the detailed mechanism of an apparently very simple reaction!

This example illustrates the point that even though the Arrhenius equation is well obeyed, detailed mechanistic studies may indicate the apparent rate constant to be composite, that is, to contain one or more equilibrium constants. In some instances the situation is obvious. Thus the reaction $2NO + O_2 \rightarrow 2NO_2$ shows a small *negative* temperature coefficient, so that the apparent activation energy is negative. The mechanism is thought to be analogous to that giving Eq. (14-44), so that the observed rate constant is really equal to kK, where K is the equilibrium constant for the preequilibrium step. One thus has

$$E_{app}^* = E_{true}^* + \Delta E_{preequil} \tag{14-87}$$

The preequilibrium involves an association, probably $2NO = N_2O_2$, and is therefore likely to be exoergic, with ΔE negative. It appears in this case that ΔE is sufficiently negative to give a net negative apparent activation energy.

In the above case, the observation of a negative apparent activation energy clearly indicates that the mechanism is at least a two-step one. In the more usual situation of a positive apparent activation energy, other information and considerations are needed to decide just what is the meaning of E_{app}^*.

In summary, if rate data do *not* obey the Arrhenius equation reasonably well, it is an indication that the apparent rate constant is a sum or some other more complicated function of rate and equilibrium constants. If the Arrhenius equation *is* obeyed, the observed rate constant may still be a product of equilibrium and true rate constants and the apparent activation energy a corresponding sum of energy quantities. The molecular interpretation of an observed activation energy can thus only be made in terms of a specific reaction mechanism.

14-7 Collision theory of gas reactions

The mass action background to the treatment of reaction rates made it natural for Arrhenius to associate E^* with a critical energy for reaction and to suggest that there might be an equilibrium between ordinary and "active" molecules, with only the latter reacting. This approach was not very fruitful for some time because it provided no detailed explanation of how molecules become activated or of the frequency factor A in the Arrhenius equation; it reappeared, however, in transition state theory (Section 14-9), but only much later. Perhaps the early emphasis on solution kinetics retarded the advance of theory because of the great difficulty in treating liquids on a molecular basis. At any rate, the first major theoretical advance was based on the kinetic molecular theory of gases.

During the period around 1920, M. Trautz, W. Lewis, C. Hinshelwood, and others developed a quantitative treatment of gas-phase reactions on the basis that only colliding molecules could react and then only if their combined kinetic energy of impact equaled or exceeded a critical energy E^*. The rate of a bimolecular reaction should then be

$$\text{rate} = (\text{collision frequency}) \begin{pmatrix} \text{fraction of impact pairs} \\ \text{expected to have } E \geqslant E^* \end{pmatrix} \qquad (14\text{-}88)$$

The first quantity is given by[†]

$$Z_{12} = 2\sqrt{2}\,\sigma_{12}^2 \left(\frac{\pi kT}{\mu_{12}}\right)^{1/2} \mathbf{n}_1 \mathbf{n}_2 \qquad [\text{Eq. (2-73)}]$$

where, we recall, σ_{12} is the average collision diameter, μ_{12} is the reduced mass, and $\mathbf{n}_1$ and $\mathbf{n}_2$ are the concentrations in molecules per cubic centimeter, so that Z_{12} is in collisions per cubic centimeter per second. Thus Z_{12} can be written as

$$Z_{AB} = (\text{constant})\ T^{1/2}(\text{A})(\text{B})$$

for a reaction between molecules A and B. The numerical example of Exercise 2-15 gives a typical Z_{AB} value of about 4×10^4 moles of collisions $\text{cm}^{-3}\ \text{s}^{-1}$ for A and B each at 1 atm pressure and 25°C, or about 3×10^{10} moles of collisions $\text{liter}^{-1}\ \text{s}^{-1}$ if A and B are each at a concentration of 1 mole liter^{-1}.

The second factor of Eq. (14-88) is arrived at as follows. We consider that E^* is to be provided by the kinetic energy of relative motion of two molecules making a head-on collision. This is equivalent to asking that a velocity c (in two dimensions) be such that $\frac{1}{2}mc^2 \geqslant E^*$, and we may therefore turn to Eq. (2-43):

[†]We write the Boltzmann constant as $\mathbf{k}$ in this chapter to avoid confusion with a rate constant.

$$\frac{dN(c)}{N_A} = \frac{m}{\mathbf{k}T} e^{-mc^2/2\mathbf{k}T} c \, dc \qquad \text{[Eq. (2-43)]}$$

That is, the probability that a single molecule will have one-dimensional velocities u and v such that $u^2 + v^2 = c^2$ is the same as the probability that two colliding molecules will have separate one-dimensional velocities such that $u_A^2 + u_B^2 = c^2$. If we make the change of variable $x = mc^2/2\mathbf{k}T$, Eq. (2-43) becomes

$$\frac{dN}{N_A} = e^{-x} \, dx$$

This is now integrated from $x = x^*$ to $x = \infty$ to give

$$\frac{\Delta N}{N_A} = e^{-E^*/\mathbf{k}T}$$

where $\Delta N/N_A$ is the fraction of molecules taken two at a time that would have an impact kinetic energy equal to or greater than E^*.

Equation (14-88) now reads

$$\text{rate} = (\text{constant}) \, T^{1/2} \, e^{-E^*/RT} \, (A)(B)$$

and the corresponding bimolecular rate constant is then

$$k = A \, e^{-E^*/RT} = A' T^{1/2} \, e^{-E^*/RT} \qquad (14\text{-}89)$$

where

$$A' = 2 \sqrt{2} \, \sigma_{12}^2 \left(\frac{\pi \mathbf{k}}{\mu_{12}}\right)^{1/2} \left(\frac{N_A}{1000}\right) \qquad (14\text{-}90)$$

(A' is reduced by a factor of 2 in the case of collisions between like molecules).

Equation (14-90) represents the initial achievement of collision theory and one which was at first very successful. It is the same as Eq. (14-78), which fits the temperature dependence data for the HI decomposition very well. Moreover, the calculated A' value is fairly close to the experimental one as such comparisons go. Thus, if one takes σ to be 3.5 Å, one obtains 3×10^{10} as compared to the experimental value of 1×10^9.

The derivation given here of Eq. (14-89) is a simplified one. As noted in connection with Eq. (2-69), the rigorous derivation of collision frequency requires detailed analysis of all possible relative velocities of the colliding pair of molecules, that is, over all possible impact parameters (see Fig. 2-15), as well as over the Boltzmann distribution of molecular velocities. Although the small forest of algebra and calculus involved is important to the theoretical kineticist, the result is still Eq. (14-89).

It remains very difficult to calculate σ_{12}. We are interested only in those collisions leading to reaction, which means only those collisions involving molecules of sufficient relative velocity to supply E^*, allowing for their impact parameter. The effect is that σ_{12} for collisions in general is not exactly the same as σ_{12} for reactive collisions. There has been some success in calculating the effective σ_{12} in the case of reactions having no activation energy; examples are $CH^+ + H_2 \rightarrow CH_2^+ + H$ and $O_2^+ + N_2 \rightarrow NO^+ + NO$. Reactions of this type are important in the upper atmosphere and in flames. Rate constants are often around 10^{11} to 10^{12} liter mol^{-1} s^{-1} (McDaniel et al., 1970).

An assumption generally embedded in collision theory treatments is that the Maxwell-

Boltzmann velocity distribution equations hold. This cannot be strictly true since chemical reaction is removing a certain selection of molecules. However, as noted in Section 2-8A, a perturbation in the velocity distribution is restored within just a few collision times, or much faster than the rate of removal of molecules by reaction. The effect must be examined in the case of very fast reactions such as the ion-molecule ones mentioned above.

Data for a number of simple bimolecular reactions are summarized in Table 14-6 and the agreement between experimental and calculated A' values can be quite good. In many cases, however, the observed rate is definitely too small. This problem of "slow" reactions was recognized by around 1925, and Eq. (14-89) was modified by the addition of an empirical factor P called the *steric factor*, where $P = A_{obs}/A_{calc}$:

$$k = PA\, e^{-E^*/RT} \tag{14-91}$$

The factor was justified qualitatively on the grounds that colliding molecules might not be suitably oriented for reaction; P then represents the fraction of energetically suitable collisions for which the orientation is also favorable.

The orientation explanation is perhaps acceptable for P values not much below about 0.1, but as shown in the last column of Table 14-6, there are many instances of much lower values, as low as 10^{-6}. There are in fact enough

TABLE 14-6
Some simple reactions[a]

Reaction	E^* (kcal mol^{-1})	log A' (liter mol^{-1} s^{-1}) Observed	Calculated	P
$2NO_2 \rightarrow 2NO + O_2$	26.6	8.42	9.85	0.038
$2NOCl \rightarrow 2NO + Cl_2$	25.8	9.51	9.47	1.1
$NO + ClNO_2 \rightarrow NOCl + NO_2$	6.6	7.73	9.76	0.01
$NO + O_3 \rightarrow NO_2 + O_2$	2.3	7.80	9.90	0.008
$NO_2 + O_3 \rightarrow NO_3 + O_2$	6.7	8.60	9.84	0.06
$NO + Cl_2 \rightarrow NOCl + Cl$	19.6	8.00	9.87	0.014
$CO + Cl_2 \rightarrow COCl + Cl$	51.3	8.5	9.87	0.04
$H + D_2 \rightarrow HD + H$	6.5	9.00	10.45	0.035
$Br + H_2 \rightarrow HBr + H$	17.6	9.31	10.23	0.12
$H + HBr \rightarrow H_2 + Br$	0.9	8.91	10.39	0.033
$H + Br_2 \rightarrow HBr + Br$	0.9	9.83	10.65	0.15
$Br + HBr \rightarrow Br_2 + H$	41.8	9.13	9.83	0.20
$CH_3 + H_2 \rightarrow CH_4 + H$	10.0	7.25	10.27	0.00095
$CH_3 + CHCl_3 \rightarrow CH_4 + CCl_3$	5.8	6.10	10.18	8.3×10^{-5}
2-Cyclopentadiene $\rightarrow$ dimer	14.5	3.39	9.91	3×10^{-7}

Termolecular reactions				
$O + O_2 + M \rightarrow O_3 + M$	-0.6[b]	7.81 [CO_2][c]	—	—
		7.42 [N_2]	—	—
$Br + Br + M \rightarrow Br_2 + M$	~ 0[b]	9.64 [N_2]	—	—
		9.14 [He]	—	—

[a]Adapted from S. W. Benson, "Foundations of Chemical Kinetics." McGraw-Hill, New York, 1960.
[b]Log A values.
[c]Species M in brackets.

of such cases to put simple collision theory into serious difficulty and to lead to the modern extensions of it discussed in Section 14-CN-1.

Table 14-6 includes some examples of termolecular reactions, discussed in Section 14-5E. It is fairly typical that these have little or no activation energy. The preexponential factor and the rate constant are the same in these cases. Note that A is in the expected range for the Br atom recombination, but rather smaller for the $O + O_2$ reaction, perhaps because of a steric factor. Note also that A varies somewhat with the nature of the inert species M, so that there is variation in the efficiency with which a collision with M can remove the energy released in the reaction.

14-8 Unimolecular reactions

The considerable success of collision theory in the period 1910–1920 seemed to establish that elementary gas-phase reactions were always bimolecular, with the colliding molecules supplying the necessary activation energy. A puzzling difficulty developed, however, with the finding that many decomposition reactions were first-order kinetically, which implied that the reaction was unimolecular. Yet the Arrhenius equation was well obeyed, giving quite respectable activation energies. How, then, could a molecule undergoing unimolecular decomposition acquire the necessary energy to react? An early proposal was that radiation absorbed and emitted between molecules provided the activation energy, but this hypothesis did not stand up to detailed inspection—molecules often would not have an absorption band in the wavelength region for which light quanta would have the requisite energy, for example.

The solution to the problem came through a recognition that there could be a lag between the time a molecule gains energy E^* and its decomposition. The decomposition reaction is essentially one of the breaking of a particular chemical bond, and if the energy E^* were distributed among various vibrational degrees of freedom, only after a number of vibrational cycles might it happen to concentrate on a particular bond vibration. The picture is difficult to present graphically, but suppose that a molecule has three vibrational degrees of freedom, that is, three independent modes of vibration, each with its characteristic frequency ν_0 [note Eq. (4-92)]. The case is shown in Fig. 14-10, where the arrows show the pattern of concerted motion of the atoms executing a given mode of vibration. The corresponding set of vibrational energy states is indicated in Fig. 14-11; in each case the first few levels are evenly spaced, as would be expected from Eq. (4-92). The spacing gets smaller and smaller, however, with increasing energy, essentially because the restoring force weakens with increasing amplitude of vibration. The corresponding situation for a diatomic molecule is illustrated in Fig. 4-13. There is in fact a covergence to a dissociation limit; bond breaking occurs at this point.

Suppose that as a result of a collision an H_2O molecule gains, in addition to kinetic energy, the vibrational state population indicated by the solid arrows in Fig. 14-11. The molecule as a whole has sufficient energy to dissociate, but cannot do so because of the way the energy is spread among the modes. While the various vibrational modes of a molecule are nearly independent of each other, there is some coupling between them; over a period of a number

FIGURE 14-10
Normal vibrational modes
for H_2O.

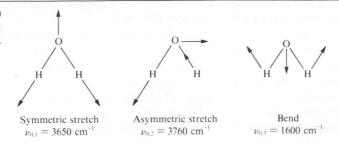

Symmetric stretch
$\nu_{0,1} = 3650$ cm^{-1}

Asymmetric stretch
$\nu_{0,2} = 3760$ cm^{-1}

Bend
$\nu_{0,3} = 1600$ cm^{-1}

of vibrations, energy can migrate between modes. Eventually, a redistribution of energy would occur, such as that shown by the dashed arrows in the figure. The asymmetric mode ν_2 has gained sufficient energy to dissociate on the next vibration, and reaction occurs. During this waiting period the molecule has a total energy E^* or more, which is much above the average molecular energy, and if a second collision occurs, it will probably take energy away rather than add to it. Thus, if reaction is to occur, it must do so between the time of the collision that has brought the energy up to E^* and that of the next collision. If the probability of this happening is small, then in effect a Boltzmann distribution of molecules of energy E^* or more is present, only slightly distorted by the disappearance of some molecules through reaction.

The kinetic scheme corresponding to this picture was proposed by F. Lindemann and J. Christiansen in 1922 and is as follows. The activating and deactivating processes are written

$$A + A \underset{k_{-2}}{\overset{k_2}{\rightleftharpoons}} A + A^* \tag{14-92}$$

where A^* is a molecule which is sufficiently energized to react, and which does so with a certain probability, and hence according to the rate law

$$A^* \overset{k_1}{\rightarrow} P \text{ (products)} \tag{14-93}$$

Equations (14-92) and (14-93) may be treated according to the stationary-state approximation described in Section 14-5C to give

$$\frac{d(P)}{dt} = \frac{k_1 k_2 (A)^2}{k_1 + k_{-2}(A)} \tag{14-94}$$

FIGURE 14-11
Qualitative energy levels
for the three vibrational
modes of H_2O. The
asymmetric stretch of ν_2
mode is assumed to be the
one most easily brought to
a dissociation limit, and
therefore to be the mode
that leads to reaction.

ν_1 $\qquad$ ν_2 $\qquad$ ν_3

If $k_{-2}(A) \gg k_1$, then Eq. (14-94) reduces to

$$\frac{d(P)}{dt} = k_1 K_2(A) = k_{app}(A) \tag{14-95}$$

where K_2 is the equilibrium constant for process (14-92), and $K_2(A)$ is therefore the equilibrium concentration of A*. This corresponds to the situation mentioned earlier, in which the rate of decomposition of energized molecules is slow enough that their distribution is the equilibrium or Boltzmann one. Equation (14-95) is pseudo-first-order, and thus accounts for the existence of first-order decomposition reactions.

A crucial test of the Lindemann–Christiansen mechanism lies in its prediction that at sufficiently small (A), Eq. (14-94) becomes

$$\frac{d(P)}{dt} = k_2(A)^2 \tag{14-96}$$

or second order. The actual test is usually made in the following way. We write Eq. (14-94) in the form

$$\frac{d(P)}{dt} = \frac{k_1 k_2(A)}{k_1 + k_{-2}(A)}(A) = k_{uni}(A) \tag{14-97}$$

That is, the initial rate of decomposition is reported as though it were a unimolecular reaction with rate constant k_{uni}. One then determines k_{uni} for various pressures and plots $k_{uni}/k_{uni,\infty}$ versus pressure, where $k_{uni,\infty}$ is the limiting value at high pressures [given by Eq. (14-95)]. This ratio should decrease toward zero at low pressures, and such behavior has indeed been found in a number of well-substantial cases. Some data on the isomerization of cyclopropane are shown in Fig. 14-12.

It is not necessary that A collide with a second molecule of its own kind in order to be energized, and Eq. (14-92) may be written in the more general form

$$A + M \underset{k_{-2}}{\overset{k_2}{\rightleftharpoons}} A^* + M$$

which leads to

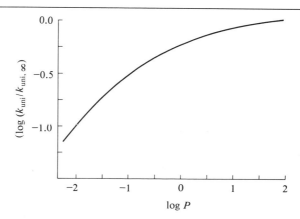

FIGURE 14-12
Variation of $k_{uni}/k_{uni,\infty}$ versus pressure for isomerization of cyclopropane. P is in cm Hg and temperature is 490°C. [From H. Pritchard, R. Sowden, and A. Trotman-Dickenson, Proc. Roy. Soc. **A217**, 563 (1953).]

$$\frac{d(\text{P})}{dt} = \frac{k_1 k_2 (\text{M})}{k_1 + k_{-2}(\text{M})}(\text{A})$$ (14-98)

where M denotes some nonreactive gas present in large excess over A, so that A–A collisions are unimportant. The decomposition will now be first order at all pressures, since (M) in Eq. (14-98) is constant, but k_{uni} will vary as the pressure of (M) is altered in successive experiments. Most studies of the Lindemann–Christiansen mechanism are made in this manner nowadays, and collisional efficiencies as given by k_2 have been studied for a large number of inert gases.

The Arrhenius parameters for the high-pressure limiting rates of some unimolecular reactions are given in Table 14-7. In the case of the cyclopropane isomerization, investigators have compared k_2 for various inert gases with that for cyclopropane itself to obtain relative efficiencies of activation. Some values are He, 0.060; N_2, 0.060; CO, 0.072; H_2O, 0.79; and toluene, 1.59. It appears that the more internal degrees of freedom a molecule has, the greater is its ability to transfer energy in a collision, which supports the idea that internal energy as well as translational energy transfers are important.

EXAMPLE   An illustration of the treatment presented here is as follows. The high-pressure limiting rate for the decomposition of azomethane is found to obey the Arrhenius equation with $A = 9.228 \times 10^{15}$ and $E^* = 51.130$ kcal mol^{-1}. At 603 K, $k_{uni}/k_{uni,\infty} = 0.10$ at $P = 0.2$ Torr. First, from Eq. (14-95),

$$k_{app} = k_{uni,\infty} = k_1 K_2 = 9.228 \times 10^{15} e^{-51,130/(1,987)(603)}$$

or $k_1 K_2 = 0.00270$ s^{-1}. Then from Eq. (14-97),

$$\frac{k_{uni}}{k_{uni,\infty}} = 0.10 = \left[\frac{k_1 k_2 P}{k_1 + k_{-2}P}\right][1/(k_1 k_2 / k_{-2}] = \frac{P}{k_1 k_{-2} + P}$$ (14-99)

or $k_1/k_{-2} = (0.2/0.10) - 0.2 = 1.8$ Torr. Alternatively, $k_2 = (k_1 K_2)(k_{-2}/k_1) = 0.00270/1.8 = 1.52 \times 10^{-3}$ Torr^{-1} s^{-1}. Thus the experimental data give values for $k_1 K_2$ and for k_2 or k_1/k_{-2}; it is not possible to obtain k_1 or K_2 alone, however, unless some assumptions are made. One could, for example, assume that k_1 corresponds to a single vibrational frequency, or to about 10^{13} s^{-1}, in which case $K_2 = 2.7 \times 10^{-16}$ and $k_{-2} = 5.6 \times 10^{12}$ Torr^{-1} s^{-1}.

An interesting point now arises. One might expect K_2 to be given simply by the Boltzmann factor for the probability of a molecule of A having energy E^*, or $\exp(-E^*/RT)$. The observed activation energy of 51 kcal mol^{-1} then leads to a theoretical value for K_2 of 3.0×10^{-19}, or much smaller than the estimate based on $k_1 = 10^{13}$ s^{-1}. It is not reasonable

TABLE 14-7 Arrhenius parameters for some unimolecular reactions[a]	Reaction	log A (s^{-1})	E^* (kcal mol^{-1})
	Cyclopropane $\rightarrow$ propylene	15.17	65.0
	cis-Isostilbene $\rightarrow$ trans-isostilbene	12.78	42.8
	$CH_3CH_2Cl \rightarrow C_2H_4 + HCl$	14.6	60.8
	CCl_3—$CH_3 \rightarrow CCl_2 = CH_2 + HCl$	12.5	47.9
	t-Butyl alcohol $\rightarrow$ isobutane + H_2O	14.68	65
	$ClCOOC_2H_5 \rightarrow C_2H_5Cl + CO_2$	10.7	29.4

[a]Adapted from S. W. Benson, "Foundations of Chemical Kinetics." McGraw-Hill, New York, 1960.

to take k_1 as any larger than this estimate, so it appears that the population of energized molecules is at least 1000 times that expected. The qualitative explanation is that the energized state is more probable than otherwise expected because of the many degrees of freedom among which E^* can be distributed; that is, its entropy as well as its energy must be taken into account. Some further discussion of this aspect is given in the Commentary and Notes section.

14-9 Absolute rate theory. The activated complex

A. The reaction profile

The material of the preceding section leads rather naturally into an alternative treatment of reaction rates. The Lindemann–Christiansen mechanism deals with molecules A* having within them enough energy to react but able to do so only at some rate k_1 determined by the speed with which E^* can localize on the particular bond that is to break. We now consider the simplest possible situation, namely a reaction of the type A' + A — A → A' — A + A, where A is an atom and the prime might denote isotopic labeling that distinguishes A' from A. There is now only one bond in the reacting molecule, and if sufficient energy is present, it should break on the next vibrational swing. There is thus the possibility of making an absolute calculation of a reaction rate.

A theory along these lines was developed by H. Eyring and M. Polanyi in the period 1930–1935. Consider the reaction

$$H' + H\!-\!H \rightarrow H'\!-\!H + H$$

where H' might in fact be a deuterium atom. Assume that colliding H' and H_2 molecules approach along a common line through the centers, so that various stages of reaction would be

$$H' + H\!-\!H \rightarrow H'\!-\!-\!-\!H\!-\!-\!H \rightarrow H'\!-\!-\!H\!-\!-\!H \rightarrow H'\!-\!-\!H\!-\!-\!-\!H$$

$$\begin{array}{ccccccccc} & 1 & & 2 & 3 & 1 & 2 & 3 & 1 & 2 & 3 \end{array}$$

$$\rightarrow H'\!-\!H + H$$

As atom 1 approaches, or the distance r_{12} decreases, the bond length between atoms 2 and 3, or r_{23}, begins to increase. The maximum potential energy must be at a point such that $r_{12} = r_{23}$, corresponding to the energized molecule which, being symmetric, might break up either to return to H' + H_2 or to give the product H'H + H. Essentially, the kinetic energy of approach of H' and H_2 is converted into potential energy, and it is possible to construct a diagram illustrating the situation as shown in Fig. 14-13. The three-dimensional model of the surface has the appearance of two valleys that interconnect through a high saddle point. The approach of H' to H_2 would be given by motion from right to left up the r_{12} valley.

Such motions or reaction trajectories are more easily seen in Fig. 14-14, in which Fig. 14-13 has been reduced to a two-dimensional plot with iso-energy contour lines. The dashed line (a) shows a nonreacting trajectory; note the oscillation in r_{23}, which shows that the rebounding H—H molecule has acquired vibrational energy. Dashed line (b) shows a possible reactive trajectory; again vibrational motion has appeared, this time in the product H'—H molecule.

FIGURE 14-13
A three-dimensional model showing the potential energy surface for the H + H_2 system as a function of r_{12} and r_{23} (see text). [See R.N. Porter and M. Karplus, J. Chem. Phys. **40,** 1105 (1964) and I. Shavitt, R.M. Stevens, F.L. Minn, and M. Karplus, J. Chem. Phys., **48,** 2700 (1968).]

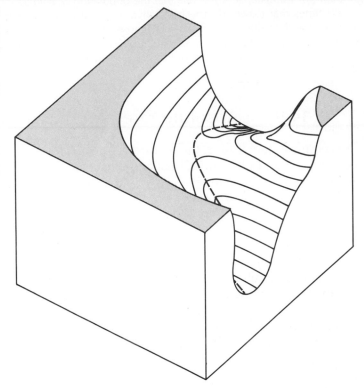

FIGURE 14-14
Vertical projection of Fig. 14-13 onto the r_{12}–r_{23} plane. Contour lines are in kcal mol^{-1} relative to H + H_2. Path (a) shows a trajectory for H' + H_2 in which there is insufficient energy for the system to surmount the saddle point, so no reaction can occur. Path (b) shows a possible trajectory leading to reaction to give products H'H + H. Note that the H'H product has vibrational energy, as indicated by the oscillation in the r_{12} distance. [See R.N. Porter and M. Karplus, J. Chem. Phys. **40,** 1105 (1964) and I. Shavitt, R.M. Stevens, F.L. Minn, and M. Karplus, J. Chem. Phys. **48,** 2700 (1968).]

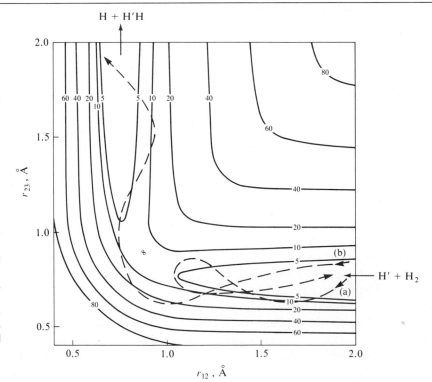

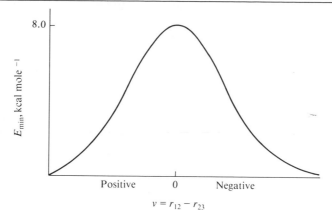

FIGURE 14-15
The reaction profile. Plot of minimum energy versus degree of advancement of reaction.

One may also compute energy profiles along a trajectory. Figure 14-15 shows the profile for the minimum energy path for reaction, and we see that the saddle point is about 8 kcal above the energy for separated H' and H—H. At this saddle point, incidentally, the symmetric species that is present, H'—H—H, has bond lengths of 0.9 Å or longer than the normal 0.74 Å H—H bond length.

Diagrams of the type of Fig. 14-14 have now been calculated for various approach angles, that is, various values of the H_1—H_2—H_3 angle. It turns out the while the saddle point is lowest for the head-on approach ($\theta = 180°$), the largest contribution to reaction is from collisions with θ around 160°. This value comes about as a trade-off between barrier energy and the angle weighting factor, $\sin \theta \, d\theta$, that is, the element of solid angle.

It has also been possible to make trajectory calculations in time. This is done by giving H' and H—H initial position and momentum values and

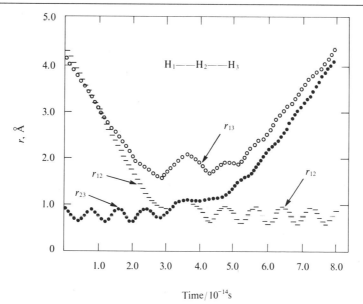

FIGURE 14-16
A reaction trajectory calculated for H + H_2. Initial conditions are: approach velocity 1.32×10^4 ms^{-1} and J = 2, $v = 0$ for H_2. [From M. Karplus, R.N. Porter, and R.D. Sharma, J. Chem. Phys. **24**, 3259 (1965)].

determining r_{12}, r_{23}, and r_{13} for successive small time increments. The result of one such calculation is shown in Fig. 14-16. We start with r_{23} small and oscillating (the H—H molecule is vibrating) and r_{12} and r_{13} large. Reaction takes place over the period 3–4×10^{-14} s, after which r_{13} stays small and oscillating, corresponding to vibrationally excited H'—H, and r_{23} and r_{12} increase. Note that the calculation was made for a specific rotational and vibrational state of the H—H molecule.

Calculations have also been made for more complex systems. Figure 14-17(a) shows the contour plot for the system $F + H_2 \rightarrow HF + H$, and Fig. 14-17(b) shows the appearance of the three-dimensional potential energy surface. The two valleys are now asymmetric, of course, with the HF + H one much deeper than the H_2 + F.

B. Transition-state theory

Theoretical calculations of the type described above have provided a framework for treating reactions generally. Species such as H'—H—H are stable with respect to vibrations involving motions against the walls of the saddle point; such species should have a set of vibrational and rotational states. They amount to a kind of compound, but a compound with an Achilles' heel—there is one mode of vibration that meets no restoring force so that the compound falls apart in this mode. Such a compound is called a *transition state* or an *activated complex.*

In transition theory it is in fact assumed that the activated complex is a distinct chemical species in equilibrium with the reactants, but a special kind of species in which one degree of vibrational freedom has become open, that is, corresponds to translation along the reaction coordinate (a parameter equivalent to $r_{12} - r_{23}$ in the H + H_2 case). The activated complex is not the same as the energized molecule of the preceding section; the latter possessed the energy E^* but might be anywhere on a reaction coordinate system, such as high on the side of a valley in Fig. 14-13, while the activated complex has the configuration corresponding to the saddle point.

The procedure is to write the intimate mechanism for an elementary reaction as

$$A + B \overset{\kappa^\ddagger}{=} [AB]^\ddagger \tag{14-100}$$

where $[AB]\ddagger$ denotes the activated complex, which then gives products

$$[AB]^\ddagger \overset{k_2}{\rightarrow} \text{products} \tag{14-101}$$

In the case of the Lindemann–Christiansen mechanism, k_2 could not be determined *a priori*, but the special nature of $[AB]^\ddagger$ now makes it possible for us to do so. The more rigorous treatment is given in the Special Topics section, but essentially k_2 is the frequency with which the open vibrational mode of the activated complex can lead to reaction and is approximated by kT/h.[†]

The rate of process (14-101) is then

$$\frac{d[AB]^\ddagger}{dt} = -\frac{kT}{h}[AB]^\ddagger \tag{14-102}$$

[†]By Eq. (4-92), $\epsilon_{vib} \simeq h\nu_0$, and if we take ϵ_{vib} to be about kT then $\nu_0 \simeq kT/h$.

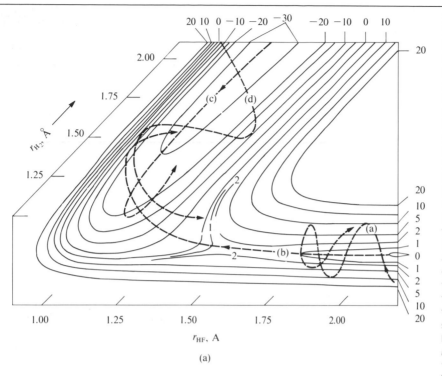

(a)

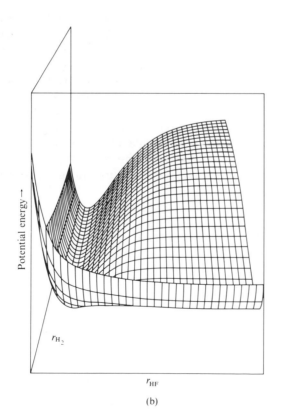

(b)

FIGURE 14-17
The system
F + H$_2$ → HF + H. (a)
Contour diagram for a
head-on approach;
contour energies are in
kcal mol^{-1} relative to
products. [From J.C.
Polanyi and J.L. Schreiber,
Chem. Phys. Lett. **29**, 319
(1974).] See Section 14-CN-
1 for a discussion of the
trajectories labelled (a),
(b), (c) and (d). Note: the
skewed coordinates are
used as a graphical device
whereby the
interconversion of kinetic
and potential energy
corresponds to that for a
frictionless ball rolling on
the surface. (b) A view of
the three-dimensional
surface looking down the
H + HF valley. [From J.W.
Moore and R.G. Pearson,
"Kinetics and Mechanism,"
3rd ed., Wiley, 1981.]

The overall reaction rate is

$$R_f = -\frac{d(A)}{dt} = \frac{\mathbf{k}T}{h} K^{\ddagger}(A)(B)$$
(14-103)

and the rate constant for an elementary reaction becomes

$$k = \frac{\mathbf{k}T}{h} K^{\ddagger}$$
(14-104)

$K^{\ddagger}$ is regarded as a normal equilibrium constant, and Eq. (7-12) therefore gives

$$k = \frac{\mathbf{k}T}{h}\exp\left(-\frac{\Delta G^{0\ddagger}}{RT}\right)$$
(14-105)

or

$$k = \frac{\mathbf{k}T}{h}\exp\left(\frac{\Delta S^{0\ddagger}}{R}\right)\exp\left(-\frac{\Delta H^{0\ddagger}}{RT}\right)^{\dagger}$$
(14-106)

EXAMPLE   The reaction $2NO_2 = 2NO + O_2$ obeys bimolecular kinetics with $k = 4.45\ M^{-1}\ s^{-1}$ at 600 K and 0.632 $M^{-1}\ s^{-1}$ at 550 K. Find $\Delta H^{0\ddagger}$, $\Delta E^{0\ddagger}$, and $\Delta S^{0\ddagger}$. First, $\ln(4.45/0.632) = (\Delta H^{0\ddagger}/R)(1/550 - 1/600)$, whence $\Delta H^{0\ddagger} = 25,600$ cal mol^{-1}. Remembering the footnote to Eq. (14-106), we obtain $\Delta E^{0\ddagger} = 27,800$ cal mol^{-1}. Then $k = 4.45 = [(1.3806 \times 10^{-16})(600)/(6.6262 \times 10^{-27})]\ \exp(\Delta S^{0\ddagger}/R)\ \exp(-25,600/600\ R)$, from which we find $\Delta S^{0\ddagger} = -14.29$ cal K^{-1} mol^{-1}.

C. Application of statistical thermodynamics to transition-state theory

The statistical mechanical methods of Chapter 6 can be applied to estimate $\Delta G^{0\ddagger}$, that is, the standard free energy of the activated complex. One needs to list the number of rotational and vibrational degrees of freedom and make a guess as to the characteristic rotational and vibrational temperatures or the moment of inertia of the activated complex and the spacings of its vibrational levels. The same can be done for the reactants, so it is possible to make reasonable calculations of $\Delta G^{0\ddagger}$ in simple cases. The procedure is developed in more detail in the Special Topics section.

More often, one uses the experimental rate constant and activation enthalpy to calculate $\Delta S^{0\ddagger}$ from Eq. (14-106). The resulting value then allows some conclusions to be drawn about the structure and general chemical nature of the activated complex.

EXAMPLE   Consider the following case. Experiment gives log $A' = 9.52$ for the reaction $H_2 + I_2 \rightarrow 2HI$, or $A = 7.94 \times 10^{10}$ mol liter^{-1} s^{-1} at 575 K, and according to transition state theory,

$$A = \frac{\mathbf{k}T}{h}\exp\left(\frac{\Delta S^{0\ddagger}}{R}\right)$$
(14-107)

where $\Delta S^{0\ddagger}$ is the standard entropy change for the reaction

$$H_2 + I_2 = [(HI)_2]^{\ddagger}$$
(14-108)

---

† For a bimolecular reaction, we can use Eq. (5-10) to replace $\Delta H^{0\ddagger}$ by $\Delta E^{0\ddagger} - RT$ and thus obtain a parallel to Eq. (14-75), with $A = (e\mathbf{k}T/h)\exp(\Delta S^{0\ddagger}/R)$.

The questions of mechanism discussed earlier do not enter into the picture at the moment; entropy being a state function, the value of $\Delta S^{0\ddagger}$ does not depend on the path whereby the transition state $[(HI)_2]^{\ddagger}$ is formed. The quantity $\mathbf{k}T/h$ has the value 1.20×10^{13} s^{-1} at 575°K and we calculate the experimental $\Delta S^{0\ddagger}$ to be -9.97 cal K^{-1} mol^{-1}.

Alternatively, $\Delta S^{0\ddagger}$ may be estimated from statistical thermodynamics as a sum of translational, rotational, and vibrational contributions. First, the translational contribution is obtained from the Sackur–Tetrode equation

$$S_{trans}^0 = R \ln(T^{5/2}M^{3/2}) - 2.31 \qquad \text{[Eq. (6-104)]}$$

Substitution of $T = 575$ K and of the appropriate molecular weights gives 31.3, 45.8, and 45.8 cal K^{-1} mol^{-1} for the entropies of H_2, I_2, and $[(HI)_2]^{\ddagger}$ at 1 atm pressure. We want, however, to use a standard state of 1 mol liter^{-1} rather than the 47.2 liter mol^{-1} for 1 atm at 575 K. The correction is then $R \ln(1/47.2) = -7.66$ cal K^{-1} mol^{-1} and yields 23.7, 38.1, and 38.1 cal K^{-1} mol^{-1} for the corrected entropies. Then $\Delta S_{trans}^{0\ddagger} = -23.7$ cal K^{-1} mol^{-1}.

We proceed to obtain the rotational entropies. The entropy S_{rot} is given by

$$S_{rot} = R + R \ln \frac{T}{\sigma \theta_{rot}} \qquad \text{[Eq. (6-108)]}$$

where, for completeness, σ is a degeneracy factor given by the number of equivalent ways of orienting the molecule in space; $\sigma = 2$ for each of the molecules involved here. From Table 6-1, $\theta_{rot} = 87.5$ K for H_2 and 0.0538 K for I_2, giving 4.4 and 19.0 cal K^{-1} mol^{-1} for the respective rotational entropies at 575 K.

We have now reached a point where some specific structure must be assumed for the activated complex $[(HI)_2]^{\ddagger}$. Equation (14-86) suggests that the structure may be the linear molecule I—H—H—I. We expect the bonds to be somewhat longer than normal and guess the I—H bond to be about 1.75 Å and the H—H bond to be 0.97 Å, so that the iodine and hydrogen atoms are 2.23 Å and 0.48 Å from the center of mass, respectively. The moment of inertia of the molecule is then

$$I^{\ddagger} = \frac{2}{N_A}[(127)(2.23 \times 10^{-8})^2 + (1)(0.48 \times 10^{-8})^2] = 2.10 \times 10^{-37} \text{ g cm}^2$$

Recalling Eq. (4-89), we obtain $\theta_{rot} = 0.0192$ K and thence $S_{rot}^{\ddagger} = 21.1$ cal K^{-1} mol^{-1}. Therefore $\Delta S_{rot}^{\ddagger} = 21.1 - 4.4 - 19.0 = -2.3$ cal K^{-1} mol^{-1}.

The vibrational entropy of a molecule is given by

$$S_{vib} = R\left[\frac{x}{e^x - 1} - \ln(1 - e^{-x})\right] \qquad \text{[Eq. (6-110)]}$$

where $x = \theta_{vib}/T$. Table 6-1 gives θ_{vib} as 5986 K and 306.8 K for H_2 and I_2, respectively. For H_2, S_{vib} is negligible in view of the large θ_{vib} and for I_2, $S_{vib} = 3.3$ cal K^{-1} mol^{-1}. The molecule I—H—H—I, being linear, has $(3)(4) - 3 - 2 = 7$ degrees of vibrational freedom, of which, however, one is open or corresponds to translation along the reaction coordinate. We guess that at least three of the remaining six vibrational degrees of freedom are low in fundamental frequency (I—H—H—I is, after all, a rather weakly bonded molecule at best) and assign them a frequency of 100 cm^{-1} each; the other three frequencies we assume to be high enough that their contribution to the entropy of the molecule is small. From Section 4-14, $\theta_{vib} = h\nu_0/\mathbf{k}$, so that for the low-energy vibrations $\theta_{vib} = 144$ K; x is then 0.25 and application of Eq. (6-110) yields $S_{vib} = (3)(4.75) = 14.2$ cal K^{-1} mol^{-1}. Then $\Delta S_{vib}^{\ddagger} = 14.2 - 3.3 = 10.9$ cal K^{-1} mol^{-1}.

We now assemble $\Delta S^{0\ddagger}$ as $-23.7 - 2.3 + 10.9 = -15.1$ cal K^{-1} mol^{-1}. We have thus calculated a value for $\Delta S^{0\ddagger}$ which is within about 5 cal K^{-1} mol^{-1} of the experimental value, which corresponds to a factor of 10 in frequency factor A.

We are able to explain experimental activation entropies theoretically. Herschbach *et al.* (1956) have carried out a number of similar calculations of $\Delta S^{0\ddagger}$ for various reactions of the type listed in Table 14-6 and again were able to obtain order-of-magnitude agreement with the experimental frequency of Arrhenius A factors in almost every case. The reactions treated included several with very low steric or P factors and the effect could be seen to arise as a natural consequence of the structure and degree of stiffness of the activated complex as compared to that of the reactants.

On the other hand, the example also illustrates that there is a considerable degree of latitude in the choice of moments of inertia and vibrational frequencies for the activated complex, so that agreement of theory with experiment can be rather subjective. For example, it was thought for many years that the activated complex in $H_2 + I_2 \rightarrow 2HI$ reaction was a four-centered molecule

and about equally successful statistical calculations were made on this assumption.

One other example might be given. In the case of a unimolecular reaction one has, in terms of transition-state theory,

$$A \overset{K}{=} [A]^{\ddagger} \qquad [A]^{\ddagger} \overset{k_2}{\rightarrow} \text{products}$$

where k_2 is still given by $\mathbf{k}T/h$. From the example of the preceding section $k = 0.00270$ s^{-1} for the decomposition of azomethane at 603 K, and the Arrhenius constant $A = 9.23 \times 10^{15}$. Equation (14-107) then gives $\Delta S^{0\ddagger} = 13.1$ cal K^{-1} mol^{-1}—a fairly positive value. From the statistical thermodynamic point of view, $\Delta S^{0\ddagger}_{trans}$ should be zero since there is no change in the number of species in the formation of the activated complex. Since A and $[A]^{\ddagger}$ must be very similar in moment of inertia, $\Delta S^{\ddagger}_{rot}$ should also be zero, or at best a small positive number if $[A]^{\ddagger}$ is somewhat expanded in size over A as a result of its high energy content. The main contribution to $\Delta S^{0\ddagger}$ must then come from $\Delta S^{\ddagger}_{vib}$. This last should be negligibly small for A, but evidently the activated complex must have one or more vibrational modes of very low θ_{vib} value, that is, some low-frequency vibrations corresponding to very weak bonds.

14-10   Collision versus transition-state theory

The collision theory for gas-phase reactions presents a very direct, indisputable physical picture of how molecules react—they must collide and must do so with the proper orientation and with a certain minimum energy E^*. One difficulty is in treating the so-called steric factor P. Not only does the simple geometric interpretation become implausible in the case of very small P values such as 10^{-6}, but this interpretation provides virtually no way to make calculational estimates or correlations.

A major further development of collision theory took place as a result of some observations of the rates of unimolecular reactions. According to the Lindemann mechanism, the maximum possible rate for a unimolecular reaction would be observed if each energized molecule decomposed on the next vibrational swing, which means that

$$k_{\text{uni},\infty} \simeq \frac{\mathbf{k}T}{h} e^{-E^*/RT} \tag{14-109}$$

Thus the maximum possible frequency factor for a unimolecular reaction should be about 10^{13} s^{-1}. However, the observed Arrhenius A parameter is over 10^{15} for cyclopropane isomerization (see Table 14-7) and is 9.23×10^{15} for the decomposition of azomethane (from the numerical illustration in Section 14-8). It appeared at first that something was seriously wrong.

The explanation that was developed by O. Rice, H. Ramsperger, and L. Kassel around 1930 is the following. G. Lewis had pointed out several years previously that it was not necessarily true that E^* came only from the kinetic energy of relative motion of the colliding molecules, since it would also be supplied by their internal energy. The resulting modification of Eq. (14-89) gives

$$k = Z' \frac{(E^*/RT)^{s-1}}{(s-1)!} e^{-E^*/RT} \tag{14-110}$$

where Z' is the bimolecular collision frequency for reactants at unit concentration and s is the number of vibrational degrees of freedom available to contribute to E^*. The equation reduces to Eq. (14-89) if $s = 1$, but has the effect of increasing the frequency factor A for larger values of s. The qualitative explanation of high frequency factors for unimolecular reactions is thus simple enough—in the cyclopropane case, for example, it would only be necessary to say that about one-third of the molecular vibrations contribute to E^*.

More precisely, however, we seek an expression for $k_{\text{uni},\infty}$. According to Eq. (14-99) $k_{\text{uni},\infty} = k_1 k_2/k_{-2}$ [cf. Eqs. (14-92) and (14-93)]. First, we use Eq. (14-110) for k_2. Second, we assume that the k_{-2} or deactivation process occurs on every collision, so that $k_{-2} = Z'$. We thus have

$$\frac{k_2}{k_{-2}} = \frac{(E^*/RT)^{s-1}}{(s-1)!} e^{-E^*/RT} \tag{14-111}$$

Finally, we need to say something about k_1, the rate constant for A^* going to products. The supposition by Rice and co-workers was that k_1 depended on the actual energy possessed by A^*. That is, greater energy in excess of the minimum amount E^* would result in greater probability of the necessary energy localizing on the particular bond that was to break before a deactivating collision occurred. If energy E was distributed among s vibrational degrees of freedom, the conclusion was that for such molecules

$$k_1 = \frac{\mathbf{k}T}{h} \left(\frac{E - E^*}{E}\right)^{s-1} \tag{14-112}$$

Note that the frequency factor has been taken to be $\mathbf{k}T/h$, which is about 10^{13} s^{-1}. This corresponds to a fairly typical vibrational frequency and is a commonly observed one for $k_{\text{uni},\infty}$.

We can now combine Eqs. (14-111) and (14-112) to obtain an expression for $k_{\text{uni},\infty}$. The actual observed value should be the sum or integral over all possible E values, so the final equation is

$$k_{\text{uni},\infty} = \left(\frac{\mathbf{k}T}{h}\right)\int_{E^*}^{\infty} \left(\frac{E - E^*}{E}\right)^{s-1} \frac{(E/RT)^{s-1}}{(s-1)!} \frac{e^{-E/RT}}{RT}\, dE \tag{14-113}$$

Eq. (14-113) integrates very simply, to give $k_{\text{uni},\infty} = (\mathbf{k}T/h)e^{-E^*/RT}$ (see Problem 14-50). Furthermore, the way in which k_{uni} should vary with inert gas concentration (M) can also be evaluated:

$$k_{\text{uni}} = A\frac{e^{-E^*/RT}}{(s-1)!}\int_{E^*}^{\infty} \frac{[(E - E^*)/RT]^{s-1}e^{(E - E^*)/RT}}{RT\{1 + [A/k_{-2}(\text{M})][(E - E^*)/E]\}^{s-1}}\, dE \tag{14-114}$$

where A may be taken to be $\mathbf{k}T/h$, and k_{-2} to be equal to Z'.

Equation (14-114) may be calculated for various s values (knowing E^* from the temperature dependence of $k_{\text{uni},\infty}$), and thus fitted to experimental data. The results have been fairly successful. In 1952, however, a further refinement was added by R.A. Marcus, and the combined Rice-Ramsperger-Kassel-Marcus or RRKM theory is now widely used. The approach is a statistical thermodynamic one, and is described in Section 14-ST-1. In effect, collision theory has approached transition-state theory, and the distinction between the two theories has become blurred when each is pursued to a high degree of refinement. The detailed treatment of the former takes into account internal vibrational and rotational energy states, and so does transition-state theory, in terms of $\Delta S^{0\ddagger}$. Thus the use of translational and rotational partition functions to calculate $\Delta S^{0\ddagger}$ is illustrated in the example of Section 14-9C. Also, a more complete statistical thermodynamic treatment of transition-state theory is given in Section 14-ST-1.

Many kineticists find it sufficient, however, to use the simple statements of the two theories, that is, Eq. (14-91) for collision theory and Eq. (14-106) for transition-state theory. One may thus report empirical steric factors or empirical $\Delta S^{0\ddagger}$ values. Collision theory as developed here is specific for gas phase reactions, but can be modified to apply to reactions in solution, as is discussed in Section 15-3. Transition-state theory is applicable to both gas and solution phase reactions, but is especially widely used for the latter. Basically, however, the choice of which simple theory to use becomes largely a matter of personal taste.

COMMENTARY AND NOTES

14-CN-1   Radicals, molecular beams, and reaction trajectories

At several points in this chapter, and especially in Section 14-5, an experimental rate law has been explained in terms of a mechanism involving reactive intermediates. These may be radicals, such as $\cdot CH_3$ or $\cdot COCH_3$, or relatively unstable compounds, such as H_2I. Although the lifetime of such species is short in a reacting system, they usually survive enough collisions to come to equilibrium with the surroundings, that is, to reach the equilibrium Boltzmann distribution of translational, rotational, and vibrational energy. They thus have definite physical and thermodynamic properties. They are perfectly "good" chemical species.

Organic free radicals have been known for some time. F. Paneth and co-workers showed around 1930 that the thermal decomposition of lead tetraethyl or of mercury dimethyl produced ethyl and methyl radicals. These could be swept along by a flow of inert gas to react downstream with a film of lead or other metal, to regenerate metal alkyl compounds. The triphenylmethyl radical, produced by the dissociation of hexaphenylethane, is stable enough in solution to be observed both spectrophotometrically and through its paramagnetism.

The world of chemistry now includes a vast number of compounds that are too reactive to be stored in bottles but whose properties can be measured by various contemporary means. For example, modern mass spectrometers routinely measure trace quantities of all kinds of molecular fragments produced either in a thermal reaction or by induced decomposition by light or by electron impact. In this last approach one determines the energy of impinging electrons needed to produce a given fragment, the threshold value being called the appearance potential. Methane produces CH_4^+ when 12 V electrons are passed through the gas, and at 14 V, CH_3^+ results. Methyl radicals generated in a chemical reaction and then electron bombarded give CH_3^+ again, but now with only 10 V electrons. The difference between 14 and 10 V must correspond to the energy for the process $CH_4 \rightarrow CH_3 + H$. Thus the energies of formation of unstable species can be estimated.

The kineticist contributes much information. For example, the activation energy for the recombination of two radicals, $\cdot R + \cdot R' \rightarrow RR'$, is often zero or very small, which means that the activation energy for the *reverse* process gives the ΔE for the overall reaction (note Fig. 14-9) and hence that bond energy. Extensive tables of bond energies and entropies make possible the fairly reliable calculation of activation energies and frequency factors for most reactions involving small molecules and radical fragments [see Benson (1968)].

Another facet of modern kinetics is the study of the intimate details of just how elementary reactions occur. An important technique is that using molecular beams, and a representative apparatus is illustrated in Fig. 14-18(a). Collimated beams of the two reactants impinge, and the angular distribution of product molecules is determined. A mass spectrometer may be used in front of the detector to select only that product desired. The beams emerge from a nozzle or an aperture in an oven with a thermal velocity distribution, but velocity selectors (see Fig. 2-13) can be used to limit the beam to a particular kinetic energy range. One of the beams may be made intermittent by means of a chopper so that the time lag before the appearance of product can be measured. It is, in brief, possible to obtain the kinetic energy as well as the angular distribution for the reaction products. One can thus calculate how much of the available energy (impact energy plus energy of reaction) appears as kinetic energy in the products, and therefore how much remains in the form of vibrational excitation. Also, ion–molecule reactions can be studied by using an ion source. Some reactions that have been studied in this manner are:

(1) $K + HBr, \rightarrow H + KBr,$
(2) $Cs + RbCl \rightarrow Rb + CsCl,$
(3) $K + I_2 \rightarrow I + KI,$
(4) $Ar^+ + D_2 \rightarrow D + ArD^+,$
(5) $K + ICH_3 \rightarrow KI + CH_3,$
(6) $F + C_2H_4 \rightarrow H + C_2H_3F,$
(7) $Cs + SF_6 \rightarrow CsF + SF_5,$
(8) $O_2^+ + D_2 \rightarrow D + O_2D^+.$

The angular distribution of the products is of great interest because it provides information about how the reaction proceeds. As one possible intimate mechanism, the reactants may form a weak association complex. This will have angular momentum and rotate in the plane defined by the two reactant beams. When the complex breaks up the products will be concentrated in this plane, but will otherwise be random in direction. This is the case for reaction (2) above. Figure 14-18(c) shows the angular distribution of the CsCl product, plotted as isoprobability contours in a polar plot, distance from the origin being proportional to velocity.† Note that there is about equal probability for the CsCl product to appear in the forward as in the backward direction.

The behavior of reaction (3) above is quite different. As shown in Fig. 14-18(b), the product KI appears in a forward directed cone. It appears as though the K atom "stripped" off an iodine atom from I_2 as it went by. The explanation is an interesting one. As the K atom approaches the I_2 molecule, a distance is reached

---

†Figure 14-18 (b,c) shows angular distributions taking the center of mass as the frame of reference. The experimental data, of course, give angular distributions in the laboratory frame of reference, and would be skewed in the direction toward which the two product beams travel. The conversion to the center-of-mass frame of reference is not a trivial process and can sometimes be done only with some approximations.

FIGURE 14-18
(a) Molecular beam apparatus used for the study of the reaction of fluorine atoms with hydrocarbons. [From J.M. Parson, K. Shobatake, Y.T. Lee, and S.A. Rice, Disc. Far. Soc. **55**, 344 (1973).] (b) The cone containing most of the KI product intensity for the reaction K + I$_2$ → I + KI. (c) Velocity-angle contour map for the CsCl product of the Cs + RbCl reaction. (From R.D. Levine and R.B. Bernstein, "Molecular Reaction Dynamics," Oxford Univ. Press, London and New York, 1974).

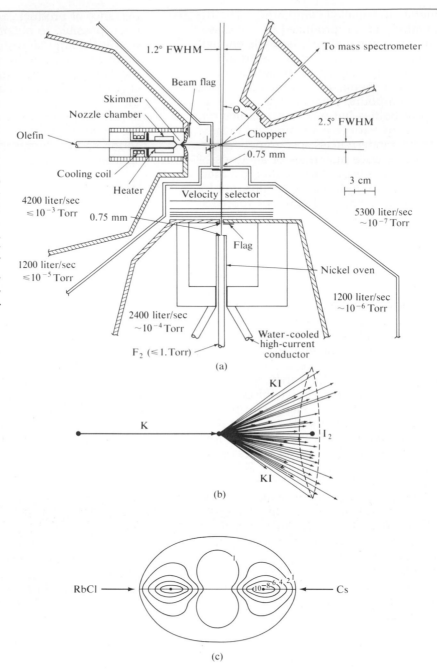

at which it becomes probable for an *electron* to transfer, giving K$^+$ and I$_2^-$. The two charged species are strongly attracted to each other, and as K$^+$ and I$^-$ form the ion pair K$^+$I$^-$, the remaining I atom is left behind. The rate constant or effective collision diameter for a reaction of this type can be quite large [127 Å^2 in the case of reaction (3)] because the transferred electron acts as a "harpoon" by which the alkali metal atom hooks the halogen atom.

Still more fundamental is the complete description of reaction trajectories. By this we mean not just the angular distribution of products but the probability of reaction from a given set of vibrational and rotational states of reactants to a given set of such states for products. We can use Fig. 14-17 to provide one kind of illustration. The figure shows that the saddle point configuration is not symmetrically located, but is more like H–H—F, that is, the H–F bond distance is greater than the H–H one. If the reaction $H_2 + F \rightarrow H + HF$ were attempted with vibrationally excited H_2, the trajectory would look something like the dashed line (a) in the figure. The system would get "bottled up" in oscillations and the chance of passing over the saddle point and hence of reacting would be small. Along the straight line path (b) in which the reactants approach with kinetic energy but little vibrational energy, reaction should be more probable. Conversely, the reaction $H + HF \rightarrow H_2 + F$ would *not* be probable along the straight line path (c). The system would "bounce" back from the wall of the potential surface with little chance of finding the saddle point. However, if the HF had some vibrational excitation, then as illustrated by path

(d), a trajectory carrying over the saddle point would be more likely.†

Similar considerations apply to the likelihood of a product having vibrational excitation. Calculations of the type illustrated in Fig. 14-16 can predict whether a product should have more or less vibrational excitation than the reactant. As an experimental illustration, in reaction (6) above, the product CsF appears with a distribution of vibrational excitation. The relative probability falls off exponentially with increasing vibrational quantum number, but CsF in, say, the $v = 4$ state, is still quite probable. The understanding of reaction kinetics at this level of detail has become a challenging and difficult exercise in wave mechanics.

†The analogy may be made to an obstacle in a miniature golf course. If one wants to putt the ball in at the lower right and have it exit at the upper left, a straight putt up the initial valley should be successful. If, however, the goal is to putt the ball in at the upper left to emerge at the lower right, then a banking shot is required.

SPECIAL TOPICS

14-ST-1   Statistical thermodynamic treatment of transition-state theory

The application of statistical thermodynamics to the calculation of $\Delta S^{0\ddagger}$ was illustrated in Section 14-9C. This in fact represents the practical approach that is the most useful. It is possible, however, to arrive at Eq. (14-102) or Eq. (14-103) on a more formal basis and one which sheds some light on the fundamental assumptions of transition-state theory. We must first develop a certain formalism for the general treatment of an equilibrium constant, which will then be applied to $K^{\ddagger}$.

A. Statistical thermodynamic treatment of equilibrium constants

The statistical thermodynamic expression for the free energy of an ideal gas can be reduced to a very simple form. Equation (6-99) reads

$$G = -RT(\ln \mathbf{Q}) + RTV\left[\frac{\partial(\ln \mathbf{Q})}{\partial V}\right]_T + RT \ln N_A - RT$$

However, only $\mathbf{Q}_{trans}$ depends on v, and for an ideal gas it is proportional to v by Eq. (4-83), so that $[\partial(\ln \mathbf{Q})/\partial V]_T$ is just $1/V$ and the second and fourth terms of Eq. (6-99) cancel. Also, Eq. (6-99) was derived on the assumption that H_0, the absolute enthalpy at 0 K, was set equal to zero. We now want to retain this reference point, and since for an ideal gas at 0 K $H_0 = E_0$, we obtain

$$G = E_0 - RT \ln \mathbf{Q} + RT \ln N_A \quad (14\text{-}115)$$

or

$$G = E_0 - RT \ln \frac{\mathbf{Q}}{N_A} \quad (14\text{-}116)$$

We now define a subsidiary partition function ϕ defined as $\mathbf{Q} = \phi V$, that is, ϕ contains all the terms of $\mathbf{Q}$ except the volume factor. If the standard state is taken to be 1 atm pressure, then $V = N_A \mathbf{k} T$, and Eq. (14-116) becomes

$$G^0 = E_0 - RT \ln(\phi \mathbf{k} T) \qquad (14\text{-}117)$$

From our standard chemical reaction

$$a\text{A} + b\text{B} + \cdots = m\text{M} + n\text{N} + \cdots$$

for which

$$\Delta G^0 = -RT \ln K_P \quad [\text{Eq. (7-12)}]$$

If the standard state is 1 atm, substitution of Eq. (14-117) gives

$$K_P = (\mathbf{k} T)^{\Delta n} K_\phi e^{-\Delta E_0 / RT} \qquad (14\text{-}118)$$

where Δn is the number of moles of products minus moles of reactants and K_ϕ is the quantity

$$K_\phi = \frac{\phi_M{}^m \phi_N{}^n \cdots}{\phi_A{}^a \phi_B{}^b \cdots}$$

[Note the resemblance to Eq. (7-64).] Since by Eq. (7-15) $K_C = K_P(RT)^{-\Delta n}$, or $K_n = K_P(\mathbf{k} T)^{-\Delta n}$, where K_n is the concentration equilibrium constant using the concentration in molecules per unit volume $\mathbf{n}$, Eq. (14-117) can be put in the final form

$$K_n = K_\phi e^{-\Delta E_0 / RT} \qquad (14\text{-}119)$$

An equilibrium constant can thus be expressed as the product and quotient of the slightly modified partition functions ϕ multiplied by the term $e^{-\Delta E_0 / RT}$, which involves the change in zero-point energy accompanying the chemical reaction.

B. Application to
transition-state
theory

We first calculate the concentration of activated complex $[\text{AB}]^\ddagger$ as

$$([\text{AB}]^\ddagger) = K_n{}^\ddagger(\text{A})(\text{B})$$
$$= K_\phi{}^\ddagger \, e^{-\Delta E_0{}^\ddagger / RT}(\text{A})(\text{B})$$

A special assumption has been made regarding Q or ϕ for $[\text{AB}]^\ddagger$, namely that one of its degrees of freedom is in translation along the reaction coordinate; this particular component of $\phi^\ddagger$ is then to be expressed separately with Eq. (4-82) for the translational partition function in one dimension, leaving $\phi^\ddagger$ containing the usual three degrees of translational freedom, two or three rotational degrees of freedom, but one less than

the normal number of vibrational contributions. We therefore write

$$([\text{AB}]^\ddagger)$$
$$= \frac{(2\pi m^\ddagger \mathbf{k} T)^{1/2}}{2h} \frac{a^\ddagger \phi^\ddagger}{\phi_A \phi_B} e^{-\Delta E_0{}^\ddagger / RT} \, (\text{A})(\text{B})$$

where $a^\ddagger$ must be some characteristic linear dimension associated with the saddle point of Fig. 14-14 and the factor of $\frac{1}{2}$ is present since only forward translation along the reaction path is being considered.

The frequency with which molecules of $[\text{AB}]^\ddagger$ will pass over the saddle is next expressed as in terms of a velocity $v^\ddagger$:

$$\text{rate} = \frac{v^\ddagger([\text{AB}]^\ddagger)}{a^\ddagger}$$
$$= v^\ddagger \frac{(2\pi m^\ddagger \mathbf{k} T)^{1/2}}{2h} \frac{\phi^\ddagger}{\phi_A \phi_B} e^{-\Delta E_0{}^\ddagger / RT} \, (\text{A})(\text{B})$$
$$(14\text{-}120)$$

The velocity $v^\ddagger$ is now supposed to have a Boltzmann distribution of values given by the treatment for a one-dimensional gas, Eq. (2-37). The average velocity for a three-dimensional gas is $(8\mathbf{k} T/\pi m)^{1/2}$ [Eq. (2-52)]; for a one-dimensional gas it turns out to be $(2\mathbf{k} T/\pi m)$, and this result is inserted in place of $v^\ddagger$ to obtain

$$\text{rate} = \frac{\mathbf{k} T}{h} \frac{\phi^\ddagger}{\phi_A \phi_B} e^{-\Delta E_0{}^\ddagger / RT}(\text{A})(\text{B})$$

from which the rate constant, expressed in molecules per cubic centimeter, is

$$k = \frac{\mathbf{k} T}{h} \frac{\phi^\ddagger}{\phi_A \phi_B} e^{-\Delta E_0{}^\ddagger / RT} \qquad (14\text{-}121)$$

One may, of course, factor the partition function ϕ in the usual manner, writing $\phi = \phi_{\text{trans}} \phi_{\text{rot}} \phi_{\text{vib}}$, and evaluate the terms from the standard expressions of Chapter 4, remembering that V is dropped out of $\mathbf{Q}_{\text{trans}}$ and $[\text{AB}]^\ddagger$ has one less than the usual number of vibrations; otherwise the ϕ's are the same as the $\mathbf{Q}$'s.

The preceding derivation has led to an equation of the same form as Eq. (14-106) but shows more specifically the assumptions that are made in transition-state theory. There is a certain artificiality to the treatment of the rate of passage over the saddle point, but the more interesting aspect is the following. Although the ϕ's are temperature-dependent, the major temperature term is $\exp(-\Delta E_0{}^{\ddagger}/RT)$. The experimental activation energy, as obtained by means of the Arrhenius equation, is thus largely identified with the difference in *zero-point energy* between $[AB]^{\ddagger}$ and the reactants A and B. Thus $[AB]^{\ddagger}$ is treated as a definite chemical species and not just as some special energized association of A and B. Its translational, rotational, and vibrational partition functions are treated as though $[AB]^{\ddagger}$ were in *ordinary thermal equilibrium* with its surroundings; that is, the ambient temperature is used in their evaluation.

The contrast with collision theory is especially striking in the case of a unimolecular reaction. Simple collision theory treats such a reaction in terms of a molecule A*, which is essentially the same molecule as A but in a highly vibrationally excited state. Transition-state theory says that [A] is a new chemical species of extra zero-point energy E_0 whose vibrations are those corresponding to equilibrium with the ambient temperature. The contradiction seems monumental, yet we see below how it has been in large measure resolved in RRKM theory.

C. Rice-Ramsperger-Kassel-Marcus or RRKM theory

The extension of RKK (Rice-Ramsperger-Kassel) theory (Section 14-10) by R.A. Marcus in the 1950's combines the concepts of the energized species A* and the transition-state species $A^{\ddagger}$. The sequence for a unimolecular reaction is now taken to be:

$$A + M \underset{k_{-2}}{\overset{k_2}{\rightleftharpoons}} A^* + M$$

$$A^* \overset{k_1}{\rightarrow} A^{\ddagger} \qquad\qquad (14\text{-}122)$$

$$A^{\ddagger} \overset{k_{\ddagger}}{\rightarrow} \text{products}$$

That is, the vibrationally excited A* species has a certain lifetime during which it may be deactivated by collision with M (the k_{-2} process), or may become the transition-state species $A^{\ddagger}$, which then goes on to product with rate constant $k_{\ddagger}$.

Figure 14-19 illustrates the model in terms of the potential energy diagram for a hypothetical XYZ molecule. The molecule, our spe-

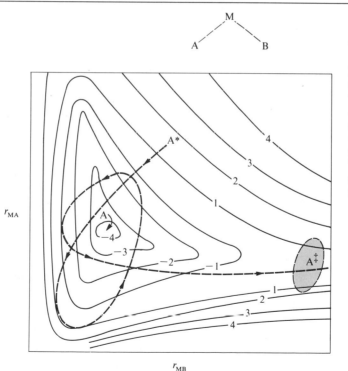

FIGURE 14-19
Potential energy diagram for a hypothetical stable molecule A of formula XYZ. The dashed line shows a possible trajectory for an energized molecule A* that eventually succeeds in entering the reaction channel, which opens in the $A^{\ddagger}$ region. The iso-energy contours are in arbitrary energy units.

cies A, is stable, as indicated by the potential well in the middle region of the figure. A collision or series of collisions produces vibrational excitation, with both the XY and the YZ bonds increasing in length. This initial energetic species is marked as A* in the figure. The A* molecule executes various vibrations, the system moving along a path such as illustrated by the dashed line. During this period A* may disappear by collisional deactivation, the k_{-2} process. It may, however, eventually wander into a trajectory leading to Y—Z bond breaking. We designate by A‡ a molecule in the shaded area; in this region there is no restoring force to further elongation of the Y—Z bond; it is now "downhill" to reaction products XY and Z.

The mathematical treatment is similar to that in Section 14-10. A* must have some minimum energy if reaction is to be possible, essentially the energy in the A‡ region of Fig. 14-19. The formation rate constant k_2 [Eq. (14-122)] is energy dependent, and so the rate of formation of A* must be averaged over the values for all energies above the minimum energy. The "strong" collision picture is assumed for the k_{-2} or deactivation process; that is, deactivation of A* back to A is taken to occur regardless of how much actual energy A* has. For any particular value of k_2, steady state analysis gives

$$(A^*) = \frac{k_2 (M)(A)}{k_{-2} (M) + k_1} \qquad (14\text{-}123)$$

and

$$k_1 (A^*) = k_‡ (A^‡) \qquad (14\text{-}124)$$

The rate of product formation is

$$\frac{d(\text{product})}{dt} = k_‡ (A^‡)$$
$$= k_1(A^*) = k(A) \qquad (14\text{-}125)$$

where

$$k = \frac{(k_1 k_2 / k_{-2})}{1 + [k_1/k_{-2} (M)]} \qquad (14\text{-}126)$$

The observed first-order rate constant, k_{uni}, is obtained by averaging k over all values of the energy of A*:

$$k_{uni} = \int_{E^*}^{\infty} \frac{k_1[dk_2/k_{-2}]}{1 + [k_1/k_{-2}(M)]} \qquad (14\text{-}127)$$

The ratio dk_2/k_{-2} gives the equilibrium constant $(A_E^*)/(A)$, where (A_E^*) is the concentration of A* molecules having energy between E and $E + dE$, $E > E^*$. Also, by Eq. (14-125), $k_1 = k_‡ (A^‡)/(A^*)$. The rate constant $k_‡$ is evaluated in terms of the rate of passage over a saddle, much in the manner used in Eq. (14-120), and the concentration ratios involving A* and A‡ are formalized in terms of the appropriate partition functions. In the high pressure limit [that is, (M) large], the result reduces to Eq. (14-121); the full Eq. (14-127) has the advantage of explaining the fall-off of k_{uni} at low (M) in terms of statistical mechanical quantities. The RRKM approach has been quite successful and is our best theory of unimolecular reactions.

14-ST-2   Heterogeneous catalysis. Chemisorption of gases

It was mentioned in Section 14-3C that zero-order reactions may be observed under conditions of heterogeneous catalysis; a brief presentation of this general subject seems worthwhile. Many commercially important gas-phase processes depend on heterogeneous catalysis to make the reaction proceed more cleanly to desired products or at a lower temperature than otherwise feasible. Often the equilibrium is favorable only at such a low temperature that the reaction rate is very slow, and it is only by catalytic means that the reaction can be made to occur on a practical basis.

Some examples of commercially important catalyzed reactions are the following. Iron oxide (with added aluminum and potassium oxides) catalyzes the formation of ammonia from hydrogen and nitrogen; mixed cobalt thorium oxides catalyze the Fischer–Tropsch reaction whereby hydrocarbons are produced from CO and H_2; nickel and platinum metals catalyze various hydrogenation reactions; various aluminas and other acidic oxides are widely used in the catalytic cracking of hydrocarbons, and so forth. In all of these cases the role of the catalyst is that of stabilizing some intermediate, which then takes part in reactions on the catalyst surface or with a molecule from the gas phase which collides with the surface. Hydrogenation catalysts may dissociate hydrogen into atoms to give M—H, where M denotes a surface metal atom. Thus a mechanism for the hydrogenation of ethylene is

(1) $C_2H_4 + M = H_2C$——CH_2
$\qquad\qquad\qquad\quad |\qquad\ |$
$\qquad\qquad\qquad\ \ M\quad\ M$

(2) $\qquad\qquad\qquad H_2\quad H$
$\qquad H_2 + M \rightarrow |\ \rightarrow 2\ |$
$\qquad\qquad\qquad\ M\qquad M$

H_2C——$CH_2 + H_2 \rightarrow CH_3 + H$
$\ \ |\qquad\quad |\qquad\ |\qquad\quad |\qquad\ |$
$\ \ M\qquad\ M\qquad M\qquad CH_2\quad M$
$\qquad\qquad\qquad\qquad\qquad\ |$
$\qquad\qquad\qquad\qquad\qquad M$

(3) $\qquad CH_3$
$\qquad\qquad |\qquad\ H$
$\qquad CH_2 + \ |\ \rightarrow C_2H_6$
$\qquad\qquad |\qquad\ M$
$\qquad\qquad M$

That for catalytic cracking probably involves the step

$\qquad\qquad\qquad\ R^+$
$RH + SH^+ \rightarrow\ |\quad + H_2$
$\qquad\qquad\qquad\ S$

where SH^+ denotes an acidic surface site, perhaps an Al—OH function, and R^+—S is a surface-adsorbed carbonium ion, which then fragments or undergoes surface reactions with other adsorbed carbonium ions.

Heterogeneous catalysis, then, is intimately related to adsorption processes in which the adsorbed species, or adsorbate, forms a chemical bond with the adsorbent. Such adsorption is called *chemisorption*, as contrasted to the much weaker adsorption of vapors on a surface due to secondary or van der Waals forces. This second kind of adsorption is more like a condensation process, and is called *physical adsorption* (or, sometimes, physisorption). It is necessary first to discuss chemisorption as a phenomenon before proceeding to the subject of the kinetics of heterogeneous catalysis.

A. Chemisorption. The
Langmuir equation

Chemisorption may be studied as a separate phenomenon from catalysis simply by using a single adsorbate species which is stable toward reaction. Examples would be the chemisorption of nitrogen on tungsten, of hydrogen on nickel, and of ammonia on carbon black. The adsorption process can be slow and temperature dependent, that is, increase in rate with increasing temperature, indicating an activation energy for adsorption; it may not always be possible to attain a reversible equilibrium. The better-behaved systems, however, show rapid adsorption and for each pressure of adsorbate gas there will be an equilibrium amount of adsorption. The data are conventionally plotted as an *adsorption isotherm*, that is, as amount adsorbed v (usually measured as cubic centimeters STP of gas adsorbed per gram of adsorbent) versus pressure, as shown in Fig. 14-20. The normal behavior is for v to approach a limiting value v_m with increasing pressure, and, as also shown in the figure, for the amount adsorbed to decrease with increasing temperature at constant pressure.

A very straightforward treatment of chemisorption was given by I. Langmuir in 1918. The picture is one of a surface having a certain number of adsorption sites S of which S_1 may be occupied by adsorbate and $S_0 = S - S_1$ are then bare. Adsorption equilibrium is treated as a dynamic state in which the rate of adsorption is equal to the rate of desorption. The former rate is taken to be nonactivated and just proportional to the surface collision frequency of the gaseous adsorbate on the bare sites, which, by Eq. (2-59), is proportional to the pressure P:

$$\text{rate of adsorption} = k_2 P S_0$$
$$= k_2 P(S - S_1)$$

The rate of desorption is proportional to the number of occupied sites S_1:

$$\text{rate of desorption} = k_1 S_1$$

The two rates are set equal, and on solving for S_1 we obtain

$$\frac{S_1}{S} = \theta = \frac{bP}{1 + bP} \qquad (14\text{-}128)$$

where θ is the fraction of surface covered and $b = k_2/k_1$. The limiting value v_m of Fig. 14-30

FIGURE 14-20
(a) Adsorption isotherms for ammonia on charcoal. (b) Adsorption isosteres for ammonia on charcoal. [From Vol. I, "Physical Adsorption" of Stephen Brunauer, "The Adsorption of Gases and Vapors" (copyright 1943 p 1971 by Princeton University Press). Reprinted by permission of Princeton University Press, Princeton, New Jersey.]

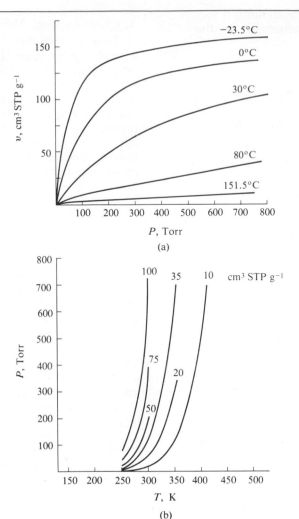

corresponds to $\theta = 1$, hence $v/v_m = \theta$, and we can write

$$v = \frac{v_m bP}{1 + bP} \tag{14-129}$$

Equations (14-128) and (14-129) are known as the Langmuir adsorption isotherm. The latter may be rearranged to the form

$$\frac{P}{v} = \frac{1}{bv_m} + \frac{P}{v_m} \tag{14-130}$$

which states that a plot of P/v versus P should be linear. Some data obeying Eq. (14-129) are plotted in this manner in Fig. 14-21.

Although the adsorption process is not con-sidered to be activated, that for desorption must be, since there will be an adsorption energy Q which the adsorbate must gain if it is to desorb; k_1 must therefore be of the form $k_1 = k_1' e^{-Q/RT}$, and the constant b consequently can be written

$$b = b_0 e^{Q/RT} \tag{14-131}$$

Equation (14-128) reduces to the form

$$\theta = bP \qquad \text{or} \qquad v = v_m bP \tag{14-132}$$

at low pressure, so that the Langmuir model requires that the adsorption isotherm be linear at low pressures; the isosteres for this region obey a Clausius–Clapeyron type of equation:

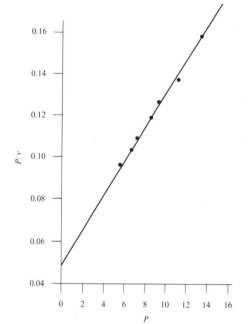

FIGURE 14-21
Adsorption of methane on mica at 90 K plotted according to Eq. (14-130), P in mm Hg and v in mm³ of methane adsorbed per gram of mica, measured at 20°C and 760 mm Hg pressure. [Data from I. Langmuir, J. Amer. Chem. Soc. **40**, 1361 (1918).]

$$\left(\frac{d \ln P}{dT}\right)_{v \text{ or } \theta} = \frac{Q}{RT^2} \qquad (14\text{-}133)$$

From its nature, Q must be a positive number, and so pressure must increase with temperature. Alternatively, at constant pressure,

$$\left(\frac{d \ln \theta}{dT}\right)_P = \frac{d \ln b}{dT} = -\frac{Q}{RT^2} \qquad (14\text{-}134)$$

Finally, if more than one species is competing for adsorption, the adsorption isotherm for some one of them is given by the Langmuir derivation as

$$\theta_i = \frac{b_i P_i}{1 + \Sigma b_j P_j} \qquad (14\text{-}135)$$

where the summation is over all species.

Relatively few chemisorption systems obey the simple Langmuir model really well; the data may not fit the Langmuir equation, especially at the extremes, or if they do, the variation of the b parameter with temperature may fail to obey the simple exponential law. The basic model seems sound, however, and such difficulties can often be attributed to nonuniformity of the surface so that Q varies with degree of surface occupancy. The adsorbate adsorbs first on the more active sites, with highest Q, and then on progressively less active ones.

Hydrogen appears to be present on the surface as individual atoms, so that the desorption process requires a recombination of two surface atoms, and is therefore bimolecular. Analysis leads to an isotherm of the form

$$\theta = \frac{bP^{1/2}}{1 + bP^{1/2}} \qquad (14\text{-}136)$$

The derivation assumes that the adsorbing gas requires two adjacent sites, which has the effect of making the rate of adsorption proportional to S_0^2.

B. Kinetics of heterogeneous catalysis

A number of simple catalytic systems can be reasonably well treated on the basis that the reactants adsorb on the catalyst surface, to react according to a mass action type of rate law.

Suppose that the reaction is $A \rightarrow C + D$ and that the surface reaction proceeds according to the rate law $dn_A/dt = -k\theta_A S$, where $\theta_A S$ is the surface covered by adsorbed A, or, by Eq. (14-135),

$$\frac{dn_A}{dt} = -kS \frac{b_A P_A}{1 + b_A P_A + b_C P_C + b_D P_D}$$ (14-137)

If the products C and D are weakly adsorbed, then Eq. (14-137) reduces to

$$\frac{dn_A}{dt} = -kS \frac{b_A P_A}{1 + b_A P_A}$$

which means that at low P_A the rate should be proportional to P_A, while at high P_A the rate levels off at the maximum value kS, and hence is independent of P_A. This last situation is then one of a zero-order reaction. Behavior of this type has been observed for the decomposition of HI on gold and platinum and of N_2O on indium sesquioxide.

If the reaction is of the type $A + B \rightarrow C + D$, and A and the products are weakly adsorbed but B is strongly adsorbed, then

$$\frac{dn_A}{dt} = -kS \frac{b_A P_A}{b_B P_B} = -\frac{k_{app} P_A}{P_B}$$ (14-138)

It is thus possible for a reactant to inhibit a reaction.

We return to Eq. (14-137) to make a final observation. At low P_A, the equation becomes

$$\frac{dn_A}{dt} = -kSb_A P_A = -k_{app} P_A$$ (14-139)

The reaction is first order, and its temperature dependence could be written as

$$\frac{d(\ln k_{app})}{dT} = \frac{E^*_{app}}{RT^2}$$ (14-140)

However, k_{app} contains the product of b_A and the true rate constant k and by Eq. (14-97), it follows that

$$E^*_{app} = E^*_{true} - Q_A$$ (14-141)

Thus the apparent activation energy is less than the true one by the heat of adsorption. There is not only this effect, of course, but the true activation energy may be less than that for the bulk-phase reaction. Thus the true activation energy for the tungsten-catalyzed decomposition of ammonia is only 39 kcal mol^{-1} as compared to the value of about 90 kcal mol^{-1} for the gas-phase reaction. One reason a catalyst is effective, then, is because of a reduction in the activation energy for reaction.

GENERAL REFERENCES

ASHMORE, P.G., ED. (1975). "Reaction Kinetics," Vol. 1. The Chemical Society, London.

BENSON, S.W. (1960). "The Foundations of Chemical Kinetics." McGraw-Hill, New York.

BENSON, S.W. (1968). "Thermochemical Kinetics: Methods for Estimation of Thermochemical Data and Rate Parameters." Wiley, New York.

BERRY, R.S., RICE, S.A., AND ROSS, J. (1980). "Physical Chemistry." Wiley, New York.

ESPENSON, J.H. (1981). "Chemical Kinetics and Reaction Mechanisms." McGraw-Hill, New York.

GARDINER, W.C., JR. (1969). "Rates and Mechanisms of Chemical Reactions." Benjamin, New York.

MOORE, J.W., AND PEARSON, R.G. (1981). "Kinetics and Mechanism," 3rd ed. Wiley, New York.

NICHOLAS, J. (1976). "Chemical Kinetics." Wiley, New York.

ROBINSON, P.J., AND HOLBROOK, K.A. (1972). "Unimolecular Reactions." Wiley (Interscience), New York.

WESTON, R.E., JR., AND SCHWARZ, H.A. (1972). "Chemical Kinetics." Prentice-Hall, Englewood Cliffs, New Jersey.

CITED REFERENCES

BENDER, C.F., O'NEILL, S.V., PEARSON, P.K., AND SCHAEFER, H.F. (1972). *Science* **176**, 1412.

BENSON, S. (1968). "Thermochemical Kinetics." Wiley, New York.

ESPENSON, J.H. (1981). "Chemical Kinetics and Reaction Mechanisms." McGraw-Hill, New York.

HARCOURT, A., AND ESSEN, W. (1865). *Proc. Roy. Soc. (London)* **14**, 470.

HARCOURT, A., AND ESSEN, W. (1866). *Phil. Trans. Roy. Soc. (London)* **156**, 193.

HARCOURT, A., AND ESSEN, W. (1867). *Phil. Trans. Roy. Soc. (London)* **157**, 117.

HERSCHBACH, D., *et al.* (1956). *J. Chem. Phys.* **25**, 737.

KUNSMAN, C.H. (1928). *J. Amer. Chem. Soc.* **50**, 2100.

MCDANIEL, E.W., CERMAK, V., FERGUSON, E.E., AND FRIEDMAN, L. (1970). "Ion-Molecule Reactions." Wiley-Interscience, New York.

SULLIVAN, (1967). *J. Chem. Phys.* **46, 73**.

EXERCISES

Take as exact numbers given to one significant figure.

Calculate k, τ, $t_{1/2}$, and $t_{1/3}$ for a first-order reaction for which $t_{1/5} = 20$ min. The coefficient a is unity. **14-1**

Ans. 8.05×10^{-2} min^{-1}, 12.43 min, 8.61 min, 13.65 min.

A reaction of stoichiometry $2A + B \rightarrow$ products is first order in (A) and in (B), and gives a linear plot of $\ln P_A$ versus time, with P_A in mm Hg. The intercept and slope are 0 and -1 hr^{-1}, respectively, and $P_B^0 = 100$ mm Hg. Calculate k_{app} and k. **14-2**

Ans. 1/2, 0.005 hr^{-1} mm Hg^{-1}.

The initial rate of the reaction $2A \rightarrow$ products is 0.5% per min. (a) Calculate $t_{1/2}$, assuming that the reaction is second order and that P_A^0 is 10 Torr. (b) The reaction would be first order if, say, the mechanism were (1) $A + B \rightarrow C$ (slow) followed by $A + C \rightarrow B$ (fast). Calculate $t_{1/2}$ for this case. **14-3**

Ans. (a) 200 min, (b) 139 min.

The reaction $A \rightarrow$ products gives a straight plot of $1/(A)$ versus t, of intercept 100 atm^{-1} and slope 3×10^{-4} s^{-1} atm^{-1}. Calculate $t_{1/2}$. **14-4**

Ans. 3.33×10^5 s.

A gas phase reaction has the stoichiometry $A + \frac{3}{2}B \rightarrow$ products and obeys the rate law $d(A)/dt = -k(A)(B)$ with $k = 3 \times 10^{-4}$ liter mol^{-1} s^{-1} at 25°C. A reaction mixture initially at 25°C initially contains 10% A and 90% B at a total pressure of 1.5 atm. Calculate the percentages of A and of B that have reacted after 1 hr. **14-5**

Ans. 5.76% A, 0.959% B.

Calculate the half-life for the reaction of Exercise 14-5 if (a) $(A)_0 = 0.005$ M and $(B)_0 = 0.5$ M, (b) $(A)_0 = 0.5$ M and $(B)_0 = 0.005$ M, and (c) $(A)_0 = 0.02$ M and $(B)_0 = 0.03$ M. We define half-life as the time for half of that reactant not in excess to have reacted. Also, assume in cases (a) and (b) that the other reactant is sufficiently in excess that the reaction is pseudo-first-order. **14-6**

Ans. (a) 4620 s, (b) 3080 s, (c) 1.11×10^5 s.

Suppose that A and B are volatile solids and that the reaction of Exercise 14-5 is carried out in a five-liter flask at 25°C in the presence of excess solid A and B. If the vapor pressures of A and B are 0.1 atm and 0.02 atm, respectively, how long should it take for 0.5 mole of A to be converted to products? **14-7**

Ans. 9.98×10^7 s.

The reaction $H_2 + D_2 \rightarrow 2HD$ was studied at 1000 K using equimolar mixtures. For initial total pressures of 4 and 8 mm Hg, the time for half reaction was 200 s and 140 s, respectively. [See A. Farkas and L. Farkas, *Proc. Roy. Soc.*, **A152,** 124, 152 (1935).] Show what the order of the reaction is. **14-8**

Ans. 3/2 order.

The rate law for the reaction $2NO + 2H_2 \rightarrow N_2 + 2H_2O$ was determined by the method of initial rates. If the initial NO pressure was 400 mm Hg, dP/dt was 1.35 mm Hg s^{-1} and 0.92 mm Hg s^{-1} for initial H_2 pressures of 290 mm Hg and 200 mm Hg, respectively. If the initial H_2 pressure was 400 mm Hg, then dP/dt was 1.50 mm Hg s^{-1} and 1.04 mm Hg s^{-1} for initial NO pressures of 360 mm Hg and 300 mm Hg, respectively. Show what a and b are in the rate law: $-dP_{NO}/dt = kP_{NO}^a P_{H_2}^b$, and calculate k. The rate dP/dt is the decrease in total pressure per unit time. **14-9**

Ans. $a = 2$, $b = 1$; $k = 5.78 \times 10^{-8}$ mm Hg^{-2} s^{-1}.

Under certain conditions the reaction $H_2 + Br_2 = 2HBr$ obeys a rate law of the form $d(HBr)/dt = k(H_2)^a(Br_2)^b(HBr)^c$. At a certain temperature T the reaction rate is found to be R when (HBr) is 2 M and (H_2) and (Br_2) are both 0.1 M. The rate varies with concentration as follows: **14-10**

	Concentration (M)		
(H_2)	(Br_2)	(HBr)	Rate
0.1	0.1	2	R
0.1	0.4	2	$8R$
0.2	0.4	2	$16R$
0.1	0.2	3	$1.88R$

Find the exponents a, b, and c.

Ans. $1, \frac{3}{2}, -1$.

14-11 Consider the reaction $A_2 + B_2 \rightarrow 2AB$. Explain what the rate law for each of the following mechanisms should be: (a) $A_2 \xrightarrow{k_1} 2A$ (slow step), $B_2 \overset{K_2}{\rightleftharpoons} 2B$ (rapid equilibrium highly displaced toward B_2) and $A + B \xrightarrow{k_3} AB$ (fast). (b) $A_2 \overset{K_1}{\underset{k_3}{\rightleftharpoons}} 2A$, $B_2 \overset{K_2}{\rightleftharpoons} 2B$ (both rapid equilibria, highly displaced toward A_2 and B_2, respectively), $A + B \xrightarrow{k_3} A_2B_2$ (slow step). (c) $A_2 + B_2 \xrightarrow{k_1} A_2B_2$ (slow step), $A_2B_2 \xrightarrow{k_2} 2AB$ (fast).

Ans. (a) $R = k_1(A_2)$; (b) $R = k_{app}(A_2)^{1/2}(B_2)^{1/2}$, where $k_{app} = K_1^{1/2}K_2^{1/2}k_3$; (c) $R = k_1(A_2)(B_2)$.

14-12 Derive the steady-state rate law for the mechanism

$$NO + H_2 \overset{k_1}{\underset{k_{-1}}{\rightleftharpoons}} NO \cdot H_2$$

$$NO \cdot H_2 + NO \xrightarrow{k_2} N_2O + H_2O.$$

Show under what condition this reduces to the form of Eq. (14-44).

Ans. $R = k_1 k_2 (NO)^2 (H_2)/[k_{-1} + k_2(NO)]$.

14-13 Verify the entry in Table 14-4 for $t = 0.1$ min.

14-14 The pyrolysis of ethyl bromide obeys first order kinetics with rate constants of 0.141 s^{-1} and 0.662 s^{-1} at 560°C and 604°C, respectively. Calculate A and E^* of Eq. (14-75) and A' and E^* of Eq. (14-78).

Ans. $A = 3.42 \times 10^{-2}$ s^{-1}, $E^*/R = 25,670$; $A' = 7.13 \times 10^{10}$ s^{-1}, $E^*/R = 25,250$.

14-15 Calculate the rate constant for the reaction $2NO_2 \rightarrow 2NO + O_2$ at 100°C.

Ans. 1.34×10^{-6} liter mol^{-1} s^{-1}.

14-16 The collision cross-section for a reaction is defined as the effective value of $\pi\sigma_{12}^2$. Calculate the collision cross-section for the reaction of Exercise 14-15.

Ans. 2.89 Å^2.

14-17 Rate constants for dilute gas reactions, such as ion molecule reactions, are often expressed as k_m in units of cm^3 molecule^{-1} s^{-1}. If k_m is 5×10^{-10} cm^3 molecule^{-1} s^{-1} for the reaction $Kr^+ + H_2 \rightarrow KrH^+ + H$, calculate k. Also calculate σ_{12} assuming $E^* = 0$ and 25°C.

Ans. $k = 3.01 \times 10^{11}$ liter mol^{-1} s^{-1}, $\sigma_{12} = 2.98$ Å.

14-18 The reaction cyclopropane $\rightarrow$ propylene conforms (approximately) to the simple Lindemann mechanism. The limiting rate constant has the parameters $\log A = 15.17$ (A in s^{-1}) and $E^* = 65$ kcal mol^{-1} and at 765 K, $k_{uni} = 10^{-4}$ s^{-1} at a pressure of 1 Torr. Calculate $k_{uni,\infty}$ at 765°C and the ratio k_1/k_{-2}.

Ans. 4.0×10^{-4} s^{-1}, 3.0 Torr.

14-19 The frequency factor A' is 7.334×10^{12} s^{-1} for the first order decomposition of ethyl iodide into HI and ethylene, and the activation energy is 221 kJ. Calculate (a) the rate constant at 300°C and (b) the value of $\Delta S^{0\ddagger}$ (assume 25°C).

Ans. (a) 1.25×10^{-6} s^{-1}, (b) 25.1 J mol^{-1} K^{-1}.

The entropy of activation is -78 kJ for the reaction $NO + Cl_2 \rightarrow NOCl + Cl$. Calculate the translational contribution to this entropy and, from this, the contribution from rotation and vibration. Assume 25°C.

14-20

Ans. $\Delta S_{tr}^{0\ddagger} = -120.2$ J mol^{-1} K^{-1}, $\Delta S_{v,r}^{0\ddagger} = 42.2$ J mol^{-1} K^{-1}.

PROBLEMS

Derive the expression for the half-life of a reaction obeying the rate law of Eq. (14-32). Explain whether your result is valid for all n values.

14-21

The half-life of a certain reaction is 10 min, 40 min, and 160 min for initial pressures of 200 mm Hg, 100 mm Hg, and 50 mm Hg, respectively. The reaction is of the type $aA \rightarrow$ products. Show what the order of the reaction is and calculate k.

14-22

The half-life for the decomposition of NH_3 as catalyzed by a hot tungsten filament (at about 1100°C) is found to vary with the initial NH_3 pressure as follows:

14-23

P(Torr)	300	100	30
$t_{1/2}$(min)	8.4	2.8	0.84

[see Kunsman (1928)]. Show what the order of the reaction is (in P_{NH_3}) and calculate the rate constant.

The decomposition of $PH_3(g)$ into $P_4(g)$ and $H_2(g)$ is studied at a certain temperature. At $t = 0$ one has 260 mm Hg of PH_3 and 500 mm Hg of inert gas; at 1 min the total pressure is 838 mm Hg and after 2 min the total pressure is 885 mm Hg. Assume that the reaction goes to completion. Show what the reaction order is and calculate k.

14-24

The rate of the reaction $2C_4H_6 \rightarrow$ cyclo-C_8H_{16} is followed at 300°C. A certain initial pressure is introduced into a flask at this temperature and after 1.94 hr, 4.16 hr, and 8.33 hr, the pressure is 0.1265 atm, 0.1073 atm, and 0.0836 atm, respectively. Find the order of the reaction and obtain k and the initial pressure.

14-25

The decomposition $ClCOOCCl_3 \rightarrow 2COCl_2$ goes to completion, and the total pressure is found to vary with time (at 280°C) as follows:

14-26

t(sec)	0	500	800	1300	1800
P(Torr)	15.0	18.9	20.7	23.0	24.8

Show what the order of the reaction is and calculate the rate constant.

A family buys a car for $8,500, paying $2500 down, so that p^0, the initial principal on their loan, is $6000. Let m be the monthly payment and i the interest rate in % per month. Derive an equation relating p, m, i, and n, where n is the number of payments to pay off the loan. What will the monthly payment be if the loan is to be paid off in three years, and the interest rate is 13% per year? What will the total amount of money paid over the three-year period? Note: this is essentially a first order rate problem, with principal p taking the place of reactant A; p decreases at the rate money is paid back and increases at the interest rate.

14-27

Integrated rate expressions can look quite different for a reaction at constant pressure. Show that for the first order reaction $A \rightarrow \nu B$, $(A)_0/(A) = \nu e^{kt} + (1 - \nu)$, assuming ideal gas behavior. Thus for $\nu = 2$, we have $(A)_0/(A) = 2e^{kt} - 1$ instead of the usual equation for constant volume conditions, $(A)/(A)_0 = e^{-kt}$. Make a plot of $(A)/(A)_0$ versus $t/t_{1/2}$ for the case of $\nu = 2$, for constant pressure and for constant volume conditions. Hint: $n_A = n_A^0 e^{-kt}$ for a first order reaction under either condition; also, at constant pressure, $v = v_0[1 + (\nu - 1)\xi]$.

14-28

Derive the integrated rate law for the autocatalytic reaction $A \rightarrow C$, where $-d(A)/dt = k(A)(C)$. An initial concentration of catalyst, $(C)_0$, is present, either deliberately or adventitiously. Construct a plot of $(x)/(A)_0$ versus kt, where (x) is the concentration of A reacted, for the case of $(C)_0 = 0.01(A)_0$.

14-29

14-30* Show that Eq. (14-28) can be put in the form

$$\ln\left[\frac{pF_A}{(p-1)+F_A}\right] = \frac{1-p}{p}(B)_0kt$$

where $F_A = (A)/(A)_0$ and $p = (B)_0/(A)_0$. Make plots of F_A versus $(B)_0kt$ for $p = 0.1, 0.5, 1.0, 2.0$, and ∞.

14-31 The following hypothetical rate data are found for the reaction $3H_2 + N_2 \rightarrow 2NH_3$ at 450°C.

	Initial Pressures (Torr)		Initial rate, $-dP_{tot}/dt$ (Torr hr^{-1})
Experiment	$P^0_{H_2}$	$P^0_{N_2}$	
1	100	1	0.01
2	200	1	0.04
3	400	0.5	0.08

(a) The rate is of the form Rate $= kP^x_{H_2}P^y_{N_3}$. Show what the values of x and y are.
(b) Calculate the rate constant. (c) Calculate the initial rate for the conditions of Experiment 1, but for 500°C, taking the activation energy to be 45 kcal mol^{-1}.

14-32 The reaction $A + B \rightarrow$ products obeys the rate law $d(A)/dt = -k(A)(B)$. In Experiment 1 the initial concentrations are 0.001 M and 2 M for A and B, respectively, and in Experiment 2 the initial concentrations are both C_0. Find the value for C_0 such that the time for A to be 90% reacted is the same for the two experiments.

14-33 Rework the Example following Eq. (14-29), assuming the reaction stoichiometry to be $2A + B \rightarrow$ products, using the same value of k.

14-34 The following rate data are obtained for $CO(g) + Cl_2(g) \rightarrow COCl_2(g)$, which goes to completion at the temperature in question. In experiment (1), $P^0_{Cl_2} = 300$ mm Hg and $P^0_{CO} = 3$ mm Hg, $P^0_{COCl_2} = 0$. After 2500 s and 5000 s, 1 mm Hg and 2.56 mm Hg of phosgene have formed, respectively. In experiment (2), $P^0_{Cl_2} = 900$ mm Hg and $P^0_{CO} = 5$ mm Hg, $P^0_{COCl_2} = 0$; after 822 s, $P_{COCl_2} = 2.5$ mm Hg. Find the reaction order with respect to P_{CO} and P_{Cl_2}, and calculate the rate constant.

14-35 The pyrolysis of cyclobutanone shows two parallel reactions, (1) to ethylene and CH_2O and (2) to cyclopropane and CO. Both are first order and in the high pressure limit the Arrhenius parameters are $A_1 = 5.0 \times 10^{14}$ s^{-1}, $A_2 = 3.4 \times 10^{14}$ s^{-1}, and $E^*_1 = 220$ kJ mol^{-1} and $E^*_2 = 245$ kJ mol^{-1}. The reaction was studied by flowing cyclobutanone (with inert carrier gas) through a high temperature chamber, the product ratio being determined by analysis of the quenched exit gases. (a) Calculate the ratio R of ethylene to cyclopropane expected at 900 K, (b) the effective dwell time in the high temperature chamber if 5% decomposition of the cyclobutanone occurred. (c) Is there a temperature at which the product ratio should be unity? [Braun et al., J. Phys. Chem. **88**, 1046 (1983).]

14-36 One mechanism for the reaction $H_2 + I_2 \rightarrow 2HI$ consists of reactions (14-83), (14-85), and (14-86). Derive the steady-state rate law for this mechanism. How, experimentally, might it be possible to distinguish this mechanism from that of reactions (14-83) plus (14-84)?

14-37 The pyrolysis of ethyl bromide to give ethylene and HBr follows first order kinetics. The following data are obtained:

T, K	800	833	877	900
k, s^{-1}	0.0361	0.141	0.662	1.410

Calculate A and E^*, and $\Delta S^{0\ddagger}$ and $\Delta H^{0\ddagger}$ from a suitable plotting of the data.

14-38 The isomerization of cyclopropane, A, follows Lindemann kinetics, that is, Eq. (14-97). Calculate the ratio of the time for 1% of the cyclopropane to isomerize when the initial pressure is 0.1 atm to the time for 1% to isomerize when the initial pressure is 1×10^{-5} atm. Use $k_1 = 1 \times 10^{11}$ min^{-1}, $k_2 = 1 \times 10^4$ atm^{-1} min^{-1}, and $k_{-2} = 1 \times 10^{14}$ atm^{-1} min^{-1}.

14-39 The rate law for the reaction A + B → products is $d(A)/dt = -k(A)^2(B)$ with $k = 1 \times 10^{-3}$ M^{-2} s^{-1}. Calculate the time for half of B to react (a) for the case of $(A)_0 = (B)_0 = 0.2$ M and (b) $(A)_0 = 0.5$ M and $(A)_0 \gg (B)_0$.

14-40 Equations (14-58) and (14-59) simplify if k_2 is much smaller than k_1 or k_{-1}. Show what this simpler form is.

14-41* Askey and Hinshelwood obtained the following data for the pyrolysis of dimethyl ether at 777 K:

Initial concentration ($M \times 10^3$)	0.58	1.20	1.89	3.12	5.01
$k_{uni} \times 10^4$ (s^{-1})	1.88	2.48	3.26	4.12	5.57
Initial concentration ($M \times 10^3$)	6.48	8.18	13.21	16.61	18.57
$k_{uni} \times 10^4$ (s^{-1})	5.58	6.29	6.90	6.64	7.45

The system was believed to conform to the general Lindemann mechanism. Plot the data in a linear form so as to obtain values for k_2 and k_{-2}/k_1 from the best least-squares slope and intercept. The observed activation energy for $k_{uni,\infty}$ is 65.5 kcal mol^{-1}. Estimate the value of s in Eq. (14-111) that will account for the experimental $k_1 K_2$.

14-42 R. Ogg proposed the following mechanism for the reaction $N_2O_5 \rightarrow N_2O_4 + \frac{1}{2}O_2$:

$$N_2O_5 \underset{k_{-1}}{\overset{k_1}{\rightleftharpoons}} NO_2 + NO_3$$

$$NO_2 + NO_3 \overset{k_2}{\rightarrow} NO + O_2 + NO_2$$

$$NO + NO_3 \overset{k_3}{\rightarrow} N_2O_4$$

Derive the stationary-state rate law for this mechanism. (Both NO_3 and NO are treated as stationary-state intermediates.)

14-43 Estimate the value of $k_1 K_2$ and k_{-2}/k_1 for the isomerization of cyclopropane (using Fig. 14-13) and also the best value of s in Eq. (14-111).

14-44 At one time the $H_2 + I_2 \rightarrow 2HI$ reaction was thought to be a simple one, the probable transition state being

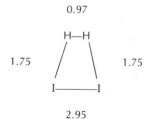

(distances in angstroms). Calculate the value of $\Delta S^{0\ddagger}$ at 575 K assuming that the three moments of inertia are 921.5, 6.9, and 928.4 (in units of 10^{-40} g cm^2) [note that $\sigma = 2$; see Eq. (4-102)], and that the vibrational frequencies are 994, 86, 1280, 1400, and 1730 (all in cm^{-1}).

14-45 The reaction $2NO + O_2 \rightarrow 2NO_2$ obeys the rate law $d(NO_2)/dt = k(NO)^2(O_2)$. Write a mechanism that will give this rate law and which avoids termolecular reactions.

14-46 The thermal decomposition of pure ozone to give O_2 obeys the rate law $d(O_2)/dt = k(O_3)^2/(O_2)$. Write a mechanism that will give this rate law.

14-47 The reaction $NO + O_3 \rightarrow NO_2 + O_2$ is second order, the rate constants showing the following temperature dependence:

T, K	203.8	222.4	272.2	307.2
k, cm^3 molec^{-1} s^{-1}	2.67	4.17	11.5	20.9

(all k values are $\times 10^{-15}$). Find A and E^*, and $\Delta S^{0\ddagger}$ and $\Delta H^{0\ddagger}$ assuming simple Arrhenius behavior. Optional: The Arrhenius plot shows some curvature, and a better fit is obtained if one takes $k = A'T^n e^{-E^*/RT}$. Find the best value for n. [R.A. Borders and J.W. Kirks, *J. Phys. Chem.* **86**, 3295 (1982).]

14-48   The following data are obtained for $k_{uni,\infty}$ for the pyrolysis of propane (to give C_2H_5 and CH_3):

T, K	900	1000	1100	1200
$k_{uni,\infty}$, s^{-1}	9.94×10^{-5}	0.0114	0.554	14.1

Find A and E^*, and $\Delta S^{0\ddagger}$ and $\Delta H^{0\ddagger}$.

14-49   Derive the rate law corresponding to the mechanism of Eq. (14-45).

14-50   Verify the integration of Eq. (14-113).

14-51   Referring to Fig. 14-7, explain to what degree the segments *ab, bc,* and *cd* should be sensitive to the amount and nature of surface (as of the reaction vessel) present.

14-52 E^* is defined by Eq. (14-75), but if k is actually given by Eq. (14-89) there is an additional temperature dependence, and $d(\ln k)/dt$ will be equal to E_a/RT^2. Derive the relationship between E^* and E_a.

SPECIAL TOPICS PROBLEMS

14-53   If the transition state for a reaction is one in which some bond M—X is stretched essentially to the point of dissociation, then the dissociation energy is regarded as making a direct contribution to the activation energy. Consider the case where M is a heavy atom and X is hydrogen or deuterium. The bond dissociation energy is less for the M—H case than for the M—D one since this energy is the difference between the potential energy of the separated atoms and the zero-point energy of the M—X vibration. This last is smaller for M—D than for M—H [note Eq. (4-93)]. Assuming the mass of M to be effectively infinite and the M—H vibrational frequency to be 3300 cm^{-1}, calculate the zero-point energies for M—H and M—D, the difference in their dissociation energies, and the resulting difference in the rate constant for the reaction. The effect is known as the *primary isotope effect.*

14-54   For the case of the reaction A + B → AB, where A and B are atoms, Eq. (14-121) reduces to

$$k = \left[\frac{8\pi(m_A + m_B)\mathbf{k}T}{m_A m_B}\right]^{1/2} r_{AB}^2 \exp\left(-\frac{\Delta E_0^{\ddagger}}{RT}\right) \; cm^3 \; molecule^{-1} s^{-1}$$

where r_{AB} is the A—B bond distance. Derive this result; also, compare the equation with the corresponding one from collision theory and comment on the similarities.

14-55   Sketch the energy versus reaction coordinate plot suggested by Fig. 14-19, that is, the plot of minimum energy versus r_{MB}. Assume that the abscissa scale is from 0 to 2 Å and that the spacing of the energy contours is 10 kcal mol^{-1}.

14-56   The Pt-catalyzed decomposition of NO (into N_2 and O_2) is found to obey the experimental rate law $dP_{NO}/dt = -kP_{NO}/P_{O_2}$. Assuming adsorbed gases obey the Langmuir equation, derive this rate law starting with some reasonable assumed mechanism for the surface reaction.

   If the heat of adsorption of NO is 20 kcal mol^{-1} and that of O_2 is 25 kcal mol^{-1}, show what the actual activation energy for the surface reaction should be, given that the apparent activation energy is 15 kcal mol^{-1} (as found from the temperature variation of k).

14-57   An adsorbent has a specific surface area of 50 m^2 g^{-1}, and the actual area occupied by either an adsorbed A or an adsorbed B molecule is 25 Å^2. In one experiment, it is found that the adsorption of gas A reaches 1/3 of its maximum amount at a pressure of 5 torr, while in a separate experiment, the adsorption of gas B reaches 1/2 of its maximum amount at 10 torr (at 50°C). (a) Calculate v_m in cm^3 STP g^{-1}. (b) Calculate b for gas A and for gas B. (c) Calculate θ_A and θ_B if adsorption occurs at 50°C from a mixture of the two gases, each at 10 torr pressure.

The following data are obtained for the chemisorption of CO on powdered Ni: **14-58**

P, torr	0.1	0.3	1	3
v, cm^3 STP g^{-1}	0.95	2.61	6.67	12.0

Obtain b and v_m, assuming that the Langmuir equation is obeyed.

Obtain from Fig. 14-21 the values of v_m, b, and the surface area per gram of mica, assuming that the   **14-59**
molecular area of methane is 18 $Å^2$.

chapter 15
Kinetics of reactions in solution

Much of the material of Chapter 14 is needed as a foundation for this chapter. Particularly important are the sections on reaction rate laws, the relationship between reaction mechanism and rate law, and the collision and transition-state theories. We proceed to consider some aspects of rate laws that were omitted before and which are often encountered in solution kinetics, as well as experimental approaches that are now more relevant. Collision and transition-state theories are discussed again in terms of their special applications to solutions, and the main portion of the chapter concludes with a discussion of a number of types of reaction mechanisms.

The flavor of this chapter differs noticeably from that of Chapter 14. There is less emphasis on detailed, quantitative theories and more on reaction mechanisms. One reason lies in the far greater variety of reactions in solution than in the gas phase. There are perhaps 30 important types of gas-phase reactions as compared to 1000 or more in solution. Gas-phase reactions are restricted to small (volatile) molecules and ions. Solution studies extend to protein and other macromolecular species. Reactions involving ions are a specialty in the gas phase, but dominate large areas of kinetics in solution.

An important and recurring theme, again special to this chapter, will be that of the role of solvent. The solvent affects the manner in which reacting molecules can come together and, further, it not only may specifically alter the structure or other properties of the reactants but often is itself a reactant. In fact, one of the difficulties of solution kinetics is the distinction between solvent as a medium and solvent as a direct participant in a rate-controlling step. Mechanistic schemes involving solvent are for this reason often more difficult to establish and more subject to controversy than those for gas-phase reactions, hence the relatively high degree of preoccupation with them by solution kineticists.

15-1   Additional comments on rate laws. Reversible reactions

It will be recalled that in Section 14-3 a detailed presentation is made of the mathematical behavior of first- and second-order rate laws. Some minor additional types of rate laws are mentioned, and the stationary-state hypothesis is developed in detail.

All of these treatments of Chapter 14 assume that the reaction goes to completion. It is entirely possible, however, for the equilibrium constant to be such that equilibrium is reached before the reactants are entirely consumed. In effect this means that the back reaction must be included in the rate equation. Although this situation can, of course, occur with either gas- or solution-phase reactions, it is perhaps more often encountered in the latter case; its consideration is therefore appropriate at this point.

A. Reversible first-order reaction

Consider the reaction

$$A + \text{other reactants} \underset{k_2}{\overset{k_1}{\rightleftharpoons}} B + \text{other products} \tag{15-1}$$

where we suppose either that A is actually the only reactant and B the only product or that the concentrations of other reactants and products are kept constant. In either case, we take the consequence to be that the forward rate is first order in A and the reverse rate is first order in B. [The more general form of Eq. (15-1) would have a moles of A going to b moles of B, but we assume as a simplification that $a = b = 1$.] If, as in Section 14-3A, only the forward reaction is considered, then the rate equation is

$$R_f = -\frac{d(A)}{dt} = k_1(A) \qquad [\text{Eq. (14-8)}]$$

and, on integration, we obtain

$$(A) = (A)_0\, e^{-k_1 t} \qquad [\text{Eq. (14-11)}]$$

where $(A)_0$ is the initial concentration, with the half-life $t_{1/2}$ given by

$$k_1 t_{1/2} = 0.693 \qquad [\text{Eq. (14-13)}]$$

The reaction thus goes to completion; (A) approaches zero as $t \to \infty$.

We next allow the reverse reaction to be significant, so that if one starts with no B present, R_f is at first large but decreases as A is consumed, and R_b, zero initially, gradually increases as B is formed. The net rate of reaction at any time is thus

$$\frac{d(A)}{dt} = -k_1(A) + k_2(B) \tag{15-2}$$

or, since $(B) = (A)_0 - (A)$,

$$\frac{d(A)}{dt} = -k_1(A) + k_2[(A)_0 - (A)]$$

Separation of the variables leads to

$$\frac{d(A)}{k_2(A)_0 - (k_1 + k_2)(A)} = dt$$

which gives, on integration,

$$-\frac{1}{k_1 + k_2} \ln[k_2(A)_0 - (k_1 + k_2)(A)] = t + \text{constant}$$

On setting $(A) = (A)_0$ at $t = 0$, we obtain

$$A = \frac{(A)_0}{k_1 + k_2} (k_2 + k_1 e^{-kt}) \tag{15-3}$$

where $k = k_1 + k_2$. Equation (15-3) may be put in a more elegant form as follows. At $t = \infty$ we have equilibrium, so that $d(A)/dt = 0$, and Eq. (15-2) gives

$$K = \frac{k_1}{k_2} = \frac{(B)_\infty}{(A)_\infty}$$

or, since $(B)_\infty = (A)_0 - (A)_\infty$, rearrangement yields

$$\frac{(A)_\infty}{(A)_0} = \frac{1}{1 + K} = \frac{k_2}{k_1 + k_2} \tag{15-4}$$

where K is the mass action law equilibrium constant. Since $(A)_0 k_2/(k_1 + k_2) = (A)_\infty$ and $(A)_0 k_1/(k_1 + k_2) = (A)_0 \{1 - [k_2/(k_1 + k_2)]\} = (A)_0 - (A)_\infty$, Eq. (15-3) becomes

$$\frac{(A) - (A)_\infty}{(A)_0 - (A)_\infty} = e^{-kt} \tag{15-5}$$

It follows from Eq. (14-40) that if $\mathscr{P}$ is any additive property, then

$$\frac{\mathscr{P} - \mathscr{P}_\infty}{\mathscr{P}_0 - \mathscr{P}_\infty} = e^{-kt} \tag{15-6}$$

Two alternative forms of Eq. (15-5) are

$$\frac{(A)}{(A)_\infty} = 1 + Ke^{-kt} \tag{15-7}$$

and

$$\frac{(A)}{(A)_0} = \frac{1}{1 + K} + \frac{K}{1 + K} e^{-kt} \tag{15-8}$$

or

$$\frac{(A)}{(A)_0} = \frac{1}{1 + K} + \frac{K}{1 + K} e^{-[(1 + K)/K]k_1 t} \tag{15-9}$$

Equation (15-5) makes clear the point that the rate law will still be first order in nature, provided that the quantity $(A) - (A)_\infty$ is used rather than just (A). Thus a plot of $\ln[(A) - (A)_\infty]$ versus t will be a straight line of slope $-k$. The usual half-life relationships are obeyed, with $kt_{1/2} = 0.693$. Notice, however, that the observed rate constant is k and not k_1. The experimental rate data thus give $k = k_1 + k_2$, and $(A)_\infty/(A)_0 = k_2/(k_1 + k_2)$, and these two pieces of information allow k_1 and k_2 to be calculated separately. If the reaction goes to completion, so that $(A)_\infty = 0$ and $k_2 \ll k_1$, Eq. (15-5) reduces to Eq. (14-10).

EXAMPLE   The reaction

$$CH_3Br + I^- \underset{k_2}{\overset{k_1}{\rightleftharpoons}} CH_3I + Br^-$$

is followed in a solvent in which KI and KBr are not very soluble. By having excess solid KI and KBr present, the I^- and Br^- concentrations are fixed and the forward and reverse reactions become pseudo-first-order; k_1 and k_2 are these first-order rate constants. In a particular experiment there is initially 100% CH_3Br; the percentage drops to 90, 71.7, and 40 after 10 min, 35 min, and at equilibrium, respectively. Find k_1 and k_2.

We can verify that the reaction is a reversible first-order one by calculating $k\ (= k_1 + k_2)$ for the 10-min and 35-min points. Thus $k = -\ln[(90 - 40)/(100 - 40)]/10 = 1.82 \times 10^{-2}$ min^{-1} and $k = -\ln[(71.7 - 40)/(100 - 40)]/35 = 1.82 \times 10^{-2}$ min^{-1}. The constancy of k shows that Eq. (15-5) is obeyed. From the equilibrium datum, $K = (CH_3I)/(CH_3Br) = 60/40 = 1.5$. Since $K = k_1/k_2$, we find $k_2 = 1.82 \times 10^{-2}/2.5 = 0.728 \times 10^{-3}$ min^{-1} and $k_1 = 1.5k_2 = 1.09 \times 10^{-3}$ min^{-1}.

The set of curves shown in Fig. 15-1 is calculated from the alternative equation (15-9), which brings out more explicitly the effect of allowing the back reaction to assume increasing importance. As K decreases from infinity, $(A)/(A)_0$ approaches a larger and larger limiting $(A)_\infty/(A)_0$ value, and the approach, while always exponential, is increasingly rapid. That is, although the initial rates are all the same, the back reaction comes in earlier and earlier to cause $(A)/(A)_0$ to level off at the equilibrium value.

Similar but more complex analyses can be made for cases where either the forward or the reverse reaction is second order or where both are. Some of these are given in the Special Topics section. However, such treatments are difficult to use and to fit accurately to data, and it is good practice to try to establish experimental conditions such that the rate law is made pseudo-first-order.

B. Kinetics of a small perturbation from equilibrium

An important type of method for studying fast reactions, discussed in Section 15-2B, consists in making a sudden change in the physical state of a system at equilibrium, such as a sudden jump in temperature, so that a different

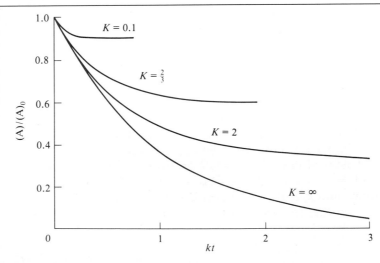

FIGURE 15-1
Rate of approach to equilibrium for the case $A \rightleftharpoons B$, for varying K. For $K = 0.1$, $(A)_\infty/(A)_0 = 0.91$; $K = \frac{2}{3}$, $(A)_\infty/(A)_0 = \frac{3}{5}$; $K = 2$, $(A)_\infty/(A)_0 = \frac{1}{3}$; $K = \infty$, $(A)_\infty/(A)_0 = 0$.

equilibrium constant applies and the system seeks its new equilibrium state. If the perturbation is relatively small, then the kinetics of this relaxation to the new equilibrium position takes on a rather simple mathematical form.

While the treatment may be generalized, it is best illustrated in terms of a specific example. Consider the dissociation of a weak acid

$$\text{HA} \underset{k_{-1}}{\overset{k_1'}{\rightleftharpoons}} \text{H}^+ + \text{A}^-, \qquad K' = \frac{k_1'}{k_{-1}'} \tag{15-10}$$

with $(\text{H}^+) = (\text{A}^-) = x_e'$ and $(\text{HA}) = a - x_e'$ initially, where a is the total formality of the acid present. As a result of a sudden temperature jump, the equilibrium constant takes on a new value K, where $K = k_1/k_{-1}$, for which the corresponding equilibrium concentrations are x_e and $a - x_e$, so that the system has been displaced from equilibrium by $\Delta x_0 = x_e' - x_e$. Net reaction will now occur with x at some subsequent time t equal to $x_e + \Delta x$. If the reaction is a simple one, that is, if Eq. (15-10) is also the mechanism for the reaction, then the reaction will take place according to the mass action rate law:

$$\frac{dx}{dt} = k_1(a - x) - k_{-1}x^2$$

or

$$\frac{d(\Delta x)}{dt} = k_1(a - x_e - \Delta x) - k_{-1}(x_e + \Delta x)^2$$

$$\frac{d(\Delta x)}{dt} = k_1(a - x_e) - k_1\Delta x - k_{-1}x_e^2 - 2k_{-1}x_e\Delta x - k_{-1}(\Delta x)^2 \tag{15-11}$$

Since x_e is the new equilibrium value, it must be true that

$$k_1(a - x_e) = k_{-1}x_e^2$$

so Eq. (15-11) reduces to

$$\frac{d(\Delta x)}{dt} = -k_1\Delta x - 2k_{-1}x_e\Delta x - k_{-1}(\Delta x)^2 \tag{15-12}$$

At this point the characteristic assumption is made that Δx is sufficiently small that square or higher-power terms in Δx can be dropped. In the present case the result is

$$\frac{d(\Delta x)}{dt} = -(k_1 + 2k_{-1}x_e)\Delta x = -k_r\Delta x \tag{15-13}$$

Equation (15-13) is first order in Δx and integrates to give

$$\Delta x = \Delta x_0 e^{-k_r t} \qquad k_r = \frac{1}{\tau} \tag{15-14}$$

where k_r, the relaxation rate constant, is often reported as its reciprocal τ, the *relaxation time*. From Eq. (15-13) we have

$$k_r = k_1 + 2k_{-1}x_e = k_1\left(1 + \frac{2x_e}{K}\right) \tag{15-15}$$

Thus if K is known, k_1 and k_{-1} may be calculated.

The general derivation is somewhat complicated, but the result is that regardless of mechanism or actual rate law, the dropping of all terms higher than first power in Δx leads to Eq. (15-14), where k_r will have some specific relationship to the actual rate constants and various equilibrium concentrations. If the mechanism is in doubt, the form of the rate law may be deduced from a study of the variation of k_r (or of τ) with system composition.

EXAMPLE

A 0.010 f solution of NH_4OH experiences a sudden temperature jump terminating at 25°C, at which temperature $K = 1.8 \times 10^{-5}$ mole liter^{-1} for the equilibrium $NH_4OH = NH_4^+ + OH^-$, so that $x_e = 4.1 \times 10^{-4}$ M. The observed relaxation time is 1.09×10^{-7}s, corresponding to $k_r = 9.2 \times 10^6$ s^{-1}. From Eq. (15-15) we have

$$k_1 = \frac{9.2 \times 10^6}{1 + [(2)(4.1 \times 10^{-4})(1.8 \times 10^{-5})]} = 2 \times 10^5 \text{ s}^{-1}$$

and $k_{-1} = 2 \times 10^5/1.8 \times 10^{-5} = 1.1 \times 10^{10}$ liter mol^{-1} s^{-1}. Note how a knowledge of k_1 has allowed the indirect determination of an extremely large k_{-1}. If there were any doubt about the correct rate law, it could be verified that k_1 and k_{-1} were in fact independent of the ammonia concentration used.

15-2 Relationship between the equilibrium constant for a reaction and its rate constants

The statement has been made several times in this and the preceding chapter that the equilibrium constant K for a chemical reaction is equal to k_f/k_b, the ratio of the forward and backward rate constants. This is strictly true only for a simple reaction, that is, one for which the mechanism is the same as the overall reaction, and it is now time to take a closer look at the situation.

If the reaction is a simple one, such as

$$A + B = C + D \qquad K = \frac{(C)(D)}{(A)(B)}$$

then $R_f = k_f(A)(B)$ and $R_b = k_b(C)(D)$, and at equilibrium $R_f = R_b$, so that

$$\frac{k_f}{k_b} = K \tag{15-16}$$

A complication arises, however, if the reaction is complex, that is, if the reaction mechanism does not coincide with the statement of the overall reaction. It turns out that the correct general statement of Eq. (15-16) is

$$\frac{k_f}{k_b} = K^n \tag{15-17}$$

where n is an integer or rational fraction. The conclusion is perhaps best demonstrated by means of a specific example.

Consider the reaction

$$2Fe^{2+} + 2Hg^{2+} = 2Fe^{3+} + Hg_2^{2+} \tag{15-18}$$

for which

$$K = \frac{(Fe^{3+})^2 \, (Hg_2^{2+})}{(Fe^{2+})^2 \, (Hg^{2+})^2} \tag{15-19}$$

The experimentally observed forward rate law is

$$R_f = k_f(Fe^{2+})(Hg^{2+}) \tag{15-20}$$

which can be explained by the mechanism

(a) $Fe^{2+} + Hg^{2+} \underset{k_{-1}}{\overset{k_1}{\rightleftharpoons}} Fe^{3+} + Hg^+$ (slow)

(b) $2Hg^+ \overset{K_2}{=} Hg_2^{2+}$ (rapid equilibrium)

Then R_f is as given by Eq. (15-20) with $k_f = k_1$, and R_b is

$$R_b = k_{-1}(Fe^{3+})(Hg^+)$$

or, on elimination of (Hg^+) by means of the expression for K_2,

$$R_b = \frac{k_{-1}}{K_2^{1/2}} (Fe^{3+})(Hg_2^{2+})^{1/2} = k_b(Fe^{3+})(Hg_2^{2+})^{1/2} \tag{15-21}$$

On equation R_f to R_b at equilibrium, we obtain

$$\frac{k_f}{k_b} = \frac{(Fe^{3+})(Hg_2^{2+})^{1/2}}{(Fe^{2+})(Hg^{2+})} \tag{15-22}$$

The ratio k_f/k_b is thus the *square root* of K as defined by Eq. (15-19).

We can now generalize somewhat. Suppose that for some reaction the mechanism is such that

$$R_f = \theta_f(k) \, \phi_f(C) \tag{15-23}$$

where $\theta_f(k)$ denotes some function of rate constants and $\phi_f(C)$ some function of concentrations, and, similarly,

$$R_b = \theta_b(k) \, \phi_b(C) \tag{15-24}$$

At equilibrium R_f must equal R_b and therefore $\theta_f(k) \, \phi_f(C) = \theta_b(k) \, \phi_b(C)$ or $\theta_f(k)/\theta_b(k) = \phi_b(C)/\phi_f(C)$. The right-hand side of this last equation is a function of C only and must be identifiable with K^n, that is,

$$K^n = \frac{\theta_f(k)}{\theta_b(k)} = \frac{\phi_b(C)}{\phi_f(C)} \tag{15-25}$$

If, now, the expression for R_f has been determined experimentally, it follows that the expression for R_b must be

$$R_b = K^n\theta_b(k) \, \phi_f(C) \tag{15-26}$$

The exponent n is not known *a priori*, and could be any rational number. Thus even though the forward rate law is known, as well as the overall reaction (and hence K), an indefinite number of possible expressions for R_b remain. Each is generated through Eq. (15-26) by making some particular choice for n.

Returning to the example, we see that R_f is given by Eq. (15-20) and K by Eq. (15-19), so that

$$R_b = \left[\frac{(Fe^{3+})^2 \, (Hg_2^{2+})}{(Fe^{2+})^2 \, (Hg^{2+})^2} \right]^n \theta_b(k)(Fe^{2+})(Hg^{2+}) \tag{15-27}$$

If n is set equal to $\frac{1}{2}$, Eq. (15-21) results, with $\theta_b(k) = k_b$, consistent with the mechanism given by reactions (a) and (b). Suppose, however, that we set $n = 1$. Now R_b must be

$$R_b = \theta_b(k) \frac{(Fe^{3+})^2 \, (Hg_2^{2+})}{(Fe^{2+}) \, (Hg^{2+})} \tag{15-28}$$

Such a reverse rate law cannot be explained by the first mechanism, but does result from the following one [where (a) is as before]:

(a) $Fe^{2+} + Hg^{2+} \overset{k_1}{\underset{k_{-1}}{\rightleftharpoons}} Fe^{3+} + Hg^+$ (slow)

(b) $Fe^{2+} + Hg^+ \overset{K_3}{=} Fe^{3+} + Hg^0$ (rapid equilibrium)

(c) $Hg^0 + Hg^{2+} \overset{K_4}{=} Hg_2^{2+}$ (rapid equilibrium)

R_b is still

$$R_b = k_{-1}(Fe^{3+})(Hg^+),$$

but on eliminating (Hg^+) by means of K_3 and K_4, we obtain

$$R_b = \frac{k_{-1}}{K_3 K_4} \frac{(Fe^{3+})^2 \, (Hg_2^{2+})}{(Fe^{2+})(Hg^{2+})} \tag{15-29}$$

which is the form required by Eq. (15-28).

Thus in the study of the rate of oxidation of ferrous by mercuric ion experimental knowledge of the forward rate law does not allow unambiguous selection among various possible mechanisms. The mass action relation between forward and reverse rate laws does serve to reduce the alternatives to a limited set. In practice, n can be expected to be a rather simple number such as 1, 2, or $\frac{1}{2}$. One therefore rarely proceeds further in considering possible mechanisms in cases where the forward rate law but not the reverse one has been determined experimentally.

15-3   Experimental methods

A. General   Most of the experimental techniques described in Section 14-4 are applicable to reactions in solution. The system may be quenched, either by cooling or by removal of a catalyst, and then analyzed chemically. Or an additive property may be used, such as absorbance at a suitable wavelength. Reactions in solution are almost universally followed at constant pressure, in contrast to the case of gas-phase reactions, which are usually studied at constant volume. In the case of dilute solutions, volume changes during reaction are typically small enough that no correction is needed (such as illustrated in Problem 14-28). Often, however, the effect of volume change on concentrations is not negligible; one example is that of reactions involving polymers.

Although volume change during a reaction in solution is ordinarily small, it usually is not zero. The partial molal volumes of the reactants and products generally will differ, with the consequence that the total volume of the system will increase or decrease a little as the reaction proceeds. Even small changes can be followed accurately if the system is in a *dilatometer*, essentially a flask capped with a capillary tube (see Fig. 15-2), and well thermostated. The level of the meniscus in the tube is measured periodically with a traveling microscope, and it usually is possible to use Eq. (14-40) to follow the degree of reaction. The dilatometric method is often resorted to if, as in the case of polymerization reactions, no very characteristic absorbance changes occur.

As noted above, one does not usually follow solution reactions by means of pressure change at constant volume. In the case of liquids the pressure necessary to maintain constant volume in the face of even a small partial molal volume change would be quite large—large enough to affect the rate constant itself (see Section 15-ST-2).

B. Fast reaction techniques

The types of methods just cited are applicable to reactions whose half times are 1 min or more. Reaction times as short as 0.1 sec can be investigated by means of rapid mixing devices such as illustrated in Fig. 14-4. The reacting solutions enter a mixing chamber and then travel down a small-bore tube; it is necessary that some color change accompany the reaction so that the light absorption at a suitable wavelength can be used in the determination of the degree of reaction at various positions down the tube and hence for various times of reaction.

A number of additional methods have come into use in recent years which allow the time scale to be shortened considerably. An important one is called the *stopped-flow method*. Reacting solutions are again delivered into a mixing chamber, now usually by means of motor-driven syringes. After an interval sufficient to establish a steady-state condition in the exit tube, the flow is stopped abruptly and the optical absorption is determined as a function of time thereafter. A schematic of such equipment is shown in Fig. 15-3. Change in absorbance is now monitored as a function of time, at a fixed point down

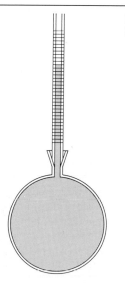

FIGURE 15-2
Dilatometer.

FIGURE 15-3
Stopped-flow apparatus.
L, light source;
G, monochromator;
C, observation cell;
M, mixer; PM, photo-
multiplier; D_1 and D_2,
driving syringes; F, flow
velocity detector;
PP, gas pressure driving
piston; EM, electro-
magnetic valve; S, pin for
stopping the flow. The
reagent solutions to be
mixed are held in the
reservoirs R_1 and R_2. [From
K. Hiromi, "Kinetics of Fast
Enzyme Reactions," Wiley,
New York, 1979.]

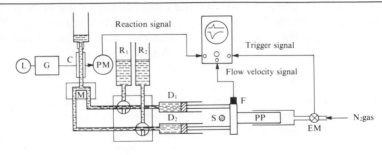

the discharge tube; the schematic oscilloscope traces show the flow velocity
profile and the change of absorbance with time in the reaction cell.

A different family of techniques is that known as *relaxation methods.* One
starts with an equilibrium mixture and subjects it to some physical perturbation such that the system must adjust or relax to a new equilibrium condition. As shown in Section 15-1B, if the departure from equilibrium is small,
the relaxation process will be first order even though the rate law is not.
Various methods have been used to produce the perturbation, the most important one perhaps being the *T-* or *temperature-jump* method. An apparatus
schematic is shown in Fig. 15-4(a). By discharging a large condensor across
electrodes placed in a conducting solution, one achieves a quick, ohmic heating. As a result of the temperature jump, the system is no longer in equilibrium, and one uses the usual monitoring light–oscilloscope technique to follow the rate of re-equilibration. Figure 15-4(b) shows the type of experimental
result than can be obtained.

Another type of relaxation method makes use of sound waves. The absorption of sound
by a medium is a consequence of various irreversible, essentially frictional processes that
take place during the compression-rarefaction cycles that occur as a sound wave passes
through the solution. If a chemical reaction system is present, its position of equilibrium
is cyclically perturbed by the sound wave, with the frequency of the sound. If the rate of
reaction is slow, then the chemical system cannot follow the changing equilibrium constant,
and the solution behaves simply as a nonreacting mixture of solutes. If the reaction rate
is fast compared to the sound frequency, then the reacting system does follow the changing
equilibrium position, with the result that energy is taken from the sound wave to supply
the heat of reaction; the absorption coefficient or rate of attenuation of the sound wave is
thereby increased. The transition between the two extremes of no effect and maximum
effect occurs fairly narrowly at a frequency corresponding to the relaxation time of the
chemical system. The variation in the absorption coefficient of the sound with its frequency—known as the *dispersion curve*—has an inflection point at the frequency equal
to $1/\tau$, where τ is the relaxation time as given by Eq. (15-4). By using high-frequency
sound, one can observe reaction times of as short as perhaps 10 microseconds.

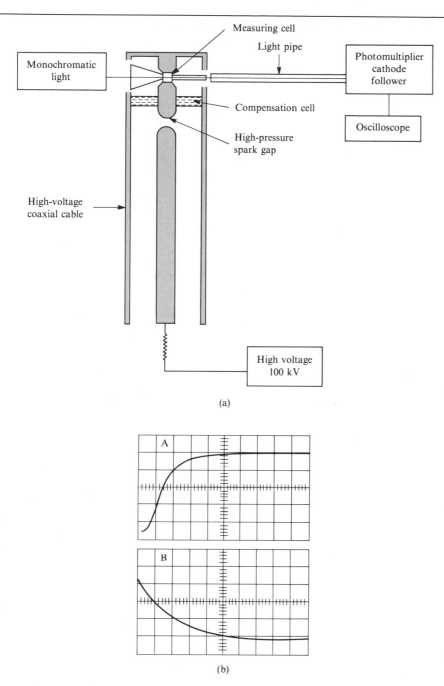

FIGURE 15-4
(a) Schematic of a nanosecond temperature-jump apparatus using the discharge of a coaxial capacitor to heat the solution. (b) Some T-jump results. Oscilloscope tracing A shows the heating curve. The ordinate is in arbitrary absorbancy units and the abscissa scale is 10 μs per division. Curve B shows the Co(II)–diglycine reaction (abscissa scale 500 μs per division). [From G.G. Hammes, ed., "Investigation of Rates and Mechanisms of Reactions," 3rd ed., Wiley-Interscience, New York, 1974).]

Reactive species may also be generated photochemically (see Section 19-4D) on a microsecond time scale by conventional flash photolysis, and in nanosecond and even picosecond times using a pulsed laser. Very fast subsequent reactions can be studied by using a monitoring beam and the photomultiplier–oscilloscope detection technique. The same is true in pulse radiolysis experiments, in which the reactive species are formed by means of a very short burst of electrons or of other ionizing radiation.

15-4   Kinetic-molecular picture of reactions in solution

A. Encounters   The physical picture of molecular motions in a liquid is rather different from that in a gas. The molecules of a liquid are about as close together as in the crystalline solid—there is usually about a 10% expansion on melting (water being an exception because of its unusually open crystal structure) which allows some looseness and randomness in the liquid structure. As illustrated in Fig. 8-6(b), it seems likely that molecules of a liquid are in a potential well, but a somewhat flattened one so that vibrations against immediate neighbors are of low energy. There is, nonetheless, a confinement that is usually referred to as the *solvent cage effect*. The physical picture is one of a molecule vibrating a number of times against the walls of its cage, that is, against its immediate neighbors, with occasional escapes to some adjacent position. The situation is illustrated in Fig. 15-5, where the molecules are shown as roughly spherical and in a somewhat expanded but essentially close-packed arrangement.

The cage model is supported by the fairly successful treatment of diffusion in liquids (see Section 10-ST-2) in which the random diffusional motion of molecules in a liquid is taken to occur as a sequence of jumps from one molecular position to the next. This elementary jump distance λ is about $2r$, where r is the radius of the molecule. By Eq. (2-89),

$$\mathscr{D} = \frac{\lambda^2}{2\tau} \tag{15-30}$$

The average time between jumps should then be

$$\tau = \frac{\lambda^2}{2\mathscr{D}} \tag{15-31}$$

Equation (2-89) assumes a continuous medium, and since we are treating a liquid as having a quasi-crystalline structure, with more or less definite molecular sites, it turns out that a somewhat more accurate statement should be

$$\mathscr{D} = \frac{\lambda^2}{6\tau} \qquad \text{or} \qquad \tau = \frac{\lambda^2}{6\mathscr{D}} \tag{15-32}$$

For small molecules a reasonable value of λ is about 4 Å, and, from Table 10-5, $\mathscr{D}$ at 25°C would be about 1×10^{-5} cm^{2}s^{-1}; τ is then about $(4 \times 10^{-8})^2/(6)(1 \times 10^{-5}) \approx 2.5 \times 10^{-11}$s. We next guess that the solvent cage is sufficiently loose that the average vibrational energy is about $\mathbf{k}T$, corresponding to a frequency $h\nu = \mathbf{k}T$ or $\nu = \mathbf{k}T/h$. Vibrations against the wall of the cage then occur at intervals of $h/\mathbf{k}T$ or about 1.5×10^{-13} s at 25°C. We conclude that a typical molecule in solution vibrates about $2.5 \times 10^{-11}/1.5 \times 10^{-13}$

FIGURE 15-5
The solvent "cage."

or about 150 times against its immediate neighbors before escaping to a new position and new neighbors.

The same analysis applies to solute molecules, and the next task is to estimate the frequency with which two solute molecules A and B will, by diffusion, accidentally become neighbors (Fig. 15-6). Such a process is called an *encounter* and the estimation of encounter frequencies is central to much of solution kinetics. The problem is a difficult one, complicated by the present impossibility of describing the structure of a liquid with any great accuracy. As an approximation, we assume the molecules of solvent and of A and B to be about the same size and to be spherical. Each A molecule should then have about 12 nearest neighbors, so that on each diffusional jump it should find 6 new ones. The chance that one of these will be a molecule of B is given by the mole fraction of B in the solution,

$$x_B = \frac{\text{molecules of B cm}^{-3}}{\text{molecules of solvent cm}^{-3}} = \frac{\mathbf{n_B}}{1/\gamma\lambda^3} \tag{15-33}$$

where γ is a geometric factor which reflects the way in which molecules are packed in the liquid and λ^3 is the molecular volume. Substitution into Eq. (15-32) gives the frequency $1/\tau_{AB}$ of encounters of A with B as

$$\frac{1}{\tau_{AB}} = \frac{6\mathscr{D}}{\lambda^2}(\mathbf{n_B}\gamma\lambda^3)(6) = 36\gamma\lambda\mathbf{n_B}\mathscr{D}$$

If we take the effective $\mathscr{D}$ to be the sum of $\mathscr{D}_A$ and $\mathscr{D}_B$, γ to be about 0.7, and λ to be the sum of the molecular radii r_{AB},

$$\frac{1}{\tau_{AB}} \simeq 25 r_{AB}\mathscr{D}_{AB}\mathbf{n_B} \tag{15-34}$$

The total number of encounters per cubic centimeter per second is this value multiplied by $\mathbf{n_A}$ or, in mole per liter units,

$$Z_{e,AB} = \frac{25 r_{AB}\mathscr{D}_{AB}N_A}{1000}(A)(B) \tag{15-35}$$

where $Z_{e,AB}$ is the A–B encounter frequency. The corresponding frequency factor is

$$A_e = \frac{25 r_{AB}\mathscr{D}_{AB}N_A}{1000} \tag{15-36}$$

For $\mathscr{D}_A = \mathscr{D}_B = 1 \times 10^{-5}$ cm^2 s^{-1} and $r_{AB} = 4$ Å, this gives $A_e = 1.2 \times 10^9$ liter mol^{-1} s^{-1}.

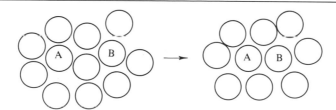

FIGURE 15-6
Diffusional encounter between A and B.

A very simple and useful approximation results if $\mathscr{D}$ is estimated by means of the Stokes–Einstein law,

$$\mathscr{D} = \frac{kT}{6\pi\eta r} \quad \text{[Eq. (10-45)]}$$

so that r_{AB} cancels out. The result is

$$A_e = 1.1 \times 10^5 \frac{T}{\eta} \text{ liter mol}^{-1} \text{ s}^{-1}. \tag{15-37}$$

In summary, for a solution 1 M in A and 1 M in B the encounter frequency at 25°C is about 4×10^9 mol liter^{-1} s^{-1}. Having made an encounter, A and B remain in their solvent cage for some 150 vibrational periods. We next contrast this picture with that provided by gas collision theory.

B. Encounters versus collisions

It would seem that collision theory, based on the kinetic molecular theory of gases, would be totally inapplicable to reactions in solution. The surprising and very significant fact is that a number of solution reactions do have an Arrhenius frequency factor A [Eq. (14-75)] close to that expected from collision theory ($\sim 5 \times 10^{11}$). Some representative data are given in Table 15-1. In other cases, although the frequency factor may be smaller than the theoretical value, it is not affected by the nature of the medium, as illustrated by the data of Table 15-2. More qualitatively, a number of other reactions have been found to have the same rate constant in the gas phase as in solution; for example, the rate of decomposition of N_2O_5 is about the same in the gas phase as in nitromethane solution.

The previous discussion about encounters does tell us, however, that although the collision frequency may be the same in the gas phase and in solution, the pattern of collisions must be quite different. Consider the case of molecules A and B both 1 M in concentration. The gas collision frequency at 25°C would be about 5×10^{11} liter^{-1} s^{-1} and, as illustrated in Fig. 15-7(a), if each A–B collision could be marked on a time scale, their pattern would be a random one.

The encounter frequency would be about 4×10^9 liter^{-1} s^{-1}, or 1/100 as often. The two molecules would stay in their mutual solvent cage for about 2.5×10^{-11} s, however, or long enough to make about 150 collisions with each other. The collision pattern is therefore as shown in Fig. 15-7(b)—occasional encounters with many collisions during each encounter.

The activation energies cited in Table 15-1 are around 20 kcal mol^{-1}, which

TABLE 15-1
Some second-order reactions in solutions[a]

Reactants	Solvent	A (liter mol^{-1} s^{-1})	E^* (kcal mol^{-1})
$CH_3Br + I^-$	CH_3OH	2.26×10^{10}	18.25
	H_2O	1.68×10^{10}	18.26
$I_2 + N_2CHCOOC_2H_5$	CCl_4	2.21×10^{11}	20.23
$C_2H_5Br + (C_2H_5)_2S$	$C_6H_5CH_2OH$	1.40×10^{11}	25.47

[a]See E. A. Moelwyn-Hughes, "Physical Chemistry," 2nd ed. Pergamon, Oxford, 1961.

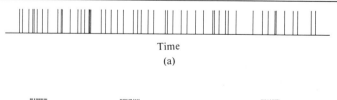

TABLE 15-2
Dimerization of
cyclopentadiene[a]

Medium	log A (liters mol^{-1} s^{-1})	E* (kcal mol^{-1})
Gas	6.1	16.7
C$_2$H$_5$OH	6.4	16.4
CS$_2$	6.2	16.9
C$_6$H$_6$	6.1	16.4

[a]Source: S. W. Benson, "The Foundations of Chemical Kinetics," McGraw-Hill, New York, 1960.

means that at 25°C the chance of a collision leading to reaction is about exp($-20,000/298R$) or about 10^{-15}. Thus only after 10^{15} collisions, or 1×10^{13} encounters, does reaction occur, on the average. The difference in pattern between Fig. 15-7(a) and Fig. 15-7(b) is thus unimportant in such cases. If, however, the activation energy is small so that reaction should occur after only a few collisions at the most, then *every* encounter should result in reaction. Further, the frequency factor for the reaction is now limited by the encounter frequency. Reactions which occur with every encounter are said to be *diffusion-controlled*. Their observed activation energy will not be zero, however, but will be determined by the temperature dependence of $\mathscr{D}_{AB}$ [in Eq. (15-36)] and hence by that of the viscosity of the solvent since by Walden's rule (Section 10-8B), $\mathscr{D}\eta$ is approximately a constant. From Table 8-5, this means that diffusion- or encounter-controlled reactions should show activation energies of 3 to 4 kcal mol^{-1} in common solvents. Reactions of this type are discussed in somewhat more detail below.

We now consider how to apply all of the above to the formulation of a rate constant for a bimolecular reaction in solution. The general picture can be expressed mechanistically as follows:

C. Encounter-collision theory of solution reactions

$$A + B \underset{k_{-d}}{\overset{k_d}{\rightleftharpoons}} [AB] \tag{15-38}$$

$$[AB] \overset{k_1}{\rightarrow} \text{products} \tag{15-39}$$

FIGURE 15-7
(a) Pattern of A–B collisions according to collision theory and (b) pattern of A–B collisions according to encounter picture.

We denote the encounter complex by [AB], and k_d and k_{-d} are the rate constants for its diffusional formation and separation back into A and B, respectively. The complex [AB] may undergo reaction to form products, a unimolecular reaction. Note that [AB] is *not* a transition state or an energized species but merely the pair of reactant molecules in a solvent cage. Reaction (15-39) may therefore require an activation energy E^*, and k_1 is given by Eq. (14-75), $k_1 = Ae^{-E^*/RT}$.

The stationary-state treatment of Eqs. (15-38) and (15-39) (Section 14-5C) gives

$$\text{Rate} = -k_1[AB] = \frac{k_1 k_d}{k_1 + k_{-d}} \text{ (A) (B)} \tag{15-40}$$

The rate constant for the reaction is thus

$$k = \frac{k_1 k_d}{k_1 + k_{-d}} \tag{15-41}$$

The two extreme cases are $k_1 \gg k_{-d}$ and $k_1 \ll k_{-d}$. First, if k_1 is large, $k_1 \gg k_{-d}$, then $k = k_d$. That is, reaction occurs on each encounter. This is the diffusion-controlled case described above.

One might expect, in this case, that k_d and hence k should be just the A_e of Eq. (15-36), but there are two kinds of corrections to make. The first of these is as follows. In deriving Eq. (15-35), it was assumed that the distribution of A and of B in solution was random. If we focus on A as the species making the encounter, the derivation also assumes that as A diffuses away after an encounter it remains available to make new encounters with the same or nearby B molecules. If, however, chemical reaction takes place with every encounter, then B acts as a sink into which A disappears and there is therefore on the average a depletion of A in the vicinity of B (and vice versa, of course). A simple treatment given by Smoluchowski in 1917 applies ordinary diffusion theory to this situation. By Fick's law [Eq. (2-87); see also Section 10-8B] the total flux of A flowing toward a B molecule is

$$J = \left(-\frac{d\mathbf{n}_A}{dr}\right)_r \mathscr{D}\mathscr{A} = -4\pi r^2 \mathscr{D}_A \left(\frac{d\mathbf{n}_A}{dr}\right)_r \tag{15-42}$$

where $\mathscr{A}$ is the area of the spherical surface around B at a distance r. We now assume a steady-state condition, that is, that $d\mathbf{n}_A/dr$ is independent of time, so that Eq. (15-42) has the form

$$\frac{d\mathbf{n}_A}{dr} = -\frac{J}{4\pi r^2 \mathscr{D}_A}$$

and integrates to give

$$\mathbf{n}_A = \frac{J}{4\pi r \mathscr{D}_A} + a$$

where a is a constant of integration. We then set $\mathbf{n}_A = 0$ at $r = r_{AB}$ and $\mathbf{n}_A = \mathbf{n}_A^\circ$, the average concentration, at $r = \infty$, which allows a to be evaluated. On solving for J, we obtain

$$J = -4\pi r_{AB} \mathscr{D}_A \mathbf{n}_A^\circ \tag{15-43}$$

We allow for the diffusion of B by using $\mathscr{D}_{AB}$ instead of just $\mathscr{D}_A$, and multiply by $\mathbf{n}_B^\circ$ to get the total encounter frequency

$$Z_{re} = 4\pi r_{AB}\mathscr{D}_{AB}\mathbf{n}_A\mathbf{n}_B \tag{15-44}$$

where the subscript "re" denotes reactive encounter, and the superscripts on the $\mathbf{n}$'s have been dropped as no longer necessary. The frequency factor for reactive encounters is thus

$$A_{re} = \frac{4\pi r_{AB}\mathscr{D}_{AB}N_A}{1000} \tag{15-45}$$

or about half that given by Eq. (15-36). Although both derivations are approximate, this comparison is probably about right. That is, the encounter frequency for a diffusion-controlled reaction is about half of that expected in the absence of reaction.

The second kind of correction provides for the possibility that A and B experience some net attraction or repulsion as they approach for an encounter. We therefore write

$$k_d = A_{re} = \frac{4\pi r_{AB}\mathscr{D}_{AB}N_A}{1000} e^{-\phi/\mathbf{k}T} \tag{15-46}$$

where ϕ is the potential (negative if attractive) and A_{re} can now be identified with k_d. If A and B are ions, ϕ will include the contribution

$$\phi = z_A z_B e^2/D \qquad \text{or} \qquad z_A z_B e^2/4\pi\epsilon_0 D \tag{15-47}$$

where z_A and z_B are the charge numbers and e is the charge on the electron. In the first formulation of Eq. (15-47), ϕ will be in ergs if e is in esu; and in the second, ϕ will be in joules if e is in coulombs. D is dielectric constant and ϵ_0 is the permittivity of vacuum (see Section 3-CN-1). Also, in the formation of [AB] the van der Waals forces of attraction (see Section 8-ST-1) between A and B will be balanced against those between A and solvent and between B and solvent, since solvent is displaced in forming [AB]. The net effect is to contribute either a positive or a negative increment to ϕ. Thus, even if A and B are uncharged, ϕ will not in general be zero.

We consider next the second extreme or limiting case of Eq. (15-41), namely that in which $k_1 \ll k_{-d}$. We now have

$$k = k_1(k_d/k_{-d}) = A(k_d/k_{-d})e^{-E^*/RT} \tag{15-48}$$

Since the k_1 process, Eq. (5-39), is a unimolecular one, we can approximate A by $\mathbf{k}T/h$; also, (k_d/k_{-d}) is just the equilibrium constant K_e for the formation of encounter complexes. We thus have an approximate expression for k:

$$k \simeq \frac{\mathbf{k}T}{h} K_e e^{-E^*/RT} \tag{15-49}$$

or

$$k \simeq \frac{\mathbf{k}T}{h} e^{\Delta S_e^0/R} e^{-\Delta H_e^0/RT} e^{-E^*/RT} \tag{15-50}$$

where ΔS_e^0 and ΔH_e^0 are the entropy and enthalpy of formation of [AB] from A and B. ΔH_e^0 can be approximately identified with ϕ, and ΔS_e^0 can be estimated as $R \ln(6/C_s)$ where C_s is the concentration of solvent in mol liter^{-1} (see Problem 15-22).

A pleasing alternative way of writing Eq. (15-49) is the following. We have $K_e = A_e\tau$; that is, we identify k_d with A_e, and k_{-d} with the lifetime τ of the encounter complex. Also, the product $\tau(\mathbf{k}T/h)$ is essentially z_e, the number A–B collisions *during* an encounter. On substituting these restatements into Eq. (15-49), we obtain

$$k = A_e z_e e^{-E^*/RT} \tag{15-51}$$

Since $A_e z_e$ is the total collision frequency factor, Eq. (15-51) is conceptually the same as Eq. (14-89)!

15-5 Diffusion-controlled reactions

A. Experimental observations

According to the treatment of the preceding section, a diffusion-controlled reaction should have a rate constant of around 10^9 liter mol^{-1} s^{-1} at 25°C in common solvents and show an activation energy corresponding to that for diffusion or for the solvent viscosity, namely of 3 to 4 kcal mol^{-1} for ordinary solvents. An implied condition is that the chemical activation energy be small enough that the reaction occurs within the first 100 collisions.

Some examples of what appear to be diffusion-controlled processes are given in Table 15-3. Note that the rate constants are around 10^9 liter mol^{-1} s^{-1}, depending somewhat on the solvent and on the size of the reactants. Equally important, the activation energies are only a few kcal mol^{-1}, or about that expected for diffusion in the solvents in question.

A number of acid–base type of reactions appear to be diffusion-controlled. Table 15-4 gives the rate parameters for several reactions of the type

$$\mathrm{HA} + \mathrm{H_2O} \rightleftharpoons \mathrm{A^-} + \mathrm{H_3O^+} \tag{15-52}$$

which are believed to be simple reactions, that is, the overall reaction also constitutes the mechanism. These reactions have been studied by one or another fast reaction techniques of the type mentioned in Section 15-3B. Notice that in all cases it is the k_2 reaction, or the transfer of a proton from $\mathrm{H_3O^+}$ to $\mathrm{A^-}$, that approaches the diffusion-controlled region of rate constant value.

These values of k_2 are distinctly larger than the 4×10^9 figure arrived at

TABLE 15-3
Some diffusion-controlled reactions[a]

Reaction	Solvent	$10^{-9}k$ (liter mol^{-1} s^{-1})	t (°C)	E^* (kcal mol^{-1})
$\mathrm{I} + \mathrm{I} \rightarrow \mathrm{I_2}$	Vapor phase	7	23	3.2
	n-Hexane	18	50	—
$2\mathrm{CCl_3} \rightarrow \mathrm{C_2Cl_6}$	Cyclohexene + $\mathrm{CCl_3Br}$	0.05	30	<6
	Vinyl acetate + $\mathrm{CCl_3Br}$	0.05	30	<6
β-Naphthylamine + $\mathrm{CCl_4} \rightarrow$ fluorescence quenching	Cyclohexane	6	20	2.5
	Isooctane	13	20	1.6

[a]Source: S. W. Benson, "The Foundations of Chemical Kinetics." McGraw-Hill, New York, 1960.

HA	pK_a	$\log k_1$	$\log k_2$
H_2O	15.7^b	-4.6	11.1
H_2S	7.0	3.9	10.9
HF	3.3	7.7	11.0
HSO_4^-	1.6	9.4	11.0
CH_3COOH	4.8	5.9	10.7
CH_3COCH_3	20	-9.3	10.7
$(CH_3)_3NH^+$	9.8	1.0	10.8
β-Naphthol^c	3.1	7.6	10.7

TABLE 15-4

Fast reactions of the type

$$HA + H_2O \underset{k_2}{\overset{k_1}{\rightleftharpoons}} A^- + H_3O^{+a}$$

^{a}Source: E. F. Caldin, "Fast Reactions in Solutions." Wiley, New York, 1964. The k values are in liter mol^{-1} s^{-1} at 25°C.

$^b K_w$ has been put on a basis consistent with other weak acids by writing it as $(H^+)(OH^-)/(H_2O)$, that is, the usual value for K_w has been divided by 55.5, the number of moles of H_2O per liter.

^{c}The reaction is one of deactivation of the first electronic excited state of β-naphthol by proton transfer, as observed by fluorescence quenching.

earlier for a diffusion-controlled reaction, and at least two possible additional factors may be present over the usual situation. It will be recalled that the mobility of H^+ ion is unusually large and that a Grotthus-type mechanism is presumably responsible. As discussed in Section 12-5C, the hydrogen-bonded structure of water makes it possible for charge to move from one end of a chain of water molecules to the other by hydrogen bond shifts, the effect being the same as if a proton moved the length of the chain. It may then be that A^- can acquire a proton from other than nearest-neighbor H_3O^+ molecules by means of a similar mechanism, so that A^- and H_3O^+ do not have to diffuse as close to each other as in a normal bimolecular reaction. Their effective encounter rate would therefore be increased. The second factor is that since the H_3O^+ and A^- ions are oppositely charged, ϕ in Eq. (15-46) should be negative, thus increasing k_d over the normal value.

A rather interesting observation is that if some particular reaction is studied for a series of related compounds, then it will often be true that $\log k$ will vary linearly with $\log K$, where k is the rate constant and K the equilibrium constant. The actual relationship, proposed by Brønsted, is

B. The Brønsted equation

$$k = (\text{constant})K_a^\alpha \tag{15-53}$$

or

$$\log k = b + \alpha \log K_a \tag{15-54}$$

where b and α are constants. The data of Table 15-4 for the reaction $HA + H_2O \xrightarrow{k_1} A^- + H_3O^+$ are plotted in this manner in Fig. 15-8, and indeed, give a reasonably straight line, whose slope (and hence α) is 1.03. All this is rather misleading, however, since for these cases k_2 is essentially constant, being at the diffusion-controlled limit; and if k_2 is constant, then k_1 *must* be proportional to K_a, so that on the $\log k_1$ versus $\log K_a$ plot a straight line of slope unity is automatically expected. The more complete picture is as shown in Fig. 15-9. Thus k_1 should increase with K_a until the diffusion-controlled limit is reached and thereafter should be constant. Conversely, k_2 should increase with de-

FIGURE 15-8
Data of Table 15-4 plotted according to the Brønsted equation (15-54).

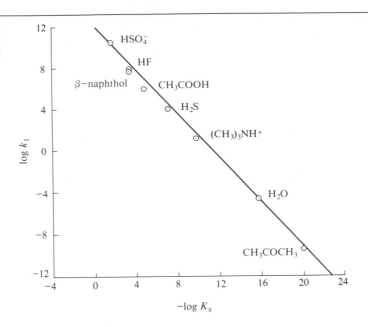

creasing K_a again until the diffusion-controlled limit is reached. When one k is at this limit, so that for it $\alpha = 0$, then the other k must obey the Brønsted equation with $\alpha = 1$. Thus α varies from 0 to 1 (or -1) along each curve. The two curves need not be symmetric, although they have to cross at log $K = 0$. Also, there can be regions where neither k is at the diffusion limit, so that there can be an intermediate region for which both α's lie between zero and unity. It happened that the early studies on rates of dissociation lay in this intermediate region and that the range of values that was experimentally accessible was small enough that α appeared to be a characteristic constant for each acid.

EXAMPLE   Estimate k_1 and k_2 for H_2CO_3. The equation of the line of Fig. 15-8 is approximately log $k_1 = 11.7 + 1.03 \log K_a$. From Table 12-8, log $K_a = -6.37$, whence log $k_1 = 5.14$ and $k_1 \simeq 1.4 \times 10^5$ s^{-1}. Then $k_2 = K_a/k_1 = 3.2 \times 10^{11}$ M^{-1} s^{-1}.

Equation (15-52) is merely a special case of an acid–base reaction

$$HA + B^- = A^- + HB \tag{15-55}$$

That is, bases other than water may be used, in which case the more general form of Eq. (15-54) is

$$\log k = b + \alpha(\log K_{HA} - \log K_{HB}) \tag{15-56}$$

and a number of systems have been studied in which B$^-$ is varied with behavior analogous to that shown in Fig. 15-9 [the abscissa now being $(\log K_{HA} - \log K_{HB})$]. The results may sometimes appear as indicated by the dashed lines in the figure, however. That is, it is not always possible to reach the diffusion limit for either k. For more detailed discussion see Eigen (1963).

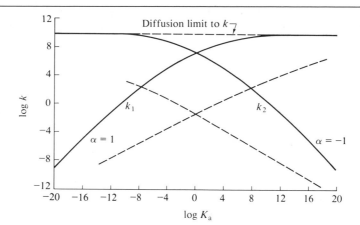

FIGURE 15-9
Brønsted plot showing both diffusion-controlled limits.

15-6   Catalysis

We take up in this section two important types of catalysis, both of which have been much studied by physical chemists. As noted in Section 7-CN-2D, a catalyst provides an alternate path for a chemical reaction—a faster one than otherwise—but, by the principal of microscopic reversibility, cannot change the equilibrium. The importance of catalysis, of course, is that the reaction mechanism and hence the reaction rate is changed.

A. Acid and base catalysis

A common type of catalysis, particularly in organic chemistry, is that by an acid or a base. In aqueous solution, the acid may be H^+ (or H_3O^+) but in general it may be any species HA capable of being a proton donor (note Section 12-CN-2). Similarly, the base may be OH^- ion but in general may be any base B capable of accepting a proton.

Our first example is the acid-catalyzed dehydration of acetaldehyde hydrate,

$$\overset{(HA)}{CH_3CH(OH)_2 \rightleftharpoons CH_3CHO + H_2O} \qquad (15\text{-}57)$$

which appears to occur through the mechanism

$$CH_3CH\overset{OH}{\underset{OH}{\big\langle}} + HA \overset{K_1}{\rightleftharpoons} CH_3\overset{H}{\underset{\underset{(HB)^+}{OH}}{C}}\!\!-\!\overset{H}{\underset{+}{O}}\!\!-\!H + A^- \qquad \text{(fast)}$$

(B)

$$CH_3\!-\!\overset{H}{\underset{\underset{OH}{+}}{C}}\!\!-\!\overset{H}{O}\!\!-\!H + A^- \overset{k_2}{\to} CH_3\!-\!\overset{H}{\underset{\underset{O^-}{+}}{C}}\!\!-\!\overset{H}{O}\!\!-\!H + HA \qquad \text{(slow)}$$

$$
\begin{array}{c}
\quad\;\; \text{H} \quad \text{H} \qquad\qquad \text{H} \\
\quad\;\; | \quad\;\; | \qquad\qquad\;\, | \\
\text{CH}_2\text{—C—OH} \rightarrow \text{CH}_3\text{—C}\!\!=\!\!\text{O} + \text{H}_2\text{O} \\
\quad\;\; | \quad\;\; + \\
\quad\;\; \text{O}^{\,-}
\end{array}
\qquad\qquad \text{(fast)}
$$

where HA is some acid. The rate law for this mechanism is

$$
\frac{dR}{dt} = -k_2 K_1 (\text{B})(\text{HA}) = -k_{\text{app}}(\text{B})(\text{HA}) \tag{15-58}
$$

where B denotes acetaldehyde hydrate [the student should verify Eq. (15-58)].

Brønsted and co-workers studied a number of reactions of this type in the 1920s and found, as an empirical observation, the Eq. (15-54) applied, with α values between zero and unity, depending on the system. Some data on reaction (15-57) are plotted according to Eq. (15-54) in Fig. 15-10; the straight line relationship is obeyed reasonably well with $\alpha = 0.53$ [see Bell and Higginson (1949)].

An alternative type of mechanism may be observed:

$$
\text{S} + \text{H}^+ \;\overset{K_1}{\rightleftharpoons}\; \text{SH}^+ \qquad\qquad \text{(fast)} \tag{15-59}
$$

$$
\text{SH}^+ + \text{R} \;\overset{k_2}{\rightarrow}\; \text{products} \qquad\qquad \text{(slow)} \tag{15-60}
$$

$$
\frac{d(\text{S})}{dt} = -k_{\text{app}}(\text{S})(\text{H}^+)(\text{R}) = -k_2 K_1 (\text{S})(\text{H}^+)(\text{R}) \tag{15-61}
$$

where S is the main reactant and R some additional one, often the solvent. The hydrolysis of esters, acetals, and ethers, as well as the inversion of sucrose, follow this type of scheme.

Various base-catalysis mechanisms occur. Thus

$$
\text{HS} + \text{B} \;\overset{k_1}{\rightleftharpoons}\; \text{S}^- + \text{BH}^+ \qquad\qquad \text{(fast)} \tag{15-62}
$$

FIGURE 15-10
Brønsted plot for the acid-catalyzed dehydration of $CH_3CH(OH)_2$. [Data from R. Bell and W. Higginson, Proc. Roy. Soc. A197, 141 (1949).]

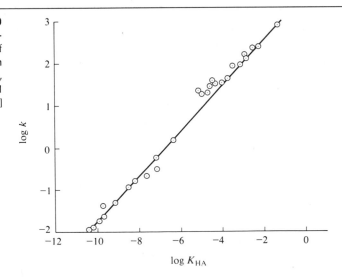

$$S^- + R \xrightarrow{k_2} \text{products} \qquad\qquad \text{(slow)} \qquad (15\text{-}63)$$

$$\frac{d(R)}{dt} = -\frac{k_2 K_1 (HS)(B)(R)}{(BH^+)} = \frac{k_2 K_1}{K_B}(HS)(R)(OH^-) \qquad (15\text{-}64)$$

where K_B is the base constant for $B + H_2O = BH^+ + OH^-$. Many of the condensation reactions of organic chemistry seem to follow this last mechanism, such as the Claisen, Michael, Perkin, and aldol condensations. The base-catalyzed reactions of many transition metal ammine complexes follow Eq. (15-64), with $B = OH^-$ and $R = H_2O$. The base-catalyzed hydrolysis of esters follows the mechanism

$$
\begin{array}{ccc}
\overset{O}{\underset{\|}{}} & & \overset{O^-}{\underset{|}{}} \\
R\!-\!C\!-\!O\!-\!R' + OH^- & \overset{K_1}{\rightleftharpoons} & R\!-\!C\!-\!O\!-\!R', \\
& & \overset{|}{OH}
\end{array}
\qquad (15\text{-}65)
$$

$$
\begin{array}{c}
\overset{O^-}{\underset{|}{}} \\
R\!-\!C\!-\!O\!-\!R' \xrightarrow{k_2} RCOO^- + R'OH \\
\overset{|}{OH}
\end{array}
\qquad (15\text{-}66)
$$

$$\frac{d(\text{ester})}{dt} = -k_2 K_1(\text{ester})(OH^-) \qquad (15\text{-}67)$$

A distinction is usually made as to whether the catalysis is a general acid or base one or is specifically by H^+ or OH^- ions. Thus Eqs. (15-61), (15-64), and (15-67) contain the specific ion H^+ or OH^-. On the other hand, Eq. (15-58) involves the acid concentration (HA), and the rate depends on this rather that on (H^+). Experimental distinction is possible since solutions can be made up which, say, vary in (H^+) at constant (HA), or vice versa. It is the cases of general acid or base catalysis to which the Brønsted equation (15-54) applies.

A very important family of catalytic reactions comprises those involving an enzyme. Enzymes are biologically developed catalysts, each usually having some one specific function in a living organism. Pepsin, secreted by the stomach, catalyzes the hydrolysis of proteins; urease, that of urea; others are involved in the mechanism of blood clotting, in the fixation of nitrogen by legumes, and so on. Enzymes are proteins, ranging in molecular weight from about 6000 to several million. Some 150 kinds have been isolated in crystalline form.

A rather widely applicable kinetic framework for enzymatic action is that known as the *Michaelis–Menten mechanism* (1913). If an enzyme E acts to catalyze the reaction of some species, called the substrate S, then the first step is an association:

$$S + E \underset{k_{-1}}{\overset{k_1}{\rightleftharpoons}} S \cdot E \qquad (15\text{-}68)$$

The enzyme–substrate complex then breaks up into products plus free enzyme,

B. Enzyme catalysis

$$S \cdot E \xrightarrow{k_2} E + P \text{ (products)} \tag{15-69}$$

The steady-state approximation (Section 14-5C) is made:

$$\frac{d(S \cdot E)}{dt} = k_1(S)(E) - k_{-1}(S \cdot E) - k_2(S \cdot E) = 0 \tag{15-70}$$

Since by material balance, $(E) + (S \cdot E) = (E)_0$, where $(E)_0$ is the total enzyme concentration, Eq. (15-77) may be solved for $(S \cdot E)$ to give

$$(S \cdot E) = \frac{k_1(S)(E)_0}{k_1(S) + k_{-1} + k_2}$$

The rate of formation of products is v, obtained from Eq. (15-69) as

$$v = \frac{d(P)}{dt} = k_2(S \cdot E) = \frac{k_2(S)(E)_0}{(S) + [(k_{-1} + k_2)/k_1]} \tag{15-71}$$

The quantity $(k_{-1} + k_2)/k_1$ is called the Michaelis constant, K_M; for a given enzyme, it will be characteristic of the substrate.

A plot of Eq. (15-71) is shown in Fig. 15-11(a). Note that v reaches a limiting

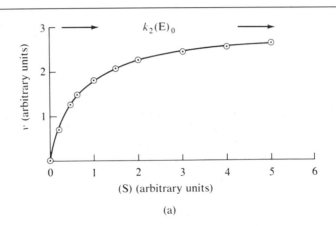

(a)

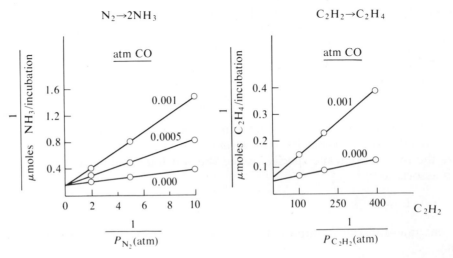

(b)

value, v_{max}, equal to $k_2(E)_0$. The quantity k_2 is shown as the *catalytic constant* or as the *turnover number* (if there is only one active site per enzyme molecule). It is sometimes given as the amount of product formed per minute (maximum rate) per unit amount of enzyme present. Catalase, for example, catalyzes the decomposition of hydrogen peroxide into water and oxygen with a k_2 value of about 10^8 min^{-1}. Further, when the rate is just half of the maximum or limiting rate, then $(k_{-1} + k_2)/k_1 = (S)_{1/2}$, so that the ratio k_1/k_{-1} can be calculated on introduction of the value for k_2.

Equation (15-71) may be put in the linear form

$$\frac{1}{v} = \frac{1}{k_2(E)_0} + \frac{k_{-1} + k_2}{k_1} \frac{1}{k_2(E)_0} \frac{1}{(S)} \tag{15-72}$$

Some early data on the reduction of nitrogen to ammonia and of acetylene to ethylene by the nitrogenase enzyme found in *Azotobacter* are shown in Fig. 15-11(b), plotted according to Eq. (15-72). See Problem 15-25 for alternative linear forms.

Many enzyme systems are more complicated kinetically than the foregoing treatment suggests. There may be more than one kind of enzyme–substrate binding site; sites within the same enzyme may interact cooperatively. Often, a cofactor is involved. That is, there may be some substance which is not an integral part of the enzyme but whose presence is necessary for enzyme activity.

As implied by the data of Fig. 15-11(b), nitrogen fixation (that is, the reduction of atmospheric nitrogen to ammonia) occurs catalytically in nature. As an example, legumes develop root nodules which harbor colonies of a bacterium capable of nitrogen fixation. The enzyme involved, nitrogenase, contains molybdenum and iron, but its exact structure is not yet known. A very schematic mechanism is shown in Fig. 15-12. Major research efforts are being directed toward elucidating this structure and toward the synthesis either of the actual enzyme or of model compounds having similar catalytic activity. The discovery of coordination compounds having dinitrogen as a ligand, such as $[(C_6H_5)_3P]_3(N_2)CoH$, seemed at first a major breakthrough,

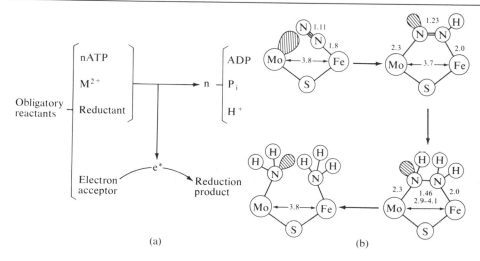

(a)   (b)

FIGURE 15-12
Mechanism of nitrogen fixation. (a) General process. (b) Possible local mechanism. [From R. W. F. Hardy, R. C. Burns, and G. W. Parshall, in "Inorganic Biochemistry" (G. L. Eichhorn, ed.), Elsevier, New York, 1973.]

but so far none has shown a significant turnover number. Not only is the problem a fascinating one in organometallic chemistry, but its solution could significantly enhance world food production.

The enzyme system is not a simple one. After all, whereas the overall reaction is downhill energetically, N_2H_2 is presumably an intermediate, and this lies some 50 kcal mol^{-1} *uphill*. The natural enzyme system indeed requires an energy source, adenosine triphosphate (ATP), and a reducing agent, ferredoxin. The ATP produces about 10 kcal mol^{-1} in its hydrolysis, and several such contributions are somehow made available to the enzyme system. In the laboratory, other reducing agents, such as hydrosulfite ($S_2O_4^{2-}$) work well (ATP is still needed). Also, the enzyme is not purely specific for N_2, but will reduce the isoelectronic compounds C_2H_2 (to C_2H_4) and HCN (to CH_4 and NH_3) as well; in the absence of other substrate, water is reduced to H_2. For reasons of analytical convenience, acetylene reduction is often used in laboratory tests of enzymatic activity and in inhibition studies. Note in Fig. 15-10(b) the inhibition by CO of both N_2 and C_2H_2 reduction. [The subject of nitrogen fixation has been reviewed by Burns and Hardy (1975).]

15-7 Transition-state theory and its applications

A. Transition-state theory for solution reactions

The formal statement of transition-state theory is the same for solution as for gas-phase reactions. An elementary bimolecular reaction is given the intimate mechanism

$$A + B \overset{K^{\ddagger}}{=} [AB]^{\ddagger} \quad \text{[Eq. (14-100)]}$$

$$[AB]^{\ddagger} \overset{k_1}{\to} \text{products [Eq. (14-101)]}$$

where $k_1 = \mathbf{k}T/h$. The rate constant is then

$$k = \frac{\mathbf{k}T}{h} K^{\ddagger} = \frac{\mathbf{k}T}{h} \exp\left(\frac{\Delta S^{0\ddagger}}{R}\right) \exp\left(\frac{\Delta H^{0\ddagger}}{RT}\right) \quad \text{[Eq. (14-106)]}$$

The equilibrium constant for forming the activated complex $[AB]^{\ddagger}$ or, alternatively, $\Delta S^{0\ddagger}$ and $\Delta H^{0\ddagger}$, must reflect not only the changes in chemical bonding that occur but also any changes in solvation. As a consequence, the complete statistical mechanical evaluation of $K^{\ddagger}$ is too complicated to be of practical use in the case of solution systems. Portions of the partition functions can be estimated, however. For example, the translational entropy change can be calculated approximately from the change in the entropy of mixing accompanying Eq. (14-100) if ideal solutions are assumed. By Eq. (9-84), the entropy of mixing of a solution of A and B for the standard state of 1 m is $-(R \ln x_A + R \ln x_B)$ and that of 1 $m[AB]^{\ddagger}$ is $-R \ln x_{[AB]}^{\ddagger}$, where each mole fraction is just $1/n_s$, n_s being the number of moles of solvent per 1000 g. We are neglecting some change in entropy of the solvent, and on this basis

$$\Delta S^{0\ddagger}_{\text{trans}} = -R \ln x_{[AB]\ddagger} + R \ln x_A + R \ln x_B = R \ln \frac{1}{ns} \quad (15\text{-}73)$$

In the case of water $n_s = 55.5$, which gives $\Delta S^{0\ddagger}_{\text{trans}} = -4R$ and $\exp(\Delta S^{0\ddagger}/R) = 0.02$.

We take up, the the subsections that follow, the use of transition-state theory as a very convenient formalism for dealing with so-called linear free energy relationships, and with the effects of nonideality on reaction rate constants. The application of transition-state theory to the effect of pressure on reaction rates in solution is presented in Section 15-ST-2.

B. Linear free energy relationships

The Brønsted equation, Eq. (15-54), is an example of what is called a *linear free energy relationship*. A related approach was used by Hammett (1940). If a series of reactions obeys Eq. (15-54) so that b is the same for each, one may take one specific reaction as a reference, with rate constant k_0 and acid constant K_0, and write

$$\log \frac{k}{k_0} = \rho \log \frac{K}{K_0} \tag{15-74}$$

where, by convention, ρ is used in place of α. The quantity $\log(K/K_0)$ is taken to be a characteristic of the system being studied and is denoted by σ, so Eq. (15-74) becomes

$$\log \frac{k}{k_0} = \rho\sigma \tag{15-75}$$

The application has been largely to reactions of substituted benzoic acids, with benzoic acid itself taken as the reference. Various types of reactions each have a characteristic ρ value, whereas σ, of course, depends only on the nature of the substituent on the benzoic acid.

The basis for the term "linear free energy relationship" is that $\log K$ is proportional to ΔG^0 for the reaction while, by transition-state theory, $\log k$ is proportional to $\Delta G^{0\ddagger}$. Equation (15-54) can thus be written

$$\Delta G^{0\ddagger} = b' + \alpha \Delta G^0 \tag{15-76}$$

This type of relationship is so often obeyed that it amounts to one of the empirical laws of kinetics, and it is important to inquire into possible explanations. It has already been pointed out that in a series of simple reactions (so that $K = k_1/k_2$) Eq. (15-56) *must* be obeyed by k_1 with $\alpha = 1$ for systems such that k_2 is at the diffusion-controlled limit, and with $\alpha = 0$ for those such that k_1 is at this limit. Intermediate values of α then arise naturally for cases lying in the transition region between these extremes. Certain sets of reactions may then appear to have an intermediate and constant value of α simply because an insufficient range of k or K values is experimentally accessible.

Data such as those of Fig. 15-10 seem to require a different explanation—the linearity extends over too large a range of values for the constancy of α to be an artifact. The alternative possibility is that there is in fact an intrinsic proportionality between $\Delta G^{0\ddagger}$ and ΔG^0 for a series of related reactions. A simple exposition is the following.

If the reactions, being related, have a constant $\Delta S^{0\ddagger}$, then the proportionality is actually between $\Delta H^{0\ddagger}$ and ΔH^0, or between the activation energy and the overall energy of reaction. Suppose, for example, that the reaction is one of proton transfer,

$$HA_1 + A_2^- \rightarrow A_1^- + HA_2$$

The reaction could proceed by the mechanism

$$HA_1 \rightarrow H^+ + A_1^- \qquad \Delta H_1^0 \quad \text{(rate determining)}$$

$$H^+ + A_2^- \rightarrow HA_2 \qquad \text{(fast)}$$

for which the activation energy would be ΔH_1^0. If, however, the reaction takes place by the route

$$HA_1 + A_2^- \rightarrow [A_1 \cdots H \cdots A_2]^- \rightarrow A_1^- + HA_2$$

then, as illustrated in Fig. 15-13, the activation energy should be smaller since complete breaking of the $H - A_1$ bond is not required. If now some second reactant HA_2' is employed and the general shape of the energy curves remains the same, as illustrated by the dashed line in the figure, then the geometry of the situation indicates that the change in activation enthalpy $\Delta(\Delta H^{0\ddagger})$ should be proportional to the change in overall reaction enthalpy, $\Delta(\Delta H^0)$, which leads to Eq. (15-56) if the entropies do not change. This last assumption may be a poor one, but it turns out that very often ΔS and ΔH quantities for a given process are linearly related to each other [Barclay and Butler (1938)], and so Eq. (15-56) is still obeyed.

C. Ionic reactions. Role of activity coefficients

Reactions in the gas phase are generally at pressures of 1 atm or less so that the species are essentially ideal in their behavior. By contrast, solution systems are often distinctly nonideal, and this is true for electrolytes even at quite low concentrations. An important question is then whether the mass action law is correctly applied to elementary reactions when concentrations rather than activities are used in the rate expression. Alternatively, if we retain the mass action law, then does the rate constant contain activity coefficient quantities?

Both transition-state theory and collision theory as modified for solutions affirm that rates should depend on activity coefficients. Considering the for-

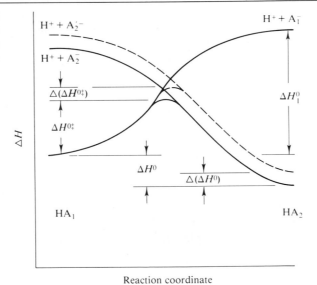

FIGURE 15-13
Possible explanation for the existence of linear free energy relationships.

Reaction coordinate

mer first, we see that the derivation of Eq. (14-103) (repeated above) should really be

$$\text{rate} = \frac{\mathbf{k}T}{h} ([AB^{\ddagger}]) = \frac{\mathbf{k}T}{h} \frac{\gamma_A \gamma_B}{\gamma_{(AB)^{\ddagger}}} K^{\ddagger} (A)(B)$$

or

$$k = \frac{\mathbf{k}T}{h} \frac{\gamma_A \gamma_B}{\gamma_{(AB)^{\ddagger}}} \exp\left(\frac{\Delta S^{0\ddagger}}{R}\right) \exp\left(-\frac{\Delta H^{0\ddagger}}{RT}\right)$$

$$= \frac{\mathbf{k}T}{h} \frac{1}{K_{\gamma}^{\ddagger}} \exp\left(\frac{\Delta S^{0\ddagger}}{R}\right) \exp\left(-\frac{\Delta H^{0\ddagger}}{RT}\right) \tag{15-77}$$

The difficulty, of course, is in the evaluation of $\gamma^{[AB]\ddagger}$, and for neutral molecules the activity coefficient factor $1/K_{\gamma}^{\ddagger}$ is generally ignored—one can argue that in dilute solutions each activity coefficient is close to unity and further that small departures from unity will tend to cancel each other.

The remaining situation is that of a reaction involving ions, and here the Debye–Hückel theory allows estimation of each γ purely on the basis of the charge of the species. Consider the elementary reaction

$$A^{z_A} + B^{z_B} \rightarrow [AB]^{\ddagger z_A + z_B} \rightarrow \text{products}$$

for which the activity coefficient factor $1/K_{\gamma}^{\ddagger}$ is given by

$$\frac{1}{K_{\gamma}^{\ddagger}} = \frac{\gamma_A \gamma_B}{\gamma_{(AB)^{\ddagger}}} \tag{15-78}$$

The Debye–Hückel limiting law is

$$\ln \gamma_i = -pz_i^2 \qquad \text{[Eq. (12-92)]}$$

where the constant $p = e^2\kappa/2\,DkT$ and κ is proportional to the square root of the ionic strength I [by Eq. (12-93)], where

$$I = \tfrac{1}{2} \sum m_i z_i^2 \qquad \text{[Eq. (12-76)]}$$

It follows that

$$\ln \frac{1}{K_{\gamma}^{\ddagger}} = -[z_A^2 + z_B^2 - (z_A + z_B)^2]\, \alpha \, \sqrt{I} = 2 z_A z_B \alpha \, \sqrt{I}$$

where for water as solvent $\alpha = 1.172$ at 25°C. Equation (15-77) may then be written

$$\ln k = \ln k_0 + 2 z_A z_B \alpha \, \sqrt{I} \tag{15-79}$$

where k_0 is the rate constant if $K_{\gamma}^{\ddagger}$ is unity—for example, at infinite dilution. Note that the same result follows from collision theory as phrased in the form of Eq. (15-49). We now deal with $\gamma_{[AB]}$, but the Debye–Hückel theory involves only the net charge on a species, and the distinction makes no difference.

Equation (15-79) predicts a linear relationship if log k is plotted against $\sqrt{I}$, with a slope proportional to $z_A z_B$. The qualitative prediction has been confirmed in that with increasing ionic strength reactions between like charged ions increase in rate constant and those between oppositely charged ions decrease in rate constant. It is questionable, however, whether Eq. (15-79) has ever been verified quantitatively. The difficulty is that the effect predicted is not very large relative to the precision of rate data until such ionic strengths are reached that serious departure from the Debye–Hückel limiting law occurs and Eq. (15-79) should not hold anyway. The reader is referred to Section 12-CN-1 for a discussion of activity coefficients of electrolytes in more concentrated solutions. There is no doubt, however, that nonparticipating electrolytes or "neutral" salts do affect reaction rate constants. Figure 15-14 shows a traditional plot of rate constants for various types of ionic reactions.

EXAMPLE   The slow step in the reduction of Hg^{2+} by Fe^{2+} is thought to be $Fe^{2+} + Hg^{2+} \rightarrow Fe^{3+} + Hg^+$, with $k = 1 \times 10^{-4} M^{-1} s^{-1}$ at 25°C. The rate constant was determined at an ionic strength of 0.2, however. Find k_0. From Eq. (15-79), $\ln k_0 = \ln k - (2)(2)(2)(1.172)(0.2)^{1/2}$, whence $\ln k_0 = -13.4$, and $k_0 = 1.5 \times 10^{-6} M^{-1} s^{-1}$.

FIGURE 15-14
Effect of ionic strength on the rates of some ionic reactions.

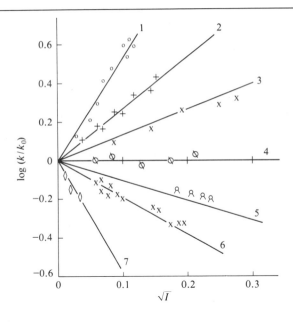

(1) $Co(NH_3)_5Br^{2+} + Hg^{2+} + H_2O \rightarrow Co(NH_3)_5 (H_2O)^{3+} + HgBr^+$.
(2) $S_2O_8^{2-} + I^- \rightarrow (SIO_4^- + SO_4^{2-}) \rightarrow I_3^- + 2SO_4^{2-}$.
(3) $O_2N—N—COOC_2H_5^- + OH^- \rightarrow N_2O + CO_3^{2-} + C_2H_5OH$.
(4)   *cane sugar* $+ OH^- \rightarrow$ *invert sugar*.
(5) $H_2O_2 + H^+ + Br^- \rightarrow H_2O + \frac{1}{2}Br_2$ (*not balanced*).
(6) $Co(NH_3)_5Br^2 + OH^- \rightarrow Co(NH_3)_5(OH)^{2+} + Br^-$.
(7) $Fe^{2+} + Co(C_2O_4)_3^{3-} \rightarrow Fe^{3+} + Co(C_2O_4)_3^{4-}$.

[Adapted from "The Foundations of Chemical Kinetics," by S. W. Benson, Copyright 1960, McGraw-Hill, New York. (Used with permission of McGraw-Hill Book Company) in which the detailed references may be found. The interested reader should also refer to A. Olson and T. Simonson, J. Chem. Phys. **17**, 1167 (1949) for some adverse comments, and also to M. Kilpatrick, Ann. Rev. Phys. Chem. **2**, 169 (1951).]

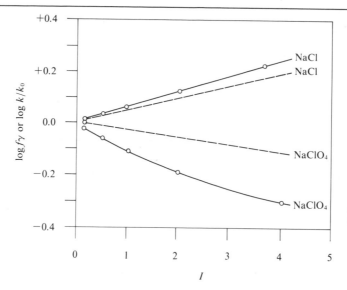

FIGURE 15-15
Effect of salt concentration on ln (k/k_0), solid lines, for the acid-catalyzed hydrolysis of γ-butyrolactone. The corresponding plots of ln $\gamma_{lactone}$ are given by the dashed lines. [From F. A. Long, W. F. McDevit, and F. B. Dunkle, *J. Phys. Colloid Chem.* **55**, 813 (1951), and F. A. Long, F. B. Dunkle, and W. F. McDevit, ibid. **55**, 829 (1951).]

The determination of a rate law usually involves changes in concentrations of reactants, as in the usual procedure of fitting data to an integrated expression, and hence changes in the ionic makeup of the solution. The consequent activity coefficient changes can be severe enough to lead to error in the determination of the actual form of the rate law. An example is the reaction

$$Co(NH_3)_5(H_2O)^{3+} + X^- \rightarrow Co(NH_3)_5X^{2+} + H_2O$$

where X^- denotes a halogen ion. Activity coefficient (or, alternatively, ion pairing) effects are so large that early studies appeared to give first-order kinetics, although the reaction is actually second order. One can reduce such effects by conducting the reaction in a medium having a high concentration of nonreactive salt such as 1 M sodium perchlorate. The activity coefficients of the reactants may then be nearly independent of their concentration.

A reaction involving neutral species should not be influenced by ionic strength, according to Eq. (15-79). In fact, however, there are salt or medium effects. An empirical observation is that for an uncharged species

$$\ln \gamma = Cm \tag{15-80}$$

[note Eq. (12-102)]. The effect of electrolyte concentration on $\ln(k/k_0)$ is shown in Fig. 15-15 for the acid-catalyzed hydrolysis of γ-butyrolactone (to γ-hydroxybutryic acid). Note that $\ln(k/k_0)$ is approximately linear in m and, moreover, tends to parallel the effect of salt concentration on the activity coefficient of the lactone.

COMMENTARY AND NOTES

15-CN-1 Comparison of collision–encounter and transition-state theories

It turns out to be quite difficult to devise a means of distinguishing experimentally between collision theory (as modified for solutions) and transition-state theory. In the case of the usual reaction of activation energy greater than about 10 kcal mol^{-1}, both treatments lead essentially to the experimentally observed Arrhenius equation. Both predict about the same frequency factor to be the "normal" one. The activity coefficient formulations are indistinguishable.

There seems no doubt, however, that the collision–encounter picture is physically correct, at least with respect to the sequence of events up to actual chemical change. There are many substantiations of the cage model for liquids, and we have seen that bimolecular reactions in solution *do* reach a diffusion-controlled limit. Further, the existence of reactions showing the same rate constant in a solvent as in the gas phase indicates that collision frequencies, as distinct from encounter frequencies, are about the same in the two phases, provided that specific molecular interactions in the encounter complex are weak.

Transition-state theory, of course, offers a most appealing formalism in ascribing essentially well-behaved thermodynamic properties to a transition state or activated complex. The formalism leads naturally to pleasing explanations of the existence of linear free energy relationships. Also, the term $\exp(\Delta S^{0\ddagger}/R)$ is amenable to some very useful structural interpretations.

For example, the denaturation of proteins—in everyday experience, the cooking of an egg—is enormously temperature-sensitive. The denaturation rate doubles with about every 1° rise in temperature (the boiling point of water has dropped by about 10°C at 14,000 ft elevation and the time to boil an egg has increased about 100-fold), corresponding to an activation energy of about 130 kcal mol^{-1}. The term $\exp(-E^*/RT)$ is then about 10^{-75} (!) at 100°C, so that the rate of denaturation should be negligible on this basis. The reaction does in fact

occur, of course, and the transition-state formalism leads to a calculated compensating entropy of activation of about 300 cal K^{-1} mol^{-1}. This is understandable since denaturation corresponds to an unraveling of the secondary and tertiary structure of the protein to a far more random structure. If the transition state is one of partial unraveling, then $\Delta S^{0\ddagger}$ should indeed be a large positive number.

Although this is an extreme example, the point is that $\Delta S^{0\ddagger}$ often lies between zero and ΔS^0 for the overall reaction. The ratio $\Delta S^{0\ddagger}/\Delta S^0$ then appears to be a useful indication of the extent to which the transition state resembles either reactants or products. These are esthetic arguments, however. Their acceptance is more a matter of taste than of scientific necessity since comparable and not much more contrived rationalizations can be made in terms of the collision–encounter model.

An important point concerning reactions in solution has to do with the manner of energy acquisition. Transition-state theory is silent on this matter since it begins with the assumption that the activated complex is in equilibrium with its surroundings. Collision theory, however, suggests the following important difference between solution- and gas-phase reactions. In the latter case activation energy E^* is gained from the collisions of the reactants, drawing on their kinetic energy of impact and on the energy distributed among various internal degrees of freedom. In solution, however, the reactants that diffuse together to form the encounter complex are essentially of normal energy content. The activation energy must then be supplied by collisions of surrounding solvent molecules. Perhaps more accurately, the Boltzmann distribution of energies in a liquid is present as regions of extra vibrational energy; these regions move around rapidly and can be thought of as soundlike waves in the liquid—called *phonons*. The activation of an encounter complex probably occurs when a confluence of phonon waves momentarily concentrates an unusually large amount of energy in the vicinity of the complex. The concentration must be a momentary one since phonon waves travel with about the velocity of sound, and would pass the region of the encounter complex in about 10^{-12} sec. The lifetime of the energized complex must thus be very small, if its energy resides in vibrational excitations.

In summary, in the collision–encounter model, the encounter complex replaces the activated complex of transition-state theory; the two "complexes" are similar in concept except that the former possesses no unusual energy. The principal question, not yet answered, is whether the energized complex may have too short a period of existence to be regarded as a species in a definite thermodynamic state.

15-CN-2 Oscillating chemical reactions

A fascinating aspect of chemical kinetics is that of the treatment of oscillating chemical reactions. A specific example of such a reaction is the cerium-ion-catalyzed oxidation of malonic acid by bromate [see Field (1972)]. The mixture might, for example, be 0.25 M in malonic acid, 0.06 M in $KBrO_3$, and 1.5 M in sulfuric acid. Cerous ammonium nitrate, $Ce(NH_4)_2(NO_3)_5$, is present in catalytic amounts, about 0.002 M, and a trace of the redox indicator Ferroin is present to make the changes more evident. After initial mixing there is an induction period; then, as shown in Fig. 15-16, oscillations occur both in Br^- concentration and in the Ce^{IV}/Ce^{III} ratio and can go on for an hour or more until the system exhausts itself. With Ferroin present, one sees periodic color changes from blue to violet and back again. The overall reaction is

$$3H^+ + 3BrO_3^-$$
$$+ 5CH_2(COOH)_2 \rightarrow 3BrCH(COOH)_2$$
$$+ 2HCOOH + 4CO_2 + 5H_2O$$

$$(15\text{-}81)$$

The detailed kinetics are fiercely complicated. It appears that three main processes are present. The first is

(A) $BrO_2^- + 2Br^- + 3CH_2(COOH)_2$
$\quad + 3H^+ \rightarrow 3BrCH(COOH)_2 + 3H_2O$

$$(15\text{-}82)$$

$$BrO_3^- + Br^- + 2H^+ \xrightarrow{k_1} HBrO_2 + HOBr \text{ (slow)}$$

$$(15\text{-}83)$$

$$HBrO_2 + Br^- + H^+ \xrightarrow{k_2} 2HOBr$$

followed by reaction of Br^- with HOBr to give Br_2 and of Br_2 with malonic acid to give $BrCH(COOH)_2$. Steady-state analysis of this subset of reactions gives

$$(HBrO_2)_A = k_1(BrO_3^-)(H^+)/k_2 \qquad (15\text{-}84)$$

The second main process that is proposed is

(B) $BrO_3^- + 4Ce^{III}$
$\quad + CH_2(COOH)_2$
$\quad + 5H^+ \rightarrow BrCH(COOH)_2$
$\quad + 4Ce^{IV} + 2H_2O$

$$(15\text{-}85)$$

for which the detailed mechanism may be

$$BrO_3^- + HBrO_2 + H^+ \xrightarrow{k_3} 2BrO_2\cdot + H_2O$$

$$\text{(slow)}$$

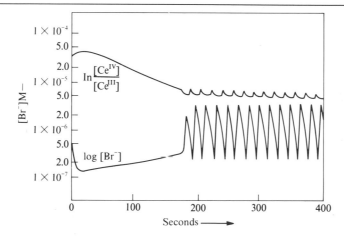

FIGURE 15-16

Oscillations of (Br^-) and of the (Ce^{IV})/(Ce^{III}) ratio in the cerium ion catalyzed oxidation of malonic acid by bromate (the Belousov reaction). Initial concentrations are: $CH_2(COOH)_2$, 0.50 M, $KBrO_3$, 0.063 M, $Ce(NH_4)_2(NO_3)_5$, 0.001 M, H_2SO_4, 0.8 M. [From R.J. Field, E. Koros, and R.M. Noyes, J. Amer. Chem. Soc., **94**, 8649 (1972).]

$$BrO_2 \cdot + Ce^{III} + H^+ \xrightarrow{k_4} HBrO_2 + Ce^{IV}$$

$$(15\text{-}86)$$

$$BrO_2 \cdot + Ce^{IV} + H_2O \xrightarrow{k_5} BrO_3^- + Ce^{III} + 2H^+$$

$$2HBrO_2 \xrightarrow{k_6} BrO_3^- + HOBr + H^+$$

followed by bromination of malonic acid by HOBr. With some approximations Eqs. (15-86) lead to a steady-state condition

$$(HBrO_2)_B = k_3(BrO_3^-)(H^+)/k_6 \qquad (15\text{-}87)$$

The third overall process is the oxidation of $BrCH(COOH)_2$ by Ce^{IV} to give Br^-, $HCOOH$, and CO_2.

Analysis of the various individual reactions suggests that (k_1/k_2) is about 10^{-5} of (k_3/k_6), so that $(HBrO_2)_A$ for process A alone should be about 10^{-5} of $(HBrO_2)_B$ for process B alone. A global steady-state analysis is required for the complete reaction system, but there is a suggestion that the system may oscillate more or less between these two separate steady-state values [see Gray (1973), Field (1972)].

The above illustrates how complicated the

FIGURE 15-17
Oscillations in a homogeneous, closed system. (a) Oscillation around equilibrium, which is not allowed. (b) Oscillation around a quasi-steady state that is approaching equilibrium. [See Degn (1972).]

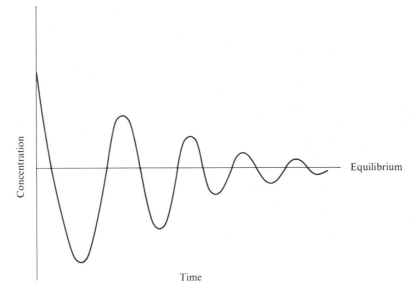

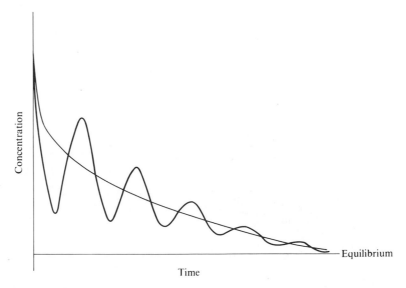

kinetics of even a simple lecture demonstration experiment can be. It also illustrates the type of requirement for an oscillating system, namely a feedback or autocatalytic system of reactions (note Problem 14-29). In the present case, $HBrO_2$ both is formed by reaction of BrO_3^- (Eq. 15-83) and reacts with it (Eq. 15-86).

The subject of chemical oscillations has fascinated physical chemists for most of this century. It can be shown (through a second law argument) that oscillations *around* equilibrium, as illustrated in Fig. 15-17(a), are not allowed. What we are discussing is the matter of an oscillatory approach to equilibrium in a reacting system, as illustrated in Fig. 15-17(b). Even this last possibility has been the subject of much debate; experimentally observed oscillations have been attributed to supersaturation effects, to adsorption effects on dust particles, and similar causes. To make matters worse, reaction systems that are potentially capable of showing oscillation are too complicated to solve exactly. Modern computers, however, make it possible to make reaction trajectory calculations in which the differential equations are applied over many successive very small time intervals, amounting to a numerical integration of the system of rate equations. Such calculations have confirmed that oscillations are possible in a reacting system.

Mathematical studies have centered on hypothetical (but experimentally conceivable) reaction schemes. Lotka showed in 1910 that the system

$$A + X \rightarrow 2X$$
$$X + Y \rightarrow 2Y \qquad (15\text{-}88)$$
$$A + Y \rightarrow \text{inert}$$

gives undamped oscillations in (X) and (Y) if (A) is constant. The mathematics is that of coupled differential equations whose solution can show a singularity for a particular (X) and (Y). Considerable progress has been made since in the theory of oscillatory systems, but the subject is still a difficult and incomplete one [see Gray (1973)].

The above case of (A) constant is essentially that of a flow reactor in which A is introduced at a constant rate. Situations of this type occur in chemical engineering processes, and there are numerous examples of reported oscillations in industrial reactors. These may include gas phase reactions, such as the oxidation of CO. Finally, systems may show *thermal* oscillations as in combustion reactions. An explosion (note Section 14-5D) is essentially an undamped oscillation!

SPECIAL TOPICS

15-ST-1   Integrated forms for some additional rate laws

There are a few additional cases that occur often enough to warrant inclusion in this text. More extensive coverage can be found in monographs such as that by Benson (1960).

A. Branching first-
order reactions

A compound may be able to react in two different ways. An example from coordination chemistry is

where "en" denotes the bidentate ligand ethylenediamine and X and Y are halogens or pseudohalogens. If both reactions are first order, then the rate of disappearance of reactant A is

$$\frac{d(A)}{dt} = -(k_1 + k_2)(A) \qquad (15\text{-}89)$$

or still first order, but with $k_{app} = k_1 + k_2$. The rates of formation of the alternative products B and C are

$$Co(en)_2(X)(Y)^+ + H_2O \underset{k_2}{\overset{k_1}{\rightleftarrows}} \begin{matrix} Co(en)_2(H_2O)(X)^{2+} + X^- \\ B \\ Co(en)_2(H_2O)(Y)^{2+} + Y^- \\ C \end{matrix} \qquad (15\text{-}90)$$

A

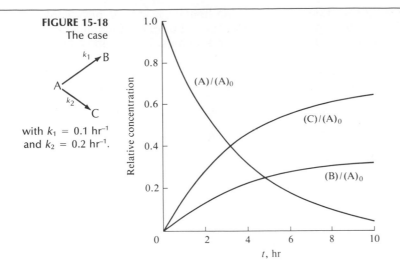

FIGURE 15-18
The case

with $k_1 = 0.1$ hr^{-1}
and $k_2 = 0.2$ hr^{-1}.

$$\frac{d(B)}{dt} = k_1(A) \qquad \frac{d(C)}{dt} = k_2(A)$$

Thus the ratio of (B) to (C) at any time must be k_1/k_2, and their sum, of course, must be $(A)_0 - (A)$. If, for example, k_1 and k_2 were 0.1 hr^{-1} and 0.2 hr^{-1}, respectively, the plots of concentrations versus time would look as shown in Fig. 15-18.

The products B and C will have some equilibrium ratio themselves, and this in general will not be the same as k_1/k_2. The example of Fig. 15-18 is based on the assumption that the B–C equilibrium is slow, so that the product ratio (B)/(C) is kinetically determined. The general situation is

$$\tag{15-91}$$

If k_3 and k_{-3} are large compared to $k_1 + k_2$, then B and C will equilibrate as they are produced, and the observed product ratio will be the equilibrium one. The distinction between these two situations is an important one if the rate data are to be used in mechanistic interpretations.

EXAMPLE   If X and Y in Eq. (15-89) are Cl$^-$ and NCS$^-$, respectively, it turns out that k_1 is much larger than k_2, so the reaction is exclusively

$$Co(en)_2(NCS)(Cl)^+ + H_2O \rightarrow Co(en)_2(H_2O)(NCS)^{2+} + Cl^- \tag{15-92}$$

The equilibration between the two possible products is slow, so that in this case the product ratio is kinetically determined. The equilibrium also strongly favors the product $Co(en)_2(H_2O)(NCS)^{2+}$, so that the rate constants parallel the equilibrium constants K_1 and K_2, that is, $k_1 \gg k_2$ and $K_1 \gg K_2$. A linear free energy relationship thus appears to be present.

This need not be the case. Continuing with the same example, the complex $Co(en)_2(NCS)(Cl)^+$, being octahedral, has *cis* and *trans* isomers as shown in the accompanying diagram. The *cis* isomer is the thermodynamically favored one, and if allowed to aquate according to Eq. (15-92), the product is entirely *cis*-$Co(en)_2(H_2O)(NCS)^{2+}$. However, if the *trans* isomer is the starting material, then about 60% *cis* and 40% *trans* product results. In this last instance, then, the reaction is definitely subject to some kinetic, stereospecific control. If X = Cl$^-$ and Y = NO$_2^-$ the chloride aquation reaction, which again

Special topics **643**

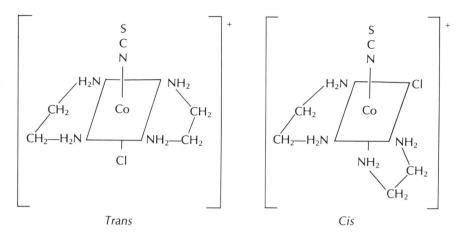

Trans

Cis

dominates, is 100% stereospecific, *cis* starting material giving *cis* product only and *trans* starting material giving *trans* product only.

The reaction scheme of Eq. (15-91) is the classic one of a triangular reaction system. The mathematics was investigated extensively in the 1920s, and for a while it appeared that if an equilibrium system were perturbed by, say, the sudden addition of more A, then the concentrations of A, B, and C could undergo oscillation. While we now know that indefinite oscillation will not occur, systems of coupled reactions *can* oscillate, although the oscillations damp out and the system finally approaches ordinary equilibrium—see Fig. 15-17 and the associated discussion.

B. Sequential first-order reactions

The reaction sequence

$$A \xrightarrow{k_1} B \xrightarrow{k_2} C \quad (15\text{-}93)$$

occurs often enough that its mathematical behavior should be explored. Each reaction is first-order or pseudo-first-order, and each goes to completion. The equations for (A) are as before,

$$\frac{d(A)}{dt} = -k_1(A) \quad [\text{Eq. (14-8)}]$$

$$(A) = (A)_0 e^{-k_1 t} \quad [\text{Eq. (14-10)}]$$

For (B), however, we have

$$\frac{d(B)}{dt} = k_1(A) - k_2(B) \quad (15\text{-}94)$$

This corresponds to the special case of Eq. (14-59) for which $k_{-1} = 0$. On putting in this condition, the result is

$$\frac{(B)}{(A)_0} = \frac{k_1}{k_2 - k_1}(e^{-k_1 t} - e^{-k_2 t}) \quad (15\text{-}95)$$

[Note that B corresponds to the intermediate I in the sequence of Eq. (14-53).]

If (C) is the final product, then, by material balance,

$$(C) = (A)_0 - (A) - (B). \quad (15\text{-}96)$$

The behavior of the reaction sequence of Eq. (15-93) is somewhat complex since, although (A) is always given by Eq. (14-10), there is a qualitative difference in the behavior of (B) and hence of (C) for $k_2 > k_1$ versus $k_2 < k_1$.

CASE 1. $k_2 > k_1$. Equation (15-95) may be put in the form

$$(B) = \frac{k_1(A)_0}{k_2 - k_1}e^{-k_1 t}$$
$$(1 - e^{-(k_2 - k_1)t}) \quad (15\text{-}97)$$
$$= \frac{k_1(A)}{k_2 - k_1}(1 - e^{-(k_2 - k_1)t})$$

The second exponential term of Eq. (15-97) goes to zero at times long compared to $1/(k_2 - k_1)$, and in this limit

$$\left(\frac{(B)}{(A)}\right)_{\text{lim}} = \frac{k_1}{k_2 - k_1} \quad \text{or}$$
$$(B)_{\text{lim}} = \frac{k_1}{k_2 - k_1}(A)_0 e^{-k_1 t} \quad (15\text{-}98)$$

Thus (B) eventually must parallel (A), differing from it by the constant factor $k_1/(k_2 - k_1)$. This limiting condition is known as one of *transient equilibrium*, a name given by early radiochemists; sequences such as Eq. (15-93) are especially common with radioactive species (see Section 21-5). Since (B) = 0 at $t = 0$ and returns toward zero as (A) goes to zero, we conclude that (B) must have a maximum value, as is indeed the case. At the maximum $d(B)/dt = 0$, and so, by Eq. (15-94), $k_1(A) = k_2(B)$. Insertion of this relationship into Eq. (15-97) gives

$$e^{-(k_2 - k_1)t} \, _{\max} = \frac{k_1}{k_2} \qquad (15\text{-}99)$$

$$t_{\max} = \frac{\ln(k_2/k_1)}{k_2 - k_1}$$

Figure 15-19(a) shows the variation of (A), (B), and (C) with time for the case of $k_1 = 0.1$

hr^{-1} and $k_2 = 0.3 \text{ hr}^{-1}$. Note that (B) never exceeds (A); in general, the mathematics of this case does not allow B ever to be the majority species. Note also that (C) shows a time lag or induction period before beginning to rise rapidly; the inflection point is at $t_{\max}$ of Eq. (15-99).

CASE 2. $k_1 > k_2$. We now write Eq. (15-95) in the form

$$(B) = \frac{k_1(A)_0}{k_1 - k_2} e^{-k_2 t}(1 - e^{-(k_1 - k_2)t}) \qquad (15\text{-}100)$$

The second exponential term goes to zero at times long compared to $1/(k_1 - k_2)$, to give

$$(B)_{\lim} = \frac{k_1}{k_1 - k_2} (A)_0 \, e^{-k_2 t} \qquad (15\text{-}101)$$

The previous condition of transient equilibrium is not present—(B) simply disappears accord-

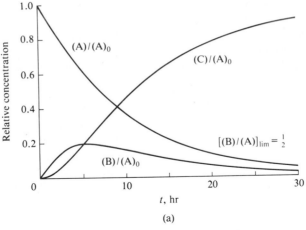

FIGURE 15-19
Sequential first-order reactions. (a) The case of transient equilibrium, $k_2 > k_1$. (Note the similarity to Fig. 14-6.)

(a)

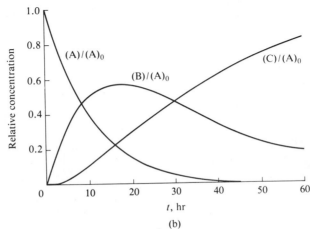

(b)

ing to its own rate constant k_2. If $k_1 \gg k_2$, (A) disappears very quickly, leaving only (B), which then reacts at its own slower rate:

$$(B)_{lim} = (A)_0 \, e^{-k_2 t}$$

An illustrative set of curves is shown in Fig. 15-19(b) for $k_1 = 0.1 \text{ hr}^{-1}$ as before, but now with $k_2 = 0.0333 \text{ hr}^{-1}$. The concentration (B) goes through a maximum as before, with t_{max} again given by Eq. (15-99), and for a time B is the dominant species. Again, there is an induction period for (C).

The above represent two important special cases. There is, however, a general solution for *any* scheme of coupled first-order reactions [see Benson (1960)].

C. Reversible second-
order reactions

We consider the simple reaction

$$A + B \underset{k_2}{\overset{k_1}{\rightleftharpoons}} C + D$$

for which the rate law is

$$\frac{d(A)}{dt} = -k_1(A)(B) + k_2(C)(D)$$

If a and b denote the initial concentrations of A and B, respectively, and x denotes the concentration of products, so that $(A) = (a - x)$ and $(B) = (b - x)$, we have

$$\frac{dx}{dt} = k_1(a - x)(b - x) - k_2 x^2$$

(supposing no C or D to be present initially), or

$$\frac{dx}{\alpha + \beta x + \gamma x^2} = dt \qquad (15\text{-}102)$$

where $\alpha = k_1 ab$, $\beta = -k_1(a + b)$, and $\gamma = k_1 - k_2$. Equation (15-102) integrates to give

$$\ln \frac{\gamma x + \frac{1}{2}(\beta - q^{1/2})}{\gamma x + \frac{1}{2}(\beta + q^{1/2})} = q^{1/2} t + \delta \qquad (15\text{-}103)$$

where $q = \beta^2 - 4\alpha\gamma$ and δ is the constant of integration, determined by setting $x = 0$ and $t = 0$. Equation (15-103) is difficult to use experimentally. A problem is that, given some experimental error in the data points and a limited range of t, it may be possible to find choices of k_1 and k_2 that appear to fit the results even though some other rate law is actually the correct one. One should either choose experimental conditions such that the system becomes a reversible first-order one (such as by having excess B and D present) or, at least, obtain an independent relation between k_1 and k_2 (such as from the equilibrium constant). Figure 15-20) illustrates the relatively uninformative shape of an $(A)/(A)_0$ versus time plot given by Eq. (15-103); it is calculated for $\alpha = 0.1 \ M$ and $b = 0.2 \ M$.

D. Rate law for
isotopic exchange

A type of kinetic approach that has been widely employed in the study of reaction mechanisms is isotopic labeling to follow the exchange of

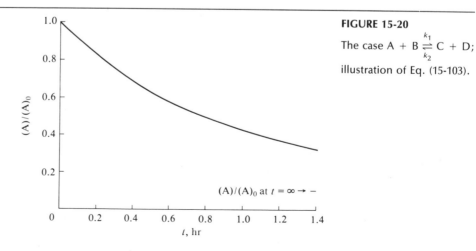

$(A)/(A)_0$ at $t = \infty \rightarrow -$

t, hr

FIGURE 15-20

The case $A + B \underset{k_2}{\overset{k_1}{\rightleftharpoons}} C + D$;

illustration of Eq. (15-103).

an atom or group between two chemical states. An example of an exchange reaction is

$$RBr + {}^{80}Br^- = R{}^{80}Br + Br^- \qquad (15\text{-}104)$$

where R might be an alkyl group, and radioactive ${}^{80}Br$ is used as a label. Other examples are

$$SO_4^{2-} + H_2{}^{18}O$$
$$= SO_3{}^{18}O^{2-} + H_2O \qquad (15\text{-}105)$$

where oxygen exchange is studied with the use of the stable isotope ${}^{18}O$ and

$$Ni(CN)_4^{2-} + {}^{14}CN^-$$
$$= Ni(CN)_3({}^{14}CN)^{2-} + CN^- \qquad (15\text{-}106)$$
$${}^{55}Fe^{3+} + Fe^{2+} = {}^{55}Fe^{2+} + Fe^{3+}$$

$$(15\text{-}107)$$

in which radiocarbon and iron are tracers.

The normal exchange procedure consists of establishing an equilibrium system containing the two chemical species AX and BX, having atom X in common and exchange between which is to be studied, and following the rate of appearance of labeled BX if the labeling was originally present in AX. That is, the reaction type is

$$AX^* + BX \rightleftharpoons AX + BX^* \qquad (15\text{-}108)$$

where the asterisk denotes the presence of a radioactive atom or an excess of some stable isotope of the atom. Samples of the system are taken periodically, compounds A and B are physically separated by some procedure (such as precipitation or solvent extraction), and the content of labeled atom in each is determined. It is ordinarily assumed that different isotopes of the same atom have essentially identical chemistry, which means that at exchange equilibrium the proportion of labeled X atoms must be the same in compound AX as in compound BX.

It turns out that these conditions imply that the kinetics of an exchange reaction will always be of the first-order reversible type, regardless of the actual mechanism whereby the exchange occurs. The demonstration of this conclusion is as follows. Since the system of Eq. (15-108) is at chemical equilibrium—that is, the amounts of AX and BX are not changing with time—it follows that R_f, the rate at which chemical species AX forms chemical species BX, is just equal to R_b, the rate of the back reaction. We denote

the total amounts of AX and of BX by a and b, respectively, and the amount of AX* and BX* by x and y, respectively,

$$\begin{array}{cccc} AX^* + & BX \rightleftharpoons & AX + & BX^* \\ x & b & a & y \end{array}$$

and suppose that x_0 is the initial amount of AX* (and $y_0 = 0$). Then the rate of appearance of BX* must be

$$\frac{dy}{dt} = R\frac{x}{a} - R\frac{y}{b} \qquad (15\text{-}109)$$

where $R = R_f = R_b$. That is, the rate of appearance of BX* is the overall chemical rate R times the fraction of atoms X in compound AX that are labeled, and the rate of back exchange is R times the fraction of atoms X in compound BX that are labeled. Since $x_0 = x + y$, Eq. (15-109) can be written

$$\frac{dy}{dt} = \frac{Rx_0}{a} - R\left(\frac{1}{a} + \frac{1}{b}\right)y$$

Separation of variables and integration gives

$$\frac{Rx_0}{a} - cy$$
$$= (\text{constant})\, e^{-ct} \qquad \text{where} \qquad (15\text{-}110)$$
$$c = R\left(\frac{1}{a} + \frac{1}{b}\right)$$

The constant of integration is evaluated from $y = 0$ at $t = 0$, and rearrangement leads to

$$cy = \frac{Rx_0}{a}(1 - e^{-ct})$$

However at exchange equilibrium, $y_\infty = x_0 b/(a + b)$, so

$$\frac{Rx_0}{a} = \frac{Ry_\infty}{ab/(a + b)} = cy_\infty$$

and the final form of the exchange rate equation is

$$1 - \frac{y}{y_\infty} = e^{-ct} \qquad (15\text{-}111)$$

which is a special case of Eq. (15-5). Thus a plot of $\ln[1 - (y/y_\infty)]$ versus t should give a straight line of slope $-c$.

Each exchange experiment provides a value of the exchange rate constant c for some particular set of values of a and b. Use of Eq. (15-110) then gives the rate R of the chemical re-

The exchange of n-C_4H_9I with I^- was followed at 50°C with the use of radioiodide ion as tracer. Samples were withdrawn periodically and the amount of radioactivity associated with the butyl iodide determined. The plots of $[1 - (y/y_\infty)]$ versus t for three sets of concentrations are shown in Fig. 15-21). Thus for (RX) = 0.1 M and (I^-) = 0.2 M, the exchange half-life is 0.588 hr, or c = 1.179 hr^{-1}. From Eq. (15-110), we find the rate of reaction R = 1.179/[1/0.1) + (1/0.2)] = 0.0786 liter mole^{-1} hr^{-1}. The same slope and R value are obtained if (RX) = 0.2 and (I^-) = 0.1. If both concentrations are 0.2 M, then the exchange half-life becomes 0.441 hr, which leads to an R value of 0.157. Thus doubling (RX) at constant (I^-) doubles the rate, indicating the rate law to be first order in (RX), and doubling (RX) while halving (I^-) does not change the rate, implying first-order dependence on (I^-) as well.

The rate then appears to be given by $R = k(RI)(I^-)$, with k = 0.0786/(0.1)(0.2) = 3.93 liter mol^{-1} hr^{-1} [or k = 0.157/(0.2)(0.2) = 3.93 liter mol^{-1} hr^{-1}]. The exchange of one halogen for another in alkyl iodides is believed to occur by the bimolecular process

$$CHR_1R_2I + X^- \rightarrow CHR_1R_2X + I^-$$

Exchange studies have, in fact, provided strong confirmatory support for this *Walden inversion mechanism*.

EXAMPLE

action responsible for the exchange. The rate law for the chemical reaction is then explored by repetition of the exchange study with varying concentrations of (AX) and (BX).

Some further points are the following. The quantities a and b refer to the amounts (or concentrations) of labeled element, so that in a case such as that of Eq. (15-106), a = 4[Ni(CN)$_4^{2-}$] and b = (CN$^-$). The rate law, or expression for R, remains, of course, in terms of the concentrations of the molecular species, [Ni(CN)$_4^{2-}$] and (CN$^-$).

It was assumed in the derivation that $x \ll a$ and $y \ll b$, that is, that the amounts of labeled elements were small compared to those of the normal ones. This is generally true in the case

of radioactive labeling but is usually not so if labeling is by means of a stable isotope. The writing of Eq. (15-109) is altered, but the final result is still Eq. (15-111).

15-ST-2 Effect of mechanical pressure on reaction rates

It is easiest to approach the effect of pressure on reaction rates through the formalism of transition-state theory, whereby we can write

$$k = \frac{kT}{h} \exp\left(-\frac{\Delta G^{0\ddagger}}{RT} \right) \qquad \text{or}$$

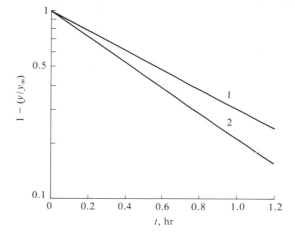

FIGURE 15-21
Exchange of n-C_4H_9I(RI) with I^-. (1) (RI) = 0.1 M, (I^-) = 0.2 M, or (RI) = 0.2 M, (I^-) = 0.1 M. (2) (RI) = (I^-) = 0.2 M.

FIGURE 15-22
Variation of some rate constants with pressure. Curve 1: rate of decomposition of benzoyl peroxide at 70°C. Curve 2: rate of dimerization of cyclopentadiene at 50°C (in monomer as solvent). Curve 3: rate for $C_2H_5I + C_2H_5O^- \rightarrow C_2H_5OC_2H_5 + I^-$ in ethanol at 25°C. (From "The Foundations of Chemical Kinetics" by S.W. Benson. Copyright 1960, McGraw-Hill, New York. Used with permission of McGraw-Hill Book Company.)

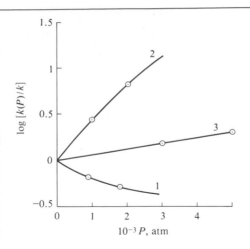

$$\ln k = \ln \frac{\mathbf{k}T}{h} - \frac{\Delta G^{0\ddagger}}{RT}$$

It follows from Eq. (6-52) that

$$\left(\frac{\partial(\ln k)}{\partial P}\right)_T = -\frac{\Delta \overline{V}^{0\ddagger}}{RT} \tag{15-112}$$

Strictly speaking, the effect of pressure on activity coefficients is being neglected, but this contribution should be small compared to the $\Delta \overline{V}^{0\ddagger}$ term if the system is dilute, so that departures from ideality are not large.

Partial molal volume changes for a reaction are generally small—of the order of 20 cm³ mol⁻¹, and pressures of thousands of atmospheres are needed to make an appreciable change in k. Experiments therefore require special high-pressure equipment. Some representative data are given in Fig. 15-22. It appears that $\Delta \overline{V}^{0\ddagger}$

values tend to lie between zero and $\Delta \overline{V}^0$, the value for the overall reaction. The results also tend to make qualitative sense in terms of structural considerations.

If a reaction involves bond breaking, it seems reasonable that $\Delta \overline{V}^{0\ddagger}$ should be positive, corresponding to an expanded or loose activated complex structure relative to that of the reactants, as in the case of the decomposition of benzoyl peroxide (curve 1 in the figure). Conversely, $\Delta \overline{V}^{0\ddagger}$ should be negative for an association reaction since the transition state should now involve incipient bond formation and hence a compaction. This case is illustrated by curve 2 in the figure. An ionization reaction also tends to give a negative $\Delta \overline{V}^{0\ddagger}$, presumably as a consequence of the electrostatic compaction of solvent around ions; that is, one assumes the transition state to correspond to incipient ionization.

EXAMPLE We read from curve 2 of Fig. 15-22 that $\log(k_P/k)$ at 50°C is about 1.0 at $P = 2000$ atm for the dimerization of cyclopentadiene, where k_P and k are the rate constants at P and at 1 atm, respectively. Log k_P is nearly linear in P and we estimate $\partial(\ln k_P)/\partial P$ to be about 9×10^{-4} atm⁻¹. From Eq. (15-112) $\Delta \overline{V}^{0\ddagger}$ is then $-(82)(323)(9 \times 10^{-4}) = -24$ cm³. For comparison, $\Delta \overline{V}^0$ for the overall reaction can be estimated to be about -30 cm³.

GENERAL REFERENCES

Basolo, F., and Pearson, R.G. (1967). "Mechanisms of Inorganic Reactions," 2nd ed. Wiley, New York.

Benson, S.W. (1960). "Foundations of Chemical Kinetics," McGraw-Hill, New York.

Berry, R.S., Rice, S.A., and Ross, J. (1980). "Physical Chemistry." Wiley, New York.

Breslow, R. (1969), "Organic Reaction Mechanisms," 2nd ed. Benjamin, New York.

CALDIN, E.F. (1964). "Fast Reactions in Solution." Wiley, New York.
CORNISH-BOWDEN, A. (1976). "Principles of Enzyme Kinetics." Butterworths, London.
ESPENSON, J.H. 1981). "Chemical Kinetics and Reaction Mechanisms." McGraw-Hill, New York.
FROST, A.A., AND PEARSON, R.G. (1961). "Kinetics and Mechanism," 2nd ed. Wiley, New York.
LANGFORD, C.H., AND GRAY, H.B. (1965). "Ligand Substitution Processes." Benjamin, New York.
MOORE, J.W., AND PEARSON, R.G. (1981). "Kinetics and Mechanism," 3rd ed. Wiley, New York.
NORTH, A.M. (1964). "The Collision Theory of Chemical Reactions in Liquids." Methuen, London.
WELLS, P.R. (1968). "Linear Free Energy Relationships." Academic Press, New York.

BARCLAY, I.M., AND BUTLER, J.A.V. (1938). *Trans. Faraday Soc.* **34,** 1445.
BELL, R., AND HIGGINSON, W. (1949). *Proc. Roy. Soc.* **A197,** 141.
BENSON, S.W. (1960), "Foundations of Chemical Kinetics." McGraw-Hill, New York.
BURNS, R.C., AND HARDY, R.W. (1975). "Molecular Biology, Biochemistry and Physics," Vol. 21 (A. Kleinzeller, G.F. Springer, and H.G. Wittmann, eds.). Springer-Verlag, Berlin and New York.
DEGN, H. (1972). *J. Chem. Ed.* **49,** 302.
EIGEN, J. (1963). *Angew. Chem.* **75,** 498.
FIELD, R.J., KOROS, E., AND NOYES, R.M. (1972). *J. Amer. Chem. Soc.* **94,** 8649.
GRAY, B.F. (1975). "Reaction Kinetics." Vol. 1 (P.G. Ashmore, ed.). The Chemical Society, Burlington House, London.
HAMMETT, L.P. (1940). "Physical Organic Chemistry." McGraw-Hill, New York.
MICHAELIS, L., AND MENTEN, M.L. (1913). *Biochem. Z.* **49,** 333.
MOORE, J.W., AND PEARSON, R.G. (1981). "Kinetics and Mechanism," 3rd ed. Wiley, New York.

CITED REFERENCES

EXERCISES

The equilibrium constant is $K = 2.5$ for the reaction

15-1

$$trans\text{-Co(en)}_2(H_2O)_2^{3+} \underset{k_2}{\overset{k_1}{\rightleftharpoons}} cis\text{-Co(en)}_2(H_2O)_2^{3+}.$$

The isomerization is first order in either direction, and the time for a solution initially 0.01 M in the *trans* complex to form half of the equilibrium concentration of the *cis* isomer, B, is 1.5 hr at 25°C and 0.3 hr at 40°C. K is essentially temperature independent. (a) Derive the expression for $(B)/(B)_\infty$ as a function of time. (b) Calculate k_1 at 25°C. (c) Calculate $\Delta H_1^{0\ddagger}$ and $\Delta S_1^{0\ddagger}$ (at 25°C).

Ans. (a) $(B)/(B)_\infty = 1 - e^{-kt}$; (b) $k_1 = 0.330$ hr^{-1}; (c) $\Delta H_1^{0\ddagger} = 19.9$ kcal mol^{-1} and $\Delta S_1^{0\ddagger} = -10.4$ cal K^{-1} mol^{-1}.

The mutarotation of glucose may be abbreviated as $G \overset{k}{\rightarrow}$ product. The rate, $d(G)/dt$, is first order in (G) and in acid concentration, (A), the acid acting as a catalyst. For A = acetic acid, and (A) = 0.02 M and $(G)_0 = 0.1$ M, $k = 1.36 \times 10^{-4}$ min^{-1} at 25°C where k is the apparent first-order rate constant. Calculate the half-life for the mutarotation if (A) = 0.04 M and $(G)_0 = 0.2$ M.

15-2

Ans. 2.55×10^3 min.

The reaction $A \underset{k_2}{\overset{k_1}{\rightleftharpoons}} B$ is first order in either direction, and at equilibrium at 25°C the ratio $(B)_\infty/(A)_\infty$ is 3. Calculate k_1 and k_2 if it takes 25 min for a system initially consisting of A only to be 50% converted into B. (Note Exericse 15-1.)

15-3

Ans. $k_1 = 3.30 \times 10^{-2}$ min^{-1}, $k_2 = 1.10 \times 10^{-2}$ min^{-1}.

A temperature-jump experiment is conducted with 0.02 M acetic acid. Calculate what the observed relaxation time should be at 25°C, using data available in the text.

15-4

Ans. 9.81×10^{-7} s.

15-5   The dissociation constant of a weak acid HA is 2×10^{-5} M and the relaxation time observed in a temperature-jump experiment is 3.5 μs, using a 2×10^{-4} M solution at 25°C. Calculate the rate constants k_1 and k_2 as defined in Table 15-4.

Ans. $k_1 = 2.48 \times 10^6$ M^{-1} s^{-1}, $k_2 = 2.23 \times 10^9$ M^{-1} s^{-1}.

15-6   A reaction 2A = 2B + D is believed to occur according to the mechanism:

$$A \underset{k_{-1}}{\overset{k_1}{\rightleftharpoons}} B + C$$

$$A + C \underset{k_{-2}}{\overset{k_2}{\rightleftharpoons}} B + D$$

Derive the rate expression for the forward reaction if (a) the first step is assumed to be a rapid reversible equilibrium, and the second reaction is rate-determining, (b) the forward direction of the first step is rate-determining, and (c) a stationary state is assumed in the intermediate C. (d) If $n = 1$ in Eq. (15-25), show what the rate law for the back reaction should be in cases (a) and (b) above, and express θ_b in terms of the various rate constants in each case.

Ans. (a) $R_f = -d(A)/dt = 2k_2K_1(A)^2/(B)$, (b) $R_f = 2k_1(A)$, (c) $R_f = 2k_1k_2(A)^2/[k_{-1}(B) + k_2(A)]$. (d) For case (a), $\theta_b = 2k_{-2}$, and for case (b), $\theta_b = 2k_{-1}/K_2$.

15-7   Consider the reaction $CH_3Br + I^-$ occurring in aqueous solution (see Table 15-1). Assume that both species are 4 Å in diameter and that Eq. (10-45) applies. Taking the solution to be 1 M in each species and at 25°C, calculate (a) the reaction rate, (b) the encounter frequency factor A_e, (c) the encounter rate, (d) the collision frequency from collision theory, and (e) the value of K_e assuming that ΔH_e^0 is zero.

Ans. (a) $R = 6.93 \times 10^{-4}$ M s^{-1}, (b) $A_e = 3.67 \times 10^9$ M^{-1} s^{-1}, (c) $Z_e = 3.67 \times 10^9$ M s^{-1}, (d) $Z_{12} = 1.03 \times 10^{11}$ M s^{-1}, (e) $K_e \simeq 6/55 = 0.11$ M^{-1}.

15-8   Estimate $\phi/\mathbf{k}T$ for the reaction $H^+ + F^- \rightarrow HF$, assuming $\mathcal{D}_{AB}$ to be 2×10^{-5} cm^2 s^{-1} and r_{AB} to be 4 Å.

Ans. 2.81 (or about 1.7 kcal mol^{-1}).

15-9   Verify Eqs. (15-58), (15-61), (15-64), and (15-48).

15-10   The equilibrium constant for the reaction $Co(NH_3)_5(H_2O)^{3+} + SO_4^{2-} = Co(NH_3)_5(SO_4)^+ + H_2O$ is 12.4 M^{-1} at 25°C, and the rate constant for the back reaction is 1.2×10^{-6} s^{-1}. Calculate the rate constant for the forward or anation reaction (a) in a low-ionic-strength medium and (b) in 0.1 M $NaClO_4$ (comment on the reliability of the answer).

Ans. (a) 1.49×10^{-5} M^{-1} s^{-1}, (b) 1.74×10^{-7} M^{-1} s^{-1}

PROBLEMS

15-11   The isomerization reaction *cis*-Pt(PEt$_3$)$_2$Cl$_2$ $\underset{k_{-1}}{\overset{k_1}{\rightleftharpoons}}$ *trans*-Pt(PEt$_3$)$_2$Cl$_2$ has been studied in benzene solution at 25°C. In the presence of a catalyst the reaction is first order in either direction and, starting with the *cis* isomer, the time of half-approach to equilibrium is 10 min; that is, half of the equilibrium amount of *trans* isomer has formed. The equilibrium constant K is 12.2, and ΔH^0 is 2.47 kcal mol^{-1}. (a) Calculate k_1 and k_{-1}, (b) find how long it should take for 90% of the *cis* isomer to undergo isomerization, and (c) obtain $\Delta H_1^{0\ddagger}$ if the time for half-isomerization is 3 min at 35°C (remember that k_1, k_{-1}, and K all change with temperature). (d) Finally, calculate $\Delta S_1^{0\ddagger}$ and ΔS^0.

15-12   Obtain the plot of (B)/(A)$_0$ vs. kt corresponding to the curve labeled $K = 2$ in Fig. 15-1. At what value of kt will the (B)/(A)$_0$ and (A)/(A)$_0$ curves cross?

15-13

A mixed ethanol–water solvent is 0.0677 M in formic acid (plus some added HCl as catalyst) and the esterification reaction is followed by periodic titration of 5 cm^3 aliquots by 0.010 m base. The following data (for 25°C) are obtained:

Time (min)	0	50	100	160	290	∞
Amount of base (cm^3)	43.52	40.40	37.75	35.10	31.09	24.29

Calculate the rate constants k_1 and k_{-1} for the formation and decomposition of the ester and the equilibrium constant K. If the rate law is written to include (H$^+$) (as catalyst), what are the values of k_1 and k_{-1}?

15-14

The first-order catalyzed decomposition of H$_2$O$_2$ in aqueous solution is followed by titration of the undecomposed H$_2$O$_2$ with KMnO$_4$ solution. By plotting the proper function, ascertain from the following data the value of the rate constant.

t (min)	0	5	10	20	30	50
cm^3 KMnO$_4$ per given amount of H$_2$O$_2$ solution	46.1	37.1	29.8	19.6	12.3	5.0

15-15

The reaction Pt(NH$_3$)$_3$Cl$^+$ + Br$^-$ = Pt(NH$_3$)$_3$Br$^+$ + Cl$^-$ was studied at 25°C using a solution initially 5 × 10^{-4} M in Pt(NH$_3$)$_3$Cl$^+$, 0.2 M in Cl$^-$ and 0.05 M in Br$^-$. The reaction was followed spectrophotometrically at a wavelength at which only the Pt complexes absorb, and the results are reported as ϵ_{app}, that is, measured absorbance divided by the path length of the cell used and by 5 × 10^{-4}. The following data were obtained. (Time, min; ϵ_{app}): (0; 90), (15; 119.2), (30; 137.6), (45; 149.1), (∞; 168.6). The extinction coefficient of the pure product is 240 M^{-1} cm^{-1} at the wavelength in question. Verify that the reaction is pseudo-first-order, and calculate the rate constant for the forward reaction and the equilibrium constant.

15-16

The rate law is found to be $R = k[\text{Ce(IV)}]^2[\text{Cr(III)}][\text{Ce(III)}]^{-1}$ for the oxidation of Cr(III) by Ce(IV) to give Cr(VI) and Ce(III), where the Roman numerals denote oxidation number. [Cr has intermediate oxidation states, Cr(IV) and Cr(V).] Find a mechanism consisting of bimolecular steps that gives the above rate law.

15-17

As an extension of Problem 15-16, find the rate law for the reverse reaction if n in Eq. (15-25) is 2/3, and now find a mechanism consistent both with this rate law and with that for the forward reaction.

15-18

The esterification of 56.5% aqueous ethanol by formic acid catalyzed by 0.026 N HCl, has been studied. The forward rate is first order in ethanol, in formic acid, and in HCl. With ethanol in great excess, the apparent forward and reverse rate constants are 1.85 × 10^{-3} min^{-1} and 1.76 × 10^{-3} min^{-1}, respectively, at 25°C. Calculate the percentage of formic acid esterified at equilibrium if the initial concentration of formic acid is 0.07 M, and the time for 45% of the formic acid to be esterified.

15-19

A solution at 25°C initially contains 0.063 M FeCl, and 0.0315 M SnCl$_2$. After the elapsed times given, the concentration of the ferrous chloride produced is determined by a titration procedure:

t (min)	1	3	7	17	40
Fe^{2+} (M)	0.0143	0.0259	0.0361	0.0450	0.0506

Determine the reaction order and the rate constant. (This one is hard!)

15-20

The simple reaction $2A + B \underset{k_b}{\overset{k_f}{\rightleftharpoons}} C$ is to be studied by the temperature-jump method. The system is initially at equilibrium, and as a result of the temperature jump it is perturbed by an amount δ so that (A) = (A)$_e$ + 2δ, (B) = (B)$_e$ + δ, and (C) = (C)$_e$ − δ. The relaxation rate constant k_r is given by $d \ln\delta/dt$. Show that

$$k_r = k_f(A)_e^2(B)_e\left[\frac{1}{(A)_e} + \frac{1}{(B)_e}\right] + k_b$$

(The derivation of Eq. (15-15) can serve as a guide.)

15-21 Calculate A_e at 25°C for the dimerization of cyclopentadiene, assuming that the molecular diameter is 4.5 Å and that Eq. (10-45) applies. Compare A_e with the experimental A factor; alternatively, calculate K_e. Assume the solvent to be benzene; obtain necessary data from a handbook and from Table 15-2.

15-22 Show that ΔS_e^0 can be approximated by $R \ln(6/C_s)$ where C_s is the solvent concentration in mol liter^{-1}. The approach can be very simple, either through considering the chance of molecule A finding B as a neighbor or by remembering that $K_e \simeq A_e\tau$.

15-23 Find the constants b and α of the Brønsted relation, given the following data for the general base-catalyzed reaction of nitramide, $H_2NNO_2 \rightarrow H_2O + N_2O$:

Base	Pyridine	Acetate ion	Formate ion	Dichloracetate ion
k (liter mol^{-1} min^{-1})	4.6	0.50	0.082	7×10^{-4}
K_b	2.3×10^{-9}	5.5×10^{-10}	4.8×10^{-11}	2.0×10^{-13}

Estimate k in water (K_w must be in liter mol^{-1} for consistency).

15-24 The following data are obtained on the kinetics of the lactonization of hydroxyvaleric acid. The overall reaction is

$$CH_3CHOHCH_2CH_2COOH \rightarrow CH_3{-}CH{-}CH_2{-}CH_2{-}\,C{=}O + H_2O$$
$$\underset{O}{\lfloor\underline{\hspace{3cm}}\rfloor}$$

	25°C				50°C	
	(H$^+$) = 0.025 M		(H$^+$) = 0.050 M		(H$^+$) = 0.025 M	
t (min)	Concentration of acid (M)	t (min)	Concentration of acid (M)	t (min)	Concentration of acid (M)	
0	0.080	0	0.100	0	0.080	
2	0.0570	2	0.0510	1	0.0408	
4	0.0408	4	0.0260	2	0.0210	
6	0.0295	6	0.0133	3	0.0107	
8	0.0210					

The rate is to be expressed by the equation $d(\text{hydroxyvaleric acid})/dt = k(\text{H}^+)^a(\text{hydroxyvaleric acid})^b$. Calculate the values of a and b and the value of the rate constant k at 25°C and 50°C, and the activation energy. Estimate the entropy of activation.

15-25 Show that the equations

$$v = V - K_M(v/S) \quad \text{and} \quad (S)/v = K_M/V + (S)/V$$

are alternative linear forms for the Michaelis–Menten mechanism. V is the maximum rate of product formation.

15-26 Estimate K_M for the two systems shown in Fig. 15-11(b) (for the cases with no CO inhibitor present).

15-27 The data of Fig. 15-11(b) show inhibition by CO. Suppose the inhibition merely modifies the Michaelis–Menten mechanism by tying up enzyme through the additional equilibrium $E + I \rightleftharpoons I \cdot E$, $K_I = (E)(I)/(I \cdot E)$. Derive the corresponding modification to Eq. (15-72), assuming that both the inhibition equilibrium and that of Eq. (15-68) are rapid. Estimate K_I for the two systems of Fig. 15-11(b).

A type of enzyme mechanism is the following: **15-28**

$$E + S \underset{}{\overset{K_1}{\rightleftharpoons}} ES$$
$$ES + S \underset{}{\overset{K_2}{\rightleftharpoons}} ES_2$$

$$ES \overset{k}{\rightarrow} products + E$$

The first two equilibria are rapid; only ES reacts to give products, regenerating E. If $(S) \gg (E)$ and the total concentration of S is varied at constant (E), find an expression for (S) when the rate of decomposition is at a maximum.

A series of solutions is made up which have the same concentration of the enzyme saccharase (E) **15-29** but different initial concentrations of substrate saccharose (S). The enzyme E is present in some small (unknown) concentration and catalyzes the hydrolysis of the saccharose, and the rate is measured from the change in optical rotation of the solutions. With increasing (S) the rate reaches a maximum R_∞; it is half of this value when (S) is 1.1% by weight. Assuming the Michaelis–Menten mechanism, calculate as much information as you can about k_1, k_{-1}, and k_2.

An alternative to the mechanism of Eqs. (15-68) and (15-69) is: **15-30**

$$E + S \overset{K_1}{\rightleftharpoons} E \cdot S$$
$$E + S \overset{k_2}{\rightarrow} E + P$$

Here, the complex $E \cdot S$ is inert, merely tying up enzyme, while the productive reaction is the direct bimolecular one. Derive the expression for $d(P)/dt$ in terms of $(E)_0$, (S), K_1 and k_2.

The use of Eq. (15-74) may be illustrated as follows. For reactions involving substituted benzoic acids, **15-31** σ values are 0.39, 0.20, and -0.25 for *meta*-Br, *para*-Cl, and *para*-OCH$_3$ substituents, respectively. In a study of the methanolysis of *meta*- and *para*-substituted $(-)$-menthylbenzoates at 40°C, log k was -2.8 and -3.3 for the first two of the above substituted analogues. Calculate log k for the *para*-OCH$_3$ substituted case.

A certain isomerization reaction can proceed by parallel first and second order reversible paths: **15-32**

$$A \underset{k_{-1}}{\overset{k_1}{\rightleftharpoons}} B \quad \text{and} \quad 2A \underset{k_{-2}}{\overset{k_2}{\rightleftharpoons}} 2B$$

Set up the expression for $R = -d(A)/dt$ [it will be convenient to express (B) as $(A)_0 - (A)$] and show that under the special condition of $K = 1$ the rate law becomes first-order. Obtain the integrated rate expression.

The kinetics of the reaction $S_2O_8^{2-} + 2I^- \rightarrow 2SO_4^{2-} + I_2$ is being studied. A solution was made up **15-33** that was 0.1 M in KI and 0.001 M in $K_2S_2O_8$, and the concentration of iodine was measured at 3 min intervals with the following results:

t (min)	0	3	6	9	12
(I_2) (mol liter^{-1})	0	0.00010	0.00019	0.00027	0.00034

The rate of the reaction will be some function of the concentrations of the species appearing in the overall equations, such as

$$d(I_2)/dt = k(S_2O_8^{2-})^a(I^-)^b(SO_4^{2-})^c(I_2)^d$$

(a) Derive from the data as much information as you can about the values of the exponents a, b, c, and d. (b) The following mechanism is proposed for the reaction:

$$S_2O_8^{2-} + I^- = SO_4I^- + SO_4^{2-} \quad \text{(rapid equilibrium)}$$

$$I^- + SO_4I^- \overset{k}{\rightarrow} I_2 + SO_4^{2-} \quad \text{(slow)}$$

Show what values of a, b, c, and d are predicted by this proposed mechanism. Discuss whether the rate data are compatible with the mechanism.

15-34 The reaction $CH_3I + C_2H_5O^- = CH_3OC_2H_5 + I^-$ is second order and k varies with temperature as follows:

$t(°C)$	0	12	18	24
k (liter mol^{-1} s^{-1})	5.60×10^{-5}	24.4×10^{-5}	48×10^{-5}	100×10^{-5}

Calcuate A, E^*, $\Delta S^{0\ddagger}$, and $\Delta^{0\ddagger}$.

15-35 The reaction $2Fe^{2+} + 2Hg^{2+} = Hg_2^{2+} + 2Fe^{3+}$ has been studied by measurement of the optical density of the solution at various times. The initial solutions contained only ferrous and mercuric ions, and at the wavelength employed the optical density D increased owing to the increased absorption of the products. The following two runs are given: Run 1, initially $(Fe^{2+}) = 0.1$, $(Hg^{2+}) = 0.1$; run 2, initially $(Fe^{2+}) = 0.1$, $(Hg^{2+}) = 0.001$.

	Run 1			Run 2	
$t(s)$	D	$(Hg^{2+})/0.1$		$t(s)$	$(Hg^{2+})/0.001$
0	0.100	—		0	1.000
1×10^5	0.400	—		0.5×10^5	0.585
2×10^5	0.500	—		1.0×10^5	0.348
3×10^5	0.550	—		1.5×10^5	0.205
∞	0.700	—		2.0×10^5	0.122
				∞	0

(a) Calculate the ratios $(Hg^{2+})/0.1$ for run 1, that is the fraction of Hg^{2+} remaining at the various times.

(b) Show what the order of the reaction is in run 1 and in run 2.

(c) If the rate equation is written in the form Rate $= k(Fe^{2+})^p(Hg^{2+})^q$, what are the values of p and q?

(d) Write a possible mechanism that would give this rate law.

15-36 The exchange reaction

$$*Ce^{4+} + Ce^{3+} = *Ce^{3+} + Ce^{4+}$$

is a simple one, with $\Delta S^{0\ddagger} = -40$ cal K^{-1} mol^{-1} and $\Delta H^{0\ddagger} = 7.7$ kcal mol^{-1}. Calculate k at (a) 25°C, at negligible ionic strength, (b) 25°C and $I = 0.1$, and (c) 35°C and $I = 0.1$.

SPECIAL TOPICS PROBLEMS

15-37 Equations (15-5) and (15-8) require the knowledge of $(A)_\infty$ or of K. Such information may be difficult to obtain accurately because of side or other reactions that perturb the system at long times. An alternative procedure is to measure $(A)/(A)_0$ at successive *equal* time intervals Δt. One thus has $(A)_1/(A)_0$, $(A)_2/(A)_0$, . . . , $(A)_i/(A)_0$, for the respective times Δt, $2\Delta t$, . . . , $i\Delta t$. Show that $[(A)_i/(A)_0 - (A)_{i+1}/(A)_0] = \alpha e^{-ik\Delta t}$ where α is a constant. Thus a plot of $\ln[(A)_i/(A)_0 - (A)_{i+1}/(A)_0]$ versus t for $i = 1,2$, etc. should give a straight line if the reaction is a reversible first-order one.

15-38 Referring to the reaction scheme of Eq. (15-91), suppose that $k_1 = 0.01$ min^{-1}, $k_{-1} = 2 \times 10^{-4}$ min^{-1}, $k_3 = 1 \times 10^{-7}$ min^{-1}, $k_{-3} = 5 \times 10^{-7}$ min^{-1}, and k_2 and k_{-2} are very small. Calculate the composition of a system consisting initially of A only after 10 min, 100 min, 10^3 min, 10^5 min, and 10^7 min. Also, find the ratio k_2/k_{-2}.

15-39 A sequential reaction, such as given by Eq. (15-93) is being studied. It is found that (B) reaches a maximum at 200 min and eventually disappears with a half time of 345 min, which is also the half-life with which (A) decreases with time. Calculate k_1 and k_2 and plot $(A)/(A)_0$ and $(B)/(A)_0$ as a function of time up to 1000 min.

15-40 Calculate k_1, k_2, and K from the plot of Fig. 15-20.

Equation (15-95) becomes indeterminate if $k_1 = k_2$. It is necessary to start with the appropriate differential equation for this case and integrate. Obtain the integrated form for this case.

15-41

Obtain the integrated rate law for the case $A \underset{k_2}{\overset{k_1}{\rightleftharpoons}} B + C$, assuming that $(B) = (C) = 0$ at $t = 0$.

15-42

A study of the kinetics of the exchange of radiocyanide ion with the hexacyanomanganate (III) ion, $Mn(CN)_6^{3-}$, gave the following values for the slope c of the exchange plot.

15-43

Concentration (M)			
CN^-	$Mn(CN)_6^{3-}$	pH	$c(min^{-1})$
0.0596	0.0199	10	1.31×10^{-2}
0.0571	0.0102	10	0.883×10^{-2}
0.104	0.0199	10	0.943×10^{-2}
0.0596	0.0199	9	1.33×10^{-2}

Determine the form of the kinetic equation for the reaction leading to exchange and calculate the rate constant. On the basis of these results suggest a possible mechanism for the exchange.

The rate of exchange between I_2^* and IO_3^- is studied with 0.00050 M I_2 and 0.00100 M HIO_3. The radioactivity of the IO_3^- at various subsequent times is given (corrected for radioactive decay). The total radioactivity of the system (or that initially present as I_2) is 1650.

15-44

t (hr)	19.1	47.3	92.8	169.2	∞
Radioactivity of IO_3^-	107	246	438	610	819

Calculate c of Eq. (15-110) and the rate R of the exchange reaction.

The rate constant for the exchange reaction

15-45

$$Cr(H_2O)_6^{3+} + H_2^{18}O \rightarrow Cr(H_2O)_5(H_2\,^{18}O)^{3+} + H_2O$$

is 5.00×10^{-5} M^{-1} s^{-1} at 25°C and 1 atm pressure, and $\overline{\Delta V}^{0\ddagger}$ is -9.3 cm^3 mol^{-1}. Calculate the rate constant for 2 kbar pressure.

chapter 16
Wave
mechanics

16-1   Introduction

The chapter on wave mechanics is probably subject to more variation in style and content than any other in a textbook of physical chemistry. Wave mechanics is of central importance to the physical chemist: Many of its results are of great utility, as in statistical thermodynamics, chemical bonding, and molecular spectroscopy. It is also the most nearly correct theory of mechanics that we have, and its very language has permeated chemistry. On the other hand, it is difficult to present. Wave mechanics is virtually unique in the history of scientific theories in that it introduces a mathematical rather than a physical model of nature—and its computational requirements become so fiercely complicated for systems of chemical interest that its practice has developed into the very specialized field of discovering and evaluating various approximation methods.

It is true, however, that the wave equation can be solved exactly for certain simple or model cases: A particle confined to a box or potential well, the hydrogen atom, and the harmonic oscillator are important illustrations. Most of the approximation methods use combinations or modifications of the exact solutions for such cases, and the procedure in this chapter will be to describe some of these exact solutions in sufficient detail that the rationale of the more advanced approaches can be appreciated. Moreover, most of the language and qualitative thinking in wave mechanics draws on the results for the simple situations. This will be true in Chapter 17 on chemical bonding and again in Chapter 19, on molecular spectroscopy.

The detailed treatment of the important model situations is thus both highly utilitarian and within the spirit of a course in physical chemistry. It might be thought, from all this, that the older quantum chemistry, as exemplified by the Bohr model, would be useless. However, the Bohr theory does give rough descriptions of the quantum states of an atom and also provides some of the terminology of modern spectroscopy. Spectroscopy, as part of the physical chemistry of electronic excited states, is taken up in Chapter 19. For the moment, we shall content ourselves with a brief review of Bohr's picture for the hydrogen atom. Also, although the origins of quantum theory lie in the development of the quantum treatment of blackbody radiation by Max Planck (around 1900), that subject itself is not ordinarily of great utility to the physical chemist and its outline is placed in a Special Topics section. It is of importance, however, to note here that Planck proposed and successfully defended the hypothesis that matter contains elementary harmonic oscillators (such as electrons) that can have only frequencies that are integral multiples of a fundamental one. Accordingly, the possible energies of an oscillator were postulated to be

$$\epsilon = h\nu, \, 2h\nu, \, 3h\nu, \, \ldots \tag{16-1}$$

where ν is the fundamental frequency and h is a proportionality constant, now called Planck's constant. Evaluation of h from experiment gives $h = 6.62618 \times 10^{-34}$ J s or 6.62618×10^{-27} erg s. Planck's ideas marked the beginning of quantum theory.

Einstein added a vital complement to Planck's hypothesis in explaining how it was that in the photoelectric effect the energy of electrons ejected from a metal by incident light had a definite energy which was independent of the flux of the light. Einstein (1905) proposed that light comes in units of energy, or quanta, of value given by the following equation, with ν now the frequency of the light radiation:

$$\epsilon = h\nu \tag{16-2}$$

Each quantum ejects one electron and since the energy of the quantum is definite, so also is the energy of the electron. Increasing the flux of incident light merely increases the number of photoelectrons, their energy being determined by the wavelength or the frequency of the radiation. Wavelength and frequency are related, of course, by the equation

$$\lambda\nu = c \tag{16-3}$$

where λ is the wavelength and c the velocity of light (3.00×10^{10} cm sec^{-1} in vacuum). Later, in the period 1908–1912, Einstein extended his law to photochemistry, saying that the absorption of each quantum of light results in the activation or excitation of a molecule by an energy corresponding to that of the quantum. This law of *photochemical equivalence* is honored by the unit called the *einstein*; an einstein is one mole of light quanta.

Equation (16-2) is sometimes called the *Planck–Einstein equation*, in reflection of its general implications. An oscillation of frequency ν, associated with atomic oscillators, electrons, or electromagnetic radiation, has energy $h\nu$. Absorption of a quantum of light changes the energy of an atom or molecule by $h\nu$, and, conversely, if an atom or molecule changes energy by an amount $\Delta\epsilon$ and thereby radiates light, that light will be of frequency given by

$$\Delta\epsilon = h\nu \tag{16-4}$$

The origins of wave mechanics trace back to Einstein's special theory of relativity, which led him to conclude that mass m and energy ϵ are interconvertible,

$$\epsilon = mc^2 \tag{16-5}$$

One of the important implications of Eq. (16-5) was perceived by L. de Broglie in 1923 as he puzzled over the fact that electromagnetic radiation sometimes appears to behave as particles or quanta, as, for example, illustrated by the law of photochemical equivalence, and yet also appears to behave as waves, as in diffraction effects. Equation (16-5) provided a connection between these two aspects; if a light quantum has energy $h\nu$, then it must have mass, given by $mc^2 = h\nu$, and momentum p,

$$p = mc = \frac{h\nu}{c} = \frac{h}{\lambda} \tag{16-6}$$

The intuitional leap made by de Broglie was to see that if radiation could have this duality, so might particles as well. Equation (16-6) provided the connecting link. Although light must always have velocity c (in vacuum, at least), a particle may have some arbitrary velocity v. The generalization of Eq. (16-6) to both radiation and particles is

$$p = \frac{h}{\lambda} \tag{16-7}$$

where, for a particle, $p = mv$. Dramatic verifications of de Broglie's equation (16-7) followed. It was found, for example, that electrons show diffraction effects and therefore a wave nature, and today, even neutrons are used in the study of crystal structures by diffraction, their wavelengths being determined by their velocities through Eq. (16-7). Many qualitative features of the wave mechanical theory that followed are derivable just from de Broglie's equation.

The remaining principle that completes the foundations of wave mechanics was formulated by Heisenberg (1927). It is called the *uncertainty principle* or, sometimes, the *principle of indeterminacy*. The principle states that neither the position nor the momentum of a particle (or of a light quantum) can be determined with unlimited accuracy, and that if the two quantities are determined simultaneously, the *minimum* uncertainty in each is such that†

$$\Delta p \, \Delta x \geqslant \frac{h}{4\pi} \tag{16-8}$$

Equation (16-8) refers to the ultimate obtainable accuracy; any actual experiments will, of course, probably be far less accurate.

The Planck–Einstein, de Broglie, and Heisenberg principles largely demolished the philosophy of classical mechanics (Newton's laws), optics, and electricity and magnetism. Neither energy nor momentum was any longer arbitrarily divisible; the results of measurements on an atomic scale could no longer be described exactly but only in terms of probabilities. It is important to remember, however, that it is the *philosophy* and not the *laws* of classical physics that are affected. The laws, being phenomenological, remain as useful as ever in the macroscopic world, although we now regard them as approx-

---

†The "uncertainty" Δx in some quantity x refers to the standard deviation of x from the mean, σ, if a very large number of determinations are made [note Eq. (2-38)]. That is, $\sigma^2 = \overline{(x - \bar{x})^2}$ or, to use a common notation, $\sigma^2 = \langle (x - \langle x \rangle)^2 \rangle$.

imations or as limiting forms derivable from the more exact theory of wave mechanics.

It is appropriate that we conclude this introduction with a review of perhaps the most dramatic early success of quantum theory, namely the model proposed by Bohr in 1913 for an atom—at first applied to the hydrogen atom, but almost immediately generalized to other atoms, especially those of the alkali metals. One of the mysteries of his day was the emission spectrum of hot hydorgen atoms, consisting of groups of sharp lines when registered on the photographic plate of a spectrograph. By contrast, a hot object or blackbody merely emits a broad, continuous spectrum of radiation. Groups of hydrogen lines were found, first in the visible (the Balmer series), then in the ultraviolet by T. Lyman, and then in the infrared, by F. Paschen and others. W. Ritz found, in 1908, geometric regularities in each series that we nowadays summarize by the equation

$$\bar{\nu} = \mathcal{R} \left(\frac{1}{n_2{}^2} - \frac{1}{n_1{}^2} \right) \tag{16-9}$$

where $\bar{\nu}$ is the reciprocal of the wavelength of the emitted light corresponding to a particular line and is called the *wavenumber*, and n_1 and n_2 are integers. The Balmer series is accurately given by $n_2 = 2$ and $n_1 = 3, 4, 5, \ldots$; the Lyman series, by $n_2 = 1$ and $n_1 = 2, 3, 4, \ldots$; and the Paschen series by $n_2 = 3$ $n_1 = 4, 5, 6, \ldots$. The same constant $\mathcal{R}$ applies to all the series, and is called the *Rydberg constant*. The modern value for $\mathcal{R}$ is 109,677.581 cm^{-1}.

Bohr had available in 1914 these very precise, very regular, and as yet mysterious spectral data, although by 1912 J. Nicholson and N. Bjerrum had arrived at the point of explaining line spectra as due to transitions between atomic or molecular states of definite energies with the frequencies of the emitted light governed by the Planck–Einstein principle, or Eq. (16-4). Bohr was the first, however, to assemble a quantitative model. He made use of the nuclear model for the atom arrived at by E. Rutherford from the then rather new phenomenon of natural radioactivity and from observations on the way in which alpha particles are scattered by matter (see Section 21-4). According to Rutherford, the mass of an atom was concentrated in a small nucleus of charge Z surrounded by electrons at a much greater distance. Bohr also retained the classical Coulomb law and the laws of mechanics but added the quantum principle of discrete, rather than continuous, energy states and the ideas of Planck and Einstein.

He then proposed the very specific picture of a hydrogen atom as consisting of an elctron orbiting around a nucleus of equal and opposite charge. To make his model work, however, he had to postulate that it was not the energy of the electron that was quantized, that is, varied in simple steps, but rather its angular momentum. By classical mechanics, the angular momentum of an orbiting electron is given by mvr, where m is the mass of the electron and r is the radius of its circular orbit. Bohr's postulate was that mvr can have only simple multiples of some minimum value, the necessary specific assumption being that

$$mvr = \frac{nh}{2\pi} \tag{16-10}$$

where n is an integer having the values $1,2,3, \cdots$. We call n the *principal quantum number*.

By classical mechanics and electrostatics, the centrifugal force on the electron mv^2/r must be balanced by the Coulomb force of attraction to the nucleus Ze^2/r^2, where e is the charge on the electron and Z is the positive charge on the nucleus expressed in units of e. Since we are using the cgs/esu system, e is in electrostatic units and force and energy are in dynes and ergs, respectively. Since the actual equations will look different, the derivation that follows will be repeated in SI units.

On balancing the centrifugal and electrostatic forces,

$$mv^2/r = Ze^2/r^2 \tag{16-11}$$

or

$$r = \frac{Ze^2}{mv^2} \tag{16-12}$$

or, in combination with Eq. (16-10) so as to eliminate v,

$$r = \frac{n^2h^2}{4\pi^2me^2Z} \tag{16-13}$$

For the lowest quantum state $n = 1$, and for the hydrogen atoms $Z = 1$, so the corresponding radius, given the special symbol a_0, is

$$a_0 = \frac{h^2}{4\pi^2me^2} \tag{16-14}$$

The total energy of an electron in its orbit is the sum of its kinetic energy $\frac{1}{2}mv^2$ and its potential energy $-Ze^2/r$, and with the use of Eq. (16-11) to eliminate mv^2, the result is

$$E = \frac{1}{2}mv^2 - \frac{Ze^2}{r} = -\frac{Ze^2}{2r} \tag{16-15}$$

The negative sign of E reflects the fact that it takes energy to remove the electron an infinite distance away from the nucleus. One next eliminates r by means of Eq. (16-13) to obtain

$$E = -\frac{2\pi^2me^4Z^2}{h^2}\frac{1}{n^2} \tag{16-16}$$

for the energies of successive quantum states.

The emission spectrum of an atom was then explained as due to the transition from one energy state E_i to a lower one E_j, the emitted radiation being of frequency given by Eq. (16-4):

$$E_i - E_j = h\nu = \frac{2\pi^2me^4Z^2}{h^2}\left(\frac{1}{n_j^2} - \frac{1}{n_i^2}\right)$$

It is customary for spectroscopists to deal with wavenumbers, $\bar{\nu} = 1/\lambda$, and from Eq. (16-3) the wavenumber of the emitted light is

$$\bar{\nu} = \frac{2\pi^2me^4Z^2}{h^3c}\left(\frac{1}{n_j^2} - \frac{1}{n_i^2}\right) \tag{16-17}$$

Equation (16-17) predicts that for the hydrogen atom, with $Z = 1$, the value of the Rydberg constant should be given by $2\pi^2 m e^4 / h^3 c$, which comes out to be 109,737 cm^{-1}. It was a major triumph for Bohr not only to have arrived at the form of Eq. (16-9), but also to have produced so accurate a value for $\mathcal{R}$ from a simple model.

As an incidental point, the system (electron plus nucleus) really revolves around its center of mass rather than the center of the nucleus. If this is taken into account, then μ, the reduced mass, replaces m in the preceding equations, where μ is given by $mM/(m + M)$, M being the mass of the nucleus. The correction is small; for example, the atomic weight of the electron is 0.00055, or 1/1837 of that of the nucleus of the hydrogen atom, so $\mu = 0.99946m$. The correct value of $\mathcal{R}$ is therefore actually 109,678 cm^{-1}.

We now repeat the above derivation in SI units. Equations (16-11) and (16-12) become

$$mv^2/r = Ze^2/4\pi\varepsilon_0 r^2 \qquad r = \frac{Ze^2}{4\pi\varepsilon_0 mv^2} \tag{16-18}$$

where ε_0 is the "permittivity of vacuum," $\varepsilon_0 = 10^7/4\pi c^2 = 8.85419 \times 10^{-12}$ C^2 J^{-1} m^{-1}, and e is now in coulombs, r in meters, and force and energy in newtons and joules, respectively, see Section 3-CN-1. Equations (16-13) and (16-14) now read

$$r = \frac{n^2 n^2 \varepsilon_0}{\pi Z e^2 m} \tag{16-19}$$

and

$$a_0 = \frac{h^2 \varepsilon_0}{\pi e^2 m} \tag{16-20}$$

Equations (16-15) and (16-16) are now

$$E = -\frac{Ze^2}{8\pi\varepsilon_0 r} = -\frac{Z^2 e^4 m}{8\varepsilon_0^2 h^2}\frac{1}{n^2} \tag{16-21}$$

Finally, the SI form of Eq. (16-17) is

$$\bar{\nu} = \frac{Z^2 e^4 m}{8\varepsilon_0^2 h^3 c}\left(\frac{1}{n_f^2} - \frac{1}{n_i^2}\right) \tag{16-22}$$

Note that the two sets of equations may be interconverted by following the instructions of Table 3-3; in this case, e in a cgs/esu equation is replaced by $e/(4\pi\varepsilon_0)^{1/2}$ to obtain the SI equation.

Example   We can calculate a_0 in both systems of units. In the cgs/esu system,

$$a_0 = \frac{(6.62618 \times 10^{-27})^2}{4\pi^2(9.10953 \times 10^{-28})(4.8032 \times 10^{-10})^2}$$

$$= 5.2918 \times 10^{-9}\text{cm or } 0.52918 \text{ Å} \tag{16-23}$$

and in SI

$$a_0 = \frac{(6.62618 \times 10^{-34})^2(8.85419 \times 10^{-12})}{\pi(1.60219 \times 10^{-19})^2(9.10953 \times 10^{-31})} = 5.2918 \times 10^{-11} \text{ m} \tag{16-24}$$

Similarly, E for the first Bohr orbit, that is, with $Z = 1$ and $n = 1$, is

$$E = -\frac{2\pi^2 (9.10953 \times 10^{-28}) (4.80324 \times 10^{-10})^4}{(6.62618 \times 10^{-27})^2}$$

$$= -2.180 \times 10^{-11} \text{ erg} \quad (16\text{-}25)$$

or

$$E = -\frac{(1.60219 \times 10^{-19})^4 (9.10953 \times 10^{-31})}{(8) (8.85419 \times 10^{-12})^2 (6.62618 \times 10^{-34})^2}$$

$$= -2.180 \times 10^{-18} \text{ J} \quad (16\text{-}26)$$

A pleasing relationship is found if one combines Eqs. (16-7) and (16-10) so as to eliminate the momentum mv. The result is

$$2\pi r = n\lambda \qquad (16\text{-}27)$$

That is, a Bohr orbit has a circumference that is an integral number of de Broglie wavelengths. The picture of the electron is thus one of a standing or stationary wave. One achievement of the Schrödinger wave equation, discussed in this chapter, is to give the amplitude of such a standing wave at every point in space and as a function of time.

16-2   Units

We pause at this point to review some of the characteristic units of atomistics and spectroscopy as well as certain simplified ways of writing the Bohr equations. First, there is a considerable variety of energy units in use. The fundamental unit is, of course, the erg in the cgs system, and the joule in the SI system. However, since charged particles are often accelerated experimentally by an electric field, it is convenient to take as a unit the energy necessary to transport 1 electronic charge across a potential difference of 1 V. The charge on the electron is 1.6021×10^{-19} C (coulomb), and from the definition of the volt, the required energy is just 1.620219×10^{-19} J. Thus

1 electron-volt or 1 eV $= 1.60219 \times 10^{-19}$ J

The energy acquired by a particle carrying Z units of electronic charge falling through a potential difference of V volts is then $1.60219 \times 10^{-19} ZV$. Chemists often prefer to consider 1 mole of electrons and to express the result in calorie units:

$$1 \text{ eV} \sim \frac{(1.60219 \times 10^{-19})(6.02205 \times 10^{23})}{4.184} \sim 23.06 \text{ kcal mol}^{-1}$$

The spectroscopist is interested in energies associated with light quanta. Although the energy of a light quantum is given by Eq. (16-2), one practice has been to use wavenumbers rather than frequency, and the energy per wavenumber is

$$1 \text{ cm}^{-1} \sim hc = (6.62618 \times 10^{-27})(2.99793 \times 10^{10})$$

$$\sim 1.9865 \times 10^{-16} \text{ erg} = 1.9865 \times 10^{-23} \text{ J}$$

Alternatively,

$$1000 \text{ cm}^{-1} \sim 0.12398 \text{ eV} \sim 2.8591 \text{ kcal mol}^{-1}$$

The unit of 1 wavenumber is sometimes called the *kayser;* 1000 cm^{-1} is then a kilokayser or 1 kK.

Frequency may be treated as an energy unit, *via* Eq. (16-2). A frequency of 1 s^{-1} is a hertz, 1 Hz. Thus 1 Hz ~ 6.62618 × 10^{-27} erg or 6.62618 × 10^{-34} J. Per mole, 1 Hz ~ 3.990 × 10^{-3} erg mol^{-1} or 3.990 × 10^{-10} J mol^{-1}.

A common practice, especially for theorists, is to express distance in units of the radius of the first Bohr orbit, a_0, and energy in units of the potential energy of the electron in the first Bohr orbit, which is e^2/a_0 or $e^2/4\pi\epsilon_0 a_0$. These are called the *atomic unit*, a.u., of length and of energy. Recalling the example of Section 16-1,

$$1 \text{ a.u. of length } = a_0 = 0.52917 \text{ Å}$$

and, by Eqs. (16-15) or (16-21),

$$1 \text{ a.u. of energy } = 4.360 \times 10^{-11} \text{ erg} = 4.360 \times 10^{-18} \text{ J}$$

Alternatively,

$$1 \text{ a.u. of energy } = 27.21 \text{ eV} \tag{16-28}$$

This unit is sometimes called the *hartree*, H, (after a famous physicist).†

Interconversions between the various energy units may be summarized as follows:

$$1 \text{ eV} = 1.6022 \times 10^{-19} \text{ J} = 1.6022 \times 10^{-12} \text{ erg} \tag{16-29}$$
$$= 8.066 \text{ kK} = 0.03675 \text{ a.u. } \sim 23.06 \text{ kcal mol}^{-1}$$

The reason that the use of atomic units is popular is that many equations then become very simple in appearance. Thus *both* Eqs. (16-15) and (16-21) become

$$E_{\text{a.u.}} = -\frac{Z^2}{2n^2} \tag{16-30}$$

The collection of physical constants is swept tidily into a closet (but not forgotten!) and questions of cgs/esu versus SI vanish. Later, it will be seen that many of the equations of wave mechanics undergo a corresponding simplification if atomic units are used.

16-3   Hydrogen and hydrogen-like atoms

The energy states of a hydrogen atom are summarized diagrammatically in Fig. 16-1; the three scales allow conversion from one energy unit to another. The energies are given by Eqs. (16-16) or (16-21), or just Eq. (16-30), and the energy change accompanying a transition between two states is given by Eqs. (16-9), (16-17), or (16-22). Several of the series observed in spectroscopy are indicated in the figure. Notice that the states become more and more closely spaced as n approaches infinity. An electron of energy greater than that corresponding to $n = \infty$ is free or is said to be dissociated. Thus the energy to dissociate an electron is 13.6 eV (or 109,678 cm^{-1} or 314 kcal mol^{-1}) if the initial state is the $n = 1$ or *ground state.* Alternatively, this energy is called

†One occasionally sees still another unit, the rydberg; 1 rydberg = $\frac{1}{2}$ a.u. = $\frac{1}{2}$ hartree.

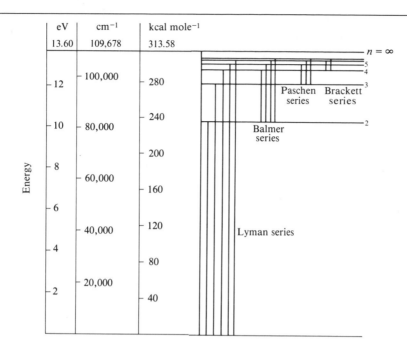

FIGURE 16-1
The hydrogen atom
spectrum (and energy
conversion scale).

the ionization energy or the ionization potential. Equation (16-30) applies to one-electron atoms other than the hydrogen atom if the appropriate value of Z is used. Thus for He^+ the ground state is still that with $n = 1$, but the ionization energy is given by Eq. (16-30) with $Z = 2$; this energy is thus 2^2 times that for a hydrogen atom, or is 2 a.u. (54.4 eV). The ionization energy for Li^{2+} is 3^2 times that for a hydrogen atom, or 4.5 a.u.

The same equations may be applied to multielectron atoms if it is recognized that the inner electrons partially shield the outer ones from the full nuclear charge. In effect, the outer electron "sees" a reduced nuclear charge Z_{eff}, which may be thought of as an empirical parameter. For example, in the case of He, the ionization energy of the first electron is only 0.92 a.u. instead of the 2 a.u. predicted by Eq. (16-30). We can account for this discrepancy by taking $Z_{eff} = [0.92 \times 2]^{1/2} = 1.36$. The difference between Z and Z_{eff} is called the *screening constant, s*. Before considering atoms beyond He it is necessary to review the additional quantum that are involved.

The rules for placing further electrons around a nucleus were developed as the Bohr theory was applied to atoms other than hydrogen. First, it was necessary to allow elliptical as well as circular orbits (of a given quantized angular momentum). The wave mechanical treatment (see Section 16-7) leads naturally to this second quantum number ℓ, called the *azimuthal* or *angular quantum number*, and further specifies that for a given n, ℓ can have the values $0, 1, ..., n − 1$. As a consequence of early descriptive assignments by spectroscopists, these ℓ states are often designated s (sharp), p (principal), d (diffuse), and f (fundamental), referring to $\ell = 0, 1, 2$, and 3, respectively (beyond this, the lettering is alphabetical: s, p, d, f, g, h, i, j, ...).

Second, the observation of splitting of spectral lines in a magnetic field (called the Zeeman effect) led to another quantum number m, called the *magnetic quantum number*, which gives the possible projections of the orbital angular momentum in the direction of the field. Again, the magnetic quantum number appears naturally in the wave mechanical treatment, which specifies that m can have the integral values $-\ell$, $-(\ell - 1)$, ..., 0, ..., $\ell - 1$, and ℓ or a total of $2\ell + 1$ values in all. Finally, the discovery of electron spin led to the introduction of a fourth quantum number, the *spin quantum number* s, whose value can be $\frac{1}{2}$ or $-\frac{1}{2}$, the angular momentum of the spining electron being $s(h/2\pi)$.

The first few possible combinations of these four quantum numbers are as follows:

$n = 1$:	$\ell = 0$ (s)	$m = 0$	$s = \frac{1}{2}, -\frac{1}{2}$
$n = 2$:	$\ell = 0$ (s)	$m = 0$	$s = \frac{1}{2}, -\frac{1}{2}$
	$\ell = 1$ (p)	$m = -1$	$s = \frac{1}{2}, -\frac{1}{2}$
		$m = 0$	$s = \frac{1}{2}, -\frac{1}{2}$
		$m = 1$	$s = \frac{1}{2}, -\frac{1}{2}$
$n = 3$:	$\ell = 0$ (s)	$m = 0$	$s = \frac{1}{2}, -\frac{1}{2}$
	$\ell = 1$ (p)	$m = -1$	$s = \frac{1}{2}, -\frac{1}{2}$
		$m = 0$	$s = \frac{1}{2}, -\frac{1}{2}$
		$m = 1$	$s = \frac{1}{2}, -\frac{1}{2}$
	$\ell = 2$ (d)	$m = -2$	$s = \frac{1}{2}, -\frac{1}{2}$
		$m = -1$	$s = \frac{1}{2}, -\frac{1}{2}$
		$m = 0$	$s = \frac{1}{2}, -\frac{1}{2}$
		$m = 1$	$s = \frac{1}{2}, -\frac{1}{2}$
		$m = 2$	$s = \frac{1}{2}, -\frac{1}{2}.$

Experimental spectroscopy led Pauli (1925) to conclude that no two electrons in the same atom may have the same four quantum numbers. By this *Pauli exclusion principle* only 2 electrons can be in the $n = 1$ state, 8 in the $n = 2$ state, 18 in the $n = 3$ state, and so on, For example, the ground- or lowest-energy state of Li has the electron configuration $1s^2 2s$, the first number denoting the principal quantum number and the superscript the number of electrons of the indicated ℓ value. Table 16-1 lists the outer electron configurations for a number of elements, along with their energies for the ionization of successive electrons.

EXAMPLE What is the electronic configuration of Na and to what Z_{eff} values do the first and second ionization potentials correspond? We see from Table 16-1 that the configuration for Na is $1s^2 2s^2 2p^6 3s$, corresponding to an atomic number of $Z = 11$. The first ionization energy is 0.19 a.u. and so from Eq. (16-30) $Z_{eff} = [(0.19)(2)(3^2)]^{1/2} = 1.85$, while the second ionization energy is 1.74 a.u., corresponding to $Z_{eff} = [(1.74)(2)(2^2)]^{1/2} = 3.73$, or almost two units of charge larger.

16-4   Wave mechanics. The Schrödinger wave equation

The Bohr model encountered increasing difficulties as attempts were made to apply it to atoms other than hydrogen; furthermore, it failed to account for the de Broglie principle that particles behave as though they had characteristic wavelengths. The de Broglie hypothesis was quantitatively confirmed by the second half of the 1920's through experiments, such as those

Element	Z	Outer electron configuration	Ionization energy I for successive electrons (a.u.)							
			I_1	I_2	I_3	I_4	I_5	I_6	I_7	I_8
H	1	1s	0.50	—	—	—	—	—	—	—
He	2	1s²	0.90	2.00	—	—	—	—	—	—
Li	3	1s²2s	0.20	2.78	4.5	—	—	—	—	—
Be	4	...2s²	0.35	0.67	5.65	8.0	—	—	—	—
B	5	...2s²2p	0.31	0.93	1.4	9.53	12.5	—	—	—
C	6	...2s²2p²	0.42	0.90	1.76	2.37	14.4	17.9	—	—
N	7	...2s²2p³	0.53	1.09	1.74	2.84	3.60	18.6	24.4	—
O	8	...2s²2p⁴	0.50	1.29	2.02	2.84	4.18	5.06	27.0	32
F	9	...2s²2p⁵	0.64	1.29	2.30	3.20	4.20	5.75	6.8	35
Ne	10	...2s²2p⁶	0.79	1.51	—	—	—	—	—	—
Na	11	...2s²2p⁶3s	0.19	1.74	2.62	—	—	—	—	—
Mg	12	...3s²	0.28	0.55	2.95	—	—	—	—	—
Al	13	...3s²3p	0.22	0.69	1.05	4.41	—	—	—	—
Si	14	...3s²3p²	0.30	0.60	1.23	1.66	6.12	—	—	—
P	15	...3s²3p³	0.41	0.73	1.11	1.89	2.39	—	—	—
S	16	...3s²3p⁴	0.38	0.86	1.29	1.74	2.66	3.22	—	—
Cl	17	...3s²3p⁵	0.48	0.88	1.47	2.00	2.50	3.56	4.18	—
Ar	18	...3s²3p⁶	0.58	1.02	1.51	2.20	—	—	—	—
K	19	...3s²3p⁶4s	0.16	1.17	1.75	—	—	—	—	—
Ca	20	...4s²	0.23	0.44	1.88	—	—	—	—	—
Sc	21	...4s²3d	0.24	0.48	0.91	2.72	—	—	—	—
Ti	22	...4s²3d²	0.25	0.50	1.04	1.59	3.67	—	—	—
V	23	...4s²3d³	0.25	0.52	1.09	1.79	2.35	4.88	—	—
Cr	24	...4s3d⁵	0.25	0.61	—	—	—	—	—	—
Mn	25	...4s3d⁵	0.28	0.58	—	—	—	—	—	—

TABLE 16-1
Ionization energies

of C. Davisson and L. Germer and G. Thomson, that showed that electrons produce x-ray-like diffraction patterns on striking a gold foil. The hypothesis is also consistent with the Heisenberg uncertainty principle that the position of a particle cannot be stated exactly but only as a probability. A new mechanics was needed that would unite all the aspects of the behavior of atoms and electrons; this was provided in 1926 by E. Schrödinger and W. Heisenberg, somewhat independently.

A. The classical wave equation

Schrödinger used the mathematical framework that had been developed for periodic motions such as those of waves. We are all familiar with water waves (ocean, lake, bathtub) and recognize a wave as a disturbance that can travel. Note that while a water wave may be moving along a surface, the water itself is not moving with the wave. A floating cork, for example, merely bobs up and down (and a little back and forth). A mathematically simpler example is that of a uniform string or rope. A person dangling a piece of rope from a balcony can give it a jerk and see a disturbance or wave travel down to the free end. This would be a travelling wave. Alternatively, the rope might be held at each end, an illustration in this case being a piano, guitar, or violin string. Upon plucking or striking the string, a vibration is set up and one may now have a standing wave. That is, there will be fixed points that do not move; these are called *nodes*. The two cases of a travelling wave and of standing waves are illustrated in Fig. 16-2.

FIGURE 16-2
Waves. (a) Travelling wave in rope held from a balcony. (b) Standing waves; left, no node; right, one node.

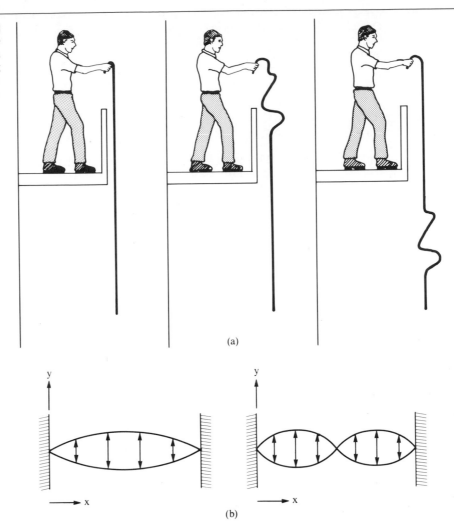

(a)

(b)

The differential equation for the uniform, elastic string is fairly easy to derive [see McQuarrie (1983)]. For the case of a one-dimensional wave, the equation is

$$\frac{\partial^2 \Psi}{\partial x^2} = \left(\frac{1}{v^2}\right) \frac{\partial^2 \Psi}{\partial t^2} \tag{16-31}$$

where t is time, x is distance along the string, and v is the wave velocity. The function $\Psi(x,t)$ gives the displacement of the string [see Fig. 16-2(b)], and is called the *amplitude*. The solution to Eq. (16-31) is

$$\Psi = A \exp\left[2\pi i\left(\frac{x}{\lambda} - vt\right)\right] \tag{16-32}$$

where λ is the wavelength and v is the frequency of the wave; this may be verified by substitution back into Eq. (16-31), recognizing that $\lambda v = v$. The constant A determines the maximum amplitude. Equations of the form of Eq.

(16-32) will appear frequently in the following material, and an important identity to keep in mind is

$$e^{\pm ix} = \cos x \pm i \sin x \qquad (16\text{-}33)$$

(as one can verify by writing out the respective series expansions). Equation (16-32) may be written as

$$\Psi = A \left\{ \cos\left[2\pi\left(\frac{x}{\lambda} - vt \right) \right] + i \sin\left[2\pi\left(\frac{x}{\lambda} - vt \right) \right] \right\} \qquad (16\text{-}34)$$

We need use only the real part of Ψ in the case of a vibrating string; this is just a cosine function oscillating between A and $-A$, with a wavelength λ if plotted against x at constant t, or of frequency v if plotted against time at constant x. Figure 16-3 shows such plots.

The intensity of the wave is given by the square of its amplitude, and if the latter is complex, by $\Psi\Psi^*$, where Ψ^* is the complex conjugate of Ψ, in this case $A \exp[-2\pi i(x/\lambda - vt)]$.†

The Schrödinger wave equation is taken by postulate—the postulates of wave mechanics are enumerated further below. We may *obtain* the wave equation, however, by applying Eq. (16-32) to a particle that obeys the de Broglie relationship, that is, has an associated wavelength. First, we restrict the situation to standing waves, or waves whose amplitude at a given x is not a function of time. This allows us to write Eq. (16-32) as

$$\Psi = \psi\, e^{-2\pi i vt} \qquad \psi = A e^{2\pi i x/\lambda} \qquad (16\text{-}35)$$

and deal only with the ψ function. The corresponding differential equation is

$$\frac{d^2\psi}{dx^2} = -\frac{4\pi^2}{\lambda^2}\, \psi \qquad (16\text{-}36)$$

[obtained by differentiating $\psi = A \exp(2\pi i x/\lambda)$ twice]. We now specifically invoke the de Broglie relationship, Eq. (16-7), to replace λ by h/p, so that the right-hand side of Eq. (16-36) becomes $-4\pi^2 p^2/h^2\psi$. The kinetic energy of a particle is $\frac{1}{2} mv^2$, or $p^2/2m$, and kinetic energy can be regarded as the total energy E minus the potential energy V. It follows that $p^2 = 2m(E - V)$, and Eq. (16-36) becomes

$$\frac{d^2\psi}{dx^2} = -\frac{8\pi^2 m}{h^2}\,(E - V)\psi$$

or

$$-\frac{h^2}{8\pi^2 m}\frac{d^2\psi}{dx^2} + V\psi = E\psi \qquad (16\text{-}37)$$

Equation (16-37) is the Schrödinger wave equation in one dimension for a particle that is present as a standing wave.

B. Obtaining the Schrödinger wave equation

---

†The complex conjugate of a complex number is obtained by replacing i by $-i$ wherever the former appears. The product of a complex number and its complex conjugate will always be real.

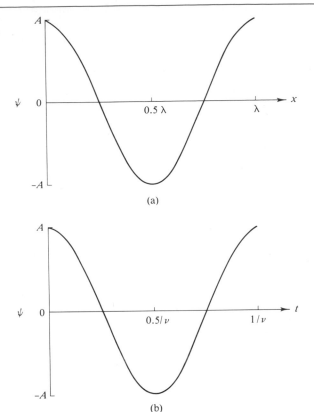

Generalization to three dimensions gives

$$-\frac{h^2}{8\pi^2 m}\left(\frac{\partial^2\psi}{\partial x^2} + \frac{\partial^2\psi}{\partial y^2} + \frac{\partial^2\psi}{\partial z^2}\right) + V\psi = E\psi \qquad (16\text{-}38)$$

If one does not wish to specify a particular coordinate system (such as the Cartesian one used here), one writes the equation in the form

$$-\frac{h^2}{8\pi^2 m}\nabla^2\psi + V\psi = E\psi \qquad (16\text{-}39)$$

where ∇^2 is known as the *Laplacian operator, del squared.* (The meaning of the term "operator" is discussed further below.) There will be some detailed expression for ∇^2 for each different coordinate system, as, for example, the polar one that is used in Section 16-7.

Schrödinger pursued the analogy to classical wave equations further by postulating the more general equation

$$-\frac{h^2}{8\pi^2 m}\nabla^2\Psi + V\Psi = \frac{h}{2\pi i}\frac{\partial\Psi}{\partial t} \qquad (16\text{-}40)$$

where Ψ is a function of both position and time. As was done with Eq. (16-32), we assume the special case in which Ψ can be written as a product $\psi\phi$, where ϕ is a function of time alone (and ψ has already been taken to be a function of position alone). If this substitution is made into Eq. (16-40), then $\nabla^2\Psi$ becomes $\phi\nabla^2\psi$ (since ϕ is not a function of coordinates) and $\partial\Psi/\partial t$ becomes $\psi\partial\phi/\partial t$ (since ψ is not a function of time). On dividing through by $\psi\phi$, we obtain

$$\frac{1}{\psi}\left(-\frac{h^2}{8\pi^2 m}\nabla^2\psi + V\psi\right) = -\frac{h}{2\pi i}\frac{1}{\phi}\frac{\partial\phi}{\partial t} \tag{16-41}$$

The left-hand side of Eq. (16-41) is a function of coordinates only [assuming that $V \neq f(t)$] and the right-hand side is a function of time only; consequently the quantity to which both sides are equal cannot be a function either of time or of coordinates—it must be a constant. Another way of phrasing one of Schrödinger's hypotheses is to say that we take this constant to be E, the total energy. Equation (16-41) then separates into two equations, Eq. (16-39) and the time-dependent equation

$$\frac{d\phi}{dt} = -\frac{2\pi i}{h}E\phi$$

which integrates to

$$\phi = (\text{constant})e^{-(2\pi iE/h)t} \tag{16-42}$$

Notice the analogy to Eq. (16-35), for which the time-dependent part is $e^{-2\pi i\nu t}$. This is the same as Eq. (16-42) if we write $E = h\nu$. Remembering relationship (16-33), ϕ is evidently a function that oscillates with time.

Nearly all of the applications of wave mechanics treated in this text will be in terms of ψ rather than Ψ.† The typical application of Eq. (16-39) will be to an electron under the influence of some attractive potential V, usually a potential well. In the case of the hydrogen atom, for example, $V = -e^2/r$, (in the cgs/esu system). The solutions to Eq. (16-39) then give the various energies that an electron can have if it is to act as a standing or stationary wave. Such a stationary state is possible only for certain wavelengths, depending on the nature of V, and hence only for certain energies. There will be additional restrictions. In each of the situations that we take up in this chapter, for example, we will require that ψ be everywhere continuous and single-valued. In addition, the solution must obey certain *boundary conditions*. Typically, ψ should go to zero at large x; we may require it to be angularly periodic. We will see that these various requirements restrict integration constants to intergral values—values that turn out to be quantum numbers. Discrete or quantized energy states will thus appear as a natural consequence of asking for stationary states and ones that must meet physically acceptable boundary conditions.

A remaining point is that while Eq. (16-39) is the wave equation for a single particle, such as an electron, obeying a potential function V and having total energy E, there is a simple recipe for constructing the wave equation for a system of particles. The first term of Eq. (16-39) plays the role of kinetic energy of the particle—this is evident from the manner in which the equation is obtained. If more than one particle is involved, an analogous term is written

---

†Rate process, such as the emission of radiation (fluorescence), require the use of the full equation, (16-40), a use which is largely beyond the scope of this text. See, however, Section 19-ST-1.

for each of them. That is, the first term becomes $\Sigma_i - [(h^2/8\pi^2 m_i)\,\nabla_i{}^2\psi]$ if i particles are present. Then V becomes the function giving all the mutual interactions between the particles; E is still the energy of the system.

C. Operators, eigenfunctions, and eigenvalues

There is a very useful shorthand notation in mathematics, called an *operator*. An operator is an instruction to carry out some mathematical procedure on the quantity or function that follows it. Thus if $\mathcal{O}$ is the general symbol for an operator, we can write

$$g(x) = \mathcal{O}f(x) \tag{16-43}$$

Equation (16-43) states that operator $\mathcal{O}$ acting on the function $f(x)$ yields the function $g(x)$. As a very simple example, let $\mathcal{O}$ be the instruction to square what follows—let's call this operator **S**. Suppose that $f(x) = (a + bx)$; then $\mathbf{S}f(x) = (a^2 + 2abx + b^2x^2) = g(x)$. More commonly, for us, an operator will be an instruction to differentiate. Thus if $\mathcal{O}$ is the instruction to differentiate with respect to x, we can call the operator **D** and have $g(x) = \mathbf{D}f(x) = df(x)/dx$. More generally, $\mathbf{D}^n$ has the meaning d^n/dx^n.

It turns out that a very important case is one where

$$g(x) = \mathcal{O}f(x) = cf(x) \tag{16-44}$$

that is, the effect of carrying out the instruction of the operator is merely to regain the original $f(x)$ but multiplied by a constant. For a given $\mathcal{O}$, there may be some functions that obey Eq. (16-44) (and some that don't). For example, if $\mathcal{O} = \mathbf{D}$ and $f(x) = ae^{bx}$, then

$$g(x) = d(ae^{bx})/dx = abe^{bx} \tag{16-45}$$

The reader can verify that if, say, $f(x) = ax^2$, then Eq. (16-44) is *not* obeyed. Functions that *do* obey Eq. (16-44) are called *eigenfunctions* of the operator $\mathcal{O}$, and the constant c is called an *eigenvalue*.

EXAMPLE   An important operator in wave mechanics is the momentum operator, **P**. In one dimension,†

$$\mathbf{P}_X = -\frac{ih}{2\pi}\frac{\partial}{\partial x} \tag{16-46}$$

Notice that $\mathbf{P}_X$ has an imaginary coefficient, and therefore has a complex conjugate, $\mathbf{P}_X^*$. Show that $f(x) = e^{ibx}$ is an eigenfunction of the operator $\mathbf{P}_X$ and find its eigenvalue. We find

$$g(x) = -\frac{ih}{2\pi}\frac{\partial e^{ibx}}{\partial x} = \frac{hb}{2\pi}e^{ibx} \tag{16-47}$$

so Eq. (16-44) is obeyed and the eigenvalue is $hb/2\pi$.

It must be apparent that the above presentation merely skims the surface of a large body of mathematics. The main purpose, however, is to present some of the language of wave mechanics in terms of fairly simple definitions. We can now proceed to the postulates of wave mechanics.

†An often used abbreviation is that of $\hbar$ in place of $h/2\pi$.

The wave equation and its meaning can be stated in a formal way by means of *postulates*. That is, we affirm certain things; we do not derive them. This is essentially what is done in the case of the laws of thermodynamics. In the formal sense, we affirm these laws as postulates—and then show that they are consistent with experience.

D. The postulates of wave mechanics

POSTULATE 1. The state of a physical system is completely described by a wave function Ψ, *which is a function of coordinates and of time.*

To fulfill this attribute, Ψ must everywhere be finite, continuous, and single-valued.

POSTULATE 2. Ψ *can be related to classical mechanics by replacing every observable by a corresponding operator.*

Some important observables and corresponding operators are as follows.

$$\text{Potential energy, } V: \quad \mathbb{O}_V = V \quad \text{(just multiply by } V) \tag{16-48}$$

$$\text{Momentum } p: \quad \mathbb{O}_p = -\frac{ih}{2\pi}\frac{\partial}{\partial x} \quad \text{[Eq. (16-46)]}$$

$$\text{Total energy } E: \quad \mathbb{O}_E = -\frac{h}{2\pi i}\frac{\partial}{\partial t} \tag{16-49}$$

We can see how this works by making the above substitutions in the classical statement:

$$\begin{array}{ccccc}
\text{total energy} & = & \text{kinetic energy} & + & \text{potential energy} \\
E & = & T & + & V
\end{array} \tag{16-50}$$

where T denotes kinetic energy. In the case of a single particle, $T = \frac{1}{2}mv^2 = p^2/2m$, so Eq. (16-50) becomes

$$E = \frac{p^2}{2m} + V \tag{16-51}$$

We now make the substitutions called for by Postulate 2, to obtain

$$-\frac{h}{2\pi i}\frac{\partial \Psi}{\partial t} = \frac{1}{2m}\left(-\frac{ih}{2\pi}\frac{\partial \Psi}{\partial x}\right)^2 + V\,\Psi \tag{16-52}$$

for the one-dimensional case. Remembering that $\mathbf{D}^n$ means d^n/dx^n, and rearranging terms, Eq. (16-52) becomes

$$-\frac{h^2}{8\pi^2 m}\frac{\partial^2 \Psi}{\partial x^2} + V\,\Psi = -\frac{h}{2\pi i}\frac{\partial \Psi}{\partial t} \tag{16-53}$$

Equation (16-53) becomes Eq. (16-40) if we generalize to three dimensions.

We can now introduce perhaps the most common operator of wave mechanics, the *Hamiltonian operator* **H**. This is defined as

$$\mathbf{H} = \frac{h^2}{8\pi^2 m}\nabla^2 + V \tag{16-54}$$

Equation (16-40) becomes

$$\mathbf{H}\Psi = -\frac{h}{2\pi i}\frac{\partial \Psi}{\partial t} \tag{16-55}$$

Furthermore, for the stationary state or standing wave case, E is not a function of time; we have Eq. (16-37), and this reduces to the statement

$$\mathbf{H}\psi = E\psi \qquad (16\text{-}56)$$

Equation (16-56) is probably the most famous statement in wave mechanics. In the sections that follow, we will be applying it to a variety of situations. Note that ψ is an *eigenfunction* [it meets the definition of Eq. (16-44)] and that the total energy E is an *eigenvalue*. We turn now to the remaining postulates of wave mechanics.

POSTULATE 3.† *Any possible measurement of an observable property $\mathscr{P}$ can yield only one of the eigenvalues of the corresponding operator $\mathscr{O}_{\mathscr{P}}$.*

The corresponding mathematical statement is: $\mathscr{O}_{\mathscr{P}}\psi = \mathscr{P}\psi$. This postulate generalizes the particular case noted above, where for the Hamiltonian operator the observable is one of the eigenvalues E.

POSTULATE 4. The probability that a particle is in the volume element $d\tau$ is given by $\psi^*\psi \, d\tau$.

In Cartesian coordinates, $d\tau$ is just $dx\, dy\, dz$. A necessary corollary of postulate 4 is that

$$\int \psi^*\psi \, d\tau \equiv \int_{-\infty}^{\infty} \int_{-\infty}^{\infty} \int_{-\infty}^{\infty} \psi^*\psi \, dx\, dy\, dz = 1 \qquad (16\text{-}57)$$

A ψ function that meets the above condition is said to be *normalized*.

POSTULATE 5. If a system is in a state described by a normalized wave function ψ, the average value of some property $\mathscr{P}$ is given by

$$\overline{\mathscr{P}} = \int \psi^*\mathscr{O}_{\mathscr{P}}\psi \, d\tau \qquad (16\text{-}58)$$

Here, $\mathscr{O}_{\mathscr{P}}$ is the operator associated with the measurable property $\mathscr{P}$; the average value $\overline{\mathscr{P}}$ is often called the *expectation value* and written $\langle \mathscr{P} \rangle$.‡

EXAMPLE   The operator corresponding to coordinate x is just $\mathscr{O}_x = x$. In a one-dimensional situation, we therefore have

$$\langle x \rangle = \int_{-\infty}^{\infty} \psi^* x \psi \, dx \qquad (16\text{-}59)$$

Similarly,

$$\langle x^2 \rangle = \int_{-\infty}^{\infty} \psi^* x^2 \psi \, dx \qquad (16\text{-}60)$$

---

†All of the postulates apply to the time-dependent wave function, but postulates 3, 4, and 5 are written in terms of ψ as being easier to discuss and closer to the applications in this text.

‡A physical paradox or contradiction develops unless it is true that $\int \psi^*\mathscr{O}_{\mathscr{P}}\,\psi \, d\tau = \int \psi\mathscr{O}_{\mathscr{P}}^*\psi^* \, d\tau$. An operator that obeys this relationship is called *Hermitian*; all of the operators of wave mechanics are Hermitian. An important further property of Hermitian operators is the following. If we have two eigenfunctions of a given operator, $\mathscr{O}\psi_m = c_m\psi_m$ and $\mathscr{O}\psi_n = c_n\psi_n$, then it can be shown that $\int \psi_m^*\psi_n \, d\tau = 0$ if $m \neq n$. Such eigenfunctions are said to be *orthogonal*. If they are also normalized, they are said to be *orthonormal*.

16-5   Some simple choices for the potential function *V*

One of the simplest applications of the Schrödinger wave equation is to a
particle for which the potential function is a constant. It is customary to take
$V = 0$; if V has some other value, it merely shifts the energy of the system
accordingly. Equation (16-38) therefore becomes

A. The free particle

$$-\frac{h^2}{8\pi^2 m}\left(\frac{\partial^2 \psi}{\partial x^2} + \frac{\partial^2 \psi}{\partial y^2} + \frac{\partial^2 \psi}{\partial z^2}\right) = E\psi \tag{16-61}$$

or

$$\mathbf{H}_x\psi + \mathbf{H}_y\psi + \mathbf{H}_z\psi = E\psi \tag{16-62}$$

where the subscripts x, y, and z mean that the operator is one of that variable
only. Any wave equation such as Eq. (16-62), in which $\mathbf{H}$ can be written as
a sum of terms that do not share coordinates, can be resolved into the set

$$\mathbf{H}_x\,\psi_x = E_1\,\psi_x \qquad \mathbf{H}_y\psi_y = E_2\psi_y \qquad \mathbf{H}_z\psi_z = E_3\psi_z \tag{16-63}$$

where $\psi = \psi_x\psi_y\psi_z$ and $E = E_1 + E_2 + E_3$. The procedure for demonstrating
this conclusion is entirely analogous to that used in obtaining Eqs. (16-41)
and (16-42). One substitutes $\psi_x\psi_y\psi_z$ for ψ in Eq. (16-61) and then divides
through by $\psi_x\psi_y\psi_z$. The left-hand side then consists of a sum of terms in
x only, y only, and z only, and therefore separates into the three equations
(16-63).

In the present case, the x, y, and z equations are identical in form, so that
only one need be solved:

$$-\frac{h^2}{8\pi^2 m}\frac{d^2\psi_x}{dx^2} = E_1\psi_x \tag{16-64}$$

A stationary-state solution to Eq. (16-64), that is, one for which E is not a
function of time, is

$$\psi_x = A_1 \sin\left[\left(\frac{8\pi^2 mE_1}{h^2}\right)^{1/2} x\right] \tag{16-65}$$

as we can verify by substituting back into Eq. (16-64): A_1 is an integration
constant (note Problem 16-39). Since the wave is not confined, there are no
restrictions on its wavelength and E_1 can have any positive value. The total
ψ is then the product

$$\psi = A \sin\left[\left(\frac{8\pi^2 mE_1}{h^2}\right)^{1/2} x\right] \sin\left[\left(\frac{8\pi^2 mE_2}{h^2}\right)^{1/2} y\right] \sin\left[\left(\frac{8\pi^2 mE_3}{h^2}\right)^{1/2} z\right] \tag{16-66}$$

and corresponds to an infinite sinusoidal wave in three dimensions.
  The wavelength in, say, the x direction is given by

$$\lambda = \frac{2\pi}{(8\pi^2 mE_1/h^2)^{1/2}} \tag{16-67}$$

since for this value of x we have $\sin 2\pi$, which completes one cycle. Equation
(16-67) simplifies to $\lambda = h/(2mE_1)^{1/2}$, but E_1 is also equal to $p_x^2/2m$, so the final

result is just the de Broglie equation, $\lambda = h/p$. This result is not too surprising in view of analogies made when writing down the wave equation.

B. The particle in a
box

We may next stipulate that the particle has zero potential energy over a region defined by $0 < x < a$, $0 < y < b$, and $0 < z < c$, but that beyond these boundaries the potential energy is infinite. That is, V_x is zero for $0 < x < a$ and is infinite for $x < 0$ and $x > a$, and similarly for V_y and V_z; hence the term "box." Equation (16-38) is now

$$-\frac{h^2}{8\pi^2 m}\left(\frac{\partial^2 \psi}{\partial x^2} + \frac{\partial^2 \psi}{\partial y^2} + \frac{\partial^2 \psi}{\partial z^2}\right) + V\psi = E\psi \tag{16-68}$$

The situation is the same as for Eq. (16-61) in that the substitution $\psi = \psi_x \psi_y \psi_z$ allows the equation to be separated into three separate ones:

$$-\frac{h^2}{8\pi^2 m}\frac{\partial^2 \psi_x}{\partial x^2} + V_x \psi_x = E_1 \psi_x \tag{16-69}$$

and similarly for y and z, where $V = V_x + V_y + V_z$ and $E = E_1 + E_2 + E_3$. Alternatively, Eq. (16-69) may be viewed as the wave question for a particle in a one-dimensional box.

The solution to Eq. (16-69) is the same as that for Eq. (16-64) for the region inside the box, since $V_x = 0$. However, the solution outside the box, where $V = \infty$, must be that $\psi_x = 0$. This conclusion can be reached on both physical and mathematical grounds. Physically, it should be impossible for the particle to escape the box, and since $\psi_x^* \psi_x \, dx$ gives the probability of finding the particle in the region dx at each x, it follows that $\psi_x^* \psi_x = 0$ and hence $\psi_x = 0$ for $x < 0$ and $x > a$. The mathematical reasoning is that the curvature of the plot of ψ_x versus x, $\partial^2 \psi_x / \partial x^2$, is proportional to $(V_x - E_1)\,\psi_x$; if $V_x = \infty$ and $\psi_x \neq 0$, then the curvature must be infinite and ψ_x must increase to infinity itself. This is an impossible catastrophe—it is not acceptable that ψ_x, and hence the probability function for the particle, be infinite anywhere. This catastrophe can be avoided only if ψ_x is zero (and hence the curvature is zero) for $x < 0$ and $x > a$.

We also do not expect ψ_x to change discontinuously, that is, to have two values at a given x. There should be only one probability for the particle being at any one position. Since ψ_x must be zero just outside the box, it must therefore also be zero just inside the box. The consequence is that only those values of E_1 in Eq. (16-65) are acceptable that make ψ_x zero at $x = 0$ and at $x = a$, which means that the wave must have a node at these boundaries. In other words, there must be an integral number n_1 of half-wavelengths in the distance a, as illustrated in Fig. 16-4. As applied to Eq. (16-67), the condition is that $a = n_1 \lambda / 2$, or, on rearrangement,

$$E_1 = \frac{h^2 n_1^2}{8ma^2} \qquad n_1 = 1, 2, 3, \ldots \tag{16-70}$$

Also, substitution of this expression for E_1 into Eq. (16-65) gives

$$\psi_x = A_1 \sin\left(\frac{n_1 \pi x}{a}\right) \tag{16-71}$$

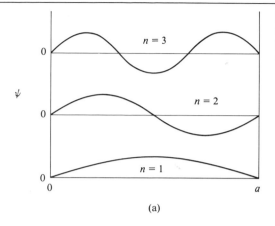

ψ

$n = 3$

$n = 2$

$n = 1$

(a)

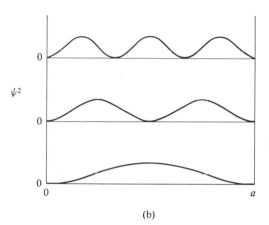

ψ^2

(b)

FIGURE 16-4
ψ and ψ^2 functions for a
particle in a one-
dimensional box.

One further step is usually taken with a solution to the wave equation. The solution will contain a constant of integration, A_1 in Eq. (16-71), and the value of A_1 is usually set so that the probability of the particle being somewhere is unity. The function is then said to be *normalized*. In the present case this condition is

$$\int \psi_x^* \psi_x \, d\tau = 1 = A_1^2 \int_0^a \left[\sin\left(\frac{n_1 \pi x}{a}\right) \right]^2 dx \qquad (16\text{-}72)$$

where the first integral is the formal statement that the integral of $\psi_x^* \psi_x$ over all space to be unity and the second integral is the specific one involved here. The integral is a standard one, equal to $a/2$ for the given integration limits. The constant A_1 is therefore $(2/a)^{1/2}$ and the normalized solution is

$$\psi_x = \left(\frac{2}{a}\right)^{1/2} \sin\left(\frac{n_1 \pi x}{a}\right) \qquad (16\text{-}73)$$

The preceding illustrates a fairly typical procedure in obtaining wave mechanical solutions. One uses the requirement that ψ be everywhere finite and single-valued if the solution is to have physical meaning, and this requirement then limits the possible number of solutions and hence of energy states. Quantization thus appears naturally and not as an *ad hoc* assumption.

To continue with the particle-in-a-box treatment, we see that the solutions for the y and z coordinates are the same as that for the x coordinate, so the complete solution for a three-dimensional box is

$$\psi = \left(\frac{8}{abc}\right)^{1/2} \sin\left(\frac{n_1\pi x}{a}\right) \sin\left(\frac{n_2\pi y}{b}\right) \sin\left(\frac{n_3\pi z}{c}\right) \tag{16-74}$$

and

$$E = \frac{h^2}{8m}\left(\frac{n_1^2}{a^2} + \frac{n_2^2}{b^2} + \frac{n_3^2}{c^2}\right) \tag{16-75}$$

where, for generality, the x, y, and z dimensions of the box are taken to be a, b, and c, respectively. Alternatively, for a two-dimensional square and a three-dimensional cubic box,

$$E_{\text{square}} = \frac{h^2}{8ma^2}\left(n_1^2 + n_1^2\right) \tag{16-76}$$

$$E_{\text{cube}} = \frac{h^2}{8ma^2}\left(n_1^2 + n_2^2 + n_3^2\right) \tag{16-77}$$

and the corresponding wave functions are

$$\psi_{\text{square}} = \frac{2}{a} \sin\left(\frac{n_1\pi x}{a}\right) \sin\left(\frac{n_2\pi y}{a}\right) \tag{16-78}$$

$$\psi_{\text{cube}} = \left(\frac{8}{a^3}\right)^{1/2} \sin\left(\frac{n_1\pi x}{a}\right) \sin\left(\frac{n_2\pi y}{a}\right) \sin\left(\left(\frac{n_3\pi z}{a}\right)\right) \tag{16-79}$$

Figure 16-5 shows how the energy levels are spaced for a square and a cubic box, expressed in units of $h^2/8ma^2$. Notice the occurrence of combinations of n_1 and n_2, and of n_1, n_2, and n_3 for which the sum of the squares is *identical*. The energies corresponding to such combinations must also be identical; such states are called *degenerate*. (The case of a square box with $n_1^2 + n_2^2 = 50$ is interesting. Why?) Such degeneracies would not occur with a rectangular box or with a three-dimensional box for which $a \neq b \neq c$. In effect, the degeneracies arise because of the symmetry introduced on going to the square or the cubic case. It is generally true in wave mechanics that degeneracies are the result of some symmetry aspect and are lifted or disappear when the symmetry is broken.

The probability of finding the particle at a given (x,y) in a square box is given by $\psi^*\psi$. That is, $\psi^2\, dx\, dy$ gives the probability of the particle being in the region $(x, x + dx)$, $(y, y + dy)$. A plot of ψ^2 is shown in Fig. 16-6 for several combinations of n_1 and n_2 values, and the resulting surface shows the corresponding standing waves. The situation is similar to that of standing waves in a square drum-head, where ψ^2 would now correspond to the square of the displacement of the parchment. Notice that in the degenerate cases (c) and (d) in the figure, the patterns are identical except for a 90° rotation.

Solutions may also be found for a circular box and for a sphere. The mathematics is more complicated but not sufficiently more revealing to warrant discussion here [see Berry, Rice, and Ross (1980)].

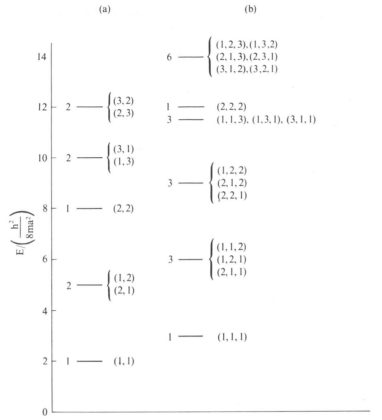

FIGURE 16-5
Energy levels for (a) a square box and (b) a cubic box. The numbers to the left of each level give the degeneracy (see text), and those in parentheses give (n_1, n_2) and (n_1, n_2, n_3) for the two cases, respectively.

What happens if the barrier height is not infinite, that is, if V does not go to infinity at the walls of the box? This is a very important and interesting situation and is discussed qualitatively in Section 16-CN-3. The particle can "leak" out in what is called *tunneling*.

C. Some applications

The equations for the particle in a box have several important applications. One of these is the use of Eq. (16-77) to supply the energy states for a gaseous molecule in a container of volume V, thereby leading to the translational partition function for an ideal gas (see Sections 4-12 and 6-12). The spacing of these energy states is quite small if the side a is of macroscopic size or if the particle is massive.

On insertion of numerical constants, Eq. (16-70) becomes

$$E = \frac{(6.6256 \times 60^{-27})^2(6.0225 \times 10^{23})(10^{16})}{8} \frac{n_1^2}{M\mathring{a}^2}$$

$$= 3.305 \times 10^{-14} \frac{n_1^2}{M\mathring{a}^2} \text{ (erg molecule}^{-1}) \tag{16-80}$$

$$= 476 \frac{n_1^2}{M\mathring{a}^2} \text{ (cal mol}^{-1})$$

FIGURE 16-6
Perspective views of ψ^2
versus x/a and y/a for a
particle in a square box
having the indicated n_1
and n_2 quantum numbers.

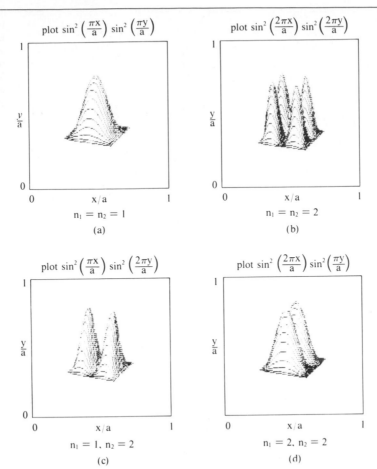

FIGURE 16-6
Perspective views of ψ^2 versus x/a and y/a for a particle in a square box having the indicated n_1 and n_2 quantum numbers.

where M is the molecular weight of the particle and $\mathring{a}$ is the length of the box in angstroms. For hydrogen gas in a 1 cm box, the first energy level would be at 2.38×10^{-14} cal mol^{-1} (!), the second at 2^2 times this, and so on.

The spacing becomes quite respectable, however, for a light particle such as an electron in a box of atomic dimensions. The atomic weight of the electron is 0.000549 g mol^{-1}, so that E is

$$E = 867 \frac{n_1^2}{\mathring{a}^2} \text{ kcal mol}^{-1} = 303 \frac{n_1^2}{\mathring{a}^2} \qquad \text{kK} = 37.6 \frac{n_1^2}{\mathring{a}^2} \text{ eV} \qquad (16\text{-}81)$$

Equation (16-81) may be applied in an approximate treatment of certain types of molecules which have electrons that are delocalized, that is, can move more or less freely over the whole structure. The electrons that form the double bond in ethylene probably behave in this way. A more interesting example is butadiene, $H_2C = CH - CH = CH_2$, in which each carbon atom contributes one electron to the double bond system, and these four electrons probably can oscillate over the entire length of the chain. That is, the double bonds are not really localized to the particular pairs of atoms shown. In effect, then, the four electrons are in a one-dimensional box equal in length to that of the chain. This last can be estimated as equal to two C=C bond lengths or 2(1.35) Å, plus one C — C bond or 1.54 Å, plus the distance of a carbon atom radius at each end or another 1.54 Å, for a total of 5.78 Å. Equation (16-81) then gives the energy states in eV as $[37.6/(5.78)^2]n_1^2 = 1.12n_1^2$. By the Pauli exclusion principle, each state can hold two electrons (with spins opposed), and so the four electrons fill the first two levels as illustrated in Fig. 16-7(a).

The energy of the first excited state of this system of four electrons is that for having one electron in the $n = 3$ level, and the energy to produce this state is $E = 1.12(3^2 - 2^2) = 5.60$ eV or 45,000 cm^{-1}. Butadiene does have an absorption band at 46,100 cm^{-1} (217 nm), and the very simple free-electron model, as this is called, has been fairly successful.

As a further example, hexatriene, $H_2C = CH — CH = CH — CH = CH_2$, should have six delocalized electrons in the double bond system, in a box of length 8.67 Å. The energy for the $n = 1$ state is then $37.6/(8.67)^2 = 0.500$ eV, and as shown in Fig. 16-7(b), the six electrons fill the first three levels. The first excited state should now have one electron in the $n = 4$ state, and ΔE for the transition should therefore be $0.500(4^2 - 3^2) = 3.50$ eV or 28,000 cm^{-1}. The experimental value is 38,500 cm^{-1}, and although the agreement is rather poor, the model has correctly predicted that the frequency of the absorption band should decrease with increasing chain length.

The free-electron model is a semiempirical one—one that does not know, for example, just how much "end space" to allow in calculating the length of the box. The uncertainty becomes less important with longer chains, and with carotene, which has alternating single and double bonds in an 11-carbon-atom chain, one estimates about 25,000 cm^{-1} for the transition to the first excited state, as compared to 22,200 cm^{-1} observed. The model may also be applied to two-dimensional boxes, such as benzene or fused aromatic ring systems, with Eq. (16-76). A more sophisticated treatment is given in Section 18-7.

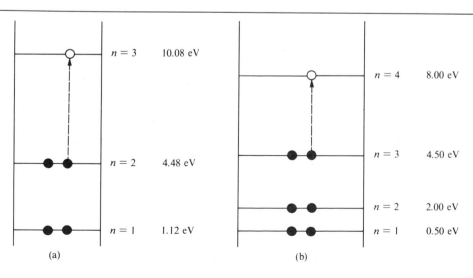

FIGURE 16-7
Particle-in-a-box energy level scheme for (a) butadiene with $\mathring{a} = 5.78$ and (b) hexatriene with $\mathring{a} = 8.67$.

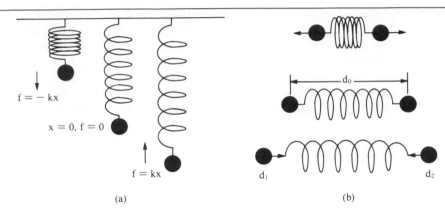

FIGURE 16-8
Harmonic oscillators. (a) A ball suspended by a spring and (b) two balls connected by a spring.

$f = -kx$

$x = 0, f = 0$

$f = kx$

(a)

d_0

d_1 d_2

(b)

16-6   The harmonic oscillator

A. The classical treatment

Mechanical oscillating systems are fairly common in everyday experience; examples are the pendulum of a clock or, as illustrated in Fig. 16-8(a), a ball supported by a spring. In the absence of friction such systems, once started, oscillate indefinitely. It is the frictionless case that we consider. A *harmonic oscillator* is one for which *Hooke's law* applies, that is, the restoring force is proportional to the displacement x. As shown in the figure, if x is positive, the force is $f = -kx$, where k is a proportionality constant, called the *force constant*. The minus sign means that the direction of the force is such as to oppose the displacement.

The potential energy V of the ball is just the negative of the integral of $f\,dx$, or $V = \frac{1}{2}kx^2$; the positive sign means that V *increases* with the displacement squared. A plot of V versus x is a parabola centered at $x = 0$. We have from Newton's law that at any point in the oscillation $f = ma$, where a is acceleration and m is the mass of the ball (the spring is taken to be massless), so that

$$m \frac{d^2x}{dt^2} = -kx \tag{16-82}$$

The solution to this differential equation is

$$x = a \cos(2\pi\nu_0 t) \tag{16-83}$$

where $\nu_0 = \dfrac{1}{2\pi}(k/m)^{1/2}$ and a is the *amplitude* or displacement at $t = 0$; that is, we have lifted the ball to a position $x = a$ and let go at zero time.† The total energy E is, of course, the sum of kinetic and potential energy, and we can write

$$E = T + V = \tfrac{1}{2}mv^2 + \tfrac{1}{2}kx^2$$

or

---

†The reader can verify that Eq. (16-83) is a solution by substitution into Eq. (16-82). The general solution to Eq. (16-82) is of the form $x = A \sin(ct) + B \cos(ct)$ where A, B, and c are constants, but we don't want the sine part (why?).

$$E = \frac{p^2}{2m} + \tfrac{1}{2}kx^2 \tag{16-84}$$

We are actually more interested in the mechanical system shown in Fig. 16-8(b), since this is the one that corresponds to the vibration of a molecule. Here two balls are connected by a spring; the balls are of masses m_1 and m_2 and the equilibrium separation is d_0 (the system at rest). If the balls are pulled apart, so that the positions are d_1 and d_2, the amount that the spring is stretched is $(d_2 - d_1 - d_0)$; this is our distance x. We again assume Hooke's law, that is, that the restoring force is given by $f = -kx$, and V is again $\tfrac{1}{2}kx^2$. It turns out from detailed analysis [see McQuarrie (1983)] that if we use the above definition of x, we regain Eq. (16-82) in the form

$$\mu \frac{d^2x}{dt^2} = -kx \tag{16-85}$$

where μ is the reduced mass, $\mu = m_1m_2/(m_1 + m_2)$. The solution is again Eq. (16-83), with $\nu_0 = \frac{1}{2\pi}(k/\mu)^{1/2}$. The two systems of Fig. 16-8 can thus be made mathematically equivalent.

In the classical case, E can have any positive value. Moreover, at $x = a$, dx/dt is zero and E is made up entirely of potential energy, while at $x = 0$, $V = 0$, and E is entirely kinetic energy; this means that the balls in Fig. 16-8(b) spend most of the time near $x = a$ (and $x = -a$), where they are moving the most slowly. We will see that the picture changes considerably when we enter the submicroscopic world in which wave mechanics must be used.

We now consider the wave mechanical harmonic oscillator. Figure 16-8(b) still applies, but the balls m_1 and m_2 are now atoms, and the spring becomes the bond between the two atoms. We still assume Hooke's law and still have a parabolic potential function. Now, however, we may change Eq. (16-84) by carrying out the instructions of postulate 2, or, more simply, we can insert $V = \tfrac{1}{2}kx^2$ into Eq. (16-39). In either case, of course, we replace m by μ since we are dealing with the situation of Fig. 16-8(b). We use the one-dimensional form of Eq. (16-39) and obtain, after a little rearranging:

B. The wave mechanical case

$$\frac{d^2\psi}{dx^2} + \frac{8\pi^2\mu}{h^2}(E - \tfrac{1}{2}kx^2)\psi = 0 \tag{16-86}$$

Equation (16-86) is not an easy one to solve because of the x^2 in the coefficient of ψ, but it turns out that it can be manipulated into a form that has been much studied by mathematicians. We let $\alpha = (8\pi^2\mu/h^2)E$ and $\beta = 2\pi\sqrt{\mu k}/h$, so that Eq. (16-86) becomes

$$\frac{d^2\psi}{dx^2} + (\alpha - \beta^2 x^2)\psi = 0 \tag{16-87}$$

and then further make the change of variable $q = \sqrt{\beta}x$, to get

$$\frac{d^2\psi}{dq^2} + \left(\frac{\alpha}{\beta} - q^2\right)\psi = 0 \tag{16-88}$$

The solution is of the form

$$\psi = \phi(q)e^{-q^2/2} \tag{16-89}$$

and substitution back into Eq. (16-88) gives the equation

$$\frac{d^2\phi}{dq^2} - 2q\frac{d\phi}{dq} + \left(\frac{\alpha}{\beta} - 1\right)\phi = 0 \tag{16-90}$$

which is identical in form to a standard differential equation known as *Hermite's equation,*

$$\frac{d^2\phi}{dq^2} - 2q\frac{d\phi}{dq} + 2n\phi = 0 \tag{16-91}$$

The admissible solutions to Eq. (16-91) are in the form of polynomials whose nature depends on the value of n. If n is an integer, the first few are as follows:

$$H_{n=0} = 1 \qquad\qquad H_{n=3} = 8q^3 - 12q$$

$$H_{n=1} = 2q \qquad\qquad H_{n=4} = 16q^4 - 48q^2 + 12$$

$$H_{n=2} = 4q^2 - 2 \qquad H_{n+1} = 2qH_n - 2nH_{n-1}$$

The last relationship above is called a *recursion formula;* it allows the next expression of a series to be obtained if the preceding two are known.

The actual solutions to Eq. (16-87) are

$$\psi_n = \left[\frac{(\beta/\pi)^{1/2}}{2^n n!}\right]^{1/2} H_n(q)e^{-q^2/2} \tag{16-92}$$

where n may be 0,1,2,3, ...This may be verified by picking an n value and substituting Eq. (16-92) back into Eq. (16-87) [and using Eq. (16-93) below)]. The factor in front of the right-hand term serves to normalize the wave function. Substitution of the expressions for q and for β then allows ψ to be plotted as a function of x for various n values.

Some decisions have been made in choosing Eq. (16-92) as the acceptable solution to Eq. (16-87). First, at large x, E drops out of Eq. (16-86) and the limiting solution is of the form $\exp(\pm q^2/2)$; however, only the *minus* sign is physically acceptable, since otherwise the catastrophe of ψ going to infinity at $x \rightarrow \infty$ would occur. This conclusion leads to Eqs. (16-89) and (16-90), the latter being Hermite's equation. Only *integral* values of n are allowed in the solution to Eq. (16-91) if the polynomials satisfying the equation are to have a finite number of terms and not themselves lead to a catastrophe. The quantization thus arises from restriction of the mathematical possibilities to *physically acceptable* solutions. As in the case of the particle in a box, n becomes a quantum number.

The energy of the various states is obtained as follows. The matching of Eqs. (16-89) and (16-90) requires that

$$\frac{\alpha}{\beta} = 2n + 1 \tag{16-93}$$

or, on substitution of the expressions for α and β and simplification,

$$E = \frac{h}{2\pi}\left(\frac{k}{\mu}\right)^{1/2}(v + \tfrac{1}{2}) = h\nu_0(v + \tfrac{1}{2}) \tag{16-94}$$

where n has been replaced by the more conventional vibrational quantum number symbol v. The quantity $(h/2\pi)(k/\mu)^{1/2}$ has the dimensions of energy, and is conventionally set equal to hv_0, since v_0 would be the natural frequency of oscillation of the classical harmonic oscillator; thus

$$v_0 = \frac{1}{2\pi}(k/\mu)^{1/2} \tag{16-95}$$

According to Eq. (16-94), the lowest energy state is *not* zero, but $\frac{1}{2}hv_0$; this is known as the *zero-point energy*. Also, ΔE for the transition from one state to another is given by

$$\Delta E = hv_0(v_2 - v_1) \tag{16-96}$$

We next examine the graphical appearance of our results, using Fig. 16-9. This figure, while a conventional one, is really a set of *three* figures. First, we see the potential function $V = \frac{1}{2}kx^2$, plotted in the heavy line as a parabola centered at $x = 0$. Next, the succession of light lines shows the spacing of the first three energy levels; note the zero point energy effect—E_0 is *not* zero. Note also that the energy levels are equally spaced, as specified by Eq. (16-94). Finally, the figure shows plots of ψ and of ψ^2. As indicated in the figure, the ordinate for each of these sets of plots is zero at the E_n line. That is, for example, ψ_0 and ψ_0^2 are plotted with the E_0 line as the base line.

It is important to look at the ψ and ψ^2 plots in a little detail. Notice that

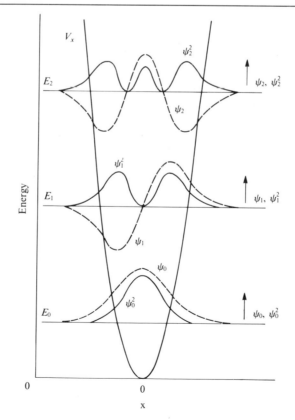

FIGURE 16-9
Energy levels for a harmonic oscillator, with superimposed plots of corresponding ψ and ψ^2. (Adapted from K. S. Pitzer, "Quantum Chemistry," © 1953. By permission of Prentice-Hall, Inc., Englewood Cliffs, New Jersey.)

they have *nodes*, or points at which the value is zero. There is an *even* number of nodes if n is even, and an *odd* number if n is odd. Notice also that the ψ plots for n even are symmetric about the vertical line through $x = 0$, while the functions of n odd are antisymmetric or inverted across the line. Mathematically, the definition of an even function is that $f(x) = f(-x)$ while, for an odd function, $f(x) = -f(-x)$. The alternating even and odd character of the ψ functions is seen later (Section 19-5) to be important in determining the probability of a transition from one vibrational level to another.

The *probability* of x being between x and $x + dx$ is given by $\psi^*\psi$ (postulate 4), or in this case by just ψ^2 since the ψ functions are not imaginary. As with the ψ functions, there are nodes—and the number of nodes is equal to n. We can at this point remark on some of the differences between the classical and the quantum mechanical harmonic oscillator. First, of course, all E values are possible for the former, while *only* the E_n values are allowed for the latter. Suppose, however, that we have a classical oscillator whose total energy happens to be exactly E_0. As shown in Fig. 16-10(a), x *cannot* exceed the value a (or $-a$). At these values, $V = E_0$ and, by Eq. (16-84), the kinetic energy must be zero. In other words, x oscillates between a and $-a$. The probability p of finding the classical oscillator at an x value between x and $x + dx$ is

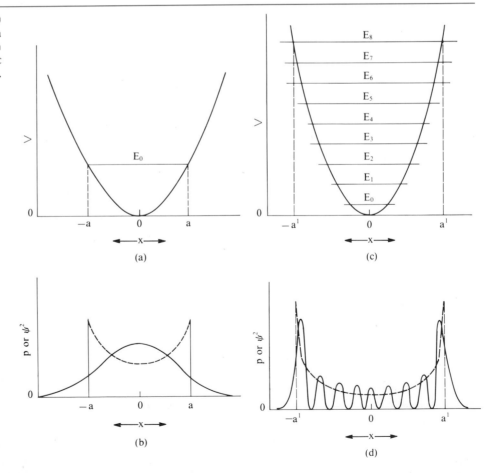

FIGURE 16-10
Comparisons between a classical and a quantum mechanical harmonic oscillator (see text).

indicated by the dashed line of Fig. 16-10(b). Notice that p is a maximum at a and $-a$, where the particle has zero kinetic energy.

The corresponding probability, ψ^2, for the quantum mechanical oscillator is shown in Fig. 16-10(b) by the solid line. Note the contrasts! First, ψ_0^2 is a *maximum* at $x = 0$, or where p is a minimum. Second, there is a finite ψ_0^2 value or probability of x *exceeding* the classical limits a and $-a$! The wave mechanical system can exist in a region where, classically, the kinetic energy is negative! This is an example of wave mechanical penetration or "leakage" through a potential energy barrier (see Section 16-CN-3).

The above picture changes if we go to a large quantum number n. Figure 16-10(c) shows, on a smaller scale, energy levels up to E_8. A classical oscillator having this total energy will have x values oscillating between a' and $-a'$, and the variation of p with x will look something like the dashed curve in Fig. 16-10(d); the situation is similar to the previous one. The picture for the quantum mechanical oscillator has changed considerably, however. The plot of ψ_8^2 versus x now has eight nodes, the intervening peaks gradually *increasing* in height as x moves toward a' and $-a'$. In fact, the classical dashed line now traces an approximate locus through the midpoints of the peaks. Also, the nonclassical or negative kinetic energy region is now relatively less important than in the $E = E_0$ case. Were we to pursue the comparison to *very* large n values, the locus of the wave mechanical ψ^2 plot would look more and more like the classical p plot. This is as it should be. For a macroscopic system, ν_0 is very small since μ is now large [Eq. (16-95)], and for any appreciable value of E, n must be very large. It is characteristic of wave mechanics that the predicted behavior approaches the classical one for macroscopic systems.

Another kind of comparison that we can make is between the harmonic oscillator and the particle in a one-dimensional box. The similarity is apparent if we compare Figs. 16-4 and 16-9, taking n_{box} as one plus n for the harmonic oscillator. There is no "leakage" past the walls of the box, however; as we see in Section 16-CN-3, this is because the potential V is infinite at the walls.

We return now to Eq. (16-94) to obtain some order-of-magnitude numbers for the $h\nu_0$ of a harmonic oscillator. This is an important quantity; according to Eq. (16-96), it gives the ΔE for a transition between adjacent vibrational energy levels. Typical force constants are about 10^5 dyn cm^{-1} or 100 N m^{-1}, and k values are often expressed in this unit. We thus have, in the cgs system,

$$h\nu_0 = \frac{h}{2\pi} \left(\frac{k}{\mu} \right)^{1/2} = \frac{6.6262 \times 10^{-27}}{2\pi} (10^5 \times 6.0221 \times 10^{23})^{1/2} \left(\frac{k'}{\mu'} \right)^{1/2}$$

or

$$h\nu_0 = 2.59 \times 10^{-13} \left(\frac{k'}{\mu'} \right)^{1/2} \text{ erg} \qquad (16\text{-}97)$$

where k' is in units of 10^5 dyn cm^{-1} and μ' is in atomic mass units (that is, calculated using the table of atomic "weights"). Alternatively,

$$h\nu_0 = 2.59 \times 10^{-20} \left(\frac{k'}{\mu'} \right)^{1/2} \text{ J} \qquad (16\text{-}98)$$

with k' in units of 100 N m^{-1} and again with μ' in atomic mass units. For $h\nu_0$ in eV the proportionality constant is 0.162, and for $h\nu_0$ in kK it is 1.30.

EXAMPLE   The first vibrational transition for H^{35}Cl is at 2886 cm^{-1}; calculate k. We have $k'/\nu' = (2886/1300)^2 = 4.928$. The reduced mass is $(1.008)(34.97)/(1.008 + 34.97) = 0.980$ (isotopic masses are found in Table 22-1), so $k' = 4.83$ or $k = 4.83 \times 10^5$ dyn cm^{-1} = 483 N m^{-1}. Note: the wavelength of light that is absorbed in this first vibrational transition is 1/2886 cm or 3.47 μm or 34,700 Å, and corresponds to the infrared region of the spectrum. Also, from the conversion table on the front cover, 2886 cm^{-1} corresponds to 8.25 kcal mol^{-1} or 34.5 kJ mol^{-1}, a quite appreciable energy. The zero point energy is $\frac{1}{2}h\nu_0$, and is therefore just half of the above values.

C. Steps beyond the harmonic oscillator

The actual potential energy curve for a diatomic molecule is like that shown in Fig. 16-11. The restoring force that keeps atoms from separating gradually weakens with higher and higher vibrational energy until a dissociation limit is reached. The harmonic oscillator model is not a bad approximation for the bottom of the correct potential well, however, and is very useful for ordinary infrared spectroscopy since the typical vibrational transition is from $v = 0$ to $v = 1$. Clearly, however, the model must fail badly for highly vibrationally excited molecules.

The wave equation has been solved for *anharmonic* (that is, not harmonic) potential functions of the shape shown in Fig. 16-11. As also illustrated in the figure, the vibrational energy levels are spaced more and more closely together with increasing vibrational quantum number. The dissociation limit is reached, however, after a finite number of quantum states.

A very convenient algebraic form for an anharmonic potential function is that due to P. Morse,

$$V = D(1 - e^{-a(r-r_e)})^2 \tag{16-99}$$

where D is the depth of the potential well at the equilibrium separation r_e. Use of this potential function in the wave equation leads to a better approximation for vibrational levels:

FIGURE 16-11
Potential energy function and vibrational level spacings for an anharmonic oscillator. The interatomic distance at the minimum in V is r_e and D is the full dissociation energy; E_0 is the actual dissociation energy, $E_0 = D - \frac{1}{2}h\nu_0$.

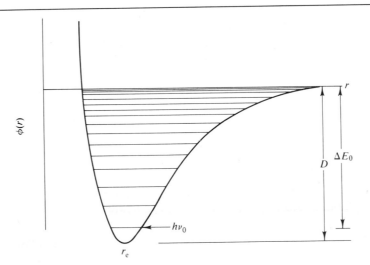

$$E = w_e(v + \tfrac{1}{2}) - x_e w_e(v + \tfrac{1}{2})^2 \tag{16-100}$$

where $w_e = (ah/\pi)(D/2\mu)^{1/2}$ and $x_e = w_e/4D$. The levels converge slowly, and the dissociation limit is reached after a finite number of states.

16-7   Solutions of the wave equation for the hydrogen atom

The most complete application of wave mechanics is undoubtedly to the case of the hydrogen atom. Not only can exact solutions be obtained, but these solutions are then widely used in approximate treatments of heavier atoms. We proceed in this section to the formal solutions for the hydrogen atom and will discuss the behavior of these solutions in detail in the following section. The derivation that follows is more frightening in appearance than actually difficult!

The potential function for the electron is now $-e^2/r$, where r is its distance from the nucleus, and Eq. (16-39) becomes

A. The Schrödinger equation for the hydrogen atom

$$-\frac{h^2}{8\pi^2 m} \nabla^2\psi - \frac{e^2}{r}\psi = E\psi \tag{16-101}$$

The presence of r as a parameter in the potential function makes it awkward for one to work in Cartesian coordinates and it turns out to be very convenient to use the polar coordinate system shown in Fig. 16-12. A point at (x,y,z) is now given by its radial distance r and the two angles θ and ϕ. In sweeping out all of space r varies from 0 to ∞, θ varies from 0 to π, and ϕ varies from 0 to 2π.

The two coordinate systems are related. The projection of r onto the z axis gives

$$z = r \cos \theta \tag{16-102}$$

and $r \sin \theta$ gives the projection onto the xy plane, so that

$$x = r \sin \theta \cos \phi \tag{16-103}$$

$$y = r \sin \theta \sin \phi \tag{16-104}$$

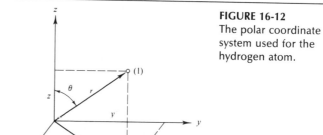

FIGURE 16-12
The polar coordinate system used for the hydrogen atom.

By inserting these relationships into the expression for ∇^2 in Cartesian coordinates, one obtains (finally) for Eq. (16-101):

$$\frac{1}{r^2}\frac{\partial}{\partial r}\left(r^2\frac{\partial\psi}{\partial r}\right) + \frac{1}{r^2\sin\theta}\frac{\partial}{\partial\theta}\left(\sin\theta\frac{\partial\psi}{\partial\theta}\right) + \frac{1}{r^2\sin^2\theta}\frac{\partial^2\psi}{\partial\phi^2}$$

$$+ \frac{8\pi^2 m}{h^2}\left(E + \frac{e^2}{r}\right)\psi = 0 \qquad (16\text{-}105)$$

[see, for example, McQuarrie (1983)].

One advantage of the new coordinate system is that it allows the variables to be separated. We do this by writing $\psi(r,\theta,\phi) = R(r)\,\Theta(\theta)\,\Phi(\phi)$, that is, as a product of three functions each having only one variable. The procedure is much the same as that used with Eq. (16-61); we substitute $\Theta(\theta)\,\Phi(\phi)[\partial R(r)/\partial r]$ for $\partial\psi/\partial r$, and so on and then divide through by $R(r)\,\Theta(\theta)\,\Phi(\phi)$ to obtain (dropping the arguments of R, Θ, and Φ for convenience)

$$\frac{1}{r^2 R}\frac{d}{dr}\left(r^2\frac{dR}{dr}\right) + \frac{1}{r^2(\sin\theta)\Theta}\frac{d}{d\theta}\left(\sin\theta\frac{d\Theta}{d\theta}\right)$$

$$+ \frac{1}{r^2(\sin^2\theta)\Phi}\frac{d^2\Phi}{d\phi^2} + \frac{8\pi^2 m}{h^2}\left(E + \frac{e^2}{r}\right) = 0$$

where the partial differentiation signs are no longer needed since each function is of a single variable only. We multiply through by $r^2\sin^2\theta$ and rearrange to obtain

$$\frac{\sin^2\theta}{R}\frac{d}{dr}\left(r^2\frac{dR}{dr}\right) + \frac{\sin\theta}{\Theta}\frac{d}{d\theta}\left(\sin\theta\frac{d\Theta}{d\theta}\right)$$

$$+ \frac{8\pi^2 m}{h^2}\left(E + \frac{e^2}{r}\right)(r^2\sin^2\theta) = -\frac{1}{\Phi}\frac{d^2\Phi}{d\phi^2} \qquad (16\text{-}106)$$

The left-hand side of Eq. (16-106) is a function of r and θ only, and the right-hand side is a function of ϕ only, so neither can depend on the other's variable(s); each side must therefore be equal to some common constant, which we take to be m^2. The right-hand side of Eq. (16-106) then gives

$$\frac{1}{\Phi}\frac{d^2\Phi}{d\phi^2} = -m^2 \qquad (16\text{-}107)$$

which we will call the Φ *equation*.

The left-hand side of Eq. (16-106) is likewise equal to m^2; we divide through by $\sin^2\theta$ and rearrange to obtain

$$\frac{1}{R}\frac{d}{dr}\left(r^2\frac{dR}{dr}\right) + \frac{8\pi^2 m}{h^2}\left(E + \frac{e^2}{r}\right)r^2$$

$$= \frac{m^2}{\sin^2\theta} - \frac{1}{\sin\theta}\frac{1}{\Theta}\frac{d}{d\theta}\left(\sin\theta\frac{d\Theta}{d\theta}\right) \qquad (16\text{-}108)$$

By the same argument as before, each side must equal some common constant, which is taken to be λ. The resulting two equations are, with some rearrangement,

$$\frac{1}{r^2}\frac{d}{dr}\left(r^2\frac{dR}{dr}\right) - \frac{\lambda}{r^2}R + \frac{8\pi^2 m}{h^2}\left(E + \frac{e^2}{r}\right)R = 0 \qquad (16\text{-}109)$$

and

$$\frac{1}{\sin\theta}\frac{d}{d\theta}\left(\sin\theta\frac{d\Theta}{d\theta}\right) - \frac{m^2\Theta}{\sin^2\theta} + \lambda\Theta = 0 \qquad (16\text{-}110)$$

These will be called the R and Θ *equations*, respectively.

The problem is now reduced to the solution of three separate equations, each in one variable only. Before proceeding to these solutions, it should be mentioned that, strictly speaking, the mass of the electron m in the $8\pi^2 m/h^2$ term should be replaced by the reduced mass μ, $\mu = (m + M)/mM$, where M is the nuclear mass. The correction is a small one, however.

Since m is a constant that can have various values, there is a family of solutions to Eq. (16-107), of the form

B. Solution of the Φ equation

$$\Phi_m = \frac{1}{\sqrt{2\pi}}e^{im\phi} = \frac{1}{\sqrt{2\pi}}(\cos m\phi + i\sin m\phi) \qquad (16\text{-}111)$$

One can easily verify that Eq. (16-111) is a solution for any m by differentiating twice and comparing with Eq. (16-107). By now it should be no surprise that the physically acceptable solutions are limited to those for which $m = 0, \pm 1, \pm 2, \ldots$. The reason is simply that the solution must be for a stationary state or a standing wave and consequently cannot depend on how many times one cycles ϕ around a circle. Φ_m must be the same for $\phi = 0$, $\phi = 2\pi$, $\phi = 4\pi$, and so on, and this is possible only if m is an integer or zero. As was anticipated, m turns out to be the magnetic quantum number of the Bohr theory. Finally, the factor $1/\sqrt{2\pi}$ is a normalization factor, such that $\int_0^{2\pi}\Phi_m^*\Phi_m\,d\phi = 1$.

The expression for Φ_m is just $1/\sqrt{2\pi}$ if $m = 0$. For other m values it is awkward to deal with an imaginary exponential. Use is therefore made of a general property of solutions of a differential equation of this type. It is that any linear combination of solutions must also be a solution. For any given m we can take the sum and difference of the solutions for m and $-m$ to obtain two new ones. Thus for $m = 1$,

$$\Phi_1 = A(\cos\phi + i\sin\phi) \qquad \Phi_{-1} = A(\cos\phi - i\sin\phi)$$

and $\Phi_1 + \Phi_{-1} = 2A\cos\phi$, and $\Phi_1 - \Phi_{-1} = 2iA\sin\phi$. It is customary to drop the factor i in the second of these combinations, since it disappears when we write $\Phi^*\Phi$. Also, the constant $2A$ must be reevaluated if the new functions are to be normalized, and is then found to be just $1/\sqrt{\pi}$. In summary, the solutions of the equation are usually written

$$\Phi_0 = \frac{1}{\sqrt{2\pi}} \qquad \Phi_m = \begin{cases} \dfrac{1}{\sqrt{\pi}}\cos(|m|\phi) \\ \dfrac{1}{\sqrt{\pi}}\sin(|m|\phi) \end{cases} \qquad (16\text{-}112)$$

Their graphical appearance is discussed in the next section.

The wave equation can be written $\mathbf{H}\psi = E\psi$, and thus conforms to Eq. (16-44). The ψ functions are therefore eigenfunctions, and the E values are eigenvalues. The eigenfunctions used in this chapter are all normalized. In addition, since $\mathbf{H}$ is an hermitian operator, it will be true that our eigenfunctions are orthogonal [see postulate 3 and the footnote following Eq. (16-58)].

In the present case this means that $\displaystyle\int_0^{2\pi} \Phi_{m_1}^* \, \Phi_{m_2} \, d\phi = 0, \; m_1 \neq m_2$

C. Solution of the Θ equation

The general solution of Eq. (16-110) requires considerations whose details are beyond the scope of this text. As with the Φ equation, the solution is in the form of a function, but now in θ, and the requirement that it be cyclic in θ restricts the values of λ to $\ell(\ell + 1)$, where ℓ is an integer. That is, in order for the solutions to represent stationary states, they must be the same for $\theta = 0, 2\pi, \dots$. The integer ℓ functions as the azimuthal or angular momentum quantum number of the Bohr theory. The requirement that the polynomial solution have a finite number of terms further requires that m in Eq. (16-110) also be an integer, but not exceeding ℓ. Once again, imposition of the physical condition of a stationary-state solution introduces integral values for constants of integration, thus converting them into quantum numbers.

The functions that are stationary-state solutions of Eq. (16-110) are known as *associated Legendre functions;* Table 16-2 lists a sufficient number of them to meet most needs. They are normalized and orthogonal. There is a somewhat tricky point here. The element of volume in polar coordinates is $r^2 \sin \theta \, dr \, d\theta \, d\phi$; this is the expression equivalent to $dx \, dy \, dz$ in Cartesian coordinates. The normalization integral for the Θ function is therefore

$$\int_0^\pi \Theta_{l,m}^* \Theta_{\ell,m} \sin \theta \, d\theta$$

or, since the solutions are real,

$$\int_0^\pi \Theta_{\ell,m}^2 \sin \theta \, d\theta = 1 \tag{16-113}$$

TABLE 16-2
Solutions to the Θ function

$\ell = 0, m = 0$ (s orbitals)		$\Theta_{0,0} = \frac{1}{2}\sqrt{2}$
$\ell = 1$ (p orbitals)	$m = 0$	$\Theta_{1,0} = \frac{1}{2}\sqrt{6}\cos\theta$
	$|m| = 1$	$\Theta_{1,1} = \frac{1}{2}\sqrt{3}\sin\theta$
$\ell = 2$ (d orbitals)	$m = 0$	$\Theta_{2,0} = \frac{1}{4}\sqrt{10}\,(3\cos^2\theta - 1)$
	$|m| = 1$	$\Theta_{2,1} = \frac{1}{2}\sqrt{15}\sin\theta\cos\theta$
	$|m| = 2$	$\Theta_{2,2} = \frac{1}{4}\sqrt{15}\sin^2\theta$
$\ell = 3$ (f orbitals)	$m = 0$	$\Theta_{3,0} = \frac{3}{4}\sqrt{14}\,(\frac{5}{3}\cos^3\theta - \cos\theta)$
	$|m| = 1$	$\Theta_{3,1} = \frac{1}{8}\sqrt{42}\,(\sin\theta)(5\cos^2\theta - 1)$
	$|m| = 2$	$\Theta_{3,2} = \frac{1}{4}\sqrt{105}\sin^2\theta\cos\theta$
	$|m| = 3$	$\Theta_{3,3} = \frac{1}{8}\sqrt{70}\sin^3\theta$

Similarly, the orthogonality integral is

$$\int_0^\pi \Theta_{\ell,m}\Theta_{\ell',m'} \sin\theta \, d\theta = 0 \tag{16-114}$$

where either ℓ and ℓ' or m and m' (or both) are different. Equations (16-113) and (16-114) might be verified by the reader for the first few entries in Table 16-2 as well as the fact that the latter are indeed solutions to Eq. (16-110).

The principal regularity in the Θ functions to be noticed at the moment is that they are all polynomials whose highest term is of the form $(\cos\theta)^a (\sin\theta)^b$, where $a + b = \ell$ and $b = |m|$. Within a given ℓ set, the progression is from an expression containing $\cos\theta$ terms only to one containing only $\sin\theta$. Also, Θ for $|m| = \ell$ is always $(\sin\theta)^\ell$ and Θ for $|m| = \ell - 1$ is always $(\sin\theta)^{\ell-1}(\cos\theta)$. These functions are very important in the applications of wave mechanics to chemical bonding and are sufficiently limited in number that the reader should familiarize himself with them.

D. Solution of the *R* equation

Equation (16-109) again can be reduced to a standard differential equation, and, as with the Θ equation, the solutions are in the form of polynomials—this time in r and known as the *associated Laguerre polynomials*. To avoid a catastrophe at $r \to \infty$, it is again necessary that the polynomials terminate after a finite number of terms; the requirement for this to happen is that λ or $\ell(\ell + 1)$ be an integer. This defines a quantum number n, the same as the principal quantum number of the Bohr theory.

It is convenient to generalize a little by replacing the potential function of Eq. (16-109), e^2/r, by Ze^2/r, to allow for a nuclear charge other than unity. Also, the polynomials are much simplified in appearance if written in terms of

$$\rho = \frac{2Z}{na_0} r \tag{16-115}$$

where a_0 is the radius of the first Bohr orbit, as given by

$$a_0 = \frac{h^2}{4\pi^2\mu e^2} \qquad [\text{Eq. (16-14)}]$$

(in the more accurate form using μ rather than m).

The most commonly needed solutions of the R equation are given in Table 16-3. These are again normalized and orthogonal. (Remember that the integrals involved are of the form $\int_0^\infty R_1^* R_2 \, r^2 \, dr$; these are equal to unity if $R_1 = R_2$ and zero otherwise). The principal regularities of these solutions are the following. Each contains the exponential $e^{-\rho/2}$ preceded by a polynomial in ρ. The highest term in the polynomial, that is, the order of the polynomial, is $n - 1$, and the polynomial is of the form

$$(a + b\rho + c\rho^2 + \cdots + \rho^{n-1-\ell})\rho^\ell$$

This general term looks complicated, and it is easier to perceive what is being said by studying the actual polynomials. Although the numerical coefficients are unimportant for our purposes, it is again very desirable to become familiar with the progression in *form* of these solutions.

TABLE 16-3 Solutions to the R equation	$n = 1$ $\ell = 0$ (1s)	$R_{1,0} = \left(\dfrac{Z}{a_0}\right)^{3/2} (2)(e^{-\rho/2})$
	$n = 2$ $\ell = 0$ (2s)	$R_{2,0} = \dfrac{(Z/a_0)^{3/2}}{2\sqrt{2}} (2 - \rho)e^{-\rho/2}$
	$\ell = 1$ (2p)	$R_{2,1} = \dfrac{(Z/a_0)^{3/2}}{2\sqrt{6}} \rho e^{-\rho/2}$
	$n = 3$ $\ell = 0$ (3s)	$R_{3,0} = \dfrac{(Z/a_0)^{3/2}}{9\sqrt{3}} (6 - 6\rho + \rho^2)e^{-\rho/2}$
	$\ell = 1$ (3p)	$R_{3,1} = \dfrac{(Z/a_0)^{3/2}}{9\sqrt{6}} (4 - \rho)\rho e^{-\rho/2}$
	$\ell = 2$ (3d)	$R_{3,2} = \dfrac{(Z/a_0)^{3/2}}{9\sqrt{30}} \rho^2 e^{-\rho/2}$
	$n = 4$ $\ell = 0$ (4s)	$R_{4,0} = \dfrac{(Z/a_0)^{3/2}}{96}(24 - 36\rho + 12\rho^2 - \rho^3)e^{-\rho/2}$
	$\ell = 1$ (4p)	$R_{4,1} = \dfrac{(Z/a_0)^{3/2}}{32\sqrt{15}} (20 - 10\rho + \rho^2)\rho e^{-\rho/2}$
	$\ell = 2$ (4d)	$R_{4,2} = \dfrac{(Z/a_0)^{3/2}}{96\sqrt{5}} (6 - \rho)\rho^2 e^{-\rho/2}$
	$\ell = 3$ (4f)	$R_{4,3} = \dfrac{(Z/a_0)^{3/2}}{96\sqrt{35}} \rho^3 e^{-\rho/2}$

E. Energy levels for the hydrogen atom

The energy corresponding to a given set of quantum numbers may be obtained by substituting the corresponding solution of the R equation back into the differential equation, that is, into Eq. (16-108), and solving for E [remembering that $\lambda = \ell(\ell + 1)$]. This procedure may be followed for each of the functions in Table 16-3, and the results are summarized by the expression

$$E = -\frac{2\pi^2 \mu e^4 Z^2}{h^2 n^2} \text{ [Eq. (16-16)]}$$

This is exactly the result obtained by the simple Bohr theory (in the more accurate version with μ replacing m)!

The more elegant approach of advanced texts is to express R in a general form and to substitute this into Eq. (16-109); one then obtains Eq. (16-16) as the general result, the quantum number ℓ canceling out. It is sufficient here to point out that the reader can verify Eq. (16-16) as the correct result for specific n and ℓ values.

16-8 The graphical appearance of hydrogen-like orbitals

The solutions of the R, Θ, and Φ equations were given in analytical form in the preceding section. Their product is ψ, and the particular function for a given n, ℓ, and m is commonly called an *orbital* when used as a one-electron wave function. We now explore their graphical appearance. The natural division at this point is into the R solutions and the product of the Θ and Φ solutions. The first contains the principal quantum number n explicitly and the nuclear charge Z, and determines the energy of the ψ wave function. That is, the energy of a given $\psi = R(r)\,\Theta(\theta)\,\Phi(\phi)$ is determined by the $R(r)$ function alone. Thus for the hydrogen atom, states of the same n but different ℓ or m are degenerate. This function is crucial in quantitative calculations. The product $\Theta(\theta)\,\Phi(\phi)$ establishes the *angular* dependence of the solution and plays a key role with respect to the symmetry or geometric appearance of the ψ function. [Sometimes the term orbital is applied to this product rather than to the complete ψ function.]

To make the point in a somewhat different way, it will be emphasized in the next chapter that a chemical bond develops if two atoms have orbitals that suitably overlap in space. Although the degree of such overlap is determined by the overall ψ function, the $R(r)$ component of ψ determines how close the atoms must come and the $\Theta(\theta)\,\Phi(\phi)$ component determines the most suitable bond angles. It is therefore this later portion of ψ that is of central importance to the qualitative understanding of molecular geometry.

The functions to be discussed are called hydrogen-like since, although they derive from the solutions for the hydrogen atom, they apply approximately to the outer electrons of any atom. That is, only the $R(r)$ functions contains Z, and for other than hydrogen it is a fairly good approximation to use a Z_{eff} or a screening constant just as was done in the Bohr model. The $\Theta(\theta)\,\Phi(\phi)$ functions do not contain Z and thus are independent of the nature of the atom. This last is true only in first approximation, of course, and a sketch of the problems encountered in more exact treatments is given in the Commentary and Notes section. Qualitative descriptions of the wave functions for atoms in general are almost invariably made in terms of the hydrogen-like functions.

Figure 16-13 presents the first few $R(r)$ functions of Table 16-3. At large r, all approach zero as a consequence of the exponential factor, but notice the

A. The R(r) functions

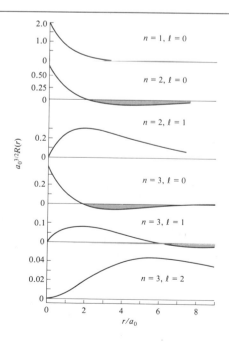

FIGURE 16-13
The R functions. (Adapted from H. Eyring, J. Walter, and G. E. Kimball, "Quantum Chemistry," Copyright 1960, Wiley, New York. Used with permission of John Wiley & Sons, Inc.)

presence of nodes or points at which $R(r)$ is zero, representing r values at which the polynomial portion is zero. The number of such nodes is simply the degree of the polynomial, and inspection of either the analytical or the graphical presentation leads to the general formula

$$\text{number of nodes} = n - \ell - 1 \tag{16-116}$$

The probability of one's finding an electron as one goes out along a particular radius line is proportional to $[R(r)]^2$. Right now, however, it is of more interest to deal with the probability of finding an electron at a distance r from

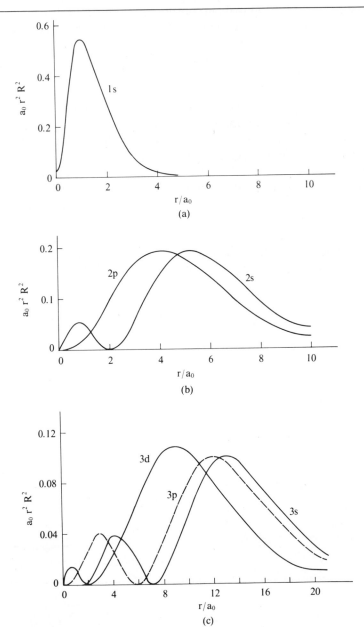

FIGURE 16-14
Radial distribution functions for the hydrogen atom. (a) $n = 1$, (b) $n = 2$, and (c) $n = 3$.

the nucleus, regardless of direction. What is now in question is the relative density of electrons in a spherical shell lying between r and $r + dr$, which means that $[R(r)]^2$ must be weighted by the area of a sphere of radius r, $4\pi r^2$. [An analogous situation was encountered in the case of the Maxwell–Boltzmann distribution of molecular speeds (see Fig. 2-4).] Figure 16-14 plots the dimensionless quantity $a_0 r^2 [R(r)]^2$, which is proportional to the probability, versus r/a_0 for various n and ℓ values. Figure 16-15 makes the same presentation in terms of a polar plot of the electron density intercepted by a plane through the nucleus.

Either of the figures makes the point that the $R(r)$ function leads to peaks in electron density, the number of which is given by $n - \ell$. There is only one such maximum for the function having the largest possible ℓ for the given n value, and within the family having the same n the number of maxima then increases stepwise with decreasing ℓ. This nodal aspect of the electron density distribution will not be referred to much henceforth but should be kept in mind. One is usually concerned mainly with the outermost maximum, which is the most important one with respect to chemical bonding. It should be noted that the radial position of the outermost maximum increases with increasing n value, and with decreasing ℓ value within a given n set, the first dependence being the more significant of the two.

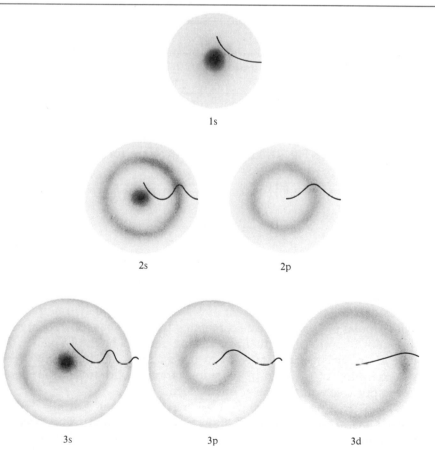

1s

2s   2p

3s   3p   3d

FIGURE 16-15
Pictorial representation of radial probability functions.

B. The angular
dependence
function

The total angular dependence function $\Theta(\theta)\,\Phi(\phi)$ modulates the $R(r)$ functions shown in Figs. 16-13 to 16-15 according to direction in space. It has become customary to show these as three-dimensional polar plots, or as two-dimensional polar plots for some particular plane. It is necessary to keep in mind that in such plots distance out from the center is proportional to the numerical value of $\Theta(\theta)\,\Phi(\phi)$ and is *not* distance from the center of the nucleus. Also, although strictly speaking the term orbital refers to a particular one-electron ψ function, it is often used to designate the $\Theta(\theta)\,\Phi(\phi)$ function for a given ℓ and will be so used in this section. An s orbital thus refers to this function for $\ell = 0$, a p orbital to the function for $\ell - 1$, and a d orbital to this function for $\ell = 2$. For simplicity of notation the product $\Theta(\theta)\,\Phi(\phi)$ will be designated as $Y(\theta, \phi)$.

Let us now examine a few of these polar plots for particular $Y(\theta, \phi)$ functions. We obtain the function for an s orbital by multiplying the $\Theta(\theta)$ function for $\ell = 0$ (from Table 16-2) by that for the $\Phi(\theta)$ function for $m = 0$ [Eq. (16-111)]:

$$Y_s = \frac{\sqrt{2}}{2\sqrt{2\pi}} = \frac{1}{2\sqrt{\pi}} \tag{16-117}$$

This is obviously independent of direction and merely modifies the appropriate radial function by the constant $1/(2\sqrt{\pi})$. The three-dimensional polar plot of an s orbital is then just a sphere of radius $1/(2\sqrt{\pi})$, and the two-dimensional plot for any plane is just a circle, again of radius $1/(2\sqrt{\pi})$, as illustrated in Fig. 16-16(a).

There are three p orbitals, corresponding to the three possible values of m for $\ell = 1$. Thus for $\ell = 1$ and $m = 0$,

$$Y_{P_z} = \frac{\sqrt{6}}{2}(\cos\theta)\frac{1}{\sqrt{2\pi}} = \frac{1}{2}\sqrt{\frac{3}{\pi}}\cos\theta \tag{16-118}$$

This is known as a p_z orbital since $\cos\theta$ gives the projection of a unit vector on the z axis (see Fig. 16-12). The other two are for $|m| = 1$:

$$Y_{P_x} = \left(\frac{\sqrt{3}}{2}\sin\theta\right)\left(\frac{1}{\sqrt{\pi}}\cos\phi\right) = \frac{1}{2}\sqrt{\frac{3}{\pi}}\sin\theta\cos\phi \tag{16-119}$$

$$Y_{P_y} = \left(\frac{\sqrt{3}}{2}\sin\theta\right)\left(\frac{1}{\sqrt{\pi}}\sin\phi\right) = \frac{1}{2}\sqrt{\frac{3}{\pi}}\sin\theta\sin\phi \tag{16-120}$$

and are designated as the p_x and p_y orbitals, respectively, since by Eqs. (16-103) and (16-104), the trigonometric functions correspond to the x and y projections of a unit vector.

The p_z orbital is independent of ϕ and is therefore symmetric about the z axis. A polar plot in any plane containing the z axis has the appearance shown in Figure 16-16(b) (the actual plane being taken to be the xz plane). The polar plot of $\cos\theta$ is a circle tangent to the xy plane at the origin, and from a consideration of the sign of $\cos\theta$ for various quadrants the upper circle or *lobe* is positive and the lower one is negative.

The p_x and p_y orbitals have their maximum values at $\theta = 90°$ and their polar plots in the xy plane are shown in Fig. 16-16(c,d). The three p orbitals are identical except for their orientation in space; each is a figure of revolution,

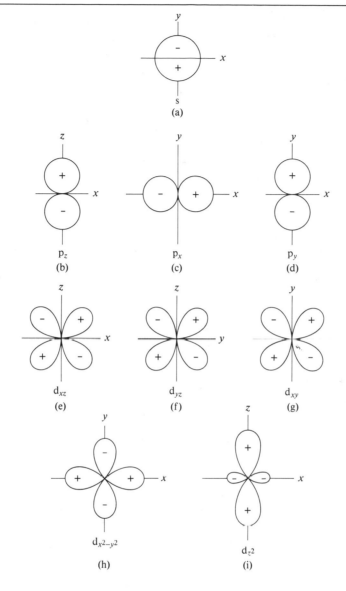

FIGURE 16-16
Polar plots for various $Y(\theta, \phi)$ functions. Each plot is in the plane of the indicated axes.

so that in a three-dimensional polar plot each consists of a pair of spheres, as shown in Fig. 16-17(b–d).

Finally (for us), there are five d orbitals, one for each of the possible m values for $\ell = 2$. Again using Eqs. (16-112) and Table 16-2, one obtains for $Y(\theta, \phi)$, when $m = 0$,

$$Y_{d_{z^2}} = \left(\frac{\sqrt{10}}{4}(3\cos^2\theta - 1)\right)\frac{1}{\sqrt{2\pi}} = \frac{1}{4}\sqrt{\frac{5}{\pi}}(3\cos^2\theta - 1) \qquad (16\text{-}121)$$

when $|m| = 1$,

$$Y_{d_{xz}} = \left(\frac{\sqrt{15}}{2}\sin\theta\cos\theta\right)\left(\frac{1}{\sqrt{\pi}}\cos\phi\right) = \frac{1}{2}\sqrt{\frac{15}{\pi}}\sin\theta\cos\theta\cos\phi \qquad (16\text{-}122)$$

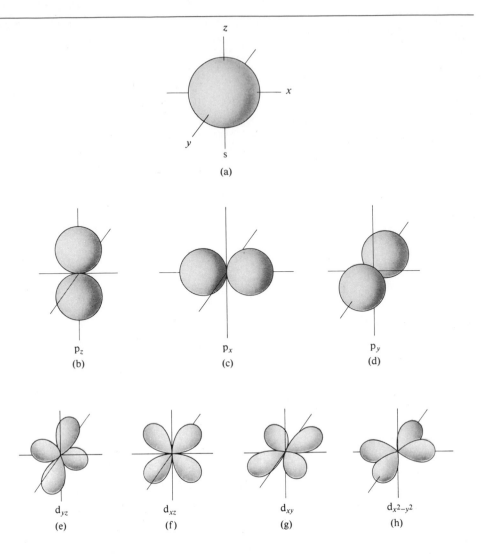

$$Y_{d_{yz}} = \left(\frac{\sqrt{15}}{2} \sin\theta\cos\theta\right)\left(\frac{1}{\sqrt{\pi}}\sin\phi\right) = \frac{1}{2}\sqrt{\frac{15}{\pi}}\sin\theta\cos\theta\sin\phi \quad (16\text{-}123)$$

and when $|m| = 2$,

$$Y_{d_{x^2-y^2}} = \left(\frac{\sqrt{15}}{4}\sin^2\theta\right)\left(\frac{1}{\sqrt{\pi}}\cos 2\phi\right) = \frac{1}{4}\sqrt{\frac{15}{\pi}}\sin^2\theta\cos 2\phi \quad (16\text{-}124)$$

$$Y_{d_{xy}} = \left(\frac{\sqrt{15}}{4}\sin^2\theta\right)\left(\frac{1}{\sqrt{\pi}}\sin 2\phi\right) = \frac{1}{4}\sqrt{\frac{15}{\pi}}\sin^2\theta\sin 2\phi \quad (16\text{-}125)$$

The designations d_{xz} and d_{yz} are fairly obvious since by Eqs. (16-102)–(16-104), the respective trigonometric functions are the product of the x and z and of the y and z projections of a unit vector. The d_{xz} orbital has its maximum value for $\phi = 0$ (or π), and, correspondingly, the polar plot in the xz plane is that shown in Fig. 16-16(e). The d_{yz} orbital has its maximum extent in the

yz plane and is shown in Fig. 16-16(f). Since the product $\sin \theta \cos \theta$ is a maximum at 45°, the lobes now point midway between the axes and carry alternate plus and minus signs, as determined from the signs of the functions for the various quadrants. They are no longer circular, but each lobe is a figure of revolution about its axis, and the three-dimensional polar plots are as illustrated in Fig. 16-17(e, f).

The d_{xy} orbital has its maximum extension in the xy plane and the $\sin 2\phi$ function produces four lobes, as shown in Figs. 16-16(g) and 16-17(g). Since $\sin 2\phi = 2 \sin \phi \cos \phi$, the trigonometric function of Eq. (16-125) corresponds to the product of the x and y projections of a unit vector, hence the name d_{xy}. The d_{xz}, d_{yz}, and d_{xy} orbitals form a group; they are identical in three-dimensional shape, differing only in their spatial orientation; each d_{ij} orbital lies in the plane defined by the i and j axes and the members of the set are thus mutually perpendicular.

There remain the $d_{x^2-y^2}$ and d_{z^2} functions. The first again has its maximum extension in the xy plane ($\sin^2 \theta = 1$) and the $\cos 2\phi$ term produces four lobes with maxima at 0°, 90°, 180°, and 270°, or as shown in Figs. 16-16(h) and 16-17(h). It further turns out that the $d_{x^2-y^2}$ orbital is identical in shape to those of the d_{ij} set. The first four so far discussed are thus equivalent except for spatial orientation.

Since an atom has no preferred direction in space, it might be thought that the situation should be symmetric, and that there should be $d_{z^2-x^2}$ and $d_{z^2-y^2}$ orbitals to complete a set of three, as with the d_{ij} ones. There can only be five independent orbitals, however, corresponding to the five possible m values; that is, the $\ell = 2$ set is just fivefold degenerate. What is done therefore is to pick the z axis as unique (with respect to the orbitals but not with respect to space) and to use a linear combination of the $d_{z^2-x^2}$ and $d_{z^2-y^2}$ pair. From Eqs. (16-102)–(16-104) these should be of the form

$$d_{z^2-x^2} \propto \cos^2\theta - \sin^2\theta \cos^2\phi \qquad d_{z^2-y^2} \propto \cos^2\theta - \sin^2 \theta \sin^2\phi$$

and their sum is then

$$d_{z^2-x^2} + d_{z^2-y^2} \propto 2 \cos^2\theta - (\sin^2 \theta)(\cos^2\phi + \sin^2\phi)$$

or, since $\cos^2 \phi + \sin^2 \phi = 1$ and $\sin^2 \theta = 1 - \cos^2 \theta$,

$$d_{z^2-x^2} + d_{z^2-y^2} \propto 3 \cos^2\theta - 1$$

Except for the numerical coefficient, this is the expression for d_{z^2} given by Eq. (16-121); because of its origin, an alternative designation sometimes seen is $d_{2z^2-x^2-y^2}$

The graphical operation corresponding to the preceding is illustrated in Fig. 16-18; what happens is that the lobes along the z axis reinforce, and the four negative lobes in the xy plane combine to give a doughnut-shaped region. The cross section shown in Fig. 16-16(i) also illustrates the point that the node between the z lobes and the doughnut lobe is at $\cos^2\theta = \frac{1}{3}$ or at a θ of about 54°.

The functions for the set of seven $\ell = 3$ or f orbitals may be assembled form Eq. (16-112) and Table 16-2 in the same manner as was done for the d orbitals. The cross section for the $m = 0$ or f_{z^3} will, for example, resemble d_{z^2} in having large lobes on the z axis, but will now have *two* small lobes in the equatorial region. The pair for $m = 3$ will have six rather than four lobes in the xy plane, and so on.

FIGURE 16-18
Illustration of the
construction of the d_{z^2}
orbital.

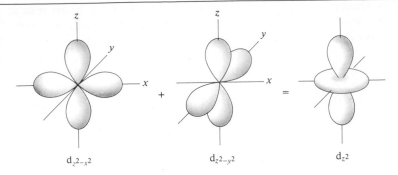

$d_{z^2-x^2}$ $d_{z^2-y^2}$ d_{z^2}

It turns out that the f orbital set obtained in this way is not very convenient in its geometric properties. Fortunately, if one has a set of solutions to a partial differential equation, any new set constructed by independent linear combinations of the old one will also constitute a set of solutions. A more useful set of f orbitals may thus be obtained, designated as f_{z^3}, f_{x^3}, f_{y^3}, f_{xyz}, $f_{x(z^2-y^2)}$, $f_{y(z^2-x^2)}$, and $f_{z(x^2-y^2)}$. The corresponding trigonometric functions follow immediately from Eqs. 16-102–16-104; the qualitative appearances of f_{xyz} and $f_{x(z^2-y^2)}$ are shown in Fig. 16-19.

16-9   Graphical appearance of the electron density around a hydrogen-like atom

A. The angular portion. Ünsold's theorem

The preceding section dealt only with $Y(\theta, \phi)$, and this function must be squared to obtain the corresponding probability or electron density functions. The effect is to elongate all lobes, and, of course, to remove their sign designation. Figure 16-20 shows the appearance of Y^2 for s, p, and d electrons as the usual polar plots in the plane of maximum electron density.

There is a useful and instructive result known as Ünsold's theorem. It states that the sum of Y^2 for the orbitals making up the set of a particular ℓ value is a constant. Thus

FIGURE 16-19
Plots of f orbitals (a) f_{xyz}
and (b) $f_{x(z^2-y^2)}$

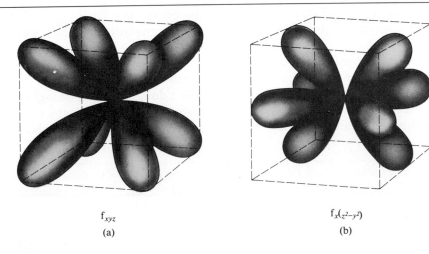

f_{xyz} $f_{x(z^2-y^2)}$
(a)   (b)

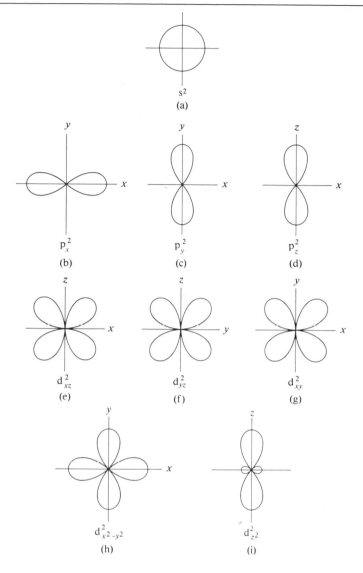

FIGURE 16-20
Polar plots of Y^2 functions.

$$Y_s^2 = \frac{1}{\pi}$$

$$Y_{P_x}^2 + Y_{P_y}^2 + Y_{P_z}^2 = \frac{3}{4\pi} \sin^2 \theta \cos^2 \phi + \frac{3}{4\pi} \sin^2 \theta \sin^2 \phi + \frac{3}{4\pi} \cos^2 \theta$$

$$= \frac{3}{4\pi} \sin^2 \theta + \frac{3}{4\pi} \cos^2 \theta = \frac{3}{4\pi}$$

The demonstration of Ünsold's theorem for the set of five d orbitals is left as an exercise.

The significance of the theorem is that with two electrons in each orbital of a set, the total electron density is independent of angle, that is, is spherically symmetric. We often speak of such a set as a closed or filled shell. [The same conclusion of spherical symmetry holds, of course, for a half-filled shell, that is, for the case of one electron in each orbital.] For example, Ne has a closed shell of six 2p and Ar a closed shell of six 3p electrons; therefore each has a spherically symmetric electron density. The outer electrons of Kr constitute closed shells of 3d and of 4p electrons, and again give a spherically symmetric total electron density. This point has been stressed to counteract the impression given by the various preceding polar plots of orbitals that an atom with a filled set would be "bumpy," with orbitals sticking out in various directions. In a sense, a set of orbitals (of a given ℓ value) represents one way of dividing a spherically symmetric electron density into a set of standing waves. Such a set of waves is known, incidentally, as one of *spherical harmonics*. Thus a sphere of electron density may be left as such, and called an s orbital. It may be divided into a set of three p orbitals or a set of five d orbitals.

B. The complete wave function

It is important to appreciate that the polar plots of the Y^2 functions are *not* representations of how the electron density actually looks; distance out from the origin of such plots is *not* distance from the nucleus but only the measure of the value of Y^2 in that direction. To obtain representations of electron density, each Y^2 function must be multiplied by the appropriate $R^2(r)$. In a polar plot of the electron density for a given ψ^2, distance out from the origin is now distance from the nucleus, and no coordinates are left to give the electron density. A qualitative graph can be constructed, however, in which the density of shading indicates the electron density. This is done in Fig. 16-21 for several cases. Thus although the 1s and 2s cases are both spherically symmetric, the radial wave function introduces a node in the latter. Similarly, the $2p_x$ and $3p_x$ differ in that the latter has a node, and so on. An alternative way of modeling the electron density around an atom, by means of *domains*, has been suggested by Adamson (1965).

16-10   The spin function

We will go into only a brief history of the concept of electron spin. The most famous experiment is that of O. Stern and W. Gerlach, who found in 1922 that a beam of silver atoms split into two components on passing through an inhomogeneous magnetic field. The only acceptable possibilities at that time were the $2\ell + 1$ components of orbital angular momentum, that is, the $2\ell + 1$ allowed values of the magnetic quantum number m, while the observed splitting implied a quantum number of $\frac{1}{2}$. The explanation was given in 1925 by G. Uhlenbeck and S. Goudsmit, namely that the electron itself possessed angular momentum and that this could be either $\frac{1}{2}(h/2\pi)$ or $-\frac{1}{2}(h/2\pi)$. Accordingly, one now has a fourth quantum number, the spin quantum number s, which can have the values $s = \frac{1}{2}$ and $s = -\frac{1}{2}$. The Schrödinger wave equation itself is silent on this matter; it was enormously satisfying when in 1928 P.A.M. Dirac constructed a quantum mechanics compatible with the requirements of relativity theory and, in so doing, found that electron spin arose in a natural way. There is, however, no classical analogue to electron spin.

We must now recognize that a complete wave function will include the spin eigenfunction. This function is essentially an algebraic mnemonic device for remembering that each electron has the two states of angular momentum $\frac{1}{2}(h/2\pi)$ and $-\frac{1}{2}(h/2\pi)$ or ω and $-\omega$. The energies of these states are unaffected

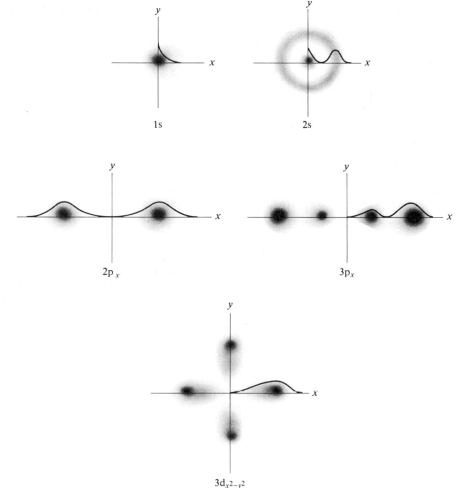

FIGURE 16-21
Pictorial representations of electron density as a function of radius and of angle, for various ψ^2 functions.

by the potential energy function V for the atom or molecule, although they are altered by an external magnetic field (see Sections 3-ST-2 and 19-CN-1).

The spin wave function is, accordingly, a rather peculiar one. There are really two functions, α and β, defined as having the values of 1 or 0 according to the convention

$$\alpha(\omega) = 1 \qquad \alpha(-\omega) = 0 \qquad \beta(-\omega) = 1 \qquad \beta(\omega) = 0 \qquad (16\text{-}126)$$

[see Daudel *et al.* (1959)]. The probability associated with the spin functions is treated in the same manner as for ordinary wave functions, and using the preceding conventions we have

$$\int \alpha(\omega)^2 \, d\omega = 1 \qquad \int \alpha(\omega) \, \beta(\omega) \, d\omega = 0 \qquad \int \beta(\omega)^2 \, d\omega = 1 \quad (16\text{-}127)$$

These integrals imply a summation over the two possible values of the ω products.

In writing a complete wave equation, one then includes $\alpha(\omega)$ or $\beta(\omega)$ as a factor. For example, ψ_{1s} becomes $\psi_{1s}\alpha$ or $\psi_{1s}\beta$. In the case of He there are two electrons to consider, and the possible combinations of the spin functions are

$$\alpha(1)\ \alpha(2) \quad \beta(1)\ \beta(2)\text{-} \quad [\alpha(1)\ \beta(2) + \alpha(2)\ \beta(1)] \quad \text{symmetric} \qquad (16\text{-}128)$$
$$[\alpha(1)\ \beta(2) - \alpha(2)\ \beta(1)] \quad \text{antisymmetric}$$

These correspond to the four different ways of assigning electrons 1 and 2 between two atoms. Three of the combinations are unchanged in sign if the coordinates of the electrons are exchanged, and are therefore said to be symmetric; one function is antisymmetric.

An alternative statement of the Pauli exclusion principle is that only those wave functions are physically significant which are antisymmetric to, that is, change sign on, the exchange of any pair of electrons. The ground-state wave function for He is just $\psi_{1s}(1)\psi_{1s}(2)$, where the numbers in parentheses denote electron one and electron two, respectively. This is symmetric, so that the spin contribution must be antisymmetric if the complete wave function is to obey the Pauli principle. Thus,

$$\psi_{\text{He}} = \psi_{1s}(1)\psi_{1s}(2)[\alpha(1)\beta(2) - \alpha(2)\beta(1)]$$

While the foregoing does serve to illustrate the use of spin function, it is approximate in that electron–electron interactions are neglected (see Section 16-ST-3).

16-11   The rigid rotator

The case of a diatomic molecule that can rotate, but whose bond length is fixed, is treated as follows. The wave equation is Eq. (16-105) but with $V = 0$ since the rotation is free. If r is distance from the center of mass, then it is a constant, so the terms involving differentiation with respect to r drop out; also, in this coordinate system, mr^2 becomes I, the moment of inertia of the molecule. Alternatively, $I = \mu R^2$ where μ is the reduced mass and R is the bond length. Equation (16-105) then reduces to

$$\frac{1}{\sin\theta}\frac{\partial}{\partial\theta}\left(\sin\theta\frac{\partial\psi}{\partial\theta}\right) + \frac{1}{\sin^2\theta}\frac{\partial^2\psi}{\partial\phi^2} + \frac{8\pi^2 IE}{h^2}\psi = 0 \qquad (16\text{-}129)$$

This separates into $\psi = \Theta(\theta)\ \Phi(\phi)$ as before, and the solution to $\Phi(\phi)$ is given by Eq. (16-111) with quantum numbers $m = 0, \pm 1, \pm 2, \ldots$. The solutions to $\Theta(\theta)$ are analogous to those for Eq. (16-110), with quantum numbers usually designated as J (but analogous to the ℓ quantum number for the hydrogen atom). As with ℓ, J can be zero or a positive interger; also $m \leqslant J$, so there are $2J + 1$ possible m values for a given J state. The actual wave functions are thus in the form of the Y spherical harmonics obtained for the hydrogen atom.

On substitution of the solutions to $\Theta(\theta)$ back into the $\Theta(\theta)$ equation, one obtains for the energy

$$E = \frac{h^2}{8\pi^2 I}J(J + 1) \qquad (16\text{-}130)$$

The degeneracy of each state is just $(2J + 1)$, corresponding to the possible m values. Unlike the case of the harmonic oscillator, E can be zero; the un-

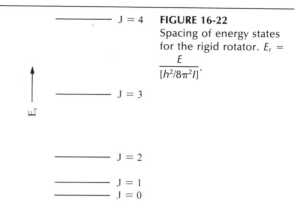

FIGURE 16-22
Spacing of energy states for the rigid rotator. $E_r = \dfrac{E}{[h^2/8\pi^2 I]}$.

certainty principle is not violated, however, because although the angular momentum is then defined exactly, the rotational *position* of the molecule is completely undefined.

EXAMPLE

The bond length for $^1H^{35}Cl$ is 1.275 Å, so that the moment of inertia is $I = (1.275 \times 10^{-8})^2[1 \times 35/36] = 2.62 \times 10^{-40}\,\text{g cm}^2$ or $2.62 \times 10^{-47}\,\text{kg m}^2$. Equation (16-130) yields, on a mole basis,

$$E = \left[\frac{(6.262 \times 10^{-27})^2}{8\pi^2 \times 2.62 \times 10^{-40}} \right] (6.022 \times 10^{23})\, J(J + 1)$$

from which

$$E = 1.28 \times 10^9\, J(J + 1)\ \text{erg mol}^{-1} = 128 J(J + 1)\ \text{J mol}^{-1}$$
$$= 10.7\, J(J + 1)\ \text{cm}^{-1}$$

Note that these energy values are small, amounting to only about 10% of RT at room temperature. Figure 16-22 illustrates how the spacing between rotational state increases with J.

COMMENTARY AND NOTES

16-CN-1   Albert Einstein

Albert Einstein is the third great man of science that we attempt to describe. Benjamin Thompson (see Section 4–CN–1) had a multifaceted brilliance and, as an individual, was effectively self-serving in a scoundrelly way. Willard Gibbs (see Section 6-CN-2) had a tremendous depth and rigor of thought; as an individual he was austere and retiring but kindly. Einstein resembled Thompson in the variety of his brilliance and Gibbs in its depth; as a person he resembled neither.

The general sequence of Einstein's life is as follows.

1879 Born on March 14 in the small Swabian Jewish community of Ulm, on the Danube. His father was the manager of a small electrochemical works and his mother an able *hausfrau* and fine pianist. The family left for Munich shortly after his birth.

1894 Completed studies at the Luitpold Humanistische Gymnasium in Munich, mainly on the basis of mathematical achievement. He then joined his family, which at this time was residing in Milan.

1895 Flunked the entrance examination of the Polytechnic Academy in Zurich but the examiners were so impressed by his mathematics that they enrolled him in a small liberal arts school in Aarau, and a year later he gained admission to the Academy.

1900 Graduated in physics but failed to receive a staff appointment. He did odd jobs, was a relief mathematics teacher in a technical school in Winterthur (and wrote an essay on capillarity), and then worked in the Swiss Patent Office in Berne until 1905. By then he was a Swiss citizen.

1902 Married Mileva, a small dark-haired girl from the Serbian town of Novi. A son, Hans Albert, was born in 1904, and then a second son, Edward. Some intense personal troubles, unknown in detail to this day, seem to have clouded the marriage. In 1902 he published a paper on the second law of thermodynamics and thermal equilibrium—apparently unaware of Gibbs' work, he essentially rediscovered statistical thermodynamics.

1905 In an amazing volume of *Annalen der Physik* Einstein published his paper on Brownian motion [note Eqs. (2-67) and (10-41)]; the Nobel-prize-winning paper on the quantum theory of light, based on photoelectric and photochemical behavior (see Section 16-1); *and* the first paper on the special theory of relativity (while still working in the patent office).

1907 Was offered an unpaid lectureship (privatdozent) at the University of Zurich. He rejected the offer but did take such a position at the University of Berne. By 1909 his accomplishments were being recognized worldwide, and, over some political infighting, the University of Zurich offered him a professorship (an *extraordinary* or nonadministrative chair), which he accepted.

1910 Was offered and accepted the chair in theoretical physics at the University of Prague—the oldest university in central Europe. At Zurich and then at Prague he continued working on relativity theory. One of his important papers on general relativity was published in 1911. That year he attended the first of several Solvay Conferences (all-expense-paid gatherings of topflight physicists and chemists, sup-

ported by a wealthy scientific hobbyist). This was Einstein's first opportunity to meet such people as Rutherford, Poincaré, Langevin, Planck, Nernst, Lorentz, and Marie Curie.

1912 Returned to Zurich as professor of theoretical physics at the Polytechnic Academy.

1913 Went to Berlin as director of a specially created research institute for physics and as a member of the Royal Prussian Academy of Science. He remained there until the rise of Hitler. In 1919 he divorced Mileva and married a second cousin, Elsa Rudolph, who had two daughters from a previous marriage.

1933 Joined the Institute for Advanced Studies at Princeton. (By this time the position of most Jews in Germany was untenable.)

1939 Wrote a letter supporting the position of Leo Szilard (and of R. Sachs, an advisor to President Roosevelt) who urged the President to begin work on an atomic bomb, pointing out that the Germans had already started.

1955 Died in Princeton Hospital on April 18 from a massive failure of the aorta.

The above factual summary cries for amplification. Einstein was perhaps the first scientist to catch the public imagination. In the 1920's he gave popular lectures to packed houses in England and on the Continent; relativity became a household word. He was a scientist's genius as well, until perhaps the Solvay Conference of 1928. Much of Einstein's contributions were in the area of statistics—statistical thermodynamics, Brownian motion, black-body radiation—and even though he was responsible for establishing the corpuscular and hence dual nature of electromagnetic radiation, he never accepted Heisenberg's uncertainty principle. He felt this to be a reflection of an inadequacy of wave mechanics rather than a fundamental limitation of nature. The whole matter came to a head at the 1928 Solvay Conference. Einstein proposed a violation of the uncertainty principle. Briefly, it should be possible to have a mechanism that would allow one photon to escape from a box at a precisely determined time. The change in mass determined by weighing the box before and after should give the exact energy of the photon that was lost. The story is that Bohr came away from that session deeply worried, but toward the end of a sleepless night perceived an answer (which had to do with the fact that the impulse given to the box by the photon would change the rate of running of the clock and hence make the precise determination of time impossible). He presented this solution the next day—and essentially defeated Einstein. This event, it appears, began the decline of Einstein as the head of the scientific world. Einstein worked until his death on an attempt to formulate a general (unified-field) theory that would contain within it both the corpuscular and wave aspects of photons and of matter. He never fully succeeded, but continued strongly to believe that wave mechanics was an incomplete model and that the search for a more complete one should continue. While the support is unneeded, this writer agrees.

As a person, Einstein seems to have had an inner reserve that prevented him from making any close personal ties. As a teacher he could be enthusiastic and inspiring, but on specific occasions and subjects. It appears that, overall, it was difficult for students to learn from him. This conclusion is supported by the fact that there was and is no "Einstein" school of physicists—he was a loner.

He could be passionate in causes, as with regard to the League of Nations. As a pacifist, he refused any war-related work during World War I. In New York in 1930, he rousingly called upon true pacifists to refuse to take up arms under any circumstances, asserting that if only 2% of the populace would do this any country's war effort would be doomed. Communists (at that time in alliance with Hitler) applauded. He was not a Communist, however. While he was "open" to the left, seeing less danger from communist than from capitalist societies, Einstein was also quite capable of walking out of an association if he perceived it to be primarily a Communist political front. His abhorrence of fascism (that is, Hitlerism) was special; it led him to promote the United States atomic bomb program. Later, in 1945, he urged the country to outlaw voluntarily use of atomic bombs and campaigned strenuously for the Emergency Committee of Atomic Scientists (along with Bethe, Pauling, and Urey).

Einstein was a Jew by descent and by culture, but not in a synagogue-attending way. His God made the universe orderly but did not inquire into the activities of individuals. He was an ardent Zionist. He toured the United States with Chaim Weizmann in 1921 to help raise money for the Palestine Foundation Fund. Israel owes much to him. [Interesting accounts of Einstein's life include the books by Schilpp (1963) and Frank (1947); see also Kerker (1985).]

Surely the trio Thompson, Gibbs, and Einstein illustrate the point that genius in science imposes no bounds on diversity in all other respects.

16-CN-2 An alternative interpretation of the uncertainty principle

The currently conventional description of the uncertainty principle is given in the Introduction. It will be recalled that Eq. (16-8) was interpreted to mean that one cannot measure both position and momentum simultaneously without errors whose product is at least as large as $h/4\pi$. This is the interpretation taken by Heisenberg and by Bohr; Einstein and others have taken strong exception to it. In his review on this subject, Ballentine (1970) describes the alternative, statistical interpretation. It is as follows. Consider an experiment in which an electron is scattered and its subsequent position (and/or momentum) is measured. Now suppose a large (in the limit, infinite) set of such experiments. Even though the experiments are identical, the results will vary, giving a distribution curve in value of position and momentum, with standard deviations δp and δx. The statistical interpretation affirms that $\delta p \, \delta x \geq h/4\pi$.

The statistical interpretation is all that is *required* by wave mechanics. That is, solutions to the wave equation, for example, do not predict the position or momentum of a *particular* particle, only the *probable* values. Thus in Fig. 16-6, $\psi^*\psi$ gives the probability of there being a given value of x; if x could be measured, the result of some one experiment could give any value. However, the collected results of a large number of identical experiments is predicted to give the distribution $\psi^*\psi$.

An intriguing aspect of the statistical interpretation is in the possibility that the supposedly identical set of experiments give distributions in p and in x may not really be identical. Might there be some as yet unknown variable that is not being controlled and whose variation leads to δp and δx? If so, wave mechanics is not a complete description of nature. As may be imagined, the *hidden variable* question has given rise to lively and as yet only partially resolved debate [see Rohrlich (1983)].

16-CN-3   Tunneling

A quite major difference between wave mechanics and classical mechanics occurs in what is known as the *tunneling effect*. The effect is a little lengthy to treat mathematically in a textbook such as this one, but can be described qualitatively in terms of a particle in a box. Figure 16-23 illustrates the new situation, in terms of the one-dimensional case. Again we have a box of length a; now, however, the potential energy V does not go to infinity outside of the box, but instead has some finite value V^0. To be more specific, we assume that V^0 applies for a distance δ, beyond which V reverts to zero. We have, in effect, a barrier of height V^0 and width δ.

We can make a qualitative prediction, based on the discussion in Section 16-6B. There it is noted that there is a finite probability for x of a one-dimensional harmonic oscillator to have values such that the total energy is *less* than the potential energy. In other words, $\psi^*\psi$ need not be zero in a region where $(E - V)$ is negative. If this is so, we can expect a finite or nonzero $\psi^*\psi$ in the region of the barriers, regions 2, and hence outside the barriers, regions 3.

The general procedure in a case such as this one [see Pitzer (1953)] is to write the wave equation for each region and the form of its solution. Thus for regions 1 and 3 we have

$$-\frac{d^2\psi}{dx^2} = \alpha^2 E\psi \qquad (16\text{-}132)$$

where $\alpha = (8\pi^2 m/h^2)^{1/2}$, which is Eq. (16-69) with $V = 0$. The general solutions for ψ_1 and ψ_3 are

$$\psi_1 = Ae^{i\alpha E^{1/2}x} + Be^{-i\alpha E^{1/2}x} \qquad (16\text{-}133)$$

and

$$\psi_3 = A'e^{i\alpha E^{1/2}x} + B'e^{-i\alpha E^{1/2}x} \qquad (16\text{-}134)$$

as may be verified by substituting back into Eq. (16-132). The wave equation for region 2 is

$$-\frac{d^2\psi}{dx^2} = \alpha^2(E - V^0)\psi \qquad (16\text{-}135)$$

and here the general solution is

$$\psi_2 = Ce^{i\alpha(E - V^0)^{1/2}x} + De^{-i\alpha(E - V^0)^{1/2}x} \qquad (16\text{-}136)$$

Since $(E - V^0)$ is a negative number, Eq. (16-136) can also be written

$$\psi_2 = Ce^{\alpha(V^0 - E)^{1/2}x} + De^{-\alpha(V^0 - E)^{1/2}x} \qquad (16\text{-}137)$$

We now apply the conditions that both ψ and $d\psi/dx$ must be continuous and single valued (a discontinuity in $d\psi/dx$ would make $d^2\psi/dx^2$ indeterminate, which is not allowable). Thus at $x = a$, $\psi_1 = \psi_2$ and $d\psi_1/dx = d\psi_2/dx$, and similarly for ψ_2 and ψ_3 at $x = a + \delta$. These requirements allow a set of simultaneous equations to be written, from which the coefficients A, B, A', B', C, and D can be evaluated.

The outcome is shown qualitatively in Fig. 16-23. Between $x = a$ and $x = -a$, ψ is an oscillating function, much like the case for the particle in a box; ψ does not go to zero, however, at the boundaries, but has a finite value in regions 2, in which it decays essentially exponentially. It resumes an oscillatory behavior

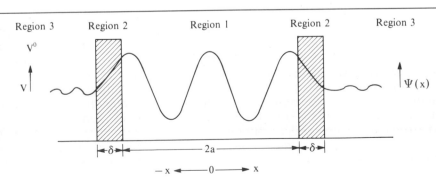

FIGURE 16-23
The particle in a box whose "walls" are of finite height and width. The heavy line shows the potential function V. Superimposed, with arbitrary base-line, is a qualitative plot of the corresponding ψ function.

in regions 3, but of much reduced amplitude. Comparison of squares of the amplitudes inside and outside the box gives the probability of the particle getting outside. This "leakage" of a particle through a barrier such that $E < V$ is a wave mechanical phenomenon called *tunneling*; it has no counterpart in classical mechanics.

A case very similar to the one illustrated in Fig. 16-23 provided one of the early triumphs of wave mechanics. This had to do with the emission of an α-particle from a nucleus (see Section 22-CN-1). The observation was that the energy of the emitted α-particle was *less* than the Coulomb barrier it had to surmount (classically) in order to escape. A treatment very similar to that outlined above yields the equation

$$P = \exp\left[-\frac{4\pi}{h}\sqrt{2M}\int \sqrt{V(r) - E}\, dr\right]$$
(16-138)

where P is the probability of a particle of mass M being outside the barrier and the integral gives a kind of area under the potential barrier $V(r)$, with E taken as the baseline. In the case of α-particle emission, $V(r)$ is just the Coulomb barrier potential, $V(r) = Z_1 Z_2/r$, where Z_1 is the charge on the α-particle and Z_2 is the nuclear charge. The form of Eq. (16-138) is a characteristic one for tunneling situations: the larger the mass of the particle and the higher and wider the potential barrier, the smaller is the tunneling probability. It was the success of this equation as applied to the probability of α-particle emission that confirmed one of the most basic nonclassical predictions of wave mechanics.

Tunneling in chemistry very often has to do with a barrier of the type shown in Fig. 16-24. Here, the particle has two regions of zero potential energy, separated by a finite barrier. Two specific examples are the following. In the first, the particle is a light atom, such as a proton or deuteron, and regions 1 and 3 correspond to equivalent but different configurations such as those of the ammonia molecule:

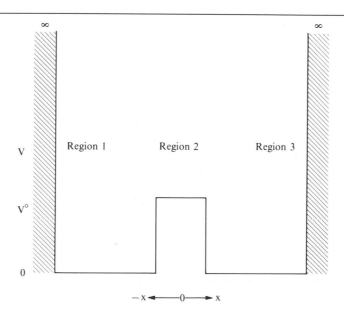

The molecule can be imagined to "flip" from the one to the other configuration not by an ordinary vibrational action but by actual tunneling. The barrier in this case would have a V^0 corresponding to the three protons being co-planar with the nitrogen atom.

The second example is rather different. Consider the specific case of the exchange of an electron between ferrocyanide and ferricyanide ions:

$$[Fe(CN)_6]^{4-}[Fe^*(CN)_6]^{3-}$$
$$\rightleftharpoons [Fe(CN)_6]^{3-}[Fe^*(CN)_6]^{4-}$$

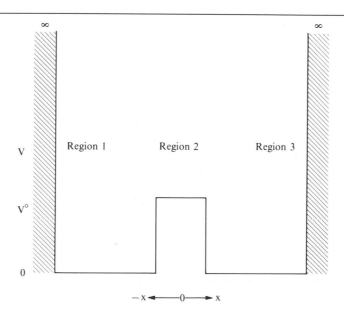

FIGURE 16-24
Particle in a two-compartment box. The outside walls are of infinite height and the middle region is occupied by a barrier of finite potential V^0.

Region 1   Region 2   Region 3

$-x \longleftarrow 0 \longrightarrow x$

Fe* is isotopically labeled iron; this allows the exchange to be observed experimentally. The picture is that the two ions have come together in a solution encounter, and an electron then tunnels from the ferrocyanide to the ferricyanide ion. This explanation of the observed exchange was an early but only partially successful one; the current treatment includes tunneling but involves a more complex analysis of nuclear motions.

16-DN-4   Steps beyond hydrogen-like wave functions

Very extensive advances have been made in the direction of obtaining more accurate solutions to the wave equation for multielectron atoms. The detailed methodology of advanced treatments must be left to subsequent elective or graduate courses. We can, however, outline the directions taken and summarize qualitatively the degree of progress made.

The difficulty of dealing with multielectron atoms becomes apparent even with the helium atom. The wave equation is now

$$-\frac{h^2}{8\pi^2 m}(\nabla_1^2 + \nabla_2^2)\psi$$
$$-\left(\frac{2e^2}{r_1} + \frac{2e^2}{r_2} - \frac{e^2}{r_{12}}\right)\psi = E\psi \qquad (16\text{-}139)$$

where the subscripts refer to electrons 1 and 2; that is, the Hamiltonian is [see Eq. (16-56)]

$$\mathbf{H} = -\frac{h^2}{8\pi^2 m}(\nabla_1^2 + \nabla_2^2)$$
$$-\frac{2e^2}{r_1} - \frac{2e^2}{r_2} + \frac{e^2}{r_{12}} \qquad (16\text{-}140)$$

One now has a second-order partial differential equation in six variables—three coordinates for each electron. The real difficulty is in the e^2/r_{12} term, which gives the interelectronic repulsion. Where it not for this, the wave equation could be separated into two, one for each electron. Put another way, the helium atom is the wave mechanical three-body problem. It has not been solved explicitly, but successive approximation methods aided by high-speed computers have led to solutions that appear to be exact theoretically and that agree well with experiment (with respect to, say, the spectrum of helium).

One approach is to start with the hydrogen-like wave functions and introduce the e^2/r_{12} term as a correction; one does this by expanding the term by means of associated Legendre polynomials and evaluating the resulting integral term by term. The first approximation for the ground-state energy is $E_{He} = (2Z^2 - \frac{5}{4}Z)\,E_H$, with $Z = 2$. The predicted ionization energy of the first electron in He is then $\frac{3}{2}E_H = 20.40$ eV, as compared to the experimental value of 24.49 eV. This type of approach makes use of what is known as *first-order perturbation theory* (Section 16-ST-3); further stages of approximation can be made, but with a cascading increase of effort.

Another approach is to consider that the electron of He which is to ionize is in a nuclear field corresponding to charge Z_{eff}, where Z_{eff} lies between 1 and 2, and to write the corresponding hydrogen-like wave function for this electron. The value of Z_{eff} makes the energy of this electron a minimum gives $E = 23.05$ eV, or much closer to the experimental value than the other method does in its first approximation. The hydrogen-like wave function with Z_{eff} can next be modified by trial and error until a yet lower or more stable energy for the electrons is found. In this manner a calculated ionization energy agreeing with experiment to within 0.002 eV has been obtained.

The alternative of the more advanced approaches has been to say that as a first approximation, one can estimate the interaction of a particular electron with all other electrons by taking the latter to obey hydrogen-like wave functions (perhaps with Slater corrections). The wave function for the particular electron is then solved, and this is done for each electron in turn. The process is repeated with the new set of wave functions. That is, one estimates the interaction of a particular electron by using the new wave functions for all the others. Eventually, a convergence is reached in which no further change in the functions occurs. The potential energy function for the electrons is now said to be *self-consistent*. This method, due to D. Hartree (and further improved by V. Fock) is the basis for many modern calculations of the energies of low-atomic-number atoms.

The struggle in all of these approximation procedures is to find wave functions that reproduce the experimental ionization and total energies for atoms. If wave functions that do this can be found, then one has some confidence in their application to other calculations.

SPECIAL TOPICS

16-ST-1 Quantum theory of blackbody radiation

The period around 1890–1900 was one of fairly intensive study of the radiation emitted by hot objects. An important, essentially thermodynamic argument is that if an object is placed in a box or other isolated system and comes to equilibrium with the ambient temperature, then it must be emitting and absorbing radiant energy at the same rate. This equivalence must apply at each wavelength, on the basis that it would otherwise be possible for the object to have a different equilibrium temperature than its surroundings. It is commonplace that different surfaces have greater or lesser ability to reflect light—some are bright, some are dull, and others are colored. They must then also have correspondingly different abilities to absorb light. Further, the most perfectly absorbing material should also be the most perfectly emitting one.

The concept of the *blackbody* emerges from such considerations. A box lined in deep black material with only a small hole for light to enter should absorb virtually all light incident on the hole. That which is not absorbed immediately should be absorbed after a few reflections around the inside of the box. Such a box should then also be an essentially perfect emitter of radiation, hence the term blackbody radiation. The achievement by 1900 was to measure accurately the spectral distribution of blackbody radiation as a function of temperature, with results as illustrated by Fig. 16-25. Several laws developed as regularities were discovered. Wien (1894) found that the short-wavelength portion of the spectral distribution could be represented by a function of the form

$$E_v \, dv = av^3 e^{-bv/T} \, dv \qquad (16\text{-}141)$$

where E_v is the emissive power (radiant energy of frequency v per unit frequency interval emitted per unit time by unit area of a blackbody) and a and b are constants. The law implies that the frequency v_{max} at the maximum of the distribution is proportional to T, or

$$\lambda_{max} = \frac{A}{T} \qquad (16\text{-}142)$$

The constant A has the observed value of 0.290 cm K.

Further, the total emissive power is $E_{tot} = \int_0^\infty E_v \, dv$, or, alternatively, $E_{tot} = \int_\infty^0 (-c/\lambda^2) E_\lambda d\lambda$, since $\lambda v = c$. That is, the emissive power per unit wavelength interval is proportional to $1/\lambda^2$ times that per unit frequency interval. If

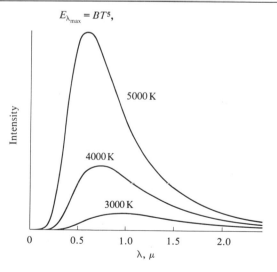

$E_{\lambda_{max}} = BT^5,$

FIGURE 16-25
Energy distribution for a blackbody radiator according to Planck's equation.

this conclusion plus Eq. (16-142) is substituted into Eq. (16-141), the result is

$$E_{\lambda_{max}} = BT^5 \qquad (16\text{-}143)$$

or the maximum in E_λ should be proportional to the fifth power of the temperature. Earlier, Boltzmann had shown by a thermodynamic argument that the total emitted energy obeyed the equation

$$E_{tot} = \sigma T^4 \qquad (16\text{-}144)$$

where σ is now generally known as the Stefan–Boltzmann constant; its value is 5.670×10^{-5} erg cm^{-2} s^{-1} K^{-4}.

Finally, Lord Rayleigh and J. Jeans were able to obtain an alternative expression to Eq. (16-141) by supposing that the equilibrium radiation in a black box must consist of standing waves. Suppose the box to be of side a, so that $n\lambda = 2a$ for standing waves, or $nc/v = 2a$. If v_0 is some observed frequency of wavelength small compared to a, then the number of possible other frequencies is of order $N = 2a/\lambda_0$. Extension of the analysis of the three-dimensional box gives

$$N = \frac{4V}{3\lambda_0{}^3} = \frac{4\pi V}{3} \frac{v_0{}^3}{c^3}$$

where $V = a^3$ and is the volume of the box. The distribution function for N is then

$$N_v \, dv = \frac{4\pi V}{c^3} v_0{}^2 \, dv$$

or, assuming equipartition so that each frequency has associated energy kT,

$$\rho_v \, dv = \frac{8\pi v^2 kT}{c^3} \, dv \qquad (16\text{-}145)$$

where ρ_v is the energy density of the radiant energy (ergs per cubic centimeter); also, a factor of two has entered to allow for the magnetic as well as electric oscillations present in electromagnetic radiation. The total emissive power E_{tot} can be shown to be

$$E_{tot} = \tfrac{1}{2}c\rho \qquad (16\text{-}146)$$

where E_{tot} is given in ergs per square centimeter per second. Equation (16-145) then becomes

$$E_v \, dv = \frac{2\pi v_0{}^2 kT \, dv}{c^2} \qquad (16\text{-}147)$$

This equation works well for the long wavelength side of the distribution function but leads to infinite emission as $\lambda \to 0$ (called the ultraviolet catastrophe). By about 1900 it was becoming necessary to question the equipartition principle itself. M. Planck suggested that the difficulty with the Rayleigh–Jeans law could be eliminated if it were supposed that an oscillating system (such as atoms in a solid) could acquire energy only in steps, rather than by infinitesimal degrees. In this way the equipartition principle would gradually come into being as kT became large compared to such steps or quanta. If v_0 were the fundamental frequency of the oscillators responsible for the radiation, then they could only emit radiation proportional to v_0 or to some integral multiple of it. This proportionality factor is, of course, now called Planck's constant h.

Planck was then able to derive the distribution law for blackbody radiation. He assumed that the Boltzmann principle applied so that the number of oscillators of energy ϵ_i was given by $N_i = N^0 e^{-\epsilon_i/kT}$ The total number becomes

$$N = N^0 + N^0 e^{-hv_0/kT} + N^0 e^{-2hv_0/kT}$$
$$+ \cdots = N^0 \sum_{i=0}^{\infty} e^{-ihv_0/kT}$$

where N^0 is the number of oscillators in the lowest energy state ($\epsilon = 0$). The total energy for oscillators in their various quantum states is therefore

$$E_{tot} = N^0(0) + N^0(hv_0)e^{-hv_0/kT}$$
$$+ N^0(2hv_0)e^{-2hv_0/kT} + \cdots$$
$$= N^0(hv_0) \sum_{i=0}^{\infty} ie^{ix}$$

where $x = hv_0/kT$, and the average energy is

$$\bar\epsilon = hv_0 \frac{\sum_{i=0}^{\infty} ie^{ix}}{\sum_{i=0}^{\infty} e^{ix}}$$

The sums are standard ones which, on evaluation, give

$$\bar\epsilon = \frac{hv_0}{e^x - 1} = \frac{hv_0}{e^{hv_0/kT} - 1} \qquad (16\text{-}148)$$

Note that if $kT \gg hv_0$, Eq. (16-148) gives $\bar\epsilon = kT$, or the classical equipartition law.

On substituting the expression for $\bar\epsilon$ into Eq.

(16-147) in place of kT, Planck obtained the distribution law†

$$E_\nu d\nu = \frac{2\pi h \nu^3}{c^2} \frac{1}{e^{h\nu/kT} - 1} d\nu \qquad (16\text{-}149)$$

or

$$E_\lambda d\lambda = \frac{2\pi h c^2}{\lambda^5} \frac{1}{e^{ch/\lambda kT} - 1} d\lambda \qquad (16\text{-}150)$$

(the subscript zero usually being dropped).

Equations (16-149) and (16-150) reduce to the Rayleigh–Jeans equation (16-147) at large wavelengths and to Wien's law, Eq. (16-141), at short wavelengths. The amazing success of Planck's treatment went far to guarantee acceptance of the then revolutionary idea of energy quanta.

16-ST-2   Atomic energy states

The treatment of atomic energy states was bypassed in the main part of the text, not because it is unimportant but because of the level of detail required. The energy states of the hydrogen atom are, of course, given by the Bohr theory and wave mechanics. Energies of states of hydrogen-like atoms may be estimated by rules such as those given in Section 16-3. Those for atoms having two or more outer electrons

†Some texts give ψ, the energy density; $\psi = 4E/c$.

require rather high-level wave mechanics for their treatment, but it is very useful at least to see how the various excited states are *named*. It is this nomenclature and the various combining rules on which it is based that we outline here.

In general the state of a multielectron atom is given by three angular momentum designations, S, L, and J. The number S gives the net spin quantum number, L is the net orbital angular momentum, and J is the vector sum of the spin and orbital angular momenta. Both S and L are zero for closed shells, and only the outer electrons have to be considered. Each individual outer electron has some particular ℓ value and its spin is $\pm\frac{1}{2}$; there are various ways in which the orbital angular momenta can combine to give a net L, and the spin angular momenta to give a net S. A particular set of individual electron ℓ values thus gives rise to a number of L and S states for the atom as a whole. For each L and S state the possible J values are $L - S$, $L - S + 1$, . . . , $L + S$, $J \geqslant 0$.†

†The scheme whereby the individual electron orbital angular momenta combine to give L and the spins to give S, which then determine the possible J values, is known as Russell–Saunders coupling. An alternative combining rule, more valid for heavy atoms, is that of first obtaining the possible net angular momenta j for each electron, where $j = \ell - \mathit{s}$, $\ell - \mathit{s} + 1$, ..., $\ell + \mathit{s}$, and then combining these j values vectorially to obtain J values.

EXAMPLE

We wish to calculate $E_{\lambda_{max}}$ for a blackbody at 1646 K and λ_{max}. Differentiating Eq. (16-150) with respect to λ and setting $dE_\lambda/d\lambda = 0$ leads to the equation $5 - x = 5e^{-x}$, where $x = ch/\lambda_{max}kT$. The solution is $x = 4.97$, whence

$$\lambda_{max} = \frac{(3.00 \times 10^{10})(6.63 \times 10^{-27})}{(4.97)(1.38 \times 10^{-16})(1646)} = 1.78 \times 10^{-4} \text{ cm} = 1.78 \text{ } \mu$$

[The same answer is obtained using Eq. (16-142) with $A = 0.290$.]

Substitution of this result into Eq. (16-150) gives

$$E_{\lambda_{max}} = \frac{2\pi(6.63 \times 10^{-27})(3.00 \times 10^{10})^2}{(1.78 \times 10^{-4})^5} \frac{1}{144 - 1} = 1.46 \times 10^{12} \text{ erg cm}^{-2} \text{ s}^{-1}$$

Our standard of luminous intensity, the *candela*, is defined in terms of blackbody radiation. Specifically, it is the luminous intensity, in the perpendicular direction, of a surface of 1/600,000 m² of a blackbody at the temperature of freezing platinum (2045 K). A source of unknown luminous intensity is rated in candelas by comparing it to the standard, using a photoelectric cell whose wavelength sensitivity is that of the human eye (a standard sensitivity–wavelength plot is used, which peaks at 550 nm).

The detailed procedure for obtaining the possible L and S states may be outlined as follows. Each outer electron has $2\ell + 1$ possible values of the magnetic quantum number, $m_\ell = 0, \pm 1, \ldots, \pm\ell$ (corresponding to the various allowed projections of the ℓ vector); the sum of a possible combination of the m_ℓ for the outer electrons then gives a particular M_L value. Likewise, a spin of $\pm\frac{1}{2}$ may be assigned to each electron, bearing in mind the Pauli exclusion principle, to yield various M_S. One thus obtains an array of all the M_L and M_S generated by vectorial combinations of the individual electron ℓ and s values. One now looks for a set of L and S states that will produce this same array. Each L state implies M_L values of 0, $\pm 1, \ldots, L$ and each S state implies M_S values of $-S, -S + 1, \ldots, S$.

An example should be helpful at this point. We consider first the case of an atom with two outer electrons of the same principal quantum number n and of the same ℓ value, namely $\ell = 1$. The configuration is thus $(np)^2$. Table 16-4 lists the various assignments of m and of s and the corresponding M_L and M_S values, and Table 16-5 summarizes the number of times each M_L and M_S combination occurs. We observe that the maximum M_L value is 2, which means that the state with $L = 2$ is present; since $M_L = 2$ occurs only with $M_S = 0$, the state is one of $S = 0$. The next step is to draw up the array for this state, given in Table 16-5(b), and to subtract it from the initial array, to give array (c).

The largest M_L value in array (c) is 1 and the largest M_S value is 1, corresponding to the state $L = 1$, $S = 1$. The array for this state is (d), which is then subtracted from (c) to give array

(e). This last corresponds to the $L = 0$, $S = 0$ state.

The naming system for L states is the same as that for ℓ ones, that is, $S, P, D, F, G, \ldots$ for $L = 0, 1, 2, 3, 4, \ldots$. It is customary to designate the S state by writing its number of possible M_S values, $2S + 1$, as an upper left superscript. Thus array (b) corresponds to a ^1D state, array (c) to a ^3P state, and array (e) to a ^1S state. The possible configurations for $(np)^2$ are thus $^1D + {}^3P + {}^1S$.

As one further example, we consider the case of $(nd)^3$. The display of possible m_ℓ, M_L, and M_S assignments is given in Table 16-6 and the tabulated array in Table 16-7. The reader can verify that this array is the sum of those for the states ^2H, ^2G, ^4F, ^2F, ^2D twice, ^4P, and ^2P. The general procedure can be extended to states involving more than one partially filled subshell, such as $(ns)^1(np)^1$.

It takes energy to pair electrons, and a rule, known as *Hund's rule*, applies here in the form of the statement that the state of lowest energy will be that of largest S value and, within this restriction, of largest L value. Thus for $(np)^2$ the ground state is ^3P, while for $(nd)^3$ it is ^4F. A further rule is that for shells less than half filled, states with lower J values are lower in energy, and the reverse is true for shells more than half filled.

16-ST-3 First-order perturbation theory

A very important method is that which treats a small perturbation to a system. The wave equation for the actual system is

TABLE 16-4 Orbital configurations for a p^2 case	m_ℓ			$M_L = \Sigma m_\ell$	$M_S = \Sigma m_s$
	1	0	−1		
	× ×	—	—	2	0^a
	—	× ×	—	0	0^a
	—	—	× ×	−2	0^a
	×	×	—	1	$-1,0,0,1^b$
	×	—	×	0	$-1,0,0,1^b$
	—	×	×	−1	$-1,0,0,1^b$

^{a}Only $M_S = 0$ is possible; since both electrons have the same three quantum numbers n, ℓ, and m_l, their spins must be opposed.

^{b}The restriction in footnote a now does not apply; each electron can have $s = \frac{1}{2}$ or $s = -\frac{1}{2}$ independently. Zero is entered twice since the first electron may have its spin up or down and the case for the second electron is reversed; since the m values are different, these are not identical configurations.

TABLE 16-5
Orbital configurations for a p² case

(a)

```
        2 |  1
        1 | 1 2 1
M_L     0 | 1 3 1
       -1 | 1 2 1
       -2 |  1
          --------
             -1 0 1
              M_s
```

or

¹D

(b)

```
        2 | 1
        1 | 1
M_L     0 | 1      +
       -1 | 1
       -2 | 1
          -----
             0
            M_s
```

(c)

```
           1 | 1 1 1
   M_L     0 | 1 2 1
          -1 | 1 1 1
             --------
               -1 0 1
                M_s
```

or

³P   ¹S

(d)

```
           1 | 1 1 1
¹D + M_L   0 | 1 1 1   +
          -1 | 1 1 1
             --------
               -1 0 1
                M_s
```

(e)

```
   M_L  0 | 1
          -----
            0
           M_s
```

or

$$p^2 \sim {}^1D + {}^3P + {}^1S$$

$$\mathbf{H}\psi - E\psi = 0 \qquad (16\text{-}151)$$

and we suppose that the Hamiltonian **H** is only slightly different from some other one, **H°**, for which the equation

$$\mathbf{H}°\psi° - E°\psi° = 0 \qquad (16\text{-}152)$$

has known solutions, the set of solutions and corresponding energy states being

$$\psi_0°, \psi_1°, \psi_2°, \ldots, \psi_k°$$
$$E_0°, E_1°, E_2°, \ldots, E_k°$$

Since the perturbation is small, we write

$$\mathbf{H} = \mathbf{H}° + \alpha\mathbf{H}'$$

where α is the measure of the perturbation.

For example, in case of the harmonic oscillator **H°** would be

$$-\frac{h^2}{8\pi^2 m}\frac{d^2}{dx^2} + \tfrac{1}{2}kx^2$$

If the oscillator is not quite harmonic, so that $V(x)$ is $\tfrac{1}{2}kx^2 + bx^4$, then **H'** is simply $-bx^4$. In the case of Eq. (16-139), the (rather large) perturbation to $V(r)$ is e^2/r_{12}. A very common situation, then, is that in which **H'** equals some additional or perturbing terms in the potential energy function.

TABLE 16-6
Orbital configurations for a d³ case

2	1	0	−1	−2	$M_L = \Sigma m_\ell$	$M_S = \Sigma m_s$
× ×	×	—	—	—	5	
× ×	—	×	—	—	4	
× ×	—	—	×	—	3	
× ×	—	—	—	×	2	
×	× ×	—	—	—	4	
—	× ×	×	—	—	2	
—	× ×	—	×	—	1	
—	× ×	—	—	×	0	
×	—	× ×	—	—	2	
—	×	× ×	—	—	1	$\dfrac{1}{2}, -\dfrac{1}{2}$
—	—	× ×	×	—	−1	
—	—	× ×	—	×	−2	
×	—	—	× ×	—	0	
—	×	—	× ×	—	−1	
—	—	×	× ×	—	−2	
—	—	—	× ×	×	−4	
×	—	—	—	× ×	−2	
—	×	—	—	× ×	−3	
—	—	×	—	× ×	−4	
—	—	—	×	× ×	−5	
×	×	×	—	—	3	
×	×	—	×	—	2	
×	×	—	—	×	1	
×	—	×	×	—	1	
×	—	×	—	×	0	$\dfrac{3}{2}, \dfrac{1}{2}, -\dfrac{1}{2}, -\dfrac{3}{2}$[a]
×	—	—	×	×	−1	
—	×	×	×	—	0	
—	×	×	—	×	−1	
—	×	—	×	×	−2	
—	—	×	×	×	−3	

The column header for columns 2–(−2) is: m_l

[a]Note that for this group of configurations there are three ways of assigning spins to give an M_S of $\frac{1}{2}$, and similarly for $M_S = -\frac{1}{2}$.

TABLE 16-7
Orbital configurations for a d³ case[a]

M_L	$-\dfrac{3}{2}$	$-\dfrac{1}{2}$	$\dfrac{1}{2}$	$\dfrac{3}{2}$
5		1	1	
4		2	2	
3	1	4	4	1
2	1	6	6	1
1	2	8	8	2
0	2	8	8	2

M_S

[a]The array is symmetric to negative M_L, so only the positive ones need be listed.

The next step is to assume that since the perturbation is to be small, the solutions to Eq. (16-151) involve only a small correction to those for the unperturbed case, Eq. (16-152), so that for the kth state

$$\psi_k = \psi_k^\circ + \alpha \psi_k' \qquad E_k = E_k^\circ + \alpha E_k'$$

If these expressions for $\mathbf{H}$, ψ_k, and E_k are substituted into Eq. (16-151), one obtains

$$(\mathbf{H}^\circ - E_k^\circ)\psi_k' = (E_k' - \mathbf{H}')\,\psi_k^\circ \qquad (16\text{-}153)$$

[Eq. (16-152) has been used to cancel some terms, and also those in α^2 have been dropped]. It is further true that the correction wave function ψ_k' can itself be expressed as some linear combination of the solutions ψ_0°, ψ_1°, and so on,

$$\psi_k' = \sum_i a_i \psi_i^\circ \qquad (16\text{-}154)$$

or

$$\mathbf{H}\,\psi_k' = \sum_i a_i \mathbf{H}^\circ \psi_i^\circ = \sum_i a_i E_i^\circ \psi_i^\circ$$

Substitution of this last equation into Eq. (16-153) gives

$$\sum_i a_i(E_i^\circ - E_k^\circ)\,\psi_i^\circ = (E_k' - \mathbf{H}')\,\psi_k^\circ \qquad (16\text{-}155)$$

Finally, Eq. (16-155) is multiplied by $\psi_k^{\circ *}$ and integrated:

$$\sum_i a_i(E_i^\circ - E_k^\circ) \int \psi_k^{\circ *}\psi_i^\circ d\tau$$
$$= \int \psi_k^{\circ *}(E_k' - \mathbf{H}')\,\psi_k^\circ d\tau \qquad (16\text{-}156)$$

where $d\tau$ represents the element of volume (and would be just dx in the case of the harmonic oscillator). The left-hand term of Eq. (16-156) is zero; the ψ° functions are all orthogonal so that $\int \psi_k^{\circ *}\psi_i^\circ d\tau = 0$ if $k \neq i$, and $E_i^\circ = E_k^\circ$ if $k = i$. The right-hand side can be separated into two integrals and solved for E_k' (remembering that $\int \psi_k^{\circ *}\psi_k^\circ \, d\tau = 1$), to give

$$E_k' = \int \psi_k^{\circ *}\mathbf{H}'\psi_k^\circ \, d\tau \qquad (16\text{-}157)$$

Equation (16-157) gives E_k' and hence $\alpha E_k'$, the first-order perturbation theory correction to the energy E°. The constant α is usually incorporated into E_k' and $\mathbf{H}'$ or, in effect, treated as equal to unity. For the kth energy state E' is obtained by averaging the perturbation func-

tion $\mathbf{H}'$ over the unperturbed state of the system. Thus for the $n = 1$ state of a harmonic oscillator and with $\mathbf{H}' = bx^4$, the correction would be

$$E_{n=1}' = b \int_{-\infty}^{\infty} \psi_{n=1}x^4\psi_{n=1}\, dx \qquad (16\text{-}158)$$

where $\psi_{n=1}$ is given by Eq. (16-92). In the case of the helium atom, for which $\mathbf{H}' = e^2/r_{12}$, the correction to the ground-state energy is

$$E' = \int \psi_{1s}^*(1)\, \psi_{1s}^*(2)\, (e^2/r_{12}) \qquad (16\text{-}159)$$
$$\psi_{1s}(1)\, \psi_{1s}(2)\, d\tau_1\, d\tau_2$$

If ψ_k° is degenerate, a complication arises in that we do not know which (if any) of the degenerate set approximates to the solution of the perturbed wave equation. The general procedure involves using a linear combination of the degenerate wave functions.

16-ST-4 The variation method. Polarizability of the hydrogren atom

A very useful means of obtaining an approximate solution to the wave equation is that known as the *variation method*. Recalling Eq. (16-56), $\mathbf{H}\psi = E\psi$, multiplication of both sides by ψ (or by ψ^* if ψ is complex) gives, on rearrangement (and recognition that E is a constant),

$$E = \frac{\int \psi^*\mathbf{H}\psi \, d\tau}{\int \psi^*\psi \, d\tau} \qquad (16\text{-}160)$$

which is a formal way of expressing the energy solution to a wave equation. The integrals are over $d\tau$, or over all space, and so if the ψ's are normalized, the denominator is unity. The basic theorem of the variation method is that if some approximate wave function ϕ is used in Eq. (16-160), the resulting E,

$$E = \frac{\int \phi^*\mathbf{H}\phi \, d\tau}{\int \phi^*\phi \, d\tau} \qquad (16\text{-}161)$$

may be greater than the correct E but never less than it. The proof is not difficult and may be found in several of the general references.

A typical situation is that in which solution of the exact wave equation, Eq. (16-56), is difficult but we can make a guess that some function ϕ will come reasonably close, where ϕ has some adjustable parameter. Alternatively, we might try a ϕ expressed as a series,

$$\phi = c_1 f_1 + c_2 f_2 + c_3 f_3 \\ + \dots + c_n f_n \quad (16\text{-}162)$$

where each function f satisfies the general conditions for a wave function and the fs have been chosen in a way that we hope will "bracket" the exact solution. In either case, one or more adjustable parameters have been introduced, and the variation method consists of finding those values of the parameters that make the calculated energy a minimum and therefore as close as possible to the correct energy.

Assuming, in the case of Eq. (16-162), that all the fs are real, substitution in Eq. (16-161) gives:

$$E = \frac{\sum_{j=1}^{n} \sum_{k=1}^{n} c_j c_k H_{jk}}{\sum_{j=1}^{n} \sum_{k=1}^{n} c_j c_k S_{jk}} \quad (16\text{-}163)$$

where

$$H_{jk} = \int f_j \mathbf{H} f_k \, d\tau \quad (16\text{-}164)$$
$$S_{jk} = \int f_j f_k \, d\tau$$

Equation (16-163) rearranges to

$$E \sum c_j c_k S_{jk} = \sum c_j c_k H_{jk} \quad (16\text{-}165)$$

One now wishes to find the coefficients c that make E a minimum. This is done by differentiating Eq. (16-165) and setting $\partial E/\partial c_i = 0$. Differentiation gives

$$\frac{\partial E}{\partial c_i} \sum_{j=1}^{n} \sum_{k=1}^{n} c_j c_k S_{jk} + E \sum_{k=1}^{n} 2 c_k S_{ik} \\ = \sum_{k=1}^{n} c_k (H_{ik} + H_{ki}) \quad (16\text{-}166)$$

It may be shown that $H_{ik} = H_{ki}$ and $S_{ik} = S_{ki}$, and the condition that $\partial E/\partial c_i$ be zero becomes

$$\sum_{k=1}^{n} c_k (H_{ik} - S_{ik} E) = 0 \quad (16\text{-}167)$$

This is a set of n simultaneous, homogeneous linear equations. If they are to have a solution (other than the trivial one of all c's equal to zero), the determinant of the coefficients of the c's must vanish. Thus

$$\begin{vmatrix} H_{11} - S_{11}E & H_{12} - S_{12}E & \cdots & H_{1n} - S_{1n}E \\ H_{21} - S_{21}E & H_{22} - S_{22}E & \cdots & H_{2n} - S_{2n}E \\ \vdots & \vdots & & \vdots \\ H_{n1} - S_{n1}E & H_{n2} - S_{n2}E & \cdots & H_{nn} - S_{nn}E \end{vmatrix} \\ = 0 \quad (16\text{-}168)$$

This determinantal equation, commonly called the *secular equation*, is then solved for E (taken to be the lowest of the roots).

A relatively simple but important application of the variation method is to the determination of E when a hydrogen atom is subjected to an electric field, $\mathbf{F}$, applied in the z direction (of Fig. 16-12). We try

$$\phi = c_1 f_1 + c_2 f_2 \quad (16\text{-}169)$$

where

$$f_1 = \psi_{1s} = (\pi a_0^3)^{-1/2} e^{-r/a_0}$$

$$f_2 = \left(\frac{z}{a_0}\right)\psi_{1s} = (\pi a_0^3)^{-1/2} e^{-r/a_0}\left(\frac{r}{a_0}\right)\cos\theta$$

(a_0 is the radius of the first Bohr orbit). The function f_2 has been selected as the simplest function that will shift the electron density in the z direction relative to f_1. We assume that the field is small enough that the Hamiltonian can be written

$$\mathbf{H} = \mathbf{H}_0 + e\mathbf{F}z = \mathbf{H}_0 + e\mathbf{F}r\cos\theta \\ (16\text{-}170)$$

where $\mathbf{H}_0$ is the Hamiltonian for the hydrogen atom with no field. The necessary integrals are easily found to be

$$H_{11} = -e^2/2a_0 \qquad S_{11} = 1$$

$$H_{12} = e\mathbf{F}a_0 \qquad S_{12} = 0$$

$$H_{22} = 0 \qquad S_{22} = 1$$

so that the secular equation is

$$\begin{vmatrix} -\dfrac{e^2}{2a_0} - E & e\mathbf{F}a_0 \\ e\mathbf{F}a_0 & -E \end{vmatrix} = 0 \quad (16\text{-}171)$$

This yields a quadratic equation in E, whence

$$2E = -\frac{e^2}{2a_0} - \frac{e^2}{2a_0} \\ \left[1 + 4e^2\mathbf{F}^2 a_0^2\left(\frac{2a_0}{e^2}\right)\right]^{1/2}$$

On using the binomial theory to expand the square root, and keeping only the term in F^2, we obtain

$$E = \frac{e^2}{2a_0} - 2a_0^3 F^2 \qquad (16\text{-}172)$$

The first term is the energy of the unperturbed atom and the second term gives the reduction in energy due to the field. Recalling Eq. (8-57),

which gives the energy of an atom in a field, F, as $-\frac{1}{2}\alpha F^2$, we thus obtain the polarizability, α, as

$$\alpha = 4(i_0)^3$$
$$= 0.59 \times 10^{-24} \text{ cm}^3 \qquad (16\text{-}173)$$

More sophisticated calculations (including more wave functions than just the ψ_{1s}) give as the exact value $\alpha = 0.670 \times 10^{-24} \text{ cm}^3$.

GENERAL REFERENCES

ANDERSON, J.M. (1969). "Introduction to Quantum Chemistry." Benjamin, New York.
BERRY, R.S., RICE, S.A., AND ROSS, J. (1980). "Physical Chemistry." Wiley, New York.
COULSON, C.A. (1961). "Valence." Oxford Univ. Press, London and New York.
DAUDEL, R., LEFEBVRE, R., AND MOSER, C., (1959). "Quantum Chemistry, Methods and Applications." Wiley (Interscience), New York.
EYRING, H., WALTER, J., AND KIMBALL, G.E. (1944). "Quantum Chemistry." Wiley, New York.
HANNA, M.W. (1969). "Quantum Mechanics in Chemistry," 2nd ed. Benjamin, New York.
McQUARRIE, D.A. (1983). "Quantum Chemistry." University Science Books, Mill Valley, California.
OFFENHARTZ, P.O'D. (1970). "Atomic and Molecular Orbital Theory." McGraw-Hill, New York.
PITZER, K.S. (1953). "Quantum Chemistry." Prentice-Hall, Englewood Cliffs, New Jersey.
TAYLOR, H.S., AND GLASSTONE, S., Eds. (1942). "A Treatise on Physical Chemistry." Van Nostrand-Reinhold, Princeton, New Jersey.

CITED REFERENCES

ADAMSON, A.W. (1965). *J. Chem. Ed.* **42**, 140.
BALLENTINE, L.E. (1970). *Rev. Mod. Phys.* **42**, 358.
BERRY, R.S, RICE, S.A., AND ROSS, J. (1980). "Physical Chemistry." Wiley, New York.
BOHR, N. (1913). *Phil. Mag.* **26**, 140.
BRIDGMAN, P.W.(1961). "The Logic of Modern Physics." Macmillan, New York.
DAUDEL, R., LEFEBVRE, R., AND MOSER, C. (1959). "Quantum Chemistry, Methods and Applications." Wiley (Interscience), New York.
EINSTEIN, A. (1905). *Ann. Phys.* (4) **17**, 132.
EYRING, H., WALTER, J., AND KIMBALL, G.E. (1944). "Quantum Chemistry." Wiley, New York.
FRANK, P. (1947). "Einstein—His Life and Times." Knopf, New York.
HEISENBERG, W. (1927). *Z.Phys.* **43**, 172.
KERKER, M. (1985). Langmuir, **1**, 531.
MICHELMORE, P. (1963). "Einstein—Profile of the Man." Muller, London.
McQUARRIE, D.A. (1983). "Quantum Chemistry." University Science Books, Mill Valley, California.
PITZER, K.S. (1953). "Quantum Chemistry." Prentice-Hall, Englewood Cliffs, New Jersey, p. 70.
ROHRLICH F. (1983). *Science* **221**, 1251.
SCHILPP, P.A. (1949). "Albert Einstein: Philospher–Scientist." Library of Living Philosophers, Evanston, Illinois.
SCHRÖDINGER, E. (1926). *Ann. Phys.* **79**, 361.
WIEN, W. (1894). *Wied. Ann.* **52**, 132.

EXERCISES

Take as exact numbers given to one significant figure.

16-1 The simple Bohr model treats the electron as a particle executing a circular orbit around the hydrogen atom. Derive the appropriate equation and calculate the velocity of the electron in the first Bohr orbit, expressed as a fraction of the velocity of light.

Ans. 0.00730.

16-2   Calculate the ratio of the radius of the outer electron orbit in Li to that of the first Bohr orbit, a_0. The outer Li electron sees a Z_{eff} reduced from Z by a screening constant of 1.74.

Ans. 3.17

16-3   As a hydrogen-like atom, Li^{2+} also has a Lyman series. Calculate the wavenumber for the first line, on the basis of Eq. (16-17), and then the more correct value assuming ^6Li and recognizing the reduced mass effect.

Ans. 740,725 cm^{-1}, 740,657 cm^{-1}.

16-4   Calculate (a) the radius of the Bohr orbit with $n = 2$, (b) the potential energy in atomic units of the electron in this orbit, (c) the energy difference in electron volts between the $n = 2$ and $n = 3$ Bohr orbits, and (d) the frequency in wavenumbers of the light emitted when a hydrogen atom in the $n = 3$ state returns to the ground state.

Ans. (a) $4a_0$ or 2.12 Å, (b) -0.25 a.u., (c) 1.89 eV, (d) 97,500 cm^{-1}.

16-5   Estimate the uncertainty in the energy (in electron-volts) of (a) a 1 eV electron confined to a 1.5 Å space and (b) a 1 eV hydrogen atom so confined.

Ans. (a) 1.30 eV, (b) 0.0304 eV.

16-6   Calculate the wavelength of (a) a 1-keV electron and (b) a 1-keV proton.

Ans. (a) 0.39 Å, (b) 9.1×10^{-11} cm.

16-7   Calculate Z_{eff} for the outer electron of (a) Ar and (b) Al^{2+}.

Ans. (a) 3.23, (b) 4.35.

16-8   Calculate the ionization energy in a.u. for (a) Na and (b) Si^{4+} if the respective screening constants are 9.16 and 7.00.

Ans. (a) 0.19 a.u., (b) 6.12 a.u.

16-9   Show that $f(x) = a \cos bx$ is an eigenfunction of the operator $\mathbf{D}^2$, and find the eigenvalue.

Ans. $-b^2$.

16-10   An operator is said to be *linear* if it is true that $\mathcal{O}[af_1(x) + bf_2(x)] = a\mathcal{O}f_1(x) + b\mathcal{O}f_2(x)$. Show whether the following operators are linear or nonlinear. (a) The operator such that $\mathcal{O}f(x) = x^3 f(x)$, (b) the operator whose instruction is to square, that is, $\mathcal{O}f(x) = [f(x)]^2$, (c) the operator whose instruction is to take the reciprocal, that is, $\mathcal{O}f(x) = 1/f(x)$.

Ans. (a) linear, (b) nonlinear, (c) nonlinear.

16-11   Show what $\langle x \rangle$ is by applying Eq. (16-59) to the case of $\psi = ae^{-bx^2}$.

Ans. $\langle x \rangle = 0$.

16-12   Benzene may be regarded as a two-dimensional box of side about 3.5 Å and containing six π electrons. What wavelength of light should be required to promote an electron from the ground to the first excited state?

Ans. 135 nm.

16-13   Calculate $\langle x \rangle$ for the $n = 1$ state of the particle in a one-dimensional box by evaluating the appropriate integral.

Ans. $\langle x \rangle = a/2$.

16-14   Suppose that a system containing 20 electrons can be likened to a cubic box. What should be the quantum numbers for the ground state and the first excited state?

Ans. (a) Any permutation of (1,1,3), (b) (2,2,2).

16-15   The force constants for the H—F and C—C bonds are about 10 and 5, respectively, in units of 10^5 dyn cm^{-1}. Show what the sequence would be if the two bonds were arranged in order of increasing wavenumber of the first vibrational absorption. Assume that each is a diatomic molecule, hypothetical in the case of C—C.

Ans. C—C, H—F.

Derive an expression for the value of q [as used in Eq. (16-88)] such that the potential energy V is equal to the total energy E.   **16-16**

Ans. $q^2 = 2(v + \frac{1}{2})$

Calculate the force constant for HBr (use Table 6-1) and, assuming it not to change, the fundamental frequency for DBr.   **16-17**

Ans. 3.84×10^5 dyn cm^{-1}, 1822 cm^{-1}.

Calculate the frequency in cm^{-1} for the first vibrational transition for CO (carbon monoxide), using the data of the example following Eq. (16-98) and taking the force constant to be 39% of that of HCl. Calculate also the ratio of the zero point energy for H^{35}Cl to that for CO.   **16-18**

Ans. (a) 2150 cm^{-1}, (b) 1.34.

The vibrational energy levels of HCl can be accurately reproduced by the equation $E(\text{cm}^{-1}) = 2988.95(v + \frac{1}{2}) - 51.65(v + \frac{1}{2})^2$. Calculate the dissociation energy of HCl in kcal mol^{-1}. What is the value of v for the last vibrational state before dissociation occurs?   **16-19**

Ans. 123.6 kcal mol^{-1} (the actual value is 103 kcal mol^{-1}), $v = 28$ or 29.

Find the value of r for which $a_0 r^2 [R(r)]^2$ is a maximum for the $n = 1$, $\ell = 0$ orbital of a hydrogen atom.   **16-20**

Ans. $r_{\text{max}} = a_0$.

Find the angle at which $[Y\theta, \phi)]^2$ is zero for the $\ell = 2$, $m = 0$ orbital of a hydrogen-like atom.   **16-21**

Ans. $\theta = 54.74°$ (independent of ϕ).

The Θ (θ) function for $\ell = 4$, $m = 4$ has the form $A \sin^4 \theta$. Find the value of the normalization constant A.   **16-22**

Ans. $A = \dfrac{3}{16} \sqrt{35}$.

Equation (16-130) can be written $E = BJ(J + 1)$ where B is called the rotational constant, and E is now given in wavenumbers. The transition from the $J = 10$ to $J = 11$ level occurs at 455 cm^{-1} in the case of HF. Calculate B for HF and the H—F bond length.   **16-23**

Ans. $B = 20.681$ cm^{-1}, $R = 0.926$ Å.

One can attempt a parallel between the wave-mechanical rigid rotator and the classical one. For the latter, the energy is given by $E = \frac{1}{2}I\omega^2$, where ω is the rotational speed in radians per second. Calculate ω for the classical dumb-bell shaped object corresponding to an H^{35}Cl molecule and having an energy equal to that of the wave mechanical case of $J = 1$. Use data from the example following Eq. (16-130). Alternatively, for what frequency would a light quantum have this energy?   **16-24**

Ans. $\omega = 9.07 \times 10^{11}$ c.p.s., $\nu = 6.42 \times 10^{11}$ s^{-1}.

PROBLEMS

The gravitational force between two masses m_1 and m_2 is given by $f = km_1 m_2/r^2$, where r is the separation and in the cgs system k has the value of 6.67×10^{-8}. Assume that the hydrogen atom is held together by gravity; that is, assume the proton and the electron to be neutral and apply the Bohr treatment to calculate the radius of the first "gravitational" orbit and the energy of an orbit in terms   **16-25** of its quantum number. Determine also if an intergral number of de Broglie wavelengths of the electron would make up the circumference of the first gravitation orbit.

16-26   Derive an expression relating the de Broglie wavelength of a particle and its kinetic energy. A tennis ball weighs about 2 oz and in a particular service, its speed is measured to be 100 mph. Calculate the de Broglie wavelength of the ball. What is the theoretical minimum uncertainty in the position of the ball if there is a 1% uncertainty in the velocity measurement?

16-27   Calculate Z_{eff} and the screening constant for Cs.

16-28   Why is the value 109,737 cm^{-1} for the Rydberg constant sometimes written as $\mathscr{R}_\infty$?

16-29   Rewrite Eq. (16-37) with x, V, and E expressed in atomic units of distance and energy. Also, verify Eq. (16-30).

16-30   Show whether the operator described by Eq. (16-46) is a linear one (see Exercise 16-10).

16-31   Find the expectation value of the momentum, $\langle p \rangle$, for a particle in a one-dimensional box, Eq. (16-73), with $n_1 = 1$. Remember Eqs. (16-46) and (16-58).

16-32   The ground-state energy for an electron in a 3-Å one-dimensional box is about 100 kcal mol^{-1}. For what size three-dimensioanl box would the ground-state energy be equal to the average kinetic energy of the electron from thermal motion at 25°C?

Ans. 54.1 Å

16-33   Show that ψ_1 and ψ_2, the wave functions for the first two levels of an electron in a one-dimensional box, are orthogonal.

16-34   Naphthalene has a pi electron transition at 260 nm. Treating naphthalene as a square box, calculate the side of the box.

16-35   Would an expression for ψ analogous to that of Eq. (16-73) but involving a cosine function also be a solution of Eq. (16-64) and, if so, why is it not used?

16-36   The case of a particle in a square box with $(n_1^2 + n_2^2) = 50$ is an interesting one. Discuss the situation.

16-37   The difference in energy ΔE between the ground and first excited state of amidinium polymethylene dyes,

$$(CH_3)_2 \overset{+}{N}=CH(-CH=CH)_n \overset{..}{N}(CH_3)_2$$

has been treated according to the one-dimensional particle-in-a-box model. Let a be the C—C bond length and d the end space or distance due to each nitrogen. Derive a general formula for ΔE in electron volts. Calculate ΔE expressed in angstroms for the case of $n = 3$; assume $a = 1.24$ Å and $d/a = 1.5$. The observed wavelength of the first absorption band is 3290 Å in this case.

16-38   Show that $\psi = Ae^{iy} + Be^{-iy}$, $y = (2\pi/h)(2mE)^{1/2}(x)$, is also a solution to Eq. (16-64). Expand this into a cos y and a sin y term and explain why the cos y portion is not used in the treatment of the particle in a box. (Consider the graphical nature of the solution involving the cos y term.)

16-39   Show that the solutions for the harmonic oscillator are orthogonal for the case of $v_2 = 1$ and $v_2 = 2$; that is, show that $\int_{-\infty}^{\infty} \psi_{v=1}\psi_{v=2}\,dx = 0$. Similarly, show that $\psi_{v=1}$ is normalized, that is $\int_{-\infty}^{\infty} \psi^2_{v=1}\,dx = 1$.

16-40*   Plot Eq. (16-99) for the case of H^{35}Cl given the information in Exercise 16-19, including the first ten vibrational levels.

16-41   Plot ψ_n for $n = 2$ as a function of x. Take $\beta = 1 \times 10^{17}$ cm.

16-42   Figure 16-9 shows plots of ψ_n^2 for the harmonic oscillator. Each shows one or more maxima. Find the value of q for the right-hand most maximum, q_{max}, for $n = 2$, 3, and 4. From these obtain the ratios q_{max}/q_v where q_v is the corresponding q value from Exercise 16-16.

The value of r_e is 1.62 Å for $H^{127}I$ and the dissociation energy is 71.4 kcal mol^{-1}; the zero point energy is 3.187 kcal mol^{-1}. (a) Calculate the constant a in Eq. (16-99). (b) Calculate the vibrational quantum number for the vibrational level just before dissociation occurs. (c) Plot Eq. (16-99), showing V from $r = 0$ to $r = 5$ Å. (d) Calculate the energy in wavenumbers for the transition for the $\nu = 0$ to the $\nu = 1$ vibrational level.

16-43

Referring to the example following Eq. (16-98), by how many wavenumbers should the frequency for the first vibrational transition in $D^{35}Cl$ shift from that for $H^{35}Cl$, assuming no change in the force constant? What would the shift be if the ^{35}Cl were bonded to some exceedingly heavy atom, again with the same force constant?

16-44

Calculate the per cent change in wavenumber for the transition from the ground to the first vibrational state for the stretching of the $C=O$ bond for free CO as compared to CO chemisorbed on a metal surface such as Ni. Assume that (a) no change in force constant occurs and (b) the $C=O$ force constant is reduced by 20% as a result of the adsorption bonding. Do each calculation assuming, first, that the adsorption bond is metal–carbon and second, that it is metal–oxygen. The effective mass of that atom which is bonded to Ni is to be regarded as infinite.

16-45

Verify that Eq. (16-111) is a solution to Eq. (16-107).

16-46

Verify that $\Theta_{1,0}$ is a solution to Eq. (16-110).

16-47

Verify that $R_{2,0}$ is a solution to Eq. (16-109) (take $\mu = m$ and $Z = 1$).

16-48

Show that the Φ_ν functions of Eq. (16-111) are normalized and orthogonal.

16-49

Show that the $\ell = 1$ Θ functions are normalized and orthogonal.

16-50

Derive the functional form of the f_{z^3} orbital and make a polar plot of this function in the xz plane.

16-51

Demonstrate Ünsold's theorem for the set of five d orbitals.

16-52

The angular part of one of the f orbitals is given by a constant times $\sin^2\theta \cos\theta \sin 2\phi$. How should this orbital be described in terms of x, y, and z projections of a unit vector (as in d_{xy}, d_{z^2}, etc.)?

16-53

Consider the radial solution of $n = 1$, $\ell = 0$, that is, $R_{1,0}$. (a) Show that $R_{1,0}$ is normalized. (b) Calculate the average radius, $\bar{r}$. (c) Calculate the most probable value of r, that is, the value of r when $4\pi^2 r^2 R^2$ is at a maximum.

16-54

In what plane does the 4f orbital ($n = 4$, $\ell = 3$, $m = 3$) lie, that is, have its maximum extension? Make a polar plot of the angular dependence of this orbital in this plane.

16-55

Show that the set of 2p hydrogen-like orbitals are mutually orthogonal.

16-56

Make a plot of the electron density for a 2s electron of Li as a function of radial distance r measured in units of a_0. What is wanted is that quantity that gives the chance of finding the electron in a spherical shell of radius between r and $r + dr$.

16-57

Verify Eq. (16-16) for the case of $n = 2$, $\ell = 1$.

16-58

The wave function for an excited state of He might be:

16-59

$$\psi_{He}^* = [\psi_{1s}(1)\,\psi_{2s}(2) - \psi_{1s}(2)\,\psi_{2s}(1)]\,\alpha(1)\,\alpha(2)$$

Explain whether this is an acceptable wave function in terms of the Pauli exclusion principle.

Show that the spacing between energy levels for a diatomic rigid rotator is just $2B(J + 1)$, where B is the rotational constant (note Exercise 16-23).

16-60

A linear triatomic molecule is treated in the same way as is a diatomic one with respect to rotation, and the same rigid rotator equation results. For example, the spectroscopic constant B (see Exercise 16-23) is 0.3903 cm^{-1} for CO_2. Calculate the carbon–oxygen bond length (assume ^{12}C and ^{16}O).

16-61

SPECIAL TOPICS PROBLEMS

16-62   Calculate from Planck's law the relative intensities of the radiation from a blackbody at 2500 K at wavelength 4500 ± 10 Å and 8500 ± 10 Å.

16-63   Calculate the emissive power E at λ_{max} for a blackbody at 4000 K.

16-64   If 2×10^4 cal min^{-1} is lost by radiation from the door of an electrically heated furnace at 950 K, how much additional electrical power must be supplied to offset the increased loss due to radiation if the furnace is heated to 1150 K?

16-65   Calculate the distribution curve for $T = 5000$ K (Fig. 16-25).

16-66   Find the set of free ion terms for the d^3 system. Explain which term should be that for the ground state.

16-67   Find the set of free ion terms (that is, the L states and their multiplicities) for the d^2 case. Explain which state should be of the lowest energy.

16-68   Find the L states and their multiplicities for a p^4 system. Explain which should be of lowest energy.

16-69   Evaluate $E'_{n=1}$ of Eq. (16-158).

16-70   Evaluate $E'_{n=1}$ for the perturbed harmonic oscillator with $\mathbf{H}' = bx^3$. Can you generalize your result to E'_n where n is any integer?

16-71   Confirm the values for the integrals H_{11}, S_{11}, H_{12}, S_{12}, and S_{22} used in Eq. (16-171) [see Pitzer (1953)].

chapter 17
Molecular
symmetry
and bonding

17-1   Introduction

It turns out that if we know the geometry of a molecule, it is much easier to set up the wave mechanical treatment; also, a number of qualitative results can be demonstrated without calculation. The emphasis of this chapter is therefore on that discipline of mathematics which allows a concise and penetrating description of the essential elements of molecular symmetry, that is, *group theory*. As a branch of mathematics, group theory requires serious and extensive study. There are numerous theorems and lemmas to be proved and many special cases to be examined in detail. We will take a very pragmatic approach here, however. Theorems will be stated rather than rigorously proved; their validity will lie in the success of their practical as opposed to their abstract demonstration. By allowing ourselves this latitude, it will be possible to present the group-theoretical approach to the symmetry properties of molecules in a rather efficient, and, this writer hopes, a stimulating way.

We proceed therefore to describe symmetry operations and to develop the concept of a set of such operations as defining a symmetry group. Some of the basic attributes of groups will be introduced. The result will be a compact description of those aspects of molecular geometry that determine the qualitative application of wave mechanics to a set of chemical bonds. The group-theoretical approach also provides a means for simplifying quantitative wave mechanical calculations. This is illustrated in the Commentary and Notes section with a discussion on the crystal field treatment of transition metal complexes.

The wave mechanics of chemical bonding, treated qualitatively and from the symmetry point of view in this chapter, is taken up more quantitatively in Chapter 18.

17-2   Symmetry and symmetry operations

The objects that we shall be dealing with, namely simple molecules, will generally have a considerable degree of symmetry. The ammonia molecule, for example, can be viewed as a triangular pyramid whose base is formed by

A. Symmetry elements and stereographic projection diagrams

the hydrogen atoms and whose apex is a nitrogen atom, as illustrated in Fig. 17-1(a). There is an axis of symmetry, labeled C_3, in recognition of the fact that if the molecule were rotated 120° about this axis, the result would be an equivalent configuration. That is, if the rotation were performed while the observer's eyes were closed, on looking again he could not tell that anything had occurred. The rotation is a symmetry operation—it gives an *equivalent configuration*. Were each hydrogen atom labeled, however, one could then tell that the rotation had been made. Only a rotation of 0°, 360°, and so on, would lead to a totally unobservable change; such an operation is called the *identity operation E*, since it results in a configuration *identical* to the starting one.

The ammonia molecule has other symmetry elements. A plane passing through the nitrogen and H_3 atoms and midway between H_1 and H_2 is a plane of symmetry. As illustrated in Fig. 17-1(b), if some molecular feature, such as H_1, occurs at a perpendicular distance x in front of the plane, then an equivalent feature is found by extending the perpendicular to a distance $-x$ to the other side of the plane, which in this case locates H_2. A plane of symmetry is designated by the symbol σ, and if it contains the principal axis of symmetry, C_3 in this case, it is called a *vertical plane* and written σ_v. There are evidently three equivalent such planes in the case of NH_3.

Three-dimensional perspective drawings are difficult to work with, and it has become conventional to use a projection down the principal axis of symmetry of the object. The ammonia molecule then looks as in Fig. 17-1(c), where the triangle imposed on the nitrogen atom stands for a C_3 rotation axis and the lines labeled σ_v locate the vertical planes of symmetry. Even Fig. 17-1(c) is too specific; quite a variety of objects would have C_3 and three σ_v's as the only symmetry elements, as illustrated in Fig. 17-1(d), and therefore belong to the same symmetry class as NH_3. A yet more abstract representation is desirable.

FIGURE 17-1
(a–c) Symmetry elements for the molecule NH_3. (d) Other objects having the same symmetry.

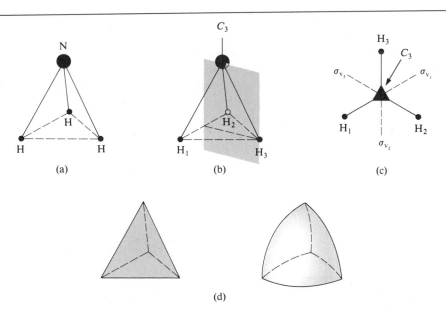

The next step is then to devise a geometric representation or pattern from which all superfluous features have been removed. It is customary to show the projection of this abstract figure as viewed down the principal axis—this is called a *stereographic projection diagram*. Figure 17-2(a) shows such a diagram for the symmetry elements C_3 and $3\sigma_v$. The triangle represents the C_3 axis and the three full lines represent the three σ_v planes. The symmetry pattern produced by these symmetry elements is generated by supposing the circle to lie in the plane of the paper and that there is initially a mathematical point lying above the plane, denoted by the plus sign. The full pattern of points is then generated by carrying out the symmetry operations. As shown in Fig. 17-2(b), the operation σ_v generates point 2 from point 1; application of C_3^1 (meaning one unit of rotation on the C_3 axis) generates point 3 from point 1 and point 4 from point 2; and application of C_3^2 similarly produces points 5 and 6. We now have obtained Fig. 17-2(a). Disregarding the labeling, no new points are produced by σ_v' or σ_v''.

A second example of a symmetric molecule is H_2O, or H—O—H. There is a twofold axis of symmetry passing through the oxygen atom, and a 180° rotation about this axis gives an equivalent configuration. In addition to the C_2 axis, there are two σ_v planes. Both contain the C_2 axis, and one is just the H—O—H plane, while the other is perpendicular to it, as illustrated in Fig. 17-3(a). The stereographic projection diagram for an object having just these symmetry elements is shown in Fig. 17-3(b), the lens denoting a twofold rotation axis. The four plus points are generated from a single initial point by application of σ_v and then C_2 (or vice versa).

A square planar molecule, such as $PtCl_4^{2-}$, possesses a number of symmetry features. There is a fourfold rotation axis C_4 passing through the platinum atom and perpendicular to the molecular plane—successive rotations of 90° give equivalent configurations. As illustrated in Fig. 17-4(a), there are also two pairs of C_2 axes, labeled C_2 and C_2'. The fourfold axis, being the highest-order one, is called the *principal axis*. The two pairs of σ_v planes contain the

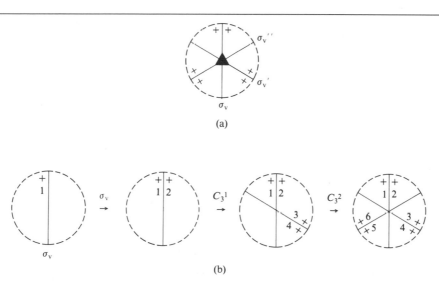

(a)

(b)

FIGURE 17-2
Stereographic projection showing the symmetry features of NH_3.

FIGURE 17-3
Stereographic projection showing the symmetry features of H_2O.

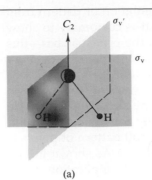

(a)

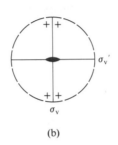

(b)

FIGURE 17-4
(a) Symmetry elements of $PtCl_4^{2-}$ and (b) corresponding stereographic projection. (c) Stereographic projection if no σ_h and no σ_v's are present.

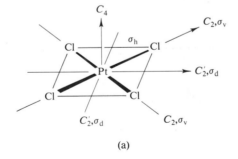

(a)

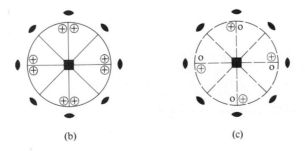

(b)　　　　　　　　　(c)

C_4 axis and either the C_2 or the C_2' ones; in addition, the molecular plane itself is a plane of symmetry. This last is a plane perpendicular to the principal axis, and is denoted σ_h.

The stereographic projection diagram generated by the symmetry features C_4, C_2, C_2', $2\sigma_v$, $2\sigma_v'$ (or $2\sigma_d$), and σ_h is shown in Fig. 17-4(b). The set of plus points is generated by carrying out σ_v on the initial one, followed by the successive C_4 operations. The circles denote mathematical points lying *below* the plane of the paper and are produced by the σ_h operation. The set of points implies all of the symmetry features—none of the other operations generate any further points. Figure 17-4(b) also illustrates some additional conventions. The presence of σ_h is signified by drawing the projection as a full rather than a dashed circle. The secondary axes C_2 and C_2' are indicated by the lenses located radially around the circle; the fact that the lines to the lenses are full means that corresponding σ_vs are also present; were they not, the lines would be dashed. Thus Fig. 17-4(c) shows the presence of the symmetry elements C_4, C_2, and C_2' only (no σ_h and no σ_v's). A few remaining conventions will be illustrated later as they come up.

A given symmetry element may generate more than one symmetry operation, and these will now be spelled out.

B. Symmetry operations

E: The identity operation has already been described; it is that which leaves the figure in an identical configuration to the initial one.

C_n: A rotation axis is one about which the molecule may be rotated to give equivalent positions, n denoting the number of positions. Thus C_4 denotes a fourfold rotation axis; it generates the individual symmetry operations C_4^1, C_4^2, C_4^3, and C_4^4, the superscript denoting the number of rotational increments of $2\pi/n$ degrees each. Since C_4^2 is the same as C_2 and $C_4^4 = E$, this sequence would usually be written C_4^1, C_2, C_4^3, E.

If a molecule has more than one rotation axis, then the one of largest n is called the *principal* axis and the others the *secondary* axes.

σ: The symbol σ denotes a plane of symmetry, that is, a plane passing through the molecule such that matching features lie on opposite sides of, or are "reflected" by, the plane. A plane perpendicular to the principal axis is denoted σ_h (for horizontal) and one which contains the principal axis is called σ_v (for vertical). If a molecule has secondary rotation axes, a vertical-type plane of symmetry that bisects the angle between two such axes is called σ_d (for dihedral). Finally, there may be two nonequivalent sets of vertical planes, and in such a case one set will be called σ_v and the other set σ_d (or σ_v'). This situation occurs if there is a C_n axis with n even *and* vertical planes of symmetry, as in the example shown in Figure 17-4(b).

i: An inversion center is denoted by i. If present, matching features lie equidistant from a point or inversion center. Alternatively, if a feature has the coordinates (x,y,z), then an equivalent feature is present at $(-x, -y, -z)$. Thus $PtCl_4^{2-}$ has an inversion center but H_2O and NH_3 do not.

S_n: This symbol stands for an *improper rotation axis*. An improper rotation may be thought of as taking place in two steps: first, a proper rotation and then, a reflection through a plane perpendicular to the rotation axis, usually a σ_h plane. The subscript n again denotes the order of the axis. By way of

FIGURE 17-5
Illustration of the
symmetry operation $S_6{}^1$.

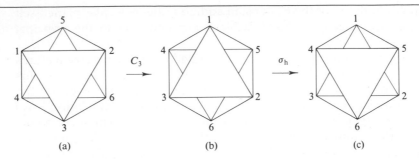

(a)   (b)   (c)

illustration, consider the triangular antiprism shown in Fig. 17-5(a). We are looking along the prism axis and see the staggered triangular ends. Rotation by 60° gives Fig. 17-5(b) and reflection in a plane perpendicular to the axis gives Fig. 17-5(c). The result is equivalent to the starting figure if the labeling is ignored, and hence is a symmetry operation. The figure therefore has the symmetry element S_6, the operation being $S_6{}^1$. The complete set of operations would be $S_6{}^1$, $C_3{}^1$, i, $C_3{}^2$, $S_6{}^5$, E (note that $S_6{}^2 = C_3{}^1$, that is, gives the identical change, and similarly, $S_6{}^4 = C_3{}^2$; also $S_6{}^2 = i$). As an exercise, the reader should examine the series $S_5{}^1$, $S_5{}^2$, ... for similar alternative statements—it is now necessary to go to S_5^{10} before the series repeats.

C. Naming of
symmetry groups

Any object or molecule has some certain set of symmetry features and this set accordingly characterizes its symmetry. Each distinctive such set is called a *point group*. As is seen in the next section, the set of symmetry operations has the property of a mathematical group and the term "point" refers to the fact that no translational motions are being considered. In the case of a crystal lattice, the addition of translations generates what are then called *space groups*.

Some easily recognizable, either very simple or highly symmetric, cases are summarized later, but the following is the general procedure for classifying a point group. The names or point group symbols used are those introduced by Schoenflies.

One first looks for rotation axes. If there is only one, then the naming is as follows:

C_n No other symmetry features are present.
C_{nh} There is also a σ_h plane.
C_{nv} There is no σ_h plane but there is a σ_v plane. (There must then be n such planes.)

If the molecule or object possesses more than one rotation axis, the one of higher order becomes the principal axis, of order n. We will only consider cases where the secondary axes are perpendicular to the principal one. The naming is now as follows:

D_n No other symmetry features are present.
D_{nh} A σ_h is present.
D_{nd} There is no σ_h but there is a σ_d (necessarily n σ_ds).

If there is an S_{2n} axis collinear with the principal C_n axis and no other symmetry features, the group is then S_{2n}. If there is only a plane of symmetry, the group is C_s, and if there is only an inversion center, it is C_i. If no symmetry elements at all are present, the group is named C_1.

There are some further point groups whose symmetry properties can usually be recognized on sight. An A—B type of diatomic molecule can be thought of as having a rotation axis of infinite order and an infinite number of σ_v planes; this combination of symmetry elements is called $C_{\infty v}$. An A—A type of molecule has also a σ_h and an infinity of twofold secondary axes, and the corresponding group name is $D_{\infty h}$.

D. Special groups

Easily recognizable but more complicated to describe in terms of symmetry are tetrahedral and octahedral molecules, for which the corresponding point group names are T_d and O_h. The tetrahedral case is best constructed by using a cubic framework, as illustrated in Fig. 17-6. The symmetry elements and operations are as follows.

1. The identity operation E.
2. Three C_2 axes coinciding with the x, y, and z axes of the figure. The operations characteristic of equivalent symmetry elements are often just totaled, and could be reported as $3C_2$ in this case.
3. Three S_4 axes collinear with the preceding C_2 ones. These generate the operations $3S_4{}^1$ and $3S_4{}^3$. There is a total of $6S_4$ operations, excluding the $S_4{}^2$ ones already accounted for (being identical to C_2).
4. Four C_3 axes, each coinciding with a body diagonal of the cube. Counting $C_3{}^1$ and $C_3{}^2$ for each, the total is $8C_3$.
5. There is a plane of symmetry containing each pair of apexes of the tetrahedron and bisecting the opposite edge, and examination of the figure shows six equivalent such planes. The symmetry operations are then 6σ, usually reported as $6\sigma_d$.

The T_d group thus has a total of 24 distinct symmetry operations.

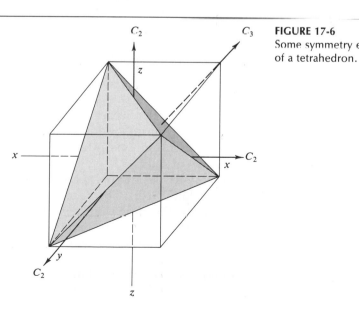

FIGURE 17-6
Some symmetry elements of a tetrahedron.

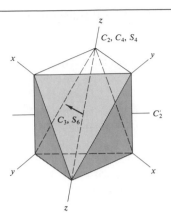

FIGURE 17-7
Symmetry elements of an octahedron.

Examination of the regular octahedron shown in Fig. 17-7 yields the following symmetry aspects.

1. The identity operation E.
2. Three C_2 axes collinear with the x, y, and z coordinates of the figure, or $3C_2$ operations.
3. Three C_4 axes collinear with the preceding C_2 axes, each generating C_4^1 and C_4^3 new operations, for a total of $6C_4$.
4. Three S_4 axes collinear with the preceding and generating S_4^1 and S_4^3 new operations, or a total of $6S_4$.
5. Six C_2' axes each bisecting opposite edges. The total is then $6C_2'$.
6. Four C_3 axes each passing through the center of opposite triangular faces, and giving $8C_3$ operations in all.
7. Four S_6 axes collinear with the C_3 ones (note Fig. 17-5!). Each generates the operations S_6^1, S_6^2 ($= C_3^1$), S_6^3($= i$), S_6^4 ($= C_3^2$), S_6^5, or two new ones, for a total of $8S_6$.
8. An inversion center i.
9. Three planes of symmetry each of which passes through four of the six apexes, or $3\sigma_h$.
10. Six planes of symmetry each of which passes through two apexes and bisects two opposite edges, or $6\sigma_d$.

The total number of symmetry operations for the octahedron is 48.

E. Some illustrations of point groups

The stereographic projection diagrams for a number of point groups are shown in Fig. 17-8. The only additional convention to be noted is that an open geometric figure in the center of the diagram indicates a corresponding improper axis of rotation.

Some molecular examples are collected in Table 17-1 [for additional ones see Adamson (1980)]. The Schoenflies designation for each can be obtained by following the procedure described in Section 17-2C. A useful point to remember is that it is not necessary to identify all the symmetry elements present in order to determine the point group designation. Thus in the case of NH_3 it is sufficient to determine that there is a C_3 axis and no secondary axes and that there is a σ_v but no σ_h, to establish the group as C_{3v}. The D_{2h}

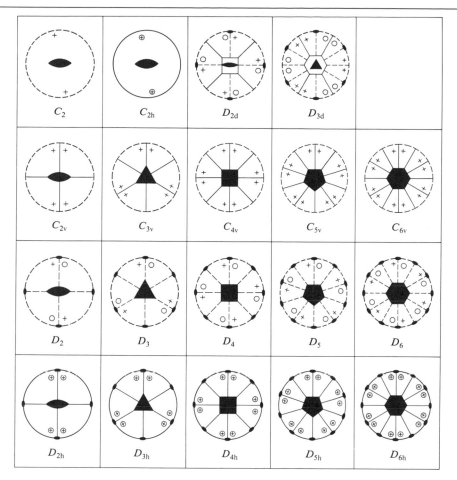

FIGURE 17-8
Stereographic projection diagrams for various point groups. (From H. Eyring, J. Walter, and G. E. Kimball, "Quantum Chemistry," Copyright 1960, Wiley, New York. Used with permission of John Wiley & Sons, Inc.)

designation for $H_2C{=}CH_2$ follows on observing that there are two mutually perpendicular C_2 axes (in addition to the principal C_2 axis) and a σ_h plane—the presence of σ_v planes is automatically assured. Dibenzene chromium is D_{6h} since the principal axis is C_6, there is at least one secondary axis, and a σ_h plane. The remaining symmetry features are implicit.

The entries for H_2O_2 illustrate another point. The molecule has C_{2v} symmetry if it is in the *cis* configuration with all atoms coplanar. Keeping this plane as a reference, one may then twist the molecule about the O—O axis, thus rotating one O—H bond forward above the plane and the other backward below the plane. A 90° rotation makes the molecule planar and *trans*; is now has C_{2h} symmetry. The actual configuration in the crystal is in between these positions, and for any such intermediate angle the symmetry is reduced to C_2. Thus the symmetry group of a molecule varies with molecular motions; one ordinarily takes the most probable positions of the atoms in the lowest vibrational energy state. In the case of $Co(NH_3)_6^{3+}$ the hydrogens are ignored in giving the symmetry as O_h; were they included, then, depending on their assumed configuration, some much lower symmetry would be reported.

The symmetry properties of a molecule are thus not absolutely invariant

TABLE 17-1
Molecular examples of various point groups

Molecule	Structure		Symmetry elements and operations	Point group
CH_4			$E, 8C_3, 3C_2, 6S_4, 6\sigma_d$	T_d
$Co(NH_3)_6^{3+\,a}$			$E, 8C_3, 6C_2', 6C_4, 3C_2, i,$ $6S_4, 8S_6, 3\sigma_h, 6\sigma_d$	O_h
HCl	H — Cl		E, C_∞, σ_v	$C_{\infty v}$
H_2	H — H		$E, C_\infty, C_2, \sigma_h, \sigma_v$	$D_{\infty h}$
H_2O_2		(planar)	$E, C_2, \sigma_v, \sigma_v'$	C_{2v}
		(planar)	E, C_2, i, σ_h	C_{2h}
			E, C_2	C_2
C_6H_6			$E, 2C_6, 2C_3, C_2, 3C_2',$ $3C_2, i, 2S_3, 2S_6, \sigma_h,$ $3\sigma_d, 3\sigma_v$	D_{6h}
$Cr(NH_3)_2(NCS)_4^-$			$E, 2C_4, C_2, 2C_2', 2C_2, i,$ $2S_4, \sigma_h, 2\sigma_v, 2\sigma_d$	D_{4h}
$(C_6H_5)_2Fe$ (ferrocene)			$E, 2C_5, 2C_5^2, 5C_2, i, 2S_{10},$ $2S_{10}^3, 5\sigma_d$	D_{5d}
$XeOF_4$			$E, 2C_4, C_2, 2\sigma_v, 2\sigma_d$	C_{4v}

[a] As is customary with such complex ions, only the central metal ion and the atoms directly bonded to it are considered.

or unique. We will be using symmetry to draw conclusions about chemical bonding and it is well to realize that assumptions as to what is important are inherent in most statements of molecular symmetry.

17-3   A set of symmetry operations as constituting a group

The term "multiplication" has the meaning in group theory of sequential performance of designated operations. Thus AB means *first* operation B, *then* operation A. With this definition in mind one can construct a "multiplication table" for the set of symmetry elements possessed by any particular object, each symmetry operation being entered separately.

Table 17-2 shows the multiplication table for the point group C_{3v}. The procedure for developing this table is as follows. It is simplest to make use of the stereographic projection diagram (Fig. 17-2 or Fig. 17-8), but with each plus identified by a number. As illustrated in Fig. 17-9(a), the effect of each separate symmetry operation is first diagrammed. Each possible symmetry operation is then performed again. As shown in Fig. 17-9(b), C_3^1 followed by C_3^1 gives the identical result as C_3^2; C_3^1 followed by σ_v gives the identical result as σ_v', and so on. In this way a table of all possible products is worked out. The convention we will follow is that in a multiplication table such as Table 17-2 the operation in the left-hand column is that performed first, followed by the one indicated by the top row. Also, of course, the multiplication AB corresponds to (row)(column).

There are several important features to notice in the table. First, each product gives a result identical to the result of some single symmetry operation of the set. Second, each symmetry operation occurs just once in each row and in each column. Each operation has a *reciprocal*, that is, for each A there exists a B such that $AB = E$ (and we then say that $A = B^{-1}$). The *associative law* of multiplication holds, that is, $A(BC) = (AB)(C)$. For example,

$$C_3^1(C_3^2\sigma_v) = C_3^1\sigma_v'' = \sigma_v \quad \text{and} \quad (C_3^1C_3^2)\,\sigma_v = E\sigma_v = \sigma_v \quad (17\text{-}1)$$

These features characterize a mathematical group, by definition; they hold for the multiplication tables of all of the point groups discussed here.

The multiplication of symmetry operations differs from ordinary multiplication in not being *commutative*, that is, the order of operation makes a difference. Thus AB is not in general the same as BA. For example, $C_3^1\sigma_v = \sigma_v''$, while $\sigma_vC_3^1 = \sigma_v'$. Finally, notice that the identity operation E either occurs along the diagonal of the multiplication table or is symmetrically disposed to it. This behavior will be a general one of symmetry point groups.

A. The multiplication table for a set of symmetry operations

	E	C_3^1	C_3^2	σ_v	σ_v'	σ_v''
E	E	C_3^1	C_3^2	σ_v	σ_v'	σ_v''
C_3^1	C_3^1	C_3^2	E	σ_v'	σ_v''	σ_v
C_3^2	C_3^2	E	C_3^1	σ_v''	σ_v	σ_v'
σ_v	σ_v	σ_v''	σ_v'	E	C_3^2	C_3^1
σ_v'	σ_v'	σ_v	σ_v''	C_3^1	E	C_3^2
σ_v''	σ_v''	σ_v'	α_v	C_3^2	C_3^1	E

TABLE 17-2
Multiplication table for the C_{3v} group

FIGURE 17-9
Illustration of symmetry operations in the C_{3v} group. (a) Effect of each symmetry operation. (b) Demonstration that $\sigma_v C_3^1 = \sigma_v'$.

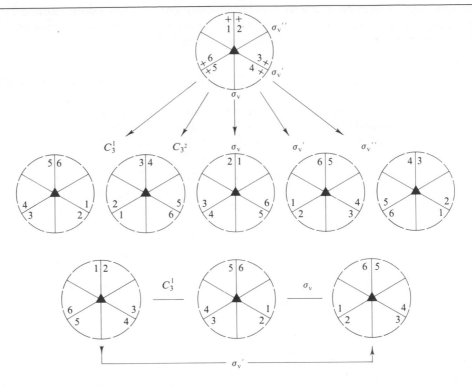

FIGURE 17-9
Illustration of symmetry operations in the C_{3v} group. (a) Effect of each symmetry operation. (b) Demonstration that $\sigma_v C_3^1 = \sigma_v'$.

B. Some properties of groups

Some general definitions and theorems for groups are now useful; we omit proofs for the latter, although in many cases the exercise is not difficult [see Cotton (1963)]. The *order h* of a symmetry group is the number of distinct symmetry operations, or captions to the rows and columns of the multiplication table; the order of the C_{3v} group is six, for example. A group may have subgroups; thus the set E, C_3^1, C_3^2 obeys the rules for a group. It can be shown that the order g of any subgroup must be an integral divisor of the group order h.

An important item is that the symmetry operations of a group will generally subdivide into *classes*. Before defining the term class we must first consider an operation known as a *similarity transformation*.

If A and X are two elements of a group and X^{-1} is the reciprocal of X, that is, $X^{-1}X = E$, then the product $X^{-1}AX$ will in general give some other element B of the group. For example, in the C_{3v} group if X is C_3^1, then X^{-1} is C_3^2; then $C_3^2 \sigma_v C_3^1 = C_3^2 \sigma_v' = \sigma_v''$. Thus if A is σ_v, its similarity transformation by C_3^1 is σ_v'.

We next take as X each element of the group in turn and apply the operation $X^{-1}AX$; if this is done for the C_{3v} group, with $A = \sigma_v$, the result is always σ_v, σ_v', or σ_v''. The same result obtains if A is σ_v' or σ_v''. The last three elements are then said to be conjugate with each other and constitute a *class* of the group. If the procedure is repeated with $A = C_3^1$ or C_3^2, the result of the similarity transformation (with X equal to *any* element of the group) is always C_3^1 or C_3^2, and these two symmetry operations constitute a second class. The identity operation always is in a class by itself.

An important theorem is that the order of any class of a group must be an integral divisor of the order h of the group. We will henceforth group symmetry operations by classes—those for C_{3v} are E, $2C_3$, and $3\sigma_v$. Note that this has been done in Table 17-1.

17-4   Representations of groups

Just as the set of symmetry elements is an abstraction of the symmetry properties of an object, there is a very useful abstraction of a group multiplication table known as a *representation*. A representation of a group is a set of numbers (or, as seen in the next section, of matrices) which if assigned to the various symmetry operations, will obey or be consistent with the group multiplication table. It turns out that all of the symmetry operations of a given class must be given the same number if the set of assignments is to work as a representation.

Two representations for the C_{3v} group are

$$(1) \quad E = 1 \qquad C_3{}^1 = C_3{}^2 = 1 \qquad \sigma_v = \sigma_v' = \sigma_v'' = 1$$

$$(2) \quad E = 1 \qquad C_3{}^1 = C_3{}^2 = 1 \qquad \sigma_v = \sigma_v' = \sigma_v'' = -1$$

If these simple designations are substituted into Table 17-2, then a self-consistent multiplication table is obtained. Considering representation 2, $\sigma_v C_3{}^2 = \sigma_v'$ and $(-1)(1) = (-1)$. It will be customary to show a representation in the more compact form given in Table 17-3, where Γ is the general symbol for a representation.

As a second example, the multiplication table for the group C_{2v} is given by Table 17-4. In this case each symmetry operation constitutes a separate class (as may be verified as an exercise). A set of simple representations for this group turns out to be as shown in Table 17-5. The designations A_1, A_2, and so on will be discussed later.

	E	$2C_3$	$3\sigma_v$
Γ_1	1	1	1
Γ_2	1	1	-1

TABLE 17-3
Two representations of C_{3v}

	E	C_2	σ_v	σ_v'
E	E	C_2	σ_v	σ_v'
C_2	C_2	E	σ_v'	σ_v
σ_v	σ_v	σ_v'	E	C_2
σ_v'	σ_v'	σ_v	C_2	E

TABLE 17-4
Multiplication table for the C_{2v} group

	E	C_2	σ_v	σ_v'
$\Gamma_1 = A_1$	1	1	1	1
$\Gamma_2 = A_2$	1	1	-1	-1
$\Gamma_3 = B_1$	1	-1	1	-1
$\Gamma_4 = B_2$	1	-1	-1	1

TABLE 17-5
Representations for C_{2v}

The concept of a representation as developed so far seems rather trivial. It provides, however, the key to using the symmetry properties of a molecule in wave mechanical treatments. To develop this application, we proceed to examine how representations may be generated.

A. Geometric transformations; matrix notation

An important method of generating a representation will be to carry out the symmetry operations of a group on some elementary object such as a point of a vector. Suppose a point is located at (x_1, y_1, z_1) and we apply the symmetry operations of the C_{2v} group to it. These consist of E, a C_2 axis collinear with the z axis, a σ_v in the xz plane, and a σ_v' in the yz plane. As illustrated in Fig. 17-10(a), application of this last changes the point to $(-x_1, y_1, z_1)$, or

$$x_2 = -x_1 + 0y_1 + 0z_1 \qquad y_2 = 0x_1 + y_1 + 0z_1 \tag{17-2}$$
$$z_2 = 0x_1 + 0y_1 + z_1$$

As shown in Fig. 17-10(b), the operation C_2 gives

$$x_2 = -x_1 + 0y_1 + 0z_1 \qquad y_2 = 0x_1 - y_1 + 0z_1 \tag{17-3}$$
$$z_2 = 0x_1 + 0y_1 + z_1$$

In these cases the new coordinates are just 1, 0, or -1 times the old ones, with no mixing—that is, x_2 does not depend on y_1 or z_1, and so on. This will not be true in general. For example, were the C_{3v} group involved, the operation C_3^1 would be as shown in Fig. 17-10(c), leading to

$$x_2 = x_1 \cos(\tfrac{2}{3}\pi) - y_1 \sin(\tfrac{2}{3}\pi) + 0z_1$$
$$y_2 = x_1 \sin(\tfrac{2}{3}\pi) + y_1 \cos(\tfrac{2}{3}\pi) + 0z_1 \tag{17-4}$$
$$z_2 = 0x_1 + 0y_1 + z_1$$

Here the x and y coordinates mix with each other but not with the z coordinate.

FIGURE 17-10
The C_{2v} point group. (a) The $\sigma_v(yz)$ operation on a general point and (b) the C_2 operation. (c) The C_3^1 operation of the C_{3v} group as applied to a general point.

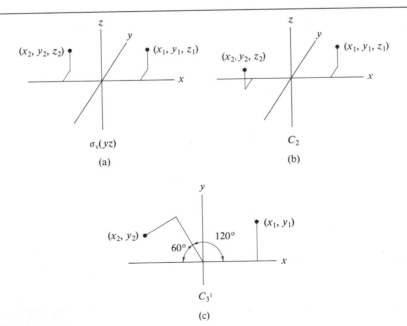

A set of transformation equations such as Eqs. (17-2), (17-3), or (17-4) may be expressed in matrix notation. Equation (17-2) becomes

$$\begin{bmatrix} -1 & 0 & 0 \\ 0 & 1 & 0 \\ 0 & 0 & 1 \end{bmatrix} \begin{bmatrix} x_1 \\ y_1 \\ z_1 \end{bmatrix} = \begin{bmatrix} x_2 \\ y_2 \\ z_2 \end{bmatrix} \tag{17-5}$$

That is, multiplication of the matrix

$$\begin{bmatrix} x_1 \\ y_1 \\ z_1 \end{bmatrix} \quad \text{by} \quad \begin{bmatrix} -1 & 0 & 0 \\ 0 & 1 & 0 \\ 0 & 0 & 1 \end{bmatrix}$$

yields the new set

$$\begin{bmatrix} x_2 \\ y_2 \\ z_2 \end{bmatrix}$$

Similarly, Eqs. (17-3) and (17-4) become

$$\begin{bmatrix} -1 & 0 & 0 \\ 0 & -1 & 0 \\ 0 & 0 & 1 \end{bmatrix} \begin{bmatrix} x_1 \\ y_1 \\ z_1 \end{bmatrix} = \begin{bmatrix} x_2 \\ y_2 \\ z_2 \end{bmatrix} \tag{17-6}$$

and

$$\begin{bmatrix} \cos(\tfrac{2}{3}\pi) & -\sin(\tfrac{2}{3}\pi) & 0 \\ \sin(\tfrac{2}{3}\pi) & \cos(\tfrac{2}{3}\pi) & 0 \\ 0 & 0 & 1 \end{bmatrix} \begin{bmatrix} x_1 \\ y_1 \\ z_1 \end{bmatrix} = \begin{bmatrix} x_2 \\ y_2 \\ z_2 \end{bmatrix} \tag{17-7}$$

The preceding statements follow from the rules for matrix multiplication. If we have the product $\mathscr{AB} = \mathscr{C}$ when a capital script letter denotes a matrix, or, for example,

$$\overset{\mathscr{A}}{\begin{bmatrix} a_{11} & a_{12} \\ a_{21} & a_{22} \\ a_{31} & a_{32} \end{bmatrix}} \overset{\mathscr{B}}{\begin{bmatrix} b_{11} & b_{12} & b_{13} & b_{14} \\ b_{21} & b_{22} & b_{23} & b_{24} \end{bmatrix}} = \overset{\mathscr{C}}{\begin{bmatrix} c_{11} & c_{12} & c_{13} & c_{14} \\ c_{21} & c_{22} & c_{23} & c_{24} \\ c_{31} & c_{32} & c_{33} & c_{34} \end{bmatrix}}$$

then the rule is

$$c_{il} = \sum_k a_{ik} b_{kl} \tag{17-8}$$

That is, the entry in row i and column l of the product matrix is the sum of the products of row i of matrix $\mathscr{A}$ with column l of matrix $\mathscr{B}$, with the restriction that the column number of a must match the row number of b. The idea of rows "into" columns may be illustrated more explicitly:

$$c_{11} = \sum_k a_{1k} b_{k1} = a_{11} b_{11} + a_{12} b_{21}$$

$$c_{21} = \sum_k a_{2k} b_{k1} = a_{21} b_{11} + a_{22} b_{21}$$

$$c_{34} = \sum_k a_{3k} b_{k4} = a_{31} b_{14} + a_{32} b_{24}$$

and so on. A requirement implicit in Eq. (17-8) is that in order to multiply two matrices, the number of columns of the left-hand one must equal the number of rows of the right-hand one.

Application of the multiplication rule to Eq. (17-7) yields

$$x_2 = c_{11} = [\cos(\tfrac{2}{3}\pi)]\, x_1 + [-\sin(\tfrac{2}{3}\pi)]\, y_1 + 0z_1$$

$$y_2 = c_{21} = [\sin(\tfrac{2}{3}\pi)\, x_1 + [\cos(\tfrac{2}{3}\pi)\, y_1 + 0z_1$$

$$z_2 = c_{31} = 0x_1 + 0y_1 + 1z_1$$

or Eq. (17-4) again.

To return to the C_{2v} case, the matrices that transform the point (x_1, y_1, z_1) into a point (x_2, y_2, z_2) are, for the various symmetry operations,

$$
\begin{array}{cccc}
E & C_2 & \sigma_v\,(xz) & \sigma_v(yz)(\text{or } \sigma_v')
\end{array}
$$

$$
\begin{bmatrix} 1 & 0 & 0 \\ 0 & 1 & 0 \\ 0 & 0 & 1 \end{bmatrix}
\quad
\begin{bmatrix} -1 & 0 & 0 \\ 0 & -1 & 0 \\ 0 & 0 & 1 \end{bmatrix}
\quad
\begin{bmatrix} 1 & 0 & 0 \\ 0 & -1 & 0 \\ 0 & 0 & 1 \end{bmatrix}
\quad
\begin{bmatrix} -1 & 0 & 0 \\ 0 & 1 & 0 \\ 0 & 0 & 1 \end{bmatrix}
$$

These matrices will turn out to be representations of the C_{2v} group. Referring to Table 17-4, $C_2\sigma_v = \sigma_v'$ and multiplication of the matrix for the σ_v' transformation by that for the σ_v one will yield the matrix for the σ_v' transformation.

A further point is that the matrices of Eq. (17-9) can all be blocked into smaller matrices, outside of which the only entries are zeros. A check on the application of Eq. (17-8) wil confirm that the products of such matrices are just the separate products of the blocked-out areas. In this case these are just matrices of dimension one. Where such blocking reduces a set of matrices to ones of lower dimension, then, necessarily, each of these last must be a representation of the group if the original matrices were. As a confirmation, representations A_1, B_1, and B_2 of Table 17-5 correspond to the three blocked-out sets of Eq. (17-9).

The set of matrices for the transformations of a point in the C_{3v} group are

$$
\begin{array}{ccc}
E & C_3{}^1 & C_3{}^2
\end{array}
$$

$$
\begin{bmatrix} 1 & 0 & 0 \\ 0 & 1 & 0 \\ 0 & 0 & 1 \end{bmatrix}
\quad
\begin{bmatrix} \cos(\tfrac{2}{3}\pi) & -\sin(\tfrac{2}{3}\pi) & 0 \\ \sin(\tfrac{2}{3}\pi) & \cos(\tfrac{2}{3}\pi) & 0 \\ 0 & 0 & 1 \end{bmatrix}
\quad
\begin{bmatrix} \cos(\tfrac{4}{3}\pi) & -\sin(\tfrac{4}{3}\pi) & 0 \\ \sin(\tfrac{4}{3}\pi) & \cos(\tfrac{4}{3}\pi) & 0 \\ 0 & 0 & 1 \end{bmatrix}
\tag{17-10}
$$

$$
\begin{array}{ccc}
\sigma_v(xz) & \sigma_v{}' & \sigma_v{}''
\end{array}
$$

$$
\begin{bmatrix} 1 & 0 & 0 \\ 0 & -1 & 0 \\ 0 & 0 & 1 \end{bmatrix}
\quad
\begin{bmatrix} -\sin(\tfrac{1}{6}\pi) & -\cos(\tfrac{1}{6}\pi) & 0 \\ -\cos(\tfrac{1}{6}\pi) & +\sin(\tfrac{1}{6}\pi) & 0 \\ 0 & 0 & 1 \end{bmatrix}
\quad
\begin{bmatrix} -\sin(\tfrac{1}{6}\pi) & \cos(\tfrac{1}{6})\pi) & 0 \\ \cos(\tfrac{1}{6}\pi) & \sin(\tfrac{1}{6}\pi) & 0 \\ 0 & 0 & 1 \end{bmatrix}
$$

Remember that in C_{3v}, σ_v' is a plane rotated 120° from $\sigma_v(xz)$ and σ_v'' is one rotated 240° from $\sigma_v(xz)$. This set of matrices is a representation of the C_{3v} group, that is, the matrices obey the group multiplication table. They may be blocked off into a 2 × 2, or two-dimensional, set and a 1 × 1, or one-dimensional, set of matrices; each set is then also a representation of the group. The 1 × 1 matrices constitute the representation Γ_1 given in Table 17-3.

We have shown that while a set of simple numbers may constitute a representation of a group, one may also have a set of matrices. This last is the more general situation. In fact, an infinite number of matrix representations may be generated for a given symmetry group merely by taking more and more complicated functions through the symmetry transformations. In the C_{2v} group, for example, the molecule H_2O (or any A_2B molecule) could be taken through the transformations by locating the oxygen atom at the origin and assigning x, y and z coordinates to each hydrogen. The original six coordinates would then transform into six others upon application of a symmetry operation, thus producing a 6×6 matrix. The set of these would again be a representation of the C_{2v} group. Such transformation matrices are always square ones, that is, they have the same number of rows as of columns.

It is unnecessarily cumbersome to write down complete matrices for a representation since it turns out that the sum of the diagonal elements sufficiently characterizes a matrix for our purposes. This sum is called the *trace* χ of a matrix. In the case of a 1×1 matrix χ is just the matrix itself, of course. For the set of 2×2 matrices of Eqs. (17-10) the traces are

$$E: 2, \quad C_3^1: \ (-0.5) + (-0.5) = -1 \quad C_3^2: \ (-0.5) + (-0.5) = -1$$
$$\sigma_v(xz): \quad (1) + (-1) = 0 \quad \sigma_v': \quad (-0.5) + (0.5) = 0$$
$$\sigma_v'': \quad (-0.5) + (0.5) = 0 \tag{17-11}$$

An important point is that while the matrices for the symmetry operations *within a given class* may differ in detail, their traces are the *same*. Also, the matrix corresponding to the identity operation must consist of just ones along the diagonal, so its trace *must* be equal to the dimension of the matrix.

It is customary to describe a representation of a group by giving just the traces of the matrices involved. Thus for C_{3v}, the result is as shown in Table 17-6. The symbols A_1, A_2, and so on used to label the representations will be described later. In the case of a representation that is higher than one-dimensional, such as that for E, it is important to remember that matrices are involved. While the traces serve to characterize them, it is the *actual matrices* that must be multiplied in verifying that they obey the group multiplication table.

The concept of a similarity transformation discussed in Section 17-3B may also be applied to square matrices. The reciprocal of a matrix $\mathcal{Q}$, denoted by $\mathcal{Q}^{-1}$, is defined by $\mathcal{Q}^{-1}\mathcal{Q} = \mathcal{E}$, where $\mathcal{E}$ is the identity matrix, that is, a matrix having ones on its diagonal and zeros elsewhere. The similarity transformation on a matrix $\mathcal{A}$ is then written

$$\mathcal{B} = \mathcal{Q}^{-1}\mathcal{A}\mathcal{Q} \tag{17-12}$$

B. The trace of a matrix

C. Reducible and irreducible representations

	E	$2C_3$	$3\sigma_v$
A_1	1	1	1
A_2	1	1	-1
E	2	-1	0

TABLE 17-6
Representations for C_{3v}

A theorem which we will not prove is that the resultant matrix $\mathscr{B}$ must have the same trace as does $\mathscr{A}$ (buy may otherwise be different). A related consequence is that if the set of matrices $\mathscr{A}_1$, $\mathscr{A}_2$, ... constitute a representation of a group, so also will the set $\mathscr{B}_1$, $\mathscr{B}_2$, ...

The importance of the similarity transformation is that it is usually possible to find one which reduces some set $\mathscr{A}_1$, $\mathscr{A}_2$, ... to a new set which has blocked-out sections and therefore decomposes into simpler matrices which are also representations. A representation which can be so simplifed is termed *reducible*. There will remain, however, certain sets of matrices which cannot be simplified further by any possible similarity transformations. Such a set is then an *irreducible representation*, hereafter abbreviated IR, of the group. It is the irreducible representations of a group which are of fundamental importance. The set of traces which a representation has for the various symmetry operations is called the *character* of the representation. Also, the tabular listing of the characters of the IR's of a group is called a *character table*.

There are several important theorems concerning the irreducible representations of a group, which we will state without derivation.

1. The sum of the squares of the dimensions l of the irreducible representations of a group is equal to the order h of the group. Thus

$$\sum_i l_i^2 = h \tag{17-13}$$

For example, the representations A_1, A_2, and E given by Table 17-6 for the C_{3v} group constitute the complete set of irreducible representations (and the display is just the character table for that group). The group is of order 6 (the total number of symmetry operations) and the dimension of each representation is given by the trace for the E symmetry operation. We see that $1^2 + 1^2 + 2^2 = 6$, as required by Eq. (17-13). A corollary is evidently that the sum of the squares of the traces of the irreducible representations for the E operation also be equal to h, or

$$\sum \chi_i^2(E) = h \tag{17-14}$$

2. The number of irreducible representations of a group is equal to the number of classes in the group. The C_{3v} group has three classes and just three irreducible representations.

3. The sum of the squares of the traces of any IR is equal to h. In applying this rule, we sum the squares of the traces over all the symmetry operations R:

$$\sum_R \chi^2(R) = h \tag{17-15}$$

Again referring to the C_{3v} group, we have

$$A_1: \quad 1^2 + (2)(1^2) + 3(1^2) \quad = 6$$

$$A_2: \quad 1^2 + (2)(1^2) + 3(-1)^2 = 6$$

$$E: \quad 2^2 + (2)(-1)^2 + (3)(0) = 6$$

A more general form of Eq. (17-15) is

$$\sum_R \chi_i(R)\,\chi_j(R) = h\delta_{ij} \tag{17-16}$$

where δ_{ij} is known as the *Kronecker* delta and has the value 1 if $i = j$ and zero if $i \neq j$.

4. It will be recalled that if a representation is reducible, there will be some similarity transformation that will convert it into a set of matrices having blocked-out areas. The ultimate case is one in which each set of blocked-out areas constitutes an irreducible representation. In general, not all of the possible irreducible representations will appear, or a given one may show up in two or more sets of blocked-out areas. There is a very useful rule which determines the number of times, a_i, the ith IR occurs in a given reducible representation:

$$a_i = \frac{1}{h}\sum_R \chi(R)\,\chi_i(R) \tag{17-17}$$

The statement is that one sums over all symmetry operations the product of the trace of the IR and that of the reducible one in question. Using the C_{3v} group as an example, Eqs. (17-10) provide an obviously reducible representation (the matrices being already blocked out). The traces for the *whole* matrices are

E	$2C_3$	$3\sigma_v$
3	0	1

Application of Eq. (17-17) gives

$$a_{A_1} = \tfrac{1}{6}[(3)(1) + 2(0)(1) + 3(1)(1)] = 1$$

$$a_{A_2} = \tfrac{1}{6}[(3)(1) + 2(0)(1) + 3(1)(-1)] = 0$$

$$a_E = \tfrac{1}{6}[(3)(2) + 2(0)(-1) + 3(1)(0)] = 1$$

which confirms that the set of matrices (17-10) consists of the irreducible representations $A_1 + E$.

17-5   Atomic orbitals as bases for representations

It was demonstrated in the preceding section that a general point (x_1, y_1, z_1) generates a representation of a symmetry group when carried through the various symmetry operations. The same is true of wave functions of an atom when all symmetry operations are about the nucleus. Since symmetry properties involve the angular rather than the radial aspect of an object, it is only necessary to consider the former portion of a wave function; we will further restrict ourselves to the hydrogen-like set of functions derived in Chapter 16.

An s function is angularly symmetric, and hence is unchanged by any symmetry operation. From Eq. (16-118), the Y_{p_z} function is proportional to $\cos\theta$, which is just the projection of a vector on the z axis; Y_{p_x} is proportional to $\sin \theta \cos \phi$ [Eq. (16-119)] or to the x projection of a vector, and Y_{p_y} is proportional to the y projection [Eq. (16-120)]. If, for example, these vectors are put through the operations of the C_{2v} group, we obtain

$$C_2: p_x = -p_x; \quad p_y = -p_y; \quad p_z = p_z$$

or the transformation matrix

$$\begin{bmatrix} -1 & 0 & 0 \\ 0 & -1 & 0 \\ 0 & 0 & 1 \end{bmatrix}$$

The operation $\sigma_v(xz)$ leads to the matrix

$$\begin{bmatrix} 1 & 0 & 0 \\ 0 & -1 & 0 \\ 0 & 0 & 1 \end{bmatrix}$$

The identity operation E leads, of course, to

$$\begin{bmatrix} 1 & 0 & 0 \\ 0 & 1 & 0 \\ 0 & 0 & 1 \end{bmatrix}$$

The set of matrices for the transformations

$$\begin{bmatrix} p_x \\ p_y \\ p_z \end{bmatrix}$$

is thus blocked out into three one-dimensional matrices, and a check with Table 17-5 identifies the matrix for p_x to be the same as B_1, that for p_y as B_2, and that for p_z as A_1. We therefore say that p_x transforms as does the B_1 irreducible representation, or forms a basis for it, p_y transforms as does B_2, and so on.

Similarly, each d wave function will transform as one or another irreducible representation of a group, or is said to form a basis for it. Thus d_{xz} has the angular nature of the product of the x and z projections of a vector. Since the z projection is unchanged by any of the operations of the C_{2v} group, d_{xz} transforms as does p_x, and thus is a basis for the B_1 irreducible representation. Similarly, d_{yz} transforms in C_{2v} as does p_y. In the cases of d_{xy} and $d_{x^2-y^2}$, however, the C_2 operation leads to no change since these orbitals are symmetric to a $180°$ rotation. In summary, the set of matrices for the transformations

$$\begin{bmatrix} d_{xz} \\ d_{yz} \\ d_{xy} \\ d_{x^2-y^2} \\ d_{z^2} \end{bmatrix}$$

are:

$$
C_2 \qquad\qquad\qquad \sigma_{xz} \qquad\qquad\qquad \sigma_{yz}
$$

$$
\begin{bmatrix}
-1 & 0 & 0 & 0 & 0 \\
0 & -1 & 0 & 0 & 0 \\
0 & 0 & 1 & 0 & 0 \\
0 & 0 & 0 & 1 & 0 \\
0 & 0 & 0 & 0 & 1
\end{bmatrix}
\begin{bmatrix}
1 & 0 & 0 & 0 & 0 \\
0 & -1 & 0 & 0 & 0 \\
0 & 0 & -1 & 0 & 0 \\
0 & 0 & 0 & 1 & 0 \\
0 & 0 & 0 & 0 & 1
\end{bmatrix}
\begin{bmatrix}
-1 & 0 & 0 & 0 & 0 \\
0 & 1 & 0 & 0 & 0 \\
0 & 0 & -1 & 0 & 0 \\
0 & 0 & 0 & 1 & 0 \\
0 & 0 & 0 & 0 & 1
\end{bmatrix}
$$

One can sum the diagonals and apply Eq. (17-17), but in this case it is easy to see by inspection that d_{z^2} and $d_{x^2-y^2}$ each transform as does the A_1 irreducible representation, while d_{xz}, d_{yx}, and d_{xy} transform as do the B_1, B_2, and A_2 irreducible representations, respectively. The conclusion is confirmed by going to Table 17-7.

17-6   Character tables

It is customary to assemble the basic information concerning each symmetry point group in what is called a character table. Several such tables are given in Table 17-7. Their organization is as follows. The first column designates each irreducible representation, the naming scheme being outlined shortly. The next group of columns lists the traces of each representation for each class of symmetry operation. The next to last column lists the types of vectors which form a basis for each representation. Thus for the C_{2v} group, the row designated A_1 shows that the z projection of a vector is appropriate, which means that the p_z wave function will form a basis for this representation. The last column lists products of vectors which are appropriate. The designation yz after the B_2 representation then means that d_{yz} orbitals will form a basis for the B_2 representation, and so on. The entry in the last column for the A_1 irreducible representation of the C_{2v} group is x^2, y^2, z^2, meaning that each separately transforms as does A_1; this means that any linear combination such as $x^2 - y^2$ also transforms as does A_1. There are occasional entries R_x, R_y, or R_z in the next to last column. These refer to a circular vector whose plane is perpendicular to the named axis. Although such vectors will not be of direct interest to us, they also form a basis for a representation.

It should be mentioned that several complexities have been side-stepped in the presentation of symmetry groups and their character tables. A number of types have been omitted from Table 17-7. The omissions are not of vital importance here, however; the omitted aspects have more the nature of complications than of new principles.

Returning to the first column, the naming scheme is one developed by R. Mulliken. Irreducible representations of a group may be one-, two- or three-dimensional (we will not encounter any of higher dimension), and may also possess various symmetry properties themselves.

One-dimensional representations are named A if they are symmetric to rotation by $2\pi/n$ about the principal or C_n axis, and B if antisymmetric to such rotation. They have subscripts 1 or 2 to designate whether they are symmetric or antisymmetric to rotation about a secondary C_2 axis, or if this is missing, to a vertical plane of symmetry.

TABLE 17-7
Character tables

C_2

C_2 :	E	C_2		
A	1	1	z, R_z	x^2, y^2, z^2, xy
B	1	-1	x, y, R_x, R_y	yz, xz

D_n

D_2 :	E	$C_2(z)$	$C_2(y)$	$C_2(x)$		
A	1	1	1	1	—	x^2, y^2, z^2
B_1	1	1	-1	-1	z, R_z	xy
B_2	1	-1	1	-1	y, R_y	xz
B_3	1	-1	-1	1	x, R_x	yz

D_3 :	E	$2C_3$	$3C_2$		
A_1	1	1	1	—	$x^2 + y^2, z^2$
A_2	1	1	-1	z, R_z	—
E	2	-1	0	$(x, y)(R_x, R_y)$	$(x^2 - y^2, xy)(xz, yz)$

D_4 :	E	$2C_4$	$C_2 (= C_4^2)$	$2C_2'$	$2C_2''$		
A_1	1	1	1	1	1	—	$x^2 + y^2, z^2$
A_2	1	1	1	-1	-1	z, R_z	—
B_1	1	-1	1	1	-1	—	$x^2 - y^2$
B_2	1	-1	1	-1	1	—	xy
E	2	0	-2	0	0	$(x, y)(R_x, R_y)$	(xz, yz)

D_5 :	E	$2C_5$	$2C_5^2$	$5C_2$		
A_1	1	1	1	1	—	$x^2 + y^2, z^2$
A_2	1	1	1	-1	z, R_z	—
E_1	2	$2\cos 72°$	$2\cos 144°$	0	$(x, y)(R_x, R_y)$	(xy, yz)
E_2	2	$2\cos 144°$	$2\cos 72°$	0	—	$(x^2 - y^2, xy)$

D_6 :	E	$2C_6$	$2C_3$	C_2	$3C_2'$	$3C_2''$		
A_1	1	1	1	1	1	1	—	$x^2 + y^2, z^2$
A_2	1	1	1	1	-1	-1	z, R_z	—
B_1	1	-1	1	-1	1	-1	—	—

B_2	1	−1	1	−1	−1	1	—	(xy, xy)
E_1	2	1	−1	−2	0	0	$(x, y)(R_x, R_y)$	—
E_2	2	−1	−1	2	0	0	—	$(x^2 − y^2, xy)$

C_{nv}

C_{2v}:	E	C_2	$\sigma_v(xz)$	$\sigma_v'(yz)$		
A_1	1	1	1	1	z	x^2, y^2, z^2
A_2	1	1	−1	−1	R_z	xy
B_1	1	−1	1	−1	x, R_y	xz
B_2	1	−1	−1	1	y, R_x	yz

C_{3v}:	E	$2C_3$	$3\sigma_v$		
A_1	1	1	1	z	$x^2 + y^2, z^2$
A_2	1	1	−1	R_z	—
E	2	−1	0	$(x, y)(R_x, R_y)$	$(x^2 − y^2, xy)(xz, yz)$

C_{4v}:	E	$2C_4$	C_2	$2\sigma_v$	$2\sigma_d$		
A_1	1	1	1	1	1	z	$x^2 + y^2, z^2$
A_2	1	1	1	−1	−1	R_z	—
B_1	1	−1	1	1	−1		$x^2 − y^2$
B_2	1	−1	1	−1	1		xy
E	2	0	−2	0	0	$(x, y)(R_x, R_y)$	(xz, yz)

C_{5v}:	E	$2C_5$	$2C_5^2$	$5\sigma_v$		
A_1	1	1	1	1	z	$x^2 + y^2, z^2$
A_2	1	1	1	−1	R_z	—
E_1	2	$2\cos 72°$	$2\cos 144°$	0	$(x, y)(R_x, R_y)$	(xz, yz)
E_2	2	$2\cos 144°$	$2\cos 72°$	0		$(x^2 − y^2, xy)$

C_{6v}:	E	$2C_6$	$2C_3$	C_2	$3\sigma_v$	$3\sigma_d$		
A_1	1	1	1	1	1	1	z	$x^2 + y^2, z^2$
A_2	1	1	1	1	−1	−1	R_z	—
B_1	1	−1	1	−1	1	−1		—
B_2	1	−1	1	−1	−1	1		—
E_1	2	1	−1	−2	0	0	$(x, y)(R_x, R_y)$	(xz, yz)
E_2	2	−1	−1	2	0	0		$(x^2 − y^2, xy)$

(cont.)

TABLE 17-7 (cont.)

C_{nh}

C_{2h}:	E	C_2	i	σ_h		
A_g	1	1	1	1	R_z	x^2, y^2, z^2, xy
B_g	1	-1	1	-1	R_x, R_y	xz, yz
A_u	1	1	-1	-1	z	—
B_u	1	-1	-1	1	x, y	—

D_{nh}

D_{2h}:	E	$C_2(z)$	$C_2(y)$	$C_2(x)$	i	$\sigma(xy)$	$\sigma(xz)$	$\sigma(yz)$		
A_g	1	1	1	1	1	1	1	1	—	x^2, y^2, z^2
B_{1g}	1	1	-1	-1	1	1	-1	-1	R_z	xy
B_{2g}	1	-1	1	-1	1	-1	1	-1	R_y	xz
B_{3g}	1	-1	-1	1	1	-1	-1	1	R_x	yz
A_u	1	1	1	1	-1	-1	-1	-1	—	—
B_{1u}	1	1	-1	-1	-1	-1	1	1	z	—
B_{2u}	1	-1	1	-1	-1	1	-1	1	y	—
B_{3u}	1	-1	-1	1	-1	1	1	-1	x	—

D_{3h}:	E	$2C_3$	$3C_2$	σ_h	$2S_3$	$3\sigma_v$		
A_1'	1	1	1	1	1	1	—	$x^2 + y^2, z^2$
A_2'	1	1	-1	1	1	-1	R_z	—
E'	2	-1	0	2	-1	0	(x, y)	$(x^2 - y^2, xy)$
A_1''	1	1	1	-1	-1	-1	—	—
A_2''	1	1	-1	-1	-1	1	z	—
E''	2	-1	0	-2	1	0	(R_x, R_y)	(xz, yz)

D_{4h}:	E	$2C_4$	C_2	$2C_2'$	$2C_2''$	i	$2S_4$	σ_h	$2\sigma_v$	$2\sigma_d$		
A_{1g}	1	1	1	1	1	1	1	1	1	1	—	$x^2 + y^2, z^2$
A_{2g}	1	1	1	-1	-1	1	1	1	-1	-1	R_z	—
B_{1g}	1	-1	1	1	-1	1	-1	1	1	-1	—	$x^2 - y^2$
B_{2g}	1	-1	1	-1	1	1	-1	1	-1	1	—	xy
E_g	2	0	-2	0	0	2	0	-2	0	0	(R_x, R_y)	(xz, yz)

(continued — D_{4h})

	E	2C4	C2	2C2′	2C2″	i	2S4	σh	2σv	2σd		
A_{1u}	1	1	1	1	1	-1	-1	-1	-1	-1	—	—
A_{2u}	1	1	1	-1	-1	-1	-1	-1	1	1	z	—
B_{1u}	1	-1	1	1	-1	-1	1	-1	-1	1	—	—
B_{2u}	1	-1	1	-1	1	-1	1	-1	1	-1	—	—
E_u	2	0	-2	0	0	-2	0	2	0	0	(x, y)	—

D_{5h}	E	2C5	2C5²	5C2	σh	2S5	2S5³	5σv		
A_1'	1	1	1	1	1	1	1	1	—	$x^2 + y^2, z^2$
A_2'	1	1	1	-1	1	1	1	-1	R_z	—
E_1'	2	2 cos 72°	2 cos 144°	0	2	2 cos 72°	2 cos 144°	0	(x, y)	—
E_2'	2	2 cos 144°	2 cos 72°	0	2	2 cos 144°	2 cos 72°	0	—	$(x^2 - y^2, xy)$
A_1''	1	1	1	1	-1	-1	-1	-1	—	—
A_2''	1	1	1	-1	-1	-1	-1	1	z	—
E_1''	2	2 cos 72°	2 cos 144°	0	-2	-2 cos 72°	-2 cos 144°	0	(R_x, R_y)	(xz, yz)
E_2''	2	2 cos 144°	2 cos 72°	0	-2	-2 cos 144°	-2 cos 72°	0	—	—

D_{6h}	E	2C6	2C3	C2	3C2′	3C2″	i	2S3	2S6	σh	3σd	3σv		
A_{1g}	1	1	1	1	1	1	1	1	1	1	1	1	—	$x^2 + y^2, z^2$
A_{2g}	1	1	1	1	-1	-1	1	1	1	1	-1	-1	R_z	—
B_{1g}	1	-1	1	-1	1	-1	1	-1	1	-1	1	-1	—	—
B_{2g}	1	-1	1	-1	-1	1	1	-1	1	-1	-1	1	—	—
E_{1g}	2	1	-1	-2	0	0	2	1	-1	-2	0	0	(R_x, R_y)	(xz, yz)
E_{2g}	2	-1	-1	2	0	0	2	-1	-1	2	0	0	—	$(x^2 - y^2, xy)$
A_{1u}	1	1	1	1	1	1	-1	-1	-1	-1	-1	-1	—	—
A_{2u}	1	1	1	1	-1	-1	-1	-1	-1	-1	1	1	z	—
B_{1u}	1	-1	1	-1	1	-1	-1	1	-1	1	-1	1	—	—
B_{2u}	1	-1	1	-1	-1	1	-1	1	-1	1	1	-1	—	—
E_{1u}	2	1	-1	-2	0	0	-2	-1	1	2	0	0	(x, y)	—
E_{2u}	2	-1	-1	2	0	0	-2	1	1	-2	0	0	—	—

D_{nd}

D_{2d}	E	2S4	C2	C2′	2σd		
A_1	1	1	1	1	1	—	$x^2 + y^2, z^2$
A_2	1	1	1	-1	-1	R_z	—
B_1	1	-1	1	1	-1	—	$x^2 - y^2$
B_2	1	-1	1	-1	1	z	xy
E	2	0	-2	0	0	$(x, y)(R_x, R_y)$	(xz, yz)

(cont.)

TABLE 17-7 (cont.)

D_{3d}	E	$2C_3$	$3C_2$	i	$2S_6$	$3\sigma_d$		
A_{1g}	1	1	1	1	1	1	—	$x^2+y^2,\ z^2$
A_{2g}	1	1	-1	1	1	-1	R_z	—
E_g	2	-1	0	2	-1	0	(R_x, R_y)	$(x^2-y^2, xy), (xz, yz)$
A_{1u}	1	1	1	-1	-1	-1	—	—
A_{2u}	1	1	-1	-1	-1	1	z	—
E_u	2	-1	0	-2	1	0	(x, y)	—

D_{4d}	E	$2S_8$	$2C_4$	$2S_8^3$	C_2	$4C_2'$	$4\sigma_d$		
A_1	1	1	1	1	1	1	1	—	$x^2+y^2,\ z^2$
A_2	1	1	1	1	1	-1	-1	R_z	—
B_1	1	-1	1	-1	1	1	-1	—	—
B_2	1	-1	1	-1	1	-1	1	z	—
E_1	2	$\sqrt{2}$	0	$-\sqrt{2}$	-2	0	0	(x, y)	—
E_2	2	0	-2	0	2	0	0	—	(x^2-y^2, xy)
E_3	2	$-\sqrt{2}$	0	$\sqrt{2}$	-2	0	0	(R_x, R_y)	(xz, yz)

D_{5d}	E	$2C_5$	$2C_5^2$	$5C_2$	i	$2S_{10}^3$	$2S_{10}$	$5\sigma_d$		
A_{1g}	1	1	1	1	1	1	1	1	—	$x^2+y^2,\ z^2$
A_{2g}	1	1	1	-1	1	1	1	-1	R_z	—
E_{1g}	2	$2\cos 72°$	$2\cos 144°$	0	2	$2\cos 72°$	$2\cos 144°$	0	(R_z, R_y)	(xz, yz)
E_{2g}	2	$2\cos 144°$	$2\cos 72°$	0	2	$2\cos 144°$	$2\cos 72°$	0	—	(x^2-y^2, xy)
A_{1u}	1	1	1	1	-1	-1	-1	-1	—	—
A_{2u}	1	1	1	-1	-1	-1	-1	1	z	—
E_{1u}	2	$2\cos 72°$	$2\cos 144°$	0	-2	$-2\cos 72°$	$-2\cos 144°$	0	(x, y)	—
E_{2u}	2	$2\cos 144°$	$2\cos 72°$	0	-2	$-2\cos 144°$	$-2\cos 72°$	0	—	—

D_{6d}	E	$2S_{12}$	$2C_6$	$2S_{12}^5$	$2C_3$	$2S_4$	C_2	$6C_2'$	$6\sigma_d$		
A_1	1	1	1	1	1	1	1	1	1	—	$x^2+y^2,\ z^2$
A_2	1	1	1	1	1	1	1	-1	-1	R_z	—
B_1	1	-1	1	-1	1	-1	1	1	-1	—	—
B_2	1	-1	1	-1	1	-1	1	-1	1	z	—

E_1	2	$\sqrt{3}$	1	0	-1	$-\sqrt{3}$	-2	0	0	(x, y)		$(x^2 - y^2, xy)$
E_2	2	1	-1	-2	-1	1	2	0	0			
E_3	2	0	-2	0	2	0	-2	0	0			
E_4	2	-1	-1	2	-1	-1	2	0	0			
E_5	2	$-\sqrt{3}$	1	0	-1	$\sqrt{3}$	-2	0	0	(R_x, R_y)		(xy, yz)

The cubic groups

T_d :	E	$8C_3$	$3C_2$	$6S_4$	$6\sigma_d$			
A_1	1	1	1	1	1			$x^2 + y^2 + z^2$
A_2	1	1	1	-1	-1			
E	2	-1	2	0	0			$(2z^2 - x^2 - y^2, x^2 - y^2)$
T_1	3	0	-1	1	-1	(R_x, R_y, R_z)		
T_2	3	0	-1	-1	1	(x, y, z)		(xy, xz, yz)

O_h :	E	$8C_3$	$6C_2$	$6C_4$	$3C_2$ $(=C_4^2)$	i	$6S_4$	$8S_6$	$3\sigma_h$	$6\sigma_d$			
A_{1g}	1	1	1	1	1	1	1	1	1	1			$x^2 + y^2 + z^2$
A_{2g}	1	1	-1	-1	1	1	-1	1	1	-1			
E_g	2	-1	0	0	2	2	0	-1	2	0			$(2z^2 - x^2 - y^2, x^2 - y^2)$
T_{1g}	3	0	-1	1	-1	3	1	0	-1	-1	(R_x, R_y, R_z)		
T_{2g}	3	0	1	-1	-1	3	-1	0	-1	1			(xz, yz, xy)
A_{1u}	1	1	1	1	1	-1	-1	-1	-1	-1			
A_{2u}	1	1	-1	-1	1	-1	1	-1	-1	1			
E_u	2	-1	0	0	2	-2	0	1	-2	0			
T_{1u}	3	0	-1	1	-1	-3	-1	0	1	1	(x, y, z)		
T_{2u}	3	0	1	-1	-1	-3	1	0	1	-1			

For example, the A_1 representation in C_{2v} [Table 17-5] is generated by a vector on the z axis, and such a vector is unchanged by C_2 and σ_v. The A_2 representation is generated by a circular vector in the xy plane, centered at the origin; it is unchanged by rotation but inverted on reflection by σ_v. The B_1 representation is like a vector on the x axis; it is inverted by C_2 but unchanged by σ_v (taken to be the xz plane), and B_2 is like a vector on the y axis and is inverted by both C_2 and σ_v.

In more complicated symmetry groups primes and double primes are attached to all letters to indicate symmetry or antisymmetry with respect to σ_h. Also the subscripts g and u (from the German *gerade* and *ungerade*) may be present to show whether the representation is symmetric or antisymmetric with respect to inversion.

17-7 Bonds as bases for reducible representations

We have pointed out that the hydrogen-like wave functions have trigonometric factors which correspond to various projections of a vector and that such projections if carried through the symmetry operations of a group, generate a representation. We may also take a set of vectors through the symmetry operations of a group, thus generating matrices which will again be representations, usually reducible ones. The detailing of such matrices is lengthy and, fortunately, it turns out that a simple rule allows the determination of the *trace* of the matrix for each symmetry operation.

The rule is that any vector of the set which remains unchanged by the symmetry operation contributes unity to the trace. A vector which is unchanged in position but reversed in sign contributes -1. These rules will always work for the cases considered here, but it should be mentioned that they have been simplified. If, for example, a set of vectors is rotated, each into the other's position, by a symmetry operation, then it may be necessary to assign appropriate fractional numbers in adding up the trace of the matrix.

A. Sigma bonds

The term *sigma bond* is conventionally used to describe a type of bond between two atoms which results from the overlap of an atomic orbital of each atom to give a cylindrically symmetric overlap. The further requirement is that the orbitals and hence their region of overlap be a figure of revolution about the line between the two nuclei. As illustrated in Fig. 17-11, two s orbitals can form a sigma bond, and if the atoms A and B lie on the x axis, then the sigma-bonding combination $s_A p_{x,B}$ is possible, as well as $p_{x,A} p_{x,B}$, $s_A d_{x^2-y^2,B}$, and $p_{x,A} d_{x^2-y^2,B}$.

The orbital pictures shown in the figure are just the polar plots of the angular portions—these must be multiplied by the radial function $R(r)$ to give the corresponding electron density. As two atoms are brought together, the product of the electron amplitudes determines what is called the overlap; the mathematical operation is the evaluation of the integral $\int \psi_1 \psi_2 \, d\tau$ for the atoms at some given distance of separation, where $d\tau$ is the element of volume (note Section 16-10). This integral normally provides a guide to the degree of bonding that is expected. For two atoms to bond, it is first necessary that their respective orbitals overlap, that is, have appreciable values in some common region of space. There must also be a reinforcement of the two orbital wave functions; their phase relation must be suitable. First, it is conventional to

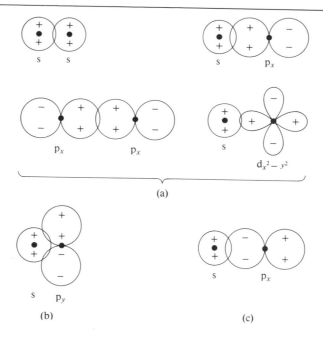

FIGURE 17-11
Arrangements of orbitals showing sigma (a) bonding, (b) nonbonding, and (c) antibonding for the case of an A—B molecule.

indicate the phase relationships between the lobes of the angular portion of a given orbital by plus and minus signs. These signs are usually chosen so that if the orbitals of two atoms overlap in regions for which their sign is the same, then a bonding interaction is indicated—the potential energy of the atoms close together should be less than when they are separated. Conversely, if the overlapping orbitals are of opposite sign, the potential energy of the system is higher than for the separated atoms. The two situations are referred to as *bonding* and *antibonding*, respectively.

It may happen alternatively that the overlap integral is small or zero either because the overlap is small or because positive and negative domains of the integral cancel. The usual consequence is that neither bonding nor repulsion is expected and the situation is called one of *nonbonding*. The three cases are illustrated in Fig. 17-11.

It should be emphasized that the overlap integral does not *in itself* give the degree of bonding; as discussed in Section 18-2, the actual integrals involve the Hamiltonian of the system. However, the preceding qualitative use of the overlap idea will usually work—it is a convenient quantum mechanical rule of thumb.

By using group theory, it is possible to determine which pairs of wave functions can give a nonzero overlap integral. The approach is the following. The symmetry properties of the set of sigma bonds form a representation of the symmetry group of the molecule and one which will contain various irreducible representations of the group, as determined by the use of Eq. (17-17). The wave functions which are bases for these irreducible representations will have the necessary symmetry properties for the overlap integral to be nonzero and either positive or negative, depending on the signs given to the orbital lobes.

As illustrated in Fig. 17-12, we draw a set of vectors corresponding to the

FIGURE 17-12
Sigma bond vectors.

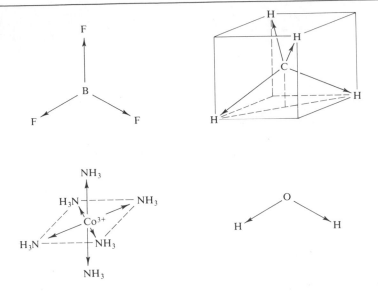

sigma bonds of the molecule. The trace of the matrix generated by each symmetry operation is then determined by applying the rule described at the beginning of this section. Equation (17-17) gives the irreducible representations contained by, or "spanned" by, the consultation of the character table for the group determines which hydrogen-like orbitals have the right symmetry for bonding.

The following examples should help to clarify matters. The planar BF_3 molecule has D_{3h} symmetry. Consultation of Table 17-7 gives the symmetry classes as E, $2C_3$, $3C_2$, σ_h, $2S_3$, and $3\sigma_v$. The operation E leaves all three bond vectors unchanged, C_3 changes all of them, C_2 leaves one unchanged, σ_h leaves all unchanged, S_3 changes all, and σ_v leaves one unchanged. The character for the reducible representation $\Gamma_{3\sigma}$ generated by the set of bonds is therefore

	E	$2C_3$	$3C_2$	σ_h	$2S_3$	$3\sigma_v$
$\Gamma_{3\sigma}$	3	0	1	3	0	1

Application of Eq. (17-17) gives

$$a_{A_1'} = \tfrac{1}{12}\,[(3)(1) + 2(0)(1) + 3(1)(1) + (3)(1) + 2(0)(1) + 3(1)(1)] = 1$$

The A_1' irreducible representation is thus contained once. On repeating the calculation for each irreducible representation of the D_{3h} group in turn, it is found that the coefficient is zero for all except E', for which it is again unity. Thus $\Gamma_{3\sigma}$ spans the $A_1' + E'$ irreducible representations. Consultation of the last two columns of the character table shows that the algebraic functions of the correct symmetry are $(x^2 + y^2)$ or z^2 for A_1' and (x,y) or $(x^2 - y^2, xy)$ for E'. The E' irreducible representation is a two-dimensional one and the functions *must* be taken together. The function $(x^2 + y^2)$ corresponds to a circle, and since the molecule is planar, this is the projection of an s orbital. Thus a combination s, p_x, p_y would do for bonding. Another possibility is s,

$d_{x^2-y^2}$, d_{xy}. The choice between these two combinations is made on the basis of whether it is the p or d orbitals that are the more stable. This is easy in the case of BF_3 since the outer electrons of boron occupy 2p orbitals, and the lowest d ones, 3d, would be very high in energy. In a less extreme case it may be appropriate to use a linear combination of the two sets of orbitals.

Notice that in this approach the fluorine atoms are really irrelevant. We are determining what orbital combinations for the central atom will have the proper symmetry to form sigma bonds in the observed bonding directions.

A second example is CH_4 (or any tetrahedral molecule), for which the point group is T_d. Turning to the character table for this group, we find

	E	$8C_3$	$3C_2$	$6S_4$	$6\sigma_d$
$\Gamma_{4\sigma}$	4	1	0	0	2

Application of Eq. (17-17) gives $\Gamma_{4\sigma} = A_1 + T_2$. The IR A_1 corresponds to $x^2 + y^2 + z^2$, or to a sphere, and hence to an s orbital. The IR T_2 is generated by (x,y,z) or (xy,xz,yz). The bonding can thus be $sp_xp_yp_z$ or $sd_{xy}d_{xz}d_{yz}$; it can be some combination of both types if the p and d orbitals are similar in energy. The first choice is the obvious one in the case of CH_4, but the tetrahedral molecule $CoCl_4^{2-}$ might use the $sd_{xy}d_{xz}d_{yz}$ combination, in view of the availability of cobalt d orbitals. Decisions of this type rest on quantitative calculations.

For a molecule having O_h symmetry, such as $Co(NH_3)_6^{3+}$ (considering only the Co and N atoms), the result is

	E	$8C_3$	$6C_2$	$6C_4$	$3C_2$	i	$6S_4$	$8S_6$	$3\sigma_h$	$6\sigma_d$
$\Gamma_{6\sigma}$	6	0	0	2	2	0	0	0	4	2

We now find $\Gamma_{6\sigma} = A_{1g} + T_{1u} + E_g$, which corresponds to the orbital set $sd_{z^2}d_{x^2-y^2}\,p_xp_yp_z$ (often reported as just d^2sp^3). A linear combination of these orbitals will indeed generate a set of new orbitals having lobes pointing to the corners of an octahedron.

As a final example, consider the molecule H_2O. This belongs to the C_{2v} group, and the procedure yields

	E	C_2	$\sigma_v(xz)$	$\sigma_v'(yz)$
$\Gamma_{2\sigma}$	2	0	2	0

with the conclusion that $\Gamma_{2\sigma} = A_1 + B_1$. From the character table the bonding might involve some combination of s, p_x, and p_z. This answer does not seem very helpful, that is, it does not suggest why the experimental bond angle in water is 104°27'. It should be recognized at this point that the oxygen atom has two additional pairs of electrons; these are not used in bonding in H_2O but are used, for example, in the hydrogen-bonded structures present in liquid water and in ice (see Section 8-CN-1). Since the treatment centers on the oxygen atom, it is more realistic to view it as having four more or less tetrahedrally disposed pairs of electrons, of which two happen to be shared in bonds. The point is that the treatment described here works best for coordinatively saturated atoms; a useful alternative approach for an AB_2 type of molecule is given in Section 18-6.

B. Hybrid bonds

The discussion above shows how one may apply group theory to determine what set or sets of atomic orbitals have the symmetry properties corresponding to the bonds formed by an atom. We now pick up the actual expressions for the $Y(\theta,\phi)$ functions given in Section 16-8 to show how the appropriate linear combinations of atomic orbitals may be formulated.

For a case such as BF_3, symmetry arguments tell us that the set sp_xp_y conforms to the situation of three coplanar bonds 120° apart. The actual algebraic expressions turn out to be the following:

$$Y_1 = \frac{1}{\sqrt{3}} Y_s + \sqrt{\frac{2}{3}} Y_{Px} \qquad Y_2 = \frac{1}{\sqrt{3}} Y_s - \frac{1}{\sqrt{6}} Y_{Px} + \frac{1}{\sqrt{2}} Y_{Py}$$

$$Y_3 = \frac{1}{\sqrt{3}} Y_s - \frac{1}{\sqrt{6}} Y_{Px} - \frac{1}{\sqrt{2}} Y_{Py} \tag{17-18}$$

A plot of these functions is shown in Fig. 17-13(a). Not only do they have the desired directional properties, but also the coefficients of Eq. (17-18) are such that Y_1, Y_2, and Y_3 are normalized and mutually orthogonal. Further, the set $Y_1^2 + Y_2^2 + Y_3^2$ is identical to the sum $Y_s^2 + Y_{Px}^2 + Y_{Py}^2$ and equal to a

FIGURE 17-13
Hybrid orbitals. (a) sp^2 and (b) sp.

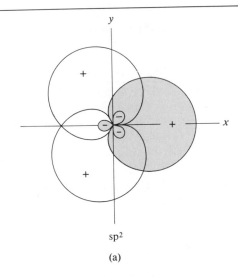

sp^2

(a)

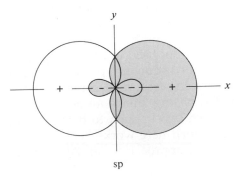

sp

(b)

constant. That is, Y_s^2 is circularly symmetric and so is $Y_{p_x}^2 + Y_{p_y}^2$, so that the sum of electron densities in the xy plane is independent of direction. The new set $Y_1 + Y_2 + Y_3$ represents an alternative way of dividing the circle. Strictly speaking, one should use the full wave functions ψ in constructing such new sets of orbitals, but the coefficients remain the same if the $R(r)$ functions are the same or can be adjusted to nearly compensate if they are not.

The combinations of s and p orbitals of Eq. (17-18) are known as *hybrid orbitals*, after L. Pauling, or as sp^2 hybrids.

Again, symmetry considerations tell us that the set of orbitals s, p_x, p_y, p_z has the symmetry of a set of tetrahedrally oriented bonds. The actual combinations of Y functions are:

$$Y_1 = \tfrac{1}{2}(Y_s + Y_{p_x} + Y_{p_y} + Y_{p_z})$$

$$Y_2 = \tfrac{1}{2}(Y_s + Y_{p_x} - Y_{p_y} - Y_{p_z})$$

$$Y_3 = \tfrac{1}{2}(Y_s - Y_{p_x} + Y_{p_y} - Y_{p_z})$$

$$Y_4 = \tfrac{1}{2}(Y_s - Y_{p_x} - Y_{p_y} + Y_{p_z})$$

(17-19)

This new set is normalized and mutually orthogonal. Again, one usually makes the assumption that $R(r)$ is the same for each wave function, and so Eqs. (17-19) can be written in terms of the complete ψ functions rather than just the angular parts. The $Y_1, \ldots, Y_4$ set is known as sp^3 hybrid orbitals.

A third and very simple case of hybrid bond formation is the sp hybrid, obtained by a linear combination of an s orbital and a p orbital. For example, the set

$$Y_1 = \sqrt{\tfrac{1}{2}}\,(Y_s + Y_{p_x}) \qquad Y_2 = \sqrt{\tfrac{1}{2}}\,(Y_s - Y_{p_x})$$

(17-20)

gives lobes lying on the x axis, as shown in Fig. 17-13(b).

The *bond strength* of an atomic orbital is taken to be proportional to the maximum value of the Y function. The usual convention is to assign Y_s the value unity, and comparison of Eqs. (16-117) and (16-118) gives the bond strength of a Y_p orbital as $\sqrt{3}$ or 1.732. Corresponding bond strength numbers may be obtained for hybrid orbitals. As an example, the maximum extension of Y_1 for the sp hybrid orbital is

$$\frac{1}{\sqrt{2}}\left[\frac{1}{2\sqrt{\pi}} + \frac{1}{2\sqrt{\pi}}\sqrt{3}\right] = \frac{1}{2\sqrt{\pi}}\left[\frac{1 + \sqrt{3}}{\sqrt{2}}\right]$$

Since Y_s is just $1/(2\sqrt{\pi})$, the maximum extension of Y_{sp} relative to this is $(1 + \sqrt{3})/\sqrt{2}$ or 1.932; this is its bond strength. Table 17-8 lists some common types of hybrid orbitals and their bond strengths. The bond strength concept is a useful one for qualitative considerations—the greater the extension of a Y function, the stronger the bond it is likely to be able to form. The idea cannot easily be made quantitative, however, since the $R(r)$ functions are not included and, of course, because the degree of spatial overlap of orbitals forming a bond is itself only a qualitative measure of actual bonding energy.

		Bond	Orbital	Bond
TABLE 17-8 Hybrid orbitals and their bond strengths	Symmetry	angle	combination	strength
	Linear	180°	sp	1.932
	Triangular	120°	sp²	1.991
	Tetrahedral	109° 28'	sp³	2.000
	Square	90°	dsp²	2.694
	Octahedral	90°	d² sp³	2.923

C. Pi bonds

The term *pi bond* applies to a bonding overlap where there is a node along the bonding axis, that is, the electron density is zero along the line of the sigma bond. Figure 17-14 illustrates some typical pi bonding orbital combinations, assuming the bond to lie on the x axis. The figures suggest, and calculation confirms, that the value of a pi bonding overlap integral is in general smaller than that of a sigma bonding one (see Section 18-2). As a consequence, pi bonding is considered not so much as providing the primary bonding holding a molecule together as supplementing an already present sigma bond.

The symmetry approach may again be used. As shown in Fig. 17-15, pi bonds must be represented by vectors since there is a plus-to-minus direction. Usually, two such vectors at right angles to each other must be considered. Various symmetry operations of the molecular point group may now leave pi bond vectors unchanged in position but reversed in sign; as noted at the beginning of this section, the rule is then that -1 is contributed to the trace of the representation generated by the pi bonding set. Also, one may consider each set of mutually perpendicular vectors separately, provided that the two sets are not mixed by any symmetry operation.

FIGURE 17-14
Arrangements of orbitals showing pi bonding for the case of an A—B molecule.

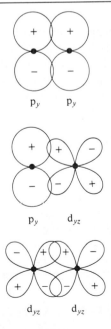

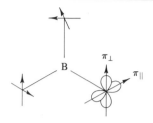

FIGURE 17-15
Pi bonding vectors.

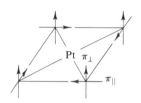

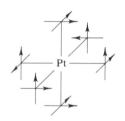

For example, a D_{3h} molecule such as BF_3 or NO_3^- generates the traces shown in Table 17-9, where $\Gamma_{3\pi\parallel}$ and $\Gamma_{3\pi\perp}$ denote the sets of pi bonding vectors parallel to and perpendicular to the molecular plane, respectively. Application of Eq. (17-17) leads to the result

$$\Gamma_{3\pi\perp} = A_2'' + E'' \qquad \Gamma_{3\pi\parallel} = A_2' + E'$$

which gives $p_x p_y$ as suitable bonding orbitals. We exclude d_{xz}, d_{yz}, listed for E'', as energetically unfavorable, and the other irreducible representations do not correspond to any hydrogen-like orbitals. However, $p_x p_y$ is also suitable for sigma bonding, and we assume that this type of overlap will take precedence over the pi bonding type. It appears that molecules such as BF_3 have mainly sigma bonding in the molecular plane. However, p_z belongs to A_2'' so $\pi_\perp$ bonding is possible.

Turning next to the D_{4h} molecule $PtCl_4^{2-}$, we find the situation shown in Table 17-10. Application of Eq. (17-17) yields

	E	$2C_3$	$3C_2$	σ_h	$2S_3$	$3\sigma_v$	
$\Gamma_{3\pi\perp}$	3	0	-1	-3	0	1	**TABLE 17-9**
$\Gamma_{3\pi\parallel}$	3	0	-1	3	0	-1	

TABLE 17-10	E	$2C_4$	C_2	$2C_2'$	$2C_2''$	i	$2S_4$	σ_h	$2\sigma_v$	$2\sigma_d$
$\Gamma_{4\sigma}$	4	0	0	2	0	0	0	4	2	0
$\Gamma_{4\pi\parallel}$	4	0	0	-2	0	0	0	4	-2	0
$\Gamma_{4\pi\perp}$	4	0	0	-2	0	0	0	-4	2	0

$$\Gamma_{4\sigma} = A_{1g} + B_{1g} + E_u \quad \text{or} \quad d_{z^2}, d_{x^2-y^2}, (p_x p_y)$$

$$\Gamma_{4\pi\parallel} = A_{2g} + B_{2g} + E_u \quad \text{or} \quad d_{xy}, (p_x p_y)$$

$$\Gamma_{4\pi\perp} = A_{2u} + B_{2u} + E_g \quad \text{or} \quad p_z, (d_{xz} d_{yz})$$

The sigma bonds are thus dsp^2 in type, leaving the d_{ij} set and p_z available for some pi bonding. We assume in the latter case that the ligands are suitably disposed, that is, have available electrons in pi bonding orbitals to enter empty metal orbitals, or vice versa. Since Pt(II) has eight d electrons of its own and chloride ion has a full octet of electrons, neither the ligands or the central metal ion has empty pi-type orbitals to accommodate the other's electrons. The conclusion, in the case of $PtCl_4^{2-}$, is that pi bonding is unimportant. Cyanide ion, however, does have some empty pi bonding type orbitals, and the $Pt(CN)_4^{2-}$ complex is considered to have both sigma and pi bonding.

Finally, and in abbreviated fashion, the situation for the combined sets of pi bonds of an O_h complex is

	E	C_3	$6C_2'$	$6C_4$	$3C_2$	i	$6S_4$	$8S_6$	$3\sigma_h$	$6\sigma_d$
$\Gamma_{12\pi}$	12	0	0	0	-4	0	0	0	0	0

This corresponds to $T_{1g} + T_{2g} + T_{1u} + T_{2u}$ or $(d_{xy} d_{xz} d_{yz})$, $(p_x p_y p_z)$. There are only six appropriate orbitals, of which the set $(p_x p_y p_z)$ is preempted by sigma bonding. We conclude that only the set d_{ij} is available for pi bonding in an O_h complex. The actual degree of such bonding will depend on the particular central metal ion and the ligands. The ion Cr(III) has only three d electrons and could accept or donate pi bonding electron density from or to ligand pi-type orbitals. If the ligand is NH_3, as in $Cr(NH_3)_6^{3+}$, the nitrogen is surrounded by hydrogen atoms, so no pi bonding at all is possible. Ligands such as Cl^- might donate some pi bonding electrons to Cr(III) and ligands such as CN^- might accept some.

These examples illustrate how symmetry plus ancillary considerations allow the physical chemist to draw qualitative conclusions as to the bonding in coordination compounds. The treatment of pi bonding in organic molecules generally involves multicenter situations, as in $CH_2 = CH_2$ or benzene, and these are usually treated by a molecular orbital approach (see Chapter 18).

17-8 The direct product

Quantum chemistry abounds in the evaluation of integrals, usually a complicated calculation. It is possible, however, to determine on symmetry grounds whether an integral of the type

$$\int f_A f_C F_B d\tau$$

should be nonzero.

We first consider the simpler integral

$$\int f_A f_B d\tau$$

where f_A and f_B might be two wave functions for a molecule. It turns out that such an integral *must* be zero unless the integrand is invariant under all operations of the symmetry group to which the molecule belongs. What this means specifically is that the product $f_A f_B$ must form the basis for the totally symmetric irreducible representation (IR) of the group—the one for which all the traces for the various symmetry operations are unity.

The procedure for determining whether or not this requirement is met is as follows. First, the traces of the representation of a product $f_A \times f_B$, called a *direct product*, are, for each symmetry operation R,

$$\chi(R) = \chi_A(R)\,\chi_B(R) \tag{17-21}$$

Consider, for example, a molecule in the C_{4v} point group. Table 17-11 gives the character table and the traces for the direct products of various IR's. Thus the direct product $A_1 \times A_2$ is just A_2, and likewise $B_1 \times B_2$. The direct product E^2 or $E \times E$ does not correspond to any of the IR's, but on application of Eq. (17-17), one finds it to contain the IR's $A_1 + A_2 + B_1 + B_2$. We conclude that an integral involving functions that are bases for the first three direct products in the table must be zero since the direct products do not contain the totally symmetric IR, in this case A_1. An integral based on the direct product E^2 would be nonzero, however, since it contains the A_1 IR.

The principle may be extended to three functions. One simply takes the direct product of the representations for which f_C and f_B are bases, and then the direct product of the result with the representation for which f_A is a basis. Again, unless the final result is or contains the totally symmetric IR of the point group, the integral must be zero. Alternatively stated, for the integral to be nonzero, the direct product of any two of the functions must have at least one IR in common with those spanned by the third function.

C_{4v}:	E	C_2	$2C_4$	$2\sigma_v$	$2\sigma_d$
A_1	1	1	1	1	1
A_2	1	1	1	-1	-1
B_1	1	-1	1	1	-1
B_2	1	-1	1	-1	1
E	2	0	-2	0	0
$A_1 \times A_2$	1	1	1	-1	-1
$B_1 \times B_2$	1	1	1	-1	-1
$B_1 \times E$	2	0	-2	0	0
$E \times E$	4	0	4	0	0
$A_1 \times E \times B_2$	2	0	-2	0	0

TABLE 17-11
Direct products in C_{4v} symmetry

COMMENTARY AND NOTES

17-CN-1 Crystal field theory

An important application of group theory is to the case of a transition metal ion that is in an environment of octahedral symmetry. The ion may be held in a crystal lattice, as is the case for Cr^{3+} in aluminum oxide (ruby is an example), or it may be a complex ion having six ligands in octahedral coordination.

We can now use a set of metal orbitals as the basis for a representation in the O_h point group. A metal s orbital is unchanged by all of the symmetry operations, and hence belongs to the A_{1g} IR (see Table 17-7). The set of three p orbitals ($4p_x$, $4p_y$, $4p_z$ for a first-row transition metal) generates the traces shown in Table 17-12. (As in Section 17-7C, 1 is entered if the orbital is unchanged, -1 if it is inverted, and 0 otherwise.) The set of traces for Γ_{p^3} is just that for the T_{1u} IR. Finally, if the process is repeated for the set of five d orbitals ($3d_{xy}$, $3d_{yz}$, $3d_{xz}$, $3d_{x^2-y^2}$, $3d_{z^2}$ for a first-row transition metal), one obtains for Γ_{d^5}: (5,1,1, -1,1,5, -1, -1,1,1). On applying Eq. (17-17), we find that this corresponds to $E_g + T_{2g}$.

We find in Section 17-7 that a set of six sigma bonds corresponds to the $A_1 + T_{1u} + E_g$ irreducible representations, which are compatible with the s, $p_x p_y p_z$, and $d_{x^2-y^2} d_{z^2}$ hydrogen-like orbital functions. The bonding is thus d^2sp^3 in type. The remaining d_{ij} orbitals belong to the T_{2g} IR; they are not suitable for sigma bonding.

From a geometric point of view, the p- and d-bonding orbitals have lobes directed toward the apexes of the octahedron. The d_{ij} set have lobes directed toward the midpoint of edges, or away from the apexes.

One of the first treatments of this situation was in terms of a transition metal ion in an ionic crystal lattice such that the ion is octahedrally surrounded by negative ions. The ef-

fect is to create an electrostatic *crystal field* of O_h symmetry, and the model used is one in which point charges are brought up to the apexes of an octahedron centered on the metal ion. As illustrated in Fig. 17-16, several work terms are involved. The point charges are attracted (a) by the nuclear charge of the metal ion and repelled (b) by its electrons. Further, the repulsion is less for electrons in the T_{2g} set (or in t_{2g} orbitals—the d_{ij} ones) than for those in the E_{2g} set. This last effect is the one of interest. It may be calculated quantitatively by means of first-order perturbation theory (Section 16-ST-3), with the mathematics much simplified by the use of symmetry considerations. One finds that the metal d orbitals are separated into a threefold degenerate set of t_{2g} types and a twofold degenerate set of e_g types [step (c)]. The energy difference, Δ (sometimes referred to as 10 Dq), is proportional to the crystal field strength. Furthermore, since the *set* of five d orbitals is spherically symmetric (Ünsold's theorem, Section 16-9A), the stabilization of the t_{2g} set must just be balanced by the destabilization of the e_g set. Allowing for the difference in degeneracy, the former is $\frac{2}{5}\Delta$, or 4 Dq, and the latter is $\frac{3}{5}\Delta$, or 6 Dq.

In the case of Cr^{3+}, the three 3d electrons would be placed in the lower-energy t_{2g} orbitals, giving a crystal field stabilization of $(3)\frac{2}{5}\Delta$ or 12 Dq. An observation, known as *Hund's rule*, is that it takes energy to pair electron spins, so that the electrons remain unpaired, as indicated by the arrows.

The ion Co^{3+} has six electrons and these all pair up in the t_{2g} set of orbitals—it evidently takes more energy to promote an electron to the e_g orbitals than to pair its spin. The resulting state is therefore diamagnetic. The *crystal field stabilization energy* is now $(6)\frac{2}{5}\Delta$ or 24 Dq.

While the original crystal field treatment due

TABLE 17-12		E	$8C_3$	$6C_2'$	$6C_4$	$3C_2$	i	$6S_4$	$8S_6$	$3\sigma_h$	$6\sigma_d$
	p_x	1	0	0	0	-1	-1	0	0	1	0
	p_y	1	0	0	0	-1	-1	0	0	1	0
	p_z	1	0	-1	1	1	-1	-1	0	-1	1
	Γ_{p^3}	3	0	-1	1	-1	-3	-1	0	1	1

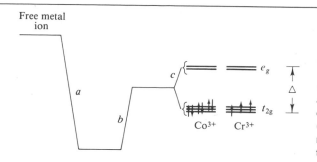

Free metal
ion

Co^{3+} Cr^{3+}

FIGURE 17-16
Schematic illustration of energy contributions as six point charges are brought up to a transition metal ion. (a) Stabilization by attraction between the charges and the metal ion, (b) destabilization due to mutual repulsion between the charges, and (c) splitting of the d-orbital energy levels in the O_h field of the charges.

to H. Bethe was applied to ions in a crystal lattice, it became evident that octahedral complexes showed much the same electronic behavior. The six ligands now furnish the octahedral crystal field. The energy Δ to promote an electron from a t_{2g} to an e_g orbital (see Fig. 17-16) can be determined from the absorption spectrum of the complex so that it is possible to obtain numerical values for the crystal field stabilization energy.

As one further example, the empirical observation is that Co^{2+} complexes exhibit a smaller crystal field strength than do Co^{3+} complexes. One consequence is that the separation Δ is now less than the spin pairing en-

ergy, so that the seven d electrons of Co^{2+} remain unpaired insofar as possible. The result is to leave three unpaired, so that the complex is paramagnetic. The contrasting situations for Co^{3+} and Co^{2+} with respect to spin pairing are known as the *strong-field* (large-Δ) and *weak-field* (small-Δ) cases.

The purely electrostatic crystal field picture has now given way largely to what is called *ligand field theory*. Perhaps the main change is that bonding is now recognized as occurring between the ligands and the central metal ion. The resulting molecular orbital treatment is discussed in Section 18-ST-1.

ADAMSON, A.W. (1980). "Understanding Physical Chemistry," 3rd ed., Chapter 18, Benjamin, New York.
COTTON, F.A. (1963). "Chemical Applications of Group Theory." Wiley (Interscience), New York.
COULSON, C.A. (1961). "Valence," 2nd ed. Oxford Univ. Press, London and New York.
FIGGIS, B.N. (1966). "Introduction to Ligand Fields." Wiley (Interscience), New York.

GENERAL REFERENCES

ADAMSON, A.W. (1980). "Understanding Physical Chemistry," 3rd ed. Benjamin, New York.
COTTON, F.A. (1963). "Chemical Applications of Group Theory." Wiley, New York.

CITED REFERENCES

EXERCISES

Write out the series of symmetry operations associated with an S_5 axis, that is, show with explanation alternative designations wherever possible.

17-1

$$Ans.\ S_5{}^1, S_5{}^2 = C_5{}^2, S_5{}^3, S_5{}^4 = C_5{}^4, S_5{}^5 = \sigma_h,$$
$$S_5{}^6 = C_5{}^1, S_5{}^7, S_5{}^8 = C_5{}^3, S_5{}^9, S_5{}^{10} = E.$$

Explain the symmetry elements present and the point group designation for the following: (a) A P. Chem. book with blank pages and cover, (b) a normally printed book, (c) a tennis ball, including the seam, (d) a two-color tennis ball, the seam dividing the two color zones, (e) an ash tray in the shape

17-2

of a round bowl and with three equally spaced grooves in the rim for holding cigarettes. (There is no printing on the tennis balls.)

Ans. (a) C_{2v}; (b) C_1; (c) D_{2d}; (d) C_{2v}; (e) C_{3v}.

17-3 Explain what symmetry elements are present and the point group designation for the following molecules: (a) pyridine, (b) pyrazine, (c) allene, (d) furan, (e) PCl_5 (a trigonal bipyramid), (f) *trans*-$Cr(NH_3)_4Cl_2^+$ (disregard the H's), (g) ruthenocene (a pentagonal prism), (h) $Pt(CN)_4^{2-}$ (square planar), (i) C_2H_6 in the staggered configuration.

Ans. (a) C_{2v}; (b) D_{2h}; (c) D_{2d}; (d) C_{2v}; (e) D_{3h};
(f) D_{4h}; (g) D_{5h}; (h) D_{4h}; (i) D_{3d}.

17-4 Work out the multiplication table for the C_{2v} group, showing each multiplication in the manner of Fig. 17-9.

Ans. See Table 17-4.

17-5 Referring to Exercise 17-4, show that $\sigma(xz)$ and $\sigma'(yz)$ belong to separate classes by carrying out the similarity transformations $X^{-1}\sigma(xz)X$ and $X^{-1}\sigma'(yz)X$, where X is each symmetry operation.

Ans. $X^{-1}\sigma(xz)X$ is $\sigma(xz)$ for $X = E(X^{-1} = E)$, $X = C_2(X^{-1} = C_2)$,
$X = \sigma(xz)$ [$X^{-1} = \sigma(xz)$], and $X = \sigma'(yz)$ [$X^{-1} = \sigma'(yz)$];
similarly $X^{-1}\sigma'(yz)X$ is $\sigma'(yz)$ for each case. Thus $\sigma(xz)$ and
$\sigma'(yz)$ do not mix and therefore belong to separate classes.

17-6 Carry out the matrix multiplication

$$\begin{bmatrix} 0 & 1 \\ 3 & 1 \end{bmatrix} \begin{bmatrix} 1 & 3 \\ 4 & 2 \end{bmatrix}.$$

Ans. $\begin{bmatrix} 4 & 2 \\ 7 & 11 \end{bmatrix}$.

17-7 Carry out the matrix multiplication

$$\begin{bmatrix} 3 & 0 & 1 \\ 1 & 0 & 1 \\ 2 & 1 & 0 \end{bmatrix} \begin{bmatrix} 2 & 1 \\ 1 & 2 \\ 0 & 1 \end{bmatrix}.$$

Ans. $\begin{bmatrix} 6 & 4 \\ 2 & 2 \\ 5 & 4 \end{bmatrix}$

17-8 Carry out the matrix multiplication

$$\begin{bmatrix} 1 & 2 & 0 \\ 0 & 2 & 0 \\ 0 & 0 & 3 \end{bmatrix} \begin{bmatrix} 3 & 2 & 0 \\ 1 & 1 & 0 \\ 0 & 0 & 2 \end{bmatrix}.$$

Ans. $\begin{bmatrix} 5 & 4 & 0 \\ 2 & 2 & 0 \\ 0 & 0 & 6 \end{bmatrix}$. (Notice that the product matrix is blocked out in the same way as are those multiplied.)

17-9 Show what irreducible representations in C_{3v} are spanned by the reducible representation

	E	$2C_3$	$3\sigma_v$
Γ	5	2	-1

Ans. $E + 2A_2 + A_1$.

Find the traces of the reducible representation $\Gamma_{5\sigma}$ generated by the set of five sigma bonds in PCl_5 (a triangular bipyramid), the irreducible representations spanned by $\Gamma_{5\sigma}$, and the types of hydrogen-like orbitals that might be used for bonding.

17-10

Ans. $\Gamma_{5\sigma} \approx 2A_1' + E' + A_2''$, corresponding
to $s(p_xp_y)p_zd_{z^2}$ or $s(d_{x^2-y^2}d_{xy})\,p_zd_{z^2}$.

Referring to Exercise 17-10 and PCl_5, find the traces for the reducible representation $\Gamma_{3\pi\perp}$ of the set of pi bonds for the three chlorines in the trigonal plane, whose lobes lie above and below the plane. Find also the irreducible representations spanned by $\Gamma_{3\pi\perp}$ and the types of hydrogen-like orbitals that might be used for pi bonding.

17-11

Ans. $\Gamma_{3\pi\perp} \sim A_2'' + E''$, corresponding
to p_z (already used in sigma bonding) and $(d_{xz}d_{yz})$.

Verify the bond strength value of 2.00 for the sp^3 hybrid bond. [In terms of the coordinate system of Fig. 17-6, the maximum extension of the first lobe of the sp^3 hybrid is at $\theta = 54°44'$ and $\phi = 45°$.]

17-12

Verify the bond strength value of 1.991 for the sp^2 hybrid bond.

17-13

Show what the symmetry elements are and give the point group for (a) a sigma bond between like atoms and (b) a pi bond between like atoms.

17-14

Ans. (a) $D_{\infty h}$; (b) D_{2h}.

Use the method of Fig. 17-9 to verify Table 17-4.

17-15

Calculate, in terms of Dq, the crystal field stabilization energy for (a) low-spin Fe(II), (b) high-spin Mn(II), and (c) Mo(IV), assuming octahedral geometry in each case.

17-16

Ans. (a) 24, (b) 0, (c) 8.

PROBLEMS

List the symmetry elements and give the point group designation for (a) a tennis racquet (no printing on it), (b) an isosceles triangle, (c) an equilateral triangle, (d) two equilateral triangles joined at an apex and in mutually perpendicular planes (spiropentane), and (e) a round pencil (no printing on it).

17-17

Show what the point group designations are for (a) ferrocene (iron sandwiched between two cyclo-pentadienyl rings, which are staggered; see the accompanying diagram), (b) ethane in the eclipsed configuration, (c) $Cr(NH_3)_5Cl^{2+}$, (d) *cis*-$Co(NH_3)_4Cl_2^+$, (e) CH_2ClCH_2Cl in the staggered configuration, (f) naphthalene, (g) cyclohexane in the chair form. (List all symmetry elements; ignore H's in the ammine complexes.)

17-18

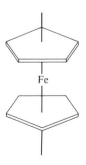

17-19   Show what the point group designations are for (a) $Fe(CN)_6^{4-}$, (b) HOCl (a bent molecule), (c) $(HNBH)_3$ (a planar six-membered ring with alternating N and B), (d) C_5H_8, spiropentane (two mutually perpendicular triangles joined apically), (e) CO, (f) $UO_2F_5^{3-}$ (a pentagonal bipyramid with apical oxygens), (g) diborane, B_2H_6 (two hydrogens are bridging and lie in a plane perpendicular to that of the other four hydrogens), (h) CH_2ClBr. (List all symmetry elements.)

17-20   Show what the point group designations are for: (a) $Cr(en)_3^{3+}$ (for present purposes, en denotes a symmetrical bidentate ligand, and the six bonds formed with Cr^{3+} lie along three mutually perpendicular axes and are equivalent); (b) the planar molecule $B(OH)_3$ with B — O bonds 120° from each other and the B — O — H angles all the same; (c) 1,3,5-trichlorobenzene.

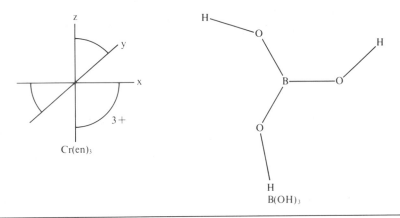

$Cr(en)_3$

$B(OH)_3$

17-21   Show what the point group designations are for (a) a triangular antiprism, (b) *trans*-$Pt(NH_3)_2Cl_2$ (ignore the H's), (c) IF_5 (four F's in a square plane and one apical F), (d) Dewar benzene. (List all the symmetry elements.)

17-22   Work out the multiplication table for the D_{3h} group. Show which are the various classes, that is, carry out the necessary similarity transformations.

17-23   Equation (17-10) gives a set of matrices that simplify to a set of two-dimensional ones and a set of one-dimensional ones. Show that the first set does in fact obey the C_{3v} multiplication table.

17-24   Carry out the matrix multiplication

$$\begin{bmatrix} 2 & 3 & 0 \\ 1 & 3 & 0 \\ 0 & 0 & 5 \end{bmatrix} \begin{bmatrix} 4 & 1 & 0 \\ 1 & 2 & 0 \\ 0 & 0 & 2 \end{bmatrix}$$

17-25   Given the matrix $\mathscr{A}$,

$$\begin{bmatrix} -1 & 0 & 0 \\ 0 & 2 & 0 \\ 0 & 0 & 3 \end{bmatrix}$$

find its reciprocal, $\mathscr{A}^{-1}$. (This one is easy because of the zeros.)

17-26   Verify Eq. (17-15) for all the IR's of the D_{3h} group.

17-27   Verify Eq. (17-16) for i and j being the A_2' and E' IR's.

17-28   Examination of the character tables shows that in the case of C_{nv} point groups there are n σ_v planes if n is odd, all belonging to the same class. If n is even, however, there are $n/2$ σ_v planes of one class and $n/2$ planes of a second class (called σ_d or sometimes σ_v'). Explain this difference in behavior. It is acceptable to take one specific set of examples, such as C_{2v} and C_{3v}, and examine what happens if one starts with a single vertical plane of symmetry and carries out the C_n operations on it, and then also look at what happens if a σ_v is followed by a C_n operation.

Complete the example in connection with Table 17-2, which showed that σ_v, σ_v', and σ_v'' belong to the same class.   **17-29**

Verify Eq. (17-10) by carrying out the transformations of a point according to the symmetry operations of C_{3v}.   **17-30**

Find the traces of the reducible representations in D_{4h} which are generated by carrying the set of (a) p orbitals and (b) d orbitals through the various symmetry operations. What irreducible representations are spanned in the two cases?   **17-31**

Show what combinations of s, p, and d orbitals are suitable for sigma bonding of n atoms to a central one if (a) $n = 2$ and the molecule is angular, (b) $n = 3$ and the symmetry is D_{3h}, (c) $n = 4$ and the n atoms lie at the corners of a square, $n = 5$ and the molecules have C_{4v} symmetry, (d) $n = 6$ and the n atoms lie at the corners of a triangular prism.   **17-32**

Show what hydrogen-like orbitals should be appropriate for sigma bonding in the case of a metal ion having five coordinated ligands, located in triangular bipyramidal geometry.   **17-33**

The allyl radical has C_{2v} symmetry, as shown: **17-34**

The p orbitals of the three carbon atoms, which are perpendicular to the plane of the atoms, can be used to form pi orbitals for the molecular as a whole. Find the characters of the reducible representation for the set of p orbitals and show which irreducible representations are spanned.

Determine what sets of orbitals should be appropriate for sigma bonding in a hypothetical $Cr(NH_3)_7^{3+}$ (a pentagonal bipyramid).   **17-35**

A reducible representation in the D_{3h} point group has the traces: $E = ?$, $C_3 = 1$, $C_2 = 0$, $i = 0$, $S_6 = 3$, $\sigma_d = 0$. The trace under the symmetry operation E is missing. Explain what is the simplest choice for this missing number and what irreducible representations are spanned.   **17-36**

A reducible representation in the T_d point group has the traces: $E = 6$, $C_3 = 0$, $C_2 = ?$, $S_4 = 0$, $\sigma_d = ?$. The traces under the symmetry operations C_2 and σ_d are missing. Explain what is the simplest set of choices for the missing numbers, and what IRs are spanned.   **17-37**

Orbital functions for an entire molecule, that is, molecular orbitals, are generally constructed from linear combinations of appropriate atomic orbitals. Any such combination *must* transform as one of the IR's of the symmetry group to which the molecule belongs. By applying group theory one can learn what symmetry types are possible for the molecular orbitals of a molecule. Illustrate the procedure by considering the water molecule, and taking the set of two 1s hydrogen orbitals plus the 2s, $2p_x$, $2p_y$, and $2p_z$ nitrogen orbitals through the operations of the C_{2v} group then finding what IR's are spanned by the resulting reducible representation.   **17-38**

Determine the hydrogen-like orbitals which are of the proper symmetry for pi bonding in a tetrahedral molecule.   **17-39**

Construct the group multiplication table for the D_4 group. Demonstrate that C_4^1 and C_4^3 belong to a different class than does $C_4^2 = C_2$.   **17-40**

Calculate the values of the three Y functions of the sp^2 set and plot them on polar coordinate paper. Show that the lobes are indeed 120° apart.   **17-41**

17-42   Calculate values for the Y_1 function of the sp^3 set to show its cross section in the plane containing the bond direction, which is $\theta = 54°44'$, $\phi = 45°$.

17-43   The crystal field splitting diagram for $Mn(CO)_5Cl$ is as shown below; this is a d^6 complex. Find the crystal field stabilization in terms of Dq for (a) maximum spin pairing and (b) minimum spin pairing.

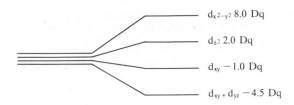

$$d_{x^2-y^2} \quad 8.0 \; Dq$$
$$d_{z^2} \quad 2.0 \; Dq$$
$$d_{xy} \quad -1.0 \; Dq$$
$$d_{xy} + d_{yz} \quad -4.5 \; Dq$$

17-44   Calculate, in terms of Dq, the crystal field stabilization energy for transition metal complexes with one through nine d electrons, assuming octahedral geometry and (a) low spin; (b) high spin.

17-45   Show how the set of five d orbitals should split in a crystal field of T_d symmetry and explain qualitatively what the energy ordering should be.

17-46   A transition metal complex, MA_6, has D_{3d} symmetry, with the ligands at the corners of a triangular antiprism. The six M—A bonds are equivalent. Find the IR's spanned by the set of six sigma bonds and infer the various hybrid orbital combinations that would be possible.

17-47   Find what irreducible representations are spanned in C_{4v} by (a) $A_2 \times E$ and (b) $E \times E$.

17-48   Find what irreducible representations are spanned in D_{3d} by (a) $A_{2g} \times E_g$ and (b) $E_g \times E_g$.

17-49   Evaluate the direct product of E_g with T_{2g} in O_h.

17-50   Show that in O_h, $E_g \times (A_{1g} \times T_{2g}) = (E_g \times A_{1g}) \times T_{2g}$.

chapter 18
Wave
mechanics
and bonding

18-1   Introduction

G. N. Lewis, perhaps the dean of American physical chemists, suggested in 1916 that the chemical bond between two atoms consists of a jointly held or shared pair of electrons. To this day, the concept of the *shared electron pair* bond has much qualitative utility for the chemist. We write, for example

$$\text{H:H} \qquad \text{H:Cl} \qquad \overset{\text{H}}{\underset{\ddot{\text{H}}}{\text{H:}\ddot{\text{C}}\text{:H}}} \qquad \overset{\text{H}}{\underset{\text{H}}{:\!\ddot{\text{C}}\text{::}\ddot{\text{C}}\!:}} \qquad \text{H:C:::C:H}$$

to show single, double, and triple bonds. By including nonparticipating electrons, the satisfying *octet rule* emerged for light-element molecules. Thus

$$:\!\ddot{\text{C}}\text{l}\!:\!\ddot{\text{C}}\text{l}\!: \qquad :\!\ddot{\text{O}}\text{::}\ddot{\text{O}}\!: \qquad \text{H:}\ddot{\text{C}}\text{l:}\ddot{\text{O}}\!: \qquad :\!\ddot{\text{O}}\text{::C::}\ddot{\text{O}}\!:$$

In each case the core of the atom "sees" an octet of electrons if the shared ones are included. In addition, the *formal charge* on each atom is obtained by again counting electrons, but now allowing only half of the shared ones, and noting the excess or deficiency over the net charge of the nucleus plus inner electrons. The formal charges are zero in the preceding examples, except for HClO, in which Cl has the formal charge $+1$ and O has -1. Thus an indication of bond polarity emerges as well.

The electron pair bond and the octet rule give chemically useful pictures of molecular bonding; they are the basis for our writing a bond as a line between atoms. The difficulty in explaining why ordinary O_2 is paramagnetic or why NO should have so little tendency to dimerize has been explained quite nicely by Linnett (1964) in a modern extension of the octet rule.

Contemporary wave mechanical treatments support much of these qualitative ideas but, of course, go far beyond them in giving detailed electron density maps, in accounting for bonding with an odd number of electrons, in explaining delocalized pi bonding as in benzene, and in giving molecular energies and other properties. In this chapter we give a brief account of some of the mathematical methods used and typical results for simple molecules. As may be imagined, the subject is a formidable one both in complexity and in the welter of detailed calculational algorithms that have been developed. Of necessity, therefore, we present here no more than an introduction to a large field of chemical physics.

The mathematical problem of dealing with the simplest common molecule, H_2, or more generally, with an AB-type molecule where the atoms are hydrogen-like, can be appreciated by looking at the complete Hamiltonian:

$$-\frac{h^2}{8\pi^2 m}(\nabla_1^2 + \nabla_2^2) - \frac{h^2}{8\pi^2 M_A}\nabla_A^2 - \frac{h^2}{8\pi^2 M_B}\nabla_B^2$$

$$- e^2\left(\frac{Z_A}{r_{A,1}} + \frac{Z_B}{r_{B,1}} + \frac{Z_A}{r_{A,2}} + \frac{Z_B}{r_{B,2}} - \frac{Z_A Z_B}{R} - \frac{1}{r_{12}}\right) \quad (18\text{-}1)$$

where 1 and 2 denote the two electrons, and A and B the two nuclei. As shown in Fig. 18-1, R is the internuclear distance; $r_{A,1}$ is that between electron 1 and nucleus A, etc.; and r_{12} is the interelectronic distance. In practice a major simplification is made at once by neglecting the ∇^2 terms for the two nuclei, yielding

$$\mathbf{H} = -\frac{h^2}{8\pi^2 m}[\nabla^2(1) + \nabla^2(2)] - e^2\left(\frac{Z_A}{r_{A,1}} + \frac{Z_B}{r_{B,1}}\right.$$

$$\left. + \frac{Z_A}{r_{A,2}} + \frac{Z_B}{r_{B,2}}\right) + e^2\left(\frac{Z_A Z_B}{R} + \frac{1}{r_{12}}\right) \quad (18\text{-}2)$$

The effect of this approximation is to say that $\psi_{\text{total}} = \psi_{\text{nuclei}}\,\psi_{\text{electronic}}$.

This approximation, known as the *Born–Oppenheimer* approximation, can be justified mathematically. A qualitative explanation is that the nuclear masses are so large compared to the electron mass that the electrons adjust instantaneously to the nuclear motions. As a consequence we treat the energy of nuclear motions separately as vibrational energy (as in Section 16-6) and

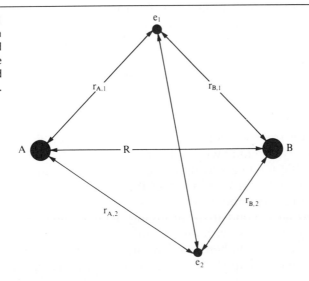

FIGURE 18-1
Coordinate scheme for an A—B type molecule. A and B are the nuclei and the two electrons are labelled e_1 and e_2.

assume that the amount of vibrational energy present simply adds to that of the electronic state. The wave equation, $\mathbf{H}\psi = E\psi$, is then to be solved for various R values, the equilibrium one being that giving the lowest energy (most negative potential energy). Explicit solutions are not possible—only more or less accurate approximations. The big difficulty is in the r_{12} term, which prevents the wave equation from being separated into two one-electron equations.

The mathematical approach is to use a linear combination of approximate wave functions combined with what is known as the *variation method*. The variation method is discussed in Section 16-ST-4 as a means of calculating the polarizability of the hydrogen atom, and we review it here in the present context. If Eq. (16-56), $\mathbf{H}\psi = E\psi$, is multiplied on both sides by ψ (or by ψ^* if ψ is complex), rearrangement gives

$$E = \frac{\int \psi^* \mathbf{H}\psi \, d\tau}{\int \psi^* \psi \, d\tau} \tag{18-3}$$

E is a constant and can be written outside of the integrals, which are over $d\tau$ or over all space. The basic theorem of the variation method is that some approximate wave function ϕ is used in Eq. (18-3), the resulting energy may be greater than the correct one but can never be *less* than it.

The formal next step is to write ϕ as

$$\phi = a\psi_1 + b\psi_2 \tag{18-4}$$

where ψ_1 and ψ_2 are alternative possibilities, chosen so that the resulting integrals *can* be solved. An example, pursued in the next section, would be to take ψ_1 as the wave function for the case of electron 1 on atom A and electron 2 on atom B, and ψ_2 as the wave function for the alternative assignment of electron 2 on atom A and electron 1 on atom B. Equation (18-3) becomes

$$E = \frac{\int (a\psi_1 + b\psi_2)^* \mathbf{H}(a\psi_1 + b\psi_2) d\tau}{\int (a\psi_1 + b\psi_2)^* (a\psi_1 + b\psi_2) d\tau} \tag{18-5}$$

It is convenient to introduce the definitions

$$H_{11} = \int \psi_1^* \mathbf{H}\psi_1 d\tau \qquad H_{12} = \int \psi_1^* \mathbf{H}\psi_2 d\tau$$
$$S_{11} = \int \psi_1^* \psi_1 d\tau \qquad S_{12} = \int \psi_1^* \psi_2 d\tau \tag{18-6}$$

so that Eq. (18-5) becomes:

$$E = \frac{a^2 H_{11} + 2ab H_{12} + b^2 H_{22}}{a^2 S_{11} + 2ab S_{12} + b^2 S_{22}} \tag{18-7}$$

(The distinction between H_{12} and H_{21} is not necessary here.) The integrals of Eq. (18-6) in general *can* be evaluated.

Now, the best possible E will be the minimum value that can be obtained using the approximate wave function ϕ, and we vary a and b in Eq. (18-4) to obtain $E_{minimum}$. Specifically, we set $(\partial E/\partial a)_b = 0$ and $(\partial E/\partial b)_a = 0$. On carrying out the first of these differentiations, one obtains

$$\left(\frac{\partial E}{\partial a}\right)_b = \frac{2aH_{11} + 2bH_{12}}{a^2S_{11} + 2abS_{12} + b^2S_{22}} - E\frac{2aS_{11} + 2bS_{12}}{a^2S_{11} + 2abS_{12} + b^2S_{22}}$$

$$= \frac{2a(H_{11} - ES_{11}) + (2b(H_{12} - ES_{12})}{a^2S_{11} + 2abS_{12} + b^2S_{22}}$$

or, on setting the derivative equal to zero,

$$a(H_{11} - ES_{11}) + b(H_{12} - ES_{12}) = 0 \qquad (18\text{-}8)$$

Similarly, on setting the derivative $(\partial E/\partial b)_a = 0$, we obtain

$$a(H_{21} - ES_{21}) + b(H_{22} - ES_{22}) = 0 \qquad (18\text{-}9)$$

As simultaneous equations in a and b, the solution is given by the determinant

$$\begin{vmatrix} (H_{12} - ES_{11}) & (H_{12} - ES_{12}) \\ (H_{12} - ES_{12}) & (H_{22} - ES_{22}) \end{vmatrix} = 0 \qquad (18\text{-}10)$$

An equation of the type of Eq. (18-10) is called a *secular equation*. It is necessary, in the variation method, to evaluate the various integrals so as to solve the secular equation for E. Substitution of the result back into Eqs. (18-8) and (18-9) then gives the values of a and b for the best possible ϕ.

Two rather different ways of setting up a ϕ function have been used. In the *valence bond* method, the ψs are constructed from those of the separated atoms, *complete with electrons* (as described above). This approach is, in a sense, the quantum mechanical outgrowth of the electron pair bond concept, hence the name. In the second approach, the *molecular orbital* method, the approximate wave function is constructed from the orbital functions of the atoms (which may be more than just two), to obtain molecular orbital functions. The requisite number of electrons are then assigned into these molecular orbitals.

We take up the valence bond approach in the next section, with emphasis on the H_2 molecule. The succeeding sections deal with the molecular orbital method, first in terms of H_2^+, for which exact solutions are possible, and then for H_2 and other diatomic molecules. Sections 18-6 and 18-7 carry aspects of the molecular orbital method to triatomic and then polyatomic molecules.

18-2   The valence bond method for an A-B type molecule

We consider in this section the simplest possible case, that of the hydrogen molecule, and then generalize somewhat to an A—B type molecule. The Hamiltonian is given by Eq. (18-2), after making the Born–Oppenheimer approximation; as noted above, the wave equation cannot be solved exactly because of the e^2/r_{12} term. The valence bond method was developed by W. Heitler and F. London around 1927 as a means of constructing approximate wave functions for use with the variation method, and was first applied to the case of the hydrogen molecule.

A. The hydrogen molecule
The approximate wave function for the variation method is set up as follows. If atoms A and B are separated, two possible wave functions are

$$\psi_1 = \psi_A(1)\psi_B(2) \qquad \psi_2 = \psi_A(2)\psi_B(1) \qquad (18\text{-}11)$$

That is, electron 1 may stay with atom A and electron 2 with atom B, or vice versa. The approximate wave function is taken to be a linear combination of the two possibilities, that is,

$$\phi = a\psi_1 + b\psi_2 \tag{18-12}$$

The approximate wave function in the valence bond method is thus assembled as a linear combination of the products of the wave functions for the separated atoms.

In the case of H_2, atoms A and B are the same, and $H_{12} = H_{21}$ and $S_{12} = S_{21}$. The integral S_{11} may be written $S_{11} = \int\int\psi_A^*(1)\psi_B^*(2)\psi_A(1)\psi_B(2)d\tau_1 d\tau_2 = \int\psi_A^*(1)\psi_A(1)d\tau_1 \int\psi_B^*(2)\psi_B(2)d\tau_2 = 1$. Each integral is just the normalization integral S, which is unity. Similarly, $S_{22} = 1$. Expansion of the secular equation, Eq. (18-10), therefore gives

$$E = \frac{(H_{11} \pm H_{12}) \mp (H_{11} \pm H_{12})S_{12}}{1 - S_{12}^2} \tag{18-13}$$

or

$$E_1 = \frac{H_{11} + H_{12}}{1 + S_{12}} \tag{18-14}$$

and

$$E_2 = \frac{H_{11} - H_{12}}{1 - S_{12}} \tag{18-15}$$

If these solutions are substituted back into Eqs. (18-8) and (18-9), to find the coefficients a and b, one obtains

$$\phi_1 = \psi_1 + \psi_2 \tag{18-16}$$

and

$$\phi_2 = \psi_1 - \psi_2 \tag{18-17}$$

(except for a normalization factor).

Integrals of the type H_{11} are known as *Coulomb integrals* (two-center in this case since **H** involves both atoms); those of the type H_{12} are called *exchange* or *resonance integrals*; and those of the type S_{12} are called *overlap integrals*. The integral S_{12} may be written S^2, where S is the one-electron overlap integral,

$$S = \int \psi_A(1)\psi_B(1)d\tau = \int \psi_A(2)\psi_B(2)d\tau \tag{18-18}$$

In the case of two hydrogen atoms, S is given by

$$S = e^{-R}(1 + R + R^2/3) \tag{18-19}$$

where R is in a_0 units. Note that $S = 1$ when $R = 0$, as it must (why?).

The evaluation of the H_{11} and especially the H_{12} integral is somewhat lengthy—one uses elliptical coordinates to simplify the algebra—and the results are shown in Fig. 18-2. Note that both H_{11} and H_{12} go through a negative minimum, corresponding to attraction between the two hydrogen atoms, that is, to chemical bond formation. It is the exchange or resonance integral H_{12}, however, that makes the biggest contribution. From its definition, Eq. (18-6),

FIGURE 18-2
Plots of the total energy. Coulomb energy, and exchange energy for H_2 as a function of internuclear distance. The H_{11} and H_{12} quantities have been divided by $(1 + S^2)$. (From C. A. Coulson, "Valence," 2nd ed. Oxford Univ. Press, London and New York, 1961.)

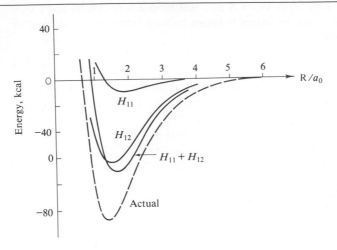

and the wave functions involved, Eq. (18-11), the integral contains the assumption of indistinguishability of electrons; it is a wave mechanical attribute that would not come up in classical mechanics. The physical explanation of the H_{12} integral is that the electrons, being indistinguishable, are not localized about the nuclei with which they were associated in the separated atoms, but expand over the system of both atoms. Without this exchange effect only a rather weak bond (given by the H_{11} contribution) would be expected.

The sum $(H_{11} + H_{12})$, divided by $(1 + S^2)$, gives E_1, Eq. (18-14), and its minimum occurs at -72.8 kcal mol^{-1} and 1.64 a_0 or 0.87 Å, as compared to the experimental value of -109.5 kcal mol^{-1} and 0.74 Å. The calculation, although extremely gratifying at the time it was first made, around 1917, is not very accurate. Clearly, some improvement in the approximate wave functions, Eq. (18-11), is needed, and the next step is to allow for the possibility that *both* electrons may be on one atom, that is, to include *ionic* wave functions. We now take ϕ to be

$$\phi = \psi_A(1)\psi_B(2) + \psi_A(2)\psi_B(1) + \alpha[\psi_A(1)\psi_A(2) + \psi_B(1)\psi_B(2)] \qquad (18\text{-}20)$$

The last two terms correspond to H^-H^+ and H^+H^-, respectively. The variation method is used to find the optimum value for α, and the best calculated value for E_1 is now -94 kcal mol^{-1} with a bond length of 0.76 Å, a considerable improvement.

The inclusion of such ionic configurations is often needed in the valence bond method. A molecule is sometimes said to resonate among various structures, the so-called *resonance* structures. It must be remembered, however, that a *single* wave function describes the state of the molecule. The so-called resonance structures are merely the components of a linear combination of wave functions that has been assembled in an effort to approximate the correct one. It is particularly important not to be misled into supposing that the molecule is somehow flickering between different electronic configurations.

The second solution, E_2, given by Eq. (18-15) involves the difference $(H_{11} - H_{12})$. This difference is positive at all internuclear distances, so ϕ_2 corresponds to a repulsion between the atoms. We speak of ϕ_1 as a *bonding* wave function and ϕ_2 as an *antibonding* wave function. The result is characteristic: For every combination of atomic wave functions that leads to bonding there is always a related antibonding combination.

The wave functions of Eqs. (18-16) and (18-17) are incomplete; they should include the spin function (Section 16-10). Recalling the combinations of Eq. (16-128), we have for the possible complete wave functions

B. The spin function

$$[\psi_A(1)\psi_B(2) + \psi_A(2)\psi_B(1)][\alpha(1)\beta(2) - \alpha(2)\beta(1)] \qquad \text{singlet} \qquad (18\text{-}21)$$

$$[\psi\alpha_A(1)\psi_B(2) - \psi_A(2)\psi_B(1)] \begin{cases} \alpha(1)\alpha(2) \\ \beta(1)\beta(2) \\ \alpha(1)\beta(2) + \alpha(2)\beta(1) \end{cases} \qquad \text{triplet} \qquad (18\text{-}22)$$

The restriction of choice to these combinations follows from an alternative statement of the *Pauli exclusion principle*: Every allowable wave function for a system of two or more electrons must be antisymmetric for the simultaneous interchange of the position and spin coordinates of any pair of electrons. This statement implies the usual one that no two electrons can have the same four quantum numbers. The function of Eq. (18-21) is symmetric in the spatial wave functions, and so must be antisymmetric in the spin function; as only one possible spin function of this type is available, there is only one complete wave function, and the state is called a *singlet* state. The *triplet* state is so called because it is triply degenerate in the spin function; since the wave function is antisymmetric, any one of the three possible symmetric spin functions may be used. In the case of H_2, the lowest energy or bonding state is singlet, while the associated antibonding state [corresponding to Eqs. (18-22)] is triplet in spin degeneracy. This is the usual situation for an A-B type molecule. An exception, discussed in Section 18-5, is the O_2 molecule, for which the lowest energy state is a triplet state.

H_{11}, H_{12}, and S_{12} integrals have been evaluated rather accurately for pairs of atoms using modern computer methods. The behavior of the overlap integral S, in particular, has been calculated for hydrogen-like wave functions for a variety of n and ℓ values [see Mulliken *et al.* (1949)], and some representative results are shown in Fig. 18-3 for the particular case of two carbon atoms (using the appropriate Z_{eff}). There is not a great deal of difference in S for the various types of orbitals beyond about 1.5 Å(the $C-C$ bond length is 1.54 Å), although that for the two pi-bonding type orbitals is distinctly lower than the rest. The $2s-2p\sigma$ and $2p\sigma-2p\sigma$ integrals go through a maximum, and the latter can actually be negative because at small distances oppositely signed lobes of the p orbitals begin to overlap (these are p_z orbitals if z denotes the bond axis, that is, they are p orbitals of sigma bonding symmetry).

C. Overlap integrals for A-B type molecules

It is clear from Fig. 18-2 that it is the Coulomb and exchange integrals that make E_1 negative for H_2 (and in general), corresponding to bonding, and we see from Eq. (18-14) that while the overlap integral acts to diminish the denominator (S and S_{12} are always less than unity), it does not determine the sign of E_1. However, as discussed in Section 17-7B, one often uses the rule of thumb that a positive overlap of orbitals is needed for a strong bond to form. The rule is a good one in spite of the comment above, since the Coulomb and exchange integrals are often roughly proportional to the overlap integral.

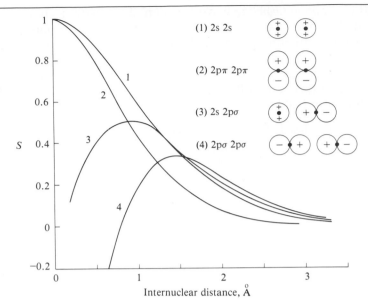

FIGURE 18-3
Overlap integrals S for
various pairs of orbitals.
[See R. S. Mulliken, C. A.
Rieke, D. Orloff, and H.
Orloff, J. Chem. Phys. **17**,
1248 (1949).]

This qualitative use of the overlap idea was extended by L. Pauling to the case of hybrid orbitals. The maximum value of the trigonometric portion gives a measure of the "reach" of an orbital in the direction of its greatest amplitude. Contrary to what might be expected from the order of hybrid orbital "bond strengths" as given in Table 17-8, one finds that for a given bond distance, S for sp³, sp², and sp hybrid orbitals *increases* in that order.

18-3   Molecular orbitals. The hydrogen molecule ion, H_2^+

The hydrogen molecule ion, H_2^+, is a relatively weakly bound species, and is not one of great chemical importance. It is of interest to us here, however, for two reasons. First, it is the one *molecule* for which exact solutions have been obtained. Second, and very important, these solutions give us a set of molecular, one-electron orbitals which are widely used in the approximate description of diatomic molecules generally.

A. Solutions to the wave equation

The wave equation for the hydrogen molecule ion is

$$\left(-\frac{\nabla^2}{2} - \frac{1}{r_A} - \frac{1}{r_B} + \frac{1}{r_{AB}}\right)\psi = E\psi \tag{18-23}$$

where energy is in a.u. (see Section 16-2), and r_A $(=r_{A,1})$, r_B $(=r_{B,1})$, and R are as defined in Fig. 18-1. Notice that, of course, the difficult r_{12} term of Eq. (18-2) is absent (and that the usual Born–Oppenheimer approximation is made). This equation has been solved exactly (not explicitly, but by series expansions) for any value of R. The algebra is much simplified if the elliptical coordinate system shown in Fig. 18-4 is used. Here, the position of the electron is given by the coordinates $\lambda = (r_A + r_B)/R$, $\mu = (r_A - r_B)/R$, and θ. It now turns out that the wave equation can be separated into three, one for each coordinate—

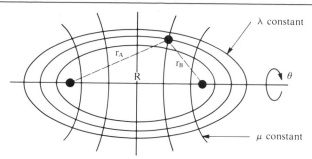

FIGURE 18-4
The elliptical coordinate
system. A point in a plane
is located by the
intersection of a pair of
curves of given λ and μ,
and the angular position of
the plane is given by θ.

much as was done for the hydrogen atom when polar coordinates are used
(Section 16-7A). On writing out the Laplacian, ∇^2, and expressing ψ as

$$\psi = L(\lambda)M(\mu)\Phi(\phi) \tag{18-24}$$

the wave equation separates into one in λ only, one in μ only, and another
in ϕ only.

The solution for the ϕ equation is just

$$\Phi = e^{i\lambda\phi} \tag{18-25}$$

where λ is now a quantum number and is restricted to the values $0, \pm 1, \pm 2,$
etc. if the solutions are to be physically acceptable. In the case of $\lambda > 1$, one
takes linear combinations so as to obtain $\cos|\lambda|\phi$ and $\sin|\lambda|\phi$ functions. The
procedure is like that for the Φ solutions for the hydrogen atom (Section 16-
7B).

In the case of H_2^+, the λ quantum number plays somewhat the same geometric role as does the ℓ quantum number for the hydrogen atom. Consistent
with this analogy, the custom is to refer to states of $\lambda = 0$ as σ states, to
those with $\lambda = \pm 1$ as π states, and to those with $\lambda = \pm 2$ as δ states. Further,
ψ can be either symmetric (sign unchanged) or antisymmetric (sign changed)
on inversion of a general point through the center of symmetry. We indicate
these two situations by the symbols g (*gerade* in German) and u (*ungerade*),
respectively. We thus have solutions designated as σ_g, σ_u, π_g, π_u, etc. The
quantum number associated with the L and M functions is not generally of
interest.

The functions $\psi(1\sigma_g)$ and $\psi(1\sigma_u)$ are plotted in Fig. 18-5 as the profile along
the H–H axis, the number 1 merely denoting the lowest energy solution.
Figure 18-5(a) is for $R = 8a_0$ (a_0 is the Bohr radius—Section 16-2), at which
separation the electron density is nearly that for two separate atoms. Each
section of each curve is approximately that of the $R(r)$ function for $n = 1$,
$\ell = 0$ of Fig. 16-13. In the case of $\psi(1\sigma_g)$, the sign of the $R(r)$-like function is
the same for both atoms, while in that of $\psi(1\sigma_u)$, it is positive for one and
negative for the other. Figure 18-5(b) is for $R = 2a_0$, or for about the equilibrium separation (see below). The $\psi(1\sigma_g)$ function now shows considerable
electron density at the midpoint between the two nuclei, corresponding to
the Lewis-type formulation H · H. Finally, in Fig. 18-5(c), the two nuclei are
merged ($R = 0$), and the ψ functions are just those for He^+. $\psi(1\sigma_g)$ has become
$\psi(1s)$ for He^+, and $\psi(1\sigma_u)$ has become $\psi(2p)$.

The potential energy for several of these molecular orbitals is shown as a

FIGURE 18-5
Normalized wave functions
for $1\sigma_g$ (left) and $1\sigma_u$ (right)
of $H_2{}^+$ along the H—H
axis. (After J. C. Slater,
"Quantum Theory of
Molecules and Solids,"
Vol. 1. McGraw-Hill, New
York, 1963.)

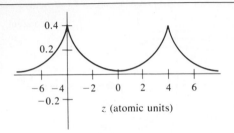
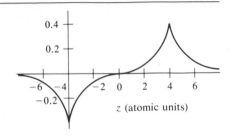

(a) $R = 8$ atomic units

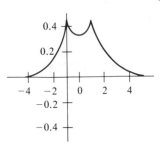
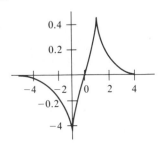

(b) $R = 2$

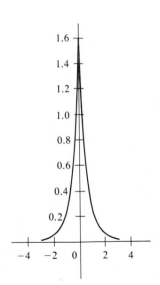
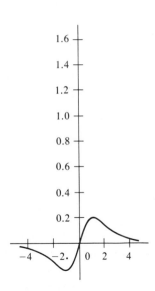

(c) $R = 0$

function of R in Fig. 18-6. Internuclear repulsion (the e^2/R term) is *not* included, and the lines terminate at the energy levels for He^+ and for separated H atoms at the left and right extremes, respectively. Again, the numbers 1 or 2 preceding the symmetry designations are just ordering numbers. The actual energy variation with R, that is, including the e^2/R term, is shown in Fig. 18-7. We now see that $\psi(1\sigma_g)$ has an energy minimum at about $R = 2a_0$ or, more exactly, at 1.06 Å. This is the equilibrium bond distance for $H_2{}^+$. The energy at the minimum is -0.6 a.u., or at -0.1 a.u. or $-(0.1)(27.21) = -2.72$ eV relative to the separated $H^+ + H$. This is the H_2^+ bond energy.

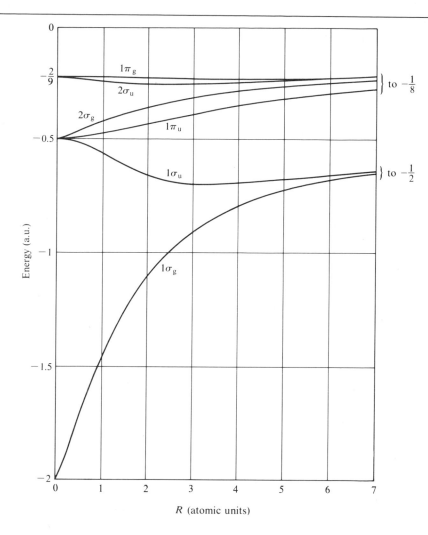

R (atomic units)

FIGURE 18-6
Energy levels of H_2^+ as a function of internuclear separation, omitting the internuclear repulsion term. (After J. C. Slater, "Quantum Theory of Molecules and Solids," Vol. 1. McGraw-Hill, New York, 1963.)

It was noted at the beginning of this section that the solutions for H_2^+ provide a set of molecular orbitals useful in the description of diatomic molecules, and, indeed, of any A—B bond. It is therefore worthwhile to take a further look at these functions. Because of the minimum in its energy (Fig. 18-7), $\psi(1\sigma_g)$ is called a *bonding* molecular orbital. In the case of $\psi(1\sigma_u)$, however, there is repulsion at all R; this is therefore called an *antibonding* orbital. Usually, such orbitals are marked with an asterisk: $1\sigma_u^*$. On the same basis, we have $1\pi_u$ (weakly bonding) and $1\pi_g^*$ (antibonding).

The various molecular orbitals can be described in other terms. It was noted earlier that $1\sigma_g$ goes over to $\psi(1s)$ for H at large R and to $\psi(1s)$ for He^+ at zero R, while $1\sigma_u$ goes from $\psi(1s)$ for H to $\psi(2p)$ for He^+. Figure 18-8 summarizes these types of behavior for a number of molecular orbitals for the general case of an A—A hydrogen-like molecule, the "*united atom*" being that resulting from the fusion of two A nuclei (He^+ in the case of H_2^+). Diagrams of this type are known as *correlation diagrams*.

B. Molecular orbitals

FIGURE 18-7
Energy levels of H₂⁺ as a
function of internuclear
separation. (After J. C.
Slater, "Quantum Theory
of Molecules and Solids,"
Vol. 1. McGraw-Hill, New
York, 1963.).

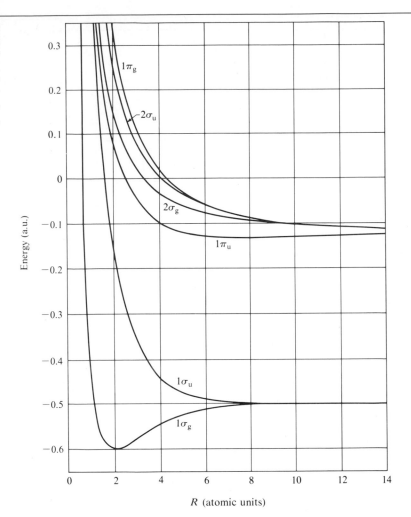

FIGURE 18-8
Correlation diagram
showing how molecular
orbital energies change
with internuclear distance
in an A—A-type molecule.
(From C. A. Coulson,
"Valence," 2nd ed. Oxford
Univ. Press, London and
New York, 1961.)

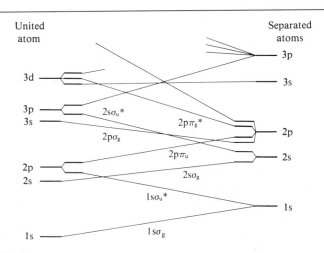

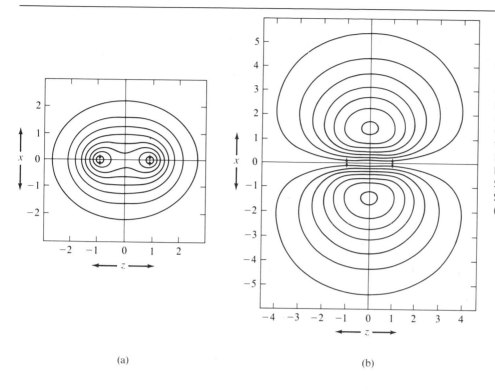

(a)   (b)

FIGURE 18-9
Contour diagrams for normalized molecular orbitals of H_2^+ with $R = 2a_0$. (a) $1\sigma_g$ (or $1s\sigma_g$); (b) $1\pi_u$ (or $2p_x\pi_u$). The small arrows mark the nuclear positions; the contours span about a tenfold range of ψ value. [Adopted from D. R. Bates, K. Ledsham, and A. L. Stewart, Phil. Trans. Roy. Soc. (London) **A246**, 215 (1953).]

A rule that we will not prove here is that states of the same orbital symmetry cannot cross. Thus a σ_g cannot cross another σ_g. A qualitative physical explanation is that at such a crossing one would have two states of the same symmetry and same energy, and some linear combination of the states must exist such that one state is raised and the other lowered in energy. The effect is that such lines avoid each other in a correlation diagram.

Notice that in Fig. 18-8 the molecular orbitals are designated not as we have been doing up to now, but rather in terms of the atomic orbital to which they go in the separated atoms. Thus our $1\sigma_g$ and $1\sigma_u$ have become $1s\sigma_g$ and $1s\sigma_u^*$.[†] The geometric meaning of the g and u designations is brought out in Fig. 18-9. Thus in Fig. 18-9(a) the 1s (or $1s\sigma_g$) molecular orbital has cylindrical symmetry around the internuclear axis; indeed, this is what the designation σ indicates. The $1\pi_u$ (or $2p_x\pi_u$, z being the axis through the nuclear centers) orbital is degenerate, with one obeying a $\cos\phi$ function and the other a $\sin\phi$ function. For one, the electron density is a maximum at $0°$ and $180°$, and for the other, at $90°$ and $270°$. Figure 18-9(b) shows the electron density contours for one of these (in the xz plane); the orbital would look like a pair of sausages lying above and below the internuclear line. From the correlation diagram, we find that the $1\pi_u$ orbital goes over to $2p_x$ orbitals on the separated atoms. Conversely, it can be viewed as resulting from the overlap of two $2p_x$ atomic orbitals. Accordingly, we now designate it as $2p_x\pi_u$ or just as $2p\pi_u$. The other of the pair would be $2p_y\pi_u$.

[†]Still another scheme, more favored by physicists, involves labeling the orbitals in terms of the atomic orbitals to which they go in the united atom.

	Bonding overlap	Antibonding overlap
ss (σ)		
sp "end-on" (σ)		
sp "sideways"		(non-bonding)
pp "end-on" (σ)		
pp "sideways" (π)		
pd "sideways" (π)		

The above way of seeing how molecular orbitals are generated is summarized in Fig. 18-10 for several combinations. Note that the signs (or phases) of the atomic orbitals used are important. Bonding overlap occurs when the overlapping regions are of the same sign, and antibonding overlap when they are of opposite sign. It can happen that the like sign and opposite sign regions balance to give zero net overlap. Such a case is called *nonbonding*. An example is the sp sideways bonding in Figure 18-10.

18-4   Variation method for obtaining molecular orbitals

The variation method may be used to obtain approximate molecular orbital functions and energies. We construct an approximate wave function for the molecule as a whole and write for a diatomic molecule

$$\phi = a\psi_A + b\psi_B = n(\psi_A + \lambda\psi_B) \tag{18-26}$$

where n is a normalization factor and λ is an adjustable constant. The ψ_A and ψ_B are suitable combinations of atomic orbitals on atoms A and B. This means of constructing ϕ is known as the *LCAO* approximation—*linear combination of atomic orbitals*. If the molecule is H_2^+ or H_2, the 1s orbitals of each hydrogen atom are the appropriate ones to use in obtaining the lowest-energy molecular orbitals (and in this case $\lambda = \pm 1$ since the atoms are the same).†

† One may include additional atomic orbitals. Thus for H_2 one may write

$$\phi = c_1(\psi_{1s_A} + \psi_{1s_B}) + c_2(\psi_{2s_A} + \psi_{2s_B}) + c_3(\psi_{2p_{z,A}} + \psi_{2p_{z,B}}) + \cdots \tag{18-27}$$

where the coefficients $c_1, c_2, c_3, \ldots$ are optimized by means of the variation method. The added terms are of rapidly diminishing importance, and the limiting result is known as the *Hartree–Fock* limit.

In general, ψ_A and ψ_B are not the same functions and the variation method is used to find an optimum value for the mixing coefficient λ. On doing this (and eliminating λ), we find

$$(E - E_A)(E - E_B) = (H_{AB} - ES_{AB})^2 \tag{18-28}$$

where E_A and E_B are the energies associated with the atomic orbitals (and are equal to the integrals $\int \psi_A^* \mathbf{H} \psi_A \, d\tau$ and $\int \psi_B^* \mathbf{H} \psi_B \, d\tau$, also called Coulomb integrals, although not the same ones as used in the valence bond method (the e^2/r_{12} term is absent); E is the energy of the best molecular orbital function ϕ. The quantity H_{AB} is the integral $\int \psi_A^* \mathbf{H} \psi_B \, d\tau$ (also called a resonance integral, although again not the same one described in Section 18-2), and S_{AB} is the overlap integral $\int \psi_A^* \psi_B \, d\tau$, which is the same as the S integral of Section 18-2. Equation (18-28) has two roots, so that two combinations of ψ_A and ψ_B in Eq. (18-26) are indicated, a bonding and an antibonding one. In the case of H_2^+, these are the $1s\sigma_g$ and $1s\sigma_u{}^*$ molecular orbitals.

As noted in the preceding section, not any combination of atomic orbitals will lead to bonding. The ψ_A and ψ_B must have the same symmetry relative to the bonding axis. Thus if z is the bonding axis and we attempt to use an s and a p_x orbital for ψ_A and ψ_B, the molecular orbital energy turns out to be just E_A or E_B; that is, no interaction results. This is a consequence of the fact that H_{AB} and S_{AB} are separable into integrals of equal magnitude and opposite sign and are therefore zero (note Fig. 18-10). As noted earlier, the situation is known as nonbonding. Pairs of orbitals that behave this way are also called orthogonal.

18-5   Molecular orbital energy levels for diatomic molecules

Molecular orbitals for A—A type molecules, where A is other than hydrogen, are qualitatively the same as for H_2^+. The energies are affected, however, by the screening of the nuclear charge by the inner electrons and a Z_{eff} must be used. At a more elaborate level, self-consistent field (Section 16-CN-4) wave functions are used as the basis, and some very extensive calculations have been made. Qualitatively, however, we order the molecular orbitals in sequence of increasing energy and proceed to populate them with the requisite number of electrons. Each orbital can hold a pair of electrons, with spins opposed, and the set of molecular orbitals is filled with pairs of electrons. This sequential placing of electrons in molecular orbitals, working upward from the lowest energy one, is known as the *aufbau* or building up procedure, and it generally works reasonably well.

Figure 18-11 shows a qualitative molecular orbital energy level diagram appropriate for light elements. Following the building-up procedure, the H_2 molecule has the configuration $(1s\sigma_g)^2$ and He_2 would have $(1s\sigma_g)^2(1s\sigma_u{}^*)^2$. In this last case there are two electrons in a bonding and two in a nonbonding orbital; the net situation is equivalent to one of essentially no bonding, as observed—He_2 is not a stable molecule. However, an electronically excited He_2 molecule of configuration $(1s\sigma_g)^2(1s\sigma_u{}^*)(2s\sigma_g)$ should have a net bonding, as is actually found to be the case.

Moving to second-row elements of the periodic table, Li_2 has the configuration $(2s\sigma_g)^2$ (the inner electrons make no net contribution and therefore are ignored); the molecule should exist, and does—its dissociation energy is

FIGURE 18-11
Molecular orbital energy
level diagram for O_2.

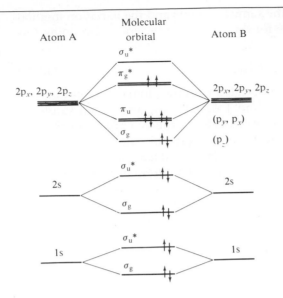

FIGURE 18-11
Molecular orbital energy
level diagram for O_2.

about 1 eV. The Be_2 molecule should be like He_2—it is, in fact, not known. The next elements now involve 2p orbitals, which may be either $2p\sigma$ or $2p\pi$ in type.

The cases of N_2 and O_2 are especially interesting. The first has the configuration $(2s\sigma_g)^2 (2s\sigma_u^*)^2 (2p\sigma)^2 (2p\pi_u)^4$. There are two $2p\pi_u$ orbitals of equal energy (or degenerate) so that four electrons may be accommodated; they are bonding, so N_2 has a net of six bonding electrons, corresponding to the triple bond in the usual Lewis formula. The O_2 molecule is $(2s\sigma_g)^2 (2s\sigma_u^*)^2 (2p\sigma_g)^2 (2p\pi_u)^4 (2p\pi_g)^2$ and therefore has a net of four bonding electrons, corresponding to a double bond ($2p\pi_g$ being antibonding). Since the $2p\pi_g$ set is only half filled, the pair of electrons in it are not required to pair their spin and, by Hund's rule (see Section 16-ST-2), do not. As a consequence, the ground state of oxygen is a paramagnetic one, called $^3\Sigma$, corresponding to two unpaired electrons. The situation is illustrated in Fig. 18-11.

These examples illustrate one important application of the molecular orbital approach. If one neglects interelectronic repulsions, a set of molecular orbital states is obtained into which the proper number of electrons are fed. One thus obtains a good qualitative understanding of the geometry of the filled orbitals and, as in the case of O_2, can predict cases of paramagnetism. In addition, it is easy to describe excited states as configurations in which one or more electrons are promoted to higher-energy molecular orbitals. This aspect will be discussed further in Chapter 19.

In the case of A—B molecules, there is no longer a center of symmetry and the g and u designations do not apply to the molecular orbitals. Also, since $\psi_A(1s)$ does not have the same energy as $\psi_B(1s)$, etc., the atomic orbital positions in a diagram such as Fig. 8-11 will lie on different energy scales. This can result in some shifting in the relative positions of the various molecular orbitals, to be determined by calculation. Given the molecular orbital energy level sequence, however, the procedure is as before—one adds the requisite number of electrons, in pairs, and beginning with the lowest-energy orbital.

18-6   Triatomic molecules. Walsh diagrams

The bonding in triatomic and in MX_n-type molecules may be treated in terms of atomic orbitals suitably hybridized so as to give a maximum overlap in the bonding directions. The use of group theory for determining what combinations of orbitals should be hybridized for the central atom is discussed in a previous chapter (Section 17-7) and the bonding can then be treated in terms of molecular orbitals formed between the hybrid orbital and an orbital on each of the X atoms.

While absolute bonding energies are very difficult to calculate, it is possible to reach qualitative conclusions about how the energies of the various molecular orbitals vary with geometry. The simplest case is that of the MH_2 molecule. If it is linear, the molecular orbitals may be of the sigma type, formed between the 1s hydrogen atomic orbitals and the sp_y hybrid orbital of the M atom (y being the H—M—H axis). As illustrated in Fig. 18-12(a),

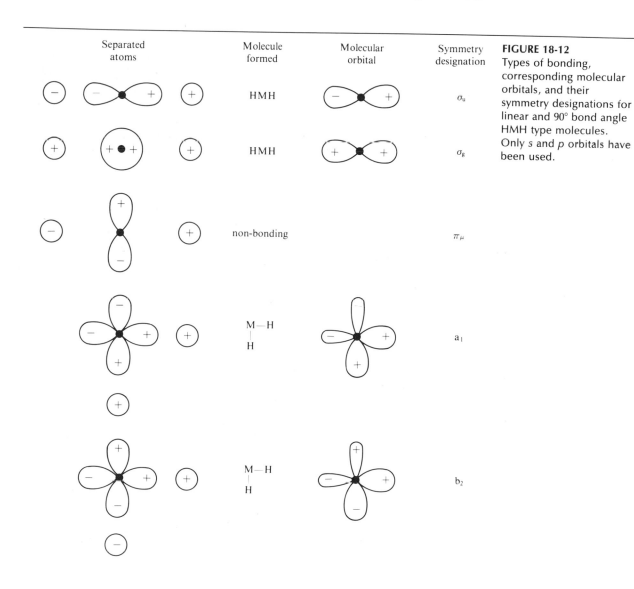

Separated atoms	Molecule formed	Molecular orbital	Symmetry designation
	HMH		σ_u
	HMH		σ_g
	non-bonding		π_μ
	M—H $\mid$ H		a_1
	M—H $\mid$ H		b_2

FIGURE 18-12
Types of bonding, corresponding molecular orbitals, and their symmetry designations for linear and 90° bond angle HMH type molecules. Only s and p orbitals have been used.

one then obtains a pair of molecular orbitals of sigma symmetry, σ_g and σ_u, the former being lower in energy. As also shown in the figure, there is no net overlap between the hydrogen 1s orbitals and either the p_x or p_z M orbitals; these molecular orbitals are nonbonding and pi in type, so the designation is π_u (two-fold degenerate).

Consider next the situation if the H—M—H angle is 90°. The molecule is now C_{2v} in symmetry, with irreducible representations A_1, A_2, B_1, and B_2. A sigma M—H bond can be formed by the overlap of a pure p orbital on M with the 1s orbital on H, and the two M—H bonds would then involve the p_z and p_y M orbitals, as shown in Figure 18-12(b). The corresponding molecular orbitals are formed by taking the in-phase and out-of-phase combinations of the two pairs of localized orbitals. As may be checked against the character table for the C_{2v} point group, the symmetry properties of these two molecular orbitals identify them as belonging to the a_1 and b_2 irreducible representations. (It is customary to use lower case letters in giving the IR to which a molecular orbital belongs.) The remaining p_x orbital is nonbonding and belongs to the b_1 irreducible representation. Finally, the s orbital on M is also nonbonding and has the symmetry designation a_1.

One may now assemble the correlation diagram shown in Fig. 18-13 [see Walsh (1953)]. As the H—M—H angle is increased from 90° the a_1 and b_2 bonding orbitals must eventually become the σ_g and σ_u orbitals, respectively, of the linear molecule. At the same time the binding energy of the orbitals must increase—one way of seeing this is that increasing s character develops in the bonds and sp hybrid orbitals give rise to a stronger bond than do those for pure p. Figure 18-13 is obtained from such considerations plus the no-crossing rule mentioned in connection with Fig. 18-8.

While the energy scale is qualitative, the figure does allow some conclusions as to molecular geometry. A given MH_2 molecule will have a certain number of outer electrons, to be placed in the lowest possible molecular orbitals; each orbital has a capacity of two electrons (with spins opposed). A molecule such as BeH_2 or HgH_2 has four outer electrons, so that the σ_g and σ_u orbitals are

FIGURE 18-13
Qualitative variation of molecular orbital energies with bond angle for an MH_2-type molecule. [From A. D. Walsh, J. Chem. Soc. **1953**, 2260 (1953).]

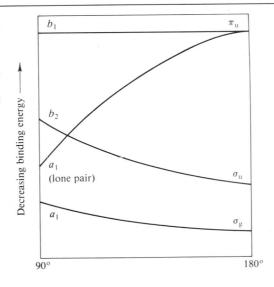

just filled. The linear configuration is the lowest-energy one, so these molecules are predicted to be linear. On the other hand, MH_2 molecules having five, six, seven, or eight valence electrons should be bent. This is the case with H_2O, for example. The actual angle cannot be obtained from Fig. 18-13; it would be determined as the optimum compromise between the opposite trends of the a_1 (s) orbital and the a_1 and b_1 orbitals. Similarly, the molecule CH_2 should be bent, but less so in that excited state which has an electron promoted to the b_1 orbital (why?). Walsh diagrams thus also allow predictions of how molecular geometry should change in excited states (see also Section 19-CN-2); more complex diagrams are available for MA_2 ($A \neq H$) molecules.

18-7 Polyatomic molecules. The Hückel method

A number of rather approximate, semiempirical methods have been developed in an effort to calculate or at least to organize in a reasonable way the experimental properties of molecules in which delocalized pi bonding seems to be present. A rather useful and widely used approach is the following.

A set of pi-bonding molecular orbitals is developed by using the variation method and taking ϕ to be a linear combination of wave functions which are of symmetry suitable for such bonding:

$$\phi = a\psi_1 + b\psi_2 + c\psi_3 \cdots \tag{18-29}$$

The condition for the optimum set of coefficients a, b, c, etc., is given by the secular equation, Eq. 18-10 (for the general form see Eq. 16-168). The problem, of course, is to evaluate the integrals!

We will now solve equations such as Eq. (18-10) by making some rather drastic-appearing approximations. In the case of molecules that are held together by a framework of sigma bonds but that also have available pi bonding orbitals, a set of pi-bonding molecular orbitals can be developed by taking ψ_1, ψ_2, and so on to be the wave functions on each atom that are of symmetry suitable for pi bonding. The usual situation is one of a chain or ring of carbon atoms, and each ψ is then a carbon p orbital perpendicular to the sigma bond. In ethylene, for example, the

A. The Hückel approximations

sigma bonding framework might be regarded as provided by sp^2 hybrid bonds of each carbon atom. The remaining carbon p orbitals then overlap to give some pi bonding.

In this case, which is typical, one then takes ψ_1, ψ_2, and so on to be the same, that is, one p orbital per carbon. The specific further approximation made in the treatment by E. Hückel is then to assume that all integrals of the type H_{ii} are the same and equal to a constant α; as mentioned earlier (Section 18-4), these are called *Coulomb integrals*. Integrals of the type H_{ij}, called *resonance integrals,* are taken to be the same and equal to a constant β if i and j refer to adjacent atoms and to be zero otherwise. The atomic p orbitals are assumed to be normalized so that all $S_{ii} = 1$; there is complete neglect of overlap integrals, that is, all S_{ij} are taken to be zero.

Equation (18-10), which applies to ethylene, then becomes

$$\begin{vmatrix} (\alpha - E) & \beta \\ \beta & (\alpha - E) \end{vmatrix} = 0 \qquad (18\text{-}30)$$

For butadiene, $H_2C\!\!=\!\!CH\!\!-\!\!CH\!\!=\!\!CH_2$, the secular equation is

$$\begin{vmatrix} (H_{11} - ES_{11}) & (H_{12} - ES_{12}) & (H_{13} - ES_{13}) & (H_{14} - ES_{14}) \\ (H_{12} - ES_{12}) & (H_{22} - ES_{22}) & (H_{23} - ES_{23}) & (H_{24} - ES_{24}) \\ (H_{13} - ES_{13}) & (H_{23} - ES_{23}) & (H_{33} - ES_{33}) & (H_{34} - ES_{34}) \\ (H_{14} - ES_{14}) & (H_{24} - ES_{24}) & (H_{34} - ES_{34}) & (H_{44} - ES_{44}) \end{vmatrix} = 0$$

$$\begin{vmatrix} (\alpha - E) & \beta & 0 & 0 \\ \beta & (\alpha - E) & \beta & 0 \\ 0 & \beta & (\alpha - E) & \beta \\ 0 & 0 & \beta & (\alpha - E) \end{vmatrix} = 0 \qquad (18\text{-}31)$$

since, for example, $H_{14} = 0$ and $S_{14} = 0$, and so on. The solution to the ethylene secular equation is simply $E = \alpha + \beta$ and $E = \alpha - \beta$ [note Eqs. (18-14) and (18-15)]. Since α corresponds to the energy for the isolated atom, it is customary to take this as a point of reference, and the solutions for ethylene are as shown in Fig. 18-14. The β integrals are negative in value, so that $\alpha + \beta$ is the lower lying of the two. There are two pi electrons to place and since they may occupy the same level if their spins are opposed, this is the assumed configuration for ethylene.

The solutions for butadiene are†

$$\alpha + \tfrac{1}{2}(\sqrt{5} + 1)\beta, \qquad \alpha + \tfrac{1}{2}(\sqrt{5} - 1)\beta, \qquad \alpha - \tfrac{1}{2}(\sqrt{5} - 1)\beta,$$

$$\alpha - \tfrac{1}{2}(\sqrt{5} + 1)\beta$$

There are now four electrons to place and these fill the first two levels in pairs. In the case of *cyclobutadiene,* integrals such as H_{14} are equal to β since these atoms are now adjacent; solution of the resulting equation gives $\alpha + 2\beta$, α, α, and $\alpha - 2\beta$. That is, there are two solutions with $E = \alpha$; such a case is called one of *degeneracy.* There are again four electrons to place, and therefore two must go into the degenerate pair of $E = \alpha$ levels. As noted in Section 18-5, an empirical rule of spectroscopy asserts that it takes energy to pair electrons, that is, to oppose their spins. The rule, known as *Hund's rule,* then

---

†Divide every row by β and let $m = (\alpha - E)/\beta$, to give the determinant

$$\begin{vmatrix} m & 1 & 0 & 0 \\ 1 & m & 1 & 0 \\ 0 & 1 & m & 1 \\ 0 & 0 & 1 & m \end{vmatrix}$$

which expands to $m^4 - 3m^2 + 1 = 0$. If now $y = m^2$, the equation becomes $y^2 - 3y + 1 = 0$, for which the solution is $y = m^2 = \tfrac{3}{2} \pm \tfrac{1}{2}\sqrt{5}$, or $m = \pm (\tfrac{1}{2} \pm \tfrac{1}{2}\sqrt{5})$.

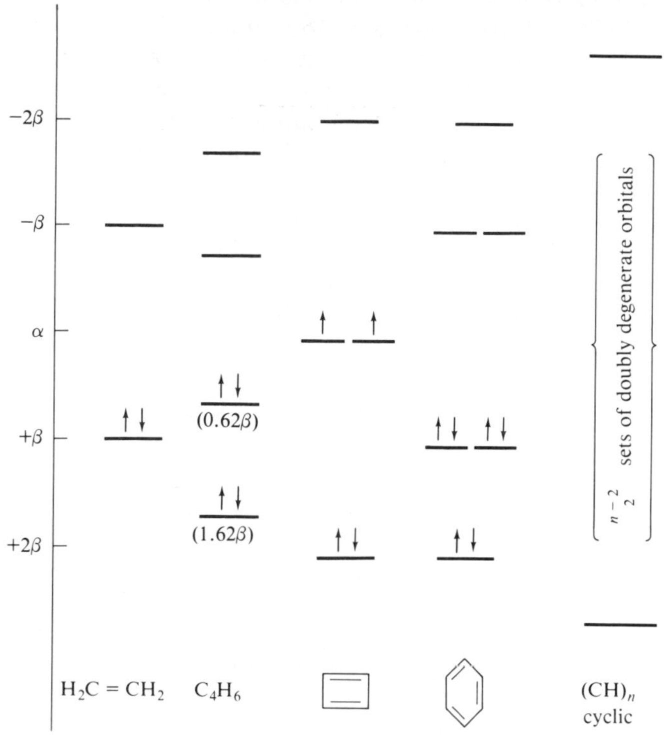

FIGURE 18-14
Pi electron molecular orbital levels according to the Hückel model.

calls for the two electrons to have their spins parallel. Cyclobutadiene should thus be a paramagnetic molecule with magnetic susceptibility corresponding to two unpaired electrons (Section 3-ST-2).

In terms of the assumptions made, the energy of the pi orbitals would be just $n\alpha$, where n is the number of pi electrons, if no mixing or linear combination is used. The difference between this and the calculated energy is then the stabilization (or destabilization) due to molecular orbital formation; it is called the *delocalization energy*. This energy is -2β for ethylene and -4.47β for butadiene. The fact that this last is more than twice the value for ethylene suggests that in butadiene the molecular orbitals extend over the whole molecule, or that one does not simply have two connected ethylene units, as the usual formula $H_2C{=}CH{-}CH{=}CH_2$ might suggest. The delocalization energy for cyclobutadiene is 4β, however, or just that of two ethylenes, which indicates that the ring structure provides no added stability. Cyclobutadiene has in fact been very difficult to synthesize; its existence seems at best to be transient, and the Hückel treatment does indicate that the molecule should be unstable since no extra delocalization energy is present to offset the strain energy of coercing the sigma bond framework to a four-membered ring.

B. Evaluation of the resonance integral β

The integrals H_{ij} can be estimated theoretically, but it is relatively easy to evaluate β experimentally, and this is what is usually done. For example, the heat of hydrogenation of ethylene is 32.8 kcal mol^{-1}, and were butadiene just two noninteracting ethylene units, the value should be 65.6 kcal mol^{-1}. The observed value is 57.1, so butadiene appears to be more stable than expected by 8.5 kcal mol^{-1}. The extra delocalization energy from the Hückel treatment is 0.47β, which makes β = 18 kcal mol^{-1}.

The case of benzene is an important one, and solution of the appropriate secular equation gives the molecular orbital levels shown in Fig. 18-14; the pi bonding energy is 8β, as compared to 6β were benzene simply three ethylenes linked together. The extra stabilization is then 2β. Either from bond energies (Section 5-ST-1) or from comparison of heats of hydrogenation of ethylene and of benzene, the experimental delocalization energy is 35 to 40 kcal mol^{-1}, again giving a β value of 18–20 kcal mol^{-1}.

If β is taken to be generally about 18 to 20 kcal mol^{-1} for carbon–carbon bonded systems, then application of the Hückel approach allows at least approximate delocalization energies to be calculated for a variety of linear and cyclic molecules. Included are rings of the type C_5H_5 and $C_5H_5^-$, C_7H_7, and other unstable species. The method has been extended to heterocyclic rings, to peroxo rings (where each carbon atom of the ring is also bonded to an oxygen), and so forth.

One may also estimate the energy to put a molecule in an excited state. That is, the lowest-lying electronic excited state is very likely that corresponding to promoting an electron to the first unoccupied pi molecular orbital. This energy should then be 2β for ethylene, 1.24β for butadiene, and so on. As shown in Fig. 18-15, a plot of the frequency of the first absorption band against the calculated energy in units of β is reasonably linear. The slope yields β = 60 kcal mol^{-1}.

C. Hückel wave functions

The coefficients of a Hückel molecular orbital

$$\phi_j = c_{1j}\psi_1 + c_{2j}\psi_2 + c_{3j}\psi_3 + \cdots \tag{18-32}$$

where ψ_1, ψ_2, . . . are the atomic orbitals, may be obtained as follows. If we take the molecular orbital ϕ to be normalized, then

---

FIGURE 18-15
Spectroscopic evaluation of β (frequency of the first π → π* transition of various polyenes against energy in units of β). (From A. Streitweiser, Jr., "Molecular Orbital Theory for Organic Chemists." Copyright 1961, Wiley, New York. Used with permission of John Wiley & Sons, Inc.)

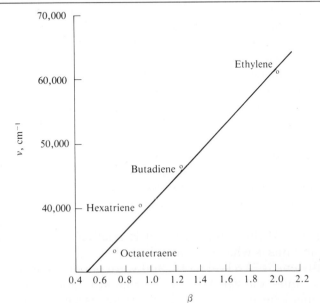

$$\int \phi_j^* \phi_j d\tau = 1 = c_{1j}^2 \int \psi_1^* \psi_1 d\tau + c_{2j}^2 \int \psi_2^* \psi_2 d\tau + \cdots$$

all the cross terms $\int \psi_i^* \psi_j d\tau$ vanishing if the ψ functions are orthogonal. If the functions are also normalized, then the remaining integrals are unity, so we have

$$c_{1j}^2 + c_{2j}^2 + c_{3j}^3 + \cdots = 1 \tag{18-33}$$

We find additional relationships between the c_{ij} as follows. The precursor equation to Eq. (18-5), Eq. (16-56), is $\mathbf{H}\phi_j = E\phi_j$, or $(\mathbf{H} - E)\phi_j = 0$. Suppose we are considering ϕ_1; then substitution of $\phi_1 = c_{11}\psi_1 + c_{21}\psi_2 + \cdots$, multiplication by ψ_1^*, and integration gives

$$c_{11} \int \psi_1^*(\mathbf{H} - E)\psi_1 d\tau + c_{21} \int \psi_1^*(\mathbf{H} - E)\psi_2 d\tau + \cdots = 0$$

or, on insertion of the Hückel assumption regarding H_{11} and H_{12},

$$c_{11}(\alpha - \epsilon) + c_{21}\beta = 0 \tag{18-34}$$

(assuming only atoms 1 and 2 are adjacent). Similar equations follow for the other coefficients on repeating the operation with ψ_2^*, etc.

Thus in the case of butadiene, Eq. (18-34) applies, and since for the lowest energy state $E = \alpha + \frac{1}{2}(\sqrt{5} + 1)\beta$, the result is $c_{21} = \frac{1}{2}(\sqrt{5} + 1)c_{11}$. Pursuing the procedure, we find that $c_{31} = \frac{1}{2}(\sqrt{5} + 1)c_{41}$ and $c_{11} = c_{41}$. All the c_{ij} may then be expressed in terms of c_{11}, and insertion of these relationships into Eq. (18-33) gives $c_{11} = 0.37$; it then follows that $c_{41} = 0.37$, $c_{21} = c_{31} = 0.60$. Thus for the lowest molecular orbital $\phi_1 - 0.37(\psi_1 + \psi_4) + 0.60(\psi_2 + \psi_3)$. The whole process can then be repeated for the other three molecular orbital states. The results are summarized in Fig. 18-16, which shows schematically how ψ_1, ψ_2, ψ_3, and ψ_4 contribute to each molecular orbital.

Somewhat of a shortcut may be taken in evaluating the ϕs if the carbon chain forms a ring. In the case of a six-membered ring (benzene) the molecular point group is D_{6h}. We can then find the traces for the reducible representation generated by a set of six $p\pi$ orbitals and then the irreducible representations

D. Application of group theory to carbon rings

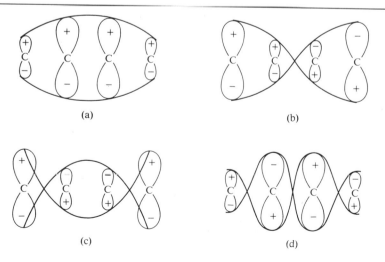

FIGURE 18-16
Hückel molecular orbitals for butadiene.

(a)

(b)

(c)

(d)

to which the set corresponds. Each irreducible representation will have the symmetry of one of the molecular orbital energy states.

If we only want to know signs of the $p\pi$ orbitals for each solution, then it is only necessary to look at the symmetry of each irreducible representation with respect to the C_6^n operations. It is therefore easier to fall back to the C_{6v} group, for which the irreducible representations are half in number, but otherwise the same as far as the traces for the C_n operations. In C_{6v}, the traces for the set of six $p\pi$ bonds are

	E	$2C_6$	$2C_3$	C_2	$3\sigma_v$	$3\sigma_d$
$p\pi$	6	0	0	0	2	0

Application of Eq. (17-17) gives $\Gamma_{6\pi p} = A_1 + B_1 + E_1 + E_2$. The A_1 IR is totally symmetric, and so must correspond to the $p\pi$ orbitals having all plus lobes on one side (and all minus lobes on the other). Thus

$$\phi_1 = (1/\sqrt{6})(\psi_1 + \psi_2 + \psi_3 + \psi_4 + \psi_5 + \psi_6)$$

the factor $1/\sqrt{6}$ is a normalization coefficient. The B_1 IR is antisymmetric to the operation C_6 (and C_6^5) and this as well as its other properties means that the signs of the ψs must alternate. Then $\phi_2 = (1/\sqrt{6})(\psi_1 - \psi_2 + \psi_3 - \psi_4 + \psi_5 - \psi_6)$. The E_1 and E_2 IR are two-dimensional, and a somewhat more lengthy procedure is needed to find the coefficients of the ϕ's; only the results are given here:

$$E_1: \quad \phi_3 = \frac{1}{\sqrt{12}}(2\psi_1 + \psi_2 - \psi_3 - 2\psi_4 - \psi_5 + \psi_6)$$

$$\phi_4 = \tfrac{1}{2}(\psi_2 + \psi_3 - \psi_5 - \psi_6)$$

$$E_2: \quad \phi_5 = \frac{1}{\sqrt{12}}(2\psi_1 - \psi_2 + 2\psi_4 - \psi_5 - \psi_6)$$

$$\phi_6 = \tfrac{1}{2}(\psi_2 - \psi_3 + \psi_4 - \psi_5)$$

FIGURE 18-17
Hückel molecular orbitals for benzene.

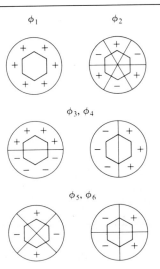

ϕ_1 ϕ_2

ϕ_3, ϕ_4

ϕ_5, ϕ_6

The various ϕ functions correspond to sets of standing molecular orbital waves as illustrated in Fig. 18-17. Note that in order to assign g and u designations to the molecular orbitals, it is necessary to go back to the full or D_{6h} point group.

The Hückel treatment leads to a simple rule, which states that a ring $(CH)_n$ will have aromatic character if $n = 4x + 2$, x being an integer. The qualitative reasoning behind this rule is the following. Referring to Fig. 18-14, molecular orbitals lower in energy than α are said to be *bonding* since electrons placed in them are lower in potential energy than in the absence of the molecular orbital formation. Orbitals equal in energy to α are said to be *nonbonding*, or neutral in this respect, and those lying above α are then *antibonding*. On placing n electrons in the set of molecular orbitals, the molecule is stabilized as a result of molecular orbital formation, that is, of delocalizing the atomic orbitals, to the extent that the total energy is less than $n\alpha$.

E. The 4x + 2 rule. Bonding, nonbonding, and antibonding molecular orbitals

It is clearly undesirable to have to place electrons in nonbonding or antibonding orbitals. If $(CH)_n$ is to be aromatic or extra stable, it is first of all necessary that n be even; since there are n pi electrons, the molecule would otherwise be a free radical. Second, the molecular orbital energy level scheme must be such that all n electrons can be accommodated, in pairs, in *bonding* orbitals. As may be inferred from Fig. 18-14 the pattern of molecular orbitals given by the Hückel treatment for n even is one of a single lowest-lying orbital and a single highest-lying one, and then one or more sets of doubly degenerate orbitals. That is, there will be $n - 2$ degenerate orbitals occurring in $(n - 2)/2$ pairs. If $(n - 2)/2$ is an odd number, then one of the sets of degenerate orbitals will be nonbonding, as in the case of $(CH)_4$. Since the $n/2$ pairs of electrons must just half fill the set of orbitals, this means that two must be nonbonding. To avoid this situation, $(n - 2)/2$ must be an even number, or $2x$, where x is an integer. It then follows that $n = 4x + 2$ if all electrons are to be paired and all to be in bonding orbitals.

According to this rule, C_4H_4 is not aromatic, C_6H_6 is, C_8H_8 is not, $C_{10}H_{10}$ is, and so on. This is the experimental observation.

COMMENTARY AND NOTES

18-CN-1   Comparison of the valence bond and molecular orbital methods

Historically speaking, the valence-bond (VB) and molecular orbital (MO) methods have been the two principal ones of wave mechanics for dealing with molecules. Both use the variation method, but with different procedures for obtaining the approximate wave function, ϕ.

In the VB method, one starts with two atoms, each with its bonding electron, but rec- ognizes that electron 1 may be on atom A and electron 2 on atom B, or vice versa. The approximate wave function whose coefficients are to be optimized is thus

$$\phi_{VB} = a\psi_A(1)\psi_B(2) + b\psi_A(2)\psi_B(1) \qquad (18\text{-}35)$$

In the MO method, one starts with two nuclei, *sans* electrons, and writes as the approximate wave function

$$\phi_1 = a_1\psi_A(1) + b_1\psi_B(1) \qquad (18\text{-}36)$$

That is, electron 1, the first electron, is taken to move in the field of the two nuclei. If we are to add a second electron to the same molecular orbital, we have

$$\phi_2 = a_2\psi_A(2) + b_2\psi_B(2) \qquad (18\text{-}37)$$

The wave function for both electrons is just the product $\phi_1\phi_2$ (neglecting interelectronic repulsion), so that

$$\phi_{MO} = a_1a_2\psi_A(1)\psi_A(2) + b_1b_2\psi_B(1)\psi_B(2)$$
$$+ a_1b_2\psi_A(1)\psi_B(2) + a_2b_1\psi_A(2)\psi_B(1)$$
$$(18\text{-}38)$$

The first two terms on the right correspond to both electrons being on atom A or both on atom B, or to the configurations A^-B^+ and A^+B^-, while the last two terms are essentially the same as ϕ_{VB}. The MO formulation thus tends to give emphasis to ionic configurations and, in fact, to overemphasize them. The VB method underestimates ionic contributions, on the other hand, and these are usually added so as to improve the results, as in Eq. (18-20).

The MO method is now the one of choice for dealing with multiatomic molecules. It gives a good picture of molecular bonding, including delocalized bonding, as in the Hückel treatment, and also provides an easy basis for the qualitative discussion of molecular excited states.

The method has been applied extensively to small molecules and increasingly to relatively large ones, with the help of simplifying approximations such as the following. The differential overlap between two atomic wave functions ψ_j and ψ_k is defined as the probability of finding an electron i in a volume element common to ψ_j and ψ_k, and is represented by

$$S_{jk} = \int\psi_j(i)\,\psi_k(i)\,d\tau$$

The approximation is that S_{jk} is zero unless j is equal to k. The abbreviated name for the procedure is CNDO (complete neglect of differential overlap).

The CNDO approximation (and its variants) allows whole categories of integrals in the secular equation to be made zero or unity. The situation is reminiscent of that of the Hückel method, except that sigma as well as pi bonding is calculated. The remaining integrals are still formidable, and to help matters, the hydrogen-like atomic wave functions are approximated by Gaussian-type functions that are similar graphically but much more tractable mathematically.

The CNDO approach, developed by Pople, Santry, and Segal (1965), has led to a whole spectrum of related calculational procedures. As a result we now have fairly good calculated bond lengths, bond energies, and dipole moments for a variety of small molecules. The method is sufficiently reliable to be used in the prediction of properties of otherwise inaccessible species such as chemical reaction intermediates.

SPECIAL TOPICS

18-ST-1 Ligand field molecular orbital diagrams

The electronic structure of an octahedral transition metal coordination compound was approached in Section 17-CN-1 in terms of the perturbing effect of the octahedral field on a set of five d electrons. As shown in Fig. 17-16, the fivefold degeneracy is partially removed, to give a set of t_{2g} orbitals and one of e_g orbitals, the former being lower in energy. One could then calculate crystal field stabilization energies by placing the requisite number of metal d electrons in these orbital sets.

The purely electrostatic crystal field picture has now given way largely to what is called *ligand field theory*. The main change is that bonding is now recognized as occurring between the ligands and the central metal ion. On making up a linear combination of six ligand sigma bonding orbitals and the s, p, $d_{x^2-y^2}$, and d_{z^2} metal orbitals, one, as usual, obtains a bonding and an antibonding set. This is illustrated in Fig. 18-18; the asterisks denote antibonding orbitals. The t_{2g} orbitals are not suited for sigma bonding and therefore do not participate, that is, are nonbonding. Since a set of bonds for the whole molecule is involved, Fig. 18-16 is usually called a *molecular orbital diagram*. The small letters a, t_{2g}, and so on are

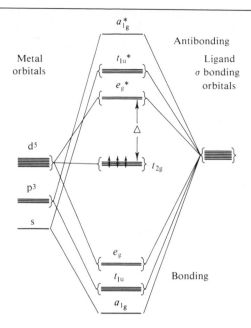

FIGURE 18-18
Sigma bonding molecular orbital diagram for an octahedral transition metal complex.

used as a reminder that the energy levels are those which a single electron could have. The bonding levels are, of course, actually occupied by 12 electrons, two from each ligand, and, in the case chosen here, Cr(III), the t_{2g} set is oc-

cupied by three electrons, thought of as belonging to the central metal ion.

A point that should be mentioned is that the *set* of nonbonding metal electrons also has a symmetry designation, in this case $^4A_{2g}$. The

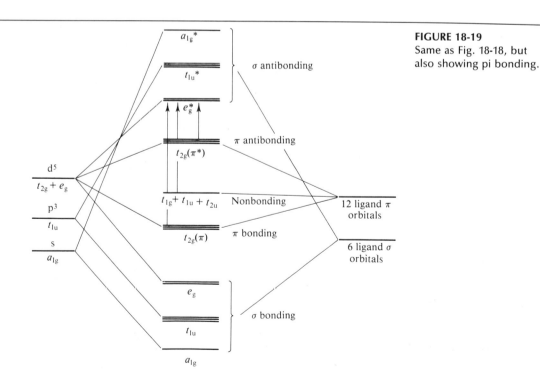

FIGURE 18-19
Same as Fig. 18-18, but also showing pi bonding.

superscript gives the spin multiplicity or number of energetically equivalent ways in which the spin $S = \frac{3}{2}$ may combine with the orbital angular momentum (see Section 16-ST-2). In the case of Co(III) complexes, the symmetry designation of the *set* of six t_{2g} nonbonding metal electrons is $^1A_{1g}$.

Returning to the figure, the quantity Δ, now called the ligand field strength, is the separation between the t_{2g} and the e_g^* sets—it functions in much the same way as in crystal field theory. Its value, however, is now treated as primarily dependent on the strength of the bonding rather than as arising from the octahedrally disposed charges of crystal field theory. We find, for example, that various ligands produce an effective ligand field strength in the order $NH_3 > H_2O > OH^- > Cl^- > Br^-$.

Recalling Section 17-CN-1, the d_{ij} or t_{2g} set of metal orbitals is suited for pi bonding, the set of 12 pi bonding ligand orbitals also generating the t_{1g}, t_{1u}, t_{2g}, and t_{2u} representations in O_h symmetry. One may next obtain bonding and corresponding antibonding combinations of pi symmetry orbitals to give the yet more complete molecular orbital diagram shown in Fig. 18-19. The pi bonding orbitals can hold up to six electrons and the nonbonding ones up to 18. It is not easy to determine the actual degree of pi bonding in a complex ion, and diagrams such as this one are necessarily rather schematic in nature. It is clear, however, that the energy separation Δ between the highest occupied level and the lowest unoccupied level must in reality be a rather complicated quantity.

GENERAL REFERENCES

ADAMSON, A.W. (1980). "Understanding Physical Chemistry," 3rd ed., Chapter 18, Benjamin, New York.

BERRY, R.S., RICE, S.A., AND ROSS, J. (1980). "Physical Chemistry." Wiley, New York.

COTTON, F.A. (1963). "Chemical Applications of Group Theory." Wiley (Interscience), New York.

COULSON, C.A. (1961). "Valence," 2nd ed. Oxford Univ. Press, London and New York.

FIGGIS, B.N. (1966). "Introduction to Ligand Fields." Wiley (Interscience), New York.

EYRING, H., WALTER, J., AND KIMBALL, G.E. (1960). "Quantum Chemistry." Wiley, New York.

HALL, L.H. (1969). "Group Theory and Symmetry in Chemistry." McGraw-Hill, New York.

HANNA, M.W. (1969). "Quantum Mechanics in Chemistry," 2nd ed. Benjamin, New York.

McQUARRIE, D.A. (1983). "Quantum Chemistry." University Science Books, Mill Valley, California.

OFFENHARTZ, P.O'D. (1970). "Atomic and Molecular Orbital Theory." McGraw-Hill, New York.

PITZER K.S. (1958). "Quantum Chemistry," Prentice-Hall, Englewood Cliffs, New Jersey.

SEGAL, G.A. (1977). "Semiempirical Methods of Electronic Structure Calculation," Part A. Plenum Press, New York.

STREITWEISER, A., JR. (1961). "Molecular Orbital Theory for Organic Chemists." Wiley, New York.

CITED REFERENCES

ADAMSON, A.W. (1980). "Understanding Physical Chemistry," 3rd ed. Benjamin, New York.

COTTON, F.A. (1963). "Chemical Applications of Group Theory," Wiley, New York.

LINNETT, J.W. (1964). "The Electronic Structure of Molecules," Methuen, New York.

MULLIKEN, R., RIEKE, C.A., ORLOFF, D., AND ORLOFF, H. (1949). *J. Chem. Phys.* **17**, 1248.

POPLE, J.A., SANTRY, D.P., AND SEGAL, G.A. (1965). *J. Chem. Phys.* **43**, S192.

WALSH, A.D. (1953). *J. Chem. Soc.* **1953**, 2260, 2266.

EXERCISES

18-1 Verify Eqs. (18-8) and (18-9).

18-2 Verify Eq. (18-13).

18-3 Make a plot of S versus R using Eq. (18-19). Find the value of S at the inflection point.

Ans. S = 0.69

Derive Eq. (18-23) from Eq. (16-56). **18-4**

Explain why the $1\sigma_g$ wave function reaches a maximum value of 0.4 in Fig. 18-5(a), and four times this, or 1.6, in Fig. 18-5(c). (The reference cited will be helpful.) **18-5**

Explain why, in Fig. 18-6, the $1\sigma_g$ molecular orbital approaches the limits -2 and $-\frac{1}{2}$, and why the $1\sigma_u$ molecular orbital approaches the limits -1 and -1. **18-6**

Sketch, in the manner of Fig. 18-10, bonding, antibonding, and nonbonding overlap situations involving an s and a d orbital. **18-7**

Referring to the elliptical coordinate system of Fig. 18-4, show that the range of λ is from 1 to ∞, and that the range of μ is from -1 to 1. **18-8**

The molecular orbitals of Fig. 18-8 are designated in terms of the atomic orbitals with which they correlate in the separated atoms. Give the alternative designations based on the united atom orbitals. **18-9**

Consider the molecule C_2. (a) Should it be stable? (b) What should the bond order be? (c) Should it be diamagnetic or paramagnetic? **18-10**

Ans. (a) Yes. (b) 2. (c) Paramagnetic.

Answer the same questions as in Exercise 18-10, but for C_2^{2-}. **18-11**

Ans. (a) Yes. (b) 3. (c) Diamagnetic.

Explain, in terms of a diagram like that of Fig. 18-11, whether the bond length in F_2 should be shorter or longer than that in F_2^+. **18-12**

Ans. The F_2^+ bond should be shorter.

Explain whether the molecule PH_2 should be bent or linear. **18-13**

Ans. Bent.

Verify the solutions for the Hückel method as applied to butadiene. Calculate the delocalization energy for butadiene in kcal mol^{-1} using spectroscopic data. **18-14**

Ans. 58 kcal mol^{-1}.

Find the energy levels for cyclobutadiene according to the Hückel treatment. **18-15**

Ans. The secular equation becomes $m^4 - 4m^2 = 0$, whence $E = \alpha$ (twice), $\alpha \pm 2\beta$.

PROBLEMS

Derive Eq. (18-19). [*Notes:* We find ψ_{1s} from Chapter 16, and substitution into Eq. (18-18) gives **18-16**

$$S = \frac{\alpha^3}{\pi} \int e^{-\alpha r_A} e^{-\alpha r_B} d\tau = \frac{\alpha^3}{\pi} I$$

where $\alpha = 1/a_0$. We use the elliptical coordinate system shown in Fig. 18-14, for which $d\tau = (R^3/8)(\lambda^2 - \mu^2)\ d\mu\ d\lambda\ d\phi$. The ψ_{1s} function is symmetric with respect to ϕ, and the integral I reduces to

$$I = \tfrac{1}{4}\pi R^3 \int_{-1}^{1} d\mu \int_{1}^{\infty} e^{-\alpha R\lambda}(\lambda^2 - \mu^2)\ d\lambda$$

The integral $\int_{x}^{\infty} t^n e^{-\alpha t}\ dt$ is known as an incomplete gamma function and is equal to

$$\frac{n!e^{-\alpha x}}{\alpha^{n+1}}\left(1 + \alpha x + \frac{1}{2!}(\alpha x)^2 + \cdots + \frac{1}{n!}(\alpha x)^n\right).]$$

18-17   Except for a normalization factor, the $1\sigma_g$ MO for H_2^+ approximates $(e^{-r_A} + e^{-r_B}$ at large R, and the $1\sigma_u$ MO approximates $(e^{-r_A} - e^{-r_B})$, with r in a_0 units. Calculate the plots of Fig. 18-5(a) using these functions. Can you explain what the normalization factor is?

18-18   Draw Fig. 18-11 for the cases of Ne_2, Na_2, and Mg_2, and show what each bond order should be.

18-19   Explain whether the molecules H_2S and MgH_2 should be bent or linear.

18-20   The MO energy level diagram for a light element AB molecule will be similar to Fig. 18-11 but with, say, the atomic orbital energies for atom B slightly lower than the corresponding ones for atom A. Sketch the diagram and assign electrons for the cases of NO, NO^+, CO, and CN^-. In each case describe the bonding and bond order.

18-21   Write the secular determinant in the Hückel treatment of cyclopropenyl, C_3H_3, and solve for its roots. Construct the molecular orbital scheme in the manner of Fig. 18-14 and place the pi electrons of the molecule.

18-22   Carry out the same procedure as in the preceding problem but now for allyl radical, C_3H_5.

18-23   Carry out the same procedure as in Problem 18-20 for trimethylene methane,

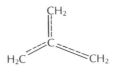

and calculate the π-electron energy. Compare this to the value for two ethylene molecules.

18-24   The π-electron energy of naphthalene is $10\alpha + 13.7\beta$. Calculate the delocalization energy in terms of β.

18-25   The Laplacian in the elliptical coordinate system is

$$\nabla^2 = \frac{4}{R^2(\lambda^2 - \mu^2)}\left\{ \frac{\partial}{\partial\lambda}\left[(\lambda^2 - 1)\frac{\partial}{\partial\lambda}\right] + \frac{\partial}{\partial\mu}\left[(1 - \mu^2)\frac{\partial}{\partial\mu}\right] \right.$$
$$\left. + \frac{\partial}{\partial\phi}\left[\frac{\lambda^2 - \mu^2}{(\lambda^2 - 1)(1 - \mu^2)}\frac{\partial}{\partial\phi}\right]\right\}$$

(18-39)

Using this and substituting ψ from Eq. (18-24), show that Eq. (18-23) separates into three equations, one in λ only, one in μ only, and one in ϕ only. Verify that Eq. (18-25) is the solution to the equation in ϕ. [*Note:* Essentially the same sequence of steps is followed as in separating the wave equation for the hydrogen atom (Section 16-7A).]

SPECIAL TOPICS PROBLEMS

18-26   Only the metal d electrons are shown in Fig. 18-18. In addition, each ligand supplies a sigma bonding pair. Add these to the diagram.

18-27   The set of d orbitals split into a t_2 and e set in tetrahedral geometry. In a transition metal complex it is the latter that are lower in energy so that the crystal field splitting is the opposite of that in octahedral geometry. Sketch the MO diagram for sigma bonding in a tetrahedral complex. Place the bonding and metal d electrons for the case of $CoCl_4^-$. Note Problem 17-45.

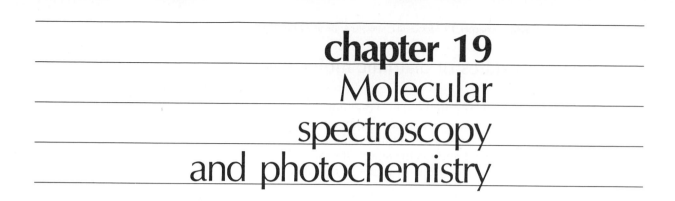

chapter 19
Molecular
spectroscopy
and photochemistry

19-1   Introduction

The field of molecular spectroscopy is a very large and a very popular one. Much of the present effort of chemical physicists is devoted to the study of the electronic, vibrational, rotational, and nuclear spin excited states of atoms and molecules. The great scope of the material is suggested by the rather lengthy list of references at the end of the chapter. The presentation here is necessarily severely limited; primary emphasis is given to electronic states of molecules—as physical chemists, we are very interested in the major changes in energy and in chemical nature that occur with electronic excitation. More than just a gain in energy is involved; an excited molecule may have a new geometry and it can undergo a variety of processes, such as emission, radiationless changes, and chemical reaction. We are beginning to see, in fact, the emergence of a distinct chemistry of excited states. We attempt therefore to present aspects of the chemical as well as of the wave mechanical approach to the subject.

The rest of molecular spectroscopy is largely concerned with phenomena whose primary interest to the physical chemist is that they provide information about the size and shape of a molecule or about the nature of the bonding in the electronic ground state. Thus analysis of vibrational and rotational spectra allows estimations of the force constants of bonds and of bond lengths. Nuclear magnetic resonance gives a special kind of information about the electronic environment of an atom, as does electron paramagnetic resonance. Such spectroscopy has become a major tool for the structural and analytical chemist. The detailed theories are of less interest, however, in a text such as this; we will present only the simplest model for each phenomenon.

The number of types of spectroscopic phenomena has grown enormously in recent years. It is impractical to discuss each of them here, but Section 19-ST-4 provides a glossary of the names in current use.

19-2   Excited states of diatomic molecules

A. The hydrogen molecule and molecular ion

The excited states of diatomic molecules are usually treated wave mechanically in terms of linear combinations of ψ functions for the separate atoms, as discussed in Sections 18-2 to 18-4. The molecular orbital method is commonly used and, as the simplest example, the ground and first excited states of H_2^+ are described in terms of the linear combinations

$$\phi_g = \psi_{1S,A} + \psi_{1S,B} \quad \text{and} \quad \phi_u = \psi_{1S,A} - \psi_{1S,B}$$

where A and B denote the two hydorgen atoms whose 1s orbitals are used; the subscripts g and u designate whether the ϕ function is symmetric or antisymmetric with respect to inversion through the center of symmetry. The energy can be calculated as a function of internuclear distance by means of the variation method (see Section 18-1) and the result is shown in Fig. 19-1. Notice that the lower curve, for ϕ_g, has a minimum, indicating that the H_2^+ molecule is stable, the depth of the minimum giving the dissociation energy. The first excited state, whose molecular orbital is σ_u, has no minimum; as a consequence, absorption of light by H_2^+ to put it in this excited state leads to prompt dissociation into H and H^+. There are further excited states, representable in first approximation by linear combinations of 2s, 2p, 3s, and so on atomic wave functions; these give a progression of states (called a *Rydberg progression*) leading eventually to ionization of the electron.

The next case, H_2, already presents a fairly difficult calculational problem. The approach is much the same as before, however, in that linear combinations of atomic orbitals are used, along with the variation method. The resulting states may be described in terms of the molecular orbitals occupied. Thus the ground state of H_2 is $(1s\sigma_g)^2$; the designation $1s\sigma_g$ refers to the molecular orbital formed from 1s atomic orbitals, or the ϕ_g orbital just given. The superscript 2 means that both electrons occupy this orbital. The above designation is often simplified to just $(1\sigma_g)^2$.

The next bound state is $1s\sigma_g 2p\sigma_u^*$; one electron is in the $1s\sigma_g$ molecular orbital and the other in the $2p\sigma_u^*$ one. This last is formed from two 2p orbitals, which must be p_z (z being the bonding axis) since the molecular orbital is of the sigma type; the asterisk means that it is an antibonding orbital. Further

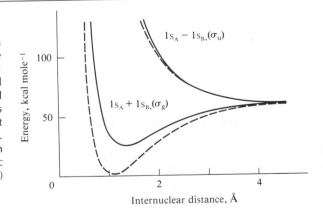

FIGURE 19-1
Energy of the H_2^+ molecular ion as a function of the internuclear distance; solid lines are calculated by the variation method and dashed lines calculated by an exact method. (From W. Kauzmann, "Quantum Chemistry." Academic Press, New York, 1957.)

Electronic configuration	Energy of the minimum (cm^{-1})^b		Internuclear distance at the minimum (Å)		TABLE 19-1 Energy states of H$_2$^{a}
	Singlet	Triplet	Singlet	Triplet	
$(1s\sigma_g)^2$ $^1\Sigma_g{}^+$	124,429	—	0.742	—	
$(1s\sigma_g 2p\sigma_u{}^*)$ $^{1,3}\Sigma_u{}^+$	32,739	~68,000^c	1.29	unstable	
$(1s\sigma_g 2p\pi_u)$ $^{1,3}\Pi_u$	24,386	28,685	1.033	1.038	
$(1s\sigma_g 2s\sigma_g)$ $^{1,3}\Sigma_g{}^+$	24,366	28,491	1.012	0.989	
$1s\sigma_g$ (H$_2$$^+$) $^2\Sigma_g{}^+$	0	—	1.06	—	

^{a}Adapted from J.G. Calvert and J.N. Pits, Jr., "Photochemistry." Wiley, New York, 1966.
^{b}Energies relative to the minimum of H$_2$$^+$.
^{c}Energy at internuclear distance of 1.29 Å (note from Fig. 19-2 that this state had no energy minimum).

excited states are listed in Table 19-1 and the calculated variation of energy with internuclear distance for several of them is shown in Fig. 19-2. Notice that this particular series of states is such that for each, one electron remains in the 1sσ_g molecular orbital. There are other excited states in which both electrons are promoted to higher energy levels.

These designations give the molecular orbitals into which each of the two electrons is placed. One may alternatively describe each state by a new set of symbols. A quantity Λ is used to indicate the component of the total electronic orbital angular momentum λ along the internuclear axis of a diatomic molecule. States of Λ number 0, 1, 2, 3, ... are called Σ, Π, Δ, Φ, ... , respectively, in capital Greek letters analogous to s, p, d, f,.... Right superscripts plus and minus denote whether or not the wave function for a Σ state changes sign on reflection in a plane through the two nuclei. If the molecule is homonuclear so that there is an inversion center, then subscripts g and u appear, to indicate whether the wave function for the state is symmetric or

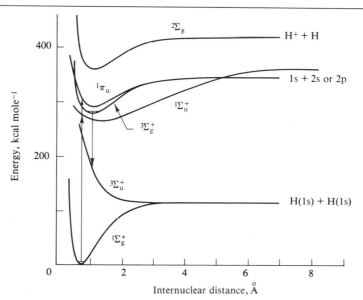

FIGURE 19-2
Potential energy curves for various electronic states of H$_2$ and H$_2$$^+$. The transitions $^1\Sigma_u{}^+ \leftarrow {}^1\Sigma_g{}^+$ and $^1\Pi_u \leftarrow {}^1\Sigma_g{}^+$ correspond to important absorption bands, while the transition $^3\Sigma_g{}^+ \rightarrow {}^3\Sigma_u{}^+$ is responsible for the continuous emission of a hydrogen arc lamp. [From E. J. Bowen, "Chemical Aspects of Light," 2nd ed. Oxford Univ. Press (Clarendon), London and New York, 1946.]

antisymmetric with respect to inversion (as noted in Section 18-3, the symbols stand for the German words *gerade* and *ungerade*). Finally, the left superscript gives the spin multiplicity, that is, whether the spin function is antisymmetric (spins paired and multiplicity 1) or symmetric (spins parallel and multiplicity 3) (see Section 18-2B).

There are certain rules, called *selection rules,* which state whether a given transition may occur or not, in first-order approximation. The wave mechanical basis for these is discussed in Section 19-3, and it is sufficient here merely to state the more important of these rules for the case of electronic transitions. The first specification is that $\Delta\Lambda = 0$ or ± 1 and that $\Delta S = 0$, or that there be no spin change. To be probable, transitions should also be ones of the type g $\leftrightarrow$ u (if the molecule has a center of symmetry so that g and u designations apply) and $+ \leftrightarrow +$ or $- \leftrightarrow -$ (but not $+ \leftrightarrow -$). Referring to Fig. 19-2, the consequence is that the ground state should undergo only the processes $^1\Sigma_u^+ \leftarrow {}^1\Sigma_g^+$ and $^1\Pi_u \leftarrow {}^1\Sigma_g^+$.† These two transitions are responsible for the important absorptions of hydorgen at 110.0 nm and 100.2 nm.

Selection rules are never absolute, however, and other states may be reached either through low-probability absorptions or by indirect means. Once formed, the $^3\Sigma_g^+$ state can emit light to drop to the $^3\Sigma_u^+$ state, for example. This last state has no potential energy minimum and therefore dissociates on the next vibration to produce two ground-state hydrogen atoms (the energy appearing as kinetic energy). As indicated in Fig. 19-2, dissociation from higher excited states may produce electronically excited hydrogen atoms. The photochemistry of hydrogen (and of diatomic molecules generally) thus consists in the production of either ground-state or excited-state atoms.

In summary, excited states of H_2 differ not just in energy but also in equilibrium internuclear distance, in dissociation energy to give atoms, and in the states of the atoms produced.

B. Other diatomic molecules

The same general theoretical approach applies to other diatomic molecules, now using hydrogen-like orbital functions. Oxygen excited states have been mentioned briefly (Section 18-5), and we show instead the somewhat analogous energy level diagram for S_2 in Fig. 19-3. The ground state, $^3\Sigma_g^-$, is paramagnetic with two unpaired spins, like that of O_2. The most probable transition is to the $^3\Sigma_u^-$ state. Notice the crossing point C in the figure where the potential energy curve for $^3\Sigma_u^-$ is intersected by that for the $^3\Pi_u$ state. If S_2 has sufficient vibrational energy in the $^3\Sigma_u^-$ state for the atoms to reach the internuclear distance corresponding to this point, then it is possible for the molecule to change to the $^3\Pi_u$ potential energy curve. Since this has no minimum, the atoms dissociate on the next vibrational swing. This type of process is called *predissociation;* it provides an explanation of how excitation to a dissociatively stable excited state can in fact lead to a prompt breakup of the molecule.

We next consider the case of a heteronuclear diatomic molecule, HI. The

---

†Note that a reverse arrow has been used in writing a transition. It is an international convention among spectroscopists that the higher or more excited state be written first regardless of the direction of the process.

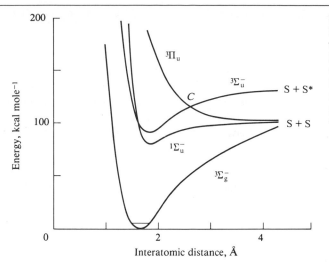

FIGURE 19-3
Potential energy curves for various electronic states of S_2. [From E. J. Bowen, " Chemical Aspects of Light," 2nd ed. Oxford Univ. Press (Clarendon), London and New York, 1946.]

potential energy curves are given in Fig. 19-4. Notice that the g and u designations have now disappeared; there is no longer an inversion center. The first few excited states are all dissociative. Irradiation of HI has with light quanta of 5 or 6 eV energy (corresponding to about 40,000 cm^{-1} or 250 nm wavelength) leads to the production of hydrogen and iodine atoms, both with considerable excess kinetic energy. Depending on the wavelength used, the iodine atoms may be in the $^2P_{1/2}$ excited state (see Section 16-ST-2 for the significance of the notation).

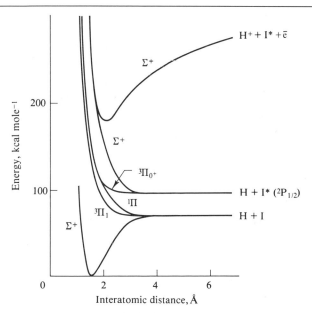

FIGURE 19-4
Potential energy curves for low-lying electronic states of HI. (From J. G. Calvert and J. N. Pitts, Jr., "Photochemistry." Copyright 1966, Wiley, New York. Used with permission of John Wiley & Sons, Inc.)

The photochemically produced hydrogen atoms may then react with HI,

$$H + HI \rightarrow H_2 + I$$

and this as well as recombination reactions

$$2I + M \rightarrow I_2 + M \qquad 2H + M \rightarrow H_2 + M$$

(where molecule M carries off the recombination energy; note Section 14-CN-1) lead to the photochemical formation of H_2 and I_2. The physical chemist makes an important distinction between *primary* photochemical processes, such as

$$HI \xrightarrow{h\nu} H + I$$

which show the immediate chemical change following excitation, and *secondary processes,* such as the other reactions just given, which the primary products undergo. These last are interesting, of course, not only for their chemistry but also in that they determine the overall photochemical change. A primary process has special importance, however, in that it describes the chemistry of a particular excited state.

19-3   Emission and absorption of radiation. Transition probability

A. Absorption absorption coefficients

The physical picture of the absorption of a light quantum is illustrated in Fig. 19-5. Electromagnetic radiation consists of an oscillating electric field (and a magnetic field at right angles to it), and absorption occurs through an interaction of the field with the electrons of the molecule. In the particular example shown an electron in an s orbital is excited to a p orbital. Note the *polarization*—the particular p orbital is the one that is aligned with the plane of the electric field.

In the theoretical treatment one assumes the train of radiation to be long enough that the atom or molecule can be regarded as immersed in an oscillating electric field. That is, we have

$$\mathbf{F} = \mathbf{F}_0 \cos 2\pi\nu t \qquad (19\text{-}1)$$

where **F** is the electric field and ν is the frequency of the radiation. If the dipole moment associated with the electron to be promoted is μ, then there will be an oscillating energy μF (note Eq. 3-27 and Section 8-ST-1) and, correspondingly, a time-dependent perturbation to the Hamiltonian:

$$\mathbf{H}(t) = \mathbf{H}^0 + \mu\mathbf{F}_0 \cos 2\pi\nu t \qquad (19\text{-}2)$$

FIGURE 19-5
Orbital representation of H atom undergoing the 1s → 2p transition (left to right). (After J. G. Calvert and J. N. Pitts, Jr., "Photochemistry." Copyright 1966, Wiley, New York. Used with permission of John Wiley & Sons, Inc.)

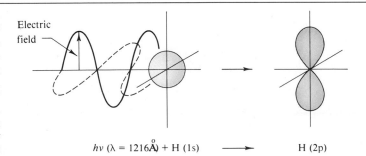

Electric field

$h\nu$ ($\lambda = 1216\overset{\circ}{A}$) + H (1s) $\longrightarrow$ H (2p)

For the case of initial state m and final state n, we write

$$\psi(t) = a(t)\psi_m + b(t)\psi_n \tag{19-3}$$

and assume that the time-dependent perturbation given by $\mathbf{H}(t)$ begins to act at $t = 0$. The problem is now one of calculating the coefficient $b(t)$; this gives the probability that the wave function for the system, $\psi(t,)$ will be just that for state n, or ψ_n. The method used is that of time-dependent perturbation theory (see Section 16-ST-3 for a discussion of ordinary perturbation theory, which is similar). The detailed treatment [see McQuarrie (1983)] leads to an expression in which the important factor is the integral

$$(\mu_x)_{nm} = \int \psi_n^* \mu_x \psi_m \, d\tau \tag{19-4}$$

Here, μ_x is the displacement of electronic charge in the direction of the oscillating field, taken to be the x direction. In the case of an atom μ_x is just $\vec{ex}$, and we have

$$(\mu_x)_{nm} = \int \psi_n^* \, (\vec{ex}) \, \mu_m \, d\tau \tag{19-5}$$

where $\vec{x}$ is the x-component of the vector that gives the position of the electron relative to the nucleus. In general, one sums over all the electrons involved. The quantity on the left-hand side is of the nature of a dipole moment, and $(\mu_x)_{nm}$, accordingly, is called the *transition dipole moment*.

As an aside, if $\psi_m = \psi_n$, Eq. (19-5) gives the actual or permanent dipole moment of the molecule (one must sum over all the electrons, of course). In the case of an atom, this intergral always turns out to be zero—an atom cannot have a permanent dipole moment.

The bottom line of the above rather "heavy" material is that the probability of absorption of radiation is proportional to $(\mu_x)_{nm}^2$; the actual expression is

$$B_{nm} = \frac{8\pi^2}{h^2} (\mu_x)_{nm}^2 \tag{19-6}$$

assuming that only the transition dipole in the x direction need be considered. Here, B_{nm} is known as the *Einstein absorption coefficient*, defined as the probability of absorption in unit time with unit radiation density. The more general expression is

$$B_{nm} = \frac{8\pi^2}{3h^2} [(\mu_x)_{nm}^2 + (\mu_y)_{nm}^2 + (\mu_z)_{nm}^2] \tag{19-7}$$

which allows for transition dipole components in the x, y, and z directions.

B_{nm} may be related to the ordinary molar extinction coefficient ϵ as defined by Eq. (3-7). since an actual absorption is spread over a band or region of wavelength, it is necessary to use the integrated intensity $\int \epsilon \, d\bar{v}$, where $\bar{v}$ is the frequency in wavenumbers. The derivation requires several steps and leads to

$$\int \epsilon \, d\bar{v} = \frac{hN_A v_0 B_{nm}}{(2.303)(1000)c^2} \tag{19-8}$$

where v_0 is the frequency at the band maximum. It is conventional to take as the "ideal" case the transition between $v = 0$ and $v = 1$ states of a harmonic oscillator of electronic mass and if the corresponding wave functions from Eq. (16-92) are substituted into Eq. (19-6), one obtains

$$B_{nm} = \frac{\pi e^2}{h m v_0}$$

Substitution of this result into Eq. (19-8) gives

$$\int \epsilon \, d\bar{v} = \frac{\pi e^2 N_A}{2303 m c^2} = 2.31 \times 10^8 \tag{19-9}$$

We take this transition probability as a reference and define the *oscillator strength* f as the actual transition probability relative to this standard. Thus

$$f = 4.33 \times 10^{-9} \int \epsilon \, d\bar{v} \tag{19-10}$$

The area under an experimental absorption band gives either B_{nm} through Eq. (19-8) or f through Eq. (19-10). For example, the area under the intense absorption band of benzene, centered at 180 nm, gives an f of about 0.7. We speak of such a transition as an allowed one. By contrast, the visible absorption band of $Co(NH_3)_6^{3+}$ due to the ligand field transition $^1T_{1g} \leftarrow {}^1A_{1g}$ has a maximum extinction coefficient of about 100 liter mol^{-1} cm^{-1} at 20,000 cm^{-1} (500 nm); the bandwidth is such that the area is about $(100)(2000) = 2 \times 10^5$, so f is about 10^{-3}. This transition is thus forbidden, that is, it is of much less intensity than that of the maximum possible.

B. Spontaneous emission

We next consider the situation in which a collection of absorbing atoms or molecules has come to equilibrium with radiation. The system is dilute enough that no collisional processes are involved, that is, no radiationless deactivations occur. Only three types of things can occur: absorption of radiation, stimulated emission of radiation, and spontaneous emission of radiation. Stimulated emission is the reverse of absorption, that is, an excited atom or molecule interacts with a radiation field with a resultant probability of undergoing a transition from excited state n to ground state m. The analysis is entirely symmetric to that for absorption, and the probability coefficient for the process, B_{mn}, is equal to B_{nm} as given by Eq. (19-6). Spontaneous emission does not depend on the presence of a radiation field, however, and has some intrinsic probability A_{mn}. The theoretical treatment of spontaneous emission requires rather advanced wave mechanics.

At equilibrium the rates of population and depopulation of the excited state have become equal, and we write

$$\text{rate of population} = B_{nm} N_m \rho_{nm}$$

$$\text{rate of depopulation} = B_{nm} N_n \rho_{nm} + A_{mn} N_n$$

where N_m and N_n are the numbers of ground- and excited-state atoms, and $\rho_{nm} = \rho_{mn} = \rho$ is the radiation density of frequency v_{nm} corresponding to the difference in energy between states n and m. Since $B_{nm} = B_{mn}$, we obtain

$$\frac{A_{mn}}{B_{nm}} = \rho_{nm}\left(\frac{N_m}{N_n} - 1\right)$$

Since the two states are in equilibrium, the Boltzmann expression applies,

$$\frac{N_n}{N_m} = e^{-h\nu_{nm}/kT}$$

Also, from Eqs. (16-146) and (16-149) the energy density of radiation, our ρ, is

$$\rho = \frac{8\pi h\nu_{nm}^3}{c^3}\frac{1}{e^{h\nu_{nm}/kT} - 1}$$

On combining these relationships, we obtain

$$\frac{A_{mn}}{B_{nm}} = \frac{8\pi h\nu_{nm}^3}{c^3} \tag{19-11}$$

The coefficient A_{mn} is mathematically equivalent to a first-order rate constant and could be written as k_e, the rate constant for spontaneous emission. The reciprocal $1/k_e$ is the (mean) lifetime for spontaneous emission; these are experimentally measurable quantities, given by the rate of decay of fluorescent or phosphorescent emission under conditions such that radiationless deactivation processes are not important (Section 19-ST-1).

Alternatively, since B_{nm} can be determined from the area under the absorption band, Eq. (19-11) can be used to calculate A_{mn} or k_e. If the observed lifetime of the excited-state emission is shorter than so calculated, one then writes $k_{e(obs)} = k_{e(natural)} + k_q$, where k_q is the sum of rates of radiationless processes. Values of k_q are often determined indirectly in this manner.

The broad absorption band for Cr(urea)$_3^{3+}$ shown in Fig. 19-12 has a maximum at about   **EXAMPLE**
16,300 cm^{-1}, and the area $\int \epsilon d\bar{\nu}$ may be estimated to be 9×10^4 with ϵ in M^{-1} cm^{-1} units and $\bar{\nu}$ in cm^{-1}. Calculate the oscillator strength and the lifetime for spontaneous emission, that is, the radiative lifetime. From Eq. (19-10), $f = (4.33 \times 10^{-9})(9 \times 10^4) = 3.90 \times 10^{-4}$; the transition must be a forbidden one. Combination of Eqs. (19-11) and (19-18) gives

$$\tau_e = 1/k_e = 1/A_{mn} = \frac{3.47 \times 10^8}{\bar{\nu}_{max}^2 \int \epsilon \, d\bar{\nu}} \tag{19-12}$$

after putting in the values of the various universal constants. Therefore $\tau_e = (3.47 \times 10^8)/[(16{,}300)^2 (9 \times 10^4)] = 1.45 \times 10^{-5}$ s, or 14.5 μs. The actual lifetime in room temperature solution is well below 1 ps, so processes other than spontaneous emission must deactivate this excited state.

The exact evaluation of the integral of Eq. (19-4) requires the use of the detailed   **C.** Selection rules
wave functions for the ground and excited states. It is possible, however, to determine on symmetry grounds whether such an integral should be nonzero.

This application of symmetry properties follows from the discussion of the direct product in Section 17-8. The integral

$$\int f_A f_C f_B \, d\tau$$

will be nonzero only if the direct product $f_A f_C f_B$ forms the basis for a representation that spans the totally symmetric IR of the point group to which the molecule belongs. In the present case, f_A and f_B are wave functions which, if they are correct for the molecule, must form bases for one or more irreducible representations (IR's) of the point group of the molecule. Further, μ_x in Eq. (19-4) is essentially a constant, e, times the x coordinate. As a consequence, the IR for which μ_x is a basis will be that listed opposite the function "x" in the character table for the point group.

As an example, for the D_{2h} group, μ_x corresponds to the B_{3u} IR (see Table 17-7). In order for the integral to be nonzero, the functions ψ_n and ψ_m must have symmetry properties such that $\psi_n \mu_x \psi_m$ contains the A_g IR of the group. For example, if ψ_n belongs to or transforms like A_g, then ψ_m must belong to B_{3u}; that is, $B_{3u} \times A_g \sim B_{3u}$ and $B_{3u} \times B_{3u} \sim A_g$. The transition is then *allowed* for radiation along the x axis. For the same ground state the transitions to states belonging to B_{2u} and B_{1u} are allowed along the y and z axes, respectively.

Notice that in this example, the IR's all have a g or u designation and that the allowed combinations are of the type g × u × u. A corollary of the general symmetry requirement is that the direct product $f_A f_C f_B$ must be g in nature (provided that the molecule does have a center of symmetry so that g and u are meaningful). Since μ_x is always u (the sign of a dipole inverts on reflection through a center of symmetry), it follows that ψ_n and ψ_m *must* be of opposite parity. This is the basis for the selection rule g ↔ u. Alternatively, a transition of type g ↔ g or u ↔ u is said to be *parity-forbidden*. The other selection rules noted in Section 19-2A can similarly be obtained from symmetry considerations.

19-4   Electronic, vibrational, and rotational transitions

A. The Franck-Condon principle

The discussion of the preceding section dealt only with electronic states, and we now consider how changes in vibrational and rotational energy may be included. The usual assumption is that electronic and vibrational–rotational energy states do not "mix," or that the total wave function can be written as

$$\psi = \psi_e \psi_{v,J} \tag{19-13}$$

where the subscripts stand for the separate wave functions. That is, we assume vibrational and rotational energies to simply superimpose on that of the electronic state. The separation of ψ_e and $\psi_{v,J}$ constitutes what is known as the *Born–Oppenheimer* approximation; the essential argument is that the nuclei, being massive, move slowly compared to electrons and may be considered at rest in solving the ψ_e. Further, rotational energies are so small that changes in them are usually neglected in considering electronic transitions, and we will do so here.

Figure 19-6 shows the hypothetical potential energy curves for the ground and first excited electronic states of a diatomic molecule. The horizontal lines indicate qualitatively the progression of vibrational states for each electronic state and, as in Fig. 16-9, the approximate appearance of the actual wave functions is included. An important implication of the Born–Oppenheimer

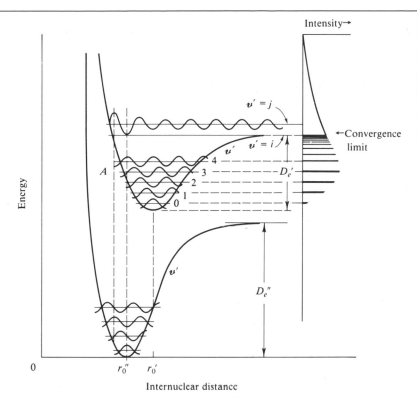

FIGURE 19-6
Illustration of the Franck–Condon principle: transitions between the ground state and excited state of a diatomic molecule. [From E. J. Bowen, " Chemical Aspects of Light," 2nd ed. Oxford Univ. Press (Clarendon), London and New York, 1946.]

approximation is that the expression (19-5) for the transition dipole moment now has the form

$$(\mu_x)_{nm} = \int \psi_e{}' \, (\overrightarrow{ex}) \, \psi_e{}'' \, d\tau_e \int \psi_{v}{}' \, \psi_{v}{}'' \, d\tau_v \tag{19-14}$$

where prime and double prime denote final and initial state, respectively, and τ is a general symbol denoting the appropriate coordinates. That is, one separates out an explicit dependence on the overlap integral of the initial and final vibrational wave functions. Now, the time for an electronic transition is about 10^{-18} sec (about that for a train of electromagnetic radiation to pass an atom), or very small compared to vibrational times, which are about 10^{-13} sec. As a consequence, nuclei do not move appreciably during an electronic excitation. The integral $\int\psi_{v}{}' \, \psi_{v}{}'' \, d\tau_v$ is therefore to be evaluated at *constant* internuclear distance, or for what is shown in Fig. 19-6 as a *vertical* transition. The idea is known as the *Franck–Condon principle*. Consider the transition from the ground electronic state of the first excited state. At ordinary temperatures most of the molecules will be in the $v'' = 0$ vibrational level. For the overlap integral to have a large value it is first of all necessary that $\psi_{v''}$ be large, which means that it is only those transitions occurring when r is near r_0'' that will be probable. It is next necessary that $\psi_{v'}$ be large for r around r_0'', which means that the most favored transition is to about the $v' = 3$ level in the figure. Transitions to various other v' levels retain some probability, however, as indicated by the satellite diagram in the right margin of the figure. In fact, the transition can be along line A, which means the excited molecule has enough energy to dissociate on the next vibration.

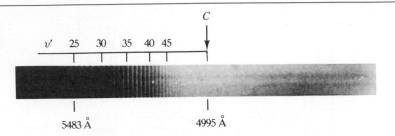

There being no barrier above D_e', the "vibrational" energy levels are essentially those of a free wave and are so close together that one sees a continuum rather than discrete energy levels.

The iodine molecule presents an example of this situation. As shown in Fig. 19-7, the first electronic transition shows as a series of lines corresponding to the spacing of vibrational states, and reaches a convergence limit or continuum. Clearly, the detailed appearance of such spectra will be sensitive to the relative shapes and positions of the potential curves for the ground and excited states. The reader might consider, for example, what the situation should be if the excited-state potential curve were very similar in shape and in r_e value to that of the ground state.

An absorption spectrum may become diffuse or blurred if a predissociation situation (see preceding section) exists. The vibrational energies of the excited state are, in effect, no longer well defined since on the first vibration the system may cross to a dissociative excited state. This is the case, for example, with the molecule S_2 (see Fig. 19-3).

Blurring also occurs often if the species is in solution or, if gaseous, is at a high pressure. Thus the absorption spectrum of I_2 in, say CCl_4, merely shows a single broad band. The qualitative reason is that collisions with gas or solvent molecules have become so frequent that a given vibrational state again has a very short life before being disturbed. Its energy is thereby made indefinite. In solution, the degree of this blurring is greater the more the molecule is solvated or highly interacting with solvent.

The above situations, in which an absorption band is broadened or made more diffuse than it should be, can be explained qualitatively in terms of a variant of the uncertainty principle (Eq. 16-8) [Pitzer (1953)]:

$$\Delta\epsilon \, \Delta t \geqslant \frac{h}{4\pi}$$

The meaning of this equation in the present context is that the shorter the lifetime of an excited state (the more certain the point in time of its existence), the greater is the range of energy value (the less certain is its energy).

B. Emission   The discussion so far has been mainly in terms of excitation, but it is, of course, also possible for an excited state to return to some lower state, ordinarily the ground state, with the emission of light. Figure 19-8 illustrates the situation with I_2. In the excitation to one component of the $^3\Pi$ state the most probable value of v' is about 26, and in the dilute gas the return is largely

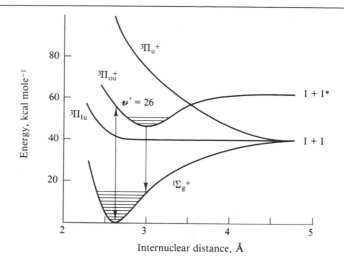

FIGURE 19-8
Potential energy diagram for various states of gaseous I_2.

from the same v' state back to the ground state. The process is known as *resonance emission*. At higher pressures or in solution what happens instead is that gas or solvent collisions remove the excess vibrational excitation so quickly that when emission does occur it is mainly from the $v' = 0$ level. This is the more usual situation and, as indicated in the figure, the Franck–Condon principle now implies that the emission will be to a high vibrational level of the ground state. The consequence is that the emission is at a *longer* wavelength than is the absorption—an observation made by Stokes in 1852.

19-5   Electronic excited states of polyatomic molecules

It is very often possible to assign features of the absorption spectrum of a polyatomic molecule to excitations that are largely localized to some particular set of atoms or bonds. A carbonyl group, for example, will usually provide rather characteristic absorption bands only secondarily modified by the rest of the molecule to which it is attached. Such groups are called *chromophores*, and a few, with their characteristic absorptions, are listed in Table 19-2. Many of these chromophores are diatomic ($R_2C{=}O$, $RC{\equiv}CR$, RCH_2X, and so on), and their theoretical treatment is that of a modified diatomic molecule. Figure 19-9 shows the set of "molecular" orbitals for formaldehyde; these are not for the molecule as a whole, but really just for the $C{=}O$ moiety. The excited states of the C—H bonds are higher enough in energy that the approximation of ignoring their mixing in with those of the $C{=}O$ portion works fairly well.

A. Localized states

There are several ways of describing the ground and excited states of a chromophore such as the $C{=}O$ group in formaldelyde. One is in terms of the molecular orbitals involved. Thus the ground state of formaldehyde is $\sigma^2\pi^2\rho_y^2$, meaning that there are two electrons in a sigma bond and two in a pi bond; the ρ_y orbitals of the oxygen are perpendicular to the $C{=}O$ axis and are not involved in bonding; their two electrons are therefore nonbonding. As indicated in the discussion of Section 18-5, for every bonding combination of atomic orbitals, there is an antibonding one, and Fig. 19-9 shows sche-

TABLE 19-2
Spectral characteristics of
organic chromophores[a]

Chromophore	Example	Absorption maximum		Approximate extinction coefficient ϵ (liter mol^{-1} cm^{-1})
		ν (kK)	λ (Å)	
C=C	$H_2C=CH_2$	55	1825	250
		57.3	1744	16,000
		58.6	1704	16,500
		62	1620	10,000
C≡C	$HC≡C—CH_2—CH_3$	58	1720	2500
C=O	H_2CO	34	2950	10
		54	1850	strong
C=S	$\underset{\parallel}{\overset{S}{CH_3C—CH_3}}$	22	4600	weak
—NO$_2$	CH_3NO_2	36	2775	15
		47.5	2100	10,000
—N=N—	$CH_3—N=N—CH_3$	28.8	3470	15
		>38.5	<2600	strong
Benzene		39	2550	200
		50	2000	6300
		55.5	1800	100,000
C—Cl	CH_3Cl	58	1725	
C—Br	CH_3Br	49	2040	1800
C—OH	CH_3OH	55	1500	1900
		67	1830	200

[a]Data from J. G. Calvert and J. N. Pitts, Jr., "Photochemistry." Wiley, New York, 1966.

matically the orbital appearance of the σ* and π* antibonding states. One then speaks of σ → σ* and π → π* transitions. In addition, an electron from the nonbonding p$_y$ orbital of the oxygen may be promoted to the σ* or a π* level of the carbonyl "molecular" orbitals; the transitions are then called $n \to \sigma^*$ and $n \to \pi^*$, respectively.

In the case of simple molecules, a formal group-theoretic designation of the ground and excited-state wave functions may be useful. That is, the wave function for a given state will form the basis for one of the irreducible representations of the symmetry group to which the molecule belongs (Section 17-5). As an example, the $n \to \pi^*$ transition in formaldehyde has the group-theoretical designation $^1A_2 \leftarrow {}^1A_1$.

Some excitations involve a substantial reorganization or transfer of electronic charge. The term *charge transfer* was introduced by R. S. Mulliken for transitions in which a significant shift in electron density occurs between atoms or groups of atoms [see Calvert and Pitts (1966)]. In simple molecules or chromophores such transitions often involve the promotion of an electron from a bonding to the corresponding antibonding orbital. The hydrogen molecule, for example, has an absorption at 110.9 nm to an excited state whose orbital picture is approximately H$^+$H$^-$. Also, the gaseous alkali halide molecules show a charge transfer absorption, as in the continuous absorption band for CsI(g) around 200 nm—the result of the absorption being to yield Cs and I atoms. The aqueous or hydrated halide ions show an absorption in which an electron is transferred to the solvent,

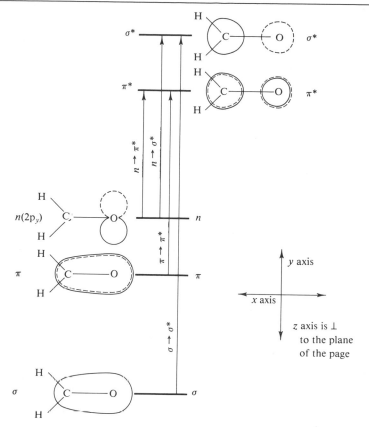

FIGURE 19-9
Molecular orbitals for
formaldehyde and their
approximate relative
energies. (From J. G.
Calvert and J. N. Pitts, Jr.,
"Photochemistry."
Copyright 1966, Wiley,
New York. Used with
permission of John Wiley
& Sons, Inc.)

$$I^-(aq) \xrightarrow{\;h\nu,232nm\;} I(aq) + e^-(aq) \qquad (19\text{-}15)$$

and similarly for metal ions such as $Fe^{2+}(aq)$ and coordination compounds such as $Fe(CN)_6^{4-}(aq)$. The $e^-(aq)$ species is a solvated electron; it reduces water in about 1 msec (millisecond) and other scavengers more quickly, but its transient absorption spectrum is well known. (There is a maximum at 680 nm and a concentrated solution of electrons would appear blue.)

The assignment of a particular absorption band as change transfer is not so easy with polyatomic molecules. Two criteria are as follows. First, charge transfer excitations are usually facile, that is, the extinction coefficient for the transition is large, and an intense band not otherwise identifiable will generally be so classed. Second, the photochemistry of a charge transfer excited state is usually one of redox decomposition, as in Eq. (19-15). As a further example, Fig. 19-10 shows the absorption spectra for acetophenone, $CH_3CO\phi$, and benzophenone, $\phi CO\phi$ ($\phi = C_6H_5$). The intense absorption around 240 nm is attributed to charge transfer and, in the former case in particular, one reason for doing so is that the products CH_3CO and C_6H_5 are observed, indicative of an electron density shift from the acetyl to the phenyl group in the excited state. Finally, the first intense absorption band of $Co(NH_3)_6^{3+}$ is classed as charge transfer (from ligand to metal), consistent with the photochemical observation that Co^{2+} and oxidized ammonia are produced.

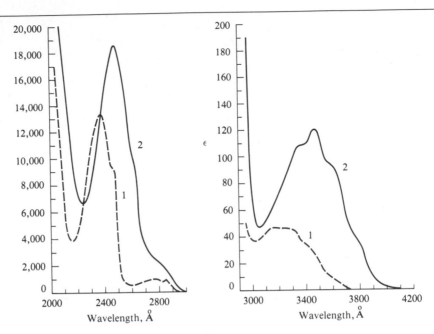

FIGURE 19-10
Absorption spectra for (1) acetophenone ($CH_3COC_6H_5$) and (2) benzophenone ($C_6H_5COC_6H_5$) (in cyclohexane at 25°C). (From J. G. Calvert and J. N. Pitts, Jr., "Photochemistry." Copyright 1966, Wiley, New York. Used with permission of John Wiley & Sons, Inc.)

B. Delocalized states   Electronic transitions may be between states whose wave functions are best described as encompassing a number of atoms. Examples are the conjugated polyenes and aromatic compounds. Both cases may be treated (rather crudely) by the particle-in-a-box model (Section 16-5) whereby the pi electrons are assigned in pairs to the successive energy levels of the set of standing waves. Such wave functions are truly molecular ones in that no use is made of atomic wave functions, the wave equation being solved for the molecule as a whole.

The Hückel treatment (Section 18-7) makes use of combinations of atomic orbitals to formulate wave functions for the whole conjugated system. The result is again a set of standing waves distributed over the entire molecule, the corresponding energy states being populated by the available pi electrons.

C. Excited-state processes   We have so far stressed the wave mechanical description of excited states in terms of their energies and electron distributions. The photochemist is also interested in the various processes which an excited state can undergo, and a generalized scheme is shown in Fig. 19-11. We suppose the molecule to be polyatomic, with several internuclear distances, so that simple potential energy diagrams such as Fig. 19-6 are no longer possible. The various families of energy levels are instead assembled in vertical arrays, the secondary lines indicating the superimposed vibrational and rotational fine structure.

The figure contains a great deal of detail which needs explanation. First, organic molecules generally have an even number of electrons, that is, they are not free radicals. Further, the spins are all paired in the ground state, which is therefore a singlet state, labeled S_0 in the figure. The more prominent absorptions are to excited singlet states, shown as S_1 and S_2. One expects a second series of states, similar to the singlet ones, but in which an electron has inverted its spin so that the molecule has a net spin of unity, and is therefore in a *triplet* state. Direct transitions such as $T_1 \leftarrow S_0$ are relatively

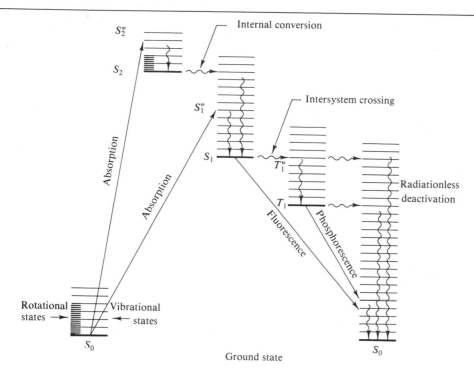

FIGURE 19-11
Various photophysical processes. Radiative transitions are given by solid lines, and radiationless ones, by wavy lines. The fine structure of lines indicates schematically vibrational and rotational excitations. (The successive displacements to the right are purely for the purpose of avoiding overlap of levels; the transitions shown are "vertical" ones.)

improbable because of the spin selection rule mentioned in Section 19-2. Typically, then, one sees various absorption bands due to singlet–singlet transitions, usually with most or all of the vibrational-rotational detail washed out if the system is in a condensed phase (solution or neat liquid or solid).

The actual absorptions are shown in the figure as $S_1^v \leftarrow S_0$ and $S_2^v \leftarrow S_0$, the superscript meaning that the transition terminates at some high vibrational level of the excited state. The situation is similar to that shown in Fig. 19-6; we assume the transitions to be vertical and that the S_1 and S_2 states are distorted relative to S_0.

A number of secondary processes may now take place. If the molecule is in solution, it is very likely that the S_2^v or S_1^v state will lose its vibrational energy or *thermally equilibrate* to the S_2 or S_1 true electronic state, the energy being dissipated into the solvent. Then S_2 may pass to the vibrationally excited state $S_1^{v'}$; the process is known as *internal conversion*. It may happen because of a crossing of the potential energy surfaces for the two states, or some interaction with solvent can be involved. The consequence is that regardless of which singlet excited state is first populated, a molecule usually ends up in the lowest excited singlet state as a result of internal conversion and rapid thermal equilibration. Thermally equilibrated excited states have been termed *thexi* states.

The S_1 state may return to the ground state S_0 by emission. We will use the term *fluorescence* for emission between states of the same spin multiplicity. Alternatively, the S_1 state may go to S_0 by *radiationless deactivation*. As the name implies, no radiation is emitted, the excess energy appearing either as vibrational excitation of S_0 or of adjacent solvent. Finally, the S_1 state may transform to a more or less vibrationally excited triplet state. The act is ra-

FIGURE 19-12
Absorption and emission spectra for Cr(urea)$_6^{3+}$ (in a glassy matrix at 77 K). [From G. B. Porter and H. L. Schläfer, Z. Phys. Chem. **37**, 110 (1963).]

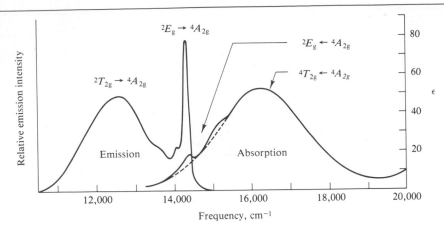

diationless, the difference in energy between S_1 and T_1 appearing in vibrational excitation of T_1 or, possibly, in solvent vibrations; we call such a transition an *intersystem crossing*. If the produced state is T_1^v, it is assumed to thermally equilibrate rapidly to T_1, which may then return to the ground state either by emission or by radiationless deactivation. Such emission now involves states of differing spin multiplicity and is called *phosphorescence*.

In general, an emission will be called a phosphorescence if it is relatively slow or long-lived, so that a strongly forbidden process appears to be involved. Figure 19-12 shows the superimposed absorption and emission spectra for an octahedral complex ion, Cr(urea)$_6^{3+}$. There is a broad absorption band corresponding to the $^4T_{2g} \leftarrow {}^4A_{2g}$ transition, and a broad emission band that is almost its mirror-image in shape. This emission is attributed to the reverse process and occurs with a very short lifetime. There are also some small bumps on the main absorption band, attributed to the $^2E_g \leftarrow {}^4A_{2g}$ transition, and a very sharp emission peak, due to the reverse process. This last emission has a relatively long lifetime. Both sets of absorption and emission features are forbidden because the transition is g ↔ g. In addition, the $^2E_g \leftrightarrow {}^4A_{2g}$ transition is spin-forbidden. The practice has been to call the spin-allowed, short-lifetime emission a fluorescence and the spin-forbidden, longer-lifetime emission a phosphorescence.

Processes of the preceding type have been termed *photophysical*, meaning that they leave the molecule intact. *Photochemical* change is, of course, that whereby an excited state undergoes isomerization, fragmentation, or reaction with solvent or a solute. A common observation with organic systems is that chemical change is largely associated with the T_1 state. For example, if a solution of *trans*-stilbene, ϕ—CH = CH—ϕ, is irradiated, one observes some fluorescence from the S_1 state, but a good deal of intersystem crossing to T_1 also occurs. The T_1 state is probably π* in type, and the weakening of the double bond allows easy isomerization. The consequence is that irradiation of the *trans*-stilbene absorption band at about 290 nm, corresponding to $S_1^v \leftarrow S_0$, results in a fairly efficient *trans* to *cis* isomerization.

A very important type of process is that of *photosensitization*, whereby an excited molecule transfers its excitation energy to some second species. Thus the T_1 state of biacetyl,

CH₃CCCH₃,

is about 55 kcal mol⁻¹ above the ground state S_0. Irradiation of biacetyl in the presence of stilbene induces isomerization of the latter, and it appears that the process

$$^3D^* + {}^1A \rightarrow {}^1D + {}^3A^* \tag{19-16}$$

has occurred, where D denotes *donor* (biacetyl) and A *acceptor* (stilbene). The energy transfer very likely occurs during an encounter between $^3D^*$ and ^1A species. Often a reaction of this type occurs on the first encounter, or with a rate constant of about 10^{10} liter mol⁻¹ sec⁻¹ (see Section 15-4).

If the $^3D^*$ state shows an observable phosphorescence, then the competition of process (19-16) leads to phosphorescence *quenching*. That is, the intensity of phosphorescence, on irradiating the system, is progressively reduced by increasing concentrations of the acceptor. The same may happen to fluorescence emission. The situation may be treated by conventional stationary-state kinetic analysis; see Section 19-ST-1.

To summarize, we have considered the following typical processes.

1. Absorption, usually $S_1^v \leftarrow S_0$ or $S_2^v \leftarrow S_0$.
2. Thermal equilibration: $S_1^v \rightarrow S_1$, $T_1^v \rightarrow T_1$, and so on.
3. Internal conversion: $S_2 \rightarrow S_1^v$, $T_2 \rightarrow T_1^v$.
4. Radiationless deactivation: $S_1 \rightarrow S_0$, $T_1 \rightarrow S_0$.
5. Intersystem crossing: $S_1 \rightarrow T_1^v$.
6. Fluorescence: $S_1 \rightarrow S_0^v + h\nu$.
7. Phosphorescence: $T_1 \rightarrow S_0^v + h\nu$.
8. Chemical reaction: S_1 or $T_1 \rightarrow$ chemical change.
9. Sensitization or energy transfer.
10. Quenching of emission.

19-6   Vibrational spectra

Most vibrational spectra are treated in terms of a set of harmonic oscillators, both for diatomic and polyatomic molecules. The case of a single harmonic oscillator was treated wave mechanically in Section 16-6, with the result given by Eq. (16-94). There is a single characteristic frequency ν_0 and the quantized energy states are

A. Diatomic molecules

$$E = h\nu_0(v + \tfrac{1}{2}) \qquad \text{[Eq. (16-94)]}$$

where v may be 0, 1, 2, … . By analogy with the equivalent mechanical system, Eq. (16-97) gives

$$h\nu_0 = 2.59 \times 10^{-13} \left(\frac{\hbar}{\mu'}\right)^{1/2} \text{ erg}$$

where $\hbar$ is the restoring force constant, in units of 10^5 dyn cm⁻¹, and μ' is the reduced mass of the diatomic molecule, in mass units.

The symmetry requirement implicit in Eq. (19-5) imposes the further condition that the probability of light absorption to produce a change in vibrational energy will be zero unless $\Delta v = \pm 1$. It further turns out that even if the symmetry requirement is met, the probability of light absorption will be zero unless some change in molecular dipole moment accompanies the transition. The consequence is that vibrational intensities are theoretically zero for homopolar or A–A-type molecules; harmonic oscillations of such a molecule may vary in amplitude but cannot produce a dipole moment.

These rules are based on the assumption of the harmonic oscillator; an actual molecule will have a potential energy versus nuclear separation curve such as shown in Fig. 19-6, and in this case the $\Delta v = \pm 1$ rule is voided. In practice, this means that while the change $\Delta v = \pm 1$ remains the most probable, other values can occur as well. The requirement that a change in dipole moment occur is a quite stringent one, however, and the consequence is that the vibrational absorption spectrum of an A–A-type molecule is very weak indeed. Heteronuclear diatomic molecules, however, show fairly intense absorptions. As noted in Section 16-6, HCl absorbs at 2886 cm^{-1}, corresponding to the $v = 0$ to $v = 1$ transition. The corresponding frequencies for HF, HBr, and HI are at 4141, 2650, and 2309 cm^{-1}, respectively. These absorptions all lie in the infrared, of course.

B. Raman spectroscopy

A molecule may reveal its vibrational energy states by an inelastic scattering of a light photon. The experimental observation is that a small fraction of the incident light is scattered and that the scatterd light may differ in energy from that of the incident light by an amount corresponding to a vibrational spacing. Usually this spacing is that for which $\Delta v = \pm 1$. This is called the *Raman effect*.

The whole effect is a second-order one, and might be rather unimportant except that its probability depends on the polarizability of the molecule rather than on its dipole moment. The relevent equation is

$$(\mu_x)_{nm} = \mathbf{F}_0 \int \psi_n^* \alpha \psi_m \, d\tau \tag{19-17}$$

where α is the molecular polarizability and $\mathbf{F}_0$ is the electric field strength of the exciting radiation. It turns out that molecules of the A—A type will show a Raman spectrum even though the usual infrared vibrational absorption spectrum is forbidden. More generally, for a molecule with a center of symmetry, infrared-active transitions are Raman-inactive, and vice versa. Raman spectroscopy is thus a valuable supplement to the usual infrared absorption measurements.

From the experimental point of view, a very intense and highly monochromatic light source is needed, the first requirement to produce appreciable scattered light and the second so that the small change in frequency can be measured accurately. For example, if light of 254 nm, or about 39,000 cm^{-1}, is used to irradiate oxygen, the Raman scattered light differs in frequency by only 1600 cm^{-1}, corresponding to the difference between the $v = 0$ and $v = 1$ vibrational levels. An important development has been the use of laser light; this provides both the high-intensity and, more importantly, the highly monochromatic light needed. To repeat, it is not necessary that the wavelength used correspond to an actual electronic transition; the physical picture given here is merely a way of visualizing the inelastic collision of the light quantum with a molecule. The intensity of Raman scattering is enhanced if the excitation wavelength of the exciting light corresponds to that of an electronic transition; one now speaks of a *resonance Raman* effect.

The discussion in Section 4-9 tells us how to count the number of vibrational degrees of freedom for any molecule, namely $3n - 6$ for a nonlinear one and $3n - 5$ for a linear one, n being the number of atoms in the molecule. We should therefore expect to observe this number of distinct vibrational frequencies, each having a value determined by the force constant and reduced mass appropriate for the particular motion involved. These frequencies bear no rational relationship to each other, and if a polyatomic molecule could be observed in some ultramicroscope, its atoms would appear to be undergoing complicated, never-repeating oscillations. These complex oscillations are, however, the result of the superposition of various primitive vibrations, and an important theoretical problem is the prediction of what these last should be.

The primitive vibrations are known as the *normal (vibrational)* modes of the molecule, and it turns out that the motions of each normal mode must form the basis for one of the irreducible representations of the point group to which the molecule belongs (Section 17-2). Consider, for example, the molecular ion CO_3^{2-}. The normal modes are depicted in Fig. 19-13. Carbonate ion, being planar, has the point group symmetry D_{3h}, for which the irreducible representations are A_1', A_2', E', A_1'', A_2'', and E'' (Table 17-7). It is easy to see that the ν_1 mode of Fig. 19-13 is totally symmetric with respect to the operations of the group, and hence belongs to the A_1' representation. The ν_2 mode is unchanged by the E, C_3, and σ_v operations but is put into the negative of itself by the C_2, S_3, and σ_h operations, thus identifying it as belonging to the A_2'' representation. The ν_{3a} and ν_{3b} modes *together* form the basis for the E'

C. Polyatomic molecules

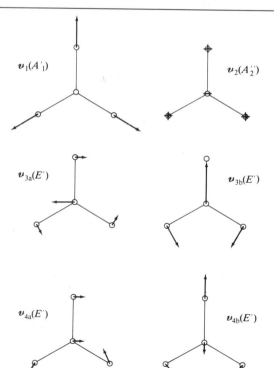

FIGURE 19-13
The normal modes of vibration of CO_3^{2-}; plus and minus denote motion in and out of the plane of the paper, respectively. [After F. A. Cotton, "Chemical Applications of Group Theory." Copyright 1963, Wiley (Interscience), New York. Used with permission of John Wiley & Sons, Inc.]

$\nu_1(A_1')$

$\nu_2(A_2'')$

$\nu_{3a}(E')$

$\nu_{3b}(E')$

$\nu_{4a}(E')$

$\nu_{4b}(E')$

representation, and so on. One expects a total of $4 \times 3 - 6 = 6$ normal modes, which is just the sum of the orders of the irreducible representations involved.

While it is relatively easy to see that the indicated normal modes for CO_3^{2-} do belong to certain representations, the reverse type of analysis is much more difficult. That is, considerable effort may be required to deduce what types of motions constitute the normal modes for some arbitrary molecule. We will not attempt to explore the problem here, except to mention that the procedure involves assigning x, y, and z vectors to each atom and carrying the set of vectors through the symmetry operations of the group. To give a brief example, the resulting reducible representation in the case of CO_3^{2-} gives

$$T = A_1' + A_2' + 3E' + 2A_2'' + E''$$

The translation of the molecule as a whole must involve representations carrying x, y, and z designations (see Table 17-7), or $A_2'' + E'$. Rotation corresponds to R_x, R_y, and R_z, so that A_2' and E'' are so assigned, leaving A_1', $2E'$, and A_2'' for vibrations.

It would be convenient if infrared absorption spectra simply showed the separate frequencies for each normal mode. Several complications enter, unfortunately. First, only those modes that involve a change in dipole moment will be infrared-active; the others will often be Raman-active. This means that certain fundamental frequencies, such as the ν_1 mode for CO_3^{2-}, will be absent. Second, the normal modes interact, so that various combinations of changes in vibrational quantum numbers may be involved. Thus CO_2 has four normal modes, of which only three are infrared-active; these have fundamental frequencies of 667 cm^{-1} (twofold degenerate) for ν_2 and 2349 cm^{-1} for ν_3. The symmetric, Raman-active vibration ν_1 is at 1384 cm^{-1}. The observed infrared spectrum shows absorptions at wavelengths corresponding to $\Delta v = 1$ for ν_2 and ν_3, but also for $\nu_3 - \nu_2$, $\nu_3 - 2\nu_2$, and other combinations, as well as for overtones, or larger Δv values. The consequence is that it can be very difficult to disentangle or assign infrared absorption frequencies to specific quantum transitions.

Fortunately, combination and overtone absorptions tend to be weaker than the fundamental ones, especially if the various normal mode frequencies are well separated in value. Furthermore, the normal modes will at least approximately correspond to simple motions of the molecular framework, with the result that to a first approximation the vibrations of a molecule may be regarded as simple stretchings or bendings of individual bonds. The consequence is that in polyatomic organic molecules characteristic functional group frequencies are found. Some of these are listed in Table 19-3. One can thus identify the presence of C—H, C—C, C—O bonds and so on in a molecule from the appearance of characteristic infrared absorption peaks. In a very real sense, the infrared spectrum of a molecule identifies or "fingerprints" it; the presence of a particular molecule in a mixture can thus be identified or even determined quantitatively. The spectra of some simple molecules are shown in Fig. 19-14; the reader can test the extent to which frequencies listed in Table 19-3 are present.

If an atom has several equivalent bonds, as in a CH_3 group, then, of course, a collection of normal modes is involved rather than just a superposition of independent C—H vibrations. The resulting spectral detail is itself a characteristic of the group of atoms, and in

TABLE 19-3

Some characteristic bond force constants and frequencies[a]

Bond	Stretching		Bending	
	Force constant ($\times 10^5$ dyn cm^{-1})	Frequency (cm^{-1})	Motion	Frequency (cm^{-1})
$\equiv$C—H	5.85	3300	$\equiv$C—H	700
$>$C—H	4.79	2960	C(H)(H)—H	1000
—C$\equiv$C—	15.59	2050	C(H)(H)	1450
$>$C=C$<$	9.6	1650	—C(H)(H)—H	1450
—C$\equiv$N	17.73	2100		
$>$C=O	12.1	1700		
—O—H	7.66	3680		
$>$C—Cl	3.64	650		

[a] Data from G. Herzberg, "Infrared and Raman Spectra of Polyatomic Molecules." Van Nostrand–Reinhold, Princeton, New Jersey, 1945.

fact, the symmetry of a group of atoms may sometimes be deduced from the splittings of the bond frequencies of the group. As an example, irradiation of the carbonyl complex $Cr(CO)_6$ produces the fragment $Cr(CO)_5$, and the splitting of the CO frequencies determines that the geometry is that of a trigonal bipyramid.

D. Fourier transform infrared spectroscopy

A very effective modification to ordinary absorption infrared spectroscopy is that known as Fourier transform infrared, FTIR, spectroscopy. In the conventional approach, a blackbody source of infrared radiation, such as a ceramic bar resistor heated to several thousand degrees, is used to produce a beam of infrared radiation, which is collimated and then passes through a grating monochromator. A particular wavelength is selected and the intensity of the beam that passes through the sample is determined, and compared to the intensity with no sample. In this way the percent transmission is obtained. The usual automatic recording instrument continuously varies the monochromator setting and the transmission is recorded as a function of wavenumber. The spectra of Fig. 19-14 were obtained in this manner.

A major disadvantage of the above approach is that only a very small fraction of the output of the infrared source is used at any one instant—namely the narrow wavelength band picked by the monochromator. In FTIR the whole output of the source is used (actually some very wide range of wavelengths corresponding to the spectral range of interest). This clever trick is accomplished by obtaining an *interferogram* based on the transmitted beam. One way of doing this is illustrated in Fig. 19-15, which shows the principle of a Michelson interferometer. The beam from the source is split by a partly reflecting–partly transmitting mirror, and the two resulting beams are reassembled to impinge on a detector. The path length of *one* of the beams is made variable by means of a movable mirror. For every one-half wavelength motion of this mirror the intensity of a monochromatic beam as seen by the detector will go through a cycle from zero to full intensity as the split beams go from destructive to constructive interference. The light intensity falling on the detector will thus

FIGURE 19-14
Infrared absorption
spectra. (a) Methyl ether in
a 10-cm cell at 10 mm Hg.
The various types of
motions shown include
stretching, ν, bending, δ,
and rocking, r. These may
be symmetric or
asymmetric, denoted by
subscripts s and as. (From
H. S. Szymanski,
"Interpreted Infrared
Spectra," Vol. 3. Plenum
Press, New York, 1967.)
(b) Methylene chloride
liquid, 0.032-mm path
length. (c) Methyl ethyl
ketone, in CCl₄ solution.
(From "Coblentz Society
Spectra," Coblentz
Society, Inc., Norwalk,
Conn.) The figures also
illustrate the various
common ways of plotting
infrared spectra.

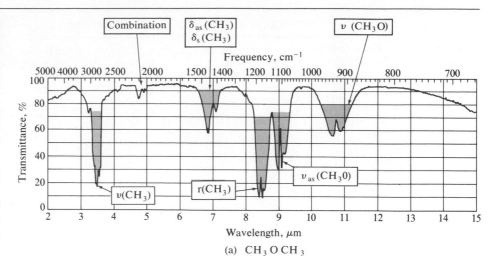

(a) CH₃OCH₃

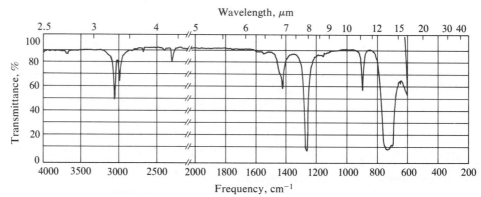

(b) CH₂Cl₂

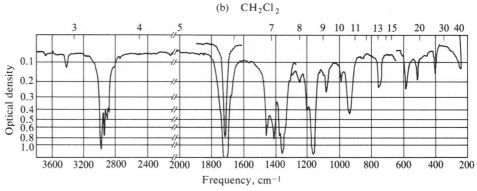

(c) CH₃C(=O)—CH₂—CH₃

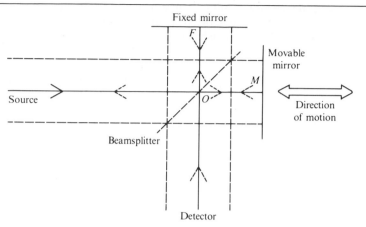

Fixed mirror

Movable mirror

Source

Direction of motion

Beamsplitter

Detector

FIGURE 19-15
Idealized diagram for a Michelson interferometer. The absorbing sample might be placed either just after the source or just in front of the detector. [See, for example, P.R. Griffiths, "Chemical Infrared/Fourier Transform Spectroscopy." Wiley, New York, 1975.]

Spectrum

Interferogram

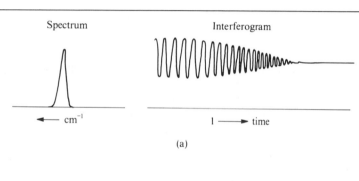

← cm⁻¹

l ⟶ time

(a)

FIGURE 19-16
Various light source spectra (left) and the corresponding interferograms. Case (d) is that of a broad-band light beam that has passed through an absorbing sample.

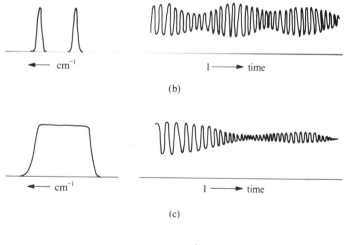

← cm⁻¹

l ⟶ time

(b)

← cm⁻¹

l ⟶ time

(c)

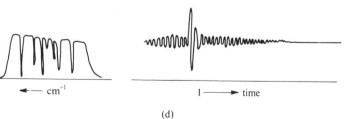

← cm⁻¹

l ⟶ time

(d)

oscillate as the movable mirror is moved over a distance large compared to the wavelength. One thus obtains an interferogram as an *oscillation* of the detector output in time, of period determined by the wavelength of the light involved and the speed of motion of the movable detector. This output might be sampled electronically at a succession of small time intervals, and the resulting data stored in a computer for later analysis.

Were the monochromatic beam of some different wavelength, λ', the path length difference between constructive and destructive interference would now be $\lambda'/2$, and for a given velocity of the movable mirror, the frequency of intensity oscillations at the detector would now be some new value. Figure 19-16(a) shows how the detector response would oscillate for a single wavelength beam, and Fig. 19-16(b) shows the resultant of super-imposing the oscillations from a beam having two wavelengths. The appearance for a broad wavelength band of radiation is shown in Fig. 19-16(c).

If, now, an absorbing sample is placed in front of the detector, certain wavelengths will be absorbed and that particular contribution to the interferogram of Fig. 19-16(c) will be missing. The FTIR instrument thus produces an interferogram in the form of a very complex oscillating output of the detector, as in Fig. 19-16(d). This oscillating output can be translated back into an intensity-versus-wavenumber spectrum by the mathematical operation known as a Fourier transform. The actual procedure is discussed in some detail in Section 19-8C, which deals with Fourier transform nuclear magnetic resonance spec-troscopy.

The important point in this qualitative discussion is that the interferogram is obtained using the entire output of the infrared source and can therefore be obtained very quickly and/or with much better signal to noise ratio than is possible in conventional absorption spectroscopy. This advantage coupled with the use of fast response time infrared detectors makes it possible to obtain infrared spectra of transient chemical species on the millisecond and even shorter time scale.

19-7   Vibrational–rotational spectra

The detailed spectrum of a molecule consists in principle of transitions be-tween states whose complete description includes the electronic, vibrational, and rotational components of the wave functions. As discussed in Section 19-3, we assume that these types of functions do not interact appreciably. Thus, as in Fig. 19-11, close-lying rotational states are superimposed on vi-brational states, which in turn are superimposed on electronic states.

We consider here the case in which both vibrational and rotational changes in state occur, and the type of absorption spectrum that results. (Pure rota-tional spectroscopy is described in Section 19-7.) The actual degree of detail that is seen in a rotation-vibration absorption spectrum depends on several factors. First, of course, is the resolution of the equipment used. Second, however, vibrational and rotational states have sufficiently long natural life-times for their energy to be made uncertain by collision processes. Thus in solution vibrational detail tends to be washed out and rotational detail dis-appears completely. This happens even with gases at high pressures.

If, however, one examines the infrared spectrum of a dilute gas, then rotational as well as vibrational detail is seen. It is such spectra that we consider briefly at this point. On combining Eqs. (16-94) and (16-130), the general expression for the vibrational–rotational energy of a diatomic molecule becomes

$$\epsilon_{vJ} = h\nu_0(v + \tfrac{1}{2}) + BhcJ(J + 1) \tag{19-18}$$

where $B = h/8\pi^2 Ic$ and is called *the rotational constant* (units are cm^{-1}).

The selection rules for a transition are that first, $\Delta v = \pm 1$ [and the molecule must have a dipole moment or else μ_x in Eq. (19-4) vanishes, so the intensity becomes zero], and second, $\Delta J = \pm 1$. We then write

$$\epsilon_{vJ} = (v' - v'')\, hv_0 + B'hcJ'(J' + 1) - B''hcJ''(J'' + 1) \tag{19-19}$$

for a transition between two vibrational–rotational states. In general the rotational constant changes on going to a different vibrational state since the vibrational amplitude is different and hence so is the moment of inertia of the molecule. If this point is ignored, Eq. (19-19) simplifies to

$$\begin{aligned}
\epsilon_{vJ} &= hv_0 + 2BhcJ' & J' - J'' &= 1 \\
\epsilon_{vJ} &= hv_0 + 2Bhc(J' + 1) & J' - J'' &= -1
\end{aligned} \tag{19-20}$$

where $v' - v''$ is taken to be unity, and ΔJ may be ± 1, primes and double primes denoting the final and initial states, respectively.

It is important to remember that at ordinary temperatures most molecules will be in the $v = 0$ vibrational state, but that the separation of rotational levels is so small that by the Boltzmann principle, an average molecule is apt to be in a $J = 10$ to 20 level. Thus the usual transition is from $v = 0$ to $v = 1$ in a diatomic molecule, but the rotational quantum number, being large, has scope to decrease as well as to increase. The consequence is that a vibrational–rotational spectrum has two branches. The R branch contains the transitions whereby the distribution of rotational states present in the collection of molecules changes by $\Delta J = 1$ and the P branch contains the transitions for which $\Delta J = -1$. These two branches are illustrated in the spectrum for the fine structure of the $v = 2$ fundamental of HCN shown in Fig. 19-17; the short-wavelength branch is, of course, that for R. In polyatomic molecules, transitions with $\Delta J = 0$ may also be observed, giving a third set of lines known as the Q branch.

Analyses of spectra of this type yield more information than analyses of pure rotational spectra since the intensities and spacings reflect the degree of interaction between the two forms of excitation. As mentioned earlier, the moment of inertia changes with the vibrational quantum number. Also, vibrational levels are affected by the rotational energy—essentially because the vibrating atoms are now in a centrifugal field.

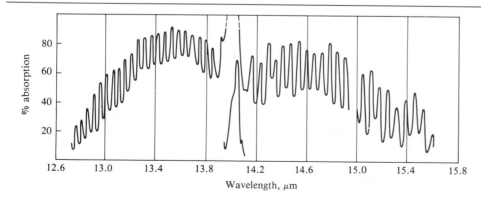

FIGURE 19-17
Fine structure of the v_2 fundamental of HCN. (See G. Herzberg, "Infrared and Raman Spectra of Polyatomic Molecules." Van Nostrand–Reinhold, Princeton, New Jersey, 1945.)

19-8   Rotational absorption spectra—microwave spectroscopy

There are several spectroscopic phenomena that yield valuable information about the geometry or the bonding of molecules in their lowest electronic or ground state. The detailed theory will not be given, but the final equations and their applications are definitely of interest. We have already considered vibrational spectra (Section 19-5) and now turn to pure rotational spectra. Mössbauer spectroscopy is mentioned in Section 22-CN-6.

Transitions from one rotational state to another may be observed directly with radiation of the appropriate wavelength (the molecule must have a permanent dipole moment). The energy levels of a diatomic rigid rotator are given by

$$E = \frac{h^2}{8\pi^2 I} J(J + 1) \qquad \text{[Eq. (16-130)]}$$

where J may be 0, 1, 2, ... The moment of inertia I is defined by $I = \Sigma m_i r_i^2$, where r_i is the distance of atom i from the center of mass, and for a diatomic molecule I values are around 10^{-40} g cm^2 (corresponding to m about 3×10^{-23} g and r about 10^{-8} cm). Rotational energies are therefore about 5×10^{-15} erg molecule^{-1}, or about $0.1 kT$ at room temperature.

The symmetry properties of rotational wave functions (see Section 19-3) lead to the rule that transitions are favored only if $\Delta J = \pm 1$, and for a transition from J to $J + 1$, Eq. (16-130) yields

$$\bar{\nu} = 2B(J + 1) \tag{19-21}$$

where $\bar{\nu}$ is frequency in wavenumbers and B, called the *rotational constant*, is $h/8\pi^2 Ic$. Usual values of B are around 1 to 100 cm^{-1}, corresponding to wavelengths of 0.5–0.005 in rotational spectra.

These numbers reveal several important qualitative aspects of rotational spectroscopy. The wavelength of the electromagnetic radiation involved approaches that of shortwave radio or radar waves. It was, in fact, the development of radar and the availability of surplus equipment after World War II that brought rotational or *microwave spectroscopy* into prominence. A second comment is that the rotational energy level spacings are so close together that a molecule at room temperature will most probably be in a fairly high rotational level; the observed spectrum is then one involving transitions from one excited state to another, unlike most other spectroscopy. It should be noted that polyatomic molecules have three moments of inertia and a correspondingly more complex set of energy states. Finally, rotational absorptions are relatively weak in intensity and the absorption bands must be quite sharp to be detected with precision. The consequence is that experiments are largely limited to rather dilute gases to minimize the line broadening that results from molecular collisions.

Figure 19-18 illustrates several of the experimental aspects. With regard to the block diagram, Fig. 19-18(a), klystron tubes were originally used as microwave generators, but new types of oscillators are used now. These may be swept automatically through the desired frequency range and are usually calibrated against a fixed reference oscillator. The microwave radiation passes through the sample cell, and its intensity is measured by a detector. The absorption spectrum is then recorded or seen on an oscilloscope. Figure 19-

18(b) shows contemporary instrumentation. Note the rectangular wave guides that carry the radiation. A single absorption line may appear as in Fig. 19-18(c); in this case the line width is only 80 kHz and since the frequency is about 40 GHz, the position of the line can be measured to within about 2 parts per million.

The *Stark* effect plays an important role. Each J state has $2J + 1$ possible orientations [recall Eq. (4-73)], corresponding to an azimuthal quantum number, M where $M = 0, \pm 1, \pm 2, \ldots, \pm J$ (note the parallel to the ℓ and m

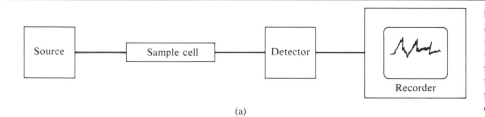

(a)

(b)

FIGURE 19-18
Microwave spectroscopy.
(a) Equipment schematic.
(b) Microwave
spectrometer. Manifolds
for introduction of
samples are in the left
console; the microwave
generator circuitry is at the
lower right. The wave
guides and associated
attenuators and wave
shapers are mounted
above the console.

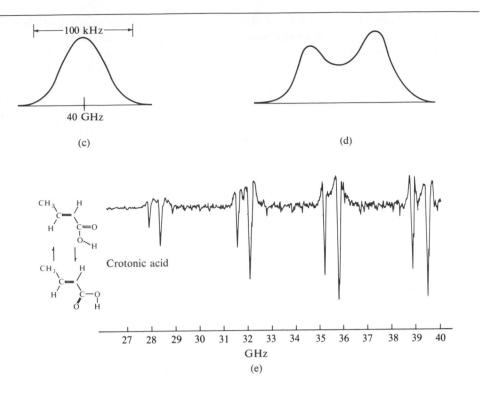

FIGURE 19-18(c)
Single absorption line. (d)
Same line as split by a
field of 1000 V cm⁻¹. (e)
Microwave spectrum of
crotonic acid. (Photograph
and spectrum courtesy of
Hewlett-Packard Co., Palo
Alto, California.)

quantum numbers for the hydrogen atom, Section 16-7B,C). This degeneracy is removed in an electric field; since the energy in the field depends on M^2, the splitting is into just $J + 1$ (rather than $2J + 1$) levels. This Stark splitting is illustrated in Fig. 19-18(d). The magnitude of the Stark effect depends on the molecular dipole moment, and microwave spectroscopy constitutes an important means of measuring dipole moments.

The effect is routinely used to enhance sensitivity. The electric field applied to the sample cell is in the form of a square-wave alternating potential (typically 0–1000 V and around 30 kHz in frequency). One now greatly reduces noise in the detector by accepting only the response having the frequency of the field.

TABLE 19-4
Geometry of some symmetric tops from microwave spectroscopy[a]

Molecules	d_{12} (Å)	d_{23} (Å)	θ
CH_3F	1.11	1.39	110°
CH_3Cl	1.113	1.781	110°31′
CH_3I	1.113	2.1392	111°14′
CCl_3H	1.767	1.073	110°24′
SiH_3Cl	1.44	2.050	110°

[a]Data from C.H. Townes and A.L. Schawlow, "Microwave Spectroscopy." McGraw-Hill, New York, 1955. The molecules are all of the C_{3v} point group; the bond lengths and angles are as shown in the diagram.

Figure 19-18(e) shows the microwave spectrum of crotonic acid. The two series of bands are for the two conformational isomers, impossible to separate chemically. Perhaps the most important application of microwave spectroscopy has been to the determination of the rotational constant B and hence of the moment (or moments) of inertia of a molecule (see Section 4-CN-2). By isotopically labeling various atoms of the molecule, individual bond lengths can be determined rather accurately, as can bond angles. A polyatomic molecule is difficult to treat theoretically unless two of its three moments of inertia are equal, so that the molecule behaves as a symmetric "top"; examples are ammonia, CH_3Cl, and $CHCl_3$. Some typical data are given in Table 19-4.

19-9   Nuclear magnetic resonance

A. Basic principles

A rather different type of spectroscopy than that of Section 19-7 is based on the splitting of otherwise degenerate *nuclear* energy states that occurs in a magnetic field. The fundamental nuclear particles, the proton and the neutron, have intrinsic angular momenta of $\frac{1}{2}h/2\pi$, usually reported as just $\frac{1}{2}$. These combine in a nucleus to give a net nuclear spin which is an even integral number of units of $\frac{1}{2}$ if there are an even number of fundamental nuclear particles, and an odd integral number of units of $\frac{1}{2}$ otherwise. For example, the even nuclei ^2H, ^{12}C, and ^{16}O have nuclear spins of 1, 0, and 0, respectively, while the odd nuclei ^1H, ^7Li, ^{15}N, and ^{19}F have spins of $\frac{1}{2}$, $\frac{3}{2}$, $\frac{1}{2}$, and $\frac{1}{2}$, respectively. This net nuclear spin is given the symbol I (not to be confused with a molecular moment of inertia).

Nuclei have, of course, a net electric charge, and a nonzero nuclear spin implies a motion of this charge or a current, hence an associated nuclear magnetic moment μ_n. The theoretical magnetic moment for a proton, treated as a spinning spherical shell of charge, is given by the nuclear magneton β_n, defined as

$$\beta_n = \frac{eh}{4\pi Mc} \tag{19-22}$$

The nuclear magneton is just m/M times the Bohr magneton (Section 3-ST-2), where m and M are the electron and proton mass, respectively, and its numerical value is 5.0508×10^{-24} erg G^{-1} (G is the abbreviation for the unit of magnetic field, gauss). Actual nuclear magnetic moments differ from this value, and it has become customary to express them as

$$\mu_n = g_n\beta_n I \tag{19-23}$$

where g_n is the nuclear g factor (sometimes called the *magnetogyric ratio* or the *gyromagnetic ratio*) and is essentially the ratio of the magnetic moment to the spin angular momentum. If μ_n is expressed in units of β_n, or as μ_n', then $g_n - \mu_n'/I$.

If a magnetic field **H** is present, then the energy of a nucleus having spin I becomes dependent on its orientation with respect to the field. The quantum restriction is that the component of μ_n in the direction of the field μ be

$$\mu = m_n g_n \beta_n \tag{19-24}$$

where m_n is a quantum number which may have the values, $I, I - 1, I - 2, \ldots,$ $-I$, or $2I + 1$ values in all. The energy of each orientation depends on the field strength,

$$E = -\mu\mathbf{H} = -m_n g_n \beta_n \mathbf{H} \tag{19-25}$$

The case of $I = \frac{1}{2}$ is illustrated in Fig. 19-19.

It turns out that the selection rule for transitions between magnetic states is $\Delta m = \pm 1$ or that $\Delta E = g_n \beta_n \mathbf{H}$. Since ΔE is also given by $h\nu$, where ν is the frequency of radiation,

$$\Delta E = h\nu = g_n \beta_n \mathbf{H} \tag{19-26}$$

$$\nu = \left(\frac{\mu_n' \beta_n}{hI}\right)\mathbf{H}$$

EXAMPLE We can calculate the resonance frequency for a proton in a field of 10,000 G (gauss) or 1 T (tesla). The proton spin is $I = \frac{1}{2}$ and μ_n' is 2.79268. We have $\Delta E = (2.79268)(5.0508 \times 10^{-24})(1 \times 10^4)/(1/2) = 2.821 \times 10^{-19}$ erg, and $\nu = \Delta E/h = 2.821 \times 10^{-19}/6.62618 \times 10^{-27} = 4.257 \times 10^7$ s^{-1} or 42.57 MHz. We may also compare ΔE to the kT value of 4.12×10^{-14} erg at 25°C, and find $\Delta E/kT = 6.847 \times 10^{-6}$.

The above example tells us that the populations of the upper and lower states typically do not differ by very much. Specifically, their ratio should be $\exp(2\Delta E/kT)$, and the exponential can be expanded to give the upper and lower state probabilities as $\frac{1}{2}[1 \mp (\Delta E/kT)]$. In the case of protons in a field of 1 T, one has $\frac{1}{2}(1 - 6.847 \times 10^{-6})$ and $\frac{1}{2}(1 + 6.847 \times 10^{-6})$, respectively, at 25°C. The natural time for nuclei to reach this equilibrium or Boltzman distribution depends on the various processes present whereby nuclei exchange energy with their surroundings and is called the *spin–lattice relaxation time T_1* (this is the reciprocal of the rate constant for the approach to the equilibrium distribution). Values of T_1 depend on the chemical (and magnetic) nature of the molecules present in the medium, but usually are in the range of 10^{-2}–10^2 s for liquids. For water T_1 is about 3.6 s and for ethanol the value is 2.2 s. Thus when the external magnetic field is turned on the protons present in a sample will adjust very quickly to the Boltzmann distribution of their two energy states.

An experimental means of measuring ΔE might be through the absorption of a quantum of electromagnetic radiation. In the above example, the frequency of 42.57 MHz is that of radiation of wavelength $2.998 \times 10^{10}/42.57 \times 10^6 = 704$ cm, corresponding to the shortwave radio region. The

FIGURE 19-19
Splitting of the proton spin states in a magnetic field.

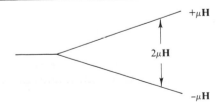

Increasing **H**

problem is that while nuclei in the lower-energy state would absorb such radiation, those in the upper state would be stimulated to emit the same wavelength radiation and return to the ground state. The theoretical probabilities for absorption and stimulated emission are identical (see Section 19-3), and since there are virtually equal numbers of nuclei in the two states, the *net* absorption of radiation will be very small. It is possible to measure it by equipment of the type shown in Fig. 19-20. First, it is easier and therefore customary to use a fixed radiofrequency source and to put the magnetic field through a small variation—one only needs perhaps 100 parts per million (ppm) change in a field of 10,000 G. When the field is such that the frequency is just right, then a minute net energy dissipation occurs in the sample around which the transmitter coil is located. If the rf circuit is delicately tuned, a drop in its output voltage will occur and can be shown on an oscilloscope or, for a single sweep of the magnetic field, on a chart recorder.

The *nuclear magnetic resonance* (nmr) effect would be no more than a somewhat obscure aspect of physics were it not that the resonance energy depends on the exact value of the *local* field H at the nucleus and that H is affected by the electron distribution in the molecule containing the nucleus. For example, the orbital electrons of each atom themselves precess in the applied field H_0 to give rise to diamagnetism (Section 3-ST-2), that is, to an induced field which opposes the applied one and is proportional to it. One then writes

$$H = H_0(1 - \sigma) \tag{19-27}$$

where σ is often called the *screening constant* since its effect is to reduce the effective field at the nucleus; its value is around 10^{-5} for protons. The effect is called a *chemical shift*.

Since H, and therefore σ, is not directly determinable, the usual procedure is to compare the value of H_0' needed to produce resonance (at a given radio frequency) in some standard compound with the value H_0 needed for the one being studied. The standard may be any liquid substance giving a simple resonance behavior—water, $CHCl_3$, and $Si(CH_3)_4$ have been used, for example, in the case of *proton magnetic resonance* (pmr). The chemical shifts involved are so small that it is convenient to report them as

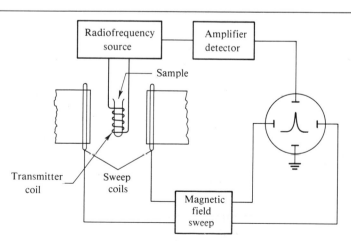

FIGURE 19-20
Schematic diagram of apparatus of an nmr experiment. (After J. A. Pople, W. G. Schneider, and H. J. Bernstein, "High Resolution Nuclear Magnetic Resonance." Copyright 1959, McGraw-Hill, New York. Used with permission of McGraw-Hill Book Company.)

FIGURE 19-21
The pmr spectrum of
liquid ethanol.

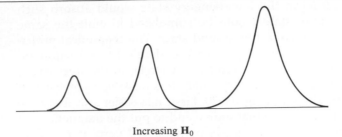

Increasing H_0

$$\delta = \frac{H_0 - H_0'}{H_0'}10^6 \qquad (19\text{-}28)$$

that is, δ is reported as the parts per million shift in the applied field needed to produce resonance.

We come now to actual pmr spectra. Each proton in some pure compound will have its own electron environment and hence chemical shift. Thus as shown schematically in Fig. 19-21, liquid ethanol shows resonances at three values of H_0, corresponding to the —OH proton, the two equivalent CH_2 protons, and the three equivalent CH_3 protons; the areas under the absorption peaks are in the ratio 1:2:3. One of the major values of pmr (and of nmr in general) is that it allows an identification of the nuclei in a molecule in terms of their various chemical environments. The chemical shifts for some compounds having only one kind of proton are given in Fig. 19-22, relative to cyclohexane. There are extensive tables of chemical shifts for protons in various chemical environments, and the pmr spectrum of a molecule not only serves to "fingerprint" it but usually allows quite detailed conclusions as to its isomeric structure. If doubts are left, deuteration of known functional groups eliminates those hydrogen atoms from pmr resonance, so that the peaks due to them in the original spectrum can be identified.

There are a large number of important effects and hence of nmr applications, two of which are illustrated in Fig. 19-23. In pure ethanol the OH, CH_2, and CH_3 peaks are seen to be split if measured with higher resolution than

FIGURE 19-22
Observed chemical shifts at room temperature of some liquids that give a single proton signal; cyclohexane is taken as an arbitrary reference point.

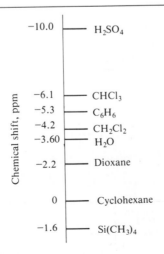

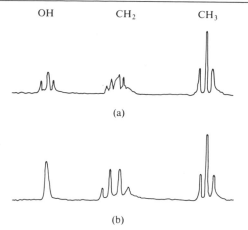

OH   CH$_2$ CH$_3$

(a)

(b)

FIGURE 19-23
The pmr spectrum of
liquid ethanol: (a) pure
dry alcohol; (b) alcohol
plus a small amount of
HCl; (c) case (b) at a lower
frequency.

used for Fig. 19-21. This is due to the mutual interactions of the proton spins on neighboring groups; thus the hydroxyl proton resonance is split into three (an unresolved central one and two satellites) by the various ways in which spin–spin interaction can occur. The second effect illustrated is that in the presence of hydrochloric acid the splitting of the —OH peak disappears and the shape of the CH$_2$ peak is greatly simplified. The reason is that the hydroxyl proton is now exchanging so rapidly with that of neighboring molecules that only its average local magnetic field is being observed. The time scale for such averaging to occur is, in simple cases, of the order of T_1.

The spin–spin interactions that split the CH$_2$ peak into four components and the CH$_3$ peak into three components (Fig. 19-23) arise as follows. The principle is that the field at a CH$_2$ proton is perturbed slightly by the net field of the CH$_3$ protons and, similarly, the field at a CH$_3$ proton is affected by the net field of the CH$_2$ protons; the number of components into which the peaks split is the number of possible values of these net fields. Thus the three CH$_3$ protons may have their spins in the relative arrangements (↑ ↑ ↑), (↑ ↑ ↓, ↑ ↓ ↑, ↓ ↑ ↑), (↑ ↓ ↓, ↓ ↑ ↓, ↓ ↓ ↑), or (↓ ↓ ↓), and a given CH$_2$ proton then "sees" one of four possible perturbing net fields in the relative probabilities 1:3:3:1. Similarly, the two CH$_2$ protons may have their spins in the relative arrangements (↑ ↑), (↑ ↓, ↓ ↑), and (↓ ↓); a CH$_3$ proton then "sees" one of three perturbing fields in the relative probabilities 1:2:1.

B. The vector model

Even a brief presentation of nmr would be inadequate without some mention of an alternative picture of the effect. The discussion has so far been in terms of energy levels, but a more physical picture is as follows. If the magnetic moment of a proton is represented by a vector, then application of an external field causes this vector to precess at an angle to the field direction that is determined by I. The two energy states then correspond to the two vector orientations shown in Fig. 19-24. The precession, known as the Larmor precession, occurs with a *frequency* that is proportional to the field and equal to that of radiation corresponding to E in Eq. (19-25).

The collection of protons in the sample will all be undergoing this precession, but not in unison as indicated schematically in Fig. 19-25(a). If, now,

FIGURE 19-24
Larmor precession of a
magnetic moment along
the z axis.

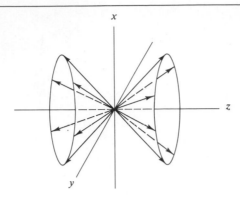

an rf field of resonance frequency is applied along the x axis, its oscillating magnetic field applies a small acceleration or retardation to the precessing magnetic moment vectors until they all come into phase, as indicated in Fig. 19-25(b). A precessing moment constitutes a source of electromagnetic radiation emitted along the y axis; since all of the nuclei are precessing together, the emission from them is in phase and so has a nonzero net amplitude. A detector coil placed perpendicular to the x axis then registers a signal. This

FIGURE 19-25
Effect of a perpendicular rf
field in bringing
precessing magnetic
moments into phase.

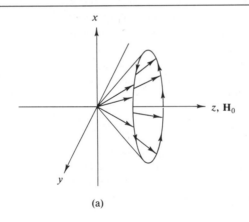

(a)

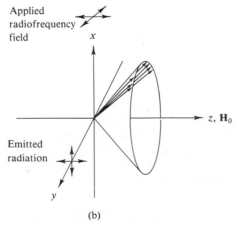

(b)

is, in fact, an alternative method for obtaining an nmr spectrum—that is, one uses an rf emitting coil and a second, receiving coil at right angles to it, the two coils directed at axes perpendicular to that of the applied magnetic field.

Some additional phenomena may now be observed. The dynamic equilibrium between the two nuclear states N and N* may be written as a balance of several rates

$$(N)(k_r + k_a) = (N^*)(k_{se} + k_e + k_r^*) \qquad (19\text{-}29)$$

where k_a and k_{se} are the rate constants for absorption and stimulated emission, respectively. These are equal to each other and proportional to the intensity of the rf field; k_e is the rate constant for spontaneous or ordinary emission. The rate constants k_r and k_r^* are those for radiationless activation and deactivation processes. If the rf intensity is zero, then the various rate constants are such as to make $(N^*)/(N)$ equal to the Boltzmann ratio, as evaluated earlier. If, however, the rf intensity is made very large, so that k_a and k_{se} dominate, then, since they are equal, $(N^*)/(N)$ becomes unity—that is, in the limit one has an equal population in the two states. There will be no nmr resonance signal at all.

If now the rf intensity is returned to its normal low value, the Boltzmann population will reestablish itself, and the nmr signal will grow back in. The reciprocal of the first-order rate constant for this return is called the *spin-lattice* or *longitudinal relaxation time* T_1. Its value is primarily a measure of that of k_r and k_r^* since k_e is generally negligibly small. The presence of paramagnetic ions in the solution will, for example, greatly reduce T_1; the magnetic susceptibility of such ions may actually be measured by this means.

There is a second relaxation time, called T_2 or the *transverse relaxation time*. This has to do with the speed with which the aligned moments of Fig. 19-25(b) would drift out of phase in the absence of an rf field, due to the different local magnetic fields that individual nuclei experience. The relaxation time T_2 can be estimated from the width of the resonance line as well as by other, somewhat more complicated experiments. In liquids the mechanism for the T_1 and T_2 relaxations are often essentially the same, so the two times are about equal. In a solid, however, T_1 may become quite large while T_2 remains small. This is because the rate constant for energy exchange with the medium, k_r or k_r^* in Eq. (19-29), has become small, but the local field inhomogeneities remain to make the different nuclear moments precess at slightly different natural rates, and thus still produce the transverse relaxation effect.

There are several special techniques that serve to improve the information from a nuclear magnetic resonance measurement, or the ease of interpretation. One very direct approach is to go to a higher radiofrequency, to, say, 300 MHz instead of the 60 MHz used in older or simpler machines. The transition energies that can be seen are now fivefold larger and, by Eq. (19-26), a correspondingly larger magnetic field is used. Since the chemical shift δ is proportional to the field, by Eq. (19-28), the spectra due to different kinds of nuclei are fivefold more widely separated. The fine structure spacing, such as shown in Fig. 19-23, is *not* affected, however. In a complicated molecule a 60 MHz spectrum might be very messy because the fine structure spacing for each type of proton overlaps the signals from different protons. This is illustrated in Fig. 19-23(c), where the use of a lower frequency (and weaker

C. Some special techniques

field) has moved the three bands close enough together that their fine structure overlaps. By going to the higher frequency, the signals from different kinds of protons are separated and the resulting spectrum is much easier to interpret.

Another technique is that known as *decoupling*. Referring again to Fig. 19-23, if a strong second source of radiation is used to stimulate CH_3 group transitions, then these nuclei are flipped back and forth rapidly. The effect is that the CH_2 protons no longer "see" the neighboring CH_3 group protons, and the quartet of lines shrinks to a single line, again much simplifying the spectrum.

A very powerful, very different approach is that known as *Fourier transform, FT, spectroscopy*. A short, say 1 μs, pulse of intense radiofrequency radiation is applied, whose frequency range or "bandwidth" covers the whole spectrum to be studied. All nuclei whose transitions fall in the frequency range are excited and, after the pulse is over, undergo their separate precessions (note Fig. 19-25). The detector of radiofrequency radiation now sees a decaying oscillation, each type of nucleus doing this at its own characteristic frequency. This is known as *free induction decay*, FID. A recording of the detector response might look as shown in Fig. 19-26. What is actually done is to store the data digitally as, say, N intensity versus time points. A dedicated computer now carries out a Fourier transform of the data for each of a succession of frequencies, ω. That is, if $I(t)$ is the signal intensity at time point t and $F(\omega)$ is the absorption at frequency ω, the Fourier transform that relates the two may be approximated by a sum over some large number of time points according to the expression

$$F(\omega) = \sum_{n=0}^{n=N} I(t)\, e^{-2\pi i \omega n\, \Delta t} \qquad (19\text{-}30)$$

Here, $n\Delta t$ is the nth time point in a series of samplings of $I(t)$ at uniformly spaced intervals Δt. As noted above, $I(t)$ is an oscillating function (and, in real situations, one that gradually damps out). If $I(t)$ has only some single frequency ω_0, $F(\omega)$ will be essentially zero except if $\omega = \omega_0$. And if two or more frequencies are superimposed in the $I(t)$ function, $F(\omega)$ will be nonzero

FIGURE 19-26
(a) Free induction decay spectrum for $(CH_2{=}CH)_4Si$ showing two principal frequencies and (b) the frequency domain spectrum after Fourier transform (the small triplet is due to the benzene-D_6 solvent). [From G.C. Levy and G.L. Nelson, "Carbon-13 Nuclear Magnetic Resonance for Organic Chemists." Wiley-Interscience, New York, 1972.]

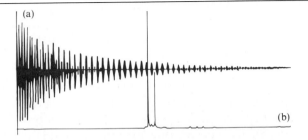

at just those frequencies. The mathematics is such that the result of the Fourier transform of a FID is to give a normal nmr spectrum. In the case shown in Fig. 19-26, there are two kinds of protons and the Fourier transform gives two peaks. See Problem 19-46 for an illustrative numerical exercise.

An important advantage of FT nmr is that data are obtained at *all* frequencies simultaneously, rather than at just one frequency at a time as is the case with conventional nmr. One can average the data from a succession of exciting pulses, and the result is that nmr spectra can be obtained that would be too weak to see by the conventional technique, such as ^{13}C spectra (the low natural abundance of ^{13}C makes it very difficult to obtain a conventional nmr spectrum).

D. NMR imaging

Some remarkable developments have occurred in recent years that allow the obtaining of two- or even three-dimensional nmr images of objects. The important object, of course, is the human body, or some portion of it such as the brain, liver, or a kidney. Figure 19-27 shows a proton nmr cross-sectional image of a liver.

Two important aspects are the following. First, by using various sequences of radio-frequency pulses, it is possible to enhance the discrimination between nuclei, such as protons, in different kinds of tissue. What is involved is that the T_1 and T_2 relaxation times (Section 19-8B) differ in different environments. One thus achieves an enhancement of contrast.

The second aspect is that of mapping the signals in space, that is, of imaging. The use of nmr to obtain the spatial distribution is sometimes called *zeugmatography*. Two types of approaches are the following. One may confine the signal to a thin disc-shaped region by using a surface coil—that is, a loop or radiofrequency coil is placed on the surface of the body. One may, for example, confine a signal to that from the muscle mass just under the skin. An alternative approach is to establish a magnetic field gradient. The chemical shift for a given nucleus thus becomes a function of distance along the field and hence of depth into the sample, and by confining the measurement to a given frequency one knows the depth of the volume element from which the signal is obtained. The actual volume element can be defined by using magnetic field gradients in different directions [see Budinger and Lauterbur (1984) and Gadian (1982)].

Finally, one is not confined to proton nmr. ^{31}P nmr, for example, yields simpler spectra and thus better resolution. It allows the specific tracking of important phosphorus-containing compounds such as ATP (adenosine triphosphate), phosphocreatine, and inorganic phosphate.

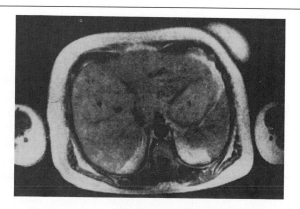

FIGURE 19-27
Proton nmr image of a liver cross-section. [From W.T. Dixon, Washington University.]

19-10   Electron spin resonance

The general principle of electron spin resonance is the same as for nmr; the relevant equation is analogous to Eq. (19-25),

$$E = -g_e m_e \beta_e \mathbf{H} \tag{19-31}$$

where g_e is called the *Landé g factor* and is 2.0023 for a free electron. The magnetic moment $m_e = \frac{1}{2}$ for a free electron and β_e is the Bohr magneton, whose value is about 2000 times that of the nuclear magneton β_n. The first consequence of these changes is that the splitting of the two spin orientations of a free electron is about a thousand times that for a nucleus. The difference in Boltzmann population of the upper and lower states is correspondingly larger, and so is the resonance signal. Thus a much lower concentration of unpaired electrons can be detected with *electron spin resonance, esr*, than of nuclei in the nmr method; smaller or more dilute samples therefore suffice. To be specific, electron spin resonance can detect about 10^{11} spins or 10^{-12} moles of unpaired electrons. Often, there is more than one spin being measured, and the more general term is *electron paramagnetic resonance* or *epr*.

Esr or epr spectra are generally measured with a fixed microwave frequency, corresponding to about 3 cm wavelength; the sample is exposed to a magnetic field that is swept through values around 10^4 gauss (or ten tesla). At resonance, the fixed microwave frequency just corresponds to

$$h\nu = g_e \beta_e \mathbf{H} \tag{19-32}$$

the equation analogous to Eq. (19-26).

EXAMPLE   A typical free radical might have a g value of 2.05. If the epr instrument uses microwave radiation of 3 cm wavelength, at what field strength will absorption occur? We have from Eq. (19-32)

$(6.626 \times 10^{-27})(2.998 \times 10^{10})/3 = (2.05)(9.2741 \times 10^{-21})\mathbf{H}$

or $\mathbf{H} = 3.483 \times 10^3$ gauss. In SI, the corresponding calculation is

$(6.626 \times 10^{-34})(2.998 \times 10^8)/0.03 = (2.05)(9.2741 \times 10^{-24})\mathbf{H}$

or $\mathbf{H} = 0.348$ T.

The absorption spectrum for a single spin will have the appearance shown in Fig. 19-28. What the instrument actually records is the *derivative* of an absorption spectrum; the absorption maximum is therefore at the field strength such that the tracing crosses the abscissa, that is, has the value zero. Note that in the above example a g value of 2.05 was used, rather than the free electron value of 2.0023. With actual species the electronic structure provides a local magnetic field, which can add to or subtract from the applied field. The effect is usually reported as an effective g value, that is, a g_{eff}. The measured g_{eff} is thus characteristic of a given radical species.

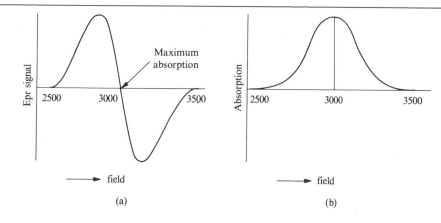

FIGURE 19-28
Hypothetical esr or epr spectrum. (a) The derivative spectrum reported by the instrument. (b) The actual absorption spectrum. The field strength corresponding to the absorption maximum is shown by the arrow in (a).

There is a further complication if the nuclei present have a magnetic moment, since this may add to or subtract from the local field. The hydrogen atom, for example, has a nuclear spin of $\frac{1}{2}$ (as discussed in the preceding subsection) and, as a consequence, the epr spectrum is split into two features, one on either side of where the absorption would be were the nuclear spin zero. The effect is known as *hyperfine splitting*. The deuterium atom has a nuclear spin of 1, and its allowed orientations relative to the electron spin permit a contribution to the local field proportional to $+1$, 0, or -1. Each possibility provides an esr absorption, so the spectrum now shows three equally spaced features. These splittings are illustrated in Fig. 19-29(a) and (b). Consider next a molecule with several protons, such as a methyl radical, $CH_3\cdot$. The three nuclear spins may combine, to give a net nuclear spin of $+\frac{3}{2}$ $+\frac{1}{2}$, $-\frac{1}{2}$, and $-\frac{3}{2}$. As discussed in the preceding subsection on nuclear magnetic resonance, the $+\frac{1}{2}$ and $-\frac{1}{2}$ cases have a relative weighting of 3. As a consequence, the epr spectrum is split into four components, equally spaced, and with the center two having three times the amplitude of the outer two. This is illustrated in Fig. 19-29(c). The protons are equivalent in the methyl radical, of course, and in cases of radicals where the protons are not equivalent, more complex splittings occur. A consequence of all this is that the detailed structure of the hyperfine splitting can give information about the geometry of a radical species, as well as indicate with which nucleus or nuclei the unpaired electron primarily resides.

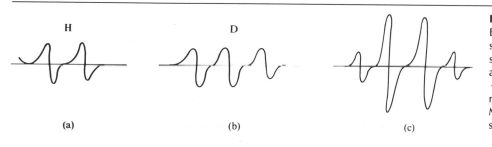

FIGURE 19-29
Examples of hyperfine splitting of esr or epr spectra. (a) Hydrogen atom, nuclear spin $+\frac{1}{2}$ or $-\frac{1}{2}$. (b) Deuterium atom, nuclear spin $+1$, 0, -1. (c) Methyl radical, net nuclear spin $+\frac{3}{2}$, $+\frac{1}{2}$, $-\frac{1}{2}$, $-\frac{3}{2}$.

COMMENTARY AND NOTES

19-CN-1 Structure and chemistry of excited states

A. Structure

Excited states do not survive long enough for conventional structure determinations, but there is no doubt that major changes from the ground state may occur. In the case of diatomic molecules, the bond length is expected to increase in an excited state and this is confirmed by calculations for H_2 (note Fig. 19-2). The bond angle in a triatomic molecule may change greatly; as examples, it has been suggested that the $^2A_1$ state of NO_2 is linear and that the first excited state of NH_3 is planar. Conversely, the first singlet excited state of formaldehyde is known to be nonlinear (deduced from an analysis of the rotational and vibrational structure of the fluorescence spectrum). In the case of states localized on a double bond, there is effectively weaker bonding in a π^* state and therefore easier rotation about the bond. It is possible that benzene is no longer planar in some of its excited states.

Considerable changes in geometry have been postulated for the ligand field excited states of coordination compounds. One of the singlet excited states of $Ni(CN)_4^{2-}$ (square planar in the ground state) is thought to have D_{2d} symmetry. Complexes such as $Cr(NH_3)_5 X^{2+}$ (where X is a halogen or pseudohalogen) may change from essentially octahedral geometry to that of a pentagonal pyramid.

One indication that an excited state is significantly different in geometry from the ground state is that the emission from the former is strongly shifted to lower energies relative to the absorption band. Recalling Fig. 19-11, the energy for the $S_1^v \leftarrow S_0$ process is shown as much greater than that for the $S_1 \rightarrow S_0^v$ fluorescent emission. As indicated in Fig. 19-6, this means that for a diatomic molecule, the bond length in the excited state is different from that in the ground state. In the case of polyatomic molecules, angle as well as bond length changes are likely.

An interesting example is that of $Cr(urea)_6^{3+}$, whose absorption and emission spectra are shown in Fig. 19-12. The main ab-

sorption band involves the process $^4T_{2g} \leftarrow {}^4A_{2g}$, and the narrower, lower-energy band involves the process $^2E_g \leftarrow {}^4A_{2g}$. The phosphorescence emission from the ^2E_g state is almost superimposed on the absorption band in its wavelength distribution, which strongly implies that the ^2E_g state has essentially the same bond lengths and the same O_h symmetry as the ground state. However, the fluorescence emission from the $^4T_{2g}$ state is broad, like the absorption band, but shifted to much longer wavelengths. The peak of the fluorescence emission is in fact at a *longer* wavelength than that of the phosphorescence emission. It seems evident that major bond length and perhaps bond angle changes have occurred in the $^4T_{2g}$ state.

As shown in Fig. 19-30, emission spectra from coordination compounds may be strongly shifted at high pressures. The effect with ruby, Fig. 19-30(a) is now widely used as a secondary calibrating standard for measuring high pressures. At sufficiently high pressure, actual inversion of energy levels may occur, and this may be one explanation of the phenomenon of *triboluminescence*. Many substances emit light when struck sharply or crushed—an old example is uranyl nitrate hexahydrate, and a newer one is Eu(acetyl acetonate)$_4$. A less esoteric example is ordinary sugar! In former times sugar brought in from the Colonies was stored in Bristol, England, in the form of large cakes. These were placed in dark storage cellars, and when men went in to break up the cakes, flashes of light were observed. The flavoring in wintergreen Life-Savers is triboluminescent as well; flashes of light may be observed on crunching one in a darkened room.

In some cases, as with sugar, sparking due to frictional electrostatic charging occurs and the observed light comes from ionized air molecules. In the other cases, however, the emission is from a fluorescent or phosphorescent excited state of the molecule in question, and a possible explanation is the following. The mechanical shock wave may produce an inversion in electronic levels so that what had been an excited state becomes the ground state. After the wave has passed, the molecules may be left in what now reverts to being an excited state, and one from which emission occurs.

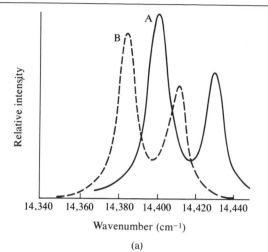

(a)

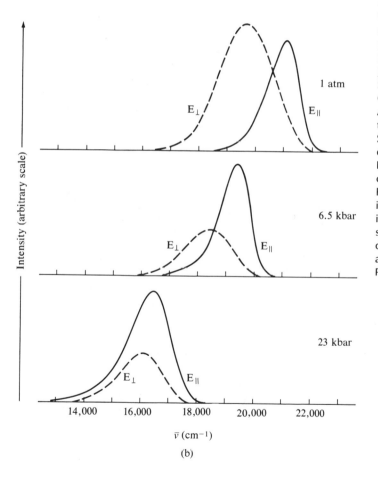

(b)

FIGURE 19-30
Effect of pressure on emission spectra. Emission from coordination compounds (and from molecules generally) is shifted in energy if pressure is applied; under high pressure the effective ligand field strength changes as a result of bond length and bond angle changes. The spectra in (a) and (b) were obtained with a cell such as shown in Fig. 8-9. (a) Emission from ruby under normal pressure (A) and under 22 kbar pressure (B). This is essentially the $^2E_g \rightarrow {}^4A_{2g}$ transition of octahedral Cr(III) (note Fig. 19-12). [From R.A. Forman, G.J. Piermarini, J.D. Barnett, and S. Block, Science **176**, 284 (1972). Copyright 1972 by the American Association for the Advancement of Science.] (b) Polarized emission spectra from Ba[Pt(CN)$_4$] • 4 H$_2$O crystals. The square planar Pt(CN)$_4^{2-}$ units are stacked in the crystal, and the increased pressure shortens the Pt-Pt distance. [From M. Stock and H. Yersin, Chem. Phys. Lett. 40, 423 (1976).]

B. Excited-state
   chemistry

The chemical nature of an excited state is in general different from that of the ground state. We refer now not to prompt molecular cleavages, but to cases where the excited state lasts long enough to function as a chemical substance, and one in thermodynamic equilibrium with its environment (except for the electronic excitation energy). This is the situation with many of the triplet states of organic molecules; these may survive one or more encounters with other solute molecules. As examples, the triplet state of coumarin,

undergoes a dimerization, and that of cyclopentadiene undergoes a Diels–Alder-type reaction with a second molecule to give endodicyclopentadiene:

A rather interesting case is that of singlet oxygen. Ordinary oxygen has a triplet ground state $^3\Sigma_g^-$ and is, perhaps for this reason, an unusually reactive molecule. It is known to photochemists for its very efficient quenching of triplet excited states—a process that usually occurs on every encounter. It is possible, however, to generate oxygen in the singlet excited state $^1\Delta_g$ lying about 22 kcal mol^{-1} above the ground state. This is a less reactive species than ground-state oxygen but shows a selective ability to add to organic dienes. Singlet oxygen may be prepared, incidentally, either by the reaction of hydrogen peroxide with metal hypochlorites or by using the triplet excited state of certain dyes such as methylene blue or eosin to sensitize the excitation $^1\Delta_g \leftarrow {}^3\Sigma_g^-$. [See Foote (1968).]

An example from coordination chemistry is that of $Cr(NH_3)_5(NCS)^{2+}$, which aquates in aqueous solution to give exclusively $Cr(NH_3)_5(H_2O)^{3+}$ and free NCS^- ion. However, on irradiation of the visible absorption bands, to produce the $^4T_{2g}$ excited state, the product is primarily $Cr(NH_3)_4(H_2O)(NCS)^{2+}$ (Zinato *et al.*, 1969).

The kinetics of excited state processes is discussed in Section 19-ST-1.

19-CN-2   Conversion of light to chemical energy

There is great contemporary interest in the conversion of solar energy to chemical energy or directly to electrical energy. One specialized process of this kind occurs with great efficiency in nature. The photosynthetic reaction in green plants amounts to

$$CO_2 + H_2O \xrightarrow{h\nu} \frac{1}{n}(CH_2O)_n + O_2$$

$$(19\text{-}33)$$

where $(CH_2O)_n$ denotes carbohydrate. The mediator for this reaction is, of course, the chlorophyll molecule. Chlorophyll absorbs in the red, however, at around 680 nm, at which wavelength a mole of light quanta has about 42 kcal of energy. Since the energy required for reaction (19-33) is about 150 kcal (it is just the reverse of the combustion of a sugar or a starch), a multistep process must be involved. The active unit in the plant cell, the *chloroplast*, contains stacked chlorophyll molecules and it appears that several, at least eight, funnel either activation energy or electrons to a central receptor which is the site of actual reaction. The full elucidation of the energy transfer apparatus and of the detailed reaction mechanism leading to Eq. (19-33) as the overall process remains one of the fascinating topics of current research.

Of equal if not greater importance is how to devise a photochemical system to supply practical energy from solar energy (other than by burning the wood produced by plants!). Charge transfer excited states of coordination compounds have been found to be good reducing agents. A possible model system is

$$RuL_3^{2+} \xrightarrow{h\nu} {}^*RuL_3^{2+}$$
$$^*RuL_3^{2+} + H_3O^+ \rightarrow RuL_3^{3+} + H_2O + \tfrac{1}{2}H_2$$
$$\underline{RuL_3^{3+} + \tfrac{3}{2}H_2O \rightarrow RuL_3^{2+} + H_3O^+ + \tfrac{1}{4}O_2}$$
$$\tfrac{1}{2}H_2O \rightarrow \tfrac{1}{2}H_2 + \tfrac{1}{4}O_2$$

where L denotes 2,2′-bipyridine (or related ligands). The oxidation potential of *RuL$_3^{2+}$ is about 1 V, or ample to reduce water, and the reduction potential of RuL$_3^{3+}$ is about 1.2 V, or enough to oxidize water. As may be seen, the net reaction is the photoinduced decomposition of water, the resulting H$_2$ then being a source of chemical energy. While the principle is clear, there are difficult problems in getting model systems to work efficiently, let alone economically.

Another approach has been to produce photoelectrons at or near an electrode surface and thus obtain a photogalvanic cell. Both semiconductor electrodes and chlorophyll-containing membranes show promise. At this writing, however, the silicon-type solar cell developed for the space program remains the most efficient, although somewhat expensive, means of generating electricity from solar energy.

Photochromic systems have attracted interest as a means of storing solar energy. An uphill reaction, often an isomerization, is driven photochemically. The reverse, thermal reaction is slow, but may be catalyzed. Thus at some later time, the chemical energy stored in the irradiated material may be recovered for heating purposes. Systems of this type that are being studied include the photoconversion of norbornadiene to quadricyclene and the photoisomerization of *trans-* to *cis*-azobenzene.

Figure 19-31 shows the energy spectrum of solar radiation reaching the earth's surface. Note that the density of solar radiation in the visible region is rather low, about 200 W m^{-2}. To supply the energy requirement of a city, even the most efficient collector would have to cover an area several times that of the city itself. Both the capital costs and the environmental impact would be considerable.

19-CN-3 Lasers

The principle of a laser can be described in terms of the two-state system discussed in Section 19-3. We can write the rate of population of excited state n as $k_a N_m$, where $k_a = B_{nm}\rho_{nm}$, and the rate of its depopulation as $k_d = k_{Sa}N_n + k_{mn}N_n$ where $k_{Sa} = B_{mn}\rho_{mn}$. If the radiation density is sufficiently high, k_d approaches $k_{Sa}N_n$ and since $B_{nm} = B_{mn}$, the consequence is that

$N_n = N_m$. Under this condition the rates of absorption and of stimulated emission are equal.

Suppose now that there exists some higher excited state n' that can undergo a conversion or crossing to excited state n. We can now populate state n *indirectly* by using radiation of frequency v'_{nm}. If A_{mn} is small enough, it will be possible to make N_n exceed N_m—after all, there is no radiation of frequency v_{nm} to depopulate state n by stimulated emission. A system having such an *inverted* population is capable of laser (light amplication by stimulated emission of radiation) action.

Suppose further that this system is established in a cavity having reflecting walls, as, for example, a cylindrical space having mirrors at each end. If some radiation of frequency v_{nm} is introduced along the cylinder axis (there will always be some from spontaneous emission), then it will stimulate further emission of the same frequency, in phase and in the same direction. Light of this frequency then reflects back and forth, gathering intensity as more and more stimulated emission occurs. The process is on the speed-of-light time scale, and the effect is that a short, intense pulse of radiation is produced. Various arrangements, such as use of a partially silvered mirror at one end, allow the escape of this pulse. Because it is in phase or coherent and accurately collimated, the beam diverges very little; laser beams can be reflected back from the moon and still be detected, for example. They may be focused down to an area comparable in dimensions to that of their wavelength to give enormous energy densities, and a focused laser beam can be used for microsurgery. Laser beams are highly monochromatic, and this has made them very useful in spectroscopy, as, for example, in Raman spectroscopy; also, their high intensity and short (nanosecond) pulse duration allows experiments in flash photolysis where a short-lived excited state is produced in sufficient amount for its absorption spectrum and other properties to be measured.

Lasing systems are now commercially available that operate in the microwave, the infrared, and the visible and ultraviolet wavelength regions. They may be pulsed or continuous; in some cases they can be tuned or varied in wavelength continuously over a region of values. A clever use of their property of coherence allows a doubling of their frequency so that lasers producing in the near

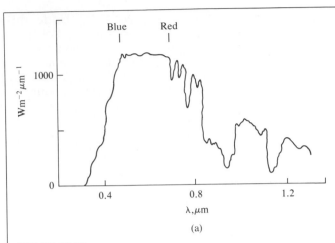

(a)

FIGURE 19-31

Solar energy. There is much interest in the use of sunlight either to produce power through solar cells, or to store energy chemically. Figure 19-31(a) shows the spectral distribution of solar energy with the sun at 60° from the zenith. The total energy is about 750 W m⁻², of which about 200 W m⁻² is in the visible region. (The dips in the spectrum are due to absorption by various atmospheric and solar atmospheric species.)

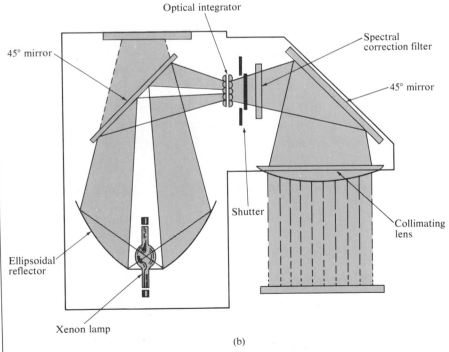

(b)

(b) A laboratory "solar" lamp. This is an Xe arc lamp collimated to give a large irradiated area, and suitably filtered so as to have about the spectral distribution shown in part (a). (Reproduction with the permission of Oriel Corporation, Stamford, Conn.)

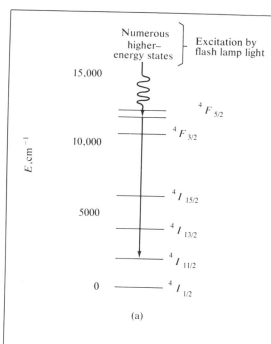

(a)

FIGURE 19-32

The Nd laser. (a) The excited-state scheme for Nd^{3+}. The ion is present as a minor constituent of yttrium aluminum garnet, $Y_3Al_5O_{12}$, "YAG," or in a glass. Flash lamp light irradiates a rod of the material, exciting various high-energy states that decay rapidly to the $^4F_{3/2}$ state. This last has a natural lifetime of 5×10^{-4} s to drop to the $^4I_{11/2}$ state, with emission of 1060-nm light. During the flash lamp excitation, a large population of $^4F_{3/2}$ states accumulates; there is nearly complete population inversion since the $^4I_{11/2}$ state is too far above the ground state to have much thermal population. Net stimulated emission may thus occur.

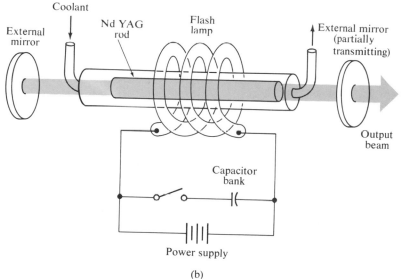

(b)

(b) A schematic of an oscillator or unit for producing stimulated emission. The flash lamp is on for about 1 ms, pumping the system to $^4F_{3/2}$ states. Light of 1060-nm wavelength (either from the flash lamp or from natural emission) is reflected back and forth between the mirrors, and stimulates further emission. An emission avalanche thus occurs, which escapes through the partially transmitting mirror as a coherent laser beam. The laser pulse may be shortened in duration and intensified if a Pockels cell is placed in the oscillator cavity. The cell is nontransmitting until polarized by a high-voltage pulse. This pulse is not applied until after the flash lamp has been on long enough for extensive population of $^4F_{3/2}$ states. On then making the Pockels cell transmitting, the stimulated emission avalanche occurs over about a 20-ns period. Such an oscillator is said to be "Q switched." (From A. Yariv, "Quantum Electronics," Wiley, New York, 1975.)

ultraviolet are possible. In brief, lasers are becoming a common and indispensable tool for the chemist, the physicist, and the engineer.

The population inversion that is crucial to laser action may be achieved in various ways. One may excite optically by means of a flash lamp, as is done with the popular ruby and Nd lasers. Some detail on the latter is given in Fig.

19-32. The widely used nitrogen, argon, and CO_2 lasers are "pumped" by an electrical discharge produced in the gas itself. Of great current interest is the use of chemical reactions that produce excited-state products. The potential high energy efficiency and portability of chemical lasers make them very attractive.

SPECIAL TOPICS

19-ST-1   Photochemical and photophysical processes

A. Photochemical processes

Some examples of excited state chemistry are noted in Section 19-CN-1B. We provide here some further illustrations, involving the more important types of organic and inorganic photochemistry.

Ketones are generally photosensitive, the primary reactions being

$$\text{RCOR}' \xrightarrow{h\nu} \begin{cases} \text{R} + \text{COR}' \\ \text{RCO} + \text{R}' \end{cases} \qquad (19\text{-}34)$$

where R and R' are alkyl groups. The RCO or R'CO fragment may then dissociate to carbon monoxide; the radicals R and R' undergo, of course, various further reactions until stable products eventually are reached. The bond that breaks tends to be the weaker of the two, the degree of discrimination being greater the longer the wavelength of the light used. Thus with gaseous $CH_3COC_2H_5$ the ratio of ethyl to methyl radicals formed in the primary dissociation reaction is about 40 with light of 313 nm but only 5.5 with light of 265 nm. The overall efficiency or quantum yield (see next subsection) varies greatly from one ketone to another, ranging from about 0.001 to nearly unity.

An alternative mode of reaction is exhibited by benzophenone, ϕ_2CO. Irradiation of the first singlet–singlet absorption band [believed to be an (n, π^*) transition] is followed by intersystem crossing to a lower-lying triplet state T_1, which now efficiently abstracts a hydrogen atom from the solvent:

$$\phi_2CO \xrightarrow{h\nu} \phi_2CO^*(S_1) \rightarrow \phi_2CO^* (T_1)$$

$$\phi_2CO^* (T_1) + RH \rightarrow \phi_2\dot{C}OH + R \qquad (19\text{-}35)$$

$$2\phi_2\dot{C}OH \rightarrow \phi_2COHCOH\phi_2$$

The resulting ketyl radicals then combine to yield benzpinacol.

The alkyl halides RX, if irradiated in their first absorption band, usually give the radicals R and X with high efficiency. Irradiation of the second, higher-energy band may, however, yield an olefin plus HX,

$$RCH_2CH_2X \xrightarrow{h\nu} RCH = CH_2 + HX \qquad (19\text{-}36)$$

Examples are ethyl and n-propyl iodides. One thus observes *spectrospecificity*, that is, wavelength-dependent photochemistry.

Photoisomerization constitutes another important class of reactions. It was mentioned earlier in this section that the stilbenes photoisomerize, mainly from the first triplet excited state. The spiropyrans form an interesting class of *photoreversible* systems. The normal form is usually colorless (depending on the ring substituents) and irradiation in the ultraviolet converts form I to colored form II. There is a return in the dark, which is accelerated photochemically on irradiation with light in the visible range.

Important primary photochemical processes in organic chemistry thus include homolytic bond breaking to yield radical species, bond weakening of a double bond to give *cis* to *trans* isomerization as well as more complex rearrangements, and excited-state reactions with solvent or solute molecules.

Coordination compounds also have a photochemistry. For example, irradiation of the first charge transfer band of a Co(III) complex such as $Co(NH_3)_6^{3+}$ leads to Co^{2+} and nitrogen, and that of $Co(NH_3)_5Br^{2+}$, to Co^{2+} and bromine atoms (which then undergo further reaction). Irradiation of a ligand field band generally leads to a substitution reaction. Thus visible light produces the reaction

$$Cr(NH_3)_5(NCS)^{2+}$$
$$+ H_2O \xrightarrow{h\nu} Cr(NH_3)_4(H_2O)NCS^{2+}$$
$$+ NH_3$$

(19-37)

with high efficiency. The ordinary thermal reaction of this complex ion is one of replacement of the NCS^- group, so in this case the photochemical and thermal reactions are distinctly different.

B. Kinetics of excited-
state processes

The overall efficiency of an excited-state process is usually described by a *quantum yield* ϕ. This may be defined as

$$n = I_a\phi$$

(19-38)

where n denotes the number of events that occur and I_a is the number of light quanta absorbed by the system in the irradiation. One may speak, for example, of a fluorescence yield ϕ_f, where the event is the emission of a light quantum from the S_1 state. Similarly, ϕ_p denotes a phosphorescence quantum yield. In the case of chemical reaction, an alternative form of Eq. (19-38) is

$$m = E\phi$$

(19-39)

where m is the number of moles of photochemical reaction that occurs and E is the number of einsteins of light absorbed (an *einstein*, abbreviated E, is one mole of light quanta). If more than one product is formed, one may assign partial quantum yields $\phi_1, \phi_2, \ldots$ to each.

A quantum yield usually refers to one or another of a set of mutually exclusive primary events such as emission or chemical reaction, and the set of quantum yields for all possible events should then total unity. However, if a photochemical quantum yield is an apparent one, being based on the amount of some final product, then values exceeding unity may be found. Thus in the case of photochemically ini-

tiated chain reactions quantum yields of several hundred or thousand may be found. Finally, it is clear from the material of this section that quantum yields are in general wavelength-dependent; they are therefore not very meaningful unless reported for at least a fairly narrow range of wavelengths.

Quantum yields refer to overall efficiencies, and one also assigns individual rate constants to separate, individual processes such as those shown in Fig. 19-11. Thus, referring to the S_1 state, the fluorescence process ($S_1 \rightarrow S_0^v$) will have a rate constant k_f; that of radiationless deactivation, k_d; and that of intersystem crossing, k_c. The total rate constant for disappearance of S_1 is then the sum $k = k_f + k_d + k_c$, and the average life of S_1 is, correspondingly, $\tau = 1/k$ [Eq. (14-17)]. Under some conditions only the emission is important, and one speaks of $1/k_f$ as the *natural lifetime* of the state.

A fairly common situation is that in which an excited state is deactivated as the result of an encounter with some solute species. Dissolved oxygen is very effective in deactivating or quenching organic triplet excited states, for example. Other substances may be effective in quenching either singlet or triplet states; the complex ion $CR(NH_3)_5(NCS)^{2+}$ quenches the fluorescence emission of acridinium ion as well as the phosphorescence emission of biacetyl.

A simple kinetic scheme for such quenching is the following:

$$S_0 \xrightarrow{h\nu} S_1 \qquad S_1 \xrightarrow{k_f} S_0 + h\nu$$
$$S_1 \xrightarrow{k_d} S_0 \qquad S_1 + M \xrightarrow{k_q} S_0 + M$$

where M is the quenching species. We apply the stationary-state hypothesis (Section 14-4C),

$$\frac{d(S_1)}{dt} = 0 = I_a - k_f(S_1)$$
$$- k_d(S_1) - k_q(S_1)(M)$$

to obtain for the rate of fluorescence

$$\frac{d(h\nu)}{dt} = k_f(S_1) = \frac{k_fI_a}{k_f + k_d + k_q(M)}$$

or

$$\phi_f = \frac{k_f}{k_f + k_d + k_q(M)}$$

(19-40)

Equation (19-40) may be put in the linear form

EXAMPLE   A 1×10^{-3} M solution of $Cr(NH_3)_5(NCS)^{2+}$ reduces the fluorescence intensity of acridinium ion in the solution by 20%, the system being irradiated at 410 nm and the emission being monitored at around 500 nm. The emission lifetime of acridinium ion in the absence of quencher is 3×10^{-8} s. Calculate k_q.

We find from Eq. (19-42) that $K_q = (1.25 - 1)/1 \times 10^{-3} = 250$ M^{-1}, whence $k_q = 250/3 \times 10^{-8} = 8.33 \times 10^9$ M^{-1} s^{-1}.

$$\frac{1}{\phi_f} = \frac{k_f + k_d}{k_f} + \frac{k_q}{k_f} \text{(M)} \qquad (19\text{-}41)$$

A particularly useful form results if we recognize that $k_f/(k_f + k_d)$ is the fluorescence yield in the absence of M, ϕ_f^0, and that τ^0, the emission lifetime in the absence of M, is just $1/(k_f + k_d)$. We now have

$$\frac{\phi_f^0}{\phi_f} = 1 + K_q \text{(M)} \qquad (19\text{-}42)$$

where K_q, called the *Stern–Volmer constant*, is given by $K_q = k_q \tau^0$. A plot of ϕ_f^0/ϕ_f vs. (M) thus gives a straight line of slope K_q; k_q follows if the emission lifetime τ^0 is known. Note that only relative emission intensities need be measured to obtain the ϕ_f^0/ϕ_f ratio. The k_q values that one obtains by this procedure often approach the diffusion limiting value of around 10^9 M^{-1} s^{-1} (see Section 15-4); that is, quenching occurs on essentially every encounter between M and an excited state species. (See the example above.)

As the example shows, K_q values can be obtained by means of ordinary emission intensity measurements. One may use, for example, a conventional spectrofluorimeter. In order to obtain *rate constants*, however, some type of rate measurement is needed. Nowadays, one generally uses a pulsed laser source, such as the Nd laser illustrated in Fig. 19-32. A collection of excited states may thus be produced on a nanosecond or even a picosecond time scale. Following the laser pulse, one may observe the decay of the emission from the excited state; this is ordinarily first order and one obtains τ, the emission lifetime from the slope of the semilogarithmic plot of emission intensity versus time. If no quencher is present, $\tau^0 = 1/(k_f + k_d)$ and there are now two ways in which k_f can be found. The deactivation of k_d process is temperature-dependent, while k_f is essentially temperature-independent. As a consequence, the former process becomes of negligible importance at a sufficiently low temperature and $\tau^0 \rightarrow \tau_f^0$. Alternatively, if the *absolute* emission yield ϕ_f^0 can be measured, then at any temperature $k_f = \phi_f^0/\tau_f^0$ (remember, the superscript zero merely means that no quencher is present—is the relationship $k_f = \phi_f/\tau_f$ also valid?). (See the example presented below.)

19-ST-2   Optical Activity

This topic is taken up here rather than in Chapter 3 because modern applications lead to useful information about excited states. The traditional aspect, however, is that of the rotation of the plane of polarization of light by an optically active substance. The optical activity may result from a crystalline arrangement of atoms or molecules in a right- or left-handed spiral,

By using a second, monitoring beam at right angles to the laser pulse, one may observe the absorption spectrum (and its decay) of excited states. Such spectra have been obtained for various organic triplet states, and for the ^2E state (see Fig. 19-12) of Cr(III) coordination compounds. The monitoring beam may also detect the change in absorption as *primary* photoproducts are formed, thus giving the rate of decay of the precursor excited state. One may, of course, also follow any subsequent reactions of the primary photoproducts.

A laser beam can be made highly monochromatic, and in the gas phase especially the vibrational structure of electronic absorption bands may be sufficiently resolved to permit the laser excitation of just one vibrational feature. In the infrared region, the isotopic shift may be sufficiently resolved to permit the selective laser excitation of molecules containing a particular isotope. This is the basis for current, intensive work on photochemical laser isotope separations, of particular significance in the possible separation of ^{235}U from ^{238}U.

as in quartz, in which case the optical activity disappears on melting. Alternatively, the individual molecules may be asymmetric, in which case the activity is retained in all physical states and in solution.

A. Rotation of plane-
polarized light

The usual experimental arrangement makes use of a *polarimeter*. Incident monochromatic light is plane-polarized by means of a special prism (as discussed later), passes through the material to be studied, and then through a second prism. The relative angular position of the two prisms for maximum (or minimum) transmission of light of a given wavelength is observed with and without the active substance, the difference in angle being the *optical rotation* α. It is customary to reduce α to *specific* rotation $[\alpha]$ by the definition

$$[\alpha]_\lambda{}^t = \frac{\alpha}{l\rho} = \frac{100\alpha}{lc} \qquad (19\text{-}43)$$

where l is the path length in decimeters, ρ is the density of the substance, if neat, and c is the number of grams of substance per 100 cm^3 of solution. The superscript and subscript give the temperature (in degrees Celsius) and wavelength; if the sodium D line is used, the subscript may be written as D. *Molar rotation* is defined as

$$[M]_\lambda{}^t = \frac{[\alpha]_\lambda{}^t M}{100} \qquad (19\text{-}44)$$

where M is the molecular weight. More rational units have been proposed: 1 biot = 10^{-3} rad cm^2 g^{-1} instead of $[\alpha]$ and 1 cotton = 0.1 rad cm^2 mol^{-1} instead of $[M]$.

If a substance rotates the plane of polarized light clockwise as viewed looking toward the light source, it is said to be *dextrorotatory* and α is reported as positive; if the rotation is to the left or counterclockwise, the substance is *levorotatory* and α is reported as negative.

Specific rotations for small organic molecules range up to about 50°; $[\alpha]_D^{25}$ is $-39°$ for L-histidine, $[\alpha]_D^{20}$ is $-12°$ for (—)-tartaric acid and $+66°$ for sucrose (all in aqueous solution). Rather larger values may be found for optically active coordination compounds; the value of $[\alpha]_{600nm}^{25°}$ is 600° for $Cr(C_2O_4)_3{}^{3-}$, for example. Optical rotation is an additive property in dilute solutions, and polarimetry is therefore quite

useful as an analytical tool. Molar rotations are to some extent constitutive (and thus resemble molar refractions, see Section 3-3) and structural conclusions may sometimes be reached on the assumption that the observed rotation is a sum of contributions from independent asymmetric centers.

B. Theory of optical
activity

A beam of plane or linearly polarized light may be represented by a wave equation such as Eq. (16-65), corresponding to a sine wave of varying electric field. The accompanying magnetic field oscillates in phase and with the same amplitude, but in the plane at right angles to that of the electric field. Considering only the electric field, if two beams are polarized at right angles to each other and both are in phase and of the same amplitude, then as illustrated in Fig. 19-33(a), the resultant will be equivalent to a plane-polarized beam at an inclination of 45° to the other two. If one beam is a quarter of a wavelength out of phase with the other, then the maximum net amplitude rotates with distance, either clockwise or counterclockwise, as shown in Fig. 19-33(b). Such a beam is said to be *circularly polarized*.

The theory of optical activity is based on the behavior of circularly polarized light. A ray of plane-polarized light may be regarded as equivalent to two circularly polarized beams that are in phase and of equal amplitude but have opposite senses of rotation [Fig. 19-33(c)]. The right- and left-handed spirals cancel except for their x components, so the resultant is plane-polarized light vibrating along the x direction. The velocities of the two circularly polarized components are the same in an inactive substance, so the angle of the equivalent plane-polarized beam does not change with distance. In an optically active material, however, the two circularly polarized components have different velocities, with the consequence that the equivalent plane-polarized beam rotates as it passes through the substance [Fig. 19-33(d)].

The velocity of light is inversely proportional to the index of refraction of the medium, and an equation due to A. Fresnel (1825) gives

$$\alpha = \frac{\pi}{\lambda} (n_l - n_r) \qquad (19\text{-}45)$$

where α is now the rotation in radians per centimeter and n_l and n_r are the indices of refrac-

FIGURE 19-33
(a) Resultant of two beams of polarized light of the same amplitude in phase and with planes of polarization perpendicular to each other. (b) Resultant if the beams are a quarter of a wavelength out of phase. (c) Plane-polarized light as the resultant of two oppositely circularly polarized components. (d) Rotation of plane of polarized ion as a consequence of two circularly polarized components having different velocities in a medium. (e) Elliptically polarized light as a consequence of two circularly polarized beams having different extinction coefficients.

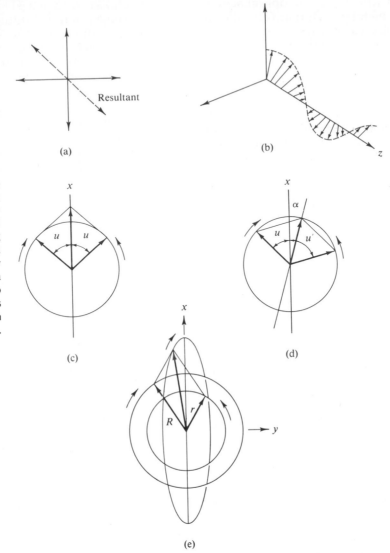

tion for left and right circularly polarized light, respectively. Since the wavelength λ is a small number in the case of visible light, an appreciable value of α results even with very small differences in the refractive indices. Thus optical rotation is a second-order effect, dependent on the small difference between relatively large numbers.

The theoretical treatment involves integrals resembling those of Eq. (19-4). No effect results, however, if only the oscillating electric field of the light is considered; it is necessary to include the oscillating magnetic field as well.

The symmetry properties of the integrals are such that the effect is still zero if the molecule possesses either a plane or a center of symmetry. It may be shown that the sufficient requirement for optical activity is that the molecule and its mirror image not be superimposable. A molecule may be transformed into its mirror image by reflection of its coordinates in any given plane; this reflection is equivalent to changing from a right-handed to a left-handed coordinate system. An optically active molecule must behave differently toward right and left circularly polarized light.

C. Rotatory dispersion and circular dichroism

An important experimental observation is that α as well as the index of refraction $n = (n_l + n_r)/2$ and the separate indices n_l and n_r vary with the wavelength of the light used. The effect may be quite dramatic as the wavelength is varied through the region of an adsorption band. This behavior is illustrated in Fig. 19-34, where the curve labeled $n_l - n_r$ is proportional to α, by Eq. (19-44). The variation of α with wavelength is known as *optical rotatory dispersion*, ORD. Note that α changes sign in the vicinity of the absorption maximum, the ordinary absorption curve being given by $(\epsilon_l + \epsilon_r)/2$, where ϵ denotes extinction coefficient (Section 3-2).

An approximate expression for this behavior was given by Drude in 1900:

$$[\alpha] = \frac{k}{\lambda^2 - \lambda_0^2} \qquad (19\text{-}46)$$

where k is a constant characteristic of the substance and λ_0 is the wavelength of the absorp-

tion maximum. Sometimes a sum of terms with different k and λ_0 values is needed to fit a rotatory dispersion curve; the implication is that two or more overlapping absorption bands are actually present. An important point is that the sign of α is not in itself a characteristic of an optically active substance; the sign depends on which side of an absorption band the measurement is made. It was perhaps fortunate for early investigators that their polarimetry was done mostly on compounds that absorb mainly in the ultraviolet, so that use of the sodium D line gives α values corresponding to the long-wavelength side of the first electronic absorption band. As a consequence, a related series of compounds, as of sugars, tend to have the same sign for α if the absolute configuration or *chirality* is the same. It is thus relatively safe to draw conclusions from how the sign of α behaves as to whether the "handedness" or chirality of an asymmetric center is retained in a chemical reaction; this rather simple approach can lead to serious errors, however.

The phenomenon of rotatory dispersion is connected with the fact that the absorption coefficients are different for right and left cir-

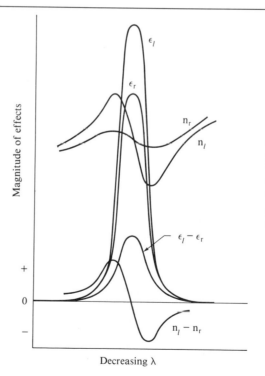

FIGURE 19-34
Schematic illustration of dispersion of index of refraction for right and left circularly polarized light and of the corresponding extinction coefficients. [After F. Woldbye, in "Technique of Inorganic Chemistry" (H.B. Jonassen and A. Weissberger, eds.), Vol. 4. Copyright 1965, Wiley (Interscience), New York. Used with permission of John Wiley & Sons, Inc.]

cularly polarized light in the case of an optically active substance. The effect is known as *circular dichroism*, CD. Both absorption coefficients are appreciable, of course, in the region of an absorption band, as illustrated in Fig. 19-34, and if they are different, the consequence is that *elliptically* polarized light results. As shown in Fig. 19-33(e), the *y* components of the amplitudes of two circularly polarized beams no longer cancel.

It is possible to measure these separate absorption coefficients to obtain the *coefficient of dichroic absorption* $\Delta\epsilon$:

$$\Delta\epsilon = \epsilon_l - \epsilon_r \qquad (19\text{-}47)$$

or the *anisotrophy* or *dissymmetry factor* $g = \Delta\epsilon/\epsilon$. The quantity $\Delta\epsilon$ varies with wavelength in the region of an absorption band, as shown in Fig. 19-34, or is said to exhibit dispersion. The dispersion of ORD and of CD constitute the *Cotton effect* (the name is French and should be pronounced accordingly).

Both ORD and CD spectra are fast becoming routine adjuncts to regular absorption spectra when dealing with optically active compounds. Rather undistinguished absorption spectra may reveal themselves as consisting of more than one absorption band by showing complicated ORD and CD behavior. Also, absorption bands not showing Cotton effects probably involve chromophores that are not themselves centers of optical activity or near such a center. Finally, if the symmetry designations of the ground and excited states are known, the CD spectrum may allow the assignment of the absolute configuration, that is, the chirality of the molecule. An example is given in Fig. 19-35; the tris-orthophenanthroline complex of Ru(II), $Ru(phen)_3^{2+}$, is basically octahedral in geometry, but the three bidentate *phen* ligands make the molecule resemble a three-bladed propeller. There are two ways for the blades to be pitched, corresponding to the two optical isomers, and the deduced absolute configuration is shown in the figure (Bosnich, 1969).

D. Instrumentation

Plane-polarized light may be produced by passing light through a suitable prism of calcite $(CaCO_3)$ or quartz. Such prisms have been cut in a plane tilted to the direction of the incident light beam and then cemented together. Or-

---

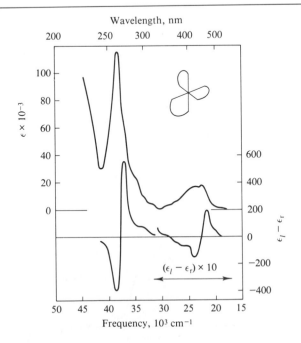

FIGURE 19-35

Circular dichroism spectrum of aqueous (+)-[Ru(phen)₃](ClO₄)₂. [From B. Bosnich, Accounts Chem. Res. 2, 266 (1969).]

dinary light can be treated as consisting of two mutually perpendicular plane-polarized beams, and the two beams will transmit differently through a properly prepared split prism and are therefore separated. Various prism constructions, such as the Nicol, Glan, and Rochan prisms, have been designed. A Polaroid sheet has a layer of oriented crystals that polarize the transmitted light.

Circularly polarized light may be obtained by passing suitably oriented plane-polarized light through a quartz prism known as a *Fresnel*

rhomb. The Fresnel rhomb is cut in such a way that the beam undergoes internal reflections before emerging, so as to cause just the right time lag between vibrations parallel and perpendicular to the plane of incidence.

Modern recording spectropolarimeters make the obtaining of ORD and CD spectra relatively easy. Just as the appearance of the recording spectrophotometer gave great stimulus to spectrophotometry, so has the appearance of ORD and CD automatic instruments led to great expansion of the study of optical activity.

GENERAL REFERENCES

ADAMSON, A.W., AND FLEISCHAUER, P.F., Eds. (1975). "Concepts of Inorganic Photochemistry." Wiley, New York.

BECKER, E.D. (1980). "High Resolution NMR," 2nd ed. Academic Press, New York.

BECKER, R.S. (1969). "Theory and Interpretation of Fluorescence and Phosphorescence." Wiley (Interscience), New York.

BERRY, R.S., RICE, S.A., AND ROSS, J. (1980). "Physical Chemistry." Wiley, New York.

CALVERT, J.G., AND PITTS, J.N., JR. (1966). "Photochemistry." Wiley, New York.

COTTON, F.A. (1963). "Chemical Applications of Group Theory." Wiley (Interscience), New York.

DAUDEL, R., LEFEBVRE, R., AND MOSER, C. (1959). "Quantum Chemistry, Methods and Applications." Wiley (Interscience), New York.

EYRING, H. (1944). "Quantum Chemistry." Wiley, New York.

GRIFFITHS, P.R. (1975). "Chemical Infrared Fourier Transform Spectroscopy." Wiley, New York.

HERZBERG, G. (1967). "Molecular Spectra and Molecular Structure. III. Electronic Spectra and Electronic Structure of Polyatomic Molecules." Van Nostrand–Reinhold, Princeton, New Jersey.

MCQUARRIE, D.A. (1983). "Quantum Chemistry." University Science Books, Mill Valley, California.

PITZER, K.S. (1953). "Quantum Chemistry." Prentice-Hall, Englewood Cliffs, New Jersey.

POPLE, J.A., SCHNEIDER, W.G., AND BERNSTEIN, H.J. (1959). "High Resolution Nuclear Magnetic Resonance." McGraw-Hill, New York.

POTTS, W.J., JR. (1963). "Chemical Infrared Spectroscopy." Wiley, New York.

STREITWEISER, A., JR. (1964). "Molecular Orbital Theory for Organic Chemists." Wiley, New York.

SZYMANSKI, H.A. (1964). "Theory and Practice of Infrared Spectroscopy." Plenum, New York.

TOWNES, C.H., AND SCHAWLOW, A.L. (1955). "Microwave Spectroscopy." McGraw-Hill, New York.

CITED REFERENCES

BOSNICH, B. (1969). *Accounts Chem. Res.* **2**, 266.

BUDINGER, T.F., AND LAUTERBUR, P.C. (1984). *Science* **226**, 288.

CALVERT, J.G., AND PITTS, J.N., Jr. (1966). "Photochemistry." Wiley, New York.

FOOTE, C.S. (1968). *Accounts Chem. Res.* **1**, 104.

GADIAN, D.G. (1982). "Nuclear Magnetic Resonance and Its Applications to Living Systems." Oxford University Press, New York.

LEVY, G.C., AND NELSON, G.L. (1972). "Carbon-13 Nuclear Magnetic Resonance for Organic Chemists." Wiley (Interscience), New York.

MCQUARRIE D.A., (1983). "Quantum Chemistry." University Science Books, Mill Valley, California.

PITZER, K.S. (1953). "Quantum Chemistry." Prentice-Hall, Englewood Cliffs, New Jersey.

ZINATO, E., LINDHOLM, R.D., AND ADAMSON, A.W. (1969). *J. Amer. Chem. Soc.* **91**, 1076.

EXERCISES

Take as exact numbers given to one significant figure.

19-1   Estimate (a) the frequency in cm^{-1} for the most probable emission from the $^3\Sigma_g^+$ to the $^3\Sigma_u^+$ state of H_2 and (b) the energy to ionize a 2s excited H atom.

Ans. (a) 34.98 kK, (b) 50 kcal.
(The student should supply a detailed explanation.)

19-2   Explain whether the $^3\Sigma_g^+ \rightarrow {}^3\Sigma_u^+$ should or should not be allowed.

Ans. Allowed (student should supply explanation).

19-3   Estimate the wavelength of light required for the transition $^1\Sigma_u^- \leftarrow {}^3\Sigma_g^-$ transition in S_2 and state whether this should be an allowed transition.

Ans. 320 nm, forbidden (student should supply explanation).

19-4   Estimate the energy difference in eV between the ground and $2P_{1/2}$ state of atomic I.

Ans. 1.1 eV.

19-5   Estimate, by means of a calculation, the oscillator strength of the absorption band of benzophenone in cyclohexane centered at about 250 nm. Calculate also the emission rate constant A_{mn} for this excited state.

Ans. $f = 0.27$, $A_{mn} = 2.9 \times 10^8$ s^{-1}.

19-6   Referring to Fig. 19-14, calculate (a) the molar extinction coefficient for the infrared absorption band of CH_2Cl_2 centered at 1275 cm^{-1} and (b) the oscillator strength.

Ans. (a) $\epsilon = 22$ M^{-1} cm^{-1}, (b) $f = 1.9 \times 10^{-6}$.

19-7   A molecule having C_{4v} symmetry has a transition from an A_1 to a B_1 state, the transition dipole moment being along the z axis. Explain whether this should be an allowed transition. How about an E to E transition? (Note Table 17-11).

Ans. Forbidden; allowed (an explanation should be provided).

19-8   A Cr(III) complex has O_h symmetry. Is the transition from the $^4A_{2g}$ ground state to a $^4T_{2g}$ excited state by light absorption allowed or forbidden?

Ans. Forbidden (explain).

19-9   What is the uncertainty in energy of an excited state whose lifetime is 1 ps? Give the answer in wavenumbers.

Ans. 2.7 cm^{-1}.

19-10   Estimate the wavelength of the absorption maximum of the first absorption band of I_2 and the wavelength of maximum intensity of the fluorescent emission.

Ans. From Fig. 19-8 the absorption maximum should be at about 20,000 cm^{-1} or 500 nm; the most probable fluorescent emission should be at about 11,000 cm^{-1} or 900 nm.

19-11   Make a guess as to the probable uv-visible absorption spectrum of benzophenone.

Ans. Use Table 19-2 for the contributions of the benzene and C=O chromophores.

19-12   The molecule CO_2 is linear and has the normal vibrational modes shown. Explain which mode or modes should be (a) infrared active and (b) Raman active.

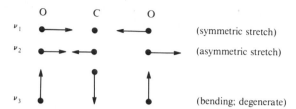

ν_1 (symmetric stretch)

ν_2 (asymmetric stretch)

ν_3 (bending; degenerate)

Ans. ν_1 is Raman active only and ν_2 and the doubly degenerate ν_3 are infrared active only. An explanation should be supplied.

The normal modes for SO_2 are shown below. Explain to which irreducible representation each belongs in the C_{2v} point group.   **19-13**

Ans. ν_1 and ν_2 transform as A_1; ν_3 transforms as B_1 if molecular plane is the xz plane.

Identify some of the characteristic bond or group frequencies in the infrared spectrum for methyl ethyl ketone [Fig. 19-14(c)].   **19-14**

Calculate the bond distance in $H^{35}Cl$ if the observed rotational constant is $B = 10.66$ cm^{-1} **19-15**

Ans. 1.275 Å.

At what magnetic field would ^1H and ^{13}C come into resonance with a radiofrequency of (a) 60 MHz and (b) 300 MHz? The respective magnetic moments are 2.7928 and 0.70203 relative to the Bohr magneton.   **19-16**

Ans. (a) 14.1 kG or 1.41 T, 56.0 kG or 5.60 T; (b) 70.5 kG, 280 kG. (A superconducting magnetic is usually needed to obtain the very high required fields for a 300 MHz nmr instrument.)

Calculate the population in the upper energy state, n^*, for a set of ^{13}C nuclei in a 10 T magnetic field. Express the result as $(n^* - n_0^*)/n_0$, where n_0^* and n_0 are the upper and lower state populations in the absence of a field, respectively. The magnetic moment of ^{13}C is 0.7023 in units of the Bohr magneton. Assume 25°C.   **19-17**

Ans. 1.723×10^{-5}.

What would be the difference in magnetic field for proton resonance in $CHCl_3$ and CH_2Cl_2 if the total field is 1×10^5 G?   **19-18**

Ans. 1.9×10^{-4} G.

We have the compound $ClCH_2-CH_2-CH_2Cl$. Show how the proton nmr spectrum of the middle H's should be split by spin-spin interaction with the protons of the CH_2Cl groups. You may assume the two CH_2Cl groups to act as four equivalent protons.   **19-19**

Ans. There should be five lines of relative intensities 1,4,6,4, and 1.

A particular esr spectrometer uses microwave radiation of 3 mm wavelength. What magnetic field is needed to bring into resonance a radical whose g_{eff} value is 3.52?   **19-20**

Ans. 3.52×10^4 G or 3.52 T.

Make the calculation analogous to the one of Exercise 19-17, but for electron spins.   **19-21**

Ans. $(n^* - n_0^*)/n_0 = 0.0461$.

19-22 A certain radical has two equivalent protons. Describe the expected hyperfine splitting of the esr line.

Ans. Three lines of relative intensity 1,2,1.

PROBLEMS

19-23 The strong Schumman–Runge absorption band of O_2 starts at about 200 nm and gradually increases in intensity to a continuum which begins at about 176 nm. Sketch the probable appearance of the ground- and excited-state potential energy curves (that is, produce a pair of plots similar to those of Fig. 19-6 but consistent with the data for O_2). Explain what happens when absorption is in the region of the continuum—is the excited state produced stable against dissociation, and if not, what might the products be? [*Note:* the energy to dissociate O_2 into two ground-state atoms is about 40,000 cm^{-1}.]

19-24 H_2 excited to the $^1\Pi_u$ state may under some conditions cross to the $^3\Sigma_g^+$ state. Light emission then occurs. Explain what happens in terms of Fig. 19-2 and estimate the range of wavelengths of the emitted light.

19-25 The data of Problem 3-17 show that ferricyanide ion has an absorption band centered around 420 nm. Calculate B_{nm} and k_e for this band. Is the transition an allowed or a forbidden one?

19-26 Assuming that the ordinate scale of Fig. 19-14(c) is for a 0.1 mm path length, estimate the oscillator strength of the absorption feature for methyl ethyl ketone at about 940 cm^{-1}.

19-27 Estimate the oscillator strength of the absorption band of Cr(urea)$_6^{3+}$ centered at (a) about 16,250 cm^{-1} and (b) 14,400 cm^{-1}. Also calculate the emission rate constant A_{mn} for these excited states and the corresponding lifetimes for emission. Note Fig. 19-12.

19-28 The ground state for an octahedral complex having two d electrons is $^3T_{1g}$. Explain whether a transition to (a) the ^1E_g and (b) to an upper $^3T_{1g}$ state is allowed either by parity or by orbital symmetry. Are the transitions spin allowed or forbidden?

19-29 A molecule has D_2 point group symmetry and a ground state of B_2 symmetry. To what symmetry excited state would a transition be allowed for radiation along the x axis?

19-30 Sketch a guessed appearance of the absorption spectrum of (a) methyl ethyl ketone, (b) azobenzene, $\phi - N = N - \phi$, and

(c) $C_6H_5-CH_2\ \overset{\overset{S}{\|}}{C}-CH_3$

Explain the basis for your spectra.

19-31 Explain what wavelength of light should be effective in the photoproduction of atoms on irradiation of (a) S_2, (b) HI, (c) the molecule of Fig. 19-6 assuming that D_e'' is 55 kcal mol^{-1}, and (d) I_2.

19-32 Sketch, with explanation, a semiquantitative plot of the likely uv-visible absorption spectrum of $CH_3-\overset{\overset{}{\underset{\|}{C}}}{}-CH_2-\overset{}{\underset{\|}{C}}-CH_3$ with S and O.

19-33 Sketch an estimated uv-visible absorption spectrum for (a) $CH_2=CH=CH_2NO_3$ and (b) ϕ-CH=CH$_2$.

19-34 The normal modes for the H_2O molecule are as shown in the accompanying diagram. Explain which should be infrared-active and -inactive and Raman-active and -inactive.

Carry the set of motion vectors through the symmetry operations of the H_2O point group and determine to what irreducible representation each mode belongs. (The answer may be arrived at by considering what vectors are left unchanged or are put into their opposites and comparing with the traces of the various irreducible representations.)

The normal modes for formaldehyde are as follows: (a) $\nu_1 = 2766$ cm^{-1}, CH$_2$(sym) stretch; (b) $\nu_2 = 1746$ cm^{-1}, C=O stretch; (c) $\nu_3 = 1501$ cm^{-1}, CH$_2$ deformation; (d) $\nu_4 = 2843$ cm^{-1}, CH$_2$(asym) stretch; (e) $\nu_5 = 1247$ cm^{-1}, CH$_2$ rock, (f) $\nu_6 = 1164$ cm^{-1}, CH$_2$ wag. One corresponds to the B_2 irreducible representation, two to the B_1, and three to the A_1. Explain which is which. In the actual infrared spectrum of formaldehyde, absorptions are observed at 3930 cm^{-1}, 2910 cm^{-1}, 2665 cm^{-1}, and 4013 cm^{-1} (among others). Explain how these frequencies arise.   **19-35**

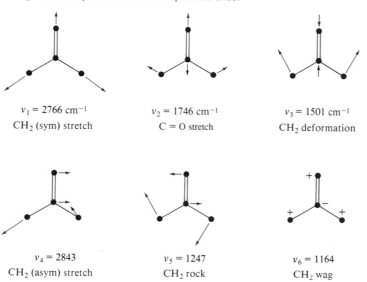

$\nu_1 = 2766$ cm^{-1}
CH$_2$ (sym) stretch

$\nu_2 = 1746$ cm^{-1}
C = O stretch

$\nu_3 = 1501$ cm^{-1}
CH$_2$ deformation

$\nu_4 = 2843$
CH$_2$ (asym) stretch

$\nu_5 = 1247$
CH$_2$ rock

$\nu_6 = 1164$
CH$_2$ wag

One of the normal modes for *trans*-$C_2H_2Cl_2$ is shown. To which IR in C_{2h} does this mode belong? Explain.   **19-36**

The fundamental absorption frequency for CN ion is about 2000 cm^{-1}. Estimate the cyanide stretching frequency that you might expect to observe in Fe(CN)$_6^{3-}$. Base your estimate on two considerations. First, assume that a change in electron distribution occurs so that the coordinated cyanide has a 25% smaller force constant. Second, suppose that the carbon is so anchored by being bonded to the iron that the effective mass of the carbon is essentially infinite.   **19-37**

One of the normal modes for formaldehyde, a planar molecule, is shown. Explain to what irreducible representation this mode belongs and whether the $\nu = 0$ to $\nu = 1$ transition in this mode should be infrared or Raman allowed (or both).   **19-38**

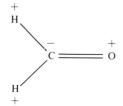

19-39   One of the normal modes for PCl_5 is shown; this is a triangular bipyramidal molecule. Show to which irreducible representation this mode belongs, and explain whether this mode should be infrared or Raman active (or both).

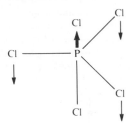

19-40   Estimate the moment of inertia of HCN from the spectrum shown in Fig. 19-15.

19-41   At what magnetic field would ^2H and ^{31}P be in resonance using (a) a 60 MHz and (b) a 300 MHz machine? The nuclear spins and g values are 1 and 0.8575 and $\frac{1}{2}$ and 2.2634, for ^2H and ^{31}P, respectively. What are the μ'_n values?

19-42   Referring to Exercise 19-17, calculate $(n^* - n_0^*)/n_0$ for ^{17}O nuclei in a 40 T magnetic field at 25°C. For ^{17}O, $I = \frac{5}{2}$ and $\mu'_n = -1.8930$.

19-43   Calculate the difference in frequency for proton resonance in benzene and cyclohexane if the magnetic field is fixed at 10,000 G.

19-44   Sketch an approximate pmr spectrum for CH_3CHO. The chemical shift of the CH_3 protons is 2.5 ppm and that of the CHO proton is 4.5 ppm; the difference is larger than the spin-spin interactions.

19-45   In the compound $C(H_a)_3$—CH_b=CH_c—CN, the H_b proton resonance is split both by the H_c proton and by the H_a protons. Assuming the splittings to be comparable in magnitude, sketch the qualitative appearance of the pmr spectrum for H_b. Do likewise for H_a. *Note:* as a first order approximation, spin-spin splittings can be applied sequentially.

19-46   (Computer or programmable calculator problem.) We can construct a simple illustration of the workings of Eq. (19-30) as follows. First, we take $I(t)$ to be just $\cos(2\pi\omega_0 t)$ and consider only the real portion of the exponential, namely $\cos(2\pi\omega t)$. We sample at successive intervals Δt, so that $t = n\Delta t$. The quantity $2\pi\omega_0\Delta t$ is a constant, which we can express in degrees, $\theta_0(\text{deg}) = 2\pi\omega_0\Delta t$. We plan to vary ω, and can write $\theta = 2\pi\omega\Delta t$, and let the actual variable be x, $\theta = x\theta_0$. The equation now reads

$$F(x) = F(2\pi\omega\Delta t) = \sum_{n=0}^{n=N} [\cos(n\theta_0)][\cos(xn\,\theta_0)]$$

As a specific case, take $\theta_0 = 30°$ and $N = 100$, and evaluate $F(x)$ for a series of x values in the vicinity of unity. We are in effect finding $F(\omega)$ when $I(t)$ is a cosine function of frequency ω_0 that is sampled for N points at intervals Δt.

Once the program is written, other calculations follow readily. One may vary θ and/or N, for example. The $I(t)$ function can be supposed to have two or more components; that is, in place of $\cos(n\theta_0)$ we can use $[a \cos \theta_{0,1} + b \cos \theta_{0,2} + \ldots]$ with x now defined by, say, $\theta = x\theta_{0,1}$. The calculated $F(x)$ will show as many maxima as there are θ_0s and will, in effect, correspond to some spectrum.

19-47   The esr spectrum of the benzene radical is split into a number of lines because of the hyperfine interaction with six equivalent hydrogen atoms. Sketch the qualitative appearance, including relative peak heights, of the spectrum.

19-48   Calculate the g_{eff} value for a radical if resonance occurs at 0.450 T using radiation of 2.5 cm wavelength.

SPECIAL TOPICS PROBLEMS

A commonly used chemical actinometer is the ferrioxalate one, for which the photochemical reaction is   **19-49**

$$Fe(C_2O_4)_3^{3-} \xrightarrow{h\nu} Fe(II) + \text{oxalate} + \text{oxidized oxalate}$$

after the irradiation the Fe(II) is complexed with 1,10-phenanthroline (neglect dilution at this point) and its concentration determined from the optical density at 510 nm, at which wavelength the extinction coefficient of the complex is 1.10×10^4. In a particular experiment 20 cm^3 of ferrioxalate solution was irradiated with light at 480 nm for 10 min and an optical density of 0.35 was subsequently found at 510 nm. The quantum yield at the irradiating wavelength is known to be 0.93. Calculate the light intensity in einsteins absorbed per second. A 1 cm cell is used.

In a photochemical experiment an intensity of 5×10^{-9} E s^{-1} is incident on a cell containing 20 cm^3 **19-50**
of 0.01 M solution of $Cr(NH_3)_5(NCS)^{2+}$. It is estimated that 90% of the incident light is absorbed and after 10 min of irradiation analysis shows that 0.6% of the complex has photoaquated to give $Cr(NH_3)_4(H_2O)(NCS)^{2+}$. Calculate the quantum yield.

Benzene vapor absorbs at 254 nm to give the first excited singlet state S_1, whose radiative lifetime is   **19-51**
6.2×10^{-7} s. S_1 disappears by intersystem crossing to the first triplet excited state T_1 with a rate constant of $k_4 = 4.7 \times 10^6$ s^{-1}, and S_1 is quenched by collisions with ground-state benzene with a rate constant of 1.00×10^9 liter mole^{-1} s^{-1}, both rate constants being for 25°C. Calculate (a) the fluorescence quantum yield at low pressures and (b) the yield at a 0.1 atm pressure (25°C).

Instead of measuring the quenching of fluorescence intensity, one may instead observe the shortening   **19-52**
of the emission lifetime due to collisions with a quencher. Derive an expression for τ^0/τ where τ is the emission lifetime and superscript zero means absence of quencher. Use the same sequence of excited state processes as for Eq. (19-42).

$Rh(NH_3)_5Cl^{2+}$ (A) exhibits emission. The emission lifetime in water solution at 5°C is 35 ns, and the   **19-53**
emission yield is 2×10^{-5}. The emission is quenched by various species and, in particular, if the solution is made 0.002 M in OH$^-$, the emission yield is reduced to one third of that in water alone. (a) Calculate k_q, the bimolecular quenching rate constant for the reaction: A* + OH$^-$ → A + OH$^-$. (b) Calculate the emission lifetime in 0.003 M OH$^-$ solution at 5°C.

The compound $W(CO)_5L$, where L = 4-cyanopyridine, phosphoresces (emission around 600 nm)   **19-54**
upon excitation with 353 nm light. Absorption of 353 nm radiation gives a first excited singlet state which, we assume, intersystem crosses with unit efficiency to a lower-lying triplet state. Emission is from this last state. The emission lifetime is 100 ns at 25°C in methylcyclohexane solution, and the emission is quenched by anthracene. Thus 2×10^{-3} M anthracene reduces the lifetime to 50 ns. The triplet state also undergoes photochemistry, to give $W(CO)_5$ plus L, with a quantum yield of 0.020 in the absence of anthracene. (a) Calculate the Stern-Volmer quenching constant for the lifetime quenching, and the value of the bimolecular quenching rate constant, k_q. (b) Show what the percent decrease in the phosphorescence emission yield should be on making a solution 3×10^{-3} M in anthracene. (c) Derive the appropriate equation, and calculate the quantum yield for $W(CO)_5$ formation if 1×10^{-3} M anthracene is present.

Absorption of light by an organic molecule A leads to phosphorescence with a quantum yield ϕ_p of   **19-55**
0.30 in a particular solvent. Show that the relation $\phi_p/\tau_p = k_p$ holds, where τ_p is the experimentally observed lifetime of the phosphorescence and k_p is the rate constant for phosphorescent emission. The value of ϕ_p is for a deaerated solution; a solution in equilibrium with air has 3.0×10^{-4} mol liter^{-1} dissolved oxygen. Encounters between dissolved oxygen and A occur with a rate constant of 5×10^9 liter mol^{-1} s^{-1}. Calculate ϕ_p' in this aerated solution; $k_p = 1.0 \times 10^5$ s^{-1}.

The specific rotation $[\alpha]_D$ of a compound in aqueous solution is 33° at 25°C. Calculate the concen-   **19-56**
tration of this compound in grams per liter in a solution that has a rotation of 3.05° when measured in a polarimeter in which the tube of solution is 20 cm long.

19-57   The specific rotation of saccharose in water at 20°C is 66.42°. Calculate the observed rotation using a polarimeter tube of 20 cm length filled with a 23.5% by weight solution of this sugar. The density of the solution is 1.108 g cm^{-3}.

19-58   A solution of 30 g of a substance of molecular weight 350 in 1 liter of water rotates the plane of polarized light by 10.5° (sodium D line, 25°C) with a 30 cm polarimeter tube. Calculate the specific and the molar rotation of the substance.

19-59   Calculate $n_l - n_r$ for the solution of Problem 19-56. The sodium D line is at 589 nm.

19-60   Read data off Fig. 19-35 to make a semiquantitative plot of g versus wavenumber for Ru(phen)$_3^{2+}$.

chapter 20
The
solid
state

We consider four more or less distinct aspects of the physical chemistry of the solid state. The emphasis is on crystalline as opposed to glassy phases, and the first topic is that of the perfect crystal as a regular, infinitely repeating structure. A special field of geometry informs us that only certain symmetry properties are possible if space is to be filled by a repeating structure or *lattice*. We then introduce some very simple examples of such lattices, involving the ways in which spherical atoms or ions can pack. This leads us to the structure of alkali metal halide and other ionic crystals and, in the Special Topics section, to calculations of the total cohesive energy of a crystal, called the *lattice energy*.

We consider next the use of x-ray diffraction as a means of determining the symmetry class of a crystal, and then its actual structure. Modern x-ray crystallography is a rather specialized subject and only the introductory aspects of it will be presented. Most crystal structures of interest today are rather complex ones, and a few examples of this type are described in the Commentary and Notes section. The main part of the chapter concludes with some material on imperfect crystals. Actual crystals have defects and dislocations, the presence of which can be quite important to their physical properties.

20-1   Space-filling lattices

There is an interesting history to the development of the awareness that the external regularity of crystals implies an inner regularity of structure. A beautifully perceptive intuition on this subject was that of Johannes Kepler. In a presentation to his benefactor, Counsellor Wackher, Kepler (1611) ponders the questions: Why should a snowflake have just six sides? Is it because this is a two-dimensional space-filling geometry? Is it significant that spheres on a flat surface pack to give a hexagonal pattern?

Topologists have since greatly developed the subject of *mosaics*, or infinitely repeating two-dimensional patterns; as illustrated in Fig. 20-1, the hexagonal unit is but one of many possibilities. The possibilities have restrictions, however. They can be shown to be limited by asking what symmetry operations

FIGURE 20-1
Mosaics of (a) modified
triangles and (b) modified
squares. (From M.
Kraitchik, "Mathematical
Recreations." Dover, New
York, 1942.)

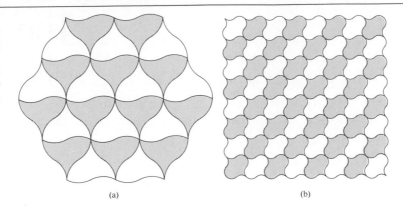

FIGURE 20-1
Mosaics of (a) modified triangles and (b) modified squares. (From M. Kraitchik, "Mathematical Recreations." Dover, New York, 1942.)

(a)   (b)

can be applied at a point that will generate a lattice. It turns out that the only types possible are (a) mirror reflections across a line and (b) one-, two-, three-, four-, or sixfold axes. A square lattice has a fourfold axis and mirror plane along each axis. An oblique lattice, such as shown in Fig. 20-2(a), has only a twofold rotation axis. There are in fact only five possible area-filling lattices: oblique, rectangular, centered rectangular, square, and hexagonal. These are known as *Bravais lattices*, after A. Bravais (1848), or just as plane lattices or plane nets.

A Bravais lattice is not in itself a crystal or, in two dimensions, a mosaic. It is a geometric construct—a repeating frame of reference. It can be described by the various symmetry operations that put the lattice into an equivalent configuration. These will be point symmetry operations, since one given lattice point remains unchanged. The *unit cell* of a lattice is the smallest portion of it exhibiting the symmetry features of the whole.

Not all of the point groups mentioned in Chapter 17 are possible for lattices. In the case of a two-dimensional lattice, we are restricted to just principal axes and vertical symmetry planes; further, as mentioned before, the order of the principal axes can only be 1, 2, 3, 4, or 6. This leaves the 10 point

FIGURE 20-2
(a) Two-dimensional oblique lattice having C_1 symmetry. (b) The same, but with C_2 symmetry as indicated by the symbols for a twofold axis (lens). (c) Lattice plus basis belonging to C_1 point group. (d) Lattice plus basis belonging to C_2 point group.

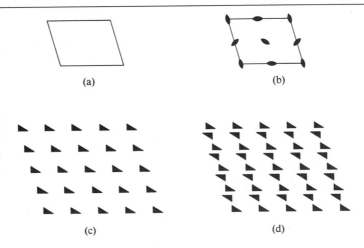

(a)   (b)

(c)   (d)

groups C_1, C_2, C_3, C_4, C_6, C_S, C_{2V}, C_{3V}, C_{4V}, and C_{6V}. (The corresponding designations in the international crystallographic notation are 1, 2, 3, 4, 6, *m*, 2*mm*, 3*m*, 4*mm*, and 6*mm*—see Section 20-ST-1.)

The unit cell of a Bravais lattice may have certain but not all of these point group symmetries. The unit cell of a two-dimensional oblique lattice may have C_1 or C_2 symmetry but not others, for example. The former is illustrated in Fig. 20-2(a) and the latter in Fig. 20-2(b), where a lens denotes a twofold rotation axis.

As an exercise the reader might convince himself that the presence of *one* twofold axis at a lattice corner implies the presence of *all* of the others shown. For example, place an asymmetric object near one corner and carry out the twofold rotation—this places a second object in the adjacent cell. The existence of the lattice implies, however, that the object can be moved about by unit translations along the lattice directions; the object produced in the adjacent cell must also appear in the original cell.

The various possible combinations of the ten point group symmetries with the five Bravais lattices generate 12 lattice types, two of which are shown in Fig. 20-2(a,b). However, because one is dealing with a lattice rather than with a single object such as a molecule, some additional symmetry operations are possible that combine rotation or reflection with translation (see Section 20-ST-1). On inclusion of possible combinations of these new operations, a final total of 17 two-dimensional lattice types are possible. These are called *plane groups*. The space group designations for the lattices of Fig. 20-2(a,b) are oblique, PC_1 and PC_2 or, in crystallographic symmetry notation, $P1$ and $P2$, where P denotes a primitive as opposed to a face-centered lattice.

In any actual crystal (or mosaic) some pattern of atoms is superimposed on a particular Bravais lattice. This pattern is called the *basis* and repeats with the unit cell of the lattice. Another way of seeing how the various space groups come about is as follows. The basis will have its own symmetry properties; if it is asymmetric, then the only symmetry of the repeating pattern is that of the lattice framework itself, as illustrated in Fig. 20-2(c). If, however, the basis has a higher symmetry, then it may conform to a more symmetric space group. Thus in Fig. 20-2(d) the basis consists of paired triangles, the pair having a C_2 axis. The presence of this axis then generates the pattern shown in the figure. The same higher lattice symmetry would be present if the basis were still more symmetric, as, for example, if it consisted of just an atom at each lattice point. Thus the space group designation implies a certain *minimum* symmetry for the basis or, conversely, one can determine the space group if the lattice and the basis are known.

The analysis of the situation in three dimensions is similar to that for two dimensions. There are 14 Bravais lattices, however, as summarized in Table 20-1. The possible point groups now include those with secondary axes and σ_h planes, but are still restricted to C_1, C_2, C_3, C_4, and C_6 principal axes. The various combinations of symmetry elements consistent with each Bravais lattice give rise to 32 lattice types or lattice point groups. If combined rotation–translation and reflection–translation operations are included, then a total of 230 space groups are obtained. A three-dimensional crystal *must* belong to one of these space groups. (Crystals belonging to only a small fraction of these space groups have actually been observed, however.) Clearly, the

TABLE 20-1 The Bravais lattices	System	Number of lattices	Lattice symbol[a]	Unit cell axes and angles	Symmetry designation[b]
	Cubic	3	P I F	$a = b = c$ $\alpha = \beta = \gamma = 90°$	$m3m$
	Tetragonal	2	P I	$a = b \neq c$ $\alpha = \beta = \gamma = 90°$	$4/mmm$
	Orthorhombic	4	P C I F	$a \neq b \neq c$ $\alpha = \beta = \gamma = 90°$	mmm
	Monoclinic	2	P C	$a \neq b \neq c$ $\alpha = \gamma = 90° \neq \beta$	$2/m$
	Rhombohedral (or trigonal)	1	P	$a = b = c$ $\alpha = \beta = \gamma \neq 90°$	$3m$
	Hexagonal	1	P	$a = b \neq c$ $\alpha = \beta = 90°; \gamma = 120°$	$6/mmm$
	Triclinic	1	P	$a \neq b \neq c$ $\alpha \neq \beta \neq \gamma \neq 90°$	$m3m$

[a]The symbols stand for; *P*, primitive, with lattice points only at the corners of the unit cell; *C*, base centered; *I*, interior or body centered; *F*, centered on all faces, or face-centered.
[b]See Section 20-ST-1.

detailed treatment of all of these symmetry combinations is too long to be practical here, and we will confine ourselves almost entirely to some important examples in the cubic system.

We will, in fact, deal mainly with rather simple crystals—ones having atoms or ions at lattice corners or in lattice faces. Figure 20-3 shows the unit cell for a simple cubic lattice whose basis consists of an atom at each lattice point. A distinction to emphasize is that the unit cell is not necessarily the repeating unit. In Fig. 20-3 each corner atom is shared with eight adjacent unit cells, and an attempt to generate the complete crystal by translations of this unit would superimpose atoms on each other. The repeating unit is in this case just the atom at the origin, that is, at position $(0, 0, 0)$. If this atom is translated

FIGURE 20-3
Simple cubic lattice having one atom at each lattice point.

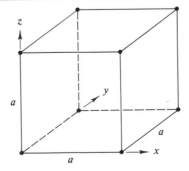

by all possible combinations of the unit lattice distance *a*, then the crystal *is* generated.

The number of atoms or ions in a unit cell is, of course, just the number present in the repeating unit. The same answer can be obtained from a diagram of the unit cell provided that corner atoms are counted as one-eighth of an atom each, face atoms as one-half, and so on. That is, each atom or ion is weighted according to the number of adjacent unit cells that share it.

Figure 20-4 shows the unit cells for the 14 Bravais lattices. These can be described by specifying the unit lengths *a, b,* and *c* and the angles α, β, and γ as indicated in Fig. 20-5. The positions of the lattice points are then given in terms of this coordinate system. Thus the simple cubic unit cell has points (or atoms in the corresponding simplest possible crystal) at $(0, 0, 0)$, $(1, 0, 0)$, $(0, 1, 0)$, $(1, 1, 0)$, $(0, 0, 1)$, $(1, 0, 1)$, $(0, 1, 1)$, and $(1, 1, 1)$, the numbers giving positions in terms of (a, b, c); similar coordinates would likewise locate the atoms in the primitive tetragonal, orthorhombic, hexagonal, and triclinic lattices, measured in units of (a, b, c) and along the required axis directions. The body-centered cubic lattice has the same positions as does the simple cubic lattice plus one at $(\frac{1}{2}, \frac{1}{2}, \frac{1}{2})$; the face-centered cubic lattice has the additional points $(\frac{1}{2}, \frac{1}{2}, 0)$, $(0, \frac{1}{2}, \frac{1}{2})$, $(1, \frac{1}{2}, \frac{1}{2})$, $(\frac{1}{2}, 1, \frac{1}{2})$, $(\frac{1}{2}, 0, \frac{1}{2})$, and $(\frac{1}{2}, \frac{1}{2}, 1)$. In these last two cases the repeating units consist of the points $(0, 0, 0)$ and $(\frac{1}{2}, \frac{1}{2}, \frac{1}{2})$,

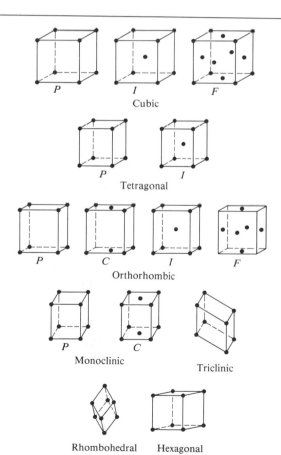

FIGURE 20-4
Unit cells for the 14 Bravais lattices.

P   I   F
Cubic

P   I
Tetragonal

P   C   I   F
Orthorhombic

P   C
Monoclinic

Triclinic

Rhombohedral   Hexagonal

FIGURE 20-5
Coordinate system for a
crystal lattice.

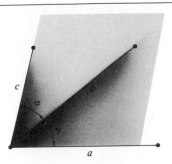

and $(0, 0, 0)$, $(\frac{1}{2}, \frac{1}{2}, 0)$, $(0, \frac{1}{2}, \frac{1}{2})$, and $(\frac{1}{2}, 0, \frac{1}{2})$, respectively. Thus the body-centered cubic repeating unit has two points and the face-centered unit has four points. The same numbers would result if the unit cell points shown in Fig. 20-4 were counted as one-eighth if at a corner, one-half if on a face, and full weight if in the interior.

It is to be remembered that there will be actual crystals such that each point in Fig. 20-4 is occupied by a single atom, as just assumed, but many other crystals where each point is occupied by some molecule or grouping of atoms—a grouping whose symmetry is reduced from that of a sphere. Thus the lattice of crystalline carbon dioxide is face-centered cubic, but the basis consists of variously oriented CO_2 molecules, as illustrated in Fig. 20-6.

20-2 Crystal planes; Miller indices

Any actual crystal is finite in extent and its surfaces are made up of planes that pass through lattice points. Thus *ideal* physical crystals are bounded by planes that are lattice or *rational* planes; the inverse reasoning was made by Haüy in 1784, namely that the regular shapes of crystals imply an inner regularity. However, the *appearance* of a crystal is very dependent on which planes bound it, as illustrated in Fig. 20-7 for a crystal whose lattice is cubic. As we truncate the corners of a simple cube progressively, surfaces corre-

FIGURE 20-6
The CO_2 unit cell.

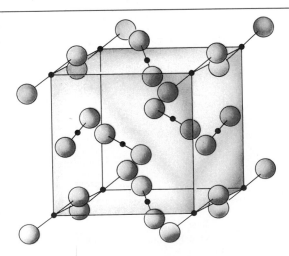

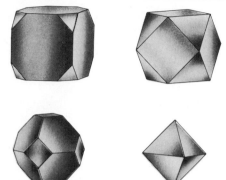

FIGURE 20-7
Some crystal habits for a
cubic crystal.

sponding to planes of the type indicated in Fig. 20-8(a) increase in extent until the crystal is bounded only by such planes. These different appearances are called crystal *habits.* It was gradually appreciated that it is not the crystal habit that is in itself fundamental, but rather its symmetry, especially the angles between crystal faces, and hence between crystal planes. One may often determine the point group to which a crystal belongs just from its external symmetry.

A system for characterizing rational planes was developed by W. H. Miller in 1839, using what are now called Miller indices. If we locate an origin at some lattice point, then any crystal plane must intercept the crystal axes originating from this point, as illustrated in Fig. 20-9. A theorem which we will not prove is that the three intercepts of any rational plane must be in the ratio of integers. This is known as the *law of rational intercepts,* due to Haüy. Parallel to any such plane there is a whole set of planes that may be generated from it by application of unit translations along the axes of the lattice.

The Miller indices of a set of planes may be obtained as follows. We consider that plane of the set that is closest to the origin (without actually containing it), then the Miller indices (hkl) are the reciprocals of the intercepts, expressed in units of the lattice distances, and with fractions cleared. Thus if the intercepts are (2, 3, 2), the reciprocals are ($\frac{1}{2}$, $\frac{1}{3}$, $\frac{1}{2}$) and the Miller indices are (3, 2, 3). In practice the indices (hkl) may refer either to this particular plane or to the whole set of equivalent planes.

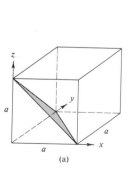

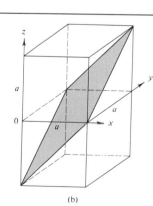

(a)   (b)

FIGURE 20-8
(a) The (111) plane of a
cubic lattice. (b) The (11$\bar{1}$)
plane of a cubic lattice.

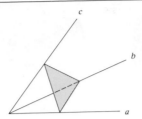

FIGURE 20-9
A crystal plane
intercepting the three
crystallographic axes.

To illustrate, the plane shown in Fig. 20-8(a) is the closest of its set to the origin and has the intercepts (1, 1, 1), expressed in units of a. The Miller indices are therefore the reciprocals, or (111). Other orientations are possible. Thus the plane shown in Fig. 20-8(b) has the intercepts (1, 1, −1), and corresponding indices (11$\bar{1}$). Similarly, the planes ($\bar{1}$11), (1$\bar{1}$1), and so on exist. We will class all of these as just (111) planes in dealing with crystals in the cubic system.

Figure 20-10(a) shows a nearest-origin plane whose intercepts are (1, 1, ∞) and corresponding Miller indices, (110). Planes parallel to one of the axes are more easily shown by means of a projection down that axis, as illustrated in Fig. 20-10(b), where the (110) planes now appear as set I. The intercepts for the nearest-origin plane of set II are (1, $\frac{1}{2}$, ∞), giving the indices (120). Those for the planes of sets III and IV are left for the reader to determine. Note, incidentally, that the equation $hx + ky + lz = 1$ is that of the plane nearest to the origin.

FIGURE 20-10
Various planes of a cubic
lattice. (a) (110) plane.
(b) Various (hk0) planes
viewed down the z axis.

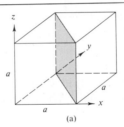

(a)

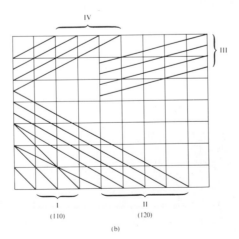

(b)

20-3   Some simple crystal structures

Many of the elements assume a crystal structure which corresponds to the closest possible packing of spheres. As illustrated in Fig. 20-11(a), a single layer of spheres forms a pattern having triangular pockets. There are two equivalent ways of locating the second layer, namely by placing the spheres either in the set of first layer pockets labeled A or in the set labeled B. If the A set is used, as in Fig. 20-11(b), then the B set is excluded. The second layer again has two kinds of triangular pockets, those labeled B, which lie above the B pockets of the first layer, and those labeled C, which lie above the centers of the first-layer spheres.

There are now two *distinguishable* ways of placing a third layer. If the spheres are put in the set of C-type pockets, they will lie directly above the first-layer spheres. The structure then repeats, the types of pockets in which successive layers rest being in the sequence ACACAC... , as shown in Fig. 20-12. The resulting symmetry is easier to visualize if the spheres are replaced by points at their centers, as in Fig. 20-13(a); the structure can now be seen to have hexagonal symmetry. This arrangement is, accordingly, called *hexagonal close packing*, or *hcp*.

Returning to Fig. 20-11(b), if the third-layer spheres are placed in the B-type pockets, this third layer will now have pockets of the Λ and C types. The next simplest repeating unit is then obtained by placing the *fourth* layer in the C pockets, so as to be directly above the first layer (Fig. 20-12). The

A. Close packing of spheres

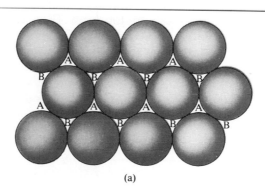

(a)

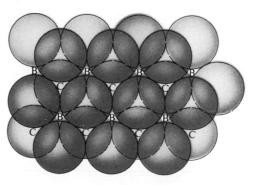

(b)

FIGURE 20-11
Close packing of spheres (a) First layer. (b) Second layer located with centers above A-type pockets of the first layer.

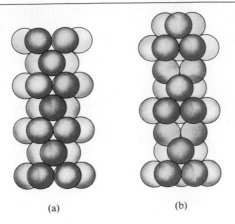

(a)   (b)

symmetry of the ABCABC arrangement is somewhat difficult to see. As shown in Fig. 20-13(b), we can trace the pattern of a face-centered cube by connecting points between the successive layers. Figure 20-14 shows the cube in a normal orientation; we now see more clearly that the successive layers of Figs. 20-12 and 20-13 correspond to (111) planes of a face-centered cubic lattice. The ABCABC arrangement is for this reason called *cubic close packing*, or *ccp*. Alternatively, of course, the structure is *face-centered cubic*, or *fcc*.

It must be remembered that it is the *symmetry* and not the density of packing that differs between the hcp and ccp structures. In both cases (but most easily seen in the hcp symmetry), each sphere in a close-packed plane has six neighbors which are touching. There is a triangle of three nearest neighbors above and one below, to give 12 neighbors in all. In the hcp structure the above and below triangles are oriented in the same way, while in the ccp structure they are rotated 120° relative to each other. It may be shown, incidentally, that the close packing of spheres leaves 26% void space.

Many metals crystallize in either the hcp or the ccp structure, the choice apparently being determined by relatively weak directional preferences in the

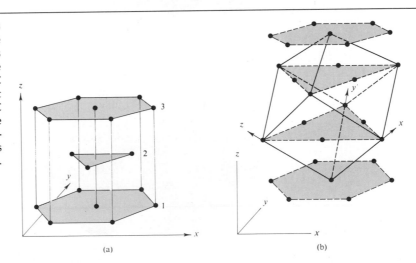

(a)   (b)

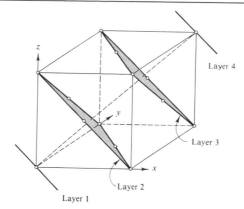

FIGURE 20-14
Cubic close packing drawn so as to show the face-centered cubic lattice.

metal–metal bonding. Metals having the hcp structure include γ-Ca, Cd, α-Co, β-Cr, Mg, and Zn, while those having the ccp structure are Cu, Ag, Au, Al, Pb, β-Ni, and γ-Fe. The rare gases Ar, Ne, and Xe also crystallize in the ccp structure.

A definite tendency toward directed bonding appears with Zn and Cd; although the structure is approximately hcp, the axial ratio c/a is 1.856 for Zn and 1.885 for Cd, as compared to 1.633 for closest packing. In other metals, this tendency leads to some other crystal structure, often *body-centered cubic*, or *bcc*. Metals showing bcc structures include Ba, α-Cr, Cs, α-Fe, δ-Fe, K, Li, Mo, Na, and β-W. Notice that a given element may show more than one crystal modification, as in the case of Cr and Fe.

While hcp and ccp structures both represent the closest packing of spheres, the physical properties of metals can be sharply dependent on which structure is present. Thus ccp metals tend to be much more ductile than hcp ones. The reason is as follows. For crystals in general it is those planes that are the most closely packed which slip past each other most easily [note Eq. (20-5)]. There is only one such set of planes in the hcp structure—namely the basal or hexagonal planes shown in Fig. 20-13(a). In the ccp structure, however, the close-packed planes are the (111) planes, and there are four of these, that is, there is a set of (111) planes, one normal to each of the four cube diagonals. In practice, one deals with metals that are polycrystalline, or have randomly oriented and mutually reinforcing crystal domains. As a consequence, the hcp metals tend to resist distortion in any given direction, since most of the crystallites will not be properly oriented for slip to occur. A much smaller fraction will be able to oppose slip in the case of a ccp metal.

One of the early triumphs of x-ray diffraction was the determination of the crystal structure of sodium chloride. It consists of two interpenetrating fcc lattices, as shown in Fig. 20-15, where, for clarity, the spherical ions are located by points marking their centers. It must be remembered, however, that in any lattice whose positions are occupied by spheres, nearest neighbors are in contact (see Special Topics section). Thus Na^+ and Cl^- ions are touching along the a axes, and the sum of the two ionic radii $r_{Na^+} + r_{Cl^-}$ is $a/2$.

The NaCl type of structure is a fairly common one for MX-type salts. It is the structure of most of the other alkali halides (M = Na^+, K^+, and Rb^+, and X = F^-, Cl^-, Br^-, and I^-, and CsF), as well as for compounds such as

B. Alkali metal halide and other MX structures

FIGURE 20-15
The NaCl structure of two
interpenetrating face
centered cubic lattices.

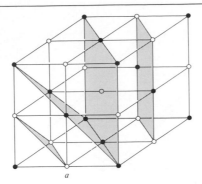

a

BaO, BaS, CaO, CaS, CdO, FeO, and other divalent metal oxides and sulfides. Ammonium ion acts like a spherical ion and NH_4I also has the NaCl structure.

The figure gives the unit cell for NaCl; this is not, however, the repeating unit, which consists of four Na^+ ions, at $(0, 0, 0)$, $(\frac{1}{2}, 0, \frac{1}{2})$, $(\frac{1}{2}, \frac{1}{2}, 0)$, and $(0, \frac{1}{2}, \frac{1}{2})$, and four Cl^- ions, at $(\frac{1}{2}, 0, 0)$, $(0, \frac{1}{2}, 0)$, $(0, 0, \frac{1}{2})$, and $(\frac{1}{2}, \frac{1}{2}, \frac{1}{2})$. If this set is translated by all possible combinations of multiples of a in the x, y, and z directions, the entire crystal is generated. The same count of ions can be obtained from Fig. 20-15.

$$
\begin{aligned}
Na^+ \text{ ions: (8 corner) } \tfrac{1}{8} &= 1 \\
\text{(6 face)} \tfrac{1}{2} &= \underline{3} \\
&\ \ 4 \\
Cl^- \text{ ions: (12 edge)} \tfrac{1}{4} &= 3 \\
\text{(1 interior)} 1 &= \underline{1} \\
&\ \ 4
\end{aligned}
$$

As a check in the counting of atoms, the proportion of each kind must always correspond to the stoichiometry of the formula of the compound. Notice also, as one characteristic of the NaCl structure, that each ion has six nearest neighbors, in an octahedral arrangement.

A second, and quite common, structure for MX salts is that of two interpenetrating simple cubic lattices, as illustrated in Fig. 20-16 for the best-known case, CsCl. The CsCl structure is also found for CsBr and CsI, as well as for other MX compounds, such as NH_4Cl and NH_4Br, and various intermetallic

FIGURE 20-16
The CsCl structure of two
interpenetrating simple
cubic lattices.

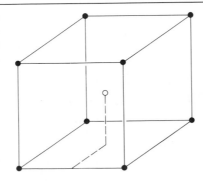

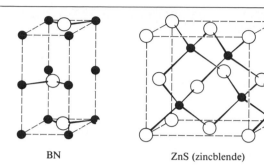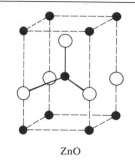

FIGURE 20-17
Typical MX lattices.

BN   ZnS (zincblende)   ZnO

compounds, such as AgCd, AuZn, and CuZn. Each ion or atom has eight nearest neighbors, in square symmetry.

The NaCl and CsCl structures are only two of a number that MX compounds have been found to exhibit. Most such compounds, however, crystallize in one or the other of the structures shown in Fig. 20-17, each labeled according to its most prominent example. Notice that not all of the structures are cubic and that the number of nearest neighbors to a given kind of atom, or its coordination number, diminishes in the sequence:

lattice type	CsCl	NaCl	ZnO	ZnS	BN
coordination number	8	6	4	4	3

It should be apparent at this point that crystal structure does not correlate with the chemistry of a compound, nor with the atomic weights of the constituents *per se*. V. Goldschmidt, who determined many of the simple crystal structures around 1925, concluded that the structure of such crystals is determined by the stoichiometry of the compound, the ratio of atomic radii, and the polarizability of the units.

C. MX$_2$ structures

Two of the important structures for MX$_2$ compounds are shown in Fig. 20-18. For example, the form of TiO$_2$ known as rutile crystallizes in the tetragonal system, with $c = 4.58$ Å and $a = b = 2.98$ Å. The unit cell contains two Ti atoms and four O atoms and is thus consistent with the formula of the compound. Each Ti has six O nearest neighbors and each O has three Ti nearest

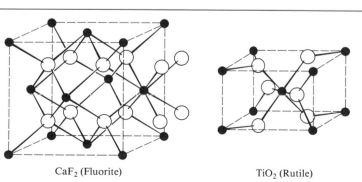

FIGURE 20-18
Common MX$_2$ structures.

CaF$_2$ (Fluorite)   TiO$_2$ (Rutile)

neighbors. If C denotes the coordination number in an M_aX_b compound, a useful rule of stoichiometry is that

$$C_M a = C_X b \tag{20-1}$$

D. Covalent crystals. Diamond and graphite

We next consider briefly two examples of covalent crystals, diamond and graphite. In the case of diamond, the C—C bonds are of the covalent, sigma-bonded type, and show the tetrahedral bond angle of 109°28′. In a sense a diamond crystal is a single large molecule; the crystal structure, however, has a cubic unit cell, displayed in Fig. 20-19. As one way of visualizing the structure, notice that there is an atom at each corner, in the center of each face, and in the centers of alternate small cubes; one set of bonds is shown (the bond distance is 1.542 Å). Germanium, silicon, and gray tin have the same structure.

The diamond structure may also occur with covalent AB-type compounds, such as ZnS (zincblende) (see Fig. 20-17), AgI, CuBr, and BN. The atoms alternate in type, so each A and B atom has four tetrahedrally disposed nearest neighbors. Again, it is not the chemistry of the compound but primarily the relative sizes of the atoms and the geometry of their bonding that determine crystal structures in these cases.

Graphite serves as an example of a layer crystal. The structure is shown in Fig. 20-20 and is seen to consist of layers having a hexagonal tile pattern. The layers are aromatic in character; the C—C distance is 1.42 Å, or not much greater than the value of 1.397 Å for benzene. The distance between layers is relatively large, 3.40 Å, and the layer–layer bonds are consequently rather weak.

The high electrical conductivity of graphite can be regarded as due to the conjugation or aromaticity of the planes, which permits electrons to move through the crystal easily. The lubricating property has been thought to stem from the ease of slippage of one layer over the next, but the actual situation has been found to be somewhat more complex. If graphite is thoroughly degassed, its coefficient of friction rises about sixfold, and it appears that gases adsorbed between layers are responsible for the very low friction ordinarily observed. The same may be true for a useful modern lubricant, MoS_2, whose structure is analogous to that of graphite, as well as for other layer crystals, such as talc.

E. Molecular crystals

The substances considered so far—metals, alkali metal halides and other ionic crystals, and covalent crystals—all have high melting points. This is a reflection of the strong bonds—metallic, ionic, or covalent—between the atoms of

---

Figure 20-19
The diamond structure. Heavy lines show the tetrahedral bonding for one of the carbon atoms.

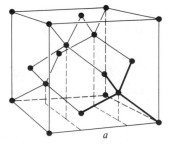

a

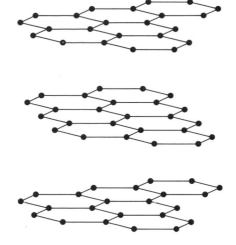

FIGURE 20-20
The graphite structure.

the lattice. A rather different situation exists in the case of crystals whose basis consists of stable molecules and which are therefore held together by secondary or van der Waals forces (note Table 8-8). Figure 20-21 shows the structure of crystalline benzene. This is a low-melting (6°C) white solid; it is soft and is not electrically conducting, and clearly is made up of only weakly interacting benzene molecules.

The crystals of most simple molecules, such as O_2, N_2, CO, CO_2, CH_4, are of this type. Most nonionic organic substances also crystallize by virtue of van der Waals forces only, except where hydrogen bonding is involved.

We conclude this section with a brief discussion of a remaining type of crystal structure, namely that in which hydrogen bonding is present. The most important example is undoubtedly that of ice, whose ordinary structure is shown in Fig. 20-22. It will be recalled from Fig. 8-8(b) that ice exhibits a number of

F. Hydrogen-bonded crystals. Ice

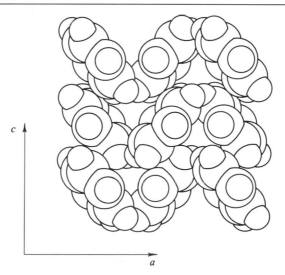

FIGURE 20-21
The benzene structure. [From E. G. Cox, D. W. J. Cruickshank, and J. A. S. Smith, Proc. Roy. Soc. **247**, 1 (1958).]

FIGURE 20-22
Hexagonal (ordinary) ice.
(Courtesy of N. H.
Fletcher.)

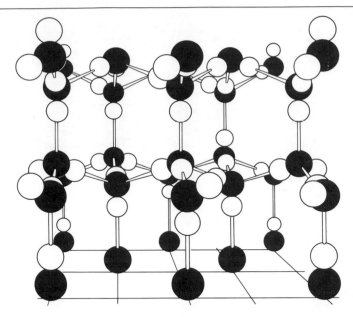

crystal modifications, but only two, tetrahedral and hexagonal ice, are stable under atmospheric pressure. It is the latter modification that is shown in Fig. 20-22. The basic structure is determined by the positions of the oxygen atoms; each is surrounded tetrahedrally by four nearest neighbors at a distance of 2.67 Å. Only recently have advanced techniques using neutron diffraction (see Section 20-CN-1) established that the hydrogen atoms are located on the lines between oxygen atoms but asymmetrically placed; each oxygen is coordinated to two close hydrogen atoms and to two that are further away. The assignment of close versus far hydrogens is random, however, and in this respect the structure is disordered.

Hydrogen bonding is important in molecular crystals in which electronegative atoms, such as fluorine, oxygen, and nitrogen, can interact with an acidic proton of a neighboring molecule. This type of interaction becomes extremely important in crystals of biologically important substances, such as proteins and nucleic acids.

20-4   Some geometric calculations

The following types of calculations can be performed on any crystal but the geometric expressions rapidly become complicated as one departs from the cubic system. The general expression for the density of a crystal is, for example,

$$\rho = \frac{\sum n_i M_i}{N_A V_c} \tag{20-2}$$

where n_i is the number of the ith kind of atom (or molecule) of molecular weight M_i in the unit cell, and V_c is the cell volume. In the case of a cubic crystal V_c is just a^3, while for a tetragonal unit cell V_c is a^2c, and so on.

Two illustrations of the use of Eq. (20-2) follow. Iron (α-Fe) crystallizes in the bcc system with $a = 2.861$ Å; the eight corner and one interior atom in the unit cell give a net of two atoms, so

$$\rho = \frac{2(55.85)}{6.02 \times 10^{23}(2.861 \times 10^{-8})^3} = 7.92 \text{ g cm}^{-3}$$

Reference to Fig. 20-4 shows that each atom in a body-centered cubic lattice has eight nearest neighbors; the center atom, for example, must therefore be in contact with the eight corner ones and the body diagonal of the cube must equal two atomic diameters. The radius of each Fe atom is therefore

$$r_{Fe} = \tfrac{1}{4}\sqrt{3}\,(2.861 \times 10^{-8}) = 1.24 \text{ Å}$$

Iron also crystallizes in the fcc structure, and a reasonable assumption is that the *radius* of the atom does not change. In the cubic close-packed arrangement each atom has twelve nearest neighbors, and examination of either Fig. 20-4 or Fig. 20-14 shows that for a face-centered atom four of these neighbors consist of the atoms at the corners of the face. The face diagonal now contains two atomic diameters, or $r_{Fe} = \tfrac{1}{4}\sqrt{2}\,a$. On setting $r_{Fe} = 1.24$ Å, we obtain $a = 3.50$ Å for the fcc structure. The unit cell has eight corner and six face atoms, or a net of four atoms, so the density of fcc iron should be

$$\rho = \frac{4(55.85)}{6.02 \times 10^{23}(3.50 \times 10^{-8})^3} = 8.66 \text{ g cm}^{-3}$$

(The experimental value is 7.86 g cm^{-3}.)

As a second illustration, the density of NaCl is 2.165 g cm^{-3}. There are four Na$^+$ and four Cl$^-$ ions per unit cell, so we have

$$2.165 = \frac{4(22.99) + 4(35.45)}{6.02 \times 10^{23}a^3}$$

whence $a = 5.64$ Å. Reference to Fig. 20-15 shows that the closest distance between ions is that along a cube edge, so that $a = 2r_{Na^+} + 2r_{Cl^-}$. The sum of the two ionic radii is then $a/2 - 2.82$ Å. As discussed in the Special Topics section, only sums of ionic radii may be determined from a crystal structure, but there are various schemes whereby individual ionic radii can be assigned.

Another important type of calculation is that of the interplanar distances in a crystal. In the case of a two-dimensional, square lattice, illustrated in Fig. 20-23, it is obvious that (10) and (01) planes are spaced just a units apart, or that $d_{10} = d_{01} = a$, and likewise that $d_{11} = \tfrac{1}{2}\sqrt{2}\,a = 0.707a$. The direction perpendicular to (12) planes is evidently that of the diagonal of a rectangle of sides 1 and 2 in the x and y directions, respectively. The particular diagonal shown is of length $\sqrt{5}\,a$ and spans five interplanar distances, so $d_{12} = (1/\sqrt{5})a$. The general formula is $d_{hk} = a/(h^2 + k^2)^{1/2}$. Extension to a three-dimensional cubic lattice gives

$$d_{hkl} = \frac{a}{(h^2 + k^2 + l^2)^{1/2}} \tag{20-3}$$

or, for a lattice with $\alpha = \beta = \gamma = 90°$ but with $a \neq b \neq c$

FIGURE 20-23
Interplanar distances for a
square lattice.

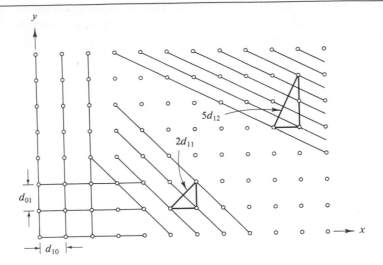

$$\frac{1}{d_{hkl}^2} = \frac{h^2}{a^2} + \frac{k^2}{b^2} + \frac{l^2}{c^2} \tag{20-4}$$

Equation (20-3) gives the interplanar distance for planes of specified Miller indices, but it is sometimes convenient to class as belonging to a given *type* all planes which are parallel to each other, regardless of actual Miller index. Thus (100) and (200) planes may be called (100)-type planes, (111) and (222) planes may be called (111)-type planes, and so on. The distinction is illustrated further as follows. The largest interplanar distances for a simple cubic crystal are, in order, d_{100}, d_{110}, and d_{111}, and their ratios are given by Eq. (20-3) as

$$\text{simple cubic:} \quad d_{100} : d_{110} : d_{111} = 1 : \frac{1}{\sqrt{2}} : \frac{1}{\sqrt{3}}$$

The largest interplanar distances for a bcc lattice are, again in order, d_{110}, d_{200}, and d_{222} (see Fig. 20-24):

$$\text{bcc:} \quad d_{110} : d_{200} : d_{222} = \frac{1}{\sqrt{2}} : \frac{1}{2} : \frac{1}{2\sqrt{3}}$$

FIGURE 20-24
Interplanar distances in a
bcc lattice.

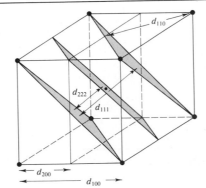

These latter two planes may be referred to as (100)- and (111)-*type* planes, however, and the ratios reported as

$$\text{bcc (type planes):} \quad d_{100} : d_{110} : d_{111} = \frac{1}{2} : \frac{1}{\sqrt{2}} : \frac{1}{2\sqrt{3}} = 1 : \frac{2}{\sqrt{2}} : \frac{1}{\sqrt{3}}$$

Similarly,

$$\text{fcc (type planes):} \quad d_{100} : d_{110} : d_{111} = 1 : \frac{1}{\sqrt{2}} : \frac{2}{\sqrt{3}}$$

(the demonstration is left to the reader). As will be seen in the next section, x-ray diffraction studies may in some cases yield an interplanar spacing where the type but not the actual Miller index of the planes is known.

A final point of geometry is the following. In the case of simple structures it will often be true that a set of planes contains all of the atoms or ions of the crystal. Thus if a metal crystallizes in the bcc structure, the (200) planes as a set pass through or contain all of the atoms of the crystal. The same is true for the sets of (110) and (222) planes. The crystal can thus be viewed as consisting of layers of (200) planes having a surface concentration σ_{200} atoms cm^{-2} and separated by d_{200} cm. The quotient σ_{200}/d_{200} must therefore equal the volume concentration of atoms in the crystal, C atoms cm^{-3}. The crystal can alternatively be regarded as consisting of layers of (110) planes separated by d_{110}, so that $\sigma_{110}/d_{110} = C$. Thus for any set of planes that contains all the atoms of the crystal, it follows that

$$\frac{\sigma_{hkl}}{d_{hkl}} = C \qquad\qquad (20\text{-}5)$$

Equation (20-5) is often quite useful. It tells us that the highest surface density of atoms will occur on that set of planes of largest interplanar spacing. These are the (110) planes in the bcc structure and the (111) planes in the fcc case, for example. Such planes tend to have the lowest surface free energy and therefore tend to be prominent faces in actual crystals. They also tend to slip or to cleave the most easily, as is noted in Section 20-3A.

20-5 Diffraction by crystals

X rays were first observed in experiments around 1890 with cathode ray tubes (see Section 22-1); by 1895 J. J. Thomson was describing the effect of this penetrating radiation on the electrical properties of gases. Later, in 1898, G. G. Stokes and G. J. Stoney suggested that x rays were electromagnetic in nature and the classic first experiment establishing x-ray diffraction was done in Munich in 1912 at the suggestion of M. von Laue.

X rays are produced by atoms which have lost a K or L electron as a result of a collision with some high-energy particle. Referring to Eq. (16-30), if a K, or $n = 1$, electron is lost from, say, a copper atom ($Z = 29$) and an L, or $n = 2$, electron falls into the vacancy, the energy of the emitted radiation is $0.5(29)^2[(1/1^2) - (1/2^2)] = 315$ a.u. or 8584 eV of energy. The wavenumber of radiation of this energy per quantum is

$$(8584)(8.066) = 6.92 \times 10^4 \quad \text{kK} = 6.92 \times 10^7 \quad \text{cm}^{-1}$$

corresponding to 1.44 Å. X-ray wavelengths are thus comparable to atomic radii. Modern equipment still consists essentially of a Crookes tube, in which a beam of perhaps 20-kV electrons impinges on a target, usually copper, iron, or tungsten. The emitted x radiation is generally filtered (see Section 22-4 on the critical absorption of x rays) and then collimated by means of slits so as to produce a narrow, monochromatic beam.

Before discussing three-dimensional diffraction it should be explained why it is that x rays act as though they are reflected specularly from each plane of a crystal. The interaction of an x-ray quantum with an atom may be regarded as an absorption followed by resonance emission (Section 19-3), that is, by emission at the same wavelength. This process is a *virtual* one; it constitutes a hypothetical mechanism yielding the same result as does the wave mechanical treatment of scattering. The effect is that each atom becomes a secondary, isotropic emitter of the x radiation, as illustrated in Fig. 20-25 for a single line of atoms.

A more detailed analysis is as follows. We suppose there to be incident radiation at angle θ to the line of atoms ABC. The radiation first hits atom A, which then emits isotropically, and by the time the wavefront reaches atom B the secondary radiation from atom A has a wavefront given by circle 1A. When the radiation reaches atom C the wavefronts of secondary radiation for atoms A and B are now given by circles 2A and 2B. Still later the reemitted radiation has reached circles 3A, 3B, and 3C. Where similarly numbered circles cross, the emitted radiation is in phase, as for example, at point a, the crossing between circles 2A and 2B, and point b, the crossing between circles 3A and 3B. As the circles move outward with time, the line along which the radiation is in phase is that given by the line ab. Alternatively, points b and c represent concurrent in-phase conditions and define the wavefront of the secondary radiation. Either reasoning leads to the conclusion that the direction of the secondary radiation is at the same angle θ to ABC as is the incident radiation, independent of the spacing between atoms, provided it is uniform.

This analysis is geometric and qualitative but serves to explain the rigorous result that the secondary radiation is mutually interfering except in the di-

FIGURE 20-25
Scattering of x rays from a single line of atoms.

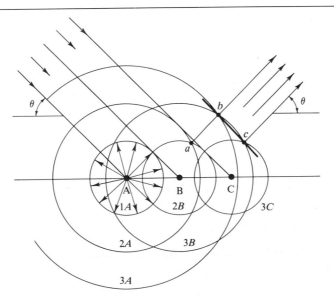

rection of specular reflection. We may now proceed to the three-dimensional situation.

The first diffraction experiment was made with the use of "white" x radiation, that is, radiation having a range of wavelengths, and a single crystal (of copper sulfate). Later, around 1912, W. Bragg and his son developed an approach which was, at that time, much easier to apply and to interpret. This consisted in using monochromatic x radiation and varying the orientation of the crystal.

The situation is illustrated in Fig. 20-26. We take each ray of incident radiation to be specularly "reflected" by a given layer of atoms, shown end-on in the figure. Reflections then occur from the successive layers, and we now ask what condition must be met for the reflected radiation to be in phase. Considering just the first two layers, the difference in path length for the respective rays is the distance abc. From the construction, this distance is $2ab$ or $2d \sin \theta$. Thus $2d \sin \theta$ must be an integral number of wavelengths if the reflected radiation is to be in phase. The *Bragg condition* is thus

$$n\lambda = 2d \sin \theta \tag{20-6}$$

where n is an integer. If $n = 1$, the reflection is called first order, if $n = 2$, second order, and so on. Note, however, that a second-order reflection for a given spacing d is at the same angle as a first-order reflection from planes of spacing $d/2$. Thus second-order reflections from (100) planes of a simple cubic crystal should be indistinguishable from first-order reflections from a hypothetical set of (200) planes. It is more convenient to treat the order of a reflection in this alternative way and to write Eq. (20-6) as simply

$$\lambda = 2d_{hkl} \sin \theta \tag{20-7}$$

For cubic crystals combination with Eq. (20-3) gives

$$\sin^2 \theta = \frac{\lambda^2}{4a^2} (h^2 + k^2 + l^2) \tag{20-8}$$

Figure 20-27 shows schematic plots of reflected intensity I versus θ for KCl and NaCl. A single crystal is oriented so that (100)-, (110)-, or (111)-type planes (or crystal faces) are perpendicular to the x-ray beam and the crystal is then

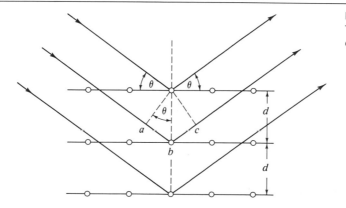

FIGURE 20-26
The Bragg scattering condition.

FIGURE 20-27
Intensity versus θ for Bragg
scattering from (100)-,
(110)-, and (111)-type
planes of (a) NaCl and
(b) KCl.

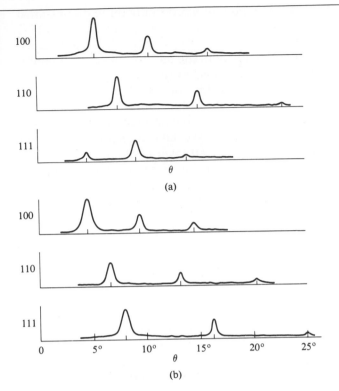

FIGURE 20-27
Intensity versus θ for Bragg scattering from (100)-, (110)-, and (111)-type planes of (a) NaCl and (b) KCl.

rotated so that the angle of incidence θ is varied. As the angles for successive Bragg conditions are met, an increase in intensity of scattered radiation is observed by means of a detector whose angular position is also rotated so that it always "sees" radiation at the specular angle. The schematic arrangement is shown in Fig. 20-28.

In the case of NaCl the first-order peaks are found in a particular experiment to occur at 5°18', 7°31', and 4°36' for reflections from (100)-, (110)-, and (111)-type planes, respectively. The interplanar distances are then in the ratio of the reciprocals of the sin θ values, or $d_{100} : d_{110} : d_{111} = 1/0.0924 : 1/0.1307 : 1/0.0801 = 1 : 0.707 : 1.154 = 1 : 1/\sqrt{2} : 2/\sqrt{3}$. This is just the ratio predicted

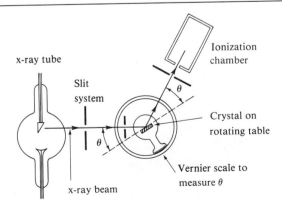

FIGURE 20-28
A Bragg x-ray spectrometer.

in the preceding section for a fcc structure. The early experimentalists knew from the symmetry of NaCl crystals that they were in the cubic system, and from the ratio test could determine which type. By reversing the density calculation of the preceding section, a could be calculated, and from this the wavelength of the x rays used.

Most other alkali halide crystals show an analogous set of intensity patterns but, as illustrated in Fig. 20-27(b), that for KCl gives $\sin \theta$ values of 0.0811, 0.1145, and 0.1405 for reflections from (100)-, (110)-, and (111)-type planes, respectively, using the same x rays. This yields a set of ratios $d_{100} : d_{110} : d_{111}$ = 1 : 0.707 : 0.577 = $1 : 1/\sqrt{2} : 1/\sqrt{3}$, or that corresponding to a simple cubic structure. Referring to Fig. 20-15, the explanation lies in the nature of the (111)-type planes. In the case of NaCl the Na$^+$ ions are contained by one set of (111) actual planes and the Cl$^-$ ions, by a second set lying midway between. The reflections from the (111) actual planes are partially but not completely canceled by the out-of-phase reflections from the planes containing the other kind of ion. Thus in Fig. 20-27 the first peak for (111)-type planes for NaCl is weaker than usual. In the case of KCl the two ions are isoelectronic and scatter x rays about equally; the effect is that K$^+$ and Cl$^-$ ions appear to be identical in the diffraction experiment. Reflections from the Cl$^-$ (111)-type planes therefore cancel those from the K$^+$ (111)-type planes. The result is that the largest interplanar distance showing a net diffraction intensity is that for (222) actual planes. Alternatively expressed, x rays do not distinguish K$^+$ from Cl$^-$ and KCl appears as a simple cubic lattice of side $a/2$, where a is the side of the actual fcc unit cell.

The example is introduced to illustrate the point that the intensity of Bragg reflections from a given (hkl) set of planes depends on the densities and nature of atoms on these and on intervening planes as well. A more formal approach is given in Section 20-ST-2.

C. Powder patterns

The Bragg method is not only historic but is also of current usefulness. The application is mainly to a crystalline powder. The individual particles are assumed to be in random orientation, so that the monochromatic x rays will always find some particle with the correct orientation for reflection from each (hkl) set of planes. The experiment consists in irradiating a thin tube filled with the powdered sample, using a collimated beam of monochromatic x rays. Those crystals whose particular (hkl) planes are at the Bragg angle θ to the beam then give a reflection at an angle θ to the crystal planes, or 2θ to the incident beam. As illustrated in Fig. 20-29, one may use a circular strip of photographic film to intercept the diffracted radiation, which then shows as curved lines on the film. Alternatively, a traveling radiation counter may be used. The $\sin \theta$ value for each line is calculated from the apparatus geometry.

The next task is to assign (hkl) values to the lines. This can be done if the crystal is known to be in the cubic, tetragonal, or orthorhombic class but is otherwise rather difficult. Often the "powder pattern" is used much as are infrared absorption spectra, namely as a kind of fingerprinting of the particular substance. A mixture of materials may then be analyzed semiquantitatively by calculating in what proportion pure substance powder patterns are present.

The procedure for assigning actual (hkl) values to individual lines can be illustrated fairly easily if the crystal is in the cubic system. From Eq. (20-8)

FIGURE 20-29
Schematic drawing of a
powder method apparatus.

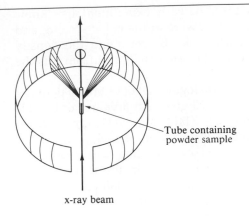

Tube containing
powder sample

x-ray beam

the $\sin^2\theta$ values are proportional to $h^2 + k^2 + l^2$. The first step is to tabulate the film data in order of increasing $\sin^2\theta$. Since $h^2 + k^2 + l^2$ must be an integer, we then look for the smallest set of integers that are in the ratios of the $\sin^2\theta$ values. Only certain integral values for $h^2 + k^2 + l^2$ are possible for each type of cubic crystal. In the case of a simple cubic crystal, the sum cannot have the values 7, 15, or 23, for example, since there are no three integers the sum of the squares of which equals one of these numbers. The sum can have only the values 2, 4, 6, 8, ... for a bcc crystal and only the values 3, 4, 8, 11, 12, ... for a fcc one, as summarized in Table 20-2. We can then determine which set the observed series of $\sin^2\theta$ values matches.

An example should be helpful at this point. We suppose that a particular sample of finely divided lead shows the following $\sin^2\theta$ values with x rays of 1.54 Å: 0.0729, 0.0972, 0.194, 0.267, 0.292, and 0.389. The procedure is shown in Table 20-3. We first divide each number by 0.0729 and observe that

TABLE 20-2
Indices of cubic crystals[a]

Indices (hkl)	$h^2 + k^2 + l^2$	Allowed reflection		
		Simple cubic	bcc	fcc
100	1	×	—	—
110	2	×	×	—
111	3	×	—	×
200	4	×	×	×
210	5	×	—	—
211	6	×	×	—
220	8	×	×	×
300, 221	9	×	—	—
310	10	×	×	—
311	11	×	—	×
222	12	×	×	×
320	13	×	—	—
321	14	×	×	—
400	16	×	×	×

[a]The sum of the squares of three integers cannot have the values $(7 + 8m)4^n$, where m and n are arbitrary integers.

Observed $\sin^2 \theta$	Result of dividing by 0.073	Integers having these ratios	(hkl)	
0.0729	1	3	(111)	**TABLE 20-3**
0.0977	1.33	4	(200)	Sample treatment of
0.194	2.65	8	(220)	powder diffraction data for
0.267	3.65	11	(311)	lead
0.291	4.01	12	(222)	
0.389	5.33	16	(400)	

the result is not a set of integers. Examination indicates that multiplying each number by three will yield integers, as shown in the third column. These integers are just those allowed by Table 20-2 for a fcc structure, and we then index each entry accordingly. In an actual experiment each $\sin^2 \theta$ value would now be divided by the assigned $h^2 + k^2 + l^2$ value to obtain a series of values of $\lambda^2/4a^2$, which would be averaged. In the present example we take this average to be $0.0729/3 = 0.0243$, and from the known λ of 1.54 Å calculate a to be 4.94 Å. We could proceed to check the result by calculating the density of lead.

COMMENTARY AND NOTES

20-CN-1 Modern crystal structure determination

The presentation in Section 20-5 emphasized the Bragg method and powder patterns as a straightforward yet actively useful approach to x-ray diffraction. It is not, however, the method used for crystal structure determination where any degree of complexity exists. The difficulty with the Bragg method is that of indexing the diffraction lines (such as shown in Fig. 20-29). If the crystal unit cell contains many atoms and especially if it is not cubic, it can be virtually impossible to find a unique set of assignments. Alternative methods are available, which avoid the loss of information in the powder method that occurs through not knowing the orientation of the crystals that give a particular reflection. These methods are described briefly below.

Once a set of reflections has been indexed and their intensities measured, the next problem is that of deducing the crystal structure. The lattice type or symmetry can often be de-

termined from an examination of a crystal under a polarizing microscope. The symmetry of the crystal habit helps to establish the Bravais lattice involved; the behavior under polarizing light may suffice to limit the space group to at least a few possibilities. The presence or absence of certain reflections is usually sufficient to complete the assignment.

A. Rotation— oscillation methods

One gains enormously in ability to index diffractions by using monochromatic x radiation *and* a single crystal that can be mounted so that one of its crystal axes is perpendicular to the x-ray beam. In the rotating crystal method, as the crystal turns various planes come into the Bragg angle with respect to the incident beam. The diffraction consists of spots generated in a series of horizontal layers on the film, as illustrated in Fig. 20-30. Those reflections at the same level as the crystal arise from planes parallel to the rotation axis, and those reflections

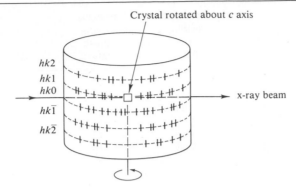

FIGURE 20-30
Rotating crystal method.

above and below arise from planes whose inclination to the axis can be calculated. Thus each spot can immediately be assigned one index. The remaining indexing can still be very difficult if the crystal has many atoms in the unit cell, and a yet better procedure is available.

The next step is to displace the film synchronously with what is usually an oscillation rather than a continuous rotation of the crystal. In this way the *orientation* of the crystal is known for each diffraction spot. The *Weissenberg method,* as it is called, thus allows each spot to be fully indexed. The method produces series of diffraction spots that lie on curved lines, however, and in the *precession method,* the motions of the crystal and the film are so regulated that the spots lie on straight lines.

Modern crystallographic equipment largely automates these procedures. Intensities are found by x-ray detectors rather than photographically, and the data are stored directly in a computer system.

B. Electron density
projections

A crystal appears to x rays as a collection of scattering centers, each of intensity approximately proportional to the number of electrons possessed by a particular atom in the unit cell. The information provided by x-ray diffraction is therefore that of the distribution of electron density $\rho(x, y, z)$ in the crystal. Since this distribution is a periodic one, it is very convenient to represent it as a set of cosine and sine terms—that is, as a Fourier series.

The formal expansion of $\rho(x, y, z)$ is

$$\rho(x, y, z)$$
$$= \frac{1}{V} \sum_h^\infty \sum_k^\infty \sum_l^\infty F_{hkl} \exp\left[-2\pi i\left(\frac{hx}{a} + \frac{ky}{b} + \frac{lz}{c}\right)\right]$$

$$(20\text{-}9)$$

where V is the volume of the unit cell and F_{hkl} is called the structure factor (see Section 20-ST-3). This form recognizes that the electron density is periodic with repeat distances of a, b, and c. If one could evaluate each term from the diffraction data, the resulting $\rho(x, y, z)$ would *be* the crystal structure. The problem is that the intensity measurements yield $|F_{hkl}|^2$ but not the sign of F_{hkl} itself.

A number of rather sophisticated procedures have been developed to help get around the problem. It is possible, for example, to determine the distribution of interatomic distances from the indexed intensity data alone. With this information and some educated guesses the crystallographer may be able to arrive at a trial structure. Even if approximate, such a structure will allow a calculation of the *sign* of each F_{hkl}, and now the experimental intensity data can be used to calculate an actual $\rho(x, y, z)$. The result usually appears as electron density contour maps calculated for one or another projection. Such a result for anthracene is shown in Fig. 20-31.

Another "trick" is to incorporate very heavy atoms (such as iodine) in the molecule. The intense scattering by such atoms dominates the diffraction pattern and it may be possible to determine the positions of the heavy atoms and thus the sign of their F factor. Sometimes a natural grouping, such as a benzene ring, can

FIGURE 20-31
Electron density contours for anthracene. [From V. L. Sinclair, J. M. Robertson, and A. McL. Mathieson, Acta Crystallogr. **3**, 254 (1950).]

$$0 \quad 1 \quad 2 \quad 3 \text{ Å}$$
Scale

be recognized in the diffraction pattern and subtracted from it.

C. Use of radiation other than X rays

The fundamental equations of this chapter are not restricted to any particular kind of radiation. The main restriction is that inherent in the Bragg equation, namely that since $\sin \theta$ cannot exceed unity, λ must be less than $d_{hkl}/2$. Thus the wavelength of the radiation used must be less than half of the interplanar spacings to be determined. This means that radiation of no more than about 1 Å can be used.

In view of the de Broglie relationship (16-7), particles have a wave nature and should also exhibit diffraction effects. Thus 40,000-eV electrons would have $(4 \times 10^4)(1.602 + 10^{-12}) =$

6.408×10^{-8} erg kinetic energy, corresponding to a momentum of 1.075×10^{-17} g cm s^{-1}. The wavelength is then $6.625 \times 10^{-27}/1.075 \times 10^{-17}$ $= 0.0616$ Å. Electrons of this energy do indeed show diffraction effects and may be used for structure determinations. Electrons, being charged, interact more strongly with matter and are more rapidly absorbed than is x radiation. As a consequence, very thin samples must be used. Standard electron microscopes may, for example, be put in a diffraction mode of operation so as to show the diffraction pattern for very thinly sliced samples. Because electrons can be focused by means of magnetic lenses, it is also possible to obtain a direct *picture*. Figure 20-32 shows a modern high-resolution electron microscope photograph of a thin flake of $Nb_{22}O_{54}$. In this photograph, the view is down the *b* axis of the crystal; the white areas are channels between columns of NbO_6 octahedra,

FIGURE 20-32
High-resolution electron micrograph of the Nb_2O_{54} lattice (see text). Arrows mark locations of crystal defects. (From S. Iijima, Department of Physics, Arizona State University.)

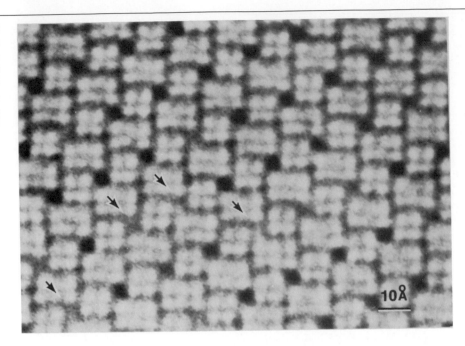

and the dark circles correspond to columns of tetrahedrally coordinated Nb atoms.

Returning to diffraction studies, we find that it is possible to obtain *molecular* structures from the diffraction of a monoenergetic beam of electrons by a molecular gas. The situation here is similar to that of the powder method since each individual molecule is in random orientation, and the analysis is similar except that one now deals with distances between atoms in a molecule rather than with interplanar distances. One finds for the intensity of the scattered beam at angle θ

$$I(\theta) = \sum_i \sum_j F_i F_j \frac{\sin x_{ij}}{x_{ij}} \qquad (20\text{-}10)$$

where F is the atomic scattering factor, and x_{ij} is given by

$$x_{ij} = \frac{4\pi r_{ij}}{\lambda} \sin \frac{\theta}{2} \qquad (20\text{-}11)$$

The distances r_{ij} are those between atoms i and j in the molecule. Equation (20-10) is known as the *Wierl* equation. It is possible to transform the $I(\theta)$ results into a radial distribution function (much as in the case of Fig. 8-1) and from this one can usually infer the molecular bond lengths and angles.

Alternatively, neutron scattering may be employed. The same calculation as the preceding applies but now with m equal to the neutron mass, about 2000 times larger than that of the electron. Neutrons of momentum 1.08×10^{-17} g cm s^{-1}, and hence wavelength 0.0616 Å, have 1/2000 the energy or about 3×10^{-11} erg molecule^{-1}, corresponding to a "temperature" of about 1000 K. Neutrons of thermal or room-temperature energy have wavelengths, then, of around 1 Å and such neutrons may be provided by nuclear reactors (see Section 22-CN-2). A velocity selector (such as illustrated in Section 2-CN-2) may then be used to provide a monoenergetic beam. The technique is a relatively difficult one, and the chief advantage in using it is that, unlike the case with x rays, the scattering of neutrons is large for light atoms. Neutron diffraction has, for example, been used to locate protons in ice and in metal hydrides, a very difficult task with x rays.

20-CN-2 The band model for solids. Semiconductors

We have so far considered a crystal to be made up of discrete atoms in a repeating structure. The wave mechanical picture, while basically not much different, does introduce an aspect of great importance with respect to the energy levels of electrons in a crystal. The point is that while completely isolated atoms would have energy levels as described in Chapter 16, when these are brought into proximity in a crystal, electron exchange begins to be possible. Suppose, for example, we had a crystal consisting of hydrogen atoms, or, more realistically, of hydrogen-like atoms, such as lithium. Considering first just one pair of atoms, then, as given by Eqs. (18-13) and (18-14), the single equal energy states of each atom combine to give a bonding and antibonding pair. In terms of the present discussion the point is that the mixing of wave functions between two atoms generates a pair of new states, one above and one below the original isolated atom state. In a crystal some number N of atoms mix their electrons wave mechanically, and the result is N states distributed in a band roughly centered at the original isolated atom state. The degree of this mixing increases as the atoms approach, and the situation is shown schematically in Fig. 20-33. In the case of a crystal of lithium, the band due to the 1s states does not span enough range of energy to overlap with that from the 2s ones. The 1s band will have N states and just $2N$ electrons, that is, two 1s electrons from each atom, to exactly fill it.

The 2s band is only half-filled, however, since only N outer electrons are present. There are thus many close-lying excited states accessible to these electrons, with the consequence that they have a Boltzmann-type distribution of kinetic energies and can move easily from one point to another. Hence the electrical conductivity of lithium.

A metal such as beryllium would exactly fill its 2s band, but it turns out that the 2p overlaps with it, so that again the outer electrons are able to move freely. The wave mechanical picture of metallic conduction is thus one of bands only partially filled with electrons. The *mobility* of the electrons is adversely affected by the thermal vibrations of the lattice, which disrupt its regularity. As a consequence, metallic conductivity tends to decrease with increasing temperature.

The profile of a band, that is, the plot of $N(E)$, the number of energy states, versus energy E, might look as in Fig. 20-34. Metals correspond to the situation shown in Fig. 20-34(a,b), that is, to an incompletely filled band or to overlapping bands. If, however, the highest energy band is completely filled and well separated from the next one, then electrons cannot move since there are no easily accessible excited states; the substance is then an insulator. An example is diamond, which has a completely filled band well separated from the next higher one, as in Fig. 20-34(c). Finally, it may happen that while a band is filled, the next higher one is so close that little additional en-

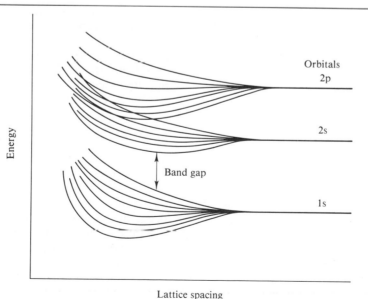

FIGURE 20-33
Splitting of energy levels as N atoms approach.

Energy

Lattice spacing

Orbitals
2p

2s

Band gap

1s

FIGURE 20-34
Distribution of energy states (shading indicates occupied states); (a, b) metals; (c) insulator; (d) semiconductor.

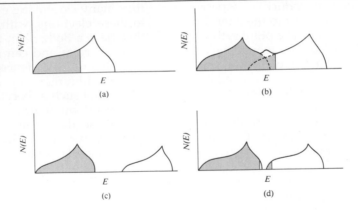

ergy is needed to promote electrons to it, as illustrated in Fig. 20-34(d). Such a substance will show an electrical conductivity that strongly increases with temperature, as the electron population of the upper band increases and leaves vacancies in the lower one. Such materials are called *intrinsic semiconductors*. Germanium has an energy gap of 0.72 eV, and gray tin one of only 0.1 eV; both behave in this manner.

It is also possible to produce electron mobility by incorporating suitable impurities in the lattice of an insulator or intrinsic semiconductor. The impurity should be able to substitute for one of the regular lattice atoms so as not to distort the lattice appreciably. Suppose, for example, a boron atom is substituted for carbon in the diamond lattice or for silicon or germanium in their crystals. The boron atoms act as C^+, Si^+, or Ge^+ ions, being one electron short, and the effect is to produce holes or vacancies in the highest energy band. The energy match is not exact, and while the boron atoms act as acceptors of electrons, some promotion energy is needed. Similarly, if P atoms are introduced, these act like C^-, Si^-, or Ge^- atoms, and with a little additional energy the extra electron can be promoted into the next higher unfilled band of the crystal. The effect of either impurity is to greatly increase the semiconductor property of the crystal. Such crystals also tend to exhibit *photoconductivity*, since light energy can also promote electrons to allow conduction.

It might be mentioned that the bands in an insulator (which do not overlap) are not uniform through the crystal but are bounded geometrically by zones called *Brillouin zones*, the shapes of which are determined by the crystal symmetry. The situation can be viewed as a confinement of electrons due to their reflection by crystal planes. In conductors, the Brillouin zones overlap, so that free motion through the crystal is possible.

20-CN-3 Crystal defects

Actual crystals do not consist of the perfect lattice array implicit in the discussion so far. First, many apparently crystalline materials are actually microcrystalline, that is, they consist of small crystalline domains or grains welded together at their boundaries. This is often true of metals and especially of alloys, which may really be eutectic mixtures. The presence of such domains affects the mechanical and electrical properties of the crystal. The x-ray diffraction pattern becomes essentially that for a powder, and if the domains are very small, the diffraction lines will also be diffuse.

This is a macroscopic, adventitious imperfection. There are two other types that are always present in an otherwise perfect crystal. First, atoms or ions may be missing from various individual lattice sites, as illustrated in Fig. 20-35(a) for the case of an ionic crystal. The crystal as a whole remains electrically neutral so that positive and negative ion vacancies must occur in pairs. (An excess of even 10^{-10} mole of one over the other would give a crystal an electrostatic charge of perhaps one million volts!) This type of imperfection is known as a *Schottky defect*. The second situation is that in which an extra atom or ion has found an abnormal or

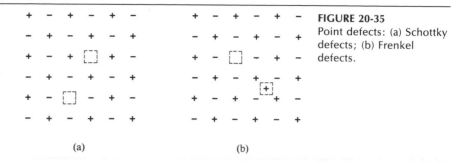

FIGURE 20-35
Point defects: (a) Schottky defects; (b) Frenkel defects.

(a)

(b)

interstitial position, as in Fig. 20-35(b); the particular situation shown is that in which a positive ion has left its lattice site, leaving a vacancy. Such defects are called *Frenkel defects.*

These two types of imperfection are known as *point defects.* Individual atom or ion displacements are involved and there is a definite energy requirement to be met. The population of point defects in a crystal is therefore governed by the Boltzmann principle, and any crystal is expected to have some equilibrium concentration of them. The fraction of such lattice vacancies will be about 10^{-5} at 1000 K, corresponding to a typical energy of 20 kcal mol^{-1}. More detailed treatments estimate the entropy as well as the energy requirement.

A second major type of imperfection is that known as a *dislocation.* This is an organized concentration of point defects and is also termed a *lattice defect.* The presence of such defects does not represent a thermodynamic equilibrium but rather is a consequence of the history of formation of the crystal and of its subsequent mechanical experience.

One important type of lattice defect is that

known as an *edge dislocation,* illustrated in Fig. 20-36. Here an extra lattice half-plane is present; the crystal below the *slip plane AB* is in tension and above it the crystal is in compression. The *dislocation line* is the slip plane at C (perpendicular to the plane of the figure). The dislocation emerges at the surface as a step or, conversely, pressure applied at the surface above *AB* has caused a slip of one lattice unit; the surface layers are in register, but the compression has localized around the dislocation line C. Continued motion of C to the left would result eventually in its emergence and the whole upper half of the solid would then have flowed one lattice unit. The process is much like moving a rug by pushing a crease down it.

This dislocation may be characterized by tracing a counterclockwise circuit around C, counting the same number of lattice points in the plus and minus directions along each axis or row. Such a circuit closes if the crystal is perfect but it does not if a dislocation is present, as illustrated in the figure. This circuit is known as a *Burgers circuit;* its failure to close distinguishes a dislocation from a point defect. The

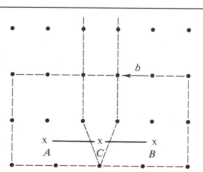

FIGURE 20-36
An edge dislocation. [From A. W. Adamson, "Physical Chemistry of Surfaces," 2nd ed. Copyright 1967, Wiley (Interscience), New York. Used with permission of John Wiley & Sons, Inc.]

FIGURE 20-37
(a) Screw dislocation. (b) The slip that produces a screw-type dislocation. Unit slip has occurred over ABCD. The screw dislocation AD is parallel to the slip vector. (From W. T. Read, Jr., "Dislocations in Crystals." Copyright 1953, McGraw-Hill, New York. Used by permission of McGraw-Hill Book Company.)

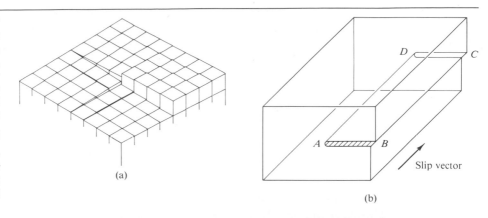

(a)

(b)

FIGURE 20-38
Screw dislocation in a carborundum crystal. [From A. R. Verma, Phil. Mag. **42**, 1005 (1951).]

ends of the circuit define a vector, the *Burgers vector b*, and the magnitude and angle of the Burgers vector are used to define the magnitude and type of dislocation.

The second major type of dislocation is the *screw dislocation*, illustrated in Fig. 20-37; each cube represents an atom or lattice site. The geometry of this may be imagined by supposing that a block of rubber has been sliced part way through and one section bent up relative to the other. A screw dislocation can be produced by slip on any plane containing the dislocation line *AB* [Fig. 20-37(b)], and the effect is that the crystal plane involved takes the form of a spiral ramp. A photomicrograph of a carborundum crystal is shown in Fig. 20-38, illustrating emergent screw dislocations on the surface of the crystal. The presence of dislocations is often made evident by an examination of the crystal surface; it may help to subject it to a mild etching.

The density of dislocations is usually stated in terms of the number of dislocation lines intersecting unit area in the crystal; it ranges from about 10^8 cm^{-2} for "good" crystals to perhaps 10^{12} cm^{-2} in cold-worked metals. Thus, dislocations are separated by 10^2 to 10^4 Å; every crystal grain larger than about 100 Å will normally have dislocations; one surface atom in 1000 is apt to be near a dislocation. Note the several dislocations evident in Fig. 20-32.

Dislocations greatly reduce the mechanical strength of a crystal; a normal specimen of metal has an experimental elastic limit about 1000 times smaller than the value for the perfect crystal. Also, the presence of emergent dislocations on crystal surfaces is very important to crystal growth. A saturated melt or solution can form new crystals only with difficulty, because of the Kelvin effect (Section 8-9) whereby the surface energy of a small crystal adds to its molar free energy and hence changes its melting temperature or solubility. It is much easier for the atoms or ions to deposit on an existing crystal surface, and especially so if they can locate at a step. Crystals often grow by successive new layers starting and then sweeping across the surface. A screw dislocation is also very effective as a site for crystal growth since crystal units depositing at an emergent screw step merely rotate it. Surface screw dislocations have been observed to be turning slowly as a crystal grows, in direct confirmation of this mechanism.

SPECIAL TOPICS

20-ST-1   Symmetry notation for crystals

The crystallographer deals with the same symmetry operations as were defined in Chapter 17 in connection with molecular symmetry. The nomenclature is somewhat different, however. A rotation axis is designated by the number giving its order, or relative to the Schoenflies symbols (Section 17-2), $1 = C_1$, $2 = C_2$, and so on. A rotation followed by an inversion through a center of symmetry is written $\bar{1}$, $\bar{2}$, and so on. The symbol m denotes a plane or symmetry or *mirror plane*; thus $2/m$ means the symmetry elements C_2 and σ_h. A succession of m's means mutually perpendicular planes of symmetry; mmm denotes three mutually perpendicular planes.

The lattice symmetries of the seven primitive lattices are given in the last column of Table 20-1. Not all the symmetry elements are generally listed, only those sufficient to define the lattice. Thus, for a primitive cubic lattice all other symmetry elements are implied by the designation $m3m$, which means a mirror plane, a threefold axis lying in the mirror plane, and a second mirror plane perpendicular to the first.

The 32 point groups may be generated by considering the various symmetry operations consistent with each type of lattice. For example, the monoclinic lattice has the symmetry elements $2/m$ and the possible point groups are 2, m (or $\bar{2}$), and $2/m$. The orthorhombic lattice has the symmetry $2/mm$ or mmm, and comprises the point groups $2m$ ($mm2$ or $\bar{2}m$), 22 (same as 222) and $2/mm$. The lattice of the triclinic system has the full symmetry 1 but need only have the symmetry $\bar{1}$. Each lattice type thus has a certain maximum possible point group symmetry, with various lesser sets of symmetry features also being consistent with it.

From the point of view of a real crystal, the

basis (see Section 20-1) may be consistent with the full lattice symmetry—a simple example being the case where an atom occupies each lattice point—or may be consistent with only some of the possible symmetry features. In this second case the crystal will belong to one of the less symmetric point groups for the lattice.

A complication is that it is not sufficient to consider only point group symmetry elements. Some new symmetry operations enter which involve translation. The first is called a *screw axis*, designated by the integers p_q, where p is the order of the axis (1, 2, 3, 4, or 6) and q/p is the fraction of a unit cell length by which the point is translated following the rotation. Points 1, 2, and 3 in Fig. 20-39(a) show the operation 3_2. Points 2 and 3 are obtained by two successive 120° rotations and advancements of two-thirds of the unit distance. Points 1', 1", 2', and 3' follow on carrying out unit translations.

The second principal space symmetry element is that called a *glide plane*, which combines reflection in a mirror plane and translation parallel to the plane. A glide plane is designated by a, b, or c if the translation is $a/2$, $b/2$, or $c/2$. An a glide operation is shown in (010) in Fig. 20-39(b). A particular combination of lattice symmetry and point group symmetry, including translational operations, is called a *space group*, and the addition of the latter operations combines the 32 point groups and the 14 Bravais lattices to yield 230 space groups.

The nomenclature for space groups cannot be pursued very far here, but we can describe one scheme briefly. The first symbol is P, C, F, or I, designating the Bravais lattice type (see Table 20-1). This is followed by the point group symbol modified by the introduction of translational symmetry elements if needed. Thus $P2_1/c$ is related to the point group $2/m$, and the space group is therefore in the monoclinic system. The designation indicates a primitive monoclinic lattice with a twofold screw axis and a glide plane perpendicular to it.

20-ST-2 X-ray diffraction intensities

We discussed in Section 20-5B how intervening (111)-type planes of Cl^- ions reduce, by interference, the intensity of scattering from the planes of Na^+ ions in NaCl, and essentially cancel the intensity for K^+ ion planes in KCl. The analysis can be put on a more general and a more quantitative basis as follows.

We consider first a simple rectangular lattice with incident radiation such that reflections from (21) planes are in phase, as shown in Fig. 20-40(a), that is, the phase difference is 2π. Recalling Eq. (16-35), the wave may be described by the mathematical form

$$\psi = A e^{2\pi i q/\lambda}$$

where, to avoid confusion, we use q to denote distance; A is the amplitude of the wave. For the set of rays shown as solid lines in Fig. 20-40, $q = \lambda$, hence the statement that the phase difference is just 2π. Alternatively, if the point of reflection of the rays moves from (1) to (2), the phase of the reflected radiation changes by 2π. We can prorate this change; the distance from (1) to (2) is a/h, where h is the Miller index of the plane. A reflection from some intermediate plane, such as the dashed one shown, would be shifted in phase by the fraction $x/(a/h)$ times 2π or $P_x = (hx/a)2\pi$, where P_x is the x component of the phase shift. The dashed plane also lies a distance y from the origin and, similarly, $P_y = (ky/b)2\pi$. The total phase shift is then $P_x + P_y = 2\pi[(hx/a) + (ky/b)]$. On extension to three dimensions, the result is

FIGURE 20-39
Illustration of (a) the screw axis operation 3_2 and (b) a glide plane operation.

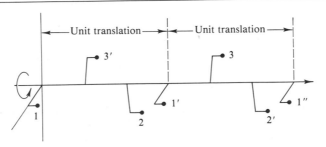

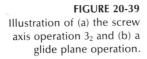

(a)

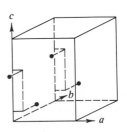

(b)

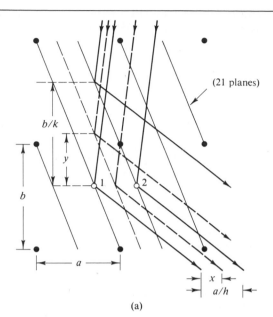

(a)

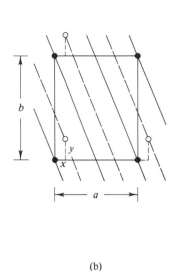

(b)

FIGURE 20-40
(a) Phase relation for scattering by a set of planes lying between (21) planes. (b) Scattering by an atom at position $(x/a, y/b)$.

$$P_{\text{tot}} = 2\pi\left(\frac{hx}{a} + \frac{ky}{b} + \frac{lz}{c}\right) \quad (20\text{-}12)$$

This turns out to be general for any lattice of unit lengths, a, b, and c and any plane displaced from the point of origin of the lattice by lengths (x/a), (y/b), and (z/c).

The actual scattering is done by individual atoms of the lattice, and a unit cell having two kinds of atoms might appear as shown in Fig. 20-40(b). A plane of index (hkl) or, in the two-dimensional case, (hk) can then be passed through each atom, and the scattering amplitude per unit cell due to *that atom* is given by its *atomic scattering factor f*. The phase shift associated with reflections from the plane is given by Eq. (20-12), and the net *amplitude* for all reflections from the unit cell is obtained by summing each contribution to ψ in Eq. (16-35),

$$F_{hkl} = \sum_j f_j e^{2\pi i[(hx/a) + (ky/b) + (lz/c)]} \quad (20\text{-}13)$$

The quantity F_{hkl} is called the *structure factor* of a crystal. The *intensity* of the diffracted radiation [due to that incident on (hkl) planes with the proper Bragg angle] is given by $|F_{hkl}|^2$. Thus if an x-ray diffraction pattern can be indexed as to hkl values for each reflection, the measured intensity gives $|F_{hkl}|^2$.

We illustrate the use of Eq. (20-13) to calculate F_{hkl} as follows. Consider first the case of

α-Fe, which crystallizes in the bcc system. The *repeating unit* (which is what must be used) consists of iron atoms at positions $(0, 0, 0)$ and $(\frac{1}{2}, \frac{1}{2}, \frac{1}{2})$, in units of the lattice unit distance a. Substitution into Eq. (20-13) gives

$$F_{hkl} = f_{\text{Fe}}\{e^{2\pi i[(h)(0) + (k)(0) + (l)(0)]} \\ + e^{2\pi i[(h)(\frac{1}{2}) + (k)(\frac{1}{2}) + (l)(\frac{1}{2})]}\}$$

We can now calculate F for particular planes, remembering that $e^{ix} = \cos x + i \sin x$. For example, the amplitude of scattering (100) planes will be

$$\begin{aligned} F_{100} = \{&f_{\text{Fe}} \cos[2\pi[(1)(0) + (0)(0) + (0)(0)]] \\ &+ i \sin[2\pi[(1)(0) + (0)(0) + (0)(0)]] \\ &+ \cos[2\pi[(1)(\tfrac{1}{2}) + (0)(\tfrac{1}{2}) + (0)(\tfrac{1}{2})]] \\ &+ i \sin[2\pi[(1)(\tfrac{1}{2}) + (0)(\tfrac{1}{2}) + (0)(\tfrac{1}{2})]]\} \\ = &f_{\text{Fe}}[1 + (i)(0) + (-1) + (i)(0)] = 0 \end{aligned}$$

Thus the intensity of reflections from (100) planes should be zero. This particular result could have been arrived at by qualitative reasoning, since in a bcc structure (100) planes would contain only half of the atoms, the rest lying on a second set of planes half way in between. Reflections from this second set would be exactly out of phase with those from the first set, so intensity cancellation should be complete.

We next consider (200) planes and obtain

$$F_{200} = f_{Fe}[1 + (i)(0) + (1) + (i)(0)] = 2f_{Fe}$$

The amplitude is just that for the two atoms of iron in the unit cell. The amplitude for any other Miller index plane can be obtained by inserting the desired (hkl) values in Eq. (20-14).

As a second example, we can use NaCl. The Na^+ ions are at the positions $(0, 0, 0)$, $(\frac{1}{2}, 0, \frac{1}{2})$, $(\frac{1}{2}, \frac{1}{2}, 0)$, and $(0, \frac{1}{2}, \frac{1}{2})$ in the repeating unit and those of Cl^- are at $(\frac{1}{2}, 0, 0)$, $(0, \frac{1}{2}, 0)$, $(0, 0, \frac{1}{2})$, and $(\frac{1}{2}, \frac{1}{2}, \frac{1}{2})$. Equation (20-13) becomes

$$F_{hkl} = f_{Na^+}\{e^{2\pi i[(h)(0) + (k)(0) + (l)(0)]}$$

$$+ e^{2\pi i[(h)(\frac{1}{2}) + (k)(0) + (l)(\frac{1}{2})]}$$

$$+ e^{2\pi i[(h)(\frac{1}{2}) + (k)(\frac{1}{2}) + (l)(0)]}$$

$$+ e^{2\pi i[(h)(0) + (k)(\frac{1}{2}) + (l)(\frac{1}{2})]}\}$$

$$+ f_{Cl^-}\{e^{2\pi i[(h)(\frac{1}{2}) + (k)(0) + (l)(0)]}$$

$$+ e^{2\pi i[(h)(0) + (k)(\frac{1}{2}) + (l)(0)]}$$

$$+ e^{2\pi i[(h)(0) + (k)(0) + (l)(\frac{1}{2})]}$$

$$+ e^{2\pi i[(h)(\frac{1}{2}) + (k)(\frac{1}{2}) + (l)(\frac{1}{2})]}\} \quad (20\text{-}15)$$

The amplitude for (111) planes becomes

$$F_{111} = f_{Na^+} [1 + (i)(0) + (1)$$

$$+ (i)(0) + 1 + (i)(0) + 1 + (i)(0)]$$

$$+ f_{Cl^-}[-1 + (i)(0)$$

$$+ (-1) + (i)(0) + (-1) + (i)(0)$$

$$+ (-1) + (i)(0)]$$

$$= 4f_{Na^+} - 4f_{Cl^-}$$

There will be a net intensity, but one proportional to $(f_{Na^+} - f_{Cl^-})^2$, and thus reduced in value. This is illustrated in Fig. 20-27, which shows the Bragg intensity pattern for reflections from (111)-type planes of NaCl. In the case of KCl, f_{K^+} is about equal to f_{Cl^-}, hence the nearly zero intensity of first-order reflections from (111)-type planes in this case.

Returning to the NaCl case, note that F_{111} could be either positive or negative, depending on the relative values of f_{Na^+} and f_{Cl^-}, but also on which set of ions is taken to be sodium. Standard tables of f values for atoms and ions allow the first point to be determined, but the sign of F_{111} is needed to establish which ions are which in the unit cell. A basic problem in crystal diffraction measurements is that intensities give only $|F_{hkl}|^2$ values and not the signs of structure factors. As discussed in the Commentary and Notes section, it is the latter that are needed to obtain the actual assignment of atoms in the unit cell.

20-ST-3 Lattice energies

A. Rare gas crystals
The total cohesive energy of a crystal may in principal be calculated if one has detailed knowledge of the forces of attraction and of repulsion between molecules. In practice, the calculation has been limited almost entirely to lattices whose points are occupied by atoms or ions.

A rare gas crystal represents about the simplest possible case. The structure is fcc and the usual assumption is that the Lennard–Jones potential function

---

It should be mentioned that the intensity of a diffraction line or spot depends not only on the F_{hkl} value for each plane but also on some essentially trivial geometric factors that have to do with the general angular dependence of diffraction. First, the intensity of scattering of electromagnetic radiation is intrinsically angle-dependent [note Eq. (10-52)]. As a consequence, the intensity will, for this reason alone, vary as $(1 + \cos^2 2\theta)$. Second, if we imagine a crystal being rotated through the Bragg angle for a given reflection, the time spent in the region of this angle is a function of θ. For powders we can speak alternatively of the fraction of randomly distributed crystals that are properly oriented for a given reflection. The factor allowing for this effect is of the form $1/(\sin^2 \theta \cos \theta)$. The general scattering intensity will fall off with increasing θ as a result of these factors, and the Bragg reflections thus appear as peaks on a descending curve of intensity. It is after correction for these geometric effects that one contains actual F_{hkl}^2 values.

$$\epsilon(r) = \alpha r^{-6} + \beta r^{-12} \qquad [\text{Eq. (1-63)}]$$

is applicable. The first term on the right gives the attractive potential due to dispersion (Section 8-ST-1), and the second amounts to a mathematically convenient way of providing a rapidly increasing potential at small distances. In the case of a cubic lattice, the distance d from an origin to some other point is just

$$d = (x^2 + y^2 + z^2)^{1/2} \quad \text{or} \qquad (20\text{-}16)$$
$$d = a(m_1^2 + m_2^2 + m_3^2)^{1/2}$$

where a is the side of the unit cell and the m's are integers. The potential energy ϵ of an atom in the interior is obtained by summing the interaction potential over all lattice sites,

$$2\epsilon = -\alpha a^{-6} \sum_{(m_1 + m_2 + m_3) \text{ even}}$$

$$\frac{1}{(m_1^2 + m_2^2 + m_3^2)^3} + \beta a^{-12} \sum_{(m_1 + m_2 + m_3) \text{even}}$$

$$\frac{1}{(m_1^2 + m_2^2 + m_3^2)^6}$$

or

$$2\epsilon = -\alpha a^{-6}A_a + \beta a^{-12}B_a \qquad (20\text{-}17)$$

The sums are restricted to even values of $(m_1 + m_2 + m_3)$ because a fcc structure corresponds to a simple cubic one in which every other site is vacant; this means, however, that a is taken to be one-half the side of the full fcc unit cell. The sums are set equal to 2ϵ to compensate for the double counting of atoms; that is, each interaction is a mutual one, only half of which should be assigned to the particular atom in question. It is apparent that ϵ is just the energy of sublimation of an atom, that is, the energy to remove one atom entirely from the lattice, the structure then closing up to eliminate the vacancy created.

The sums A_a and B_a in Eq. (20-17) are geometric ones whose values are independent of a, and the coefficients α and β may be estimated from the nonideality of the behavior of the corresponding gas (Section 1-ST). It has been possible to make fairly good calculations of energies of vaporization in this way.

B. Ionic crystals

The application of the foregoing procedure to ionic crystals has been of much more interest and importance. One now usually neglects the dispersion term, considering that the Coulombic attraction between unlike ions dominates the attractive part of the potential, which may be written as

$$\epsilon(r) = \frac{z_1 z_2 e^2}{r} + \frac{be^2}{r^n} \qquad (20\text{-}18)$$

where z_1 and z_2 are the charges on the ions in question and the repulsion term is left open as to the exponent of r. The constant for this latter term is written as be^2 purely as a matter of algebraic convenience.

In the case of NaCl, a sodium ion experiences repulsions from other sodium ions as given by

$$2\epsilon_{\text{Na}^+ - \text{Na}^+} = e^2 a^{-1} \sum_{(m_1 + m_2 + m_3) \text{ even}}$$

$$\frac{1}{(m_1^2 + m_2^2 + m_3^2)^{1/2}} + be^2 a^{-n} \sum_{(m_1 + m_2 + m_3) \text{ even}}$$

$$\frac{1}{(m_1^2 + m_2^2 + m_3^2)^{n/2}} \qquad (20\text{-}19)$$

The attraction between Na^+ and Cl^- ions is given by a similar sum, but now restricted to $(m_1 + m_2 + m_3)_{\text{odd}}$ (why?). Again the sums are geometric ones that can be evaluated, and the total potential energy for a pair of Na^+, Cl^- ions is usually written in the form

$$\epsilon = -\frac{Ae^2}{a} + \frac{Be^2}{a^n} \qquad (20\text{-}20)$$

where A and B are essentially these geometric sums, A being known as the *Madelung constant* and B also containing the constant b. Unlike the case with a rare gas crystal, it is difficult to evaluate b directly, and we therefore treat a as a parameter, which is at the equilibrium value when $d\epsilon/da = 0$. On carrying out the former differentiation, an expression for B in terms of A is found, whereby the former may be eliminated from Eq. (20-20) to give

$$\epsilon_0 = -\frac{Ae^2}{a_0}\left(1 - \frac{1}{n}\right) \qquad (20\text{-}21)$$

where ϵ_0 and a_0 are now the equilibrium energy and distance (again a_0 is half the side of the unit cell). The lattice energy E_0 is defined as the energy released in the formation of one mole of the crystal from the gaseous ions, $E_0 = -N_A\epsilon_0$.

This approach may be extended to any ionic crystal, A now being the geometric sum appropriate for the lattice type,

$$E_0 = \frac{N_A A e^2 Z^2}{a_0} \left(1 - \frac{1}{n}\right) \qquad (20\text{-}22)$$

where Z is defined as the highest common factor of the ionic charges (one for NaCl, Na_2O, Al_2O_3, ... , and two for MgO, TiO_2, ...). Table 20-4 gives the Madelung constants of several common minerals. The repulsion exponent n may be estimated from the compressibility of the crystal; values range from 6 to 10 for various substances. A slightly better treatment appears to result if the repulsion term is written as the exponential $be^{-r/\rho}$, in which case the procedure yields

$$E_0 = \frac{N_A A e^2 Z^2}{a_0} \left(1 - \frac{r}{\rho}\right) \qquad (20\text{-}23)$$

Of course, a_0 is obtainable from x-ray diffraction studies, so the lattice energies of simple crystals may be calculated absolutely. Strictly speaking, the result is for 0 K; the differential dG/da rather than $d\epsilon/da$ is needed otherwise to give the equilibrium condition, G being the lattice free energy.

C. The Born-Haber cycle

Lattice energies can be related to other thermodynamic quantities by means of a cycle known as the *Born–Haber cycle*. The formation of a solid salt MX from the elements may be formulated in two alternative ways:

$$\begin{array}{ccc}
MX(c) & \xleftarrow{\quad -E_0 \quad} & M^+(g) + X^-(g) \\
\Delta H_f \Big\uparrow & & \Big\uparrow I - A \\
M(c) + \tfrac{1}{2}X_2(g) & \xrightarrow{\ S + \frac{1}{2}D\ } M(g) & + X(g)
\end{array}$$

$$(20\text{-}24)$$

where I is the ionization potential of the gaseous metal atom, A the electron affinity of the gaseous halogen atom, D the dissociation energy of $X_2(g)$, S the sublimation energy of the metal, and ΔH_f the heat of formation MX(c) from the elements. The change in energy must be independent of path, so ΔH_f is equal to the algebraic sum of the other quantities:

$$\Delta H_f = S + \tfrac{1}{2}D + I - A - E_0 \qquad (20\text{-}25)$$

Equation (20-25) may be used in various ways. Since the electron affinity A is the least accurately known, one use of the equation is to obtain an indirect value for it. The various quantities are given in Table 20-5 for several alkali metal halides. Note that all of the quantities make an appreciable contribution to E_0.

20-ST-4 Ionic radii

It was pointed out in Section 20-4 that nearest-neighbor atoms or ions in a crystal are regarded as being in contact; the more correct statement is that they have approached to the point of being at the potential minimum or balance between attraction and repulsion. In effect, we *define* the crystal size of atoms or ions on this basis. Thus in NaCl the side of the unit cell is, by definition, equal to $2r_{Na^+} + 2r_{Cl^-}$; the body diagonal in CsCl is, again by definition, equal to $2r_{Cs^+} + 2r_{Cl^-}$.

Crystal lattice dimensions thus in general give a sum of radii for a pair of oppositely charged ions, and some additional information is needed if individual ionic radii are to be obtained. Since molar refraction is a measure of atomic or ionic volume (Section 3-3), one method has been to use this as a basis for dividing the internuclear distance. The currently accepted procedure, however, is one due to Pauling whereby isoelectronic ions (such as K^+ and Cl^-,

TABLE 20-4 Madelung constants	Structure	Madelung constant	Structure	Madelung constant
	NaCl	1.7476	TiO_2 (rutile)	4.816
	CsCl	1.7627	TiO_2 (anatase)	4.800
	ZnS (zincblende)	1.6381	CdI_2	4.71
	ZnS (wurzite)	1.641	SiO_2 (β quartz)	4.4394
	CaF_2 (fluorite)	5.0388	Al_2O_3 (corundum)	25.0312

TABLE 20-5

Thermochemical data and electron affinities at 0 K[a,b]

Salt	$-\Delta H_f$ (298 K)	S (298 K)	$\frac{1}{2}D$ (298 K)	$H_1{}^c$	$H_2{}^d$	I	E_0 (Theory)	A [from Eq. (19-22)]
NaF	136.3	25.9	18.9	3.0	1.9	118.4	218.7	79.7
KF	134.5	21.5	18.9	3.0	2.3	100.0	194.4	79.8
LiCl	96.0	38.4	28.9	3.0	1.8	124.4	202.0	84.5
NaCl	98.2	25.9	28.9	3.0	2.5	118.4	185.9	85.1
KCl	104.2	21.5	28.9	3.0	2.8	100.0	169.4	85.1
RbCl	103.4	19.5	28.9	3.0	3.0	96.3	164.0	84.2
CsCl	106.9	18.7	28.9	3.0	3.3	89.7	155.9	88.7
NaBr	86.0	25.9	26.8	3.0	2.8	118.4	176.7	80.2
KBr	93.7	21.5	26.8	3.0	3.0	100.0	162.4	79.6
NaI	68.8	25.9	25.5	3.0	3.1	118.4	165.4	73.3
KI	78.3	21.5	25.5	3.0	3.2	100.0	153.0	72.5

[a]Data given in kilocalories per mole.

[b]Adapted from B. E. Douglas and D. H. McDaniel, "Concepts and Models of Inorganic Chemistry." Ginn (Blaisdell), Boston, Massachusetts, 1965.

[c]Enthalpy to take M(g) and X(g) from 298 K to 0 K.

[d]Enthalpy to take MX(c) from 298 K to 0 K.

or Na^+ and F^-) are taken to have radii inversely proportional to their effective nuclear charge (Section 16-3). Thus for NaF, $r_{Na^+} + r_{F^-} = 2.31$Å and the screening constant σ is taken from spectroscopy to be 4.5, so that the respective Z_{eff} values are $11 - 4.5 = 6.5$ and $9 - 4.5 = 4.5$. We write $(C/6.5) + (C/4.5) = 2.31$, whence the constant $C = 6.14$ for ions of the neon configuration. The radius of Na^+ is then $6.14/6.5 = 0.95$ Å and that of F^- is $6.14/4.5 = 1.36$ Å. One may proceed to estimate radii for a hypothetical ion, such as O^-; Z_{eff} is now $8 - 4.5 = 3.5$, so $r_{O^-} = 6.14/3.5 = 1.75$ Å.

By applying these assumptions to crystallographic data for sums of radii, Pauling has calculated a number of crystal radii for ions of various charges, a selection of which is given in Table 20-6. It is to be remembered that these rules are semi-empirical, so that particular sums of radii may come close to but will not in general give the measured crystallographic sum exactly.

Points to notice are that in an isoelectronic series of ions the radius decreases steadily with increasing net positive charge, attributed to the increasing mutual attraction of the electrons. Conversely, ions increase in size with increasing net negative charge. The comparison between M^{5+} and M^{3-} for the Group V ions dramatically illustrates this charge effect. Note, too, that the radius of O^{2-} is 1.40 Å as compared to 1.75 Å for O^-.

Knowledge of the individual ionic radii helps to understand why certain MX crystals have the NaCl structure, others the ZnS one, and

TABLE 20-6

Crystal radii[a,b]

Li^+ 0.60	Be^{2+} 0.31	B^{3+} 0.20	C^{4+} 0.15	N^{5+} 0.11	N^{3-} 1.71	O^{2-} 1.40	F^- 1.36	Ne 1.22
Na^+ 0.95	Mg^{2+} 0.65	Al^{3+} 0.50	Si^{4+} 0.41	P^{5+} 0.34	P^{3-} 2.12	S^{2-} 1.84	Cl^- 1.81	Ar 1.54
K^+ 1.33	Ca^{2+} 0.99	Sc^{3+} 0.81	Ti^{4+} 0.68	V^{5+} 0.59	—	Cr^{6+} 0.52	Mn^{7+} 0.46	—
Cu^+ 0.96	Zn^{2+} 0.74	Ga^{3+} 0.62	Ge^{4+} 0.53	As^{5+} 0.47	As^{3-} 2.22	Se^{2-} 1.98	Br^- 1.95	Kr 1.69

[a]Data given in angstroms.

[b]See L. Pauling, "The Nature of the Chemical Bond," 3rd ed. Cornell Univ. Press, Ithaca, New York, 1960.

FIGURE 20-41
(a) Face of the unit cell of the NaCl structure. (b) Condition of double repulsion—negative ion–negative ion contact just occurs.

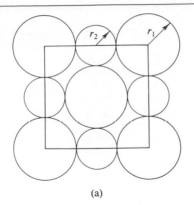

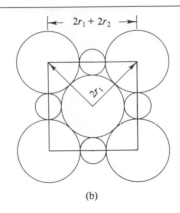

(a)

(b)

still others that of CsCl. Considering first the NaCl structure, the face of the unit cell appears as shown in Fig. 20-41(a), where r_2 is the radius of the smaller of the two ions, ordinarily that of the cation. The oppositely charged ions are in contact, but not the like charged ones. However, as r_2 is decreased relative to r_1 a point is reached, shown in Fig. 20-41(b), such that the larger, liked charged ions have just come in contact. This condition is known as one of *double repulsion*, meaning that further approach will be resisted not only by Coulombic repulsion but also by the general strong repulsion of the electronic clouds [as given by the $1/r^n$ term in Eq. (20-18)]. One would expect the lattice energy of the crystal to decrease dramatically from this point on. The radius ratio r_2/r_1 for this critical condition can be calculated from the geometry of the situation. The right angle triangle shown in the figure yields the relationship

$$(2r_1)^2 + (2r_1)^2 = (2r_1 + 2r_2)^2, \quad \text{whence} \quad r_2/r_1 = 0.41.$$

The condition for double repulsion in the CsCl structure may similarly be calculated to be $r_2/r_1 = 0.73$, and that for the ZnS structure to be 0.22. The energetics of the situation is illustrated in Figure 20-42. In the absence of double repulsion, the CsCl structure should have the largest lattice energy since each ion has eight nearest neighbors. However, when the radius ratio drops to 0.73, double repulsion sets in, and the CsCl structure becomes unstable relative to the NaCl one, with six nearest neighbors. This in turn yields to the ZnS structure with four nearest neighbors when r_2/r_1 drops below 0.22.

These radius ratio effects, as they are called, can be involved in explanation of a number of the shifts in structure that occur in the various series of MX, MX$_2$, ... ionic lattices.

FIGURE 20-42
Qualitative variation of lattice energy with radius ratio for CsCl, NaCl, and ZnS structures.

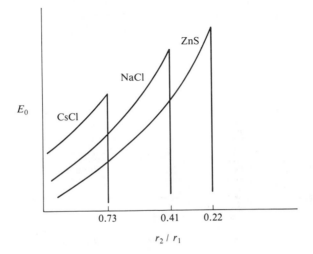

BUERGER, M.J. (1963). "Elementary Crystallography." Wiley, New York.
BUERGER, M.J. (1964). "X-ray Crystallography," Wiley, New York.
EITEL, W. (1954). "The Physical Chemistry of the Silicates." Univ. of Chicago Press, Chicago, Illinois.
PHILLIPS, F.C. (1946). "An Introduction to Crystallography." Longmans Green, New York.
STOUT, G.H., AND JENSEN, L.H. (1968). "X-ray Structure Determination." Macmillan, New York.
WELLS, A.F. (1950). "Structural Inorganic Chemistry," 2nd ed. Oxford Univ. Press, London and New York.
WYCKOFF, R.W.G. (1931). "The Structure of Crystals," 2nd ed. Chem. Catalog Co. (Tudor), New York.

GENERAL
REFERENCES

KEPLER, J. (1611). "A New Year's Gift, Or On the Six-Cornered Snowflake" (translated from the Latin by C. Hardie), Oxford Univ. Press (Clarendon), London and New York, 1966.
WYCKOFF, R.W.G. (1931). "The Structure of Crystals," 2nd ed. Chem. Catalog Co. (Tudor), New York.

CITED
REFERENCES

EXERCISES

Explain what the Miller indices are for the planes of set IV in Fig. 20-10.

20-1

Ans. $(1\bar{2}0)$.

Explain what the Miller indices are for the planes of set III in Fig. 20-10.

20-2

Ans. $(1\bar{4}0)$.

Explain how many ions of each kind are present in the unit cell of (a) BN, (b) ZnS, and (c) ZnO.

20-3

Ans. (a) two of each, (b) four of each, (c) two of each.

Explain how many ions of each kind are present in the unit cell of (a) CaF_2, (b) TiO_2, and (c) diamond.

20-4

Ans. (a) four Ca, eight F; (b) two Ti, four O; (c) eight.

The α crystalline modification of Cr has a body centered cubic structure of side $a = 2.878$ Å. Calculate the density.

20-5

Ans. 7.246 g cm^{-3}.

The density of solid Xe is 2.7 g cm^{-3} at −140°C, the unit cell being ccp. Assuming no change in the radius of Xe, calculate the density of a bcc crystalline form.

20-6

Ans. 2.48 g cm^{-3}.

Show what symmetry designation should be given the mosaic of Fig. 20-1(b).

20-7

Ans. C_{2v}.

Show that $d_{100} : d_{110} : d_{111} = 1 : 1/\sqrt{2} : 2/\sqrt{3}$, referring to type planes of an fcc structure.

20-8

Calculate the surface density of ions of either kind in (100) type planes of NaCl; $a = 5.627$ Å.

20-9

Ans. 12.6×10^{14} ions cm^{-2}.

An fcc crystal of an element gives a diffration peak at $\theta = 6°30'$ for (240) planes (actual). There is another peak at $10°55'$; to what Miller index planes might this correspond to?

20-10

Ans. (264) (or any permutation).

20-11 Calculate the density of Pb from the data of Section 20-5C.

Ans. 11.4 g cm⁻³.

20-12 Extend the listing of $h^2 + k^2 + l^2$ values of Table 20-2 up to 35, giving for each the possible (hkl) value(s).

20-13 Show that the close packing of spheres leaves 26% void space.

20-14 Show that the diamond structure given in Fig. 20-19 does indeed lead to a C—C—C angle of 109°28′.

20-15 The density of NbN is 8.4 g cm⁻³. The compound crystallizes in one of the cubic systems, with $a = 4.41$ Å. How many molecules of NbN are in the unit cell and which type of cubic system is it?

Ans. 4; NaCl type.

20-16 The element Mo crystallizes in one of the cubic systems. A diffraction experiment using 1.089 Å x rays and a powdered sample showed reflections at 9°38′, 14°14′, 17°31′, 20°18′, 22°50′, 25°31′, 27°49′ (and further ones at higher angles). Show which cubic system is involved and calculate the value of a (the side of the unit cell) and the density of Mo.

Ans. bcc, 3.14 Å, 10.3 g cm⁻³.

20-17 The element W crystallizes in the bcc system. If d_{120} (actual planes) is 1.412 Å and if 0.35 Å x rays are used, calculate (a) the density of W, and (b) the angle of incidence at which there should be the first order Bragg reflection from (120)-*type* planes (careful!).

Ans. (a) 19.30 g cm⁻³, (b) 14° 21′.

PROBLEMS

20-18 The molar volume of a crystal always has the form $V = cN_Ar^3$, where r is a characteristic interparticle distance. If r is taken to be the positive ion–positive ion distance, calculate the value of c for (a) the NaCl structure, (b) the CsCl structure, and (c) the zincblende structure.

20-19 Calculate the Ni—Ni distances in NiO and Ni at 25°C, given that NiO crystallizes with the NaCl structure (approximately), with $a = 4.18$ Å and Ni crystallizes in the fcc structure, with $a = 3.53$ Å. Comment on the result.

20-20 Cerium metal crystallizes in one of the cubic systems. The side of the unit cell is 5.12 Å, and the density is 7.0 g cm⁻³. (a) show whether the system is simple cubic, bcc, or fcc. (b) Assuming the structure to be bcc, calculate the radius of the Ce atom.

20-21 A powder diffraction photograph of KCl gave lines at the following distances from the center spot when Mo x rays (wavelength 0.708 Å) were used in a camera of radius 5.74 cm: 13.2, 18.4, 22.8, 26.2, 29.4, 32.2, 37.2, 39.6, 41.8, 43.8, 46.0, all in mm. Index the lines, identify the kind of unit cell, and determine its size. (Careful!)

20-22 Make a minimum modification in the mosaic of Fig. 20-1(b) to reduce the symmetry from C_{2v} to C_2 (note Exercise 20-7).

20-23 Find the Miller indices of sets of planes labeled III and IV in Fig. 20-10 if each set is rotated 90° counterclockwise.

20-24 Calculate the surface density of ions of either kind in (110)-type planes of NaCl; $a = 5.627$ Å.

20-25 Calculate the scattering pattern from a single body centered cubic crystal of Ag, with $a = 4.078$ Å and using Cu K$_\alpha$ x rays of wavelength 1.54 Å. Suppose that the crystal is oriented to give Bragg reflections from (100) planes and then from (111) planes. Give the value of a suitable orientation angle for the crystal face relative to the incident x rays (the crystal habit is such that there are faces parallel to (100) and to (111) planes). Identify reflections by their order.

A hypothetical salt, MX, crystallizes in the cubic system. The structure is of two interpenetrating body-centered lattices with the *repeating unit* consisting of M ions at $(0,0,0)$ and $(\frac{1}{2},\frac{1}{2},\frac{1}{2})$ and X ions at $(\frac{1}{4},\frac{1}{4},\frac{1}{4})$ and $(\frac{3}{4},\frac{3}{4},\frac{3}{4})$. (a) Locate the M and X ions in a sketch of the unit cell. (b) Calculate the molecular weight of MX if a is 6 Å and the density of crystalline MX is 3.5 g cm⁻³. (c) Explain what the coordination numbers of M and of X are, that is, what and how many nearest neighbors are there. (d) Calculate the smallest sin θ values for reflections of nonzero intensity from (100) type planes; 0.4 Å x rays are used. (e) Explain briefly why this particular crystal structure seems improbable.   **20-26**

Li_2Te crystallizes in the CaF_2 system. The unit cell contains Te atoms at $(0,0,0)$, $(1,0,0)$, $(1,1,0)$, $(0,0,1)$, $(0,1,0)$, $(0,1,1)$, $(1,1,0)$, $(1,1,1)$, $(\frac{1}{2},0,\frac{1}{2})$, $(\frac{1}{2},\frac{1}{2},0)$, $(0,\frac{1}{2},\frac{1}{2})$, $(\frac{1}{2},\frac{1}{2},1)$, $(\frac{1}{2},1,\frac{1}{2})$, and $(1,\frac{1}{2},\frac{1}{2})$. The Li atoms are located in the positions given by all possible combinations of $\frac{1}{4}$ and $\frac{3}{4}$. (a) Draw a perspective view of the unit cell and locate all atoms in it. (b) Calculate the density of Li_2Te. The a value is 5.47 Å. (c) Sketch the relative intensities of the first four orders of (100) type reflections in a Bragg experiment, that is, make a qualitative plot of I vs. sin θ.   **20-27**

BeS is found to be cubic from microscopic examination. A powder pattern obtained with Cu x rays (1.539 Å) gives lines at the following values of sin^2 θ: 0.0746; 0.0992; 0.2011; 0.2767; 0.3019; 0.4030; 0.4786; 0.5027; 0.6038; 0.6789. Show which type of cubic lattice is present (index the sin^2 θ values; that is, assign values of hkl to each). Calculate the side of the unit cell and the number of atoms per unit cell (the density is 2.36).   **20-28**

The mineral spinel contains 37.9% Al, 17.1% Mg, and 45% oxygen. The density is 3.57 g cm⁻³. The smallest unit (unit cell) in the crystal is a cube of edge 8.09 Å. How many atoms of each kind are in the unit cell?   **20-29**

Calculate the size of the sphere that can be accommodated in the octahedral hole of the fcc structure; cube edge $= a$, atom radius $= r$.   **20-30**

What is the highest order diffraction line of (100) that can be observed from a CsCl crystal with x radiation of 1.54 Å? (Remember that sin θ cannot exceed unity.)   **20-31**

The sin^2 θ values observed on a sample of MgO powder with 0.710 Å x rays are as follows: 0.02134, 0.02857, 0.05734, 0.07846, 0.08613, 0.11437, 0.13671, 0.14358, 0.17219, 0.22939, 0.25836. Show to which type of MX cubic lattice the data correspond and the side of the unit cell. The density of MgO is 3.58 g cm⁻³. [Data from Wyckoff (1931).]   **20-32**

Using Eq. (20-4), calculate d_{hkl}/a as a function of c/a for a tetragonal lattice. Cover the range $c/a = 2$ to $c/a = 0.2$, and $h^2 + k^2 + l^2$ values up to 10. Plot the results as log (d_{hkl}/a) versus c/a, using semilogarithmic graph paper. Graphs of this kind are useful in fitting powder diffraction data. As an example, the data of Problem 20-32 can be converted to a series of numbers proportional to the corresponding d's. If these numbers are marked on a strip of the same semilogarithmic scale, the strip can be slid up and down along the $c/a = 1$ line until a match is obtained; the (hkl) values can then be assigned directly.   **20-33**

Spheres of 1.5 Å diameter are in a close-packed arrangement. Calculate the side of the unit cell if the arrangement is ccp and the values of a and c if it is hcp.   **20-34**

SPECIAL TOPICS PROBLEMS

Calculate $F(331)/f_{Ce}$ for crystalline Ce. See Problem 20-20; assume the structure to be bcc.   **20-35**

The zincblende structure is cubic Zn in an fcc structure and S at $(\frac{1}{4},\frac{1}{4},\frac{1}{4})$, $(\frac{1}{4},\frac{3}{4},\frac{3}{4})$, $(\frac{3}{4},\frac{1}{4},\frac{3}{4})$, and $(\frac{3}{4},\frac{3}{4},\frac{1}{4})$. Calculate the structure factor of the lowest angle reflection from a ZnS crystal. The atomic scattering   **20-36** factors are $f_{Zn} = 24.7$ and $f_S = 12.3$. The lowest angle reflection is observed at $\theta = 14.3°$ when the wavelength of the x rays used is 0.1542 nm. Calculate the side a of the unit cell.

20-37 Referring to Problem 20-26 and assuming that $f_M = f_X$ (M and X isoelectronic), calculate F/f for reflections from (300) planes.

20-38 The repeating unit for an MX$_2$-type cubic crystal is: M at $(0,0,0)$ and $(\frac{1}{2},\frac{1}{2},\frac{1}{2})$; X at $(\frac{1}{4},\frac{1}{4},0)$, $(\frac{3}{4},\frac{3}{4},0)$, $(\frac{1}{4},\frac{3}{4},\frac{1}{2})$, and $(\frac{3}{4},\frac{1}{4},\frac{1}{2})$. Calculate F for (111) reflections, that is, the coefficients a and b in $F(111) = af_M + bf_X$.

20-39 Calculate the lattice energy for NaCl, assuming $n = 12$.

20-40 Calculate the lattice energy for CaF$_2$, assuming $n = 8$. The parameter a_0 is taken to be the Ca-F distance.

20-41 Calculate the lattice energy of AgCl from the following data. Heat of vaporization to give AgCl(g) is 54; heat of reaction Ag(g) + Cl(g) = AgCl(g) is -72; electron affinity of Cl is 84; ionization energy of Ag(g) is 174 (all values in kilocalories).

20-42 Calculate the proton affinity for ammonia, that is, E for the process NH$_3$(g) + H$^+$(g) = NH$_4^+$(g), from the following data. Heat of vaporization of NH$_4$Cl(s) to NH$_4^+$(g) and Cl$^-$(g) is 153; proton affinity of Cl$^-$(g) is 327; heat of formation of NH$_4$Cl(s) from HCl(g) and NH$_3$(g) is -42 (values in kilocalories).

20-43 Show that the value of the radius ratio r_2/r_1 for onset of double repulsion in CsCl is 0.73.

20-44 Estimate the data of Table 20-6 the screening constant that is used in proportioning ionic radii between ions isoelectronic with argon.

20-45 Explain the alternative symmetry notation for the point groups (a) C$_{2h}$, (b) 2 mm, (c) 2/m, 2/m, 2/m, (d) C$_{4v}$.

chapter 21
Colloids and macromolecules

The twin topics of this chapter have vast extensions into applied areas, ranging from the behavior of detergents, emulsions, suspensions, and gels, to the properties of synthetic and biological polymers. We will be dealing with particles small enough or molecules large enough that their size, shape, and interfacial properties play a major role in determining behavior. Particle–particle interactions may be of great importance; these will be of the van der Waals type, including hydrogen bonding. One of the very significant contributions of colloid chemistry to fundamental science has been to the elucidations of the nature and manner of propagation of such secondary interactions. Modern colloid chemistry is in fact far more interesting and important than the meaning of the word *colloid*—gluelike—would imply!

The physical chemistry of macromolecules overlaps with that of colloidal systems. There are many experimental techniques that are common to the two subjects. In addition, however, polymer chemistry is concerned with reaction kinetics and with the stereochemistry of the primary or chemical bonding of the molecule.

The plan of the chapter is as follows. We take up first the topic of lyophobic colloids—a more complex subject than might at first be imagined—then lyophilic colloids and an excursion into rheology. Because of the difficulty in covering the field in any really adequate way, only a few rather descriptive aspects of polymer chemistry are presented. We conclude with a Special Topics section on electrokinetic phenomena, a part of surface chemistry not so far considered.

21-1 Lyophobic colloids

The term "lyophobic" means solvent-hating, and the class of lyophobic colloids includes all dispersed systems in which the dispersed material is neither in true solution nor aggregated. A gold or silver iodide *sol* (a sol being a suspension of solid particles in a liquid medium) is an example of a lyophobic system. The particles tend to stay separated from each other; they are not strongly solvated nor do they otherwise interact with the solvent, in contrast to gel-forming substances (see Section 21-3). We consider here only one aspect of the subject, namely the theoretical explanation of why lyophobic colloids are fundamentally unstable toward flocculation on the one hand, and why, on the other hand, the rate of flocculation may be very slow. As an example of this last point, some of Michael Faraday's gold sols—not yet flocculated— are still to be seen in the British Museum. There must evidently be present forces both of attraction and of repulsion.

The attraction is due to van der Waals forces, often mainly dispersion in type (see Section 8-ST-1), and the repulsion is due to the electric field around each particle. Both forces can be long-range, that is, extend over hundreds of angstroms. Being quite general in nature, these forces play an important role in any nonideal condensed system, which, incidentally, means virtually all biological ones. The importance of lyophobic colloids is partly that their study has served as a means for the experimental characterization of long-range forces.

A. Long-range dispersion forces

The dispersion attraction between two like atoms is given by Eq. (8-61), which we will write in the simplified form

$$\epsilon(x) = -\frac{A}{x^6} \tag{21-1}$$

where $A = \frac{3}{4}h\nu_0\alpha^2$, α being the polarizability of the atom and $h\nu_0$, its ionization potential. This is a general force of attraction, largely independent of the chemical nature of the atom; it arises from small, mutually induced perturbations in the electron clouds and is additive (to a first approximation). A representative value of A would be 10^{-60} erg cm^6.

A consequence of the additive nature of dispersion interactions is that an atom near a large body of matter experiences a total attraction given by integrating Eq. (21-1) over the length, breadth, and depth of the body. For a semi-infinite slab, the result is

$$\epsilon(x)_{\text{atom-slab}} = -\frac{\pi \mathbf{n}A}{6x^3} \tag{21-2}$$

where $\mathbf{n}$ is the number of atoms in the slab per cubic centimeter. To get the attraction force between two slabs, we must consider a column of atoms in the second one, as illustrated in Fig. 21-1, and integrate over its depth. The result for two infinitely thick slabs is

$$\epsilon(x)_{\text{slab-slab}} = -\frac{1}{12\pi}\frac{H}{x^2} \tag{21-3}$$

where H is called the *Hamaker* constant (alternatively given the symbol A), and is equal to $\pi^2\mathbf{n}^2A$. For two closely approaching spheres,

$$\epsilon(x) = \frac{aH}{12x} \tag{21-4}$$

where a is the radius and x is the surface-to-surface distance, assumed to be small compared to a. A typical value for H is about 10^{-13} erg. These interaction potentials are long-range in the sense that they fall off only slowly with distance, that is, as $1/x^2$ or as $1/x$.

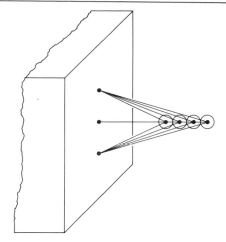

FIGURE 21-1
Van der Waals forces between a surface and a column of molecules. [From A. W. Adamson, "Physical Chemistry of Surfaces," 3rd ed. Copyright 1976, Wiley (Interscience), New York. Used by permission of John Wiley & Sons, Inc.]

Consider two spherical colloidal particles of 1μ (10^{-4} cm) radius, for which H is 10^{-18} erg, so that from Eq. (21-4) $\epsilon(x) = -10^{-18}/x$ erg. At 100 Å or 10^{-6} cm separation, $\epsilon(x)$ is therefore -10^{-12} erg. This is a small quantity, but much larger than the kinetic energy of the particles kT, which is 4×10^{-14} erg at 25°C.

As the example demonstrates, once two colloidal particles are in each other's vicinity, either as a result of diffusion or of the action of convection currents, they should drift together and seize. The repulsive potential that prevents this is now discussed.

We consider first a single particle, large enough for its surface to be treated as flat, and having a surface charge σ_0 per square centimeter and corresponding potential ψ_0. The particle is in a liquid medium containing an electrolyte of concentration n_0, which we take to be uni-univalent in type. We further assume ψ_0 to be positive so that negative ions of the electrolyte are attracted to the surface and positive ions are repelled. The potential energy of an ion in a potential ψ is just $e\psi$, so by the Boltzmann principle the concentration of ions in a region near the surface is

B. Electrostatic repulsion. The diffuse double layer

$$\mathbf{n}^- = \mathbf{n}_0 e^{e\psi/kT} \qquad \mathbf{n}^+ = \mathbf{n}_0 e^{-e\psi/kT} \tag{21-5}$$

where ψ is the potential at some distance x away from the surface. The net charge density at this point is then

$$p = e(\mathbf{n}^+ - \mathbf{n}^-) = -2\mathbf{n}_0 e \sinh \frac{e\psi}{kT} \tag{21-6}$$

[remembering that $\sinh x = \frac{1}{2}(e^x - e^{-x})$]. The derivation is so far entirely analogous to that of the Debye–Hückel theory [compare Eq. (12-82)], and we proceed to invoke the Poisson equation (12-83), thus obtaining

$$\nabla^2\psi = \frac{8\pi\mathbf{n}_0 e}{D} \sinh \frac{e\psi}{kT} \qquad \text{[Eq. (12-86)]}$$

where D is the dielectric constant of the medium.

The derivation now takes a turn form the Debye–Hückel one, since in the present case $\nabla^2\psi$ is just $d^2\psi/dx^2$ and Eq. (12-86) can be integrated directly. The result is somewhat complicated, but if $e\psi/kT$ (for ions of charge z, $ze\psi/kT$) is small compared to unity, one obtains the simple result

$$\psi = \psi_0 e^{-\kappa x} \tag{21-7}$$

where κ is the Debye–Hückel constant, defined for a uni-univalent electrolyte by Eq. (12-88),

$$\kappa^2 = \frac{8\pi n_0 e^2}{DkT} \tag{21-8}$$

The potential thus decays exponentially with distance into the solution.

The behavior of Eq. (21-7) is illustrated in Fig. 21-2 for $\psi_0 = 25$ mV (1 mV = 0.001 V). For a singly charged ion the corresponding potential energy is 23 cal mol^{-1}; 25 mV therefore

FIGURE 21-2
The diffuse double layer.
[Courtesy K. J. Mysels.]

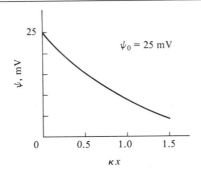

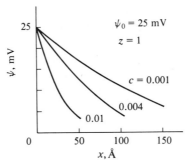

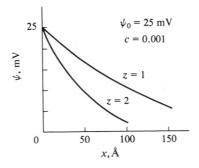

corresponds to about kT at 25°C. Notice that the range of ψ is about 100 Å for fairly typical electrolyte concentrations, and that ψ drops off more rapidly the higher the concentration and the higher the valence of the electrolyte ions.

The physical situation is that for a surface of potential ψ_0; the nearby solution has an excess of negative over positive ions, or a net charge density ρ which, along with ψ, diminishes approximately exponentially with distance. The system as a whole is electrically neutral, so that the surface of the solid must have a surface charge density as illustrated in Fig. 21-3,

$$\sigma_0 = -\int_0^\infty \rho\, dx \qquad (21\text{-}9)$$

The picture is one of a positively charged surface and a counterbalancing net excess of negative charge in the neighboring solution. This last is diffuse, and the system is called a *diffuse double layer*. The potential ψ has decayed to $1/e$ of ψ_0 at a distance $1/\kappa$, and consequently the electrical "center of gravity" of charge in the solution may be regarded as located at a distance $1/\kappa$ from the surface. In the case of the Debye–Hückel theory, $1/\kappa$ is the radius of the ionic atmosphere.

If two charged surfaces, such as those of two colloidal particles, approach, the repulsion depends on the value of ψ at the midpoint. The derivation will not be given here [see Adamson (1982)] but only an approximate result:

$$\epsilon(x) = \frac{64\mathbf{n}_0 kT}{\kappa}\gamma^2 e^{-2\kappa x} \qquad (21\text{-}10)$$

where $\gamma = (e^{y_0/2} - 1)/(e^{y_0/2} + 1)$, $y_0 = ze\psi_0/kT$, $\mathbf{n}_0$ is in molecules per cubic centimeter, and $\epsilon(x)$ is the repulsion potential in ergs per square centimeter.

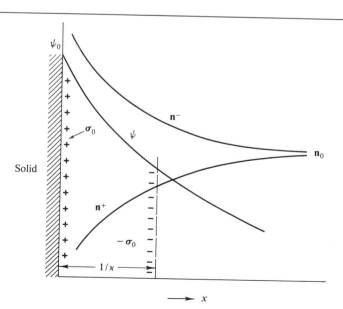

FIGURE 21-3
The diffuse double layer as equivalent to a plane of excess charge in solution at distance $1/\kappa$ from a surface of equal but opposite charge.

Suppose that $y_0 = 1$, $x = 10^{-6}$ cm, and the concentration of uni-univalent electrolyte is 0.001 M, so that κ is about 10^6 cm^{-1} (and $\kappa x = 1$). We obtain

$$\epsilon(x) = \frac{64(0.001)(10^{-3})6.02 \times 10^{23}kT}{10^6}\left(\frac{1.65 - 1}{1.65 + 1}\right)^2 0.135 = 3.13 \times 10^{11}kT \text{ erg cm}^{-2}$$

C. Flocculation of colloidal suspensions

The flocculation or coming together of colloidal particles can be examined in the light of the preceding estimates of the attraction and repulsion potentials. The approximate net potential is given by the sum of Eq. (21-3) or Eq. (21-4) and Eq. (21-8),

$$\epsilon(x) = \frac{64\mathbf{n}_0 kT}{\kappa}\gamma^2 e^{-2\kappa x} - \frac{1}{12\pi}\frac{H}{x^2} \qquad (21\text{-}11)$$

The function $\epsilon(x)$ is illustrated in Fig. 21-4. We assume particles of radius about 10^{-5} cm or area about 3×10^{-10} cm^2 in applying Eq. (21-3). At small κ or low electrolyte concentration the electrostatic repulsion is strong and the colloidal particles will be virtually unable to approach each other; the suspension should be relatively stable. The barrier diminishes with increasing electrolyte concentration and eventually disappears.

One may, in fact, calculate the critical electrolyte concentration such that the barrier is just zero from the condition that $\epsilon(x) = 0$ and $d\epsilon(x)/dx = 0$. The resulting equation gives $\mathbf{n}_{0(\text{crit})}$ as proportional to $1/z^6$. Thus for a z–z

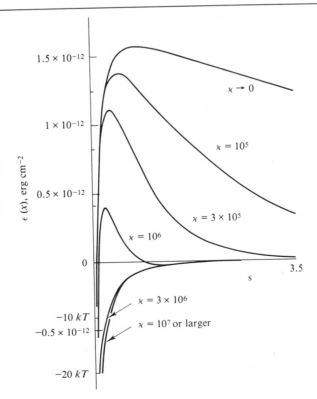

FIGURE 21-4
The effect of electrolyte concentration on the interaction energy between two spheres; s is distance measured as x/a, where x is the distance between the surfaces of the spheres and a is their radius. (From E.J.W. Verwey and J.Th.G. Overbeek, "Theory of Lyophobic Colloids," Elsevier, Amsterdam, 1948.)

electrolyte, equivalent conditions with respect to producing flocculation should be in the order $1 : (\frac{1}{2})^6 : (\frac{1}{3})^6$ or $100 : 1.6 : 0.13$ for 1–1, 2–2, and 3–3 electrolytes, respectively. This is essentially the experimental observation as embodied in what is known as the *Schulze–Hardy rule*.

Another factor, not explicitly mentioned so far, is that while ψ_0 reflects the intrinsic nature of the particle–medium interface, it may be modified by adsorbed ions. The potential of colloidal AgI particles is, for example, very dependent on the concentration of Ag^+ or I^- ions in the solution and in general would be affected by the presence of any other adsorbable ions. One may thus vary ψ_0 as well as electrolyte concentration, and studies of this type have allowed approximate experimental values of the Hamaker constant H to be calculated from flocculation rates.

The basis of the calculation is that the flocculation rate should be just the encounter rate given by Eq. (15-44) modfied by the factor $\exp(-\epsilon^*/kT)$, where ϵ^* is the height of the barrier (such as shown in Fig. 21-4). Thus ϵ^* may be calculated from the measured flocculation rate and, by means of Eq. (21-11), related to H.

It should be mentioned that H has been measured directly as a weak force of attraction between a spherical surface and a plate. The force is very small, of the order of 0.01 dyn for a separation of about 10^{-4} cm, and the required apparatus is quite ingenious in design—the measurements come from the laboratories of J. Th. Overbeek in Holland and B. V. Derjaguin in the USSR. It is quite a triumph that the approximate wave mechanical theory, the direct measurements, and the indirect calculation through flocculation rates agree in value of H to within at least an order of magnitude. See also Israelachvili (1985).

21-2   Association colloids. Colloidal electrolytes

A solution of a soap or in general of a detergent will exhibit colloidal behavior under certain conditions. The qualitative phase diagram shown in Fig. 21-5 is typical for ordinary soaps (that is, sodium or potassium salts of long-chain fatty acids such as stearic or oleic acid). Below a certain temperature T_K, known

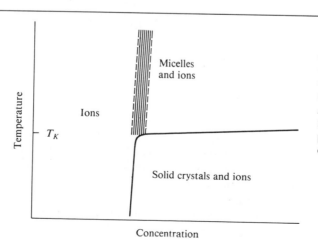

FIGURE 21-5
Phase map for a colloidal electrolyte. [From W.C. Preston, J. Phys. Colloid Chem. **52**, 84 (1948). Copyright 1948 by the American Chemical Society. Reproduced by permission of the copyright owner.]

FIGURE 21-6
Schematic representation
of a spherical micelle.

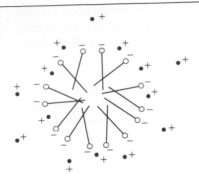

as the *Kraft temperature*, the soap exhibits a fairly normal solubility behavior, but above T_K, there is a critical concentration beyond which the solution consists of colloidal aggregates plus free ions. These aggregates are called *micelles* and consist of 50 to 100 monomer molecules in a more or less spherical unit, with about half of the cations bound and the rest present in an electrical double layer. A schematic illustration of a micelle is shown in Fig. 21-6.

Micelles are in thermodynamic equilibrium with the monomer electrolyte and their formation occurs over a rather narrow range of concentration (as shown in Fig. 21-5); one therefore speaks of the *critical micelle concentration*. The presence of micelles seems to be necessary for, or at least to correlate with, detergent action. A soap below its Kraft temperature does not function well, for example. One explanation is that micelles are able to absorb or "solubilize" oily material such as is present in dirt and one function of a soap in washing is in detaching dirt particles from the fabric and then keeping them in suspension.

21-3   Gels

A semisolid system having either a yield value (see Section 21-4) or a very high viscosity is called a gel, jelly, or paste. It is easily understandable that very concentrated suspensions should be semisolid or very viscous in behavior. It is possible, however, for quite dilute colloidal systems to show such properties also; a few percent by weight or gelatin or agar in water can give a gel. Such systems are properly termed lyophilic in that the colloidal material is strongly solvated and individual units interact with each other through their solvation sheaths. Weakly bound aggregates result, which form a loose three-dimensional network.

21-4   Rheology

Rheology is the science of the deformation and flow of matter. It is that branch of physics which is concerned with the mechanics of deformable bodies, primarily those that are roughly describable as liquids or solids. We considered a simple rheological situation in the treatment of Newtonian viscosity (see Sections 2-9 and 8-10) defined by Eq. (2-78). An ideal or Newtonian fluid is one whose viscosity coefficient η is independent of the value of dv/dx, that is, independent of *shear*.

Another type of limiting behavior is that of a *ideal elastic* body. Such a body deforms or shows *strain* which is proportional to the applied unbalanced force or *stress*. The ratio of stress over strain is called the *elastic modulus*, and a crystal of cubic symmetry has three kinds of elastic modulus: Young's modulus Y(change in normal stress divided by the resulting relative change in length), shear or rigidity modulus G (change in tangential stress divided by change in the resulting angle of extension), and bulk modulus K (change in hydrostatic pressure divided by resulting change in volume). These deformations are illustrated in Fig. 21-7. A less isotropic solid has moduli that are different for different directions of stress—up to a maximum of 21 in all.

The subject of rheology is made considerably more complicated by nonideal behavior. A liquid whose viscosity decreases with increasing stress (such as increasing rate of flow or of stirring) is called *pseudoplastic*; if the viscosity increases with stress, the liquid is *dilatant*. These cases are illustrated in Fig. 21-8, in which *rate of shear* is plotted against the *shearing stress*. Referring to Eq. (2-78), $f = \eta \mathcal{A} \, dv/dx$, the quantity dv/dx is the rate of shear, and $f/\mathcal{A}$, the force differential per unit area, is the shearing stress.

An ideal elastic solid shows no shear at any shearing stress, but actual solids have a *yield point* beyond which flow begins to occur. A solid is known as a *Bingham plastic* if, once flow takes place, the rate of shear is proportional to the shearing stress in excess of the yield value, illustrated by curve 4 in figure. Curves 5 and 6 are those for pseudoplastic and dilatant materials showing a yield value.

The measured viscosity of a system is given by the ratio $(f/\mathcal{A})/(dv/dx)$. Thus Bingham and pseudoplastic solids as well as non-Newtonian liquids have viscosities that are dependent on the rate of shear.

The situation is yet worse than that described since the measured viscosity can vary with time as well as with shearing stress. A liquid that becomes more fluid with increasing time of flow is said to be *thixotropic*, while if the opposite is true, we say it exhibits *rheopexy*.

Examples of these types of behavior are as follows. Gases and pure single-phase liquids exhibit Newtonian viscosity, while suspensions, slurries, and

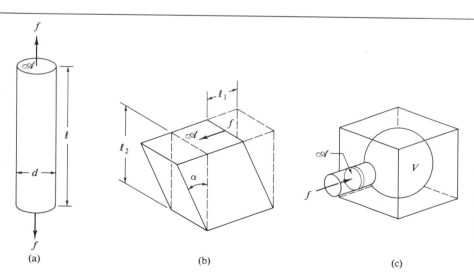

(a) (b) (c)

FIGURE 21-7
Diagrams explaining the definitions of the elastic constants. (a) Young's modulus Y; (b) shear modulus G; and (c) bulk modulus K. ($\mathcal{A}$ is area, d is diameter, l is length, V is volume, and f is force.) [From J.R. Van Wazer, J.W. Lyons, K.Y. Kim, and R.E. Colwell, "Viscosity and Flow Measurement." Copyright 1963, Wiley (Interscience), New York. Used by permission of John Wiley & Sons, Inc.]

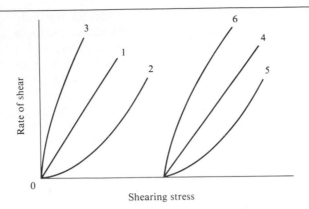

FIGURE 21-8
Shear versus stress plots.
(1) Newtonian fluid;
(2) pseudoplastic fluid;
(3) dilatant fluid; (4)
Bingham plastic; (5)
pseudoplastic with a yield
value; and (6) dilatant
material with a yield value.

emulsions are apt to show dilatant behavior—a common example being that of a thick starch paste. It is often difficult to distinguish betwen behaviors 2 and 5 of Fig. 21-8, that is, between a pseudoplastic fluid and a pseudoplastic solid with yield value; the term pseudoplastic is therefore often applied without attempting to make the distinction. Materials in this general category include melts or solutions of high molecular weight solutes. Household paints are often pseudoplastic so as to brush easily, yet not run; the same is true for printing inks.

Ordinary crystalline solids will always have a yield value, but often a very high one. That for metals is around 10 to 50 kg mm^{-2}, for example—flow begins to occur once this yield pressure is exceeded.

Some gels are thixotropic—they will liquify on shaking. The same is true of certain types of suspensions, such as of bentonite clay. They settle very slowly on standing, but rapidly if tapped gently.

21-5   Liquid crystals. Mesohases of matter

There are a number of crystalline substances that show a very peculiar behavior on warming. For example, the compound 5-chloro-6-*n*-heptyloxy-2-naphthoic acid,

$$HOOC-\underset{\displaystyle Cl}{\bigcirc\!\!\bigcirc}-O-n\text{-}C_7H_{15}$$

exists in ordinary crystalline form. On heating to 165.5°C, the crystals collapse to a turbid, viscous melt, which adheres to the walls of the container. If this is spread on a flat plate, one may see steps or ridges; it is birefringent, showing colored areas under polarized light and having different optical properties parallel and perpendicular to the plate. Further heating of the compound produces a second change, at 176.5°C, to a fluid but turbid liquid. Only at 201°C does final true melting to a clear liquid occur.

This type of behavior is called *liquid crystalline* because of the combination

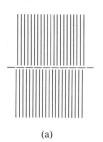 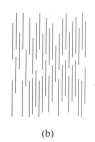 

(a) (b) (c)

FIGURE 21-9
Molecular arrangements for (a) smectic mesophase, (b) nematic mesophase, and (c) isotropic liquid.

of properties exhibited, and the phases involved are called *mesophases* (lying between crystalline and liquid). Liquid crystals tend to occur with compounds that are highly asymmetric in shape, often having well-separated polar and nonpolar portions to the molecule. The two types of mesophases described here are called *smectic* and *nematic*, and the type of order present is illustrated in Fig. 21-9. In the smectic phase the molecules have retained their parallel orientation but have lost the crystalline regularity in the spacing between them so that successive layers no longer match. In the nematic phase the molecules, while still parallel, have further lost their planar array.

A great many biological structures might be called liquid crystalline—there is partial but not complete ordering between molecular units. Muscle fibers show double refraction; sperm cells may possess a truly liquid crystalline state; a solution of tobacco mosaic virus contains nematic-type micellar units. At a simpler level, cholesterol and its derivatives may show both a smectic and a so-called cholesteric mesophase, the transition from the latter to isotropic melt occurring at around body temperature. The change from isotropic to cloudy liquid crystalline phase is very sensitive to electric fields and, in thin layers, to how the molecules orient in adsorbing at the boundary interfaces. Some phenomenal practical applications have resulted; one example is in the display of time on watches.

21-6 Polymers

The great importance of polymers hardly needs to be stressed. Biological polymers such as proteins and nucleic acids are central to the functioning of living things. Natural polymers such as rubber and the host of synthetic ones such as polytetrafluoroethylene ("Teflon"), polyethylene, and polystyrene are vital materials to our technology.

There are perhaps three major aspects to the general subject. The first is that of the chemistry of polymers—their synthesis and degradation reactions; the second important subject is the physical chemistry of polymer solutions; and the third is that of the overall configurational structure of polymers. The material that follows is intended merely to be illustrative of each of these very large fields.

The characteristic feature of a polymer is that it is composed of a large number of identical or at least similar units which are chemically bonded together. The compound which forms the "repeating" unit is called the *monomer*. Thus in polystyrene,

A. Polymerization chemistry and kinetics

$$- CH - CH_2 - (CH - CH_2)_n - CH - CH_2 -$$
$$\quad\phi \qquad\qquad \phi \qquad\qquad \phi$$

the monomer unit is styrene, $C_6H_5CH = CH_2$. Polymers of this type are called *linear polymers.* Other examples include polyethylene, polyvinyl chloride, polyacrylic esters and biological polymers such as proteins and polysaccharides. If more than two polymer-forming functions are present, as with divinyl benzene ($CH_2 = CH - C_6H_4 - CH = CH_2$), then cross-linked chains form, yielding a three-dimensional network. Cross-linking may also be induced to occur in an already formed linear polymer. Thus in the case of rubber,

$$—CH_2—CH{=}CH—CH_2—CH_2—C{=}CH—CH_2—$$
$$\qquad\qquad |\qquad\qquad\qquad\qquad\quad |$$
$$\qquad\qquad CH_3 \qquad\qquad\qquad\quad CH_3$$

the natural linear polymer chains may be tied at random positions by mixing in sulfur and heating to form $C—S_x—C$ bonds bridging two chains. Irradiation of polyethylene with gamma rays knocks off hydrogen atoms at random points and the resulting active carbon atoms can then bond to a neighboring chain.

Polymerization reactions generally occur by one of two types of mechanism. Vinyl monomers are polymerized by a free radical process such as that for ethylene,

$$R_0\cdot \ + \ CH_2{=}CH_2 \rightarrow R_0—CH_2—CH_2\cdot$$

$$R_0—CH_2—CH_2\cdot \ + \ CH_2{=}CH_2 \rightarrow R_0—CH_2—CH_2—CH_2—CH_2 \quad (21\text{-}12)$$

where $R_0\cdot$ is an initiating radical produced by some catalyst such as a peroxy acid,

$$\overset{\displaystyle 0}{\underset{\displaystyle RCOOH}{\|}}$$

or photochemically. Styrene will polymerize just on heating in the presence of oxygen.

The formal kinetic scheme for the radical mechanism is as follows:

$$R_0 \ + \ M \overset{k_i}{\rightarrow} M_1\cdot \qquad \text{(initiation)} \qquad (21\text{-}13)$$

$$M_1\cdot \ + \ M \overset{k_1}{\rightarrow} M_2\cdot \qquad M_2\cdot \ + \ M \overset{k_2}{\rightarrow} M_3\cdot \qquad \text{(propagation)} \qquad (21\text{-}14)$$

$$M_n\cdot \ + \ M_t\cdot \overset{k_i}{\rightarrow} M_n \ + \ M_s \ (\text{or } M_{n+s}) \qquad \text{(termination)} \qquad (21\text{-}15)$$

Application of the stationary-state hypothesis (Section 14-4C) yields

$$R_i \ = \ R_t \ = \ k_t \left[\sum_{n=1}^{\infty} (M_n\cdot) \right]^2 \ = \ k_t(M\cdot)^2 \qquad (21\text{-}16)$$

where R_i and R_t are the initiation and termination rates, respectively, it being assumed that any two radical chains react to terminate with the same rate constant k_t; $(M\cdot)$ denotes the total radical concentration, $\Sigma_n (M_n\cdot)$. If we further assume that $k_1 = k_2 = k_n = k_p$ for the reactions of Eq. (21-14), where k_p is the common propagation rate constant, then the rate of disappearance of monomer is

$$-\frac{d(M)}{dt} = (M)\, k_p \sum_n (M_n\cdot) \tag{21-17}$$

On substitution for $(M\cdot)$ from Eq. (21-16), we obtain

$$-\frac{d(M)}{dt} = k_p \left(\frac{R_i}{k_t}\right)^{1/2} (M) \tag{21-18}$$

In general, each polymerization system must be subjected to separate kinetic analysis since the appropriate set of approximations may vary. A particularly simple special case of Eq. (21-18), however, is that in which R_i is photochemical and therefore given by ϕI_a, where ϕ is the quantum yield for radical production and I_a is the number of quanta of light absorbed per unit volume and time. A point implicit in schemes such as this is that after some time t there will be a distribution of polymer molecular weights. A more detailed analysis would predict both the average molecular weight and the breadth of the distribution.

A second type of mechanism is that in which a Lewis acid such as $AlCl_3$ or an ordinary strong protonic acid such as H_2SO_4 adds a proton to the monomer. A typical sequence is that for the production of butyl rubber by the polymerization of isobutylene,

$$\text{H}^+ \overset{\frown}{} \text{CH}_2 = \underset{\underset{\text{CH}_3}{|}}{\overset{\overset{\text{CH}_3}{|}}{\text{C}}} \overset{\frown}{} \text{CH}_2 = \underset{\underset{\text{CH}_3}{|}}{\overset{\overset{\text{CH}_3}{|}}{\text{C}}} \overset{\frown}{} \text{CH}_2 = \underset{\underset{\text{CH}_3}{|}}{\overset{\overset{\text{CH}_3}{|}}{\text{C}}} \rightarrow \text{CH}_3 - \underset{\underset{\text{CH}_3}{|}}{\overset{\overset{\text{CH}_3}{|}}{\text{C}}} - \text{CH}_2 - \underset{\underset{\text{CH}_3}{|}}{\overset{\overset{\text{CII}_3}{|}}{\text{C}}} - \text{CH}_2 - \underset{\underset{\text{CH}_3}{|}}{\overset{\overset{\text{CH}_3}{|}}{\text{C}}} \cdots \tag{21-19}$$

More recently, transition metal ions have been found to be excellent catalysts for polymerization reactions, often giving highly stereoregular or *isotactic* polymers. (A polymer such as polystyrene is isotactic if, when the chain is stretched out, all of the phenyl groups are on one side.)

A related mechanism is that by which water is eliminated between two functional groups. Thus a protein chain,

$$\text{H} - \underset{\underset{\text{H}}{|}}{\overset{\overset{\text{R}}{|}}{\text{C}}} - \underset{\underset{\text{H}}{|}}{\overset{}{\text{N}}} - \overset{\overset{\text{O}}{\|}}{\text{C}} - (\underset{\underset{\text{H}}{|}}{\overset{\overset{\text{R}}{|}}{\text{C}}} - \underset{\underset{\text{H}}{|}}{\overset{}{\text{N}}} - \overset{\overset{\text{O}}{\|}}{\text{C}} -)_n - \underset{\underset{\text{H}}{|}}{\overset{\overset{\text{R}}{|}}{\text{C}}} - \underset{\underset{\text{H}}{|}}{\overset{}{\text{N}}} - \overset{\overset{\text{O}}{\|}}{\text{C}} - \text{H}$$

forms on elimination of water between RCOOH and RNH_2 groups. The R's vary down the chain in natural proteins but may be the same in synthetic polypeptides. As another illustration, nylon results from the condensation of adipic acid $(HOOC(CH_2)_4—COOH)$ with hexamethylenediamine $(H_2N(CH_2)_6NH_2)$. These various polymerizations follow kinetic schemes similar to that of Eqs. (21-13) to (21-15).

B. Polymer solutions

The study of polymer solutions can be regarded as a branch of colloid chemistry. As with colloidal suspensions, one is interested in the distribution of particle sizes (in this case, polymer molecular weights), in the shape of the particles, in their interaction with the solvent medium, and in the general physical properties of the system.

1. MOLECULAR WEIGHT. The general methods for molecular weight determination are discussed in Section 10-7. For polymers these include measurement of osmotic pressure, diffusion rate, sedimentation equilibrium and sedimentation rate, and light scattering. If a range of molecular weights is present, a given method will yield either a number-average or a weight-average molecular weight; if both types of average molecular weight can be determined, their ratio provides a measure of the broadness of the molecular weight distribution.

2. VISCOSITY. There is an additional method of molecular weight estimation, which makes use of viscosity measurements. Einstein showed in 1906 that for a dilute suspension of spheres in a viscous medium

$$\lim_{\phi \to 0} \frac{(\eta/\eta_0) - 1}{\phi} = 2.5 \tag{21-20}$$

where η and η_0 are the viscosities of the solution or suspension and pure solvent, respectively, and ϕ is the volume fraction of the solution occupied by the spheres. The fraction η/η_0 is called the viscosity ratio or *relative viscosity* η_r, and $[(\eta/\eta_0) - 1]$ is called the *specific viscosity* η_{sp}. In the case of a polymer solution the concentration C is usually given as grams per cubic centimeter, and $\eta_{sp}/100\,C$ is called the *reduced viscosity* and its limiting value as C approaches zero is the *intrinsic viscosity* $[\eta]$.† Since $\phi = v_2 C$, where v_2 is the specific volume (strictly, the partial specific volume) of the polymer in cubic centimeters per gram, it follows from Eq. (21-20) that $[\eta] = 0.025 v_2$. The specific volume as determined from $[\eta]$ will in general be larger than that obtained from the density of the dry polymer, the difference being attributed to bound solvent.

The factor 2.5 in Eq. (21-20) applies to the case of spheres, and linear polymers are, of course, not spherical, in the sense of the molecule being a long chain. On the other hand, the completely extended configuration is a very improbable one, and the situation is more like that illustrated in Fig. 21-10, which shows various *random coils*. The problem is essentially the statistical one of calculating the most probable end-to-end distance d assuming the polymer to consist of n flexible links; d, as might be expected, turns out to be proportional to the square root of the chain length. The proportionality constant depends, however, on the balance between solvent–chain and chain–chain interactions. In a "good" solvent, the random coil will be very extended, while in a "poor" solvent, the chain prefers its own environment and therefore assumes a compact configuration. While none of these configurations is exactly spherical, Eq. (21-20) can still be used, but with an empirical

---

†The definitions above are the traditional ones. The current recommendation is that η/η_0 be called the *viscosity ratio* and not relative viscosity; that $[(\eta/\eta_0) - 1]/C$ be called the *viscosity number*; and that the limiting value as C approaches zero, or *limiting viscosity number*, be designated $[\eta]$.

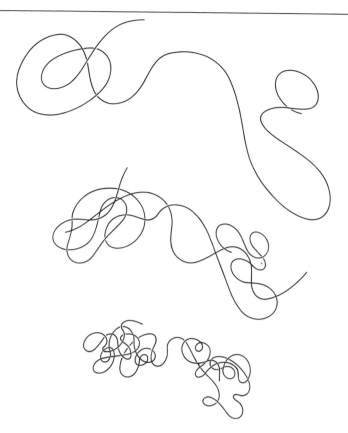

parameter v in place of the Einstein coefficient 2.5, and v evidently will depend on the nature of the solvent as well as on that of the polymer.

An alternative development of Eq. (21-20) is the following. If ϕ_0 denotes the volume fraction calculated using v_2 for the dry polymer, then the correct ϕ is given by $\phi_0\beta$, where β is the ratio of the hydrated to dry specific volumes. The dry volume of a polymer is, of course, proportional to its molecular weight M. The hydrated volume will be proportional to d^3, or to $M^{3/2}$, so β should be proportional to $M^{1/2}$. Thus Eq. (21-20) can be put in the form

$$[\eta] = KM^{1/2} \tag{21-21}$$

If, however, the polymer were a rigid rod of length d, its effective volume would be that swept out by the tumbling of the rod. The "hydrated" volume is now proportional to M^3, and β is proportional to M^2.

The exponent in Eq. (21-21) can thus be expected to vary depending on the stiffness of the polymer, and in an equation proposed by M. Staudinger in 1932 it is treated as an empirical parameter:

$$[\eta] = KM^a \tag{21-22}$$

The experimental observation is that K and a are approximately constant for a given type of polymer and a given solvent. Polymers of about the same chemical composition may differ greatly in molecular weight, and Eq. (21-22) allows the relative molecular weights of different preparations to be determined. If two preparations of different known molecular weight can be obtained, then K and a can be evaluated, and absolute molecular weights may then be found for any other sample. It might be mentioned that if a range of molecular weights is present in the sample, then the value calculated by means of Eq. (21-22) will be an average lying between the number- and weight-average values.

This type of study has been made for a number of types of polymer and values of a are usually in the range 0.5–1.1. As expected, a decreases if one goes from a good to a poor solvent. That is, the relatively compact configuration present in a poor solvent behaves more like a sphere than a rod.

3. THERMODYNAMICS. One approach to the statistical treatment of polymer solutions is to consider the solution as consisting of a number of cells or sites. Each site may be occupied by a solvent molecule or by one link or monomer unit of the polymer; as illustrated in Fig. 21-11, the requirement is that successive links occupy adjacent sites. One then treats the "solution" of links as a regular one (Section 9-CN-2) or essentially according to the model of Section 9-2E, but with the added aspect that the entropy of the links is reduced because of their having to remain connected.

The vapor pressure or activity of the polymer is then given by an equation similar to Eq. (9-17) and varies with solution composition in a way similar to that shown in Fig. 9-9. Thus for $\alpha > 2$, there is a maximum and a minimum, corresponding to partial miscibility; however, the effect of the entropy complication is to distort the behavior in the direction of making intermediate compositions less probable than for an ordinary regular solution. As a consequence, one finds that for $\alpha > 2$, the dilute solution tends to be very dilute and the concentrated solution very concentrated, corresponding to slightly solvated polymer in equilibrium with a dilute solution. This explains the experimental observation that polymers tend to be either slightly soluble or quite soluble in a given solvent.

The quantity α in Eq. (9-17) is actually an interaction energy ω divided by kT. The critical condition is therefore that $\alpha = \omega/kT_c = 2$, or $T_c = \omega/2k$. This relationship accounts for the common observation that below a critical temperature a polymer will not be very soluble, whereas above this temperature it is almost completely miscible with the solvent.

FIGURE 21-11
Illustration of the statistical fitting of solvent molecules (open circles) and monomer units (shaded circles) on a two-dimensional lattice.

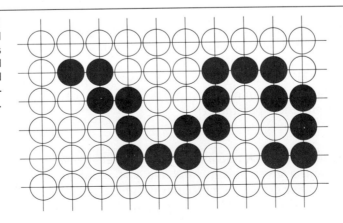

4. SECONDARY STRUCTURE. The primary structure of a polymer is that of the chemically bonded chain itself, including its stereochemistry. Since we are dealing with a macromolecule, there is a second level of structure, namely that of the polymer chain as a whole. We saw in the preceding section that linear polymers in solution assume a random coil configuration, for example. The coil becomes tighter the poorer the solvent or the more strongly interacting the polymer chain is with itself.

As a limiting case of this last situation, amino acid polymers, that is, proteins, can form hydrogen bonds between various portions of the chain. L. Pauling proposed a now widely accepted structure known as the *α-helix*. The helix makes one turn for every 3.7 amino acid residues, thus allowing hydrogen bonding between the CO and NH groups of adjacent coils.

In the case of the nucleic acids, the primary structure is that of a linear

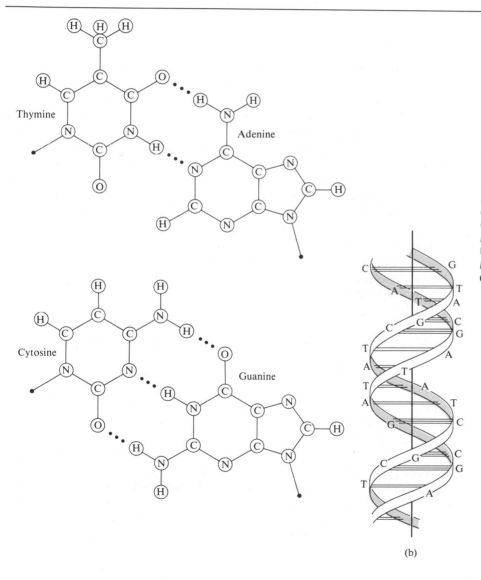

(b)

FIGURE 21-12
Structure of DNA. (a) Hydrogen bonding between adenine (A) and thymine (T) and between cytosine (C) and guanine (G). (b) Two strands held in a double helix by C–G and T–A hydrogen bonds. [See M.H.F. Wilkins and S. Arnott, J. Molec. Biol. **11**, 391 (1965).] [Part (b) from A. White, P. Handler, and E.L. Smith, "Principles of Biochemistry," 4th ed. Copyright © 1968, McGraw-Hill, New York. Used with permission of McGraw-Hill Book Company.]

polymer in which adenine (A), guanine (G), cytosine (C), and thymine (T) rings are linked by phosphate groups; adjacent chains are hydrogen bonded as shown in Fig. 21-12. The secondary structure is that of a double helix in which C–G and A–T pairs form hydrogen bonds. The detailed configuration of these hydrogen bonds is at present subject to some controversy.

The ordinary denaturation of proteins and nucleic acids generally consists of a loss of the secondary structure but not of the primary structure. Intermediate situations are possible, too, in which part of the polymer has the helical configuration and part that of random coils. It has been possible in some cases to observe helix–coil transitions.

Finally, one may speak of a tertiary structure. A protein helix is still a long unit and may fold in on itself. Regions that have few polar groups find water a poor solvent, and hence tend to approach each other—the effect has been rather misnamed as *hydrophobic bonding*. The consequence is that native proteins tend to form a tertiary structure that is more or less globular and with the nonpolar groups in the interior.

21-7   Electrokinetic effects

Several very interesting phenomena may occur if there is relative motion between a charged surface and an electrolyte solution; these are called *electrokinetic effects*. The surface is ordinarily either that of a solid particle suspended in the solution or the solid wall of tube down which the solution flows. What happens is essentially that a charged surface experiences a force in an electric field or, conversely, that a field is induced by the motion of such a surface relative to the solution. Thus if a field is applied to a suspension of colloidal particles, these will move or, alternatively, if a solution is made to flow down a tube an electric field develops. In each case there is a plane of shear between the surface and the solution, a plane which lies within the electrical double layer. The potential at this plane will not be ψ_0, but some lower value called the *zeta potential* ζ. The zeta potential does give a measure of ψ_0, however, and one important application of electrokinetic phenomena is to the measurement of ζ and hence estimation of ψ_0.

There are four types of electrokinetic effect, depending on whether the solid surface is stationary (as a wall) or moves (as a particle) and whether the field is applied or induced. In *electrophoresis* an applied field causes suspended particles to move, while in the *sedimentation potential effect* the motion of the particles in a centrifugal field induces a field. If the solid surface is that of the wall of a tube, application of a potential gradient or field induces a flow of solution, or *electroosmosis*, while a forced flow of solution induces a potential gradient or *streaming potential*. The four effects are summarized in Table 21-1. We will consider only two of them, however, namely electrophoresis and the streaming potential effect.

A. Electrophoresis   The most familiar type of electrokinetic experiment consists of setting up a potential gradient in a solution containing charged particles and determining their rate of motion. If the particles consist of ordinary small ions, the phenomenon is one of *ionic conductance*, treated in Chapter 12. As shown in Section 12-5, the velocity of an ion in a field $\mathbf{F}$ is

	Nature of solid surface	
Potential	Stationary[a]	Moving[b]
Applied	Electroosmosis	Electrophoresis
Induced	Streaming potential	Sedimentation potential

TABLE 21-1
Electrokinetic effects

[a]For example, a wall or apparatus surface.
[b]For example, a colloidal particle.

$$v_i = z_i e \omega_i \mathbf{F} \qquad [\text{Eq. (12-34)}]$$

where ω_i is the mobility and v_i will be in centimeters per second if e is in esu and $\mathbf{F}$ in esu volts per centimeter. Alternatively, by Eq. (12-35), $v_i = u_i\mathbf{F}$, where u_i is the electrochemical mobility and $\mathbf{F}$ is now given in ordinary volts per centimeter; also, by Eq. (12-38), $\lambda_i = u_i\mathbf{F}$.

In the case of a macromolecule, the total charge z_i is not known, and it is customary to refer to the motion of such molecules in an electric field as *electrophoresis.*

However, if the molecular weight is known, then the radius can be calculated assuming the shape to be spherical. This allows estimation of the friction factor f from Stokes' law, Eq. (10-36), and hence of $\omega = 1/f$. Alternatively, if the diffusion coefficient can be measured, then f can be obtained without any geometric assumptions from Eq. (10-44). The charge on the polymer molecule may then be calculated from the electrophoretic velocity using Eq. (12-34).

As would be expected from the discussion of the acid–base equilibria of amino acids (Section 12-CN-3), the net charge on a protein is very pH-dependent. For each protein there will be some one pH at which z and hence the electrophoretic velocity is zero. Conversely, at a given pH each protein will have a characteristic velocity in an electric field, depending both on its molecular weight and on its net charge. A mixture of proteins will therefore separate into different velocity groups in an electrophoresis experiment.

We turn now to colloidal particles. As with macromolecules, the charge is not known; in addition, the size cannot be determined very accurately, although it may be possible to estimate particle radii from microscopic (optical or electron) examination. Ordinary molecular weight determination is, of course, not possible (why?) nor is the diffusion rate fast enough for a determination of f. As noted in the introduction to this section, we assume instead that the particle is large enough for its interface with the solution to be treated as planar, and apply electrical double layer theory.

The complication mentioned at the beginning of this section is that, as illustrated in Fig. 21-13, there is undoubtedly a layer of solvent molecules and electrolyte ions that is strongly enough held that it moves with the particle rather than with the solution. We must deal, therefore, with the plane of *shear* rather than with the actual interface; the potential ψ_0 is reduced to some value $\psi = \zeta$ at this plane, called the zeta potential, and $\boldsymbol{\sigma}_0$ to some value $\boldsymbol{\sigma}_s$. The remainder of the diffuse double layer must still balance $\boldsymbol{\sigma}_s$, or, by Eq. (21-9),

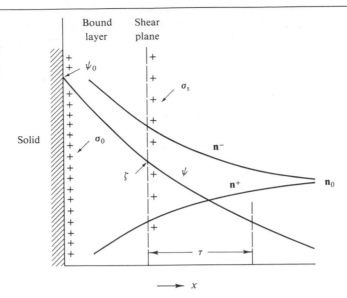

$\sigma_s = \int_{x=s}^{\infty} \rho \, dx$. The "center of gravity" of this portion of the double layer is taken to be at the point where ψ has dropped to $1/e$ of ζ, at distance τ from the plane of shear.

As a first approximation, the situation is likened to that of a parallel plate condenser, for which the standard formula gives $\Delta V = 4\pi q d/D\mathcal{A}$, where ΔV is the potential difference across the plates, d is their separation, $\mathcal{A}$ is their area, and D is the dielectric constant. One plate carries charge $+q$ and the other carries $-q$. We now equate ΔV to ζ, d to τ, and $q/\mathcal{A}$ to σ_s to obtain

$$\sigma_s = \frac{D\zeta}{4\pi\tau} \tag{21-23}$$

We next proceed to balance the electrical force acting per square centimeter of the particle with that due to viscous drag. The first is just $\sigma_s\mathbf{F}$, where $\mathbf{F}$ is the field, and the second may be obtained from the defining equation for viscosity, Eq. (2-78). We assume that the relative velocity between the particle and the solution drops linearly with distance from v at $x = s$ to 0 at $x = \tau$ and therefore write $dv/dx = v/\tau$, which gives the viscous force or drag per square centimeter as $\eta v/\tau$. On equating the two forces, we obtain

$$v = \frac{\sigma_s\mathbf{F}\tau}{\eta} = \frac{\sigma_s\mathbf{F}\tau}{300\eta} \tag{21-24}$$

where in the second form, $\mathbf{F}$ is expressed in volts per centimeter. An alternative form results on combination with Eq. (21-23):

$$v = \frac{\zeta D\mathbf{F}}{4\pi\eta} \tag{21-25}$$

All of the fundamental statements have been written in the esu system insofar as electrical quantities are involved, but one usually reports ζ in volts (actually, millivolts), having measured $\mathbf{F}$ in volts per centimeter. If one converts ζ to millivolts by means of the factor 1000/300 and $\mathbf{F}$ to volts per centimeter by the factor 1/300, one obtains a more practical form of Eq. (21-25):

$$\zeta = 12.9 \frac{v}{\mathbf{F}} \qquad (21\text{-}26)$$

where, in addition, v is in microns per second and D and η have been taken to be for water at 25°C. Thus if a colloidal particle moves 50 μ s^{-1} in a field of 10 V cm^{-1}, its zeta potential is 65 mV, positive or negative depending on the direction of motion relative to the field. The field may be applied by means of nongassing electrodes at each end of a tube containing the colloidal suspension and the rate of motion of individual particles observed under a microscope; the instrument is called a *zetameter*.

Most measured zeta potentials are in the range of ± 100 mV for lyophobic colloids, but with actual values dependent on the nature and concentration of the electrolyte. Table 21-2 gives some data on a gold sol, which show that the charge decreases and then reverses with increasing Al^{3+} ion concentration. Note that the sol is stable if the particles are charged but flocculates rapidly if the zeta potential is small.

If an electrolyte solution flows through a tube, one observes a potential difference between the inflowing and outflowing solution. The physical basis for the effect is that the flowing solution carries with it the charge density σ_s in the diffuse double layer, so that there is in effect an electrical current at the wall. The current is just $i = 2\pi r \sigma_s v_\tau$, where v_τ is the streamline velocity at distance τ from the wall; it is thus proportional to the surface area and, by Eq. (21-23), to the zeta potential. A potential drop $\mathbf{E}$ develops in the electrolyte solution until a counter current, $i = \mathbf{E}/R$, develops, where R is the resistance of the solution, such as to just balance the wall current. It is this steady-state potential that is called the *streaming potential*.

The actual derivation is somewhat in the same vein as that of the Poiseuille equation given in Section 8-10. Assuming streamline flow, the velocity at a radius x from the center of the tube is

$$v = \frac{P(r^2 - x^2)}{4\eta l} \qquad \text{[Eq. (8-64)]}$$

where P is the pressure drop and r the radius of the tube, which is of length l. The double layer is centered at $x = r - \tau$, and substitution into Eq. (8-64) gives

B. Streaming potential

TABLE 21-2
Flocculation of gold sol

Concentration of Al^{3+} (equiv liter^{-1} × 10^6)	Electrophoretic velocity (cm V^{-1} s^{-1} × 10^6)	Stability
0	3.30 (toward anode)	Indefinitely stable
21	1.71	Flocculates in 4 hr
—	0	Flocculates spontaneously
42	0.17 (toward cathode)	Flocculates in 4 hr
70	1.35	Incompletely flocculated after 4 days

$$v_d = \frac{\tau r P}{2\eta l} \tag{21-27}$$

if the term in τ^2 is neglected. The current due to the motion of the double layer is then

$$i = 2\pi r \boldsymbol{\sigma}_s v_d \tag{21-28}$$

or

$$i = \frac{\pi r^2 \boldsymbol{\sigma}_s P}{\eta l} \tag{21-29}$$

If κ is the specific conductivity of the solution, then the actual conductivity of the liquid in the tube is $C = \pi r^2 \kappa / l$ and, by Ohm's law, the streaming potential is just $\mathbf{E} = i/C$. Combining these equations, we have

$$\mathbf{E} = \frac{\tau \boldsymbol{\sigma}_s P}{\eta \kappa} \tag{21-30}$$

or, using Eq. (21-23),

$$\mathbf{E} = \frac{\zeta P D}{4\pi \eta \kappa} \tag{21-31}$$

A practical form of Eq. (21-31) is

$$\mathbf{E} = 8.0 \times 10^{-5} \frac{\zeta P}{C\Lambda} \tag{21-32}$$

where ζ is now in millivolts and P in atmospheres; C is the electrolyte concentration in equivalents per liter and Λ is the equivalent conductivity. Water at 25°C is assumed. Thus for a pressure drop of 10 atm, a zeta potential of 50 mV, and 10^{-5} N aqueous NaCl at 25°C (equivalent conductivity 126), $\mathbf{E} = 32$ V. As this example illustrates, the electrolyte must be quite dilute if an appreciable effect is to occur.

Streaming potentials may cause trouble in the filtering of dilute electrolyte solutions, especially nonaqueous ones whose electrolyte concentration and hence specific conductivity may be quite small. A quite real hazard developed in the early days of jet aircraft—the high pumping rate of the kerosene fuel led to streaming potentials large enough to cause sparks. The problem has since been eliminated by incorporating small amounts of conducting solutes in the fuel.

GENERAL REFERENCES

ADAMSON, A.W. (1982). "The Physical Chemistry of Surfaces," 4th ed. Wiley (Interscience), New York.

BILLMEYER, F.W., JR. (1962). "Textbook of Polymer Science." Wiley (Interscience), New York.

GRAY, G.W. (1962). "Molecular Structure and the Properties of Liquid Crystals." Academic Press, New York.

KRUYT, H.R. (1952). "Colloid Science." Elsevier, New York.

McBAIN, J.W. (1950). "Colloid Science." Heath, Indianapolis, Indiana.

MYSELS, K.J. (1959). "Introduction to Colloid Chemistry." Wiley (Interscience), New York.

VAN WAZER, J.R., LYONS, J.W., KIM, K.Y., AND COLWELL, R.E. (1963). "Viscosity and Flow Measurement." Wiley (Interscience), New York.

Verwey, E.J.W., and Overbeek, J.Th.G. (1948). "Theory of the Stability of Lyophobic Colloids." Elsevier, New York.

White, A., Handler, P., and Smith, E.L. (1968). "Principles of Biochemistry," 4th ed. McGraw-Hill, New York.

Adamson, A.W. (1982). "The Physical Chemistry of Surfaces," 4th ed. Wiley (Interscience), New York.

Israelachvili, J.N. (1985). "Intermolecular and Surface Forces," Academic, New York.

CITED REFERENCES

EXERCISES

Take as exact numbers given to one significant figure.

For what value of ψ is $e\psi/kT = 1$ at 25°C?

21-1

Ans. 25.7 mV.

A surface of potential 100 mV is in contact with 0.2 M NaCl solution at 25°C. Calculate the concentrations of Na$^+$ and of Cl$^-$ ions at the surface. Comment on the result.

21-2

Ans. (Na$^+$) = 4.08×10^{-3} M, (Cl$^-$) = 9.79 M.

Referring to Exercise 21-2, calculate ψ and (Na$^+$), (Cl$^-$), and the charge density ρ at a distance of 20 Å from the surface.

21-3

Ans. 5.30 mV, 0.163 M, 0.246 M, 0.083 mole of charge per liter or 2.40×10^{10} esu cm^{-3}.

Still referring to Exercises 21-2 and 21-3, calculate the repulsion energy per square centimeter between two charged surfaces in 0.2 M NaCl and 20 Å apart (at 25°C) for $\psi = 100$ mV.

21-4

Ans. 0.0341 erg cm^{-2}.

Calculate the attractive potential due to dispersion forces between two platelets (say, of a colloidal suspension of a mica, vermiculite) for which $H = 4.5 \times 10^{-13}$ erg and the surface-to-surface distance is 20 Å. Assume the platelets to be oriented parallel to each other, and express your result in units of kT at 25°C (a) as ϵ/kt per cm^2 and (b) as ϵ/kT if the platelets are squares 0.5 μ on the side.

21-5

Ans. (a) 7.25×10^{12} cm^{-2}, (b) 1.81×10^4.

Calculate the attractive potential due to dispersion forces between two spherical colloidal particles of radius 0.5 μ and 20 Å apart, taking H to be 4.5×10^{-13} erg. Express the result in units of kT at 25°C; comment on the comparison with the answer to Exercise 21-5.

21-6

Ans. 228.

Combine the results of Exercises 21-4 and 21-5 to obtain the net attraction or repulsion between platelets 0.5 μ on the side, again in units of kT at 25°C.

21-7

Ans. 1.60×10^4 (attractive).

A rheological study of a certain fluid gives $y = 0.1x + 2x^2$, where y is the rate of strain in second^{-1} and x is the shear stress in dynes per square centimeter. Calculate the viscosity at a shear stress of 10 dyn cm^{-2}. What type of rheological behavior is being exhibited?

21-8

Ans. 0.0498 P.

Suppose that the polymerization kinetics of Eqs. (21-14) and (21-15) is followed, but that the initiation step is photochemical, with radicals produced by the process R $\xrightarrow{h\nu}$ 2M. Show that Eq. (21-18) becomes $-d(M)/dt = k_P(2\phi I/k_t)^{1/2}(M)$, where ϕ is the quantum yield and I the rate of absorption of light quanta.

21-9

The relative viscosity of a solution of polystyrene has the following values:

21-10

C'(g per 100 cm^3)	0.08	0.12	0.16	0.20
η_r	1.213	1.330	1.452	1.582

Find $[\eta]$ and estimate v_2 for the polymer. Calculate K of Eq. (21-21) if the molecular weight of the polymer is 250,000 g mol^{-1}, and calculate the relative viscosity of a 0.10 g per 100 cm^3 of solution of polymer of molecular weight 500,000. (The slope of the plot of $\eta_{sp}/100C$ should vary as $[\eta]^2$.)

Ans. $[\eta] = 2.49$, $v_2 = 100$ cm^3 g^{-1}, $K = 4.98 \times 10^{-3}$, $\eta_r = 1.395$.

21-11   While the result should not have very exact physical meaning, as an exercise, calculate the zeta potential of sodium ion, knowing that its equivalent conductivity is 50 cm^2 equiv^{-1} ohm^{-1} in water at 25°C.

Ans. 67 mV.

21-12   Calculate the expected streaming potential **E** for pure water at 25°C when flowing through a quartz tube under 10 atm pressure. Take ζ to be 150 mV.

Ans. 2190 V.

21-13   A quartz particle of 1.5×10^{-4} cm diameter moves through water at a velocity of 2.50×10^{-3} cm s^{-1} at 25°C under a potential gradient of 10.0 V cm^{-1}. Calculate the zeta potential of the particle in this medium. Assuming Stokes's law to hold, what total charge does the particle appear to carry? What is the equivalent conductivity of the particle?

Ans. $z = 1963$; $\lambda = 24$ cm^2 equiv^{-1} ohm^{-1}.

PROBLEMS

21-14   Show that Eq. (21-9) can be phrased as $\sigma = -(D/4\pi)(d\psi/dx)_{x=0}$ and further show that an approximate value for σ is $D\kappa\psi_0/4\pi$.

21-15   The value of x for which ψ/ψ_0 is $1/e$ is called the double layer "thickness" τ. Calculate τ as a function of concentration of NaCl at 25°C up to 0.5 M and plot the result.

21-16   A fourth plot that might have been shown in Fig. 21-2 is that of surface charge density (electron charges per 100 Å^2) versus ψ^0. Calculate and plot several points on this curve, for the range ψ^0 from zero up to about 100 mV. Assume a univalent electrolyte at 0.01 M and 25°C. Note that Eq. (12-86) becomes

$$\frac{d^2y}{dx^2} = \kappa^2 \sinh y$$

where $y = e\psi/kT$. It can be shown that integration gives $dy/dx = -2\kappa \sinh(y/2)$. Remember also that from Eq. (12-83)

$$\rho = -\left(\frac{D}{4\pi}\right)\frac{d^2\psi}{dx^2} = -\left(\frac{D}{4\pi}\right)\left(\frac{kT}{e}\right)\frac{d^2y}{dx^2}$$

21-17   Derive Eq. (21-2) from Eq. (21-1) by carrying out the required triple integration.

21-18   Calculate the interaction energy in calories per mole for an argon atom adsorbed on a hypothetical argon surface; assume the atom–surface distance to be 3.5 Å. (Note Table 8-8.)

21-19*   Calculate, using Eqs. (21-3) and 21-10), the net interaction energy in units of kT for two colloidal platelets as a function of their distance of separation and for $\kappa = 10^5$ cm^{-1}, 10^6 cm^{-1}, and 10^7 cm^{-1}. Assume that $H = 3 \times 10^{-13}$ erg, $\psi_0 = 25.6$ mV, $T = 25$°C, and assume an aqueous NaCl medium. For each κ value, plot your results as ϵ/kT versus x in angstroms up to about 200 Å. The area of each platelet is 10^{-9} cm^2.

21-20   Two argon atoms have an attractive potential due to dispersion interaction of 6×10^{-12} erg when 2 Å apart. Calculate the attractive potential for an argon atom positioned 2 Å away from the surface of an argon crystal. The molar volume of the crystal is 24 cm^3 mol^{-1}. The result may seem surprising; comment on it.

The following data were obtained for the viscosity of nitrocellulose in butyl acetate:

C(g per 100 cm³)	0.032	0.075	0.135	0.180
η_r	1.540	2.323	3.964	8.025

Calculate the intrinsic viscosity of this polymer and its specific volume.

The following relative viscosities were measured for a polyisobutylene in cyclohexane.

C(g per 100 cm³)	0.050	0.152	0.271	0.441
η_r	1.2895	2.067	3.312	6.579

Calculate the molecular weight of the polymer if the constants in Eq. (21-22) are $\alpha = 0.73$ and $K = 5.1 \times 10^{-4}$.

The viscosity of a globular protein is investigated at 5°C using an Ubbelohde viscometer (based on the Poiseuille equation, Section 8-10; note that $P_1 - P_2$ is proportional to the density of the liquid). The following data are obtained:

C (g per 100 cm³)	0	2.10	4.20	6.30	8.36	10.45
Density (g cm⁻³)	—	1.0052	1.0106	1.0160	1.0213	1.0268
Flow time (sec)	42.1	44.0	46.3	49.2	53.7	60.3

The true density of the protein is 1.310 g cm⁻³. Making appropriate calculations, decide whether this protein is hydrated or not and if it is, about what volume of solvent is bound to (moves with) a gram of protein.

Verify Eq. (21-26). What number replaces the 12.9 if ζ is in volts, v in m s⁻¹ and **F** in V m⁻¹?

It was mentioned that streaming potentials could be a real problem in jet aircraft. Suppose that a hydrocarbon fuel (such as kerosene) of dielectric constant 8 and viscosity 0.03 poise is being pumped under a driving pressure of 30 atm. The ζ potential between the pipe and the fuel is, say, 125 mV, and the fuel has a low ion concentration in it, equivalent to 10^{-8} M NaCl. Making and stating any necessary and reasonable assumptions, calculate the streaming potential that should be developed. [Consider carefully the handling of units.]

A colloidal particle suspended in an aqueous solution has a zeta potential of 65 mV. Calculate the radius of the particle, assuming Stokes's law, if the particle carries 2500 units of electron charge.

Micelle formation can be treated as a mass action equilibrium, for example,

$$40 \text{ Na}^+ + 80 \text{ R}^- = (\text{Na}_{40}\text{R}_{80})^{-40}$$

The cmc (critical micelle concentration) for sodium dodecylbenzenesulfonate is about 10^{-3} M at 25°C. Calculate K for the above reaction, assuming that it is the only process involved in micelle formation and that the cmc corresponds to 10% of the surfactant present in micelle form. [It is worthwhile to invest a little time in reflecting on how to proceed with your calculation.] Optional: make a plot of fraction present as micelles as a function of overall surfactant concentration.

chapter 22
Nuclear chemistry
and radiochemistry

22-1   Introduction

The subjects of nuclear chemistry and radiochemistry are marked by many major, far-reaching discoveries. Three early lines of development are those relating to the electrical nature of matter, to radioactivity, and to the nuclear model for the atom. The modern subject has branched into the separate fields of radiochemistry and nuclear chemistry, radiation chemistry, and the physics of fundamental particles. The historical outline that follows will serve as a foundation for the more detailed and modern aspects taken up in subsequent sections.

A. Electrical nature of matter

Various studies on the emission of electrified particles into the evacuated space around a hot electrode were made in the 19th century; the device was called a Crookes tube. It was not until about 1898, however, that J.J. Thomson carried out a series of experiments that allowed the determination of the ratio of charge to mass of these particles (now known as electrons). As illustrated in Fig. 22-1, a collimated beam from a hot cathode (called a *cathode ray beam*) is deflected by an electric field $\mathbf{E}$ or by a magnetic field $\mathbf{H}$. Application of the latter alone produces a force on the electrons in the beam given by $\mathbf{H}ev$, where e is the charge on the electron and v is its velocity; the particle follows a curved path such that this force is just balanced by the centrifugal force mv^2/r, where r is the radius of the path as determined from the deflection of the beam. If an electrical field is also applied, just sufficient to annul the deflection, then $\mathbf{E}e = \mathbf{H}ev$. Elimination of v between these two relationships gives $e/m = \mathbf{E}/\mathbf{H}^2 r$. The modern value of e/m for the electron is 5.273×10^{17} esu g^{-1}.

Thomson also obtained an approximate value for e. Later, in 1909, R. Millikan obtained a precise value by measuring the rate of fall of individual oil droplets in air and the electric field needed to just prevent their falling. The limiting velocity of fall v is such that the frictional force $6\pi\eta rv$ [as given by Stoke's law, Eq. (10-36)] is just equal to that due to gravity, where η is the viscosity of air. If an electric field is now applied that is just sufficient to keep

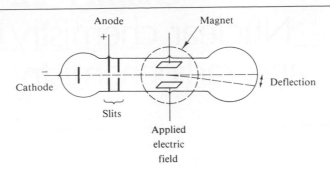

the drop from falling, it follows that $\mathbf{E}q = 6\pi\eta rv$, where q is the charge carried by the drop. Millikan determined the drop radius by microscopic examination and measured v and $\mathbf{E}$ to obtain q. The measured q values for various drops were integral multiples of a unit charge, that of the electron. Millikan's value for e of 4.77×10^{-10} esu differs from the modern one of 4.803×10^{-10} esu mainly due to an error in the then available value for the viscosity of air.

The procedure for determining e/m for the electron may be applied to positive ions. That is, positive ions generated by a hot filament or by collisions with a beam of electrons may be accelerated electrostatically, the accelerated ions collimated into a beam by means of slits, and the bending of the beam by a magnetic field then determined. The procedure was first applied by J. Aston in 1919, with the arrangement of Fig. 22-2. The figure shows, in exaggerated fashion, how the positive ion beam is first deflected by a small angle θ and then focused by a magnetic field. Aston's work led to the discovery of stable isotopes and then to the precise measurement of isotopic atomic weights.

Modern mass spectrometry is a highly instrumented science, now largely used in the analysis of the isotopic content of samples and in the study of the fragments produced in the electron bombardment of polyatomic molecules. Detailed, computer-assisted analysis of the distribution of the fragments from a sample consisting of a mixture of organic compounds allows its composition to be determined.

FIGURE 22-2
One of Aston's early mass spectrographs. The beam of positive ions is collimated by slits S_1 and S_2, then bent by an angle θ by the electric field of the charged plates at a. A magnetic field at b bends the beam back, the geometry working out so that the ions come to a focus at c.

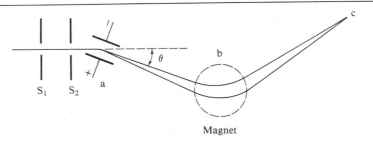

We turn now to a second line of discovery, that of radioactivity. W. Roentgen reported in 1895 that cathode rays (that is, an electron beam) generated a penetrating radiation on hitting a solid target. X rays, as they were called, could penetrate matter and were not deflected by electric or magnetic fields. The electromagnetic nature of x rays was later established, and in 1912 their wavelength was measured by crystal diffraction experiments (see Section 20-5).

It was at first thought that x rays were a kind of fluorescence, and it was for this reason that H. Becquerel looked for x radiation from naturally fluorescing minerals. Among others he tried potassium uranyl sulfate, $K_2UO_2(SO_4)_2 \cdot 2H_2O$. His great discovery, in 1896, was that this mineral emitted penetrating radiation without any prior exposure to light. He called the phenomenon *radioactivity*.

Chemical fractionation of UO_2^{2+} solutions led Marie Curie (working with her husband, Pierre Curie) to the discovery first of *polonium*, and then, in 1898, of *radium*. By 1902, pure radium had been isolated and its atomic weight determined.

Three types of radiation were characterized in due course. These are known as α, β, and γ radiation, in order of their penetrating power. We now know that they consist, respectively, of high-speed He^{2+} ions, electrons, and short-wavelength electromagnetic radiation. It was recognized very early that these radiations are emitted from individual atoms. The early radiochemists also appreciated that in the process the atom disintegrates or is converted to some other kind of atom. The development of the nuclear model for the atom led to the realization that α, β, and γ radiation came from the nucleus itself, and permitted a precise formulation of the radioactive decay process. For example, the first two steps of the uranium decay series are written

$$^{238}_{92}U \rightarrow {}^{4}_{2}He(\alpha \text{ particle}) + {}^{234}_{90}Th \tag{22-1}$$

$$^{234}_{90}Th \rightarrow {}^{0}_{-1}\beta(\beta^- \text{ particle}) + {}^{234}_{91}Pa \tag{22-2}$$

It is conventional to write the atomic number (or nuclear charge) Z as a lower left subscript and the *mass number A* (atomic weight to the nearest whole number) as a left superscript. Nuclear processes are balanced with respect to mass number and with respect to atomic number.

The energy of nuclear radiations is usually expressed in electron volts (eV): 1 eV is the energy gained by a particle of unit charge falling through a potential difference of 1 V; nuclear radiations may have energies as high as several million electron volts (MeV). It is not surprising that they ionize air or other matter in their path. A very useful application of this behavior is the *cloud chamber*, which was developed by C.T.R. Wilson in 1912. The principle is that if a vapor is supersaturated, then the ionization caused by high-speed particles induces condensation to occur along the track of each particle. A spray of α particle tracks is illustrated in Fig. 22-3; notice that each particle travels a definite length before stopping.

The cloud chamber helped E. Rutherford to make the discovery of nuclear transmutation; he observed tracks in a cloud chamber that terminated abruptly, with the appearance of a new track which could only be that of a proton. Such an event is shown in Fig. 22-3. The reaction is

FIGURE 22-3
Spray of α particle tracks
in a cloud chamber. Each
particle travels an
essentially straight line
path until it has lost nearly
all of its energy; the
bending that then occurs
is known as straggling.
Note that one particle has
undergone a large angular
deflection as a result of a
nuclear collision.

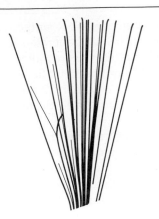

$$^{14}_{7}N + {}^{4}_{2}He \rightarrow {}^{1}_{1}H + {}^{17}_{8}O \tag{22-3}$$

A number of such transmutations were studied during the period 1919–1930 and eventually a new type of penetrating radiation was noticed. This was identified in 1932 by J. Chadwick as a neutral particle, which he named the *neutron*. A typical neutron-producing reaction is

$$^{10}_{5}B + {}^{4}_{2}He \rightarrow {}^{1}_{0}n + {}^{13}_{7}N \tag{22-4}$$

The actual atomic weight of the neutron is 1.0086654 (see Table 22-1). About the same time C. Anderson discovered another new particle, the *positron*, a particle identical to the electron except for having the opposite charge. A further major discovery made in this very active period was that of artificial radioactivity. Irene Curie (daughter of M. and P. Curie) and her husband F. Joliot (both known as Joliot-Curie) observed reactions such as the following:

$$^{24}_{12}Mg + {}^{4}_{2}He \rightarrow {}^{27}_{14}Si + {}^{1}_{0}n \tag{22-5}$$

The isotope produced, $^{27}_{14}Si$, is radioactive, emitting positrons,

$$^{27}_{14}Si \rightarrow {}^{27}_{13}Al + {}^{0}_{1}\beta \tag{22-6}$$

Up to this point the study of transmutation reactions was limited to the use of natural α particle emitters. Various electrostatic accelerating devices were developed, but the major advance came in 1932 with the invention of the *cyclotron* by E. Lawrence. The very clever principle involved was the following. If a charged particle is injected into the region of a magnetic field, it will follow a circular orbit of radius $r = mv/He$. The circumference of the orbit is $2\pi r$, so that the frequency of revolution of the particle is $\omega = v/2\pi r$, or $\omega = He/2\pi m$. The important point is that this frequency is independent of the velocity and hence of the energy of the particle. As illustrated in Fig. 22-4, the particles move inside hollow electrodes called "dees" (after their shape), crossing the gap between them twice on each revolution. Since the frequency of crossing is constant, one may impose an alternating potential on the dees such that the particle is accelerated each time it crosses the gap. Its actual path is therefore the spiral one shown in the figure, and the particle eventually emerges through a window (such as one of thin mica) with very high energy.

TABLE 22-1
Selected isotopic masses[a]

Species	Abundance (%)	Isotopic mass	Spin	Half-life	Decay mode[b]
e	—	0.0005486	$\frac{1}{2}$	—	—
$^1_0$n	—	1.0086654	$\frac{1}{2}$	12.8 min	β^- 0.782
$^1_1$H	99.985	1.0078252	$\frac{1}{2}$	—	—
^{2}H	0.015	2.0141022	1	—	—
^{3}H	—	3.0160494	$\frac{1}{2}$	12.26 yr	β^- 0.0186
$^3_2$He	~1 × 10^{-4}	3.0160299	$\frac{1}{2}$	—	—
^{4}He	100	4.0026036	0	—	—
^{6}He	—	6.01890	0	0.82 s	β^- 3.51
$^6_3$Li	7.42	6.015126	1	—	—
^{7}Li	92.58	7.016005	$\frac{3}{2}$	—	—
^{8}Li	—	8.022488	—	0.84 s	β^- 13
$^7_4$Be	—	7.016931	—	53.6 day	EC; γ 0.477
^{8}Be	—	8.005308	—	10^{-16} s	2 α's
^{9}Be	100	9.01218	$\frac{3}{2}$	—	—
^{10}Be	—	10.013535	—	2.5 × 10^6 yr	β^- 0.555
$^{10}_5$B	18	10.012939	3	—	—
^{11}B	80.39	11.0093051	$\frac{3}{2}$	—	—
^{12}B	—	12.0143535	—	0.020 s	β^- 13.37
$^{11}_6$C	—	11.011433	$\frac{3}{2}$	20.4 min	β^+ 2.1, γ 0.72
^{12}C	98.893	12.0000000	0	—	—
^{13}C	1.107	13.003354	$\frac{1}{2}$	—	—
^{14}C	—	14.0032419	0	5720 yr	β^- 0.155
$^{13}_7$N	—	13.005739	$\frac{1}{2}$	10.0 min	β^+ 1.19
^{14}N	99.634	14.0030744	1	—	—
^{15}N	0.366	15.000108	$\frac{1}{2}$	—	—
^{16}N	—	16.00609	—	7.38 s	β^- 4.26, 10.4, 3.3; γ 6.13 (others)
$^{15}_8$O	—	15.003072	$\frac{1}{2}$	2.0 min	β^+ 1.72
^{16}O	99.759	15.9949149	0	—	—
^{17}O	0.0374	16.999133	$\frac{5}{2}$	—	—
^{18}O	0.2039	17.9991598	0	—	—
^{19}O	—	19.003577	—	29.4 s	β^- 3.25, 4.60; γ 0.20 (m)
$^{20}_{10}$Ne	90.92	19.9923304	0	—	—
^{21}Ne	0.257	20.993849	$\frac{3}{2}$	—	—
^{22}Ne	8.82	21.991385	0	—	—
$^{22}_{11}$Na	—	21.994435	3	2.58 yr	β^+ 0.544; EC; γ 1.274
^{23}Na	100	22.989773	$\frac{3}{2}$	—	—
^{24}Na	—	23.990967	4	15.0 hr	β^- 1.39; γ 1.368, 2.753
$^{23}_{12}$Mg	—	23.99414	0	—	—
^{24}Mg	78.79	23.985045	0	—	—
^{25}Mg	10.13	24.985840	$\frac{5}{2}$	—	—
^{26}Mg	11.17	25.982591	0	—	—
^{27}Mg	—	26.984354	—	9.5 min	β^- 1.75, 1.59; γ 0.834, 1.015 (others)
$^{27}_{13}$Al	100	26.981535	$\frac{5}{2}$	—	—
^{28}Al	—	27.981908	—	2.3 min	β^- 2.87; γ 1.78
$^{27}_{14}$Si	—	26.98670	—	4.2 s	β^+ 3.85; (γ)
^{28}Si	92.21	27.976927	0	—	—
^{29}Si	4.70	28.976491	$\frac{1}{2}$	—	—

(continued)

TABLE 22-1
Selected isotopic masses[a]
(continued)

Species	Abundance (%)	Isotopic mass	Spin	Half-life	Decay mode[b]
^{30}Si	3.09	29.973761	0	—	—
$^{34}_{17}$Cl	—	33.97376	—	1.6 s	β^+ 4.4
^{35}Cl	75.53	34.968854	$\frac{3}{2}$	—	—
^{36}Cl	—	35.96831	2	3.0×10^5 yr	β^- 0.714; (EC); (β^+)
^{37}Cl	24.47	36.965896	$\frac{3}{2}$	—	—
^{38}Cl	—	37.96800	—	37.3 min	β^- 4.81, 1.11, 2.77; γ 2.15, 1.60
$^{39}_{19}$K	93.10	38.963714	$\frac{3}{2}$	—	—
^{40}K	0.0118	39.964008	4	1.27×10^9 yr	β^- 1.32; EC; γ 1.46; (β^+)
^{41}K	6.88	40.961835	$\frac{3}{2}$	—	—
^{42}K	—	41.96242	2	12.36 hr	β^- 3.55, 1.98; γ 1.52
$^{56}_{26}$Fe	91.66	55.93493	—	—	—
$^{57}_{27}$Co	—	56.93629	$\frac{7}{2}$	270 days	EC; γ (e$^-$) 0.122, e$^-$ (γ) 0.0144 (m), (γ 0.136)
^{59}Co	100	58.933189	$\frac{7}{2}$	—	—
^{60}Co	—	59.93381	5	5.26 yr	β^- 0.32; γ 1.173, 1.333
$^{88}_{38}$Sr	82.56	87.9056	0	—	—
$^{111}_{48}$Cd	12.75	109.90300	—	—	—
^{113}Cd^m	—	—	—	14 yr	β^- 0.57
$^{140}_{56}$Ba	—	—	—	12.8 days	β^- 1.02, 0.48; γ's
$^{140}_{57}$La	—	—	3	40.2 hr	β^- 1.34; γ's
$^{140}_{59}$Pr	—	139.90878	—	3.5 min	EC; β^+ 2.4; γ 1.2
$^{197}_{79}$Au	100	196.96655	$\frac{3}{2}$	—	—
$^{232}_{90}$Th	100	232.03821	—	1.39×10^{10} yr	γ 4.01, 3.95; e$^-$ (γ) 0.059
$^{238}_{92}$U	99.274	238.0508	—	4.51×10^9 yr	α 4.19 (others); (γ 0.045): (SF)

[a]From G. Friedlander, J.W. Kennedy, and J.M. Miller, "Nuclear and Radiochemistry," 2nd ed. Wiley, New York, 1964. Beyond oxygen only selected stable isotope masses are given.

[b]Energies of the indicated radiations are given in MeV; SF denotes spontaneous fission; the designation m stands for metastable; EC is electron capture.

FIGURE 22-4
Schematic arrangement of a cyclotron.

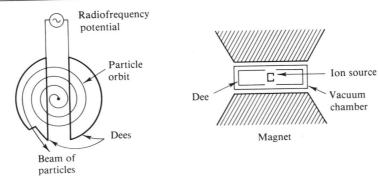

By using alternating voltages of 50,000–100,000 V, early cyclotrons quickly achieved the production of 20-MeV particles (protons and deuterons). Later models reached energies of several hundred MeV. Today we have very large doughnut-shaped cyclotron-type accelerators as well as linear accelerators in which the particle is accelerated through successive high-voltage gaps. Attainable energies are now in the GeV (giga-electron volt = 10^9 eV) range.

With the advent of artificially produced high-energy particles it became practical to generate high-intensity beams of neutrons and to study neutron-induced transmutations. Such studies culminated in 1939 with the discovery of uranium fission. Interestingly, it was a pair of chemists, O. Hahn and F. Strassmann, who made the discovery. The first nuclear chain reaction was demonstrated at the University of Chicago in 1942 and the first fission or "atomic" bomb in 1945. In 1952, an inventive contribution of E. Teller led to the first fusion or "hydrogen" bomb; the energy release comes from a reaction of the type

$$\begin{matrix} {}^2_1\text{H} + {}^2_1\text{H} \rightarrow {}^3_2\text{He} + {}^1_0\text{n} \end{matrix} \tag{22-7}$$

The period from 1950 to date has been marked more by a steady exploitation of past discoveries than by major new ones. Nuclear weaponry and nuclear power have become rather exact technologies. Many new transuranic elements have been discovered; accelerating devices have become incredibly powerful machines capable of generating a host of transient new particles, including mesons and baryons.

22-2 Nuclear energetics and existence rules

We consider nuclei to be made up of protons and neutrons moving in a mutual potential field. The number of protons must be just the nuclear charge Z and, since the proton and neutron both have a mass number of unity, the total number of *nucleons* must be A, the mass number of the nucleus. The number of neutrons N is then $A - Z$. Nuclei of the same Z but differing N are called *isotopes*; those of the same N but differing Z are called *isotones*.

Figure 22-5 shows a plot of Z versus N for the stable isotopes. It is evident that the existence of stable nuclei is not random but is confined to a rather narrow zone. An important corollary is that one can predict the type of radioactivity an unstable nucleus should show. Figure 22-6 shows a portion of the isotope chart; the shaded squares mark stable isotopes. Consider the nucleus ^{24}Na. Possible modes of disintegration are β^- emission, β^+ emission, electron capture (EC), α emission, and spontaneous fission. The last two modes are important only for heavy elements. The first three processes may be written for a general isotope ^A_ZI:

A. Existence rules for nuclei

$$^A_Z\text{I} \rightarrow {}_{Z+1}^A\text{I}' + \beta^- \tag{22-8}$$

$$^A_Z\text{I} \rightarrow {}_{Z-1}^A\text{I}' + \beta^+ \tag{22-9}$$

$$^A_Z\text{I} \rightarrow {}_{Z-1}^A\text{I}' \text{ (EC)} \tag{22-10}$$

Electron capture is a process whereby a K- or an L-orbital electron is acquired by the nucleus. As in positron emission, Z decreases by one unit.

FIGURE 22-5
A plot of Z versus N for stable nuclei. (From G. Friedlander, J.W. Kennedy, and J.M. Miller, "Nuclear and Radiochemistry," 2nd ed. Copyright 1964, Wiley, New York. Used with permission of John Wiley & Sons, Inc.)

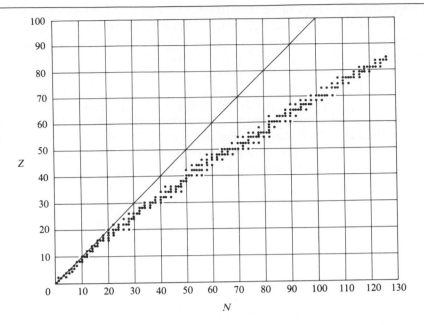

It is evident that $^{24}_{11}$Na is most unlikely to decay by either β^+ emission or EC, since the consequence would be to produce $^{24}_{10}$Ne, or an isotope yet further removed from the line of stability. We conclude (correctly) that the most likely process is that of β^- emission to give stable $^{24}_{12}$Mg. Similar reasoning leads to the conclusion that $^{22}_{11}$Na should decay by either positron emission or EC; the former is the observed process.

B. Nuclear energetics

The energy change in a nuclear reaction is large enough to be measured as a gain or loss of mass using the Einstein relativity relationship $E = mc^2$, where c is the velocity of light.† Accordingly, the table of isotopic masses (or "weights") (see Table 22-1) may be used to calculate energy changes in a transmutation process. One mass unit corresponds to $(1)(2.9979 \times 10^{10})^2/(6.0225 \times 10^{23}) = 1.492 \times 10^{-3}$ erg. This may be related to the electron volt unit of energy: 1 eV corresponds to 1.602×10^{-19} J; hence 1 MeV corresponds to 1.602×10^{-6} erg and the energy equivalent of one mass unit corresponds to 932 MeV.

Consider reaction (22-4). We calculate the energy release Q by adding the masses of the reactants and subtracting from the masses of the products:

$$\begin{array}{ccccccc}
^{10}_{5}\text{B} & + & ^{4}_{2}\text{He} & \rightarrow & ^{1}_{0}\text{n} & + & ^{13}_{7}\text{N} \\
10.012939 & & 4.002604 & & 1.008665 & & 13.005739 \\
& 14.015543 & & & & 14.014404 & \quad\text{[Eq. (22-4)]}
\end{array}$$

so that $Q = 14.015543 - 14.014404 = 0.001139$ mass units or $(1.139)(0.932) = 1.06$ MeV. Thus about 1 MeV of energy is released, in the form of kinetic energy of the products.

† A related consequence is that the mass of a particle increases with its velocity v and approaches infinity as $v \rightarrow c$. The equation obtained by Einstein is $m = m_0/[1 - (v/c)^2]^{1/2}$, where m_0 is the mass at $v = 0$, called the *rest mass*.

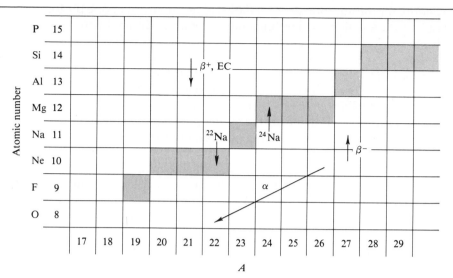

FIGURE 22-6
Illustration of how the atomic and mass numbers change for various types of disintegration process.

A similar calculation may be made for a disintegration process. Consider

$$^{14}_{6}C \rightarrow {}^{14}_{7}N + \beta^- \qquad (22\text{-}11)$$

From Table 22-1, $Q = 14.0032419 - 14.0030744 = 0.0001675$ mass units or 0.156 MeV. Most of this energy appears as kinetic energy of the β^- particle, although there is a small recoil energy of the $^{14}_{7}N$. Notice that the mass of the β^- particle is not used. This is because isotopic masses are given for the *neutral atoms*; the value for $^{14}_{6}C$ thus includes the mass of the six outer electrons, and that for $^{14}_{7}N$ includes the mass of its seven outer electrons. The extra electron in Eq. (22-11) is automatically taken care of. In the case of the positron emission, however, the mass of *two* electrons must be included in the products in order for the bookkeeping to be correct.

A useful quantity is the *mass defect* Δ, defined as $W - A$, where W is the atomic weight of the isotope. The demonstration that $Q = \Sigma\Delta_{\text{reactants}} - \Sigma\Delta_{\text{products}}$ is left as an exercise. It may be noticed from Table 22-1 that the mass defects are initially positive, decrease to a minimum of about -0.07 near iron, and then increase to about 0.05 near uranium. The initial decrease can be explained as reflecting the mutual attraction between protons and neutrons. The minimum and subsequent rise are due to the increasing mutual electrostatic repulsion of the protons. Also, as Z increases it becomes energetically favorable for nuclei to take on additional neutrons over the otherwise preferred 1:1 ratio, thus increasing the N/Z ratio and "diluting" the charge.

The energetics of nuclei can be treated in terms of the *nuclear binding energy* Q_B. This is defined as the energy released when a given isotope is assembled from the requisite number of protons and neutrons (actually, the neutral atom is assembled from hydrogen atoms and neutrons). Thus we write

$$3\,{}^{1}_{1}H + 4\,{}^{1}_{0}n \rightarrow {}^{7}_{3}Li \qquad (22\text{-}12)$$

$Q_B = (3)(1.007852) + (4)(1.0086654) - 7.016005 = 0.04221$ mass units or 39.3 MeV. This last corresponds to about 6 MeV per nucleon, a figure that is roughly constant for the light elements. We may also speak of the binding energy for the addition or the removal of one neutron or one proton. A general observation is that the binding energy for an additional neutron is again 6 to 7 MeV.

Empirical observation indicates that nuclei having 2, 8, 20, 28, 50, 82, or 126 protons or neutrons are especially stable. These numbers have been called "magic" or closed-shell numbers. The natural isotopic abundance of stable closed-shell nuclei is unusually high, as, for example, ^4_2He, $^{16}_8O$, and $^{40}_{20}Ca$, for which both proton and neutron numbers correspond to closed-shell values, and $^{88}_{38}Sr$ and $^{208}_{82}Pb$, for which the neutron number is magic. Tin, with $Z = 50$, has no less than ten stable isotopes, which is taken as an indication of the stabilizing effect of the presence of a closed shell of protons. The two pips on the fission yield plot of Fig. 22-7 correspond to neutron numbers of 50 and 82—again closed-shell numbers.

The presence of magic numbers has been explained in terms of a relatively simple wave mechanical model of the nucleus which assumes that protons and neutrons move in a spherically symmetric potential. The energy states are given by quantum numbers n and ℓ, defined similarly to those for the hydrogen atom, although n does not now limit the possible ℓ values. One can have 1s, 1p, 1d, 1f, ... states. In this scheme protons and neutrons have separate sets of energy levels, and the magic numbers are found to correspond to certain groupings of filled n and ℓ shells. More elaborate theories permit the nucleus to be deformed from a spherical shape and allow yet more detailed calculations of nuclear properties.

22-3   Nuclear reactions

A. Types of reactions   Observable nuclear transmutation reactions were at first confined to the Rutherford type, illustrated by Eq. (22-3). A given element was bombarded with α particles, or, with the advent of accelerators, with protons or deuterons.

FIGURE 22-7
Mass distribution of the fission products of ^{235}U. Note the two small peaks at "magic number" values of A (Section 22-2B). (From G. Friedlander, J.W. Kennedy, and J.M. Miller, "Nuclear and Radiochemistry," 2nd ed. Copyright 1964, Wiley, New York. Used with permission of John Wiley & Sons, Inc.)

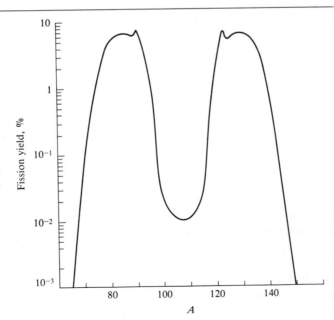

The products typically consist of another light particle and a new element. Such reactions can be written $^A_Z I(x,y)^{A'}_{Z'} I'$, where x might be α, p (proton), or d (deuteron) and y might be α, p, or n. With the development of neutron sources, such as from (α, n) reactions, x could be n and y could be α, p, or γ. In this last case the neutron simply adds to the initial isotope and the binding energy appears as a γ photon, usually of 6 to 7 MeV energy. A further possibility for neutrons is the $(n, 2n)$ reaction, for which Q is usually -6 to -7 MeV.

The energy of the bombarding particle x must of course be at least equal to the Q of the reaction if Q is negative. However, if x is a charged particle, its energy must be at least several MeV even though Q is positive. The reason is that the particle x must have sufficient energy to surmount the Coulomb repulsion between it and the nucleus. An empirical formula gives

$$E_{\text{barrier}} = \frac{0.9zZ}{A_x^{1/3} + A_I^{1/3}} \qquad (22\text{-}13)$$

where z is the charge number of x and E is in MeV. Thus the Coulomb barrier for Eq. (22-3) is 3.15 MeV. The Q for the reaction is -1.19 MeV, so that in this case the barrier determines the minimum or *threshold* energy of the α particles.

With the development of the cyclotron and of other high-energy accelerators it became possible to make bombardments with particles of much higher energy than the 5 to 10 MeV needed for simple (x,y) reactions. Roughly speaking, each additional 7 to 10 MeV allows the escape of one additional small particle. With 30-MeV αs, for example, reactions of the type $(\alpha, 3n)$ and/or $(\alpha, 2p)$ become possible. If yet higher energies are used, the number of multiple small product nuclei increases. The process is now called a *spallation*. The mechanism is thought to be one in which the x particle passes through the nucleus, knocking out bits and pieces on the way.

B. Fission

Nuclei of very large atomic number show a new type of process: fission. The mutual Coulomb repulsion of the protons in such nuclei is so large that a spontaneous breakup into two approximately equal fragments occurs. For example, an isotope of californium $^{254}_{98}\text{Cf}$ has a half-life for spontaneous fission of about 60 days.

It turns out that $^{235}_{92}\text{U}$ is close to the point of being able to undergo spontaneous fission. The binding energy of about 7 MeV gained by the addition of a neutron is sufficient to make the product nucleus, $^{236}_{92}\text{U}$, unstable. A typical fission process is

$$^{235}_{92}\text{U} + ^1_0\text{n} \rightarrow [^{236}_{92}\text{U}] \rightarrow ^{140}_{56}\text{Ba} + ^{93}_{36}\text{Kr} + 3\,^1_0\text{n} \qquad (22\text{-}14)$$

The Q for the process is about 200 MeV; uranium has a large positive Δ while the two fission products have negative Δs. Equation (22-14) is merely typical; the fission may occur in various ways, and the observed distribution of fission products is shown in Fig. 22-7. Since fission products tend to be neutron-rich, they lie below the line of stable isotopes and hence are β^- emitters. The fission product ^{140}Xe undergoes a succession of β^- decays, for example, until stable ^{140}Ce is finally reached.

Equation (22-14) makes clear the stoichiometric basis for a fission chain reaction. Each neutron that induces fission leads to the production of two or three new neutrons, or to one or two extra ones. A brief discussion of nuclear reactors is given in the Commentary and Notes section.

C. Reaction probabilities

There are two distinct situations involved in the treatment of the probability of a nuclear transmutation reaction. The first is that dealing with charged particles. As discussed in Section 22-4, such particles steadily lose energy through Coulomb interactions with the orbital electrons of the atoms of the absorbing medium. As a consequence, charged particles such as α particles possess a definite range or path length. If a target thicker than this range is used, one observes that a certain fraction of the particles produce a transmutation. This fraction, called the *yield*, is typically around 10^{-6}.

In the case of neutrons, however, there is no Coulomb interaction with orbital electrons of the absorber, and neutrons disappear only by nuclear collisions (their natural decay time is long enough not to compete appreciably with transmutation processes). We write for each layer dx of absorbing medium

$$-dI = In\sigma\, dx \qquad (22\text{-}15)$$

where I is the number of neutrons incident per square centimeter per second, n is the number of target nuclei per cubic centimeter, and σ is the probability of a neutron capture process. The quantity σ is called a *cross section* and is usually expressed in units of 10^{-24} cm^2, called a *barn*† (see also Section 3-2). Cross sections for transmutation reactions are often around 1 b so that σ corresponds roughly to the physical target area of a nucleus. However, σ for some (n, γ) reactions is hundreds to thousands of barns. Integration of Eq. (22-15) gives

$$k = I_0 - I = I_0(1 - e^{n\sigma x}) \qquad (22\text{-}16)$$

where k is the rate of reaction per square centimeter of target bombarded.

EXAMPLE

Suppose that a neutron beam of 10^8 neutrons cm^{-2} s^{-1} strikes a gold sheet 0.2 mm thick and 2 cm^2 in area. The capture cross section for the reaction $^{197}_{79}$Au(n, γ)$^{198}_{79}$Au is 100 b. The density of Au is 19.3 g cm^{-3} and its atomic weight is 197.2. Thus $n = (19.3)(6.02 \times 10^{23})/197.2 = 5.89 \times 10^{22}$ cm^{-3} and application of Eq. (22-16) gives $(10^8)(2)\{1 - \exp[-(5.89 \times 10^{22})(100)(10^{-24})(0.02)]\} = 2.22 \times 10^7$ ^{198}Au nuclei formed per second.

Neutrons may lose their kinetic energy by collisions with nuclei, and if this occurs in a medium for which the capture cross section is small (as in graphite or heavy water), the neutrons may reach thermal energy, that is, an average energy corresponding to the ambient temperature. Nuclear reactors, for example, are often constructed so that most of the fission product neutrons are reduced to thermal energy before they undergo any capture process. A sample inserted in a reactor is therefore immersed in some concentration n of thermal neutrons. The number hitting a target of 1 cm^2 cross section of sample is then

† The story, probably not apocryphal, is that in the early days of the atomic energy project, boron was a serious impurity because of its large capture cross section for neutrons. E. Fermi is said to have exclaimed at one point that boron had a cross section as big as a barn—and the term became the unit of cross section.

nv per second, where v is the average neutron velocity. The quantity **n**v is called the *neutron flux*; a typical value for an experimental nuclear reactor is 10^{12} to 10^{14} neutrons cm^{-2} s^{-1}. This quantity then replaces I in Eq. (22-15).

Suppose that 2 g of NaCl is placed in a nuclear reactor whose slow neutron flux is 10^{13} neutrons cm^{-2} s^{-1}. The cross section of ^{23}Na for the reaction ^{23}Na(n,γ)^{24}Na is 0.536 b. The rate of production of ^{24}Na is then $k = [2(6.02 \times 10^{23})/58.5]10^{13}$ $(0.536 \times 10^{-24}) = 1.10 \times 10^{11}$ atoms s^{-1}.

EXAMPLE

22-4   Absorption of radiation

Radiation, such as α particles, β particles, and γ rays, is absorbed by matter due to interactions with orbital electrons. All such particles dissipate their energy to form ions, that is, positive and negative ion pairs, in the absorbing medium. On the average, 35 eV is expended per ion pair; a 1-MeV particle produces about 30,000 ion pairs before its energy is lost. However, while the overall process is the same, the rate and the mechanism vary considerably.

Charged particles lose energy continuously as they pass through matter, as a result of Coulomb interactions with the orbital electrons of atoms. As a consequence, a charged particle exhibits a definite range or distance it can traverse before coming to rest. In the case of heavy particles, such as high-speed α particles, protons, deuterons, and so on, the path is nearly straight, as illustrated in Fig. 22-3. Heavy charged particles are not very penetrating. The range of a 5-MeV α particle is only about 3 cm in air, for example; such particles cannot penetrate skin and would not be dangerous externally.

A. Charged particles

Beta particles also have a definite range. An empirical equation by N. Feather gives

$$R = 0.543E - 0.160 \tag{22-17}$$

where R is the range in aluminum in grams per square centimeter and E is in million electron volts. Note that R is about 100 times that for an α particle of the same energy. The range of β particles is hard to measure, however, since, being very light, they are easily deflected by the atoms of the absorber and therefore pursue a tortuous path. If the radiation from a β^- emitter is measured through successive increases in thickness of absorber, the intensity is found to diminish approximately exponentially with distance for two or perhaps three half-thicknesses before beginning its rapid drop to zero as the range is approached.

There is a second reason for the approximately exponential absorption curve for beta emitters. It turns out that the energy of a β^- particle emitted from a given nucleus may have a value ranging from near zero up to a maximum, as illustrated in Fig. 22-8. It is this maximum energy that appears in Eq. (22-17). The reason for the energy distribution is that the total decay energy is shared with a *neutrino* that is also emitted. The neutrino is a neutral particle of zero or near zero rest mass and spin $\frac{1}{2}$; its interaction with matter, although very weak, has been detected.

The situation with positrons is very similar to that with β^- particles; similar absorption curves are found and a similar range–energy relationship. The

FIGURE 22-8
Shapes of β-particle
energy distributions.

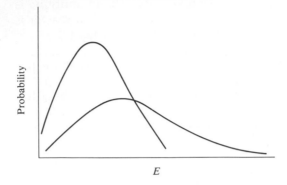

FIGURE 22-8
Shapes of β-particle
energy distributions.

positron, however, being "antimatter" in nature, eventually fuses with an electron, and the combined mass energy is converted into two γ quanta. The atomic weight of an electron or of a positron is 0.00055 mass units, corresponding to 0.5 MeV. Thus positron emission is always accompanied by 0.5-MeV γ radiation, known as *annihilation* radiation.

B. Electromagnetic radiation

Gamma and x rays, being electromagnetic in nature, obey the same Beer–Lambert absorption law as does ordinary light (Section 3-2). The equation

$$I = I_0 e^{-\mu x} \tag{22-18}$$

applies, where μ is the linear absorption coefficient. The corresponding half-thickness is $0.693/\mu$. Figure 22-9 shows the variation of half-thickness with energy for various absorbers. For example, 1-MeV γ radiation has a half-thickness of about 10 g cm^{-2} in aluminum.

The absorption process is primarily one of ionizing orbital electrons of the atoms in the absorbing material. The gamma quantum may impart all of its energy to the electron, in the *photoelectric* effect, or it may make an elastic collision with it, resulting in an accelerated electron and a gamma quantum of reduced energy, in the *Compton effect*. Gamma quanta of above 1.1 MeV energy may also, by interaction with the field of a nucleus, create an electron–positron pair; the effect is known as *pair formation.*

If we consider gamma quanta of successively lower energy, we reach the x-ray region. The energy becomes comparable to that required to ionize a K or an L electron of the absorbing medium. As Fig. 22-9 shows, the half-thickness suddenly rises when the energy falls to just *below* that necessary to ionize a particular type of electron. This happens first at the ionization energy of the K electrons of the absorber, then at that of the L electrons, and so on. The effect is known as *critical absorption;* it allows one to select combinations of absorbers that will preferentially pass a particular energy range of x rays.

X-ray emission occurs, of course, when an L electron falls into a K vacancy, or an M electron into an L vacancy. An alternative to such emission is the ejection of an outer electron in an intra-atomic photoelectric process known as the *Auger* effect.

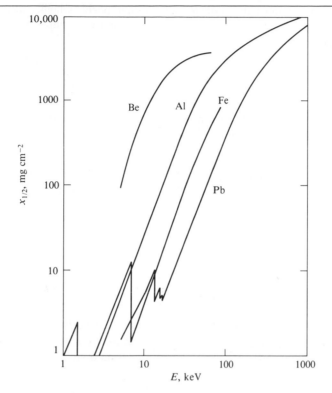

FIGURE 22-9
Half-thickness for the absorption of γ or x radiation in various substances. [Data from "Handbook of Chemistry and Physics" (C. D. Hodgman, ed.), 44th ed. Chem. Rubber Publ., Cleveland, Ohio, 1963.]

Radiation *dosimetry* is the measurement of the amount of energy expended in absorbing material; 1 *rad* is that radiation dose which deposits 100 erg in 1 g of material. An earlier, more complicated unit is the *roentgen* (R), defined as the quantity of γ or x radiation that produces ions carrying 1 esu of electric charge per cubic centimeter of dry air at STP; 1 R corresponds to the absorption of about 84 erg of energy. The roentgen and the rad are thus roughly equivalent; 1 g of radium gives a dose of about 3 rad hr^{-1} at a point 1 m away.

C. Dosage

The biological hazard of radiation varies with the type. Neutrons cause more damage per rad than does γ radiation, for example. The dosage in rads is therefore scaled accordingly to give dosage in *rems* (roentgen equivalent man). One rad of γ or of β$^-$ radiation is equivalent to about 1 rem, but 1 rad of neutron radiation corresponds to 10 rem.

The allowed industrial radiation level is about 3 mrem hr^{-1} (mrem = millirem) for whole body exposure and about 30 mrem hr^{-1} for hands or feet only. The total *accumulated* dose should not exceed, in rem, about five times the number of working years or the person's age minus 20. Acute and chronic exposure have different effects. A person might receive 150 rem over a working career with no harm, but 150 rem received all at once would cause some radiation sickness. The median lethal *acute* dose is about 500 rem (whole body exposure).

The dosage from cosmic radiation, natural radioactivity in the earth and in bricks, and so on amounts to about 0.1 rem yr^{-1} or about 0.003 of the maximum industrial exposure. Radiation due to fallout from nuclear testing is even less than this. The matter has been much debated, of course, but it is questionable whether such low levels of radiation have even a statistical effect on human longevity or health [see, for example, Holcomb (1970)].

Radiation chemistry is the study of chemical change induced by high-energy radiation. It is customary to measure efficiencies in terms of the number of molecules destroyed or reacted per 100 eV of energy absorbed; this is called the G value. For example, irradiation of aqueous solutions leads primarily to ionization and dissociation of water as the primary processes. A typical sequence of events is that assumed for dilute, air-saturated, and acidic ferrous sulfate:

$$H_2O \rightarrow H + OH \qquad OH + Fe^{2+} \rightarrow Fe^{3+} + OH^-$$
$$H + O_2 \rightarrow HO_2 \qquad H^+ + HO_2 + Fe^{2+} \rightarrow H_2O_2 + Fe^{3+}$$
$$H_2O_2 + Fe^{2+} \rightarrow OH + Fe(OH)^{2+} \qquad \text{(and so on)}$$

Thus for every H_2O molecule decomposed in the primary reaction, four Fe^{2+} ions are oxidized.

22-5   Kinetics of radioactive decay

Nuclear disintegration or decay is a statistical event. With a large enough sample, however, the observed decay rate approximates the most probable one and we can therefore treat it as a simple first-order rate process. Thus

$$D = -\frac{dN}{dt} = \lambda N \tag{22-19}$$

where D is the disintegration rate, N is the number of atoms present, and λ is the decay constant. This is the same as Eq. (14-8), and the integrated forms, Eqs. (14-9) and (14-10), also apply. Equation (14-13) gives the half-life as $t_{1/2} = 0.6931/\lambda$.

One ordinarily measures a radioactive material by its disintegration rate rather than by the number of atoms present. The *curie* (Ci) is defined as 3.700×10^{10} disintegrations s^{-1}. As an example, the disintegration rate of 1 g of ^{14}C may be calculated by means of Eq. (22-19). The number of atoms present is $(1)(6.02 \times 10^{23})/14 = 4.30 \times 10^{22}$; $t_{1/2} = 5720$ yr, so that

$$\lambda = 0.6931/(5720)(365.3)(24)(60)(60) = 3.8 \times 10^{-12} \ s^{-1}$$

Then $D = (4.30 \times 10^{22})(3.8 \times 10^{-12}) = 1.65 \times 10^{11}$ dis s^{-1} g^{-1} or 4.46 Ci g^{-1}.

An important special case is that in which a radioactive species is produced at some constant rate k. The differential equation is

$$\frac{dN}{dt} = k - \lambda N \tag{22-20}$$

which integrates to give

$$D = k(1 - e^{-\lambda t}) \tag{22-21}$$

In the example of Section 22-3C, the rate of production of ^{24}Na is 1.10×10^{11} s^{-1}. The half-life of ^{24}Na is 15.0 hr, and, for example, a 15 hr irradiation would yield $D = k/2$ or 5.5×10^{10} dis s^{-1}. With a sufficiently prolonged irradiation, D approaches k. The *saturation* activity is then 1.10×10^{11} dis s^{-1}.

   A second common situation is that in which a mixture of radioactive species is present. The case of two species, 1 and 2, is illustrated in Fig. 22-10. Each decays independently, and $D_{tot} = D_1 + D_2$. If species 1 is the longer-lived, then $D_{tot} \rightarrow D_1$ at large times. As illustrated in the figure, subtraction of the D_1 line from D_{tot} gives D_2; we thus find $D_1{}^0 = 2000$ dis s^{-1}, $t_{1/2}(1) = 3$ hr, and $D_2{}^0 = 4000$ dis s^{-1}, $t_{1/2}(2) = 0.5$ hr.

   The third case to be considered is that in which a parent radioactive species 1 decays to a daughter 2, which is also radioactive,

$$\text{species } 1 \overset{\lambda_1}{\rightarrow} \text{ species } 2 \overset{\lambda_2}{\rightarrow} \text{ stable product}$$

This situation is discussed in Selection 15-ST-1B; Eq. (15-95) may be written

$$D_2 = \frac{\lambda_2 D_1{}^0}{\lambda_2 - \lambda_1}\left(e^{-\lambda_1 t} - e^{-\lambda_2 t}\right) \tag{22-22}$$

If $\lambda_2 > \lambda_1$, then at large times Eq. (22-22) reduces to that for *transient equilibrium*

$$D_2 = \frac{\lambda_2}{\lambda_2 - \lambda_1}D_1 \tag{22-23}$$

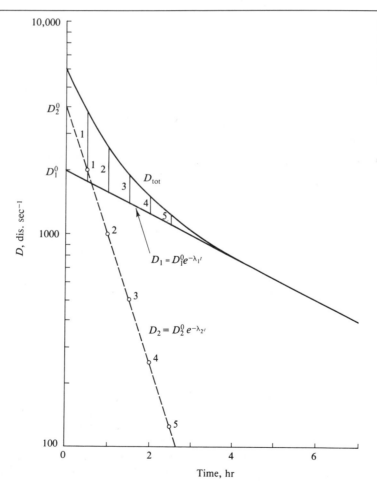

FIGURE 22-10
Analysis of the composite decay curve for two independently decaying species.

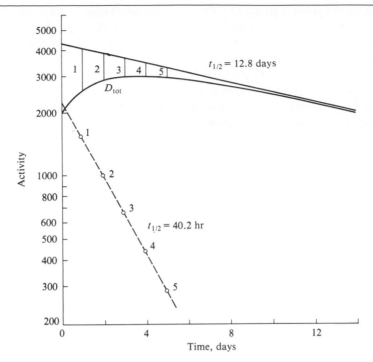

FIGURE 22-11
Analysis of the composite decay curve for the parent–daughter sequence
$$^{140}Ba \xrightarrow{12.8 \text{ day}} {}^{140}La \xrightarrow{40.2 \text{ hr}} {}^{140}Ce.$$

If $\lambda_2 \gg \lambda_1$, then at times large compared to $1/\lambda_2$, species 1 has still not appreciably decayed and $D_2 = D_1^0$. This situation is known as one of *secular equilibrium*.

The experimental application of Eq. (22-22) may be illustrated as follows. We suppose that a sample of pure ^{140}Ba has been isolated. It decays with a half-life of 12.8 days into daughter ^{140}La, whose half-life is 40.2 hr; both emit β^- particles. The plot of D_{tot} versus time is shown in Fig. 22-11; D_{tot} goes through a maximum due to the growth of the daughter ^{140}La but eventually becomes linear with time in the semilogarithmic plot. That is,

$$D_{tot}(t \text{ large}) \rightarrow D_1^0 \, e^{-\lambda_1 t} \left(1 + \frac{\lambda_2}{\lambda_2 - \lambda_1} \right) \tag{22-24}$$

The slope of the limiting line thus gives λ_1, corresponding in this case to $t_{1/2} = 12.8$ days. Subtraction of the limiting line from D_{tot} yields a difference line whose slope gives λ_2, corresponding to $t_{1/2} = 40.2$ hr.

COMMENTARY AND NOTES

22-CN-1 Theories of radioactive decay

The theoretical treatment of radioactive decay is in some ways similar to that of emission from a molecular excited state. Thus the emission of a gamma quantum from a nucleus is essentially a nuclear fluorescence or phosphorescence (see Section 19-4, although it is usually called an *isomeric transition* (IT). The emission is subject to selection rules; its probability depends on the change in nuclear angular momentum Δl

that occurs, and on whether or not a change occurs in the electric charge distribution in the nucleus. An allowed transition occurs in about 10^{-13} s, but if Δl is large, the half-life can be quite large. Thus, referring to Table 22-1, ^{113}Cd^m decays to ground-state ^{113}Cd with a half-life of 14 yr.

A complication is that an alternative to actual gamma emission is the ejection of an orbital electron, which carries off the energy of the isomeric transition. The effect is called *internal conversion* (IC), and the probability that the decay will take this route increases with increasing Δl. Thus in the case of ^{113}Cd^m the actual emission is by IC, with monoenergetic 0.58-MeV electrons produced.

Beta decay is also treated as an emission process, but with the complication that there is a simultaneous emission of two particles from the nucleus: an electron and a neutrino. The theoretical treatment of the lifetime of beta decay is somewhat complicated, but again the half-life increases with increasing Δl for the transition.

The treatment of α emission is quite different from the preceding two cases sine the particle emitted is essentially a piece of the nucleus itself. The situation is sketched in Fig. 22-12. If we consider the process $^A_Z I \rightarrow ^{A-4}_{Z-2} I' + ^4_2 He$, the reverse reaction has a $Q_{barrier}$ as given by Eq. (22-13). In the case of $^{234}_{90}$Th this barrier amounts to about 20 MeV, so that it would require a 20-MeV α particle to enter the $^{234}_{90}$Th nucleus to give $^{238}_{90}$U. Yet the reverse process is spontaneous and the emitted α particle has *only* 4 MeV energy. Clearly the emitted particles are never at the potential energy of the top of the Coulomb barrier.

The explanation of this energy paradox con-stituted one of the first great triumphs of wave mechanics. The theory is that of a particle in a box with a finite barrier (note Section 16-5B). As illustrated in Fig. 16-23, it turns out that the wave function for the particle (the α particle in this case) has a finite value outside of the box—the walls of the box being the Coulomb barrier of Fig. 22-12. Integration of the square of the wave function over the region of the barrier gives the probability of the α particle being outside of the box or nucleus, and hence of being emitted. The actual equation is of the form of Eq. (16-138).

22-CN-2 Nuclear reactors and "Atomic" bombs

A nuclear reactor is a device that allows a nuclear fission reaction to be self-sustaining. We consider first the problem of arranging this situation in the case of natural uranium, which contains 99.27% ^{238}U and 0.72% ^{235}U. According to Eq. (22-14), the slow neutron fission of ^{235}U produces additional neutrons that could, in turn, lead to further fissions. The problem is that ^{238}U also captures neutrons:

$$^{238}_{82}U + ^1_0 n \rightarrow ^{239}_{92}U \xrightarrow[\text{23.5 min}]{\beta^-} ^{239}_{94}Np \xrightarrow[\text{23.5 day}]{\beta^-} ^{229}_{94}Pu$$

$$(22\text{-}25)$$

The result is the production of plutonium (half-life 2.44×10^4 yr). The capture cross section of ^{238}U for neutrons goes through a peak or *resonance* absorption region as the neutrons lose their energy, and the consequence is that a sample of natural uranium is unable to sustain

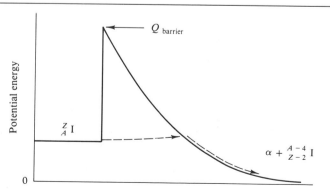

FIGURE 22-12
Quantum mechanical model for α emission.

a chain reaction; too many neutrons go into reaction (22-25).

The solution to this problem that was found during World War II was to place the uranium in a lattice of highly purified graphite. Carbon does not absorb neutrons strongly, and the neutrons produced by the fission of ^{235}U wander through the graphite until they are slowed down by collisions to thermal energies before again encountering uranium atoms. Thus the dangerous resonance absorption region is spent in graphite and away from ^{238}U. A sufficient fraction of the neutrons produced by each act of ^{235}U fission now survives to carry the chain reaction. This fraction is called the *reproduction constant k*.

If k is greater than unity, a chain reaction should occur. In any actual reactor, however, there is a loss of neutrons from the surface into surrounding space, so that the finite reactor has a practical reproduction constant $k' < k$. The importance of the surface effect diminishes with increasing size of the reactor, so that if a particular infinite reactor has $k > 1$, then there will be some critical size of an actual structure such that k' is just equal to unity.

The lifetime of a neutron is about 10^{-5} s; this is the time from its formation in a fission event to its eventual capture by another ^{235}U nucleus. Since the concentration of neutrons in a reactor should increase by the factor k' with each generation, this means that the neutron concentration and hence the power output of the reactor should increase by $(k')^{10^5}$ per second. If this were actually the case, the situation would be extremely dangerous. As soon as construction of the reactor exceeded its critical size, the power would rise catastrophically in a fraction of a second. No actual explosion would occur in ordinary reactors; the reactor would be ruined, however.

There is a most fortunate natural situation that prevents this catastrophe from happening. It so happens that about 1% of the neutrons produced in fission are tied up in isotopes called delayed neutron emitters. These are energetic β^- emitters decaying to excited nuclear states that then promptly emit a neutron. The half-lives of the β^- emitters range from milliseconds to seconds. Thus if k' is less than about 1.01, the effective generation time is not 10^{-5} but about 1 s. A reactor that is just critical in size thus increases its power only slowly and may easily be controlled by moving neutron-ab-

sorbing control rods inward so as to bring k' back to unity whenever the power level is to be stabilized.

The original reactors of World War II were intended not to produce power, but to produce ^{239}Pu by reaction (22-25). The reason was that ^{239}Pu was expected to fission easily, like ^{235}U, and being chemically different from uranium, could be concentrated into the pure element. The alternative approach used was to fractionate natural uranium into relatively pure ^{235}U by isotopic fractionation procedures.

Atomic bombs (strictly speaking, nuclear bombs) consist of relatively pure ^{239}Pu or ^{235}U. One modern arrangement is to have a sphere of the material that has *just* too much surface area for k' to exceed unity. The sphere is surrounded by a conventional high explosive and when this is detonated it compresses the sphere, thereby reducing its area and making k' greater than unity. The ensuing chain reaction is very rapid in this case since the neutrons need not be slowed down; the generation time is much less than 10^{-5} s and the result is the incredible nuclear explosion.

Contemporary power-producing nuclear reactors may use plutonium or ^{235}U enriched uranium. Uranium- or plutonium-containing ceramic rods may, for example, be spaced in a bath of heavy water, or, if sufficiently enriched, in ordinary water. The fission energy appears ultimately as heat, of course, and heat exchangers transfer this heat to a working fluid that is then used for power generation.

22-CN-3 Nuclear chemistry

The term nuclear chemistry applies to the study of the properties of nuclei and of "fundamental" particles. It is possible, for example, to draw inferences about the size, shape, and ease of deformation of nuclei from a study of how they scatter high-energy particles such as protons or neutrons. It appears that most nuclei are not spherical but are prolate ellipsoids. Scattering experiments also allow a mapping of nuclear energy states. These states are quantized, and charts of them look much like those from atomic spectroscopy.

We have considered nuclei to be made up of neutrons and protons, with the implication that these last are fundamental particles. Actually, the experiments of high-energy physics have disclosed some 80 "fundamental" particles. These may be classified as *baryons*, which are heavy particles and include the proton and the neutron, *mesons*, whose rest masses range down to about 0.1 mass units, and *leptons*, which comprise the electron, the *mu* particle, and the neutrino. The study of the properties of these particles has led to the formulation of several quantities that are conserved in nuclear processes. In addition to charge, mass number, spin angular momentum, and parity, quantities called hypercharge, isotopic spin, and strangeness have been defined.

It also appears that for every elementary particle there is an antiparticle; the two annihilate each other in a collision. Thus the positron is the antimatter equivalent of the electron. Antiprotons and antineutrons as well as various other antibaryons have been discovered. It is possible, then, to imagine a hydrogen atom consisting of an antiproton nucleus and an orbital positron—or, indeed, an antimatter universe. The detection of astronomical antimatter seems impossible at present. Stellar spectra cannot be used, for example, since these should be invariant with respect to the type of matter involved.

22-CN-4   Quantum statistics

The usual Boltzmann expression is actually a limiting case of more general expressions. Particles whose wave function is antisymmetric, such as the electron or proton, obey the Pauli exclusion principle in that no two can have the same wave function. In an assemblage of such particles, detailed analysis gives the probable number N_i or particles having energy ϵ_i as

$$N_i = \frac{g_i}{\alpha e^{\epsilon_i/kT} + 1} \qquad (22\text{-}26)$$

where g_i is the degeneracy of the state and α is the normalization factor required to make $\Sigma N_i = 1$. Equation (22-26) reduces to the Boltzmann expression if $\alpha e^{\epsilon_i/kT} \gg 1$. It turns out that α can be identified with the reciprocal of the absolute or rational activity (Section 9-CN-3), and the condition for Eq. (22-26) to reduce to the Boltzmann law is therefore either that $\epsilon_i \gg kT$ or that there are many particles, so that α is

large. The analysis giving Eq. (22-26) is known as *Fermi–Dirac* statistics. Particles to which it applies are called *fermions*.

If a particle contains an even number of fermions, such as deuterium or a helium nucleus, then its wave function is symmetric and the Pauli exclusion principle does not apply; any number of such particles may be in a given state. The change puts -1 instead of $+1$ in the denominator of Eq. (22-26). The analysis is known as *Bose–Einstein* statistics and the particles obeying these statistics are called *bosons*. Interestingly, it is found that photons obey Bose-Einstein statistics (Berry *et al.*, 1980).

22-CN-5   Experimental detection methods

Most detection devices are based on the ability of high-energy radiation to ionize matter. A quartz fiber may be electrostatically charged, for example, so as to be repelled from its support. Radiation ionizes the air, rendering it somewhat conducting so that the charge leaks away and the fiber slowly returns to its rest position. The rate of such return is a measure of the intensity of the radiation. Such *electroscopes* are widely used for personal dosimeters; they can be constructed in the size and shape of a fountain pen.

Photographic film is exposed by high-energy radiation and is much used, of course, in medical x-ray examinations. Film has the two advantages of providing a cumulative measure of radiation and of mapping its intensity distribution. However, the photographic method is inconvenient to use for accurate measurements or where discrimination among different kinds of radiation is desired.

The ionization produced by radiation may be measured electronically. A potential of several hundred volts is applied through a high resistance to a pair of electrodes separated by an air gap. The ions produced by the radiation collect on the electrodes to generate a small current, which drops the voltage, and this voltage change can be amplified by a dc amplifier. Since each high-speed particle produces a burst of ions, the current actually occurs in pulses, and each pulse can be amplified by means of an ac amplifier. Discrimination between different kinds of particles is now possible since the pulse sizes will be different. It is thus possible to "count" α particle pulses in the presence of a heavy background of β⁻ or of λ radiation.

If the potential between the electrodes is raised to about 1000 V, the ions produced by the radiation are so strongly accelerated as to produce further ionization. The result is a cascade of ions and a very large pulse—one that can be detected with little amplification. The device is known as a *Geiger* counter. It does not, of course, discriminate among kinds of particles since all pulses are now the same.

Perhaps the most common detection device is the *scintillation counter*. Materials such as crystalline anthracene and sodium iodide [with a trace of Tl(I)] emit light under irradiation; each high-speed particle produces a quick burst of photons and the light pulse is detected by means of a photomultiplier. The use of NaI(Tl) is particularly advantageous for the detection of γ rays since the heavy I atoms absorb electromagnetic radiation strongly. Moreover, each photoelectron has the energy of the γ quantum, and the light pulse resulting from such electrons is proportional to their energy. It is thus possible to measure the energy spectrum of incident γ radiation. Semiconductors have also been used. Each ionizing particle produces a current pulse very nearly proportional to its energy.

22-CN-6   The Mössbauer effect

Gamma rays emitted in an isomeric transition ordinarily have less energy than the true difference between the two nuclear states involved. Conservation of momentum demands that some energy go into a recoil of the emitting nucleus. The discovery by R. Mössbauer in 1958 was that if the emitting isotope is embedded in a fairly stiff crystal lattice, then part of the time the emission is recoilless. In effect, the recoil momentum is absorbed by the crystal as a whole, and the recoil *energy* loss of the emitted γ ray becomes negligible.

The energy of such γ quanta is now exactly that between the two nuclear states, and such gammas show a strong resonance absorption by ground-state nuclei. The wavelength spread of the recoilless radiation is so narrow that if the source is moving even a few centimeters per second, the Doppler effect is enough to destroy the resonance absorption by a stationary absorber.

Much as the nmr, the Mössbauer effect, although remarkable, would have a limited application except that gamma energies are measurably affected by the density of orbital electrons in the vicinity of the nucleus, and hence by the chemical environment of the atom. There is thus a chemical shift in the energy of the recoilless γ emission. If the source and the absorber atoms are in different chemical states, resonance is destroyed.

A typical experimental procedure is as follows. Some ^{57}Co (270 day half-life) is deposited on the surface of some cobalt metal. This isotope decays to an excited state of iron ^{57}Fe^m, which then emits a 14.4-keV γ quantum. The absorber consists of iron in the chemical state to be studied, and one looks for resonance absorption by stable ^{57}Fe. With source and absorber stationary there will in general be no resonance because their chemical states are different, and one therefore puts the source through a periodic velocity cycle (it may be driven by a loudspeaker cone, for example). At those velocities at which the Doppler shift restores resonance, absorption is strong, and the intensity of radiation passing through the absorber decreases. A Mössbauer spectrum is shown typically as a plot of radiation intensity through the absorber versus source velocity. The general flavor of Mössbauer spectrometry is similar to that of nmr, and the derived information is somewhat analogous.

SPECIAL TOPIC

22-ST-1   Statistical fluctuations in radioactive decay

Consider N atoms of a radioactive substance. The probability that any arbitrarily selected set of exactly m nuclei will decay during an interval t is

$$(1 - e^{-\lambda t})^m (e^{-\lambda t})^{N-m} = p^m q^{N-m} \quad (22\text{-}27)$$

where p^m is the probability that the m nuclei will decay and q^{N-m} is the probability that the other $N-m$ nuclei will not. The probability given by (22-27) must be multiplied by the number of ways of picking m nuclei out of N total ones. This is $N!/(N-m)!$. However, since the order of picking the m nuclei is immaterial, we must divide by the ways of permuting m objects, $m!$. The final equation for the probability $W(m)$ is

$$W(m) = \frac{N!}{(N-m)!\, m!}\, p^m q^{N-m} \quad (22\text{-}28)$$

We can relate this result to the standard deviation of a measurement of m. In general, the standard deviation σ of a series of n measurements of a quantity x is given by

$$\sigma^2 = \frac{1}{n} \sum_{i=1}^{n} (x_i - \overline{x})^2 \quad (22\text{-}29)$$

Expansion of the sums leads to the alternative form

$$\sigma^2 = \overline{x^2} - \overline{x}^2 \quad (22\text{-}30)$$

That is, σ^2 is given by the difference between the average value of x^2 and the square of the average value of x.

The quantities $\overline{m}$ and $\overline{m^2}$ may be evaluated from Eq. (22-28). It turns out that $\overline{m} = N(1 - e^{-\lambda t})$, which reduces to Eq. (22-19) if t is small. Also, $(N-1)Np^2 = \overline{m^2} - \overline{m}$, and insertion of these results into Eq. (22-30) gives $\sigma^2 = \overline{m}q$. If $\overline{m}$ is small compared to N, the q is essentially unity, and $\sigma = \overline{m}^{1/2}$. Thus the standard deviation of a measurement of the disintegrations occurring in time t is just the square root of the average value.

A particular sample of radioactive material registers 1000 disintegrations in 10 min; we assume that the figure of 1000 is close to the true $\overline{m}$, and estimate σ to be $(1000)^{1/2}$ or 31.6. The activity is then reported as 100 ± 3.2 dis min^{-1}, or $\pm 3.2\%$. Were 10,000 disintegrations observed over 100 min, σ would be 100 and we would now report 100 ± 1 dis min^{-1}, or an uncertainty of 1%. Thus the more total disintegrations or emitted particles counted, the smaller is the percentage of error in the measurement.   **EXAMPLE**

FRIEDLANDER, G., KENNEDY, J.W., AND MILLER, J.M. (1964). "Nuclear and Radiochemistry," 2nd ed. Wiley, New York.
SIEGBAHN, K., Ed. (1965). " Alpha-, Beta-, and Gamma-Ray Spectroscopy." North-Holland Publ., Amsterdam.   **GENERAL REFERENCES**

BERRY, R.S., RICE, S.A., AND ROSS, J. (1980). "Physical Chemistry." Wiley, New York.
HOLCOMB, R. (1970). *Science* **168**, 853.   **CITED REFERENCES**

EXERCISES

Take as exact numbers given to one significant figure.

An electric field of 10^4 V cm^{-1} is just sufficient to balance the effect of a magnetic field of 200 G on an electron. What is the velocity of the electron? SI units are convenient here.   **22-1**

Ans. 5×10^7 m s^{-1}.

22-2 Use the data of Exercise 22-1 to calculate the radius of curvature of the path of the electron if the electric field is turned off, and also the value of e/m.

Ans. 1.422×10^{-2} m, 1.758×10^{11} C/kg.

22-3 Calculate **E** in volts per centimeter required to maintain from falling a water droplet which, in the absence of the field, falls at the rate of 10^{-4} cm s^{-1}. Assume the drop carries three units of electronic charge and the viscosity of air is 1.85×10^{-4} P.

Ans. 0.67 V cm^{-1}.

22-4 Calculate the specific activity of ^{14}C in disintegrations per second per gram.

Ans. 1.65×10^{11} dis s^{-1} g^{-1}.

22-5 A piece of wood found at an archeological site shows 1.20 dis s^{-1} per gram of carbon. Contemporary wood would show 2.69 dis s^{-1} per gram. How old is the ancient wood? (The radioactivity is due to ambient ^{14}C in the atmosphere from natural sources; it is assumed that the ^{14}C level has remained constant over geological time. The isotope is incorporated into trees from ^{14}CO$_2$ and then ceases to exchange with the atmosphere.)

Ans. 6380 yr.

22-6 The magnetic field of a cyclotron is such that the operating frequency for protons is 10 MHz. Calculate this field and also the terminal radius of the proton orbit if the exit energy is to be 20 MeV (neglect relativistic effects).

Ans. 6570 G; 98 cm.

22-7 What is the relativistic increase in mass of a 10-MeV deuteron, of a 10-MeV electron? What is the velocity of a 1-MeV electron?

Ans. 0.533%, 19.6-fold, 0.94 c.

22-8 Explain how ^{26}Al should decay; likewise for ^{20}F.

22-9 Calculate Q for the overall decay of the uranium series,

$$^{238}_{92}\text{U} \rightarrow \,^{206}_{82}\text{Pb} + 8\,^{4}_{2}\text{He} + 6\beta^-$$

The isotopic mass of ^{206}Pb is 205.9745.

Ans. 51.7 MeV.

22-10 Calculation Q for (a) $^{10}_{5}$ B $+ \,^{4}_{2}$He $\rightarrow \,^{1}_{0}$n $+ \,^{13}_{7}$N and (b) $^{22}_{11}$Na $\rightarrow \,^{22}_{11}$Ne $+ \beta^+$.

Ans. (a) 1.06 MeV; (b) 1.82 MeV.

22-11 Calculate the binding energy of the last neutron added to (a) ^{16}O, (b) ^{17}O. Comment on the difference between the two values.

Ans. (a) 15.7 MeV, (b) 4.14 MeV.

22-12 Calculate the Coulomb barrier in the bombardment of ^{197}Au with alpha particles.

Ans. 19.2 MeV.

22-13 The cross section for the reaction ^{59}Co(n, γ)^{60}Co is 20 b. Calculate the number of ^{60}Co atoms formed if 2 g of metal foil is exposed to a neutron flux of 1.5×10^{13} neutrons cm^{-2} s^{-1} for 10 min, and the resulting radioactivity in millicuries.

Ans. 0.415 mCi.

22-14 Calculate the dosage in roentgens per hour 1 m away from a 1-Ci source of 1-MeV γ radiation. [*Note:* Estimate the absorption coefficient per centimeter of air from Fig. 22-9 to determine the fraction of the γ radiation absorbed per centimeter and hence the energy dissipated; this gives the number of ions produced. The number of ions per cubic centimeter 1 m away can then be calculated, and thence the dosage.]

Ans. 1.30 R hr^{-1}.

^{38}Cl is formed by the neutron irradiation of NaCl at the rate of 5×10^9 atoms per second. How many millicuries will be present (a) immediately after a 10 min irradiation, (b) 10 min after the end of the irradiation? (c) What is the saturation activity?   **22-15**

Ans. (a) 22.8 mCi; (b) 19.0 mCi; (c) 135 mCi.

The total activity is followed as a function of time for a sample suspected to consist of two or more   **22-16** independently decaying radioisotopes. Find the half-life of each component and the number of disintegrations per minute of each component present at zero time.

t (hr)	0	0.5	1.0	1.5	2.0	2.5
Dis min^{-1}	7300	4500	2900	1950	1350	990
t (hr)	3.0	3.5	4.0	5.0	6.0	7.0
Dis min^{-1}	740	580	480	370	310	280
t (hr)	8	10	12	14		
Dis min^{-1}	225	210	180	155		

Ans. (a)$t_{1/2} = 0.7$ hr, $D^0 = 6800$;
(b)$t_{1/2} = 8.2$ hr, $D^0 = 470$.

^{146}Ce decays by beta emission to ^{146}Pr, with $t_{1/2} = 14$ min; the ^{146}Pr decays in turn to stable ^{146}Nd   **22-17** with $t_{1/2} = 25$ min. A sample consists initially of 0.5 mCi of pure ^{146}Ce. Calculate the activity of ^{146}Ce and of ^{146}Pr present 30 min later. Is this a case of secular equilibrium?

Ans. $D_1 = 0.113$ mCi, $D_2 = 0.385$ mCi.

PROBLEMS

The specific activity of pure ^{244}Cf is 30.8 Ci per microgram. Calculate the half-life.   **22-18**

Calculate the average binding energy per nucleon for (a) ^{16}O and (b) ^{27}Al.   **22-19**

Derive the equation for the total binding energy of an isotope in terms of its mass number for isotopes   **22-20** of optimum charge. Use the equation

$$E(\text{MeV}) = 11.6A - \frac{20}{A}(A - 2Z)^2 - 0.06Z^2 \qquad (22\text{-}31)$$

to find $Z_A = f(A)$ such that $\partial E/(\partial Z)_A = 0$. By means of this result make a plot of the mass defect against A; also compare your result with the plot for stable isotopes as actually found (compare with Fig. 22-5).

Calculate the isotopic mass of ^{59}Co from Eq. (22-31). Calculate also the Q for adding one neutron to   **22-21** obtain ^{60}Co.

By means of the equations developed in Problem 22-20, calculate the energy of α emissions for   **22-22**

$$^A_Z\text{I} \rightarrow {}^{A-4}_{Z-2}\text{I} + {}^4_2\text{He}$$

when $A = 200$. Estimate the average number of betas per alpha for a decay series around $A = 200$.

When a magnetic field of 8000 G is imposed in a cloud chamber, the electrons and positrons formed   **22-23** by pair production from incident γ radiation are found to follow a path of 3 cm radius of curvature. What is the energy in MeV of the γ ray?

RaE(^{210}Bi) emits a beta particle whose range in aluminum is 0.470 g cm^{-2}. Write the balanced equation   **22-24** for the decay process and calculate the difference between the mass defect of ^{210}Bi and the product isotope.

22-25   A source emits equal numbers per second of two kinds of gamma rays. The first kind is of 1 keV energy, and the second of 1 MeV energy. By interpolation using Fig. 22-9, calculate the fraction of the initial total intensity that would penetrate 10 cm of lead.

22-26   What is the dosage in roentgens per hour 1 m away from a 100-Ci point source of 1-MeV gamma rays of absorption coefficient $\mu = 7 \times 10^{-5} cm^{-1}$ in air? Assume that only singly charged ions are formed and that 30 eV are required to produce each ion pair.

22-27   The high voltage across the dees of a cyclotron has a frequency of 12 MHz. What must the magnetic field be if deuterons are to be accelerated?

22-28   Calculate Q for the reaction $^{14}N(p,\alpha)^{11}C$.

22-29   A certain radioisotope may be made by proton bombardment of Ne, by neutron irradiation of NaCl, and by proton bombardment of $MgSO_4$. Explain what this radioisotope is most likely to be.

22-30   A radioelement of half-life T_1 yields a daughter of half-life T_2. Assuming that at zero time no daughter is present, derive the expression for the time of maximum combined activity (that is, disintegrations per unit time) of parent plus daughter.

22-31   What is the cross section for the formation of ^{14}C by (n, p) reaction if, on irradiation of 100 liter of 0.1 M ammonium nitrate for 1 wk at an average pile power of 3000 kW, 0.1 mCi of ^{14}C is obtained? (Assume 10^5 neutrons cm^{-2} s^{-1} W^{-1}.)

22-32   The cross section for the reaction $^{26}Mg (n, \gamma)^{27}Mg$ is 0.027 b. Calculate the activity of ^{27}Mg produced by a 20 min irradiation of 2 g of $MgSO_4$ in a neutron flux of 3×10^{12} neutrons cm^{-2} s^{-1}.

22-33   The following decay data are obtained on a sample suspected of containing several radioelements:

t (hr)	0.0	0.5	1.0	1.5	2.0	3.0	4.0	5.0	6.0	8.0
Dis min^{-1}	8000	1875	850	543	410	288	211	168	139	108

t (hr)	10	12	14
Dis min^{-1}	90	80	73

Analyze the data graphically to obtain the half-lives of all the radioelements present, and the percentage of the initial total activity due to each.

22-34   A sample of radioactive zirconium is obtained in pure form by an appropriate processing of a fission product mixture. The activity of the sample increased with time, as shown by the tabulated activity (total disintegrations per minute), due to the growth of a niobium daughter. Analyze the data to determine the half-lives of the parent and the daughter.

t (days)	0	20	40	60	100	150	200	300	400	500
D_{tot} (dis min^{-1})	2000	2250	2200	2050	1600	1050	666	240	82	28

22-35   Uranium-containing minerals may be age-dated on the assumption that all of the helium produced by the ^{238}U series remains in the sample (the overall series is given in Exercise 22-9). As an example, a meteorite contains 1.5×10^{-3}% U, and 0.066 cm^3 STP of He can be extracted from 10 g of the meteorite. Calculate its minimum age.

SPECIAL TOPICS PROBLEMS

22-36   A counter has a background of about 35 counts min^{-1}. How long should a sample of approximately 1000 counts min^{-1} be counted and how long a background count should be taken in order that the net count can be determined in a minimum total counting time with a probable standard deviation of less than 1%?

A sample of radioactive material is counted for 1 min and, from the result, the standard deviation is estimated to be 5.0%. Calculate the standard deviation had the sample been counted for 10 min. **22-37**

A counter has a measured background rate of 600 counts in 25 min. With a sample in place the total measured rate is 1050 counts in 20 min. Give the net counting rate per minute for the sample and its standard deviation. **22-38**

Index

Tables are indicated by "t" following the page number.